Library of Congress Cataloging-in-Publication Data

Montgomery, John H. (John Harold), 1955-
    Groundwater chemicals desk reference / John H. Montgomery, --2nd ed.
      p. cm.
    Includes bibliographical references and index.
    ISBN 1-56670-165-1 (alk. paper)
    1. Groundwater--Pollution--Handbooks, manuals, etc. 2. Pollutants--Handbooks, manuals, etc. I.
    Title.
    TD426.M66   1996
    628.1'61--dc20   95-45365
                                                         CIP

© 1996 by CRC Press, Inc.
Lewis Publishers is an imprint of CRC Press

No claim to original U.S. Government works
International Standard Book Number 1-56670-165-1
Library of Congress Card Number 95-45365
Printed in the United States of America 1 2 3 4 5 6 7 8 9 0
Printed on acid-free paper

Second Edition

GROUNDWATER CHEMICALS

Desk Reference

John H. Montgomery

LEWIS PUBLISHERS

Boca Raton   New York   London   Tokyo

Preface

The continuation of research publications dealing with the fate, transport, and remediation of hazardous substances in the environment has stimulated this new edition.

The compounds in this book were previously presented in two earlier versions - Groundwater Chemical Desk Reference and Groundwater Chemicals Desk Reference - Volume 2. The following data fields have been added: odor thresholds in air, heat of fusion, interfacial tension with water, diffusivity in water, bioconcentration factors, toxicity to aquatic organisms and mammals, symptoms of chemical exposure, and federal drinking water standards. This data should prove useful by the environmental regulated community and environmental consultants especially in conducting risk-based contamination assessments. Moreover, four appendices have been added. These include conversion factors between different concentration units, bulk density values of soils and selected rocks, porosity values of selected rocks and unconsolidated sediments, and solubility of miscellaneous organic compounds not profiled in the text.

The presentation of data has been revised to make it easier for the reader to locate information. The Environmental Fate section has been expanded and includes much more information than in the previous edition. This section is subdivided into the following categories: Biological, Soil, Plant, Surface Water, Groundwater, Photolytic, and Chemical/Physical. When available, the half-lives of chemicals in various media are included. These include photolytic, hydrolysis, biodegradation, and volatilization half-lives. To conserve space, references are no longer given at the end of each chemical profile. All references now appear at the end of the text, before the appendices.

The book is based on more than 1,400 references. Most of the citations reviewed from the documented literature and included in this book pertain to the fate and transport of chemicals in the subsurface environment (about 400 new references). Every effort has been made to select the most accurate information and present it factually and accurately. The publisher and author would appreciate hearing from readers regarding corrections and suggestions for material that might be included for use in future editions.

The author is grateful to the staff of CRC Press, in particular, Mr. Bob Esposito and Mr. James K. Brody for their invaluable contributions and suggestions during the preparation of this book. The author also extends thanks to the many anonymous reviewers for their comments and suggestions on draft proofs.

Introduction

The compounds profiled include solvents, herbicides, insecticides, fumigants, and other hazardous substances most commonly found in the groundwater environment. Compounds profiled in this book include the organic Priority Pollutants promulgated by the U.S. Environmental Protection Agency (U.S. EPA) under the Clean Water Act of 1977 (40 CFR 136, 1977) and compounds most commonly found in the work environment.

The compound headings are those commonly used by the U.S. EPA and many agricultural organizations. Positional and/or structural prefixes set in italic type are not an integral part of the chemical name and are disregarded in alphabetizing. These include *asym-*, *sym-*, *n-*, *sec-*, *cis-*, *trans-*, α-, β-, γ-, *o-*, *m-*, *p-*, *N-*, *S-*, etc.

Synonyms: These are listed alphabetically following the convention used for the compound headings. Compounds in boldface type are the Chemical Abstracts Service (CAS) names listed in the eighth or ninth Collective Index. If no synonym appears in boldface type, then the compound heading is the CAS assigned name. Synonyms include chemical names, common or generic names, trade names, registered trademarks, government codes and acronyms. All synonyms found in the literature are listed.

Although synonyms were retrieved from several references, most of them were retrieved from the Registry of Toxic Effects of Chemical Substances (RTECS, 1985).

Beneath the synonyms is the structural formula. This is given for every compound regardless of its complexity. The structural formula is a graphic representation of atoms or group(s) of atoms relative to each other. Clearly, the limitation of structural formulas is that they depict these relationships in two dimensions.

Chemical Abstracts Service (CAS) Registry Number: This is a unique identifier assigned by the American Chemical Society to chemicals recorded in the CAS Registry System. This number is used to access various chemical databases such as the Hazardous Substances Data Bank (HSDB), CAS Online, Chemical Substances Information Network and many others. This entry is also useful to conclusively identify a substance regardless of the assigned name.

Department of Transportation (DOT) Designation: This is a four-digit number assigned by the U.S. Department of Transportation (DOT) for hazardous materials and is identical to the United Nations identification number (which is preceded by the letters UN). This number is required on shipping papers, on placards or orange panels on tanks and on a label or package containing the material. These numbers are widely used for personnel responding to emergency situations, e.g., overturned tractor trailers, in which the identification of the transported material is quickly and easily determined. The DOT designations and appropriate responses to each chemical or compounds are cross-referenced in the Emergency Response Guidebook. Additional information is provided in this book and may be obtained through the U.S. Department of Transportation, Research and Special Programs Administration, Materials Transportation Bureau, Washington, DC 20590.

DOT label: The label is the hazard classification assigned by the U.S. Department of Transportation and is required by regulation on all containers being shipped.

Molecular formula: This is arranged by carbon, hydrogen and remaining elements in alphabetical order in accordance with the system developed by Hill (1900). Molecular formulas are useful in identifying isomers (i.e. compounds with identical molecular formulas) and are required if one wishes to calculate the formula weight of a substance.

Formula weight: This is calculated to the nearest hundredth using the empirical formula and the 1981 Table of Standard Atomic Weights as reported in Weast (1986). Formula weights are required for many calculations, such as converting weight/volume units, e.g., mg/L or g/L, to molar units (mol/L); with density for calculating molar volumes; and for estimating Henry's law constants.

Registry of Toxic Effects of Chemical Substances (RTECS) Number: Many compounds are assigned a unique accession number consisting of two letters followed by seven numerals. This number is needed to quickly and easily locate additional toxicity and health-based data which are cross-referenced in the RTECS (1985). Contact the National Institute for Occupational Safety and Health (NIOSH), U.S. Department of Health and Human Services, Mail Stop C-13, 4676 Columbia Parkway, Cincinnati, OH 45226-1998 (toll free: 1-800-35-NIOSH; fax: 513-533-8573) for additional information.

Physical state, color, and odor: The appearance, including the color and physical state (solid, liquid, or gas) of a chemical at room temperature (20-25 °C) is provided. If the compound can be detected by the olfactory sense, the odor is noted. Unless noted otherwise, the information provided in this category is for the pure substance and was obtained from many sources (CHRIS Hazardous Chemical Data, 1984; Hazardous Substances Data Bank, 1989; Hawley, 1981; Keith and Walters, 1992; Sax, 1984; Sax and Lewis, 1987; Toxic and Hazardous Industrial Chemicals Safety Manual for Handling and Disposal with Toxicity and Hazard Data, 1986; Sittig, 1985; Verschueren, 1983; Windholz et al., 1983). If available, odor thresholds are given. The odor threshold is the lowest compound concentration in air that can be detected by the olfactory sense and are usually reported as parts per billion (ppb) as micrograms per cubic meter ($\mu g/m^3$). Sometimes, the odor threshold in water is given and is reported as $\mu g/kg$.

Boiling point: This is defined as the temperature at which the vapor pressure of a liquid equals the atmospheric pressure. Unless otherwise noted, all boiling points are reported at 1.0 atmosphere pressure (760 mmHg). Although not used in environmental assessments, boiling points for aromatic compounds have been found to be linearly correlated with aqueous solubility (Almgren et al., 1979). Boiling points are also useful in assessing entry of toxic substances into the body. Body contact with high-boiling liquids is the most common means of entry into the body whereas the inhalation route is the most common for low-boiling liquids (Shafer, 1987).

Diffusivity in water: Molecular diffusion is defined as the transport of molecules (e.g., organic compounds) in either liquid or gaseous states. Typically, molecular diffusion is not a major factor under most environmental conditions. However, in saturated aquifers having low pore water velocities (i.e., <0.002 cm/sec), diffusion can be a contributing factor in the transport of organic compounds.

Very few experimentally determined diffusivities of organic substances in water are available. If experimentally determined diffusivity values are not available, Hayduk and Laudie (1974) recommends the following equation for estimating this parameter:

$$D = (1.326 \times 10^4)/\mu^{1.14}V^{0.589}) \qquad [1]$$

where D is the diffusivity of the substance in water (cm^2/sec), μ is the viscosity of water (cps) and V is the molar volume of the solute (cm^3/mol). The molar volume is easily determined if the liquid density of the solute is available. Molar volume may also be determined using the LeBas incremental method as described in Lyman et al. (1982). The latter method is required to determine molar volumes of substances that are solids at ordinary temperatures.

Dissociation constant: In an aqueous solution, an acid (HA) will dissociate into the carboxylate anion (A$^-$) and hydrogen ion (H$^+$) and may be represented by the general equation:

$$HA_{(aq)} \rightleftharpoons H^+ + A^- \qquad [2]$$

At equilibrium, the ratio of the products (ions) to the reactant (non-ionized electrolyte) is related by the equation:

$$K_a = ([H^+][A^-]/[HA]) \qquad [3]$$

where K_a is the dissociation constant. This expression shows that K_a increases if there is increased ionization and vice versa. A strong acid (weak base) such as hydrochloric acid ionizes readily and has a large K_a, whereas a weak acid (or stronger base) such as benzoic acid ionizes to a lesser extent and has a lower K_a. The dissociation constants for weak acids are sometimes expressed as K_b, the dissociation constant for the base and both are related to the dissociation constant for water by the expression:

$$K_w = K_a + K_b \qquad [4]$$

where K_w is the dissociation constant for water (10^{-14} at 25 °C), K_a is the acid dissociation constant, K_b is the base dissociation constant.

The dissociation constant is usually expressed as $pK_a = -\log_{10}K_a$. Equation [4] becomes:

$$pK_w = pK_a + pK_b \qquad [5]$$

When the pH of the solution and the pK_a are equal, 50% of the acid will have dissociated into ions. The percent dissociation of an acid or base can be calculated if the pH of the solution and the pK_a of the compound are known (Guswa et al., 1984):

For organic acids: $\alpha_a = [100/(1 + 10^{(pH-pKa)})]$ [6]

For organic bases: $\alpha_b = [100/(1 + 10^{(pKw-pKb-pH)})]$ [7]

where α_a is the percent of the organic acid that is nondissociated, α_b is the percent of the organic base that is nondissociated, pK_a is the $-\log_{10}$ dissociation constant for an acid, pK_w is the $-\log_{10}$ dissociation constant for water (14.00 at 25 °C), pK_b is the $-\log_{10}$ dissociation constant for base ($pK_b = pK_w-pK_a$) and pH is the $-\log_{10}$ hydrogen ion activity (concentration) of the solution.

Since ions tend to remain in solution, the degree of dissociation will affect processes such as volatilization, photolysis, adsorption and bioconcentration (Howard, 1989).

Heat of fusion: This value, normally reported in kcal/mol, is also referred to as the heat of melting. For solids, the heat of fusion is required in estimating the solubility of the solute to account for crystal lattice interactions. The theoretical basis for introducing this value into the estimation of aqueous solubility of organic solids is explained by Irmann (1965) and Yalkowsky and Valvani (1979). Heat of fusion data is available in many texts including Dean (1987), Weast (1986), and CHRIS (1984).

Henry's law constant: Sometimes referred to as the air-water partition coefficient, the Henry's law constant is defined as the ratio of the partial pressure of a compound in air to the concentration of the compound in water at a given temperature under equilibrium conditions. If the vapor pressure and solubility of a compound are known, this parameter can be calculated at 1.0 atm (760 mmHg) as follows:

$$K_H = Pfw/760S \qquad [8]$$

where K_H is Henry's law constant (atm·m^3/mol), P is the vapor pressure (mmHg), S is the solubility in water (mg/L) and fw is the formula weight (g/mol).

Henry's law constant can also be expressed in dimensionless form and may be calculated using one of the following equations:

$$K_H' = K_H/RK \qquad or \qquad K_H' = S_a/S \qquad [9]$$

where K_H' is Henry's law constant (dimensionless), R is the ideal gas constant (8.20575 x 10^{-5} atm·m^3/mol·K), K is the temperature of water (degrees Kelvin), S_a is the solute concentration in air (mol/L), S is the aqueous solute concentration (mol/L).

It should be noted that estimating Henry's law constant assumes that the gas obeys the ideal gas law and the aqueous solution behaves as an ideally dilute solution. The solubility and vapor pressure data inputted into the equations are valid only for the pure compound and must be in the same standard state at the same temperature.

The major drawback in estimating Henry's law constant is that both the solubility and the vapor pressure of the compound are needed in equation [8]. If one or both these parameters are unknown, an empirical equation based on quantitative structure-activity relationships (QSAR) may be used to estimate Henry's law constants (Nirmalakhandan and Speece, 1988). In this QSAR model, only the structure of the compound is needed. From this, connectivity indexes (based on molecular topology), polarizability (based on atomic contributions) and the propensity of the compound to form hydrogen bonds can easily be determined. These parameters, when regressed against known Henry's law constants for 180 organic compounds, yielded an empirical equation that explained more than 98% of the variance in the data set having an average standard error of only 0.262 logarithm units.

Henry's law constant may also be estimated using the bond or group contribution method developed by Hine and Mookerjee (1975). The constants for the bond and group contributions were determined using experimentally determined Henry's law constants for 292 compounds. The authors found that those estimated values significantly deviating from observed values (particularly for compounds containing halogen, nitrogen, oxygen and sulfur substituents) could be explained by "distant polar interactions", i.e., interactions between polar bonds or structural groups.

A more recent study for estimating Henry's law constants using the bond contribution method was provided by Meylan and Howard (1991). In this study, the authors updated and revised the method developed by Hine and Mookerjee (1975) due to new experimental data that have become available since 1975. Bond contribution values were determined for 59 chemical bonds based on known Henry's law constants for 345 organic compounds. A good statistical fit [correlation coefficient (r^2) = 0.94] was obtained when the bond contribution values were regressed against known Henry's law constants for all compounds. For selected chemicals classes, r^2 increased slightly to 0.97.

Russell et al. (1992) conducted a similar study using the same data set from Hine and Mookerjee (1975). They developed a computer-assisted model that was based on five molecular descriptors which was related to the compound's bulk, lipophilicity and polarity. They found that 63 molecular structures was highly correlative with the log of Henry's law constants (r^2 = 0.96).

Henry's law constants provided an indication of the relative volatility of a substance. According to Lyman et al. (1982), if K_H <10^{-7} atm·m^3/mol, the substance has a low volatility. If K_H is >10^{-7} but <10^{-5} atm·m^3/mol, the substance will volatilize slowly. Volatilization becomes an important transfer mechanism in the range 10^{-5}< H <10^{-3} atm·m^3/mol. Values of K_H >10^{-3} atm·m^3/mol indicate volatilization will proceed rapidly.

The rate of volatilization will also increase with an increase in temperature. ten Hulscher et al. (1992) studied the temperature dependence of Henry's law constants for three chlorobenzenes, three chlorinated biphenyls and six polynuclear aromatic hydrocarbons. They observed that over the temperature range of 10 to 55 °C, Henry's law constant was doubled for every 10 °C increase in temperature. This temperature relationship should be considered when assessing the role of chemical volatilization from large surface water bodies whose temperatures are generally higher than those typically observed in groundwater.

Interfacial tension with water: With few exceptions, most organic compounds entering the aqueous environment are non-miscible liquids. The interfacial tension between the compound and the water (i.e., groundwater, surface water bodies, etc.) is numerically equivalent to the free surface energy that is formed at the interface. Compounds with high interfacial tension values relative to water are easy to separate after mixing and are not likely to form emulsions (Lyman et al., 1982). The interfacial tension of organic compounds can be used to calculate the spreading coefficient to determine whether it form a lens or macromolecular film with water (Demond and Lindner, 1993):

$$s_{OW} = \gamma_{W(O)} - \gamma_{O(W)} - \gamma_{OW} \qquad [10]$$

where s_{OW} is the spreading coefficient of the organic liquid at the air-water interface, $\gamma_{W(O)}$ is the surface tension of water saturated with the organic liquid, $\gamma_{O(W)}$ is the surface tension of the organic liquid saturated with water and γ_{OW} is the interfacial tension between organic liquid and water. Organic liquids with spreading coefficients more than zero will form a thin layer on the water surface. Conversely, the organic liquid will forms a lens if the spreading coefficient is negative.

The units of interfacial tension are identical for surface tension, i.e., dyn/cm. Interfacial tension values of organic compounds range from zero for completely miscible liquids (e.g., acetone, methanol, ethanol) up to the surface tension of water at 25 °C which is 72 dyn/cm (Lyman et al., 1982). Interfacial tension values may be affected by pH, surface-active agents and gas in solution (Schowalter, 1979). Most of the interfacial tension values reported in this book were obtained from Dean (1987), Demond and Lindner (1993) and references cited therein.

Ionization potential: The ionization potential of a compound is defined as the energy required to remove a given electron from the molecule's atomic orbit (outermost shell) and is expressed in electron volts (eV). One electron volt is equivalent to 23,053 cal/mol.

Knowing the ionization potential of a contaminant is required in determining the appropriate photoionization lamp for detecting that contaminant or family of contaminants. Photoionization instruments are equipped with a radiation source (ultraviolet lamp), pump, ionization chamber, an amplifier and a recorder (either digital or meter). Generally, compounds with ionization potentials smaller than the radiation source (UV lamp rating) being used will readily ionize and will be detected by the instrument. Conversely, compounds with ionization potentials higher than the lamp rating will not ionize and will not be detected by the instrument.

Bioconcentration factor, log BCF: The bioconcentration factor is defined as the ratio of the chemical accumulated in tissue to the concentration in water. Generally, high bioconcentration factors tend to be associated with very lipophilic compounds. Conversely, low bioconcentration factors are associated with compounds having high aqueous solubilities.

Bioconcentration factors have been shown to be correlated with the octanol/water partition coefficient in aquatic organisms (Davies and Dobbs, 1984; de Wolf et al., 1992; Isnard and Lambert, 1988) and fish (Davies and Dobbs, 1984; Kenaga, 1980; Isnard and Lambert, 1988; Neely et al., 1974; Ogata et al., 1984; Oliver and Niimi, 1985).

Soil sorption coefficient, log K_{oc}: The soil/sediment partition or sorption coefficient is defined as the ratio of adsorbed chemical per unit weight of organic carbon to the aqueous solute concentration. This value provides an indication of the tendency of a chemical to partition between particles containing organic carbon and water. Compounds that bind strongly to organic carbon have characteristically low solubilities, whereas compounds with low tendencies to adsorb onto organic particles have high solubilities.

Nonionizable chemicals that sorb onto organic materials in an aquifer (i.e., organic carbon) are retarded in their movement in groundwater. The sorbing solute travels at linear velocity that is lower than the groundwater flow velocity by a factor of R_d, the retardation factor. If the K_{oc} of a compound is known, the retardation factor may be calculated using the following equation from Freeze and Cherry (1974) for unconsolidated sediments:

$$R_d = V_w/V_c = [1 + (BK_d/n_e)] \qquad [11]$$

where R_d is the retardation factor (unitless), V_w is the average linear velocity of groundwater (e.g., ft/day), V_c is the average linear velocity of contaminant (e.g., ft/day), B is the average soil bulk density (g/cm^3), n_e is the effective porosity (unitless), K_d is the distribution (sorption) coefficient (cm^3/g).

By definition, K_d is defined as the ratio of the concentration of the solute on the solid to the concentration of the solute in solution. This can be represented by the Freundlich equation:

$$K_d = VM_S/MM_L = C_S/C_L^n \qquad [12]$$

where V is the volume of the solution (cm^3), M_S is the mass of the sorbed solute (g), M is the mass of the porous medium (g), M_L is mass of the solute in solution (g), C_S is the concentration of the sorbed solute (g/cm^3), C_L is the concentration of the solute in the solution (g/cm^3) and n is a constant.

Values of n are normally between 0.7 and 1.1 although values of 1.6 have been reported (Lyman et al., 1982). If n is unknown, it is assumed to be unity and a plot of C_S versus C_L will be linear. The distribution coefficient is related to K_{oc} by the equation:

$$K_{oc} = K_d/f_{oc} \qquad [13]$$

where f_{oc} is the fraction of naturally occurring organic carbon in soil.

Sometimes K_d is expressed on an organic-matter basis and is defined as:

$$K_{om} = K_d/f_{om} \qquad [14]$$

where f_{om} is the fraction of naturally occurring organic matter in soil. The relationship between K_{oc} and K_{om} is defined as:

$$K_{om} = 0.58K_{oc} \qquad [15]$$

where the constant 0.58 is assumed to represent the fraction of carbon present in the soil or sediment organic matter (Allison, 1965).

For fractured rock aquifers in which the porosity of the solid mass between fractures is insignificant, Freeze and Cherry (1974) report the retardation equation as:

$$R_d = V_w/V_c = [1 + (2K_A/b)] \tag{16}$$

where K_A is the distribution coefficient (cm) and b is the aperture of fracture (cm).

To calculate the retardation factors for ionizable compounds such as acids and bases, the fraction of un-ionized acid (α_a) or base (α_b) needs to be determined (see **Dissociation constant**). According to Guswa et al. (1984), if it is assumed only the un-ionized portion of the acid is adsorbed onto the soil, the retardation factor for the acid becomes:

$$R_a = [1 + (\alpha_a BK_d/n_e)] \tag{17}$$

However, for a base they assume that the ionized portion is exchanged with a monovalent ion and the un-ionized portion of the base is adsorbed hydrophobically. Therefore, the retardation factor for the base is:

$$R_b = \{1 + [(\alpha_b BK_d)/n_e] + [CECB(1-\alpha_b)]/(100\Sigma z^+ n_e)\} \tag{18}$$

where CEC is the cation exchange capacity of the soil (cm^3/g) and Σz^+ is the sum of all positively charged particles in the soil (milliequivalents/cm^3). Guswa et al. (1984) report that the term Σz^+ is approximately 0.001 for most agricultural soils.

Correlations between K_{oc} and bioconcentration factors in fish and beef have shown a log-log linear relationship (Kenaga, 1980) as well as solubility of organic compounds in water (Abdul et al., 1987; Means et al., 1980). Moreover, the log K_{oc} has been shown to be related to molecular connectivity indices (Govers et al., 1984; Gerstl and Helling, 1987; Koch, 1983; Meylan et al., 1992; Sabljić and Protić, 1982; Sabljić, 1984, 1987) and high performance liquid chromatography (HPLC) capacity factors (Haky and Young, 1984; Hodson and Williams, 1988; Szabo et al., 1990, 1990a).

In instances where experimentally determined K_{oc} values are not available, they can be estimated using recommended regression equations as cited in Lyman et al. (1982) or Meylan et al. (1992). All the K_{oc} estimations are based on regression equations in which the aqueous solubility or the K_{ow} of the substance is known.

Octanol/water partition coefficient, log K_{ow}: The K_{ow} of a substance is the *n*-octanol/water partition coefficient and is defined as the ratio of the solute concentration in the water-saturated *n*-octanol phase to the solute concentration in the *n*-octanol-saturated water phase. Values of K_{ow} are therefore unitless.

The partition coefficient has been recognized as a key parameter in predicting the environmental fate of organic compounds. The log K_{ow} has been shown to be linearly correlated with log bioconcentration factors (BCF) in aquatic organisms (Davies and Dobbs, 1984; de Wolf et al., 1992; Isnard and Lambert, 1988), in fish (Davies and Dobbs, 1984; Kenaga, 1980; Isnard and Lambert, 1988; Neely et al., 1974; Ogata et al., 1984; Oliver and Niimi, 1985), log soil/sediment partition coefficients (K_{oc}) (Chiou et al., 1979; Kenaga and Goring, 1980), log of the solubility of organic compounds in water (Banerjee et al., 1980; Chiou et al., 1977, 1982; Hansch et al., 1968; Isnard and Lambert, 1988; Miller et al., 1984, 1985; Tewari et al., 1982; Yalkowsky and Valvani, 1979, 1980), molecular surface area (Camilleri et al., 1988; Funasaki et al., 1985; Miller et al., 1984; Woodburn et al., 1992; Yalkowsky and Valvani, 1979, 1980), molar refraction (Yoshida et al., 1983), molecular connectivity indices (Govers et al., 1984; Patil, 1991; Woodburn et al., 1992), reversed-phase liquid chromatography (RPLC) retention factors (Khaledi and Breyer, 1989; Woodburn et al., 1992), RPLC capacity factors (Braumann, 1986; Minick et al., 1989), isocratic RPLC capacity factors (Hafkenscheid and Tomlinson, 1983); HPLC capacity factors (Brooke et al., 1986; Carlson et al., 1975; DeKock and Lord, 1987; Eadsforth, 1986; Hammers et al., 1982; Harnisch et al., 1983; Kraak et al., 1986; Miyake et al., 1982, 1987, 1988; Szabo et al., 1990), HPLC retention times (Burkhard and Kuehl, 1986; Mirrlees et al., 1976; Sarna et al., 1984; Veith and Morris, 1978; Webster et al., 1985), reversed-phase thin layer chromatography retention parameters (Bruggeman et al., 1982), gas chromatography retention indices (Valkó et al., 1984), distribution coefficients (Campbell et al., 1983), solvatochromic parameters (Sadek et al., 1985), biological responses (Schultz et al., 1989), log of the n-hexane/water and L-a-phosphatidycholine dimyristol partitioning coefficients (Gobas et al., 1988), molecular descriptors and physicochemical properties (Bodor et al., 1989; Warne et al., 1990), molecular structure (Suzuki, 1991), linear solvation energy relationships (LSER) (Kamlet et al., 1988) and substituent constants which are based on empirically derived atomic or group constants and structural factors (Hansch and Anderson, 1967; Hansch et al., 1972). Variables needed for employing the LSER method have recently been presented by Hickey and Passino-Reader (1991).

For ionizable compounds (e.g., acids, amines and phenols), K_{ow} values are a function of pH. Unfortunately, many investigators have neglected to report the pH of the solution at which the K_{ow} was determined. If a K_{ow} value is used for an ionizable compound for which the pH is known, both values should be noted.

Melting point: The melting point of a substance is defined as the temperature at which a solid substance undergoes a phase change to a liquid. The reverse process, the temperature at which a liquid freezes to a solid, is called the freezing point. For a given substance, the melting point is identical to the freezing point.

Unless noted otherwise, all melting points are reported at the standard pressure of 1.0 atm (760 mmHg). Although the melting point of a substance is not directly used in predicting its behavior in the environment, it is useful in determining the phase in which the substance would be found under typical conditions.

Solubility in organics: The presence of small quantities of solvents can enhance a compound's solubility in water (Nyssen et al., 1987). Consequently, its fate and transport in soils, sediments and groundwater will be changed due to the presence of these cosolvents. For example, soils contaminated with compounds having low water solubilities tend to remain bound to the soil by adsorbing onto organic carbon and/or by interfacial tension with water. A solvent introduced to an unsaturated soil environment (e.g., a surface spill, leaking aboveground tank, etc.) may come in contact with existing soil contaminants. As the solvent interacts with the existing contamination, it may mobilize it, thereby facilitating its migration. Consequently, the organic solvent can facilitate the leaching of contaminants from the soil to the water table. Therefore, the presence of cosolutes must be considered when predicting the fate and transport of contaminants in the unsaturated zone, the water table and surface water bodies.

Solubility in water: The water solubility of a compound is defined as the saturated concentration of the compound in water at a given temperature and pressure. This parameter is perhaps the most important factor in estimating a chemical's fate and transport in the aquatic environment. Compounds with high water solubilities tend to desorb from soils and sediments (i.e., they have low K_{oc} values), are less likely to volatilize from water, and are susceptible to biodegradation. Conversely, compounds with low solubilities tend to adsorb onto soils and sediments (have high K_{oc}), volatilize more readily from water and bioconcentrate in aquatic organisms. The more soluble compounds commonly enter the water table more readily than their less soluble counterparts.

The water solubility of a compound varies with temperature, pH (particularly, ionizable compounds such as acids and bases) and other dissolved constituents, e.g., inorganic salts (electrolytes) and organic chemicals including naturally occurring organic carbon, such as humic and fulvic acids. At a given temperature, the variability/discrepancy of water-solubility measurements documented by investigators may be attributed to one or more of the following: (1) purity of the compound, (2) analytical method employed, (3) particle size (for solid solubility determinations only), (4) adsorption onto the container and/or suspended solids, (5) time allowed for equilibrium conditions to be reached, (6) losses due to volatilization and (7) chemical transformations (e.g., hydrolysis).

The water solubility of chemical substances has been related to bioconcentration factors (BCF), soil/sediment partition coefficients (K_{oc}) (Abdul et al., 1987; Means et al., 1980), n-octanol/water partition coefficients (K_{ow}) (Chiou et al., 1977, 1982; Hansch et al., 1968; Miller et al., 1984, 1985; Yalkowsky and Valvani, 1979, 1980), HPLC capacity factors (Hafkenscheid and Tomlinson, 1981; Whitehouse and Cooke, 1982), molecular descriptors and physicochemical properties (Warne et al., 1990), soil organic matter K_{om} (Chiou et al., 1983), total molecular surface area (Amidon et al., 1975; Hermann, 1972; Lande and Banerjee, 1981; Lande et al., 1985; Valvani et al., 1976), the compound's molecular structure (Nirmalakhandan and Speece, 1988a, 1989; Patil, 1991; Suzuki, 1989), boiling points (Almgren et al., 1979) and for homologous series of hydrocarbons or classes of organic compounds-carbon number (Bell, 1973; Krzyzanowska and Szeliga, 1978; Mitra et al., 1977; Robb, 1966) and molar volumes (Lande and Banerjee, 1981). With the exception of the molecular structure-solubility relationship, regression equations generated from 2the other relationships have demonstrated a log-log linear relationship for these properties. The reported regression equations are useful in estimating the solubility of a compound in water if experimental values are not available. In addition, the solubility of a compound may be estimated from experimentally determined Henry's law constants (Kamlet et al., 1987) or from measured infinite dilution activity coefficients (Wright et al., 1992).

Unless otherwise noted, all reported solubilities were determined using distilled and/or deionized water. For some compounds, solubilities were determined using groundwater, natural seawater or artificial seawater.

Solubility concentrations can be expressed many ways. These include, molarity (mol/L), molality (mol/kg), mole fraction, weight percent, mass per unit volume (e.g. g/L), etc. The conversion formulas for solutions having different concentration units are presented in Appendix A.

Methods for estimating the aqueous solubilities of organic solutes can be found in Lyman et al. (1982) and Yalkowsky and Banerjee (1992).

Specific density: The specific density, also known as relative density, is defined as:

$$\rho = d_s/d_w \qquad [19]$$

where d_s is the density of a substance (g/mL or g/cm^3) and d_w is the density of distilled water (g/mL or g/cm^3). Values of specific density are unitless and are reported in the form ρ at T_s/T_w where ρ is the specific density of the substance, T_s is the temperature of substance at the time of measurement (°C) and T_w is the water temperature (°C).

For example, the value 1.1750 at 20/4 °C indicates a specific density of 1.1750 for the substance at 20 °C with respect to water at 4 °C. At 4 °C, the density of water is exactly 1.0000 g/mL (g/cm^3). Therefore, the specific density of a substance is equivalent to the density of the substance relative to the density of water at 4 °C.

The density of a hydrophobic substance enables it to sink or float in water. Density values are especially important for liquids migrating through the unsaturated zone and encountering the water table as "free product." Generally, liquids that are less dense than water "float" on the water table. Conversely, organic liquids that are more dense than water commonly "sink" through the water table, e.g., dense nonaqueous phase liquids (DNAPLs) such as chloroform, dichloroethane and tetrachloroethylene.

Hydrophilic substances, on the other hand, behave differently. Acetone, which is less dense than water, does not float on water because it is freely miscible with it in all proportions. Therefore, the solubility of a substance must be considered in assessing its behavior in the subsurface.

Environmental Fate: Chemicals released in the environment are susceptible to several degradation pathways. These include chemical (i.e., hydrolysis, oxidation, reduction, dealkylation, dealkoxylation, decarboxylation, methylation, isomerization and conjugation), photolysis or photooxidation and biodegradation. Compounds transformed by one or more of these processes may result in the formation of more toxic or less toxic substances. In addition, the transformed product(s) will behave differently than the parent compound due to changes in their physicochemical properties. Many researchers focus their attention on transformation rates rather than the transformation products. Consequently, only limited data exist on the transitional and resultant end products. Where available, compounds that are transformed into identified products as well as environmental fate rate constants and/or half-lives are listed.

In addition to chemical transformations occurring under normal environmental conditions, abiotic degradation products are also included. Types of abiotic transformation processes or treatment technologies fall into two categories -- physical and chemical. Types of physical processes used in removing or eliminating hazardous wastes include sedimentation, centrifugation, flocculation, oil/water separation, dissolved air flotation, heavy media separation, evaporation, air stripping, steam stripping, distillation, soil flushing, chelation, liquid-liquid extraction, supercritical extraction, filtration, carbon adsorption, absorption, reverse osmosis, ion exchange and electrodialysis. This information can be useful in evaluating abiotic degradation as a possible remedial measure. Chemical processes include neutralization, precipitation, hydrolysis (acid or base catalyzed), photolysis or irradiation, oxidation-reduction, oxidation by hydrogen peroxide, alkaline chlorination, electrolytic oxidation, catalytic dehydrochlorination and alkali metal dechlorination.

If available, experimentally determined hydrolysis and photolytic half-lives of chemicals are provided. The half-life of a chemical is the time required for the parent chemical to reach one-half or 50% of its original concentration.

Chemicals will undergo photolysis if they can absorb sunlight. Photolysis can occur in air, soil, water and plants. The rate of photolysis is dependent upon the pH, temperature, presence of sensitizers, sorption to soil, depth of the compound in soil and water. Lyman et al. (1982) present an excellent overview of the photolysis process.

The rate of chemical hydrolysis is highly dependent upon the compound's solubility, temperature and pH. Since other environmental factors such as photolysis, adsorption, volatility (i.e., Henry's law constants) and adsorption can affect the rate of hydrolysis, these factors are virtually eliminated by performing hydrolysis experiments under carefully controlled laboratory conditions. The hydrolysis half-lives reported in the literature were calculated using experimentally determined hydrolysis rate constants.

Most of the abiotic chemical transformation products reported in this book are limited to only three processes: hydrolysis, photooxidation and chemical oxidation-reduction. These processes are the most widely studied and reported in the literature. Detailed information describing the above technologies, their availability/limitation and company sources is available (U.S. EPA, 1987).

Vapor density: The vapor density of a substance is defined as the ratio of the mass of vapor per unit volume. An equation for estimating vapor density is readily derived from a varied form of the ideal gas law:

$$PV = MRK/fw \qquad [20]$$

where P is the vapor pressure (atm), V is the volume (L), M is the mass (g), R is the ideal gas constant (8.20575×10^{-2} atm·L/mol·K) and K is the temperature (degrees Kelvin). Recognizing that the density of a substance is defined as:

$$d_s = M/V \qquad [21]$$

Substituting this equation into the equation [20], rearranging and simplifying results in an expression to determine the vapor density (g/L):

$$d_s = Pfw/RK \qquad [22]$$

At standard temperature (293.15K) and pressure (1 atm), equation [22] simplifies to:

$$d_s = fw/24.47 \qquad [23]$$

The specific vapor density of a substance relative to air is determined using:

$$p_v = fw/24.47p_{air} \qquad [24]$$

where p_v is the specific vapor density of a substance (unitless) and p_{air} is the vapor density of air (g/L).

The specific vapor density, p_v, is simply the ratio of the vapor density of the substance to that of air under the same pressure and temperature. According to Weast (1986), the vapor density of dry air at 20 °C and 760 mmHg is 1.204 g/L. At 25 °C, the vapor density of air decreases slightly to 1.184 g/L. Calculated specific vapor densities are reported relative to air (set equal to 1) only for compounds which are liquids at room temperature (i.e., 25 °C).

Vapor pressure: The vapor pressure of a substance is defined as the pressure exerted by the vapor (gas) of a substance when it is under equilibrium conditions. It provides a semi-quantitative rate at which it will volatilize from soil and/or water. The vapor pressure of a substance is a required input parameter for calculating the air-water partition coefficient (see **Henry's Law Constant**), which in turn is used in estimating the volatilization rate of compounds from groundwater to the unsaturated zone and from surface water bodies to the atmosphere.

FIRE HAZARDS

Flash point: The flash point is defined as the minimum temperature at which a substance releases ignitable flammable vapors in the presence of an ignition source. e.g., spark or flame. Flash points may be determined by two methods --- Tag closed cup (ASTM method D56) or Cleveland open cup (ASTM method D93). Unless otherwise noted, all flash point values represent closed cup method determinations. Flash point values determined by the open cup method are slightly higher (about 10-15 °C) than those determined by the closed cup method; however, the open cup method is more representative of actual environmental conditions.

According to Sax (1984), a material with a flash point of 100 °F or less is considered dangerous, whereas a material having a flash point greater than 200 °F is considered to have a low flammability. Substances with flash points within this temperature range are considered to have moderate flammabilities.

Lower explosive limit: The minimum concentration (vol % in air) of a flammable gas or vapor required for ignition or explosion to occur in the presence of an ignition source (see also **Flash point**).

Upper explosive limit: The maximum concentration (vol % in air) of a flammable gas or vapor required for ignition or explosion to occur in the presence of an ignition source (see also **Flash point**).

HEALTH HAZARD DATA

Immediately Dangerous to Life or Health (IDLH): According to the National Institute of Occupational Safety and Health (1994), the IDLH level ". . . for the purpose of respirator selection represents a maximum concentration from which, in the event of respirator failure, one could escape within 30 minutes without experiencing any escape-impairing or irreversible health effects." Concentrations are typically reported in parts per million (ppm) or milligrams per cubic meter (mg/m^3).

Exposure Limits: The permissible exposure limits (PELs) in air, set by the Occupational Health and Safety Administration (OSHA), can be found in the Code of Federal Regulations (General Industry Standards for Toxic and Hazardous Substances, 1977). Unless noted otherwise, the PELs are 8-hour time-weighted average (TWA) concentrations.

If NIOSH (1994) recommended exposure limits (RELs) and/or the American Conference of Governmental Industrial Hygienists (ACGIH) threshold limit values (TLVs) has published recommended exposure limits, these are also included. Unless noted otherwise, the NIOSH RELs are TWA concentrations for up to a 10-hour workday during a 40-hour workweek. The short-term exposure limit (STEL) is a 15-minute TWA that should be exceeded at any time during the workday. Recommended ceiling values are concentrations that should never be exceeded at any time during the day.

The ACGIH's TLVs, are subdivided into three exposure classes (Threshold Limit Values and Biological Exposure Indices for 1987-1988). The TLVs, which are updated annually, are defined as follows:

Threshold Limit Value-Time Weighted Average (TLV-TWA) - the TWA concentration for a normal 8-hour workday and a 40-hour workweek, to which nearly all workers may be repeatedly exposed, day after day, without adverse effect.

Threshold Limit Value-Short Term Exposure Limit (TLV-STEL) - the concentration to which workers can be exposed continuously for a short period of time without suffering from 1) irritation, 2) chronic or irreversible tissue damage, or 3) narcosis of sufficient degree to increase the likelihood of accidental injury, impair self-rescue or materially reduce work efficiency and provided that the daily TLV-TWA is not exceeded. It is not a separate independent exposure limit; rather, it supplements the TWA limit where there are recognized acute toxic effects from a substance whose toxic effects are primarily of a chronic nature. STELs are recommended only where toxic effects have been reported from high short-term exposures in either humans or animals.

A STEL is defined as a 15-minute time-weighted average exposure which should not be exceeded at any time during a workday even if the 8-hour TWA is within the TLV. Exposures at the STEL should not be longer than 15 minutes and should not be repeated more than four times per day. There should be at least 60 minutes between successive exposures at the STEL. An averaging period other than 15 minutes may be recommended when this is warranted by observed biological effects.

Threshold Limit Value-Ceiling - the concentration that should not be exceeded during any part of the working exposure.

For additional information from OSHA, write to Technical Data Center, U.S. Department of Labor, Washington, DC 20210. The NIOSH Pocket Guide to Chemical Hazards is available in electronic formats (e.g., CD-ROM and diskettes) and is available from Industrial Hygiene Services, Inc., 941 Gardenview Office Parkway, St. Louis, MO 63141 (toll free: 1-800-732-3015; fax: 314-993-3193), Micromedex, Inc., 6200 South Syracuse Way, Suite 300, Englewood, CO (toll free: 1-800-525-9083; fax: 303-486-6464) or Praxis Environmental Systems, Inc., 251 Nortontown Road, Guilford, CT (telephone 203-458-7111; fax: 203-458-7121). The ACGIH's address is 6500 Glenway Ave., Building D-7, Cincinnati, OH 45211-7881.

Symptoms of Exposure: Effects of exposure caused by inhalation of gases, ingestion of liquids or solids, contact with eyes or skin are provided. This information should only be used as a guide to identify potential effects of exposure. If an exposure to a chemical is suspected or known, seek immediate medical attention. Additional information on the symptoms and effects of chemical exposure can be obtained from Patnaik (1992), Sax and Lewis (1987) and CHRIS (1984).

Toxicology: Information on toxicity to aquatic life was obtained primarily from the Royal Society of Chemistry (Hartley and Kidd, 1987) and the Chemical Hazard Response Information System (CHRIS) Manual (1984). Information on toxicity to rats and/or mice were obtained from Ashton and Monaco (1991), Hartley and Kidd (1987) and RTECS (1985). The absence of toxicity data does not imply that toxic effects do not exist.

Drinking Water Standard: Drinking water standards established by the U.S. EPA are given in as maximum contaminant level goals (MCLGs) and/or maximum contaminant levels (MCLs). The MCLG is a non-enforceable concentration of a drinking water contaminant that is protective of adverse human health effects and allows an adequate margin of safety. The MCL is the maximum permissible concentration of a drinking water contaminant that is delivered to any user of a public water supply system. The reader should be aware that many states may have adopted more stringent drinking water standards than those promulgated and/or regulated by the U.S. EPA. For additional information on drinking water regulations, contact the U.S. EPA, Office of Water, Washington, DC. Their phone number is (202) 260-7571. The Safe Drinking Water Hotline is 1-800-426-4791 and is open Monday through Friday, 8:30 am to 5:00 pm Eastern Standard Time.

Uses: Descriptions of specific uses are based on one or more of the following sources - HSDB (1989), CHRIS Manual (1984), Sittig (1985) and Verschueren

(1983). This information is useful in attempting to identify potential sources of the industrial and environmental contamination.

REFERENCES

Abdul, S.A., T.L. Gibson, and D.N. Rai. "Statistical Correlations for Predicting the Partition Coefficient for Nonpolar Organic Contaminants between Aquifer Organic Carbon and Water," *Haz. Waste Haz. Mater.*, 4(3):211-222 (1987).

Allison, L.E. "Organic Carbon" in *Methods of Soil Analysis, Part 2.*, Black, C., Evans, D., White, J., Ensminger, L., and F. Clark, eds. (Madison, WI: American Society of Agronomy, 1965), pp. 1367-1378.

Almgren, M., F. Grieser, J.R. Powell, and J.K. Thomas. "A Correlation between the Solubility of Aromatic Hydrocarbons in Water and Micellar Solutions, with Their Normal Boiling Points," *J. Chem. Eng. Data*, 24(4):285-287 (1979).

Amidon, G.L., S.H. Yalkowsky, S.T. Anik, and S.C. Valvani. "Solubility of Nonelectrolytes in Polar Solvents. V. Estimation of the Solubility of Aliphatic Monofunctional Compounds in Water Using a Molecular Surface Area Approach," *J. Phys. Chem.*, 79(21):2239-2246 (1975).

Ashton, F.M. and T.J. Monaco. *Weed Science* (New York: John Wiley & Sons, Inc., 1991), 466 p.

Banerjee, S., S.H. Yalkowsky, and S.C. Valvani. "Water Solubility and Octanol/Water Partition Coefficients of Organics. Limitations of the Solubility-Partition Coefficient Correlation," *Environ. Sci. Technol.*, 14(10):1227-1229 (1980).

Bell, G.H. "Solubilities of Normal Aliphatic Acids, Alcohols and Alkanes in Water," *Chem. Phys. Lipids*, 10:1-10 (1973).

Bodor, N., Z. Gabanyi, and C.-K. Wong. "A New Method for the Estimation of Partition Coefficient," *J. Am. Chem. Soc.*, 111(11):3783-3786 (1989).

Braumann, T. "Determination of Hydrophobic Parameters By Reversed-Phase Liquid Chromatography: Theory, Experimental Techniques, and Application in Studies on Quantitative Structure-Activity Relationships," *J. Chromatogr.*, 373:191-225 (1986).

Brooke, D.N., A.J. Dobbs, and N. Williams. "Octanol:Water Partition Coefficients (P): Measurement, Estimation, and Interpretation, Particularly for Chemicals with P >10^5," *Ecotoxicol. Environ. Saf.*, 11(3):251-260 (1986).

Bruggeman, W.A., J. Van Der Steen, and O. Hutzinger. "Reversed-Phase Thin-Layer Chromatography of Polynuclear Aromatic Hydrocarbons and Chlorinated Biphenyls. Relationship with Hydrophobicity as Measured by Aqueous Solubility and Octanol-Water Partition Coefficient," *J. Chromatogr.*, 238:335-346 (1982).

Burkhard, L.P. and D.W. Kuehl. "*n*-Octanol/Water Partition Coefficients by Reverse Phase Liquid Chromatography/Mass Spectrometry for Eight

Tetrachlorinated Planar Molecules," *Chemosphere*, 15(2):163-167 (1986).

Camilleri, P., S.A. Watts, and J.A. Boraston. "A Surface Area Approach to Determination of Partition Coefficients," *J. Chem. Soc., Perkin Trans. 2*, (September 1988), pp 1699-1707.

Campbell, J.R., R.G. Luthy, and M.J.T. Carrondo. "Measurement and Prediction of Distribution Coefficients for Wastewater Aromatic Solutes," *Environ. Sci. Technol.*, 17(10):582-590 (1983).

Carlson, R.M., R.E. Carlson, and H.L. Kopperman. "Determination of Partition Coefficients by Liquid Chromatography," *J. Chromatogr.*, 107:219-223 (1975).

Chiou, C.T., V.H. Freed, D.W. Schmedding, and R.L. Kohnert. "Partition Coefficients and Bioaccumulation of Selected Organic Chemicals," *Environ. Sci. Technol.*, 11(5):475-478 (1977).

Chiou, C.T., L.J. Peters, and V.H. Freed. "A Physical Concept of Soil-Water Equilibria for Nonionic Organic Compounds," *Science (Washington, DC)*, 206(4420):831-832 (1979).

Chiou, C.T., P.E. Porter, and D.W. Schmedding. "Partition Equilibria of Nonionic Organic Compounds between Organic Matter and Water," *Environ. Sci. Technol.*, 17(4):227-231 (1983).

Chiou, C.T., D.W. Schmedding, and M. Manes. "Partitioning of Organic Compounds in Octanol-Water Systems," *Environ. Sci. Technol.*, 16(1):4-10 (1982).

"CHRIS Hazardous Chemical Data, Vol. 2," U.S. Department of Transportation, U.S. Coast Guard, U.S. Government Printing Office (November, 1984).

Davies, R.P. and A.J. Dobbs. "The Prediction of Bioconcentration in Fish," *Water Res.*, 18(10):1253-1262 (1984).

DeKock, A.C. and D.A. Lord. "A Simple Procedure for Determining Octanol-Water Partition Coefficients using Reverse Phase High Performance Liquid Chromatography (RPHPLC)," *Chemosphere*, 16(1):133-142 (1987).

Demond, A.H. and A.S. Lindner. "Estimation of Interfacial Tension between Organic Liquids and Water," *Environ. Sci. Technol.*, 27(12):2318-2331 (1993).

de Wolf, W., J.H.M. de Bruijn, W. Sienen, and J.L.M. Hermens. "Influence of Biotransformation on the Relationship between Bioconcentration Factors and Octanol-Water Partition Coefficients," *Environ. Sci. Technol.*, 26(6):1197-1201 (1992).

Eadsforth, C.V. "Application of Reverse-Phase H.P.L.C. for the Determination of Partition Coefficients," *Pestic. Sci.*, 17(3):311-325 (1986).

Freeze, R.A. and J.A. Cherry. *Groundwater* (Englewood Cliffs, NJ: Prentice-Hall, Inc., 1974), 604 p.

Funasaki, N., S. Hada, and S. Neya. "Partition Coefficients of Aliphatic Ethers - Molecular Surface Area Approach," *J. Phys. Chem.*, 89(14):3046-3049 (1985).

Gerstl, Z. and C.S. Helling. "Evaluation of Molecular Connectivity as a Predictive Method for the Adsorption of Pesticides in Soils," *J. Environ. Sci. Health*, B22(1):55-69 (1987).

Gobas, F.A.P.C., J.M. Lahittete, G. Garofalo, W.Y. Shiu, and D. Mackay. "A

Novel Method for Measuring Membrane-Water Partition Coefficients of Hydrophobic Organic Chemicals: Comparison with 1-Octanol-Water Partitioning," *J. Pharm. Sci.*, 77(3):265-272 (1988).

Govers, H., C. Ruepert, and H. Aiking. "Quantitative Structure-Activity Relationships for Polycyclic Aromatic Hydrocarbons: Correlation between Molecular Connectivity, Physico-Chemical Properties, Bioconcentration and Toxicity in *Daphnia Pulex*," *Chemosphere*, 13(2):227-236 (1984).

"Guidelines Establishing Test Procedures for the Analysis of Pollutants," U.S. Code of Federal Regulations, 40 CFR 136, 44(233):69464-69575.

Guswa, J.H., W.J. Lyman, A.S. Donigan, Jr., T.Y.R. Lo, and E.W. Shanahan. *Groundwater Contamination and Emergency Response Guide* (Park Ridge, NJ: Noyes Publications, 1984), 490 p.

Hafkenscheid, T.L. and E. Tomlinson. "Estimation of Aqueous Solubilities of Organic Non-Electrolytes Using Liquid Chromatographic Retention Data," *J. Chromatogr.*, 218:409-425 (1981).

Hafkenscheid, T.L. and E. Tomlinson. "Correlations between Alkane/Water and Octan-1-ol/Water Distribution Coefficients and Isocratic Reversed-Phase Liquid Chromatographic Capacity Factors of Acids, Bases and Neutrals," *Int. J. Pharm.*, 16:225-239 (1983).

Haky, J.E. and A.M. Young. "Evaluation of a Simple HPLC Correlation Method for the Estimation of the Octanol-Water Partition Coefficients of Organic Compounds," *J. Liq. Chromatogr.*, 7(4):675-689 (1984).

Hammers, W.E., G.J. Meurs, and C.L. De Ligny. "Correlations between Chromatographic Capacity Ratio Data on Lichrosorb RP-18 and Partition Coefficients in the Octanol-Water System," *J. Chromatogr.*, 247:1-13 (1982).

Hansch, C. and S.M. Anderson. "The Effect of Intramolecular Hydrophobic Bonding on Partition Coefficients," *J. Org. Chem.*, 32:2853-2586 (1967).

Hansch, C., A. Leo, and D. Nikaitani. "On the Additive-Constitutive Character of Partition Coefficients," *J. Org. Chem.*, 37(20):3090-3092 (1972).

Hansch, C., J.E. Quinlan, and G.L. Lawrence. "The Linear Free-Energy Relationship between Partition Coefficients and Aqueous Solubility of Organic Liquids," *J. Org. Chem.*, 33(1):347-350 (1968).

Harnisch, M., H.J. Mockel, and G. Schulze. "Relationship between Log P_{ow} Shake-Flask Values and Capacity Factors Derived from Reversed-Phase High Performance Liquid Chromatography for *n*-Alkylbenzene and Some OECD Reference Substances," *J. Chromatogr.*, 282:315-332 (1983).

Hartley, D. and H. Kidd, Eds. *The Agrochemicals Handbook*, 2nd ed. (England: Royal Society of Chemistry, 1987).

Hawley, G.G. *The Condensed Chemical Dictionary* (New York: Van Nostrand Reinhold Co., 1981), 1135 p.

Hayduk, W. and H. Laudie. Prediction of Difusion Coefficients for Nonelectrolytes in Dilute Aqueous Solution," *Am. Inst. Chem. Eng.*, 20(3):611-615 (1974).

Hazardous Substances Data Bank. National Library of Medicine, Toxicology

Information Program (1989).

Hermann, R.B. "Theory of Hydrophobic Bonding. II. The Correlation of Hydrocarbon Solubility in Water with Solvent Cavity Surface Area," *J. Phys. Chem.*, 76(19):2754-2759 (1972).

Hickey, J.P. and D.R. Passino-Reader. "Linear Solvation Energy Relationships: "Rules of Thumb" for Estimation of Variable Values," *Environ. Sci. Technol.*, 25(10):1753-1760 (1991).

Hill, E.A. "On a System of Indexing Chemical Literature; Adopted by the Classification Division of the U.S. Patent Office," *J. Am. Chem. Soc.*, 22(8):478-494 (1900).

Hine, J. and P.K. Mookerjee. "The Intrinsic Hydrophobic Character of Organic Compounds. Correlations in Terms of Structural Contributions," *J. Org. Chem.*, 40(3):292-298 (1975).

Hodson, J. and N.A. Williams. "The Estimation of the Adsorption Coefficient (K_{oc}) for Soils by High Performance Liquid Chromatography," *Chemosphere*, 19(1):67-77 (1988).

Howard, P.H. *Handbook of Environmental Fate and Exposure Data for Organic Chemicals - Volume I. Large Production and Priority Pollutants* (Chelsea, MI: Lewis Publishers, Inc., 1989), 574 p.

Irmann, F. "Eine einfache Korrelation zwischen Wasserloslichkeit und Struktur von Kohlenwasserstoffen und Halogenkohlenwasserstoffen," *Chemie. Ing. Techn.*, 37:789-798 (1965).

Isnard, S. and S. Lambert. "Estimating Bioconcentration Factors from Octanol-Water Partition Coefficient and Aqueous Solubility," *Chemosphere*, 17(1):21-34 (1988).

Kamlet, M.J., R.M. Doherty, M.H. Abraham, P.W. Carr, R.F. Doherty, and R.W. Taft. "Linear Solvation Energy Relationships. 41. Important Differences between Aqueous Solubility Relationships for Aliphatic and Aromatic Solutes," *J. Phys. Chem.*, 91(7):1996-2004 (1987).

Kamlet, M.J., R.M. Doherty, P.W. Carr, D. Mackay, M.H. Abraham, and R.W. Taft. "Linear Solvation Energy Relationships. 44. Parameter Estimation Rules That Allow Accurate Prediction of Octanol/Water Partition Coefficients and Other Solubility and Toxicity Properties of Polychlorinated Biphenyls and Polycyclic Aromatic Hydrocarbons," *Environ. Sci. Technol.*, 22(5):503-509 (1988).

Kenaga, E.E. "Correlation of Bioconcentration Factors of Chemicals in Aquatic and Terrestrial Organisms with Their Physical and Chemical Properties," *Environ. Sci. Technol.*, 14(5):553-556 (1980).

Kenaga, E.E. and C.A.I. Goring. "Relationship between Water Solubility, Soil Sorption, Octanol-Water Partitioning and Concentration of Chemicals in Biota," in *Aquatic Toxicology, ASTM STP 707*, Eaton, J.G., P.R. Parrish, and A.C. Hendricks, Eds. (Philadelphia, PA: American Society for Testing and Materials, 1980), pp. 78-115.

Khaledi, M.G. and E.D. Breyer. "Quantitation of Hydrophobicity with Micellar Liquid Chromatography," *Anal. Chem.*, 61(9):1040-1047 (1989).

Koch, R. "Molecular Connectivity Index for Assessing Ecotoxicological Behaviour of Organic Compounds," *Toxicol. Environ. Chem.*, 6(2):87-96 (1983).

Kraak, J.C., H.H. Van Rooij, and J.L.G. Thus. "Reversed-Phase Ion-Pair Systems for the Prediction of *n*-Octanol-Water Partition Coefficients of Basic Compounds by High-Performance Liquid Chromatography," *J. Chromatogr.*, 352:455-463 (1986).

Krzyzanowska, T. and J. Szeliga. "A Method for Determining the Solubility of Individual Hydrocarbons," *Nafta*, 28:414-417 (1978).

Lande, S.S. and S. Banerjee. "Predicting Aqueous Solubility of Organic Nonelectrolytes from Molar Volume," *Chemosphere*, 10(7):751-759 (1981).

Lande, S.S., D.F. Hagen, and A.E. Seaver. "Computation of Total Molecular Surface Area from Gas Phase Ion Mobility Data and its Correlation with Aqueous Solubilities of Hydrocarbons," *Environ. Toxicol. Chem.*, 4(3):325-334 (1985).

Lyman, W.J., W.F. Reehl, and D.H. Rosenblatt. *Handbook of Chemical Property Estimation Methods: Environmental Behavior of Organic Compounds* (New York: McGraw-Hill, Inc., 1982).

Means, J.C., S.G. Wood, J.J. Hassett, and W.L. Banwart. "Sorption of Polynuclear Aromatic Hydrocarbons by Sediments and Soils," *Environ. Sci. Technol.*, 14(2):1524-1528 (1980).

Meylan, W. and P.H. Howard. "Bond Contribution Method for Estimating Henry's Law Constants," *Environ. Toxicol. Chem.*, 10(10):1283-1293 (1991).

Meylan, W., P.H. Howard, and R.S. Boethling. "Molecular Topology/Fragment Contribution Method for Predicting Soil Sorption Coefficients," *Environ. Sci. Technol.*, 26(8):1560-1567 (1992).

Miller, M.M., S. Ghodbane, S.P. Wasik, Y.B. Tewari, and D.E. Martire. "Aqueous Solubilities, Octanol/Water Partition Coefficients, and Entropies of Melting of Chlorinated Benzenes and Biphenyls," *J. Chem. Eng. Data*, 29(2):184-190 (1984).

Miller, M.M., S.P. Wasik, G.-L. Huang, W.-Y. Shiu, and D. Mackay. "Relationships between Octanol-Water Partition Coefficient and Aqueous Solubility," *Environ. Sci. Technol.*, 19(6):522-529 (1985).

Minick, D.J., D.A. Brent, and J. Frenz. "Modeling Octanol-Water Partition Coefficients by Reversed-Phase Liquid Chromatography," *J. Chromatogr.*, 461:177-191 (1989).

Mirrlees, M.S., S.J. Moulton, C.T. Murphy, and P.J. Taylor. "Direct Measurement of Octanol-Water Partition Coefficient by High-Pressure Liquid Chromatography," *J. Med. Chem.*, 19(5):615-619 (1976).

Mitra, A., R.K. Saksena, and C.R. Mitra. "A Prediction Plot for Unknown Water Solubilities of Some Hydrocarbons and Their Mixtures," *Chem. Petro-Chem. J.*, 8:16-17 (1977).

Miyake, K., F. Kitaura, N. Mizuno, and H. Terada. "Phosphatidylcholine-Coated

Silica as a Useful Stationary Phase for High-Performance Liquid Chromatographic Determination of Partition Coefficients between Octanol and Water," *J. Chromatogr.*, 389(1):47-56 (1987).

Miyake, K., N. Mizuno, and H. Terada. "Effect of Hydrogen Bonding on the High-Performance Liquid Chromatographic Behaviour of Organic Compounds. Relationship between Capacity Factors and Partition Coefficients," *J. Chromatogr.*, 439:227-235 (1988).

Miyake, K. and H. Terada. "Determination of Partition Coefficients of Very Hydrophobic Compounds by High-Performance Liquid Chromatography on Glyceryl-Coated Controlled-Pore Glass," *J. Chromatogr.*, 240(1):9-20 (1982).

Neely, W.B., D.R. Branson, and G.E. Blau. "Partition Coefficient to Measure Bioconcentration Potential of Organic Chemicals in Fish," *Environ. Sci. Technol.*, 8(13):1113-1115 (1974).

NIOSH. 1994. "NIOSH Pocket Guide to Chemical Hazards," U.S. Department of Health and Human Services, U.S. Government Printing Office, 398 p.

Nirmalakhandan, N.N. and R.E. Speece. "QSAR Model for Predicting Henry's Constant," *Environ. Sci. Technol.*, 22(11):1349-1357 (1988).

Nirmalakhandan, N.N. and R.E. Speece. "Prediction of Aqueous Solubility of Organic Compounds Based on Molecular Structure," *Environ. Sci. Technol.*, 22(3):328-338 (1988a).

Nirmalakhandan, N.N. and R.E. Speece. "Prediction of Aqueous Solubility of Organic Compounds Based on Molecular Structure. 2. Application to PNAs, PCBs, PCDDs, etc.," *Environ. Sci. Technol.*, 23(6):708-713 (1989).

Nyssen, G.A., E.T. Miller, T.F. Glass, C.R. Quinn II, J. Underwood, J., and D.J. Wilson. "Solubilities of Hydrophobic Compounds in Aqueous-Organic Solvent Mixtures," *Environ. Monit. Assess.*, 9(1):1-11 (1987).

Ogata, M., K. Fujisawa, Y. Ogino, and E. Mano. "Partition Coefficients as a Measure of Bioconcentration Potential of Crude Oil in Fish and Sunfish," *Bull. Environ. Contam. Toxicol.*, 33(5):561-567 (1984).

Oliver, B.G. and A.J. Niimi. "Bioconcentration Factors of Some Halogenated Organics for Rainbow Trout: Limitations in Their Use for Prediction of Environmental Residues," *Environ. Sci. Technol.*, 19(9):842-849 (1985).

Patil, G.S. "Correlation of Aqueous Solubility and Octanol-Water PArtition Coefficient Based on Molecular Structure," *Chemosphere*, 22(8):723-738 (1991).

Patnaik, P. *A Comprehensive Guide to the Hazardous Properties of Chemical Substances* (New York: Van Nostrand Reinhold, 1992), 763 p.

RTECS. 1985. "Registry of Toxic Effects of Chemical Substances," U.S. Department of Health and Human Services, National Institute for Occupational Safety and Health, 2050 p.

Robb, I.D. "Determination of the Aqueous Solubility of Fatty Acids and Alcohols," *Aust. J. Chem.*, 18:2281-2285 (1966).

Russell, C.J., S.L. Dixon, and P.C. Jurs. "Computer-Assisted Study of the Relationship between Molecular Structure and Henry's Law Constant," *Anal.*

Chem., 64(13):1350-1355 (1992).

Sabljić, A. "On the Prediction of Soil Sorption Coefficients of Organic Pollutants from Molecular Structure: Application of Molecular Topology Model," *Environ. Sci. Technol.*, 21(4):358-366 (1987).

Sabljić, A. "Predictions of the Nature and Strength of Soil Sorption of Organic Pollutants by Molecular Topology," *J. Agric. Food Chem.*, 32(2):243-246 (1984).

Sabljić, A. and M. Protić. "Relationship between Molecular Connectivity Indices and Soil Sorption Coefficients of Polycyclic Aromatic Hydrocarbons," *Bull. Environ. Contam. Toxicol.*, 28(2):162-165 (1982).

Sadek, P.C., P.W. Carr, R.M. Doherty, M.J. Kamlet, R.W. Taft, and M.H. Abraham. "Study of Retention Processes in Reversed-Phase High-Performance Liquid Chromatography by the Use of the Solvatochromic Comparison Method," *Anal. Chem.*, 57(14):2971-2978 (1985).

Sarna, L.P., P.E. Hodge, and G.R.B. Webster. "Octanol-Water Partition Coefficients of Chlorinated Dioxins and Dibenzofurans by Reversed-Phase HPLC Using Several C_{18} Columns," *Chemosphere*, 13(9):975-983 (1984).

Sax, N.I. *Dangerous Properties of Industrial Materials* (New York: Van Nostrand Reinhold Co., 1984), 3124 p.

Sax, N.I. and R.J. Lewis, Sr. *Hazardous Chemicals Desk Reference* (New York: Van Nostrand Reinhold Co., 1987), 1084 p.

Schowalter, T.T. "Mechanics of Secondary Hydrocarbon Migration and Entrapment," *Assoc. Pet. Geol. Bull.*, 63(5):723-760 (1979).

Schultz, T.W., S.K. Wesley, and L.L. Baker. "Structure-Activity Relationships for Di and Tri Alkyl and/or Halogen Substituted Phenols," *Bull. Environ. Contam. Toxicol.*, 43(2):192-198 (1989).

Shafer, D. *Hazardous Materials Training Handbook* (Madison, CT: Bureau of Law and Business, Inc., 1987), 206 p.

Sittig, M. *Handbook of Toxic and Hazardous Chemicals and Carcinogens* (Park Ridge, NJ: Noyes Publications, 1985), 950 p.

Suzuki, T. "Development of an Automatic Estimation System for both the Partition Coefficient and Aqueous Solubility," *J. Comput.-Aided Mol. Des.*, 5:149-166 (1991).

Szabo, G., S.L. Prosser, and R.A. Bulman. "Adsorption Coefficient (K_{oc}) and HPLC Retention Factors of Aromatic Hydrocarbons," *Chemosphere*, 21(4/5):495-505 (1990).

Szabo, G., S.L. Prosser, and R.A. Bulman. "Determination of the Adsorption Coefficient (K_{oc}) of Some Aromatics for Soil by RP-HPLC on Two Immobilized Humic Acid Phases," *Chemosphere*, 21(6):777-788 (1990a).

Szabo, G., S.L. Prosser, and R.A. Bulman. "Prediction of the Adsorption Coefficient (K_{oc}) for Soil by a Chemically Immobilized Humic Acid Column using RP-HPLC," *Chemosphere*, 21(6):729-739 (1990).

ten Hulscher, Th.E.M., L.E. van der Velde, and W.A. Bruggeman. "Temperature Dependence of Henry's Law Constants for Selected Chlorobenzenes,

Polychlorinated Biphenyls and Polycyclic Aromatic Hydrocarbons," *Environ. Toxicol. Chem.*, 11(11):1595-1603 (1992).

Tewari, Y.B., M.M. Miller, S.P. Wasik, and D.E. Martire. "Aqueous Solubility of Octanol/Water Partition Coefficient of Organic Compounds at 25.0 °C," *J. Chem. Eng. Data*, 27(4):451-454 (1982).

Threshold Limit Values and Biological Exposure Indices for 1987-1988 (Cincinnati, OH: American Conference of Governmental Industrial Hygienists, 1987), 114 p.

Toxic and Hazardous Industrial Chemicals Safety Manual for Handling and Disposal with Toxicity and Hazard Data (Tokyo, Japan: International Technical Information Institute, 1986), 700 p.

U.S. EPA. "A Compendium of Technologies Used in the Treatment of Hazardous Wastes," Office of Research and Development, U.S. EPA Report-625/8-87-014 (1987), 49 p.

Valkó, K., O. Papp, and F. Darvas. "Selection of Gas Chromatographic Stationary Phase Pairs for Characterization of the 1-Octanol-Water Partition Coefficient," *J. Chromatogr.*, 301:355-364 (1984).

Valvani, S.C., S.H. Yalkowsky, and G.L. Amidon. "Solubility of Nonelectrolytes in Polar Solvents. VI. Refinements in Molecular Surface Area Computations," *J. Phys. Chem.*, 80(8):829-835 (1976).

Veith, G.D. and R.T. Morris. "A Rapid Method for Estimating Log P for Organic Chemicals," U.S. EPA Report-600/3-78-049 (1978), 15 p.

Verschueren, K. *Handbook of Environmental Data on Organic Chemicals* (New York: Van Nostrand Reinhold Co., 1983), 1310 p.

Warne, M. St.J., D.W. Connell, D.W. Hawker, and G. Schüürmann. "Prediction of Aqueous Solubility and the Octanol-Water Partition Coefficient for Lipophilic Organic Compounds Using Molecular Descriptors and Physicochemical Properties," *Chemosphere*, 21(7):877-888 (1990).

Weast, R.C., Ed. *CRC Handbook of Chemistry and Physics*, 67th ed. (Boca Raton, FL: CRC Press, Inc., 1986), 2406 p.

Webster, G.R.B., K.J. Friesen, L.P. Sarna, and D.C.G. Muir. "Environmental Fate Modeling of Chlorodioxins: Determination of Physical Constants," *Chemosphere*, 14(6/7):609-622 (1985).

Whitehouse, B.G. and R.C. Cooke. "Estimating the Aqueous Solubility of Aromatic Hydrocarbons by High Performance Liquid Chromatography," *Chemosphere*, 11(8):689-699 (1982).

Windholz, M., S. Budavari, R.F. Blumetti, and E.S. Otterbein, Eds., *The Merck Index*, 10th ed. (Rahway, NJ: Merck and Co., 1983), 1463 p.

Woodburn, K.B., J.J. Delfino, and Rao, P.S.C. "Retention of Hydrophobic Solutes on Reversed-Phase Liquid Chromatography Supports: Correlation with Solute Topology and Hydrophobicity Indices," *Chemosphere*, 24(8):1037-1046 (1992).

Worthing, C.R. and S.B. Walker, Eds. *The Pesticide Manual - A World Compendium*, 9th ed. (Great Britain: British Crop Protection Council, 1991),

1141 p.

Wright, D.A., S.I. Sandler, and D. DeVoll. "Infinite Dilution Activity Coefficients and Solubilities of Halogenated Hydrocarbons in Water at Ambient Temperature," *Environ. Sci. Technol.*, 26(9):1828–1831 (1992).

Yalkowsky, S.H. and S. Banerjee. *Aqueous Solubility-Methods of Estimation for Organic Compounds*, (New York: Marcel Dekker, Inc., 1992), 264 p.

Yalkowsky, S.H. and S.C. Valvani. "Solubilities and Partitioning 2. Relationships between Aqueous Solubilities, Partition Coefficients, and Molecular Surface Areas of Rigid Aromatic Hydrocarbons," *J. Chem. Eng. Data*, 24(2):127–129 (1979).

Yalkowsky, S.H. and S.C. Valvani. "Solubility and Partitioning I: Solubility of Nonelectrolytes in Water," *J. Pharm. Sci.*, 69(8):912–922 (1980).

Yoshida, K., S. Tadayoshi, and F. Yamauchi. "Relationship between Molar Refraction and *n*-Octanol/Water Partition Coefficient," *Ecotoxicol. Environ. Saf.*, 7(6):558–565 (1983).

Abbreviations and Symbols

A	Amp
Å	angstrom
α	alpha
α_a	percent of acid that is nondissociated
α_b	percent of base that is nondissociated
$\approx$	approximately equal to
ACGIH	American Conference of Governmental Industrial Hygienists
ASTM	American Society for Testing and Materials
asym-	asymmetric
atm	atmosphere
b	aperture of fracture
B	average soil bulk density (g/cm^3)
β	beta
BCF	bioconcentration factor
C	ceiling
°C	degrees Centigrade (Celsius)
cal	calorie
CAS	Chemical Abstracts Service
CEC	cation exchange capacity (meq/L unless noted otherwise)
CERCLA	Comprehensive Environmental Response, Compensation and Liability Act
CHRIS	Chemical Hazard Response Information System
cm	centimeter
C_L	concentration of solute in solution
CO_2	carbon dioxide
cps	centipoise
C_S	concentration of sorbed solute
D	diffusivity
DOT	Department of Transportation (U.S.)
d_s	density of a substance
d_w	density of water
dyn	dyne
δ	delta
EC_{50}	concentration necessary for 50% of the aquatic species tested showing abnormal behavior
et al.	and others
eV	electron volts
°F	degrees Fahrenheit
f_{oc}	fraction of organic carbon
fw	formula weight
γ	gamma
g	gram

gal	gallon
GC/MS	gas chromatography/mass spectrometry
>	greater than
≥	greater than or equal to
HPLC	high performance liquid chromatography
HSDB	Hazardous Substances Data Bank
Hz	Hertz
IDLH	immediately dangerous to life or health
IP	ionization potential
K	Kelvin (°C + 273.15)
K_a	acid dissociation constant
K_A	distribution coefficient (cm)
K_b	base dissociation constant
kcal	kilocalories
K_d	distribution coefficient (cm^3/g)
kg	kilogram
K_H	Henry's law constant ($atm \cdot m^3/mol \cdot K$)
$K_{H'}$	Henry's law constant (dimensionless)
K_{oc}	soil/sediment partition coefficient (organic carbon basis)
K_{om}	soil/sediment partition coefficient (organic matter basis)
K_{ow}	n-octanol/water partition coefficient
kPa	kilopascal
K_w	dissociation constant for water (10^{-14} at 25 °C)
<	less than
≤	less than or equal to
L	liter
lb	pound
LC_{50}	lethal concentration necessary to kill 50% of the aquatic species tested
LC_{100}	lethal concentration necessary to kill 100% of the aquatic species tested
LD_{50}	lethal dose necessary to kill 50% of the mammals tested
m	meter
m-	meta (as in m-dichlorobenzene)
M	molarity (moles/liter)
M	mass
MCL	maximum contaminant level
MCLG	maximum contaminant level goal
meq	milliequivalents
mg	milligram
min	minute(s)
mL	milliliter
M_L	mass of sorbed solute

mmHg	millimeters of mercury
mmol	millimole
mol	mole
M_S	mass of solute in solution
mV	millivolt
N	normality (equivalents/liter)
n-, N-	normal (as in n-propyl, N-nitroso)
n_e	effective porosity
ng	nanogram
NIOSH	National Institute for Occupational Safety and Health
nm	nanometer
o-	ortho (as in o-dichlorobenzene)
OSHA	Occupational Safety and Health Administration
ρ	specific density (unitless)
p-	para (as in p-dichlorobenzene)
P	pressure
Pa	pascal
p_{air}	vapor density of air
PEL	permissible exposure limit
pH	$-\log_{10}$ hydrogen ion activity (concentration)
pK_a	$-\log_{10}$ dissociation constant of an acid
pK_b	$-\log_{10}$ dissociation constant of a base
pK_w	$-\log_{10}$ dissociation constant of water
ppb	parts per billion (μg/L)
pph	parts per hundred
ppm	parts per million (mg/L)
p_v	specific vapor density
QSAR	quantitative structure-activity relationships
R	ideal gas constant (8.20575×10^{-5} atm·m^3/mol)
R_a	retardation factor for an acid
R_b	retardation factor for a base
RCRA	Resource Conservation and Recovery Act
R_d	retardation factor
REL	recommended exposure limit
rpm	revolutions per minute
RTECS	Registry of Toxic Effects of Chemical Substances
S	solubility
S_a	solute concentration in air (mol/L)
SARA	Superfund Amendments and Reauthorization Act
sec-	secondary (as in sec-butyl)
s_{OW}	spreading coefficient of organic liquid at air-water table interface (dyn/cm)
sp.	species

spp.	species (plural)
STEL	short-term exposure limit
sym	symmetric
t-	tertiary (as in *t*-butyl; but *tert*-butyl)
TLV	threshold limit value
TOC	total organic carbon (mg/L)
T_s	temperature of a substance
T_w	temperature of water
TWA	time-weighted average
μ	micro (10^{-6})
μ	viscosity
μg	microgram
unsym	unsymmetric
U.S. EPA	U.S. Environmental Protection Agency
UV	ultraviolet
V, vol	volume
V	molar volume
V_c	average linear velocity of contaminant (e.g., ft/day)
V_w	average linear velocity of groundwater (e.g., ft/day)
W	watt
λ	wavelength
γ_{OW}	interfacial tension between organic liquid and water (dyn/cm)
$\gamma_{O(W)}$	surface tension of organic liquid saturated with water (dyn/cm)
$\gamma_{W(O)}$	surface tension of water saturated with organic liquid (dyn/cm)
wt	weight
z^+	positively charged species (milliequivalents/cm^3)

Contents

xl   **CONTENTS**

ACENAPHTHENE

Synonyms: 1,2-Dihydroacenaphthylene; Ethylenenaphthalene; 1,8-Ethylenenaph-thalene; 1,8-Hydroacenaphthylene; Periethylenenaphthalene.

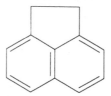

CAS Registry No.: 83-32-9
Molecular formula: $C_{12}H_{10}$
Formula weight: 154.21
RTECS: AB1000000

Physical state, color and odor
White crystalline solid. Odor threshold in air ranges from 0.02 to 0.22 ppm (Keith and Walters, 1992).

Melting point (°C):
96.2 (Weast, 1986)
93.0 (Pearlman et al., 1984)
89.9 (Casellato et al., 1973)

Boiling point (°C):
279 (Weast, 1986)

Density (g/cm^3):
1.0242 at 90/4 °C (Weast, 1986)

Diffusivity in water (10^5 cm^2/sec):
0.64 at 20 °C using method of Hayduk and Laudie (1974).

Dissociation constant:
>15 (Christensen et al., 1975)

Heat of fusion (kcal/mol):
4.95 (Tsonopoulos and Prausnitz, 1971)
5.13 (Osborn and Douslin, 1975)
5.23 (Wauchope and Getzen, 1972)

Henry's law constant (10^5 atm·m^3/mol):
14.6 (Mackay et al., 1979)

1

15.5 (Mackay and Shiu, 1981)
24.1 (Warner et al., 1987)
19 (Petrasek et al., 1983)
6.36 at 25 °C (Fendinger and Glotfelty, 1990)

Bioconcentration factor, log BCF:
2.58 (bluegill sunfish, Veith et al., 1980)

Soil sorption coefficient, log K_{oc}:
1.25 (Mihelcic and Luthy, 1988)
5.38 (average, Kayal and Connell, 1990)

Octanol/water partition coefficient, log K_{ow}:
3.92 (Banerjee et al., 1980; Veith et al., 1980)

Solubility in organics (g/L):
Methanol (17.9), ethanol (32.3), propanol (40.0), chloroform (400.0), benzene or toluene (200.0), glacial acetic acid (32) (Windholz et al., 1983)

Solubility in water:
3.47 mg/kg at 25 °C (Eganhouse and Calder, 1976)
3.93 mg/L at 25 °C (Mackay and Shiu, 1977)
4.16 mg/L at 25 °C (Walters and Luthy, 1984)
2.42 mg/kg at 25 °C. In seawater (salinity = 35 g/kg): 0.214, 0.55 and 1.84 mg/kg at 15, 20 and 25 °C, respectively (Rossi and Thomas, 1981)
47.8 μmol/L at 25 °C (Banerjee et al., 1980)
In mg/kg (°C): 3.57 (22.0), 4.60, 4.72 and 4.76 (30.0), 5.68, 5.73 and 6.00 (34.5), 6.8, 7.0 and 7.1 (39.3), 9.3, 9.4 and 9.4 (44.7), 12.4, 12.4 and 12.5 (50.1), 15.8, 15.9 and 16.3 (55.6), 25.9 and 27.8 (64.5), 22.8, 23.4 and 23.7 (65.2), 30.1, 33.6 and 34.3 (69.8), 35.2 (71.9), 39.1 and 40.1 (73.4), 39.3 and 40.8 (74.7) (Wauchope and Getzen, 1972)
23.3 μmol/kg at 25.0 °C (Vesala, 1974)

Vapor pressure (10^3 mmHg):
2.30 at 25 °C (Banerjee et al., 1990)
149 at 65 °C, 231 at 70 °C, 351 at 75 °C, 529 at 80 °C, 787 at 85 °C, 1,151 at 90 °C, 1,388 at 92.5 °C, 1,463 at 93.195 °C (Osborn and Douslin, 1975)
2.15 at 25 °C (Sonnefeld et al., 1983)

Environmental Fate
Biological. When acenaphthene was statically incubated in the dark at 25 °C with yeast extract and settled domestic wastewater inoculum, significant biodegradation with rapid adaptation was observed. At concentrations of 5 and 10 mg/L, 95 and

100% biodegradation, respectively, were observed after 7 days (Tabak et al., 1981).

Chemical/Physical. Ozonation in water at 60 °C produced 7-formyl-1-inda-none, 1-indanone, 7-hydroxy-1-indanone, 1-indanone-7-carboxylic acid, indane-1,7-dicarboxylic acid and indane-1-formyl-7-carboxylic acid (Chen et al., 1979). Wet oxidation of acenaphthene at 320 °C yielded formic and acetic acids (Randall and Knopp, 1980).

Toxicity: Carcinogenicity in animals has not been determined. Mutagenicity results were inconclusive (Patnaik, 1992).

Drinking Water Standard: As of May 1994, no MCLGs or MCLs have been proposed (U.S. EPA, 1994).

Uses: Manufacture of dye intermediates, pharmaceuticals, insecticides, fungicides and plastics; chemical research. Derived from coal tar and petroleum refining.

ACENAPHTHYLENE

Synonym: Cyclopenta[*d,e*]naphthalene.

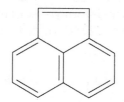

CAS Registry No.: 208-96-8
Molecular formula: $C_{12}H_8$
Formula weight: 152.20
RTECS: AB1254000

Physical state
Solid.

Melting point (°C):
92-93 (Weast, 1986)
96.2 (Lande et al., 1985)

Boiling point (°C):
280 (Aldrich, 1988)

Density (g/cm³):
0.8988 at 16/4 °C (Keith and Walters, 1992)

Diffusivity in water (10^5 cm²/sec):
0.66 at 20 °C using method of Hayduk and Laudie (1974).

Dissociation constant:
>15 (Christensen et al., 1975)

Henry's law constant (10^4 atm·m³/mol at 25 °C):
1.14 (Warner et al., 1987)
1.125 (Fendinger and Glotfelty, 1990)

Ionization potential (eV):
8.73 (Franklin et al., 1969)
8.29 (Rav-Acha and Choshen, 1987)

Bioconcentration factor, log BCF:
2.58 (Isnard and Lambert, 1988)

Soil sorption coefficient, log K_{oc}:
3.68 using method of Karickhoff et al. (1979).

Octanol/water partition coefficient, log K_{ow}:
3.94 (Yalkowsky and Valvani, 1979)
4.07 (Yoshida et al., 1983)

Solubility in organics:
Soluble in ethanol, ether and benzene (U.S. EPA, 1985).

Solubility in water (mg/L at 25 °C):
3.93 (Verschueren, 1983)
16.1 (Walters and Luthy, 1984)

Vapor pressure (10^3 mmHg):
29.0 at 20 °C (Sims et al., 1988)
6.68 at 25 °C (Sonnefeld et al., 1983)

Environmental Fate
   Biological. When acenaphthylene was statically incubated in the dark at 25 °C with yeast extract and settled domestic wastewater inoculum, significant bio-degradation with rapid adaptation was observed. At concentrations of 5 and 10 mg/L, 100 and 94% biodegradation, respectively, were observed after 7 days (Tabak et al., 1981).
   Photolytic. Based on data for structurally similar compounds, acenaphthylene may undergo photolysis to yield quinones (U.S. EPA, 1985). In a toluene solution, irradiation of acenaphthylene at various temperatures and concentrations all resulted in the formation of dimers (Chen et al., 1979).
   Chemical/Physical. Ozonation in water at 60 °C produced 1,8-naphthalene dial-dehyde, 1,8-naphthalene anhydride, 1,2-epoxyacenaphthylene, 1-naphthoic acid and 1,8-naphthaldehydic acid (Calvert and Pitts, 1966).

Exposure Limits: No individual standards have been set. As a constituent in coal tar pitch volatiles, the following exposure limits have been established (mg/m^3): NIOSH REL: TWA 0.1 (cyclohexane-extractable fraction), IDLH 80; OSHA PEL: TWA 0.2 (benzene-soluble fraction); ACGIH TLV: TWA 0.2 (benzene solubles).

Toxicity: Carcinogenicity in animals and humans is not known (Patnaik, 1992).

Uses: Research chemical. Derived from industrial and experimental coal gasifi-cation operations where maximum concentration detected in gas, liquid and coal tar streams were 28, 4.1 and 18 mg/m^3, respectively (Cleland, 1981).

ACETALDEHYDE

Synonyms: Acetic aldehyde; Aldehyde; Ethanal; Ethylaldehyde; NCI-C56326; RCRA waste number U001; UN 1089.

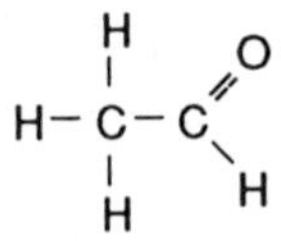

$$H-\overset{\displaystyle \overset{H}{|}}{\underset{\displaystyle \underset{H}{|}}{C}}-\overset{\displaystyle O}{C}\diagdown_H$$

CAS Registry No.: 75-07-0
DOT: 1089
DOT label: Flammable liquid
Molecular formula: C_2H_4O
Formula weight: 44.05
RTECS: AB1925000

Physical state, color and odor
Colorless, mobile, fuming liquid or gas with a penetrating, pungent odor; fruity odor when diluted. Odor threshold in air is 0.05 ppm (Amoore and Hautala, 1983).

Melting point (°C):
-121 (Weast, 1986)
-123.5 (Windholz et al., 1983)

Boiling point (°C):
20.8 (Weast, 1986)

Density (g/cm^3):
0.7834 at 18/4 °C (Weast, 1986)
0.788 at 16/4 °C (Windholz et al., 1983)

Diffusivity in water (10^5 cm^2/sec):
1.23 at 20 °C using method of Hayduk and Laudie (1974).

Dissociation constant:
14.15 at 0 °C (HSDB, 1989)

Flash point (°C):
-37.8 (NIOSH, 1994)
-40 (open cup, Hawley, 1981)

Lower explosive limit (%):
4.0 (NIOSH, 1994)

6

Upper explosive limit (%):
57 (Sax and Lewis, 1987)
60 (NIOSH, 1994)

Heat of fusion (kcal/mol):
0.770 (Dean, 1987)

Henry's law constant (10^5 atm·m^3/mol at 25 °C):
6.58 (Hine and Mookerjee, 1975)
6.61 (Buttery et al., 1969)
6.67 (Gaffney et al., 1987)
6.71 (Zhou and Mopper, 1990)
8.87 (Betterton and Hoffmann, 1988)

Ionization potential (eV):
10.21 (Franklin et al., 1969)

Soil sorption coefficient, log K_{oc}:
Unavailable because experimental methods for estimation of this parameter for aldehydes are lacking in the documented literature. However, its miscibility in water suggests its adsorption to soil will be nominal (Lyman et al., 1982).

Octanol/water partition coefficient, log K_{ow}:
0.52 (Sangster, 1989)

Solubility in organics:
Miscible with acetone, alcohol, benzene, ether, gasoline, solvent naphtha, toluene, turpentine and xylene (Hawley, 1981).

Solubility in water:
Miscible (Palit, 1947).

Vapor density:
1.80 g/L at 25 °C, 1.52 (air = 1)

Vapor pressure (mmHg):
740 at 20 °C (NIOSH, 1994)
900 at 25 °C (Lide, 1990)

Environmental Fate
Photolytic. Photooxidation of acetaldehyde in nitrogen oxide-free air using radiation between 2900-3500 Å yielded hydrogen peroxide, alkylhydroperoxides, carbon monoxide and lower molecular weight aldehydes. In the presence of nitro-

gen oxides, photooxidation products include ozone, hydrogen peroxide and peroxyacyl nitrates (Kopczynski et al., 1974). Anticipated products from the reaction of acetaldehyde with ozone or hydroxyl radicals in the atmosphere are formaldehyde and carbon dioxide (Cupitt, 1980). Reacts with nitrogen dioxide forming peroxyacyl nitrates, formaldehyde and methyl nitrate (Altshuller, 1983). Irradiation in the presence of chlorine yielded peroxyacetic acid, carbon monoxide and carbon dioxide (Hanst and Gay, 1983). Synthetic air containing gaseous nitrous acid and exposed to artificial sunlight (λ = 300-450 nm) photooxidized acetaldehyde into formic acid, methyl nitrate and peroxyacetal nitrate (Cox et al., 1980).

The room-temperature photooxidation of acetaldehyde in the presence of oxygen with continuous irradiation (λ >2200 Å) resulted in the following by-products: methanol, carbon monoxide, carbon dioxide, water, formaldehyde, formic acid, acetic acids, CH_3OOCH_3 and probably $CH_3C(O)OOH$ (Johnston and Heicklen, 1964).

Chemical/Physical. Oxidation in air yields acetic acid (Windholz et al., 1983). In the presence of sulfuric, hydrochloric or phosphoric acids, polymerizes explosively forming trimeric paraldehyde (Huntress and Mulliken, 1941; Patnaik, 1992). In an aqueous solution at 25 °C, acetaldehyde is partially hydrated, i.e., 0.60 expressed as a mole fraction, forming a gem-diol (Bell and McDougall, 1960). Acetaldehyde decomposes above 400 °C, forming carbon monoxide and methane (Patnaik, 1992).

Exposure Limits: Potential occupational carcinogen. NIOSH REL: IDLH 2,000 ppm; OSHA PEL: TWA 200 ppm (360 mg/m³); ACGIH TLV: TWA 100 ppm (180 mg/m³), STEL 150 ppm (270 mg/m³).

Symptoms of Exposure: Conjunctivitis, central nervous system, eye and skin burns and dermatitis are symptoms of ingestion. Inhalation may cause irritation of the eyes, nose and throat. At high concentrations headache, sore throat and paralysis of respiratory muscles may occur (Patnaik, 1992).

Toxicity: LC_{50} (48-hour) for red killifish 1,820 mg/L (Yoshioka et al., 1986), hamsters 17,000 ppm/4-hour, mice 1,400 ppm/4-hour, rats 37 gm/m³/30-minute; acute oral LD_{50} for rats 1,930 mg/kg (RTECS, 1985).

Uses: Manufacture of acetic acid, acetic anhydride, aldol, aniline dyes, 1-butanol, 1,3-butylene glycol, cellulose acetate, chloral, 2-ethylhexanol, paraldehyde, pentaerythritol, peracetic acid, pyridine derivatives, terephthalic acid, trimethylolpropane, flavors, perfumes, plastics, synthetic rubbers, disinfectants, drugs, explosives, antioxidants, yeast; silvering mirrors; hardening gelatin fibers.

ACETIC ACID

Synonyms: Acetic acid (aqueous solution); Ethanoic acid; Ethylic acid; Glacial acetic acid; Methanecarboxylic acid; Pyroligneous acid; UN 2789; UN 2790; Vinegar acid.

CAS Registry No.: 64-19-7
DOT: 2789 (glacial, >80 wt %), 2790 (10-80% solution)
DOT label: Corrosive
Molecular formula: $C_2H_4O_2$
Formula weight: 60.05
RTECS: AF1225000

Physical state, color and odor
Clear, colorless, corrosive liquid with a strong vinegar-like odor. Lower and upper odor thresholds in air are 5 and 80 ppm, respectively (Keith and Walters, 1992). Very sour taste.

Melting point (°C):
16.6 (Weast, 1986)

Boiling point (°C):
117.9 (Weast, 1986)

Density (g/cm³):
1.0492 at 20/4 °C (Weast, 1986)

Diffusivity in water (10^5 cm²/sec):
1.19 at 25 °C (Hayduk and Laudie, 1974)

Dissociation constant:
4.74 (Windholz et al., 1983)

Flash point (°C):
39.5 (NIOSH, 1994)

Lower explosive limit (%):
4.0 (NFPA, 1984)

9

Upper explosive limit (%):
19.9 at 93–94 °C (NFPA, 1984)

Heat of fusion (kcal/mol):
2.80 (Dean, 1987)

Henry's law constant (10^7 atm·m^3/mol):
1,230 at 25 °C (Hine and Mookerjee, 1975)
1 at pH 4 (Gaffney et al., 1987)

Ionization potential (eV):
10.69 ± 0.03 (Franklin et al., 1969)
10.37 (Gibson et al., 1977)

Soil sorption coefficient, log K_{oc}:
0.00 (Meylan et al., 1992)

Octanol/water partition coefficient, log K_{ow}:
-0.29, -0.30 (Sangster, 1989)
-0.17 (Leo et al., 1971)
-0.53 (Onitsuka et al., 1989)
-0.31 (Collander, 1951)

Solubility in organics:
Miscible with alcohol, carbon tetrachloride, glycerol (Windholz et al., 1983).

Solubility in water:
Miscible (NIOSH, 1994).

Vapor density:
2.45 g/L at 25 °C, 2.07 (air = 1)

Vapor pressure (mmHg):
11.4 at 20 °C, 20 at 30 °C (Verschueren, 1983)

Environmental Fate
Biological. Near Wilmington, NC, organic wastes containing acetic acid (representing 52.6% of total dissolved organic carbon) were injected into an aquifer containing saline water to a depth of about 1,000 feet. The generation of gaseous components (hydrogen, nitrogen, hydrogen sulfide, carbon dioxide and methane) suggests acetic acid and possibly other waste constituents, were anaerobically degraded by microorganisms (Leenheer et al., 1976).
Chemical/Physical. Ozonolysis of acetic acid in distilled water at 25 °C yielded

glyoxylic acid which was oxidized readily to oxalic acid. Oxalic acid was further oxidized to carbon dioxide. Ozonolysis accompanied by UV irradiation enhanced the removal of acetic acid (Kuo et al., 1977).

Exposure Limits: NIOSH REL: TWA 10 ppm (25 mg/m^3), STEL 15 ppm (37 mg/m^3), IDLH 50 ppm; OSHA PEL: TWA 10 ppm; ACGIH TLV: TWA 10 ppm, STEL 15 ppm.

Symptoms of Exposure: Produces skin burns. Causes eye irritation on contact. Inhalation may cause irritation of the respiratory tract. Acute toxic effects following ingestion may include corrosion of mouth and gastrointestinal tract, vomiting, diarrhea, ulceration, bleeding from intestines and circulatory collapse (Patnaik, 1992; Windholz et al., 1983).

Toxicity: Acute oral LD$_{50}$ for rats 3,530 mg/kg; LC$_{50}$ (inhalation) for mice 5,620 ppm/1 hour (RTECS, 1985); LC$_{50}$ (24-hour) for fathead minnows in Lake Superior water 122 mg/L (Mattson et al., 1976).

Uses: Manufacture of acetate rayon, acetic anhydride, acetone, acetyl compounds, cellulose acetates, chloroacetic acid, ethyl alcohol, ketene, methyl ethyl ketone, vinyl acetate, plastics and rubbers in tanning; laundry sour; acidulant and preservative in foods; printing calico and dyeing silk; solvent for gums, resins, volatile oils and other substances; manufacture of nylon and fiber, vitamins, antibiotics and hormones; production of insecticides, dyes, photographic chemicals, stain removers; latex coagulant; textile printing.

ACETIC ANHYDRIDE

Synonyms: Acetic acid anhydride; Acetic oxide; Acetyl anhydride; Acetyl ether; Acetyl oxide; Ethanoic anhydrate; Ethanoic anhydride; UN 1715.

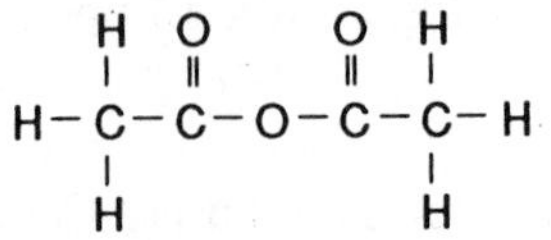

$$
\begin{array}{ccccccc}
 & H & O & & O & H & \\
 & | & \| & & \| & | & \\
H- & C- & C- & O- & C- & C- & H \\
 & | & & & & | & \\
 & H & & & & H &
\end{array}
$$

CAS Registry No.: 108-24-7
DOT: 1715
DOT label: Corrosive
Molecular formula: $C_4H_6O_3$
Formula weight: 102.09
RTECS: AK1925000

Physical state, color and odor
Colorless, very mobile liquid with a very strong acetic acid-like odor. Odor threshold in air is 0.13 ppm (Amoore and Hautala, 1983).

Melting point (°C):
-68 to -73 (Verschueren, 1983)

Boiling point (°C):
139.55 (Weast, 1986)

Density (g/cm^3):
1.0820 at 20/4 °C (Weast, 1986)
1.080 at 15/4 °C (Windholz et al., 1983)

Flash point (°C):
48.9 (NIOSH, 1994)
54.4 (Windholz et al., 1983)

Lower explosive limit (%):
2.7 (NIOSH, 1994)

Upper explosive limit (%):
10.3 (NIOSH, 1994)

Heat of fusion (kcal/mol):
2.51 (Dean, 1987)

Ionization potential (eV):
10.00 (NIOSH, 1994)

Soil sorption coefficient, log K_{oc}:
Unavailable because experimental methods for estimation of this parameter for anhydrides and its acetic acid (hydrolysis product) are lacking in the documented literature. However, its high solubility in water and low K_{ow} for its hydrolysis product (acetic acid) suggest its adsorption to soil will be nominal (Lyman et al., 1982).

Octanol/water partition coefficient, log K_{ow}:
-0.31, -0.17 (acetic acid, Leo et al., 1971)

Solubility in organics:
Miscible with acetic acid, alcohol and ether (Hawley, 1981).

Solubility in water:
12 wt % at 20 °C (NIOSH, 1994)

Vapor density:
4.17 g/L at 25 °C, 3.52 (air = 1)

Vapor pressure (mmHg):
3.5 at 20 °C, 5 at 25 °C, 7 at 30 °C (Verschueren, 1983)

Environmental Fate
Chemical/Physical. Slowly dissolves in water forming acetic acid. In ethanol, ethyl acetate is formed (Windholz et al., 1983).

Exposure Limits: NIOSH REL: ceiling 5 ppm (20 mg/m^3), IDLH 200 ppm; OSHA PEL: 5 ppm; ACGIH TLV: ceiling 5 ppm.

Symptoms of Exposure: Severe eye and skin irritant (NIOSH, 1994).

Toxicity: Acute oral LD_{50} for rats 1,780 mg/kg; LC_{50} (inhalation) 1,000 ppm/4-hour (RTECS, 1985).

Uses: Preparation of acetyl compounds and cellulose acetates; detection of rosin; dehydrating and acetylating agent in the production of pharmaceuticals, dyes, explosives, perfumes and pesticides; organic synthesis.

ACETONE

Synonyms: Chevron acetone; Dimethylformaldehyde; Dimethylketal; Dimethyl ketone; DMK; Ketone propane; β-Ketopropane; Methyl ketone; Propanone; **2-Propanone**; Pyroacetic acid; Pyroacetic ether; RCRA waste number U002; UN 1090.

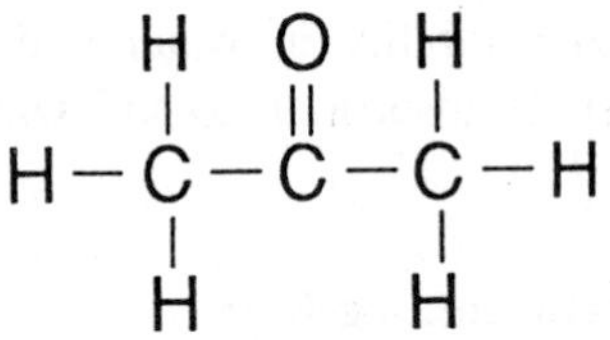

$$\begin{array}{ccccccc} & H & & O & & H & \\ & | & & \| & & | & \\ H - & C & - & C & - & C & - H \\ & | & & & & | & \\ & H & & & & H & \end{array}$$

CAS Registry No.: 67-64-1
DOT: 1090
DOT label: Flammable liquid
Molecular formula: C_3H_6O
Formula weight: 58.08
RTECS: AL3150000

Physical state, color and odor
Clear, colorless, volatile, liquid with a sweet, fragrant odor. Odor threshold in air is 13 ppm (Amoore and Hautala, 1983). Sweetish taste.

Melting point (°C):
-95.35 (Weast, 1986)

Boiling point (°C):
56.2 (Weast, 1986)

Density (g/cm³ at 20/4 °C):
0.7899 (Weast, 1986)

Diffusivity in water (10^5 cm²/sec):
1.28 at 25 °C (Hayduk and Laudie, 1974)

Flash point (°C):
-18 (NIOSH, 1994)

Lower explosive limit (%):
2.5 (NIOSH, 1994)

Upper explosive limit (%):
12.8 (NIOSH, 1994)

Dissociation constant:
≈ 20 (Gordon and Ford, 1972)

Heat of fusion (kcal/mol):
1.366 (Dean, 1987)

Henry's law constant (10^5 atm·m^3/mol at 25 °C):
3.97 (Hine and Mookerjee, 1975)
3.30 (Butler and Ramchandani, 1935)
3.67 (Buttery et al., 1969)
3.33 (Gaffney et al., 1987)
3.46 (Zhou and Mopper, 1990)

Ionization potential (eV):
9.68 (Gibson, 1977)

Soil sorption coefficient, log K_{oc}:
Unavailable because experimental methods for estimation of this parameter for ketones are lacking in the documented literature. However, its low K_{ow} and miscibility in water suggest its adsorption to soil will be nominal (Lyman et al., 1982).

Octanol/water partition coefficient, log K_{ow}:
-0.24, -0.48 (Sangster, 1989)
-0.23 (Collander, 1951)

Solubility in organics:
Soluble in ethanol, benzene and chloroform (U.S. EPA, 1985). Miscible with dimethylformaldehyde, chloroform, ether and most oils (Windholz et al., 1983).

Solubility in water:
Miscible (Palit, 1947). A saturated solution in equilibrium with its own vapor had a concentration of 440.6 g/L at 25 °C (Kamlet et al., 1987).

Vapor density:
2.37 g/L at 25 °C, 2.01 (air = 1)

Vapor pressure (mmHg):
180 at 20 °C (ACGIH, 1986; NIOSH, 1994)
235 at 25 °C (Howard, 1990)

Environmental Fate
   Photolytic. Photolysis in air yields carbon monoxide and free radicals, but in iso-

propanol, pinacol is formed (Calvert and Pitts, 1966). Photolysis of acetone vapor with nitrogen dioxide via a mercury lamp gave peroxyacetyl nitrate as the major product with smaller quantities of methyl nitrate (Warneck and Zerbach, 1992).

Chemical/Physical. Hypochlorite ions, formed by the chlorination of water for disinfection purposes, may react with acetone to form chloroform. This reaction is expected to be significant within the pH range of 6-7 (Stevens et al., 1976).

Exposure Limits: NIOSH REL: TWA 250 ppm (590 mg/m^3), IDLH 2,500 ppm; OSHA PEL: TWA 1,000 ppm (2,400 mg/m^3); ACGIH TLV: TWA 750 ppm (1,780 mg/m^3), STEL 1,000 ppm.

Symptoms of Exposure: Inhalation of acetone at high concentrations may produce headache, mouth dryness, fatigue, nausea, dizziness, muscle weakness, speech impairment and dermatitis. Ingestion causes headache, dizziness and drowsiness (Patnaik, 1992). Prolonged contact with skin may produce erythema and dryness (Windholz et al., 1983).

Toxicity: Acute oral LD$_{50}$ for rats 5,800 mg/kg, mice 3,000 mg/kg (RTECS, 1985).

Uses: Intermediate in the manufacture of many chemicals including acetic acid, chloroform, methyl isobutyl ketone, methyl isobutyl carbinol, methyl methacrylate, bisphenol-A; paint, varnish and lacquer solvent; spinning solvent for cellulose acetate; to clean and dry parts for precision equipment; solvent for potassium iodide, potassium permanganate, cellulose acetate, nitrocellulose, acetylene; delustrant for cellulose acetate fibers; specification testing for vulcanized rubber products; extraction of principals from animal and plant substances; ingredient in nail polish remover; manufacture of rayon, photographic films, explosives; sealants and adhesives; pharmaceutical manufacturing; production of lubricating oils; organic synthesis.

ACETONITRILE

Synonyms: Cyanomethane; Ethanenitrile; Ethyl nitrile; Methanecarbonitrile; Methyl cyanide; NA 1648; NCI-C60822; RCRA waste number U003; UN 1648; USAF EK-488.

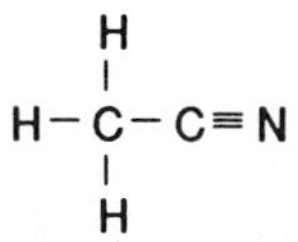

$$H-\underset{\underset{H}{|}}{\overset{\overset{H}{|}}{C}}-C\equiv N$$

CAS Registry No.: 75-05-8
DOT: 1648
DOT label: Flammable liquid and poison
Molecular formula: C_2H_3N
Formula weight: 41.05
RTECS: AL7700000

Physical state, color and odor
Colorless liquid with an ether-like or pungent odor of vinegar. Odor threshold in air is 40 ppm (Keith and Walters, 1992).

Melting point (°C):
-41.0 (Stull, 1947)
-45.7 (Weast, 1986)

Boiling point (°C):
81.6 (Weast, 1986)

Density (g/cm³):
0.7857 at 20/4 °C (Weast, 1986)
0.7138 at 30/4 °C (Windholz et al., 1983)

Diffusivity in water (10^5 cm²/sec):
1.23 at 20 °C using method of Hayduk and Laudie (1974).

Dissociation constant:
29.1 (Riddick et al., 1986)

Flash point (°C):
5.6 (open cup, NIOSH, 1994)
12.8 (Windholz et al., 1983)
6 (open cup, NFPA, 1984)

17

Lower explosive limit (%):
3.0 (NFPA, 1984)

Upper explosive limit (%):
16 (NIOSH, 1994)

Heat of fusion (kcal/mol):
1.952 (Riddick et al., 1986)

Henry's law constant (10^5 atm·m^3/mol at 25 °C):
2.93 (Snider and Dawson, 1985)
3.46 (Hine and Mookerjee, 1975)

Ionization potential (eV):
12.22 (Franklin et al., 1969)

Soil sorption coefficient, log K_{oc}:
0.34 (calculated, Mercer et al., 1990)

Octanol/water partition coefficient, log K_{ow}:
-0.34 (Hansch and Anderson, 1967)
-0.54 (Tanni and Hashimoto, 1984)

Solubility in organics:
Miscible with acetamide solutions, acetone, carbon tetrachloride, chloroform, 1,2-dichloroethane, ether, ethyl acetate, methanol, methyl acetate and many unsaturated hydrocarbons (Windholz et al., 1983). Immiscible with many saturated hydrocarbons (Keith and Walters, 1992).

Solubility in water:
Miscible (NIOSH, 1994). A saturated solution in equilibrium with its own vapor had a concentration of 139.1 g/L at 25 °C (Kamlet et al., 1987).

Vapor density:
1.68 g/L at 25 °C, 1.42 (air = 1)

Vapor pressure (mmHg):
73 at 20 °C (NIOSH, 1994)
88.8 at 25 °C (Banerjee et al., 1990)
87.02 at 25.00 °C (Hussam and Carr, 1985)
86.6 at 25 °C (Hoy, 1970)
100 at 27.0 (Stull, 1947)
115 at 30 °C (Verschueren, 1983)

Environmental Fate

Chemical/Physical. The estimated hydrolysis half-life at 25 °C and pH 7 is >150,000 years (Ellington et al., 1988). A by-product in the synthesis of acrylo-nitrile (Patnaik, 1992).

Exposure Limits: NIOSH REL: TWA 20 ppm (34 mg/m^3), IDLH 500 ppm; OSHA PEL: TWA 40 ppm (70 mg/m^3); ACGIH TLV: TWA 40 ppm, STEL 60 ppm.

Symptoms of Exposure: Inhalation may cause nausea, vomiting, asphyxia and tightness of the chest. Symptoms of ingestion may include gastrointestinal pain, vomiting, nausea, stupor, convulsions and weakness (Patnaik, 1992).

Toxicity: Acute oral LD$_{50}$ in guinea pigs 177 mg/kg, rats 2,730 mg/kg, mice 269 mg/kg; LC$_{50}$ (inhalation) for cats 18 gm/cm^3, guinea pigs 5,655 ppm/4-hour, mice 2,693 ppm/1 hour, rabbits 2,828 ppm/4-hour (RTECS, 1985).

Uses: Preparation of acetamidine, acetophenone, α-naphthaleneacetic acid, thi-amine; dyeing and coating textiles; extracting fish liver oils, fatty acids and other animal and vegetable oils; recrystallizing steroids; solvent for polymers, spinning fibers, casting and molding plastics; manufacture of pharmaceuticals; chemical intermediate for pesticide manufacture; catalyst.

2-ACETYLAMINOFLUORENE

Synonyms: AAF; 2-AAF; 2-Acetamidofluorene; 2-Acetaminofluorene; Acetoaminofluorene; 2-(Acetylamino)fluorene; *N*-Acetyl-2-aminofluorene; FAA; 2-FAA; *N*-9*H*-Fluoren-2-ylacetamide; 2-Fluorenylacetamide; *N*-2-Fluorenylacetamide; *N*-Fluoren-2-acetylacetamide; *N*-Fluorenyl-2-acetamide; RCRA waste number U005.

CAS Registry No.: 53-96-3
Molecular formula: $C_{15}H_{13}NO$
Formula weight: 223.27
RTECS: AB9450000

Physical state and color
Tan crystalline solid or needles.

Melting point (°C):
194 (Weast, 1986)

Diffusivity in water (10^5 cm^2/sec):
0.52 at 20 °C using method of Hayduk and Laudie (1974).

Soil sorption coefficient, log K_{oc}:
3.20 (calculated, Mercer et al., 1990)

Octanol/water partition coefficient, log K_{ow}:
3.28 (Mercer et al., 1990)

Solubility in organics:
Soluble in acetone, acetic acid, alcohol (Weast, 1986), glycols and fat solvents (Windholz et al., 1983).

Solubility in water:
10.13 mg/L at 26.3 °C (Ellington et al., 1987)

Environmental Fate
Chemical/Physical. Releases toxic nitrogen oxides when heated to decomposition (Sax and Lewis, 1987).

Exposure Limits: Potential occupational carcinogen. Since no standards have been established, exposure should be kept at lowest feasible limit.

Toxicity: Acute oral LD_{50} for mice 1,020 mg/kg (RTECS, 1985).

Uses: Biochemical research.

ACROLEIN

Synonyms: Acraldehyde; Acrylaldehyde; Acrylic aldehyde; Allyl aldehyde; Aqualin; Aqualine; Biocide; Crolean; Ethylene aldehyde; Magnacide; NSC 8819; Propenal; **2-Propenal**; Prop-2-en-1-al; 2-Propen-1-one; RCRA waste number P003; Slimicide; UN 1092.

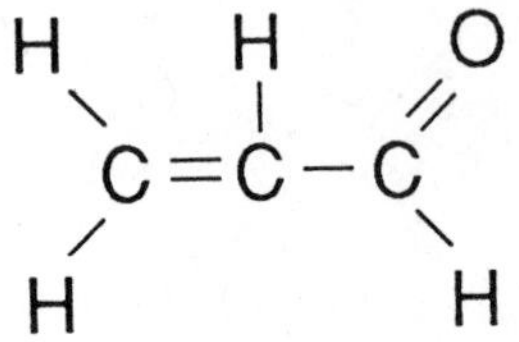

Note: Normally inhibited to prevent polymerization (Keith and Walters, 1992).

CAS Registry No.: 107-02-8
DOT: 1092 (inhibited), 2607 (stabilized dimer)
DOT label: Flammable liquid and poison
Molecular formula: C_3H_4O
Formula weight: 56.06
RTECS: AS1050000

Physical state, color and odor
Colorless to yellow, watery liquid with a very sharp, pungent, irritating odor. Odor threshold in air is 0.16 ppm (Amoore and Hautala, 1983).

Melting point (°C):
-86.9 (Weast, 1986)

Boiling point (°C):
52.7 (Standen, 1963)

Density (g/cm³):
0.8410 at 20/4 °C (Weast, 1986)

Diffusivity in water (10^5 cm²/sec):
1.12 at 20 °C using method of Hayduk and Laudie (1974).

Flash point (°C):
-25 (Weiss, 1986)
-18 (open cup, Aldrich, 1988)

Lower explosive limit (%):
2.8 (NIOSH, 1994)

Upper explosive limit (%):
31 (NIOSH, 1994)

Henry's law constant (10^6 atm·m^3/mol):
4.4 at 25 °C (Howard, 1989)
122 (Gaffney et al., 1987)

Ionization potential (eV):
10.10 ± 0.01 (Franklin et al., 1969)

Bioconcentration factor, log BCF:
2.54 (bluegill sunfish, Veith et al., 1980)

Soil sorption coefficient, log K_{oc}:
Unavailable because experimental methods for estimation of this parameter for unsaturated ketones are lacking in the documented literature. However, its low K_{ow} and high solubility in water suggest its adsorption to soil will be low (Lyman et al., 1982).

Octanol/water partition coefficient, log K_{ow}:
-0.01 (Sangster, 1989)
0.90 (Veith et al., 1980)

Solubility in organics:
Soluble in ethanol, ether and acetone (U.S. EPA, 1985).

Solubility in water:
208 g/L at 20 °C (Lide, 1990)

Vapor density:
2.29 g/L at 25 °C, 1.94 (air = 1)

Vapor pressure (mmHg):
210 at 20 °C (NIOSH, 1994)
265 at 25 °C (Howard, 1989)
273 at 25 °C (Banerjee et al., 1990)

Environmental Fate

Biological. Microbes in site water degraded acrolein to β-hydroxypropionaldehyde (Kobayashi and Rittman, 1982). This product also forms when acrolein is hydrated in distilled water (Burczyk et al., 1968). When 5 and 10 mg/L of acrolein were statically incubated in the dark at 25 °C with yeast extract and settled domestic wastewater inoculum, complete degradation was observed after 7 days

(Tabak et al., 1981).

Photolytic. Photolysis products include carbon monoxide, ethylene, free radicals and a polymer (Calvert and Pitts, 1966). Anticipated products from the reaction of acrylonitrile with ozone or hydroxyl radicals in the atmosphere are glyoxal, formaldehyde, formic acid and carbon dioxide (Cupitt, 1980). The major product reported from the photooxidation of acrolein with nitrogen oxides is formaldehyde with a trace of glyoxal (Altshuller, 1983).

Chemical/Physical. Wet oxidation of acrolein at 320 °C yielded formic and acetic acids (Randall and Knopp, 1980). May polymerize in the presence of light and explosively in the presence of concentrated acids (Worthing and Hance, 1991) forming disacryl, a white plastic solid (Humburg et al., 1989; Windholz et al., 1983). In distilled water, acrolein hydrolyzed to β-hydroxypropionaldehyde (Burczyk et al., 1968; Reinert and Rodgers, 1987).

Exposure Limits: NIOSH REL: TWA 0.1 ppm (0.25 mg/m^3), STEL 0.3 ppm (0.8 mg/m^3), IDLH 2 ppm; OSHA PEL: TWA 0.1 ppm; ACGIH TLV: 0.1 ppm, STEL 0.3 ppm.

Symptoms of Exposure: Strong lachrymator and nasal irritant. Eye contact may damage cornea. Skin contact may cause delayed pulmonary edema (Patnaik, 1992).

Toxicity: EC_{50} (96-hour) for oysters 55 μg/L (salt water); EC_{50} (24-hour) for salmon 80 μg/L (fresh water); LC_{50} (24-hour): for bluegill sunfish 79 μg/L, mosquito fish 0.39 mg/L, rainbow trout 0.15 mg/L and shiners 0.04 mg/L; LC_{50} (48-hour): for oysters 0.56 mg/L and shrimp 0.10 mg/L (Worthing and Hance, 1991); acute oral LD_{50} for rats 46 mg/kg (Ashton and Monaco, 1991), 25.1 mg/kg (RTECS, 1985), mice 40 mg/kg, rabbits 7 mg/kg; LD_{50} (inhalation) for mice 66 ppm/6-hour, rats 300 mg/m^3/30-minute (RTECS, 1985).

Uses: Intermediate in the manufacture of many chemicals (glycerine, 1,3,6-hexanediol, β-chloropropionaldehyde, 1,2,3,6-tetrahydrobenzaldehyde, β-picoline, nicotinic acid), pharmaceuticals, polyurethane, polyester resins, liquid fuel, slimicide; herbicide; antimicrobial agent; control of aquatic weeds; warning agent in gases.

ACRYLAMIDE

Synonyms: Acrylamide monomer; Acylic amide; Ethylene carboxamide; Propenamide; **2-Propenamide**; RCRA waste number U007; UN 2074.

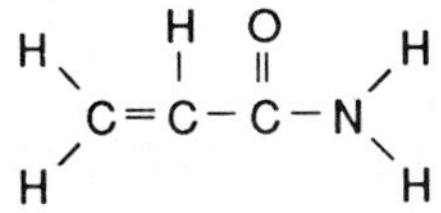

CAS Registry No.: 79-06-1
DOT: 2074
DOT label: Poison
Molecular formula: C_3H_5NO
Formula weight: 71.08
RTECS: AS3325000

Physical state, color and odor
Colorless, odorless solid or flake-like crystals.

Melting point (°C):
84-85 (Weast, 1986)

Boiling point (°C):
125 at 25 mmHg (Windholz et al., 1983)

Density (g/cm^3):
1.122 at 30/4 °C (Windholz et al., 1983)

Diffusivity in water (10^5 cm^2/sec):
1.49 at 30 °C using method of Hayduk and Laudie (1974).

Flash point (°C):
137 (NIOSH, 1994)

Henry's law constant (10^9 atm·m^3/mol):
3.03 at 20 °C (approximate - calculated from water solubility and vapor pressure)

Ionization potential (eV):
9.50 (NIOSH, 1994)

Soil sorption coefficient, log K_{oc}:
Unavailable because experimental methods for estimation of this parameter for

25

amides are lacking in the documented literature. However, its miscibility in water suggests its adsorption to soil will be nominal (Lyman et al., 1982).

Octanol/water partition coefficient, log K_{ow}:
-0.78 (Lide, 1990)

Solubility in organics:
At 30 °C (g/L): acetone (631), benzene (3.46), chloroform (26.6), ethanol (862), ethyl acetate (126), heptane (0.068), methanol (1,550) (Windholz et al., 1983).

Solubility in water (g/L):
2,155 at 30 °C (Windholz et al., 1983)
2,050 (Verschueren, 1983)

Vapor pressure (10^3 mmHg):
7 at 20 °C (NIOSH, 1994)

Environmental Fate
    Soil. Under aerobic conditions, ammonium ions is oxidized to nitrite ions and nitrate ions. The ammonium ions produced in soil may volatilize as ammonia or accumulate as nitrite ions in sandy or calcareous soils (Abdelmagid and Tabatabai, 1982).
    Chemical/Physical. Readily polymerizes at the melting point or under UV light (Windholz et al., 1983).
    Acrylic acid (Abdelmagid and Tabatabai, 1982; Brown and Rhead, 1979) and ammonium ions were reported as hydrolysis products. The hydrolysis rate constant for acrylamide at pH 7 and 25 °C was determined to be <2.1 x 10^{-6}/hour, resulting in a half-life of <37.7 years (Ellington et al., 1988). The hydrolysis half-lives are reduced significantly at varying pHs and temperature. At 88.0 °C and pH values of 2.99 and 7.04, the half-lives were 2.3 and 6.0 days, respectively (Ellington et al., 1986).
    Decomposes between 175 and 300 °C (NIOSH, 1994).

Exposure Limits (mg/m^3): Potential occupational carcinogen. NIOSH REL: TWA 0.03, IDLH: 60; OSHA PEL: TWA 0.3; ACGIH TLV: TWA 0.03.

Toxicity: Acute oral LD_{50} for mice 107 mg/kg, quail 186 mg/kg, rats 124 mg/kg (RTECS, 1985).

Drinking Water Standard (final): MCLG: zero; MCL: lowest concentration obtained using conventional treatment techniques (U.S. EPA, 1984).

Uses: Synthesis of dyes; flocculants; polymers or copolymers as plastics, adhesives, soil conditioning agents; sewage and waste treatment; ore processing; permanent press fabrics.

ACRYLONITRILE

Synonyms: Acritet; Acrylon; Acrylonitrile monomer; An; Carbacryl; Cyanoethylene; ENT 54; Fumigrain; Miller's fumigrain; Nitrile; Propenenitrile; **2-Propenenitrile**; RCRA waste number U009; TL 314; UN 1093; VCN; Ventox; Vinyl cyanide.

$$\begin{array}{c} H \quad\quad H \\ \backslash \quad\quad | \\ C=C-C\equiv N \\ / \\ H \end{array}$$

Note: Inhibited with hydroquinone monomethyl ether to prevent polymerization (Aldrich, 1990).

CAS Registry No.: 107-13-1
DOT: 1093 (inhibited)
DOT label: Flammable liquid and poison
Molecular formula: C_3H_3N
Formula weight: 53.06
RTECS: AT5250000

Physical state, color and odor
Clear, colorless, watery, volatile liquid with a sweet irritating odor resembling peach pits. Odor threshold in air is 17 ppm (Amoore and Hautala, 1983).

Melting point (°C):
-83 (Verschueren, 1983)

Boiling point (°C):
77.5-79 (Weast, 1986)

Density (g/cm^3):
0.8060 at 20/4 °C (Weast, 1986)
0.8004 at 25/4 °C (Standen, 1963)

Diffusivity in water (10^5 cm^2/sec):
1.12 at 20 °C using method of Hayduk and Laudie (1974).

Flash point (°C):
-1 (NIOSH, 1994)

Lower explosive limit (%):
3.05 ± 0.5 (Standen, 1963)

Upper explosive limit (%):
17.0 ± 0.5 (Standen, 1963)

Henry's law constant (10^4 atm·m^3/mol):
1.10 at 25 °C (Howard, 1989)

Ionization potential (eV):
10.91 ± 0.01 (Franklin et al., 1969)

Bioconcentration factor, log BCF:
1.68 (bluegill sunfish, Veith et al., 1980)

Soil sorption coefficient, log K_{oc}:
1.10 (Captina silt loam), 1.01 (McLaurin sandy loam) (Walton et al., 1992)

Octanol/water partition coefficient, log K_{ow}:
0.00, 0.09, 0.25 (Sangster, 1989)
1.20 (Veith et al., 1980)

Solubility in organics:
Soluble in ether, acetone and benzene (U.S. EPA, 1985), carbon tetrachloride and toluene (Yoshida et al., 1983). Miscible with alcohol and chloroform (Meites, 1963).

Solubility in water:
7.2% at 0 °C, 7.35% at 20 °C, 7.9% at 40 °C (WHO, 1983)
7.5% at 25 °C (Gunther et al., 1968)
15.6 wt % at 20 °C (Riddick et al., 1986)
79,000 mg/L at 25 °C (Mabey et al., 1982)
110 g/kg at 60.3 °C (Fordyce and Chapin, 1947)

Vapor density:
2.17 g/L at 25 °C, 1.83 (air = 1)

Vapor pressure (mmHg):
83 at 20 °C (NIOSH, 1994)
137 at 30 °C (Verschueren, 1983)
107.8 at 25 °C (Howard, 1989)

Environmental Fate
   Biological. Degradation by the microorganism *Nocardia rhodochrous* yielded ammonium ion and propionic acid, the latter being oxidized to carbon dioxide and water (DiGeronimo and Antoine, 1976). When 5 and 10 mg/L of acrylonitrile were statically incubated in the dark at 25 °C with yeast extract and settled domestic

wastewater inoculum, complete degradation was observed after 7 days (Tabak et al., 1981).

Photolytic. In an aqueous solution at 50 °C, UV light photooxidized acrylonitrile to carbon dioxide. After 24 hours, the concentration of acrylonitrile was reduced 24.2% (Knoevenagel and Himmelreich, 1976).

Chemical/Physical. Ozonolysis of acrylonitrile in the liquid phase yielded form-aldehyde and the tentatively identified compounds glyoxal, an epoxide of acrylo-nitrile and acetamide (Munshi et al., 1989). In the gas phase, cyanoethylene oxide was reported as an ozonolysis product (Munshi et al., 1989). Anticipated products from the reaction of acrylonitrile with ozone or hydroxyl radicals in the atmos-phere are formaldehyde, formic acid, HC(O)CN and cyanide ions (Cupitt, 1980).

Incineration or heating to decomposition releases toxic nitrogen oxides (Sittig, 1985) and cyanides (Lewis, 1990). Wet oxidation of acrylonitrile at 320 °C yielded formic and acetic acids (Randall and Knopp, 1980). Polymerizes readily in the absence of oxygen or on exposure to visible light (Windholz et al., 1983).

The hydrolysis rate constant for acrylonitrile at pH 2.87 and 68 °C was deter-mined to be 6.4×10^{-3}/hour, resulting in a half-life of 4.5 days. At 68.0 °C and pH 7.19, no hydrolysis/disappearance was observed after 2 days. However, when the pH was raised to 10.76, the hydrolysis half-life was calculated to be 1.7 hours (Ellington et al., 1986).

Exposure Limits (ppm): Potential occupational carcinogen. NIOSH REL: TWA 1, 15-minute ceiling 1, IDLH 85; OSHA PEL: TWA 2, 15-minute ceiling 10; ACGIH TLV: TWA 2.

Symptoms of Exposure: Eye and skin irritant. Inhalation may cause asphyxia and headache. Ingestion and skin absorption may cause headache, lightheadedness, sneezing, weakness, nausea and vomiting (Patnaik, 1992).

Toxicity: LC_{50} (48-hour) for red killifish 600 mg/L (Yoshioka et al., 1986); LC_{100} (24-hour) for all fish 100 mg/L (fresh water); acute oral LD_{50} for mice 27 mg/kg, guinea pigs 50 mg/kg, rats 78 mg/kg, rabbits 93 mg/kg (RTECS, 1985).

Drinking Water Standard (tentative): MCLG: zero (U.S. EPA, 1984).

Uses: Copolymerized with methyl acrylate, methyl methacrylate, vinyl acetate, vinyl chloride, or 1,1-dichloroethylene to produce acrylic and modacrylic fibers and high-strength fibers; ABS (acrylonitrile-butadiene-styrene) and acrylonitrile-styrene copolymers; nitrile rubber; cyanoethylation of cotton; synthetic soil block (acrylonitrile polymerized in wood pulp); manufacture of adhesives; organic syn-thesis; grain fumigant; pesticide; monomer for a semi-conductive polymer that can be used similar to inorganic oxide catalysts in dehydrogenation of *t*-butyl alcohol to isobutylene and water; pharmaceuticals; antioxidants; dyes and surfactants.

ALDRIN

Synonyms: Aldrec; Aldrex; Aldrex 30; Aldrite; Aldrosol; Altox; Compound 118; Drinox; ENT 15949; Hexachlorohexahydro-*endo,exo*-dimethanonaphthalene; **1,2,3,4,10,10-Hexachloro-1,4,4a,5,8,8a-hexahydro-1,4:5,8-dimethanonaph-thalene;** 1,2,3,4,10,10-Hexachloro-1,4,4a,5,8,8a-hexahydro-1,4-*endo,exo*-5,8-di-methanonaphthalene; 1,2,3,4,10,10-Hexachloro-1,4,4a,5,8,8a-hexahydro-*exo*-1,4-*endo*-5,8-dimethanonaphthalene; 1,4,4a,5,8,8a-Hexahydro-1,4-*endo,exo*-5,8-di-methanonaphthalene; HHDN; NA 2761; NA 2762; NCI-C00044; Octalene; RCRA waste number P004; Seedrin; Seedrin liquid.

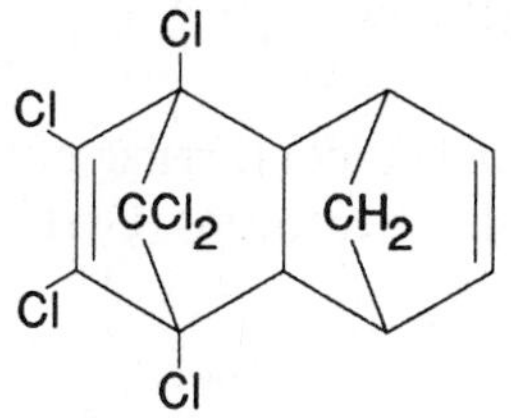

CAS Registry No.: 309-00-2
DOT: 2761 (additional numbers may exist and may be provided by the supplier)
DOT label: Poison
Molecular formula: $C_{12}H_8Cl_6$
Formula weight: 364.92
RTECS: IO2100000

Physical state, color and odor
White, odorless crystals when pure; technical grades are tan to dark brown with a mild chemical odor.

Melting point (°C):
104 (Weast, 1986)
107 (Sims et al., 1988)

Boiling point (°C):
145 at 2 mmHg (Hayes, 1982)

Density (g/cm³):
1.70 at 20/4 °C (Hayes, 1982)

Diffusivity in water (10^5 cm²/sec):
0.45 at 20 °C using method of Hayduk and Laudie (1974).

Flash point (°C):
Not applicable (NIOSH, 1994)

31

Lower explosive limit (%):
Not applicable (NIOSH, 1994)

Upper explosive limit (%):
Not applicable (NIOSH, 1994)

Henry's law constant (10^6 atm·m^3/mol):
1.4 (Eisenreich et al., 1981)
496 at 25 °C (Warner et al., 1987)

Bioconcentration factor, log BCF:
4.09 (algae, Geyer et al., 1984)
4.26 (activated sludge), 3.44 (golden ide) (Freitag et al., 1985)
3.50 (freshwater fish), 3.50 (fish, microcosm) (Garten and Trabalka, 1983)

Soil sorption coefficient, log K_{oc}:
2.61 (Kenaga, 1980)
4.69 (Batcombe silt loam, Briggs, 1981)

Octanol/water partition coefficient, log K_{ow}:
5.52 (Travis and Arms, 1988)
6.496 (de Bruijn et al., 1989)
7.4 (Briggs, 1981)

Solubility in organics:
50 g/L in alcohol at 25 °C (Meites, 1963)

Solubility in water (μg/L):
27 at 25-29 °C (Park and Bruce, 1968)
17 at 25 °C (Weil et al., 1974)
105 at 15 °C, 180 at 25 °C, 350 at 35 °C, 600 at 45 °C (particle size ≤5 μ, Biggar and Riggs, 1974)

Vapor pressure (10^5 mmHg):
7.5 at 20 °C (Windholz et al., 1983)
2.31 at 20 °C (Martin, 1972)
15.2, 17.3 at 25 °C (Hinckley et al., 1990)
24.9 at 25 °C (Bidleman, 1984)

Environmental Fate
   Biological. Dieldrin is the major metabolite from the microbial degradation of aldrin by oxidation or epoxidation. Dieldrin may further degrade to photodieldrin (Kearney and Kaufman, 1976). A pure culture of the marine alga, namely *Duna-*

liella sp. degraded aldrin to dieldrin and the diol 23.2 and 5.2%, respectively (Patil et al., 1972). In four successive 7-day incubation periods, aldrin (5 and 10 mg/L) was recalcitrant to degradation in a settled domestic wastewater inoculum (Tabak et al., 1981). Incubation with a mixed anaerobic population resulted in a degradation yield of 87% in 4 days. Two dechlorinated products were reported (Maule et al., 1987).

Soil. Aldrin was found to be very persistent in an agricultural soil. Fifteen years after application at a rate of 20 lb/acre, 5.8% of the applied amount was recovered as dieldrin and 0.2% was recovered as photodieldrin (Lichtenstein et al., 1971). Patil and Matsumura (1970) reported 13 of 20 soil microorganisms were able to degrade aldrin to dieldrin under laboratory conditions.

Plant. Photoaldrin and photodieldrin formed when aldrin was codeposited on bean leaves and exposed to sunlight (Ivie and Casida, 1971a). Dieldrin and 1,2,3,4,7,8-hexachloro-1,4,4a,6,7,7a-hexahydro-1,4-*endo*-methyleneindene-5,7-dicarboxylic acid were identified in aldrin-treated soil on which potatoes were grown (Flein et al., 1973).

Surface Water. Under oceanic conditions, aldrin may undergo dihydroxylation at the chlorine free double bond to produce aldrin diol (Verschueren, 1983). In a river die-away test using raw water from the Little Miami River in Ohio, 26, 60 and 80% of aldrin present degraded after 2, 4 and 6 weeks, respectively (Eichelberger and Lichtenberg, 1971).

Photolytic. Photolysis of 0.33 ppb aldrin in San Francisco Bay water by sunlight produced photodieldrin with a reported photolysis half-life of 1.1 days (Singmaster, 1975). Aldrin on silica gel plates in the presence of photosensitizers and exposed to sunlight produced photoaldrin (Ivie and Casida, 1971). Photoaldrin also formed when a benzene solution containing aldrin and benzophenone as a sensitizer was exposed to UV light (λ = 268-356 nm) (Rosen and Carey, 1968). Photodegradation of aldrin by sunlight yielded the following products after 1 month: dieldrin, photodieldrin, photoaldrin and a polymeric substance (Rosen and Sutherland, 1967). Photolysis of solid aldrin using a high pressure mercury lamp with a pyrex filter (λ >300 nm) yielded a polymeric substance with small amounts of photoaldrin, dieldrin, hydrochloric acid and carbon dioxide (Gäb et al., 1974).

Sunlight and UV light can convert aldrin to photoaldrin (Georgacakis and Khan, 1971). When an aqueous solution containing aldrin was photooxidized by UV light at 90-95 °C, 25, 50 and 75% degraded to carbon dioxide after 14.1, 28.2 and 109.7 hours, respectively (Knoevenagel and Himmelreich, 1976). Aldrin in a hydrogen peroxide solution (5 μM) was irradiated by UV light (λ = 290 nm). After 12 hours, the aldrin concentration was reduced 79.5%. Dieldrin, photoaldrin and an unidentified compound were reported as metabolites (Draper and Crosby, 1984). After a short-term exposure to sunlight (<1 hour), aldrin on silica gel chromatoplates was converted to photoaldrin. Photodecomposition was accelerated by several photosensitizing agents (Ivie and Casida, 1971).

When aldrin vapor in a reaction vessel was irradiated by a sunlamp for 45 hours,

14-34% degraded to dieldrin (50-60 μg) and photodieldrin (20-30 μg). However, when the aldrin vapor concentration was reduced to μg and irradiation time extended to 14 days, 60% degraded to dieldrin (0.63 μg), photodieldrin (0.02 μg) and photoaldrin (0.02 μg) (Crosby and Moilanen, 1974). When an aqueous solution of aldrin (0.07 μM) in natural water samples collected from California and Hawaii were irradiated (λ <220 nm) for 36 hours, 25% was photooxidized to dieldrin (Ross and Crosby, 1985). In an aqueous solution containing peracetic acid maintained in the dark, aldrin was transformed to dieldrin (Ross and Crosby, 1975).

Chemical/Physical. In an aqueous solution containing peracetic acid, aldrin was transformed to dieldrin in the dark (Ross and Crosby, 1975). Oxidation of aldrin by oxygen atoms yielded dieldrin (Saravanja-Bozanic et al., 1977). The hydrolysis rate constant for aldrin at pH 7 and 25 °C was determined to be 3.8 x 10^{-5}/hour, resulting in a half-life of 760 days (Ellington et al., 1988). At higher temperatures, the hydrolysis half-lives decreased significantly. At 68 °C and pH values of 3.03, 6.99 and 10.70, the calculated hydrolysis half-lives were 1.9, 2.9 and 2.5 days, respectively (Ellington et al., 1986).

Exposure Limits (mg/m^3): Potential occupational carcinogen. NIOSH REL: 0.25, IDLH 25; OSHA PEL: TWA 0.25; ACGIH TLV: TWA 0.25.

Symptoms of Exposure: Inhalation may cause nausea, vomiting, asphyxia and tightness of the chest. Symptoms of ingestion may include gastrointestinal pain, vomiting, nausea, stupor, convulsions and weakness (Patnaik, 1992).

Toxicity: Acute oral LD_{50} for male and female rats: 39, 60 mg/kg (Windholz et al., 1983), wild birds 7.2 mg/kg, cats 10 mg/kg, chickens 10 mg/kg, ducks 520 mg/kg, dogs 65 mg/kg, guinea pigs 33 mg/kg, hamsters 100 mg/kg, mice 44 mg/kg, pigeons 56.2 mg/kg, quail 42.1 mg/kg, rabbits 50 mg/kg (RTECS, 1985); LC_{50} (96-hour) for American eel 5 ppb, mummichog 4-8 ppb, striped killifish 17 ppb, Atlantic silverside 13 ppb, striped mullet 100 ppb, bluehead 12 ppb, northern puffer 36 ppb, fathead minnow 28 μg/L, bluegill sunfish 13 μg/L, rainbow trout 17.7 μg/L, coho salmon 45.9 μg/L, chinook 7.5 μg/L, striped bass 10 μg/L, pumpkinseed 20 μg/L and white perch 42 μg/L; LC_{50} (48-hour) for mos-quito fish 36 ppb (Verschueren, 1985); LC_{50} (48-hour) for red killifish 220 μg/L (Yoshioka et al., 1986); LC_{50} (24-hour) for bluegill sunfish 260 ppb (Verschueren, 1983).

Drinking Water Standard: As of May 1994, no MCLGs or MCLs have been proposed (U.S. EPA, 1994).

Uses: Insecticide and fumigant.

ALLYL ALCOHOL

Synonyms: AA; Allyl al; Allylic alcohol; 3-Hydroxypropene; Orvinylcarbinol; Propenol; Propenol-3; Propen-1-ol-3; 1-Propenol-3; 1-Propen-3-ol; 2-Propenol; **2-Propen-1-ol**; Propenyl alcohol; 2-Propenyl alcohol; RCRA waste number P005; UN 1098; Vinyl carbinol.

$$\begin{array}{ccccc} H & & H & H & \\ \diagdown & & | & | & \\ & C = C & - & C & - OH \\ \diagup & & & | & \\ H & & & H & \end{array}$$

CAS Registry No.: 107-18-6
DOT: 1098
DOT label: Flammable liquid; poison
Molecular formula: C_3H_6O
Formula weight: 58.08
RTECS: BA5075000

Physical state, color and odor
Colorless liquid with a pungent, mustard-like odor. Odor threshold in air is 1.1 ppm (Amoore and Hautala, 1983).

Melting point (°C):
-129 (Weast, 1986)

Boiling point (°C):
97.1 (Weast, 1986)

Density (g/cm^3 at 20/4 °C):
0.8540 (Weast, 1986)
0.8250 (Verschueren, 1983)

Diffusivity in water (10^5 cm^2/sec):
1.10 at 20 °C using method of Hayduk and Laudie (1974).

Flash point (°C):
23.9, 21.1 (open cup, Windholz et al., 1983)

Lower explosive limit (%):
2.5 (NIOSH, 1994)

Upper explosive limit (%):
18.0 (NIOSH, 1994)

Henry's law constant (10^6 atm·m^3/mol):
5.00 at 25 °C (Hine and Mookerjee, 1975)

Ionization potential (eV):
9.67 ± 0.05 (Franklin et al., 1969)

Soil sorption coefficient, log K_{oc}:
0.51 (calculated, Mercer et al., 1990)

Octanol/water partition coefficient, log K_{ow}:
0.17 (Leo et al., 1971)

Solubility in organics:
Miscible with alcohol, chloroform, ether and petroleum ether (Windholz et al., 1983).

Solubility in water:
Miscible (Gunther et al., 1968).

Vapor density:
2.37 g/L at 25 °C, 2.01 (air = 1)

Vapor pressure (mmHg):
20 at 20 °C, 32 at 30 °C (Verschueren, 1983)
28.1 at 25 °C (Banerjee et al., 1990)

Environmental Fate
   Chemical/Physical. Will slowly polymerize over time into a viscous liquid (Windholz et al., 1983).

Exposure Limits: NIOSH REL: TWA 2 ppm (5 mg/m^3), STEL 4 ppm (10 mg/m^3), IDLH 20 ppm; OSHA PEL: TWA 2 ppm; ACGIH TLV: TWA 2 ppm, STEL 4 ppm.

Symptoms of Exposure: Inhalation may cause severe irritation of mucous membranes. Ingestion may cause irritation of intestinal tract (Patnaik, 1992).

Toxicity: LD$_{50}$ for rats 64 mg/kg, mice 96 mg/kg, rabbits 71 mg/kg; LC$_{50}$ (inhalation) for mice 500 mg/m^3/2-hour, rats 76 ppm/8-hour (RTECS, 1985).

Uses: Manufacture of acrolein, allyl compounds, glycerol, plasticizers, resins, military poison gas; contact pesticide for weed seeds and certain fungi; intermediate for pharmaceuticals and other organic compounds; herbicide.

ALLYL CHLORIDE

Synonyms: Chlorallylene; 3-Chloroprene; 1-Chloropropene-2; 1-Chloro-2-propene; 3-Chloropropene; 3-Chloroprene-1; **3-Chloro-1-propene**; 1-Chloro-2-propylene; 1-Chloropropylene-2; 3-Chloropropylene; α-Chloropropylene; 3-Chloro-1-propylene; NCI-C04615; UN 1100.

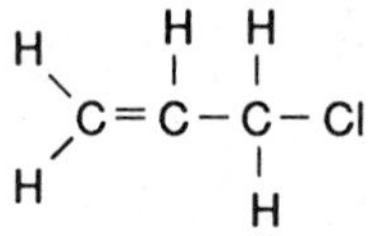

$$H_2C=CH-CH_2-Cl$$

CAS Registry No.: 107-05-1
DOT: 1100
DOT label: Flammable liquid
Molecular formula: C_3H_5Cl
Formula weight: 76.53
RTECS: UC7350000

Physical state, color and odor
Colorless or yellow liquid with a pungent, unpleasant odor. Odor threshold in air is 1.2 ppm (Amoore and Hautala, 1983).

Melting point (°C):
-134.5 (Weast, 1986)

Boiling point (°C):
45 (Weast, 1986)

Density (g/cm^3):
0.9376 at 20/4 °C (Weast, 1986)

Diffusivity in water (10^5 cm^2/sec):
0.99 at 20 °C using method of Hayduk and Laudie (1974).

Flash point (°C):
-31.7 (NIOSH, 1994)
-29 (open cup, Hawley, 1981)

Lower explosive limit (%):
2.9 (NFPA, 1984)

Upper explosive limit (%):
11.1 (NFPA, 1984)

37

Henry's law constant (10^2 atm·m^3/mol):
1.08 at 25 °C (calculated, Dilling, 1977)

Ionization potential (eV):
10.05 (NIOSH, 1994)

Soil sorption coefficient, log K_{oc}:
1.68 using method of Kenaga and Goring (1980).

Octanol/water partition coefficient, log K_{ow}:
1.79 using method of Hansch et al. (1968).

Solubility in organics:
Miscible with alcohol, chloroform, ether and petroleum ether (Windholz et al., 1983).

Solubility in water:
3,600 mg/L at 20 °C (Krijgsheld and van der Gen, 1986)

Vapor density:
3.13 g/L at 25 °C, 2.64 (air = 1)

Vapor pressure (mmHg):
295 at 20 °C (NIOSH, 1994)
340 at 20 °C, 440 at 30 °C (Verschueren, 1983)
360 at 25 °C (Nathan, 1978)

Environmental Fate
  Photolytic. Anticipated products from the reaction of allyl chloride with ozone or hydroxyl radicals in the atmosphere are formaldehyde, formic acid, chloroacetaldehyde, chloroacetic acid and chlorinated hydroxy carbonyls (Cupitt, 1980).
  Chemical/Physical. Hydrolysis under alkaline conditions will yield allyl alcohol (Hawley, 1981). The estimated hydrolysis half-life in water at 25 °C and pH 7 is 69 days (Mabey and Mill, 1978).
  The evaporation half-life of allyl chloride (1 mg/L) from water at 25 °C using a shallow-pitch propeller stirrer at 200 rpm at an average depth of 6.5 cm is 26.6 minutes (Dilling, 1977).

Exposure Limits: NIOSH REL: TWA 1 ppm (3 mg/m^3), STEL 2 ppm (6 mg/m^3), IDLH 250 ppm; OSHA PEL: TWA 1 ppm; ACGIH TLV: STEL 2 ppm.

Symptoms of Exposure: Irritation of eyes and respiratory passages (Windholz et al., 1983).

Toxicity: Acute oral LD_{50} for rats 64 mg/kg; LD_{50} (skin) for rabbits 2,066 mg/kg (RTECS, 1985).

Uses: Preparation of epichlorohydrin, glycerol, allyl compounds, pharmaceuticals; thermosetting resins for adhesives, plastics, varnishes; glycerol and insecticides.

ALLYL GLYCIDYL ETHER

Synonyms: AGE; Allyl-2,3-epoxypropyl ether; 1-Allyloxy-2,3-epoxypropane; 1,2-Epoxy-3-allyloxypropane; Glycidyl allyl ether; NCI-C56666; [(2-Propenyloxy)-methyl]oxirane; UN 2219.

CAS Registry No.: 106-92-3
DOT: 2219
Molecular formula: $C_6H_{10}O_2$
Formula weight: 114.14
RTECS: RR0875000

Physical state, color and odor
Colorless liquid with a strong, pleasant odor.

Melting point (°C):
-100 (Verschueren, 1983)

Boiling point (°C):
154 (NIOSH, 1994)

Density (g/cm^3):
0.9698 at 20/4 °C (Sax and Lewis, 1987)

Diffusivity in water (10^5 cm^2/sec):
0.80 at 20 °C using method of Hayduk and Laudie (1974).

Flash point (°C):
57.2 (NIOSH, 1994)

Henry's law constant (10^6 atm·m^3/mol):
3.83 at 20 °C (approximate - calculated from water solubility and vapor pressure)

Soil sorption coefficient, log K_{oc}:
Unavailable because experimental methods for estimation of this parameter for amides are lacking in the documented literature. However, its very high solubility in water suggests its adsorption to soil will be nominal (Lyman et al., 1982).

40

Octanol/water partition coefficient, log K_{ow}:
0.63 using method of Hansch et al. (1968).

Solubility in organics:
Miscible with toluene (Keith and Walters, 1992).

Solubility in water:
141 g/L (Verschueren, 1983)

Vapor density:
4.67 g/L at 25 °C, 3.94 (air = 1)

Vapor pressure (mmHg):
2 at 20 °C (NIOSH, 1994)
3.6 at 20 °C, 5.8 at 30 °C (Verschueren, 1983)

Exposure Limits: NIOSH REL: TWA 5 ppm (22 mg/m^3), STEL 10 ppm (44 mg/m^3), IDLH 50 ppm; OSHA PEL: ceiling 10 ppm; ACGIH TLV: TWA 5 ppm, STEL 10 ppm.

Symptoms of Exposure: Irritation of eyes, nose, skin and respiratory system (NIOSH, 1994).

Toxicity: Acute oral LD_{50} for mice 390 mg/kg, rats 922 mg/kg; LC_{50} (96-hour) for goldfish 30 mg/L, LC_{50} (24-hour) for goldfish 78 mg/L (Verschueren, 1983).

Uses: Ingredient in epoxy resins.

4-AMINOBIPHENYL

Synonyms: *p*-Aminobiphenyl; 4-Aminodiphenyl; *p*-Aminodiphenyl; Anilinobenzene; Biphenylamine; 4-Biphenylamine; *p*-Biphenylamine; **[1,1'-Biphenyl]-4-
amine**; Paraminodiphenyl; 4-Phenylaniline; *p*-Phenylaniline; Xenylamine.

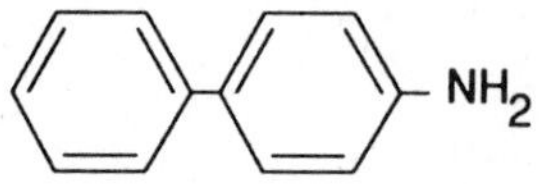

CAS Registry No.: 92-67-1
Molecular formula: $C_{12}H_{11}N$
Formula weight: 169.23
RTECS: DU8925000

Physical state, color and odor
Colorless to yellowish-brown crystals with a floral-like odor. Becomes purple on
exposure to air.

Melting point (°C):
53-54 (Weast, 1986)
50-52 (Sittig, 1985)

Boiling point (°C):
302 (Weast, 1986)

Density (g/cm^3):
1.160 at 20/20 °C (Sax and Lewis, 1987)

Diffusivity in water (10^5 cm^2/sec):
0.59 at 20 °C using method of Hayduk and Laudie (1974).

Dissociation constant:
4.27 at 25 °C (Dean, 1987)

Henry's law constant (10^{10} atm·m^3/mol):
3.89 at 25 °C (calculated, Mercer et al., 1990)

Soil sorption coefficient, log K_{oc}:
2.03 (calculated, Mercer et al., 1990)

Octanol/water partition coefficient, log K_{ow}:
2.86 (Sangster, 1989)

Solubility in organics:
Soluble in alcohol, chloroform and ether (Weast, 1986).

Solubility in water:
842 mg/L at 20-30 °C (Mercer et al., 1990)

Vapor pressure (mmHg):
6×10^{-5} at 20-30 °C (Mercer et al., 1990)

Exposure Limits: Potential occupational carcinogen.

Symptoms of Exposure: Headache, dizziness, lethargy, dyspnea, ataxia, weakness, urinary burning (NIOSH, 1994).

Toxicity: Acute oral LD_{50} for rats 200 mg/kg, for mice 50 mg/kg (Verschueren, 1983).

Uses: Detecting sulfates; formerly used as a rubber antioxidant; cancer research.

2-AMINOPYRIDINE

Synonyms: Amino-2-pyridine; α-Aminopyridine; *o*-Aminopyridine; **2-Pyridin-amine**; α-Pyridinamine; α-Pyridylamine; 2-Pyridylamine; UN 2671.

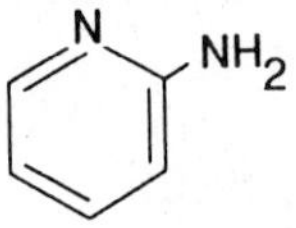

CAS Registry No.: 504-29-0
DOT: 2671
DOT label: Poison
Molecular formula: $C_5H_6N_2$
Formula weight: 94.12
RTECS: US1575000

Physical state, color and odor
Colorless solid or crystals with a characteristic odor.

Melting point (°C):
57-58 (Weast, 1986)

Boiling point (°C):
210.6 (Windholz et al., 1983)

Density (g/cm^3):
1.073 at 20/4 °C (calculated, Lyman et al., 1982)

Diffusivity in water (10^5 cm^2/sec):
0.84 at 20 °C using method of Hayduk and Laudie (1974).

Dissociation constant:
6.86 at 25 °C (Dean, 1973)

Flash point (°C):
68 (NIOSH, 1994)
92 (Dean, 1987)

Ionization potential (eV):
8.00 (NIOSH, 1994)

Soil sorption coefficient, log K_{oc}:
Unavailable because experimental methods for estimation of this parameter for

amides are lacking in the documented literature. However, its high solubility in water and low K_{ow} suggest its adsorption to soil will be nominal (Lyman et al., 1982).

Octanol/water partition coefficient, log K_{ow}:
-0.22 (Verschueren, 1983)

Solubility in organics:
Soluble in acetone, alcohol, benzene and ether (Weast, 1986).

Solubility in water:
Miscible (NIOSH, 1994).

Vapor pressure (mmHg):
25 at 25 °C (NIOSH, 1994)

Environmental Fate
  Chemical/Physical. Releases toxic nitrogen oxides when heated to decomposition (Sax and Lewis, 1987).

Exposure Limits: NIOSH REL: TWA 0.5 ppm (2 mg/m^3), IDLH 5 ppm; OSHA PEL: 0.5 ppm; ACGIH TLV: TWA 0.5 ppm.

Toxicity: LC_{50} (48-hour) for red killifish 63 mg/L (Yoshioka et al., 1986).

Uses: Manufacture of pharmaceuticals, especially antihistamines.

AMMONIA

Synonyms: Am-fol; Ammonia anhydrous; Ammonia gas; Anhydrous ammonia; Nitro-sil; R 717; Spirit of Hartshorn; UN 1005.

CAS Registry No.: 7664-41-7
DOT: 1005 (anhydrous), 2672 (12-44% solution), 2073 (>44% solution)
DOT label: Liquefied compressed gas
Molecular formula: H_3N
Formula weight: 17.04
RTECS: BO0875000

Physical state, color and odor
Colorless gas with a penetrating, pungent, suffocating odor. Odor threshold in air is 5.2 ppm (Amoore and Hautala, 1983).

Melting point (°C):
-77.8 (NIOSH, 1994)

Boiling point (°C):
-33.3 (NIOSH, 1994)

Density (g/cm^3):
0.77 at 0/4 °C (Hawley, 1981)

Diffusivity in water (10^5 cm^2/sec):
1.10 at 0 °C using method of Hayduk and Laudie (1974).

Dissociation constant:
9.247 at 25 °C (as ammonium hydroxide, Gordon and Ford, 1972)

Flash point (°C):
Not applicable (NIOSH, 1994)

Lower explosive limit (%):
15 (NFPA, 1984)

Upper explosive limit (%):
28 (NFPA, 1984)

46

Henry's law constant (10^4 atm·m^3/mol):
2.91 at 20 °C (approximate - calculated from water solubility and vapor pressure)

Ionization potential (eV):
10.15 (Gibson et al., 1977)
10.2 (Franklin et al., 1969)

Soil sorption coefficient, log K_{oc}:
0.49 (calculated, Mercer et al., 1990)

Octanol/water partition coefficient, log K_{ow}:
0.00 (Mercer et al., 1990)

Solubility in organics:
Soluble in chloroform, ether, methanol (16 wt % at 25 °C) and ethanol (10 and 20 wt % at 0 and 25 °C, respectively) (Windholz et al., 1983).

Solubility in water:
In wt %: 47 at 0 °C, 38 at 15 °C, 34 at 20 °C, 31 at 25 °C, 28 at 30 °C, 18 at 50 °C
  (Windholz et al., 1983)
895 g/L at 0 °C, 531 g/L at 20 °C, 440 g/L at 28 °C (Verschueren, 1983)

Vapor density:
0.7714 g/L, 0.5967 (air = 1) (Windholz et al., 1983)

Vapor pressure (mmHg):
7,600 at 25.7 °C (Sax and Lewis, 1987)
6,460 at 20 °C (NIOSH, 1994)

Environmental Fate
   Chemical/Physical. Reacts violently with acetaldehyde, ethylene oxide, ethylene dichloride (Patnaik, 1992). Reacts with acids forming water-soluble salts.

Exposure Limits: NIOSH REL: TWA 25 ppm (18 mg/m^3), STEL 35 ppm (27 mg/m^3), IDLH 300 ppm; OSHA PEL: STEL 50 ppm; ACGIH TLV: TWA 25 ppm, STEL 35 ppm.

Symptoms of Exposure: Very irritating to eyes, nose and respiratory tract. Exposure to 3,000 ppm for several minutes may result in serious blistering of skin, lung edema and asphyxia leading to death (Patnaik, 1992). Ingestion may cause bronchospasm, difficulty in breathing, chest pain and pulmonary edema. Contact with liquid ammonia or aqueous solutions may cause vesiculation or frostbite (NIOSH, 1994).

Toxicity: Acute oral LD_{50} for rats 250 mg/kg; LC_{50} (inhalation) for mice 4,230 ppm/1 hour, rats 2,000 ppm/4-hour, rabbits 7 $gm/m^3/4$-hour (RTECS, 1985); LC_{50} (96-hour) for guppy fry 1.26-74 mg/L (Verschueren, 1983).

Uses: Manufacture of acrylonitrile, hydrazine hydrate, hydrogen cyanide, nitric acid, sodium carbonate, urethane, explosives, synthetic fibers, fertilizers; refrigerant; condensation catalyst; dyeing; neutralizing agent; synthetic fibers; latex preservative; fuel cells, rocket fuel; nitrocellulose; nitroparaffins; ethylenediamine, melamine; sulfite cooking liquors; developing diazo films; yeast nutrient.

n-AMYL ACETATE

Synonyms: Acetic acid amyl ester; **Acetic acid pentyl ester**; Amyl acetate; Amyl acetic ester; Amyl acetic ether; Banana oil; Birnenoel; Pear oil; Pentacetate; Pentacetate 28; 1-Pentanol acetate; 1-Pentyl acetate; *n*-Pentyl acetate; Primary amyl acetate; UN 1104.

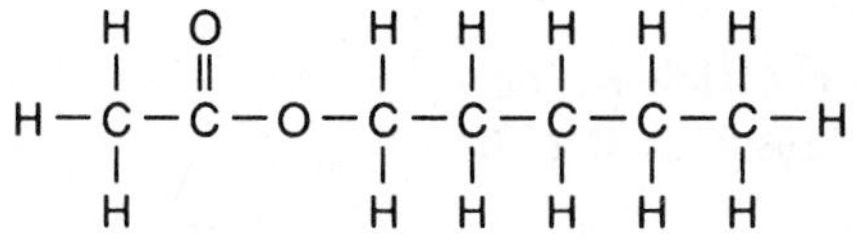

```
        H   O       H   H   H   H   H
        |   ||      |   |   |   |   |
    H — C — C — O — C — C — C — C — C — H
        |           |   |   |   |   |
        H           H   H   H   H   H
```

CAS Registry No.: 628-63-7
DOT: 1104
DOT label: Flammable liquid
Molecular formula: $C_7H_{14}O_2$
Formula weight: 130.19
RTECS: AJ1925000

Physical state, color and odor
Colorless liquid with a banana-like odor. Odor threshold in air is 54 ppb (Amoore and Hautala, 1983).

Melting point (°C):
-70.8 (Weast, 1986)

Boiling point (°C):
149.25 (Weast, 1986)

Density (g/cm³):
0.8756 at 20/4 °C (Weast, 1986)

Diffusivity in water (10^5 cm²/sec):
0.70 at 20 °C using method of Hayduk and Laudie (1974).

Flash point (°C):
25 (NIOSH, 1994)
16-21 (NFPA, 1984)

Lower explosive limit (%):
1.1 (NIOSH, 1994)

Upper explosive limit (%):
7.5 (NIOSH, 1994)

Henry's law constant (10^4 atm·m^3/mol):
3.88 at 25 °C (Hine and Mookerjee, 1975)

Soil sorption coefficient, log K_{oc}:
Unavailable because experimental methods for estimation of this parameter for aliphatic esters are lacking in the documented literature.

Octanol/water partition coefficient, log K_{ow}:
2.349 using method of Hansch et al. (1968).

Solubility in organics:
Miscible with alcohol and ether (Hawley, 1981).

Solubility in water:
1.8 g/L at 20 °C (Verschueren, 1983)
In wt %: 0.29 at 0 °C, 0.22 at 19.7 °C, 0.16 at 30.6 °C, 0.16 at 39.5 °C, 0.10 at 50.0 °C, 0.10 at 60.3 °C, 0.17 at 70.2 °C, 0.17 at 80.1 °C (Stephenson and Stuart, 1986)

Vapor density:
5.32 g/L at 25 °C, 4.49 (air = 1)

Vapor pressure (mmHg):
4.1 at 25 °C (Abraham, 1984)

Environmental Fate
   Chemical/Physical. Hydrolyzes in water forming acetic acid and 1-pentanol.

Exposure Limits: NIOSH REL: TWA 100 ppm (525 mg/m^3), IDLH 1,000 ppm; OSHA PEL: TWA 100 ppm; ACGIH TLV: TWA 100 ppm.

Symptoms of Exposure: Irritating to eyes and respiratory tract. At concentrations of 1,000 ppm, inhalation may cause headache, somnolence and narcotic effects (Patnaik, 1992).

Toxicity: Acute oral LD$_{50}$ for rats 6,500 mg/kg (RTECS, 1985).

Uses: Solvent for lacquers and paints; leather polishes; flavoring agent; photographic film; extraction of penicillin; nail polish; printing and finishing fabrics; odorant.

sec-AMYL ACETATE

Synonyms: 2-Acetoxypentane; 1-Methylbutyl acetate; **2-Pentanol acetate**; 2-Pentyl acetate; *sec*-Pentyl acetate.

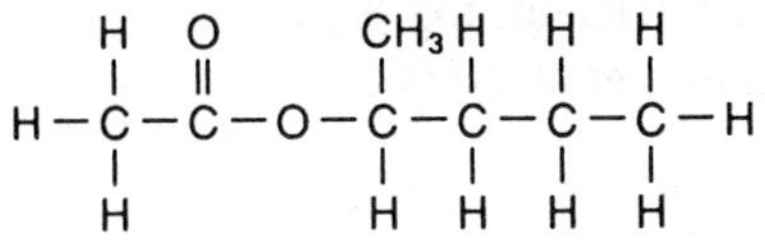

$$H-\underset{\underset{H}{|}}{\overset{\overset{H}{|}}{C}}-\underset{}{\overset{\overset{O}{\|}}{C}}-O-\underset{\underset{H}{|}}{\overset{\overset{CH_3}{|}}{C}}-\underset{\underset{H}{|}}{\overset{\overset{H}{|}}{C}}-\underset{\underset{H}{|}}{\overset{\overset{H}{|}}{C}}-\underset{\underset{H}{|}}{\overset{\overset{H}{|}}{C}}-H$$

CAS Registry No.: 626-38-0
DOT: 1104
DOT label: Flammable liquid
Molecular formula: $C_7H_{14}O_2$
Formula weight: 130.19
RTECS: AJ2100000

Physical state, color and odor
Clear, colorless liquid with a fruity odor. Odor threshold in air is 2.0 ppb (Amoore and Hautala, 1983).

Melting point (°C):
-78.4 (NIOSH, 1994)

Boiling point (°C):
134 (Weast, 1986)

Density (g/cm³):
0.862-0.866 at 20/20 °C (Hawley, 1981)

Diffusivity in water (10^5 cm²/sec):
0.69 at 20 °C using method of Hayduk and Laudie (1974).

Flash point (°C):
31.7 (NIOSH, 1994)

Lower explosive limit (%):
1 (NIOSH, 1994)

Upper explosive limit (%):
7.5 (NIOSH, 1994)

Henry's law constant (10^4 atm·m³/mol):
7.7 at 25 °C (approximate - calculated from water solubility and vapor pressure)

Soil sorption coefficient, log K_{oc}:
Unavailable because experimental methods for estimation of this parameter for aliphatic esters are lacking in the documented literature.

Octanol/water partition coefficient, log K_{ow}:
5.26 using method of Hansch et al. (1968).

Solubility in organics:
Soluble in alcohol and ether (Weast, 1986).

Solubility in water:
2.2 g/L at 25 °C (Montgomery, 1989)

Vapor density:
5.32 g/L at 25 °C, 4.49 (air = 1)

Vapor pressure (mmHg):
10 at 35.2 °C (estimated, Weast, 1986)

Environmental Fate
   Chemical/Physical. Slowly hydrolyzes in water forming acetic acid and 2-pentanol.

Exposure Limits: NIOSH REL: TWA 125 ppm (650 mg/m^3), IDLH 1,000 ppm; OSHA PEL: TWA 125 ppm; ACGIH TLV: TWA 125 ppm.

Symptoms of Exposure: Irritating to eyes, nose and respiratory tract (NIOSH, 1994).

Uses: Solvent for nitrocellulose and ethyl cellulose; coated paper, lacquers; cements; nail enamels, leather finishes; textile sizing and printing compounds; plastic wood.

ANILINE

Synonyms: Aminobenzene; Aminophen; Aniline oil; Anyvim; **Benzenamine**; Benzidam; Blue oil; C.I. 76000; C.I. oxidation base 1; Cyanol; Krystallin; Kyanol; NCI-CO3736; Phenylamine; RCRA waste number U012; UN 1547.

CAS Registry No.: 62-53-3
DOT: 1547
DOT label: Poison
Molecular formula: C_6H_7N
Formula weight: 93.13
RTECS: BW6650000

Physical state, color and odor
Colorless, oily liquid with a faint ammonia-like odor and burning taste. Darkens on exposure to air or light. The lower and upper odor thresholds are 2 and 128 ppm, respectively (Keith and Walters, 1992).

Melting point (°C):
-6.3 (Weast, 1986)

Boiling point (°C):
184 (Weast, 1986)

Density (g/cm³):
1.02173 at 20/4 °C (Weast, 1986)

Diffusivity in water (10^5 cm²/sec):
1.05 at 25 °C (Hayduk and Laudie, 1974)

Dissociation constant:
4.630 at 25 °C (Gordon and Ford, 1972)

Flash point (°C):
70 (NFPA, 1984)
76 (Windholz et al., 1983)

Lower explosive limit (%):
1.3 (NFPA, 1984)

Upper explosive limit (%):
11 (NFPA, 1984)

Heat of fusion (kcal/mol):
2.519 (Dean, 1987)

Henry's law constant (10^6 atm·m^3/mol at 25 °C):
136,000 at pH 7.3 (Hakuta et al., 1977)
1.99 (Jayasinghe et al., 1992)

Interfacial tension with water (dyn/cm at 20 °C):
5.77 (Demond and Lindner, 1993)

Ionization potential (eV):
7.70 (Franklin et al., 1969)

Bioconcentration factor, log BCF:
0.60 (algae, Geyer et al., 1984)
2.70 (activated sludge, Freitag et al., 1985)

Soil sorption coefficient, log K_{oc}:
1.96 (river sediment), 3.4 (coal wastewater sediment) (Kopinke et al., 1995)
1.41 (Briggs, 1981)
2.11 (Hagerstown clay loam), 2.61 (Palouse silt loam) (Pillai et al., 1982)

Octanol/water partition coefficient, log K_{ow}:
0.90 (Campbell and Luthy, 1985; Fujita et al., 1964; Hammers et al., 1982; Mirrlees
    et al., 1976)
1.09 (Geyer et al., 1984)
0.781 (Klein et al., 1988)
0.940 (Brooke et al., 1990; de Bruijn et al., 1989)
0.942 (Brooke et al., 1990)
0.93 (Könemann et al., 1979)
0.79, 0.96 (Garst and Wilson, 1984)

Solubility in organics:
Miscible with alcohol, benzene and chloroform (Windholz et al., 1983).

Solubility in water:
35 g/L at 20 °C (Patnaik, 1992)
34.1 g/L (Fu et al., 1986)
In g/L: 27.2 at 4 °C, 35.4 at 25 °C, 47.8 at 40 °C (Moreale and Van Bladel, 1979)
36.65 mg/L at 25 °C (Hill and Macy, 1924)

Vapor density:
3.81 g/L at 25 °C, 3.22 (air = 1)

Vapor pressure (mmHg at 25 °C):
0.6 (Sonnefeld et al., 1983)
0.49 (Banerjee et al., 1990)

Environmental Fate

Biological. Under anaerobic conditions using a sewage inoculum, 10% of the aniline present degraded to acetanilide and 2-methylquinoline (Hallas and Alexander, 1983). In a 56-day experiment, [^{14}C]aniline applied to soil-water suspensions under aerobic and anaerobic conditions gave $^{14}CO_2$ yields of 26.5 and 11.9%, respectively (Scheunert et al., 1987). A bacterial culture isolated from the Oconee River in North Georgia degraded aniline to the intermediate catechol (Paris and Wolfe, 1987).

Silage samples (chopped corn plants) containing aniline were incubated in an anaerobic chamber for 2 weeks at 28 °C. After 3 days, aniline was biologically metabolized to formanilide, propioanilide, 3,4-dichloroaniline, 3- and 4-chloroaniline (Lyons et al., 1985). Various microorganisms isolated from soil degraded aniline to acetanilide, 2-hydroxyacetanilide, 4-hydroxyaniline and two unidentified phenols (Smith and Rosazza, 1974). In activated sludge, 20.5% mineralized to carbon dioxide after 5 days (Freitag et al., 1985).

Soil. A reversible equilibrium is quickly established when aniline covalently bonds with humates in soils forming imine linkages. These quinodal structures may oxidize to give nitrogen-substituted quinoid rings (Parris, 1980). In sterile soil, aniline partially degraded to azobenzene, phenazine, formanilide and acetanilide and the tentatively identified compounds nitrobenzene and *p*-benzoquinone (Pillai et al., 1982).

Surface Water. Aniline degraded in pond water containing sewage sludge to catechol, which further degraded to carbon dioxide. Intermediate compounds identified in minor degradative pathways include acetanilide, phenylhydroxylamine, *cis,cis*-muconic acid, β-ketoadipic acid, levulinic acid and succinic acid (Lyons et al., 1984).

Photolytic. A carbon dioxide yield of 46.5% was achieved when aniline adsorbed on silica gel was irradiated with light (λ >290 nm) for 17 hours (Freitag et al., 1985). Products identified from the gas-phase reaction of ozone with aniline in synthetic air at 23 °C were nitrobenzene, formic acid, hydrogen peroxide and a nitrated salt having the formula: $[C_6H_5NH_3]^+NO_3^-$ (Atkinson et al., 1987).

Irradiation of an aqueous solution at 50 °C for 24 hours resulted in a 28.5% yield of carbon dioxide (Knoevenagel and Himmelreich, 1976).

Chemical/Physical. Alkali or alkaline earth metals dissolve with hydrogen evolution and the formation of anilides (Windholz et al., 1983). Laha and Luthy (1990) investigated the redox reaction between aniline and a synthetic manganese dioxide in aqueous suspensions at the pH range 3.7-6.5. They postulated that aniline undergoes oxidation by loss of one electron forming cation radicals. These radicals may undergo head-to-tail, tail-to-tail and head-to-head couplings forming 4-aminophenylamine, benzidine and hydrazobenzene, respectively. These compounds were further oxidized, in particular, hydrazobenzene to azobenzene at pH 4 (Laha and Luthy, 1990).

Kanno et al. (1982) studied the aqueous reaction of aniline and other substituted aromatic hydrocarbons (toluidine, 1-naphthylamine, phenol, cresol, pyrocatechol, resorcinol, hydroquinone and 1-naphthol) with hypochlorous acid in the presence of ammonium ion. They reported that the aromatic ring was not chlorinated as expected but was cleaved by chloramine forming cyanogen chloride (Kanno et al., 1982). The amount of cyanogen chloride formed increased at lower pHs. At pH 6, the greatest amount of cyanogen chloride was formed when the reaction mixture contained ammonium ion and hypochlorous acid at a ratio of 2:3 (Kanno et al., 1982). When aniline in an aqueous solution containing nitrite ion was ozonated, nitrosobenzene, nitrobenzene, 4-aminodiphenylamine, azobenzene, azoxybenzene, benzidine, phenazine (Chan and Larson, 1991), 2-, 3- and 4-nitroaniline formed as products (Chan and Larson, 1991a). The yields of nitroanilines were higher at a low pH (6.25) than at high pH (10.65) and the presence of carbonates inhibited their formation (Chan and Larson, 1991a).

Exposure Limits: Potential occupational carcinogen. NIOSH REL: IDLH 100 ppm; OSHA PEL: TWA 5 ppm (19 mg/m^3); ACGIH TLV: TWA 2 ppm (10 mg/m^3).

Symptoms of Exposure: Absorption through skin may cause headache, weakness, dizziness, ataxia and cyanosis (Patnaik, 1992).

Toxicity: Acute oral LD$_{50}$ for mice is 464 mg/kg, wild birds 562 mg/kg, dogs 195 mg/kg, quail 750 mg/kg, rats 250 mg/kg (RTECS, 1985); LC$_{50}$ (inhalation) for mice 175 ppm/7-hour; LC$_{50}$ (48-hour) for red killifish 1,820 mg/L (Yoshioka et al., 1986).

Uses: Manufacture of dyes, resins, varnishes, medicinals, perfumes, photographic chemicals, shoe blacks, chemical intermediates; solvent; vulcanizing rubber; isocyanates for urethane foams; explosives; petroleum refining; diphenylamine; phenolics; fungicides; herbicides.

o-ANISIDINE

Synonyms: 2-Aminoanisole; *o*-Aminoanisole; 1-Amino-2-methoxybenzene; 2-Anisylamine; 2-Methoxy-1-aminobenzene; 2-Methoxyaniline; *o*-Methoxyaniline; **2-Methoxybenzenamine**; *o*-Methoxyphenylamine; UN 2431.

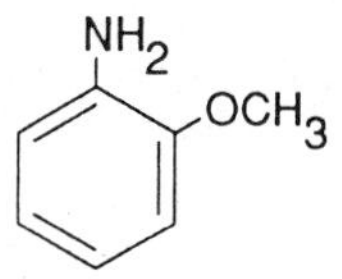

CAS Registry No.: 90-04-0
DOT: 2431
DOT label: Poison
Molecular formula: C_7H_9NO
Formula weight: 123.15
RTECS: BZ5410000

Physical state, color and odor
Colorless, yellow or pink to reddish liquid with an amine-like odor. Becomes brown on exposure to air.

Melting point (°C):
6.2 (Weast, 1986)
5 (Windholz et al., 1983)

Boiling point (°C):
224 (Weast, 1986)

Density (g/cm^3):
1.0923 at 20/4 °C (Weast, 1986)

Diffusivity in water (10^5 cm^2/sec):
0.82 at 20 °C using method of Hayduk and Laudie (1974).

Dissociation constant:
4.09 at 25 °C (Dean, 1973)

Flash point (°C):
118 (open cup, NFPA, 1984)

Henry's law constant (10^6 atm·m^3/mol):
1.25 at 25 °C (approximate - calculated from water solubility and vapor pressure)

Soil sorption coefficient, log K_{oc}:
Unavailable because experimental methods for estimation of this parameter for anilines are lacking in the documented literature.

Octanol/water partition coefficient, log K_{ow}:
0.95 (Leo et al., 1971)
1.18 (Camilleri et al., 1988)
1.23 (Unger et al., 1978)

Solubility in organics:
Soluble in acetone and benzene (Weast, 1986). Miscible with alcohol and ether (Windholz et al., 1983).

Solubility in water:
1 wt % at 20 °C (NIOSH, 1994)

Vapor density:
5.03 g/L at 25 °C, 4.25 (air = 1)

Vapor pressure (mmHg):
<0.1 at 25 °C (NIOSH, 1994)

Exposure Limits (mg/m^3): Potential occupational carcinogen. NIOSH REL: TWA 0.5, IDLH 50; OSHA PEL: TWA 0.5; ACGIH TLV: TWA 0.5.

Symptoms of Exposure: Absorption and inhalation may cause headache and dizziness (Patnaik, 1992).

Toxicity: Acute oral LD_{50} for rats 2,000 mg/kg, wild birds 422 mg/kg, mice 1,400 mg/kg, rabbits 870 mg/kg (RTECS, 1985).

Uses: Manufacture of azo dyes.

p-ANISIDINE

Synonyms: 4-Aminoanisole; *p*-Aminoanisole; 1-Amino-4-methoxybenzene; *p*-Aminomethoxybenzene; *p*-Aminomethylphenyl ether; 4-Anisidine; *p*-Anisylamine; 4-Methoxy-1-aminobenzene; 4-Methoxyaniline; *p*-Methoxyaniline; **4-Methoxybenzenamine**; *p*-Methoxybenzenamine; 4-Methoxyphenylamine; *p*-Methoxyphenylamine.

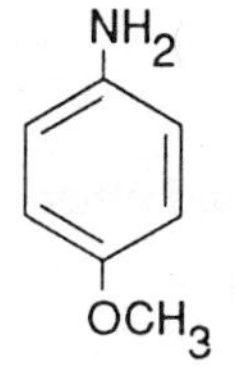

CAS Registry No.: 104-94-9
DOT: 2431
DOT label: Poison
Molecular formula: C_7H_9NO
Formula weight: 123.15
RTECS: BZ5450000

Physical state, color and odor
Light brown solid or crystals with a characteristic amine-like odor.

Melting point (°C):
57.2 (Weast, 1986)
57-60 (Aldrich, 1988)

Boiling point (°C):
243 (Weast, 1986)
246 (Windholz et al., 1983)
240-243 (Aldrich, 1988)

Density (g/cm³):
1.096 at 20/4 °C (Aldrich, 1988)

Diffusivity in water (10^5 cm²/sec):
0.73 at 20 °C using method of Hayduk and Laudie (1974).

Dissociation constant:
4.49 at 25 °C (Dean, 1973)

Ionization potential (eV):
7.44 (NIOSH, 1994)
7.82 (Franklin et al., 1969)

Soil sorption coefficient, log K_{oc}:
Unavailable because experimental methods for estimation of this parameter for anilines are lacking in the documented literature.

Octanol/water partition coefficient, log K_{ow}:
0.95 (Leo et al., 1971)
0.83 (Garst and Wilson, 1984)

Solubility in organics:
Soluble in acetone, alcohol, benzene and ether (Weast, 1986).

Solubility in water:
3.3 mg/L at 20-25 °C using method of Kenaga and Goring (1980).

Vapor pressure (mmHg):
6×10^{-3} at 25 °C (NIOSH, 1994)

Environmental Fate
   Chemical/Physical. Releases toxic nitrogen oxides when heated to decomposition (Sax and Lewis, 1987).

Exposure Limits (mg/m^3): NIOSH REL: TWA 0.5, IDLH 50; OSHA PEL: TWA 0.5; ACGIH TLV: 0.5.

Symptoms of Exposure: Anemia and cyanosis (Patnaik, 1992).

Toxicity: Acute oral LD_{50} for rats 1,400 mg/kg, mice 810 mg/kg, rabbits 2,900 mg/kg (RTECS, 1985).

Uses: Azo dyestuffs; chemical intermediate.

ANTHRACENE

Synonyms: Anthracin; Green oil; Paranaphthalene; Tetra olive N2G.

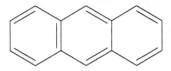

CAS Registry No.: 120-12-7
Molecular formula: $C_{14}H_{10}$
Formula weight: 178.24
RTECS: CA9350000

Physical state, color and odor
White to yellow crystalline flakes with a bluish or violet fluorescence and a weak aromatic odor.

Melting point (°C):
216.2-216.4 (Aldrich, 1988)
219.5 (Casellato et al., 1973)

Boiling point (°C):
339.9 (Dean, 1973)

Density (g/cm³):
1.283 at 25/4 °C (Weast, 1986)
1.24 at 20/4 °C (Weiss, 1986)

Diffusivity in water (10^5 cm²/sec):
0.59 at 20 °C using method of Hayduk and Laudie (1974).

Flash point (°C):
121.1 (Weiss, 1986)

Lower explosive limit (%):
0.6 (Weiss, 1986)

Dissociation constant:
>15 (Christensen et al., 1975)

Heat of fusion (kcal/mol):
6.9 (Tsonopoulos and Prausnitz, 1971)

Henry's law constant (10^5 atm·m³/mol):
140 (Petrasek et al., 1983)
6.51 at 25 °C (Southworth, 1979)
1.77 at 25 °C (Hine and Mookerjee, 1975)
1.93 at 25 °C (Fendinger and Glotfelty, 1990)

Ionization potential (eV):
7.55 (Franklin et al., 1969)
7.58 (Yoshida et al., 1983)
7.43 (Cavalieri and Rogan, 1985)

Bioconcentration factor, log BCF:
3.83 (activated sludge), 2.96 (golden ide) (Freitag et al., 1985)
2.58 (*Daphnia pulex*, Southworth et al., 1978)
2.95 (bluegill sunfish, Spacie et al., 1983)
2.21 (goldfish, Ogata et al., 1984)
3.89 (algae, Geyer et al., 1984)

Soil sorption coefficient, log K_{oc}:
4.27 (aquifer sands, Abdul et al., 1987)
4.41 (Karickhoff et al., 1979)
4.205 (Nkedi-Kizza et., 1985)
4.93 (Gauthier et al., 1986)
5.76 (average, Kayal and Connell, 1990)
4.11 (fine sand, Enfield et al., 1989)

Octanol/water partition coefficient, log K_{ow}:
4.45 (Hansch and Fujita, 1964; DeKock and Lord, 1987)
4.54 (Miller et al., 1985)
4.34 (Mackay, 1982)
4.63 (Bruggeman et al., 1982)

Solubility in organics:
Ethanol (14.9 g/L), methanol (14.3 g/L), benzene (16.1 g/L), carbon disulfide (32.3 g/L), carbon tetrachloride (11.6 g/L), chloroform (11.8 g/L) and toluene (8.0 g/L) (Windholz et al., 1983)

6.6, 21 and 16 mmol/L at 25 °C in isooctane, butyl ether and pentyl ether, respectively (Anderson et al., 1980)

In *N,N*-dimethylformamide, g/kg (°C): 13.3414 (29.8), 16.9352 (34.8), 19.9337 (39.6), 22.5539 (44.2), 27.1358 (49.6). In 1,4-dioxane, g/kg (°C): 2.0787 (29.8), 3.7332 (34.8), 5.4112 (39.6), 8.3659 (44.2), 13.6541 (49.6). In ethylene glycol, g/kg (°C): 0.4384 (64.8), 0.7955 (78.8), 1.0680 (86.8), 1.5346 (97.8), 2.3934 (110.8), 3.3500 (123.8), 6.5100 (146.2), 8.5000 (159.8) (Cepeda et al., 1989)

Solubility in water:

At 20 °C: 180, 118, 107 and 126 nmol/L in distilled water, Pacific seawater, artificial seawater and 35% NaCl, respectively (Hashimoto et al., 1984)

44.6 μg/kg at 25 °C, 57.0 μg/kg at 29 °C. In seawater (salinity = 35.0 g/kg): 31.1 μg/kg at 25 °C (May et al., 1978a)

75 μg/L at 27 °C (Davis et al., 1942; Klevens, 1950)

73 μg/L at 25 °C (Mackay and Shiu, 1977)

70 μg/L at 23 °C (Pinal et al., 1991)

30 μg/L at 25 °C (Schwarz and Wasik, 1976)

112.5 μg/L at 25 °C (Sahyun, 1966)

In mg/kg: 0.119-0.125 at 35.4 °C, 0.148-0.152 at 39.3 °C, 0.206-0.210 at 44.7 °C, 0.279 at 47.5 °C, 0.297-0.302 at 50.1 °C, 0.389-0.402 at 54.7 °C, 0.480-0.525 at 59.2 °C, 0.62-0.72 at 64.5 °C, 0.64-0.67 at 65.1 °C, 0.92 at 69.8 °C, 0.90-0.97 at 70.7 °C, 0.91 at 71.9 °C, 1.13-1.26 at 74.7 °C (Wauchope and Getzen, 1972)

In μg/kg: 12.7 at 5.2 °C, 17.5 at 10.0 °C, 22.2 at 14.1 °C, 29.1 at 18.3 °C, 37.2 at 22.4 °C, 43.4 at 24.6 °C, 55.7 at 28.7 °C (May et al., 1978)

In nmol/L: 131 at 8.6 °C, 137 at 11.1 °C, 144 at 12.2 °C, 154 at 14 °C, 166 at 15.5 °C, 181 at 18.2 °C, 222 at 20.3 °C, 234 at 23.0 °C, 230 at 25.0 °C, 267 at 26.2 °C, 325 at 28.5 °C, 390 at 31.3 °C. In 0.5 M NaCl (nmol/L): 93 at 8.6 °C, 101 at 8.6 °C, 122 at 11.7 °C, 147 at 19.2 °C, 168 at 21.5 °C, 204 at 25.0 °C, 192 at 25.3 °C, 202 at 27.1 °C, 246 at 30.2 °C (Schwarz, 1977)

41 μg/L at 20 °C (Kishi and Hashimoto, 1989)

250 nmol/L at 25 °C (Akiyoshi et al., 1987; Wasik et al., 1983)

70 μg/L at 23 °C (Pinal et al., 1990)

At pH 9 containing humic acids (wt %) derived from Sagami Bay: 470 μg/L (0.02), 172 μg/L (0.04), 157 μg/L (0.06), 154 μg/L (0.09), 343 μg/L (0.12) (Shinozuka et al., 1987)

69.8 μg/L at 25 °C (Walters and Luthy, 1984)

In nmol/L: 53.9, 72.4, 99.3, 133, 181 and 248 at 4.6, 8.8, 12.9, 17.0, 21.1 and 25.3 °C, respectively. In seawater (salinity = 36.5 g/kg): 37.9, 50.7, 68.4, 97.6, 131 and 182 at 4.6, 8.8, 12.9, 17.0, 21.1 and 25.3 °C, respectively (Whitehouse, 1984)

58 μg/L at 25 °C (Vadas et al., 1991)

44.3 and 34 μg/L at 25 °C (Billington et al., 1988)

Vapor pressure (10^5 mmHg):

0.60 at 25 °C (Wasik et al., 1983)

19.5 at 25 °C (Radding et al., 1976)

51.7, 75 at 25 °C (Hinckley et al., 1990)

669 at 85.25 °C, 1,020 at 90.15 °C (Macknick and Prausnitz, 1979)

8.6 at 65.7 °C, 10.5 at 67.10 °C, 11.8 at 68.75 °C (Bradley and Cleasby, 1953)

0.43 at 25 °C (McVeety and Hites, 1982)

33,000 at 127 °C (Eiceman and Vandiver, 1983)

0.56 at 25 °C (de Kruif, 1980)

48 at 25 °C (Bidleman, 1984)

1.25, 1.91 and 2.91 at 95, 100 and 105 °C, respectively (Kelley and Rice, 1964)

Environmental Fate

Biological. Catechol is the central metabolite in the bacterial degradation of anthracene. Intermediate byproducts included 3-hydroxy-2-naphthoic acid and salicylic acid (Chapman, 1972). Anthracene was statically incubated in the dark at 25 °C with yeast extract and settled domestic wastewater inoculum. Significant biodegradation with gradual adaptation was observed. At concentrations of 5 and 10 mg/L, biodegradation yields at the end of 4 weeks of incubation were 92 and 51%, respectively (Tabak et al., 1981). A mixed bacterial community isolated from seawater foam degraded anthraquinone, a photodegradation product of anthracene, to traces of benzoic and phthalic acids (Rontani et al., 1975). In activated sludge, only 0.3% mineralized to carbon dioxide after 5 days (Freitag et al., 1985).

Soil. In a 14-day experiment, [^{14}C]anthracene applied to soil-water suspensions under aerobic and anaerobic conditions gave $^{14}CO_2$ yields of 1.3 and 1.8%, respectively (Scheunert et al., 1987).

Photolytic. Oxidation of anthracene adsorbed on silica gel or alumina by oxygen in the presence of UV-light yielded anthraquinone. This compound further oxidized to 1,4-dihydroxy-9,10-anthraquinone. Anthraquinone was also formed by the oxidation of anthracene in diluted nitric acid or nitrogen oxides (Nikolaou et al., 1984) and in the dark when adsorbed on fly ash (Korfmacher et al., 1980). Irradiation of anthracene (2.6 mmol/L) in cyclohexanone solutions gave 9,10-anthraquinone as the principal product (Korfmacher et al., 1980). Photocatalysis of anthracene and sulfur dioxide at -25 °C in various solvents yielded anthracene-9-sulfonic acid (Nielsen et al., 1983).

A carbon dioxide yield of 16.0% was achieved when anthracene adsorbed on silica gel was irradiated with light (λ >290 nm) for 17 hours (Freitag et al., 1985).

Chemical/Physical. In urban air from St. Louis, MO, anthracene reacted with NO_x to form 9-nitroanthracene (Ramdahl et al., 1982).

Exposure Limits: Potential occupational carcinogen. No individual standards have been set, however, as a constituent in coal tar pitch volatiles, the following exposure limits have been established (mg/m^3): NIOSH REL: TWA 0.1 (cyclohexane-extractable fraction), IDLH 80; OSHA PEL: TWA 0.2 (benzene-soluble fraction); ACGIH TLV: TWA 0.2 (benzene solubles).

Toxicity: Intraperitoneal LD$_{50}$ for mice is 430 mg/kg (RTECS, 1985).

Drinking Water Standard: As of May 1994, no MCLGs or MCLs have been proposed (U.S. EPA, 1994).

Uses: Dyes; starting material for the preparation of alizarin, phenanthrene, carbazole, 9,10-anthraquinone, 9,10-dihydroanthracene and insecticides; in calico printing; as component of smoke screens; scintillation counter crystals; organic semiconductor research; wood preservative.

ANTU

Synonyms: Anturat; Bantu; Chemical 109; Krysid; **1-Naphthalenylthiourea**; 1-(1-Naphthyl)-2-thiourea; α-Naphthylthiourea; *N*-1-Naphthylthiourea; α-Naphthyl-thiocarbamide; Rattrack.

$$\underset{\text{NHCNH}_2}{\overset{\overset{\textstyle S}{\|}}{}}$$

CAS Registry No.: 86-88-4
DOT: 1651
Molecular formula: $C_{11}H_{10}N_2S$
Formula weight: 202.27
RTECS: YT9275000

Physical state, color and odor
Colorless to gray, odorless solid.

Melting point (°C):
198 (Weast, 1986)

Boiling point (°C):
Decomposes (NIOSH, 1994)

Density (g/cm³):
1.895 using method of Lyman et al. (1982).

Diffusivity in water (10^5 cm²/sec):
0.56 at 20 °C using method of Hayduk and Laudie (1974).

Flash point (°C):
Not applicable (NIOSH, 1994)

Lower explosive limit (%):
Not applicable (NIOSH, 1994)

Upper explosive limit (%):
Not applicable (NIOSH, 1994)

Solubility in organics:
Solubility (g/L): acetone (24.3), triethylene glycol (86) (Windholz et al., 1983).

66

Solubility in water:
600 mg/L at 20 °C (Windholz et al., 1983)

Environmental Fate
   Chemical/Physical. The hydrolysis rate constant for ANTU at pH 7 and 25 °C was determined to be 8 x 10^{-5}/hour, resulting in a half-life of 361 days (Ellington et al., 1988).
   Emits toxic fumes of nitrogen and sulfur oxides when heated to decomposition (Lewis, 1990).

Exposure Limits (mg/m^3): NIOSH REL: TWA 0.3, IDLH 100; OSHA PEL: TWA 0.3.

Symptoms of Exposure: Vomiting, dyspnea, cyanosis, course pulmonary rales after ingestion of large doses (NIOSH, 1994).

Toxicity: Acute oral LD_{50} for Norwegian rats 6-8 mg/kg (Hartley and Kidd, 1987).

Use: Rodenticide.

BENZENE

Synonyms: Annulene; Benxole; Benzol; Benzole; Benzolene; Bicarburet of hydrogen; Carbon oil; Coal naphtha; Coal tar naphtha; Cyclohexatriene; Mineral naphthalene; Motor benzol; NCI-C55276; Nitration benzene; Phene; Phenyl hydride; Pyrobenzol; Pyrobenzole; RCRA waste number U019; UN 1114.

CAS Registry No.: 71-43-2
DOT: 1114
DOT label: Flammable liquid
Molecular formula: C_6H_6
Formula weight: 78.11
RTECS: CY1400000

Physical state, color and odor
Clear, colorless to light yellow watery liquid with an aromatic or gasoline-like odor. Odor threshold in air is 0.84 ppm (Keith and Walters, 1992).

Melting point (°C):
5.533 (Standen, 1964)

Boiling point (°C):
80.100 (Standen, 1964)

Density (g/cm³):
0.8765 at 20/4 °C (Weast, 1986)
0.8784 at 20/4 °C, 0.8680 at 30/4 °C, 0.8572 at 40/4 °C (Sumer et al., 1968)
0.87378 at 25/4 °C (Kirchnerová and Cave, 1976)

Diffusivity in water (10^5 cm²/sec):
1.02 at 20 °C (Witherspoon and Bonoli, 1969)
1.09 at 25 °C (Hayduk and Laudie, 1974)

Flash point (°C):
-11 (NIOSH, 1994)

Lower explosive limit (%):
1.2 (NIOSH, 1994)

Upper explosive limit (%):
7.8 (NIOSH, 1994)

Dissociation constant:
≈ 37 (Gordon and Ford, 1972)

Heat of fusion (kcal/mol):
2.370 (Tsonopoulos and Prausnitz, 1971)

Henry's law constant (10^3 atm·m^3/mol):
5.48 at 25 °C (Hine and Mookerjee, 1975)
5.43 at 25 °C (Mackay and Shiu, 1981)
3.30, 3.88, 4.52, 5.28 and 7.20 at 10, 15, 20, 25 and 30 °C, respectively (Ashworth et al., 1988)
2.86, 3.75, 4.54, 5.96 and 7.31 at 10, 15, 20, 25 and 30 °C, respectively (Perlinger et al., 1993)
1.73, 2.20, 2.38, 3.70 and 4.75 at 2.0, 6.0, 10.0, 18.2 and 25.0 °C, respectively (Dewulf et al., 1995)

Interfacial tension with water (dyn/cm at 25 °C):
34.1 (Donahue and Bartell, 1952)

Ionization potential (eV):
9.25 (Lo et al., 1986)
9.56 (Yoshida et al., 1983)

Bioconcentration factor, log BCF:
0.54 (eels, Ogata and Miyake, 1978)
1.10 (fathead minnow, Veith et al., 1980)
0.63 (goldfish, Ogata et al., 1984)
3.23 (activated sludge, Freitag et al., 1985)
1.48 (algae, Geyer et al., 1984)

Soil sorption coefficient, log K_{oc}:
1.69 (aquifer sands, Abdul et al., 1987)
1.92 (Schwarzenbach and Westall, 1981)
1.96, 2.00 (Rogers et al., 1980)
1.58, 1.64, 1.73 (various Norwegian soils, Seip et al., 1986)
1.50 (Woodburn silt loam soil, Chiou et al., 1983)
1.42 (estuarine sediment, Vowles and Mantoura, 1987)
2.10, 2.40 (Allerod), 2.30 (Borris, Brande), 2.55 (Brande), 2.15 (Finderup), 2.65, 2.68 (Gunderup), 2.48 (Herborg), 2.92 (Rabis), 2.40, 2.50 (Tirstrup), 2.05 (Tylstrup), 2.70 (Vasby), 2.38, 2.78 (Vejen), 2.85, 2.95, 2.28 (Vorbasse) (Larsen

et al., 1992)

1.74 (Captina silt loam), 1.81 (McLaurin sandy loam) (Walton et al., 1992)

From crude oil: 0.68 (Grimsby silt loam), 1.00 (Vaudreil sandy loam), 1.54 (Wendover silty clay), -0.53 (Rideau silty clay) (Nathwani and Phillips, 1977)

1.73 (river sediment), 1.70 (coal wastewater sediment) (Kopinke et al., 1995)

Octanol/water partition coefficient, log K_{ow}:
2.13 (DeKock and Lord, 1987; Hansch and Fujita, 1964)
2.11 (Mackay, 1982)
1.56 (Rogers and Cammarata, 1969)
2.15 (Campbell and Luthy, 1985; Leo et al., 1971)
2.12 (Veith et al., 1980)
1.95 (Eadsforth, 1986)
2.186 (de Bruijn et al., 1989)
2.20 (Hammers et al., 1982)
2.16, 2.28 (Suntio et al., 1988)

Solubility in organics:
Miscible with ethanol, ether, glacial acetic acid, acetone, chloroform, carbon tetrachloride (U.S. EPA, 1985), carbon disulfide, oils (Windholz et al., 1983) and hexane (Corby and Elworthy, 1971).

Solubility in water:
0.181 wt % at 25 °C (Lo et al., 1986)
1,820 mg/L at 22 °C (Chiou et al., 1977)
1,800 mg/L at 25 °C (Howard and Durkin, 1974; Klevens, 1950)
1,790 mg/L at 25 °C (Bohon and Claussen, 1951; Wasik et al., 1983)
1,750 mg/L at 25 °C (Banerjee et al., 1980)
1,850 mg/kg at 30 °C (Gross and Saylor, 1931)
1,755 mg/L at 25 °C (McDevit and Long, 1952)
1,740 mg/L at 25 °C (Andrews and Keefer, 1949)
1,780 mg/kg at 25 °C (McAuliffe, 1963, 1966)
1,791 µg/kg at 25 °C (May et al., 1978a)
In wt %: 0.153 at 0 °C, 0.163 at 10 °C, 0.175 at 20 °C, 0.180 at 25 °C, 0.190 at 30 °C, 0.206 at 40 °C, 0.225 at 50 °C, 0.250 at 60 °C, 0.277 at 70 °C, 0.344 at 80 °C, 0.393 at 90 °C, 0.504 at 107.4 °C (Stephen and Stephen, 1963)
1,755 mg/kg at 25 °C (Polak and Lu, 1973)
1,740 mg/kg at 25 °C, 1,391 mg/kg in artificial seawater (34.472 mg NaCl/kg) at 25 °C (Price, 1976)
1,710 mg/L at 20 °C (Freed et al., 1977)
21.7 mmol/L at 25.00 °C (Keely et al., 1988)
1,860 mg/kg at 25 °C (Stearns et al., 1947)
1,000 mg/L in fresh water at 25 °C), 1,030 mg/L in salt water at 25 °C

(Krasnoshchekova and Gubergrits, 1975)

0.18775 wt % at 23.5 °C (Schwarz, 1980)

23.3 mmol/L at 25 °C (Ben-Naim and Wilf, 1980)

17.76 mmol/L in 0.5 M NaCl at 25 °C (Wasik et al., 1984)

24.2 mmol/L at 35 °C (Hine et al., 1963)

1.74 g/kg at 25 °C (Chey and Calder, 1972)

20.7, 20.2, 20.7, 21.8 and 22.8 mmol/L at 5, 15, 25, 35 and 45 °C, respectively (Sanemasa et al., 1982)

32 mmol/L at 25 °C (Hogfeldt and Bolander, 1963)

22 mmol/L at 25 °C (Taha et al., 1966)

23.6 and 24.3 mmol/kg at 30 and 35 °C, respectively (Saylor et al., 1938)

1,510 mg/L at 25 °C (McBain and Lissant, 1951)

1,779.5 mg/L at 25 °C (Mackay and Shiu, 1973)

2,170 mg/L at 25 °C (Worley, 1967)

1,650 mg/L (Coutant and Keigley, 1988)

22.0 mmol/kg at 25 °C (Morrison and Billett, 1952)

In g/kg: 1.84, 1.85, 1.81, 1.81, 1.77, 1.77, 1.79, 1.79 and 1.76 at 4.5, 6.3, 7.1, 9.0, 11.8, 12.1, 15.1, 17.9 and 20.1 °C, respectively. In artificial seawater (g/kg): 1.323, 1.376, 1.347, 1.318 and 1.296 at 0.19, 5.32, 10.05, 14.96 and 20.04 °C, respectively (Brown and Wasik, 1974)

1.76 g/L at 25 °C (Brady and Huff, 1958)

In g/kg: 1.79, 1.77, 1.80, 1.83, 1.92, 2.03, 2.14, 2.34 and 2.57 at 9.4, 16.8, 24.0, 31.0, 38.0, 44.7, 51.5, 58.8 and 65.4 °C, respectively (Alexander, 1959)

10^4 mole fraction (°C): 3.95 (17.0), 3.97 (22.0), 3.99 (26.0), 4.02 (29.0), 4.12 (32.0), 4.20 (35.0), 4.39 (40.5), 4.40 (42.0), 4.45 (44.0), 4.57 (46.0), 4.78 (51.0), 5.03 (56.0), 5.31 (61.0), 5.42 (63.0) (Franks et al., 1963)

21.8 mmol/L at 20 °C (Corby and Elworthy, 1971)

In wt % (°C): 1.283 (153), 1.913 (178), 2.902 (204), 3.790 (225), 4.471 (241), 5.073 (154) (Guseva and Parnov, 1963).

1,765 mg/L at 25 °C (Leinonen and Mackay, 1973)

2.0403 mL/L at 25 °C (Sada et al., 1975)

24.4 mmol/kg at 25.0 °C (Vesala, 1974)

15.4 mmol/L at 25.00 °C (Sanemasa et al., 1985)

4.03×10^{-4} at 25 °C (mole fraction, Li et al., 1993)

Vapor density:
3.19 g/L at 25 °C, 2.70 (air = 1)

Vapor pressure (mmHg):
60 at 15 °C, 76 at 20 °C, 118 at 30 °C (Verschueren, 1983)
95.2 at 25 °C (Mackay and Leinonen, 1975)
397 at 60.3 °C, 556 at 70.3 °C, 764 at 80.3 °C, 1,031 at 90.3 °C (Eon et al., 1971)
146.8 at 35 °C (Hine et al., 1963)

95 at 25 °C (Milligan, 1924)
93.56 at 25.00 °C (Hussam and Carr, 1985)

Environmental Fate

Biological. A mutant of *Pseudomonas putida* dihydroxylyzed benzene into *cis*-benzene glycol, accompanied by partial dehydrogenation, yielding catechol (Dagley, 1972). Bacterial dioxygenases can cleave catechol at the *ortho* and *meta* positions to yield *cis,cis*-muconic acid and α-hydroxymuconic semialdehyde, respectively (Chapman, 1972). Pure microbial cultures hydroxylated benzene to phenol and two unidentified phenols (Smith and Rosazza, 1974). Muconic acid was reported to be the biooxidation product of benzene by *Nocardia corallina* V-49 using hexadecane as the substrate (Keck et al., 1989). In activated sludge, 29.2% of the applied benzene mineralized to carbon dioxide after 5 days (Freitag et al., 1985). In anoxic groundwater near Bemidji, MI, benzene was anaerobically biodegraded to phenol (Cozzarelli et al., 1990). When benzene was statically incubated in the dark at 25 °C with yeast extract and settled domestic wastewater inoculum, significant biodegradation with rapid adaptation was observed. At concentrations of 5 and 10 mg/L, 49 and 37% biodegradation, respectively, were observed after 7 days. After 14 days of incubation, benzene demonstrated complete dissimilation (Tabak et al., 1981).

Chemical/Physical. Titanium dioxide suspended in an aqueous solution and irradiated with UV light (λ = 365 nm) converted benzene to carbon dioxide at a significant rate (Matthews, 1986). Irradiation of benzene in an aqueous solution yields mucondialdehyde. Photolysis of benzene vapor at 1849-2000 Å yields ethylene, hydrogen, methane, ethane, toluene and a polymer resembling cuprene. Other photolysis products reported under different conditions include fulvene, acetylene, substituted trienes (Howard, 1990), phenol, 2-nitrophenol, 4-nitrophenol, 2,4-dinitrophenol, 2,6-dinitrophenol, nitrobenzene, formic acid and peroxyacetyl nitrate (Calvert and Pitts, 1966). Under atmospheric conditions, the gas-phase reaction with hydroxyl radicals and nitrogen oxides resulted in the formation of phenol and nitrobenzene (Atkinson, 1990). A carbon dioxide yield of 40.8% was achieved when benzene adsorbed on silica gel was irradiated with light (λ >290 nm) for 17 hours.

Kanno et al. (1982) studied the aqueous reaction of benzene and other aromatic hydrocarbons (toluene, xylene and naphthalene) with hypochlorous acid in the presence of ammonium ion. They reported that the aromatic ring was not chlorinated as expected (forming chlorobenzene) but was cleaved by chloramine forming cyanogen chloride (Kanno et al., 1982). The amount of cyanogen chloride formed was inversely proportional to the pH of the solution. At pH 6, the greatest amount of cyanogen chloride was formed when the reaction mixture contained ammonium ion and hypochlorous acid at a ratio of 2:3 (Kanno et al., 1982). In the gas phase, benzene reacted with nitrate radicals in purified air forming nitrobenzene (Chiodini et al., 1993).

Exposure Limits (ppm): Potential occupational carcinogen. NIOSH REL: TWA 0.1, STEL 1, IDLH 500; OSHA PEL: TWA 1, STEL 5; ACGIH TLV: TWA 10.

Symptoms of Exposure: Hallucination, distorted perception, euphoria, somnolence, nausea, vomiting and headache. Narcotic at air concentrations of 200 ppm. At higher concentrations, convulsions may occur. Eye, nose and respiratory irritant (Patnaik, 1992).

Toxicity: Acute oral LD_{50} for mice 4,700 ppm, rats 3,306 mg/kg; LC_{50} (inhalation) for mice 9,980 ppm, rats 3,306 ppm/7-hour (RTECS, 1985).

Drinking Water Standard (final): MCLG: zero; MCL: 5 μg/L (U.S. EPA, 1994).

Uses: Manufacture of ethylbenzene (preparation of styrene monomer), dodecylbenzene (for detergents), cyclohexane (for nylon), nitrobenzene, aniline, maleic anhydride, biphenyl, benzene hexachloride, benzene sulfonic acid, phenol, dichlorobenzenes, insecticides, pesticides, fumigants, explosives, aviation fuel, flavors, perfume, medicine, dyes and many other organic chemicals; paints, coatings, plastics and resins; food processing; photographic chemicals; nylon intermediates; paint removers; rubber cement; antiknock gasoline; solvent for fats, waxes, resins, inks, oils, paints, plastics and rubber.

BENZIDINE

Synonyms: Azoic diazo component 112; Benzidine base; *p*-Benzidine; 4,4'-Bi-aniline; *p,p'*-Bianiline; **(1,1'-Biphenyl)-4,4'-diamine;** 4,4'-Biphenyldiamine; *p,p'*-Biphenyldiamine; 4,4'-Biphenylenediamine; *p,p'*-Biphenylenediamine; C.I. 37225; C.I. azoic diazo component 112; 4,4'-Diaminobiphenyl; *p,p'*-Diaminobi-phenyl; 4,4'-Diamino-1,1'-biphenyl; 4,4'-Diaminodiphenyl; *p*-Diaminodiphenyl; *p,p'*-Diaminodiphenyl; 4,4'-Dianiline; *p,p'*-Dianiline; 4,4'-Diphenylenediamine; *p,p'*-Diphenylenediamine; Fast corinth base B; NCI-C03361; RCRA waste number U021; UN 1885.

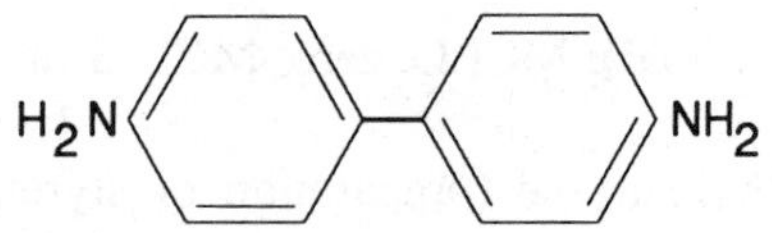

CAS Registry No.: 92-87-5
DOT: 1885
DOT label: Poison
Molecular formula: $C_{12}H_{12}N_2$
Formula weight: 184.24
RTECS: DC9625000

Physical state, color and odor
Grayish-yellow powder or white to pale reddish, odorless crystals. Darkens on exposure to air or light.

Melting point (°C):
128 (Weast, 1986)
117.2 (ACGIH, 1986)

Boiling point (°C):
401.7 (Sax, 1984)
400 (Patnaik, 1992)

Density (g/cm³):
1.250 at 20/4 °C (Shriner et al., 1978)

Diffusivity in water (10^5 cm²/sec):
0.57 at 20 °C using method of Hayduk and Laudie (1974).

Flash point (°C):
Combustible solid, but difficult to burn (NIOSH, 1994).

Dissociation constant (25 °C):
pK_1 = 3.63, pK_2 = 4.70 (Dean, 1973)

Henry's law constant (10^{11} atm·m^3/mol):
3.88 at 25 °C (estimated, Howard, 1989)

Soil sorption coefficient, log K_{oc}:
5.72 (average using 4 soils, Graveel et al., 1986)
3.46 (Meylan et al., 1992)

Octanol/water partition coefficient, log K_{ow}:
1.34 (Mabey et al., 1982)
1.63 (Hassett et al., 1980)

Solubility in organics:
Soluble in ethanol (U.S. EPA, 1985) and ether (1 g/50 mL) (Windholz et al., 1983).

Solubility in water (mg/L):
400 at 12 °C, 9,400 at 100 °C (Verschueren, 1983)
500 at 25 °C (Morrison and Boyd, 1971)
360 at 24 °C (Hassett et al., 1980)
520 at 25 °C (Shriner et al., 1978)

Vapor density:
7.50 g/L presumably at 20 °C (Sims et al., 1988)

Vapor pressure (mmHg):
Based on the specific vapor density value of 6.36 (Sims et al., 1988), the vapor pressure was calculated to be 0.83 at 20 °C.

Environmental Fate
Biological. In activated sludge, <0.1% mineralized to carbon dioxide after 5 days (Freitag et al., 1985).

Soil. Benzidine was added to different soils and incubated in the dark at 23 °C under a carbon dioxide-free atmosphere. After 1 year, 8.3 to 11.6% of the added benzidine degraded to carbon dioxide primarily by microbial metabolism and partially by hydrolysis (Graveel et al., 1986). Tentatively identified biooxidation compounds using GC/MS include hydroxybenzidine, 3-hydroxybenzidine, 4-amino-4'-nitrobiphenyl, N,N'-dihydroxybenzidine, 3,3'-dihydroxybenzidine and 4,4'-dinitrobiphenyl (Baird et al., 1977).

In the presence of hydrogen peroxide and acetylcholine at pH 11 and 20 °C, benzidine oxidized to 4-amino-4'-nitrobiphenyl (Aksnes and Sandberg, 1957).

Photolytic. A carbon dioxide yield of 40.8% was achieved when benzene adsorbed on silica gel was irradiated with light (λ >290 nm) for 17 hours (Freitag et al., 1985).

Exposure Limits: Potential occupational carcinogen.

Toxicity: Acute oral LD_{50} for mice 214 mg/kg, rats 309 mg/kg (RTECS, 1985); LC_{50} (48-hour) for red killifish 57.5 mg/L (Yoshioka et al., 1986).

Uses: Organic synthesis; manufacture of azo dyes, especially Congo Red; detection of blood stains; stain in microscopy; laboratory reagent in determining cyanide, sulfate, nicotine and some sugars; stiffening agent in rubber compounding.

BENZO[a]ANTHRACENE

Synonyms: BA; B(a)A; Benzanthracene; **Benz[a]anthracene**; 1,2-Benzanthracene; 1,2-Benz[a]anthracene; 2,3-Benzanthracene; Benzanthrene; 1,2-Benzanthrene; Benzoanthracene; 1,2-Benzoanthracene; Benzo[a]phenanthrene; Benzo[b]phenanthrene; 2,3-Benzophenanthrene; Naphthanthracene; RCRA waste number U018; Tetraphene.

CAS Registry No.: 56-55-3
Molecular formula: $C_{18}H_{12}$
Formula weight: 228.30
RTECS: CV9275000

Physical state and color
Colorless leaflets or plates with a greenish-yellow fluorescence.

Melting point (°C):
162 (Weast, 1986)
161.1 (Casellato et al., 1973)
156.9 (Murray et al., 1974)

Boiling point (°C):
435 with sublimation (Weast, 1986)
437.6 (Aldrich, 1988)
400 (Sims et al., 1988)

Density (g/cm^3 at 20 °C):
1.274 (HSDB, 1989)
1.2544 (Mailhot and Peters, 1988)

Diffusivity in water (10^5 cm^2/sec):
0.52 at 20 °C using method of Hayduk and Laudie (1974).

Dissociation constant:
>15 (Christensen et al., 1975)

Henry's law constant (10^7 atm·m^3/mol):
80 (Southworth, 1979)

Ionization potential (eV):
8.01 (Franklin et al., 1969)
7.45 (Yoshida et al., 1983)
7.54 (Cavalieri and Rogan, 1985)

Bioconcentration factor, log BCF:
4.00 (*Daphnia pulex*, Southworth et al., 1978)
4.00 (fathead minnow, Veith et al., 1979)
4.01 (*Daphnia magna*, Newsted and Giesy, 1987)
4.39 (activated sludge), 3.50 (algae), 2.54 (golden ide) (Freitag et al., 1985)

Soil sorption coefficient, log K_{oc}:
5.0 (Meylan et al., 1992)
6.30 (average, Kayal and Connell, 1990)

Octanol/water partition coefficient, log K_{ow}:
5.61 (Radding et al., 1976)
5.91 (Yoshida et al., 1983)

Solubility in organics:
Soluble in ethanol, ether, acetone and benzene (U.S. EPA, 1985).

Solubility in water:
10 μg/L at 25 °C (Klevens, 1950)
14 μg/L at 25 °C (Mackay and Shiu, 1977)
9.4 and 12.2 μg/kg at 25 and 29 °C, respectively. In seawater (salinity = 35.0 g/kg):
    5.6 μg/kg at 25 °C (May et al., 1978a)
11 μg/L at 27 °C (Davis et al., 1942)
44 μg/L at 24 °C (practical grade, Hollifield, 1979)
5.7 μg/L at 20 °C (Smith et al., 1978)
16.8 μg/L at 25 °C (Walters and Luthy, 1984)
In nmol/L: 21.3, 18.9, 18.8, 20.6, 27.6 and 37.4 at 3.7, 8.0, 12.4, 16.7, 20.9 and 25.0
    °C, respectively. In seawater (salinity = 32.1 g/kg): 118, 65.4, 57.2, 38.5, 36.7 and
    40.9 at 3.7, 8.0, 12.4, 16.7, 20.9 and 25.0 °C, respectively (Whitehouse, 1984)
94.4 μg/L at 25 °C (Billington et al., 1988)

Vapor pressure (10^9 mmHg):
5 at 20 °C (Pupp et al., 1974)
210 at 25 °C (Sonnefeld et al., 1983)
2,250 at 25 °C (Bidleman, 1984)
30.5 at 25 °C (Banerjee et al., 1990)
110.3 at 25 °C (Murray et al., 1974)
54.8 at 25 °C (de Kruif, 1980)

Environmental Fate

Biological. In an enclosed marine ecosystem containing planktonic primary production and heterotrophic benthos, the major metabolites were water soluble and could not be extracted with organic solvents. The only degradation product identified was benzo[a]anthracene-7,12-dione (Hinga and Pilson, 1987). Under aerobic conditions, *Cunninghanella elegans* degraded benzo[a]anthracene to 3,4-, 8,9- and 10,11-dihydrols (Kobayashi and Rittman, 1982; Riser-Roberts, 1992).

A strain of *Beijerinckia* was able to oxidize benzo[a]anthracene producing 1-hydroxy-2-anthranoic acid as the major product. Two other metabolites identified were 2-hydroxy-3-phenanthroic acid, 3-hydroxy-2-phenanthroic acid and *cis*-1,1-dihydroxy-1,2-dihydrobenzo[a]anthracene (Gibson et al., 1975; Mahaffey et al., 1988).

In a marine microcosm containing Narragansett Bay sediments, the polychaete *Mediomastis ambesita* and the bivalve *Nucula anulata*, benzo[a]anthracene degraded to carbon dioxide, phenols and quinones (Hinga et al., 1980).

In activated sludge, <0.1% mineralized to carbon dioxide after 5 days (Freitag et al., 1985). When benzo[a]anthracene (5 and 10 mg/L) was statically incubated in the dark at 25 °C with yeast extract and settled domestic wastewater inoculum, no significant biodegradation was observed (Tabak et al., 1981).

Photolytic. Benzo[a]anthracene-7,12-dione formed from the photolysis of benzo[a]anthracene (λ = 366 nm) in an air-saturated, acetonitrile-water solvent (Smith et al., 1978).

A carbon dioxide yield of 25.3% was achieved when benzo[a]anthracene adsorbed on silica gel was irradiated with light (λ >290 nm) for 17 hours (Freitag et al., 1985).

Chemical/Physical. Benzo[a]anthracene-7,12-dione and a monochlorinated product were formed during the chlorination of benzo[a]anthracene. At pH 4, the reported half-lives at chlorine concentrations of 0.6 and 10 mg/L were 2.3 and <0.2 hours, respectively (Mori et al., 1991).

Exposure Limits: Potential occupational carcinogen. No individual standards have been set, however, as a constituent in coal tar pitch volatiles, the following exposure limits have been established (mg/m^3): NIOSH REL: TWA 0.1 (cyclohexane-extractable fraction), IDLH 80; OSHA PEL: TWA 0.2 (benzene-soluble fraction); ACGIH TLV: TWA 0.2 (benzene solubles).

Toxicity: LD$_{50}$ for mice by intravenous injection 10 mg/kg (Patnaik, 1992).

Drinking Water Standard (proposed): MCLG: zero; MCL: 0.1 μg/L (U.S. EPA, 1984).

Uses: Organic synthesis. Not manufactured commercially but is derived from industrial and experimental coal gasification operations where maximum concentration detected in gas, liquid and coal tar streams were 28, 4.1 and 18 mg/m^3, respectively (Cleland, 1981).

BENZO[*b*]FLUORANTHENE

Synonyms: **Benz[*e*]acephenanthrylene**; 3,4-Benz[*e*]acephenanthrylene; 2,3-Benz-fluoranthene; 3,4-Benzfluoranthene; Benzo[*b*]fluoranthene; Benzo[*e*]fluoranthene; 2,3-Benzofluoranthene; 3,4-Benzofluoranthene; 3,4-Benzo[*b*]fluoranthene; B(*b*)F.

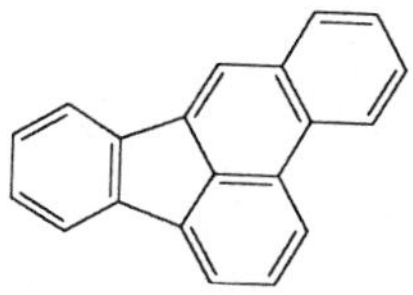

CAS Registry No.: 205-99-2
Molecular formula: $C_{20}H_{12}$
Formula weight: 252.32
RTECS: CU1400000

Physical state
Solid.

Melting point (°C):
168 (Weast, 1986)
163-165 (Aldrich, 1988)
161.6 (Casellato et al., 1973)

Diffusivity in water (10^5 cm²/sec):
0.49 at 20 °C using method of Hayduk and Laudie (1974).

Dissociation constant:
>15 (Christensen et al., 1975)

Henry's law constant (10^7 atm·m³/mol):
2.47, 5.03, 11.74, 14.90, 20.53 and 36.52 at 10.0, 20.0, 35.0, 40.1, 45.0 and 55.0 °C, respectively (ten Hulscher et al., 1992)

Bioconcentration factor, log BCF:
4.00 (*Daphnia magna*, Newsted and Giesy, 1987)
0.96 (*Polychaete* sp.), 0.23 (*Capitella capitata*) (Bayona et al., 1991)

Soil sorption coefficient, log K_{oc}:
5.74 (U.S. EPA, 1982)

Octanol/water partition coefficient, log K_{ow}:
6.06 (Mabey et al., 1982)
6.40 (Bayona et al., 1991)

Solubility in organics:
Soluble in most solvents (U.S. EPA, 1985).

Solubility in water (μg/L):
14 (U.S. EPA, 1982)
1.2 at 25 °C (U.S. EPA, 1980)

Vapor pressure (mmHg):
5 x 10^{-7} at 20 °C (U.S. EPA, 1982)

Exposure Limits: Potential occupational carcinogen. No individual standards have been set, however, as a constituent in coal tar pitch volatiles, the following exposure limits have been established (mg/m^3): NIOSH REL: TWA 0.1 (cyclohexane-extractable fraction), IDLH 80; OSHA PEL: TWA 0.2 (benzene-soluble fraction); ACGIH TLV: TWA 0.2 (benzene solubles).

Drinking Water Standard (proposed): MCLG: zero; MCL: 0.2 μg/L (U.S. EPA, 1984).

Use: Research chemical. Derived from industrial and experimental coal gasification operations where maximum concentration detected in gas, liquid and coal tar streams were 0.38, 0.033 and 3.2 mg/m^3, respectively (Cleland, 1981).

BENZO[*k*]FLUORANTHENE

Synonyms: 8,9-Benzfluoranthene; 8,9-Benzofluoranthene; 11,12-Benzofluor-anthene; 11,12-Benzo[*k*]fluoranthene; B(*k*)F; 2,3,1′,8′-Binaphthylene; Di-benzo[*b,jk*]fluorene.

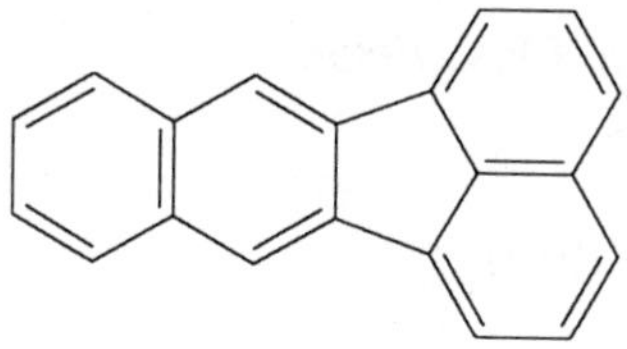

CAS Registry No.: 207-08-9
Molecular formula: $C_{20}H_{12}$
Formula weight: 252.32
RTECS: DF6350000

Physical state and color
Pale yellow needles.

Melting point (°C):
198-217 (Murray et al., 1974)
217 (Weast, 1986)

Boiling point (°C):
480 (Pearlman et al., 1984)
481 (Bjørseth, 1983)

Diffusivity in water (10^5 cm^2/sec):
0.49 at 20 °C using method of Hayduk and Laudie (1974).

Dissociation constant:
>15 (Christensen et al., 1975)

Henry's law constant (10^7 atm·m^3/mol):
2.17, 4.24, 10.56, 13.62, 19.54 and 39.77 at 10.0, 20.0, 35.0, 40.1, 45.0 and 55.0 °C, respectively (ten Hulscher et al., 1992)

Bioconcentration factor, log BCF:
4.00 (*Daphnia magna*, Newsted and Giesy, 1987)
1.15 (*Polychaete* sp.), 0.26 (*Capitella capitata*) (Bayona et al., 1991)

Soil sorption coefficient, log K_{oc}:
5.99 (average, Kayal and Connell, 1990)

Octanol/water partition coefficient, log K_{ow}:
6.85 (Mills et al., 1985)
6.40 (Bayona et al., 1991)

Solubility in organics:
Soluble in most solvents (U.S. EPA, 1985).

Solubility in water:
0.55 μg/L at 25 °C (Walton, 1985)

Vapor pressure (mmHg):
9.59 x 10^{-11} at 25 °C (Radding et al., 1976)

Exposure Limits: Potential occupational carcinogen. No individual standards have been set, however, as a constituent in coal tar pitch volatiles, the following exposure limits have been established (mg/m^3): NIOSH REL: TWA 0.1 (cyclohexane-extractable fraction), IDLH 80; OSHA PEL: TWA 0.2 (benzene-soluble fraction); ACGIH TLV: TWA 0.2 (benzene solubles).

Drinking Water Standard (proposed): MCLG: zero; MCL: 0.2 μg/L (U.S. EPA, 1984).

Use: Research chemical. Derived from industrial and experimental coal gasification operations where maximum concentration detected in liquid and coal tar streams were 0.017 and 1.6 mg/m^3, respectively (Cleland, 1981).

BENZOIC ACID

Synonyms: Benzenecarboxylic acid; Benzeneformic acid; Benzenemethanoic acid; Benzoate; Carboxybenzene; Dracylic acid; NA 9094; Phenylcarboxylic acid; Phenylformic acid; Retarder BA; Retardex; Salvo liquid; Salvo powder; Tenn-plas.

CAS Registry No.: 65-85-0
DOT: 9094
Molecular formula: $C_7H_6O_2$
Formula weight: 122.12
RTECS: DG0875000

Physical state, color and odor
Colorless to white needles, scales or powder with a faint benzoin or benzaldehyde-like odor.

Melting point (°C):
122.13 (Weast, 1986)

Boiling point (°C):
249.2 (Weast, 1986)

Density (g/cm^3):
1.2659 at 15/4 °C (Weast, 1986)
1.316 at 28/4 °C (Standen, 1964)

Diffusivity in water (10^5 cm^2/sec):
0.79 at 20 °C using method of Hayduk and Laudie (1974).

Flash point (°C):
121 (Aldrich, 1988)

Dissociation constant:
4.21 at 25 °C (Dean, 1973)

Heat of fusion (kcal/mol):
4.32 (Dean, 1987)

Henry's law constant (10^8 atm·m^3/mol):
7.02 (calculated, U.S. EPA, 1980)

Ionization potential (eV):
9.73 ± 0.09 (Franklin et al., 1969)

Bioconcentration factor, log BCF:
0.48 (algae, Geyer et al., 1984)
3.11 (activated sludge, Freitag et al., 1985)

Soil sorption coefficient, log K_{oc}:
1.48-2.70 (average = 2.26 for 10 Danish soils, Løkke, 1984)

Octanol/water partition coefficient, log K_{ow}:
1.68, 1.94 (Sangster, 1989)
1.87 (Leo et al., 1971)
2.03 (Lu and Metcalf, 1975)
1.81-1.88 (Fujita et al., 1964)
2.18 (Garst and Wilson, 1984)

Solubility in organics:
Soluble in chloroform (222 g/L), ether (333 g/L), acetone (333 g/L), carbon tetrachloride (33 g/L), benzene (100 g/L), carbon disulfide (33 g/L), volatile and fixed oils (Windholz et al., 1983).

Solubility in water (g/L):
3.00, 3.20 and 3.40 at 18, 23.5 and 25 °C, respectively (Stephen and Stephen, 1963)
1.70 at 0 °C, 2.10 at 10 °C, 2.90 at 20 °C, 3.40 at 25 °C, 4.20 at 30 °C, 6.00 at 40 °C, 8.50 at 50 °C, 12.00 at 60 °C, 17.70 at 70 °C, 27.50 at 80 °C, 45.50 at 90 °C, 68.00 at 95 °C (Standen, 1964)
1.80 at 4 °C, 2.70 at 18 °C, 22.00 at 75 °C (Hodgman et al., 1961)

Vapor pressure (10^4 mmHg):
45 at 25 °C (Howard, 1989)
8.18 at 25 °C (Sonnefeld et al., 1983)
4.5 at 20 °C (Klöpffer et al., 1988)

Environmental Fate
   Biological. Benzoic acid may degrade to catechol if it is the central metabolite whereas, if protocatechuic acid (3,4-dihydroxybenzoic acid) is the central metabolite, the precursor is 3-hydroxybenzoic acid (Chapman, 1972). Other compounds identified following degradation of benzoic acid to catechol include *cis,cis*-muconic acid, (+)-muconolactone, 3-oxoadipate enol lactone and 3-oxo-

adipate (Verschueren, 1983). Pure microbial cultures hydroxylated benzoic acid to 3,4-dihydroxybenzoic acid, 2- and 4-hydroxybenzoic acid (Smith and Rosazza, 1974). In a methanogenic enrichment culture, 91% of the added benzoic acid anaerobically biodegraded to carbon dioxide and methane (Healy and Young, 1979). In activated sludge, 65.5% mineralized to carbon dioxide after 5 days (Freitag et al., 1985).

Photolytic. Titanium dioxide suspended in an aqueous solution and irradiated with UV light (λ = 365 nm) converted benzoic acid to carbon dioxide at a significant rate (Matthews, 1986). An aqueous solution containing chlorine and irradiated with UV light (λ = 350 nm) converted benzoic acid to salicylaldehyde and other unidentified chlorinated compounds (Oliver and Carey, 1977). A carbon dioxide yield of 10.2% was achieved when benzoic acid adsorbed on silica gel was irradiated with light (λ >290 nm) for 17 hours (Freitag et al., 1985).

Toxicity: Acute oral LD_{50} for mice 1,940 mg/kg, cats 2,000 mg/kg, dogs 2,000 mg/kg, rats 2,530 mg/kg (RTECS, 1985); LC_{50} (48-hour) for red killifish 910 mg/L (Yoshioka et al., 1986).

Uses: Preparation of sodium and butyl benzoates, benzoyl chloride, phenol, caprolactum and esters for perfume and flavor industry; plasticizers; manufacture of alkyl resins; preservative for food, fats and fatty oils; seasoning; tobacco; dentifrices; standard in analytical chemistry; antifungal agent; synthetic resins and coatings; pharmaceutical and cosmetic preparations; plasticizer manufacturing (to modify resins such as polyvinyl chloride, polyvinyl acetate, phenol-formaldehyde).

BENZO[ghi]PERYLENE

Synonyms: 1,12-Benzoperylene; 1,12-Benzperylene; B(*ghi*)P.

CAS Registry No.: 191-24-2
Molecular formula: $C_{22}H_{12}$
Formula weight: 276.34
RTECS: DI6200500

Physical state
Solid.

Melting point (°C):
222 (Cleland and Kingsbury, 1977)
278-280 (Fluka, 1988)
275-277 (Murray et al., 1974)

Boiling point (°C):
>500 (Aldrich, 1988)

Diffusivity in water (10^5 cm^2/sec):
0.49 at 20 °C using method of Hayduk and Laudie (1974).

Dissociation constant:
>15 (Christensen et al., 1975)

Henry's law constant (10^7 atm·m^3/mol):
1.88, 2.66, 5.13, 5.33, 6.51 and 8.59 at 10.0, 20.0, 35.0, 40.1, 45.0 and 55.0 °C, respectively (ten Hulscher et al., 1992)

Ionization potential (eV):
7.24 (Franklin et al., 1969)

Bioconcentration factor, log BCF:
4.45 (*Daphnia magna*, Newsted and Giesy, 1987)

Soil sorption coefficient, log K_{oc}:
6.89 using method of Karickhoff et al. (1979).

Octanol/water partition coefficient, log K_{ow}:
7.10 (Bruggeman et al., 1982; Mackay et al., 1980)

Solubility in organics:
Soluble in most solvents (U.S. EPA, 1985).

Solubility in water:
0.26 μg/L at 25 °C (Mackay and Shiu, 1977)

Vapor pressure (10^{10} mmHg at 25 °C):
1.01 (Radding et al., 1976)
1.04 (Murray et al., 1974)

Exposure Limits: Potential occupational carcinogen. No individual standards have been set, however, as a constituent in coal tar pitch volatiles, the following exposure limits have been established (mg/m^3): NIOSH REL: TWA 0.1 (cyclohexane-extractable fraction), IDLH 80; OSHA PEL: TWA 0.2 (benzene-soluble fraction); ACGIH TLV: TWA 0.2 (benzene solubles).

Drinking Water Standard: As of May 1994, no MCLGs or MCLs have been proposed (U.S. EPA, 1994).

Use: Research chemical. Derived from industrial and experimental coal gasification operations where the maximum concentration detected in coal tar streams was 2.7 mg/m^3 (Cleland, 1981).

BENZO[a]PYRENE

Synonyms: Benzo[d,e,f]chrysene; 1,2-Benzopyrene; 3,4-Benzopyrene; 6,7-Benzo-pyrene; Benz[a]pyrene; Benzo[α]pyrene; 1,2-Benzpyrene; 3,4-Benzpyrene; 3,4-Benz[a]pyrene; 3,4-Benzypyrene; BP; B(a)P; 3,4-BP; RCRA waste number U022.

CAS Registry No.: 50-32-8
Molecular formula: $C_{20}H_{12}$
Formula weight: 252.32
RTECS: DJ3675000

Physical state, color and odor
Odorless, yellow, orthorhombic or monoclinic crystals. Solution in concentrated sulfuric acid is orange-red with a green fluorescence (Keith and Walters, 1992).

Melting point (°C):
179-179.3 (Weast, 1986)
181.3 (Casellato et al., 1973)
176.4 (Murray et al., 1974)

Boiling point (°C):
495 (Aldrich, 1988)

Density (g/cm^3):
1.351 (Kronberger and Weiss, 1944)

Diffusivity in water (10^5 cm^2/sec):
0.50 at 20 °C using method of Hayduk and Laudie (1974).

Dissociation constant:
>15 (Christensen et al., 1975)

Henry's law constant (10^7 atm·m^3/mol):
2.17, 3.36, 7.30, 9.08, 10.86 and 23.89 at 10.0, 20.0, 35.0, 40.1, 45.0 and 55.0 °C, respectively (ten Hulscher et al., 1992)

Ionization potential (eV):
7.23 (Cavalieri and Rogan, 1985)

Bioconcentration factor, log BCF:
2.69 (bluegill sunfish), 3.69 (*Lepomis macrochirus*) (Spacie et al., 1983)
3.90 (*Daphnia magna*, McCarthy, 1983)
4.00 (activated sludge), 3.53 (algae), 2.68 (golden ide) (Freitag et al., 1985)

Soil sorption coefficient, log K_{oc}:
5.95 (Meylan et al., 1992)
6.26 (average, Kayal and Connell, 1990)
7.01 (Rotterdam Harbor sediment, Hageman et al., 1995)

Octanol/water partition coefficient, log K_{ow}:
6.00 (Sangster, 1989)
5.99 (Mallon and Harrison, 1984)
6.04 (Radding et al., 1976)
6.50 (Bruggeman et al., 1982; Landrum et al., 1984)
5.81 (Zepp and Scholtzhauer, 1979)

Solubility in organics:
Soluble in benzene, toluene and xylene; sparingly soluble in ethanol and methanol (Windholz et al., 1983).

Solubility in water:
3.8 µg/L at 25 °C (Mackay and Shiu, 1977)
4 µg/L at 25 °C (Schwarz and Wasik, 1976)
1.2 ng/L at 22 °C (Smith et al., 1978)
0.505 µg/L in Lake Michigan water at ≈ 25 °C (Eadie et al., 1990)
In nmol/L: 2.64, 3.00, 3.74, 4.61 and 6.09 at 8.0, 12.4, 16.7, 20.9 and 25.0 °C, respectively. In seawater (salinity = 36.7 g/kg): 1.40, 2.07, 2.50, 3.43 and 4.45 at 8.0, 12.4, 16.7, 20.9 and 25.0 °C, respectively (Whitehouse, 1984)
4.8 nmol/L (Mill et al., 1981)
3.0, 3.5, 4.0, 4.0 and 4.5 µg/L at 27 °C (Davis et al., 1942)
4.0 µg/L at 27 °C (Davis and Parke, 1942)
19 nmol/L at 25 °C (Barone et al., 1967)
1.6 µg/L at 25 °C (Billington et al., 1988)

Vapor pressure (10^9 mmHg):
5.49 at 25 °C (Radding et al., 1976)
5.6 at 25 °C (Murray et al., 1984)
5 at 25 °C (Smith et al., 1978)
509 at 20 °C (Sims et al., 1988)
2.4 at 25 °C (McVeety and Hites, 1988)
113 at 25 °C (Bidleman, 1984)
543, 840 at 25 °C (Hinckley et al., 1990)

Environmental Fate

Biological. Benzo[a]pyrene was biooxidized by *Beijerinckia* B836 to *cis*-9,10-dihydroxy-9,10-dihydrobenzo[a]pyrene. Under nonenzymatic conditions, this metabolite monodehydroxylated to form 9-hydroxybenzo[a]pyrene (Verschueren, 1983). Under aerobic conditions, *Cunninghanella elegans* degraded benzo[a]pyrene to *trans*-7,8-dihydroxy-7,8-dihydrobenzo[a]pyrene (Kobayashi and Rittman, 1982), 3-hydroxy-benzo[a]pyrene, 9-hydroxybenzo[a]pyrene and vicinal dihydrols including *trans*-9,10-dihydroxy-9,10-dihydrobenzo[a]pyrene (Cerniglia and Gibson, 1980; Gibson et al., 1975). The microorganisms *Candida lipolytica* and *Saccharomyces cerevisiae* oxidized benzo[a]pyrene to *trans*-7,8-dihydroxy-7,8-dihydrobenzo[a]pyrene, 3- and 9-hydroxybenzo[a]pyrene (Cerniglia and Crow, 1980; Wiseman et al., 1978) whereas 3-hydroxybenzo[a]pyrene was the main degradation product by the microbe *Neurospora crassa* (Lin and Kapoor, 1979).

After a 30-day incubation period, the white rot fungus *Phanerochaete chrysosporium* converted benzo[a]pyrene to carbon dioxide. Mineralization began between the third and sixth day of incubation. The production of carbon dioxide was highest between 3-18 days of incubation, after which the rate of carbon dioxide produced decreased until the 30th day. It was suggested that the metabolism of benzo[a]pyrene and other compounds including *p,p'*-DDT, TCDD and lindane, was dependent on the extracellular lignin-degrading enzyme system of this fungus (Bumpus et al., 1985). In activated sludge, <0.1% mineralized to carbon dioxide after 5 days (Freitag et al., 1985).

Soil. Lu et al. (1977) studied the degradation of benzo[a]pyrene in a model ecosystem containing Drummer silty clay loam. Samples were incubated at 27.6 °C for 1, 2 and 4 weeks before extraction with acetone for TLC analysis. After 4 weeks, only 8.05% of benzo[a]pyrene degraded forming one polar compound and two unidentified compounds.

Photolytic. Coated glass fibers exposed to air containing 100-200 ppb ozone yielded benzo[a]pyrene-4,5-oxide. At 200 ppb ozone, conversion yields of 50 and 80% were observed after 1 and 4 hours, respectively (Pitts et al., 1990). Free radical oxidation and photolysis of benzo[a]pyrene at a wavelength of 366 nm yielded the following tentatively identified products: benzo[a]pyrene-1,6-quinone, benzo[a]pyrene-3,6-quinone and benzo[a]pyrene-6,12-quinone (Smith et al., 1978).

In a solution containing oxygen, photolysis yields a mixture of 6,12-, 1,6- and 3,6-diones. Nitration by nitrogen dioxide forms 6-nitro-, 1-nitro- and 3-nitrobenzo[a]pyrene. When benzo[a]pyrene in methanol (1 g/L) was irradiated at 254 nm in a quartz flask for 1 hour, the solution turned pale yellow. After 2 hours, the solution turned yellow and back to clear after 4 hours of irradiation. After 4 hours, 99.67% of benzo[a]pyrene was converted to polar compounds. One of these compounds was identified as a methoxylated benzo[a]pyrene (Lu et al., 1977). A carbon dioxide yield of 26.5% was achieved when benzo[a]pyrene adsorbed on silica gel was irradiated with light (λ >290 nm) for 17 hours (Freitag et al., 1985).

Chemical/Physical. Ozonolysis to benzo[a]pyrene-1,6-quinone or benzo[a]pyrene-3,6-quinone followed by further oxidation to benzanthrone dicarboxylic anhydride was reported (IARC, 1983).

In a simulated atmosphere, direct epoxidation by ozone led to the formation of benzo[a]pyrene-4,5-oxide. Benzo[a]pyrene reacted with benzoyl peroxide to form the 6-benzoyloxy derivative (Nikolaou et al., 1984). It was reported that benzo[a]pyrene adsorbed on fly ash and alumina reacted with sulfur dioxide (10%) in air to form benzo[a]pyrene sulfonic acid (Nielsen et al., 1983). Benzo[a]pyrene coated on a quartz surface was subjected to ozone and natural sunlight for 4 and 2 hours, respectively. The compounds 1,6-quinone, 3,6-quinone and the 6,12-quinone of benzo[a]pyrene were formed in both instances (Rajagopalan et al., 1983).

When benzo[a]pyrene adsorbed from the vapor phase onto coal fly ash, silica and alumina was exposed to nitrogen dioxide, no reaction occurred. However, in the presence of nitric acid, nitrated compounds were produced (Yokley et al., 1985). Chlorination of benzo[a]pyrene in polluted humus poor lake water gave 11,12-dichlorobenzo[a]pyrene and 1,11,12-, 3,11,12- or 3,6,11-trichlorobenzo[a]pyrene, representing 99% of the chlorinated products formed (Johnsen et al., 1989).

Exposure Limits: Potential occupational carcinogen. No individual standards have been set, however, as a constituent in coal tar pitch volatiles, the following exposure limits have been established (mg/m^3): NIOSH REL: TWA 0.1 (cyclohexane-extractable fraction), IDLH 80; OSHA PEL: TWA 0.2 (benzene-soluble fraction); ACGIH TLV: TWA 0.2 (benzene solubles).

Toxicity: LD$_{50}$ for mice (subcutaneous) 50 mg/kg (RTECS, 1985).

Drinking Water Standard (final): MCLG: zero; MCL: 0.2 μg/L (U.S. EPA, 1984).

Use: Research chemical. Derived from industrial and experimental coal gasification operations where maximum concentration detected in gas, liquid and coal tar streams were 5.0, 0.036 and 3.5 mg/m^3, respectively (Cleland, 1981).

BENZO[e]PYRENE

Synonyms: 1,2-Benzopyrene; 1,2-Benzpyrene; 4,5-Benzopyrene; 4,5-Benzo[e]pyrene; B(e)P.

CAS Registry No.: 192-97-2
Molecular formula: $C_{20}H_{12}$
Formula weight: 252.32
RTECS: DJ4200000

Physical state
Crystalline, solid, prisms or plates.

Melting point (°C):
178-179 (Windholz et al., 1983)
178.8 (Murray et al., 1974)
179 (Bjørseth, 1983)

Boiling point (°C):
493 (Bjørseth, 1983)

Density (g/cm^3):
0.8769 at 20/4 °C (Windholz et al., 1983)

Diffusivity in water (10^5 cm^2/sec):
0.50 at 20 °C using method of Hayduk and Laudie (1974).

Dissociation constant:
>14 (Schwarzenbach et al., 1993)

Henry's law constant (10^7 atm·m^3/mol):
4.84 at 25 °C (approximate - calculated from water solubility and vapor pressure)

Soil sorption coefficient, log K_{oc}:
5.60 using method of Karickhoff et al. (1979).

Octanol/water partition coefficient, log K_{ow}:
6.75 using method of Yalkowsky and Valvani (1979).

Solubility in organics:
Slightly soluble in methanol (Patnaik, 1992).

Solubility in water:
In ng/kg: 325 at 8.6 °C, 358 at 14.0 °C, 444 at 17.0 °C, 394 at 17.5 °C, 459 at 20.0
°C, 479 at 20.2 °C, 535 at 23.2 °C, 507 at 23.0 °C, 643 at 29.2 °C, 681 at 31.7 °C.
In seawater: 3.32 ng/kg at 25 °C. In NaCl solution (salinity = 30 g/kg), nmol/L:
0.082 at 8.9 °C, 0.088 at 10.8 °C, 0.101 at 15.6 °C, 0.104 at 19.2 °C, 0.113 at 21.7
°C, 0.135 at 25.3 °C, 0.142 at 27.1 °C, 0.166 at 30.2 °C (Schwarz, 1977)
29 nmol/L at 25 °C (Barone et al., 1967)

Vapor pressure (10^9 mmHg):
2.4 (McVeety and Hites, 1988)
5.54 at 25 °C (Pupp et al., 1974)
5.55 at 25 °C (Murray et al., 1974)
644 at 25 °C (Hinckley et al., 1990)

Uses: Chemical research.

BENZYL ALCOHOL

Synonyms: Benzal alcohol; Benzene carbinol; **Benzene methanol**; Benzoyl alcohol; α-Hydroxytoluene; NCI-C06111; Phenol carbinol; Phenyl carbinol; Phenyl methanol; Phenyl methyl alcohol; α-Toluenol.

CAS Registry No.: 100-51-6
Molecular formula: C_7H_8O
Formula weight: 108.14
RTECS: DN3150000

Physical state, color and odor
Colorless, hygroscopic liquid with a faint, pleasant, aromatic odor. Odor threshold in air is 5.5 ppm (Keith and Walters, 1992).

Melting point (°C):
-15.3 (Stull, 1947)
-11 to -9 (Fluka, 1988)

Boiling point (°C):
204.7 (Stull, 1947)
205.3 (Weast, 1986)

Density (g/cm³):
1.044 at 20/4 °C (Fluka, 1988)
1.0424 at 25/4 °C, 1.0383 at 30/4 °C (Abraham et al., 1971)

Diffusivity in water (10^5 cm²/sec):
0.93 at 25 °C (Hayduk and Laudie, 1974)

Flash point (°C):
93 (NFPA, 1984)

Interfacial tension with water (dyn/cm at 22.5 °C):
4.75 (Demond and Lindner, 1993)

Ionization potential (eV):
9.14 ± 0.05 (Franklin et al., 1969)

Soil sorption coefficient, log K$_{oc}$:
<0.70 (Apison, Fullerton and Dormant soils, Southworth and Keller, 1986)

Octanol/water partition coefficient, log K$_{ow}$:
1.00 (Sangster, 1989)
1.10 (Fujita et al., 1964)
1.16 (Garst and Wilson, 1984)

Solubility in organics:
Miscible with ether and absolute alcohol (Windholz et al., 1983).

Solubility in water:
42,900 mg/L at 25 °C (Banerjee, 1984)
3.66 wt % at 20 °C (Stephen and Stephen, 1963)
In wt %: 4.8 at 0 °C, 4.7 at 9.8 °C, 4.3 at 20.1 °C, 4.3 at 29.6 °C, 4.6 at 40.2 °C, 5.2
    at 50.0 °C (Stephenson and Stuart, 1986)

Vapor density:
4.42 g/L at 25 °C, 3.73 (air = 1)

Vapor pressure (mmHg):
1 at 58 °C (estimated, Weast, 1986)

Environmental Fate
   Chemical/Physical. Slowly oxidizes in air to benzaldehyde (Huntress and
Mulliken, 1941).

Toxicity: Acute oral LD$_{50}$ for wild bird 100 mg/kg, mouse 1,580 mg/kg, rat 1,230
mg/kg and rabbit 1,040 mg/kg (RTECS, 1985).

Uses: Manufacture of esters for use in perfumes, flavors, soaps, lotions, ointments;
photographic developer for color movie films; dying nylon filament, textiles and
sheet plastics; solvent for dyestuffs, cellulose, esters, casein, waxes, etc.;
heat-sealing polyethylene films; bacteriostat, insect repellant; emulsions; ballpoint
pen inks and stencil inks; surfactant.

BENZYL BUTYL PHTHALATE

Synonyms: BBP; **1,2-Benzenedicarboxylic acid butyl phenylmethyl ester;** Benzyl *n*-butyl phthalate; Butyl benzyl phthalate; *n*-Butyl benzyl phthalate; Butyl phenylmethyl 1,2-benzenedicarboxylate; NCI-C54375; Palatinol BB; Santicizer 160; Sicol 160; Unimoll BB.

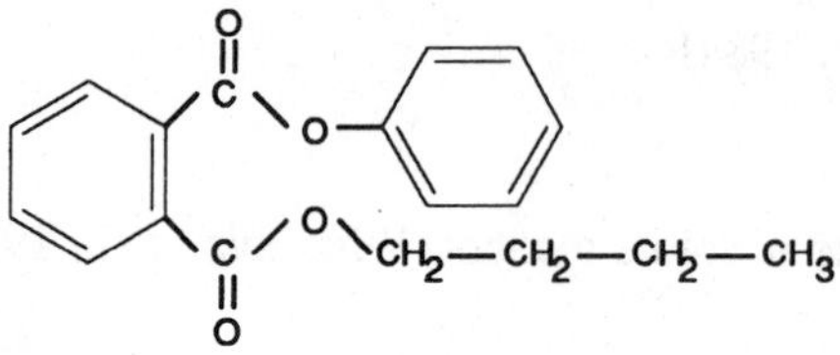

CAS Registry No.: 85-68-7
Molecular formula: $C_{19}H_{20}O_4$
Formula weight: 312.37
RTECS: TH9990000

Physical state and odor
Clear, oily liquid with a faint odor.

Melting point (°C):
-35 (Fishbein and Albro, 1972)

Boiling point (°C):
370 (Verschueren, 1983)
377 (Fishbein and Albro, 1972)

Density (g/cm³):
1.12 at 20/4 °C (Weiss, 1986)
1.111 at 25/4 °C (Standen, 1968)

Diffusivity in water (10^5 cm²/sec):
0.48 at 20 °C using method of Hayduk and Laudie (1974).

Flash point (°C):
197 (Weiss, 1986)
110 (Aldrich, 1988)

Henry's law constant (10^6 atm·m³/mol):
1.3 at 25 °C (calculated, Howard, 1989)

Bioconcentration factor, log BCF:
2.89 (bluegill sunfish, Veith et al., 1980)

Soil sorption coefficient, log K_{oc}:
4.23 (Meylan et al., 1992)

Octanol/water partition coefficient, log K_{ow}:
4.05 (Veith et al., 1980)
4.91 (Leyder and Boulanger, 1983)
4.77 (Gledhill et al., 1980)

Solubility in water (mg/L):
2.82 at 20 °C (Leyder and Boulanger, 1983)
2.69 at 25 °C (Howard et al., 1985)
0.710 at 24 °C (practical grade, Hollifield, 1979)
2.0 at 25 °C (Russell and McDuffie, 1986)

Vapor density:
12.76 g/L at 25 °C, 10.78 (air = 1)

Vapor pressure (10^6 mmHg):
8.6 at 20 °C (Verschueren, 1983)
8.25 at 25 °C (Banerjee et al., 1990)

Environmental Fate
   Biological. In anaerobic sludge diluted to 10%, benzyl butyl phthalate bio-degraded to monobutyl phthalate, which subsequently degraded to phthalic acid (Shelton et al., 1984). When benzyl butyl phthalate (5 and 10 mg/L) was statically incubated in the dark at 25 °C with yeast extract and settled domestic wastewater inoculum, complete biodegradation with rapid adaptation was observed after 7 days (Tabak et al., 1981).
   Chemical/Physical. Slowly hydrolyzes in water forming mono-*n*-butyl phthalate, monobenzyl phthalate, phthalic acid, benzyl alcohol and 1-butanol.

Symptoms of Exposure: Toxic symptoms include nausea, somnolence, hallucination and dizziness (Patnaik, 1992).

Toxicity: Acute oral LD_{50} for guinea pigs 13,750 mg/kg, mice 4,170 mg/kg, rats 2,330 mg/kg (RTECS, 1985).

Drinking Water Standard (proposed): MCLG: zero; MCL: 0.1 mg/L (U.S. EPA, 1984).

Uses: Plasticizer used in polyvinyl chloride formulations; additive in polyvinyl acetate emulsions, ethylene glycol and ethyl cellulose; organic synthesis.

BENZYL CHLORIDE

Synonyms: Benzyl chloride anhydrous; **(Chloromethyl)benzene;** Chlorophenyl-methane; α-Chlorotoluene; Ω-Chlorotoluene; NCI-C06360; RCRA waste number P028; Tolyl chloride; UN 1738.

CAS Registry No.: 100-44-7
DOT: 1738
DOT label: Corrosive material
Molecular formula: C_7H_7Cl
Formula weight: 126.59
RTECS: XS8925000

Physical state, color and odor
Colorless to pale yellowish-brown liquid with a pungent, aromatic, irritating odor. Odor threshold in air is 47 ppb (Keith and Walters, 1992).

Melting point (°C):
-39 (Weast, 1986)
-43 to -48 (Windholz et al., 1983)

Boiling point (°C):
179.3 (Weast, 1986)

Density (g/cm³):
1.1002 at 20/4 °C (Weast, 1986)

Diffusivity in water (10^5 cm²/sec):
0.81 at 20 °C using method of Hayduk and Laudie (1974).

Flash point (°C):
67.3 (NIOSH, 1994)

Lower explosive limit (%):
1.1 (NIOSH, 1994)

Henry's law constant (10^4 atm·m³/mol):
3.04 at 20 °C (approximate - calculated from water solubility and vapor pressure)

Soil sorption coefficient, log K_{oc}:
2.28 using method of Chiou et al. (1979).

Octanol/water partition coefficient, log K_{ow}:
2.30 (Leo et al., 1971)

Solubility in organics:
Miscible with alcohol, chloroform and ether (Windholz et al., 1983).

Solubility in water:
493 mg/L at 20 °C (Howard, 1989)

Vapor density:
5.17 g/L at 25 °C, 4.37 (air = 1)

Vapor pressure (mmHg):
1.3 at 25 °C (Banerjee et al., 1990)

Environmental Fate
    Biological. When incubated with raw sewage and raw sewage acclimated with hydrocarbons, benzyl chloride degraded forming non-chlorinated products (Jacobson and Alexander, 1981).
    Chemical/Physical. Anticipated products from the reaction of benzyl chloride with ozone or hydroxyl radicals in the atmosphere are chloromethylphenols, benzaldehyde and chlorine radicals (Cupitt, 1980).
    Slowly hydrolyzes in water forming hydrochloric acid and benzyl alcohol. The estimated hydrolysis half-life in water at 25 °C and pH 7 is 15 hours (Mabey and Mill, 1978). The hydrolysis rate constant for benzyl chloride at pH 7 and 59.2 °C was determined to be 0.0204/minute, resulting in a half-life of 34 minutes (Ellington et al., 1986).
    May polymerize in contact with metals except nickel and lead (NIOSH, 1994).

Exposure Limits: NIOSH REL: 15-minute ceiling 1 ppm (5 mg/m^3), IDLH 10 ppm; OSHA PEL: TWA 1 ppm; ACGIH TLV: TWA 1 ppm.

Symptoms of Exposure: Eye contact may cause corneal injury. Exposure to fumes may cause irritation to eyes, nose, skin and throat (NIOSH, 1994; Patnaik, 1992).

Toxicity: Acute oral LD_{50} for mice 1,624 mg/kg, rats 1,231 mg/kg; LD_{50} (subcutaneous) for rats 1 gm/kg; LC_{50} (inhalation) for mice 80 ppm/2-hour, rats 150 ppm/2-hour (RTECS, 1985).

Uses: Manufacture of perfumes, benzyl compounds, pharmaceutical products, resins, dyes, photographic developers; gasoline gum inhibitors; quaternary ammonium compounds; penicillin precursors; chemical intermediate.

α-BHC

Synonyms: Benzene hexachloride-α-isomer; α-Benzene hexachloride; ENT 9232; α-HCH; α-Hexachloran; α-Hexachlorane; α-Hexachlorcyclohexane; α-Hexachlorocyclohexane; 1,2,3,4,5,6-Hexachloro-α-cyclohexane; **1α,2α,3β,4α,5β,6β-Hexachlorocyclohexane**; α-1,2,3,4,5,6-Hexachlorocyclohexane; α-Lindane; TBH.

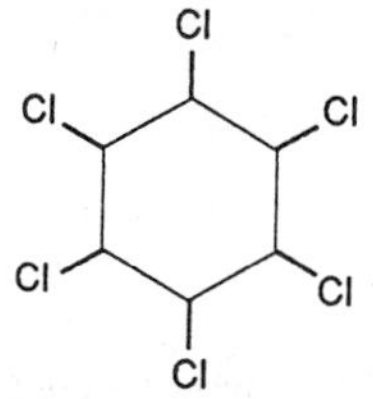

CAS Registry No.: 319-84-6
DOT: 2761
DOT label: Poison
Molecular formula: $C_6H_6Cl_6$
Formula weight: 290.83
RTECS: GV3500000

Physical state, color and odor
Brownish to white crystalline solid or powder with a phosgene-like odor (technical grade).

Melting point (°C):
156-161 (Aldrich, 1988)
159.8 (Horvath, 1982)

Boiling point (°C):
288 (Weast, 1986)

Density (g/cm³):
1.870 at 20/4 °C (Horvath, 1982)

Diffusivity in water (10^5 cm²/sec):
0.50 at 20 °C using method of Hayduk and Laudie (1974).

Henry's law constant (10^6 atm·m³/mol):
5.3 at 20 °C (approximate - calculated from water solubility and vapor pressure)

Bioconcentration factor, log BCF:
3.20-3.38 (fish tank), 2.85 (Lake Ontario) (rainbow trout, Oliver and Niimi, 1985)
2.49 (*Chlamydomonas*, Canton et al., 1977)
2.79 (freshwater fish), 0.00 (fish, microcosm) (Garten and Trabalka, 1983)

Soil sorption coefficient, log K$_{oc}$:
3.279 (Karickhoff, 1981)
3.30 (Meylan et al., 1992)

Octanol/water partition coefficient, log K$_{ow}$:
3.81 (Kurihara et al., 1973)
3.46 (Geyer et al., 1987)
3.72 (Schwarzenbach et al., 1983)
3.776 (de Bruijn et al., 1989)

Solubility in organics:
Soluble in ethanol, benzene, chloroform (Weast, 1986), cod liver oil and octanol (Montgomery, 1993).

Solubility in water (mg/L):
1.4 (salt water, Verschueren, 1983)
At 28 °C: 1.48, 1.77 (0.05 μ particle size), 1.21, 2.03 (0.1 μ particle size) (Kurihara et al., 1973)
2.00 at 25 °C (Weil et al., 1974)
1.63 at 20 °C (Brooks, 1974)

Vapor pressure (10^5 mmHg):
2.5 at 20 °C (Balson, 1947)
2.15 at 20 °C (Sims et al., 1988)
173 at 25 °C (Hinckley et al., 1990)

Environmental Fate
Biological. Clostridium sphenoides biodegraded α-BHC to δ-3,4,5,6-tetrachloro-1-cyclohexane (Heritage and MacRae, 1977). In four successive 7-day incubation periods, α-BHC (5 and 10 mg/L) was recalcitrant to degradation in a settled domestic wastewater inoculum (Tabak et al., 1981).
Soil. Under aerobic conditions, indigenous microbes in contaminated soil produced pentachlorocyclohexane. However, under methanogenic conditions, α-BHC was converted to chlorobenzene, 3,5-dichlorophenol and the tentatively identified compound 2,4,5-trichlorophenol (Bachmann et al., 1988). *Clostridium sphenoides* biodegraded α-BHC to δ-3,4,5,6-tetrachloro-1-cyclohexane (Heritage and MacRae, 1977).
Photolytic. When an aqueous solution containing α-BHC was photooxidized by UV light at 90-95 °C, 25, 50 and 75% degraded to carbon dioxide after 4.2, 24.2 and 40.0 hours, respectively (Knoevenagel and Himmelreich, 1976). In basic, aqueous solutions, α-BHC dehydrochlorinates forming pentachlorocyclohexene, which is further transformed to trichlorobenzenes. In a buffered aqueous solution at pH 8 and 5 °C, the calculated hydrolysis half-life is 26 years (Ngabe et al.,

1993).

Chemical/Physical. Emits very toxic chloride fumes when heated to decomposition (Lewis, 1990).

Toxicity: Acute oral LD_{50} for rats 177 mg/kg (RTECS, 1985); LC_{50} (96-hour) for guppies >1.4 mg/L (Verschueren, 1983).

Use: Not produced commercially in the U.S. and its sale is prohibited by the U.S. EPA.

β-BHC

Synonyms: *trans*-α-Benzene hexachloride; β-Benzene hexachloride; Benzene-*cis*-hexachloride; ENT 9233; β-HCH; β-Hexachlorobenzene; 1α,2β,3α,4β,5α,6β-Hexachlorocyclohexane; β-Hexachlorocyclohexane; 1,2,3,4,5,6-Hexachloro-β-cyclohexane; 1,2,3,4,5,6-Hexachloro-*trans*-cyclohexane; β-1,2,3,4,5,6-Hexachlorocyclohexane; β-Isomer; β-Lindane; TBH.

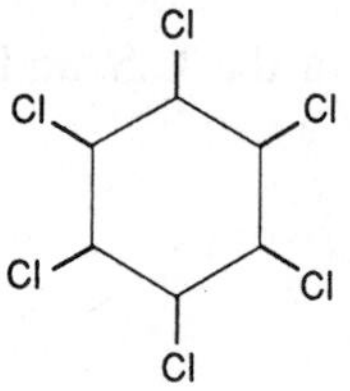

CAS Registry No.: 319-85-7
DOT: 2761
DOT label: Poison
Molecular formula: $C_6H_6Cl_6$
Formula weight: 290.83
RTECS: GV4375000

Physical state
Solid.

Melting point (°C):
314.5 (Horvath, 1982)
311.7 (Standen, 1964)

Boiling point (°C):
60 at 0.58 mmHg (Horvath, 1982)
Sublimes at 760 mmHg (U.S. EPA, 1980)

Density (g/cm^3):
1.89 at 19/4 °C (Weast, 1986)

Diffusivity in water (10^5 cm^2/sec):
0.50 at 20 °C using method of Hayduk and Laudie (1974).

Henry's law constant (10^7 atm·m^3/mol):
2.3 at 20 °C (calculated)

Bioconcentration factor, log BCF:
2.82 (brown trout, Sugiura et al., 1980)
3.08 (activated sludge), 2.26 (algae), 2.65 (golden ide) (Freitag et al., 1985)
2.86 (freshwater fish), 2.97 (fish, microcosm) (Garten and Trabalka, 1983)

106

Soil sorption coefficient, log K$_{oc}$:
3.462 (silt loam soil, Chiou et al., 1979)
3.322 (Karickhoff, 1981)
3.553 (Reinbold et al., 1979)
3.50 (Meylan et al., 1992)
3.55 (Ca-Staten peaty muck at 20 °C, Mills and Biggar, 1969)

Octanol/water partition coefficient, log K$_{ow}$:
3.80 (Kurihara et al., 1973)
4.50 (Geyer et al., 1987)
3.842 (de Bruijn et al., 1989)

Solubility in organics:
Soluble in ethanol, benzene and chloroform (Weast, 1986).

Solubility in water:
130, 200 ppb at 28 °C (0.1 μm particle size, Kurihara et al., 1973)
240 μg/L at 25 °C (Weil et al., 1974)
700 ppb at 25 °C (Brooks, 1974)
5 mg/L at 20 °C (Chiou et al., 1979)
2.7 mg/L at 20 °C (Mills and Biggar, 1969)

Vapor pressure (10^7 mmHg):
2.8 at 20 °C (Balson, 1947)
4.66 at 25 °C (Banerjee et al., 1990)

Environmental Fate
Biological. No biodegradation of β-BHC was observed under denitrifying and sulfate-reducing conditions in a contaminated soil collected from the Netherlands (Bachmann et al., 1988). In four successive 7-day incubation periods, β-BHC (5 and 10 mg/L) was recalcitrant to degradation in a settled domestic wastewater inoculum (Tabak et al., 1981).
Chemical/Physical. Emits very toxic fumes of chloride, hydrochloric acid and phosgene when heated to decomposition (Lewis, 1990).

Toxicity: Acute oral LD$_{50}$ for rats 6,000 mg/kg (RTECS, 1985).

Use: Insecticide.

δ-BHC

Synonyms: δ-Benzene hexachloride; ENT 9234; δ-HCH; δ-Hexachlorocyclohexane; δ-1,2,3,4,5,6-Hexachlorocyclohexane; δ-(aeeeee)-1,2,3,4,5,6-Hexachlorocyclohexane; 1α,2α,3α,4β,5β,6β-Hexachlorocyclohexane; 1,2,3,4,5,6-Hexachloro-δ-cyclohexane; δ-Lindane; TBH.

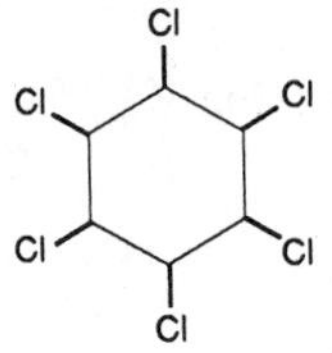

CAS Registry No.: 319-86-8
DOT: 2761
DOT label: Poison
Molecular formula: $C_6H_6Cl_6$
Formula weight: 290.83
RTECS: GV4550000

Physical state and odor
Solid with a faint musty-like odor.

Melting point (°C):
141.8 (Horvath, 1982)

Boiling point (°C):
60 at 0.34 mmHg (Horvath, 1982)

Density (g/cm³):
≈ 1.87 (Hawley, 1981)

Diffusivity in water (10^5 cm²/sec):
0.50 at 20 °C using method of Hayduk and Laudie (1974).

Henry's law constant (10^7 atm·m³/mol):
2.5 at 20 °C (approximate - calculated from water solubility and vapor pressure)

Soil sorption coefficient, log K_{oc}:
3.279 (Karickhoff, 1981)

Octanol/water partition coefficient, log K_{ow}:
4.14 (Kurihara et al., 1973)
2.80 (Geyer et al., 1987)

Solubility in organics:
Soluble in ethanol, benzene and chloroform (Weast, 1986).

Solubility in water (ppm):
At 28 °C: 8.64, 11.6 (0.05 μ particle size), 10.7, 15.7 (0.1 μ particle size) (Kurihara et al., 1973)
31.4 at 25 °C (Weil et al., 1974)
21.3 at 25 °C (Brooks, 1974)
7 at 20 °C (Worthing and Hance, 1991)

Vapor pressure (10^5 mmHg):
1.7 at 20 °C (Balson, 1947)
3.52 at 25 °C (Banerjee et al., 1990)

Environmental Fate
Biological. Dehydrochlorination of δ-BHC by a *Pseudomonas* sp. under aerobic conditions was reported by Sahu et al. (1992). They also reported that when deionized water containing δ-BHC was inoculated with this species, the concentration of δ-BHC decreased to undetectable levels after 8 days with concomitant formation of chloride ions and δ-pentachlorocyclohexane. In four successive 7-day incubation periods, δ-BHC (5 and 10 mg/L) was recalcitrant to degradation in a settled domestic wastewater inoculum (Tabak et al., 1981).

Chemical/Physical. δ-BHC dehydrochlorinates in the presence of alkalies. Though no products were reported, the hydrolysis half-lives at pH values of 7 and 9 are 191 days and 11 hours, respectively (Worthing and Hance, 1991).

Toxicity: Acute oral LD_{50} for rats 1,000 mg/kg (RTECS, 1985).

Use: Insecticide.

BIPHENYL

Synonyms: Bibenzene; **1,1′-Biphenyl**; Diphenyl; Lemonene; Phenylbenzene.

CAS Registry No.: 92-52-4
Molecular formula: $C_{12}H_{10}$
Formula weight: 154.21
RTECS: NU8050000

Physical state, color and odor
White scales with a pleasant but peculiar odor. Odor threshold in air is 0.83 ppb (Amoore and Hautala, 1983).

Melting point (°C):
71 (Weast, 1986)
68.6 (Parks and Huffman, 1931)

Boiling point (°C):
255.9 (Weast, 1986)
254-255 (Windholz et al., 1983)

Density (g/cm³):
0.8660 at 20/4 °C (Weast, 1986)
1.18 at 0/4 °C (Verschueren, 1983)

Diffusivity in water (10^5 cm²/sec):
0.62 at 20 °C using method of Hayduk and Laudie (1974).

Dissociation constant:
>14 (Schwarzenbach et al., 1993)

Flash point (°C):
113 (NIOSH, 1994)

Lower explosive limit (%):
0.6 at 111 °C (NIOSH, 1994)

Upper explosive limit (%):
5.8 at 155 °C (NIOSH, 1994)

Heat of fusion (kcal/mol):
4.18 (Miller et al., 1984)
4.44 (Dean, 1987)

Henry's law constant (10^4 atm·m^3/mol at 25 °C):
1.93 (Fendinger and Glotfelty, 1990)
4.08 (Mackay et al., 1979)
4.15 (Hine and Mookerjee, 1975)

Ionization potential (eV):
7.95 (NIOSH, 1994)
8.27 ± 0.01 (Franklin et al., 1969)
8.30 (Yoshida et al., 1983)

Bioconcentration factor, log BCF:
3.12 (rainbow trout, Veith et al., 1979)
2.73 (algae, Geyer et al., 1984)
3.41 (activated sludge), 2.45 (golden ide) (Freitag et al., 1985)

Soil sorption coefficient, log K_{oc}:
3.03 (Kishi et al., 1990)
3.52 (Apison soil), 2.95 (Fullerton soil), 2.94 (Dormont soil) (Southworth and
   Keller, 1986)

Octanol/water partition coefficient, log K_{ow}:
3.16 (Rogers and Cammarata, 1969)
4.09 (Rogers and Cammarata, 1969; Johnsen et al., 1989)
4.04 (Banerjee et al., 1980; Rogers and Cammarata, 1969)
3.76 (Geyer et al., 1984; Miller et al., 1984)
3.95 (Ruepert et al., 1985)
4.00 (DeKock and Lord, 1987)
4.008 (de Bruijn et al., 1989)
3.89 (Camilleri et al., 1988; Eadsforth, 1986; Woodburn et al., 1984)
4.11 (Garst and Wilson, 1984)
3.91 (Tipker et al., 1988)
3.79 (Rapaport and Eisenreich, 1984)
4.10 (Bruggeman et al., 1982)

Solubility in organics:
Soluble in alcohol, benzene and ether (Weast, 1986).

Solubility in water:
39.1 μmol/L at 25 °C (Banerjee et al., 1980)

43.5 μmol/L at 25 °C (Miller et al., 1984)

7.45 mg/kg at 25 °C, 4.76 mg/kg in artificial seawater (salinity = 35 g/kg) at 25 °C (Eganhouse and Calder, 1976)

In mg/L: 2.83 at 0.4 °C, 2.97 at 2.4 °C, 3.38 at 5.2 °C, 3.64 at 7.6 °C, 4.06 at 10.0 °C, 4.58 at 12.6 °C, 5.11 at 14.9 °C, 5.27 at 15.9 °C, 7.48 at 25.0 °C, 7.78 at 25.6 °C, 9.64 at 30.2 °C, 9.58 at 30.4 °C, 11.0 at 33.3 °C, 11.9 at 34.9 °C, 12.5 at 36.0 °C, 17.2 at 42.8 °C (Bohon and Claussen, 1951)

5.94 mg/L at 25 °C (Andrews and Keefer, 1949)

7.0 mg/L at 25 °C (Mackay and Shiu, 1977)

2.64, 7.08, 8.88, 13.8, 22.1 and 37.2 mg/kg at 0.0, 25.0, 30.3, 40.1, 50.1, 60.5 °C, respectively (Wauchope and Getzen, 1972)

42 μmol/L at 29 °C (Stucki and Alexander, 1987)

6.92 mg/L (Morehead et al., 1986)

At 20 °C: 38, 23.9, 23.9 and 26.3 μmol/L in doubly distilled water, Pacific seawater, artificial seawater and 35% NaCl, respectively (Hashimoto et al., 1984)

In NaCl (g/kg) at 25 °C, μg/kg: 6.08 (13.25), 5.46 (26.24), 4.62 (39.05), 4.16 (46.28), 4.13 (51.62), 3.54 (63.97), 3.45 (63.97) (Paul, 1952)

7.2 mg/L at 25 °C (Billington et al., 1988)

45.7 μmol/L at 25 °C (Akiyoshi et al., 1987)

45.70 μmol/L at 30 °C (Yalkowsky et al., 1983)

49.1 μmol/kg at 25.0 °C (Vesala, 1974)

Vapor pressure (10^4 mmHg):
145,500 at 127 °C (Eiceman and Vandiver, 1983)

100 at 25 °C (Mackay et al., 1982)

5.84 at 20.70 °C, 8.87 at 24.00 °C, 15.4 at 29.15 °C (Bradley and Cleasby, 1953)

530 at 25 °C (Bidleman, 1984)

1,200 at 53.0 °C, 2,600 at 61.0 °C, 6,900 at 71.0 °C (Sharma and Palmer, 1974)

497 at 25 °C (Foreman and Bidleman, 1985)

97.5 at 25 °C (Bright, 1951)

251 at 25 °C (Hinckley et al., 1990)

Environmental Fate

Biological. Reported biodegradation products include 2,3-dihydro-2,3-dihydroxybiphenyl, 2,3-dihydroxybiphenyl, 2-hydroxy-6-oxo-6-phenylhexa-2,4-dienoate, 2-hydroxy-3-phenyl-6-oxohexa-2,4-dienoate, 2-oxopenta-4-enoate, phenylpyruvic acid (Verschueren, 1983), 2-hydroxybiphenyl, 4-hydroxybiphenyl and 4,4'-hydroxybiphenyl (Smith and Rosazza, 1974). The microbe *Candida lipolytica* degraded biphenyl into the following products: 2-, 3- and 4-hydroxybiphenyl, 4,4'-dihydroxybiphenyl and 3-methoxy-4-hydroxybiphenyl (Cerniglia and Crow, 1981). With the exception of 3-methoxy-4-hydroxybiphenyl, these products were also identified as metabolites by *Cunninghanella elegans* (Dodge et al., 1979). In activated sludge, 15.2% mineralized to carbon dioxide after 5 days

(Freitag et al., 1985).

Under aerobic conditions, *Beijerinckia* sp. degraded biphenyl to *cis*-2,3-dihy-dro-2,3-dihydroxybiphenyl. In addition, *Oscillatoria* sp. and *Pseudomonas putida* degraded biphenyl to 4-hydroxybiphenyl and benzoic acid, respectively (Kobayashi and Rittman, 1982).

Photolytic. A carbon dioxide yield of 9.5% was achieved when biphenyl adsorbed on silica gel was irradiated with light (λ >290 nm) for 17 hours (Freitag et al., 1985). Irradiation of biphenyl (λ >300 nm) in the presence of nitrogen monoxide resulted in the formation of 2- and 4-nitrobiphenyl (Fukui et al., 1980).

Chemical/Physical. The aqueous chlorination of biphenyl at 40 °C over a pH range of 6.2 to 9.0 yielded 2-chlorobiphenyl and 3-chlorobiphenyl (Snider and Alley, 1979). In an acidic aqueous solution (pH = 4.5) containing bromide ions and a chlorinating agent (sodium hypochlorite), 4-bromobiphenyl formed as the major product. Minor products identified include 2-bromobiphenyl, 2,4- and 4,4'-di-bromobiphenyl (Lin et al., 1984).

Exposure Limits: NIOSH REL: TWA 1 mg/m^3 (0.2 ppm), IDLH 100 mg/m^3; OSHA PEL: TWA 1 mg/m^3; ACGIH TLV: TWA 0.2 ppm.

Symptoms of Exposure: Irritation of throat, eyes; headache, nausea, fatigue, numbness in limbs (Hamburg et al., 1989).

Toxicity: Acute oral LD$_{50}$ for rats 3,280 mg/kg (RTECS, 1985).

Uses: Heat transfer liquid; fungistat for oranges; plant disease control; manufacture of benzidine; organic synthesis.

BIS(2-CHLOROETHOXY)METHANE

Synonyms: BCEXM; Bis(2-chloroethyl)formal; Bis(β-chloroethyl)formal; Dichloro-diethyl formal; Dichlorodiethyl methylal; Dichloroethyl formal; Di-2-chloroethyl formal; **1,1'-[Methylenebis(oxy)]bis(2-chloroethane)**; 1,1'-[Methylenebis(oxy)]-bis(2-chloroformaldehyde); Bis(β-chloroethyl)acetal ethane; Formaldehyde bis(β-chloroethylacetal); RCRA waste number U024.

$$Cl - \underset{\underset{\textstyle H}{|}}{\overset{\overset{\textstyle H}{|}}{C}} - \underset{\underset{\textstyle H}{|}}{\overset{\overset{\textstyle H}{|}}{C}} - O - \underset{\underset{\textstyle H}{|}}{\overset{\overset{\textstyle H}{|}}{C}} - O - \underset{\underset{\textstyle H}{|}}{\overset{\overset{\textstyle H}{|}}{C}} - \underset{\underset{\textstyle H}{|}}{\overset{\overset{\textstyle H}{|}}{C}} - Cl$$

CAS Registry No.: 111-91-1
DOT: 1916
DOT label: Poison
Molecular formula: $C_5H_{10}Cl_2O_2$
Formula weight: 173.04
RTECS: PA3675000

Physical state and color
Colorless liquid.

Melting point (°C):
-32.8 (Hawley, 1981)

Boiling point (°C):
218.1 (Webb et al., 1962)

Density (g/cm^3):
1.2339 at 20/20 °C (Hawley, 1981)

Diffusivity in water (10^5 cm^2/sec):
0.72 at 20 °C using method of Hayduk and Laudie (1974).

Flash point (°C):
110 (open cup, Hawley, 1981)

Henry's law constant (10^7 atm·m^3/mol):
3.78 (calculated, U.S. EPA, 1980)

Soil sorption coefficient, log K_{oc}:
2.06 using method of Kenaga and Goring (1980).

Octanol/water partition coefficient, log K_{ow}:
1.26 (calculated, Leo et al., 1971)

Solubility in water:
81,000 mg/L at 25 °C using method of Moriguchi (1975).

Vapor density:
7.07 g/L at 25 °C, 5.97 (air = 1)

Vapor pressure (mmHg):
1 at 53 °C (Weast, 1986)

Environmental Fate
    Biological. Using settled domestic wastewater inoculum, bis(2-chloroethoxy)methane (5 and 10 mg/L) did not degrade after 28 days of incubation at 25 °C (Tabak et al., 1981).

Toxicity: Acute oral LD_{50} for rats 65 mg/kg (RTECS, 1985).

Uses: Manufacture of insecticides, polymers; degreasing solvent; intermediate for polysulfide rubber.

BIS(2-CHLOROETHYL)ETHER

Synonyms: Bis(β-chloroethyl)ether; Chlorex; 1-Chloro-2-(β-chloroethoxy)ethane; Chloroethyl ether; 2-Chloroethyl ether; (β-Chloroethyl)ether; DCEE; Dichlorodiethyl ether; 2,2'-Dichlorodiethyl ether; β,β'-Dichlorodiethyl ether; Dichloroether; Dichloroethyl ether; α,α'-Dichloroethyl ether; Di(β-chloroethyl)-ether; Di(2-chloroethyl)ether; *sym*-Dichloroethyl ether; 2,2'-Dichloroethyl ether; Dichloroethyl oxide; ENT 4504; **1,1'-Oxybis(2-chloroethane)**; RCRA waste number U025; UN 1916.

$$Cl-\underset{\underset{H}{|}}{\overset{\overset{H}{|}}{C}}-\underset{\underset{H}{|}}{\overset{\overset{H}{|}}{C}}-O-\underset{\underset{H}{|}}{\overset{\overset{H}{|}}{C}}-\underset{\underset{H}{|}}{\overset{\overset{H}{|}}{C}}-Cl$$

CAS Registry No.: 111-44-4
DOT: 1916
DOT label: Poison
Molecular formula: $C_4H_8Cl_2O$
Formula weight: 143.01
RTECS: KN0875000

Physical state, color and odor
Clear, colorless liquid with a strong, fruity odor.

Melting point (°C):
-47 (Aldrich, 1988)
-50 (NIOSH, 1994)

Boiling point (°C):
178.5 (Dean, 1973)

Density (g/cm^3):
1.2199 at 20/4 °C (Weast, 1986)

Diffusivity in water (10^5 cm^2/sec):
0.80 at 20 °C using method of Hayduk and Laudie (1974).

Flash point (°C):
55 (NFPA, 1984)

Henry's law constant (10^5 atm·m^3/mol):
1.3 (Schwille, 1988)

Ionization potential (eV):
9.85 (Franklin et al., 1969)

Bioconcentration factor, log BCF:
1.04 (bluegill sunfish, Veith et al., 1980)

Soil sorption coefficient, log K_{oc}:
1.15 (Schwille, 1988)

Octanol/water partition coefficient, log K_{ow}:
1.12 (Veith et al., 1980)
1.29 (Lide, 1990)

Solubility in organics:
Soluble in acetone, ethanol, benzene and ether (Weast, 1986)

Solubility in water (mg/L):
10,700 at 20 °C (Dean, 1973)
10,200 at 20 °C (Du Pont, 1966)
10,400 at 20 °C, 10,300 at 31 °C (Stephenson, 1992)
17,195 at 25 °C (Veith et al., 1975)

Vapor density:
5.84 g/L at 25 °C, 4.94 (air = 1)

Vapor pressure (mmHg):
0.71 at 20 °C, 1.4 at 25 °C (Verschueren, 1983)
1.55 at 25 °C (Howard, 1989)

Environmental Fate
  Biological. When 5 and 10 mg/L of bis(2-chloroethyl)ether were statically incubated in the dark at 25 °C with yeast extract and settled domestic wastewater inoculum, complete degradation was observed after 7 days (Tabak et al., 1981).
  Chemical/Physical. The hydrolysis rate constant for bis(2-chloroethyl) ether at pH 7 and 25 °C was determined to be 2.6 x 10^{-5}/hour, resulting in a half-life of 3.0 years. Products of hydrolysis include 2-(2-chloroethoxy)ethanol and bis(2-hydroxyethyl) ether (Ellington et al., 1988).
  Decomposes in the presence of moisture forming hydrochloric acid (NIOSH, 1994).

Exposure Limits: Potential occupational carcinogen. NIOSH REL: TWA 5 ppm (30 mg/m^3), STEL 10 ppm (60 mg/m^3), IDLH 100 ppm; OSHA PEL: ceiling 15 ppm (90 mg/m^3); ACGIH TLV: TWA 30 mg/m^3, STEL 60 mg/m^3.

Symptoms of Exposure: Eye contact may cause conjunctival irritation and injury to the corneal injury. Ingestion of low concentrations may cause nausea and vomiting (Patnaik, 1992). Symptoms of inhalation include irritation of nose and throat (NIOSH, 1994).

Toxicity: Acute oral LD_{50} for mice 112 mg/kg, rats 75 mg/kg (RTECS, 1985).

Uses: Scouring and cleaning textiles; fumigants; processing fats, waxes, greases, cellulose esters; dewaxing agent for lubricating oils; preparation of insecticides, butadiene, pharmaceuticals; solvent in paints, varnishes and lacquers; selective solvent for production of high-grade lubricating oils; fulling, wetting and penetrating compounds; finish removers; spotting and dry cleaning; soil fumigant; acaricide; organic synthesis.

BIS(2-CHLOROISOPROPYL)ETHER

Synonyms: BCIE; BCMEE; Bis(β-chloroisopropyl)ether; Bis(2-chloro-1-methyl-ethyl)ether; 1-Chloro-2-(β-chloroisopropoxy)propane; 2-Chloroisopropyl ether; β-Chloroisopropylether; (2-Chloro-1-methylethyl)ether; Dichlorodiisopropyl ether; Dichloroisopropyl ether; 2,2'-Dichloroisopropyl ether; NCI-C50044; **2,2'-Oxybis-(1-chloropropane)**; RCRA waste number U027; UN 2490.

CAS Registry No.: 108-60-1
DOT label: Poison and corrosive material
Molecular formula: $C_6H_{12}Cl_2O$
Formula weight: 171.07
RTECS: KN1750000

Physical state and color
Colorless to brown oily liquid.

Melting point (°C):
-20 (Sax and Lewis, 1987)

Boiling point (°C):
187 (Weast, 1986)

Density (g/cm³):
1.103 at 20/4 °C (Weast, 1986)
1.1127 at 25/4 °C (Standen, 1964)

Diffusivity in water (10^5 cm²/sec):
0.68 at 20 °C using method of Hayduk and Laudie (1974).

Flash point (°C):
85 (Hawley, 1981)

Henry's law constant (10^4 atm·m³/mol):
1.1 (Pankow and Rosen, 1988)

Soil sorption coefficient, log K_{oc}:
1.79 (Schwille, 1988)

120   Bis(2-chloroisopropyl)ether

Octanol/water partition coefficient, log K_{ow}:
2.58 (U.S. EPA, 1980)

Solubility in organics:
Soluble in acetone, ethanol, benzene and ether (Weast, 1986).

Solubility in water (mg/L):
1,700 at 20 °C (Dean, 1973)
2,450 at 19.1 °C, 2,370 at 31.0 °C (Stephenson, 1992)

Vapor density:
6.99 g/L at 25 °C, 5.91 (air = 1)

Vapor pressure (mmHg at 20 °C):
0.85 (Verschueren, 1983)
0.56 (Worthing and Hance, 1991)

Environmental Fate
   Biological. When bis(2-chloroisopropyl)ether (5 and 10 mg/L) was statically incubated in the dark at 25 °C with yeast extract and settled domestic wastewater inoculum, complete biodegradation was achieved after 14 days (Tabak et al., 1981).

Symptoms of Exposure: Exposure to vapors may cause eye and respiratory tract irritation (Patnaik, 1992).

Toxicity: Acute oral LD_{50} for rats 240 mg/kg (RTECS, 1985); LC_{50} (48-hour) for carp >40 mg/L (Hartley and Kidd, 1987).

Drinking Water Standard: As of May 1994, no MCLGs or MCLs have been proposed (U.S. EPA, 1994).

Uses: Chemical intermediate in the manufacturing of dyes, resins and pharmaceuticals; solvent and extractant for fats, waxes and greases; textile manufacturing; agent in paint and varnish removers, spotting and cleaning agents; a combatant in liver fluke infections; preparation of glycol esters in fungicidal preparations and as an insecticidal wood preservative; apparently used as a nematocide in Japan but is not registered in the U.S. for use as a pesticide.

BIS(2-ETHYLHEXYL)PHTHALATE

Synonyms: 1,2–Benzenedicarboxylic acid bis(2–ethylhexyl)ester; Bioflex 81; Bioflex DOP; Bis(2-ethylhexyl)-1,2-benzenedicarboxylate; Compound 889; DAF 68; DEHP; Di(2-ethylhexyl)orthophthalate; Di(2-ethylhexyl)phthalate; Dioctyl phthalate; Di-*sec*-octyl phthalate; DOP; Ergoplast FDO; Ethylhexyl phthalate; 2-Ethylhexyl phthalate; Eviplast 80; Eviplast 81; Fleximel; Flexol DOP; Flexol plasticizer DOP; Good-rite GP 264; Hatcol DOP; Hercoflex 260; Kodaflex DOP; Mollan 0; NCI-C52733; Nuoplaz DOP; Octoil; Octyl phthalate; Palatinol AH; Phthalic acid bis(2-ethylhexyl) ester; Phthalic acid dioctyl ester; Pittsburgh PX-138; Platinol AH; Platinol DOP; RC plasticizer DOP; RCRA waste number U028; Reomol D 79P; Reomol DOP; Sicol 150; Staflex DOP; Truflex DOP; Vestinol 80; Witicizer 312.

$$COOCH_2CH(C_2H_5)(CH_2)_3CH_3$$
$$COOCH_2CH(C_2H_5)(CH_2)_3CH_3$$

CAS Registry No.: 117-81-7
Molecular formula: $C_{24}H_{38}O_4$
Formula weight: 390.57
RTECS: TI0350000

Physical state, color and odor
Colorless, oily liquid with a very faint odor.

Melting point (°C):
-55 (Verschueren, 1983)
-46 (Standen, 1968)

Boiling point (°C at 5 mmHg):
386.9 (Fishbein and Albro, 1972)
230 (Howard, 1989)

Density (g/cm³):
0.985 at 20/4 °C (Fluka, 1988)

Diffusivity in water (10^5 cm²/sec):
0.39 at 20 °C using method of Hayduk and Laudie (1974).

Flash point (°C):
207 (Aldrich, 1988)

196 (open cup, Broadhurst, 1972)
215 (open cup, NFPA, 1984)

Lower explosive limit (%):
0.3 at 245 °C (NFPA, 1984)

Henry's law constant (10^5 atm·m^3/mol):
1.1 at 25 °C (calculated, Howard, 1989)

Bioconcentration factor, log BCF:
3.73 (algae, Geyer et al., 1984)
3.48 (activated sludge), 1.60 (golden ide) (Freitag et al., 1985)
2.49 (freshwater fish), 2.11 (fish, microcosm) (Garten and Trabalka, 1983)

Soil sorption coefficient, log K_{oc}:
5.0 (Neely and Blau, 1985)
4.94 (Meylan et al., 1992)

Octanol/water partition coefficient, log K_{ow}:
4.20 (Mackay, 1982)
5.11 (Geyer et al., 1984)
5.0 (Klein et al., 1988)
7.453 (de Bruijn et al., 1989)
7.137 (Brooke et al., 1990)
4.88 (Wams, 1987)

Solubility in organics:
Miscible with mineral oil and hexane (U.S. EPA, 1985).

Solubility in water (μg/L):
400 at 25 °C (Wolfe et al., 1980)
41 at 20 °C (Leyder and Boulanger, 1983)
At 25 °C: 340, 300 (well water), 160 (natural seawater) (Howard et al., 1985)
285 at 24 °C (technical grade, Hollifield, 1979)
47 at 25 °C (Klöpffer et al., 1982)
360 (DeFoe et al., 1990)
300 at 25 °C (Wams, 1987)

Vapor density:
15.96 g/L at 25 °C, 13.48 (air = 1)

Vapor pressure (10^9 mmHg):
200 at 20 °C (Hirzy et al., 1978)

50 at 68 °C, 5,000 at 120 °C (Gross and Colony, 1973)
6,450 at 25 °C (Howard et al., 1985)
100 at 20 °C (Broadhurst, 1972)
62 at 25 °C (Giam et al., 1980)
8.25 at 20 °C (Riederer, 1990)
340 at 25 °C (Wams, 1987)
143 at 25 °C (Hinckley et al., 1990)

Environmental Fate
   Biological. Bis(2-ethylhexyl)phthalate degraded in both amended and unamended calcareous soils from New Mexico. After 146 days, 76 to 93% degraded to carbon dioxide. No other metabolites were detected (Fairbanks et al., 1985). In a 56-day experiment, [^{14}C]bis(2-ethylhexyl)phthalate applied to soil-water suspensions under aerobic and anaerobic conditions gave $^{14}CO_2$ yields of 11.6 and 8.1%, respectively (Scheunert et al., 1987). When bis(2-ethylhexyl)-phthalate was statically incubated in the dark at 25 °C with yeast extract and settled domestic wastewater inoculum, no degradation was observed after 7 days. Over a 21-day period, however, gradual adaptation did occur resulting in 95 and 93% losses at concentrations of 5 and 10 mg/L, respectively (Tabak et al., 1981).
   Chemical/Physical. Hydrolyzes in water to phthalic acid and 2-ethylhexyl alcohol (Wolfe et al., 1980). Pyrolysis of bis(2-ethylhexyl)phthalate in the presence of polyvinyl chloride at 600 °C for 10 minutes gave the following compounds: methylindene, naphthalene, 1-methylnaphthalene, 2-methylnaphthalene, biphenyl, dimethylnaphthalene, acenaphthene, fluorene, methylacenaphthene, methyl-fluorene, phenanthrene, anthracene, methylphenanthrene, methylanthracene, methylpyrene or fluoranthene and 17 unidentified compounds (Bove and Dalven, 1984).

Exposure Limits (mg/m^3): Potential occupational carcinogen. NIOSH REL: TWA 5, STEL 10, IDLH 5,000; OSHA PEL: TWA 5; ACGIH TLV: TWA 5, STEL 10.

Symptoms of Exposure: Ingestion of 10 mL may cause gastrointestinal pain, hypermobility and diarrhea (Patnaik, 1992).

Toxicity: Acute oral LD$_{50}$ for guinea pigs 26 gm/kg, mice 30 gm/kg, rats 30,600 mg/kg, rabbits 34 gm/kg (RTECS, 1985).

Uses: Plasticizer; in vacuum pumps.

BROMOBENZENE

Synonyms: Monobromobenzene; NCI-C55492; Phenyl bromide; UN 2514.

CAS Registry No.: 108-86-1
DOT: 2514
DOT label: Flammable liquid
Molecular formula: C_6H_5Br
Formula weight: 157.01
RTECS: CY9000000

Physical state, color and odor
Clear, colorless liquid with an aromatic odor.

Melting point (°C):
-30.8 (Weast, 1986)

Boiling point (°C):
156 (Weast, 1986)

Density (g/cm^3):
1.5017 at 15/4 °C, 1.4952 at 20/4 °C, 1.4815 at 30/4 °C (Windholz et al., 1983)

Diffusivity in water (10^5 cm^2/sec):
0.85 at 20 °C using method of Hayduk and Laudie (1974).

Flash point (°C):
51 (Windholz et al., 1983)

Heat of fusion (kcal/mol):
2.54 (Dean, 1987)

Henry's law constant (10^3 atm·m^3/mol at 25 °C):
2.40 (Valsaraj, 1988)
2.08 (Hine and Mookerjee, 1975)

Interfacial tension with water (dyn/cm at 25 °C):
38.1 (Donahue and Bartell, 1952)

Ionization potential (eV):
8.98 ± 0.02 (Franklin et al., 1969)
9.41 (Yoshida et al., 1983)

Bioconcentration factor, log BCF:
3.18 (activated sludge), 2.28 (algae), 1.70 (golden ide) (Freitag et al., 1985)

Soil sorption coefficient, log K_{oc}:
2.33 using method of Chiou et al. (1979).

Octanol/water partition coefficient, log K_{ow}:
3.01 (Garst and Wilson, 1984; Watarai et al., 1982)
2.99 (Hansch and Anderson, 1967)
2.96 (Kenaga and Goring, 1980)
2.98 (Wasik et al., 1981, 1983)

Solubility in organics:
At 25 °C (wt %): alcohol (10.4) and ether (71.3). Miscible with benzene, chloroform, petroleum ethers (Windholz et al., 1983) and many other organic solvents (Patnaik, 1992).

Solubility in water:
500 mg/L at 20 °C (Verschueren, 1983)
446 mg/L at 30 °C (Chiou et al., 1977)
409 mg/L at 25 °C (Valsaraj, 1988)
446 mg/kg at 30 °C (Gross and Saylor, 1931)
3.62 mmol/L at 25 °C (Andrews and Keefer, 1950)
2.92 mmol/L at 35 °C (Hine et al., 1963)
2.3 mmol/L at 25 °C (Yalkowsky et al., 1979)
2.62 mmol/L at 25 °C (Wasik et al., 1981, 1983)
2.84 mmol/kg at 25.0 °C (Vesala, 1974)

Vapor density:
6.42 g/L at 25 °C, 5.41 (air = 1)

Vapor pressure (mmHg):
3.3 at 20 °C (Verschueren, 1983)
3.8 at 25 °C (Valsaraj, 1988)
4.14 at 25 °C (Mackay et al., 1982)
7.48 at 35 °C (Hine et al., 1963)

Environmental Fate
Biological. In activated sludge, 34.8% of the applied bromobenzene mineralized

to carbon dioxide after 5 days (Freitag et al., 1985).

Photolytic. A carbon dioxide yield of 19.7% was achieved when bromobenzene adsorbed on silica gel was irradiated with light (λ >290 nm) for 17 hours (Freitag et al., 1985). Irradiation of bromobenzene in air containing nitrogen oxides gave phenol, 4-nitrophenol, 2,4-dinitrophenol, 4-bromophenol, 3-bromonitrobenzene, 3-bromo-2-nitrophenol, 3-bromo-4-nitrophenol, 3-bromo-6-nitrophenol, 2-bromo-4-nitrophenol and 2,6-dibromo-4-nitrophenol (Nojima et al., 1980).

Toxicity: Acute oral LD_{50} for rats 2,699 mg/kg, mice 2,700 mg/kg (RTECS, 1985).

Drinking Water Standard: As of May 1994, bromobenzene was listed for regulation but no MCLGs or MCLs have been proposed (U.S. EPA, 1994).

Uses: Preparation of phenyl magnesium bromide used in organic synthesis; solvent for fats, waxes and oils; motor oil additive; crystallizing solvent; chemical intermediate.

BROMOCHLOROMETHANE

Synonyms: CB; CBM; Chlorobromomethane; Halon 1011; Methylene chlorobromide; Mil-B-4394-B; UN 1887.

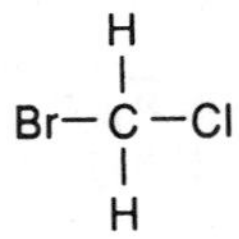

$$\begin{array}{c} H \\ | \\ Br-C-Cl \\ | \\ H \end{array}$$

CAS Registry No.: 74-97-5
DOT: 1887
Molecular formula: CH_2BrCl
Formula weight: 129.39
RTECS: PA5250000

Physical state, color and odor
Clear, colorless liquid with a sweet, chloroform-like odor.

Melting point (°C):
-87.5 (Riddick et al., 1986)

Boiling point (°C):
68.1 (Weast, 1986)

Density (g/cm^3 at 20/4 °C):
1.9344 (Weast, 1986)
1.991 (Aldrich, 1988)

Diffusivity in water (10^5 cm^2/sec):
1.13 at 20 °C using method of Hayduk and Laudie (1974).

Flash point (°C):
Non applicable (NIOSH, 1994)

Lower explosive limit (%):
Not applicable (NIOSH, 1994)

Upper explosive limit (%):
Not applicable (NIOSH, 1994)

Henry's law constant (10^3 atm·m^3/mol):
1.44 at 25 °C (approximate - calculated from water solubility and vapor pressure)

Ionization potential (eV):
10.77 ± 0.01 (Franklin et al., 1969)

Soil sorption coefficient, log K_{oc}:
1.43 using method of Chiou et al. (1979)

Octanol/water partition coefficient, log K_{ow}:
1.41 (Tewari et al., 1982; Wasik et al., 1981)

Solubility in organics:
Soluble in acetone, alcohol, benzene and ether (Weast, 1986).

Solubility in water (25 °C):
129 mmol/L (Tewari et al., 1982; Wasik et al., 1981)
15,000 mg/L (O'Connell, 1963)

Vapor density:
5.29 g/L at 25 °C, 4.47 (air = 1)

Vapor pressure (mmHg):
115 at 20 °C (NIOSH, 1994)
141.07 at 24.05 °C (Kudchadker et al., 1979)

Environmental Fate
   Biological. When bromochloromethane (5 and 10 mg/L) was statically incubated in the dark at 25 °C with yeast extract and settled domestic wastewater inoculum for 7 days, 100% biodegradation with rapid adaptation was observed (Tabak et al., 1981).
   Chemical/Physical. Though no products were identified, the estimated hydrolysis half-life in water at 25 °C and pH 7 is 44 years (Mabey and Mill, 1978).

Exposure Limits: NIOSH REL: TWA 200 ppm (1,050 mg/m^3), IDLH 2,000 ppm; OSHA PEL: TWA 200 ppm; ACGIH TLV: TWA 200 ppm, STEL 250 ppm (1,300 mg/m^3).

Toxicity: Acute oral LD$_{50}$ for rats 5,000 mg/kg, mice 4,300 mg/kg; LC$_{50}$ (inhalation) for mice 15,850 mg/m^3/8-hour, rats 28,800 ppm/15-minute (RTECS, 1985).

Drinking Water Standard: As of May 1994, no MCLGs or MCLs have been proposed (U.S. EPA, 1994).

Uses: Fire extinguishing agent; organic synthesis.

BROMODICHLOROMETHANE

Synonyms: BDCM; Dichlorobromomethane; NCI-C55243.

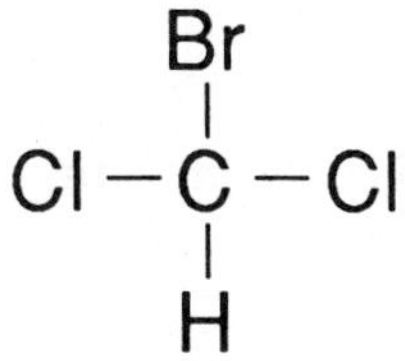

$$Cl - \underset{\underset{H}{|}}{\overset{\overset{Br}{|}}{C}} - Cl$$

CAS Registry No.: 75-27-4
Molecular formula: $CHBrCl_2$
Formula weight: 163.83
RTECS: PA5310000

Physical state and color
Clear, colorless liquid.

Melting point (°C):
-57.1 (Weast, 1986)

Boiling point (°C):
90.1 (Dean, 1973)

Density (g/cm^3 at 20/4 °C):
1.980 (Weast, 1986)

Diffusivity in water (10^5 cm^2/sec):
0.98 at 20 °C using method of Hayduk and Laudie (1974).

Flash point (°C):
None (Dean, 1987)

Henry's law constant (10^4 atm·m^3/mol):
2.12 at 25 °C (Warner et al., 1987)
16 (Nicholson et al., 1984)
16, 26 and 40 at 20, 30 and 40 °C, respectively (Tse et al., 1992)
In seawater: 5.5, 10.5 and 19 at 0, 10 and 20 °C, respectively (Moore et al., 1995)

Ionization potential (eV):
10.88 ± 0.05 (Franklin et al., 1969)

Soil sorption coefficient, log K_{oc}:
1.79 (Schwille, 1988)

Octanol/water partition coefficient, log K_{ow}:
1.88 (Mills et al., 1985)
2.10 (Hansch and Leo, 1979; Mabey et al., 1982)

Solubility in organics:
Soluble in acetone, ethanol, benzene, chloroform and ether (Weast, 1986).

Solubility in water (mg/L):
4,500 at 0 °C (Schwille, 1988)
4,700 at 22 °C (Mabey et al., 1982)
2,968 at 30 °C (McNally and Grob, 1984)
3,031.9 at 30 °C (McNally and Grob, 1983)

Vapor density:
6.70 g/L at 25 °C, 5.66 (air = 1)

Vapor pressure (mmHg):
50 at 20 °C (Mills et al., 1985)

Environmental Fate
   Biological. Bromodichloromethane showed significant degradation with gradual adaptation in a static-culture flask-screening test (settled domestic wastewater inoculum) conducted at 25 °C. At concentrations of 5 and 10 mg/L, percent losses after 4 weeks of incubation were 59 and 51, respectively. At a substrate concentration of 5 mg/L, 8% was lost due to volatilization after 10 days (Tabak et al., 1981).
   Chemical/Physical. Though no products were identified, the estimated hydrolysis half-life in water at 25 °C and pH 7 is 137 years (Mabey and Mill, 1978).

Toxicity: Acute oral LD_{50} for rats 916 mg/kg, mice 450 mg/kg (RTECS, 1985).

Drinking Water Standard (tentative): MCLG: zero; MCL: 0.1 mg/L. Total for all trihalomethanes cannot exceed a concentration of 0.08 mg/L (U.S. EPA, 1984).

Uses: Component of fire extinguisher fluids; solvent for waxes, fats and resins; degreaser; flame retardant; heavy liquid for mineral and salt separations; chemical intermediate; laboratory use.

BROMOFORM

Synonyms: Methenyl tribromide; Methyl tribromide; NCI-C55130; RCRA waste number U225; **Tribromomethane**; UN 2515.

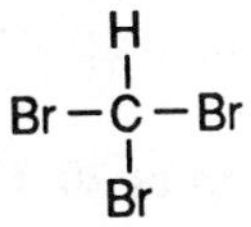

$$Br-\overset{\displaystyle H}{\underset{\displaystyle Br}{C}}-Br$$

CAS Registry No.: 75-25-2
DOT: 2515
DOT label: Poison
Molecular formula: $CHBr_3$
Formula weight: 252.73
RTECS: PB5600000

Physical state, color and odor
Clear, colorless to yellow liquid with a chloroform-like odor.

Melting point (°C):
8.05 (Riddick et al., 1986)
8.3 (Weast, 1986)

Boiling point (°C):
149.5 (Weast, 1986)

Density (g/cm^3 at 20/4 °C):
2.9031 (Dean, 1987)
2.8899 (Weast, 1986)
2.89165 (Kudchadker et al., 1979)

Diffusivity in water (10^5 cm^2/sec):
0.95 at 20 °C using method of Hayduk and Laudie (1974).

Flash point (°C):
Noncombustible liquid (NIOSH, 1994)

Lower explosive limit (%):
Not applicable (NIOSH, 1994)

Upper explosive limit (%):
Not applicable (NIOSH, 1994)

131

Henry's law constant (10^4 atm·m^3/mol):
5.6 (Pankow and Rosen, 1988)
5.32 at 25 °C (Warner et al., 1987)
4.3 at 20 °C (Nicholson et al., 1984)
4, 7 and 12 at 20, 30 and 40 °C, respectively (Tse et al., 1992)
In seawater: 0.14, 2.9 and 5.2 at 0, 10 and 20 °C, respectively (Moore et al., 1995)

Interfacial tension with water (dyn/cm at 20 °C):
40.85 (Demond and Lindner, 1993)

Ionization potential (eV):
10.51 ± 0.02 (Franklin et al., 1969)

Soil sorption coefficient, log K_{oc}:
2.45 (Abdul et al., 1987)
2.06 (Schwille, 1988)

Octanol/water partition coefficient, log K_{ow}:
2.30 (Mills et al., 1985)
2.38 (Valsaraj, 1988)

Solubility in organics:
Soluble in ligroin (Weast, 1986). Miscible with benzene, chloroform, ether, petroleum ether, acetone and oils (Windholz et al., 1983).

Solubility in water:
3,010 mg/kg at 15 °C, 3,190 mg/kg at 30 °C (Gross and Saylor, 1931)
3,130 mg/L at 25 °C (Valsaraj, 1988)
3,050 mg/L 20 °C (Munz and Roberts, 1987)
0.318 wt % at 30 °C (Riddick et al., 1986)
3,931 mg/L at 30 °C (McNally and Grob, 1984)
12.6 mmol/L at 30 °C (Horvath, 1982)

Vapor density:
10.33 g/L at 25 °C, 8.72 (air = 1)

Vapor pressure (mmHg):
4 at 20 °C (Munz and Roberts, 1987)
5.4 at 25 °C (Mackay et al., 1982)

Environmental Fate
   Biological. Bromoform showed significant degradation with gradual adaptation in a static-culture flask-screening test (settled domestic wastewater inoculum)

conducted at 25 °C. At concentrations of 5 and 10 mg/L, percent losses after 4 weeks of incubation were 48 and 35, respectively (Tabak et al., 1981).

Chemical/Physical. Though no products were identified, the estimated hydrolysis half-life in water at 25 °C and pH 7 is 686 years (Mabey and Mill, 1978). To an aqueous solution containing 9.08 μmol bromoform was bubbled hydrogen. After 24 hours, only 5% of the bromoform reacted to form methane and minor traces of ethane. In the presence of colloidal platinum catalyst, the reaction proceeded at a much faster rate forming the same end products (Wang et al., 1988). In an earlier study, water containing 2,000 ng/μL of bromoform and colloidal platinum catalyst was irradiated with UV light. After 20 hours, about 50% of the bromoform had reacted. A duplicate experiment was performed but the concentration of bromoform was increased to 3,000 ng/μL and 0.1 g zinc was added. After 14 hours, only 0.1 ng/μL bromoform remained. Anticipated transformation products include methane and bromide ions (Wang and Tan, 1988).

Photolysis of an aqueous solution containing bromoform (989 μmol) and a catalyst [Pt(colloid)/Ru(bpy)$^{2+}$/MV/EDTA] yielded the following products after 25 hours (μmol detected): bromide ions (250), methylene bromide (475) and unreacted bromoform (421) (Tan and Wang, 1987).

Exposure Limits: NIOSH REL: TWA 0.5 ppm (5 mg/m^3), IDLH 850 ppm; OSHA PEL: 0.5 ppm; ACGIH TLV: STEL 0.5 ppm.

Symptoms of Exposure: Inhalation may cause respiratory irritation (Patnaik, 1992).

Toxicity: Acute oral LD$_{50}$ for rats 1,147 mg/kg, mice 1,400 mg/kg (RTECS, 1985).

Drinking Water Standard (tentative): MCLG: zero; MCL: 0.1 mg/L. Total for all trihalomethanes cannot exceed a concentration of 0.08 mg/L (U.S. EPA, 1984).

Uses: Solvent for waxes, greases and oils; separating solids with lower densities; component of fire-resistant chemicals; geological assaying; medicine (sedative); gauge fluid; intermediate in organic synthesis.

4-BROMOPHENYL PHENYL ETHER

Synonyms: 4-Bromodiphenyl ether; *p*-Bromodiphenyl ether; **1-Bromo-4-phen-oxybenzene**; 1-Bromo-*p*-phenoxybenzene; 4-Bromophenyl ether; *p*-Bromophenyl ether; *p*-Bromophenyl phenyl ether; Phenyl-4-bromophenyl ether; Phenyl-*p*-bromophenyl ether.

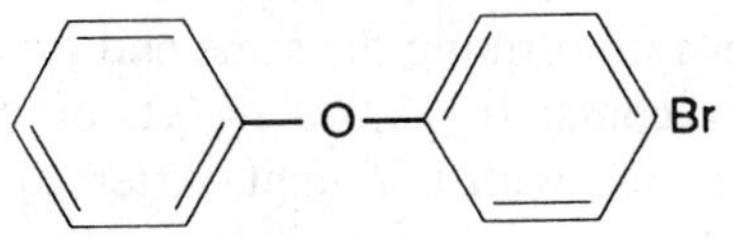

CAS Registry No.: 101-55-3
Molecular formula: $C_{12}H_9BrO$
Formula weight: 249.20

Physical state
Liquid.

Melting point (°C):
18.7 (Weast, 1986)

Boiling point (°C):
310.1 (Weast, 1986)
305 (Aldrich, 1988)

Density (g/cm^3):
1.4208 at 20/4 °C (Weast, 1986)

Diffusivity in water (10^5 cm^2/sec):
0.63 at 20 °C using method of Hayduk and Laudie (1974).

Flash point (°C):
>110 (Aldrich, 1988)

Henry's law constant (10^4 atm·m^3/mol):
1 (Pankow and Rosen, 1988)

Soil sorption coefficient, log K_{oc}:
4.94 using method of Karickhoff et al. (1979).

Octanol/water partition coefficient, log K_{ow}:
5.15 (Walton, 1985)

Solubility in organics:
Soluble in ether (Weast, 1986).

Vapor density:
10.19 g/L at 25 °C, 8.60 (air = 1)

Vapor pressure (mmHg):
1.5 x 10^{-3} at 20 °C (calculated, Dreisbach, 1952)

Environmental Fate
   Biological. Using settled domestic wastewater inoculum, 4-bromophenyl phenyl ether (5 and 10 mg/L) did not degrade after 28 days of incubation at 25 °C (Tabak et al., 1981).

Use: Research chemical.

BROMOTRIFLUOROMETHANE

Synonyms: Bromofluoroform; F 13B1; Freon 13-B1; Halocarbon 13B1; Halon 1301; Monobromotrifluoromethane; Refrigerant 13B1; Trifluorobromomethane; Trifluoromonobromomethane; UN 1009.

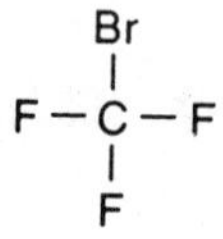

$$\begin{array}{c} Br \\ | \\ F-C-F \\ | \\ F \end{array}$$

CAS Registry No.: 75-63-8
DOT: 1009
Molecular formula: $CBrF_3$
Formula weight: 148.91
RTECS: PA5425000

Physical state, color and odor
Colorless gas with an ether-like odor.

Melting point (°C):
-168 (Horvath, 1982)

Boiling point (°C):
-58 to -57 (Aldrich, 1988)

Density (g/cm³):
1.538 at 20/4 °C (Horvath, 1982)

Diffusivity in water (10^5 cm²/sec):
0.90 at 20 °C using method of Hayduk and Laudie (1974).

Flash point (°C):
Nonflammable gas (NIOSH, 1994)

Lower explosive limit (%):
Not applicable (NIOSH, 1994)

Upper explosive limit (%):
Not applicable (NIOSH, 1994)

Henry's law constant (atm·m³/mol):
0.500 at 25 °C (Hine and Mookerjee, 1975)

Ionization potential (eV):
11.78 (NIOSH, 1994)

Soil sorption coefficient, log K_{oc}:
2.44 using method of Chiou et al. (1979).

Octanol/water partition coefficient, log K_{ow}:
1.54 (Hansch and Leo, 1979)

Solubility in organics:
Soluble in chloroform (Weast, 1986).

Solubility in water:
0.03 wt % at 20 °C (NIOSH, 1994)

Vapor density:
6.09 g/L at 25 °C, 5.14 (air = 1)

Vapor pressure (mmHg):
>760 at 20 °C (NIOSH, 1994)

Exposure Limits: NIOSH REL: TWA 1,000 ppm (6,100 mg/m^3), IDLH 40,000 ppm; OSHA PEL: TWA 1,000 ppm.

Symptoms of Exposure: Exposure to 10% in air for 3 minutes caused lightheadedness and paresthesia (Patnaik, 1992).

Toxicity: LC_{50} (inhalation) for mice 381 gm/m^3, rats 416 gm/m^3 (RTECS, 1985).

Uses: Chemical intermediate; metal hardening; refrigerant; fire extinguishers.

1,3-BUTADIENE

Synonyms: Biethylene; Bivinyl; Butadiene; Buta-1,3-diene; α,γ-Butadiene; Divinyl erythrene; NCI-C50602; Pyrrolylene; Vinylethylene.

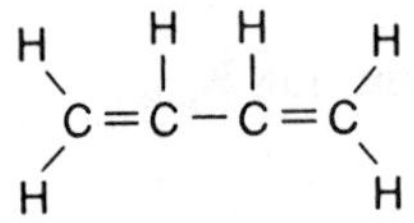

CAS Registry No.: 106-99-0
DOT: 1010
DOT label: Flammable liquid
Molecular formula: C_4H_6
Formula weight: 54.09
RTECS: EI9275000

Physical state, color and odor
Colorless gas with a mild, aromatic odor. Odor threshold in air is 160 ppb (4 mg/m^3) (Keith and Walters, 1992).

Melting point (°C):
-108.9 (Weast, 1986)

Boiling point (°C):
-4.4 (Weast, 1986)
-4.50 (Howard, 1989)

Density (g/cm^3):
0.6211 at 20/4 °C and 0.6149 at 25/4 °C at saturation pressure (Dreisbach, 1959).
0.6789 at 25/4 °C (Hayduk and Minhas, 1987)

Diffusivity in water (10^5 cm^2/sec):
0.95 at 20 °C using method of Hayduk and Laudie (1974).

Dissociation constant:
>14 (Schwarzenbach et al., 1993)

Flash point (°C):
-76 (NIOSH, 1994)

Lower explosive limit (%):
2.0 (NIOSH, 1994)

Upper explosive limit (%):
12.0 (NIOSH, 1994)

Heat of fusion (kcal/mol):
1.908 (Dean, 1987)

Henry's law constant (10^2 atm·m^3/mol):
6.3 at 25 °C (Hine and Mookerjee, 1975)

Ionization potential (eV):
9.07 (Franklin et al., 1969)

Soil sorption coefficient, log K_{oc}:
2.08 (calculated, Mercer et al., 1990)

Octanol/water partition coefficient, log K_{ow}:
1.99 (Hansch and Leo, 1979)

Solubility in organics:
In *n*-heptane (mole fraction): 0.668, 0.360 and 0.210 at 4.00, 25.00 and 50.00 °C, respectively (Hayduk and Minhas, 1987).

Solubility in water:
735 ppm at 25 °C (McAuliffe, 1966)
At 37.8 °C, the reported mole fraction solubilities at 517 and 1,034 mmHg are 8 x 10^{-5} and 1.6 x 10^{-4}, respectively (Reed and McKetta, 1959)
11 mmol/L at 25 °C and 520 mmHg (Fischer and Ehrenberg, 1948)
6.8 x 10^{-4} at 4.00 °C (mole fraction, Hayduk and Minhas, 1987)

Vapor density:
2.29 g/L at 25 °C, 1.87 (air = 1)

Vapor pressure (mmHg):
1,840 at 21 °C (Sax and Lewis, 1987)
2,105 at 25 °C (Wilhoit and Zwolinski, 1971)

Environmental Fate
   Chemical/Physical. Will polymerize in the presence of oxygen if no inhibitor is present (Hawley, 1981).

Exposure Limits: Potential occupational carcinogen. NIOSH REL: IDLH 2,000 ppm; OSHA PEL: TWA 1,000 ppm (2,200 mg/m^3); ACGIH TLV: TWA 10 ppm (22 mg/m^3).

Symptoms of Exposure: An asphyxiant. Inhalation may cause hallucinations, distorted perception, eye, nose and throat irritation. At high concentrations drowsiness, lightheadedness and narcosis may occur (Patnaik, 1992). Contact of liquid with skin may result in frostbite (NIOSH, 1994).

Toxicity: LC_{50} (inhalation) for mice 270 gm/m^3/2-hour, rats 285 gm/m^3/4-hour (RTECS, 1985).

Uses: Synthetic rubbers and elastomers (styrene-butadiene, polybutadiene, neoprene); organic synthesis (Diels-Alder reactions); latex paints; resins; chemical intermediate.

BUTANE

Synonyms: *n*-Butane; Diethyl; Methylethylmethane; UN 1011.

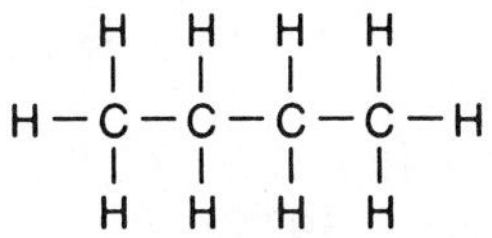

```
      H   H   H   H
      |   |   |   |
  H — C — C — C — C — H
      |   |   |   |
      H   H   H   H
```

CAS Registry No.: 106-97-8
DOT: 1011
DOT label: Flammable gas
Molecular formula: C_4H_{10}
Formula weight: 58.12
RTECS: EJ4200000

Physical state, color and odor
Colorless gas with a natural gas-like odor. Odor threshold in air is 2,700 ppm (Amoore and Hautala, 1983).

Melting point (°C):
-138.4 (Weast, 1986)
-135.0 (Stull, 1947)

Boiling point (°C):
-0.5 (Weast, 1986)

Density (g/cm^3):
0.6012 at 0/4 °C, 0.5788 at 20/4 °C (Weast, 1986)
0.57287 at 25/4 °C (Riddick et al., 1986)

Diffusivity in water (10^5 cm^2/sec):
0.89 at 20 °C (Witherspoon and Bonoli, 1969)
0.97 at 25 °C (Hayduk and Laudie, 1974)

Dissociation constant:
>14 (Schwarzenbach et al., 1993)

Flash point (°C):
-138 (Windholz et al., 1983)

Lower explosive limit (%):
1.6 (NFPA, 1984)

142   Butane

Upper explosive limit (%):
8.4 (NFPA, 1984)

Heat of fusion (kcal/mol):
1.114 (Dean, 1987)

Henry's law constant (atm·m³/mol):
0.930 at 25 °C (Hine and Mookerjee, 1975)

Ionization potential (eV):
10.63 ± 0.03 (Franklin et al., 1969)

Soil sorption coefficient, log K_{oc}:
Unavailable because experimental methods for estimation of this parameter for aliphatic hydrocarbons are lacking in the documented literature.

Octanol/water partition coefficient, log K_{ow}:
2.89 (Hansch and Leo, 1979)

Solubility in organics:
At 17 °C (mL/L): alcohol (150 at 770 mmHg), chloroform (25,000), ether (30,000) (Windholz et al., 1983)
At 10 °C (mole fraction): acetone (0.2276), aniline (0.04886), benzene (0.5904), 2-butanone (0.3885), cyclohexane (0.6712), ethanol (0.1647), methanol (0.04457), 1-propanol (0.2346), 1-butanol (0.2817). At 25 °C (mole fraction): acetone (0.1108), aniline (0.03241), benzene (0.2851), 2-butanone (0.1824), cyclohexane (0.3962), ethanol (0.07825), methanol (0.03763), 1-propanol (0.1138), 1-butanol (0.1401) (Miyano and Hayduk, 1986)
Mole fraction solubility in 1-butanol: 0.140, 0.0692 and 0.0397 at 25, 30 and 70 °C, respectively; in chlorobenzene: 0.274, 0.129 and 0.0800 at 25, 30 and 70 °C, respectively and in octane: 0.423, 0.233 and 0.152 at 25, 30 and 70 °C, respectively (Hayduk et al., 1988)

Solubility in water:
61.4 mg/kg at 25 °C (McAuliffe, 1963, 1966)
At 0 °C, 0.0327 and 0.0233 volume of gas dissolved in a unit volume of water at 19.8 and 29.8 °C, respectively (Claussen and Polglase, 1952)
7 mmol/L at 17 °C and 772 mmHg (Fischer and Ehrenberg, 1948)
1.09 mmol/L at 25 °C (Barone et al., 1966)
3.21, 1.26 and 0.66 mM at 4, 25 and 50 °C, respectively (Kresheck et al., 1965)

Vapor density:
2.38 g/L at 25 °C, 2.046 (air = 1)

Vapor pressure (mmHg):
1,820 at 25 °C (Wilhoit and Zwolinski, 1971)

Environmental Fate

Biological. In the presence of methane, *Pseudomonas methanica* degraded butane to 1-butanol, butyric acid and 2-butanone (Leadbetter and Foster, 1959). 2-Butanone was also reported as a degradation product of butane by the microorganism *Mycobacterium smegmatis* (Riser-Roberts, 1992). Butane may biodegrade in two ways. The first is the formation of butyl hydroperoxide which decomposes to 1-butanol followed by oxidation to butyric acid. The other pathway involves dehydrogenation yielding 1-butene, which may react with water forming 1-butanol (Dugan, 1972). Microorganisms can oxidize alkanes under aerobic conditions (Singer and Finnerty, 1984). The most common degradative pathway involves the oxidation of the terminal methyl group forming the corresponding alcohol (1-butanol). The alcohol may undergo a series of dehydrogenation steps forming butanal followed by oxidation forming butyric acid. The fatty acid may then be metabolized by β-oxidation to form the mineralization products, carbon dioxide and water (Singer and Finnerty, 1984).

Photolytic. Major products reported from the photooxidation of butane with nitrogen oxides are acetaldehyde, formaldehyde and 2-butanone. Minor products included peroxyacyl nitrates and methyl, ethyl and propyl nitrates, carbon monoxide and carbon dioxide. Biacetyl, *tert*-butyl nitrate, ethanol and acetone were reported as trace products (Altshuller, 1983; Bufalini et al., 1971). Irradiation of butane in the presence of chlorine yielded carbon monoxide, carbon dioxide, hydroperoxides, peroxyacid and other carbonyl compounds (Hanst and Gay, 1983). Nitrous acid vapor and butane in a "smog chamber" were irradiated with UV light. Major oxidation products identified included 2-butanone, acetaldehyde and butyraldehyde. Minor products included peroxyacetyl nitrate, methyl nitrate and other unidentified compounds (Cox et al., 1981).

Chemical/Physical. Complete combustion in air gives carbon dioxide and water.

Exposure Limits: NIOSH REL: TWA 800 ppm (1,900 mg/m^3); ACGIH TLV: TWA 800 ppm (1,900 mg/m^3).

Symptoms of Exposure: High concentrations may cause narcosis (Patnaik, 1992).

Toxicity: LC$_{50}$ (inhalation) for mice 680 gm/m^3/2-hour, rats 658 gm/m^3/4-hour (RTECS, 1985).

Uses: Manufacture of synthetic rubbers, ethylene; raw material for high octane motor fuels; solvent; refrigerant; propellant in aerosols; calibrating instruments; organic synthesis.

1-BUTANOL

Synonyms: Butan-1-ol; *n*-Butanol; Butyl alcohol; *n*-Butyl alcohol; Butyl hydroxide; Butyric alcohol; CCS 203; 1-Hydroxybutane; Methylolpropane; NA 1120; NBA; Propylcarbinol; Propylmethanol; RCRA waste number U031; UN 1120.

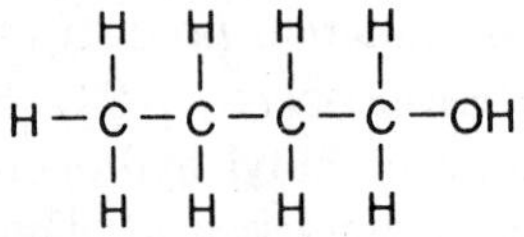

$$H-\underset{\underset{H}{|}}{\overset{\overset{H}{|}}{C}}-\underset{\underset{H}{|}}{\overset{\overset{H}{|}}{C}}-\underset{\underset{H}{|}}{\overset{\overset{H}{|}}{C}}-\underset{\underset{H}{|}}{\overset{\overset{H}{|}}{C}}-OH$$

CAS Registry No.: 71-36-3
DOT: 1120
DOT label: Flammable liquid
Molecular formula: $C_4H_{10}O$
Formula weight: 74.12
RTECS: EO1400000

Physical state, color and odor
Colorless liquid with a characteristic odor similar to fusel oil. Odor threshold in air is 0.83 ppm (Amoore and Hautala, 1983).

Melting point (°C):
-89.5 (Weast, 1986)

Boiling point (°C):
117.2 (Weast, 1986)

Density (g/cm³):
0.8098 at 20/4 °C (Weast, 1986)
0.8057 at 25/4 °C (Huntress and Mulliken, 1941)

Diffusivity in water (10^5 cm²/sec):
0.97 at 25 °C (Hayduk and Laudie, 1974)

Dissociation constant:
>14 (Schwarzenbach et al., 1993)

Flash point (°C):
28.9 (NIOSH, 1994)
36-38 (Windholz et al., 1983)

Lower explosive limit (%):
1.4 (NIOSH, 1994)

144

Upper explosive limit (%):
11.2 (NIOSH, 1994)

Heat of fusion (kcal/mol):
2.24 (Dean, 1987)

Henry's law constant (10^6 atm·m^3/mol at 25 °C):
7.90 (Snider and Dawson, 1985)
8.48 (Hine and Mookerjee, 1975)
8.81 (Buttery et al., 1969)

Interfacial tension with water (dyn/cm at 25 °C):
1.8 (Donahue and Bartell, 1952)

Ionization potential (eV):
10.04 (Franklin et al., 1969)

Soil sorption coefficient, log K_{oc}:
0.50 (Gerstl and Helling, 1987)

Octanol/water partition coefficient, log K_{ow}:
0.88 (Leo et al., 1971)
0.785 (Tewari et al., 1982; Wasik et al., 1981)

Solubility in organics:
Miscible with alcohol, ether and many other organic solvents (Windholz et al., 1983).

Solubility in water:
77,085 mg/L at 20 °C (Mackay and Yeun, 1983)
74,700 mg/L at 25 °C in Lake Superior water having a hardness and alkalinity of 45.5 and 42.2 mg/L as $CaCO_3$, respectively (Veith et al., 1983)
0.854 M at 25.0 °C (Tewari et al., 1982; Wasik et al., 1981)
6.4 wt % at 20 °C (Palit, 1947)
7.31 wt % at 25 °C (Butler et al., 1933)
77 g/L (Price et al., 1974)
0.9919 M at 18 °C (Fühner, 1924)
In wt %: 10.33 at 0 °C, 8.98 at 9.6 °C, 8.03 at 20.0 °C, 7.07 at 30.8 °C, 6.77 at 40.1 °C, 6.54 at 50.0 °C, 6.35 at 60.1 °C, 6.73 at 70.2 °C, 7.04 at 80.1 °C, 7.26 at 90.6 °C (Stephenson and Stuart, 1986)

Vapor density:
3.03 g/L at 25 °C, 2.56 (air = 1)

Vapor pressure (mmHg):
6 at 20 °C (NIOSH, 1994)
4.4 at 20 °C, 6.5 at 25 °C, 10 at 30 °C (Verschueren, 1983)

Environmental Fate
Biological. 1-Butanol degraded rapidly, presumably by microbes, in New Mexico soils releasing carbon dioxide (Fairbanks et al., 1985).

Photolytic. An aqueous solution containing chlorine and irradiated with UV light (λ = 350 nm) converted 1-butanol into numerous chlorinated compounds which were not identified (Oliver and Carey, 1977).

Exposure Limits: NIOSH REL: ceiling 50 ppm (150 mg/m^3), IDLH 1,400 ppm; OSHA PEL: TWA 100 ppm (300 mg/m^3); ACGIH TLV: ceiling 50 ppm (150 mg/m^3).

Symptoms of Exposure: Inhalation may cause irritation to eyes, nose and throat. Chronic exposure to high concentrations may cause photophobia, blurred vision and lacrimation (Patnaik, 1992).

Toxicity: Acute oral LD$_{50}$ for wild birds 2,500 mg/kg, mice 5,200 mg/kg, rats 790 mg/kg, rabbits 384 mg/kg (RTECS, 1985).

Uses: Preparation of butyl esters, glycol ethers and di-*n*-butyl phthalate; solvent for resins and coatings; hydraulic fluid; ingredient in perfumes and flavors; gasoline additive.

2-BUTANONE

Synonyms: Butanone; Ethyl methyl ketone; Meetco; MEK; Methyl acetone; Methyl ethyl ketone; RCRA waste number U159; UN 1193; UN 1232.

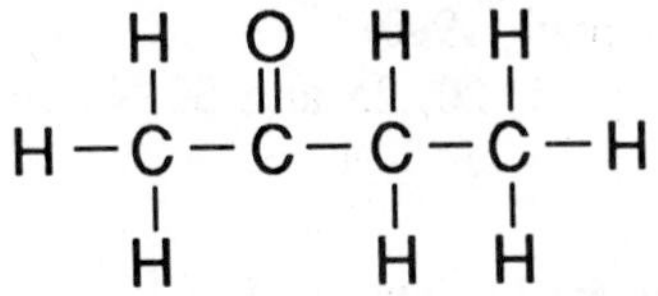

$$H - \underset{\underset{H}{|}}{\overset{\overset{H}{|}}{C}} - \underset{}{\overset{\overset{O}{\|}}{C}} - \underset{\underset{H}{|}}{\overset{\overset{H}{|}}{C}} - \underset{\underset{H}{|}}{\overset{\overset{H}{|}}{C}} - H$$

CAS Registry No.: 78-93-3
DOT: 1193
DOT label: Flammable liquid
Molecular formula: C_4H_8O
Formula weight: 72.11
RTECS: EL6475000

Physical state, color and odor
Clear, colorless liquid with a sweet, mint-like odor. Odor threshold in air is 2.0 ppm (Keith and Walters, 1992).

Melting point (°C):
-86.9 (Dean, 1973)

Boiling point (°C):
79.6 (Weast, 1986)

Density (g/cm^3):
0.8054 at 20/4 °C (Weast, 1986)

Diffusivity in water (10^5 cm^2/sec):
0.94 at 20 °C using method of Hayduk and Laudie (1974).

Flash point (°C):
-9 (NFPA, 1984)

Lower explosive limit (%):
1.4 at 93 °C (NIOSH, 1994)

Upper explosive limit (%):
11.4 at 93 °C (NIOSH, 1994)

Heat of fusion (kcal/mol):
2.017 (Dean, 1987)

147

Henry's law constant (10^5 atm·m^3/mol):
4.65 at 25 °C (Buttery et al., 1971)
1.05 at 25 °C (Snider and Dawson, 1985)
7.0 at 25 °C (Hawthorne et al., 1985)
1.98 at 25 °C (Zhou and Mopper, 1990)
28, 39, 19, 13 and 11 at 10, 15, 20, 25 and 30 °C, respectively (Ashworth et al., 1988)

Interfacial tension with water (dyn/cm at 25 °C):
1.0 (Demond and Lindner, 1993)

Ionization potential (eV):
9.54 (NIOSH, 1994)
9.53 (Gibson, 1977)

Soil sorption coefficient, log K_{oc}:
1.47 (Captina silt loam), 1.53 (McLaurin sandy loam) (Walton et al., 1992)

Octanol/water partition coefficient, log K_{ow}:
0.26, 0.50 (Sangster, 1989)
0.29 (Leo et al., 1971)
0.69 (Wasik et al., 1981)
0.25 (Collander, 1951)

Solubility in organics:
Miscible with acetone, ethanol, benzene and ether (U.S. EPA, 1985).

Solubility in water:
353 g/L at 10 °C, 190 g/L at 90 °C (Verschueren, 1983)
24.00 wt % at 20 °C (Palit, 1947; Riddick et al., 1986)
26.7 wt % at 20 °C (Stephen and Stephen, 1963)
In wt %: 27.33 at 20 °C, 25.57 at 25 °C, 24.07 at 30 °C (Ginnings et al., 1940)
268 g/L (Price et al., 1974)
1.89 mol/L at 25.0 °C (Wasik et al., 1981)

Vapor density:
2.94 g/L at 25 °C, 2.49 (air = 1)

Vapor pressure (mmHg):
77.5 at 20 °C (Verschueren, 1983)
71.2 at 20 °C (Standen, 1967)
90.6 at 25 °C (Ambrose et al., 1975)
92.64 at 25.00 °C (Hussam and Carr, 1985)

Environmental Fate

Photolytic. Synthetic air containing gaseous nitrous acid and exposed to artificial sunlight (λ = 300-450 nm) photooxidized 2-butanone into peroxyacetyl nitrate and methyl nitrate (Cox et al., 1980). The hydroxyl-initiated photooxidation of 2-butanone in a smog chamber produced peroxyacetyl nitrate and acetaldehyde (Cox et al., 1981).

Exposure Limits: NIOSH REL: TWA 200 ppm (590 mg/m^3), STEL 300 ppm (885 mg/m^3), IDLH 3,000 ppm; OSHA PEL: TWA 200 ppm; ACGIH TLV: TWA 200 ppm, STEL 300 ppm.

Symptoms of Exposure: Inhalation may cause irritation of eyes and nose and headache. Narcotic at high concentrations (Patnaik, 1992).

Toxicity: Acute oral LD$_{50}$ for rats 2,737 mg/kg, mouse 4,050 mg/kg; LD$_{50}$ (skin) for rabbit 13 gm/kg (RTECS, 1985).

Drinking Water Standard (proposed): As of May 1994, no MCLGs or MCLs have been proposed (U.S. EPA, 1994).

Uses: Solvent in nitrocellulose coatings, vinyl films and "Glyptal" resins; paint removers; cements and adhesives; organic synthesis; manufacture of smokeless powders and colorless synthetic resins; preparation of 2-butanol, butane and amines; cleaning fluids; printing; catalyst carrier; acrylic coatings.

1-BUTENE

Synonyms: α-Butylene; Ethylethylene.

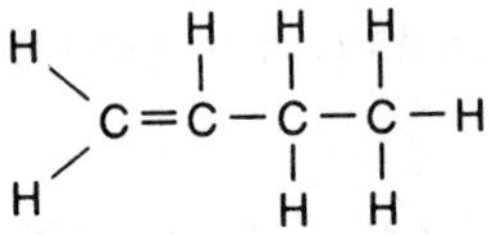

$$H \diagdown \atop H \diagup C = C - \underset{\underset{H}{|}}{\overset{\overset{H}{|}}{C}} - \underset{\underset{H}{|}}{\overset{\overset{H}{|}}{C}} - H$$

CAS Registry No.: 106-98-9
DOT: 1012
Molecular formula: C_4H_8
Formula weight: 56.11

Physical state, color and odor
Colorless gas with a weak, aromatic odor.

Melting point (°C):
-185.3 (Weast, 1986)

Boiling point (°C):
-6.3 (Weast, 1986)

Density (g/cm^3):
0.5951 at 20/4 °C (Weast, 1986)

Diffusivity in water (10^5 cm^2/sec):
0.91 at 20 °C using method of Hayduk and Laudie (1974).

Dissociation constant:
>14 (Schwarzenbach et al., 1993)

Flash point (°C):
-79 (Hawley, 1981)

Lower explosive limit (%):
1.6 (NFPA, 1984)

Upper explosive limit (%):
10.0 (NFPA, 1984)

Heat of fusion (kcal/mol):
0.92 (Dean, 1987)

Henry's law constant (atm·m^3/mol):
0.25 at 25 °C (Hine and Mookerjee, 1975)

Ionization potential (eV):
9.6 (Franklin et al., 1969)

Soil sorption coefficient, log K$_{oc}$:
Unavailable because experimental methods for estimation of this parameter for aliphatic hydrocarbons are lacking in the documented literature.

Octanol/water partition coefficient, log K$_{ow}$:
2.40 (Hansch and Leo, 1979)

Solubility in organics:
Soluble in alcohol, benzene and ether (Weast, 1986).

Solubility in water:
222 ppm at 25 °C (McAuliffe, 1966)

Vapor density:
2.29 g/L at 25 °C, 1.94 (air = 1)

Vapor pressure (mmHg at 25 °C):
2,708 at 25 °C (estimated from Anoine equation, Stull, 1947)
2,230 at 25 °C (Wilhoit and Zwolinski, 1971)

Environmental Fate
   Biological. Biooxidation of 1-butene may occur yielding 3-buten-1-ol, which may further oxidize to give 3-butenoic acid (Dugan, 1972). Washed cell suspensions of bacteria belonging to the genera *Mycobacterium*, *Nocardia*, *Xanthobacter* and *Pseudomonas* and growing on selected alkenes metabolized 1-butene to 1,2-epoxybutane (van Ginkel et al., 1987).
   Photolytic. Products identified from the photoirradiation of 1-butene with nitrogen dioxide in air are epoxybutane, 2-butanone, propanal, ethanol, ethyl nitrate, carbon monoxide, carbon dioxide, methanol and nitric acid (Takeuchi et al., 1983).
   Chemical/Physical. Complete combustion in air yields carbon dioxide and water.

Symptoms of Exposure: Narcotic at high concentrations (Patnaik, 1992).

Uses: Polybutylenes; polymer and alkylate gasoline; intermediate for butyl and pentyl aldehydes, alcohols, maleic acid and other organic compounds.

2-BUTOXYETHANOL

Synonyms: Butyl cellosolve; Butyl glycol; Butyl glycol ether; Butyl oxitol; Dowanol EB; Ektasolve; Ethylene glycol monobutyl ether; Ethylene glycol mono-*n*-butyl ether; Glycol monobutyl ether; Jeffersol EB.

$$H-\underset{\underset{H}{|}}{\overset{\overset{H}{|}}{C}}-\underset{\underset{H}{|}}{\overset{\overset{H}{|}}{C}}-\underset{\underset{H}{|}}{\overset{\overset{H}{|}}{C}}-\underset{\underset{H}{|}}{\overset{\overset{H}{|}}{C}}-O-\underset{\underset{H}{|}}{\overset{\overset{H}{|}}{C}}-\underset{\underset{H}{|}}{\overset{\overset{H}{|}}{C}}-OH$$

CAS Registry No.: 111-76-2
DOT: 2369
DOT label: Poison and combustible liquid
Molecular formula: $C_6H_{14}O_2$
Formula weight: 118.18
RTECS: KJ8575000

Physical state, color and odor
Clear, colorless, oily liquid with a mild, ether-like odor. Odor threshold in air is 0.10 ppm (Amoore and Hautala, 1983).

Melting point (°C):
-77 (NIOSH, 1994)

Boiling point (°C):
171 (Weast, 1986)
171 at 743 mmHg (Aldrich, 1988)

Density (g/cm³):
0.9015 at 20/4 °C (Weast, 1986)

Diffusivity in water (10^5 cm²/sec):
0.75 at 20 °C using method of Hayduk and Laudie (1974).

Flash point (°C):
62 (NIOSH, 1994)

Lower explosive limit (%):
1.1 at 93 °C (NIOSH, 1994)

Upper explosive limit (%):
12.7 at 93 °C (NIOSH, 1994)

Henry's law constant (10^6 atm·m^3/mol):
2.36 (approximate - calculated from water solubility and vapor pressure)

Ionization potential (eV):
10.00 (NIOSH, 1994)

Soil sorption coefficient, log K_{oc}:
Unavailable because experimental methods for estimation of this parameter for cellosolves are lacking in the documented literature.

Octanol/water partition coefficient, log K_{ow}:
0.45 using method of Hansch et al. (1968).

Solubility in organics:
Soluble in alcohol, ether (Weast, 1986) and mineral oil (Windholz et al., 1983).

Solubility in water:
Miscible (Price et al., 1974).

Vapor density:
4.83 g/L at 25 °C, 4.08 (air = 1)

Vapor pressure (mmHg):
0.8 at 20 °C (NIOSH, 1994)

Exposure Limits: NIOSH REL: TWA 5 ppm (24 g/m^3), IDLH 700 ppm; OSHA PEL: TWA 50 ppm (240 mg/m^3); ACGIH TLV: TWA 25 ppm (120 mg/m^3).

Symptoms of Exposure: An 8-hour exposure to 200 ppm may cause nausea, vomiting and headache (Patnaik, 1992).

Toxicity: Acute oral LD_{50} for rats is 450 mg/kg (Patnaik, 1992).

Uses: Dry cleaning; solvent for nitrocellulose, cellulose acetate, resins, oil, grease, albumin; perfume fixative; coating compositions for paper, cloth, leather; lacquers.

n-BUTYL ACETATE

Synonyms: Acetic acid butyl ester; 1-Butanol acetate; Butyl acetate; 1-Butyl acetate; Butyl ethanoate; *n*-Butyl ethanoate; UN 1123.

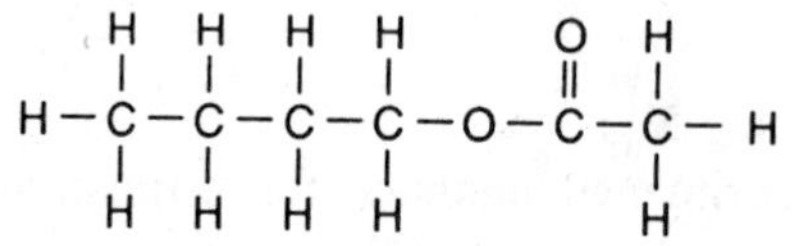

$$H-\underset{\underset{H}{|}}{\overset{\overset{H}{|}}{C}}-\underset{\underset{H}{|}}{\overset{\overset{H}{|}}{C}}-\underset{\underset{H}{|}}{\overset{\overset{H}{|}}{C}}-\underset{\underset{H}{|}}{\overset{\overset{H}{|}}{C}}-O-\overset{\overset{O}{\|}}{C}-\underset{\underset{H}{|}}{\overset{\overset{H}{|}}{C}}-H$$

CAS Registry No.: 123-86-4
DOT: 1123
DOT label: Flammable liquid
Molecular formula: $C_6H_{12}O_2$
Formula weight: 116.16
RTECS: AF7350000

Physical state, color and odor
Colorless liquid with a fruity odor. Odor thresholds in air ranges from 7 to 20 ppm (Keith and Walters, 1992).

Melting point (°C):
-77.9 (Weast, 1986)

Boiling point (°C):
126.5 (Weast, 1986)
124-126 (Aldrich, 1988)

Density (g/cm³):
0.8825 at 20/4 °C (Weast, 1986)

Diffusivity in water (10^5 cm²/sec):
0.75 at 20 °C using method of Hayduk and Laudie (1974).

Flash point (°C):
22.2 (NIOSH, 1994)
36.6 (open cup, Hawley, 1981)

Lower explosive limit (%):
1.7 (NIOSH, 1994)

Upper explosive limit (%):
7.6 (NIOSH, 1994)

154

Henry's law constant (10^4 atm·m^3/mol):
3.3 at 25 °C (Hine and Mookerjee, 1975)

Interfacial tension with water (dyn/cm at 25 °C):
14.5 (Donahue and Bartell, 1952)

Ionization potential (eV):
9.56 ± 0.03 (Franklin et al., 1969)
10.01 (HNU, 1986)

Soil sorption coefficient, log K_{oc}:
Unavailable because experimental methods for estimation of this parameter for aliphatic esters are lacking in the documented literature.

Octanol/water partition coefficient, log K_{ow}:
1.82 (Tewari et al., 1982; Wasik et al., 1981)

Solubility in organics:
Soluble in benzene (Weast, 1986) and most hydrocarbons (Windholz et al., 1983). Miscible with alcohol and ether (Windholz et al., 1983).

Solubility in water:
14,000 mg/L at 20 °C, 5,000 mg/L at 25 °C (Verschueren, 1983)
57.7 mmol/L at 25.0 °C (Tewari et al., 1982; Wasik et al., 1981)
6.8 g/L (Price et al., 1974)
0.84 wt % at 25 °C (Lo et al., 1986)
In wt %: 0.96 at 0 °C, 0.76 at 9.1 °C, 0.64 at 19.7 °C, 0.52 at 30.3 °C, 0.50 at 39.6 °C, 0.50 at 50.0 °C, 0.50 at 60.2 °C, 0.47 at 70.2 °C, 0.48 at 80.1 °C, 0.48 at 90.5 °C (Stephenson and Stuart, 1986)

Vapor density:
4.75 g/L at 25 °C, 4.01 (air = 1)

Vapor pressure (mmHg):
10 at 20 °C (NIOSH, 1994)
11.0 at 25 °C (Abraham, 1984)

Environmental Fate
Chemical/Physical. Hydrolyzes in water forming 1-butanol and acetic acid.

Exposure Limits: NIOSH REL: TWA 150 (710 mg/m^3), STEL 200 ppm (950 mg/m^3), IDLH 1,700 ppm; OSHA PEL: TWA 150 ppm; ACGIH TLV: TWA 150 ppm, STEL 200 ppm.

Symptoms of Exposure: Exposure to 300-400 ppm may cause moderate irritation of the eyes and throat and headache. May be narcotic at higher concentrations (Patnaik, 1992).

Toxicity: Acute oral LD_{50} for rats 14 mg/kg, mice 7,060 mg/kg, rabbits 7,400 mg/kg; LC_{50} (inhalation) for mice 6 gm/m^3/2-hour, rats 2,000 ppm/4-hour (RTECS, 1985).

Uses: Manufacture of artificial leathers, plastics, safety glass, photographic films, lacquers; as a solvent in the production of perfumes, natural gums and synthetic resins; solvent for nitrocellulose lacquers; dehydrating agent.

sec-BUTYL ACETATE

Synonyms: Acetic acid 2-butoxy ester; Acetic acid *sec*-butyl ester; **Acetic acid 1-methylpropyl ester;** 2-Butanol acetate; 2-Butyl acetate; *sec*-Butyl alcohol acetate; 1-Methylpropyl acetate; UN 1123.

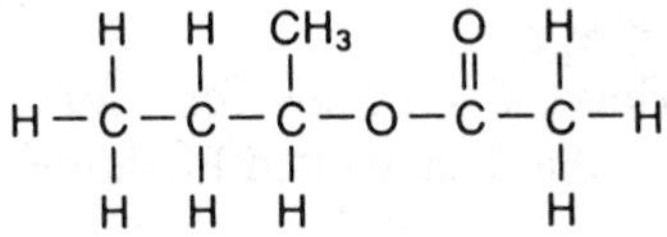

```
        H   H   CH₃    O   H
        |   |   |      ||  |
   H — C — C — C — O — C — C — H
        |   |   |          |
        H   H   H          H
```

CAS Registry No.: 105-46-4
DOT: 1123
DOT label: Flammable liquid
Molecular formula: $C_6H_{12}O_2$
Formula weight: 116.16
RTECS: AF7380000

Physical state, color and odor
Colorless liquid with a pleasant odor.

Melting point (°C):
-37.8 (NIOSH, 1994)

Boiling point (°C):
112 (Weast, 1986)

Density (g/cm³):
0.8758 at 20/4 °C (Weast, 1986)
0.865 at 25/4 °C (Windholz et al., 1983)

Diffusivity in water (10^5 cm²/sec):
0.74 at 20 °C using method of Hayduk and Laudie (1974).

Flash point (°C):
16.7 (NIOSH, 1994)
31 (open cup, Windholz et al., 1983)

Lower explosive limit (%):
1.7 (NIOSH, 1994)

Upper explosive limit (%):
9.8 (NIOSH, 1994)

157

Henry's law constant (10^4 atm·m^3/mol):
1.91 at 20 °C (approximate - calculated from water solubility and vapor pressure)

Ionization potential (eV):
9.91 ± 0.03 (Franklin et al., 1969)

Soil sorption coefficient, log K_{oc}:
Unavailable because experimental methods for estimation of this parameter for aliphatic esters are lacking in the documented literature.

Octanol/water partition coefficient, log K_{ow}:
1.66 using method of Hansch et al. (1968).

Solubility in organics:
Soluble in acetone, alcohol and ether (Weast, 1986).

Solubility in water (wt %):
1.330 at 0 °C, 0.879 at 9.6 °C, 0.869 at 19.5 °C, 0.753 at 29.7 °C, 0.663 at 39.9 °C, 0.629 at 50.0 °C, 0.613 at 60.1 °C, 0.605 at 70.5 °C, 0.622 at 80.2 °C, 0.604 at 90.5 °C (Stephenson and Stuart, 1986)

Vapor density:
4.75 g/L at 25 °C, 4.01 (air = 1)

Vapor pressure (mmHg):
10 at 20 °C (NIOSH, 1994)

Environmental Fate
   Chemical/Physical. Slowly hydrolyzes in water forming *sec*-butyl alcohol and acetic acid.

Exposure Limits: NIOSH REL: TWA 200 ppm (950 mg/m^3), IDLH 1,700 ppm; OSHA PEL: TWA 200 ppm; ACGIH TLV: TWA 200 ppm.

Symptoms of Exposure: Exposure to vapors may cause irritation to eyes and respiratory passages (Patnaik, 1992).

Uses: Solvent for nitrocellulose lacquers; nail enamels, thinners; leather finishers.

tert-BUTYL ACETATE

Synonyms: Acetic acid *tert*-butyl ester; **Acetic acid 1,1-dimethylethyl ester;** *t*-Butyl acetate; Texaco lead appreciator; TLA; UN 1123.

$$H-\overset{\displaystyle H}{\underset{\displaystyle H}{\overset{\displaystyle |}{\underset{\displaystyle |}{C}}}}-\overset{\displaystyle CH_3}{\underset{\displaystyle CH_3}{\overset{\displaystyle |}{\underset{\displaystyle |}{C}}}}-O-\overset{\displaystyle O}{\overset{\displaystyle ||}{C}}-\overset{\displaystyle H}{\underset{\displaystyle H}{\overset{\displaystyle |}{\underset{\displaystyle |}{C}}}}-H$$

CAS Registry No.: 540-88-5
DOT: 1123
DOT label: Flammable liquid
Molecular formula: $C_6H_{12}O_2$
Formula weight: 116.16
RTECS: AF7400000

Physical state, color and odor
Colorless liquid with a fruity odor.

Boiling point (°C):
97-98 (Weast, 1986)

Density (g/cm^3):
0.8665 at 20/4 °C (Weast, 1986)
0.8593 at 25/4 °C (Windholz et al., 1983)

Diffusivity in water (10^5 cm^2/sec):
0.74 at 20 °C using method of Hayduk and Laudie (1974).

Flash point (°C):
15 (Dean, 1987)
22.2 (NIOSH, 1994)

Lower explosive limit (%):
1.5 (NIOSH, 1994)

Soil sorption coefficient, log K_{oc}:
Unavailable because experimental methods for estimation of this parameter for aliphatic esters are lacking in the documented literature.

Solubility in organics:
Miscible with alcohol and ether (Windholz et al., 1983).

159

Solubility in water (wt %):
1.170 at 0 °C, 1.000 at 9.2 °C, 0.803 at 19.2 °C, 0.703 at 29.6 °C, 0.620 at 40.0 °C, 0.573 at 50.0 °C, 0.526 at 60.5 °C, 0.538 at 70.5 °C, 0.499 at 80.5 °C (Stephenson and Stuart, 1986)

Vapor density:
4.75 g/L at 25 °C, 4.01 (air = 1)

Environmental Fate
   Chemical/Physical. Hydrolyzes in water to *tert*-butyl alcohol and acetic acid. The estimated hydrolysis half-life at 25 °C and pH 7 is 140 years (Mabey and Mill, 1978).

Exposure Limits: NIOSH REL: TWA 200 ppm (950 mg/m^3), IDLH 1,500 ppm; OSHA PEL: TWA 200 ppm; ACGIH TLV: TWA 200 ppm.

Symptoms of Exposure: Exposure to vapors may cause irritation to eyes and respiratory passages; narcotic at high concentrations (Patnaik, 1992).

Uses: Gasoline additive; solvent.

sec-BUTYL ALCOHOL

Synonyms: Butanol-2; Butan-2-ol; **2-Butanol**; *sec*-Butanol; 2-Butyl alcohol; Butyl-ene hydrate; CCS 301; Ethylmethyl carbinol; 2-Hydroxybutane; Methylethylcar-binol; SBA.

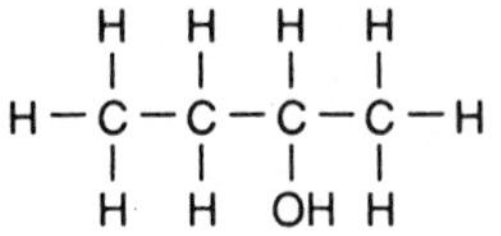

$$
\begin{array}{ccccccc}
 & H & H & H & H & \\
 & | & | & | & | & \\
H & - C & - C & - C & - C & - H \\
 & | & | & | & | & \\
 & H & H & OH & H &
\end{array}
$$

CAS Registry No.: 78-92-2
DOT: 1120
DOT label: Flammable liquid
Molecular formula: $C_4H_{10}O$
Formula weight: 74.12
RTECS: EO1750000

Physical state, color and odor
Colorless liquid with a pleasant odor. Odor threshold in air is 2.6 ppm (Amoore and Hautala, 1983).

Melting point (°C):
-114.7 (Windholz et al., 1983)

Boiling point (°C):
99.5 (Weast, 1986)

Density (g/cm³):
0.8063 at 20/4 °C (Weast, 1986)
0.80235 at 25/4 °C (Huntress and Mulliken, 1941)

Diffusivity in water (10^5 cm²/sec):
0.92 at 20 °C using method of Hayduk and Laudie (1974).

Dissociation constant:
>14 (Schwarzenbach et al., 1993)

Flash point (°C):
23.9 (NIOSH, 1994)
31 (open cup, Windholz et al., 1983)

Lower explosive limit (%):
1.7 at 100 °C (NFPA, 1984)

Upper explosive limit (%):
9.8 at 100 °C (NFPA, 1984)

Henry's law constant (10^5 atm·m^3/mol):
1.02 at 25 °C (Hine and Mookerjee, 1975)

Ionization potential (eV):
10.10 (NIOSH, 1994)

Soil sorption coefficient, log K_{oc}:
Unavailable because experimental methods for estimation of this parameter for aliphatic alcohols are lacking in the documented literature.

Octanol/water partition coefficient, log K_{ow}:
0.61 (Hansch and Anderson, 1967)

Solubility in organics:
Miscible with acetone, alcohol, benzene and ether (Windholz et al., 1983).

Solubility in water (wt %):
20 at 20 °C (Palit, 1947)
26.0 at 0 °C, 23.5 at 10.0 °C, 19.6 at 20.0 °C, 17.0 at 29.9 °C, 15.1 at 40.0 °C, 14.0 at 50.0 °C, 13.4 at 60.3 °C, 13.3 at 70.1 °C, 13.6 at 80.1 °C, 14.5 at 90.2 °C (Stephenson and Stuart, 1986)

Vapor density:
3.03 g/L at 25 °C, 2.56 (air = 1)

Vapor pressure (mmHg):
10 at 20 °C (Sax and Lewis, 1987)
12 at 20 °C, 24 at 30 °C (Verschueren, 1983)

Exposure Limits: NIOSH REL: TWA 100 ppm (305 mg/m^3), STEL 150 ppm (455 mg/m^3), IDLH 2,000 ppm; OSHA PEL: TWA 150 ppm; ACGIH TLV: TWA 100 ppm, STEL 150 ppm.

Symptoms of Exposure: Inhalation may cause irritation to eyes. Narcotic at high concentrations. May irritate skin on contact (Patnaik, 1992).

Toxicity: Acute oral LD$_{50}$ for rats 6,480 mg/kg, rabbits 4,893 mg/kg (RTECS, 1985).

Uses: Manufacture of flotation agents, esters (perfumes and flavors), dyestuffs, wetting agents; ingredient in industrial cleaners and paint removers; preparation of methyl ethyl ketone; solvent in lacquers; in hydraulic brake fluids; organic synthesis.

tert-BUTYL ALCOHOL

Synonyms: *t*-Butanol; *tert*-Butanol; *t*-Butyl alcohol; *t*-Butyl hydroxide; 1,1-Di-methylethanol; **2-Methyl-2-propanol**; NCI-C55367; TBA; Trimethylcarbinol; Trimethyl methanol; UN 1120.

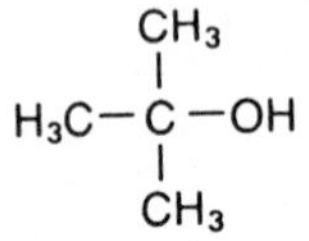

$$H_3C - \underset{\underset{CH_3}{|}}{\overset{\overset{CH_3}{|}}{C}} - OH$$

CAS Registry No.: 75-65-0
DOT: 1120
DOT label: Flammable liquid
Molecular formula: $C_4H_{10}O$
Formula weight: 74.12
RTECS: EO1925000

Physical state, color and odor
Colorless liquid or crystals with a camphor-like odor. Odor threshold in air is 47 ppm (Amoore and Hautala, 1983).

Melting point (°C):
25.5 (Weast, 1986)

Boiling point (°C):
82.3 (Weast, 1986)

Density (g/cm³):
0.78581 at 20/4 °C, 0.78086 at 25/4 °C (Windholz et al., 1983)

Diffusivity in water (10^5 cm²/sec):
0.91 at 20 °C using method of Hayduk and Laudie (1974).

Dissociation constant:
≈ 19 (Gordon and Ford, 1972)

Flash point (°C):
11.1 (NIOSH, 1994)
4 (Aldrich, 1988)

Lower explosive limit (%):
2.4 (NIOSH, 1994)

Upper explosive limit (%):
8.0 (NIOSH, 1994)

Heat of fusion (kcal/mol):
1.06 (Dean, 1987)

Henry's law constant (10^5 atm·m³/mol at 25 °C):
1.19 (Butler et al., 1935)
1.44 (Snider and Dawson, 1985)

Ionization potential (eV):
9.70 (NIOSH, 1994)

Soil sorption coefficient, log K_{oc}:
Unavailable because experimental methods for estimation of this parameter for alcohols are lacking in the documented literature. However, its miscibility in water and low K_{ow} suggest its adsorption to soil will be nominal (Lyman et al., 1982).

Octanol/water partition coefficient, log K_{ow}:
0.37 (Hansch and Anderson, 1967)
0.59 (Youezawa and Urushi, 1979)

Solubility in organics:
Miscible with alcohol and ether (Windholz et al., 1983).

Solubility in water:
Miscible (NIOSH, 1994; Palit, 1947). A saturated solution in equilibrium with its own vapor had a concentration of 316.2 g/L at 25 °C (Kamlet et al., 1987).

Vapor pressure (mmHg):
42 at 25 °C, 56 at 30 °C (Verschueren, 1983)

Environmental Fate
   Chemical/Physical. May react with strong mineral acids (e.g., hydrochloric, sulfuric) or oxidizers releasing isobutylene (HSDB, 1989; NIOSH, 1994). Complete combustion in air yields carbon dioxide and water vapor.

Exposure Limits (ppm): NIOSH REL: TWA 100, STEL 150, IDLH 1,600; OSHA PEL: TWA 100; ACGIH TLV: TWA 100, STEL 150.

Symptoms of Exposure: Ingestion may cause headache, dizziness, dry skin and narcosis. Inhalation may cause drowsiness and mild irritation to eyes, nose and skin (NIOSH, 1994; Patnaik, 1992).

Toxicity: Acute oral LD_{50} for rats 3,500 mg/kg, rabbits 3,559 mg/kg (RTECS, 1985); LD_{50} (24-hour) for goldfish 4,300 mg/L (Verschueren, 1983).

Uses: Denaturant for ethyl alcohol; manufacturing flavors, perfumes (artificial musk), flotation agents; solvent; paint removers; octane booster for unleaded gasoline; dehydrating agent; chemical intermediate.

n-BUTYLAMINE

Synonyms: 1-Aminobutane; Butylamine; **1-Butanamine**; Monobutylamine; Mono-*n*-butylamine; Norvalamine; UN 1125.

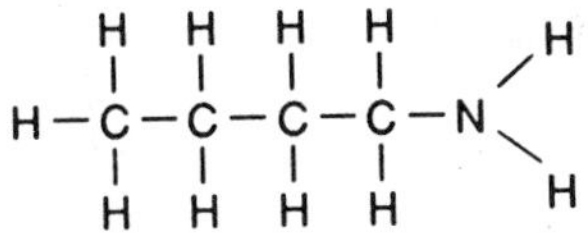

$$H-\underset{\underset{H}{|}}{\overset{\overset{H}{|}}{C}}-\underset{\underset{H}{|}}{\overset{\overset{H}{|}}{C}}-\underset{\underset{H}{|}}{\overset{\overset{H}{|}}{C}}-\underset{\underset{H}{|}}{\overset{\overset{H}{|}}{C}}-N\overset{\diagup H}{\diagdown H}$$

CAS Registry No.: 109-73-9
DOT: 1125
DOT label: Flammable liquid
Molecular formula: $C_4H_{11}N$
Formula weight: 73.14
RTECS: EO2975000

Physical state, color and odor
Clear, colorless liquid with an ammonia-like odor. Odor threshold in air is 1.8 ppm (Amoore and Hautala, 1983).

Melting point (°C):
-49.1 (Weast, 1986)

Boiling point (°C):
77.8 (Weast, 1986)

Density (g/cm^3):
0.7414 at 20/4 °C (Weast, 1986)
0.7327 at 25/4 °C (Windholz et al., 1983)

Diffusivity in water (10^5 cm^2/sec):
0.89 at 20 °C using method of Hayduk and Laudie (1974).

Dissociation constant:
9.33 at 20 °C (Gordon and Ford, 1972)

Flash point (°C):
-12.2 (NIOSH, 1994)
-1 (open cup, Windholz et al., 1983)

Lower explosive limit (%):
1.7 (NIOSH, 1994)

167

Upper explosive limit (%):
9.8 (NIOSH, 1994)

Henry's law constant (10^5 atm·m^3/mol at 25 °C):
1.51 (Hine and Mookerjee, 1975)
2.18 (Amoore and Buttery, 1978)

Ionization potential (eV):
8.71 ± 0.03 (Franklin et al., 1969)

Soil sorption coefficient, log K_{oc}:
1.88 (Meylan et al., 1992)

Octanol/water partition coefficient, log K_{ow}:
0.81, 0.88 (Leo et al., 1971)
0.68 (Collander, 1951)

Solubility in organics:
Soluble in alcohol and ether (Weast, 1986).

Solubility in water:
Miscible (NIOSH, 1994). A saturated solution in equilibrium with its own vapor had a concentration of 667.0 g/L at 25 °C (Kamlet et al., 1987).

Vapor density:
2.99 g/L at 25 °C, 2.52 (air = 1)

Vapor pressure (mmHg at 20 °C):
82 (NIOSH, 1994)
72 (Verschueren, 1983)

Environmental Fate
   Chemical/Physical. Reacts with mineral acids forming water-soluble salts.

Exposure Limits: NIOSH REL: ceiling 5 ppm (15 mg/m^3), IDLH 300 ppm; ACGIH TLV: ceiling 5 ppm.

Symptoms of Exposure: Severe irritant to eyes and respiratory tract. Contact may cause severe burns (Patnaik, 1992).

Toxicity: Acute oral LD_{50} for guinea pigs 430 mg/kg, mice 430 mg/kg, rats 366 mg/kg; LC_{50} (inhalation) for mice 800 gm/m^3/2-hour, rats 4,000 ppm/4-hour (RTECS, 1985).

Uses: Intermediate for dyestuffs, emulsifying agents, pharmaceuticals, rubber chemicals, insecticides and synthetic tanning agents; preparation of isocyanates for coatings.

n-BUTYLBENZENE

Synonyms: Butylbenzene; 1-Phenylbutane.

$$CH_2CH_2CH_2CH_3$$

CAS Registry No.: 104-51-8
DOT: 2709
Molecular formula: $C_{10}H_{14}$
Formula weight: 134.22
RTECS: CY9070000

Physical state and color
Colorless liquid.

Melting point (°C):
-88 (Weast, 1986)

Boiling point (°C):
183.31 (Wilhoit and Zwolinski, 1971)

Density (g/cm^3):
0.8601 at 20/4 °C (Weast, 1986)

Diffusivity in water (10^5 cm^2/sec):
0.68 at 20 °C using method of Hayduk and Laudie (1974).

Dissociation constant:
>14 (Schwarzenbach et al., 1993)

Flash point (°C):
71 (open cup, Windholz et al., 1983)
59 (Aldrich, 1988)

Lower explosive limit (%):
0.8 (Sax and Lewis, 1987)

Upper explosive limit (%):
5.8 (Sax and Lewis, 1987)

Henry's law constant (10^3 atm·m^3/mol):
12.5 at 25 °C (Hine and Mookerjee, 1975)

5.35, 8.17, 11.0, 16.7 and 21.4 at 10, 15, 20, 25 and 30 °C, respectively (Perlinger et al., 1993)

Interfacial tension with water (dyn/cm at 20 °C):
39.6 (Demond and Lindner, 1993)

Ionization potential (eV):
8.69 ± 0.01 (Franklin et al., 1969)

Soil sorption coefficient, log K_{oc}:
3.39 (Schwarzenbach and Westall, 1981)
3.40 (estuarine sediment, Vowles and Mantoura, 1987)

Octanol/water partition coefficient, log K_{ow}:
4.643 (Klein et al., 1988)
4.29 (Brooke et al., 1990; Schantz and Martire, 1987)
4.377 (Brooke et al., 1990; de Bruijn et al., 1989)
4.26 (Camilleri et al., 1988; Hammers et al., 1982; Hansch and Leo, 1979)
4.28 (Tewari et al., 1982; Wasik et al., 1981, 1983)
4.18 (Bruggeman et al., 1982)
4.44 (Hammers et al., 1982)

Solubility in organics:
Miscible with alcohol, benzene and ether (Windholz et al., 1983) and many other organic solvents.

Solubility in water:
In mg/kg at 25 °C: 11.8 (distilled water), 7.09 (artificial seawater) (Sutton and Calder, 1975)
73.0 μmol/L at 25.0 °C (Andrews and Keefer, 1950a)
In μmol/L (°C): 96.6 (15.0), 101.8 (20.0), 102.5 (25.0) (Owens et al., 1986)
434 μmol/L at 25.0 °C (Tewari et al., 1982)
15.4 and 17.7 mg/L at 25 °C (Mackay and Shiu, 1977)
64.2 μmol/L in 0.5 M NaCl at 25 °C (Wasik et al., 1984)
103 μmol/L at 25 °C (Wasik et al., 1981, 1983)
50.0 mg/L at 25 °C (Klevens, 1950)
1.77 x 10^{-6} at 25 °C (mole fraction, Li et al., 1993)

Vapor density:
5.49 g/L at 25 °C, 4.63 (air = 1)

Vapor pressure (mmHg):
1.03 at 25 °C (Mackay et al., 1982)

Drinking Water Standard: As of May 1994, no MCLGs or MCLs have been proposed (U.S. EPA, 1994).

Uses: Pesticide manufacturing; plasticizer; solvent for coatings; surface active agents; polymer linking agent; ingredient in naphtha; asphalt component; organic synthesis.

sec-BUTYLBENZENE

Synonyms: (1-Methylpropyl)benzene; 2-Phenylbutane; UN 2709.

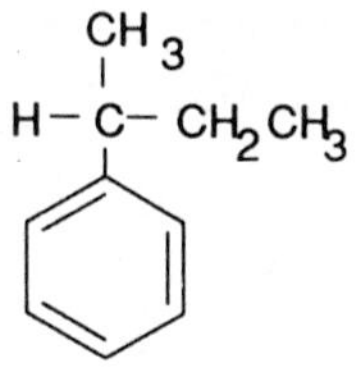

$$H-\underset{\underset{\displaystyle C_6H_5}{|}}{\overset{\overset{\displaystyle CH_3}{|}}{C}}-CH_2CH_3$$

CAS Registry No.: 135-98-8
DOT: 2709
Molecular formula: $C_{10}H_{14}$
Formula weight: 134.22
RTECS: CY9100000

Physical state and color
Colorless liquid.

Melting point (°C):
-75.5 (Weast, 1986)
-82.7 (Windholz et al., 1983)

Boiling point (°C):
173 (Weast, 1986)
173.34 (Wilhoit and Zwolinski, 1971)

Density (g/cm³ at 20/4 °C):
0.8621 (Weast, 1986)
0.8608 (Windholz et al., 1983)

Diffusivity in water (10^5 cm²/sec):
0.68 at 20 °C using method of Hayduk and Laudie (1974).

Dissociation constant:
>14 (Schwarzenbach et al., 1993)

Flash point (°C):
52 (Windholz et al., 1983)
62.7 (open cup, Hawley, 1981)
45 (Aldrich, 1988)

Lower explosive limit (%):
0.8 (Sax and Lewis, 1987)

173

Upper explosive limit (%):
6.9 (Sax and Lewis, 1987)

Henry's law constant (10^2 atm·m^3/mol):
1.14 at 25 °C (Hine and Mookerjee, 1975)

Ionization potential (eV):
8.68 ± 0.01 (Franklin et al., 1969)

Soil sorption coefficient, log K_{oc}:
2.95 using method of Kenaga and Goring (1980).

Octanol/water partition coefficient, log K_{ow}:
4.24 using method of Hansch et al. (1968).

Solubility in organics:
Miscible with alcohol, benzene and ether (Windholz et al., 1983).

Solubility in water (25 °C):
17.6 mg/kg (distilled water), 11.9 mg/kg (artificial seawater) (Sutton and Calder, 1975)
171 mg/L at (Andrews and Keefer, 1950a)
87.6 μmol/L in 0.5 M NaCl (Wasik et al., 1984)

Vapor density:
5.49 g/L at 25 °C, 4.63 (air = 1)

Vapor pressure (mmHg):
1.1 at 20 °C (Verschueren, 1983)
1.81 at 25 °C (Mackay et al., 1982)

Toxicity: Acute oral LD_{50} for rats 2,240 mg/kg (RTECS, 1985).

Drinking Water Standard: As of May 1994, no MCLGs or MCLs have been proposed (U.S. EPA, 1994).

Uses: Solvent for coating compositions; plasticizer; surface-active agents; organic synthesis.

tert-BUTYLBENZENE

Synonyms: *t*-Butylbenzene; **(1,1-Dimethylethyl)benzene**; 2-Methyl-2-phenyl-propane; Pseudobutylbenzene; Trimethylphenylmethane; UN 2709.

$$H_3C-\underset{\underset{\displaystyle}{\overset{\displaystyle CH_3}{|}}}{C}-CH_3$$

CAS Registry No.: 98-06-6
DOT: 2709
Molecular formula: $C_{10}H_{14}$
Formula weight: 134.22
RTECS: CY9120000

Physical state and color
Colorless liquid.

Melting point (°C):
-57.8 (Weast, 1986)

Boiling point (°C):
169 (Weast, 1986)
169.15 (Wilhoit and Zwolinski, 1971)

Density (g/cm³):
0.8665 at 20/4 °C (Weast, 1986)

Diffusivity in water (10^5 cm²/sec):
0.68 at 20 °C using method of Hayduk and Laudie (1974).

Dissociation constant:
>14 (Schwarzenbach et al., 1993)

Flash point (°C):
60 (open cup, Windholz et al., 1983)
34 (Aldrich, 1988)

Lower explosive limit (%):
0.7 at 100 °C (Sax and Lewis, 1987)

Upper explosive limit (%):
5.7 at 100 °C (Sax and Lewis, 1987)

Henry's law constant (10^2 atm·m^3/mol):
1.17 at 25 °C (Hine and Mookerjee, 1975)

Ionization potential (eV):
8.68 ± 0.01 (Franklin et al., 1969)

Soil sorption coefficient, log K_{oc}:
2.83 using method of Kenaga and Goring (1980).

Octanol/water partition coefficient, log K_{ow}:
4.11 (Leo et al., 1971)
4.07 (Nahum and Horvath, 1980)

Solubility in organics:
Soluble in acetone (Weast, 1986) but miscible with alcohol, benzene and ether (Windholz et al., 1983).

Solubility in water:
29.5 mg/kg at 25 °C, 21.2 mg/kg in seawater at 25 °C (Sutton and Calder, 1975)
196 μmol/L at 25.0 °C (Andrews and Keefer, 1950a)
134 μmol/L in 0.5 M NaCl at 25 °C (Wasik et al., 1984)

Vapor density:
5.49 g/L at 25 °C, 4.63 (air = 1)

Vapor pressure (mmHg):
2.14 at 25 °C (Mackay et al., 1982)
1.5 at 20 °C (Verschueren, 1983)

Drinking Water Standard: As of May 1994, no MCLGs or MCLs have been proposed (U.S. EPA, 1994).

Uses: Polymerization solvent; polymer linking agent; organic synthesis.

n-BUTYL MERCAPTAN

Synonyms: Butanethiol; **1-Butanethiol;** *n*-Butanethiol; Butyl mercaptan; *n*-Butyl thioalcohol; 1-Mercaptobutane; NCI-C60866; Thiobutyl alcohol; UN 2347.

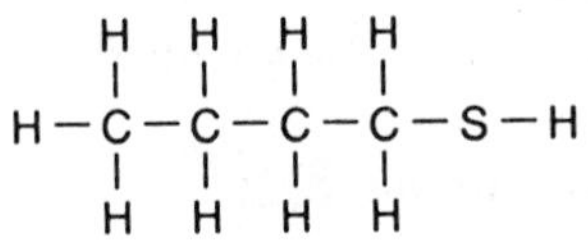

$$H-\overset{\displaystyle H}{\underset{\displaystyle H}{C}}-\overset{\displaystyle H}{\underset{\displaystyle H}{C}}-\overset{\displaystyle H}{\underset{\displaystyle H}{C}}-\overset{\displaystyle H}{\underset{\displaystyle H}{C}}-S-H$$

CAS Registry No.: 109-79-5
DOT: 2347
Molecular formula: $C_4H_{10}S$
Formula weight: 90.18
RTECS: EK6300000

Physical state, color and odor
Colorless liquid with a strong garlic, cabbage or heavy skunk-like odor. Odor threshold in air is 0.97 ppm (Amoore and Hautala, 1983).

Melting point (°C):
-115.7 (Weast, 1986)

Boiling point (°C):
98.4 (Weast, 1986)

Density (g/cm³):
0.8337 at 20/4 °C (Weast, 1986)
0.83679 at 25/4 °C (Windholz et al., 1983)
0.8412 at 20/4 °C (Hawley, 1981)

Diffusivity in water (10⁵ cm²/sec):
0.84 at 20 °C using method of Hayduk and Laudie (1974).

Flash point (°C):
1.7 (NIOSH, 1994)
12 (Aldrich, 1988)

Heat of fusion (kcal/mol):
2.500 (Dean, 1987)

Henry's law constant (10³ atm·m³/mol):
7.04 at 20 °C (approximate - calculated from water solubility and vapor pressure)

Ionization potential (eV):
9.14 ± 0.02 (Franklin et al., 1969)

Soil sorption coefficient, log K_{oc}:
Unavailable because experimental methods for estimation of this parameter for aliphatic mercaptans are lacking in the documented literature.

Octanol/water partition coefficient, log K_{ow}:
2.28 (Sangster, 1989)

Solubility in organics:
Soluble in alcohol and ether (Weast, 1986).

Solubility in water:
590 mg/L at 22 °C (Verschueren, 1983)

Vapor density:
3.69 g/L at 25 °C, 3.11 (air = 1)

Vapor pressure (mmHg):
35 at 20 °C (NIOSH, 1994)
55.5 at 25 °C (Wilhoit and Zwolinski, 1971)

Environmental Fate
Chemical/Physical. Releases toxic sulfur oxide fumes when heated to decomposition (Sax and Lewis, 1987).

Exposure Limits: NIOSH REL: 15-minute ceiling 0.5 ppm (1.8 mg/m^3), IDLH 500 ppm; OSHA PEL: TWA 10 ppm (35 mg/m^3); ACGIH TLV: TWA 0.5 ppm.

Toxicity: Acute oral LD_{50} for rats 1,500 mg/kg (RTECS, 1985).

Uses: Chemical intermediate; solvent.

CAMPHOR

Synonyms: 2-Bornanone; 2-Camphanone; Camphor-natural; Camphor-synthetic; Formosa camphor; Gum camphor; Japan camphor; 2-Keto-1,7,7-trimethylnor-camphane; Laurel camphor; Matricaria camphor; Norcamphor; 2-Oxobornane; Synthetic camphor; **1,7,7-Trimethylbicyclo[2.2.1]heptan-2-one**; UN 2717.

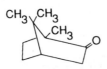

CAS Registry No.: 76-22-2
DOT: 2717
DOT label: Flammable solid
Molecular formula: $C_{10}H_{16}O$
Formula weight: 152.24
RTECS: EX1225000

Physical state, color and odor
Colorless to white crystalline semi-solid with a penetrating, fragrant or aromatic odor. Odor threshold in air is 0.27 ppm (Amoore and Hautala, 1983).

Melting point (°C):
174-179 (Verschueren, 1983)

Boiling point (°C):
Sublimes at 204 (Weast, 1986)

Density (g/cm³):
0.990 at 25/4 °C (Weast, 1986)

Diffusivity in water (10^5 cm²/sec):
0.78 at 25 °C using method of Hayduk and Laudie (1974).

Flash point (°C):
65.6 (NIOSH, 1994)

Lower explosive limit (%):
0.6 (NIOSH, 1994)

Upper explosive limit (%):
3.5 (NIOSH, 1994)

Heat of fusion (kcal/mol):
1.635 (Dean, 1987)

Henry's law constant (10^5 atm·m^3/mol):
3.00 at 20 °C (approximate - calculated from water solubility and vapor pressure)

Ionization potential (eV):
8.76 (NIOSH, 1994)

Soil sorption coefficient, log K_{oc}:
Unavailable because experimental methods for estimation of this parameter for cyclic ketones are lacking in the documented literature.

Octanol/water partition coefficient, log K_{ow}:
2.42 using method of Hansch et al. (1968).

Solubility in organics:
At 25 °C (g/L): acetone (2,500), alcohol (1,000), benzene (2,500), chloroform (2,000), ether (1,000), glacial acetic acid (2,500), oil of turpentine (667). Also soluble in aniline, carbon disulfide, decalin, methylhexalin, nitrobenzene, petroleum ether, tetralin, higher alcohols, in fixed and volatile oils (Windholz et al., 1983).

Solubility in water:
≈ 1.25 g/L at 25 °C (Windholz et al., 1983)

Vapor pressure (mmHg):
0.2 at 20 °C (NIOSH, 1994)

Exposure Limits (mg/m^3): NIOSH REL: TWA 2, IDLH 200; OSHA PEL: TWA 2.

Symptoms of Exposure: Vapors may irritate eyes, mucous membranes and throat. Ingestion may cause headache, nausea, vomiting and diarrhea (NIOSH, 1994; Patnaik, 1992).

Toxicity: LD_{50} (intraperitoneal) for mice 3,000 mg/kg (RTECS, 1985).

Uses: Plasticizer for cellulose esters and ethers; manufacture of plastics, cymene, incense, celluloid; in lacquers, explosives and embalming fluids; pyrotechnics; moth repellent; preservative in pharmaceuticals and cosmetics; odorant/flavorant in household, pharmaceutical and industrial products; tooth powders.

CARBARYL

Synonyms: Arylam; Carbamine; Carbatox; Carbatox 60; Carbatox 75; Carpolin; Carylderm; Cekubaryl; Crag sevin; Denapon; Devicarb; Dicarbam; ENT 23969; Experimental insecticide 7744; Gamonil; Germain's; Hexavin; Karbaspray; Karbatox; Karbosep; Methyl carbamate-1-naphthalenol; Methyl carbamate-1-naphthol; Methyl carbamic acid 1-naphthyl ester; *N*-Methyl-1-naphthylcarbamate; *N*-Methyl-α-naphthyl carbamate; *N*-Methyl-α-naphthylurethan; NA 2757; NAC; **1-Naphthalenol methyl carbamate**; 1-Naphthol-*N*-methyl carbamate; 1-Naphthyl methyl carbamate; 1-Naphthyl-*N*-methyl carbamate; α-Naphthyl-*N*-methyl carbamate; OMS-29; Panam; Ravyon; Rylam; Seffein; Septene; Sevimol; Sevin; Sok; Tercyl; Toxan; Tricarnam; UC 7744; Union Carbide 7744.

CAS Registry No.: 63-25-2
DOT: 2757
DOT label: Poison
Molecular formula: $C_{12}H_{11}NO_2$
Formula weight: 201.22
RTECS: FC5950000

Physical state and color
Colorless solid or white crystals.

Melting point (°C):
142 (Worthing and Hance, 1991)
145.1 (NIOSH, 1994)

Density (g/cm³):
1.232 at 20/20 °C (Windholz et al., 1983)

Diffusivity in water (10^5 cm²/sec):
0.56 at 20 °C using method of Hayduk and Laudie (1974).

Flash point (°C):
195 (ACGIH, 1986)

Henry's law constant (10^5 atm·m³/mol):
1.27 at 20 °C (approximate - calculated from water solubility and vapor pressure)

Bioconcentration factor, log BCF:
1.86 (algae, Geyer et al., 1984)
1.95 (activated sludge), 1.48 (golden ide) (Freitag et al., 1985)
0.00 (fish, microcosm) (Garten and Trabalka, 1983)

Soil sorption coefficient, log K_{oc}:
2.42 (Commerce soil), 2.55 (Tracy soil), 2.59 (Catlin soil) (McCall et al., 1981)
2.36 (Kenaga and Goring, 1980)
2.02 (Batcombe silt loam, Briggs, 1981)

Octanol/water partition coefficient, log K_{ow}:
2.36, 2.56 (Leo et al., 1971)
2.32 (Briggs, 1981; Geyer et al., 1984)
2.31 (Bowman and Sans, 1983a)
2.34 (Swann et al., 1983)

Solubility in organics:
Moderately soluble in acetone, cyclohexanone, N,N-dimethylformamide and isophorone (Windholz et al., 1983).

Solubility in water:
0.4105 mol/m^3 at 25 °C (Swann et al., 1983)
120 mg/L at 30 °C (Windholz et al., 1983)
40 mg/L at 30 °C (Gunther et al., 1968)
In mg/L: 72.4 at 10 °C, 104 at 20 °C, 130 at 30 °C (Bowman and Sans, 1985)
350 μmol/L at 25 °C (LaFleur, 1979)
67 at 12 °C, 114 ppm at 24 °C (Wauchope and Haque, 1973)

Vapor pressure (10^3 mmHg):
5 at 25 °C (Swann et al., 1983)
2 at 40 °C (Meister, 1988)

Environmental Fate
   Biological. Carbaryl degraded completely after 4 weeks of incubation in the dark in Holland March canal water (Sharom et al., 1980). Fourteen soil fungi metabolized [methyl-^{14}C]carbaryl via hydroxylation to 1-naphthyl-N-hydroxymethyl carbamate, 4-hydroxy-1-naphthylmethyl carbamate and 5-hydroxy-1-naphthylmethyl carbamate (Bollag and Liu, 1972). Carbaryl was degraded by a culture of *Aspergillus terreus* to 1-naphthyl carbamate. The half-life was determined to be 8 days (Liu and Bollag, 1971). When ^{14}C-carbonyl-labeled carbaryl (200 ppm) was added to five different soils and incubated at 25 °C for 32 days, evolution of $^{14}CO_2$ varied from 2.2-37.4% (Kazano et al., 1972).
   Soil. The rate of hydrolysis of carbaryl in flooded soil increased when the soil

was pretreated with the hydrolysis product, 1-naphthol (Rajagopal et al., 1986). Carbaryl is hydrolyzed in both flooded and nonflooded soils but the rate is slightly higher under flooded conditions (Rajagopal et al., 1983). When ^{14}C-carbonyl-labeled carbaryl (200 ppm) was added to five different soils and incubated at 25 °C for 32 days, evolution of $^{14}CO_2$ varied from 2.2-37.4% (Kazano et al., 1972). Metabolites identified in soil included 1-naphthol (hydrolysis product) (Ramanand et al., 1988a; Sud et al., 1972), hydroquinone, catechol, pyruvate (Sud et al., 1972), coumarin, carbon dioxide (Kazano et al., 1972), 1-naphthyl carbamate, 1-naphthyl *N*-hydroxymethyl carbamate, 5-hydroxy-1-naphthylmethyl carbamate, 4-hydroxy-1-naphthylmethyl carbamate and 1-naphthylhydroxymethyl carbamate (Liu and Bollag, 1971, 1971a). Sud et al. (1972) discovered that a strain of *Achromobacter* sp. utilized carbaryl as the sole source of carbon in a salt medium. The organism grew on the degradation products 1-naphthol, hydroquinone and catechol. 1-Naphthol, a metabolite of carbaryl in soil, was recalcitrant to further degradation by a bacterium tentatively identified as an *Arthrobacter* sp. under anaerobic conditions (Ramanand et al., 1988a). Carbaryl or its metabolite 1-naphthol at normal and ten times the field application rate had no effect on the growth of *Rhizobium* sp. or *Azotobacter chroococcum* (Kale et al., 1989). The half-lives of carbaryl under flooded and non-flooded conditions were 13-14 and 23-28 days, respectively (Venkateswarlu et al., 1980).

Rajagopal et al. (1984) proposed degradation pathways of carbaryl in soil and in microbial cultures included the following compounds: 5,6-dihydrodihydroxy carbaryl, 2-hydroxy carbaryl, 4-hydroxy carbaryl, 5-hydroxy carbaryl, 1-naphthol, *N*-hydroxymethyl carbaryl, 1-naphthyl carbamate, 1,2-dihydroxynaphthalene, 1,4-dihydroxynaphthalene, *o*-coumaric acid, *o*-hydroxybenzalpyruvate, 1,4-naphthoquinone, 2-hydroxy-1,4-naphthoquinone, coumarin, γ-hydroxy-γ-*o*-hydroxyphenyl-α-oxobutyrate, 4-hydroxy-1-tetralone, 3,4-dihydroxy-1-tetralone, pyruvic acid, salicylaldehyde, salicylic acid, phenol, hydroquinone, catechol, carbon dioxide and water. When carbaryl was incubated at room temperature in a mineral salts medium by soil-enrichment cultures for 30 days, 26.8 and 31.5% of the applied insecticide remained in flooded and nonflooded soils, respectively (Rajagopal et al., 1984a). A *Bacillus* sp. and the enrichment cultures both degraded carbaryl to 1-naphthol. Mineralization to carbon dioxide was negligible (Rajagopal et al., 1984a).

Plant. In plants, the *N*-methyl group may be subject to oxidation or hydroxylation (Kuhr, 1968).

Surface Water. In a laboratory aquaria containing estuarine water, 43% of dissolved carbaryl was converted to 1-naphthol in 17 days at 20 °C (pH = 7.5-8.1). The half-life of carbaryl in estuarine water without mud at 8 °C was 38 days. When mud was present, both carbaryl and 1-naphthol decreased to less than 10% in the estuarine water after 10 days. Based on a total recovery of only 40%, it was postulated that the remainder was evolved as methane (Karinen et al., 1967). The rate of hydrolysis of carbaryl increased with an increase in temperature (Karinen et al., 1967) and in increases of pH values above 7.0 (Rajagopal et al., 1984). The presence of a micelle [hexadecyltrimethylammonium bromide (HDATB), 3 x 10^{-3} M] in natural waters greatly enhanced the hydrolysis rate. The hydrolysis half-lives in natural water samples with and without HDATB were 0.12-0.67 and 9.7-138.6 hours, respectively (González et al., 1992). In the dark, carbaryl was incubated in 21 °C water obtained from the Holland Marsh drainage canal. Degradation was complete after 4 weeks (Sharom et al., 1980).

In pond water, carbaryl rapidly degraded to 1-naphthol. The latter was further degraded, presumably by *Flavobacterium* sp., into hydroxycinnamic acid, salicylic acid and an unidentified compound (HSDB, 1989). Four days after carbaryl (30 mg/L and 300 μg/L) was added to Fall Creek water, >60% was mineralized to carbon dioxide. At pH 3, however, <10% was converted to carbon dioxide (Boethling and Alexander, 1979). Under these conditions, hydrolysis of carbaryl to 1-naphthol was rapid. The authors could not determine how much carbon dioxide was attributed to biodegradation of carbaryl and how much was due to the bio-degradation of 1-naphthol (Boethling and Alexander, 1979). Hydrolysis half-lives of carbaryl in filtered and sterilized Hickory Hills (pH 6.7) and U.S. Department of Agriculture Number 1 pond water (pH 7.2) were 30 and 12 days, respectively (Wolfe et al., 1978).

Photolytic. Based on data for phenol, a structurally related compound, an aqueous solution containing the 1-naphthoxide ion (3 x 10^{-4} M) in room light would be expected to photooxidize to give 2-hydroxy-1,4-naphthoquinone (Tomkiewicz et al., 1971). 1-Naphthol, methyl isocyanate and other unidentified cholinesterase inhibitors were reported as products formed from the direct photolysis of carbaryl by sunlight (Wolfe et al., 1976). In an aqueous solution at 25 °C, the photolysis half-life of carbaryl by natural sunlight or UV light (λ = 313 nm) is 6.6 days (Wolfe et al., 1978a). A photolysis half-life of 1.88 days was found when carbaryl in an acidic (pH 5.5), buffered aqueous solution was irradiated with UV light (λ >290 nm) (Wolfe et al., 1976).

Chemical/Physical. Ozonation of carbaryl in water yielded 1-naphthol, naph-thoquinone, phthalic anhydride, N-formyl carbamate of 1-naphthol (Martin et al., 1983), naphthoquinones and acidic compounds (Shevchenko et al., 1982). Hydrolysis and photolysis of carbaryl forms 1-naphthol (Wauchope and Haque, 1973; Rajagopal et al., 1984, 1986; Miles et al., 1988; MacRae, 1989; Ramanand et al., 1988a; Lewis, 1989; Somasundaram et al., 1991) and 2-hydroxy-1,4-naphtho-quinone (Wauchope and Haque, 1973), respectively. In aqueous solutions, carbaryl hydrolyzes to 1-naphthol (Boethling and Alexander, 1979; Vontor et al., 1972), methylamine and carbon dioxide (Vontor et al., 1972), especially under alkaline conditions (Wolfe et al., 1978). At pH values of 5, 7 and 9, the hydrolysis half-lives at 27 °C are 1,500, 15 and 0.15 days, respectively (Wolfe et al., 1978).

Miles et al. (1988) studied the rate of hydrolysis of carbaryl in phosphate-buffered water (0.01 M) at 26 °C with and without a chlorinating agent (10 mg/L hypochlorite solution). The hydrolysis half-lives at pH 7 and 8 with and without chlorine were 3.5 and 10.3 days and 0.05 and 1.2 days, respectively (Miles et al., 1988). The reported hydrolysis half-lives of carbaryl in water at pH values of 7, 8, 9 and 10 were 10.5 days, 1.3 days, 2.5 hours and 15.0 minutes, respectively (Aly and El-Dib, 1971). 1-Naphthol, the major product of carbaryl hydrolysis, dissociates in water to the 1-naphthoxide ion (Wauchope and Haque, 1973). The hydrolysis half-lives of carbaryl in a sterile 1% ethanol/water solution at 25 °C and pH values of 4.5, 6.0, 7.0 and 8.0, were 300, 58, 2.0 and 0.27 weeks, respectively (Chapman and Cole, 1982).

Releases toxic nitrogen oxides when heated to decomposition (Sax and Lewis, 1987; Lewis, 1990). Products reported from the combustion of carbaryl at 900 °C include carbon monoxide, carbon dioxide, ammonia and oxygen (Kennedy et al., 1972).

Exposure Limits (mg/m^3): NIOSH REL: TWA 5, IDLH 100; OSHA PEL: TWA 5; ACGIH TLV: TWA 5.

Symptoms of Exposure: Symptoms may include nausea, vomiting, diarrhea, abdominal cramps, miosis, lachrymation, excessive salivation, nasal discharge, sweating, muscle twitching, convulsions and coma (Patnaik, 1992).

Toxicity: Acute oral LD_{50} for wild birds 56 mg/kg, cats 150 mg/kg, guinea pigs 250 mg/kg, gerbils 491 mg/kg, hamsters, 250 mg/kg, mice, 212 mg/kg, rats 250 mg/kg, rabbits 710 mg/kg (RTECS, 1985); LC_{50} (48-hour) for red killifish 32 mg/L (Yoshioka et al., 1986).

Drinking Water Standard: As of May 1994, no MCLGs or MCLs have been proposed (U.S. EPA, 1994).

Uses: Contact insecticide.

CARBOFURAN

Synonyms: Bay 70143; **2,3-Dihydro-2,2-dimethyl-7-benzofuranol methyl carbamate**; 2,2-Dimethyl-7-coumaranyl *N*-methyl carbamate; 2,2-Dimethyl-2,3-dihydro-7-benzofuranyl-*N*-methyl carbamate; ENT 27164; Furadan; Methyl carbamic acid 2,3-dihydro-2,2-dimethyl-7-benzofuranyl ester; NIA 10242.

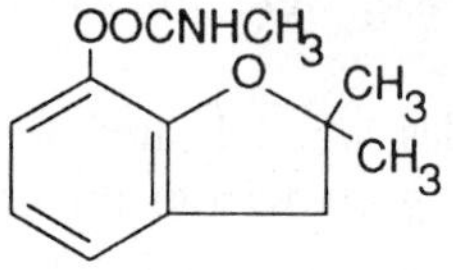

CAS Registry No.: 1563-66-2
DOT: 2757
DOT label: Poison
Molecular formula: $C_{12}H_{15}NO$
Formula weight: 221.26
RTECS: FB9450000

Physical state, color and odor
Odorless, white to grayish crystalline solid.

Melting point (°C):
150-153 (Windholz et al., 1983)
151 (Bowman and Sans, 1983a)

Boiling point (°C):
152 (NIOSH, 1994)

Density (g/cm^3):
1.18 at 20/20 °C (Verschueren, 1983)

Diffusivity in water (10^5 cm^2/sec):
0.54 at 20 °C using method of Hayduk and Laudie (1974).

Henry's law constant (10^8 atm·m^3/mol):
3.88 at 30 °C (approximate - calculated from water solubility and vapor pressure)

Soil sorption coefficient, log K_{oc}:
2.20 (Commerce soil), 1.98 (Tracy soil), 2.02 (Catlin soil) (McCall et al., 1981)

Octanol/water partition coefficient, log K_{ow}:
1.60 (Kenaga and Goring, 1980)

186

1.63 (Bowman and Sans, 1983a)
1.23-1.41 (Worthing and Hance, 1991)

Solubility in organics (g/L):
Methylene chloride (>200), 2-propanol (20-50) (Worthing and Hance, 1991)

Solubility in water:
415 ppm (Kenaga and Goring, 1980)
700 ppm at 25 °C (Windholz et al., 1983)
In mg/L: 291 at 10 °C, 320 at 20 °C, 375 at 30 °C (Bowman and Sans, 1985)
320 mg/L at 19.0 °C (Bowman and Sans, 1979)

Vapor pressure (mmHg):
2×10^{-5} at 33 °C (Verschueren, 1983)

Environmental Fate
   Biological. Carbofuran or their metabolites (3-hydroxycarbofuran and 3-keto-carbofuran) at normal and ten times the field application rate had no effect on *Rhizobium* sp. However, in a nitrogen-free culture medium, *Azotobacter chroococcum* growth was inhibited by carbofuran, 3-hydroxycarbofuran and 3-ketocarbofuran (Kale et al., 1989). Under *in vitro* conditions, 15 of 20 soil fungi degraded carbofuran to one or more of the following compounds: 3-hydroxycar-bofuran, 3-ketocarbofuran, carbofuran phenol and 3-hydroxyphenol (Arunachalam and Lakshmanan, 1988). *Pseudomonas* sp. and *Achromobacter* sp. were also capable of degrading carbofuran (Felsot et al., 1981). Derbyshire et al. (1987) reported that an enzyme, isolated from the microorganism *Achromobacter* sp., hydrolyzed carbofuran via the carbamate linkage forming 2,3-dihydro-2,2-dimethyl-7-benzofuranol. The optimum pH and temperature for the degradation of carbofuran and two other pesticides (aldicarb and carbaryl) was 9.0-10.5 and 45 and 53 °C, respectively.
   Soil. Carbofuran is relatively persistent in soil, especially in dry, acidic and low temperature soils (Ahmad et al., 1979; Caro et al., 1973; Fuhremann and Lichtenstein, 1980; Gorder et al., 1982; Greenhalgh and Belanger, 1981; Ou et al., 1982). In alkaline soil with high moisture content, microbial degradation of carbofuran was more important than leaching and chemical degradation (Gorder et al., 1982; Greenhalf and Balanger, 1981).
   Carbofuran phenol is formed from the hydrolysis of carbofuran at pH 7.0. Carbofuran phenol was also found to be the major biodegradation product by *Azospirillum lipoferum* and *Streptomyces* spp. isolated from a flooded alluvial soil (Venkateswarlu and Sethunathan, 1984). The hydrolysis of carbofuran to carbo-furan phenol was catalyzed by the addition of rice straw in an anaerobic flooded soil where it accumulated (Venkateswarlu and Sethunathan, 1979). The rate of transformation of carbofuran in soil increased with repeated applications (Harris et

al., 1984). In an alluvial soil, carbaryl and its analog carbosulfan 2,3-dihydro-2,2-dimethyl-7-benzofuranyl [(di-n-butyl)aminosulfenyl methyl carbamate] both degraded faster at 35 °C than at 25 °C with carbosulfan degrading to carbofuran (Sahoo et al., 1990). An enrichment culture isolated from a flooded alluvial soil (Ramanand et al., 1988) and a bacterium tentatively identified as an *Arthrobacter* sp. (Ramanand et al., 1988a) readily mineralized carbofuran to carbon dioxide at 35 °C. Mineralization was slower at lower temperatures (20-28 °C). Under anaerobic conditions, carbofuran did not degrade (Ramanand et al., 1988a). The reported half-lives in soil are 1-2 months (Hartley and Kidd, 1987); 11-13 days at a pH value of 6.5 and 60-75 days for a granular formulation (Ahmad et al., 1979).

Rajagopal et al. (1984) proposed degradation pathways of carbofuran in both soils and microbial cultures included the following compounds: 3-hydroxycarbofuran, 3-ketocarbofuran, carbofuran phenol, 3-hydroxycarbofuran phenol, 3-ketocarbofuran phenol, 6,7-dihydroxycarbofuran phenol, 3,6,7-trihydroxycarbofuran phenol, 3-keto-6,7-dihydroxycarbofuran phenol and carbon dioxide. In soils, microorganisms degraded carbofuran to carbofuran phenol (Ou et al., 1982), then to 3-hydroxycarbofuran and 3-ketocarbofuran (Kale et al., 1989; Ou et al., 1982).

Hydrolyzes in soil and water to carbofuran phenol, carbon dioxide and methylamine (Rajagopal et al., 1986; Seiber et al., 1978; Somasundaram et al., 1989, 1991). Hydrolysis of carbofuran occurs in both flooded and nonflooded soils, but the rate is slightly higher under flooded conditions (Venkateswarlu et al., 1977), especially when the soil is pretreated with the hydrolysis product, carbofuran phenol (Rajagopal et al., 1986). In addition, the hydrolysis of carbofuran was found to be pH dependent in both deionized water and rice paddy water. At pH values of 7.0, 8.7 and 10.0, the hydrolysis half-lives in deionized water were 864, 19.4 and 1.2 hours, respectively. In paddy water, the hydrolysis half-lives at pH values of 7.0, 8.7 and 10.0 were and 240, 13.9 and 1.3, respectively (Seiber et al., 1978).

Plant. Carbofuran is rapidly metabolized in plants to nontoxic products (Cremlyn, 1991). Metcalf et al. (1968) reported that carbofuran undergoes hydroxylation and hydrolysis in plants, insects and mice. Hydroxylation of the benzylic carbon gives 3-hydroxycarbofuran, which is subsequently oxidized to 3-ketocarbofuran. In carrots, carbofuran initially degraded to 3-hydroxycarbofuran. This compound reacted with naturally occurring angelic acid in carrots forming a conjugated metabolite identified as 2,3-dihydro-2,2-dimethyl-7-(((methylamino)carbonyl)oxy)-3-benzofuranyl (Z)-2-methyl-2-butenoic acid (Sonobe et al., 1981). Metabolites identified in three types of strawberries (Day-Neutral, Tioga and Tufts) were 2,3-dihydro-2,2-dimethyl-3-hydroxy-7-benzofuranyl-*N*-methyl carbamate, 2,3-dihydro-2,2-dimethyl-3-oxo-7-benzofuranyl-*N*-methyl carbamate, 2,3-dihydro-2,2-dimethyl-3-benzofuranol, 2,3-dihydro-2,2-dimethyl-3,7-benzofuranol and 2,3-dihydro-2,2-dimethyl-3-oxo-7-benzofuranol (Archer et al., 1977). Oat plants were grown in two soils treated with [^{14}C]carbofuran. Most of the residues recovered in oat leaves were in the form of carbofuran and 3-hydroxy-carbofuran. Other metabolites identified were 3-ketocarbofuran, a 3-keto-7-phenol and 3-hydroxy-7-phenol (Fuhremann and Lichtenstein, 1980).

Photolytic. 2,3-Dihydro-2,2-dimethylbenzofuran-4,7-diol and 2,3-dihydro-3-keto-2,2-dimethylbenzofuran-7-yl carbamate were formed when carbofuran dissolved in water was irradiated by sunlight for 5 days (Raha and Das, 1990).

Chemical/Physical. Releases toxic nitrogen oxides when heated to decomposition (Sax and Lewis, 1987; Lewis, 1990). The hydrolysis half-lives of carbofuran in a sterile 1% ethanol/water solution at 25 °C and pH values of 4.5, 5.0, 6.0, 7.0 and 8.0 were 170, 690, 690, 8.2 and 1.0 weeks, respectively (Chapman and Cole, 1982).

Exposure Limits (mg/m^3): NIOSH REL: TWA 0.1; ACGIH TLV: TWA 0.1.

Drinking Water Standard (final): MCLG: 40 μg/L; MCL: 40 μg/L (U.S. EPA, 1984).

Uses: Systematic insecticide, nematocide and acaricide.

CARBON DISULFIDE

Synonyms: Carbon bisulfide; Carbon bisulphide; Carbon disulphide; Carbon sulfide; Carbon sulphide; Dithiocarbonic anhydride; NCI-C04591; RCRA waste number P022; Sulphocarbonic anhydride; UN 1131; Weeviltox.

$$S=C=S$$

CAS Registry No.: 75-15-0
DOT: 1131
DOT label: Flammable liquid
Molecular formula: CS_2
Formula weight: 76.13
RTECS: FF6650000

Physical state, color and odor
Clear, water-white to pale yellow liquid; ethereal odor when pure; technical and reagent grades have strong, foul odors. Odor threshold in air is 0.11 ppm (Amoore and Hautala, 1983).

Melting point (°C):
-111.5 (Weast, 1986)

Boiling point (°C):
46.2 (Weast, 1986)

Density (g/cm³):
1.2632 at 20/4 °C (Weast, 1986)
1.2632 at 15/4 °C, 1.2559 at 25/4 °C, 1.2500 at 30/4 °C (Standen, 1964)

Diffusivity in water (10^5 cm²/sec):
1.18 at 20 °C using method of Hayduk and Laudie (1974).

Flash point (°C):
-30 (NIOSH, 1994)

Lower explosive limit (%):
1.3 (NIOSH, 1994)

Upper explosive limit (%):
50.0 (NIOSH, 1994)

Heat of fusion (kcal/mol):
1.049 (Dean, 1987)

Henry's law constant (atm·m³/mol):
24.25 at 24 °C (Elliott, 1989)

Interfacial tension with water (dyn/cm at 25 °C):
48.1 (Donahue and Bartell, 1952)

Ionization potential (eV):
10.08 (Franklin et al., 1969)

Soil sorption coefficient, log K_{oc}:
2.38–2.55 using method of Kenaga and Goring (1980).

Octanol/water partition coefficient, log K_{ow}:
1.84, 2.16 (calculated, Leo et al., 1971)

Solubility in organics:
Soluble in ethanol, chloroform and ether (Weast, 1986).

Solubility in water:
2,300 mg/L at 22 °C (Verschueren, 1983)
2,200 mg/L at 22 °C (ACGIH, 1986)
2,000 mg/L at 20 °C (WHO, 1979)
0.1185 wt % at 25 °C (Stephen and Stephen, 1963)
0.210 wt % at 20 °C (Riddick et al., 1986)
0.294% at 20 °C (Windholz et al., 1983)
0.286 wt % at 25 °C (Lo et al., 1986)

Vapor density:
3.11 g/L at 25 °C, 2.63 (air = 1)

Vapor pressure (mmHg):
297 at 20 °C (NIOSH, 1994)
360 at 25 °C (Lide, 1990)
430 at 30 °C (Verschueren, 1983)

Environmental Fate
Chemical/Physical. In alkaline solutions, carbon disulfide hydrolyzes to carbon dioxide and hydrogen sulfide (Peyton et al., 1976). In an aqueous alkaline solution containing hydrogen peroxide, dithiopercarbonate, sulfide, elemental sulfur and polysulfides may be expected (Elliott, 1990). In an aqueous, alkaline solution (pH

≥8), carbon disulfide reacted with hydrogen peroxide forming sulfate and carbonate ions. However, when the pH is lowered to 7 and 7.4, colloidal sulfur is formed (Adewuyi and Carmichael, 1987). Forms a hemihydrate which decomposes at -3 °C (Keith and Walters, 1992). An aqueous solution containing carbon disulfide reacts with sodium hypochlorite forming carbon dioxide, sulfuric acid and sodium chloride (Patnaik, 1992). Burns with a blue flame releasing carbon dioxide and sulfur dioxide (Windholz et al., 1983).

Oxidizes in the troposphere forming carbonyl sulfide. The atmospheric half-lives of carbon disulfide and carbonyl sulfide were estimated to be approximately 2 years and 13 days, respectively (Khalil and Rasmussen, 1984).

Exposure Limits: NIOSH REL: TWA 1 ppm (3 mg/m^3), STEL 10 ppm, IDLH 500 ppm; OSHA PEL: TWA 20 ppm, ceiling 30 ppm, 100 ppm peak for 30 minutes; ACGIH TLV: TWA 10 ppm.

Symptoms of Exposure: Dizziness, headache, poor sleep, fatigue, nervousness; anorexia, weight loss; polyneuropathy, burns, dermatitis. Contact with skin causes burning pain, erythema and exfoliation (NIOSH, 1994).

Toxicity: Acute oral LD$_{50}$ for rats 3,188 mg/kg, guinea pigs 2,125 mg/kg, mice 2,780 mg/kg, rabbits 2,550 mg/kg (RTECS, 1985).

Uses: Manufacture of viscose rayon, cellophane, flotation agents, ammonium salts, carbon tetrachloride, carbanilide, paints, enamels, paint removers, varnishes, tallow, textiles, rocket fuel, soil disinfectants, electronic vacuum tubes, herbicides; grain fumigants; solvent for fats, resins, phosphorus, sulfur, bromine, iodine and rubber; petroleum and coal tar refining; solvent and eluant for organics adsorbed on charcoal for air analysis.

CARBON TETRACHLORIDE

Synonyms: Benzinoform; Carbona; Carbon chloride; Carbon tet; ENT 4705; Fasciolin; Flukoids; Freon 10; Halon 104; Methane tetrachloride; Necatorina; Necatorine; Perchloromethane; R 10; RCRA waste number U211; Tetrachloor-metaan; Tetrachlorocarbon; **Tetrachloromethane**; Tetrafinol; Tetraform; Tetrasol; UN 1846; Univerm; Vermoestricid.

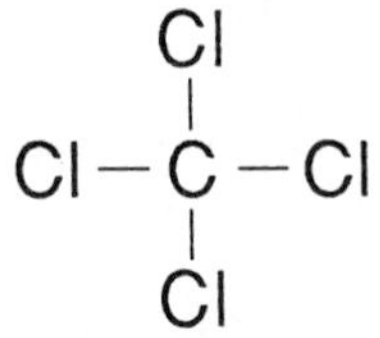

$$Cl - \overset{\displaystyle Cl}{\underset{\displaystyle Cl}{\overset{|}{\underset{|}{C}}}} - Cl$$

CAS Registry No.: 56-23-5
DOT: 1846
DOT label: Poison
Molecular formula: CCl_4
Formula weight: 153.82
RTECS: FG4900000

Physical state, color and odor
Clear, colorless, heavy, watery liquid with a strong, sweetish, distinctive odor resembling ether. Odor threshold in air is 21.4 ppm (Keith and Walters, 1992).

Melting point (°C):
-22.99 (Horvath, 1982)

Boiling point (°C):
76.54 (Horvath, 1982)

Density (g/cm³):
1.59472 at 20/4 °C (Standen, 1964)
1.5844 at 25/4 °C (Kirchnerová and Cave, 1976)

Diffusivity in water (10^5 cm²/sec):
0.90 at 20 °C using method of Hayduk and Laudie (1974).

Flash point (°C):
Noncombustible (Rogers and MacFarlane, 1981)

Heat of fusion (kcal/mol):
0.601 (Dean, 1987)

Henry's law constant (10^2 atm·m³/mol):
3.02 at 25 °C (Warner et al., 1987)

2.4 at 20 °C (Roberts and Dändliker, 1983)
10.2 at 37 °C (Sato and Nakajima, 1979)
3.04 at 24.8 °C (Gossett, 1987)
0.2226 at 20 °C (Roberts et al., 1985)
3.78 at 30 °C (Jeffers et al., 1989)
2.64 at 25 °C (Liss and Slater, 1974)
1.48, 1.91, 2.32, 2.95 and 3.78 at 10, 15, 20, 25 and 30 °C, respectively (Ashworth et al., 1988)
2.04, 3.37, 3.82 and 4.52 at 20, 30, 35 and 40 °C, respectively (Tse et al., 1992)
At 25 °C: 2.40 and 3.68 in distilled water and seawater, respectively (Hunter-Smith et al., 1983)
0.894, 0.963, 1.098, 1.946 and 2.568 at 2.0, 6.0, 10.0, 18.2 and 25.0 °C, respectively (Dewulf et al., 1995)

Interfacial tension with water (dyn/cm at 25 °C):
43.7 (Donahue and Bartell, 1952)

Ionization potential (eV):
11.47 ± 0.01 (Franklin et al., 1969)

Bioconcentration factor, log BCF:
1.48 (bluegill sunfish, Veith et al., 1980)
2.48 (algae, Geyer et al., 1984)
2.68 (activated sludge, Freitag et al., 1985)

Soil sorption coefficient, log K_{oc}:
2.35 (Abdul et al., 1987)
2.62 (Chin et al., 1988)
2.16 (Captina silt loam), 1.69 (McLaurin sandy loam) (Walton et al., 1992)
1.78 (normal soils), 2.01 (suspended bed sediments) (Kile et al., 1995)

Octanol/water partition coefficient, log K_{ow}:
2.83 (Chou and Jurs, 1979)
2.73 (Banerjee et al., 1980)

Solubility in organics:
Miscible with ethanol, benzene, chloroform, ether, carbon disulfide (U.S. EPA, 1985), petroleum ether, solvent naphtha and volatile oils (Yoshida et al., 1983).

Solubility in water:
785 mg/L at 20 °C (Pearson and McConnell, 1975)
970 mg/L at 0 °C (Dean, 1973)
800 mg/L at 25 °C (Valsaraj, 1988)

757 mg/L at 25 °C (Banerjee et al., 1980)
770 mg/kg at 15 °C, 810 mg/kg at 30 °C (Gross and Saylor, 1931)
803 mg/L at 20 °C (Munz and Roberts, 1987)
10^2 wt %: 8.3 at 10 °C, 8.0 at 20 °C, 8.5 at 30 °C (Stephen and Stephen, 1963)
0.077 wt % at 25 °C (Riddick et al., 1986)
770 mg/kg at 25 °C (Gross, 1929)
805 mg/L at 20 °C (Howard, 1990)
0.16 wt % at 25 °C (Nathan, 1978)
8 g/kg at 25 °C (McGovern, 1943)
0.080 wt % at 25 °C (Lo et al., 1986)
780 mg/L at 23-24 °C (Broholm and Feenstra, 1995)

Vapor density:
6.29 g/L at 25 °C, 5.31 (air = 1)

Vapor pressure (mmHg):
56 at 10 °C, 90 at 20 °C, 113 at 25 °C, 137 at 30 °C (Verschueren, 1983)
115 at 25 °C (Rogers and MacFarlane, 1981)

Environmental Fate

Biological. An anaerobic species of *Clostridium* biodegraded carbon tetrachloride by reductive dechlorination yielding trichloromethane (chloroform), dichloromethane and unidentified products (Gälli and McCarty, 1989).

Carbon tetrachloride (5 and 10 mg/L) showed significant degradation with rapid adaptation in a static-culture flask-screening test (settled domestic wastewater inoculum) conducted at 25 °C. Complete degradation was observed after 14 days of incubation (Tabak et al., 1981).

Chemical/Physical. Under laboratory conditions, carbon tetrachloride in aqueous solutions partially hydrolyzed forming chloroform and carbon dioxide. Complete hydrolysis yielded carbon dioxide and hydrochloric acid. Chloroform also formed by microbial degradation of carbon tetrachloride using denitrifying bacteria (Smith and Dragun, 1984). Though no products were identified, the estimated hydrolysis half-life in water at 25 °C and pH 7 is 7,000 years (Mabey and Mill, 1978) and 40.5 years (Jeffers et al., 1989).

Anticipated products from the reaction of carbon tetrachloride with ozone or hydroxyl radicals in the atmosphere are phosgene and chlorine radicals (Cupitt, 1980). Phosgene is hydrolyzed readily to hydrogen chloride and carbon dioxide (Morrison and Boyd, 1971).

Carbon tetrachloride slowly reacts with hydrogen sulfide in aqueous solution yielding carbon dioxide via the intermediate carbon disulfide. However, in the presence of two micaceous minerals (biotite and vermiculite) and amorphous silica, the rate transformation increases. At 25 °C and a hydrogen sulfide concentration of 0.001 M, the half-lives of carbon tetrachloride were calculated to be 2,600, 160 and 50 days for the silica, vermiculite and biotite studies, respectively. In all three studies, the major transformation pathway is the formation of carbon disulfide. This compound is then hydrolyzed to carbon dioxide (81-86% yield) and HS⁻. Minor intermediates detected include chloroform (5-15% yield), carbon monoxide (102% yield) and a non-volatile compound tentatively identified as formic acid (3-6% yield) (Kriegman-King and Reinhard, 1992).

Matheson and Tratnyek (1994) studied the reaction of fine-grained iron metal in an anaerobic aqueous solution (15 °C) containing carbon tetrachloride (151 μM). Initially, carbon tetrachloride underwent rapid dehydrochlorination forming chloroform, which further degraded to methylene chloride and chloride ions. The rate of reaction decreased with each dehydrochlorination step. However, after 1 hour of mixing, the concentration of carbon tetrachloride decreased from 151 to approximately 15 μM. No additional products were identified although the authors concluded that environmental circumstances may exist where degradation of methylene chloride may occur. They also reported that reductive dehalogenation of carbon tetrachloride and other chlorinated hydrocarbons used in this study appears to take place in conjunction with the oxidative dissolution or corrosion of the iron metal through a diffusion-limited surface reaction.

The evaporation half-life of carbon tetrachloride (1 mg/L) from water at 25 °C using a shallow-pitch propeller stirrer at 200 rpm at an average depth of 6.5 cm is 29 minutes (Dilling et al., 1977).

Emits toxic chloride and phosgene fumes when heated to decomposition (Lewis, 1990).

Exposure Limits: Potential occupational carcinogen. NIOSH REL: STEL (1 hour) 2 ppm (12.6 mg/m^3), IDLH 200 ppm; OSHA PEL: TWA 10 ppm, ceiling 25 ppm, 5-minute/4-hour peak 200 ppm; ACGIH TLV: TWA 5 ppm (30 mg/m^3).

Symptoms of Exposure: Central nervous system depression, nausea, vomiting, skin irritation (NIOSH, 1994).

Toxicity: LC_{50} (14-day) for guppies 67 ppm; LC_{50} (48-hour) for red killifish 617 mg/L (Yoshioka et al., 1986); acute oral LD_{50} for rats 2,800 mg/kg, guinea pigs 5,760 mg/kg, mice 8,263 mg/kg, rabbits 5,760 mg/kg (RTECS, 1985).

Drinking Water Standard (final): MCLG: zero; MCL: 5 μg/L (U.S. EPA, 1984).

Uses: Preparation of dichlorodifluoromethane, refrigerants, aerosols and propellants; metal degreasing; agricultural fumigant; chlorinating unsaturated organic compounds; production of semiconductors; solvent for fats, oils, rubber, etc; dry cleaning operations; industrial extractant; spot remover; in fire extinguishers; veterinary medicine; organic synthesis.

CHLORDANE

Synonyms: A 1068; Aspon-chlordane; Belt; CD 68; Chlordan; γ-Chlordan; Chloridan; Chlorindan; Chlor kil; Chlorodane; Chlortox; Corodane; Cortilan-neu; Dichlorochlordene; Dowklor; ENT 9932; ENT 25552-X; HCS 3,260; Kypchlor; M 140; M 410; NA 2762; NCI-C00099; Niran; Octachlor; 1,2,4,5,6,7,8,8-Octachlor-2,3,3a,4,7,7a-hexahydro-4,7-methanoindane; Octachlorodihydrodicyclopentadiene; 1,2,4,5,6,7,8,8-Octachloro-2,3,3a,4,7,7a-hexahydro-4,7-methanoindene; **1,2,4,5,6,7,8,8-Octachloro-2,3,3a,4,7,7a-hexahydro-4,7-methano-1H-indene**; 1,2,4,5,6,7,8,8-Octachloro-3a,4,7,7a-hexahydro-4,7-methyleneindane; Octachloro-4,7-methanohydroindane; Octachloro-4,7-methanotetrahydroindane; 1,2,4,5,6,7,8,8-Octachloro-4,7-methano-3a,4,7,7a-tetrahydroindane; 1,2,4,5,6,7,8,8-Octachloro-3a,4,7,7a-tetrahydro-4,7-methanoindan; 1,2,4,5,6,7,8,8-Octachloro-3a,4,7,7a-tetrahydro-4,7-methanoindane; Octaklor; Octaterr; Orthoklor; RCRA waste number U036; SD 5532; Shell SD-5532; Synklor; Tat chlor 4; Topichlor 20; Topiclor; Topiclor 20; Toxichlor; Velsicol 1068.

Note: Chlordane is a mixture of *cis-* and *trans*-chlordane and other complex chlorinated hydrocarbons including heptachlor and nonachlor. According to Brooks (1974), technical chlordane has the approximate composition: *trans*-chlordane (24%), four chlordene isomers ($C_{10}H_6Cl_8$) (21.5%), *cis*-chlordane (19%), heptachlor (10%), nonachlor (7%), pentachlorocyclopentadiene (2%), hexachlorocyclopentadiene (>1%), octachlorocyclopentene (1%), $C_{10}H_{7-8}Cl_{6-7}$ (8.5%) and other unidentified compounds (6%).

CAS Registry No.: 57-74-9
DOT: 2762
DOT label: Poison
Molecular formula: $C_{10}H_6Cl_8$
Formula weight: 409.78
RTECS: PB9800000

Physical state, color and odor
Colorless to amber to yellowish-brown, viscous liquid with an aromatic, slight pungent odor similar to chlorine.

Melting point (°C):
103-109 (NIOSH, 1994)

Boiling point (°C):
175 at 2 mmHg (Roark, 1951)

Density (g/cm^3):
1.59–1.63 at 20/4 °C (Melnikov, 1971)

Diffusivity in water (10^5 cm^2/sec):
0.43 at 20 °C using method of Hayduk and Laudie (1974).

Flash point (°C):
The solid is noncombustible but may be utilized in flammable solutions (NIOSH, 1994).

Lower explosive limit (%):
0.7 (kerosene solution, Weiss, 1986)

Upper explosive limit (%):
5 (kerosene solution, Weiss, 1986)

Henry's law constant (10^5 atm·m^3/mol):
4.8 at 25 °C (Warner et al., 1987)

Bioconcentration factor, log BCF:
4.58 (freshwater fish), 3.92 (fish, microcosm) (Garten and Trabalka, 1983)

Soil sorption coefficient, log K_{oc}:
5.15, 5.57 (Chin et al., 1988)
4.85, 4.88 (sand), 4.72, 4.78 (silt) (Johnson-Logan et al., 1992)

Octanol/water partition coefficient, log K_{ow}:
6.00 (Travis and Arms, 1988)

Solubility in organics:
Miscible with aliphatic and aromatic solvents (U.S. EPA, 1985).

Solubility in water:
56 ppb at 25 °C (Sanborn et al., 1976)
1.85 ppm at 25 °C (Weil et al., 1974)
9 ppb at 25 °C (NAS, 1977)
34 μg/L at 24 °C (Johnson-Logan et al., 1992)

Vapor density:
16.75 g/L at 25 °C, 14.15 (air = 1)

Vapor pressure (10^6 mmHg):
10 at 25 °C (Sunshine, 1969)

Environmental Fate

Biological. In four successive 7-day incubation periods, chlordane (5 and 10 mg/L) was recalcitrant to degradation in a settled domestic wastewater inoculum (Tabak et al., 1981).

Soil. The actinomycete *Nocardiopsis* sp. isolated from soil extensively degraded pure *cis*- and *trans*-chlordane to dichlorochlordene, oxychlordane, heptachlor, heptachlor *endo*-epoxide, chlordene chlorohydrin and 3-hydroxy-*trans*-chlordane. Oxychlordane slowly degraded to 1-hydroxy-2-chlorochlordene (Beeman and Matsumura, 1981). The reported half-life in soil is approximately 1 year (Hartley and Kidd, 1987).

Chemical/Physical. In an alkaline medium or solvent, carrier, diluent or emulsifier having an alkaline reaction, chlorine will be released (Windholz et al., 1983). Technical grade chlordane that was passed over a 5% platinum catalyst at 200 °C resulted in the formation of tetrahydrodicyclopentadiene (Musoke et al., 1982). The hydrolysis half-life at pH 7 and 25 °C was estimated to be >197,000 years (Ellington et al., 1988).

Emits very toxic chloride fumes when heated to decomposition (Lewis, 1990).

Exposure Limits (mg/m^3): Potential occupational carcinogen. NIOSH REL: TWA 0.5, IDLH 100; OSHA PEL: TWA 0.5; ACGIH TLV: TWA 0.5, STEL 2.

Symptoms of Exposure: Blurred vision, confusion, ataxia, delirium, coughing, abdominal pain, nausea, vomiting, diarrhea, irritability, tremor, convulsions, anuria (NIOSH, 1994).

Toxicity: LC_{50} (96-hour) for rainbow trout 90 μg/L and bluegill sunfish 70 μg/L (Hartley and Kidd, 1987); LC_{50} (48-hour) for red killifish 245 μg/L (Yoshioka et al., 1986); LC_{50} (inhalation) for cats 100 mg/m^3/4-hour; EC_{50} (96-hour) for goldfish 0.5 mg/L; acute oral LD_{50} for rats 283 mg/kg, chickens 220 mg/kg, ducks 1,200 mg/kg, hamsters 1,720 mg/kg, mice 145 mg/kg, rabbits 100 mg/kg (RTECS, 1985).

Drinking Water Standard (final): MCLG: zero; MCL: 2 μg/L (U.S. EPA, 1984).

Uses: Insecticide and fumigant.

cis-CHLORDANE

Synonyms: α-Chlordane; β-Chlordane; α-1,2,4,5,6,7,8,8-Octachloro-3a,4,7,7a-tetrahydro-4,7-methanoindan.

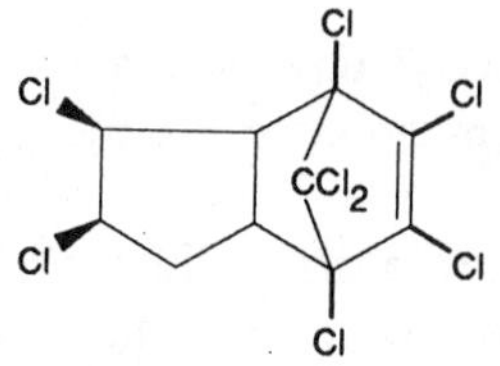

CAS Registry No.: 5103-74-2
DOT: 2762
DOT label: Flammable liquid
Molecular formula: $C_{10}H_6Cl_8$
Formula weight: 409.78
RTECS: PC0175000

Physical state
Solid.

Melting point (°C):
107.0-108.8 (Callahan et al., 1979)

Boiling point (°C):
175 (technical grade containing both *cis* and *trans* isomers, Sims et al., 1988)

Diffusivity in water (10^5 cm²/sec):
0.43 at 20 °C using method of Hayduk and Laudie (1974).

Henry's law constant (10^4 atm·m³/mol):
8.75 and 41.3 at 23 °C in distilled water and seawater, respectively (Atlas et al., 1982)

Bioconcentration factor, log BCF:
3.62-4.34 (fish tank), 6.15 (Lake Ontario) (rainbow trout, Oliver and Niimi, 1985)
3.30 (*B. subtilis*, Grimes and Morrison, 1975)

Soil sorption coefficient, log K_{oc}:
5.40 (river sediments, Oliver and Charlton, 1984)
5.57 (Chin et al., 1988)
4.77 (Meylan et al., 1992)

Octanol/water partition coefficient, log K_{ow}:
5.93 using method of Kenaga and Goring (1980).

Solubility in organics:
Miscible with aliphatic and aromatic solvents (Windholz et al., 1983).

Solubility in water:
51 µg/L at 20–25 °C (Geyer et al., 1984)

Vapor pressure (mmHg):
3.6 x 10^{-5} at 25 °C (Hinckley et al., 1990).

Environmental Fate
   Photolytic. Irradiation of *cis*-chlordane by a 450-W high-pressure mercury lamp gave photo-*cis*-chlordane (Ivie et al., 1972).
   Chemical/Physical. In an alkaline medium or solvent, carrier, diluent or emulsifier having an alkaline reaction, chlorine will be released (U.S. EPA, 1985). The hydrolysis rate constant for *cis*-chlordane at pH 11 and 25 °C was determined to be 4.3 x 10^{-3}/hour. The calculated half-life at pH 7 is >197,000 years (Ellington et al., 1987).
   Emits toxic chloride fumes when heated to decomposition (Lewis, 1990).

Symptoms of Exposure: Blurred vision, confusion, ataxia, delirium, coughing, abdominal pain, nausea, vomiting, diarrhea, irritability, tremor, convulsions, anuria (NIOSH, 1994).

Toxicity: LC$_{50}$ (96-hour) for rainbow trout 90 µg/L and bluegill sunfish 70 µg/L (Hartley and Kidd, 1987); EC$_{50}$ (96-hour) for goldfish 0.5 mg/L; acute oral LD$_{50}$ for rats 365–590 mg/kg (Hartley and Kidd, 1987).

Drinking Water Standard: See chlordane.

Use: Insecticide.

trans-CHLORDANE

Synonyms: α-Chlordan; *cis*-Chlordan; α-Chlordane; $\alpha(cis)$-Chlordane; γ-Chlordane; 1,2,4,5,6,7,8,8-Octachloro-3a,4,7,7a-tetrahydro-4,7-methanoindan.

CAS Registry No.: 5103-71-9
DOT: 2762
DOT label: Flammable liquid
Molecular formula: $C_{10}H_6Cl_8$
Formula weight: 409.78
RTECS: PB9705000

Physical state
Solid.

Melting point (°C):
103.0-105.0 (Callahan et al., 1979)

Boiling point (°C):
175 (technical grade containing both *cis* and *trans* isomers, Sims et al., 1988)

Diffusivity in water (10^5 cm^2/sec):
0.43 at 20 °C using method of Hayduk and Laudie (1974).

Henry's law constant (10^3 atm·m^3/mol at 23 °C):
1.34 and 5.59 in distilled water and seawater, respectively (Atlas et al., 1982).

Bioconcentration factor, log BCF:
4.18-4.30 (fish tank), 4.88 (Lake Ontario) (rainbow trout, Oliver and Niimi, 1985)
3.36 (*B. subtilis*, Grimes and Morrison, 1975)

Soil sorption coefficient, log K_{oc}:
5.48 (river sediments, Oliver and Charlton, 1984)
An average value of 6.3 was experimentally determined using 10 suspended sediment samples collected from the St. Clair and Detroit Rivers (Lau et al., 1989).

Octanol/water partition coefficient, log K_{ow}:
8.69, 9.65 using method of Kenaga and Goring (1980).

Solubility in organics:
Miscible with aliphatic and aromatic solvents (U.S. EPA, 1985).

Vapor pressure (mmHg):
5.03 x 10^{-5} at 25 °C (Hinckley et al., 1990)

Environmental Fate
   Photolytic. Irradiation of *trans*-chlordane by a 450-W high-pressure mercury lamp gave photo-*trans*-chlordane (Ivie et al., 1972).
   Chemical/Physical. In an alkaline medium or solvent, carrier, diluent or emulsifier having an alkaline reaction, chlorine will be released (Windholz et al., 1983).
   Emits toxic chloride fumes when heated to decomposition (Lewis, 1990).

Drinking Water Standard: See chlordane.

Use: Insecticide.

CHLOROACETALDEHYDE

Synonyms: 2-Chloroacetaldehyde; Chloroacetaldehyde monomer; 2-Chloroethanal; 2-Chloro-1-ethanal; Monochloroacetaldehyde; RCRA waste number P023; UN 2232.

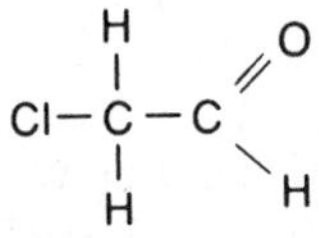

CAS Registry No.: 107-20-0
DOT: 2232
DOT label: Poison
Molecular formula: C_2H_3ClO
Formula weight: 78.50
RTECS: AB2450000

Physical state, color and odor
Clear, colorless liquid with an irritating, acrid odor.

Melting point (°C):
-19.5 (40% aqueous solution, NIOSH, 1994)

Boiling point (°C):
85-85.5 at 748 mmHg (Weast, 1986)
85 (40% solution, Hawley, 1981)

Density (g/cm³ at 20/4 °C):
1.19 (40% solution, NIOSH, 1994)
1.236 (Aldrich, 1988)

Diffusivity in water (10^5 cm²/sec):
1.15 at 20 °C using method of Hayduk and Laudie (1974).

Flash point (°C):
87.8 (40% aqueous solution, NIOSH, 1994)
53 (Aldrich, 1988)

Ionization potential (eV):
10.61 (NIOSH, 1994)

Soil sorption coefficient, log K_{oc}:
Unavailable because experimental methods for estimation of this parameter for

halogenated aldehydes are lacking in the documented literature. However, its miscibility in water suggests its adsorption to soil will be nominal (Lyman et al., 1982).

Octanol/water partition coefficient, log K_{ow}:
Unavailable because experimental methods for estimation of this parameter for halogenated aldehydes are lacking in the documented literature.

Solubility in organics:
Soluble in ether (Weast, 1986), acetone and methanol (Hawley, 1981).

Solubility in water:
>50 wt %, forms a hemihydrate (Hawley, 1981).

Vapor density:
3.21 g/L at 25 °C, 2.71 (air = 1)

Vapor pressure (mmHg):
100 at 20 °C (NIOSH, 1994)

Environmental Fate
  Chemical/Physical. Polymerizes on standing (Windholz et al., 1983).

Exposure Limits: NIOSH REL: ceiling 1 ppm (3 mg/m^3), IDLH 45 ppm; OSHA PEL: ceiling 1 ppm.

Symptoms of Exposure: Severe irritation and blurred vision on exposure to vapors. Skin contact with 40% aqueous solution can cause skin burn and tissue damage (Patnaik, 1992).

Toxicity: Acute oral LD_{50} for rats 75 mg/kg, mice 69 mg/kg (RTECS, 1985).

Uses: Removing barks from trees; manufacture of 2-aminothiazole; chemical intermediate; organic synthesis.

α-CHLOROACETOPHENONE

Synonyms: CAF; CAP; Chemical mace; 2-Chloroacetophenone; Ω-Chloroaceto-phenone; Chloromethyl phenyl ketone; **2-Chloro-1-phenylethanone**; CN; Mace; NCI-C55107; Phenacyl chloride; Phenyl chloromethyl ketone; Tear gas; UN 1697.

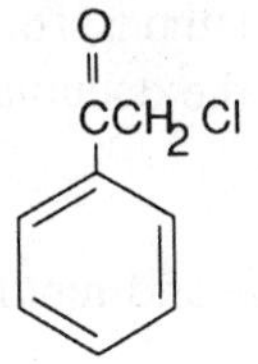

CAS Registry No.: 532-27-4
DOT: 1697
DOT label: Poison
Molecular formula: C_8H_7ClO
Formula weight: 154.60
RTECS: AM6300000

Physical state, color and odor
Colorless to gray crystalline solid with a sharp, penetrating, irritating odor. Odor threshold in air is 0.035 ppm (Amoore and Hautala, 1983).

Melting point (°C):
56.5 (Weast, 1986)
58–59 (Windholz et al., 1983)
54–56 (Aldrich, 1988)

Boiling point (°C):
247 (Weast, 1986)
244–245 (Windholz et al., 1983)

Density (g/cm³):
1.324 at 15/4 °C (Weast, 1986)

Diffusivity in water (10^5 cm²/sec):
0.69 at 15 °C using method of Hayduk and Laudie (1974).

Flash point (°C):
117.9 (NIOSH, 1994)

Ionization potential (eV):
9.5 (Franklin et al., 1969)
9.44 (NIOSH, 1994)

Soil sorption coefficient, log K_{oc}:
Unavailable because experimental methods for estimation of this parameter for halogenated aromatic ketones are lacking in the documented literature. However, its miscibility in water suggests its adsorption to soil will be nominal (Lyman et al., 1982).

Octanol/water partition coefficient, log K_{ow}:
Unavailable because experimental methods for estimation of this parameter for halogenated aromatic ketones are lacking in the documented literature.

Solubility in organics:
Soluble in acetone (Weast, 1986). Freely soluble in alcohol, benzene and ether (Windholz et al., 1983).

Vapor pressure (10^3 mmHg):
5.4 at 20 °C (Windholz et al., 1983)
4 at 20 °C, 14 at 30 °C (Verschueren, 1983)

Environmental Fate
   Chemical/Physical. Releases toxic chloride fumes when heated to decomposition (Sax and Lewis, 1987).

Exposure Limits: NIOSH REL: TWA 0.3 mg/m^3 (0.05 ppm), IDLH 15 mg/m^3; OSHA PEL: TWA 0.05 ppm (0.3 mg/m^3).

Toxicity: Acute oral LD_{50} for rats 50 mg/kg, rabbits 118 mg/kg, mice 139 mg/kg, guinea pigs 158 mg/kg (RTECS, 1985).

Uses: Riot control agent.

4-CHLOROANILINE

Synonyms: 1-Amino-4-chlorobenzene; 1-Amino-*p*-chlorobenzene; 4-Amino-chlorobenzene; *p*-Aminochlorobenzene; 4-Chloraniline; *p*-Chloraniline; *p*-Chloroaniline; **4-Chlorobenzamine**; *p*-Chlorobenzamine; 4-Chlorophenylamine; *p*-Chlorophenylamine; NCI-C02039; RCRA waste number P024; UN 2018; UN 2019.

CAS Registry No.: 106-47-8
DOT: 2018 (solid); 2019 (liquid)
DOT label: Poison
Molecular formula: C_6H_6ClN
Formula weight: 127.57
RTECS: BX0700000

Physical state, color and odor
Yellowish-white solid with a mild, sweetish odor. Odor threshold in air is 287 ppm (Keith and Walters, 1992).

Melting point (°C):
72.50 (Martin et al., 1979)
70 (Morrison and Boyd, 1971)

Boiling point (°C):
232 (Weast, 1986)

Density (g/cm³):
1.429 at 19/4 °C (Weast, 1986)

Diffusivity in water (10^5 cm²/sec):
0.75 at 20 °C using method of Hayduk and Laudie (1974).

Dissociation constant:
4.15 (Weast, 1986)
3.98 (Howard, 1989)

Flash point (°C):
>1,205 (Weiss, 1986)

Lower explosive limit (%):
Not pertinent (Weiss, 1986)

Upper explosive limit (%):
Not pertinent (Weiss, 1986)

Henry's law constant (10^5 atm·m^3/mol):
1.07 at 25 °C (calculated, Howard, 1989)

Bioconcentration factor, log BCF:
2.41 (algae, Geyer et al., 1984)

Soil sorption coefficient, log K_{oc}:
2.42 (Hodson and Williams, 1988)
1.98, 2.05, 3.10, 3.13, 3.18 (Rippen et al., 1982)
2.75 (Rao and Davidson, 1980)
1.96 (Meylan et al., 1992)
2.60, 2.70, 2.82, 2.91 (Van Bladel and Moreale, 1977)

Octanol/water partition coefficient, log K_{ow}:
1.83 (Leo et al., 1971)
2.78 (Hammers et al., 1982)
1.88 (Garst and Wilson, 1984; Hammers et al., 1982)
2.02 (Geyer et al., 1984)

Solubility in organics:
Soluble in ethanol and ether (Weast, 1986). Freely soluble in acetone and carbon disulfide (Windholz et al., 1983)

Solubility in water (g/L):
3.9 at 20-25 °C (Kilzer et al., 1979)
3.1 at 4 °C, 4.7 at 25 °C, 7.8 at 40 °C (Moreale and Van Bladel, 1979)
2.8 at 20 °C (Hafkenscheid and Tomlinson, 1983)

Vapor pressure (10^2 mmHg):
1.5 at 20 °C, 5 at 30 °C (Verschueren, 1983)
2.5 at 25 °C (Kilzer et al., 1979; Piacente et al., 1985)

Environmental Fate
Biological. In an anaerobic medium, the bacteria of the *Paracoccus* sp. converted 4-chloroaniline to 1,3-bis(*p*-chlorophenyl)triazene and 4-chloroacetanilide with product yields of 80 and 5%, respectively (Minard et al., 1977). In a field experiment, [^{14}C]4-chloroaniline was applied to a soil at a depth of 10 cm. After 20

weeks, 32.4% of the applied amount was recovered in soil. Metabolites identified include 4-chloroformanilide, 4-chloroacetanilide, 4-chloronitrobenzene, 4-chloronitrosobenzene, 4,4′-dichloroazoxybenzene and 4,4′-dichloroazobenzene (Freitag et al., 1984).

In a 56-day experiment, [^{14}C]4-chloroaniline applied to soil-water suspensions under aerobic and anaerobic conditions gave $^{14}CO_2$ yields of 3.0 and 2.3%, respectively (Scheunert et al., 1987). Silage samples (chopped corn plants) containing 4-chloroaniline were incubated in an anaerobic chamber for 2 weeks at 28 °C. After 3 days, 4-chloroaniline was biologically metabolized to 4-chloroacetanilide and another compound, tentatively identified as 4-chloroformanilide (Lyons et al., 1985). In activated sludge, 22.7% mineralized to carbon dioxide after 5 days (Freitag et al., 1985).

Soil. 4-Chloroaniline covalently bonds with humates in soils to form quinoidal structures followed by oxidation to yield a nitrogen-substituted quinoid ring (Parris, 1980). Catechol, a humic acid monomer, reacted with 4-chloroaniline yielding 4,5-bis(4-chlorophenylamino)-3,5-cyclohexadiene-1,2-dione (Adrian et al., 1989).

Photolytic. Under artificial sunlight, river water containing 2-5 ppm 4-chloroaniline photodegraded to 4-aminophenol and unidentified polymers (Mansour et al., 1989). Photooxidation of 4-chloroaniline (100 μM) in air-saturated water using UV light (λ >290 nm) produced 4-chloronitrobenzene and 4-chloronitrosobenzene. About 6 hours later, 4-chloroaniline completely reacted leaving dark purple condensation products (Miller and Crosby, 1983). A carbon dioxide yield of 27.7% was achieved when 4-chloroaniline adsorbed on silica gel was irradiated with light (λ >290 nm) for 17 hours (Freitag et al., 1985).

Toxicity: Acute oral LD_{50} for wild birds 100 mg/kg, guinea pigs 350 mg/kg, mice 100 mg/kg, quail 237 mg/kg, rats 310 mg/kg; LD_{50} (inhalation) for mice 250 mg/m^3/6-hour; LC_{50} (48-hour) for red killifish 219 mg/L (Yoshioka et al., 1986).

Uses: Dye intermediate; pharmaceuticals; agricultural chemicals.

CHLOROBENZENE

Synonyms: Benzene chloride; Chlorbenzene; Chlorbenzol; Chlorobenzol; MCB; Monochlorbenzene; Monochlorobenzene; NCI-C54886; Phenyl chloride; RCRA waste number U037; UN 1134.

CAS Registry No.: 108-90-7
DOT: 1134
DOT label: Flammable liquid
Molecular formula: C_6H_5Cl
Formula weight: 112.56
RTECS: CZ0175000

Physical state, color and odor
Clear, colorless, watery liquid with a sweet almond odor. Odor threshold in air is 0.21 ppm (Keith and Walters, 1992).

Melting point (°C):
-45.6 (Weast, 1986)

Boiling point (°C):
132 (Weast, 1986)

Density (g/cm^3):
1.1058 at 20/4 °C (Weast, 1986)
1.1009 at 25/4 °C (Kirchnerová and Cave, 1976)

Diffusivity in water (10^5 cm^2/sec):
0.87 at 20 °C using method of Hayduk and Laudie (1974).

Flash point (°C):
27.8 (NIOSH, 1994)

Lower explosive limit (%):
1.3 (NIOSH, 1994)

Upper explosive limit (%):
9.6 (NIOSH, 1994)

Heat of fusion (kcal/mol):
2.28 (Dean, 1987)

Henry's law constant (10^3 atm·m^3/mol):
3.93 at 25 °C (Warner et al., 1987)
3.6 (Pankow and Rosen, 1988)
3.7 (Valsaraj, 1988)
4.45 at 25 °C (Hine and Mookerjee, 1975)
6.21 at 37 °C (Sato and Nakajima, 1979)
2.44, 2.81, 3.41, 3.60 and 4.73 at 10, 15, 20, 25 and 30 °C, respectively (Ashworth et al., 1988)

Interfacial tension with water (dyn/cm at 20 °C):
37.41 (Demond and Lindner, 1993)

Ionization potential (eV):
9.07 (Horvath, 1982)
9.14 (Yoshida et al., 1983)

Bioconcentration factor, log BCF:
3.23 (activated sludge), 1.70 (algae), 1.88 (golden ide) (Freitag et al., 1985)
2.65 (fathead minnow, Veith et al., 1979)

Soil sorption coefficient, log K_{oc}:
1.92 (Woodburn silt loam soil, Chiou et al., 1983)
2.50 (Captina silt loam), 2.17 (McLaurin sandy loam) (Walton et al., 1992)

Octanol/water partition coefficient, log K_{ow}:
2.84 (Fujita et al., 1964; Garst and Wilson, 1984; Mirrlees et al., 1976)
2.81 (Mirrless et al., 1976)
2.98 (Tewari et al., 1982; Wasik et al., 1981, 1983)
2.71 (Schwarzenbach and Westall, 1981)
2.898 (Brooke et al., 1990; de Bruijn et al., 1989)
2.784 (Brooke et al., 1990)
2.83 (Hammers et al., 1982; Yoshida et al., 1983)
2.65 (Campbell and Luthy, 1985)

Solubility in organics:
Soluble in ethanol, benzene, chloroform, ether (U.S. EPA, 1985) and many other organic solvents.

Solubility in water:
502 mg/L at 25 °C (Banerjee, 1984)

295 mg/L at 25 °C (Tewari et al., 1982; Wasik et al., 1981)
4.43 mmol/L at 25 °C (Wasik et al., 1983)
471.7 mg/L at 25 °C (Aquan-Yuen et al., 1979)
500 mg/L at 25 °C, 488 mg/kg at 30 °C (Andrews and Keefer, 1950)
446 mg/kg at 30 °C (Gross and Saylor, 1931)
448 mg/L at 30 °C (Freed et al., 1977)
503 mg/L at 25 °C (Yalkowsky et al., 1979)
534 mg/kg at 25 °C (Chey and Calder, 1972)
0.0512 wt % at 25 °C (Nathan, 1978)
474.0 mg/L at 30 °C (McNally and Grob, 1983, 1984)
0.035 wt % at 25 °C (Lo et al., 1986)
4.11 mmol/kg at 25.0 °C (Vesala, 1974)
3.78 mM at 25.0 °C (Sanemasa et al., 1987)

Vapor density:
4.60 g/L at 25 °C, 3.88 (air = 1)

Vapor pressure (mmHg):
9 at 20 °C (NIOSH, 1994)
11.86 at 25 °C (Mackay et al., 1982)

Environmental Fate
Biological. In activated sludge, 31.5% of the applied chlorobenzene mineralized to carbon dioxide after 5 days (Freitag et al., 1985). A mixed culture of soil bacteria or a *Pseudomonas* sp. transformed chlorobenzene to chlorophenol (Ballschiter and Scholz, 1980). Pure microbial cultures isolated from soil hydroxylated chlorobenzene to 2- and 4-hydroxychlorobenzene (Smith and Rosazza, 1974). Chlorobenzene was statically incubated in the dark at 25 °C with yeast extract and settled domestic wastewater inoculum. At a concentration of 5 mg/L, biodegradation yields at the end of 1 and 2 weeks were 89 and 100%, respectively. At a concentration of 10 mg/L, significant degradation with gradual adaptation was observed. Complete degradation was not observed until after the 3rd week of incubation (Tabak et al., 1981).
Photolytic. Under artificial sunlight, river water containing 2-5 ppm chlorobenzene photodegraded to phenol and chlorophenol. The lifetimes of chloro-benzene in distilled water and river water were 17.5 and 3.8 hours, respectively (Mansour et al., 1989). In distilled water containing 1% acetonitrile exposed to artificial sunlight, 28% of chlorobenzene in solution photolyzed to phenol, chloride ion and acetanilide with reported product yields of 55, 112 and 2%, respectively (Dulin et al., 1986).
Titanium dioxide suspended in an aqueous solution and irradiated with UV light (λ = 365 nm) converted chlorobenzene to carbon dioxide at a significant rate (Matthews, 1986). Products identified as intermediates in this reaction include

three monochlorophenols, chlorohydroquinone and hydroxyhydroquinone as intermediates (Kawaguchi and Furuya, 1990).

Photooxidation of chlorobenzene in air containing nitric oxide in a Pyrex glass
vessel and a quartz vessel gave 3-chloronitrobenzene, 2-chloro-6-nitrophenol, 2-
chloro-4-nitrophenol, 4-chloro-2-nitrophenol, 4-nitrophenol, 3-chloro-4-nitro-
phenol, 3-chloro-6-nitrophenol and 3-chloro-2-nitrophenol (Kanno and Nojima,
1979). A carbon dioxide yield of 18.5% was achieved when chlorobenzene
adsorbed on silica gel was irradiated with light (λ >290 nm) for 17 hours. The
sunlight irradiation of chlorobenzene (20 g) in a 100-mL borosilicate glass-
stoppered Erlenmeyer flask for 28 days yielded 1,060 ppm monochlorobiphenyl
(Uyeta et al., 1976).

When an aqueous solution containing chlorobenzene (190 μM) and a nonionic
surfactant micelle (Brij 58, a polyoxyethylene cetyl ether) was illuminated by a
photoreactor equipped with 253.7-nm monochromatic ultraviolet lamps, phenol,
hydrogen and chloride ions formed as the major products. Although not measured,
it was reported that aromatic aldehydes, organic acids and carbon dioxide would
form from the photoreaction of benzene in water under similar conditions. A
duplicate experiment was conducted using an ionic micelle (triethylamine, 5 mM),
which serves as a hydrogen source. Products identified were phenol and benzene
(Chu and Jafvert, 1994).

Chemical/Physical. Anticipated products from the reaction of chlorobenzene
with ozone or hydroxyl radicals in the atmosphere are chlorophenols and ring
cleavage compounds (Cupitt, 1980).

In the absence of oxygen, chlorobenzene reacted with Fenton's reagent forming
chlorophenols, dichlorobiphenyls and phenolic polymers as major intermediates.
With oxygen, chlorobenzoquinone, chlorinated and nonchlorinated diols formed
(Sedlak and Andren, 1991).

Based on an assumed base mediated 1% disappearance after 16 days at 85 °C and
pH 9.70 (pH 11.26 at 25 °C), the hydrolysis half-life was estimated to be >900
years (Ellington et al., 1988).

Exposure Limits: NIOSH REL: Awaiting OSHA ruling to determine if the recommended TWA exposure limit of 75 ppm is protective of human health, IDLH
1,000 ppm; OSHA PEL: TWA 75 ppm (350 mg/m^3); ACGIH TLV: TWA 75 ppm.

Symptoms of Exposure: Inhalation of vapors may cause drowsiness, incoordination
and liver damage. May irritate eyes and skin (Patnaik, 1992).

Toxicity: Acute oral LD$_{50}$ for rats 2,910 mg/kg, guinea pigs 5,060 mg/kg, rabbits
2,250 mg/kg (RTECS, 1985); LC$_{50}$ for goldfish 1.8 mL/kg (Verschueren, 1983).

Drinking Water Standard (final): MCLG: 0.1 mg/L; MCL: 0.1 mg/L (U.S. EPA,
1994).

Uses: Preparation of phenol, 4-chlorophenol, chloronitrobenzene, aniline, 2-, 3- and 4-nitrochlorobenzenes; solvent carrier for methylene diisocyanate; solvent for paints; insecticide, pesticide and dyestuffs intermediate; heat transfer agent.

o-CHLOROBENZYLIDENEMALONONITRILE

Synonyms: 2-Chlorobenzalmalononitrile; *o*-Chlorobenzalmalononitrile; 2-Chloro-benzylidenemalononitrile; 2-Chlorobnm; **[(2-Chlorophenyl)methylene]propanedi-nitrile**; CS; *β,β*-Dicyano-*o*-chlorostyrene; NCI-C55118; OCBM; USAF KF-11.

CAS Registry No.: 2698-41-1
Molecular formula: $C_{10}H_5ClN_2$
Formula weight: 188.61
RTECS: OO3675000

Physical state, color and odor
White crystalline solid with a pepper-like odor.

Melting point (°C):
95-96 (Windholz et al., 1983)

Boiling point (°C):
310-315 (Windholz et al., 1983)

Density (g/cm^3):
1.472 using method of Lyman et al. (1982).

Diffusivity in water:
Not applicable - reacts with water.

Lower explosive limit:
For a dust, 25 mg/m^3 is the minimum explosive concentration in air (NIOSH, 1994).

Henry's law constant (atm·m^3/mol):
Not applicable - reacts with water.

Soil sorption coefficient, log K$_{oc}$:
Not applicable - reacts with water.

Octanol/water partition coefficient, log K$_{ow}$:
Not applicable - reacts with water.

Solubility in organics:
Soluble in acetone, benzene, 1,4-dioxane, ethyl acetate and methylene chloride (Windholz et al., 1983).

Solubility in water:
Not applicable - reacts with water.

Vapor pressure (mmHg):
3×10^{-5} at 20 °C (NIOSH, 1994)

Environmental Fate
   Chemical/Physical. Hydrolyzes in water forming 2-chlorobenzaldehyde and malononitrile (Verschueren, 1983).

Exposure Limits: NIOSH REL: ceiling 0.05 ppm (0.4 mg/m^3), IDLH 2 mg/m^3; OSHA PEL: TWA 0.05 ppm; ACGIH TLV: ceiling 0.05 ppm.

Toxicity: Acute oral LD$_{50}$ for rats 178 mg/kg, guinea pigs 212 mg/kg, rabbits 143 mg/kg, mice 282 mg/kg (RTECS, 1985).

Uses: Riot control agent.

p-CHLORO-*m*-CRESOL

Synonyms: Aptal; Baktol; Baktolan; Candaseptic; *p*-Chlor-*m*-cresol; Chlorocresol; 4-Chlorocresol; *p*-Chlorocresol; 4-Chloro-*m*-cresol; 6-Chloro-*m*-cresol; 4-Chloro-1-hydroxy-3-methylbenzene; 2-Chlorohydroxytoluene; 2-Chloro-5-hydroxytoluene; 4-Chloro-3-hydroxytoluene; 6-Chloro-3-hydroxytoluene; **4-Chloro-3-methylphenol**; *p*-Chloro-3-methylphenol; 3-Methyl-4-chlorophenol; Ottafact; Parmetol; Parol; PCMC; Peritonan; Preventol CMK; Raschit; Raschit K; Rasenanicon; RCRA waste number U039.

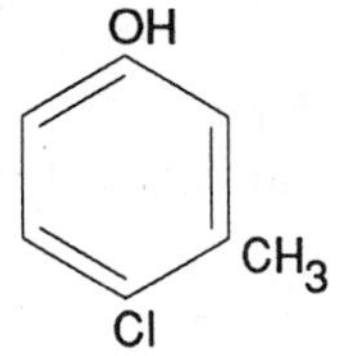

CAS Registry No.: 59-50-7
DOT: 2669
DOT label: Poison
Molecular formula: C_7H_7ClO
Formula weight: 142.59
RTECS: GO7100000

Physical state, color and odor
Colorless, white or pinkish crystals with a slight phenolic odor. On exposure to air it slowly becomes light brown.

Melting point (°C):
66-68 (Weast, 1986)
63-65 (Fluka, 1988)

Boiling point (°C):
235 (Weast, 1986)

Diffusivity in water (10^5 cm²/sec):
0.71 at 20 °C using method of Hayduk and Laudie (1974).

Dissociation constant:
9.549 at 25 °C (Dean, 1973)

Henry's law constant (10^6 atm·m³/mol):
2.5 at 20 °C (calculated, Mabey et al., 1982)

Soil sorption coefficient, log K_{oc}:
2.89 using method of Karickhoff et al. (1979).

Octanol/water partition coefficient, log K_{ow}:
2.18 (Hansch and Leo, 1979)
3.10 (Leo et al., 1971)

Solubility in organics:
Soluble in ethanol, ether (Weast, 1986), benzene, chloroform, acetone, petroleum ether, fixed oils, terpenes and aqueous alkaline solutions (Windholz et al., 1983).

Solubility in water (mg/L):
3,846 at 20 °C (Windholz et al., 1983)
3,990 at 25 °C and pH 5.1 (Blackman et al., 1955)

Vapor pressure (mmHg):
No data was found, however, a value of 5×10^{-2} at 20 °C was assigned by analogy (Mabey et al., 1982).

Environmental Fate
Biological. When *p*-chloro-*m*-cresol was statically incubated in the dark at 25 °C with yeast extract and settled domestic wastewater inoculum, significant bio-degradation with rapid adaptation was observed. At concentrations of 5 and 10 mg/L, 78 and 76% biodegradation respectively were observed after 7 days (Tabak et al., 1981).

Toxicity: LD_{50} for rats (subcutaneous) 400 mg/kg, acute oral LD_{50} for rats 1,830 mg/kg (RTECS, 1985).

Uses: External germicide; preservative for gums, glues, paints, inks, textile and leather products; topical antiseptic (veterinarian).

CHLOROETHANE

Synonyms: Aethylis; Aethylis chloridum; Anodynon; Chelen; Chlorethyl; Chloridum; Chloryl; Chloryl anesthetic; Ether chloratus; Ether hydrochloric; Ether muriatic; Ethyl chloride; Hydrochloric ether; Kelene; Monochlorethane; Monochloroethane; Muriatic ether; Narcotile; NCI-C06224; UN 1037.

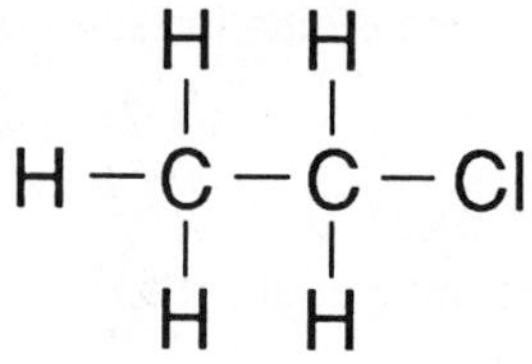

$$H-\underset{\underset{H}{|}}{\overset{\overset{H}{|}}{C}}-\underset{\underset{H}{|}}{\overset{\overset{H}{|}}{C}}-Cl$$

CAS Registry No.: 75-00-3
DOT: 1037
DOT label: Flammable gas/liquid
Molecular formula: C_2H_5Cl
Formula weight: 64.52
RTECS: KH75250000

Physical state, color and odor
Colorless gas or liquid with a pungent ether-like odor. Odor threshold in air is 4.2 ppm (Amoore and Hautala, 1983).

Melting point (°C):
-136.4 (Weast, 1986)
-139.0 (Stull, 1947)

Boiling point (°C):
12.3 (Weast, 1986)

Density (g/cm³):
0.9028 at 15/4 °C, 0.8970 at 20/4 °C (Standen, 1964)
0.8706 (Dreisbach, 1959)

Diffusivity in water (10^5 cm²/sec):
1.07 at 20 °C using method of Hayduk and Laudie (1974).

Flash point (°C):
-50 (NIOSH, 1994)
-43 (open cup, Windholz et al., 1983)

Lower explosive limit (%):
3.8 (NIOSH, 1994)

Upper explosive limit (%):
15.4 (NIOSH, 1994)

Heat of fusion (kcal/mol):
1.064 (Dean, 1987)

Henry's law constant (10^3 atm·m^3/mol):
11.1 (Gossett, 1987)
9.3 (Pankow and Rosen, 1988)
8.5 at 25 °C (Hine and Mookerjee, 1975)
7.59, 9.58, 11.0, 12.1 and 14.3 at 10, 15, 20, 25 and 30 °C, respectively (Ashworth
  et al., 1988)

Ionization potential (eV):
10.97, 11.01 (Horvath, 1982)

Soil sorption coefficient, log K_{oc}:
0.51 using method of Chiou et al. (1979).

Octanol/water partition coefficient, log K_{ow}:
1.43 (Hansch et al., 1975; Valvani et al., 1981)

Solubility in organics:
Soluble in ethanol and ether (U.S. EPA, 1985)

Solubility in water:
3,330 mg/L at 0 °C (Verschueren, 1983)
4,470 mg/L at 0 °C (Standen, 1964)
0.57 wt % at 17.5 °C (Stephen and Stephen, 1963)
0.455 wt % at 0 °C (Konietzko, 1984)
5,710 mg/L at 20 °C (Mackay and Shiu, 1981)
89 mmol/L at 20 °C (Fischer and Ehrenberg, 1948)
0.574 wt % at 20 °C (Nathan, 1978)
89.0 mmol/L (Fühner, 1924)

Vapor density:
2.76 kg/m^3 at 20 °C (Konietzko, 1984)

Vapor pressure (mmHg):
1,011 at 20 °C (Standen, 1964)
1,444 at 30 °C (Verschueren, 1983)
766 at 12.5 °C (Howard, 1990)
1,199 at 25 °C (Nathan, 1978)

Environmental Fate

Chemical/Physical. Under laboratory conditions, chloroethane hydrolyzed to ethanol (Smith and Dragun, 1984). An estimated hydrolysis half-life in water at 25 °C and pH 7 is 38 days, with ethanol and hydrochloric acid being the expected end-products (Mabey and Mill, 1978). In the atmosphere, formyl chloride is the initial photooxidation product (U.S. EPA, 1985). In the presence of water, formyl chloride hydrolyzes to hydrochloric acid and carbon monoxide (Morrison and Boyd, 1971). Burns with a smoky, greenish flame releasing hydrogen chloride (Windholz et al., 1983).

The evaporation half-life of chloroethane (1 mg/L) from water at 25 °C using a shallow-pitch propeller stirrer at 200 rpm at an average depth of 6.5 cm is 23.1 minutes (Dilling, 1977).

Exposure Limits: Potential occupational carcinogen. NIOSH REL: IDLH 3,800 ppm; OSHA PEL: TWA 1,000 ppm (2,600 mg/m^3); ACGIH TLV: TWA 1,000 ppm.

Symptoms of Exposure: May cause stupor, eye irritation, incoordination, abdominal cramps, anesthetic effects, cardiac arrest and unconsciousness (Patnaik, 1992).

Toxicity: LC$_{50}$ (inhalation) for mice 146 gm/m^3/2-hour, rats is 160 gm/m^3/2-hour (RTECS, 1985).

Drinking Water Standard: As of May 1994, chloroethane has been listed for regulation but no MCLGs or MCLs have been proposed (U.S. EPA, 1994).

Uses: Intermediate for tetraethyl lead and ethyl cellulose; topical anesthetic; organic synthesis; alkylating agent; refrigeration; analytical reagent; solvent for phosphorus, sulfur, fats, oils, resins and waxes; insecticides.

2-CHLOROETHYL VINYL ETHER

Synonyms: 2-Chlorethyl vinyl ether; **(2-Chloroethoxy)ethene**; RCRA waste number U042; Vinyl 2-chloroethyl ether; Vinyl β-chloroethyl ether.

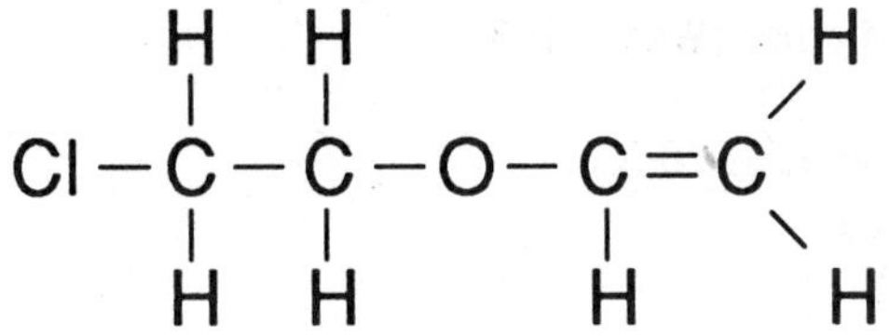

$$Cl-\underset{\underset{H}{|}}{\overset{\overset{H}{|}}{C}}-\underset{\underset{H}{|}}{\overset{\overset{H}{|}}{C}}-O-\underset{\underset{H}{|}}{\overset{}{C}}=\overset{\overset{\diagup H}{}}{\underset{\diagdown H}{C}}$$

CAS Registry No.: 110-75-8
Molecular formula: C_4H_7ClO
Formula weight: 106.55
RTECS: KN6300000

Physical state and color
Colorless liquid.

Melting point (°C):
-70.3 (Sax, 1984)
-69.7 (Dean, 1987)

Boiling point (°C):
108 (Weast, 1986)
109 (Standen, 1970)
110 (Dean, 1987)

Density (g/cm^3 at 20/4 °C):
1.0475 (Weast, 1986)
1.0493 (Standen, 1970)

Diffusivity in water (10^5 cm^2/sec):
0.87 at 20 °C using method of Hayduk and Laudie (1974).

Flash point (°C):
16 (Aldrich, 1988)
27 (open cup, NFPA, 1984)

Henry's law constant (10^4 atm·m^3/mol):
2.5 (Pankow and Rosen, 1988)

Soil sorption coefficient, log K_{oc}:
0.82 (Schwille, 1988)

Octanol/water partition coefficient, log K_{ow}:
1.28 (calculated, Leo et al., 1971)

Solubility in organics:
Soluble in ethanol and ether (Weast, 1986).

Solubility in water (mg/L at 20 °C):
15,000 (Schwille, 1988)
6,000 (Standen, 1970)

Vapor density:
4.36 g/L at 25 °C, 3.68 (air = 1)

Vapor pressure (mmHg):
26.75 at 20 °C (U.S. EPA, 1980)

Environmental Fate
  Biological. When 2-chloroethyl vinyl ether was statically incubated in the dark at 25 °C with yeast extract and settled domestic wastewater inoculum, significant biodegradation with rapid adaptation was observed. At concentrations of 5 and 10 mg/L, complete degradation was observed after 21 days (Tabak et al., 1981).
  Chemical/Physical. Chlorination of 2-chloroethyl vinyl ether to α-chloroethyl ethyl ether or β-chloroethyl ethyl ether may occur in water treatment facilities. The *alpha* compound is very unstable in water and decomposes almost as fast as it is formed (Summers, 1955; Tou and Kallos, 1974). Though stable in sodium hydroxide solutions, in dilute acid solutions hydrolysis yields acetaldehyde and chlorohydrin (Windholz et al., 1983).

Symptoms of Exposure: Exposure to vapors may cause irritation of eyes, nose and lungs (Patnaik, 1992).

Toxicity: Acute oral LD_{50} for rats 250 mg/kg (RTECS, 1985).

Uses: Anesthetics, sedatives and cellulose ethers; copolymer of 95% ethyl acrylate with 5% 2-chloroethyl vinyl ether is used to produce an acrylic elastomer.

CHLOROFORM

Synonyms: Formyl trichloride; Freon 20; Methane trichloride; Methenyl chloride; Methenyl trichloride; Methyl trichloride; NCI-C02686; R 20; R 20 (refrigerant); RCRA waste number U044; TCM; Trichloroform; **Trichloromethane**; UN 1888.

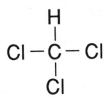

CAS Registry No.: 67-66-3
DOT: 1888
DOT label: Poison
Molecular formula: $CHCl_3$
Formula weight: 119.38
RTECS: FS9100000

Physical state, color and odor
Clear, colorless, volatile liquid with a strong, sweet, pleasant ether-like odor. Odor threshold in air is 0.3 mg/m^3 (Keith and Walters, 1992).

Melting point (°C):
-63.5 (McGovern, 1986)

Boiling point (°C):
61.7 (Weast, 1986)

Density (g/cm^3):
1.4832 at 20/4 °C (Weast, 1986)
1.4890 at 20/4 °C, 1.48069 at 25/4 °C (Standen, 1964)

Diffusivity in water (10^5 cm^2/sec):
1.00 at 20 °C using method of Hayduk and Laudie (1974).

Flash point (°C):
Noncombustible (NIOSH, 1994)

Heat of fusion (kcal/mol):
2.28 (Dean, 1987)

Henry's law constant (10^3 atm·m^3/mol):
3.39 at 25 °C (Warner et al., 1987)

3.23 (Valsaraj, 1988)
5.3 at 20 °C (Roberts and Dändliker, 1983)
3.18 at 25 °C (Dilling, 1977)
7.27 at 37 °C (Sato and Nakajima, 1979)
4.35 at 25 °C (Hine and Mookerjee, 1975)
5.76 at 20 °C (Roberts et al., 1985)
3 at 20 °C (Nicholson et al., 1984)
5.54 at 30 °C (Jeffers et al., 1989)
1.720, 2.33, 33.2, 42.1 and 55.4 at 10, 15, 20, 25 and 30 °C, respectively (Ashworth et al., 1988)
4.90 at 25 °C in seawater (Hunter-Smith et al., 1983)
In seawater: 1.3, 2.1 and 3.5 at 0, 10 and 20 °C, respectively (Moore et al., 1995)
1.24, 1.43, 1.72, 2.79 and 3.75 at 2.0, 6.0, 10.0, 18.2 and 25.0 °C, respectively (Dewulf et al., 1995)
2.80 (McConnell et al., 1975)

Interfacial tension with water (dyn/cm at 25 °C):
31.6 (Donahue and Bartell, 1952)

Ionization potential (eV):
11.42 ± 0.03 (Franklin et al., 1969)
11.50 (Horvath, 1982)

Bioconcentration factor, log BCF:
0.78 (bluegill sunfish, Veith et al., 1980)

Soil sorption coefficient, log K_{oc}:
1.80 (Potomac-Raritan-Magothy aquifer sand), 1.94 (Cohansey aquifer sand) (Uchrin and Michaels, 1986)
1.57 (Captina silt loam), 1.46 (McLaurin sandy loam) (Walton et al., 1992)

Octanol/water partition coefficient, log K_{ow}:
1.97 (Hansch and Anderson, 1967; Hansch et al., 1975)
1.90 (Veith et al., 1980)
1.95 (Mackay, 1982)

Solubility in organics:
Miscible with ethanol, ether, benzene and ligroin (U.S. EPA, 1985).

Solubility in water:
8,200 mg/L at 20 °C (Pearson and McConnell, 1975)
7,950 mg/L at 25 °C (Kenaga, 1975)
7,222 mg/L at 25 °C (Banerjee et al., 1980)

7,840 mg/L at 25 °C (Dilling, 1977)
8,520 mg/kg at 15 °C, 7,710 mg/kg at 30 °C (Gross and Saylor, 1931)
0.815 wt % at 20 °C (Riddick et al., 1986)
8.950 g/L at 10 °C, 8.220 g/L at 20 °C, 7.760 g/L at 30 °C (Standen, 1964)
2,524 mg/L at 30 °C (McNally and Grob, 1984)
0.82 wt % at 25 °C (Lo et al., 1986)
7.9 g/kg at 25 °C (McGovern, 1943)
8,700 mg/L at 23–24 °C (Broholm and Feenstra, 1995)
1.11×10^{-3} at 25 °C (mole fraction, Li et al., 1993)

Vapor density:
4.88 g/L at 25 °C, 4.12 (air = 1)

Vapor pressure (mmHg):
160 at 20 °C, 245 at 30 °C (Verschueren, 1983)
150.5 at 20 °C (McConnell et al., 1975)
198 at 25 °C (Warner et al., 1987)
246 at 25 °C (Howard, 1990)

Environmental Fate
Biological. An anaerobic species of *Clostridium* biodegraded chloroform (a metabolite of carbon tetrachloride) by reductive dechlorination yielding methylene chloride and unidentified products (Gälli and McCarty, 1989). Chloroform showed significant degradation with gradual adaptation in a static-culture flask-screening test (settled domestic wastewater inoculum) conducted at 25 °C. At concentrations of 5 and 10 mg/L, complete degradation was observed at the end of the third subculture period (28 days). The amount lost due to volatilization after 10 days was 6–24% (Tabak et al., 1981).

Photolytic. Complete mineralization was reported when distilled deionized water containing trichloromethane (118 ppm) and 0.1 wt % titanium dioxide as a catalyst was irradiated with UV light. Mineralization products included carbon dioxide and hydrochloric acid (Pruden and Ollis, 1983). In a similar experiment, titanium dioxide suspended in an aqueous solution and irradiated with UV light ($\lambda = 365$ nm) converted chloroform to carbon dioxide at a significant rate. Intermediate compounds were not identified (Matthews, 1986). The estimated hydrolysis half-lives in water at 25 °C and pH 7 are 3,500 years (Mabey and Mill, 1978) and 1,849.6 years (Jeffers et al., 1989).

An aqueous solution containing 300 ng/μL chloroform and colloidal platinum catalyst was irradiated with UV light. After 15 hours, only 10 ng/μL chloroform remained. A duplicate experiment was performed but 0.1 g zinc was added to the system. At approximately 2 hours, 10 ng/μL chloroform remained and 210 ng/μL methane was produced (Wang and Tan, 1988).

Photolysis of an aqueous solution containing chloroform (314 μmol) and the catalyst [Pt(colloid)/Ru(bpy)$^{2+}$/MV/EDTA] yielded the following products after 15 hours (μmol detected): chloride ions (852), methane (265), ethylene (0.05), ethane (0.52) and unreacted chloroform (10.5) (Tan and Wang, 1987). In the troposphere, photolysis of chloroform via hydroxyl radicals may yield formyl chloride, carbon monoxide, hydrogen chloride and phosgene as the principal products (Spence et al., 1976). Phosgene is hydrolyzed readily to hydrogen chloride and carbon dioxide (Morrison and Boyd, 1971).

Chemical/Physical. Matheson and Tratnyek (1994) studied the reaction of fine-grained iron metal in an anaerobic aqueous solution (15 °C) containing chloroform (107 μM). Initially, chloroform underwent rapid dehydrochlorination forming methylene chloride and chloride ions. As the concentration of methylene chloride increased, the rate of reaction appeared to decrease. After 140 hours, no additional products were identified. The authors reported that reductive dehalogenation of chloroform and other chlorinated hydrocarbons used in this study appears to take place in conjunction with the oxidative dissolution or corrosion of the iron metal through a diffusion-limited surface reaction.

The evaporation half-life of chloroform (1 mg/L) from water at 25 °C using a shallow-pitch propeller stirrer at 200 rpm at an average depth of 6.5 cm is 20.2 minutes (Dilling, 1977). The experimental half-life for hydrolysis of chloroform in water at 25 °C is approximately 15 months (Dilling et al., 1975).

When chloroform is heated to decomposition, phosgene gas is formed (NIOSH, 1994).

Exposure Limits: Potential occupational carcinogen. NIOSH REL: STEL (1 hour) 2 ppm (9.78 mg/m^3), IDLH 500 ppm; OSHA PEL: ceiling 50 ppm (240 mg/m^3); ACGIH TLV: TWA 10 ppm (50 mg/m^3).

Symptoms of Exposure: Dizziness, lightheadedness, dullness, hallucination, nausea, headache, fatigue and anesthesia (Patnaik, 1992).

Toxicity: LC$_{50}$ (48-hour) for red killifish 1,260 mg/L (Yoshioka et al., 1986); acute oral LD$_{50}$ for rats 23 mg/kg (Patnaik, 1992), 908 mg/kg, mice 36 mg/kg, guinea pigs 820 mg/kg (RTECS, 1985).

Drinking Water Standard (tentative): MCLG: zero; MCL: 0.1 mg/L. Total for all trihalomethanes cannot exceed a concentration of 0.08 mg/L (U.S. EPA, 1984).

Uses: Manufacture of fluorocarbon refrigerants, fluorocarbon plastics and propellants; solvent for natural products; analytical chemistry; cleansing agent; soil fumigant; insecticides; preparation of chlorodifluoromethane, methyl fluoride, salicylaldehyde; cleaning electronic circuit boards; in fire extinguishers.

2-CHLORONAPHTHALENE

Synonyms: β-Chloronaphthalene; RCRA waste number U047.

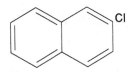

CAS Registry No.: 91-58-7
Molecular formula: $C_{10}H_7Cl$
Formula weight: 162.62
RTECS: QJ2275000

Physical state and color
Off-white monoclinic plates or leaflets.

Melting point (°C):
61 (Weast, 1986)

Boiling point (°C):
256 (Weast, 1986)

Density (g/cm³):
1.1377 at 71/4 °C (Dean, 1987)

Diffusivity in water (10^5 cm²/sec):
0.65 at 20 °C using method of Hayduk and Laudie (1974).

Flash point (°C):
Nonflammable (Sittig, 1985).

Henry's law constant (10^4 atm·m³/mol):
6.12 (calculated, U.S. EPA, 1980)

Soil sorption coefficient, log K_{oc}:
3.93 using method of Karickhoff et al. (1979).

Octanol/water partition coefficient, log K_{ow}:
3.98 (Sangster, 1989)

Solubility in organics:
Soluble in chloroform and carbon disulfide (Windholz et al., 1983).

Solubility in water:
<1 mg/L at 30 °C (McNally and Grob, 1984)

Vapor pressure (mmHg):
1.7 x 10^{-2} at 25 °C (calculated, U.S. EPA, 1980)

Environmental Fate

Biological. Reported biodegradation products include 8-chloro-1,2-dihydro-1,2-dihydroxynaphthalene and 3-chlorosalicylic acid (Callahan et al., 1979). When 2-chloronaphthalene was statically incubated in the dark at 25 °C with yeast extract and settled domestic wastewater inoculum, complete biodegradation was observed after 7 days (Tabak et al., 1981).

Chemical/Physical. The hydrolysis rate constant for 2-chloronaphthalene at pH 7 and 25 °C was determined to be 9.5 x 10^{-6}/hour, resulting in a half-life of 8.3 years (Ellington et al., 1988).

Toxicity: Acute oral LD_{50} for rats 2,078 mg/kg, mice 886 mg/kg (RTECS, 1985).

Uses: Chlorinated naphthalenes were formerly used in the production of electric condensers, insulating electric condensers, electric cables and wires; additive for high pressure lubricants.

p-CHLORONITROBENZENE

Synonyms: **1-Chloro-4-nitrobenzene**; 4-Chloronitrobenzene; 4-Chloro-1-nitro-benzene; 4-Nitrochlorobenzene; *p*-Nitrochlorobenzene; PCNB; PNCB; UN 1578.

CAS Registry No.: 100-00-5
DOT: 1578
DOT label: Poison
Molecular formula: $C_6H_4ClNO_2$
Formula weight: 157.56
RTECS: CZ1050000

Physical state, color and odor
Yellow, crystalline solid with a sweet odor.

Melting point (°C):
83.6 (Weast, 1986)

Boiling point (°C):
242 (Weast, 1986)

Density (g/cm^3):
1.520 at 18/4 °C (Verschueren, 1983)

Diffusivity in water (10^5 cm^2/sec):
0.76 at 20 °C using method of Hayduk and Laudie (1974).

Flash point (°C):
127.3 (NIOSH, 1994)

Henry's law constant (10^5 atm·m^3/mol):
9.2 at 20-30 °C (approximate - calculated from water solubility and vapor pressure)

Ionization potential (eV):
9.96 (NIOSH, 1994)

Soil sorption coefficient, log K_{oc}:
K_d = 44 mL/g on a Cs^+-kaolinite (Haderlein and Schwarzenbach, 1993)

Octanol/water partition coefficient, log K$_{ow}$:
2.39 (Fujita et al., 1964)

Solubility in organics:
Soluble in acetone and alcohol (Weast, 1986).

Solubility in water (20 °C):
2,877 μmol/L (Eckert, 1962)
3.40 g/L (Hafkenscheid and Tomlinson, 1983)

Vapor pressure (mmHg):
0.2 at 30 °C (NIOSH, 1994)

Environmental Fate

Biological. Under aerobic conditions, the yeast *Rhodosporidium* sp. metabolized 4-chloronitrobenzene to 4-chloroacetanilide and 4-chloro-2-hydroxyacetanilide as final major metabolites. Intermediate compounds identified include 4-chloronitrosobenzene, 4-chlorophenylhydroxylamine and 4-chloroaniline (Corbett and Corbett, 1981).

Under continuous flow conditions involving feeding, aeration, settling and reflux, a mixture of *p*-chloronitrobenzene and 2,4-dinitrochlorobenzene was reduced 61-70% after 8-13 days by *Arthrobacter simplex*, a microorganism isolated from industrial waste. A similar experiment was conducted using two aeration columns. One column contained *A. simplex*, the other a mixture of *A. simplex* and microorganisms isolated from soil (*Streptomyces coelicolor*, *Fusarium* sp., probably *aquaeductum* and *Trichoderma viride*). After 10 days, 89.5-91% of the nitro compounds was reduced. *p*-Chloronitrobenzene was reduced to 4-chloroaniline and six unidentified compounds (Bielaszczyk et al., 1967).

Photolytic. An aqueous solution containing *p*-chloronitrobenzene and a titanium dioxide (catalyst) suspension was irradiated with UV light (λ >290 nm). 2-Chloro-5-nitrophenol was the only compound identified as a minor degradation product. Continued irradiation caused further degradation yielding carbon dioxide, water, hydrochloric and nitric acids (Hustert et al., 1987).

Irradiation of *p*-chloronitrobenzene in air and nitrogen produced 4-chloro-2-nitrophenol and *p*-chlorophenol, respectively (Kanno and Nojima, 1979).

Exposure Limits (mg/m^3): Potential occupational carcinogen. NIOSH REL: IDLH 100; OSHA PEL: TWA 1.

Symptoms of Exposure: Anoxia, unpleasant taste, anemia (NIOSH, 1994).

Toxicity: Acute oral LD$_{50}$ for mice 650 mg/kg, rats 420 mg/kg; LD$_{50}$ (skin) for rats 16 gm/kg (RTECS, 1985).

Uses: Intermediate for dyes; rubber and agricultural chemicals; manufacture of *p*-nitrophenol.

1-CHLORO-1-NITROPROPANE

Synonyms: Chloronitropropane; Korax; Lanstan.

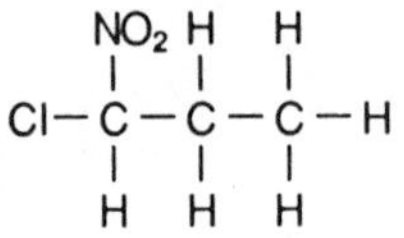

$$
\begin{array}{cccc}
 & NO_2 & H & H \\
 & | & | & | \\
Cl- & C - & C - & C -H \\
 & | & | & | \\
 & H & H & H
\end{array}
$$

CAS Registry No.: 600-25-9
Molecular formula: $C_3H_6ClNO_2$
Formula weight: 123.54
RTECS: TX5075000

Physical state, color and odor
Colorless liquid with an unpleasant odor.

Boiling point (°C):
170.6 at 745 mmHg (Hawley, 1981)

Density (g/cm^3):
1.209 at 20/20 °C (Weast, 1986)

Diffusivity in water (10^5 cm^2/sec):
0.87 at 20 °C using method of Hayduk and Laudie (1974).

Flash point (°C):
62.3 (open cup, NIOSH, 1994)

Henry's law constant (atm·m^3/mol):
0.157 at 20 °C (approximate - calculated from water solubility and vapor pressure)

Ionization potential (eV):
9.90 (NIOSH, 1994)

Soil sorption coefficient, log K_{oc}:
3.34 using method of Chiou et al. (1979).

Octanol/water partition coefficient, log K_{ow}:
4.25 using method of Kenaga and Goring (1980).

Solubility in organics:
Soluble in alcohol and ether (Weast, 1986).

Solubility in water:
6 mg/L at 20 °C (Verschueren, 1983)

Vapor density:
5.05 g/L at 25 °C, 4.26 (air = 1)

Vapor pressure (mmHg):
6 at 25 °C (NIOSH, 1994)

Exposure Limits: NIOSH REL: TWA 2 ppm (10 mg/m^3), IDLH 100 ppm; OSHA PEL: TWA 20 ppm (100 mg/m^3); ACGIH TLV: TWA 2 ppm.

Toxicity: Acute oral LD$_{50}$ for mice 510 mg/kg (RTECS, 1985).

Use: Fungicide.

2-CHLOROPHENOL

Synonyms: 1-Chloro-2-hydroxybenzene; 1-Chloro-*o*-hydroxybenzene; 2-Chloro-hydroxybenzene; *o*-**Chlorophenol**; 1-Hydroxy-2-chlorobenzene; 1-Hydroxy-*o*-chlorobenzene; 2-Hydroxychlorobenzene; RCRA waste number U048; UN 2020; UN 2021.

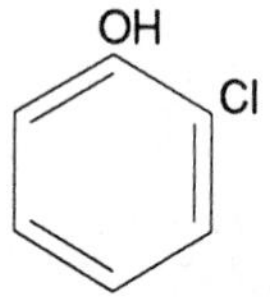

CAS Registry No.: 95-57-8
DOT: 2020 (liquid); 2021 (solid)
DOT label: Poison
Molecular formula: C_6H_5ClO
Formula weight: 128.56
RTECS: SK2625000

Physical state, color and odor
Pale amber liquid with a slight phenolic odor. Odor threshold in air is 20 ppb (Keith and Walters, 1992).

Melting point (°C):
9.00 (Martin et al., 1979)
7.0 (Stull, 1947)

Boiling point (°C):
174.9 (Weast, 1986)
174.5 (Stull, 1947)

Density (g/cm³):
1.2634 at 20/4 °C (Weast, 1986)
1.257 at 25/4 °C (Krijgsheld and van der Gen, 1986)

Diffusivity in water (10^5 cm²/sec):
0.87 at 20 °C using method of Hayduk and Laudie (1974).

Dissociation constant:
8.48 at 25 °C (Dean, 1973)

Flash point (°C):
64 (NFPA, 1984)

Henry's law constant (10^7 atm·m^3/mol):
5.6 at 25 °C (estimated, Howard, 1989)

Ionization potential (eV):
9.28 (Franklin et al., 1969)

Bioconcentration factor, log BCF:
2.33 (bluegill sunfish, Veith et al., 1980)

Soil sorption coefficient, log K_{oc}:
1.71 (Brookstone clay loam, Boyd, 1982)
3.69 (fine sediments), 3.60 (coarse sediments) (Isaacson and Frink, 1984)
2.60 (Meylan et al., 1992)
2.00 (coarse sand), 1.36 (loamy sand) (Kjeldsen et al., 1990)
K_d = <0.1 mL/g on a Cs^+-kaolinite (Haderlein and Schwarzenbach, 1993)

Octanol/water partition coefficient, log K_{ow}:
2.16 (Banerjee et al., 1980; Veith et al., 1980)
2.25 (Menges et al., 1990)
2.19 (Leo et al., 1971)
2.15 (Fujita et al., 1964)

Solubility in organics:
Soluble in ethanol, benzene and ether (Weast, 1986).

Solubility in water:
28,500 mg/L at 20 °C (Verschueren, 1983)
24,650 mg/L at 20 °C (Mulley and Metcalf, 1966)
22,000 mg/L at 25 °C (Roberts et al., 1977)
28,000 mg/L at 25 °C (Morrison and Boyd, 1971)
11,350 mg/L at 25 °C (Banerjee et al., 1980)
0.2 M at 25 °C (Caturla et al., 1988)
2.77 wt % (Nathan, 1978)

Vapor density:
5.25 g/L at 25 °C, 4.44 (air = 1)

Vapor pressure (mmHg at 25 °C):
1.42 (Howard, 1989)
2.25 (Nathan, 1978)

Environmental Fate
  Biological. Chloroperoxidase, a fungal enzyme isolated from *Caldariomyces*

fumago, reacted with 2-chlorophenol yielding traces of 2,4,6-trichlorophenol, 2,4-
and 2,6-dichlorophenols (Wannstedt et al., 1990). When 2-chlorophenol was
statically incubated in the dark at 25 °C with yeast extract and settled domestic
wastewater inoculum, significant biodegradation with rapid adaptation was
observed. At concentrations of 5 and 10 mg/L, 86 and 83% biodegradation
respectively were observed after 7 days (Tabak et al., 1981).

Photolytic. Monochlorophenols exposed to sunlight (UV radiation) produced
catechol and other hydroxybenzenes (Hwang et al., 1986). Titanium dioxide
suspended in an aqueous solution and irradiated with UV light (λ = 365 nm)
converted 2-chlorophenol to carbon dioxide at a significant rate (Matthews, 1986).
In a similar experiment, irradiation of an aqueous solution containing 2-
chlorophenol and titanium dioxide with UV light (λ >340 nm) resulted in the
formation of chlorohydroquinone and trace amounts of catechol. Hydroxylation of
both of these compounds forms the intermediate hydroxyhydroquinone, which
degrades quickly to unidentified carboxylic acids and carbonyl compounds
(D'Oliveira et al., 1990).

Irradiation of an aqueous solution at 296 nm and pH values from 8 to 13 yielded
different products. Photolysis at a pH nearly equal to the dissociation constant
(undissociated form) yielded pyrocatechol. At an elevated pH, 2-chlorophenol is
almost completely ionized; photolysis yielded cyclopentadienic acid (Boule et al.,
1982). Irradiation of an aqueous solution at 296 nm containing hydrogen peroxide
converted 2-chlorophenol to catechol and 2-chlorohydroquinone (Moza et al.,
1988). In the dark, nitric oxide (10^{-3} vol %) reacted with 2-chlorophenol forming
4-nitro-2-chlorophenol and 6-nitro-2-chlorophenol yields of 36 and 30%,
respectively (Kanno and Nojima, 1979).

Chemical/Physical. Wet oxidation of 2-chlorophenol at 320 °C yielded formic
and acetic acids (Randall and Knopp, 1980). Wet oxidation of 2-chlorophenol at
elevated pressure and temperature gave the following products: acetone, acetaldehyde, formic, acetic, maleic, oxalic, muconic and succinic acids (Keen and
Baillod, 1985).

Toxicity: Acute oral LD_{50} for mice 345 mg/kg, rats 670 mg/kg (RTECS, 1985).

Drinking Water Standard: As of May 1994, no MCLGs or MCLs have been
proposed (U.S. EPA, 1994).

Uses: Component of disinfectant formulations; chemical intermediate for phenolic
resins; solvent for polyester fibers, antiseptic (veterinarian); preparation of 4-
nitroso-2-methylphenol and other compounds.

4-CHLOROPHENYL PHENYL ETHER

Synonyms: 4-Chlorodiphenyl ether; *p*-Chlorodiphenyl ether; **1-Chloro-4-phenoxybenzene**; 1-Chloro-*p*-phenoxybenzene; 4-Chlorophenyl ether; *p*-Chlorophenyl ether; *p*-Chlorophenyl phenyl ether; Monochlorodiphenyl oxide.

CAS Registry No.: 7005-72-3
Molecular formula: $C_{12}H_9ClO$
Formula weight: 204.66

Physical state
Liquid.

Melting point (°C):
-8 (U.S. EPA, 1980)

Boiling point (°C):
284-285 (Weast, 1986)

Density (g/cm^3):
1.2026 at 15/4 °C (Weast, 1986)
1.193 at 20/4 °C (Aldrich, 1988)

Diffusivity in water (10^5 cm^2/sec):
0.64 at 20 °C using method of Hayduk and Laudie (1974).

Flash point (°C):
110 (Aldrich, 1988)

Henry's law constant (10^4 atm·m^3/mol):
2.2 (Pankow and Rosen, 1988)

Soil sorption coefficient, log K_{oc}:
3.60 using method of Kenaga and Goring (1980).

Octanol/water partition coefficient, log K_{ow}:
4.08 (Branson, 1978)

Solubility in water:
3.3 mg/L at 25 °C (Branson, 1978)

Vapor density:
8.36 g/L at 25 °C, 7.06 (air = 1)

Vapor pressure (mmHg):
2.7×10^{-3} at 25 °C (calculated, Branson, 1978)

Environmental Fate

Biological. 4-Chlorophenyl phenyl ether (5 and 10 mg/L) did not significantly biodegrade following incubation in settled domestic wastewater inoculum at 25 °C. Percent losses reached a maximum after 2-3 weeks but decreased thereafter suggesting a deadaptive process was occurring (Tabak et al., 1981).

Photolytic. In a methanolic solution irradiated with UV light (λ >290 nm), dechlorination of 4-chlorophenyl phenyl ether resulted in the formation of diphenyl ether (Choudhry et al., 1977). Photolysis of an aqueous solution containing 10% acetonitrile with UV light (λ = 230-400 nm) yielded 4-hydroxybiphenyl ether and chloride ion (Dulin et al., 1986).

Use: Research chemical.

CHLOROPICRIN

Synonyms: Acquinite; Chlor-o-pic; Dolochlor; G 25; Larvacide 100; Microlysin; NA 1583; NCI-C00533; Nitrochloroform; Nitrotrichloromethane; Pic-clor; Picfume; Picride; Profume A; PS; S 1; **Trichloronitromethane**; Tri-clor; UN 1580.

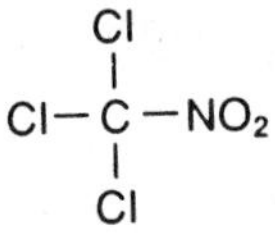

$$Cl-\underset{\underset{Cl}{|}}{\overset{\overset{Cl}{|}}{C}}-NO_2$$

CAS Registry No.: 76-06-2
DOT: 1580
DOT label: Poison
Molecular formula: CCl_3NO_2
Formula weight: 164.38
RTECS: PB6300000

Physical state, color and odor
Colorless to pale yellow, oily liquid with a sharp, penetrating odor. Odor threshold in air is 0.78 ppm (Amoore and Hautala, 1983).

Melting point (°C):
-64.5 (Weast, 1986)
-64 (Worthing and Hance, 1991)

Boiling point (°C):
111.8 (Weast, 1986)

Density (g/cm^3):
1.6558 at 20/4 °C, 1.6483 at 25/4 °C (Windholz et al., 1983)

Diffusivity in water (10^5 cm^2/sec):
0.89 at 20 °C using method of Hayduk and Laudie (1974).

Henry's law constant (10^2 atm·m^3/mol):
8.4 (Kawamoto and Urano, 1989)

Soil sorption coefficient, log K_{oc}:
0.82 using method of Chiou et al. (1979).

Octanol/water partition coefficient, log K_{ow}:
1.03 (Kawamoto and Urano, 1989)

Solubility in organics:
Miscible with benzene, carbon disulfide and ethanol (Windholz et al., 1983).

Solubility in water:
2,270 mg/L at 0 °C (Gunther et al., 1968)
1.621 g/L at 25 °C (Windholz et al., 1983)
2,300 mg/L (Kawamoto and Urano, 1989)

Vapor density:
6.72 g/L at 25 °C, 5.67 (air = 1)

Vapor pressure (mmHg):
16.9 at 20 °C, 33 at 30 °C (Verschueren, 1983)
23.8 at 25 °C (Kawamoto and Urano, 1989)
18.3 at 20 °C (Meister, 1988)

Environmental Fate
   Photolytic. Photodegrades under simulated atmospheric conditions to phosgene and nitrosyl chloride. Photolysis of nitrosyl chloride yields chlorine and nitric oxide (Moilanen et al., 1978).
   Chemical/Physical. Releases very toxic fumes of chlorides and nitrogen oxides when heated to decomposition (Sax and Lewis, 1987).

Exposure Limits: NIOSH REL: TWA 0.1 ppm (0.7 mg/m^3), IDLH 2 ppm; OSHA PEL: 0.1 ppm.

Toxicity: Acute oral LD$_{50}$ for rats 250 mg/kg; LC$_{50}$ (inhalation) for mice 1,500 mg/m^3/10-minute (RTECS, 1985).

Drinking Water Standard: As of May 1994, no MCLGs or MCLs have been proposed although chloropicrin has been listed for regulation (U.S. EPA, 1994).

Uses: Disinfecting cereals and grains; fumigant and soil insecticide; dyestuffs; fungicide; rat exterminator; organic synthesis; war gas.

CHLOROPRENE

Synonyms: Chlorobutadiene; 2-Chlorobutadiene-1,3; 2-Chlorobuta-1,3-diene; 2-Chloro-1,3-butadiene; β-Chloroprene; Neoprene; UN 1991.

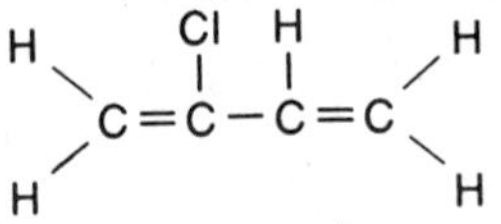

$$H \diagdown \quad \overset{\displaystyle Cl}{\underset{\displaystyle |}{C}} \quad \overset{\displaystyle H}{\underset{\displaystyle |}{} } \qquad H$$

C=C—C=C

CAS Registry No.: 126-99-8
DOT: 1991
DOT label: Flammable liquid
Molecular formula: C_4H_5Cl
Formula weight: 88.54
RTECS: EI9625000

Physical state, color and odor
Colorless liquid with a pungent, ether-like odor.

Melting point (°C):
-103 (NIOSH, 1994)

Boiling point (°C):
59.4 (Weast, 1986)

Density (g/cm^3):
0.9583 at 20/4 °C (Weast, 1986)

Diffusivity in water (10^5 cm^2/sec):
0.93 at 20 °C using method of Hayduk and Laudie (1974).

Flash point (°C):
-20 (NIOSH, 1994)

Lower explosive limit (%):
4.0 (NIOSH, 1994)

Upper explosive limit (%):
20.0 (NIOSH, 1994)

Henry's law constant (10^2 atm·m^3/mol):
3.20 using method of Hine and Mookerjee (1975).

243

Ionization potential (eV):
8.79 (NIOSH, 1994)

Solubility in organics:
Soluble in acetone, benzene and ether (Weast, 1986).

Vapor density:
3.62 g/L at 25 °C, 3.06 (air = 1)

Vapor pressure (mmHg):
188 at 20 °C (NIOSH, 1994)
174 at 25 °C (Boublik et al., 1984)
118 at 10 °C, 275 at 30 °C, 200 at 20 °C (Verschueren, 1983)

Environmental Fate
   Chemical/Physical. Anticipated products from the reaction of chloroprene with ozone or hydroxyl radicals in the atmosphere are formaldehyde, 2-chloroacrolein, OHCCHO, ClCOCHO, $H_2CCHCClO$, chlorohydroxy acids and aldehydes (Cupitt, 1980).
   Chloroprene will polymerize at room temperature unless inhibited with anitioxidants (NIOSH, 1994).

Exposure Limits: Potential occupational carcinogen. NIOSH REL: 15-minute ceiling 1 ppm (3.6 mg/m^3), IDLH 300 ppm; OSHA PEL: TWA 25 ppm (90 mg/m^3); ACGIH TLV: TWA 10 ppm.

Symptoms of Exposure: Irritation of eyes, skin and respiratory system; dermatitis; nervousness (NIOSH, 1994).

Toxicity: Acute oral LD$_{50}$ for mice 260 mg/kg, rats 900 mg/kg (RTECS, 1985).

Use: Manufacture of neoprene.

CHLORPYRIFOS

Synonyms: Brodan; Chlorpyrifos-ethyl; Detmol U.A.; *O,O*-Diethyl-*O*-3,5,6-tri-chloro-2-pyridyl phosphorothioate; Dowco-179; Dursban; Dursban F; ENT 27311; Eradex; Lorsban; NA 2783; OMS-971; **Phosphorothionic acid *O,O*-diethyl *O*-(3,5,6-trichloro-2-pyridyl) ester**; Pyrinex.

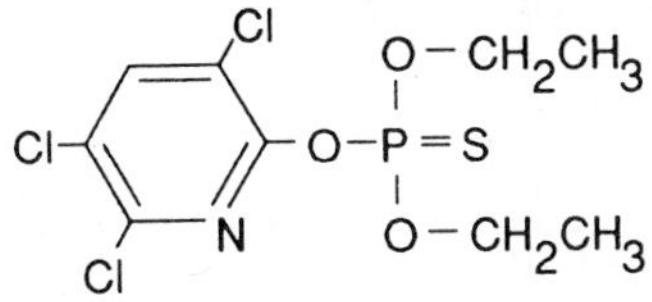

CAS Registry No.: 2921-88-2
DOT: 2783
DOT label: Poison
Molecular formula: $C_9H_{11}Cl_3NO_3PS$
Formula weight: 350.59
RTECS: TF6300000

Physical state, color and odor
Colorless to white granular crystals or amber-colored oil with a mercaptan-like odor.

Melting point (°C):
42.3 (NIOSH, 1994)

Boiling point (°C):
Decomposes at 160 (NIOSH, 1994)

Density (g/cm³):
1.398 at 43.5/4 °C (Verschueren, 1983)

Henry's law constant (10^6 atm·m³/mol):
4.16 at 25 °C (Fendinger and Glotfelty, 1990)

Bioconcentration factor, log BCF:
2.67 (freshwater fish), 2.50 (fish, microcosm) (Garten and Trabalka, 1983)

Soil sorption coefficient, log K_{oc}:
3.86 (Commerce soil), 3.77 (Tracy soil), 3.78 (Catlin soil) (McCall et al., 1981)
4.13 (Kenaga and Goring, 1980)
3.27 (average of 2 soil types, Kanazawa, 1989)
3.34-3.79 (average = 3.66 in 4 sterilized Iowa soils, Felsot and Dahm, 1979)

Octanol/water partition coefficient, log K_{ow}:
5.11 (Chiou et al., 1977)
4.80 (DeKock and Lord, 1987)
5.267 (de Bruijn et al., 1989)

Solubility in organics (kg/kg):
Soluble in acetone (6.5), benzene (7.9), chloroform (6.3) and methanol (0.450) (Worthing and Hance, 1991).

Solubility in water (ppm):
1.12 at 24 °C (Felsot and Dahm, 1979)
0.4 at 23 °C (Chiou et al., 1977)
0.45 at 10 °C, 0.73 at 20 °C, 1.3 at 30 °C (Bowman and Sans, 1985)
0.7 at 19.0 °C (Bowman and Sans, 1979)

Vapor pressure (mmHg):
5.03 x 10^{-5} at 25 °C (Hinckley et al., 1990)

Environmental Fate
   Photolytic. 3,5,6-Trichloro-2-pyridinol formed by the photolysis of chlorpyrifos in water. Continued photolysis yielded chloride ions, carbon dioxide, ammonia and possibly polyhydroxychloropyridines. The following photolytic half-lives in water at north 40° latitude were reported: 31 days during midsummer at a depth of 10^{-3} cm; 345 days during midwinter at a depth of 10^{-3} cm; 43 days at a depth of 1 m; 2.7 years during midsummer at a depth of 1 m in river water (Dilling et al., 1984).
   Chemical/Physical. Hydrolysis products include 3,5,6-trichloro-2-pyridinol, *O*-ethyl *O*-hydrogen-*O*-(3,5,6-trichloro-2-pyridyl)phosphorthioate and *O,O*-di-hydrogen-*O*-(3,5,6-trichloro-2-pyridyl)phosphorothioate. Reported half-lives in buffered distilled water at 25 °C at pH values of 8.1, 6.9 and 4.7 are 22.8, 35.3 and 62.7 days, respectively (Meikle and Youngson, 1978).

Exposure Limits (mg/m^3): NIOSH REL: TWA 0.2, STEL 0.6; ACGIH TLV: TWA 0.2, STEL 0.6.

Toxicity: Acute oral LD_{50} for chickens 25.4 mg/kg, ducks 76 mg/kg, guinea pigs 504 mg/kg, mice 60 mg/kg, quail 13.3 mg/kg, rats 82 mg/kg, rabbits 1,000 mg/kg (RTECS, 1985); LD_{50} (skin) for rats 202 mg/kg, rabbits 2,000 mg/kg; LC_{50} (48-hour) for red killifish 1,350 mg/L (Yoshioka et al., 1986).

Drinking Water Standard: As of May 1994, no MCLGs or MCLs have been proposed (U.S. EPA, 1994).

Use: Insecticide.

CHRYSENE

Synonyms: Benz[*a*]phenanthrene; Benzo[*a*]phenanthrene; Benzo[α]phenanthrene; 1,2-Benzophenanthrene; 1,2-Benzphenanthrene; 1,2-Dibenzonaphthalene; 1,2,5,6-Dibenzonaphthalene; RCRA waste number U050.

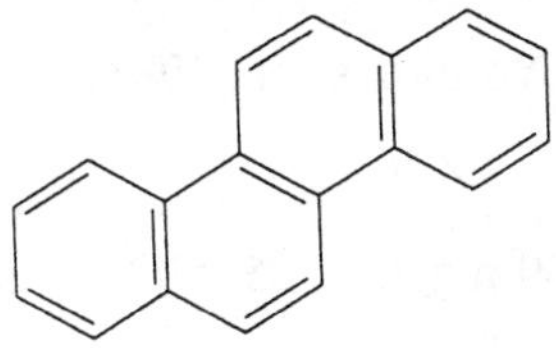

CAS Registry No.: 218-01-9
Molecular formula: $C_{18}H_{12}$
Formula weight: 228.30
RTECS: GC0700000

Physical state
Orthorhombic, bipyramidal plates from benzene exhibiting strong fluorescence under UV light.

Melting point (°C):
255-256 (Weast, 1986)
258.2 (Casellato, 1973)

Boiling point (°C):
448 (Weast, 1986)

Density (g/cm^3):
1.274 at 20/4 °C (Weast, 1986)

Diffusivity in water (10^5 cm^2/sec):
0.63 at 20 °C using method of Hayduk and Laudie (1974).

Dissociation constant:
>15 (Christensen et al., 1975)

Henry's law constant (10^{20} atm·m^3/mol):
7.26 at 20 °C (approximate - calculated from water solubility and vapor pressure)

Ionization potential (eV):
7.85 ± 0.15 (Franklin et al., 1969)

Bioconcentration factor, log BCF:
3.79 (*Daphnia magna*, Newsted and Giesy, 1987)
1.17 (*Polychaete* sp.), 0.79 (*Capitella capitata*) (Bayona et al., 1991)

Soil sorption coefficient, log K$_{oc}$:
6.27 (average, Kayal and Connell, 1990)

Octanol/water partition coefficient, log K$_{ow}$:
5.60 (Mills et al., 1985)
5.91 (Bruggeman et al., 1982; Yoshida et al., 1983)

Solubility in organics:
Soluble in absolute alcohol (769 mg/L at 25 °C) (Windholz et al., 1983).

Solubility in water:
2.0 µg/L at 25 °C (Mackay and Shiu, 1977)
1.8 µg/kg at 25 °C, 2.2 µg/kg at 29 °C (May et al., 1978a)
1.5 µg/L at 27 °C (Davis et al., 1942)
17 µg/L at 24 °C (practical grade, Hollifield, 1979)
6 µg/L at 25 °C (Klevens, 1950)
3.27 µg/L at 25 °C (Walters and Luthy, 1984)
1.6 µg/L at 25 °C (Vadas et al., 1991)
1.02 and µg/L at 25 °C (Billington et al., 1988)
2.1 µg/L at 23 °C (Pinal et al., 1991)

Vapor pressure (10^9 mmHg):
6.3 at 25 °C (Mabey et al., 1982)
630 at 20 °C (Sims et al., 1988)
4.3 at 25 °C (de Kruif, 1980)

Environmental Fate
   Biological. When chrysene was statically incubated in the dark at 25 °C with yeast extract and settled domestic wastewater inoculum, significant biodegradation with varied adaptation rates was observed. At concentrations of 5 and 10 mg/L, 59 and 38% biodegradation, respectively, were observed after 28 days (Tabak et al., 1981).
   Photolytic. Based on structurally related compounds, chrysene may undergo photolysis to yield quinones (U.S. EPA, 1985) and/or hydroxy derivatives (Nielsen et al., 1983).
   Chemical/Physical. A monochlorochrysene product was formed during the chlorination of chrysene in aqueous solutions at pH 4. The reported half-lives at chlorine concentrations of 0.6 and 10 mg/L were >24 and 0.45 hours, respectively (Mori et al., 1991).

Exposure Limits: Potential occupational carcinogen. No individual standards have been set, however, as a constituent in coal tar pitch volatiles, the following exposure limits have been established (mg/m^3): NIOSH REL: TWA 0.1 (cyclohexane-extractable fraction), IDLH 80; OSHA PEL: TWA 0.2 (benzene-soluble fraction); ACGIH TLV: TWA 0.2 (benzene solubles).

Drinking Water Standard (proposed): MCLG: zero; MCL: 0.2 μg/L (U.S. EPA, 1984).

Use: Organic synthesis. Derived from industrial and experimental coal gasification operations where maximum concentration detected in gas, liquid and coal tar streams were 7.3, 0.16 and 8.6 mg/m^3, respectively (Cleland, 1981).

CROTONALDEHYDE

Synonyms: 2-Butenal; Crotenaldehyde; Crotonal; Crotonic aldehyde; **1,2-Ethane-diol dipropanoate**; Ethylene glycol dipropionate; Ethylene dipropionate; Ethylene propionate; β-Methylacrolein; NCI-C56279; Propylene aldehyde; RCRA waste number U053; Topanel.

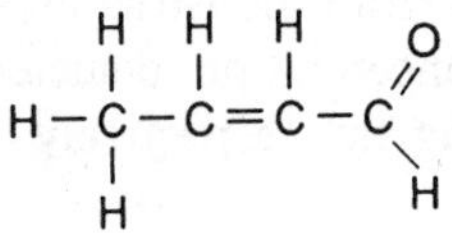

CAS Registry No.: 4170-30-3
DOT: 1143
DOT label: Flammable liquid
Molecular formula: C_4H_6O
Formula weight: 70.09
RTECS: GP9499000

Physical state, color and odor
Clear, colorless to straw-colored liquid with a pungent, irritating, suffocating odor. Odor threshold in air is 7 ppm (Keith and Walters, 1992).

Melting point (°C):
-74 (Weast, 1986)
-76.5 (Windholz et al., 1983)

Boiling point (°C):
104-105 (Weast, 1986)
99 (Verschueren, 1983)

Density (g/cm³):
0.853 at 20/20 °C (Weast, 1986)
0.8477 at 20.5/4 °C (Huntress and Mulliken, 1941)

Diffusivity in water (10^5 cm²/sec):
0.99 at 20 °C using method of Hayduk and Laudie (1974).

Flash point (°C):
7.2 (NIOSH, 1994)
13 (open cup, Windholz et al., 1983)

Lower explosive limit (%):
2.1 (NIOSH, 1994)

Upper explosive limit (%):
15.5 (NIOSH, 1994)

Henry's law constant (10^5 atm·m^3/mol):
1.96 (Gaffney et al., 1987)

Ionization potential (eV):
9.73 ± 0.01 (Franklin et al., 1969)

Soil sorption coefficient, log K_{oc}:
Unavailable because experimental methods for estimation of this parameter for aldehydes are lacking in the documented literature. However, its high solubility in water suggests its adsorption to soil will be nominal (Lyman et al., 1982).

Octanol/water partition coefficient, log K_{ow}:
Unavailable because experimental methods for estimation of this parameter for aliphatic aldehydes are lacking in the documented literature.

Solubility in organics:
Miscible with alcohol, benzene, gasoline, kerosene, solvent naphtha and toluene (Hawley, 1981).

Solubility in water (wt %):
19.2 at 5 °C, 18.1 at 20 °C (Windholz et al., 1983)

Vapor density:
2.86 g/L at 25 °C, 2.41 (air = 1)

Vapor pressure (mmHg):
19 at 20 °C (NIOSH, 1994)

Environmental Fate
   Chemical/Physical. Slowly oxidizes in air forming crotonic acid (Windholz et al., 1983). At elevated temperatures, crotonaldehyde may polymerize (NIOSH, 1994).

Exposure Limits: NIOSH REL: TWA 2 ppm (6 mg/m^3), IDLH 50 ppm; OSHA PEL: TWA 2 ppm; ACGIH TLV: TWA 2 ppm.

Symptoms of Exposure: Irritation of eyes and respiratory system (NIOSH, 1994).

Toxicity: Acute oral LD_{50} for mice 104 mg/kg, rats 206 mg/kg (RTECS, 1985).

Uses: Preparation of 1-butanol, butyraldehyde, 2-ethylhexanol, quinaldine;

chemical warfare; insecticides; leather tanning; alcohol denaturant; solvent; warning agent in fuel gases; purification of lubricating oils; organic synthesis.

CYCLOHEPTANE

Synonyms: Heptamethylene; Suberane; UN 2241.

CAS Registry No.: 291-64-5
DOT: 2241
Molecular formula: C_7H_{14}
Formula weight: 98.19
RTECS: GU3140000

Physical state
Oily liquid.

Melting point (°C):
-12 (Weast, 1986)

Boiling point (°C):
118.5 (Weast, 1986)

Density (g/cm^3 at 20/4 °C):
0.8098 (Weast, 1986)
0.8011 (Riddick et al., 1986)

Diffusivity in water (10^5 cm^2/sec):
0.79 at 20 °C using method of Hayduk and Laudie (1974).

Dissociation constant:
>14 (Schwarzenbach et al., 1993)

Flash point (°C):
15 (Sax and Lewis, 1987)
6 (Aldrich, 1988)

Lower explosive limit (%):
1.1 (NFPA, 1984)

Upper explosive limit (%):
6.7 (NFPA, 1984)

Heat of fusion (kcal/mol):
0.450 (Dean, 1987)

Henry's law constant (10^2 atm·m^3/mol):
9.35 (approximate – calculated from water solubility and vapor pressure)

Soil sorption coefficient, log K_{oc}:
Unavailable because experimental methods for estimation of this parameter for alicyclic hydrocarbons are lacking in the documented literature.

Octanol/water partition coefficient, log K_{ow}:
2.64 using method of Hansch et al. (1968).

Solubility in organics:
Soluble in alcohol, benzene, chloroform, ether and lignoin (Weast, 1986).

Solubility in water:
30 mg/kg at 25 °C (McAuliffe, 1966)
27.1 mg/L at 30 °C, 17.0 mg/L in artificial seawater (34.5 parts NaCl per 1,000 parts water) at 30 °C (Groves, 1988)

Vapor density:
4.01 g/L at 25 °C, 3.39 (air = 1)

Vapor pressure (mmHg):
21.7 (extrapolated, Boublik et al., 1986)

Environmental Fate
Biological. Cycloheptane may be oxidized by microbes to cycloheptanol, which may further oxidize to give cycloheptanone (Dugan, 1972).

Uses: Organic synthesis; gasoline component.

CYCLOHEXANE

Synonyms: Benzene hexahydride; Hexahydrobenzene; Hexamethylene; Hexa-naphthene; RCRA waste number U056; UN 1145.

CAS Registry No.: 110-82-7
DOT: 1145
DOT label: Flammable liquid
Molecular formula: C_6H_{12}
Formula weight: 84.16
RTECS: GU6300000

Physical state, color and odor
Colorless liquid with a sweet, chloroform-like odor. Odor threshold in air is 410 ppb (Keith and Walters, 1992).

Melting point (°C):
6.5 (Weast, 1986)
6.6 (Stull, 1947)

Boiling point (°C):
80.7 (Weast, 1986)

Density (g/cm³):
0.7785 at 20/4 °C (Weast, 1986)

Diffusivity in water (10^5 cm²/sec):
0.84 at 20 °C (Witherspoon and Bonoli, 1969)
0.90 at 25 °C (Hayduk and Laudie, 1974)

Dissociation constant:
≈ 45 (Gordon and Ford, 1972)

Flash point (°C):
-18 (NIOSH, 1994)

Lower explosive limit (%):
1.3 (NIOSH, 1994)

Upper explosive limit (%):
8.0 (NFPA, 1984)

Heat of fusion (kcal/mol):
0.640 (Dean, 1987)

Henry's law constant (atm·m^3/mol):
0.194 at 25 °C (Hine and Mookerjee, 1975)
0.103, 0.126, 0.140, 0.177 and 0.223 at 10, 15, 20, 25 and 30 °C, respectively (Ashworth et al., 1988)

Interfacial tension with water (dyn/cm at 25 °C):
50.0 (Donahue and Bartell, 1952)

Ionization potential (eV):
9.88 (Franklin et al., 1969)
11.00 (Yoshida et al., 1983)

Soil sorption coefficient, log K_{oc}:
Unavailable because experimental methods for estimation of this parameter for alicyclic hydrocarbons are lacking in the documented literature.

Octanol/water partition coefficient, log K_{ow}:
3.44 (Yoshida et al., 1983)

Solubility in organics:
Miscible with acetone, benzene, chloroform, carbon tetrachloride, ethanol and ethyl ether (Windholz et al., 1983).
In methanol, g/L: 344 at 15 °C, 384 at 20 °C, 435 at 25 °C, 503 at 30 °C, 600 at 35 °C, 740 at 40 °C (Kiser et al., 1961).

Solubility in water:
66.5 mg/kg at 25 °C (Price, 1976)
55.0 mg/kg at 25 °C (McAuliffe, 1963, 1966)
57.5 mg/L at 25 °C (Mackay and Shiu, 1975)
52 mg/kg at 23.5 °C (Schwarz, 1980)
At 25 °C: 58.4 mg/L, 40.1 mg/L in 3.3% NaCl solution (Groves, 1988)
100 mg/L at 20 °C (Korenman and Aref'eva, 1977)
80 mg/L at 25 °C (McBain and Lissant, 1951)
In mg/kg: 0.008 at 25 °C, 0.017 at 56 °C, 0.028 at 94 °C, 0.0517 at 127 °C, 0.146 at 162 °C, 1.785 at 220.5 °C (Guseva and Parnov, 1963a).
56.7 mg/L at 25 °C (Leinonen and Mackay, 1973)
627 μmmol/L at 25 °C (Sanemasa et al., 1987)

Vapor density:
3.44 g/L at 25 °C, 2.91 (air = 1)

Vapor pressure (mmHg):
78 at 20 °C (NIOSH, 1994)
97.61 at 25 °C (Wilhoit and Zwolinski, 1971)

Environmental Fate
  Biological. Microbial degradation products reported include cyclohexanol (Dugan, 1972; Verschueren, 1983), 1-oxa-2-oxocycloheptane, 6-hydroxy-heptanoate, 6-oxohexanoate, adipic acid, acetyl-CoA, succinyl-CoA (Verschueren, 1983) and cyclohexanone (Dugan, 1972; Keck et al., 1989).

Exposure Limits: NIOSH REL: TWA 300 ppm (1,050 mg/m^3), IDLH 1,300 ppm; OSHA PEL: TWA 300 ppm; ACGIH TLV: TWA 300 ppm.

Symptoms of Exposure: Irritation of the eyes and respiratory system (Patnaik, 1992).

Toxicity: Acute oral LD_{50} for mice 813 mg/kg, rats 12,705 mg/kg (RTECS, 1985).

Uses: Manufacture of nylon; solvent for cellulose ethers, fats, oils, waxes, resins, bitumens, crude rubber; paint and varnish removers; extracting essential oils; glass substitutes; solid fuels; fungicides; gasoline and coal tar component; organic synthesis.

CYCLOHEXANOL

Synonyms: Adronal; Anol; Cyclohexyl alcohol; Hexahydrophenol; Hexalin; Hydralin; Hydrophenol; Hydroxycyclohexane; Naxol.

CAS Registry No.: 108-93-0
Molecular formula: $C_6H_{12}O$
Formula weight: 100.16
RTECS: GV7875000

Physical state, color and odor
Colorless to pale yellow, viscous, hygroscopic liquid with a camphor-like odor. Odor threshold in air is 0.15 ppm (Amoore and Hautala, 1983).

Melting point (°C):
25.1 (Weast, 1986)
23 (Hawley, 1981)
20-22 (Aldrich, 1988)

Boiling point (°C):
161.1 (Weast, 1986)

Density (g/cm³):
0.9624 at 20/4 °C (Weast, 1986)
0.9449 at 25/4 °C (Sax and Lewis, 1987)
0.937 at 37/4 °C (Hawley, 1981)

Diffusivity in water (10^5 cm²/sec):
0.86 at 20 °C using method of Hayduk and Laudie (1974).

Dissociation constant:
>14 (Schwarzenbach et al., 1993)

Flash point (°C):
67.8 (NIOSH, 1994)

Heat of fusion (kcal/mol):
0.406 (Dean, 1987)

Henry's law constant (10^6 atm·m^3/mol):
5.74 at 25 °C (Hine and Mookerjee, 1975)

Interfacial tension with water (dyn/cm at 16.2 °C):
3.92 (Demond and Lindner, 1993)

Ionization potential (eV):
10.00 (NIOSH, 1994)

Soil sorption coefficient, log K_{oc}:
Unavailable because experimental methods for estimation of this parameter for alicyclic alcohols are lacking in the documented literature.

Octanol/water partition coefficient, log K_{ow}:
1.23 (Hansch and Anderson, 1967)

Solubility in organics:
Miscible with aromatic hydrocarbons, ethanol, ethyl acetate, linseed oil and petroleum solvents (Windholz et al., 1983).

Solubility in water:
56,700 mg/L at 15 °C, 36,000 mg/L at 20 °C (Verschueren, 1983)
3.82 wt % at 20 °C (Palit, 1947)
43 g/kg at 30 °C (Patnaik, 1992)

Vapor density:
4.09 g/L at 25 °C, 3.46 (air = 1)

Vapor pressure (mmHg):
1 at 20 °C (NIOSH, 1994)
3.5 at 34 °C (Verschueren, 1983)

Environmental Fate
   Biological. Reported biodegradation products include cyclohexanone (Dugan, 1972; Verschueren, 1983), 2-hydroxyhexanone, 1-oxa-2-oxocycloheptane, 6-hydroxyheptanonate, 6-oxohexanoate and adipate (Verschueren, 1983).

Exposure Limits: NIOSH REL: TWA 50 ppm (200 mg/m^3), IDLH 400 ppm; OSHA PEL: TWA 50 ppm; ACGIH TLV: TWA 50 ppm.

Symptoms of Exposure: May irritate eyes, nose and throat. Ingestion can cause nausea, trembling, gastrointestinal disturbances (Patnaik, 1992) and narcosis (NIOSH, 1994).

Toxicity: Acute oral LD_{50} for rats 2,060 mg/kg; LD_{50} (subcutaneous) for mice 2,480 mg/kg (RTECS, 1985).

Uses: In paint and varnish removers; solvent for lacquers, shellacs and resins; manufacture of adipic acid, caprolactum, benzene, cyclohexene, chlorocyclohexane, cyclohexanone, nitrocyclohexane and solid fuel for camp stoves; fungicidal formulations; polishes; plasticizers; soap and detergent manufacturing (stabilizer); emulsified products; blending agent; recrystallizing steroids; germicides; plastics; organic synthesis.

CYCLOHEXANONE

Synonyms: Anone; Cyclohexyl ketone; Hexanon; Hytrol O; Ketohexamethylene; Nadone; NCI-C55005; Pimelic ketone; Pimelin ketone; RCRA waste number U057; Sextone; UN 1915.

CAS Registry No.: 108-94-1
DOT: 1915
DOT label: Flammable liquid
Molecular formula: $C_6H_{10}O$
Formula weight: 98.14
RTECS: GW1050000

Physical state, color and odor
Clear, colorless to pale yellow, oily liquid with a peppermint-like odor. Odor threshold in air is 120 ppb (Keith and Walters, 1992).

Melting point (°C):
-16.4 (Weast, 1986)
-32.1 (Windholz et al., 1983)
-26 (Verschueren, 1983)
-47 (Aldrich, 1988)

Boiling point (°C):
155.6 (Weast, 1986)

Density (g/cm³):
0.9478 at 20/4 °C (Weast, 1986)
0.9421 at 25/4 °C (Windholz et al., 1983)

Diffusivity in water (10^5 cm²/sec):
0.87 at 20 °C using method of Hayduk and Laudie (1974).

Flash point (°C):
35.2 (NIOSH, 1994)
46 (Aldrich, 1988)

Lower explosive limit (%):
1.1 at 100 °C (NIOSH, 1994)

Upper explosive limit (%):
9.4 (NFPA, 1984)

Henry's law constant (10^5 atm·m^3/mol):
1.2 at 25 °C (Hawthorne et al., 1985)

Ionization potential (eV):
9.14 ± 0.01 (Franklin et al., 1969)

Soil sorption coefficient, log K_{oc}:
Unavailable because experimental methods for estimation of this parameter for alicyclic ketones are lacking in the documented literature. However, its high solubility in water suggests its adsorption to soil will be nominal (Lyman et al., 1982).

Octanol/water partition coefficient, log K_{ow}:
0.81 (Leo et al., 1971)

Solubility in organics:
Soluble in acetone, alcohol, benzene, chloroform and ether (Weast, 1986).

Solubility in water (g/L):
150 at 10 °C, 50 at 30 °C (Windholz et al., 1983)
23 g/L at 20 °C, 24 g/L at 31 °C (Verschueren, 1983)

Vapor density:
4.01 g/L at 25 °C, 3.39 (air = 1)

Vapor pressure (mmHg):
4 at 20 °C, 6.2 at 30 °C (Verschueren, 1983)
5 at 20 °C (NIOSH, 1994)
4.5 at 25 °C (Banerjee et al., 1990)

Exposure Limits: NIOSH REL: TWA 25 ppm (100 mg/m^3), IDLH 700 ppm; OSHA PEL: TWA 50 ppm (200 mg/m^3); ACGIH TLV: TWA 25 ppm.

Symptoms of Exposure: May irritate eyes and throat. Contact with eyes may cause cornea damage (Patnaik, 1992).

Toxicity: Acute oral LD$_{50}$ for mice 1,400 mg/kg, rats 1,535 mg/kg (RTECS, 1985).

Uses: Solvent for cellulose acetate, crude rubber, natural resins, nitrocellulose, vinyl resins, waxes, fats, oils, shellac, rubber, DDT and other pesticides; preparation of adipic acid and caprolactum; wood stains; paint and varnish removers; degreasing of metals; spot remover; lube oil additive; leveling agent in dyeing and delustering silk.

CYCLOHEXENE

Synonyms: Benzene tetrahydride; Tetrahydrobenzene; 1,2,3,4-Tetrahydrobenzene; UN 2256.

CAS Registry No.: 110-83-8
DOT: 2256
DOT label: Flammable liquid
Molecular formula: C_6H_{10}
Formula weight: 82.15
RTECS: GW2500000

Physical state, color and odor
Colorless liquid with a sweet odor. Odor threshold in air is 0.18 ppm (Amoore and Hautala, 1983).

Melting point (°C):
-103.5 (Weast, 1986)

Boiling point (°C):
83 (Weast, 1986)

Density (g/cm³):
0.8102 at 20/4 °C (Weast, 1986)
0.7823 at 50/4 °C (Windholz et al., 1983)
0.81096 at 20/4 °C, 0.80609 at 25/4 °C (Dreisbach, 1959)

Diffusivity in water (10^5 cm²/sec):
0.88 at 20 °C using method of Hayduk and Laudie (1974).

Dissociation constant:
>14 (Schwarzenbach et al., 1993)

Flash point (°C):
-11.7 (NIOSH, 1994)

Heat of fusion (kcal/mol):
0.787 (Dean, 1987)

Henry's law constant (10^2 atm·m^3/mol):
4.6 at 25 °C (Hine and Mookerjee, 1975)

Ionization potential (eV):
8.72 (Franklin et al., 1969)
8.945 (HNU, 1986)
9.20 (Rav-Acha et al., 1987)

Soil sorption coefficient, log K_{oc}:
Unavailable because experimental methods for estimation of this parameter for alicyclic hydrocarbons are lacking in the documented literature.

Octanol/water partition coefficient, log K_{ow}:
2.86 (Hansch and Leo, 1979)

Solubility in organics:
Soluble in acetone, alcohol, benzene and ether (Weast, 1986).

Solubility in water:
130 mg/L at 25 °C (McBain and Lissant, 1951)
213 mg/kg at 25 °C (McAuliffe, 1966)
281, 286 mg/kg at 23.5 °C (Schwarz, 1980)
In 1 mM nitric acid: 4.95, 3.46 and 2.49 mmol/L at 30, 35 and 40 °C, respectively
    (Natarajan and Venkatachalam, 1972).

Vapor density:
3.36 g/L at 25 °C, 2.84 (air = 1)

Vapor pressure (mmHg):
67 at 20 °C (NIOSH, 1994)
88.8 at 25 °C (estimated using Antoine equation, Dreisbach, 1959)

Environmental Fate
   Biological. Cyclohexene reportedly biodegrades to cyclohexanone (Dugan, 1972; Verschueren, 1983).
   Chemical/Physical. Gaseous products formed from the reaction of cyclohexene with ozone were (% yield): formic acid (12), carbon monoxide (18), carbon dioxide (42), ethylene (1) and valeraldehyde (17) (Hatakeyama et al., 1987).
   Cyclohexene reacts with chlorine dioxide in water forming 2-cyclohexen-1-one (Rav-Acha et al., 1987).

Exposure Limits: NIOSH REL: TWA 300 ppm (1,015 mg/m^3), IDLH 2,000 ppm; OSHA PEL: TWA 300 ppm; ACGIH TLV: TWA 300 ppm.

Symptoms of Exposure: Irritation of eyes, skin and respiratory tract. Inhalation of high concentrations may cause drowsiness (NIOSH, 1994; Patnaik, 1992).

Uses: Manufacture of adipic acid, hexahydrobenzoic acid, maleic acid, 1,3-butadiene; catalyst solvent; oil extraction; component of coal tar; stabilizer for high octane gasoline; organic synthesis.

CYCLOPENTADIENE

Synonyms: 1,3-Cyclopentadiene; Pentole; R-pentine; Pyropentylene.

CAS Registry No.: 542-92-7
Molecular formula: C_5H_6
Formula weight: 66.10
RTECS: GY1000000

Physical state, color and odor
Colorless liquid with a turpentine-like odor. Odor threshold in air is 1.9 ppm (Amoore and Hautala, 1983).

Melting point (°C):
-97.2 (Weast, 1986)
-85 (Windholz et al., 1983)

Boiling point (°C):
40.0 (Weast, 1986)
41.5-42 (Windholz et al., 1983)

Density (g/cm³):
0.8021 at 20/4 °C (Weast, 1986)
0.7966 at 25/4 °C (Windholz et al., 1983)

Diffusivity in water (10^5 cm²/sec):
0.99 at 20 °C using method of Hayduk and Laudie (1974).

Dissociation constant:
15 (Gordon and Ford, 1972)
16.0 (Streitwieser and Nebenzahl, 1976)

Flash point (°C):
25.0 (open cup, NIOSH, 1994)

Henry's law constant (10^2 atm·m³/mol):
5.1 at 25 °C (approximate - calculated from water solubility and vapor pressure)

Ionization potential (eV):
8.56 (NIOSH, 1994)
8.97 (Franklin et al., 1969)

Soil sorption coefficient, log K_{oc}:
Unavailable because experimental methods for estimation of this parameter for alicyclic hydrocarbons are lacking in the documented literature.

Octanol/water partition coefficient, log K_{ow}:
2.34 using method of Hansch et al. (1968).

Solubility in organics:
Miscible with acetone, benzene, carbon tetrachloride and ether. Soluble in acetic acid, aniline and carbon disulfide (Windholz et al., 1983).

Solubility in water:
10.3 mmol/L at 20-25 °C (Streitwieser and Nebenzahl, 1976)

Vapor pressure (mmHg):
400 at 20 °C (NIOSH, 1994)

Vapor density:
2.70 g/L at 25 °C, 2.28 (air = 1)

Environmental Fate
   Biological. Cyclopentadiene may be oxidized by microbes to cyclopentanone (Dugan, 1972).
   Chemical/Physical. Dimerizes to dicyclopentadiene on standing (NIOSH, 1994; Windholz et al., 1983).

Exposure Limits: NIOSH REL: TWA 75 ppm (200 mg/m³), IDLH 750 ppm; OSHA PEL: TWA 75 ppm; ACGIH TLV: TWA 75 ppm.

Symptoms of Exposure: Irritation of eyes and nose (Patnaik, 1992).

Toxicity: Acute oral LD_{50} (dimeric form) for rats is 820 mg/kg (Patnaik, 1992).

Uses: Manufacture of resins; chlorinated insecticides; organic synthesis (Diels-Alder reaction).

CYCLOPENTANE

Synonyms: Pentamethylene; UN 1146.

CAS Registry No.: 287-92-3
DOT: 1146
DOT label: Flammable liquid
Molecular formula: C_5H_{10}
Formula weight: 70.13
RTECS: GY2390000

Physical state and color
Colorless, mobile liquid.

Melting point (°C):
-93.9 (Weast, 1986)
-95 (Huntress and Mulliken, 1941)

Boiling point (°C):
49.2 (Weast, 1986)

Density (g/cm³):
0.7457 at 20/4 °C (Weast, 1986)
0.74059 at 15/4 °C (Huntress and Mulliken, 1941)
0.74394 at 25/4 °C (Riddick et al., 1986)

Diffusivity in water (10^5 cm²/sec):
0.93 at 20 °C (Witherspoon and Bonoli, 1969)

Dissociation constant:
≈ 44 (Gordon and Ford, 1972)

Flash point (°C):
-7 (Sax and Lewis, 1987)
-37.2 (NIOSH, 1994)

Lower explosive limit (%):
1.1 (NIOSH, 1994)

Upper explosive limit (%):
8.7% (NIOSH, 1994)

Heat of fusion (kcal/mol):
0.1455 (Dean, 1987)

Henry's law constant (atm·m^3/mol):
0.186 at 25 °C (Hine and Mookerjee, 1975)

Ionization potential (eV):
10.53 ± 0.05 (Franklin et al., 1969)

Soil sorption coefficient, log K_{oc}:
Unavailable because experimental methods for estimation of this parameter for alicyclic hydrocarbons are lacking in the documented literature.

Octanol/water partition coefficient, log K_{ow}:
3.00 (Leo et al., 1975)

Solubility in organics:
Miscible with ether and other hydrocarbon solvents (Windholz et al., 1983).
In methanol, g/L: 680 at 5 °C, 860 at 10 °C, 1,400 at 15 °C. Miscible at higher temperatures (Kiser et al., 1961).

Solubility in water:
In mg/kg: 160 at 25 °C, 163 at 40.1 °C, 180 at 55.7 °C, 296 at 99.1 °C, 372 at 118.0 °C, 611 at 137.3 °C, 792 at 153.1 °C (Price, 1976)
156 mg/kg at 25 °C (McAuliffe, 1963, 1966)
164 mg/L at 25 °C, 128 mg/L in artificial seawater (34.5 parts NaCl per 1,000 parts water) at 25 °C (Groves, 1988)

Vapor density:
2.87 g/L at 25 °C, 2.42 (air = 1)

Vapor pressure (mmHg):
400 at 31.0 °C (estimated, Weast, 1986)

Environmental Fate
   Biological. Cyclopentane may be oxidized by microbes to cyclopentanol, which may further oxidize to cyclopentanone (Dugan, 1972).

Exposure Limits: NIOSH REL: 600 ppm (1,720 mg/m^3); ACGIH TLV: TWA 600 ppm.

Symptoms of Exposure: Exposure to high concentrations may produce depression of central nervous system. Symptoms include excitement, loss of equilibrium, stupor and coma (Patnaik, 1992).

Uses: Solvent for cellulose ethers and paints; azeotropic distillation agent; motor fuel; extractions of fats and wax; shoe industry; organic synthesis.

CYCLOPENTENE

Synonym: UN 2246.

CAS Registry No.: 142-29-0
DOT: 2246
Molecular formula: C_5H_8
Formula weight: 68.12
RTECS: GY5950000

Physical state and color
Colorless liquid.

Melting point (°C):
-135 (Weast, 1986)
-135.08 (Dreisbach, 1959)

Boiling point (°C):
44.2 (Weast, 1986)

Density (g/cm³):
0.77199 at 20/4 °C, 0.76653 at 25/4 °C (Dreisbach, 1959)

Diffusivity in water (10^5 cm²/sec):
0.95 at 20 °C using method of Hayduk and Laudie (1974).

Dissociation constant:
>14 (Schwarzenbach et al., 1993)

Flash point (°C):
-28.9 (Sax and Lewis, 1987)
-34 (Aldrich, 1988)

Heat of fusion (kcal/mol):
0.804 (Dean, 1987)

Henry's law constant (10^2 atm·m³/mol):
6.3 at 25 °C (Hine and Mookerjee, 1975)

Ionization potential (eV):
9.01 ± 0.01 (Franklin et al., 1969)

Soil sorption coefficient, log K_{oc}:
Unavailable because experimental methods for estimation of this parameter for alicyclic hydrocarbons are lacking in the documented literature.

Octanol/water partition coefficient, log K_{ow}:
2.45 using method of Hansch et al. (1968).

Solubility in organics:
Soluble in alcohol, benzene, ether and petroleum (Weast, 1986).

Solubility in water:
535 mg/kg at 25 °C (McAuliffe, 1966)
In 1 mM nitric acid: 9.21, 8.97 and 8.71 mmol/L at 30, 35 and 40 °C, respectively (Natarajan and Venkatachalam, 1972)

Vapor density:
2.78 g/L at 25 °C, 2.35 (air = 1)

Vapor pressure (mmHg):
380 at 25 °C (estimated using Antoine equation, Dreisbach, 1959)

Environmental Fate
   Biological. Cyclopentene may be oxidized by microbes to cyclopentanol, which may further oxidize to cycloheptanone (Dugan, 1972).
   Chemical/Physical. Gaseous products formed from the reaction of cyclopentene with ozone were (% yield): formic acid (11), carbon monoxide (35), carbon dioxide (42), ethylene (12), formaldehyde (13) and butyraldehyde (11). Particulate products identified include succinic acid, glutaraldehyde, 5-oxopentanoic acid and glutaric acid (Hatakeyama et al., 1987).

Toxicity: Acute oral LD_{50} for rats is 1,656 mg/kg (RTECS, 1985).

Uses: Cross-linking agent; organic synthesis.

2,4-D

Synonyms: Agrotect; Amidox; Amoxone; Aqua-kleen; BH 2,4-D; Brush-rhap; B-Selektonon; Chipco turf herbicide "D"; Chloroxone; Crop rider; Crotilin; D 50; 2,4-D acid; Dacamine; Debroussaillant 600; Decamine; Ded-weed; Ded-weed LV-69; Desormone; Dichlorophenoxyacetic acid; **(2,4-Dichlorophenoxy)acetic acid**; Dicopur; Dicotox; Dinoxol; DMA-4; Dormone; Emulsamine BK; Emulsamine E-3; ENT 8538; Envert 171; Envert DT; Esteron; Esteron 76 BE; Esteron 44 weed killer; Esteron 99; Esteron 99 concentrate; Esteron brush killer; Esterone 4; Estone; Farmco; Fernesta; Fernimine; Fernoxone; Ferxone; Foredex 75; Formula 40; Hedonal; Herbidal; Ipaner; Krotiline; Lawn-keep; Macrondray; Miracle; Monosan; Moxone; NA 2765; Netagrone; Netagrone 600; NSC 423; Pennamine; Pennamine D; Phenox; Pielik; Planotox; Plantgard; RCRA waste number U240; Rhodia; Salvo; Spritz-hormin/2,4-D; Spritz-hormit/2,4-D; Super D weedone; Transamine; Tributon; Trinoxol; U 46; U-5043; U 46DP; Vergemaster; Verton; Verton D; Verton 2D; Vertron 2D; Vidon 638; Visko-rhap; Visko-rhap drift herbicides; Visko-rhap low volatile 4L; Weedar; Weddar-64; Weddatul; Weed-b-gon; Weedez wonder bar; Weedone; Weedone LV4; Weed-rhap; Weed tox; Weedtrol.

CAS Registry No.: 94-75-7
DOT: 2765
DOT label: Poison
Molecular formula: $C_8H_6Cl_2O_3$
Formula weight: 221.04
RTECS: AG6825000

Physical state, color and odor
Odorless, white to pale yellow, powder or prismatic crystals.

Melting point (°C):
140-141 (Weast, 1986)
140.5 (Worthing and Hance, 1991)
138.2-138.8 (Crosby and Tutass, 1966)
136 (Riederer, 1990)
138 (Windholz et al., 1983)

Boiling point (°C):
160 at 0.4 mmHg (Weast, 1986)

Density (g/cm^3):
1.416 at 25/4 °C (Verschueren, 1983)
1.57 at 30/4 °C (Bailey and White, 1965)

Diffusivity in water (10^5 cm^2/sec):
0.57 at 20 °C using method of Hayduk and Laudie (1974).

Dissociation constant:
2.73 (Nelson and Faust, 1969)
2.64, 2.80, 3.22 at 60 °C (Bailey and White, 1965)
2.90 (Jafvert et al., 1990)

Flash point:
Noncombustible solid (NIOSH, 1994)

Henry's law constant (10^2 atm·m^3/mol):
1.95 at 20 °C (approximate - calculated from water solubility and vapor pressure)

Bioconcentration factor, log BCF:
0.78 (algae, Geyer et al., 1984)
1.30 (activated sludge, Freitag et al., 1985)
0.00 (fish, microcosm) (Garten and Trabalka, 1983)

Soil sorption coefficient, log K_{oc}:
1.68 (Commerce soil), 1.88 (Tracy soil), 1.76 (Catlin soil) (McCall et al., 1981)
1.30 (includes salts, Kenaga and Goring, 1980)
2.04-2.35 (Hodson and Williams, 1988)
1.70-2.73 (average = 2.18 for 10 Danish soils, Løkke, 1984)
2.05-2.16 (average = 2.11 for 3 soil types, Rippen et al., 1982)

Octanol/water partition coefficient, log K_{ow}:
2.81 (Leo et al., 1971)
1.57, 4.88 (Geyer et al., 1984)
2.50 (Riederer, 1990)
2.14, 2.16 (Jafvert et al., 1990)

Solubility in organics:
At 25 °C (g/L): carbon tetrachloride (1), ethyl ether (270), acetone (850) and ethyl alcohol (1,300) (Bailey and White, 1965).

Solubility in water:
890 ppm at 25 °C (Chiou et al., 1977)
400 mg/L at 20 °C (Riderer, 1990)

725 ppm at 25 °C (Bailey and White, 1965)
530 mg/L at 17 °C, 2.36 mM at 25 °C (Gunther et al., 1968)
2,940 μmol/L at 25 °C (LaFleur, 1979)
620 mg/L at 25 °C (Worthing and Hance, 1991)

Vapor pressure (10^3 mmHg):
4.7 at 20 °C (Riderer, 1990)

Environmental Fate
   Biological. 2,4-D degraded in anaerobic sewage sludge to 4-chlorophenol
(Mikesell and Boyd, 1985). In clear water and muddy water, hydrolysis half-lives
of 18->50 days and 10-25 days, respectively, were reported (Nesbitt and Watson,
1980).
   Soil. In moist soils, 2,4-D degraded to 2,4-dichlorophenol and 2,4-dichloro-
anisole as intermediates followed by complete mineralization to carbon dioxide
(Smith, 1985; Stott, 1983). 2,4-Dichlorophenol was reported as a hydrolysis
metabolite (Somasundaram et al., 1989, 1991; Somasundaram and Coats, 1991). In a
soil pretreated with its hydrolysis metabolite, 80% of the applied [^{14}C]2,4-D
mineralized to $^{14}CO_2$ within 4 days. In soils not treated with the hydrolysis product
(2,4-dichlorophenol), only 6% of the applied [^{14}C]2,4-D degraded to $^{14}CO_2$ after 4
days (Somasundaram et al., 1989). Steenson and Walker (1957) reported that the soil
microorganisms *Flavobacterium peregrinum* and *Achromobacter* both degraded 2,4-
D yielding 2,4-dichlorophenol and 4-chlorocatechol as metabolites. The microorganisms *Gloeosporium olivarium*, *Gloeosporium kaki* and *Schisophyllum com-
muns* also degraded 2,4-D in soil forming 2-(2,4-dichlorophenoxy)ethanol as the
major metabolite (Nakajima et al., 1973). Microbial degradation of 2,4-D was
more rapid under aerobic conditions (half-life = 1.8-3.1 days) than under
anaerobic conditions (half-life = 69-135 days) (Liu et al., 1981). In a 5-day
experiment, [^{14}C]2,4-D applied to soil water suspensions under aerobic and
anaerobic conditions gave $^{14}CO_2$ yields of 0.5 and 0.7%, respectively (Scheunert et
al., 1987). Degradation was observed to be lowest at low redox potentials (Gambrell
et al., 1984). The reported half-lives in soil are 15 days (Jury et al., 1987) and <7
days (Hartley and Kidd, 1987). Residual activity in soil is limited to approximately
6 weeks (Hartley and Kidd, 1987).
   Plant. Reported metabolic products in bean and soybean plants include 4-*O*-β-
glucosides of 4-hydroxy-2,5-dichlorophenoxyacetic acid, 4-hydroxy-2,3-dichlo-
rophenoxyacetic acid, *N*-(2,4-dichlorophenoxyacetyl)-L-aspartic acid and *N*-(2,4-
dichlorophenocyacetyl)-L-glutamic acid. Metabolites identified in cereals and
strawberries include 1-*O*-(2,4-dichlorophenoxyacetyl)-β-D-glucose and 2,4-di-
chlorophenol, respectively (Verschueren, 1983). In alfalfa, the side chain in the
2,4-D molecule was found to be lengthened by two and four methylene groups
resulting in the formation of (2,4-dichlorophenoxy)butyric acid and (2,4-dichloro-
phenoxy)hexanoic acid, respectively. In several resistant grasses, however, the side

chain increased by one methylene group forming (2,4-dichlorophenoxy)propionic acid (Hagin and Linscott, 1970).

2,4-D was metabolized by soybean cultures forming 2,4-dichlorophenoxyacetyl derivatives of alanine, leucine, phenylalanine, tryptophan, valine, aspartic and glutamic acids (Feung et al., 1971, 1972, 1973). On bean plants, 2,4-D degraded via β-oxidation and ring hydroxylation to form 2,4-dichloro-4-hydroxyphenoxy-acetic acid, 2,3-dichloro-4-hydroxyphenoxyacetic acid (Hamilton et al., 1971) and 2-chloro-4-hydroxyphenoxyacetic acid. 2,5-Dichloro-4-hydroxyphenoxyacetic acid was the predominant product identified in several weed species, as well as in smaller quantities of 2-chloro-4-hydroxyphenoxyacetic acid in wild buckwheat, yellow foxtail and wild oats (Fleeker and Steen, 1971).

Photolytic. Photolysis of 2,4-D in distilled water using mercury arc lamps (λ = 254 nm) or by natural sunlight yielded 2,4-dichlorophenol, 4-chlorocatechol, 2-hydroxy-4-chlorophenoxyacetic acid, 1,2,4-benzenetriol and polymeric humic acids. The half-life for this reaction is 50 minutes (Crosby and Tutass, 1966). A half-life of 2-4 days was reported for 2,4-D in water irradiated at 356 nm (Baur and Bovey, 1974).

Bell (1956) reported that the composition of photodegradation products formed were dependent upon the initial 2,4-D concentration and pH of the solutions. 2,4-D undergoes reductive dechlorination when various polar solvents (methanol, butanol, isobutyl alcohol, *t*-butyl alcohol, octanol, ethylene glycol) are irradiated at wavelengths between 254-420 nm. Photoproducts formed included 2,4-dichloro-phenol, 2,4-dichloroanisole, 4-chlorophenol, 2- and 4-chlorophenoxyacetic acid (Que Hee and Sutherland, 1981).

Surface Water. In filtered lake water at 29 °C, 90% of 2,4-D (1 mg/L) mineralized to carbon dioxide. The half-life was <5 days. At low concentrations (0.2 mg/L), no mineralization was observed (Wang et al., 1984). Subba-Rao et al. (1982) reported that 2,4-D in very low concentrations mineralized in one of three lakes tested. Mineralization did not occur when concentrations were at the pico-gram level.

Chemical/Physical. In a helium pressurized reactor containing ammonium nitrate and polyphosphoric acid at temperatures of 121 and 232 °C, 2,4-D was oxidized to carbon dioxide, water and hydrochloric acid (Leavitt and Abraham, 1990). Carbon dioxide, chloride, aldehydes, oxalic and glycolic acids, were reported as ozonation products of 2,4-D in water at pH 8 (Struif et al., 1978). Reacts with alkalies, metals and amines forming water soluble salts (Hartley and Kidd, 1987).

When 2,4-D was heated at 900 °C, carbon monoxide, carbon dioxide, chlorine, hydrochloric acid and oxygen were produced (Kennedy et al., 1972, 1972a). Total mineralization of 2,4-D was observed when a solution containing the herbicide and Fenton's reagent (ferrous ions and hydrogen peroxide) was subjected to UV light (λ = 300-400 nm). One intermediate compound identified was oxalic acid (Sun and Pignatello, 1993).

Exposure Limits (mg/m^3): NIOSH REL: TWA 10, IDLH 100; OSHA PEL: TWA 10; ACGIH TLV: TWA 10.

Toxicity: Acute oral LD$_{50}$ for chickens 541 mg/kg, dogs 100 mg/kg, guinea pigs 469 mg/kg, hamsters 500 mg/kg, mice 368 mg/kg, rats 370 mg/kg (RTECS, 1985).

Drinking Water Standard (final): MCLG: 70 μg/L; MCL: 70 μg/L (U.S. EPA, 1984).

Uses: Herbicide; weed killer and defoliant.

p,p′-DDD

Synonyms: 1,1-Bis(4-chlorophenyl)-2,2-dichloroethane; 1,1-Bis(*p*-chlorophenyl)-2,2-dichloroethane; 2,2-Bis(4-chlorophenyl)-1,1-dichloroethane; 2,2-Bis-(*p*-chlorophenyl)-1,1-dichloroethane; DDD; 4,4′-DDD; 1,1-Dichloro-2,2-bis(*p*-chlorophenyl)ethane; 1,1-Dichloro-2,2-di(4-chlorophenyl)ethane; 1,1-Dichloro-2,2-di-(*p*-chlorophenyl)ethane; Dichlorodiphenyldichloroethane; 4,4′-Dichlorodiphenyldichloroethane; *p,p′*-Dichlorodiphenyldichloroethane; **1,1′-(2,2-Dichloroethylidene)bis[4-chlorobenzene]**; Dilene; ENT 4225; ME-1700; NA 2761; NCI-C00475; RCRA waste number U060; Rhothane; Rhothane D-3; Rothane; TDE; 4,4′-TDE; *p,p′*-TDE; Tetrachlorodiphenylethane.

CAS Registry No.: 72-54-8
DOT: 2761
DOT label: Poison
Molecular formula: $C_{14}H_{10}Cl_4$
Formula weight: 320.05
RTECS: KI0700000

Physical state and color
Crystalline white solid.

Melting point (°C):
112 (Verschueren, 1983)
107-109 (Aldrich, 1988)

Boiling point (°C):
193 (Sax, 1985)

Density (g/cm^3):
1.476 at 20/4 °C (Weiss, 1986)

Diffusivity in water (10^5 cm^2/sec):
0.45 at 20 °C using method of Hayduk and Laudie (1974).

Flash point (°C):
Not pertinent (Weiss, 1986)

Lower explosive limit (%):
Not pertinent (Weiss, 1986)

Upper explosive limit (%):
Not pertinent (Weiss, 1986)

Henry's law constant (10^5 atm·m^3/mol):
2.16 (calculated, U.S. EPA, 1980)

Bioconcentration factor, log BCF:
4.92 (fish, microcosm) (Garten and Trabalka, 1983)

Soil sorption coefficient, log K_{oc}:
5.38 (Jury et al., 1987)
6.6 (average using 9 suspended sediment samples from the St. Clair and Detroit Rivers, Lau et al., 1989)

Octanol/water partition coefficient, log K_{ow}:
6.02 (Jury et al., 1987)
5.99 (Callahan et al., 1979)
5.061 (Rao and Davidson, 1980)
5.80 (DeKock and Lord, 1987)
6.217 (de Bruijn et al., 1989)

Solubility in water (ppb):
160 at 24 °C (Verschueren, 1983)
20 at 25 °C (Weil et al., 1974)
5 (Jury et al., 1987)
50 at 15 °C, 90 at 25 °C, 150 at 35 °C, 240 at 45 °C (particle size ≤5 μ) (Biggar and Riggs, 1974)

Vapor density:
17.2 ng/L at 30 °C (Spencer and Cliath, 1972)

Vapor pressure (10^6 mmHg):
1.02 at 30 °C (Spencer and Cliath, 1972)
4.68 at 25 °C (Bidleman, 1984)
8.25, 12.2, at 25 °C (Hinckley et al., 1990)

Environmental Fate
    Biological. It was reported that *p,p'*-DDD, a major biodegradation product of *p,p'*-DDT, was degraded by *Aerobacter aerogenes* under aerobic conditions to yield 1-chloro-2,2-bis(*p*-chlorophenyl)ethylene, 1-chloro-2,2-bis(*p*-chlorophenyl)ethane

and 1,1-bis(*p*-chlorophenyl)ethylene. Under anaerobic conditions, however, four additional compounds were identified: bis(*p*-chlorophenyl)acetic acid, *p,p′*-dichlorodiphenylmethane, *p,p′*-dichlorobenzhydrol and *p,p′*-dichlorobenzophenone (Fries, 1972). Under reducing conditions, indigenous microbes in Lake Michigan sediments degraded DDD to 2,2-bis(*p*-chlorophenyl)ethane and 2,2-bis(*p*-chlorophenyl)ethanol (Leland et al., 1973). Incubation of *p,p′*-DDD with ·hematin and ammonia gave 4,4′-dichlorobenzophenone, 1-chloro-2,2-bis(*p*-chlorophenyl)ethylene and bis(*p*-chlorophenyl)acetic acid methyl ester (Quirke et al., 1979). Using settled domestic wastewater inoculum, *p,p′*-DDD (5 and 10 mg/L) did not degrade after 28 days of incubation at 25 °C (Tabak et al., 1981).

Chemical/Physical. The hydrolysis rate constant for *p,p′*-DDD at pH 7 and 25 °C was determined to be 2.8×10^{-6}/hour, resulting in a half-life of 28.2 years (Ellington et al., 1987).

Toxicity: Acute oral LD_{50} for rats 113 mg/kg (RTECS, 1985).

Uses: Dusts, emulsions and wettable powders for contact control of leaf rollers and other insects on vegetables and tobacco.

p,p'-DDE

Synonyms: 2,2-Bis(4-chlorophenyl)-1,1-dichloroethene; 2,2-Bis(*p*-chlorophenyl)-1,1-dichloroethene; 1,1-Bis(4-chlorophenyl)-2,2-dichloroethylene; 1,1-Bis(*p*-chlorophenyl)-2,2-dichloroethylene; DDE; 4,4'-DDE; DDT dehydrochloride; 1,1-Dichloro-2,2-bis(*p*-chlorophenyl)ethylene; Dichlorodiphenyldichloroethylene; *p,p'*-Dichlorodiphenyldichloroethylene; **1,1'-(Dichloroethenylidene)bis(4-chlorobenzene)**; NCI-C00555.

CAS Registry No.: 72–55–9
DOT: 2761
DOT label: Poison
Molecular formula: $C_{14}H_8Cl_4$
Formula weight: 319.03
RTECS: KV9450000

Physical state and color
White crystalline powder.

Melting point (°C):
88–90 (Leffingwell, 1975)
112 (Melnikov, 1971)

Diffusivity in water (10^5 cm^2/sec):
0.46 at 20 °C using method of Hayduk and Laudie (1974).

Henry's law constant (10^3 atm·m^3/mol at 23 °C):
1.22 and 3.65 in distilled water and seawater, respectively (Atlas et al., 1982).

Bioconcentration factor, log BCF:
4.00–4.15 (fish tank), 7.26 (Lake Ontario) (rainbow trout, Oliver and Niimi, 1985)
4.08 (fish, Metcalf et al., 1975)
4.71 (freshwater fish), 4.44 (fish, microcosm) (Garten and Trabalka, 1983)

Soil sorption coefficient, log K$_{oc}$:
5.386 (Reinbold et al., 1979)
6.6 (average using 10 suspended sediment samples collected from the St. Clair and Detroit Rivers, Lau et al., 1989)
K_d = 2.5 (Barcelona coastal sediments, Bayona et al., 1991).

Octanol/water partition coefficient, log K_{ow}:
5.83 (Travis and Arms, 1988)
5.69 (Freed et al., 1977, 1979a)
5.766 (Kenaga and Goring, 1980)
5.89 (Burkhard et al., 1985a)
6.20 (DeKock and Lord, 1987)
6.956 (de Bruijn et al., 1989)

Solubility in organics:
Soluble in fats and most solvents (IARC, 1974).

Solubility in water (ppb):
40 at 20 °C, 1.3 μg/L at 25 °C (Metcalf et al., 1973)
14 at 20 °C (Weil et al., 1974)
40 at 20 °C (Chiou et al., 1977)
55 at 15 °C, 120 at 25 °C, 235 at 35 °C, 450 at 45 °C (particle sizes ≤5 μ) (Biggar and Riggs, 1974)
65 at 24 °C (Hollifield, 1979)

Vapor density:
109 ng/L at 30 °C (Spencer and Cliath, 1972)

Vapor pressure (10^6 mmHg):
6.49 at 30 °C (Spencer and Cliath, 1972)
13 at 30 °C (Wescott et al., 1981)
15.7 at 25 °C (Bidleman, 1984)
7.4 at 25 °C (Wescott and Bidleman, 1981)
14.0, 20.3 at 25 °C (Hinckley et al., 1990)

Environmental Fate

Biological. In four successive 7-day incubation periods, *p,p'*-DDE (5 and 10 mg/L) was recalcitrant to degradation in a settled domestic wastewater inoculum (Tabak et al., 1981).

Photolytic. When an aqueous solution of *p,p'*-DDE (0.004 μM) in natural water samples from California and Hawaii were irradiated (maximum λ = 240 nm) for 120 hours, 62% was photooxidized to *p,p'*-dichlorobenzophenone (Ross and Crosby, 1985). In an air-saturated distilled water medium irradiated with monochromic light (λ = 313 nm), *p,p'*-DDE degraded to *p,p'*-dichlorobenzophenone, 1,1-bis(*p*-chlorophenyl)-2-chloroethylene (DDMU) and 1-(4-chloro-phenyl)-1-(2,4-dichlorophenyl)-2-chloroethylene (*o*-chloro DDMU). Identical photoproducts were also observed using tap water containing Mississippi River sediments (Miller and Zepp, 1979). The photolysis half-life under sunlight irradiation was reported to be 1.5 days (Mansour et al., 1989).

Chemical/Physical. May degrade to bis(chlorophenyl)acetic acid in water (Verschueren, 1983), or oxidize to *p,p'*-dichlorobenzophenone using UV light as a catalyst (HSDB, 1989).

Toxicity: Acute oral LD_{50} for rats 880 mg/kg (RTECS, 1985).

Uses: Military product; chemical research.

p,p'-DDT

Synonyms: Agritan; Anofex; Arkotine; Azotox; 2,2-Bis(4-chlorophenyl)-1,1,1-trichloroethane; 2,2-Bis(*p*-chlorophenyl)-1,1,1-trichloroethane; α,α-Bis(*p*-chlorophenyl)-β,β,β-trichloroethane; 1,1-Bis(*p*-chlorophenyl)-2,2,2-trichloroethane; Bosan Supra; Bovidermol; Chlorophenothan; Chlorophenothane; Chlorophenotoxum; Citox; Clofenotane; DDT; 4,4′-DDT; Dedelo; Deoval; Detox; Detoxan; Dibovan; Dichlorodiphenyltrichloroethane; *p,p'*-Dichlorodiphenyltrichloroethane; 4,4′-Dichlorodiphenyltrichloroethane; Dicophane; Didigam; Didimac; Diphenyltrichloroethane; Dodat; Dykol; ENT 1506; Estonate; Genitox; Gesafid; Gesapon; Gesarex; Gesarol; Guesapon; Gyron; Havero-extra; Ivoran; Ixodex; Kopsol; Mutoxin; NCI-C00464; Néocid; Parachlorocidum; PEB1; Pentachlorin; Pentech; PPzeidan; RCRA waste number U061; Rukseam; Santobane; Trichlorobis(4-chlorophenyl)ethane; Trichlorobis(*p*-chlorophenyl)ethane; 1,1,1-Trichloro-2,2-bis(*p*-chlorophenyl)ethane; 1,1,1-Trichloro-2,2-di(4-chlorophenyl)ethane; 1,1,1-Trichloro-2,2-di(*p*-chlorophenyl)ethane; **1,1′-(2,2,2-Trichloroethylidene)bis[4-chlorobenzene]**; Zeidane; Zerdane.

CAS Registry No.: 50-29-3
DOT: 2761
DOT label: Poison
Molecular formula: $C_{14}H_9Cl_5$
Formula weight: 354.49
RTECS: KJ3325000

Physical state, color and odor
Colorless crystals or white powder, odorless to slightly fragrant powder. Odor threshold in air is 200 ppb (Keith and Walters, 1992).

Melting point (°C):
108.5 (Bowman et al., 1960)
108-109 (Weast, 1986)

Boiling point (°C):
260 (Weast, 1986)
185 (U.S. EPA, 1980)

Density (g/cm^3):
1.56 at 15/4 °C (Weiss, 1986)

Diffusivity in water (10^5 cm^2/sec):
0.37 at 20 °C using method of Hayduk and Laudie (1974).

Flash point (°C):
72.3–77.3 (NIOSH, 1994)

Henry's law constant (10^5 atm·m^3/mol):
3.8 (Eisenreich et al., 1983)
48.9 (Jury et al., 1984)
10.3 (Jury et al., 1984a)
5.2 (Jury et al., 1983)
1.29 at 23 °C (Fendinger et al., 1989)

Bioconcentration factor, log BCF:
3.97 (algae, Geyer et al., 1984)
4.15 (activated sludge), 3.28 (golden ide) (Freitag et al., 1985)
4.53 (freshwater fish), 4.49 (fish, microcosm) (Garten and Trabalka, 1983)

Soil sorption coefficient, log K_{oc}:
5.146 (silt soil loam, Chiou et al., 1979)
5.14 (Schwarzenbach and Westall, 1981)
5.20 (Commerce soil), 5.17 (Tracy soil), 5.18 (Catlin soil) (McCall et al., 1981)
6.26 (marine sediments, Pierce et al., 1974)
5.39 (Rao and Davidson, 1980)
6.7 (average using 7 suspended sediment samples from the St. Clair and Detroit Rivers, Lau et al., 1989)
5.77 (Mivtahim soil), 5.48 (Gilat soil), 5.22 (Neve Yaar soil), 4.98 (Malkiya soil), 5.40 (Kinneret sediment), 5.30 (Kinneret-G sediment) (Gerstl and Mingelgrin, 1984)

Octanol/water partition coefficient, log K_{ow}:
6.36 (Chiou et al., 1982)
6.19 (DeKock and Lord, 1987; Freed et al., 1977, 1979; Johnsen et al., 1989)
5.76 (Travis and Arms, 1988)
5.98 (Mackay and Paterson, 1981)
5.38 (Kenaga, 1980)
6.16, 6.17, 6.22, 6.44 (Brooke et al., 1986)
6.28 (Geyer et al., 1984)
4.89 (Wolfe et al., 1977)
5.44 (Gerstl and Mingelgrin, 1984; Burkhard et al., 1985a)
6.914 (Brooke et al., 1990; de Bruijn et al., 1989)
6.307 (Brooke et al., 1990)
6.38 (Hammers et al., 1982)

Solubility in organics:
In g/L: acetone (580), benzene (780), benzyl benzoate (420), carbon tetrachloride (450), chlorobenzene (740), cyclohexanone (1,160), ethyl ether (280), gasoline (100), isopropanol (30), kerosene (80-100), morpholine (750), peanut oil (110), pine oil (100-160), tetralin (610), tributyl phosphate (500) (Windholz et al., 1983).

Solubility in water:
5.5 ppb at 25 °C (Weil et al., 1974)
1.2 μg/L at 25 °C (Bowman et al., 1960)
7 μg/L at 20 °C (Nisbet and Sarofim, 1972)
5.4 μg/L at 24 °C (Chiou et al., 1986)
4 μg/L at 24-25 °C (Hollifield, 1979; Chiou et al., 1979)
5.9 ppb at 2 °C, 37.4 ppb at 25 °C, 45 ppb at 37.5 °C (Babers, 1955)
10-100 ppb at 22 °C (Roeder and Weiant, 1946)
2 ppb (Kapoor et al., 1973)
In ppb: 17 at 15 °C, 25 at 25 °C, 37 at 35 °C, 45 at 45 °C (particle size ≤5 μ)
   (Biggar and Riggs, 1974)
260 ppb at 25 °C (NAS, 1977)
At 20-25 °C: 40 ppb (particle size ≤5 μ), 16 ppb (particle size ≤0.05 μ) (Robeck et
   al., 1965)
7.7 μg/L at 20 °C (Friesen et al., 1985)
40 μg/L at 20 °C (Ellgehausen et al., 1981)
3 nmol/L at 25 °C (LaFleur, 1979)
3.54 μg/L in Lake Michigan water at ≈ 25 °C (Eadie et al., 1990)
4.5 ppb at 25 °C (Gerstl and Mingelgrin, 1984)

Vapor density:
13.6 ng/L at 30 °C (Spencer and Cliath, 1972)

Vapor pressure (10^7 mmHg):
1 at 25 °C (Mackay and Wolkoff, 1973)
1.5 at 20 °C (Balson, 1947)
7.26 at 30 °C (Spencer and Cliath, 1972)
1.29 at 20 °C, 4.71 at 50 °C, 6.76 at 100 °C (Webster et al., 1985)
1.40 at 30 °C (Wescott and Bidleman, 1981)
2.2, 4.3, 9.3, 40, 150, 480, 1,500 and 4,500 at 20, 25, 30, 40, 50, 60, 70 and 80 °C,
   respectively (Rothman, 1980).
150, 533, 5,850 and 14,900 at 50.1, 60.1, 80.4 and 90.2 °C, respectively (Dickenson,
   1956)
62.3 at 25 °C (Hinckley et al., 1990)

Environmental Fate
   Biological. In four successive 7-day incubation periods, *p,p′*-DDT (5 and 10

mg/L) was recalcitrant to degradation in a settled domestic wastewater inoculum (Tabak et al., 1981).

Castro (1964) reported that iron(II) porphyrins in dilute aqueous solution was rapidly oxidized by DDT to form the corresponding iron(III) chloride complex (hematin) and DDE, respectively. Incubation of *p,p'*-DDT with hematin and ammonia gave *p,p'*-DDD, *p,p'*-DDE, bis(*p*-chlorophenyl)acetonitrile, 1-chloro-2,2-bis(*p*-chlorophenyl)ethylene, 4,4'-dichlorobenzophenone and the methyl ester of bis(*p*-chlorophenyl)acetic acid (Quirke et al., 1979).

In 1 day, *p,p'*-DDT reacted rapidly with reduced hematin forming *p,p'*-DDD and unidentified products (Baxter, 1990). The white rot fungus *Phanerochaete chrysosporium* degraded *p,p'*-DDT yielding the following metabolites: 1,1-dichloro-2,2-bis(4-chlorophenyl)ethane (*p,p'*-DDD), 2,2,2-trichloro-1,1-bis(4-chlorophenyl)ethanol (dicofol), 2,2-dichloro-1,1-bis(4-chlorophenyl)ethanol and 4,4'-dichlorobenzophenone. Mineralization of *p,p'*-DDT by the white rot fungi *Pleurotus ostreatus*, *Phellinus weirri* and *Polyporus versicolor* was also demonstrated (Bumpus and Aust, 1987). Fries (1972) reported that *Aerobacter aerogenes* degraded *p,p'*-DDT under aerobic conditions forming *p,p'*-DDD, *p,p'*-DDE, 1-chloro-2,2-bis(*p*-chlorophenyl)ethylene, 1-chloro-2,2-bis(*p*-chlorophenyl)ethane and 1,1-bis(*p*-chlorophenyl)ethylene. Under anaerobic conditions the same organism produced four additional compounds. These were bis(*p*-chlorophenyl)-acetic acid, *p,p'*-dichlorodiphenylmethane, *p,p'*-dichlorobenzhydrol and *p,p'*-dichlorobenzophenone. Other degradation products of *p,p'*-DDT under aerobic and anaerobic conditions in soils using various cultures not previously mentioned include 1,1-bis(*p*-chlorophenyl)-2,2,2-trichloroethanol (Kelthane) and 4-chlorobenzoic acid (Fries, 1972).

Under aerobic conditions, the amoeba *Acanthamoeba castellanii* (Neff strain ATCC 30.010) degraded *p,p'*-DDT to *p,p'*-DDE, *p,p'*-DDD and dibenzophenone (Pollero and dePollero, 1978). Incubation of *p,p'*-DDT with hematin and ammonia gave *p,p'*-DDD, *p,p'*-DDE, bis(*p*-chlorophenyl)acetonitrile, 1-chloro-2,2-bis(*p*-chlorophenyl)ethylene, 4,4'-dichlorobenzophenone and the methyl ester of bis(*p*-chlorophenyl)acetic acid (Quirke et al., 1979).

Thirty-five microorganisms isolated from marine sediment and marine water samples taken from Hawaii and Houston, TX were capable of degrading *p,p'*-DDT. *p,p'*-DDD was identified as the major metabolite. Minor transformation products included 2,2-bis(*p*-chlorophenyl)ethanol, 2,2-bis(*p*-chlorophenyl)ethane and *p,p'*-DDE (Patil et al., 1972).

In a 42-day experiment, [^{14}C]*p,p'*-DDT applied to soil water suspensions under aerobic and anaerobic conditions gave $^{14}CO_2$ yields of 0.8 and 0.7%, respectively (Scheunert et al., 1987). Similarly, Matsumura et al. (1971) found that *p,p'*-DDT was degraded by numerous aquatic microorganisms isolated from water and silt samples collected from Lake Michigan and its tributaries in Wisconsin. The major metabolites identified were TDE, DDNS and DDE. *p,p'*-DDT was metabolized by the following microorganisms under laboratory conditions to *p,p'*-DDD:

Actinomycetes (Chacko et al., 1966), *Escherichia coli* (Langlois, 1967), *Aerobacter aerogenes* (Plimmer et al., 1968; Wedemeyer, 1966) and *Proteus vulgaris* (Wedemeyer, 1966). In addition, *p,p'*-DDT was degraded to *p,p'*-DDD, *p,p'*-DDE and dicofol by *Trichoderma viride* (Matsumura and Boush, 1968) and to *p,p'*-DDD and *p,p'*-DDE by *Ankistrodemus amalloides* (Neudorf and Khan, 1975).

Jensen et al. (1972) studied the anaerobic degradation of *p,p'*-DDT (100 mg) in 1 L of sewage sludge containing *p,p'*-DDD (4.0%) and *p,p'*-DDE (3.1%) as contaminants. The sludge was incubated at 20 °C for 8 days under a nitrogen atmosphere. The parent compound degraded rapidly (half-life = 7 hours) forming *p,p'*-DDD, *p,p'*-dichlorodiphenylbenzophenone (DBP), 1,1-bis(*p*-chlorophenyl)-2-chloroethylene (DDMU) and bis(*p*-chlorophenyl)acetonitrile. After 48 hours, the original amount of *p,p'*-DDD added to the sewage sludge had completely reacted. In a similar study, Pfaender and Alexander (1973) observed the cometabolic conversion of DDT (0.005%) in unamended sewage sludge to give DDD, DDE and DBP. When the sewage sludge was amended with glucose (0.10%), the rate of DDD formation was enhanced. However, with the addition of diphenylmethane, the rate of formation of both DDD and DBP was reduced. The diphenylmethane-amended sewage sludge showed the greatest abundance of bacteria capable of cometabolizing DDT, whereas the unamended sewage showed the fewest number of bacteria. Zoro et al. (1974) also reported that *p,p'*-DDT in untreated sewage sludge was converted to *p,p'*-DDD, especially in the presence of sodium dithionate, a widely used reducing agent.

In an *in vitro* fermentation study, rumen microorganisms metabolized both isomers of [^{14}C]DDT (*o,p'*- and *p,p'*-) to the corresponding DDD isomers at a rate of 12%/hour. With *p,p'*-DDT, 11% of the ^{14}C detected was an unidentified polar product associated with microbial and substrate residues (Fries et al., 1969). In another *in vitro* study, extracts of *Hydrogenomonas* sp. cultures degraded DDT to DDD, 1-chloro-2,2-bis(*p*-chlorophenyl)ethane (DDMS), DBP and several other products under anaerobic conditions. Under aerobic conditions containing whole cells, one of the rings is cleaved and *p*-chlorophenylacetic acid is formed (Pfaender and Alexander, 1972).

Soil. *p,p'*-DDD and *p,p'*-DDE are the major metabolites of *p,p'*-DDT in the environment (Metcalf, 1973). In soils under anaerobic conditions, *p,p'*-DDT is rapidly converted to *p,p'*-DDD via reductive dechlorination (Johnsen, 1976) and very slowly to *p,p'*-DDE under aerobic conditions (Guenzi and Beard, 1967; Kearney and Kaufman, 1976). The aerobic degradation of *p,p'*-DDT under flooded conditions is very slow with *p,p'*-DDE forming as the major metabolite. Dicofol was also detected in minor amounts (Lichtenstein et al., 1971). In addition to *p,p'*-DDD and *p,p'*-DDE, 2,2-bis(*p*-chlorophenyl)acetic acid (DDA), bis(*p*-chlorophenyl)methane (DDM), *p,p'*-dichlorobenzhydrol (DBH), DBP and *p*-chlorophenylacetic acid (PCPA) were also reported as metabolites of *p,p'*-DDT in soil under aerobic conditions (Subba-Rao and Alexander, 1980).

The anaerobic conversion of *p,p'*-DDT to *p,p'*-DDD in soil was catalyzed by the presence of ground alfalfa or glucose (Burge, 1971). Under flooded conditions, *p,p*-DDT was rapidly converted to TDE via reductive dehalogenation and other metabolites (Castro and Yoshida, 1971; Guenzi and Beard, 1967). Degradation was faster in flooded soil than an upland soil and was faster in soils containing high organic matter (Castro and Yoshida, 1971). Other reported degradation products under aerobic and anaerobic conditions by various soil microbes include 1,1'-bis(*p*-chlorophenyl)-2-chloroethane, 1,1'-bis(*p*-chlorophenyl)-2-hydroxyethane and *p*-chlorophenyl acetic acid (Kobayashi and Rittman, 1982). It was also reported that *p,p'*-DDE formed by hydrolyzing *p,p'*-DDT (Wolfe et al., 1977). The clay-catalyzed reaction of DDT to form DDE was reported by Lopez-Gonzales and Valenzuela-Calahorro (1970). They observed that DDT adsorbed of sodium bentonite clay surfaces was transformed more rapidly than on the corresponding hydrogen-bentonite clay. In one day, *p,p'*-DDT reacted rapidly with reduced hematin forming *p,p'*-DDD and unidentified products (Baxter, 1990). In an Everglades muck, *p,p'*-DDT was slowly converted to *p,p'*-DDD and *p,p'*-DDE (Parr and Smith, 1974). The reported half-life in soil is 3,800 days (Jury et al., 1987).

Oat plants were grown in two soils treated with [^{14}C]*p,p'*-DDT. Most of the residues remained bound to the soil. Metabolites identified were *p,p'*-DDE, *o,p'*-DDT, TDE, DBP, dicofol and DDA (Fuhremann and Lichtenstein, 1980).

In a 42-day experiment, [^{14}C]*p,p'*-DDT applied to soil water suspensions under aerobic and anaerobic conditions gave $^{14}CO_2$ yields of 0.8 and 0.7%, respectively (Scheunert et al., 1987).

Photolytic. Photolysis of *p,p*-DDT in nitrogen-sparged methanol solvent by UV light (λ = 260 nm) produced DDD and DDMU. But photolysis of *p,p'*-DDT at 280 nm in an oxygenated methanol solution yielded a complex mixture containing the methyl ester of 2,2-bis(*p*-chlorophenyl)acetic acid (Plimmer et al., 1970). *p,p'*-DDT in an aqueous solution containing suspended titanium dioxide as a catalyst and irradiated with UV light (λ >340 nm) formed chloride ions. Based on the amount of chloride ions generated, carbon dioxide and hydrochloric acid were reported as the end products (Borello et al., 1989). When an aqueous solution containing *p,p'*-DDT was photooxidized by UV light at 90-95 °C, 25, 50 and 75% degraded to carbon dioxide after 25.9, 66.5 and 120.0 hours, respectively (Knoevenagel and Himmelreich, 1976).

When *p,p'*-DDT on quartz was subjected to UV radiation (2537 Å) for 2 days, 80% of *p,p'*-DDT degraded to 4,4'-dichlorobenzophenone, 1,1-dichloro-2,2-bis-(*p*-chlorophenyl)ethane and 1,1-dichloro-2,2-bis(*p*-chlorophenyl)ethene. Irradiation of a hexane solution yielded 1,1-dichloro-2,2-bis(*p*-chlorophenyl)-ethane and hydrochloric (Mosier et al., 1969).

Chemical/Physical. In alkaline solutions and temperatures >108.5 °C, *p,p'*-DDT undergoes dehydrochlorination releasing hydrochloric acid to give the noninsecticidal *p,p'*-DDE (Hartley and Kidd, 1987; Worthing and Hance, 1991). This reaction is also catalyzed by ferric and aluminum chlorides and UV light (Worthing and Hance, 1991).

When *p,p′*-DDT was heated at 900 °C, carbon monoxide, carbon dioxide, chlorine, hydrochloric acid and other unidentified substances were produced (Kennedy et al., 1972, 1972a). Emits hydrochloric acid and chlorine when incinerated (Sittig, 1985).

Exposure Limits (mg/m^3): Potential occupational carcinogen. NIOSH REL: TWA 0.5, IDLH 500; OSHA PEL: TWA 1; ACGIH TLV: TWA 1.

Toxicity: Acute oral LD_{50} for dogs 150 mg/kg, frogs 7,600 μg/mg, guinea pigs 150 mg/kg, monkeys 200 mg/kg, mice 135 mg/kg, rats 87 mg/kg, rabbits 250 mg/kg (RTECS, 1985); LC_{50} (48-hour) for red killifish 100 μg/L (Yoshioka et al., 1986).

Uses: Use as an insecticide is now prohibited; chemical research; nonsystemic stomach and contact insecticide.

DECAHYDRONAPHTHALENE

Synonyms: Bicyclo[4.4.0]decane; Dec; Decalin; Decalin solvent; Dekalin; Naphthalane; Naphthane; Perhydronaphthalene; UN 1147.

CAS Registry No.: 91-17-8
DOT: 1147
Molecular formula: $C_{10}H_{18}$
Formula weight: 138.25
RTECS: QJ3150000

Physical state, color and odor
Water-white liquid with a methanol-like odor.

Melting point (°C):
-43 (*cis*), -30.4 (*trans*) (Weast, 1986)

Boiling point (°C):
195.6 (*cis*), 187.2 (*trans*) (Weast, 1986)

Density (g/cm^3):
0.8965 at 20/4 °C (*cis*), 0.8699 at 20/4 °C (*trans*) (Weast, 1986)

Diffusivity in water (10^5 cm^2/sec at 20 °C):
0.68 and 0.67 for *cis* and *trans* isomers, respectively, using method of Hayduk and Laudie (1974).

Dissociation constant:
>14 (Schwarzenbach et al., 1993)

Flash point (°C):
58 (commercial mixture, Windholz et al., 1983)
58 (*cis*), 52 (*trans*) (Dean, 1987)

Lower explosive limit (%):
0.7 at 100 °C (Sax and Lewis, 1987)

Upper explosive limit (%):
4.9 at 100 °C (Sax and Lewis, 1987)

Henry's law constant (10^2 atm·m^3/mol):
7.00, 8.37, 10.6, 11.7 and 19.9 at 10, 15, 20, 25 and 30 °C, respectively (Ashworth et al., 1988).

Interfacial tension with water (dyn/cm at 20 °C):
51.24 (*cis*), 50.7 (*trans*) (Demond and Lindner, 1993)

Soil sorption coefficient, log K_{oc}:
Unavailable because experimental methods for estimation of this parameter for alicyclic hydrocarbons are lacking in the documented literature.

Octanol/water partition coefficient, log K_{ow}:
4.00 using method of Hansch et al. (1968).

Solubility in organics:
Soluble in acetone, alcohol, benzene, ether and chloroform (Weast, 1986). Miscible with propanol, isopropanol and most ketones and ethers (Windholz et al., 1983).

Solubility in water (25 °C):
889 µg/kg (Price, 1976)
<0.2 mL/L (Booth and Everson, 1948)

Vapor density:
5.65 g/L at 25 °C, 4.77 (air = 1)

Vapor pressure (mmHg):
195.77 and 187.27 at 25 °C for *cis* and *trans* isomers, respectively (Wilhoit and Zwolinski, 1971)

Toxicity: Acute oral LD_{50} for rats 4,170 mg/kg (RTECS, 1985).

Uses: Solvent for naphthalene, waxes, fats, oils, resins, rubbers; motor fuel and lubricants; cleaning machinery; substitute for turpentine; shoe-creams; stain remover.

DECANE

Synonyms: *n*-Decane; Decyl hydride; UN 2247.

$$
\begin{array}{c}
\text{H}\ \ \text{H}\ \ \text{H}\ \ \text{H}\ \ \text{H}\ \ \text{H}\ \ \text{H}\ \ \text{H}\ \ \text{H}\ \ \text{H} \\
|\ \ \ |\ \ \ |\ \ \ |\ \ \ |\ \ \ |\ \ \ |\ \ \ |\ \ \ |\ \ \ | \\
\text{H}-\text{C}-\text{C}-\text{C}-\text{C}-\text{C}-\text{C}-\text{C}-\text{C}-\text{C}-\text{C}-\text{H} \\
|\ \ \ |\ \ \ |\ \ \ |\ \ \ |\ \ \ |\ \ \ |\ \ \ |\ \ \ |\ \ \ | \\
\text{H}\ \ \text{H}\ \ \text{H}\ \ \text{H}\ \ \text{H}\ \ \text{H}\ \ \text{H}\ \ \text{H}\ \ \text{H}\ \ \text{H}
\end{array}
$$

CAS Registry No.: 124-18-5
DOT: 2247
Molecular formula: $C_{10}H_{22}$
Formula weight: 142.28
RTECS: HD6550000

Physical state and color
Clear, colorless liquid.

Melting point (°C):
-29.7 (Weast, 1986)
-30.0 (Stephenson and Malanowski, 1987)

Boiling point (°C):
174.1 (Dreisbach, 1959)

Density (g/cm³):
0.73005 at 20/4 °C (Dreisbach, 1959)
0.72625 at 25/4 °C (Riddick et al., 1986)

Dissociation constant:
>14 (Schwarzenbach et al., 1993)

Diffusivity in water (10^5 cm²/sec):
0.60 at 20 °C using method of Hayduk and Laudie (1974).

Flash point (°C):
46.1 (Sax and Lewis, 1987)

Lower explosive limit (%):
0.8 (Sax and Lewis, 1987)

Upper explosive limit (%):
5.4 (Sax and Lewis, 1987).

Heat of fusion (kcal/mol):
6.863 (Dean, 1987)

Henry's law constant (atm·m³/mol):
0.187 at 25 °C (approximate - calculated from water solubility and vapor pressure)

Interfacial tension with water (dyn/cm at 20 °C):
51.2 (Girifalco and Good, 1957)

Soil sorption coefficient, log K_{oc}:
Unavailable because experimental methods for estimation of this parameter for aliphatic hydrocarbons are lacking in the documented literature.

Octanol/water partition coefficient, log K_{ow}:
6.69 (Burkhard et al., 1985a)

Solubility in organics:
Miscible with hexane (Corby and Elworthy, 1971).
In methanol, g/L: 62 at 5 °C, 68 at 10 °C, 74 at 15 °C, 81 at 20 °C, 89 at 25 °C, 98 at 30 °C, 109 at 35 °C, 120 at 40 °C (Kiser et al., 1961).

Solubility in water:
2.20 x 10^{-5} mL/L at 25 °C (Baker, 1959)
52 μg/kg at 25 °C (McAuliffe, 1969)
19.8 μg/kg at 25 °C (Franks, 1966)
9 μg/L at 20 °C (distilled water), 0.087 mg/L at 20 °C (seawater) (Verschueren, 1983)

Vapor density:
5.82 g/L at 25 °C, 4.91 (air = 1)

Vapor pressure (mmHg):
2.7 at 20 °C (Verschueren, 1983)
1.35 at 25 °C (Wilhoit and Zwolinski, 1971)

Environmental Fate
   Biological. Decane may biodegrade in two ways. The first is the formation of decyl hydroperoxide, which decomposes to 1-decanol, followed by oxidation to decanoic acid. The other pathway involves dehydrogenation to 1-decene, which may react with water giving 1-decanol (Dugan, 1972). Microorganisms can oxidize alkanes under aerobic conditions (Singer and Finnerty, 1984). The most common degradative pathway involves the oxidation of the terminal methyl group forming the corresponding alcohol (1-decanol). The alcohol may undergo a series of

dehydrogenation steps, forming decanal, followed by oxidation forming decanoic acid. The fatty acid may then be metabolized by β-oxidation to form the mineralization products, carbon dioxide and water (Singer and Finnerty, 1984). Hou (1982) reported 1-decanol and 1,10-decanediol as degradation products by the microorganism *Corynebacterium*.

Toxicity: LC_{50} (inhalation) for mice 72,300 gm/kg/2-hour (RTECS, 1985).

Uses: Solvent; standardized hydrocarbon; manufacturing paraffin products; jet fuel research; paper processing industry; rubber industry; organic synthesis.

DIACETONE ALCOHOL

Synonyms: DAA; Diacetone; Diacetonyl alcohol; Diketone alcohol; Dimethyl-acetonylcarbinol; 4-Hydroxy-2-keto-4-methylpentane; 4-Hydroxy-4-methylpen-tanone-2; 4-Hydroxy-4-methylpentan-2-one; **4-Hydroxy-4-methyl-2-penta-none**; 2-Methyl-2-pentanol-4-one; Pyranton; Pyranton A; UN 1148.

$$\begin{array}{c} H_3C \\ \diagdown \\ \diagup \\ H_3C \end{array} C - \overset{\overset{\displaystyle OH}{|}}{\underset{\underset{\displaystyle H}{|}}{C}} - \overset{\overset{\displaystyle H}{|}}{\underset{\underset{\displaystyle H}{|}}{C}} - \overset{\overset{\displaystyle O}{||}}{C} - \overset{\overset{\displaystyle H}{|}}{\underset{\underset{\displaystyle H}{|}}{C}} - H$$

CAS Registry No.: 123-42-2
DOT: 1148
DOT label: Flammable liquid
Molecular formula: $C_6H_{12}O_2$
Formula weight: 116.16
RTECS: SA9100000

Physical state, color and odor
Colorless liquid with a mild, pleasant odor.

Melting point (°C):
-44 (Weast, 1986)

Boiling point (°C):
164 (Weast, 1986)
167.9 (Windholz et al., 1983)

Density (g/cm^3):
0.9387 at 20/4 °C (Weast, 1986)
0.9306 at 25/4 °C (Windholz et al., 1983)

Diffusivity in water (10^5 cm^2/sec):
0.78 at 20 °C using method of Hayduk and Laudie (1974).

Flash point (°C):
51.7 (NIOSH, 1994)
66 (reagent grade), 48 (commercial grade), 13 (commercial grade - open cup) (Windholz et al., 1983)

Lower explosive limit (%):
1.8 (NIOSH, 1994)

Upper explosive limit (%):
6.9 (NIOSH, 1994)

Soil sorption coefficient, log K_{oc}:
Unavailable because experimental methods for estimation of this parameter for ketones are lacking in the documented literature. However, its miscibility in water suggests its adsorption to soil will be nominal (Lyman et al., 1982).

Solubility in organics:
Soluble in alcohol and ether (Weast, 1986).

Solubility in water:
Miscible (NIOSH, 1994).

Vapor density:
4.75 g/L at 25 °C, 4.01 (air = 1)

Vapor pressure (mmHg):
1 at 20 °C (NIOSH, 1994)
1.7 at 30 °C (Verschueren, 1983)

Exposure Limits: NIOSH REL: TWA 50 ppm (240 mg/m^3), IDLH 1,800 ppm; OSHA PEL: TWA 50 ppm.

Symptoms of Exposure: May cause irritation of eyes, nose, throat and skin (Patnaik, 1992).

Toxicity: Acute oral LD_{50} for rats 4,000 mg/kg, mice 3,950 mg/kg (RTECS, 1985).

Uses: Solvent for celluloid, cellulose acetate, fats, oils, waxes, nitrocellulose and resins; wood preservatives; rayon and artificial leather; imitation gold leaf; extraction of resins and waxes; in antifreeze mixtures and hydraulic fluids; laboratory reagent; preservative for animal tissue; dyeing mixtures; stripping agent for textiles.

DIBENZ[*a,h*]ANTHRACENE

Synonyms: 1,2:5,6-Benzanthracene; DBA; 1,2,5,6-DBA; DB[*a,h*]A; 1,2:5,6-Dibenz-anthracene; 1,2:5,6-Dibenz[*a,h*]anthracene; 1,2:5,6-Dibenzoanthracene; Di-benzo[*a,h*]anthracene; RCRA waste number U063.

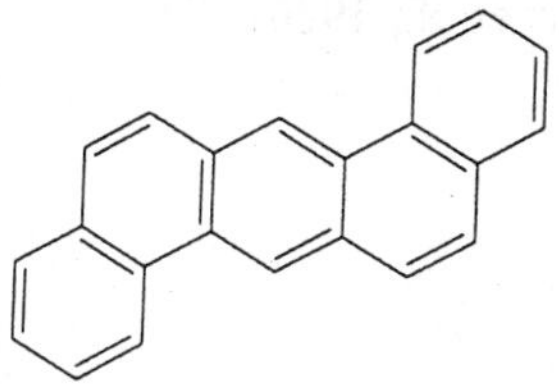

CAS Registry No.: 53-70-3
Molecular formula: $C_{22}H_{14}$
Formula weight: 278.36
RTECS: HN2625000

Physical state and color
White, monoclinic or orthorhombic crystals or leaflets.

Melting point (°C):
271 (Casellato et al., 1973)
262-265 (Fluka, 1988)

Boiling point (°C):
524 (Verschueren, 1983)

Density (g/cm³):
1.282 (IARC, 1973)

Diffusivity in water (10^5 cm²/sec):
0.46 at 20 °C using method of Hayduk and Laudie (1974).

Dissociation constant:
>15 (Christensen et al., 1975)

Henry's law constant (10^6 atm·m³/mol):
1.70 at 25 °C (approximate - calculated from water solubility and vapor pressure)

Ionization potential (eV):
7.28 ± 0.29 (Franklin et al., 1969)

Bioconcentration factor, log BCF:
4.00 (*Daphnia magna*, Newsted and Giesy, 1987)
4.63 (activated sludge), 3.39 (algae), 1.00 (fish) (Freitag et al., 1985)

Soil sorption coefficient, log K_{oc}:
6.22 (Abdul et al., 1987)

Octanol/water partition coefficient, log K_{ow}:
6.50 (Abdul et al., 1987; Means et al., 1980)
6.36 (Chiou et al., 1982)
5.97 (Sims et al., 1988)
6.58 (Burkhard et al., 1985a)

Solubility in organics:
Soluble in petroleum ether, benzene, toluene, xylene and oils. Slightly soluble in alcohol and ether (Windholz et al., 1983).

Solubility in water:
0.5 μg/L at 27 °C (Davis et al., 1942)
2.49 μg/L at 25 °C (Means et al., 1980)
2.15 nmol/L at 25 °C (Klevens, 1950)

Vapor pressure (mmHg):
2.78×10^{-12} at 25 °C (de Kruif, 1980)

Environmental Fate
   Biological. In activated sludge, <0.1% of the applied dibenz[a,h]anthracene mineralized to carbon dioxide after 5 days (Freitag et al., 1985).
   Photolytic. A carbon dioxide yield of 45.3% was achieved when dibenz[a,h]anthracene adsorbed on silica gel was irradiated with light (λ >290 nm) for 17 hours (Freitag et al., 1985).

Exposure Limits: Potential occupational carcinogen. No individual standards have been set, however, as a constituent in coal tar pitch volatiles, the following exposure limits have been established (mg/m^3): NIOSH REL: TWA 0.1 (cyclohexane-extractable fraction), IDLH 80; OSHA PEL: TWA 0.2 (benzene-soluble fraction); ACGIH TLV: TWA 0.2 (benzene solubles).

Toxicity: LD_{50} (intravenous) for rats is 10 mg/kg (Patnaik, 1992).

Drinking Water Standard (proposed): MCLG: zero; MCL: 0.3 μg/L (U.S. EPA, 1984).

Use: Research chemical. Though not produced commercially in the U.S., dibenz[a,h]anthracene is derived from industrial and experimental coal gasification operations where maximum concentration detected in gas and coal tar streams were 0.0061 and 3.4 mg/m^3, respectively (Cleland, 1981).

DIBENZOFURAN

Synonyms: Biphenylene oxide; Diphenylene oxide.

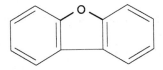

CAS Registry No.: 132-64-9
Molecular formula: $C_{12}H_8O$
Formula weight: 168.20

Physical state and color
Colorless crystals.

Melting point (°C):
86–87 (Weast, 1986)
82 (Banerjee et al., 1980)

Boiling point (°C):
287 (Weast, 1986)

Density (g/cm³):
1.0886 at 99/4 °C (Weast, 1986)

Diffusivity in water (10^5 cm²/sec):
0.63 at 20 °C using method of Hayduk and Laudie (1974).

Heat of fusion (kcal/mol):
4.6845 (Rordorf, 1989)

Henry's law constant (10^5 atm·m³/mol):
5.82 at 25 °C (approximate - calculated from water solubility and vapor pressure)

Ionization potential (eV):
8.59 (Franklin et al., 1969)

Soil sorption coefficient, log K_{oc}:
3.91–4.10 using method of Karickhoff et al. (1979).

Octanol/water partition coefficient, log K_{ow}:
4.17 (Banerjee et al., 1980)

4.12 (Leo et al., 1971)
4.31 (Doucette and Andren, 1988)

Solubility in organics:
Soluble in acetic acid, acetone, ethanol and ether (Weast, 1986).

Solubility in water (mg/L at 25 °C):
10.03 (Banerjee et al., 1980)
4.22 (Doucette and Andren, 1988a)

Vapor pressure (mmHg):
2.63 x 10^{-3} at 25 °C (Rordorf, 1989)

Environmental Fate
   Chemical/Physical. It was suggested that the chlorination of dibenzofuran in tap water accounted for the presence of chlorodibenzofuran (Shiraishi et al., 1985).

Use: Research chemical. Derived from industrial and experimental coal gasification operations where the maximum concentration detected in coal gas tar streams was 12 mg/m^3 (Cleland, 1981).

1,4-DIBROMOBENZENE

Synonym: *p*-Dibromobenzene.

CAS Registry No.: 106-37-6
DOT: 2711
DOT label: Combustible liquid
Molecular formula: $C_6H_4Br_2$
Formula weight: 235.91
RTECS: CZ1791000

Physical state, color and odor
Colorless liquid with a pleasant, aromatic odor.

Melting point (°C):
87.3 (Weast, 1986)

Boiling point (°C):
218-219 (Weast, 1986)
219 (Dean, 1987)
220.4 (Windholz et al., 1983)
225 (Hawley, 1981)

Density (g/cm³):
1.9767 at 25/4 °C (Hawley, 1981)
1.841 at 87-89 °C (Aldrich, 1988)

Diffusivity in water (10^5 cm²/sec):
0.71 at 20 °C using method of Hayduk and Laudie (1974).

Flash point (°C):
None (Dean, 1987)

Henry's law constant (10^4 atm·m³/mol):
5.0 at 25 °C (Hine and Mookerjee, 1975)

Soil sorption coefficient, log K_{oc}:
3.20 using method of Chiou et al. (1979).

303

Octanol/water partition coefficient, log K_{ow}:
3.79 (Watarai et al., 1982)
3.89 (Gobas et al., 1988)

Solubility in organics:
Miscible with acetone, alcohol, benzene, carbon tetrachloride, ether and heptane (Hawley, 1981).

Solubility in water:
16.5 mg/L at 25 °C (Andrews and Keefer, 1950)
0.112 mmol/L at 35 °C (Hine et al., 1963)
20.0 mg/L at 25 °C (Mackay and Shiu, 1977)

Vapor pressure (mmHg):
0.161 at 25 °C (Mackay et al., 1982)
0.134 at 35 °C (Hine et al., 1963)

Toxicity: Acute oral LD_{50} for mice 120 mg/kg (RTECS, 1985).

Uses: Solvent for oils; ore flotation; motor fuels; organic synthesis.

DIBROMOCHLOROMETHANE

Synonyms: Chlorodibromomethane; CDBM; NCI-C55254.

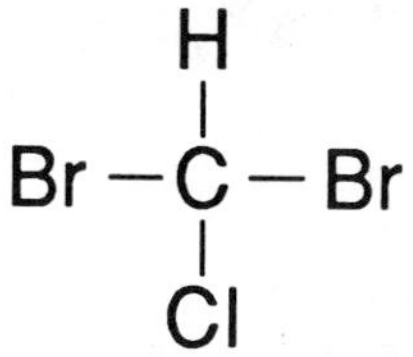

$$\begin{array}{c} H \\ | \\ Br-C-Br \\ | \\ Cl \end{array}$$

CAS Registry No.: 124-48-1
Molecular formula: $CHBr_2Cl$
Formula weight: 208.28
RTECS: PA6360000

Physical state and color
Clear, colorless to pale yellow, heavy liquid.

Melting point (°C):
-23 to -21 (Dean, 1973)

Boiling point (°C):
116 (Hawley, 1981)
122 (Horvath, 1982)

Density (g/cm^3):
2.451 at 20/4 °C (Weast, 1986)

Diffusivity in water (10^5 cm^2/sec):
0.97 at 20 °C using method of Hayduk and Laudie (1974).

Flash point (°C):
Noncombustible (Aldrich, 1988)

Henry's law constant (10^4 atm·m^3/mol):
78.3 at 25 °C (Warner et al., 1987)
8.7 at 20 °C (Nicholson et al., 1984)
3.8, 4.5, 10.3, 11.8 and 15.2 at 10, 15, 20, 25 and 30 °C, respectively (Ashworth et al., 1988)
8, 14 and 22 at 20, 30 and 40 °C, respectively (Tse et al., 1992)
In seawater: 2.8, 5.5 and 10.2 at 0, 10 and 20 °C, respectively (Moore et al., 1995)

Ionization potential (eV):
10.59 (HNU, 1986)

305

Soil sorption coefficient, log K_{oc}:
1.92 (Schwille, 1988)

Octanol/water partition coefficient, log K_{ow}:
2.24 (Hansch and Leo, 1979)
2.08 (Mills et al., 1985)

Solubility in organics:
Miscible with oils, dichloropropane and isopropanol (U.S. EPA, 1985).

Solubility in water (mg/L at 30 °C):
2,509 (McNally and Grob, 1984)
1,049.9 (McNally and Grob, 1983)

Vapor density:
8.51 g/L at 25 °C, 7.19 (air = 1)

Vapor pressure (mmHg):
76 at 20 °C (Schwille, 1988)

Environmental Fate

Biological. Dibromochloromethane showed significant degradation with gradual adaptation in a static-culture flask-screening test (settled domestic wastewater inoculum) conducted at 25 °C. At concentrations of 5 and 10 mg/L, percent losses after 4 weeks of incubation were 39 and 25, respectively. At a substrate concentration of 5 mg/L, 16% was lost due to volatilization after 10 days (Tabak et al., 1981).

Photolytic. Water containing 2,000 ng/μL of dibromochloromethane and colloidal platinum catalyst was irradiated with UV light. After 20 hours, dibromochloromethane degraded to 80 ng/μL bromochloromethane, 22 ng/μL methyl chloride and 1,050 ng/μL methane. A duplicate experiment was performed but 1 g zinc was added. After about 1 hour, total degradation was achieved. Presumed transformation products include methane, bromide and chloride ions (Wang and Tan, 1988).

Chemical/Physical. The estimated hydrolysis half-life in water at 25 °C and pH 7 is 274 years (Mabey and Mill, 1978). Hydrogen gas was bubbled in an aqueous solution containing 18.8 μmol bromodichloromethane. After 24 hours, only 18% of the bromodichloromethane reacted to form methane and minor traces of ethane. In the presence of colloidal platinum catalyst, the reaction proceeded at a much faster rate forming the same end products (Wang et al., 1988).

Drinking Water Standard (tentative): MCLG: zero; MCL: 0.1 mg/L. Total for all trihalomethanes cannot exceed a concentration of 0.08 mg/L (U.S. EPA, 1984).

Toxicity: Acute oral LD_{50} for rats 848 mg/kg, mice 800 mg/kg (RTECS, 1985).

Uses: Manufacture of fire extinguishing agents, propellants, refrigerants and pesticides; organic synthesis.

1,2-DIBROMO-3-CHLOROPROPANE

Synonyms: BBC 12; 1-Chloro-2,3-dibromopropane; 3-Chloro-1,2-dibromopropane; DBCP; Dibromochloropropane; Fumagon; Fumazone; Fumazone 86; Fumazone 86E; NCI-C00500; Nemabrom; Nemafume; Nemagon; Nemagon 20; Nemagon 20G; Nemagon 90; Nemagon 206; Nemagon soil fumigant; Nemanax; Nemapaz; Nemaset; Nematocide; Nematox; Nemazon; OS 1987; Oxy DBCP; RCRA waste number U066; SD 1897; UN 2872.

$$
\begin{array}{ccc}
\text{H} & \text{H} & \text{H} \\
| & | & | \\
\text{H}-\text{C}-\text{C}-\text{C}-\text{H} \\
| & | & | \\
\text{Br} & \text{Br} & \text{Cl}
\end{array}
$$

CAS Registry No.: 96-12-8
DOT: 2872
Molecular formula: $C_3H_5Br_2Cl$
Formula weight: 236.36
RTECS: TX8750000

Physical state, color and odor
Yellow to brown liquid with a pungent odor at high concentrations.

Melting point (°C):
5 (NIOSH, 1994)

Boiling point (°C):
196 (Windholz et al., 1983)

Density (g/cm³):
2.093 at 14/4 °C (Windholz et al., 1983)
2.05 at 20/4 °C (Hawley, 1981)

Diffusivity in water (10^5 cm²/sec):
0.81 at 20 °C using method of Hayduk and Laudie (1974).

Flash point (°C):
76.7 (open cup, NIOSH, 1994)

Henry's law constant (10^4 atm·m³/mol):
2.49 at 20 °C (approximate - calculated from water solubility and vapor pressure)

Soil sorption coefficient, log K_{oc}:
2.11 (Kenaga and Goring, 1980)
1.49-2.16 (Panoche clay loam, Biggar et al., 1984)

Octanol/water partition coefficient, log K_{ow}:
2.63 using method of Hansch et al. (1968).

Solubility in organics:
Miscible with oils, dichloropropane and isopropanol (Windholz et al., 1983).

Solubility in water:
1,270 ppm (Kenaga and Goring, 1980)
1,000 mg/L at 20-25 °C (Verschueren, 1983)

Vapor density:
9.66 g/L at 25 °C, 8.16 (air = 1)

Vapor pressure (mmHg):
0.8 at 21 °C (Verschueren, 1983)

Environmental Fate
Biological. Soil water cultures converted 1,2-dibromo-3-chloropropane to 1-propanol, bromide and chloride ions. Precursors to the alcohol formation include allyl chloride and allyl alcohol (Castro and Belser, 1968). The reported half-life in soil is 6 months (Jury et al., 1987).
Chemical/Physical. Hydrolysis of 1,2-dibromo-3-chloropropane yielded the intermediates 2-bromo-3-chloropropene and 2,3-dibromopropene. Both were readily converted to 2-bromoallyl alcohol. The hydrolysis half-life at pH 7 and 25 °C was calculated to be 38 years (Burlinson et al., 1982).

Exposure Limits: Potential occupational carcinogen. OSHA PEL: TWA 1 ppb.

Toxicity: Acute oral LD_{50} for chickens 60 mg/kg, guinea pigs 150 mg/kg, mice 257 mg/kg, rats 170 mg/kg, rabbits 180 mg/kg (RTECS, 1985).

Drinking Water Standard (final): MCLG: zero; MCL: 0.2 μg/L (U.S. EPA, 1984).

Uses: Soil fumigant, nematocide; pesticide; organic synthesis.

DIBROMODIFLUOROMETHANE

Synonyms: Difluorodibromomethane; Freon 12-B2; Halon 1202; UN 1941.

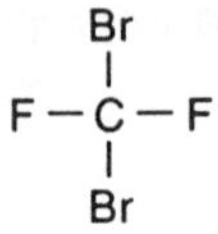

$$\begin{array}{c} Br \\ | \\ F-C-F \\ | \\ Br \end{array}$$

CAS Registry No.: 75-61-6
DOT: 1941
Molecular formula: CBr_2F_2
Formula weight: 209.82
RTECS: PA7525000

Physical state, color and odor
Colorless liquid or gas with a characteristic odor.

Melting point (°C):
-146.2 (NIOSH, 1994)
-142 to -141 (Aldrich, 1988)

Boiling point (°C):
24.5 (Weast, 1986)
22-23 (Aldrich, 1988)
23-24 (Dean, 1987)

Density (g/cm^3):
2.3063 at 15/4 °C (Horvath, 1982)
2.288 at 15/4 °C (Hawley, 1981)
2.297 at 20/4 °C (Aldrich, 1988)

Diffusivity in water (10^5 cm^2/sec):
0.93 at 20 °C using method of Hayduk and Laudie (1974).

Flash point (°C):
Noncombustible (NIOSH, 1994)

Ionization potential (eV):
11.07 (NIOSH, 1994)

Solubility in organics:
Soluble in acetone, alcohol, benzene and ether (Weast, 1986).

Vapor density:
8.58 g/L at 25 °C, 7.24 (air = 1)

Vapor pressure (mmHg):
688 at 20 °C (calculated using the reported constants for the Antoine equation) (Kudchadker et al., 1979).

Exposure Limits: NIOSH REL: TWA 100 ppm (860 mg/m^3), IDLH 2,000 ppm; OSHA PEL: TWA 100 ppm.

Symptoms of Exposure: May cause headache drowsiness and excitement (Patnaik, 1992).

Toxicity: Fifteen minute exposure to 6,400 and 8,000 ppm were fatal to rats and mice, respectively (Patnaik, 1992).

Uses: Synthesis of dyes; quaternary ammonium compounds; pharmaceuticals; fire-extinguishing agent.

DI-*n*-BUTYL PHTHALATE

Synonyms: 1,2-Benzenedicarboxylate; **1,2-Benzenedicarboxylic acid dibutyl ester;** *o*-Benzenedicarboxylic acid dibutyl ester; Benzene-*o*-dicarboxylic acid di-*n*-butyl ester; Butyl phthalate; *n*-Butyl phthalate; Celluflex DPB; DBP; Dibutyl-1,2-benzenedicarboxylate; Dibutyl phthalate; Elaol; Hexaplas M/B; Palatinol C; Phthalic acid dibutyl ester; Polycizer DP; PX 104; RCRA waste number U069; Staflex DBP; Witicizer 300.

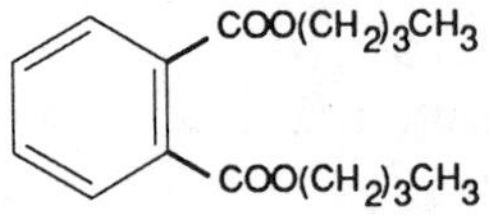

CAS Registry No.: 84-74-2
DOT: 9095
Molecular formula: $C_{16}H_{22}O_4$
Formula weight: 278.35
RTECS: TI0875000

Physical state, color and odor
Colorless to pale yellow, oily, viscous liquid with a mild, aromatic odor.

Melting point (°C):
-35 (Verschueren, 1983)
-40 (Standen, 1968)

Boiling point (°C):
340 (Weast, 1986)
335 (Weiss, 1986)

Density (g/cm^3):
1.042 at 25/4 °C (Standen, 1968)

Diffusivity in water (10^5 cm^2/sec):
0.49 at 20 °C using method of Hayduk and Laudie (1974).

Flash point (°C):
157 (NFPA, 1984)
159 (open cup, Broadhurst, 1972)

Lower explosive limit (%):
0.5 at 235 °C (NFPA, 1984)

Upper explosive limit (%):
2.5 (calculated, Weiss, 1986)

Henry's law constant (10^5 atm·m^3/mol):
6.3 (Petrasek et al., 1983)

Soil sorption coefficient, log K_{oc}:
3.14 (Russell and McDuffie, 1986)
3.00 (Apison soil), 2.98 (Fullerton soil), 2.60 (Dormont soil) (Southworth and Keller, 1986)

Octanol/water partition coefficient, log K_{ow}:
4.31 (Doucette and Andren, 1988)
4.57 (Leyder and Boulanger, 1983)
4.79 (Howard et al., 1985)
4.72 (DeFoe et al., 1990)
5.20 (Wang et al., 1992)

Solubility in organics:
Soluble in ethanol, benzene, ether (Weast, 1986). Very soluble in acetone (Windholz et al., 1983).

Solubility in water:
9.40 mg/L (DeFoe et al., 1990)
13 ppm at 25 °C (Fukano and Obata, 1976)
10.1 mg/L at 20 °C (Leyder and Boulanger, 1983)
11.2 mg/L at 25 °C (Howard et al., 1985)
100 mg/L at 22 °C (Nyssen et al., 1987)
1.300 mg/L at 20-25 °C (Narragansett Bay water, 1.8 mg/L dissolved organic carbon, Boehm and Quinn, 1973)
1.83×10^{-3} wt % at 23.5 °C (Schwarz, 1980)
9.2 mg/L at 25 °C (Russell and McDuffie, 1986)

Vapor density:
11.38 g/L at 25 °C, 9.61 (air = 1)

Vapor pressure (10^5 mmHg at 25 °C):
1.4 (Giam et al., 1980)
7.3 (Banerjee et al., 1990; Howard et al., 1985)
4.2 (Hinckley et al., 1990)

Environmental Fate
   Biological. Under aerobic conditions using a freshwater hydrosoil, mono-*n*-butyl

phthalate and phthalic acid were produced. Under anaerobic conditions, phthalic acid was not present (Verschueren, 1983). In anaerobic sludge, di-*n*-butyl phthalate degraded as follows: monobutyl phthalate to phthalic acid to proto-catechuic acid followed by ring cleavage and mineralization (Shelton et al., 1984). Engelhardt et al. (1975) reported that a variety of microorganisms were capable of degrading of di-*n*-butyl phthalate and suggested the following degradation scheme: di-*n*-butyl phthalate to mono-*n*-butyl phthalate to phthalic acid to 3,4-dihy-droxybenzoic acid and other unidentified products. Di-*n*-butyl phthalate was degraded to benzoic acid by tomato cell suspension cultures (*Lycopericon lycopersicum*) (Pogány et al., 1990).

In a static-culture-flask screening test, di-*n*-butyl phthalate showed significant biodegradation with rapid adaptation. The ester (5 and 10 mg/L) was statically incubated in the dark at 25 °C with yeast extract and settled domestic wastewater inoculum. After 7 days, 100% biodegradation was achieved (Tabak et al., 1981).

Photolytic. An aqueous solution containing titanium dioxide and subjected to UV radiation (λ >290 nm) produced hydroxyphthalates and dihydroxyphthalates as intermediates (Hustert and Moza, 1988).

Chemical/Physical. Pyrolysis of di-*n*-butyl phthalate in the presence of polyvinyl chloride at 600 °C gave the following compounds: indene, methylindene, naphthalene, 1-methylnaphthalene, 2-methylnaphthalene, biphenyl, dimethylnaph-thalene, acenaphthene, fluorene, methylacenaphthene, methylfluorene and six unidentified compounds (Bove and Dalven, 1984). Hydrolyzes in water to phthalic acid and 1-butanol (Wolfe et al., 1980).

Exposure Limits (mg/m³): NIOSH REL: TWA 5, IDLH 4,000; OSHA PEL: TWA 5; ACGIH TLV: TWA 5.

Symptoms of Exposure: Ingestion at a dose level of 150 mg/kg may cause nausea, vomiting, hallucination, dizziness, distorted vision, lacrimation and conjunctivitis (Patnaik, 1992).

Toxicity: Acute oral LD_{50} for guinea pigs 10 gm/kg, mice 5,289 mg/kg, rats 8,000 mg/kg (RTECS, 1985); LC_{50} (48-hour) for red killifish 15 mg/L (Yoshioka et al., 1986).

Drinking Water Standard: As of May 1994, no MCLGs or MCLs have been proposed (U.S. EPA, 1984).

Uses: Manufacture of plasticizers, insect repellents, printing inks, paper coatings, explosives, adhesives, safety glass; organic synthesis.

1,2-DICHLOROBENZENE

Synonyms: Chloroben; Chloroden; Cloroben; DCB; 1,2-DCB; o-DCB; 1,2-Dichlorbenzene; o-Dichlorbenzene; 1,2-Dichlorbenzol; o-Dichlorbenzol; o-Dichlorobenzene; 1,2-Dichlorobenzol; o-Dichlorobenzol; Dilantin DB; Dilatin DB; Dizene; Dowtherm E; NCI-C54944; ODB; ODCB; Orthodichlorobenzene; Orthodichlorobenzol; RCRA waste number U070; Special termite fluid; Termitkil; UN 1591.

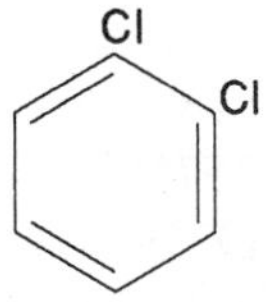

CAS Registry No.: 95-50-1
DOT: 1591
Molecular formula: $C_6H_4Cl_2$
Formula weight: 147.00
RTECS: CZ4500000

Physical state, color and odor
Clear, colorless to pale yellow liquid with a pleasant, aromatic odor. Odor thresholds in air ranges from 2.0 to 4.0 ppm (Keith and Walters, 1992).

Melting point (°C):
-16 to -14 (Fluka, 1988)
-17.00 (Martin et al., 1979)

Boiling point (°C):
180.5 (Weast, 1986)
179.5 (Standen, 1964)

Density (g/cm³):
1.3048 at 20/4 °C (Weast, 1986)
1.30024 at 25/4 °C (Kirchnerová and Cave, 1976)

Diffusivity in water (10^5 cm²/sec):
0.82 at 20 °C using method of Hayduk and Laudie (1974).

Flash point (°C):
66 (NIOSH, 1994)

Lower explosive limit (%):
2.2 (NIOSH, 1994)

Upper explosive limit (%):
9.2 (NIOSH, 1994)

Heat of fusion (kcal/mol):
3.09 (Weast, 1986)
3.19 (Dean, 1987)

Henry's law constant (10^3 atm·m^3/mol):
1.9 (Pankow and Rosen, 1988)
1.2 at 20 °C (Oliver, 1985)
2.4 at 25 °C (Hine and Mookerjee, 1975)
2.83 at 37 °C (Sato and Nakajima, 1979)
1.63, 1.43, 1.68, 1.57 and 2.37 at 10, 15, 20, 25 and 30 °C, respectively (Ashworth et al., 1988)

Ionization potential (eV):
9.06 (NIOSH, 1994)

Bioconcentration factor, log BCF:
1.95 (bluegill sunfish, Veith et al., 1980)
2.43-2.75 (Oliver and Niimi, 1983)

Soil sorption coefficient, log K_{oc}:
2.27 (log K_{om} for a Woodburn silt loam soil, Chiou et al., 1983)
2.255 (silt loam soil, Chiou et al., 1979)
2.59 (Appalachee soil, Stauffer and MacIntyre, 1986)
3.10 (Captina silt loam), 2.90 (McLaurin sandy loam) (Walton et al., 1992)
3.02 (Tinker), 2.83 (Carswell), 2.45 (Barksdale), 2.91 (Blytheville), 3.51 (Traverse City), 3.29 (Borden), 2.85 (Lula) (Stauffer et al., 1989)
2.46 (normal soils), 2.70 (suspended bed sediments) (Kile et al., 1995)

Octanol/water partition coefficient, log K_{ow}:
3.38 (Leo et al., 1971; Wasik et al., 1981)
3.40 (Banerjee et al., 1980)
3.55 (Könemann et al., 1979)
3.433 (de Bruijn et al., 1989)
3.34 (Hammers et al., 1982)

Solubility in organics:
Miscible with alcohol, ether and benzene (Windholz et al., 1983).

Solubility in water:
100 mg/L at 20 °C (Verschueren, 1983)

137 mg/L at 25 °C (Banerjee, 1984)
156 mg/L at 25 °C (Banerjee et al., 1980)
0.0309 vol % at 25 °C (Stephen and Stephen, 1963)
0.0156 wt % at 25 °C (Riddick et al., 1986)
148 mg/L at 20 °C (Chiou et al., 1979)
92.7 mg/L at 25 °C (Yalkowsky et al., 1979)
0.628 mmol/L at 25 °C (Miller et al., 1984)
145 mg/L at 25 °C (Bailey and White, 1965)
0.0145 wt % at 25 °C (Nathan, 1978)
142.3 mg/L at 30 °C (McNally and Grob, 1984)
149.4 mg/L at 30 °C (McNally and Grob, 1983)
0.017 wt % at 25 °C (Lo et al., 1986)

Vapor density:
6.01 g/L at 25 °C, 5.07 (air = 1)

Vapor pressure (mmHg):
1.9 at 30 °C (Verschueren, 1983)
62 at 100 °C (Bailey and White, 1965)
1.282 at 25 °C (Nathan, 1978)
1.5 at 25 °C (Mackay et al., 1982)

Environmental Fate

Biological. Pseudomonas sp. isolated from sewage samples produced 3,4-dichloro-*cis*-1,2-dihydroxycyclohexa-3,5-diene. Subsequent degradation of this metabolite yielded 3,4-dichlorocatechol, which underwent ring cleavage to form 2,3-dichloro-*cis,cis*-muconate, followed by hydrolysis to form 5-chloromaleylacetic acid (Haigler et al., 1988). When 1,2-dichlorobenzene was statically incubated in the dark at 25 °C with yeast extract and settled domestic wastewater inoculum, significant biodegradation with gradual acclimation was followed by a deadaptive process in subsequent subcultures. At a concentration of 5 mg/L, 45, 66, 48 and 29% losses were observed after 7, 14, 21 and 28-day incubation periods, respectively. At a concentration of 10 mg/L, only 20, 59, 32 and 18% losses were observed after 7, 14, 21 and 28-day incubation periods, respectively (Tabak et al., 1981).

Photolytic. Titanium dioxide suspended in an aqueous solution and irradiated with UV light (λ = 365 nm) converted 1,2-dichlorobenzene to carbon dioxide at a significant rate (Matthews, 1986). The sunlight irradiation of 1,2-dichlorobenzene (20 g) in a 100-mL borosilicate glass-stoppered Erlenmeyer flask for 56 days yielded 2,270 ppm 2,3',4'-trichlorobiphenyl (Uyeta et al., 1976).

When an aqueous solution containing 1,2-dichlorobenzene (190 μM) and a nonionic surfactant micelle (Brij 58, a polyoxyethylene cetyl ether) was illuminated by a photoreactor equipped with 253.7-nm monochromatic ultraviolet lamps, photoisomerization took place yielding 1,3- and 1,4-dichlorobenzene as the principal products. The half-life for this reaction, based on the first-order photodecomposition rate of 1.35×10^{-3}/sec, is 8.6 minutes (Chu and Jafvert, 1994).

Chemical/Physical. Anticipated products from the reaction of 1,2-dichlorobenzene with ozone or hydroxyl radicals in the atmosphere are chlorinated phenols, ring cleavage products and nitro compounds (Cupitt, 1980). Based on an assumed base-mediated 1% disappearance after 16 days at 85 °C and pH 9.70 (pH 11.26 at 25 °C), the hydrolysis half-life was estimated to be >900 years (Ellington et al., 1988).

Exposure Limits: NIOSH REL: ceiling 50 ppm (300 mg/m^3), IDLH 200 ppm; OSHA PEL: ceiling 50 ppm; ACGIH TLV: ceiling 50 ppm.

Symptoms of Exposure: Lacrimation, depression of central nervous system, anesthesia and liver damage (Patnaik, 1992).

Toxicity: Acute oral LD_{50} for mice 4,386 mg/kg, rats 500 mg/kg, rabbits 500 mg/kg (RTECS, 1985); LC_{50} (48-hour) for red killifish 68 mg/L (Yoshioka et al., 1986).

Drinking Water Standard (final): MCLG: 0.6 mg/L; MCL: 0.6 mg/L (U.S. EPA, 1984).

Uses: Preparation of 3,4-dichloroaniline; solvent for a wide variety of organic compounds and for oxides of nonferrous metals; solvent carrier in products of toluene diisocyanate; intermediate for dyes; fumigant; insecticide for termites; degreasing hides and wool; metal polishes; degreasing agent for metals, wood and leather; industrial air control; disinfectant; heat transfer medium.

1,3-DICHLOROBENZENE

Synonyms: 1,3-DCB; *m*-DCB; 1,3-Dichlorbenzene; *m*-Dichlorbenzene; 1,3-Dichlorbenzol; *m*-Dichlorbenzol; *m*-Dichlorobenzene; 1,3-Dichlorobenzol; *m*-Dichlorobenzol; RCRA waste number U071; UN 1591.

CAS Registry No.: 541-73-1
Molecular formula: $C_6H_4Cl_2$
Formula weight: 147.00
RTECS: CZ4499000

Physical state and color
Colorless liquid. Odor threshold in air is 20 ppb (Keith and Walters, 1992).

Melting point (°C):
-24.70 (Martin et al., 1979)

Boiling point (°C):
174 (Miller et al., 1984)
173 (Weast, 1986)

Density (g/cm^3):
1.2881 at 20/4 °C, 1.2799 at 25/4 °C (Standen, 1964)

Diffusivity in water (10^5 cm^2/sec):
0.82 at 20 °C using method of Hayduk and Laudie (1974).

Flash point (°C):
63 (Aldrich, 1988)

Lower explosive limit (%):
2.02 (estimated, Weiss, 1986)

Upper explosive limit (%):
9.2 (estimated, Weiss, 1986)

Heat of fusion (kcal/mol):
3.021 (Weast, 1986)

Henry's law constant (10^3 atm·m^3/mol):
3.6 (Pankow and Rosen, 1988)
2.63 at 25 °C (Warner et al., 1987)
4.7 at 25 °C (Hine and Mookerjee, 1975)
4.63 at 37 °C (Sato and Nakajima, 1979)
1.8 at 20 °C (Oliver, 1985)
2.21, 2.31, 2.94, 2.85 and 4.22 at 10, 15, 20, 25 and 30 °C, respectively (Ashworth et al., 1988)

Ionization potential (eV):
9.12 (Franklin et al., 1969)

Bioconcentration factor, log BCF:
1.82 (bluegill sunfish, Veith et al., 1980)
2.62–2.87 (rainbow trout, Oliver and Niimi, 1983)
1.99 (fathead minnow, Carlson and Kosian, 1987)

Soil sorption coefficient, log K_{oc}:
2.23 (log K_{om} for a Woodburn silt loam soil, Chiou et al., 1983)
K_d = 1.4 mL/g on a Cs^+-kaolinite (Haderlein and Schwarzenbach, 1993)

Octanol/water partition coefficient, log K_{ow}:
3.38 (Leo et al., 1971)
3.43 (Miller et al., 1985)
3.48 (Miller et al., 1984; Wasik et al., 1981)
3.44 (Banerjee, 1984); 3.53 (Watarai et al., 1982)
3.72 at 13 °C, 3.55 at 19 °C, 3.48 at 28 °C, 3.42 at 33 °C (Opperhuizen et al., 1988)
3.60 (Könemann et al., 1979)
3.525 (de Bruijn et al., 1989)
3.46 (Hammers et al., 1982)

Solubility in organics:
Soluble in ethanol, acetone, ether, benzene, carbon tetrachloride and ligroin (U.S. EPA, 1985).

Solubility in water:
69 mg/L at 22 °C (Verschueren, 1983)
143 mg/L at 25 °C (Banerjee, 1984)
133 mg/L at 25 °C (Banerjee et al., 1980)
0.0111 wt % at 20 °C (Riddick et al., 1986)
0.847 mmol/L at 25 °C (Miller et al., 1984)
0.01465 wt % at 23.5 °C (Schwarz, 1980)
0.0123 wt % at 25 °C (Nathan, 1978)

125.5 mg/L at 30 °C (McNally and Grob, 1983, 1984)
0.700 μmol/kg at 25.0 °C (Vesala, 1974)

Vapor density:
6.01 g/L at 25 °C, 5.07 (air = 1)

Vapor pressure (mmHg at 25 °C):
1.9 (Warner et al., 1987)
2.3 (Mackay et al., 1982)
2.15 (Banerjee et al., 1990)

Environmental Fate

Biological. When 1,3-dichlorobenzene was statically incubated in the dark at 25 °C with yeast extract and settled domestic wastewater inoculum, significant biodegradation with gradual acclimation was followed by a deadaptive process in subsequent subcultures. At a concentration of 5 mg/L, 59, 69, 39 and 35% losses were observed after 7, 14, 21 and 28-day incubation periods, respectively. At a concentration of 10 mg/L, percent losses were virtually unchanged. After 7, 14, 21 and 28-day incubation periods, percent losses were 58, 67, 31 and 33, respectively (Tabak et al., 1981).

Photolytic. The sunlight irradiation of 1,3-dichlorobenzene (20 g) in a 100-mL borosilicate glass-stoppered Erlenmeyer flask for 56 days yielded 520 ppm trichlorobiphenyl (Uyeta et al., 1976).

When an aqueous solution containing 1,3-dichlorobenzene (190 μM) and a nonionic surfactant micelle (Brij 58, a polyoxyethylene cetyl ether) was illuminated by a photoreactor equipped with 253.7-nm monochromatic ultraviolet lamps, photoisomerization took place yielding 1,2- and 1,4-dichlorobenzene as the principal products. The half-life for this reaction, based on the first-order photodecomposition rate of 1.40×10^{-3}/sec, is 8.3 minutes (Chu and Jafvert, 1994).

Chemical/Physical. Anticipated products from the reaction of 1,3-dichlorobenzene with atmospheric ozone or hydroxyl radicals are chlorinated phenols, ring cleavage products and nitro compounds (Cupitt, 1980). Based on an assumed base-mediated 1% disappearance after 16 days at 85 °C and pH 9.70 (pH 11.26 at 25 °C), the hydrolysis half-life was estimated to be >900 years (Ellington et al., 1988).

Drinking Water Standard (final): MCLG: 0.6 mg/L; MCL: 0.6 mg/L (U.S. EPA, 1994).

Uses: Fumigant and insecticide; organic synthesis.

1,4-DICHLOROBENZENE

Synonyms: 4-Chlorophenyl chloride; *p*-Chlorophenyl chloride; 1,4-DCB; *p*-DCB; Di-chloricide; 4-Dichlorobenzene; *p*-Dichlorobenzene; 4-Dichlorobenzol; *p*-Dichlorobenzol; Evola; NCI-C54955; Paracide; Para crystals; Paradi; Paradichlorobenzene; Paradichlorobenzol; Paradow; Paramoth; Paranuggetts; Parazene; Parodi; PDB; PDCB; Persia-Perazol; RCRA waste number U072; Santochlor; UN 1592.

CAS Registry No.: 106-46-7
DOT: 1592
Molecular formula: $C_6H_4Cl_2$
Formula weight: 147.00
RTECS: CZ4550000

Physical state, color and odor
Colorless to white crystals with a penetrating, mothball-like odor. Odor thresholds in air ranges from 15 to 30 ppm (Keith and Walters, 1992).

Melting point (°C):
53.10 (Martin et al., 1979)

Boiling point (°C):
174.4 (Dean, 1973)

Density (g/cm³):
1.2475 at 20/4 °C (Weast, 1986)

Diffusivity in water (10^5 cm²/sec):
0.82 at 20 °C using method of Hayduk and Laudie (1974).

Flash point (°C):
65.6 (NIOSH, 1994)

Lower explosive limit (%):
2.5 (NIOSH, 1994)

Heat of fusion (kcal/mol):
4.35 (Tsonopoulos and Prausnitz, 1971)
4.54 (Miller et al., 1984)

Henry's law constant (10^3 atm·m³/mol):
3.1 (Pankow and Rosen, 1988)
2.72 at 25 °C (Warner et al., 1987)
4.45 at 25 °C (Hine and Mookerjee, 1975)
1.5 at 20 °C (Oliver, 1985)
2.12, 2.17, 2.59, 3.17 and 3.89 at 10, 15, 20, 25 and 30 °C, respectively (Ashworth et al., 1988)

Ionization potential (eV):
8.98 (NIOSH, 1994)
8.95 (Horvath, 1982)
9.07 (Yoshida et al., 1983)

Bioconcentration factor, log BCF:
2.71-2.95 (rainbow trout, Oliver and Niimi, 1985)
1.78 (bluegill sunfish, Veith et al., 1980)
2.33 (rainbow trout, Neely et al., 1974)
2.00 (algae), 1.70 (fish), 2.75 (activated sludge) (Freitag et al., 1985)

Soil sorption coefficient, log K_{oc}:
2.20 (log K_{om} for a Woodburn silt loam soil, Chiou et al., 1983)
2.82 (Apison soil), 2.93 (Fullerton soil), 2.45 (Dormont soil) (Southworth and Keller, 1986)

Octanol/water partition coefficient, log K_{ow}:
3.39 (Leo et al., 1971)
3.37 (Banerjee et al., 1980; Garst and Wilson, 1984; Wasik et al., 1981)
3.38 (Chiou et al., 1977)
3.53 (Mackay, 1982)
3.62 (Könemann et al., 1979)
3.444 (Brooke et al., 1990; de Bruijn et al., 1989)
3.355 (Brooke et al., 1990)
3.41 (Hammers et al., 1982)
3.40 (Campbell and Luthy, 1985)

Solubility in organics:
Soluble in ethanol, acetone, ether, benzene, carbon tetrachloride, ligroin (U.S. EPA, 1985), carbon disulfide and chloroform (Windholz et al., 1983).

Solubility in water:
65.3 mg/L at 25 °C (Banerjee, 1984)
74 mg/L at 25 °C (Banerjee et al., 1980)
77 mg/kg at 30 °C (Gross and Saylor, 1931)

76 mg/L at 25 °C, 79.1 mg/kg at 25 °C (Andrews and Keefer, 1950)
87.15 mg/L at 25 °C (Aquan-Yuen et al., 1979)
80 mg/L at 25 °C (Gunther et al., 1968)
90.6 mg/L at 25 °C (Yalkowsky et al., 1979)
0.21 mmol/L at 25 °C (Miller et al., 1984)
94.4 mg/L at 30 °C (McNally and Grob, 1984)
92.13 mg/L at 30 °C (McNally and Grob, 1983)
In mg/kg: 78.0, 77.5 at 22.2 °C, 83.6, 83.3 and 83.4 at 24.6 °C, 87.8, 85.6 at 25.5 °C,
    92.1, 93.1 at 30.0 °C, 101, 102 and 103 at 34.5 °C, 121 at 38.4 °C, 159 at 47.5 °C,
    172, 175 at 50.1 (Wauchope and Getzen, 1972)
0.580 mmol/kg at 25.0 °C (Vesala, 1974)

Vapor pressure (mmHg at 25 °C):
0.4 (Standen, 1964)
0.7 (Mackay et al., 1982)

Environmental Fate
   Biological. In activated sludge, <0.1% mineralized to carbon dioxide after 5 days
(Freitag et al., 1985). When 1,4-dichlorobenzene was statically incubated in the
dark at 25 °C with yeast extract and settled domestic wastewater inoculum, significant biodegradation with gradual acclimation was followed by a deadaptive
process in subsequent subcultures. At a concentration of 5 mg/L, 55, 61, 34 and
16% losses were observed after 7, 14, 21 and 28-day incubation periods,
respectively. At a concentration of 10 mg/L, only 37, 54, 29 and 0% losses were
observed after 7, 14, 21 and 28-day incubation periods, respectively (Tabak et al.,
1981).
   Photolytic. Under artificial sunlight, river water containing 2-5 ppm of 1,4-di-
chlorobenzene photodegraded to chlorophenol and phenol (Mansour et al., 1989). A
carbon dioxide yield of 5.1% was achieved when 1,4-dichlorobenzene adsorbed on
silica gel was irradiated with light (λ >290 nm) for 17 hours (Freitag et al., 1985).
Irradiation of 1,4-dichlorophenol in air containing nitrogen oxides gave 2,5-di-
chloro-6-phenol (major product), 2,5-dichloronitrobenzene, 2,5-dichlorophenol
and 2,5-dichloro-4-nitrophenol (Nojima and Kanno, 1980). The sunlight irradiation of 1,4-dichlorobenzene (20 g) in a 100-mL borosilicate glass-stoppered
Erlenmeyer flask for 56 days yielded 1,860 ppm 4,2',5'-trichlorobiphenyl (Uyeta et
al., 1976).
   When an aqueous solution containing 1,2-dichlorobenzene (190 μM) and a
nonionic surfactant micelle (Brij 58, a polyoxyethylene cetyl ether) was illuminated
by a photoreactor equipped with 253.7-nm monochromatic ultraviolet lamps,
photoisomerization took place, yielding 1,3- and 1,4-dichlorobenzene as the
principal products. The half-life for this reaction, based on the first-order
photodecomposition rate of 1.34 x 10^{-3}/sec, is 8.6 minutes (Chu and Jafvert, 1994).
   Chemical/Physical. Anticipated products from the reaction of 1,4-dichloro-

benzene with ozone or hydroxyl radicals in the atmosphere are chlorinated phenols, ring cleavage products and nitro compounds (Cupitt, 1980). Based on an assumed base-mediated 1% disappearance after 16 days at 85 °C and pH 9.70 (pH 11.26 at 25 °C), the hydrolysis half-life was estimated to be >900 years (Ellington et al., 1988).

Exposure Limits: Potential occupational carcinogen. NIOSH REL: IDLH 150 ppm; OSHA PEL: TWA 75 ppm (450 mg/m^3); ACGIH TLV: TWA 75 ppm, STEL 110 ppm (675 mg/m^3).

Symptoms of Exposure: Repeated inhalation of high concentrations of vapors may cause headache, weakness, dizziness, nausea, vomiting, diarrhea, loss of weight and injury to kidney and liver (Patnaik, 1992).

Toxicity: Acute oral LD_{50} for mice 2,950 mg/kg, rats 500 mg/kg, rabbits 2,830 mg/kg (RTECS, 1985).

Drinking Water Standard (final): MCLG: 75 μg/L; MCL: 75 μg/L (U.S. EPA, 1994).

Uses: Moth repellent; general insecticide, fumigant and germicide; space odorant; manufacture of 2,5-dichloroaniline and dyes; pharmacy; agriculture (fumigating soil); disinfectant and chemical intermediate.

3,3'-DICHLOROBENZIDINE

Synonyms: C.I. 23060; Curithane C126; DCB; 4,4'-Diamino-3,3'-dichlorobiphenyl; 4,4'-Diamino-3,3'-dichlorodiphenyl; Dichlorobenzidine; Dichlorobenzidine base; *m,m'*-Dichlorobenzidine; **3,3'-Dichloro-1,1'-(biphenyl)-4,4'-diamine**; 3,3'-Dichlorobiphenyl-4,4'-diamine; 3,3'-Dichloro-4,4'-biphenyldiamine; 3,3'-Dichloro-4,4'-diaminobiphenyl; 3,3'-Dichloro-4,4'-diamino-(1,1-biphenyl); RCRA waste number U073.

Note: Normally found as the dihydrochloride.

CAS Registry No.: 91-94-1
Molecular formula: $C_{12}H_{10}Cl_2N_2$
Formula weight: 253.13
RTECS: DD0525000

Physical state, color and odor
Colorless to grayish-purple crystals with a mild odor.

Melting point (°C):
132-133 (Windholz et al., 1983)

Boiling point (°C):
420 (NIOSH, 1994)

Dissociation constant:
<4 (Boyd et al., 1984)

Diffusivity in water (10^5 cm^2/sec):
0.51 at 20 °C using method of Hayduk and Laudie (1974).

Henry's law constant (10^8 atm·m^3/mol):
4.5 at 25 °C (estimated, Howard, 1989)

Bioconcentration factor, log BCF:
3.49 (activated sludge), 2.97 (algae), 2.79 (golden ide) (Freitag et al., 1985)

Soil sorption coefficient, log K_{oc}:
4.35 (Meylan et al., 1992)

Octanol/water partition coefficient, log K_{ow}:
3.51 (Banerjee et al., 1980)

Solubility in organics:
Soluble in ethanol, benzene and glacial acetic acid (Windholz et al., 1983).

Solubility in water:
3.11 mg/L at 25 °C (Banerjee et al., 1980)
4.0 mg/L at 22 °C (dihydrochloride, U.S. EPA, 1980)
3.99 ppm at pH 6.9 (Appleton and Sikka, 1980)

Vapor pressure (mmHg):
10^{-5} at 22 °C (assigned by analogy, Mabey et al., 1982)
4.2×10^{-7} at 25 °C (estimated, Howard, 1989)

Environmental Fate
 Biological. In activated sludge, 2.7% mineralized to carbon dioxide after 5 days (Freitag et al., 1985).
 Photolytic. An aqueous solution subjected to UV radiation caused a rapid degradation to monochlorobenzidine, benzidine and several unidentified chromophores (Banerjee et al., 1978). A carbon dioxide yield of 41.2% was achieved when 3,3'-dichlorobenzidine adsorbed on silica gel was irradiated with light (λ >290 nm) for 17 hours (Freitag et al., 1985).

Exposure Limits: Potential occupational carcinogen.

Symptoms of Exposure: May cause irritation of eyes, nose, throat and skin (Patnaik, 1992).

Toxicity: Acute oral LD_{50} for rats 5,250 mg/kg (RTECS, 1985).

Uses: Intermediate for azo dyes and pigments; curing agent for isocyanate-terminated polymers and resins; rubber and plastic compounding ingredient; formerly used as chemical intermediate for direct red 61 dye.

DICHLORODIFLUOROMETHANE

Synonyms: Algofrene type 2; Arcton 6; Difluorodichloromethane; Electro-CF 12; Eskimon 12; F 12; FC 12; Fluorocarbon 12; Freon 12; Freon F-12; Frigen 12; Genetron 12; Halon; Halon 122; Isceon 122; Isotron 2; Kaiser chemicals 12; Ledon 12; Propellant 12; R 12; RCRA waste number U075; Refrigerant 12; Ucon 12; Ucon 12/halocarbon 12; UN 1028.

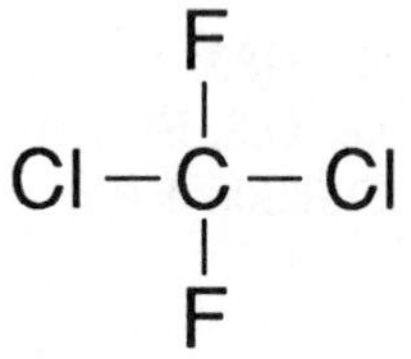

$$Cl - \underset{\underset{F}{|}}{\overset{\overset{F}{|}}{C}} - Cl$$

CAS Registry No.: 75-71-8
DOT: 1028
DOT label: Nonflammable gas
Molecular formula: CCl_2F_2
Formula weight: 120.91
RTECS: PA8200000

Physical state, color and odor
Colorless gas with an ethereal odor.

Melting point (°C):
-158 (Weast, 1986)

Boiling point (°C):
-29.8 (Weast, 1986)

Density (g/cm³):
1.329 at 20/4 °C (Verschueren, 1983)
1.311 at 25/4 °C (Horvath, 1982)

Diffusivity in water (10^5 cm²/sec):
0.93 at 20 °C using method of Hayduk and Laudie (1974).

Flash point (°C):
Nonflammable (Weiss, 1986)

Lower explosive limit (%):
Nonflammable (Weiss, 1986)

Upper explosive limit (%):
Nonflammable (Weiss, 1986)

Henry's law constant (atm·m³/mol):
3.0 (Pankow and Rosen, 1988)
0.425 at 25 °C (Hine and Mookerjee, 1975)

Ionization potential (eV):
11.75 (NIOSH, 1994)
12.31 ± 0.05 (Franklin et al., 1969)

Soil sorption coefficient, log K_{oc}:
2.56 using method of Kenaga and Goring (1980).

Octanol/water partition coefficient, log K_{ow}:
2.16 (Hansch et al., 1975)

Solubility in organics:
Soluble in acetic acid, acetone, chloroform, ether (Weast, 1986) and ethanol (ITII, 1986).

Solubility in water:
280 mg/L at 25 °C (Pearson and McConnell, 1975)
301 mg/L at 25 °C (Munz and Roberts, 1987)

Vapor density:
4.94 g/L at 25 °C, 4.17 (air = 1)

Vapor pressure (mmHg):
4,250 at 20 °C, 5,776 at 30 °C (Verschueren, 1983)
4,870 at 25 °C (Jordan, 1954)
4,306 at 20 °C (McConnell et al., 1975)

Exposure Limits: NIOSH REL: TWA 1,000 ppm (4,950 mg/m³), IDLH 15,000 ppm; OSHA PEL: TWA 1,000 ppm; ACGIH TLV: TWA 1,000 ppm.

Toxicity: LC_{50} (inhalation) for guinea pigs 80 pph/30-minute, mice 76 pph/30-minute, rats, 80 pph/30-minute, rabbits 80 pph/30-minute (RTECS, 1985).

Drinking Water Standard: As of May 1994, no MCLGs or MCLs have been proposed although dichlorodifluoromethane has been listed for regulation (U.S. EPA, 1994).

Uses: Refrigerant; aerosol propellant; plastics; blowing agent; low temperature solvent; chilling cocktail glasses; freezing foods by direct contact; leak-detecting agent.

1,3-DICHLORO-5,5-DIMETHYLHYDANTOIN

Synonyms: Dactin; DCA; DDH; **1,3-Dichloro-5,5-dimethyl-2,4-imidazolidinedione**; Dichlorodimethylhydantoin; Halane; NCI-C03054; Omchlor.

CAS Registry No.: 118-52-5
Molecular formula: $C_5H_6Cl_2N_2O_2$
Formula weight: 197.03
RTECS: MU0700000

Physical state, color and odor
White powder or solid with a chlorine-like odor.

Melting point (°C):
132 (NIOSH, 1994)

Boiling point:
Sublimes at 100 °C (Keith and Walters, 1992).

Density (g/cm³):
1.5 at 20/20 °C (Windholz et al., 1983)

Diffusivity in water:
Not applicable - reacts with water.

Flash point (°C):
174.6 (NIOSH, 1994)

Henry's law constant (atm·m³/mol):
Not applicable - reacts with water.

Soil sorption coefficient, log K_{oc}:
Not applicable - reacts with water.

Octanol/water partition coefficient, log K_{ow}:
Not applicable - reacts with water.

Solubility in organics (wt % at 25 °C):
benzene (9.2), carbon tetrachloride (12.5), chloroform (14), 1,2-dichloroethane

(32.0), methylene chloride (30.0), 1,1,2,2-tetrachloroethane (17.0) (Windholz et al., 1983).

Solubility in water (wt %):
0.21 at 25 °C, 0.60 at 60 °C (Windholz et al., 1983)

Environmental Fate
   Chemical/Physical. Reacts with water (pH = 7.0) releasing hypochlorous acid. At pH 9, nitrogen chloride is formed (Windholz et al., 1983).

Exposure Limits (mg/m^3): NIOSH REL: TWA 0.2, STEL 0.4, IDLH 5; OSHA PEL: TWA 0.2.

Uses: Chlorinating agent; industrial deodorant, disinfectant; intermediate for amino acids, drugs and insecticides; polymerization catalyst; stabilizer for vinyl chloride polymers; household laundry bleach; water treatment; organic synthesis.

1,1-DICHLOROETHANE

Synonyms: Chlorinated hydrochloric ether; 1,1-Dichlorethane; *asym*-Dichloroethane; Ethylidene chloride; Ethylidene dichloride; 1,1-Ethylidene dichloride; NCI-C04535; RCRA waste number U076; UN 2362.

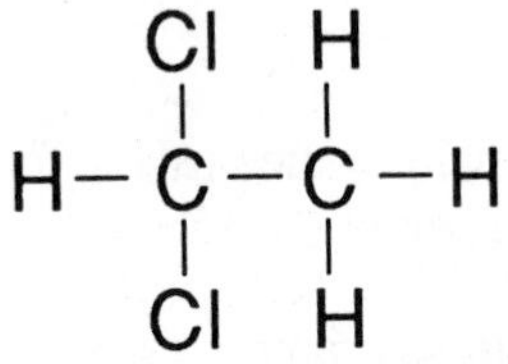

$$\begin{array}{ccc} Cl & & H \\ | & & | \\ H-C&-&C-H \\ | & & | \\ Cl & & H \end{array}$$

CAS Registry No.: 75-34-3
DOT: 2362
DOT label: Flammable liquid
Molecular formula: $C_2H_4Cl_2$
Formula weight: 98.96
RTECS: KI0175000

Physical state, color and odor
Clear, colorless, oily liquid with a chloroform-like odor.

Melting point (°C):
-97.4 (Dean, 1973)
-96.98 (Howard, 1990)

Boiling point (°C):
57.3 (Weast, 1986)

Density (g/cm^3):
1.1757 at 20/4 °C (Weast, 1986)
1.1830 at 15/4 °C, 1.60010 at 30/4 °C (Standen, 1964)

Diffusivity in water (10^5 cm^2/sec):
0.98 at 20 °C using method of Hayduk and Laudie (1974).

Flash point (°C):
-16.7 (NIOSH, 1994)

Lower explosive limit (%):
5.4 (NIOSH, 1994)

Upper explosive limit (%):
11.4 (NIOSH, 1994)

332

Heat of fusion (kcal/mol):
1.881 (Dean, 1987)

Henry's law constant (10^3 atm·m^3/mol):
4.3 (Pankow and Rosen, 1988)
5.45 at 25 °C (Warner et al., 1987)
5.87 at 25 °C (Hine and Mookerjee, 1975)
9.43 at 37 °C (Sato and Nakajima, 1979)
7.76 at 30 °C (Jeffers et al., 1989)
3.68, 4.54, 5.63, 6.25 and 7.76 at 10, 15, 20, 25 and 30 °C, respectively (Ashworth et al., 1988)
4.6, 7.0 and 10.2 at 20, 30 and 40 °C, respectively (Tse et al., 1992)
1.63, 2.06, 2.03, 3.75 and 5.05 at 2.0, 6.0, 10.0, 18.2 and 25.0 °C, respectively (Dewulf et al., 1995)

Ionization potential (eV):
11.06 (NIOSH, 1994)

Soil sorption coefficient, log K_{oc}:
1.48 (Schwille, 1988)

Octanol/water partition coefficient, log K_{ow}:
1.78 (Mills et al., 1985)
1.79 (Hansch and Leo, 1979)

Solubility in organics:
Miscible with ethanol (U.S. EPA, 1985).

Solubility in water:
5,500 mg/L at 20 °C (Verschueren, 1983)
5,060 mg/kg at 25 °C (Gross, 1929)
4,589 mg/L at 30 °C (McNally and Grob, 1984)
4,834.4 mg/L at 30 °C (McNally and Grob, 1983)

Vapor density:
4.04 g/L at 25 °C, 3.42 (air = 1)

Vapor pressure (mmHg):
234 at 25 °C, 270 at 30 °C (Verschueren, 1983)
227 at 25 °C (Howard, 1990)

Environmental Fate
  Biological. 1,1-Dichloroethane showed significant degradation with gradual

adaptation in a static-culture flask-screening test (settled domestic wastewater inoculum) conducted at 25 °C. At concentrations of 5 and 10 mg/L, percent losses after 4 weeks of incubation were 91 and 83, respectively. At a substrate concentration of 5 mg/L, 19% was lost due to volatilization after 10 days (Tabak et al., 1981). Under anoxic conditions, indigenous microbes in uncontaminated sediments produced vinyl chloride (Barrio-Lage et al., 1986).

Photolytic. Titanium dioxide suspended in an aqueous solution and irradiated with UV light (λ = 365 nm) converted dichloroethane to carbon dioxide at a significant rate (Matthews, 1986). The initial photodissociation product of 1,1-dichloroethane was reported to be chloroacetyl chloride (U.S. EPA, 1975). This compound is readily hydrolyzed to hydrochloric acid and chloroacetic acid (Morrison and Boyd, 1971).

Chemical/Physical. A glass bulb containing air and 1,1-dichloroethane degraded outdoors to carbon dioxide and hydrochloric acid. The half-life for this reaction was 17 weeks (Pearson and McConnell, 1975). Hydrolysis of 1,1-dichloroethane under alkaline conditions yielded vinyl chloride. The reported hydrolysis half-life at 25 °C and pH 7 is 61.3 years (Jeffers et al., 1989).

The evaporation half-life of 1,1-dichloroethane (1 mg/L) from water at 25 °C using a shallow-pitch propeller stirrer at 200 rpm at an average depth of 6.5 cm is 32.2 minutes (Dilling, 1977).

Exposure Limits: NIOSH REL: TWA 100 ppm (400 mg/m^3), IDLH 3,000 ppm; OSHA PEL: TWA 100 ppm; ACGIH TLV: TWA 200 ppm (810 mg/m^3), STEL 250 ppm (1,010 mg/m^3).

Symptoms of Exposure: May cause irritation of eyes, nose, throat and skin (Patnaik, 1992).

Toxicity: Acute oral LD$_{50}$ for rats 725 mg/kg (RTECS, 1985).

Drinking Water Standard: As of May 1994, no MCLGs or MCLs have been proposed although 1,1-dichloroethane has been listed for regulation (U.S. EPA, 1994).

Uses: Extraction solvent; insecticide and fumigant; preparation of vinyl chloride; paint, varnish and finish removers; degreasing and drying metal parts; ore flotation; solvent for plastics, oils and fats; chemical intermediate for 1,1,1-trichloroethane; in rubber cementing, fabric spreading and fire extinguishers; formerly used as an anesthetic; organic synthesis.

1,2-DICHLOROETHANE

Synonyms: 1,2-Bichloroethane; Borer sol; Brocide; 1,2-DCA; 1,2-DCE; Destruxol borer-sol; Dichloremulsion; 1,2-Dichlorethane; Dichlormulsion; α,β-Dichloroethane; *sym*-Dichloroethane; Dichloroethylene; Dutch liquid; Dutch oil; EDC; ENT 1656; Ethane dichloride; Ethene dichloride; Ethylene chloride; Ethylene dichloride; 1,2-Ethylene dichloride; Freon 150; Glycol dichloride; NCI-C00511; RCRA waste number U077; UN 1184.

$$Cl-\underset{\underset{H}{|}}{\overset{\overset{H}{|}}{C}}-\underset{\underset{H}{|}}{\overset{\overset{H}{|}}{C}}-Cl$$

CAS Registry No.: 107-06-2
DOT: 2362
DOT label: Flammable liquid
Molecular formula: $C_2H_4Cl_2$
Formula weight: 98.96
RTECS: KI0525000

Physical state, color and odor
Clear, colorless, oily liquid with a pleasant, chloroform-like odor. Odor thresholds in air ranges from 6 to 40 ppm (Keith and Walters, 1992).

Melting point (°C):
-35.3 (Weast, 1986)

Boiling point (°C):
83.5 (Weast, 1986)

Density (g/cm³):
1.2351 at 20/4 °C (Weast, 1986)
1.26000 at 15/4 °C, 1.25280 at 20/4 °C, 1.24530 at 25/4 °C (Standen, 1964)

Diffusivity in water (10^5 cm²/sec):
1.01 at 20 °C using method of Hayduk and Laudie (1974).

Flash point (°C):
13.3 (Fordyce and Meyer, 1940; NIOSH, 1994)

Lower explosive limit (%):
6.2 (NIOSH, 1994)

335

Upper explosive limit (%):
16 (NIOSH, 1994)

Heat of fusion (kcal/mol):
2.112 (Dean, 1987)

Henry's law constant (10^4 atm·m^3/mol):
11.1 at 25 °C (Warner et al., 1987)
9.8 at 25 °C (Dilling, 1977)
13.1 at 25 °C (Hine and Mookerjee, 1975)
22.5 at 37 °C (Sato and Nakajima, 1979)
17.4 at 30 °C (Jeffers et al., 1989)
11.7, 13.0, 14.7, 14.1 and 17.4 at 10, 15, 20, 25 and 30 °C, respectively (Ashworth et al., 1988)
10, 15, 18 and 22 at 20, 30, 35 and 40 °C, respectively (Tse et al., 1992)
3.43, 4.48, 4.12, 7.47 and 10.09 at 2.0, 6.0, 10.0, 18.2 and 25.0 °C, respectively (Dewulf et al., 1995)

Ionization potential (eV):
11.05 (NIOSH, 1994)
11.12 ± 0.05 (Franklin et al., 1969)

Bioconcentration factor, log BCF:
0.30 (bluegill sunfish, Veith et al., 1980)

Soil sorption coefficient, log K_{oc}:
1.279 (silt loam soil, Chiou et al., 1979)

Octanol/water partition coefficient, log K_{ow}:
1.48 (Konietzko, 1984)
1.45 (Banerjee et al., 1980)

Solubility in organics:
Miscible with ethanol, chloroform and ether (U.S. EPA, 1985).

Solubility in water:
9,200 mg/L at 0 °C, 8,690 mg/L at 20 °C (Verschueren, 1983)
8,300 mg/L at 25 °C (Warner et al., 1987)
7,986 mg/L at 25 °C (Banerjee et al., 1980)
8,650 mg/kg at 25 °C (Gross, 1929)
8,800 mg/L at 20 °C (McConnell et al., 1975)
0.81 wt % at 20 °C (Riddick et al., 1986)
8,450 mg/L at 20 °C (Chiou et al., 1979)

0.873 wt % at 0 °C (Konietzko, 1984)
7,200 mg/L at 19.7 °C, 8,100 mg/L at 29.7 °C (Stephenson, 1992)
8,720 mg/kg at 15 °C, 9,000 mg/kg at 30 °C (Gross and Saylor, 1931)
8,524 mg/L at 25 °C (Howard, 1990)
8,100 mg/L (Price et al., 1974)
3,506 mg/L at 30 °C (McNally and Grob, 1984)
1.56 x 10^{-3} at 25 °C (mole fraction, Li et al., 1993)
8,690 mg/L at 25 °C (Cowen and Baynes, 1980)

Vapor density:
4.04 g/L at 25 °C, 3.42 (air = 1)

Vapor pressure (mmHg):
87 at 25 °C (ACGIH, 1986)
78.7 at 20 °C (Howard, 1990)
82 at 25 °C (Nathan, 1978)

Environmental Fate

Biological. Methanococcus thermolithotrophicus, Methanococcus deltae and *Methanobacterium thermoautotrophicum* metabolized 1,2-dichloroethane releasing methane and ethylene (Belay and Daniels, 1987). 1,2-Dichloroethane showed slow to moderate biodegradative activity with concomitant rate of volatilization in a static-culture flask-screening test (settled domestic wastewater inoculum) conducted at 25 °C. At concentrations of 5 and 10 mg/L, percent losses after 4 weeks of incubation were 63 and 53, respectively. At a substrate concentration of 5 mg/L, 27% was lost due to volatilization after 10 days (Tabak et al., 1981).

Photolytic. Titanium dioxide suspended in an aqueous solution and irradiated with UV light (λ = 365 nm) converted dichloroethane to carbon dioxide at a significant rate (Matthews, 1986).

Chemical/Physical. Anticipated products from the reaction of 1,2-dichloroethane with ozone or hydroxyl radicals in the atmosphere are chloroacetaldehyde, chloroacetyl chloride, formaldehyde and ClHCHO (Cupitt, 1980).

Hydrolysis of 1,2-dichloroethane under alkaline and neutral conditions yielded vinyl chloride and ethylene glycol, respectively, with 2-chloroethanol forming as the intermediate (Ellington et al., 1988; Jeffers et al., 1989). The reported hydrolysis half-life in distilled water at 25 °C and pH 7 is 72.0 years (Jeffers et al., 1989), but in a 0.05 M phosphate buffer solution the hydrolysis half-life is 37 years (Barbash and Reinhard, 1989).

In an aqueous solution, 1,2-dichloroethane reacted with hydrogen sulfide ions forming 1,2-dithioethane (Barbash and Reinhard, 1989).

The evaporation half-life of 1,2-dichloroethane (1 mg/L) from water at 25 °C using a shallow-pitch propeller stirrer at 200 rpm at an average depth of 6.5 cm is 28.0 minutes (Dilling, 1977).

When heated to 600 °C, 1,2-dichloroethane decomposes to vinyl chloride and hydrogen chloride (NIOSH, 1994).

Exposure Limits: Potential occupational carcinogen. NIOSH REL: TWA 1 ppm (4 mg/m^3), STEL 2 ppm, IDLH 50 ppm; OSHA PEL: TWA 50 ppm, ceiling 100 ppm, 5-minute/3-hour peak 200 ppm; ACGIH TLV: TWA 10 ppm (40 mg/m^3).

Symptoms of Exposure: Depression of central nervous system, irritation of eyes, corneal opacity, nausea, vomiting, diarrhea, ulceration, somnolence, cyanosis, pulmonary edema, coma. Ingestion of liquid may cause death (Patnaik, 1992).

Toxicity: Acute oral LD_{50} for mice 489 mg/kg, rats 670 mg/kg, rabbits 860 mg/kg (RTECS, 1985).

Drinking Water Standard (final): MCLG: zero; MCL: 5 μg/L (U.S. EPA, 1994).

Uses: Manufacture of acetyl cellulose, vinyl chloride and ethylenediamine; vinyl chloride solvent; lead scavenger in antiknock unleaded gasoline; paint, varnish and finish remover; metal degreasers; soap and scouring compounds; wetting and penetrating agents; ore flotation; tobacco flavoring; soil and foodstuff fumigant; solvent for oils, fats, waxes, resins, gums and rubber.

1,1-DICHLOROETHYLENE

Synonyms: 1,1-DCE; **1,1-Dichloroethene**; *asym*-Dichloroethylene; NCI-C54262; RCRA waste number U078; Sconatex; VDC; Vinylidene chloride; Vinylidene chloride (II); Vinylidene dichloride; Vinylidine chloride.

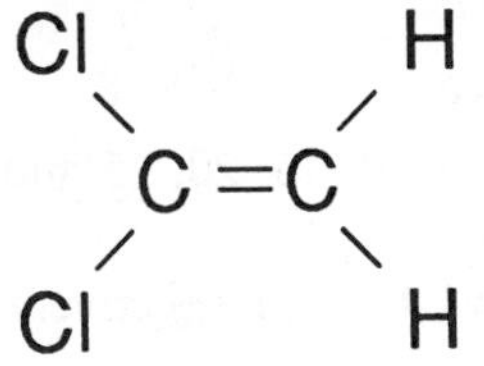

$$\begin{array}{ccc} Cl & & H \\ \diagdown & & \diagup \\ & C = C & \\ \diagup & & \diagdown \\ Cl & & H \end{array}$$

CAS Registry No.: 75-35-4
DOT: 1303
DOT label: Combustible liquid
Molecular formula: $C_2H_2Cl_2$
Formula weight: 96.94
RTECS: KV9275000

Physical state, color and odor
Colorless liquid or gas with a mild, sweet, chloroform-like odor. Odor threshold in air is 190 ppm (Amoore and Hautala, 1983).

Melting point (°C):
-122.1 (Weast, 1986)

Boiling point (°C):
31.56 (Boublik et al., 1986)

Density (g/cm³ at 20/4 °C):
1.218 (Weast, 1986)
1.2132 (Riddick et al., 1986)

Diffusivity in water (10^5 cm²/sec):
1.01 at 20 °C using method of Hayduk and Laudie (1974).

Flash point (°C):
-19 (NIOSH, 1994)

Lower explosive limit (%):
6.5 (NFPA, 1984)

Upper explosive limit (%):
15.5 (NFPA, 1984)

Heat of fusion (kcal/mol):
1.557 (Dean, 1987)

Henry's law constant (10^2 atm·m^3/mol):
1.5 at 25 °C (Warner et al., 1987)
19 (Pankow and Rosen, 1988)
3.18 at 30 °C (Jeffers et al., 1989)
1.54, 2.03, 2.18, 2.59 and 3.18 at 10, 15, 20, 25 and 30 °C, respectively (Ashworth et al., 1988)
2.29, 3.37 and 4.75 at 20, 30 and 40 °C, respectively (Tse et al., 1992)

Ionization potential (eV):
9.81 ± 0.35 (Franklin et al., 1969)
9.46 (Horvath, 1982)
10.00 (NIOSH, 1994)

Soil sorption coefficient, log K_{oc}:
1.81 (Schwille, 1988)

Octanol/water partition coefficient, log K_{ow}:
2.13 (Mabey et al., 1982)
1.48 (HSDB, 1989)

Solubility in organics:
Slightly soluble in ethanol, acetone, benzene and chloroform (U.S. EPA, 1985).

Solubility in water:
400 mg/L at 20 °C (Pearson and McConnell, 1975)
0.021 wt % at 25 °C (Riddick et al., 1986)
In wt % (°C): 0.24 (15), 0.255 (17), 0.25 (20), 0.225 (25), 0.24 (28.5), 0.255 (29.5), 0.22 (38.5), 0.21 (45), 0.23 (51), 0.24 (60), 0.225 (65), 0.295 (71), 0.25 (74.5), 0.295 (81), 0.37 (85.5), 0.35 (90.5) (DeLassus and Schmidt, 1981)
2,232 mg/L at 30 °C (McNally and Grob, 1984)

Vapor density:
3.96 g/L at 25 °C, 3.35 (air = 1)

Vapor pressure (mmHg):
591 at 25 °C, 720 at 30 °C (Verschueren, 1983)
495 at 20 °C, 760 at 31.8 °C (Standen, 1964)

Environmental Fate
Biological. 1,1-Dichloroethylene significantly degraded with rapid adaptation in

a static-culture flask-screening test (settled domestic wastewater inoculum) conducted at 25 °C. Complete degradation was observed after 14 days. At concentrations of 5 and 10 mg/L, the amount lost due to volatilization at the end of 10 days was 24 and 15%, respectively. (Tabak et al., 1981).

Soil. In a methanogenic aquifer material, 1,1-dichloroethylene biodegraded to vinyl chloride (Wilson et al., 1986). Under anoxic conditions, indigenous microbes in uncontaminated sediments degraded 1,1-dichloroethylene to vinyl chloride (Barrio-Lage et al., 1986).

Photolytic. Photooxidation of 1,1-dichloroethylene in the presence of nitrogen dioxide and air yielded phosgene, chloroacetyl chloride, formic acid, hydrochloric acid, carbon monoxide, formaldehyde and ozone (Gay et al., 1976).

Chemical/Physical. Above 0 °C in the presence of oxygen or other catalysts, 1,1-dichloroethylene will polymerize to a plastic (Windholz et al., 1983). The alkaline hydrolysis of 1,1-dichloroethylene yielded chloro-acetylene. The reported hydrolysis half-life at 25 °C and pH 7 is 1.2×10^8 years (Jeffers et al., 1989).

The evaporation half-life of 1,1-dichloroethylene (1 mg/L) from water at 25 °C using a shallow-pitch propeller stirrer at 200 rpm at an average depth of 6.5 cm is 27.2 minutes (Dilling, 1977).

Exposure Limits: Potential occupational carcinogen. ACGIH TLV: TWA 5 ppm (20 mg/m^3), STEL 20 ppm (80 mg/m^3).

Symptoms of Exposure: Irritation of mucous membranes. Narcotic at high concentrations (Patnaik, 1992).

Toxicity: LC$_{50}$ (inhalation) for rats 6,350 ppm/4-hour, mice 98 ppm/22-hour (RTECS, 1985).

Drinking Water Standard (final): MCLG: 7 μg/L; MCL: 7 μg/L (U.S. EPA, 1994).

Uses: Synthetic fibers and adhesives; chemical intermediate in vinylidene fluoride synthesis; comonomer for food packaging, coating resins and modacrylic fibers.

trans-1,2-DICHLOROETHYLENE

Synonyms: Acetylene dichloride; *trans*-Acetylene dichloride; 1,2-Dichloroethene; **(E)-1,2-Dichloroethene**; *trans*-Dichloroethylene; 1,2-*trans*-Dichloroethene; 1,2-*trans*-Dichloroethylene; *sym*-Dichloroethylene; Dioform.

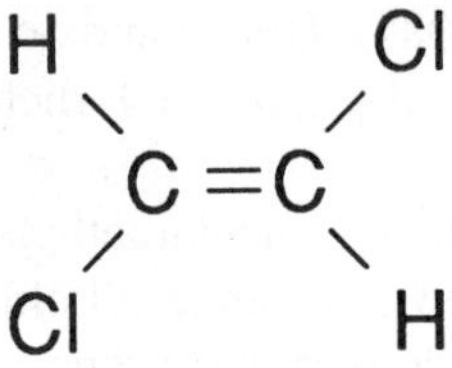

$$H \diagdown \qquad\qquad Cl \diagup$$
$$C = C$$
$$Cl \diagup \qquad\qquad \diagdown H$$

CAS Registry No.: 156-60-5
DOT: 1150 (isomeric mixture)
DOT label: Flammable liquid
Molecular formula: $C_2H_2Cl_2$
Formula weight: 96.94
RTECS: KV9400000

Physical state, color and odor
Colorless, viscous liquid with a sweet, pleasant odor. Odor threshold in air is 17 ppm (Amoore and Hautala, 1983).

Melting point (°C):
-50.0 (McGovern, 1943)

Boiling point (°C):
47.5 (Weast, 1986)

Density (g/cm³):
1.2565 at 20/4 °C (Horvath, 1982)
1.27 at 25/4 °C (isomeric mixture, Weiss, 1986)
1.2631 at 10/4 °C (Standen, 1964)
1.2546 at 25/4 °C (Dean, 1987)

Diffusivity in water (10^5 cm²/sec):
1.03 at 20 °C using method of Hayduk and Laudie (1974).

Flash point (°C):
2 (Sax, 1984)
4 (Fordyce and Meyer, 1940)

Lower explosive limit (%):
9.7 (Sax, 1984)

Upper explosive limit (%):
12.8 (Sax, 1984)

Heat of fusion (kcal/mol):
1.72 (Dean, 1987)

Henry's law constant (10^4 atm·m^3/mol):
3,840 (Gossett, 1987)
72 (Pankow and Rosen, 1988)
67.4 at 25 °C (Hine and Mookerjee, 1975)
121 at 37 °C (Sato and Nakajima, 1979)
57.5 at 30 °C (Jeffers et al., 1989)
5.9, 70.5, 85.7, 94.5 and 121 at 10, 15, 20, 25 and 30 °C, respectively (Ashworth et al., 1988)
79, 118 and 117 at 20, 30 and 40 °C, respectively (Tse et al., 1992)

Ionization potential (eV):
9.64 (Franklin et al., 1969)
9.95 (Horvath, 1982)

Soil sorption coefficient, log K_{oc}:
1.77 (Schwille, 1988)

Octanol/water partition coefficient, log K_{ow}:
2.09 (Mabey et al., 1982)
2.06 (Howard, 1990)

Solubility in organics:
Miscible with acetone, ethanol, ether and very soluble in benzene and chloroform (U.S. EPA, 1985).

Solubility in water (25 °C):
6.3 g/kg at 25 °C (McGovern, 1943)
6,300 mg/L (Dilling, 1977)
0.63 wt % (Riddick et al., 1986)
6,260 mg/L (Kamlet et al., 1987)

Vapor density:
3.96 g/L at 25 °C, 3.35 (air = 1)

Vapor pressure (mmHg):
185 at 10 °C, 265 at 20 °C, 410 at 30 °C (Standen, 1964)
340 at 25 °C (Howard, 1990)

Environmental Fate

Soil. In a methanogenic aquifer material, *trans*-1,2-dichloroethylene biodegraded to vinyl chloride (Wilson et al., 1986). Under anoxic conditions *trans*-1,2-dichloroethylene, when subjected to indigenous microbes in uncontaminated sediments, degraded to vinyl chloride (Barrio-Lage et al., 1986). *trans*-1,2-dichloroethylene showed slow to moderate degradation concomitant with the rate of volatilization in a static-culture flask-screening test (settled domestic wastewater inoculum) conducted at 25 °C. At concentrations of 5 and 10 mg/L, percent losses after 4 weeks of incubation were 95 and 93, respectively. The amount lost due to volatilization was 26-33% after 10 days (Tabak et al., 1981).

Photolytic. Carbon monoxide, formic and hydrochloric acids were reported to be photooxidation products (Gay et al., 1976).

Chemical/Physical. Slowly decomposes in the presence of air, light and moisture releasing hydrogen chloride (Windholz et al., 1983). The reported hydrolysis half-life at 25 °C and pH 7 is 2.1×10^{10} years (Jeffers et al., 1989).

The evaporation half-life of *trans*-1,2-dichloroethane (1 mg/L) from water at 25 °C using a shallow-pitch propeller stirrer at 200 rpm at an average depth of 6.5 cm is 24.0 minutes (Dilling, 1977).

Exposure Limits (isomeric mixture): OSHA PEL: TWA 200 ppm (790 mg/m^3); ACGIH TLV: TWA 200 ppm.

Symptoms of Exposure: Vapor inhalation may cause somnolence and ataxia. Narcotic at high concentrations (Patnaik, 1992).

Toxicity: Acute oral LD_{50} for mice 2,122 mg/kg (RTECS, 1985).

Drinking Water Standard (final): MCLG: 0.1 mg/L; MCL: 0.1 mg/L (U.S. EPA, 1994).

Uses: A mixture of *cis* and *trans* isomers is used as a solvent for fats, phenols, camphor; ingredient in perfumes; low temperature solvent for sensitive substances such as caffeine; refrigerant; organic synthesis.

DICHLOROFLUOROMETHANE

Synonyms: Algofrene type 5; Arcton 7; Dichloromonofluoromethane; Fluorocarbon 21; Fluorodichloromethane; Freon 21; Genetron 21; Halon 21; Refrigerant 21; UN 1029.

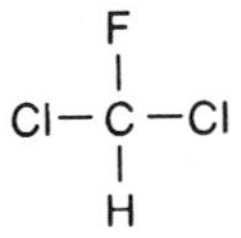

$$Cl-\underset{\underset{H}{|}}{\overset{\overset{F}{|}}{C}}-Cl$$

CAS Registry No.: 75-43-4
DOT: 1029
Molecular formula: $CHCl_2F$
Formula weight: 120.91
RTECS: PA8400000

Physical state, color and odor
Colorless liquid or gas with an ether-like odor.

Melting point (°C):
-135 (Horvath, 1982)

Boiling point (°C):
8.92 (Horvath, 1982)

Density (g/cm^3):
1.366 at 25/4 °C (Horvath, 1982)

Diffusivity in water (10^5 cm^2/sec):
1.08 at 25 °C using method of Hayduk and Laudie (1974).

Flash point (°C):
Nonflammable gas (NIOSH, 1994)

Ionization potential (eV):
12.39 ± 0.20 (Franklin et al., 1969)

Soil sorption coefficient, log K_{oc}:
1.54 using method of Chiou et al. (1979).

Octanol/water partition coefficient, log K_{ow}:
1.55 (Hansch and Leo, 1979)

Solubility in organics:
Soluble in acetic acid, alcohol and ether (Weast, 1986).

Solubility in water:
0.7 wt % at 30 °C (NIOSH, 1994)

Vapor density:
4.94 g/L at 25 °C, 4.17 (air = 1)

Vapor pressure (mmHg):
1,216 at 11.8 °C (NIOSH, 1994)

Exposure Limits: NIOSH REL: TWA 10 ppm (40 mg/m^3), IDLH 5,000 ppm; OSHA PEL: TWA 1,000 ppm (4,200 mg/m^3); ACGIH TLV: TWA 10 ppm.

Toxicity: LC$_{50}$ (inhalation) for rats 49,900 ppm/4-hour (RTECS, 1985).

Uses: Fire extinguishers; solvent; refrigerant.

sym-DICHLOROMETHYL ETHER

Synonyms: BCME; Bis(chloromethyl)ether; Bis-cme; Chloro(chloromethoxy)methane; Chloromethyl ether; Dimethyl-1,1'-dichloroether; *sym*-Dichlorodimethyl ether; Dichloromethyl ether; **Oxybis(chloromethane)**; RCRA waste number P016; UN 2249.

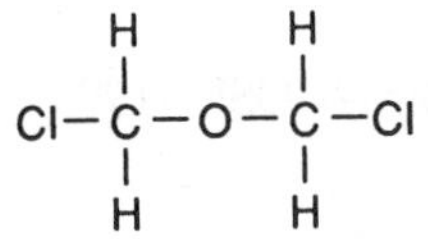

$$Cl-\underset{\underset{\displaystyle H}{|}}{\overset{\overset{\displaystyle H}{|}}{C}}-O-\underset{\underset{\displaystyle H}{|}}{\overset{\overset{\displaystyle H}{|}}{C}}-Cl$$

CAS Registry No.: 542-88-1
DOT: 2249
DOT label: Poison and flammable liquid
Molecular formula: $C_2H_4Cl_2O$
Formula weight: 114.96
RTECS: KN1575000

Physical state, color and odor
Colorless liquid with a suffocating odor.

Melting point (°C):
-41.5 (Weast, 1986)

Boiling point (°C):
104 (Weast, 1986)
106 (Fishbein, 1979)

Density (g/cm³):
1.328 at 15/4 °C (Weast, 1986)
1.315 at 20/4 °C (Fishbein, 1979)

Diffusivity in water:
Not applicable - reacts with water.

Flash point (°C):
<19 (Sax and Lewis, 1987)

Henry's law constant (atm·m³/mol):
Not applicable - reacts with water.

Soil sorption coefficient, log K_{oc}:
Not applicable - reacts with water.

Octanol/water partition coefficient, log K_{ow}:
Not applicable - reacts with water.

Solubility in organics:
Soluble in alcohol, ether (Weast, 1986) and benzene (Hawley, 1981).

Solubility in water:
Not applicable - reacts with water (NIOSH, 1994).

Vapor density:
4.70 g/L at 25 °C, 3.97 (air = 1)

Environmental Fate
   Chemical/Physical. Reacts with water forming hydrochloric acid and form-aldehyde (Fishbein, 1979; NIOSH, 1994). Anticipated products from the reaction of *sym*-dichloromethyl ether with ozone or hydroxyl radicals in the atmosphere, excluding the decomposition products formaldehyde and hydrochloric acid, are chloromethyl formate and formyl chloride (Cupitt, 1980).

Exposure Limits: Potential occupational carcinogen. ACGIH TLV: TWA 1 ppb (5 $\mu g/m^3$).

Symptoms of Exposure: Irritation of eyes, nose and throat (Patnaik, 1992).

Toxicity: Acute oral LD_{50} for rats is 210 mg/kg (RTECS, 1985).

Uses: Intermediate in anionic-exchange quaternary resins; chloromethylating agent.

2,4-DICHLOROPHENOL

Synonyms: 3-Chloro-4-hydroxychlorobenzene; DCP; 2,4-DCP; 2,4-Dichlorohy-droxybenzene; 4,6-Dichlorohydroxybenzene; NCI-C55345; RCRA waste number U081.

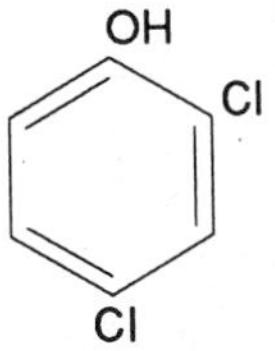

CAS Registry No.: 120-83-2
Molecular formula: $C_6H_4Cl_2O$
Formula weight: 163.00
RTECS: SK8575000

Physical state, color and odor
Colorless to yellow crystals with a sweet, musty or medicinal odor. Odor threshold in air is 210 ppb (Keith and Walters, 1992).

Melting point (°C):
45.0 (Stull, 1947)
43.0 (Renner, 1990)

Boiling point (°C):
210 (Weast, 1986)
216 (Weiss, 1986)

Density (g/cm^3):
1.40 at 15/4 °C (Weiss, 1986)

Dissociation constant:
7.65 (Keith and Walters, 1992)
7.70 (Xie, 1983)
7.80 (Blackman et al., 1980)
7.85 (Dean, 1973)

Flash point (°C):
113.9, 93.3 (open cup, Weiss, 1986)

Henry's law constant (10^6 atm·m^3/mol):
6.66 (calculated, U.S. EPA, 1980)
3.23 at 25 °C (estimated, Leuenberger et al., 1985)

Bioconcentration factor, log BCF:
2.53 (activated sludge), 2.41 (algae), 2.00 (golden ide) (Freitag et al., 1985)
1.0 (brown trout, Hattula et al., 1981)

Soil sorption coefficient, log K_{oc}:
3.60 (fine sediments), 3.50 (coarse sediments) (Isaacson and Frink, 1984)
2.10 (Brookston clay loam, Boyd, 1982)
2.81 (river sediment, Eder and Weber, 1980)
2.17 (loamy sand, Kjeldsen et al., 1990)
2.22, 2.26 and 2.32 in aerobic, anaerobic and autoclaved Brookston clay loam soil,
    respectively (Boyd and King, 1984)

Octanol/water partition coefficient, log K_{ow}:
3.15 (Roberts, 1981)
3.06 (Banerjee et al., 1984; Hansch and Leo, 1979)
3.08 (Krijgsheld and van der Gen, 1986; Leo et al., 1971)
3.23 (Schellenberg et al., 1984)
3.20 (Kishi and Kobayashi, 1994)

Solubility in organics:
Soluble in ethanol, benzene, ether, chloroform (U.S. EPA, 1985) and carbon
tetrachloride (ITII, 1986).

Solubility in water:
4,600 mg/L at 20 °C, 4,500 mg/L at 25 °C (Verschueren, 1983)
4,500 mg/L at 20 °C (Krijgsheld and van der Gen, 1986)
5,000 mg/L at 25 °C (Roberts et al., 1977)
6,194 mg/L at 25 °C and pH 5.1 (Blackman, 1955)
92 mmol/L at 25 °C (Caturla, 1988)

Vapor pressure (10^2 mmHg):
1.5 at 8 °C, 8.9 at 25 °C (Leuenberger et al., 1985)

Environmental Fate
   Biological. In activated sludge, 2.8% mineralized to carbon dioxide after 5 days
(Freitag et al., 1985). In freshwater lake sediments, anaerobic reductive
dechlorination produced 4-chlorophenol (Kohring et al., 1989). Chloroperoxidase,
a fungal enzyme isolated from *Caldariomyces fumago*, converted 9-12% of 2,4-
dichlorophenol to 2,4,6-trichlorophenol (Wannstedt et al., 1990). When 2,4-di-
chlorophenol was statically incubated in the dark at 25 °C with yeast extract and
settled domestic wastewater inoculum, significant biodegradation with rapid
adaptation was observed. At concentrations of 5 and 10 mg/L, 100 and 99%
biodegradation, respectively, were observed after 7 days (Tabak et al., 1981).

Photolytic. In distilled water, photolysis occurs at a slower rate than in estuarine waters containing humic substances. Photolysis products identified in distilled water were the three isomers of chlorocyclopentadienic acid (Hwang et al., 1986). In a similar experiment, titanium dioxide suspended in an aqueous solution and irradiated with UV light (λ = 365 nm) converted 2,4-dichlorophenol to carbon dioxide at a significant rate (Matthews, 1986). An aqueous solution containing hydrogen peroxide and irradiated by UV light (λ = 296 nm) converted 2,4-dichlorophenol to chlorohydroquinone and 1,4-dihydroquinone (Moza et al., 1988). A carbon dioxide yield of 50.4% was achieved when 2,4-dichlorophenol adsorbed on silica gel was irradiated with light (λ >290 nm) for 17 hours (Freitag et al., 1985).

Toxicity: Acute oral LD_{50} for mice 1,276 mg/kg, rats 580 mg/kg (RTECS, 1985).

Drinking Water Standard: As of May 1994, no MCLGs or MCLs have been proposed (U.S. EPA, 1994).

Uses: A chemical intermediate in the manufacture of the pesticide 2,4-dichlorophenoxyacetic acid (2,4-D) and other compounds for use as germicides, antiseptics and seed disinfectants.

1,2-DICHLOROPROPANE

Synonyms: α,β-Dichloropropane; ENT 15406; NCI-C55141; Propylene chloride; Propylene dichloride; α,β-Propylene dichloride; RCRA waste number U083.

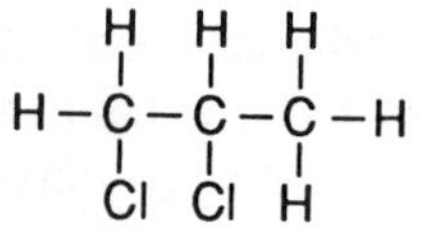

$$H-\underset{\underset{Cl}{|}}{\overset{\overset{H}{|}}{C}}-\underset{\underset{Cl}{|}}{\overset{\overset{H}{|}}{C}}-\underset{\underset{H}{|}}{\overset{\overset{H}{|}}{C}}-H$$

CAS Registry No.: 78-87-5
DOT: 1279
DOT label: Flammable liquid
Molecular formula: $C_3H_6Cl_2$
Formula weight: 112.99
RTECS: TX9625000

Physical state, color and odor
Clear, colorless liquid with a sweet, chloroform-like odor. Odor threshold in air is 50 ppm (Keith and Walters, 1992).

Melting point (°C):
-100.4 (Dreisbach, 1959)

Boiling point (°C):
96.22 (Boublik et al., 1973)
96.0 (Banerjee et al., 1990)

Density (g/cm³ at 20/4 °C):
1.15597 (Riddick et al., 1986)
1.1560 (Dreisbach, 1959)

Diffusivity in water (10^5 cm²/sec):
0.90 at 20 °C using method of Hayduk and Laudie (1974).

Flash point (°C):
15.6 (NIOSH, 1994)

Lower explosive limit (%):
3.4 (NIOSH, 1994)

Upper explosive limit (%):
14.5 (NIOSH, 1994)

Henry's law constant (10^3 atm·m^3/mol):
2.3 (Pankow and Rosen, 1988)
2.94 at 25 °C (Hine and Mookerjee, 1975)
4.71 at 37 °C (Sato and Nakajima, 1979)
2.07 at 25 °C (Howard, 1990)
1.22, 1.26, 1.90, 3.57 and 2.86 at 10, 15, 20, 25 and 30 °C, respectively (Ashworth et al., 1988)
2.1, 3.2 and 4.8 at 20, 30 and 40 °C, respectively (Tse et al., 1992)

Ionization potential (eV):
10.87 (NIOSH, 1994)

Soil sorption coefficient, log K_{oc}:
1.71 (Schwille, 1988)
1.431 (silt loam soil, Chiou et al., 1979)

Octanol/water partition coefficient, log K_{ow}:
2.28 (Mills et al., 1985)
2.00 (Hansch and Leo, 1979)

Solubility in organics:
Miscible with organic solvents (U.S. EPA, 1985).

Solubility in water:
2,700 mg/L at 20 °C (Gunther et al., 1968)
0.280 wt % at 25 °C (Stephen and Stephen, 1963)
2,800 mg/kg at 25 °C (Gross, 1929)
2,740 mg/L at 25 °C (Howard, 1990)
2,069 mg/L at 30 °C (McNally and Grob, 1984)
2,420.4 mg/L at 30 °C (McNally and Grob, 1983)
2,096 mg/L at 25 °C (Jones et al., 1977)

Vapor density:
4.62 g/L at 25 °C, 3.90 (air = 1)

Vapor pressure (mmHg):
42 at 20 °C, 50 at 25 °C, 66 at 30 °C (Verschueren, 1983)
53.3 at 25 °C (Banerjee et al., 1990)

Environmental Fate
Biological. 1,2-Dichloropropane showed significant degradation with gradual adaptation in a static-culture flask-screening test (settled domestic wastewater inoculum) conducted at 25 °C. At concentrations of 5 and 10 mg/L, percent losses

after 4 weeks of incubation were 89 and 81, respectively. The amount lost due to volatilization was only 0-3% (Tabak et al., 1981).

Soil. Boesten et al. (1992) investigated the transformation of [^{14}C]1,2-dichloropropane under laboratory conditions of three subsoils collected from the Netherlands (Wassenaar low-humic sand, Kibbelveen peat, Noord-Sleen humic sand podsoil). The groundwater saturated soils were incubated in the dark at 9.5-10.5 °C. In the Wassenaar soil, no transformation of 1,2-dichloropropane was observed after 156 days of incubation. After 608 and 712 days, however, more than 90% was degraded to nonhalogenated volatile compounds, which were detected in the headspace above the soil. These investigators postulated that these compounds could be propylene and propane in a ratio of 8:1. Degradation of 1,2-dichloropropane in the Kibbelveen peat and Noord-Sleen humic sand podsoil was not observed, possibly because the soil redox potentials in both soils (50-180 and 650-670 mV, respectively) were higher than the redox potential in the Wassenaar soil (10-20 mV).

Photolytic. Distilled water irradiated with UV light (λ = 290 nm) yielded the following photolysis products: 2-chloro-1-propanol, allyl chloride, allyl alcohol and acetone. The photolysis half-life in distilled water is 50 minutes, but in distilled water containing hydrogen peroxide, the half-life decreased to less than 30 minutes (Milano et al., 1988).

Chemical/Physical. Hydrolysis in distilled water at 25 °C produced 1-chloro-2-propanol and hydrochloric acid. The reported half-life for this reaction is 23.6 years (Milano et al., 1988). The hydrolysis rate constant for 1,2-dichloropropane at pH 7 and 25 °C was determined to be 5 x 10^{-6}/hour, resulting in a half-life of 15.8 years (Ellington et al., 1987). Ozonolysis yielded carbon dioxide at low ozone concentrations (Medley and Stover, 1983). Emits toxic chloride fumes when heated to decomposition (Lewis, 1990).

Exposure Limits: Potential occupational carcinogen. NIOSH REL: IDLH 400 ppm; OSHA PEL: TWA 75 ppm (350 mg/m^3); ACGIH TLV: TWA 75 ppm, STEL 110 ppm (510 mg/m^3).

Toxicity: Acute oral LD_{50} for guinea pigs 2,000 mg/kg, rats 2,196 mg/kg (RTECS, 1985).

Drinking Water Standard (final): MCLG: zero; MCL: 5 μg/L (U.S. EPA, 1994).

Uses: Preparation of tetrachloroethylene and carbon tetrachloride; lead scavenger for antiknock fluids; metal cleanser; soil fumigant for nematodes; solvent for oils, fats, gums, waxes and resins; spotting agent.

cis-1,3-DICHLOROPROPYLENE

Synonyms: *cis*-1,3-Dichloropropene; *cis*-1,3-Dichloro-1-propene; (Z)-1,3-Di-chloropropene; **(Z)-1,3-Dichloro-1-propene**; 1,3-Dichloroprop-1-ene; *cis*-1,3-Dichloro-1-propylene.

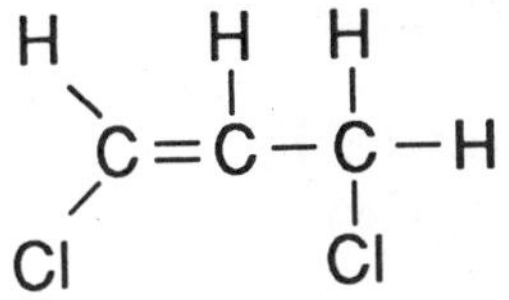

$$\underset{Cl}{\overset{H}{\diagdown}}C = \underset{}{\overset{H}{\underset{|}{C}}} - \underset{Cl}{\overset{H}{\underset{|}{C}}} - H$$

CAS Registry No.: 10061-01-5
DOT: 2047 (isomeric mixture)
DOT label: Flammable liquid
Molecular formula: $C_3H_4Cl_2$
Formula weight: 110.97
RTECS: UC8325000

Physical state, color and odor
Colorless to amber-colored liquid with a chloroform-like odor.

Melting point (°C):
-84 (isomeric mixture, Krijgsheld and van der Gen, 1986)

Boiling point (°C):
104.3 (Horvath, 1982)

Density (g/cm³ at 20/4 °C):
1.224 (Melnikov, 1971)
1.217 (Horvath, 1982)

Diffusivity in water (10^5 cm²/sec):
0.94 at 20 °C using method of Hayduk and Laudie (1974).

Flash point (°C):
35 (isomeric mixture, NFPA, 1984)

Lower explosive limit (%):
5.3 (isomeric mixture, NFPA, 1984)

Upper explosive limit (%):
14.5 (isomeric mixture, NFPA, 1984)

Henry's law constant (10^3 atm·m^3/mol):
1.3 (Pankow and Rosen, 1988)

Soil sorption coefficient, log K$_{oc}$:
1.36 (Kenaga, 1980)
1.75 (isomeric mixture, Meylan et al., 1992)

Octanol/water partition coefficient, log K$_{ow}$:
1.41 (Krijgsheld and van der Gen, 1986)

Solubility in organics:
Soluble in benzene, chloroform and ether (U.S. EPA, 1985).

Solubility in water (mg/L):
2,700 at 20 °C (Dilling, 1977)
911.2 at 30 °C (McNally and Grob, 1984)
1,071.0 at 30 °C (McNally and Grob, 1983)

Vapor density:
4.54 g/L at 25 °C, 3.83 (air = 1)

Vapor pressure (mmHg):
43 at 25 °C (Verschueren, 1983)
25 at 20 °C (Schwille, 1988)

Environmental Fate
    Biological. cis-1,3-Dichloropropylene was reported to hydrolyze to 3-chloro-2-propen-1-ol and can be biologically oxidized to 3-chloropropenoic acid, which is further oxidized to formylacetic acid. Decarboxylation of this compound yields carbon dioxide (Connors et al., 1990). The isomeric mixture showed significant degradation with gradual adaptation in a static-culture flask-screening test (settled domestic wastewater inoculum) conducted at 25 °C. At concentrations of 5 and 10 mg/L, percent losses after 4 weeks of incubation were 85 and 84, respectively. Ten days into the incubation study, 7-19% was lost due to volatilization (Tabak et al., 1981).
    Chemical/Physical. Hydrolysis in distilled water at 25 °C produced 2-chloro-3-propenol and hydrochloric acid. The reported half-life for this reaction is 1 day (Milano et al., 1988). Hydrolysis in wet soil resulted in the formation of *cis*-3-chloroallyl alcohol (Castro and Belser, 1966).
    Chloroacetaldehyde, formyl chloride and chloroacetic acid were formed from the ozonation of dichloropropylene at approximately 23 °C and 730 mmHg. Chloroacetaldehyde and formyl chloride also formed from the reaction of dichloropropylene and hydroxyl radicals (Tuazon et al., 1984).

Emits chlorinated acids when incinerated. Incomplete combustion may release toxic phosgene (Sittig, 1985).

The evaporation half-life of *cis*-1,3-dichloropropylene (1 mg/L) from water at 25 °C using a shallow-pitch propeller stirrer at 200 rpm at an average depth of 6.5 cm is 29.6 minutes (Dilling, 1977).

Drinking Water Standard (tentative): MCLG: zero; MCL: none proposed (U.S. EPA, 1994).

Uses: A mixture containing *cis* and *trans* isomers is used as a soil fumigant and a nematocide.

trans-1,3-DICHLOROPROPYLENE

Synonyms: (*E*)-1,3-Dichloropropene; *trans*-1,3-Dichloropropene; **(*E*)-1,3-Di-chloro-1-propene**; *trans*-1,3-Dichloro-1-propene; 1,3-Dichloroprop-1-ene; *trans*-1,3-Dichloro-1-propylene.

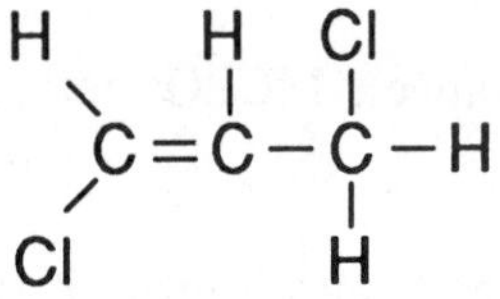

$$\begin{array}{ccccc} H & & H & Cl & \\ \backslash & & | & | & \\ & C = C & - & C & - H \\ / & & & | & \\ Cl & & & H & \end{array}$$

CAS Registry No.: 10061-02-6
DOT: 2047 (isomeric mixture)
DOT label: Flammable liquid
Molecular formula: $C_3H_4Cl_2$
Formula weight: 110.97
RTECS: UC8320000

Physical state, color and odor
Clear, colorless liquid with a chloroform-like odor.

Melting point (°C):
-84 (isomeric mixture, Krijgsheld and van der Gen, 1986)

Boiling point (°C):
112.0 (Horvath, 1982)
112.1 (Melnikov, 1971)

Density (g/cm^3 at 20/4 °C):
1.224 (Horvath, 1982)
1.217 (Krijgsheld and van der Gen, 1986)

Diffusivity in water (10^5 cm^2/sec):
0.92 at 20 °C using method of Hayduk and Laudie (1974).

Flash point (°C):
5.3 (isomeric mixture, NFPA, 1984)

Lower explosive limit (%):
5.3 (isomeric mixture, NFPA, 1984)

Upper explosive limit (%):
14.5 (isomeric mixture, NFPA, 1984)

Henry's law constant (10^3 atm·m³/mol):
1.3 (Pankow and Rosen, 1988)

Soil sorption coefficient, log K_{oc}:
1.68 (Schwille, 1988)
1.415 (Kenaga, 1980)
1.75 (isomeric mixture, Meylan et al., 1992)

Octanol/water partition coefficient, log K_{ow}:
1.41 (Krijgsheld and van der Gen, 1986)

Solubility in organics:
Soluble in benzene, chloroform and ether (U.S. EPA, 1985).

Solubility in water (mg/L):
1,000 at 20 °C (Schwille, 1988)
2,800 at 20 °C (Dilling, 1977)
1,019.9 at 30 °C (McNally and Grob, 1984)
1,188.1 at 30 °C (McNally and Grob, 1983)

Vapor density:
4.54 g/L at 25 °C, 3.83 (air = 1)

Vapor pressure (mmHg):
34 at 25 °C (Verschueren, 1983)
25 at 20 °C (Schwille, 1988)

Environmental Fate

Biological. The isomeric mixture showed significant degradation with gradual adaptation in a static-culture flask-screening test (settled domestic wastewater inoculum) conducted at 25 °C. At concentrations of 5 and 10 mg/L, percent losses after 4 weeks of incubation were 85 and 84, respectively. Ten days into the incubation study, 7-19% was lost due to volatilization (Tabak et al., 1981).

Chemical/Physical. Hydrolysis in distilled water at 25 °C produced 2-chloro-3-propenol and hydrochloric acid. The reported half-life for this reaction is only 2 days (Milano et al., 1988). Hydrolysis in wet soil resulted in the formation of *trans*-3-chloroallyl alcohol (Castro and Belser, 1966). *trans*-1,3-Dichloropropylene was reported to hydrolyze to 3-chloro-2-propen-1-ol and can be biologically oxidized to 3-chloropropenoic acid which is further oxidized to formylacetic acid. Decarboxylation of this compound yields carbon dioxide (Connors et al., 1990). Chloroacetaldehyde, formyl chloride and chloroacetic acid were formed from the ozonation of dichloropropylene at approximately 23 °C and 730 mmHg. Chloroacetaldehyde and formyl chloride also formed from the reaction of dichloropropylene and hydroxyl radicals (Tuazon et al., 1984). Emits chlorinated acids when incinerated. Incomplete combustion may release toxic phosgene (Sittig, 1985).

The evaporation half-life of *trans*-1,3-dichloropropylene (1 mg/L) from water at 25 °C using a shallow-pitch propeller stirrer at 200 rpm at an average depth of 6.5 cm is 24.6 minutes (Dilling, 1977).

Drinking Water Standard (tentative): MCLG: zero; MCL: none proposed (U.S. EPA, 1994).

Uses: A mixture containing *cis* and *trans* isomers is used as a soil fumigant and a nematocide.

DICHLORVOS

Synonyms: Apavap; Astrobot; Atgard; Atgard C; Atgard V; Bay 19149; Benfos; Bibesol; Brevinyl; Brevinyl E50; Canogard; Cekusan; Chlorvinphos; Cyanophos; Cypona; DDVF; DDVP; Dedevap; Deriban; Derribante; Devikol; Dichlorman; 2,2-Dichloroethenyl dimethyl phosphate; 2,2-Dichloroethenyl phosphoric acid dimethyl ester; Dichlorophos; 2,2-Dichlorovinyl dimethyl phosphate; 2,2-Dichlorovinyl dimethyl phosphoric acid ester; Dichlorovos; Dimethyl 2,2-dichloroethenyl phosphate; Dimethyl dichlorovinyl phosphate; Dimethyl 2,2-dichlorovinyl phosphate; *O,O*-Dimethyl *O*-(2,2-dichlorovinyl)phosphate; Divipan; Duo-kill; Duravos; ENT 20738; Equigard; Equigel; Estrosel; Estrosol; Fecama; Fly-die; Fly fighter; Herkal; Herkol; Krecalvin; Lindan; Mafu; Mafu strip; Marvex; Mopari; NA 2783; NCI-C00113; Nerkol; Nogos; Nogos 50; Nogos G; No-pest; No-pest strip; NSC 6738; Nuva; Nuvan; Nuvan 100EC; Oko; OMS 14; **Phosphoric acid 2,2-dichloroethenyl dimethyl ester;** Phosphoric acid 2,2-dichlorovinyl dimethyl ester; Phosvit; SD 1750; Szklarniak; Tap 9VP; Task; Task tabs; Tenac; Tetravos; UDVF; Unifos; Unifos 50 EC; Vapona; Vaponite; Vapora II; Verdican; Verdipor; Vinyl alcohol 2,2-dichlorodimethyl phosphate; Vinylofos; Vinylophos.

CAS Registry No.: 62-73-7
DOT: 2783
DOT label: Poison
Molecular formula: $C_4H_7Cl_2O_4P$
Formula weight: 220.98
RTECS: TC0350000

Physical state, color and odor
Colorless to yellow liquid with an aromatic odor.

Boiling point (°C):
74 at 1 mmHg (Worthing and Hance, 1991)
140 at 20 mmHg (Windholz et al., 1983)

Density (g/cm^3):
1.415 at 25/4 °C (Windholz et al., 1983)

Diffusivity in water (10^5 cm^2/sec):
0.78 at 25 °C using method of Hayduk and Laudie (1974).

Flash point (°C):
>80 (NIOSH, 1994)

Henry's law constant (10^3 atm·m^3/mol):
5.0 (Kawamoto and Urano, 1989)

Soil sorption coefficient, log K_{oc}:
9.57 using method of Saeger et al. (1979).

Octanol/water partition coefficient, log K_{ow}:
1.40 (Leo et al., 1971)

Solubility in organics:
Miscible with alcohol and most non-polar solvents (Windholz et al., 1983).

Solubility in water:
16,000 mg/L (Kawamoto and Urano, 1989)

Vapor density:
9.03 g/L at 25 °C, 7.63 (air = 1)

Vapor pressure (10^2 mmHg):
1.2 at 20 °C (Kawamoto and Urano, 1989; Windholz et al., 1983)
5.27 at 25 °C (Kim et al., 1984)

Environmental Fate

Biological. Dichlorvos incubated with sewage sludge for 1 week at 29 °C degraded to dichloroethanol, dichloroacetic acid, ethyl dichloroacetate and an inorganic phosphate. In addition, dimethyl phosphate formed in the presence or absence of microorganisms (Lieberman and Alexander, 1983).

Plant. Metabolites identified in cotton leaves include dimethyl phosphate, phosphoric acid, methyl phosphate and *O*-dimethyl dichlorvos (Bull and Ridgway, 1969).

Chemical/Physical. Releases very toxic fumes of chlorides and phosphorous oxides when heated to decomposition (Sax and Lewis, 1987). Slowly hydrolyzes in water to dimethyl hydrogen phosphate and dichloroacetaldehyde (Worthing and Hance, 1991).

Exposure Limits (mg/m^3): NIOSH REL: TWA 1, IDLH 100; OSHA PEL: TWA 1.

Symptoms of Exposure: Miosis, eye ache, headache, rhinorrhea, salivation, wheezing, cyanosis, anorexia, vomiting, diarrhea, sweating, muscle fasiculation, paralysis, ataxia, convulsions, low blood pressure (NIOSH, 1994).

Toxicity: Acute oral LD_{50} for wild birds 12 mg/kg, chickens 6.45 mg/kg, ducks 7.8 mg/kg, dogs 1,090 mg/kg, mice 101 mg/kg, pigeons 23.7 mg/kg, pigs 157 mg/kg, quail 23.7 mg/kg, rats 25 mg/kg, rabbits 10 gm/kg (RTECS, 1985); LC_{50} (96-hour) for bluegill sunfish 869 μg/L (Verschueren, 1983); LC_{50} (24-hour) for bluegill sunfish 1.0 mg/L (Hartley and Kidd, 1987); LC_{50} (48-hour) for red killifish 81 mg/L (Yoshioka et al., 1986).

Uses: Insecticide and fumigant.

DIELDRIN

Synonyms: Alvit; Compound 497; Dieldrite; Dieldrix; ENT 16225; HEOD; Hexa-chloroepoxyoctahydro-*endo,exo*-dimethanonaphthalene; 1,2,3,4,10,10-Hexachloro-6,7-epoxy-1,4,4a,5,6,7,8,8a-octahydro-1,4-*endo,exo*-5,8-dimethanonaphthalene; **3,4,5,6,9,9-Hexachloro-1a,2,2a,3,6,6a,7,7a-octahydro-2,7:3,6-dimethanonaphth[2,3-b]oxirene;** Illoxol; Insecticide 497; NA 2761; NCI-C00124; Octalox; Panoram D-31; Quintox; RCRA waste number P037.

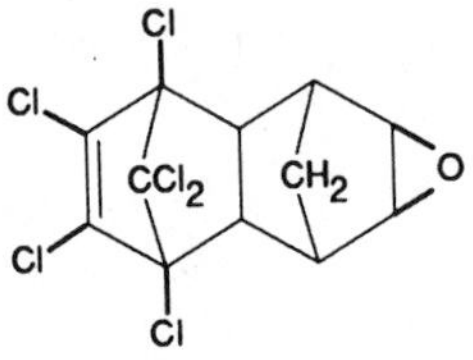

CAS Registry No.: 60-57-1
DOT: 2761
DOT label: Poison
Molecular formula: $C_{12}H_8Cl_6O$
Formula weight: 380.91
RTECS: IO1750000

Physical state, color and odor
White crystals to pale tan flakes with an odorless to mild chemical odor. Odor threshold in air is 41 μg/L (Keith and Walters, 1992).

Melting point (°C):
175-176 (Weast, 1986)
143-144 (technical grade ≈ 90%, Aldrich, 1988)

Boiling point (°C):
Decomposes (Weast, 1986)

Density (g/cm^3):
1.75 at 20/4 °C (Weiss, 1986)

Diffusivity in water (10^5 cm^2/sec):
0.44 at 20 °C using method of Hayduk and Laudie (1974).

Flash point (°C):
Nonflammable (Weiss, 1986)

Henry's law constant (10^7 atm·m^3/mol):
2 (Eisenreich et al., 1981)

580 at 25 °C (Warner et al., 1987)
290 at 20 °C (Slater and Spedding, 1981)

Bioconcentration factor, log BCF:
3.53 (*B. subtilis*, Grimes and Morrison, 1975)
4.25 (activated sludge), 3.36 (algae), 3.48 (golden ide) (Freitag et al., 1985)
4.15 (freshwater fish), 3.61 (fish, microcosm) (Garten and Trabalka, 1983)

Soil sorption coefficient, log K_{oc}:
4.11 (Batcombe silt loam, Briggs, 1981)
4.15 (clay loam, Travis and Arms, 1988)

Octanol/water partition coefficient, log K_{ow}:
5.16 (Kishi et al., 1990)
6.2 (Briggs, 1981)
5.48 (Mackay, 1982)
4.32 (Geyer et al., 1987)
4.49, 4.51, 4.55, 4.66 (Brooke et al., 1986)
3.692 (Rao and Davidson, 1980)
5.401 (de Bruijn et al., 1989)
5.30 (Hammers et al., 1982)

Solubility in organics:
Soluble in ethanol and benzene (Weast, 1986).

Solubility in water (μg/L):
200 at 20 °C (Weil et al., 1974)
186 at 25-29 °C (Park and Bruce, 1968)
50 at 26 °C (Melnikov, 1971)
90 at 15 °C, 195 at 25 °C, 400 at 35 °C, 650 at 45 °C (particle size $\leq$5 μ, Biggar and Riggs, 1974)
20-25 °C: 180 (particle size $\leq$5 μ), 140 (particle size $\leq$0.04 μ) (Robeck et al., 1965)
50 (Gile and Gillett, 1979)
250 at 20-25 °C (Herzel and Murty, 1984)

Vapor density (ng/L):
54 at 20 °C, 202 at 30 °C, 676 at 40 °C (Spencer and Cliath, 1969)

Vapor pressure (10^7 mmHg):
31 at 20 °C (Windholz et al., 1983)
28 at 20 °C (Spencer and Cliath, 1969)
100 at 30 °C (Tinsley, 1979)
7.78 at 20.25 °C (Gile and Gillett, 1979)

448 at 25 °C (Bidleman, 1984)
239, 399 at 25 °C (Hinckley et al., 1990)

Environmental Fate

Biological. Identified metabolites of dieldrin from solution cultures containing *Pseudomonas* sp. in soils include aldrin and dihydroxydihydroaldrin. Other unidentified byproducts included a ketone, an aldehyde and an acid (Matsumura et al., 1968; Kearney and Kaufman, 1976). A pure culture of the marine alga, namely *Dunaliella* sp., degraded dieldrin to photodieldrin and an unknown metabolite at yields of 8.5 and 3.2%, respectively. Photodieldrin and the diol were also identified as metabolites in field-collected samples of marine water, sediments and associated biological materials (Patil et al., 1972). At least 10 different types of bacteria comprising a mixed anaerobic population degraded dieldrin, via monodechlorination at the methylene bridge carbon, to give *syn*- and *anti*-monodechlorodieldrin. Three isolates, *Clostridium bifermentans*, *Clostridium glycolium* and *Clostridium* sp., were capable of dieldrin dechlorination but the rate was much lower than that of the mixed population (Maule et al., 1987). Using settled domestic wastewater inoculum, dieldrin (5 and 10 mg/L) did not degrade after 28 days of incubation at 25 °C in four successive 7-day incubation periods (Tabak et al., 1981).

Soil. Dieldrin is very persistent in soil under both aerobic and anaerobic conditions (Castro and Yoshida, 1971). Reported half-lives in soil ranged from 175 days to 3 years (Howard et al., 1991).

Photolytic. Photolysis of an aqueous solution by sunlight for 3 months resulted in a 70% yield of photodieldrin (Henderson and Crosby, 1968). A solid film of dieldrin exposed to sunlight for 2 months resulted in a 25% yield of photodieldrin (Benson, 1971). In addition to sunlight, UV light converts dieldrin to photodieldrin (Georgacakis and Khan, 1971). Solid dieldrin exposed to UV light (λ <300 nm) under a stream of oxygen yielded small amounts of photodieldrin (Gäb et al., 1974). Many other investigators reported photodieldrin as a photolysis product of dieldrin under various conditions (Crosby and Moilanen, 1974; Ivie and Casida, 1970, 1971, 1971a; Rosen and Carey, 1968; Robinson et al., 1976; Rosen et al., 1966). One of the photoproducts identified besides photodieldrin was photoaldrin chlorohydrin [1,1,2,3,3a,5(or 6),7a-heptachloro-6(or 5)-hydroxydecahydro-2,4,7-metheno-1*H*-cyclopenta[*a*]pentalene] (Lombardo et al., 1972). After a 1 hour exposure to sunlight, dieldrin was converted to photodieldrin. Photodecomposition was accelerated by a number of photosensitizing agents (Ivie and Casida, 1971). When an aqueous solution containing dieldrin was photooxidized by UV light at 90-95 °C, 25, 50 and 75% degraded to carbon dioxide after 2.9, 4.8 and 12.5 hours, respectively (Knoevenagel and Himmelreich, 1976).

Chemical/Physical. The hydrolysis rate constant for dieldrin at pH 7 and 25 °C was determined to be 7.5 x 10^{-6}/hour, resulting in a half-life of 10.5 years (Ellington et al., 1987). At higher temperatures, the hydrolysis half-lives decreased significantly. At 69 °C and pH values of 3.13, 7.22 and 10.45, the calculated hydrolysis half-lives were 19.5, 39.5 and 29.2 days, respectively (Ellington et al., 1986). Products reported from the combustion of dieldrin at 900 °C include carbon monoxide, carbon dioxide, hydrochloric acid, chlorine and unidentified compounds (Kennedy et al., 1972). Emits very toxic chloride fumes when heated to decomposition (Lewis, 1990).

Exposure Limits (mg/m^3): Potential occupational carcinogen. NIOSH REL: TWA 0.25, IDLH 50; OSHA PEL: TWA 0.25; ACGIH TLV: TWA 0.25.

Symptoms of Exposure: Headache, dizziness, nausea, vomiting, malaise, sweating, myoclonic limb jerks, clonic and tonic convulsions, coma, respiratory failure (NIOSH, 1994).

Toxicity: Acute oral LD_{50} for wild birds 13.3 mg/kg, chickens 20 mg/kg, ducks 381 mg/kg, dogs 65 mg/kg, guinea pigs 49 mg/kg, hamsters 60 mg/kg, monkeys 3 mg/kg, mice 38 mg/kg, pigeons 23,700 mg/kg, pigs 38 mg/kg, quail 10.78 mg/kg, rats 38.3 mg/kg, rabbits 45 mg/kg (RTECS, 1985); LC_{50} (96-hour) for goldfish 37 μg/L (Hartley and Kidd, 1987), bluegill sunfish 8 μg/L, fathead minnow 16 μg/L (Henderson et al., 1959), rainbow trout 10 μg/L, coho salmon 11 μg/L, chinook 6 μg/L (Katz, 1961), pumpkinseed 6.7 μg/L, channel catfish 4.5 μg/L (Verschueren, 1983); LC_{50} (48-hour) for mosquito fish 8 ppb (Verschueren, 1983), for red killifish 11 μg/L (Yoshioka et al., 1986); LC_{50} (24-hour) for bluegill sunfish 170 ppb and fathead minnow 24 ppb (Verschueren, 1983).

Drinking Water Standard: As of May 1994, no MCLGs or MCLs have been proposed (U.S. EPA, 1994).

Uses: Insecticide; wool processing industry.

DIETHYLAMINE

Synonyms: Diethamine; *N,N*-Diethylamine; *N*-**Ethylethanamine**; UN 1154.

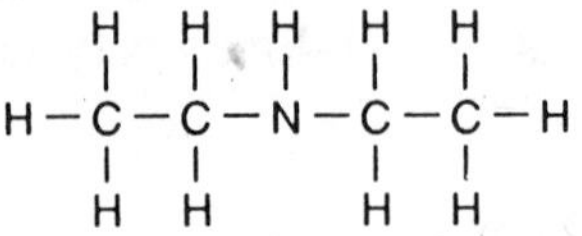

$$H-\overset{\displaystyle H}{\underset{\displaystyle H}{C}}-\overset{\displaystyle H}{\underset{\displaystyle H}{C}}-\overset{\displaystyle H}{N}-\overset{\displaystyle H}{\underset{\displaystyle H}{C}}-\overset{\displaystyle H}{\underset{\displaystyle H}{C}}-H$$

CAS Registry No.: 109-89-7
DOT: 1154
DOT label: Flammable liquid
Molecular formula: $C_4H_{11}N$
Formula weight: 73.14
RTECS: HZ8750000

Physical state, color and odor
Colorless liquid with a fishy, ammonia-like odor. Odor threshold in air is 140 ppb (Keith and Walters, 1992).

Melting point (°C):
-48 (Weast, 1986)
-50 (Windholz et al., 1983)

Boiling point (°C):
56.3 (Weast, 1986)
55.5 (Windholz et al., 1983)

Density (g/cm³):
0.7056 at 20/4 °C (Weast, 1986)
0.711 at 18/4 °C (Verschueren, 1983)

Diffusivity in water (10^5 cm²/sec at 25 °C):
1.11 (Hayduk and Laudie, 1974)

Dissociation constant:
11.090 at 20 °C (Gordon and Ford, 1972)
10.93 at 25 °C (Dean, 1973)

Flash point (°C):
-28 (Aldrich, 1988)

Lower explosive limit (%):
1.8 (NIOSH, 1994)

Upper explosive limit (%):
10.1 (NIOSH, 1994)

Henry's law constant (10^5 atm·m³/mol):
2.56 at 25 °C (Hine and Mookerjee, 1975)

Ionization potential (eV):
8.01 ± 0.01 (Franklin et al., 1969)

Soil sorption coefficient, log K_{oc}:
Unavailable because experimental methods for estimation of this parameter for aliphatic amines are lacking in the documented literature. However, its high solubility in water suggests its adsorption to soil will be nominal (Lyman et al., 1982).

Octanol/water partition coefficient, log K_{ow}:
0.43, 0.58 (Sangster, 1989)
0.44 (Collander, 1951)
0.57 (Leo et al., 1971)
0.81 (Eadsforth, 1986)

Solubility in organics:
Miscible with alcohol (Windholz et al., 1983).

Solubility in water:
815,000 mg/L at 14 °C (Verschueren, 1983)

Vapor density:
2.99 g/L at 25 °C, 2.52 (air = 1)

Vapor pressure (mmHg):
192 at 20 °C (NIOSH, 1994)
200 at 20 °C, 290 at 30 °C (Verschueren, 1983)

Environmental Fate
   Chemical/Physical. Diethylamine reacted with NO_x in the dark forming diethylnitrosamine. In an outdoor chamber, photooxidation by natural sunlight yielded the following products: diethylnitramine, diethylformamide, diethylacetamide, ethylacetamide, ozone, acetaldehyde and peroxyacetylnitrate (Pitts et al., 1978).
   Reacts with acids forming water-soluble salts.

Exposure Limits: NIOSH REL: TWA 10 ppm (30 mg/m³), STEL 25 ppm (75 mg/m^3), IDLH 200 ppm; OSHA PEL: TWA 25 ppm; ACGIH TLV: TWA 10 ppm, STEL 25 ppm.

Symptoms of Exposure: Strong irritant to the eyes, skin and mucous membranes. Eye contact may cause corneal damage (Patnaik, 1992).

Toxicity: Acute oral LD$_{50}$ for mice 500 mg/kg, rats 540 mg/kg (RTECS, 1985).

Uses: In flotation agents, resins, dyes, resins, pesticides, rubber chemicals and pharmaceuticals; selective solvent; polymerization and corrosion inhibitors; petroleum chemicals; electroplating; organic synthesis.

2-DIETHYLAMINOETHANOL

Synonyms: DEAE; Diethylaminoethanol; β-Diethylaminoethanol; *N*-Diethylamino-ethanol; 2-*N*-Diethylaminoethanol; 2-Diethylaminoethyl alcohol; β-Diethylamino-ethyl alcohol; Diethylethanolamine; *N,N*-Diethylethanolamine; Diethyl(2-hy-droxyethyl)amine; *N,N*-Diethyl-*N*-(β-hydroxyethyl)amine; 2-Hydroxytriethyl-amine; UN 2686.

$$CH_3CH_2 \diagdown \\ \quad\quad N-\overset{\overset{\displaystyle H}{|}}{\underset{\underset{\displaystyle H}{|}}{C}}-\overset{\overset{\displaystyle H}{|}}{\underset{\underset{\displaystyle H}{|}}{C}}-OH \\ CH_3CH_2 \diagup$$

CAS Registry No.: 100-37-8
DOT: 2686
Molecular formula: $C_6H_{15}NO$
Formula weight: 117.19
RTECS: KK5075000

Physical state, color and odor
Colorless, hygroscopic liquid with a nauseating, ammonia-like odor. Odor threshold in air is 11 ppb (Amoore and Hautala, 1983).

Melting point (°C):
-70 (Dean, 1987)

Boiling point (°C):
163 (Windholz et al., 1983)
161 (Aldrich, 1988)

Density (g/cm³):
0.8800 at 25/4 °C (Windholz et al., 1983)
0.884 at 20/4 °C (Aldrich, 1988)

Diffusivity in water (10^5 cm²/sec):
0.75 at 20 °C using method of Hayduk and Laudie (1974).

Flash point (°C):
52 (NIOSH, 1994)
60 (open cup, Sax and Lewis, 1987)
48 (Aldrich, 1988)

Soil sorption coefficient, log K_{oc}:
Unavailable because experimental methods for estimation of this parameter for

371

aliphatic alcohols are lacking in the documented literature. However, its high solubility in water suggest its adsorption to soil will be nominal (Lyman et al., 1982).

Solubility in organics:
Soluble in alcohol, benzene and ether (Windholz et al., 1983).

Solubility in water:
Miscible (NIOSH, 1994).

Vapor density:
4.79 g/L at 25 °C, 4.05 (air = 1)

Vapor pressure (mmHg):
1.4 at 20 °C (Sax and Lewis, 1987)

Exposure Limits: NIOSH REL: TWA 10 ppm (50 mg/m^3), IDLH 100 ppm; OSHA PEL: TWA 10 ppm.

Toxicity: Acute oral LD_{50} for rats 1,300 mg/kg (RTECS, 1985).

Uses: Water-soluble salts; textile softeners; antirust formulations; fatty acid derivatives; pharmaceuticals; curing agent for resins; emulsifying agents in acid media; organic synthesis.

DIETHYL PHTHALATE

Synonyms: Anozol; **1,2-Benzenedicarboxylic acid diethyl ester;** DEP; Diethyl-*o*-phthalate; Estol 1550; Ethyl phthalate; NCI-C60048; Neantine; Palatinol A; Phthalol; Placidol E; RCRA waste number U088; Solvanol.

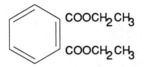

CAS Registry No.: 84-66-2
Molecular formula: $C_{12}H_{14}O_4$
Formula weight: 222.24
RTECS: TI1050000

Physical state, color and odor
Clear, colorless, oily liquid with a mild, chemical odor. Bitter taste.

Melting point (°C):
-40.5 (Verschueren, 1983)

Boiling point (°C):
298 (Weast, 1986)
296 (Standen, 1968)
302 (Sax, 1984)

Density (g/cm³):
1.1175 at 20/4 °C (Weast, 1986)
1.123 at 25/4 °C (Fishbein and Albro, 1972)

Diffusivity in water (10^5 cm²/sec):
0.59 at 20 °C using method of Hayduk and Laudie (1974).

Flash point (°C):
140 (Windholz et al., 1983)
163 (open cup, Sax, 1984)

Lower explosive limit (%):
0.7 at 186 °C (NFPA, 1984)

Henry's law constant (10^7 atm·m³/mol):
8.46 (U.S. EPA, 1980)

Interfacial tension with water (dyn/cm at 20.5 °C):
16.27 (Donahue and Bartell, 1952)

Bioconcentration factor, log BCF:
2.07 (bluegill sunfish, Veith et al., 1980)

Soil sorption coefficient, log K_{oc}:
1.84 (Russell and McDuffie, 1986)

Octanol/water partition coefficient, log K_{ow}:
2.35 (Leyder and Boulanger, 1983)
2.47 (Mabey et al., 1982)
2.24 (Howard et al., 1985)
1.40 (Veith et al., 1980)
2.82 (DeKock and Lord, 1987)

Solubility in organics:
Soluble in acetone and benzene; miscible with ethanol, ether, esters and ketones (U.S. EPA, 1985).

Solubility in water:
896 mg/L at 25 °C (Callahan et al., 1979)
928 mg/L at 20 °C (Leyder and Boulanger, 1983)
1,080 mg/L at 25 °C (Howard et al., 1985)
1,200 ppm at 25 °C (Fukano and Obata, 1976)
0.1 wt % at 20 °C (Fishbein and Albro, 1972)
680 mg/L at 25 °C (Russell and McDuffie, 1986)

Vapor density:
9.08 g/L at 25 °C, 7.67 (air = 1)

Vapor pressure (10^3 mmHg):
50 at 70 °C (Fishbein and Albro, 1972)
1.65 at 25 °C (Banerjee et al., 1980; Howard et al., 1985)
2.1 at 25 °C (Hinckley et al., 1990)

Environmental Fate

Biological. A proposed microbial degradation mechanism is as follows: 4-hydroxy-3-methylbenzyl alcohol to 4-hydroxy-3-methylbenzaldehyde to 3-methyl-4-hydroxybenzoic acid to 4-hydroxyisophthalic acid to protocatechuic acid to β-ketoadipic acid (Chapman, 1972). In anaerobic sludge, diethyl phthalate degraded as follows: monoethyl phthalate to phthalic acid to protocatechuic acid followed by ring cleavage and mineralization (Shelton et al., 1984).

In a static-culture-flask screening test, diethyl phthalate showed significant biodegradation with rapid adaptation. The ester (5 and 10 mg/L) was statically incubated in the dark at 25 °C with yeast extract and settled domestic wastewater inoculum. After 7 days, 100% biodegradation was achieved (Tabak et al., 1981).

Photolytic. An aqueous solution containing titanium dioxide and subjected to UV radiation (λ >290 nm) produced hydroxyphthalates and dihydroxyphthalates as intermediates (Hustert and Moza, 1988).

Chemical/Physical. Hydrolyzes in water to phthalic acid and ethanol (Wolfe et al., 1980). Pyrolysis of diethyl phthalate in a flow reactor at 700 °C yielded the following products: ethanol, ethylene, benzene, naphthalene, phthalic anhydride and 2-phenylenaphthalene (Bove and Arrigo, 1985).

Exposure Limits (mg/m^3): NIOSH REL: TWA 5; ACGIH TLV: TWA 5.

Symptoms of Exposure: Vapor inhalation may cause lacrimation, coughing and throat irritation (Patnaik, 1992).

Toxicity: Acute oral LD_{50} for guinea pigs 8,600 mg/kg, mice 6,172 mg/kg, rats 8,600 mg/kg (RTECS, 1985); LC_{50} (48-hour) for red killifish 98 mg/L (Yoshioka et al., 1986).

Drinking Water Standard: As of May 1994, no MCLGs or MCLs have been proposed (U.S. EPA, 1994).

Uses: Plasticizer; plastic manufacturing and processing; denaturant for ethyl alcohol; ingredient in insecticidal sprays and explosives (propellant); dye application agent; wetting agent; perfumery as fixative and solvent; solvent for nitrocellulose and cellulose acetate; camphor substitute.

1,1-DIFLUOROTETRACHLOROETHANE

Synonyms: 1,1-Difluoro-1,2,2,2-tetrachloroethane; 2,2-Difluoro-1,1,1,2-tetrachloroethane; Freon 112a; Halocarbon 112a; Refrigerant 112a; **1,1,1,2-Tetrachloro-2,2-difluoroethane.**

```
        F   Cl
        |   |
  Cl— C — C —Cl
        |   |
        F   Cl
```

CAS Registry No.: 76-11-9
DOT: 1078
Molecular formula: $C_2Cl_4F_2$
Formula weight: 203.83
RTECS: KI1425000

Physical state, color and odor
Colorless solid with a faint ether-like odor.

Melting point (°C):
40.6 (Weast, 1986)

Boiling point (°C):
91.5 (Weast, 1986)

Density (g/cm^3):
1.65 at 20/4 °C (NIOSH, 1994)

Diffusivity in water (10^5 cm^2/sec):
0.71 at 20 °C using method of Hayduk and Laudie (1974).

Flash point (°C):
Noncombustible solid (NIOSH, 1994)

Solubility in organics:
Soluble in alcohol, ether and chloroform (Weast, 1986).

Vapor pressure (mmHg):
40 at 20 °C (NIOSH, 1994)

Exposure Limits: NIOSH REL: TWA 500 ppm (4,170 mg/m^3), IDLH 2,000 ppm; OSHA PEL: TWA 500 ppm; ACGIH TLV: TWA 500 ppm.

Symptoms of Exposure: Eye and skin irritation, drowsiness, central nervous system depression (NIOSH, 1994).

Use: Organic synthesis.

1,2-DIFLUOROTETRACHLOROETHANE

Synonyms: 1,2-Difluoro-1,1,2,2-tetrachloroethane; F 112; Freon 112; Genetron 112; Halocarbon 112; Refrigerant 112; **1,1,2,2-Tetrachloro-1,2-difluoroethane.**

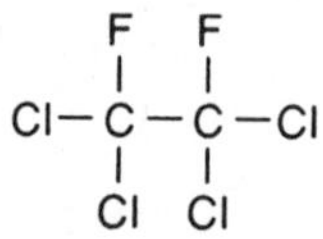

$$\begin{array}{ccc} & F & F \\ & | & | \\ Cl- & C-C & -Cl \\ & | & | \\ & Cl & Cl \end{array}$$

CAS Registry No.: 76-12-0
DOT: 1078
Molecular formula: $C_2Cl_4F_2$
Formula weight: 203.83
RTECS: KI1420000

Physical state, color and odor
Colorless liquid or solid with a faint, ether-like odor.

Melting point (°C):
25 (Weast, 1986)

Boiling point (°C):
93 (Weast, 1986)
91.58 (Boublik et al., 1984)

Density (g/cm^3):
1.6447 at 25/4 °C (Weast, 1986)

Diffusivity in water (10^5 cm^2/sec):
0.89 at 25 °C using method of Hayduk and Laudie (1974).

Flash point (°C):
Noncombustible (NIOSH, 1994)

Henry's law constant (atm·m^3/mol):
0.102 at 25 °C (approximate - calculated from water solubility and vapor pressure)

Ionization potential (eV):
11.30 (NIOSH, 1994)

Soil sorption coefficient, log K_{oc}:
2.78 using method of Chiou et al. (1979).

378

Octanol/water partition coefficient, log K_{ow}:
3.39 using method of Hansch et al. (1968).

Solubility in organics:
Soluble in alcohol, chloroform and ether (Weast, 1986).

Solubility in water:
120 mg/L at 25 °C (Du Pont, 1966)

Vapor density:
8.33 g/L at 25 °C, 7.04 (air = 1)

Vapor pressure (mmHg):
40 at 20 °C (NIOSH, 1994)
45.8 at 25 °C (Boublik et al., 1984)

Exposure Limits: NIOSH REL: TWA 500 ppm (4,170 mg/m^3), IDLH 2,000 ppm; OSHA PEL: TWA 500 ppm.

Toxicity: Acute oral LD_{50} for mice 800 mg/kg; LD_{50} (inhalation) for mice 123 gm/m^3/2-hour (RTECS, 1985).

Uses: Organic synthesis.

DIISOBUTYL KETONE

Synonyms: DIBK; *sym*-Diisopropylacetone; 2,6-Dimethylheptan-4-one; **2,6-Di-methyl-4-heptanone;** Isobutyl ketone; Isovalerone; UN 1157; Valerone.

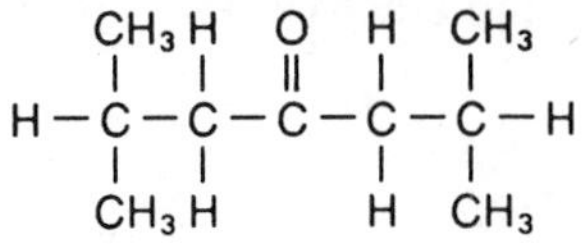

$$
\begin{array}{ccccc}
CH_3\,H & O & H & CH_3 \\
| \quad | & \| & | & | \\
H-C-C-C-C-C-H \\
| \quad | & & | & | \\
CH_3\,H & & H & CH_3 \\
\end{array}
$$

CAS Registry No.: 108-83-8
DOT: 1157
DOT label: Combustible liquid
Molecular formula: $C_9H_{18}O$
Formula weight: 142.24
RTECS: MJ5775000

Physical state, color and odor
Clear, colorless liquid with a mild, sweet, ether-like odor. Odor threshold in air is 0.11 ppm (Amoore and Hautala, 1983).

Melting point (°C):
-46 to -42 (Verschueren, 1983)

Boiling point (°C):
168 (Weast, 1986)

Density (g/cm^3):
0.8053 at 20/4 °C (Weast, 1986)

Diffusivity in water (10^5 cm^2/sec):
0.63 at 20 °C using method of Hayduk and Laudie (1974).

Flash point (°C):
49 (NIOSH, 1994)

Lower explosive limit (%):
0.8 at 93 °C (NFPA, 1984)

Upper explosive limit (%):
7.1 at 93 °C (NOSH, 1994)

Henry's law constant (10^4 atm·m^3/mol):
6.36 at 20 °C (approximate - calculated from water solubility and vapor pressure)

Ionization potential (eV):
9.04 (NIOSH, 1994)

Soil sorption coefficient, log K_{oc}:
Unavailable because experimental methods for estimation of this parameter for aliphatic ketones are lacking in the documented literature.

Octanol/water partition coefficient, log K_{ow}:
2.58 using method of Hansch et al. (1968).

Solubility in organics:
Soluble in alcohol and ether (Weast, 1986).

Solubility in water:
0.05 wt % at 20 °C (NIOSH, 1994)

Vapor density:
5.81 g/L at 25 °C, 4.91 (air = 1)

Vapor pressure (mmHg):
1.7 at 20 °C, 2.3 at 30 °C (Verschueren, 1983)

Exposure Limits: NIOSH REL: TWA 25 ppm (150 mg/m^3), IDLH 500 ppm; OSHA PEL: TWA 50 ppm (290 mg/m^3); ACGIH TLV: TWA 25 ppm.

Symptoms of Exposure: Vapor inhalation may cause irritation of eyes, nose and throat (Patnaik, 1992).

Toxicity: Acute oral LD$_{50}$ for rats 5,750 mg/kg, mice 1,416 mg/kg (RTECS, 1985).

Uses: Solvent for nitrocellulose, synthetic resins, rubber, lacquers; coating compositions; inks and stains; organic synthesis.

DIISOPROPYLAMINE

Synonyms: DIPA; *N*-(1-Methylethyl)-2-propanamine; UN 1158.

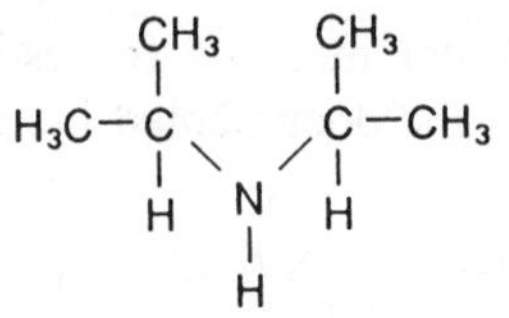

$$\begin{array}{ccccc} & CH_3 & & CH_3 & \\ & | & & | & \\ H_3C-C & & & C-CH_3 \\ & | & \diagdown \quad \diagup & | & \\ & H & N & H & \\ & & | & & \\ & & H & & \end{array}$$

CAS Registry No.: 108-18-9
DOT: 1158
DOT label: Flammable liquid
Molecular formula: $C_6H_{15}N$
Formula weight: 101.19
RTECS: IM4025000

Physical state, color and odor
Colorless liquid with an ammonia-like odor. Odor threshold in air is 1.8 ppm (Amoore and Hautala, 1983).

Melting point (°C):
-61 (Weast, 1986)
-96.3 (Verschueren, 1983)

Boiling point (°C):
84 (Weast, 1986)

Density (g/cm^3):
0.7169 at 20/4 °C (Weast, 1986)
0.722 at 22/4 °C (Windholz et al., 1983)

Diffusivity in water (10^5 cm^2/sec):
0.72 at 20 °C using method of Hayduk and Laudie (1974).

Dissociation constant:
11.13 at 21 °C (Gordon and Ford, 1972)

Flash point (°C):
-6.7 (NIOSH, 1994)
-1.11 (open cup, Hawley, 1981)

Lower explosive limit (%):
1.1 (NIOSH, 1994)

Upper explosive limit (%):
7.1 (NIOSH, 1994)

Ionization potential (eV):
7.73 ± 0.03 (Franklin et al., 1969)

Soil sorption coefficient, log K_{oc}:
Unavailable because experimental methods for estimation of this parameter for aliphatic amines are lacking in the documented literature. However, its miscibility in water suggests its adsorption to soil will be nominal (Lyman et al., 1982).

Solubility in organics:
Soluble in acetone, alcohol, benzene and ether (Weast, 1986).

Solubility in water:
Miscible (NIOSH, 1994).

Vapor density:
4.14 g/L at 25 °C, 3.49 (air = 1)

Vapor pressure (mmHg):
70 at 20 °C (NIOSH, 1994)

Environmental Fate
   Chemical/Physical. Reacts with acids forming water-soluble salts.

Exposure Limits: NIOSH REL: TWA 5 ppm (20 mg/m^3), IDLH 200 ppm; OSHA PEL: TWA 5 ppm; ACGIH TLV: TWA 5 ppm.

Symptoms of Exposure: Severe irritation of eyes, skin and respiratory tract. Contact with skin causes burns. Visual disturbance and cloudy swelling of cornea accompanied by partial or total loss of vision (Patnaik, 1992).

Toxicity: Acute oral LD_{50} for guinea pigs, 2,800 mg/kg, mice 2,120 mg/kg, rats 770 mg/kg, rabbits 4,700 mg/kg (RTECS, 1985).

Uses: Intermediate; catalyst; organic synthesis.

N,N-DIMETHYLACETAMIDE

Synonyms: Acetdimethylamide; Acetic acid dimethylamide; Dimethylacetamide; Dimethylacetone amide; Dimethylamide acetate; DMA; DMAC; Hallucinogen; NSC 3138; U 5954.

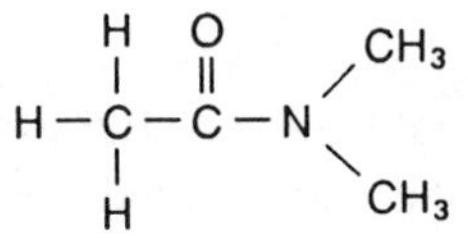

$$H-\overset{\overset{\displaystyle H}{|}}{\underset{\underset{\displaystyle H}{|}}{C}}-\overset{\overset{\displaystyle O}{\|}}{C}-N\overset{\diagup CH_3}{\diagdown CH_3}$$

CAS Registry No.: 127-19-5
Molecular formula: C_4H_9NO
Formula weight: 115.18
RTECS: AB7700000

Physical state, color and odor
Clear, colorless liquid with a weak, ammonia-like odor. Odor threshold in air is 47 ppm (Amoore and Hautala, 1983).

Melting point (°C):
–20 (Weast, 1986)
27 (Verschueren, 1983)

Boiling point (°C):
165 at 758 mmHg (Weast, 1986)
165.5 (Dean, 1987)

Density (g/cm³):
0.9366 at 25/4 °C (Weast, 1986)

Diffusivity in water (10^5 cm²/sec):
0.89 at 20 °C using method of Hayduk and Laudie (1974).

Flash point (°C):
77.2 (open cup, Sax and Lewis, 1987)
70 (NFPA, 1984)

Lower explosive limit (%):
1.8 at 100 °C (NFPA, 1984)

Upper explosive limit (%):
11.5 at 160 °C (NFPA, 1984)

384

Ionization potential (eV):
8.81 ± 0.03 (Franklin et al., 1969)
8.60 (HNU, 1986)

Soil sorption coefficient, log K_{oc}:
Unavailable because experimental methods for estimation of this parameter for aliphatic amides are lacking in the documented literature. However, its miscibility in water and low K_{ow} suggest its adsorption to soil will be nominal (Lyman et al., 1982).

Octanol/water partition coefficient, log K_{ow}:
-0.77 (Sangster, 1989)

Solubility in organics:
Miscible with aromatics, esters, ketones and ethers (Hawley, 1981).

Solubility in water:
Miscible (NIOSH, 1994).

Vapor pressure (mmHg):
2 at 20 °C (NIOSH, 1994)
1.3 at 25 °C, 9 at 60 °C (Verschueren, 1983)

Environmental Fate
Chemical/Physical. Releases toxic fumes of nitrogen oxides when heated to decomposition (Sax and Lewis, 1987).

Exposure Limits: NIOSH REL: TWA 10 ppm (35 mg/m^3), IDLH 300 ppm; OSHA PEL: TWA 10 ppm.

Toxicity: Acute oral LD_{50} for mice 4,620 mg/kg, rats 5,000 mg/kg (RTECS, 1985).

Uses: Solvent used in organic synthesis; paint removers; solvent for plastics, resins, gums and electrolytes; intermediate; catalyst.

DIMETHYLAMINE

Synonyms: DMA; *N*-Methylmethanamine; RCRA waste number U092; UN 1032; UN 1160.

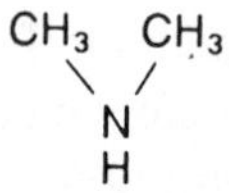

$$CH_3 \quad CH_3$$
$$\diagdown \quad \diagup$$
$$N$$
$$H$$

CAS Registry No.: 124-40-3
DOT: 1032 (anhydrous), 1160 (aqueous solution)
DOT label: Flammable gas/flammable liquid (aqueous)
Molecular formula: C_2H_7N
Formula weight: 45.08
RTECS: IP8750000

Physical state, color and odor
Clear, colorless liquid or gas with a strong, ammonia-like odor. Odor threshold in air is 0.34 ppm (Amoore and Hautala, 1983).

Melting point (°C):
-93 (Weast, 1986)
-96 (Windholz et al., 1983)

Boiling point (°C):
7.4 (Weast, 1986)
6.9 (Dean, 1987)

Density (g/cm³):
0.6804 at 0/4 °C (Weast, 1986)

Diffusivity in water (10^5 cm²/sec):
1.11 at 20 °C using method of Hayduk and Laudie (1974).

Dissociation constant:
10.732 at 25 °C (Gordon and Ford, 1972)

Flash point (°C):
-17.7 (25% solution, Hawley, 1981)

Lower explosive limit (%):
2.8 (NIOSH, 1994)

Upper explosive limit (%):
14.4 (NIOSH, 1994)

Heat of fusion (kcal/mol):
1.420 (Dean, 1987)

Henry's law constant (10^5 atm·m³/mol):
1.77 at 25 °C (Hine and Mookerjee, 1975)

Ionization potential (eV):
8.24 ± 0.02 (Franklin et al., 1969)
8.36 (Gibson et al., 1977)

Soil sorption coefficient, log K_{oc}:
Unavailable because experimental methods for estimation of this parameter for aliphatic amines are lacking in the documented literature. However, its miscibility in water suggests its adsorption to soil will be nominal (Lyman et al., 1982).

Octanol/water partition coefficient, log K_{ow}:
-0.38 (Sangster, 1989)

Solubility in organics:
Soluble in alcohol and ether (Weast, 1986).

Solubility in water:
24 wt % at 60 °C (NIOSH, 1994)

Vapor density:
1.84 g/L at 25 °C, 1.56 (air = 1)

Vapor pressure (mmHg):
1,292 at 20 °C (Verschueren, 1983)

Environmental Fate
Photolytic. Dimethylnitramine, nitrous acid, formaldehyde, *N,N*-dimethyl-formamide and carbon monoxide were reported as photooxidation products of dimethylamine with NO_x. An additional compound was tentatively identified as tetramethylhydrazine (Tuazon et al., 1978).

Soil. After 2 days, degradation yields in an Arkport fine sandy loam (Varna, NY) and sandy soil (Lake George, NY) amended with sewage and nitrite-N were 50 and 20%, respectively. *N*-Nitrosodimethylamine was identified as the major metabolite (Greene et al., 1981). Mills and Alexander (1976) reported that *N*-nitrosodimethylamine also formed in soil, municipal sewage and lake water supplemented with dimethylamine (ppm) and nitrite-N (100 ppm). They found that nitrosation occurred under nonenzymatic conditions at neutral pHs.

Chemical/Physical. In an aqueous solution, chloramine reacted with dimethylamine to form *N*-chlorodimethylamine (Isaac and Morris, 1983).

In the atmosphere, reacts with hydroxyl radicals forming formaldehyde and/or amides (Atkinson et al., 1978).

Exposure Limits: NIOSH REL: TWA 10 ppm (18 mg/m^3), IDLH 500 ppm; OSHA PEL: TWA 10 ppm; ACGIH TLV: TWA 10 ppm.

Symptoms of Exposure: Strong irritation of eyes, skin and mucous membranes. Contact with skin may cause necrosis. Eye contact with liquid can cause corneal damage and loss of vision (Patnaik, 1992).

Toxicity: Acute oral LD_{50} for guinea pigs 340 mg/kg, mice 316 mg/kg, rats 698 mg/kg, rabbits 240 mg/kg (RTECS, 1985).

Uses: Detergent soaps; accelerator for vulcanizing rubber; detection of magnesium; tanning; acid gas absorbent solvent; gasoline stabilizers; textile chemicals; pharmaceuticals; surfactants; manufacture of *N,N*-dimethylformamide and *N,N*-dimethylacetamide; rocket propellants; missile fuels; dehairing agent; electroplating.

p-DIMETHYLAMINOAZOBENZENE

Synonyms: Atul fast yellow R; Benzeneazodimethylaniline; Brilliant fast oil yellow; Brilliant fast spirit yellow; Brilliant fast yellow; Brilliant oil yellow; Butter yellow; Cerasine yellow CG; C.I. 11020; C.I. solvent yellow 2; DAB; Dimethylaminobenzene; 4-Dimethylaminoazobenzene; 4-(*N,N*-Dimethylamino)azobenzene; *N,N*-Dimethyl-4-aminoazobenzene; *N,N*-Dimethyl-*p*-aminoazobenzene; Dimethylaminoazobenzol; 4-Dimethylaminoazobenzol; 4-Dimethylaminophenylazobenzene; *N,N*-Dimethyl-*p*-azoaniline; *N,N*-Dimethyl-4-(phenylazo)benzamine; *N,N*-Dimethyl-*p*-(phenylazo)benzamine; **N,N-Dimethyl-4-(phenylazo)benzenamine**; *N,N*-Dimethyl-*p*-(phenylazo)benzenamine; Dimethyl yellow; Dimethyl yellow analar; Dimethyl yellow *N,N*-dimethylaniline; DMAB; Enial yellow 2G; Fast oil yellow B; Fast yellow; Fat yellow; Fat yellow A; Fat yellow AD OO; Fat yellow ES; Fat yellow ES extra; Fat yellow extra conc.; Fat yellow R; Fat yellow R (8186); Grasal brilliant yellow; Methyl yellow; Oil yellow; Oil yellow 20; Oil yellow 2625; Oil yellow 7463; Oil yellow BB; Oil yellow D; Oil yellow DN; Oil yellow FF; Oil yellow FN; Oil yellow G; Oil yellow G-2; Oil yellow 2G; Oil yellow GG; Oil yellow GR; Oil yellow II; Oil yellow N; Oil yellow PEL; Oleal yellow 2G; Organol yellow ADM; Orient oil yellow GG; PDAB; Petrol yellow WT; RCRA waste number U093; Resinol yellow GR; Resoform yellow GGA; Silotras yellow T2G; Somalia yellow A; Stear yellow JB; Sudan GG; Sudan yellow; Sudan yellow 2G; Sudan yellow 2GA; Toyo oil yellow G; USAF EK-338; Waxoline yellow AD; Waxoline yellow ADS; Yellow G soluble in grease.

CAS Registry No.: 60-11-7
Molecular formula: $C_{14}H_{15}N_3$
Formula weight: 225.30
RTECS: BX7350000

Physical state and color
Yellow leaflets or crystals.

Melting point (°C):
114-117 (Windholz et al., 1983)

Boiling point (°C):
Sublimes (Weast, 1986).

Density (g/cm^3):
1.212 using method of Lyman et al. (1982).

Diffusivity in water (10^5 cm^2/sec):
0.50 at 20 °C using method of Hayduk and Laudie (1974).

Soil sorption coefficient, log K$_{oc}$:
3.00 (calculated, Mercer et al., 1990)

Octanol/water partition coefficient, log K$_{ow}$:
4.58 (Verschueren, 1983)

Solubility in organics:
Room temperature (g/L): dimethyl sulfoxide (5-10), acetone (50-100), toluene (12-30) (Keith and Walters, 1992).

Solubility in water:
10^{-3} wt % at 20 °C (NIOSH, 1994)
13.6 mg/L at 20-30 °C (Mercer et al., 1990)

Vapor pressure (mmHg):
3 x 10^{-7} (estimated, NIOSH, 1994)

Environmental Fate
   Chemical/Physical. Releases toxic nitrogen oxides when heated to decomposition (Sax and Lewis, 1987).

Exposure Limits: Potential occupational carcinogen.

Toxicity: Acute oral LD$_{50}$ for mice 300 mg/kg, rats 200 mg/kg (RTECS, 1985).

Uses: Not commercially produced in the United States. pH indicator; determining hydrochloric acid in gastric juice; coloring agent; organic research.

DIMETHYLANILINE

Synonyms: Dimethylaminobenzene; *N,N*-Dimethylaniline; *N,N*-**Dimethylbenzenamine**; Dimethylphenylamine; *N,N*-Dimethylphenylamine; NCI-C56428; UN 2253; Versneller NL 63/10.

$$H_3C \diagdown \diagup CH_3$$
$$N$$

CAS Registry No.: 121-69-7
DOT: 2253
Molecular formula: $C_8H_{11}N$
Formula weight: 121.18
RTECS: BX4725000

Physical state, color and odor
Straw to brown-colored liquid with an amine-like odor. Odor threshold in air is 13 ppb (Amoore and Hautala, 1983).

Melting point (°C):
2.45 (Weast, 1986)

Boiling point (°C):
194 (Weast, 1986)

Density (g/cm³):
0.9557 at 20/4 °C (Weast, 1986)

Diffusivity in water (10^5 cm²/sec):
0.77 at 20 °C using method of Hayduk and Laudie (1974).

Dissociation constant:
5.21 at 25 °C (Dean, 1973)

Flash point (°C):
61 (NIOSH, 1994)

Henry's law constant (10^6 atm·m³/mol):
4.98 at 20 °C (approximate - calculated from water solubility and vapor pressure)

Ionization potential (eV):
7.12 (Franklin et al., 1969)

Soil sorption coefficient, log K_{oc}:
2.26 (Meylan et al., 1992)

Octanol/water partition coefficient, log K_{ow}:
2.29 (Sangster, 1989)
2.31 (Leo et al., 1971)
2.62 (Rogers and Cammarata, 1969)

Solubility in organics:
Soluble in acetone, alcohol, benzene, chloroform and ether (Weast, 1986).

Solubility in water:
1,105.2 mg/L at 25 °C (Chiou et al., 1982)

Vapor density:
4.95 g/L at 25 °C, 4.18 (air = 1)

Vapor pressure (mmHg):
0.5 at 20 °C, 1.1 at 30 °C (Verschueren, 1983)
0.52 at 25 °C (Banerjee et al., 1990)

Environmental Fate
  Chemical/Physical. Products identified from the gas-phase reaction of ozone with *N,N*-dimethylaniline in synthetic air at 23 °C were: *N*-methylformanilide, formaldehyde, formic acid, hydrogen peroxide and a nitrated salt having the formula: $[C_6H_6NH(CH_3)_2]^+NO_3^-$ (Atkinson et al., 1987). Reacts with acids forming water-soluble salts.

Exposure Limits: NIOSH REL: TWA 5 ppm (25 mg/m^3), STEL 10 ppm (50 mg/m^3), IDLH 100 ppm; OSHA PEL: TWA 5 ppm; ACGIH TLV: TWA 5 ppm, STEL 10 ppm.

Toxicity: Acute oral LD_{50} for rats 1,410 mg/kg; LC_{50} (48-hour) for red killifish 275 mg/L (Yoshioka et al., 1986).

Uses: Manufacture of vanillin, Michler's ketone, methyl violet and other dyes; solvent; reagent for methyl alcohol, hydrogen peroxide, methyl furfural, nitrate and formaldehyde; chemical intermediate; stabilizer; reagent.

2,2-DIMETHYLBUTANE

Synonyms: Neohexane; UN 1208.

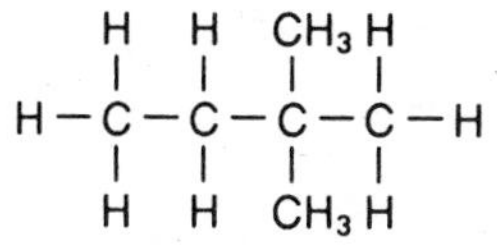

$$H-\underset{\underset{H}{|}}{\overset{\overset{H}{|}}{C}}-\underset{\underset{H}{|}}{\overset{\overset{H}{|}}{C}}-\underset{\underset{CH_3}{|}}{\overset{\overset{CH_3}{|}}{C}}-\underset{\underset{H}{|}}{\overset{\overset{H}{|}}{C}}-H$$

CAS Registry No.: 75-83-2
DOT: 2457
Molecular formula: C_6H_{14}
Formula weight: 86.18
RTECS: EJ9300000

Physical state, color and odor
Colorless liquid with a mild gasoline-like odor.

Melting point (°C):
-99.9 (Weast, 1986)

Boiling point (°C):
49.7 (Weast, 1986)

Density (g/cm³):
0.6485 at 20/4 °C (Weast, 1986)
0.6570 at 25/4 °C (Hawley, 1981)

Diffusivity in water (10^5 cm²/sec):
0.75 at 20 °C using method of Hayduk and Laudie (1974).

Dissociation constant:
>14 (Schwarzenbach et al., 1993)

Flash point (°C):
-47.8 (Sax and Lewis, 1987)
-34 (Aldrich, 1988)

Lower explosive limit (%):
1.2 (Sax and Lewis, 1987)

Upper explosive limit (%):
7.0 (Sax and Lewis, 1987)

393

Heat of fusion (kcal/mol):
0.138 (Dean, 1987)

Henry's law constant (atm·m^3/mol):
1.943 at 25 °C (Hine and Mookerjee, 1975)

Ionization potential (eV):
10.06 (HNU, 1986)

Soil sorption coefficient, log K_{oc}:
Unavailable because experimental methods for estimation of this parameter for aliphatic hydrocarbons are lacking in the documented literature.

Octanol/water partition coefficient, log K_{ow}:
3.82 (Hansch and Leo, 1979)

Solubility in organics:
In methanol: 590 and 800 g/L at 5 and 10 °C, respectively. Miscible at higher temperatures (Kiser et al., 1961).

Solubility in water (mg/kg):
21.2 at 25 °C (Price, 1976)
18.4 at 25 °C (McAuliffe, 1963, 1966)
39.4 at 0 °C, 23.8 at 25 °C (Polak and Lu, 1973)

Vapor density:
3.52 g/L at 25 °C, 2.98 (air = 1)

Vapor pressure (mmHg):
325 at 24.47 °C (Willingham et al., 1945)
319.1 at 25 °C (Wilhoit and Zwolinski, 1971)

Use: Intermediate for agricultural chemicals; in high octane fuels.

2,3-DIMETHYLBUTANE

Synonyms: Biisopropyl; Diisopropyl; Isopropyldimethylmethane; UN 2457.

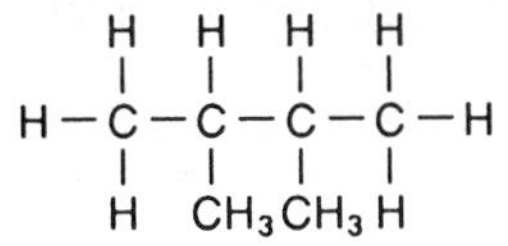

```
      H   H   H   H
      |   |   |   |
  H − C − C − C − C − H
      |   |   |   |
      H  CH₃ CH₃ H
```

CAS Registry No.: 79-29-8
DOT: 2457
Molecular formula: C_6H_{14}
Formula weight: 86.18
RTECS: EJ9350000

Physical state, color and odor
Colorless liquid with a mild gasoline-like odor.

Melting point (°C):
-128.5 (Weast, 1986)
-135 (Verschueren, 1983)

Boiling point (°C):
58 (Weast, 1986)

Density (g/cm³):
0.66164 at 20/4 °C, 0.65702 at 25/4 °C (Dreisbach, 1959)

Diffusivity in water (10^5 cm²/sec):
0.43 at 20 °C using method of Hayduk and Laudie (1974).

Dissociation constant:
>14 (Schwarzenbach et al., 1993)

Flash point (°C):
-28.9 (Hawley, 1981)
-33 (Aldrich, 1988)

Lower explosive limit (%):
1.2 (NFPA, 1984)

Upper explosive limit (%):
7.0 (NFPA, 1984)

Heat of fusion (kcal/mol):
0.194 (Dean, 1987)

Henry's law constant (atm·m^3/mol):
1.18 at 25 °C (approximate - calculated from water solubility and vapor pressure)

Ionization potential (eV):
10.02 (HNU, 1986)

Soil sorption coefficient, log K_{oc}:
Unavailable because experimental methods for estimation of this parameter for aliphatic hydrocarbons are lacking in the documented literature.

Octanol/water partition coefficient, log K_{ow}:
3.85 (Hansch and Leo, 1979)

Solubility in organics:
In methanol: 495, 593, 760 and 1,700 g/L at 5, 10, 15 and 20 °C, respectively. Miscible at higher temperatures (Kiser et al., 1961).

Solubility in water (mg/kg):
19.1 at 25 °C, 19.2 at 40.1 °C, 23.7 at 55.1 °C, 40.1 at 99.1 °C, 56.8 at 121.3 °C, 97.9 at 137.3 °C, 171.0 at 149.5 °C (Price, 1976)
32.9 at 0 °C, 22.5 at 25 °C (Polak and Lu, 1973)

Vapor density:
3.52 g/L at 25 °C, 2.98 (air = 1)

Vapor pressure (mmHg):
200 at 20 °C (Verschueren, 1983)
217 at 23.10 °C (Willingham et al., 1945)
234.6 at 25 °C (Wilhoit and Zwolinski, 1971)

Environmental Fate
   Photolytic. Major products reported from the photooxidation of 2,3-dimethylbutane with nitrogen oxides are carbon monoxide and acetone. Minor products included formaldehyde, acetaldehyde and peroxyacyl nitrates (Altshuller, 1983). Synthetic air containing gaseous nitrous acid and exposed to artificial sunlight (λ = 300-450 nm) photooxidized 2,3-dimethylbutane into acetone, hexyl nitrate, peroxyacetal nitrate and a nitro aromatic compound tentatively identified as a propyl nitrate (Cox et al., 1980).

Uses: Organic synthesis; gasoline component.

cis-1,2-DIMETHYLCYCLOHEXANE

Synonyms: *cis-o-*Dimethylcyclohexane; *cis*-1,2-Hexahydroxylene.

CAS Registry No.: 2207-01-4
DOT: 2263
Molecular formula: C_8H_{16}
Formula weight: 112.22

Physical state and color
Colorless liquid.

Melting point (°C):
-50.1 (Weast, 1986)

Boiling point (°C):
129.7 (Weast, 1986)

Density (g/cm³):
0.79627 at 20/4 °C, 0.79222 at 25/4 °C (Riddick et al., 1986)

Diffusivity in water (10^5 cm²/sec):
0.72 at 20 °C using method of Hayduk and Laudie (1974).

Dissociation constant:
>14 (Schwarzenbach et al., 1993)

Flash point (°C):
-12 (Aldrich, 1988)

Heat of fusion (kcal/mol):
0.393 (Dean, 1987)

Henry's law constant (atm·m³/mol):
0.354 at 25 °C (Hine and Mookerjee, 1975)

Ionization potential (eV):
10.08 ± 0.02 (Franklin et al., 1969)

Soil sorption coefficient, log K_{oc}:
Unavailable because experimental methods for estimation of this parameter for alicyclic hydrocarbons are lacking in the documented literature.

Octanol/water partition coefficient, log K_{ow}:
3.26 using method of Hansch et al. (1968).

Solubility in organics:
Soluble in acetone, alcohol, benzene, ether and lignoin (Weast, 1986).

Solubility in water:
6.0 mg/kg at 25 °C (McAuliffe, 1966)

Vapor density:
4.59 g/L at 25 °C, 3.87 (air = 1)

Vapor pressure (mmHg):
14.5 at 25 °C (Wilhoit and Zwolinski, 1971)

Use: Organic synthesis.

trans-1,4-DIMETHYLCYCLOHEXANE

Synonym: *trans*-*p*-Dimethylcyclohexane.

CAS Registry No.: 6876-23-9
DOT: 2263
Molecular formula: C_8H_{16}
Formula weight: 112.22

Physical state and color
Colorless liquid.

Melting point (°C):
-37 (Weast, 1986)

Boiling point (°C):
119.3 (Weast, 1986)

Density (g/cm^3):
0.76255 at 20/4 °C, 0.75835 at 25/4 °C (Dresibach, 1959)

Diffusivity in water (10^5 cm^2/sec):
0.70 at 20 °C using method of Hayduk and Laudie (1974).

Dissociation constant:
>14 (Schwarzenbach et al., 1993)

Flash point (°C):
≈ 10 (isomeric mixture, Hawley, 1981)

Heat of fusion (kcal/mol):
2.491-2.508 (Dean, 1987)

Henry's law constant (atm·m^3/mol):
0.870 at 25 °C (approximate - calculated from water solubility and vapor pressure)

Ionization potential (eV):
10.08 ± 0.03 (Franklin et al., 1969)

Soil sorption coefficient, log K_{oc}:
Unavailable because experimental methods for estimation of this parameter for alicyclic hydrocarbons are lacking in the documented literature.

Octanol/water partition coefficient, log K_{ow}:
3.41 using method of Hansch et al. (1968).

Solubility in organics:
Soluble in acetone, alcohol, benzene, ether and lignoin (Weast, 1986).

Solubility in water:
3.84 mg/kg at 25 °C (Price, 1976)

Vapor density:
4.59 g/L at 25 °C, 3.87 (air = 1)

Vapor pressure (mmHg):
22.65 at 25 °C (Mackay et al., 1982)

Use: Organic synthesis.

N,N-DIMETHYLFORMAMIDE

Synonyms: Dimethylformamide; DMF; DMFA; *N*–Formyldimethylamine; NCI–C60913; NSC 5536; U 4224; UN 2265.

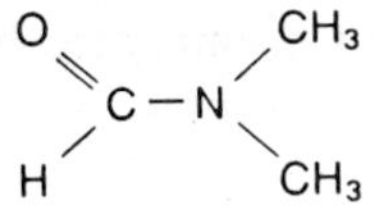

CAS Registry No.: 68–12–2
DOT: 2265
DOT label: Combustible liquid
Molecular formula: C_3H_7NO
Formula weight: 73.09
RTECS: LQ2100000

Physical state, color and odor
Colorless to light yellow, mobile liquid with a faint, ammonia-like odor. Odor threshold in air is 2.2 ppm (Amoore and Hautala, 1983).

Melting point (°C):
–60.5 (Weast, 1986)

Boiling point (°C):
149–156 (Weast, 1986)

Density (g/cm^3):
0.9487 at 20/4 °C (Weast, 1986)
0.9445 at 25/4 °C (Windholz et al., 1983)

Diffusivity in water (10^5 cm^2/sec):
1.03 at 20 °C using method of Hayduk and Laudie (1974).

Flash point (°C):
57.8 (NIOSH, 1994)
67 (open cup, Windholz et al., 1983)

Lower explosive limit (%):
2.2 at 100 °C (NFPA, 1984)

Upper explosive limit (%):
15.2 (NFPA, 1984)

Ionization potential (eV):
9.12 ± 0.02 (Franklin et al., 1969)

Soil sorption coefficient, log K_{oc}:
Unavailable because experimental methods for estimation of this parameter for aliphatic amines are lacking in the documented literature. However, its miscibility in water and low K_{ow} suggest its adsorption to soil will be nominal (Lyman et al., 1982).

Octanol/water partition coefficient, log K_{ow}:
−1.01 (Sangster, 1989)

Solubility in organics:
Miscible with most organic solvents (Windholz et al., 1983).

Solubility in water:
Miscible (NIOSH, 1994). A saturated solution in equilibrium with its own vapor had a concentration of 5,294 g/L at 25 °C (Kamlet et al., 1987).

Vapor density:
2.99 g/L at 25 °C, 2.52 (air = 1)

Vapor pressure (mmHg):
3 at 20 °C (NIOSH, 1994)
3.7 at 25 °C (Sax and Lewis, 1987)

Environmental Fate
  Biological. Incubation of [^{14}C]N,N-dimethylformamide (0.1–100 µg/L) in natural seawater resulted in the compound mineralizing to carbon dioxide. The rate of carbon dioxide formation was inversely proportional to the initial concentration (Ursin, 1985).

Exposure Limits: NIOSH REL: TWA 10 ppm (30 mg/m^3), IDLH 500 ppm; OSHA PEL: TWA 10 ppm; ACGIH TLV: TWA 10 ppm.

Toxicity: Acute oral LD_{50} for mice 3,750 mg/kg, rats 2,800 mg/kg (RTECS, 1985).

Uses: Solvent for liquids, gases and vinyl resins; polyacrylic fibers; gas carrier; catalyst in carboxylation reactions; organic synthesis.

1,1-DIMETHYLHYDRAZINE

Synonyms: UDMH; Dimazine; *asym*-Dimethylhydrazine; *unsym*-Dimethylhydrazine; *N,N*-Dimethylhydrazine.

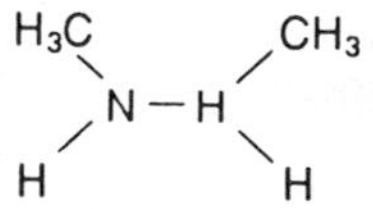

$$H_3C \quad\quad CH_3$$
$$N-H$$
$$H \quad\quad\quad H$$

CAS Registry No.: 57-14-7
DOT: 1163 (*asym*), 2382 (*sym*)
DOT label: Flammable liquid and poison
Molecular formula: $C_2H_8N_2$
Formula weight: 60.10
RTECS: MV2450000

Physical state, color and odor
Clear, colorless to yellow, fuming liquid with an amine-like odor. Odor threshold in air is 1.7 ppm (Amoore and Hautala, 1983).

Melting point (°C):
-57.8 (NIOSH, 1994)

Boiling point (°C):
63.9 (Windholz et al., 1983)

Density (g/cm^3):
0.7914 at 22/4 °C (Weast, 1986)

Diffusivity in water (10^5 cm^2/sec):
1.04 at 20 °C using method of Hayduk and Laudie (1974).

Flash point (°C):
-15.1 (NIOSH, 1994)
1 (Aldrich, 1988)

Lower explosive limit (%):
2 (NIOSH, 1994)

Upper explosive limit (%):
95 (NIOSH, 1994)

Henry's law constant (10^9 atm·m^3/mol):
2.45 at 25 °C (Mercer et al., 1990)

403

Ionization potential (eV):
7.67 ± 0.05 (Franklin et al., 1969)
8.05 (NIOSH, 1994)

Soil sorption coefficient, log K_{oc}:
-0.70 (calculated, Mercer et al., 1990)

Octanol/water partition coefficient, log K_{ow}:
-2.42 (Mercer et al., 1990)

Solubility in organics:
Miscible with alcohol, *N,N*-dimethylformamide, ether and hydrocarbons (Windholz et al., 1983).

Solubility in water:
Miscible (NIOSH, 1994).

Vapor density:
2.46 g/L at 25 °C, 2.07 (air = 1)

Vapor pressure (mmHg):
103 at 20 °C (NIOSH, 1994)
157 at 25 °C (Verschueren, 1983)

Environmental Fate
   Chemical/Physical. Releases toxic nitrogen oxides when heated to decomposition (Sax and Lewis, 1987). Ignites spontaneously in air or in contact with hydrogen peroxide, nitric acid, or other oxidizers (Patnaik, 1992).
   N-Nitrosodimethylamine was the major product of ozonation of 1,1-dimethyl-hydrazine in the dark. Hydrogen peroxide, methyl hydroperoxide and methyl diazene were also identified (HSDB, 1989).

Exposure Limits: Potential occupational carcinogen. NIOSH REL: 2-hour ceiling 0.6 ppm (1.2 mg/m^3), IDLH 15 ppm; OSHA PEL: TWA 0.5 ppm (1 mg/m^3); ACGIH TLV: TWA 0.5 ppm.

Symptoms of Exposure: Irritation of eyes, nose and throat. May cause diarrhea, stimulation of central nervous system, tremor and convulsions (Patnaik, 1992).

Toxicity: Acute oral LD_{50} for rats 122 mg/kg, mice 265 mg/kg (RTECS, 1985).

Uses: Rocket fuel formulations; stabilizer for organic peroxide fuel additives; absorbent for acid gases; plant control agent; photography; in organic synthesis.

2,3-DIMETHYLPENTANE

Synonym: 3,4-Dimethylpentane.

```
        H   H   H   H   H
        |   |   |   |   |
   H — C — C — C — C — C — H
        |   |   |   |   |
        H  CH₃ CH₃ H   H
```

CAS Registry No.: 565-59-3
Molecular formula: C_7H_{16}
Formula weight: 100.20

Physical state and color
Clear, colorless liquid.

Boiling point (°C):
89.8 (Weast, 1986)

Density (g/cm³):
0.6951 at 20/4 °C (Weast, 1986)

Diffusivity in water (10^5 cm²/sec):
0.71 at 20 °C using method of Hayduk and Laudie (1974).

Dissociation constant:
>14 (Schwarzenbach et al., 1993)

Flash point (°C):
-6 (Aldrich, 1988)

Lower explosive limit (%):
1.1 (NFPA, 1984)

Upper explosive limit (%):
6.7 (NFPA, 1984)

Henry's law constant (atm·m³/mol):
1.73 at 25 °C (approximate - calculated from water solubility and vapor pressure)

Soil sorption coefficient, log K_{oc}:
Unavailable because experimental methods for estimation of this parameter for aliphatic hydrocarbons are lacking in the documented literature.

Octanol/water partition coefficient, log K_{ow}:
3.26 using method of Hansch et al. (1968).

Solubility in organics:
Soluble in acetone, alcohol, benzene, chloroform and ether (Weast, 1986).

Solubility in water:
5.25 mg/kg at 25 °C (Price, 1976)

Vapor density:
4.10 g/L at 25 °C, 3.46 (air = 1)

Vapor pressure (mmHg):
68.9 at 25 °C (Wilhoit and Zwolinski, 1971)
24.937 at 4.999 °C (Osborn and Douslin, 1974)

Uses: Organic synthesis; gasoline component.

2,4-DIMETHYLPENTANE

Synonym: Diisopropylmethane.

$$H-\underset{\underset{\displaystyle H}{|}}{\overset{\overset{\displaystyle H}{|}}{C}}-\underset{\underset{\displaystyle CH_3}{|}}{\overset{\overset{\displaystyle H}{|}}{C}}-\underset{\underset{\displaystyle H}{|}}{\overset{\overset{\displaystyle H}{|}}{C}}-\underset{\underset{\displaystyle CH_3}{|}}{\overset{\overset{\displaystyle H}{|}}{C}}-\underset{\underset{\displaystyle H}{|}}{\overset{\overset{\displaystyle H}{|}}{C}}-H$$

CAS Registry No.: 108-08-7
Molecular formula: C_7H_{16}
Formula weight: 100.20

Physical state and color
Colorless liquid.

Melting point (°C):
-119.2 (Weast, 1986)
-123 (Aldrich, 1988)

Boiling point (°C):
80.5 (Weast, 1986)

Density (g/cm³):
0.6727 at 20/4 °C, 0.66832 at 25/4 °C (Riddick et al., 1986)

Diffusivity in water (10⁵ cm²/sec):
0.70 at 20 °C using method of Hayduk and Laudie (1974).

Dissociation constant:
>14 (Schwarzenbach et al., 1993)

Flash point (°C):
-12.1 (Hawley, 1981)

Heat of fusion (kcal/mol):
1.636 (Dean, 1987)

Henry's law constant (atm·m³/mol):
3.152 at 25 °C (Hine and Mookerjee, 1975)

Soil sorption coefficient, log K_{oc}:
Unavailable because experimental methods for estimation of this parameter for aliphatic hydrocarbons are lacking in the documented literature.

Octanol/water partition coefficient, log K_{ow}:
3.24 using method of Hansch et al. (1968).

Solubility in organics:
Soluble in acetone, alcohol, benzene, chloroform and ether (Weast, 1986).

Solubility in water (mg/kg):
4.41 at 25 °C (Price, 1976)
3.62 at 25 °C (McAuliffe, 1963)
4.06 at 25 °C (McAuliffe, 1966)
6.50 at 0 °C, 5.50 at 25 °C (Polak and Lu, 1973)

Vapor density:
4.10 g/L at 25 °C, 3.46 (air = 1)

Vapor pressure (mmHg):
98.4 at 25 °C (Wilhoit and Zwolinski, 1971)

Uses: Organic synthesis; gasoline component.

3,3-DIMETHYLPENTANE

Synonyms: None.

CAS Registry No.: 562-49-2
Molecular formula: C_7H_{16}
Formula weight: 100.20

Physical state and color
Colorless liquid.

Melting point (°C):
-134.4 (Weast, 1986)

Boiling point (°C):
86.1 (Weast, 1986)

Density (g/cm^3):
0.69327 at 20/4 °C, 0.68908 at 25/4 °C (Dreisbach, 1959)

Diffusivity in water (10^5 cm^2/sec):
0.71 at 20 °C using method of Hayduk and Laudie (1974).

Dissociation constant:
>14 (Schwarzenbach et al., 1993)

Flash point (°C):
-6 (Aldrich, 1988)

Heat of fusion (kcal/mol):
1.689 (Dean, 1987)

Henry's law constant (atm·m^3/mol):
1.84 at 25 °C (approximate - calculated from water solubility and vapor pressure)

Soil sorption coefficient, log K_{oc}:
Unavailable because experimental methods for estimation of this parameter for aliphatic hydrocarbons are lacking in the documented literature.

Octanol/water partition coefficient, log K_{ow}:
3.22 using method of Hansch et al. (1968).

Solubility in organics:
Soluble in acetone, alcohol, benzene, chloroform and ether (Weast, 1986).

Solubility in water (mg/kg):
5.92 at 25 °C, 6.78 at 40.1 °C, 8.17 at 55.7 °C, 10.3 at 69.7 °C, 15.8 at 99.1 °C, 27.3 at 118.0 °C, 67.3 at 120.4 °C, 86.1 at 150.4 °C (Price, 1976)

Vapor density:
4.10 g/L at 25 °C, 3.46 (air = 1)

Vapor pressure (mmHg):
82.8 at 25 °C (Wilhoit and Zwolinski, 1971)

Uses: Organic synthesis; gasoline component.

2,4-DIMETHYLPHENOL

Synonyms: 4,6-Dimethylphenol; 2,4-DMP; 1-Hydroxy-2,4-dimethylbenzene; 4-Hydroxy-1,3-dimethylbenzene; RCRA waste number U101; 1,3,4-Xylenol; 2,4-Xylenol; *m*-Xylenol.

CAS Registry No.: 105-67-9
Molecular formula: $C_8H_{10}O$
Formula weight: 122.17
RTECS: ZE5600000

Physical state and color
Colorless solid, slowly turning brown on exposure to air.

Melting point (°C):
24.5 (Andon et al., 1960)
27 (Dean, 1987)

Boiling point (°C):
210 (Weast, 1986)
210-212 (Dean, 1987)

Density (g/cm³ at 20/4 °C):
0.9650 (Weast, 1986)
1.02017 (Andon et al., 1960)

Diffusivity in water (10^5 cm²/sec):
0.77 at 20 °C using method of Hayduk and Laudie (1974).

Dissociation constant:
10.63 (Howard, 1989)

Flash point (°C):
>110 (Aldrich, 1988)

Henry's law constant (10^3 atm·m³/mol):
8.29, 6.74, 10.1, 4.93 and 3.75 at 10, 15, 20, 25 and 30 °C, respectively (Ashworth et al., 1988).

411

Bioconcentration factor, log BCF:
2.18 (bluegill sunfish, Veith et al., 1980)

Soil sorption coefficient, log K_{oc}:
2.08 (river sediment), 2.02 (coal wastewater sediment) (Kopinke et al., 1995)

Octanol/water partition coefficient, log K_{ow}:
2.54 (Sangster, 1989)
2.42 (Veith et al., 1980)
2.30 (Mabey et al., 1982)
2.47 (Garst and Wilson, 1984)
2.34 (Wasik et al., 1981)

Solubility in organics:
Freely soluble in ethanol, chloroform, ether and benzene (U.S. EPA, 1985).

Solubility in water (mg/L):
4,200 at 20 °C (Verschueren, 1983)
7,868 at 25 °C (Banerjee et al., 1980)
6,200 at 25 °C (Leunberger et al., 1985)
7,819 at 25.0 °C (Wasik et al., 1981)
8,795 at 25 °C and pH 5.1 (Blackman, 1955)
7,888 at 25 °C (Veith et al., 1980)

Vapor pressure (10^2 mmHg):
6.21 at 20 °C (supercooled liquid, Andon et al., 1960)
9.8 at 25.0 °C (Leuenberger et al., 1985)

Environmental Fate
   Biological. When 2,4-dimethylphenol was statically incubated in the dark at 25 °C with yeast extract and settled domestic wastewater inoculum, significant biodegradation with rapid adaptation was observed. At concentrations of 5 and 10 mg/L, 100 and 99% biodegradation, respectively, were observed after 7 days (Tabak et al., 1981).
   Chemical/Physical. Wet oxidation of 2,4-dimethylphenol at 320 °C yielded formic and acetic acids (Randall and Knopp, 1980).

Toxicity: Acute oral LD_{50} for mice 809 mg/kg, rats 3,200 mg/kg (RTECS, 1985).

Uses: Wetting agent; dyestuffs; preparation of phenolic antioxidants; plastics, resins, solvent, disinfectant, pharmaceuticals, insecticides, fungicides and rubber chemicals manufacturing; lubricant and gasoline additive; possibly used a pesticide; plasticizers.

DIMETHYL PHTHALATE

Synonyms: Avolin; **1,2-Benzenedicarboxylic acid dimethyl ester;** Dimethyl-1,2-benzenedicarboxylate; Dimethylbenzene-*o*-dicarboxylate; DMP; ENT 262; Fermine; Methyl phthalate; Mipax; NTM; Palatinol M; Phthalic acid dimethyl ester; Phthalic acid methyl ester; RCRA waste number U102; Solvanom; Solvarone.

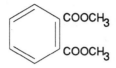

CAS Registry No.: 131-11-3
Molecular formula: $C_{10}H_{10}O_4$
Formula weight: 194.19
RTECS: TI1575000

Physical state, color and odor
Clear, colorless, odorless, moderately viscous, oily liquid.

Melting point (°C):
0 (Fishbein and Albro, 1972)
5.5 (U.S. EPA, 1980)

Boiling point (°C):
283.8 (Weast, 1986)

Density (g/cm³):
1.1905 at 20/4 °C (Weast, 1986)

Diffusivity in water (10^5 cm²/sec):
0.66 at 20 °C using method of Hayduk and Laudie (1974).

Flash point (°C):
146 (NIOSH, 1994)

Lower explosive limit (%):
0.9 at 180 °C (NFPA, 1984)

Henry's law constant (10^7 atm·m³/mol):
4.2 (Petrasek et al., 1983)

Ionization potential (eV):
9.64 (NIOSH, 1994)

Bioconcentration factor, log BCF:
1.76 (bluegill sunfish, Veith et al., 1980)

Soil sorption coefficient, log K_{oc}:
0.88, 1.63, 1.84 (various Norwegian soils, Seip et al., 1986)
2.28 (Banerjee et al., 1985)

Octanol/water partition coefficient, log K_{ow}:
1.53 (Leyder and Boulanger, 1983)
1.56 (Mabey et al., 1982)
1.47 (Howard et al., 1985)
1.61 (Veith et al., 1980)
1.86 (Eadsforth, 1986)
1.80 (Johnsen et al., 1989)

Solubility in organics:
Soluble in ethanol, ether and benzene (Weast, 1986).

Solubility in water:
4,320 mg/L at 25 °C (Wolfe et al., 1980)
4,290 mg/L at 20 °C (Leyder and Boulanger, 1983)
At 25 °C (mg/L): 4,000 (distilled water), 3,960 (well water), 3,160 (natural seawater) (Howard et al., 1985)
4,500 ppm at 25 °C (Fukano and Obata, 1976)
0.5 wt % at 20 °C (Fishbein and Albro, 1972)

Vapor density:
7.94 g/L at 25 °C, 6.70 (air = 1)

Vapor pressure (10^3 mmHg at 25 °C):
8.93 (Hinckley et al., 1990)
1.65 (Banerjee et al., 1990; Howard et al., 1985)

Environmental Fate
   Biological. In anaerobic sludge, degradation occurred as follows: monomethyl phthalate to phthalic acid to protocatechuic acid followed by ring cleavage and mineralization (Shelton et al., 1984). In a static-culture-flask screening test, dimethyl phthalate showed significant biodegradation with rapid adaptation. The ester (5 and 10 mg/L) was statically incubated in the dark at 25 °C with yeast extract and settled domestic wastewater inoculum. After 7 days, 100% bio-degradation was achieved (Tabak et al., 1981).
   Photolytic. An aqueous solution containing titanium dioxide and subjected to UV radiation (λ >290 nm) produced hydroxyphthalates and dihydroxyphthalates as

intermediates (Hustert and Moza, 1988).

Chemical/Physical. Hydrolyzes in water to phthalic acid and methyl alcohol (Wolfe et al., 1980).

Exposure Limits (mg/m^3): NIOSH REL: TWA 5, IDLH 2,000; OSHA PEL: TWA 5; ACGIH TLV: TWA 5.

Symptoms of Exposure: Irritates nasal passages, upper respiratory system, stomach; eye ache. Ingestion may cause central nervous system depression (NIOSH, 1994).

Toxicity: Acute oral LD_{50} for chickens 8,500 mg/kg, guinea pigs 2,400 mg/kg, mice 6,800 mg/kg, rats 6,800 mg/kg, rabbits 4,400 mg/kg (RTECS, 1985); LC_{50} (8-day) for grass shrimp larvae 100 ppm (Verschueren, 1983).

Drinking Water Standard: As of May 1994, no MCLGs or MCLs have been proposed (U.S. EPA, 1984).

Uses: Plasticizer for cellulose acetate, nitrocellulose, resins, rubber, elastomers; ingredient in lacquers; coating agents; safety glass; insect repellant; molding powders; perfumes.

2,2-DIMETHYLPROPANE

Synonyms: Neopentaene; Neopentane; *tert*-Pentane; Tetramethylmethane; UN 1265; UN 2044.

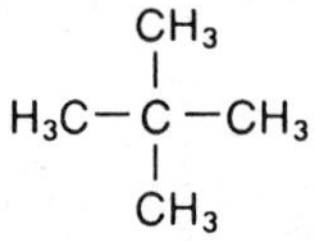

$$H_3C-\underset{\underset{CH_3}{|}}{\overset{\overset{CH_3}{|}}{C}}-CH_3$$

CAS Registry No.: 463-82-1
DOT: 2044
Molecular formula: C_5H_{12}
Formula weight: 72.15
RTECS: TY1190000

Physical state
Gas.

Melting point (°C):
-16.5 (Weast, 1986)
-19.8 (Windholz et al., 1983)

Boiling point (°C):
9.5 (Weast, 1986)

Density (g/cm³):
0.591 at 20/4 °C, 0.5852 at 25/4 °C (Dreisbach, 1959)

Diffusivity in water (10^5 cm²/sec):
0.80 at 20 °C using method of Hayduk and Laudie (1974).

Dissociation constant:
>14 (Schwarzenbach et al., 1993)

Flash point (°C):
-65.0 (Hawley, 1981)

Lower explosive limit (%):
1.4 (NFPA, 1984)

Upper explosive limit (%):
7.5 (NFPA, 1984)

Heat of fusion (kcal/mol):
0.752 (Dean, 1987)

Henry's law constant (atm·m^3/mol):
2.18 at 25 °C (Hine and Mookerjee, 1975)

Ionization potential (eV):
10.35 (Franklin et al., 1969)

Soil sorption coefficient, log K_{oc}:
Unavailable because experimental methods for estimation of this parameter for aliphatic hydrocarbons are lacking in the documented literature.

Octanol/water partition coefficient, log K_{ow}:
3.11 (Hansch and Leo, 1979)

Solubility in organics:
Soluble in alcohol and ether (Weast, 1986).

Solubility in water:
33.2 mg/kg at 25 °C (McAuliffe, 1966)

Vapor density:
2.95 g/L at 25 °C, 2.49 (air = 1)

Vapor pressure (mmHg):
1,287 at 25 °C (Wilhoit and Zwolinski, 1971)
1,074.43 at 19.492 °C, 1,267.75 at 24.560 °C (Osborn and Douslin, 1974)

Toxicity: Acute oral LD_{50} (intraperitoneal) for mice 100 mg/kg (RTECS, 1985).

Uses: Butyl rubber; organic synthesis.

2,7-DIMETHYLQUINOLINE

Synonyms: None.

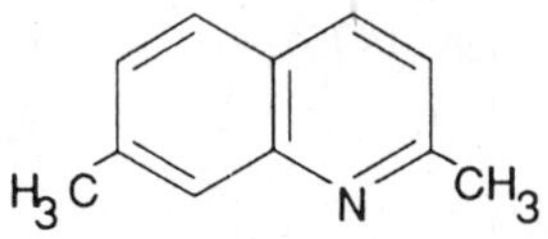

CAS Registry No.: 93-37-8
Molecular formula: $C_{11}H_{11}N$
Formula weight: 157.22

Physical state
Liquid.

Melting point (°C):
61 (Weast, 1986)
57-59 (Verschueren, 1983)

Boiling point (°C):
262-265 (Weast, 1986)

Density (g/cm³):
1.054 using method of Lyman et al. (1982).

Solubility in organics:
Soluble in alcohol, ether, chloroform (Weast, 1986) and benzene (Hawley, 1981).

Solubility in water:
1,795 mg/kg at 25 °C (Price, 1976)

Uses: Organic synthesis; dye intermediate.

DIMETHYL SULFATE

Synonyms: Dimethyl monosulfate; DMS; Methyl sulfate; RCRA waste number U103; **Sulfuric acid dimethyl ester;** UN 1595.

CAS Registry No.: 77-78-1
DOT: 1595
DOT label: Corrosive
Molecular formula: $C_2H_6O_4S$
Formula weight: 126.13
RTECS: WS8225000

Physical state, color and odor
Colorless, oily liquid with an onion-like odor.

Melting point (°C):
-31.7 (Weast, 1986)
-26.8 (Hawley, 1981)

Boiling point (°C):
Decomposes at 188.5 (Weast, 1986).

Density (g/cm^3):
1.3283 at 20/4 °C (Weast, 1986)

Diffusivity in water (10^5 cm^2/sec):
0.91 at 20 °C using method of Hayduk and Laudie (1974).

Flash point (°C):
83.4 (NIOSH, 1994)

Henry's law constant (10^6 atm·m^3/mol):
2.96 at 20 °C (approximate - calculated from water solubility and vapor pressure)

Soil sorption coefficient, log K_{oc}:
0.61 (calculated, Mercer et al., 1990)

Octanol/water partition coefficient, log K_{ow}:
-1.24 (Mercer et al., 1990)

Solubility in organics:
Soluble in alcohol, benzene, ether (Weast, 1986), 1,4-dioxane and aromatic hydrocarbons (Windholz et al., 1983).

Solubility in water:
28 g/L at 18 °C (Windholz et al., 1983)

Vapor density:
4.35 (air = 1) (Windholz et al., 1983)
5.16 g/L at 25 °C

Vapor pressure (mmHg):
0.5 at 20 °C (Weast, 1986)

Environmental Fate
   Chemical/Physical. Hydrolyzes in water (half-life = 1.2 hours) to methyl alcohol and sulfuric acid (Robertson and Sugamori, 1966).

Exposure Limits: Potential occupational carcinogen. NIOSH REL: TWA 0.1 ppm (0.5 mg/m^3), IDLH 7 ppm; OSHA PEL: TWA 1 ppm (5 mg/m^3); ACGIH TLV: 0.1 ppm.

Toxicity: Acute oral LD_{50} for mice 140 mg/kg, rats 205 mg/kg (RTECS, 1985).

Uses: In organic synthesis as a methylating agent.

1,2-DINITROBENZENE

Synonyms: *o*-Dinitrobenzene; *o*-Dinitrobenzol; UN 1597.

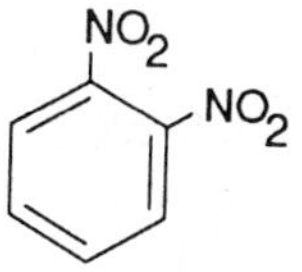

CAS Registry No.: 528-29-0
DOT: 1597
DOT label: Poison
Molecular formula: $C_6H_4N_2O_4$
Formula weight: 168.11
RTECS: CZ7450000

Physical state and color
Colorless to yellow needles.

Melting point (°C):
118.50 (Martin et al., 1979)

Boiling point (°C):
319 at 775 mmHg (Weast, 1986)

Density (g/cm^3):
1.565 at 17/4 °C (Weast, 1986)

Diffusivity in water (10^5 cm^2/sec):
0.79 at 20 °C using method of Hayduk and Laudie (1974).

Flash point (°C):
150 (NIOSH, 1994)

Ionization potential (eV):
10.71 (NIOSH, 1994)

Soil sorption coefficient, log K_{oc}:
Unavailable because experimental methods for estimation of this parameter for nitroaromatic hydrocarbons are lacking in the documented literature.
K_d = 1.7 mL/g on a Cs^+-kaolinite (Haderlein and Schwarzenbach, 1993)

Octanol/water partition coefficient, log K_{ow}:
1.58 (Leo et al., 1971)

421

Solubility in organics:
Soluble in alcohol ($\approx$ 16.7 g/L), benzene (50 g/L); freely soluble in chloroform and ethyl acetate (Windholz et al., 1983).

Solubility in water (mg/L):
151.5 (cold water), 370 at 100 °C (Windholz et al., 1983)
100 (cold water), 3,800 at 100 °C (Verschueren, 1983)

Environmental Fate
   Biological. Under anaerobic and aerobic conditions using a sewage inoculum, 1,2-dinitrobenzene degraded to nitroaniline (Hallas and Alexander, 1983).
   Chemical/Physical. Releases toxic nitrogen oxides when heated to decomposition (Sax and Lewis, 1987).

Exposure Limits (mg/m^3): NIOSH REL: TWA 1, IDLH 50; OSHA PEL: TWA 1.

Uses: Organic synthesis; dyes.

1,3-DINITROBENZENE

Synonyms: Binitrobenzene; 2,4-Dinitrobenzene; *m*-Dinitrobenzene; 1,3-Dinitrobenzol; UN 1597.

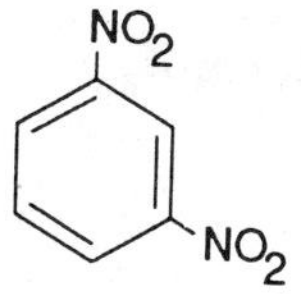

CAS Registry No.: 99-65-0
DOT: 1597
DOT label: Poison
Molecular formula: $C_6H_4N_2O_4$
Formula weight: 168.11
RTECS: CZ7350000

Physical state and color
White to yellowish crystals.

Melting point (°C):
90 (Weast, 1986)
90.20 (Martin et al., 1979)

Boiling point (°C):
291 at 756 mmHg (Weast, 1986)
300–303 (Windholz et al., 1983)

Density (g/cm³):
1.5751 at 18/4 °C (Weast, 1986)
1.368 at 89/4 °C (Aldrich, 1988)

Diffusivity in water (10^5 cm²/sec):
0.79 at 20 °C using method of Hayduk and Laudie (1974).

Flash point (°C):
150 (NIOSH, 1994)

Henry's law constant (10^7 atm·m³/mol):
2.75 at 35 °C (approximate – calculated from water solubility and vapor pressure)

Ionization potential (eV):
10.43 (NIOSH, 1994)

Soil sorption coefficient, log K_{oc}:
Unavailable because experimental methods for estimation of this parameter for nitroaromatic hydrocarbons are lacking in the documented literature.
K_d = 1,800 mL/g on a Cs^+-kaolinite (Haderlein and Schwarzenbach, 1993)

Octanol/water partition coefficient, log K_{ow}:
1.49 (Leo et al., 1971)

Solubility in organics:
Soluble in acetone, ether, pyrimidine (Weast, 1986), alcohol (27 g/L), pyridine (3,940 g/kg at 20-25 °C) (Dehn, 1917); freely soluble in benzene, chloroform and ethyl acetate (Windholz et al., 1983).

Solubility in water:
500 mg/L in cold water, 3.13 g/L at 100 °C (Windholz et al., 1983)
469 mg/L at 15 °C, 3,200 mg/L at 100 °C (Verschueren, 1983)
654 mg/kg at 30 °C (Gross and Saylor, 1931)
4.67 mmol/L at 35 °C (Hine et al., 1963)
>21.4 g/kg at 20-25 °C (Dehn, 1917)
5.12 mmol/kg at 25.0 °C (Vesala, 1974)

Vapor pressure (mmHg):
8.15 x 10^{-4} at 35 °C (Hine et al., 1963)

Environmental Fate
  Biological. Under anaerobic and aerobic conditions using a sewage inoculum, 1,3-dinitrobenzene degraded to nitroaniline (Hallas and Alexander, 1983).
  Chemical/Physical. Releases toxic nitrogen oxides when heated to decomposition (Sax and Lewis, 1987).

Exposure Limits (mg/m^3): NIOSH REL: TWA 1, IDLH 50; OSHA PEL: TWA 1.

Toxicity: Acute oral LD_{50} for rats 83 mg/kg, wild birds 42 mg/kg (RTECS, 1985).

Drinking Water Standard: As of May 1994, no MCLGs or MCLs have been proposed (U.S. EPA, 1984).

Uses: Organic synthesis; dyes.

1,4-DINITROBENZENE

Synonyms: *p*-Dinitrobenzene; Dithane A-4; UN 1597.

CAS Registry No.: 100-25-4
DOT: 1597
DOT label: Poison
Molecular formula: $C_6H_4N_2O_4$
Formula weight: 168.11
RTECS: CZ7525000

Physical state and color
White to yellow crystalline solid.

Melting point (°C):
174.00 (Martin et al., 1979)

Boiling point (°C):
299 (NIOSH, 1994)

Density (g/cm³):
1.625 at 18/4 °C (Weast, 1986)

Diffusivity in water (10^5 cm²/sec):
0.79 at 20 °C using method of Hayduk and Laudie (1974).

Henry's law constant (10^7 atm·m³/mol):
4.79 at 35 °C (approximate - calculated from water solubility and vapor pressure)

Ionization potential (eV):
10.50 (NIOSH, 1994)

Soil sorption coefficient, log K_{oc}:
Unavailable because experimental methods for estimation of this parameter for aliphatic hydrocarbons are lacking in the documented literature.
$K_d \approx 4,000$ mL/g on a Cs^+-kaolinite (Haderlein and Schwarzenbach, 1993)

Octanol/water partition coefficient, log K_{ow}:
1.46, 1.49 (Leo et al., 1971)

425

Solubility in organics:
Soluble in acetone, acetic acid, benzene, toluene (Weast, 1986) and alcohol (3.3 g/L) (Windholz et al., 1983).

Solubility in water:
0.01 wt % at 20 °C (NIOSH, 1994)
80 mg/L in cold water, 1.8 g/L at 100 °C (Windholz et al., 1983)
0.617 mmol/L at 35 °C (Hine et al., 1963)

Vapor pressure (mmHg):
2.25×10^{-4} at 35 °C (Hine et al., 1963)

Environmental Fate
   Chemical/Physical. Releases toxic nitrogen oxides when heated to decomposition (Sax and Lewis, 1987).

Exposure Limits (mg/m^3): NIOSH REL: TWA 1, IDLH 50; OSHA PEL: TWA 1.

Uses: Organic synthesis; dyes.

4,6-DINITRO-*o*-CRESOL

Synonyms: Antinonin; Antinonnon; Arborol; Capsine; Chemsect DNOC; Degrassan; Dekrysil; Detal; Dinitrocresol; Dinitro-*o*-cresol; 2,4-Dinitro-*o*-cresol; 3,5-Dinitro-*o*-cresol; Dinitrodendtroxal; 3,5-Dinitro-2-hydroxytoluene; Dinitrol; Dinitromethyl cyclohexyltrienol; 2,4-Dinitro-2-methylphenol; 2,4-Dinitro-6-methylphenol; 4,6-Dinitro-2-methylphenol; Dinitrosol; Dinoc; Dinurania; DN; DNC; DN-dry mix no. 2; DNOC; Effusan; Effusan 3436; Elgetol; Elgetol 30; Elipol; ENT 154; Extrar; Hedolit; Hedolite; K III; K IV; Kresamone; Krezotol 50; Lipan; **2-Methyl-4,6-dinitrophenol**; 6-Methyl-2,4-dinitrophenol; Nitrador; Nitrofan; Prokarbol; Rafex; Rafex 35; Raphatox; RCRA waste number P047; Sandolin; Sandolin A; Selinon; Sinox; Trifina; Trifocide; Winterwash.

CAS Registry No.: 534-52-1
DOT: 1598
Molecular formula: $C_7H_6N_2O_5$
Formula weight: 198.14
RTECS: GO9625000

Physical state, color and odor
Yellow, odorless crystals.

Melting point (°C):
86.5 (Weast, 1986)

Boiling point (°C):
312 (ACGIH, 1986)

Diffusivity in water (10^5 cm^2/sec):
0.69 at 20 °C using method of Hayduk and Laudie (1974).

Dissociation constant:
4.35 at 25 °C (Dean, 1973)
4.39, 4.46 (Jafvert et al., 1990)
4.31 at 21.5 °C (Schwarzenbach et al., 1988)

Flash point (°C):
Noncombustible solid (NIOSH, 1994)

Lower explosive limit:
For a dust, 30 mg/m^3 is the minimum explosive concentration in air (NIOSH, 1994).

Henry's law constant (10^6 atm·m^3/mol):
1.4 at 25 °C (Warner et al., 1987)

Soil sorption coefficient, log K$_{oc}$:
2.41 (Meylan et al., 1992)

Octanol/water partition coefficient, log K$_{ow}$:
2.14, 2.16 (Jafvert et al., 1990)
2.12 (Schwarzenbach et al., 1988)
2.85 (Mills et al., 1985)

Solubility in organics:
At 15 °C (mg/L): methanol (7.33), ethanol (9.12), chloroform (37.2), acetone (100.6) (Bailey and White, 1965).

Solubility in water:
0.013 wt % at 15 °C (Berg, 1983)
128 mg/L at 20 °C (Meites, 1963)
198 mg/L (aqueous buffer solution pH 1.5 and 20 °C, Schwarzenbach et al., 1988)

Vapor pressure (10^5 mmHg):
5.2 at 25 °C (Melnikov, 1971)
5 at 20 °C (ACGIH, 1986)
32 at 20 °C (Schwarzenbach et al., 1988)

Environmental Fate
   Biological. In plants and soils, the nitro groups are reduced to amino groups (Hartley and Kidd, 1987). When 4,6-dinitro-*o*-cresol was statically incubated in the dark at 25 °C with yeast extract and settled domestic wastewater inoculum, no significant biodegradation and necessary acclimation for optimum biooxidation within the 4-week incubation period was observed (Tabak et al., 1981).
   Chemical/Physical. Reacts with alkalies and amines forming water-soluble salts which are indicative of phenols (Morrison and Boyd, 1971).

Exposure Limits (mg/m^3): NIOSH REL: TWA 0.2, IDLH 5; OSHA PEL: TWA 0.2.

Symptoms of Exposure: Headache, fever, profuse sweating, rapid pulse and respiration, cough, shortness of breath and coma (Patnaik, 1992).

Toxicity: Acute oral LD_{50} for mice 47 mg/kg, rats 10 mg/kg; LD_{50} (skin) for rats 200 mg/kg (RTECS, 1985).

Uses: Dormant ovicidal spray for fruit trees (highly phototoxic and cannot be used successfully on actively growing plants); selective herbicide and insecticide.

2,4-DINITROPHENOL

Synonyms: Aldifen; Chemox PE; α-Dinitrophenol; DNP; 2,4-DNP; Fenoxyl carbon n; 1-Hydroxy-2,4-dinitrobenzene; Maroxol-50; Nitro kleenup; NSC 1532; RCRA waste number P048; Solfo black B; Solfo black BB; Solfo black 2B supra; Solfo black G; Solfo black SB; Tetrasulphur black PB; Tetrosulphur PBR.

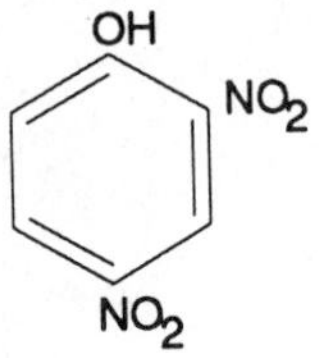

CAS Registry No.: 51-28-5
DOT: 0076
DOT label: Poison
Molecular formula: $C_6H_4N_2O_5$
Formula weight: 184.11
RTECS: SL2800000

Physical state, color and odor
Yellow crystals with a sweet, musty odor.

Melting point (°C):
115–116 (Weast, 1986)
106–108 (Aldrich, 1988)
111–113 (Fluka, 1988)

Boiling point (°C):
Sublimes (Weast, 1986)

Density (g/cm³):
1.683 at 24/4 °C (Weast, 1986)
1.68 at 20/4 °C (Weiss, 1986)

Diffusivity in water (10^5 cm²/sec):
0.76 at 20 °C using method of Hayduk and Laudie (1974).

Dissociation constant:
4.09 at 25 °C (Dean, 1973)
3.94 at 21.5 °C (Schwarzenbach et al., 1988)

Henry's law constant (10^8 atm·m³/mol):
1.57 at 20 °C (approximate - calculated from water solubility and vapor pressure)

Soil sorption coefficient, log K_{oc}:
1.25 (estimated, Montgomery, 1989)

Octanol/water partition coefficient, log K_{ow}:
1.51, 1.54 (Leo et al., 1971)
1.50 (Stockdale and Selwyn, 1971)
1.56 (Korenman et al., 1977)
1.67 (Schwarzenbach et al., 1988)

Solubility in organics:
At 15 °C (wt %): 13.46 in ethyl acetate, 26.42 in acetone, 5.11 in chloroform, 16.72 in pyridine, 0.42 in carbon tetrachloride, 5.98 in toluene (Windholz et al., 1983); 30.5 g/L in alcohol (Meites, 1963).

Solubility in water:
5,600 mg/L at 18 °C, 43,000 mg/L at 100 °C (Verschueren, 1983)
6,000 mg/L at 25 °C (Morrison and Boyd, 1971)
At pH 1.5 (mg/L): 172 at 5 °C, 207 at 10 °C, 335 at 20 °C, 473 at 30 °C (Schwarzenbach et al., 1988)
3.46 mmol/L at 25 °C (Caturla et al., 1988)
In wt %: 0.137 at 54.5 °C, 0.301 at 75.8 °C, 0.587 at 87.4 °C, 1.22 at 96.2 °C (Windholz et al., 1983)
202 mg/L at 12.5 °C (Meites, 1963)

Vapor pressure (10^5 mmHg):
39 at 20 °C (Schwarzenbach et al., 1988)

Environmental Fate
Biological. When 2,4-dinitrophenol was statically incubated in the dark at 25 °C with yeast extract and settled domestic wastewater inoculum, significant biodegradation with rapid adaptation was observed. At concentrations of 5 and 10 mg/L, 60 and 68% biodegradation, respectively, were observed after 7 days (Tabak et al., 1981).

Photolytic. When an aqueous 2,4-dinitrophenol solution containing titanium dioxide was illuminated by UV light, ammonium and nitrate ions formed as the major products (Low et al., 1991).

Chemical/Physical. Ozonation of an aqueous solution containing 2,4-dinitrophenol (100 mg/L) yielded formic, acetic, glyoxylic and oxalic acids (Wang, 1990).

Symptoms of Exposure: Heavy sweating, nausea, vomiting, collapse and death (Patnaik, 1992).

Toxicity: Acute oral LD_{50} for mice 45 mg/kg, guinea pigs 81 mg/kg, rats 30 mg/kg, rabbits 30 mg/kg, wild birds 13 mg/kg; LD_{50} (subcutaneous) for rats 25 mg/kg (RTECS, 1985).

Uses: Organic synthesis; photographic agent; manufacture of pesticides, herbicides, explosives and wood preservatives; yellow dyes; preparation of picric acid and diaminophenol (photographic developer); indicator; analytical reagent for potassium and ammonium ions; insecticide.

2,4-DINITROTOLUENE

Synonyms: 2,4-Dinitromethylbenzene; Dinitrotoluol; 2,4-Dinitrotoluol; DNT; 2,4-DNT; 1-Methyl-2,4-dinitrobenzene; NCI-C01865; RCRA waste number U105.

CAS Registry No.: 121-14-2
DOT: 1600 (liquid); 2038 (solid)
Molecular formula: $C_7H_6N_2O_4$
Formula weight: 182.14
RTECS: XT1575000

Physical state, color and odor
Yellow to red needles or yellow liquid with a faint, characteristic odor.

Melting point (°C):
71.1 (Lenchitz and Velicky, 1970)
67-70 (Aldrich, 1988)

Boiling point (°C):
300 with slight decomposition (Howard, 1989)

Density (g/cm^3):
1.521 at 15/4 °C (Sax, 1984)
1.379 at 20/4 °C (Weiss, 1986)
1.3208 at 71/4 °C (Keith and Walters, 1992)

Diffusivity in water (10^5 cm^2/sec):
0.76 at 20 °C using method of Hayduk and Laudie (1974).

Flash point (°C):
206.7 (Weiss, 1986)

Henry's law constant (10^7 atm·m^3/mol):
8.67 (Howard, 1989)

Soil sorption coefficient, log K_{oc}:
1.79 using method of Karickhoff et al. (1979).

Octanol/water partition coefficient, log K_{ow}:
1.98 (Mabey et al., 1982)

Solubility in organics:
Soluble in acetone, ethanol, benzene, ether and pyrimidine (Weast, 1986).

Solubility in water:
270 mg/L at 22 °C (Verschueren, 1983)

Vapor pressure (10^4 mmHg):
12.98 at 58.8 °C (Lenchitz and Velicky, 1970)
1.1 at 20 °C (Howard, 1989)

Environmental Fate
Biological. When 2,4-dinitrotoluene was statically incubated in the dark at 25 °C with yeast extract and settled domestic wastewater inoculum, significant biodegradation with gradual acclimation was followed by deadaptive process in subsequent subcultures. At a concentration of 5 mg/L, 77, 61, 50 and 27% losses were observed after 7, 14, 21 and 28-day incubation periods, respectively. At a concentration of 10 mg/L, only 50, 49, 44 and 23% were observed after 7, 14, 21 and 28-day incubation periods, respectively (Tabak et al., 1981).

Chemical/Physical. Wet oxidation of 2,4-dinitrotoluene at 320 °C yielded formic and acetic acids (Randall and Knopp, 1980).

Toxicity: Acute oral LD_{50} for mice 790 mg/kg, rats 268 mg/kg, guinea pigs 1,300 mg/kg (RTECS, 1985).

Drinking Water Standard: As of May 1994, no MCLGs or MCLs have been proposed although 2,4-dichlorotoluene has been listed for regulation (U.S. EPA, 1994).

Uses: Organic synthesis; intermediate for toluidine, dyes and explosives.

2,6-DINITROTOLUENE

Synonyms: 2,6-Dinitromethylbenzene; 2,6-Dinitrotoluol; 2,6-DNT; **2-Methyl-1,3-dinitrobenzene;** RCRA waste number U106.

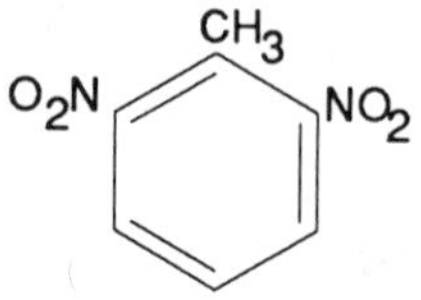

CAS Registry No.: 606-20-2
DOT: 1600 (liquid); 2038 (solid)
Molecular formula: $C_7H_6N_2O_4$
Formula weight: 182.14
RTECS: XT1925000

Physical state and color
Yellow crystals. Odor threshold in water is 100 ppb (Keith and Walters, 1992).

Melting point (°C):
66 (Weast, 1986)
60.5 (Weiss, 1986)

Boiling point (°C):
285 (Maksimov, 1968)

Density (g/cm³):
1.2833 at 111/4 °C (Weast, 1986)

Diffusivity in water (10^5 cm²/sec):
0.76 at 20 °C using method of Hayduk and Laudie (1974).

Flash point (°C):
206.7 (calculated, Weiss, 1986)

Henry's law constant (10^7 atm·m³/mol):
2.17 (Howard, 1989)

Soil sorption coefficient, log K_{oc}:
1.79 using method of Karickhoff et al. (1979).

Octanol/water partition coefficient, log K_{ow}:
2.00 (Mills et al., 1985)

435

Solubility in organics:
Soluble in ethanol (Weast, 1986).

Solubility in water:
≈ 300 mg/L (Mills et al., 1985)

Vapor pressure (10^4 mmHg):
3.5 at 20 °C (Howard, 1989)
5.67 at 25 °C (Banerjee et al., 1990)

Environmental Fate
   Biological. When 2,6-dinitrotoluene was statically incubated in the dark at 25 °C with yeast extract and settled domestic wastewater inoculum, significant biodegradation with gradual acclimation was followed by deadaptive process in subsequent subcultures. At a concentration of 5 mg/L, 82, 55, 47 and 29% losses were observed after 7, 14, 21 and 28-day incubation periods, respectively. At a concentration of 10 mg/L, only 57, 49, 35 and 13% were observed after 7, 14, 21 and 28-day incubation periods, respectively (Tabak et al., 1981). Under anaerobic and aerobic conditions, a sewage inoculum degraded 2,6-dinitrotoluene to aminonitrotoluene (Hallas and Alexander, 1983).

Exposure Limits (mg/m^3): TWA: 1.5, IDLH: 200 (Weiss, 1986).

Toxicity: Acute oral LD_{50} for mice 621 mg/kg, rats 177 mg/kg (RTECS, 1985).

Drinking Water Standard: As of May 1994, no MCLGs or MCLs have been proposed although 2,6-dinitrotoluene has been listed for regulation (U.S. EPA, 1994).

Uses: Organic synthesis; manufacture of explosives.

DI-*n*-OCTYL PHTHALATE

Synonyms: 1,2-Benzenedicarboxylic acid dioctyl ester; 1,2-Benzenedicarboxylic acid di-*n*-octyl ester; *o*-Benzenedicarboxylic acid dioctyl ester; Celluflex DOP; Dinopol NOP; Dioctyl-*o*-benzenedicarboxylate; Dioctyl phthalate; *n*-Dioctyl phthalate; DNOP; DOP; Octyl phthalate; *n*-Octyl phthalate; Polycizer 162; PX 138; RCRA waste number U107; Vinicizer 85.

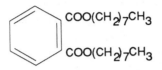

CAS Registry No.: 117-84-0
Molecular formula: $C_{24}H_{38}O_4$
Formula weight: 390.57
RTECS: TI1925000

Physical state, color and odor
Clear, light colored, viscous, oily liquid with a slight odor.

Melting point (°C):
-30 (Clayton and Clayton, 1981)
-25 (Fishbein and Albro, 1972)

Boiling point (°C):
386 (Weiss, 1986)

Density (g/cm³):
0.978 at 25/4 °C (Standen, 1968)
0.978 at 20/4 °C (Fishbein and Albro, 1972)

Diffusivity in water (10^5 cm²/sec):
0.39 at 20 °C using method of Hayduk and Laudie (1974).

Flash point (°C):
218.3 (Weiss, 1986)

Lower explosive limit (%):
Not pertinent (Weiss, 1986)

Upper explosive limit (%):
Not pertinent (Weiss, 1986)

Henry's law constant (10^{12} atm·m^3/mol):
1.41 at 25 °C (approximate - calculated from water solubility and vapor pressure)

Soil sorption coefficient, log K_{oc}:
8.99 using method of Karickhoff et al. (1979).

Octanol/water partition coefficient, log K_{ow}:
9.2 (calculated, Leo et al., 1971)

Solubility in water (mg/L):
0.285 at 24 °C (Verschueren, 1983)
3.0 at 25 °C (Wolfe et al., 1980)

Vapor density:
16.00 g/L at 25 °C, 13.52 (air = 1)

Vapor pressure (10^8 mmHg):
5 at 82 °C (Gross and Colony, 1973)
14,000 at 25 °C (assigned by analogy, Mabey et al., 1982)

Environmental Fate
Biological. o-Phthalic acid was tentatively identified as the major degradation product of di-*n*-octyl phthalate produced by the bacterium *Serratia marcescens* (Mathur and Rouatt, 1975). When di-*n*-octyl phthalate was statically incubated in the dark at 25 °C with yeast extract and settled domestic wastewater inoculum, no degradation was observed after 7 days. Over a 21-day period, however, gradual adaptation did occur, resulting in 94 and 93% losses at concentrations of 5 and 10 mg/L, respectively (Tabak et al., 1981).

Chemical/Physical. Hydrolyzes in water to phthalic acid and 1-octanol (Wolfe et al., 1980). The hydrolysis half-life at pH 7 and 25 °C was estimated to be 107 years (Ellington et al., 1988).

Symptoms of Exposure: Ingestion may cause nausea, somnolence, hallucination and lacrimation (Patnaik, 1992).

Toxicity: Acute oral LD_{50} for mice 6,513 mg/kg (RTECS, 1985).

Use: Plasticizer for poly(vinyl chloride) (PVC) and other vinyl polymers.

1,4-DIOXANE

Synonyms: Diethylene dioxide; 1,4-Diethylene dioxide; Diethylene ether; Diethylene oxide; Diokan; 1,4-Dioxacyclohexane; Dioxan; Dioxane-1,4; **1,4-Dioxane**; *p*-Dioxane; Dioxyethylene ether; Glycol ethylene ether; NCI-C03689; RCRA waste number U108; Tetrahydro-1,4-dioxin; Tetrahydro-*p*-dioxin; UN 1165.

CAS Registry No.: 123-91-1
DOT: 1165
DOT label: Flammable liquid
Molecular formula: $C_4H_8O_2$
Formula weight: 88.11
RTECS: JG8225000

Physical state, color and odor
Clear, colorless liquid with a pleasant, ether-like odor. Odor threshold in air is 24 ppm (Amoore and Hautala, 1983).

Melting point (°C):
11.8 (Weast, 1986)

Boiling point (°C):
101.32 (Riddick et al., 1986)
101.1 (Windholz et al., 1983)

Density (g/cm³):
1.0337 at 20/4 °C (Weast, 1986)
1.02797 at 25/4 °C (Riddick et al., 1986)

Diffusivity in water (10^5 cm²/sec):
0.97 at 20 °C using method of Hayduk and Laudie (1974).

Flash point (°C):
12.8 (NIOSH, 1994)
18.3 (open cup, Hawley, 1981)

Lower explosive limit (%):
2.0 (NIOSH, 1994)

Upper explosive limit (%):
22 (NIOSH, 1994)

Heat of fusion (kcal/mol):
3.07 (Dean, 1987)

Henry's law constant (10^6 atm·m^3/mol at 25 °C):
4.88 (Hine and Mookerjee, 1975)
9.05 (Amoore and Buttery, 1978)

Ionization potential (eV):
9.13 ± 0.03 (Franklin et al., 1969)

Soil sorption coefficient, log K_{oc}:
0.54 (calculated, Mercer et al., 1990)

Octanol/water partition coefficient, log K_{ow}:
-0.42 (Collander, 1951)
-0.27 (Hansch and Leo, 1979)

Solubility in organics:
Soluble in acetone, alcohol, benzene and ether (Weast, 1986). Miscible with most organic solvents (Huntress and Mulliken, 1941).

Solubility in water:
Miscible (Palit, 1947).

Vapor density:
3.60 g/L at 25 °C, 3.04 (air = 1)

Vapor pressure (mmHg):
29 at 20 °C (NIOSH, 1994)
37 at 25 °C, 50 at 30 °C (Verschueren, 1983)
38.1 at 25 °C (Banerjee et al., 1990)
34.28 at 25.00 °C (Hussam and Carr, 1985)

Environmental Fate
Photolytic. Irradiation of pure 1,4-dioxane through quartz using a 450-W medium-pressure mercury lamp gave meso and racemic forms of 1-hydroxyethyldioxane, a pair of diastereomeric dioxane dimers (Mazzocchi and Bowen, 1975), dioxanone, dioxanol, hydroxymethyldioxane and hydroxyethylidenedioxane (Houser and Sibbio, 1977). When 1,4-dioxane is subjected to a megawatt ruby laser, 4% was decomposed yielding ethylene, carbon monoxide, hydrogen and a

trace of formaldehyde (Watson and Parrish, 1971).

Chemical/Physical. Anticipated products from the reaction of 1,4-dioxane with ozone or hydroxyl radicals in the atmosphere are glyoxylic acid, oxygenated formates and $OHCOCH_2CH_2OCHO$ (Cupitt, 1980). Storage of 1,4-dioxane in the presence of air resulted in the formation of 1,2-ethanediol monoformate and 1,2-ethane diformate (Jewett and Lawless, 1980).

Exposure Limits: Potential occupational carcinogen. NIOSH REL: 30-minute ceiling 1 ppm (3.6 mg/m^3), IDLH 500 ppm; OSHA PEL: TWA 100 ppm; ACGIH TLV: TWA 25 ppm (90 mg/m^3).

Symptoms of Exposure: Ingestion or skin absorption may cause drowsiness, headache, respiratory distress, nausea and vomiting (Patnaik, 1992).

Toxicity: Acute oral LD_{50} for mice 5,700 mg/kg, cats 2,000 mg/kg, guinea pigs 3,150 mg/kg, rats 4,200 mg/kg, rabbits 2,000 mg/kg (RTECS, 1985); LC_{50} (48-hour) for red killifish 81,280 mg/L (Yoshioka et al., 1986).

Drinking Water Standard: As of May 1994, no MCLGs or MCLs have been proposed although 1,4-dioxane has been listed for regulation (U.S. EPA, 1994).

Uses: Solvent for cellulose acetate, benzyl cellulose, ethyl cellulose, waxes, resins, oils, resins and many organic compounds; cements, cosmetics, deodorants; fumigants; paint and varnish removers, cleaning and detergent preparations; wetting and dispersing agent in textile processing, dyes baths, stain and printing compositions; polishing compositions; stabilizer for chlorinated solvents; scintillation counter; organic synthesis.

1,2-DIPHENYLHYDRAZINE

Synonyms: *N,N′*-Bianiline; *N,N′*-Diphenylhydrazine; *sym*-Diphenylhydrazine; DPH; Hydrazobenzene; 1,1′-Hydrazobenzene; Hydrazodibenzene; NCI-C01854; RCRA waste number U109.

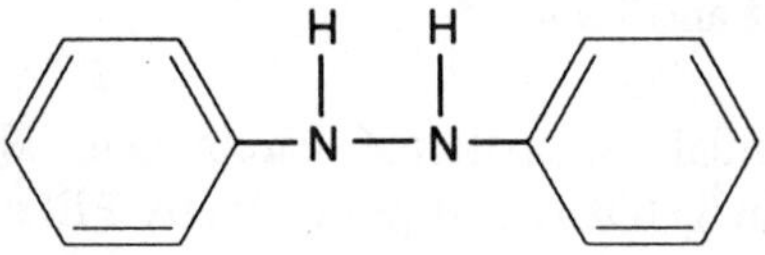

CAS Registry No.: 122-66-7
Molecular formula: $C_{12}H_{12}N_2$
Formula weight: 184.24
RTECS: MW2625000

Physical state and color
Colorless to pale yellow crystals.

Melting point (°C):
131 (Weast, 1986)
126-128 (Fluka, 1988)

Boiling point (°C):
Decomposes near the melting point (U.S. EPA, 1980).

Density (g/cm³):
1.158 at 16/4 °C (Weast, 1986)

Diffusivity in water (10^5 cm²/sec):
0.57 at 20 °C using method of Hayduk and Laudie (1974).

Henry's law constant (10^{11} atm·m³/mol):
4.11 at 25 °C (approximate - calculated from water solubility and vapor pressure)

Soil sorption coefficient, log K_{oc}:
2.82 using method of Karickhoff et al. (1979).

Octanol/water partition coefficient, log K_{ow}:
2.94 (Mabey et al., 1982)

Solubility in organics:
Soluble in ethanol (Weast, 1986).

Solubility in water:
221 mg/L at 25 °C (U.S. EPA, 1980)

Vapor pressure (mmHg):
2.6 x 10^{-5} at 25 °C (Mabey et al., 1982)

Environmental Fate
   Biological. When 5 and 10 mg/L of diphenylhydrazine was statically incubated in the dark at 25 °C with yeast extract and settled domestic wastewater inoculum, 80 and 72% biodegradation, respectively, were observed after 7 days (Tabak et al., 1981).
   Chemical/Physical. Wet oxidation of 1,2-diphenylhydrazine at 320 °C yielded formic and acetic acids (randall and Knopp, 1980).

Toxicity: Acute oral LD_{50} for rats 301 mg/kg (RTECS, 1985).

Uses: Manufacture of benzidine and starting material for pharmaceutical drugs.

DIURON

Synonyms: AF 101; Cekiuron; Crisuron; Dailon; DCMU; Diater; Dichlorfenidim; 3-(3,4-Dichlorophenol)-1,1-dimethylurea; 3-(3,4-Dichlorophenyl)-1,1-dimethylurea; **N'-(3,4-Dichlorophenyl)-N,N-dimethylurea**; 1,1-Dimethyl-3-(3,4-dichlorophenyl)urea; Di-on; Direx 4L; Diurex; Diurol; DMU; Dynex; Farmco diuron; Herbatox; HW 920; Karmex; Karmex diuron herbicide; Karmex DW; Marmer; NA 2767; Sup'r flo; Telvar; Telvar diuron weed killer; Unidron; Urox D; USAF P-7; USAF XR-42; Vonduron.

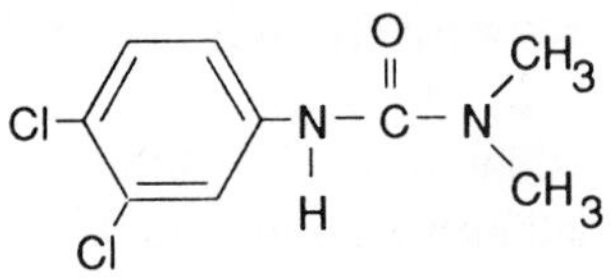

CAS Registry No.: 330-54-1
DOT: 2767
DOT label: Poison
Molecular formula: $C_9H_{10}Cl_2N_2O$
Formula weight: 233.11
RTECS: YS8925000

Physical state and color
White crystalline solid.

Melting point (°C):
158-159 (Windholz et al., 1983)
150-155 (Bailey and White, 1965)

Boiling point (°C):
Decomposes at 180 (Hawley, 1981)

Density (g/cm³):
1.385 using method of Lyman et al. (1982).

Diffusivity in water (10^5 cm²/sec):
0.53 at 20 °C using method of Hayduk and Laudie (1974).

Dissociation constant:
-1 to -2 (Bailey and White, 1965)

Flash point (°C):
Noncombustible solid (NIOSH, 1994)

Henry's law constant (10^9 atm·m^3/mol):
1.46 at 25 °C (approximate - calculated from water solubility and vapor pressure)

Soil sorption coefficient, log K_{oc}:
2.51 (Commerce soil), 2.62 (Tracy soil), 2.60 (Catlin soil) (McCall et al., 1981)
2.87 (Webster soil, Nkedi-Kizza et al., 1983)
2.21 (Briggs, 1981)
2.95 (Eustis fine sand, Wood et al., 1990)

Octanol/water partition coefficient, log K_{ow}:
1.97 (Kenaga and Goring, 1980)
2.68 (Briggs, 1981)
2.60 (Ellgehausen et al., 1981)

Solubility in organics:
In acetone: 5.3 wt % at 27 °C (Meister, 1988).

Solubility in water (ppm):
40 at 20 °C (Gunther et al., 1968)
42 at 25 °C (Bailey and White, 1965)
22 at 20 °C (Ellgehausen et al., 1981)

Vapor pressure (10^7 mmHg):
31 at 50 °C (Bailey and White, 1965)
2 at 30 °C (Hawley, 1981)

Environmental Fate
  Biological. Radiolabeled diuron degraded in aerobic soils to 3-(3,4-dichlorophenyl)-1-methylurea and 3-(3,4-dichlorophenyl)urea. 3,4-Dichloroaniline was reported as a minor degradation product of diuron (Lewis, 1989). Incubation of diuron in soil releases carbon dioxide. The rate of carbon dioxide formation nearly tripled when the soil temperature was increased from 25 to 35 °C. Half-lives (soil temperature, °C) of diuron in an Adkins loamy sand were: 705 (25), 414 (30) and 225 days (35). However, in a Semiahoo mucky peat, the half-lives were considerable higher: 3,991, 2,164 and 1,165 days at 25, 30 and 35 °C, respectively (Madhun and Freed, 1987).

Exposure Limits: NIOSH REL: 10 mg/m^3.

Symptoms of Exposure: May irritate eyes, skin, nose and throat (NIOSH, 1994).

Toxicity: Acute oral LD$_{50}$ for rats 3,400 mg/kg (Hartley and Kidd, 1987), 1,017 mg/kg (RTECS, 1985), 437 mg/kg (Windholz et al., 1983); LC$_{50}$ (96-hour) for rainbow trout 5.6 mg/L, bluegill sunfish 5.9 mg/L and guppies 25 mg/L (Hartley and Kidd, 1987); LC_{50} (48-hour) for bluegill sunfish 7.4 ppm, rainbow trout 4.3 ppm and coho salmon 16.0 mg/L (Verschueren, 1983).

Drinking Water Standard: As of May 1994, no MCLGs or MCLs have been proposed although diuron has been listed for regulation (U.S. EPA, 1994).

Uses: Pre-emergence herbicide used in soil to control germinating broadleaf grasses and weeds in crops such as apples, cotton, grapes, pears, pineapples and alfalfa; sugar cane flowering depressant.

DODECANE

Synonyms: Adakane 12; Bihexyl; Dihexyl; *n*-Dodecane; Duodecane.

$$H-\overset{\overset{\displaystyle H}{|}}{\underset{\underset{\displaystyle H}{|}}{C}}-\overset{\overset{\displaystyle H}{|}}{\underset{\underset{\displaystyle H}{|}}{C}}-\overset{\overset{\displaystyle H}{|}}{\underset{\underset{\displaystyle H}{|}}{C}}-\overset{\overset{\displaystyle H}{|}}{\underset{\underset{\displaystyle H}{|}}{C}}-\overset{\overset{\displaystyle H}{|}}{\underset{\underset{\displaystyle H}{|}}{C}}-\overset{\overset{\displaystyle H}{|}}{\underset{\underset{\displaystyle H}{|}}{C}}-\overset{\overset{\displaystyle H}{|}}{\underset{\underset{\displaystyle H}{|}}{C}}-\overset{\overset{\displaystyle H}{|}}{\underset{\underset{\displaystyle H}{|}}{C}}-\overset{\overset{\displaystyle H}{|}}{\underset{\underset{\displaystyle H}{|}}{C}}-\overset{\overset{\displaystyle H}{|}}{\underset{\underset{\displaystyle H}{|}}{C}}-\overset{\overset{\displaystyle H}{|}}{\underset{\underset{\displaystyle H}{|}}{C}}-\overset{\overset{\displaystyle H}{|}}{\underset{\underset{\displaystyle H}{|}}{C}}-H$$

CAS Registry No.: 112-40-3
Molecular formula: $C_{12}H_{26}$
Formula weight: 174.34
RTECS: JR2125000

Physical state and color
Colorless liquid.

Melting point (°C):
-9.6 (Weast, 1986)
-12 (Sax and Lewis, 1987)

Boiling point (°C):
216.3 (Weast, 1986)
213 (Hawley, 1981)

Density (g/cm^3):
0.74869 at 20/4 °C, 0.74516 at 25/4 °C (Dreisbach, 1959)

Diffusivity in water (10^5 cm^2/sec):
0.54 at 20 °C using method of Hayduk and Laudie (1974).

Dissociation constant:
>14 (Schwarzenbach et al., 1993)

Flash point (°C):
71.1 (Hawley, 1981)
79 (Affens and McLaren, 1972)

Lower explosive limit (%):
0.6 (Sax and Lewis, 1987)

Heat of fusion (kcal/mol):
8.57 (Dean, 1987)

447

Henry's law constant (atm·m^3/mol at 25 °C):
24.2 (approximate – calculated from water solubility and vapor pressure)

Interfacial tension with water (dyn/cm at 24.5 °C):
52.8 (Demond and Lindner, 1993)

Bioconcentration factor, log BCF:
3.80 (algae, Geyer et al., 1984)
3.11 (activated sludge), 1.70 (golden ide) (Freitag et al., 1985)

Soil sorption coefficient, log K_{oc}:
Unavailable because experimental methods for estimation of this parameter for aliphatic hydrocarbons are lacking in the documented literature.

Octanol/water partition coefficient, log K_{ow}:
5.64 (Geyer et al., 1984)
6.10 (Coates et al., 1985)
7.24 (Burkhard et al., 1985a)

Solubility in organics:
Soluble in acetone, alcohol, chloroform, ether (Weast, 1986) and many hydrocarbons.

Solubility in water (25 °C):
3.7 μg/kg (distilled water), 2.9 μg/kg (seawater) (Sutton and Calder, 1974)
3.4 μg/L (Mackay and Shiu, 1981)
8.42 μg/kg (Franks, 1966)
5 x 10^{-10} in seawater (mole fraction, Krasnoshchekova and Gubergrits, 1973)

Vapor density:
7.13 g/L at 25 °C, 6.02 (air = 1)

Vapor pressure (mmHg):
0.3 at 20 °C, 1 at 48 °C (Verschueren, 1983)
0.39 at 25 °C (Mackay et al., 1982)

Environmental Fate

Biological. Dodecane may biodegrade in two ways. The first is the formation of dodecyl hydroperoxide which decomposes to 1-dodecanol. The alcohol is further oxidized forming dodecanoic acid. The other pathway involves dehydrogenation to 1-dodecene, which may react with water, giving 1-dodecanol (Dugan, 1972).

Chemical/Physical. Complete combustion in air yields carbon dioxide and water.

Uses: Solvent; jet fuel research; rubber industry; manufacturing paraffin products; paper processing industry; standardized hydrocarbon; distillation chaser; gasoline component; organic synthesis.

α-ENDOSULFAN

Synonyms: Benzoepin; Beosit; Bio 5462; Chlorthiepin; Crisulfan; Cyclodan; Endocel; Endosol; Endosulfan; Endosulfan I; Endosulphan; ENT 23979; FMC 5462; 1,2,3,7,7-Hexachlorobicyclo[2.2.1]-2-heptene-5,6-bisoxymethylene sulfite; α,β-1,2,3,7,7-Hexachlorobicyclo[2.2.1]-2-heptene-5,6-bisoxymethylene sulfite; Hexachlorohexahydromethano-2,4,3-benzodioxathiepin-3-oxide; (3α,5aβ,6α,9α,9aβ)-6,7,8,9,10,10-Hexachloro-1,5,5a,6,9,9a-hexahydro-6,9-methano-2,4,3-benzodioxathiepin-3-oxide; 1,4,5,6,7,7-Hexachloro-5-norbor-ene-2,3-dimethanol cyclic sulfite; Hildan; HOE 2671; Insectophene; KOP-thiodan; Malix; NCI-C00566; NIA 5462; Niagara 5462; OMS-570; RCRA waste number P050; Thifor; Thimul; Thiodan; Thiofor; Thiomul; Thionex; Thiosulfan; Tionel; Tiovel.

CAS Registry No.: 959-98-8
DOT: 2761
DOT label: Poison
Molecular formula: $C_9H_6Cl_6O_3S$
Formula weight: 406.92
RTECS: RB9275000

Physical state, color and odor
Colorless to brown crystals with a sulfur dioxide odor.

Melting point (°C):
108-110 (Ali, 1978)

Density (g/cm³):
1.745 at 20/4 °C (Sax, 1984)

Diffusivity in water (10^5 cm²/sec):
0.45 at 20 °C using method of Hayduk and Laudie (1974).

Henry's law constant (10^4 atm·m³/mol):
1.01 at 25 °C (approximate - calculated from water solubility and vapor pressure)

Soil sorption coefficient, log K_{oc}:
3.31 using method of Kenaga and Goring (1980).

Octanol/water partition coefficient, log K$_{ow}$:
3.55 (Ali, 1978)

Solubility in water:
530 ppb at 25 °C (Weil et al., 1974)
0.32 mg/L at 22 °C (Worthing and Hance, 1991)
0.51 mg/L at 20 °C (Bowman and Sans, 1983)

Vapor pressure (mmHg):
4.58 x 10^{-5} at 25 °C (Hinckley et al., 1990)

Environmental Fate

Soil. Metabolites of endosulfan identified in seven soils were: endosulfandiol, endosulfanhydroxy ether, endosulfan lactone and endosulfan sulfate (Dreher and Podratzki, 1988; Martens, 1977). Endosulfan sulfate was the major biodegradation product in soils under aerobic, anaerobic and flooded conditions. In flooded soils, endolactone was detected only once, whereas endodiol and endohydroxy ether were identified in all soils under these conditions. Under anaerobic conditions, endodiol formed in low amounts in two soils (Martens, 1977). Indigenous microorganisms obtained from a sandy loam degraded endosulfan to endosulfan diol. This diol was converted to endosulfanhydroxy ether and trace amounts of endosulfan ether and both were degraded to endosulfan lactone (Miles and Moy, 1979). Using settled domestic wastewater inoculum, α-endosulfan (5 and 10 mg/L) did not degrade after 28 days of incubation at 25 °C (Tabak et al., 1981).

Plant. Endosulfan sulfate was formed when endosulfan was translocated from the leaves to roots in both bean and sugar beet plants (Beard and Ware, 1969). In tobacco leaves, α-endosulfan is hydrolyzed to endosulfandiol (Chopra and Mahfouz, 1977). Stewart and Cairns (1974) reported the metabolite endosulfan sulfate was identified in potato peels and pulp at concentrations of 0.3 and 0.03 ppm, respectively. They also reported that the half-life for the conversion of α-endosulfan to β-endosulfan was 60 days.

Surface Water. Endosulfan sulfate was identified as a metabolite in a survey of 11 agricultural watersheds located in southern Ontario, Canada (Frank et al., 1982). When endosulfan (α- and β- isomers, 10 µg/L) was added to Little Miami River water, sealed and exposed to sunlight and UV light for 1 week, a degradation yield of 70% was observed. After 2 and 4 weeks, 95% and 100% of the applied amount, respectively, degraded. The major degradation product was identified as endosulfan alcohol by IR spectrometry (Eichelberger and Lichtenberg, 1971).

Photolytic. Thin films of endosulfan on glass and irradiated by UV light (λ >300 nm) produced endosulfan diol with minor amounts of endosulfan ether, lactone, α-hydroxyether and other unidentified compounds (Archer et al., 1972). When an aqueous solution containing endosulfan was photooxidized by UV light at 90-95 °C, 25, 50 and 75% degraded to carbon dioxide after 5.0, 9.5 and 31.0 hours, respectively (Knoevenagel and Himmelreich, 1976). Slowly hydrolyzes to the diol and sulfur dioxide (Worthing and Hance, 1991).

Chemical/Physical. Endosulfan slowly hydrolyzes forming endosulfandiol and endosulfan sulfate (Worthing and Hance, 1991). The hydrolysis rate constant for α-endosulfan at pH 7 and 25 °C was determined to be 3.2×10^{-3}/hour, resulting in a half-life of 9.0 days (Ellington et al., 1988). The hydrolysis half-lives are reduced significantly at varying pHs and temperature. At temperatures (pH) of 87.0 (3.12), 68.0 (6.89) and 38.0 °C (8.69), the half-lives were 4.3, 0.10 and 0.08 days, respectively (Ellington et al., 1986).

Emits toxic fumes of chlorides and sulfur oxides when heated to decomposition (Lewis, 1990).

Exposure Limits: 0.1 mg/m^3 on skin (Weiss, 1986).

Toxicity: LC$_{50}$ (96-hour) for golden orfe 2 μg/L (Hartley and Kidd, 1987), rainbow trout 0.3 μg/L, white sucker 3.0 μg/L (Verschueren, 1983).

Use: Insecticide for vegetable crops.

β-ENDOSULFAN

Synonyms: Benzoepin; Beosit; Bio 5462; Chlorthiepin; Crisulfan; Cyclodan; Endocel; Endosol; Endosulfan; Endosulfan II; Endosulphan; ENT 23979; FMC 5462; 1,2,3,7,7-Hexachlorobicyclo[2.2.1]-2-heptene-5,6-bisoxymethylene sulfite; α,β-1,2,3,7,7-Hexachlorobicyclo[2.2.1]-2-heptene-5,6-bisoxymethylene sulfite; Hexachlorohexahydromethano-2,4,3-benzodioxathiepin-3-oxide; (3α,5aα,6β,9β,9aα)-6,7,8,9,10,10-Hexachloro-1,5,5a,6,9,9a-hexahydro-6,9-methano-2,4,3-benzodioxathiepin-3-oxide; 1,4,5,6,7,7-Hexachloro-5-norborene-2,3-dimethanol cyclic sulfite; Hildan; HOE 2671; Insectophene; KOP-thiodan; Malix; NCI-C00566; NIA 5,462; Niagara 5462; OMS-570; RCRA waste number P050; Thifor; Thimul; Thiomul; Thiodan; Thiofor; Thionex; Thiosulfan; Tionel; Tiovel.

CAS Registry No.: 33213-65-9
DOT: 2761
DOT label: Poison
Molecular formula: $C_9H_6Cl_6O_3S$
Formula weight: 406.92
RTECS: RB9275000

Physical state, color and odor
Colorless to brown crystals with a sulfur dioxide odor.

Melting point (°C):
207-209 (Ali, 1978)

Density (g/cm³):
1.745 at 20/20 °C (Sax, 1984)

Diffusivity in water (10^5 cm²/sec):
0.45 at 20 °C using method of Hayduk and Laudie (1974).

Henry's law constant (10^5 atm·m³/mol):
1.91 at 25 °C (approximate - calculated from water solubility and vapor pressure)

Soil sorption coefficient, log K_{oc}:
3.37 using method of Kenaga and Goring (1980).

Octanol/water partition coefficient, log K_{ow}:
3.62 (Ali, 1978)

Solubility in water (μg/L):
280 at 25 °C (Weil et al., 1974)
330 at 20 °C (Worthing and Hance, 1991)
450 at 20 °C (Bowman and Sans, 1983)

Vapor pressure (mmHg):
2.40 x 10^{-5} at 25 °C (Hinckley et al., 1990)

Environmental Fate

Soil. In aerobic soils, β-endosulfan is converted to the corresponding alcohol and ether (Perscheid et al., 1973). Metabolites of endosulfan identified in seven soils were: endosulfandiol, endosulfanhydroxy ether, endosulfan lactone and endosulfan sulfate (Dreher and Podratzki, 1988; Martens, 1977). Endosulfan sulfate was the major biodegradation product in soils under aerobic, anaerobic and flooded conditions. In flooded soils, endosulfan lactone was detected only once, whereas endosulfan diol and endosulfanhydroxy ether were identified in all soils under these conditions. Under anaerobic conditions, endosulfandiol formed in low amounts in two soils (Martens, 1977). Indigenous microorganisms in a sandy loam degraded β-endosulfan to endosulfan diol. This diol was converted to endosulfan α-hydroxy ether and trace amounts of endosulfan ether and both were degraded to endosulfan lactone (Miles and Moy, 1979).

In four successive 7-day incubation periods, β-endosulfan (5 and 10 mg/L) was recalcitrant to degradation in a settled domestic wastewater inoculum (Tabak et al., 1981).

Plant. Endosulfan sulfate was formed when endosulfan was translocated from the leaves to roots in both bean and sugar beet plants (Beard and Ware, 1969). In tobacco leaves, β-endosulfan hydrolyzed into endosulfandiol (Chopra and Mahfouz, 1977). Stewart and Cairns (1974) reported the metabolite endosulfan sulfate was identified in potato peels and pulp at concentrations of 0.3 and 0.03 ppm, respectively. They also reported that the half-life for the oxidative conversion of β-endosulfan to endosulfan sulfate was 800 days.

Surface Water. Endosulfan sulfate was identified as a metabolite in a survey of 11 agricultural watersheds located in southern Ontario, Canada (Frank et al., 1982). When endosulfan (α- and β- isomers, 10 μg/L) was added to Little Miami River water, sealed and exposed to sunlight and UV light for 1 week, a degradation yield of 70% was observed. After 2 and 4 weeks, 95% and 100% of the applied amount, respectively, degraded. The major degradation product was identified as endosulfan alcohol by IR spectrometry (Eichelberger and Lichtenberg, 1971).

Photolytic. Thin films of endosulfan on glass and irradiated by UV light (λ >300 nm) produced endosulfan diol with minor amounts of endosulfan ether, lactone, αhydroxyether and other unidentified compounds (Archer et al., 1972). Gaseous β-endosulfan subjected to UV light (λ >300 nm) produced endosulfan ether, endosulfan diol, endosulfan sulfate, endosulfan lactone, α-endosulfan and a dechlorinated ether (Schumacher et al., 1974). Irradiation of β-endosulfan in hexane by UV light produced the photoisomer α-endosulfan (Putnam et al., 1975). When an aqueous solution containing endosulfan was photooxidized by UV light at 90-95 °C, 25, 50 and 75% degraded to carbon dioxide after 5.0, 9.5 and 31.0 hours, respectively (Knoevenagel and Himmelreich, 1976).

Chemical/Physical. Endosulfan detected in the Little Miami River, OH was readily hydrolyzed and tentatively identified as endosulfan diol (Eichelberger and Lichtenberg, 1971). The hydrolysis rate constant for β-endosulfan at pH 7 and 25 °C was determined to be 3.7×10^{-3}/hour, resulting in a half-life of 7.8 days (Ellington et al., 1988). The hydrolysis half-lives were determined to be a function of both pH and temperature. At temperatures (pH) of 87.0 (3.32), 68.0 (6.89) and 38.0 °C (8.69), the half-lives were 2.7, 0.07 and 0.04 days, respectively (Ellington et al., 1986).

Toxicity: LC_{50} (96-hour) for golden orfe 2 μg/L (Hartley and Kidd, 1987), rainbow trout 0.3 μg/L, white sucker 3.0 μg/L (Verschueren, 1983).

Use: Insecticide for vegetable crops.

ENDOSULFAN SULFATE

Synonyms: 6,7,8,9,10,10-Hexachloro-1,5,5a,6,9,9a-hexahydro-3,3-dioxide; 6,9-Methano-2,4,3-benzodioxathiepin.

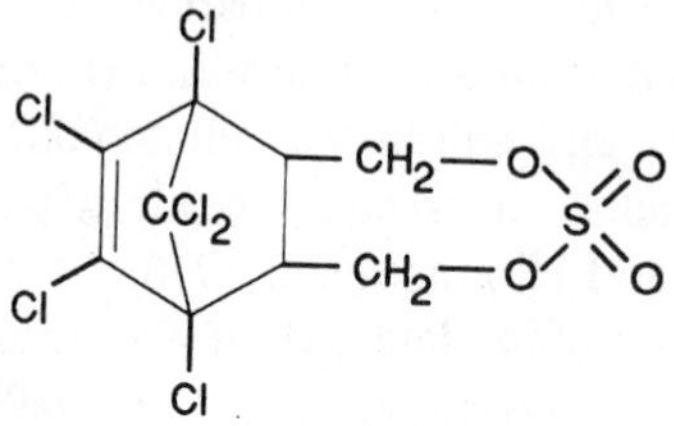

CAS Registry No.: 1031-07-8
DOT: 2761
DOT label: Poison
Molecular formula: $C_9H_6Cl_6O_4S$
Formula weight: 422.92

Physical state
Solid.

Melting point (°C):
181 (Keith and Walters, 1992)
198–201 (Ali, 1978)

Diffusivity in water (10^5 cm^2/sec):
0.44 at 20 °C using method of Hayduk and Laudie (1974).

Henry's law constant (10^5 atm·m^3/mol):
4.64 at 25 °C (approximate - calculated from water solubility and vapor pressure)

Soil sorption coefficient, log K_{oc}:
3.37 using method of Kenaga and Goring (1980).

Octanol/water partition coefficient, log K_{ow}:
3.66 (Ali, 1978)

Solubility in water:
117 ppb (Ali, 1978)

Vapor pressure (mmHg):
9.75 x 10^{-6} at 25 °C (Hinckley et al., 1990)

Environmental Fate
   Biological. A mixed culture of soil microorganisms biodegraded endosulfan

456

sulfate to endosulfan ether, endosulfan-α-hydroxy ether and endosulfan lactone (Verschueren, 1983). Indigenous microorganisms obtained from a sandy loam degraded endosulfan sulfate (a metabolite of α- and β-endosulfan) to endosulfan diol. This diol was converted to endosulfan α-hydroxy ether and trace amounts of endosulfan ether and both were degraded to endosulfan lactone (Miles and Moy, 1979). Using settled domestic wastewater inoculum, endosulfan sulfate (5 and 10 mg/L) did not degrade after 28 days of incubation at 25 °C (Tabak et al., 1981).

Plant. In tobacco leaves, endosulfan sulfate was formed to α-endosulfan, which subsequently hydrolyzed to endosulfandiol (Chopra and Mahfouz, 1977).

Uses: Not known. Based on data obtained from photolysis studies of tobacco, this compound, α- and β-endosulfan and other products were identified in tobacco smoke (Chopra et al., 1977).

ENDRIN

Synonyms: Compound 269; Endrex; ENT 17251; Experimental insecticide 269; Hexachloroepoxyoctahydro-*endo,endo*-dimethanonaphthalene; 1,2,3,4,10,10-Hexachloro-6,7-epoxy-1,4,4a,5,6,7,8,8a-octahydro-*endo,endo*-1,4:5,8-dimethanonaphthalene; **3,4,5,6,9,9-Hexachloro-1a,2,2a,3,6,6a,7,7a-octahydro-2,7:3,6-dimethanonaphth[2,3-*b*]oxirene;** Hexadrin; Isodrin epoxide; Mendrin; NA 2761; NCI-C00157; Nendrin; RCRA waste number P051.

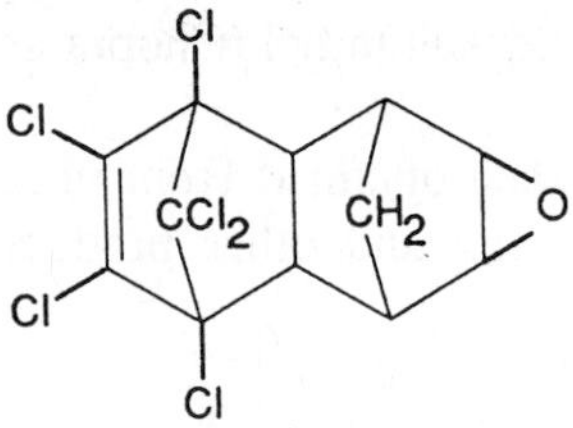

CAS Registry No.: 72-20-8
DOT: 2761
DOT label: Poison
Molecular formula: $C_{12}H_8Cl_6O$
Formula weight: 380.92
RTECS: IO1575000

Physical state, color and odor
White, odorless, crystalline solid when pure; light tan color with faint chemical odor for technical grades. Odor thresholds in air ranges from 18 to 41 ppb (Keith and Walters, 1992).

Melting point (°C):
200 (Caswell et al., 1981)

Boiling point (°C):
Decomposes at 245 (ACGIH, 1986)

Density (g/cm³):
1.65 at 25/4 °C (Weiss, 1986)

Diffusivity in water (10^5 cm²/sec):
0.44 at 20 °C using method of Hayduk and Laudie (1974).

Flash point (°C):
Noncombustible solid but may be dissolved in flammable liquids (NIOSH, 1994).

Lower explosive limit (%):
1.1 in xylene (Weiss, 1986)

Upper explosive limit (%):
7.0 in xylene (Weiss, 1986)

Henry's law constant (10^7 atm·m^3/mol):
5.0 (U.S. EPA, 1980)

Bioconcentration factor, log BCF:
2.15-2.35 (algae, Grant, 1976)
3.42 (freshwater fish), 3.13 (fish, microcosm) (Garten and Trabalka, 1983)

Soil sorption coefficient, log K_{oc}:
4.16 (Log K_{om} for sand, Sharom et al., 1980a)

Octanol/water partition coefficient, log K_{ow}:
5.16 (Travis and Arms, 1988)
5.339 (Kenaga and Goring, 1980)
3.209 (Rao and Davidson, 1980)
5.195 (de Bruijn et al., 1989)

Solubility in organics:
At 25 °C (g/L): acetone (170), benzene (138), carbon tetrachloride (33), hexane (71), xylene (183) (Windholz et al., 1983). Soluble in aromatic hydrocarbons, esters and ketones (ITII, 1986).

Solubility in water (ppb):
230 at 20-25 °C (Geyer et al., 1980)
130 at 15 °C, 250 at 25 °C, 420 at 35 °C, 625 at 45 °C (particle size ≤5 μ, Biggar and Riggs, 1974)
At 20-25 °C: 260 (particle size ≤5 μ), 190 (particle size ≤0.06 μ) (Robeck et al., 1965)

Vapor pressure (mmHg):
2 x 10^{-7} at 25 °C (ACGIH, 1986)

Environmental Fate
Biological. Microbial degradation of endrin in soil formed several ketones and aldehydes of which *keto*-endrin was the only metabolite identified (Kearney and Kaufman, 1976). In four successive 7-day incubation periods, endrin (5 and 10 mg/L) was recalcitrant to degradation in a settled domestic wastewater inoculum (Tabak et al., 1981).
Soil. In eight Indian rice soils, endrin degraded rapidly to low concentrations after 55 days. Degradation was highest in a pokkali soil and lowest in a sandy soil (Gowda and Sethunathan, 1976).

Surface Water. Algae isolated from a stagnant fish pond degraded 24.4% of the applied endrin to ketoendrin (Patil et al., 1972).

Photolytic. Photolysis of thin films of solid endrin using UV light (λ = 254 nm) produced δ-ketoendrin and endrin aldehyde and other compounds (Rosen et al., 1966). Endrin exposed to sunlight for 17 days completely isomerized to δ-keto-endrin or 1,8-*exo*-9,10,11,11-hexachlorocyclo-6.2.1.1^{3,6}.0^{2,7}.0^{4,10}-dodecan-5-one (Burton and Pollard, 1974). Irradiation of endrin by UV light (λ = 253.7 and 300 nm) or by natural sunlight in cyclohexane and hexane solution resulted in an 80% yield of 1,8-*exo*-9,11,11-pentachloropentacyclo[6.2.1.1^{3,6}.0^{2,7}.0^{4,10}]-dodecan-5-one (Zabik et al., 1971). When an aqueous solution containing endrin was photo-oxidized by UV light at 90-95 °C, 25, 50 and 75% degraded to carbon dioxide after 15.0, 41.0 and 172.0 hours, respectively (Knoevenagel and Himmelreich, 1976).

Chemical/Physical. When heated to decomposition, hydrogen chloride and phosgene may be released (NIOSH, 1994).

Exposure Limits (mg/m^3): NIOSH REL: TWA 0.1, IDLH 2; OSHA PEL: TWA 0.1; ACGIH TLV: TWA 0.1.

Symptoms of Exposure: Epileptiform, convulsions, stupor, headache, dizziness, abdominal discomfort, nausea, vomiting, insomnia, aggressive confusion, lethargy, weakness, anorexia (NIOSH, 1994).

Toxicity: LC$_{50}$ (96-hour) for bluegill sunfish 0.6 μg/L, fathead minnow 1.0 μg/L (Henderson et al., 1959), rainbow trout 0.6 μg/L, coho salmon 0.5 μg/L and chinook 1.2 μg/L (Katz, 1961); LC$_{50}$ (48-hour) for red killifish 21 μg/L (Yoshioka et al., 1986); acute oral LD$_{50}$ for rats 39 mg/kg, wild birds 13.3 mg/kg, chickens 20 mg/kg, ducks 381 mg/kg, dogs 65 mg/kg, guinea pigs 49 mg/kg, hamsters 60 mg/kg, monkeys 3 mg/kg, mice 38 mg/kg, pigeons 23.7 mg/kg, pigs 38 mg/kg, quail 10.78 mg/kg, rats 38.3 mg/kg, rabbits 45 mg/kg (RTECS, 1985), male and female rats 18 and 7.5 mg/kg, respectively (Windholz et al., 1983).

Drinking Water Standard (final): MCLG: 2 μg/L; MCL: 2 μg/L (U.S. EPA, 1994).

Use: Insecticide.

ENDRIN ALDEHYDE

Synonym: 2,2a,3,3,4,7-Hexachlorodecahydro-1,2,4-methenocyclopenta[c,d]pen-talene-5-carboxaldehyde.

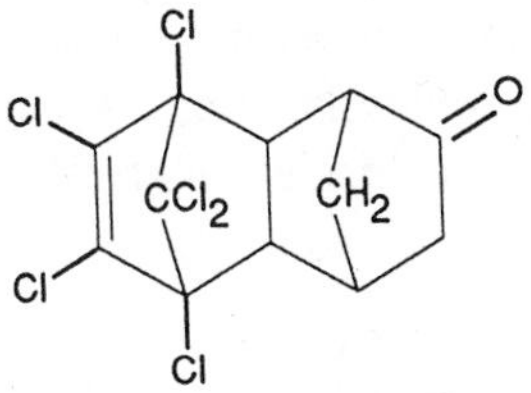

CAS Registry No.: 7421-93-4
DOT: 2761
DOT label: Poison
Molecular formula: $C_{12}H_8Cl_6O$
Formula weight: 380.92

Physical state
Solid.

Melting point (°C):
145-149 (U.S. EPA, 1980)

Boiling point (°C):
Decomposes at 235 (Callahan et al., 1979)

Diffusivity in water (10^5 cm^2/sec):
0.43 at 20 °C using method of Hayduk and Laudie (1974).

Henry's law constant (10^7 atm·m^3/mol):
3.86 at 25 °C (approximate - calculated from water solubility and vapor pressure)

Soil sorption coefficient, log K_{oc}:
4.43 using method of Kenaga and Goring (1980).

Octanol/water partition coefficient, log K_{ow}:
5.6 (calculated, Neely et al., 1974)

Solubility in water:
260 ppb at 25 °C (Weil et al., 1974)

Vapor pressure (mmHg):
2 x 10^{-7} at 25 °C (Martin, 1972)

Uses: Not known.

EPICHLOROHYDRIN

Synonyms: 1-Chloro-2,3-epoxypropane; 3-Chloro-1,2-epoxypropane; (Chloro-methyl)ethylene oxide; **(Chloromethyl)oxirane**; 2-(Chloromethyl)oxirane; 2-Chloropropylene oxide; γ-Chloropropylene oxide; 3-Chloro-1,2-propylene oxide; ECH; α-Epichlorohydrin; (*dl*)-α-Epichlorohydrin; Epichlorophydrin; 1,2-Epoxy-3-chloropropane; 2,3-Epoxypropyl chloride; Glycerol epichlorohydrin; RCRA waste number U041; UN 2023.

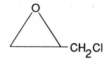

CAS Registry No.: 106-89-8
DOT: 2023
DOT label: Flammable liquid and poison
Molecular formula: C_3H_5ClO
Formula weight: 92.53
RTECS: TX4900000

Physical state, color and odor
Clear, colorless, mobile liquid with a strong, irritating, chloroform-like odor. Odor threshold in air is 0.93 ppm (Amoore and Hautala, 1983).

Melting point (°C):
-48 (Weast, 1986)
-57 (Verschueren, 1983)

Boiling point (°C):
116.5 (Weast, 1986)
117.9 (Sax and Lewis, 1987)

Density (g/cm³):
1.1801 at 20/4 °C (Weast, 1986)
1.1750 at 25/4 °C (Windholz et al., 1983)

Diffusivity in water (10^5 cm²/sec):
1.02 at 20 °C using method of Hayduk and Laudie (1974).

Flash point (°C):
40 (open cup, Windholz et al., 1983)
33.9 (NIOSH, 1994)
31 (open cup, NFPA, 1984)

Lower explosive limit (%):
3.8 (NIOSH, 1994)

Upper explosive limit (%):
21.0 (NIOSH, 1994)

Henry's law constant (10^5 atm·m^3/mol):
2.4 at 20 °C (approximate - calculated from water solubility and vapor pressure)

Ionization potential (eV):
10.60 (NIOSH, 1994)

Soil sorption coefficient, log K_{oc}:
1.00 (calculated, Mercer et al., 1990)

Octanol/water partition coefficient, log K_{ow}:
0.30 (Lide, 1990)
0.45 (Deneer et al., 1988)

Solubility in organics:
Soluble in benzene (Weast, 1986). Miscible with alcohol, carbon tetrachloride, chloroform, ether and tetrachloroethylene (Windholz et al., 1983).

Solubility in water (g/L at 20 °C):
60 (Verschueren, 1983)
66 (Lide, 1990)

Vapor density:
3.78 g/L at 25 °C, 3.19 (air = 1)

Vapor pressure (mmHg):
12.5 at 20 °C (Hawley, 1981)
16.5 at 25 °C (Lide, 1990)

Environmental Fate
   Chemical/Physical. Anticipated products from the reaction of epichlorohydrin with ozone or hydroxyl radicals in the atmosphere are formaldehyde, glyoxylic acid and $ClCH_2O(O)OHCHO$ (Cupitt, 1980).
   Hydrolyzes in water forming 1-chloro-2,3-hydroxypropane. The estimated half-life for this reaction at 20 °C and pH 7 is 8.2 days (Mabey and Mill, 1978).

Exposure Limits: Potential occupational carcinogen. NIOSH REL: IDLH 75 ppm; OSHA PEL: TWA 5 ppm (19 mg/m^3); ACGIH TLV: TWA 2 ppm (10 mg/m^3).

Symptoms of Exposure: Irritation of eyes, skin and respiratory tract (Patnaik, 1992).

Toxicity: Acute oral LD_{50} for guinea pigs 280 mg/kg, mice 194 mg/kg, rats 90 mg/kg, rabbits 345 mg/kg; LC_{50} (48-hour) for harlequin fish 36 mg/L (Verschueren, 1983).

Drinking Water Standard (final): MCLG: zero; MCL: lowest feasible limit following conventional treatment (U.S. EPA, 1994).

Uses: Solvent for natural and synthetic resins, paints, cellulose esters and ethers, gums, paints, varnishes, lacquers and nail enamels; manufacturing of glycerol, epoxy resins, surface active agents, pharmaceuticals, insecticides, adhesives, coatings, plasticizers, glycidyl ethers, ion-exchange resins and fatty acid derivatives; organic synthesis.

EPN

Synonyms: ENT 17798; EPN 300; Ethoxy-4-nitrophenoxy phenylphosphine sulfide; Ethyl *p*-nitrophenyl benzenethionophosphate; Ethyl *p*-nitrophenyl benzenethiophosphonate; Ethyl *p*-nitrophenyl ester; *O*-Ethyl *O*-4-nitrophenyl phenylphosphonothioate; Ethyl *p*-nitrophenyl phenylphosphonothioate; *O*-Ethyl *O*-*p*-nitrophenyl phenylphosphonothioate; Ethyl *p*-nitrophenyl thionobenzenephosphate; Ethyl *p*-nitrophenyl thionobenzenephosphonate; *O*-Ethyl phenyl *p*-nitrophenyl phenylphosphorothioate; Ethyl *p*-nitrophenyl thionobenzenephosphate; *O*-Ethyl phenyl *p*-nitrophenyl thiophosphonate; Phenylphosphonothioic acid *O*-ethyl *O*-*p*-nitrophenyl ester; **Phosphonothioic acid 0,0-diethyl *O*-(3,5,6-trichloro-2-pyridinyl) ester**; Pin; Santox.

CAS Registry No.: 2104-64-5
Molecular formula: $C_{14}H_{14}NO_4PS$
Formula weight: 323.31
RTECS: TB1925000

Physical state, color and odor
Yellow solid or brown liquid with an aromatic odor.

Melting point (°C):
36 (Weast, 1986)

Boiling point (°C):
215 at 5 mmHg (Worthing and Hance, 1991)

Density (g/cm³):
1.27 at 25/4 °C (Weast, 1986)

Flash point (°C):
Noncombustible solid (NIOSH, 1994)

Soil sorption coefficient, log K_{oc}:
3.12 (average of 2 soil types, Kanazawa, 1989)

Octanol/water partition coefficient, log K_{ow}:
3.85 (Kanazawa, 1989)

Solubility in organics:
Miscible with acetone, benzene, methanol, isopropanol, toluene and xylene (Windholz et al., 1983).

Vapor pressure (10^7 mmHg):
9.4 at 25 °C (Worthing and Hance, 1991)
3,000 at 100 °C (Verschueren, 1983)

Environmental Fate
   Chemical/Physical. On heating, EPN is converted to the *S*-ethyl isomer (Worthing and Hance, 1991). EPN is rapidly hydrolyzed in alkaline solutions to *p*-nitrophenol, alcohol and benzene thiophosphoric acid (Sittig, 1985). Releases toxic sulfur, phosphorous and nitrogen oxides when heated to decomposition (Sax and Lewis, 1987).

Exposure Limits (mg/m^3): NIOSH REL: TWA 0.5, IDLH 5; OSHA PEL: TWA 0.5.

Toxicity: Acute oral LD_{50} for wild birds 2.37 mg/kg, chickens 5 mg/kg, ducks 3 mg/kg, dogs 20 mg/kg, mice 14.5 mg/kg, pigeons 4.21 mg/kg, quail 5 mg/kg, rats 7 mg/kg, rabbits 45 mg/kg (RTECS, 1985); LC_{50} for rainbow trout 0.21 mg/L (Worthing and Hance, 1987), bluegill sunfish 100 μg/L (Sanders and Cope, 1968), fathead minnow 110 mg/L (Solon and Nair, 1970).

Uses: Insecticide; acaricide.

ETHANOLAMINE

Synonyms: 2-Aminoethanol; β-Aminoethyl alcohol; Colamine; β-Ethanolamine; Ethylolamine; Glycinol; 2-Hydroxyethylamine; β-Hydroxyethylamine; MEA; Monoethanolamine; Olamine; Thiofalco M-50; UN 2491; USAF EK-1597.

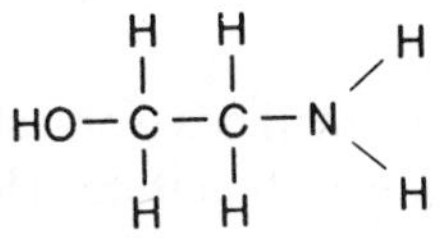

$$HO-\underset{\underset{H}{|}}{\overset{\overset{H}{|}}{C}}-\underset{\underset{H}{|}}{\overset{\overset{H}{|}}{C}}-N\overset{\diagup H}{\diagdown H}$$

CAS Registry No.: 141-43-5
DOT: 2491
DOT label: Corrosive material
Molecular formula: C_2H_7NO
Formula weight: 61.08
RTECS: KJ5775000

Physical state, color and odor
Colorless, viscous, hygroscopic liquid with an unpleasant, mild, ammonia-like odor. Odor threshold in air is 2.6 ppm (Amoore and Hautala, 1983).

Melting point (°C):
10.3 (Weast, 1986)

Boiling point (°C):
170 (Weast, 1986)

Density (g/cm³):
1.0180 at 20/4 °C (Weast, 1986)
1.0117 at 25/4 °C (Windholz et al., 1983)

Diffusivity in water (10^5 cm²/sec):
1.19 at 20 °C using method of Hayduk and Laudie (1974).

Dissociation constant:
9.50 at 25 °C (Dean, 1973)

Flash point (°C):
86 (NIOSH, 1994)
90.6 (Windholz et al., 1983)

Lower explosive limit (%):
3.0 at 140 °C (NIOSH, 1994)

467

Upper explosive limit (%):
23.5 (NIOSH, 1994)

Ionization potential (eV):
8.96 (NIOSH, 1994)

Soil sorption coefficient, log K_{oc}:
Unavailable because experimental methods for estimation of this parameter for amines are lacking in the documented literature. However, its miscibility in water and low K_{ow} suggests its adsorption to soil will be nominal (Lyman et al., 1982).

Octanol/water partition coefficient, log K_{ow}:
-1.31 (Collander, 1951)

Solubility in organics:
Miscible with acetone and methanol; soluble in benzene (1.4 wt %), carbon tetrachloride (0.2 wt %) and ether (2.1 wt %) (Windholz et al., 1983).

Solubility in water:
Miscible (NIOSH, 1994).

Vapor density:
2.50 g/L at 25 °C, 2.11 (air = 1)

Vapor pressure (mmHg):
0.4 at 20 °C, 6 at 60 °C (Verschueren, 1983)
0.48 at 20 °C (Hawley, 1981)

Environmental Fate
   Chemical/Physical. Aqueous chlorination of ethanolamine at high pH produced N-chloroethanolamine, which slowly degraded to unidentified products (Antelo et al., 1981).

Exposure Limits: NIOSH REL: TWA 3 ppm (8 mg/m^3), STEL 6 ppm (15 mg/m^3), IDLH 30 ppm; OSHA PEL: TWA 3 ppm.

Symptoms of Exposure: Severe irritation of eyes and moderate irritation of the skin (Patnaik, 1992).

Toxicity: Acute oral LD_{50} for guinea pigs 620 mg/kg, mice 700 mg/kg, rats 2,050 mg/kg, rabbits 1,000 mg/kg (Patnaik, 1992).

Uses: Removing carbon dioxide and hydrogen sulfide from natural gas; in emulsifiers, hair waving solutions, polishes; softening agent for hides; agricultural sprays; pharmaceuticals, chemical intermediates; corrosion inhibitor; rubber accelerator; non-ionic detergents used in dry cleaning; wool treatment.

2-ETHOXYETHANOL

Synonyms: Cellosolve; Cellosolve solvent; Dowanol EE; Ektasolve EE; Ethyl cellosolve; Ethylene glycol ethyl ether; Ethylene glycol monoethyl ether; Glycol ether EE; Glycol monoethyl ether; Hydroxy ether; Jeffersol EE; NCI-C54853; Oxitol; Poly-solv EE; UN 1171.

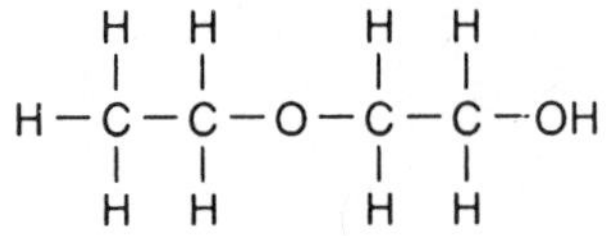

$$H-\overset{\displaystyle H}{\underset{\displaystyle H}{C}}-\overset{\displaystyle H}{\underset{\displaystyle H}{C}}-O-\overset{\displaystyle H}{\underset{\displaystyle H}{C}}-\overset{\displaystyle H}{\underset{\displaystyle H}{C}}-OH$$

CAS Registry No.: 110-80-5
DOT: 1171
DOT label: Combustible liquid
Molecular formula: $C_4H_{10}O_2$
Formula weight: 90.12
RTECS: KK8050000

Physical state, color and odor
Clear, colorless liquid with a sweetish odor. Odor threshold in air is 2.7 ppm (Amoore and Hautala, 1983).

Melting point (°C):
-70 (Keith and Walters, 1992)

Boiling point (°C):
135 (Weast, 1986)

Density (g/cm³):
0.9297 at 20/4 °C (Weast, 1986)

Diffusivity in water (10^5 cm²/sec):
0.90 at 20 °C using method of Hayduk and Laudie (1974).

Flash point (°C):
43 (NIOSH, 1994)
44 (Windholz et al., 1983)

Lower explosive limit (%):
1.7 at 93 °C (NIOSH, 1994)

Upper explosive limit (%):
15.6 at 93 °C (NIOSH, 1994)

Soil sorption coefficient, log K_{oc}:
Unavailable because experimental methods for estimation of this parameter for aliphatic amines are lacking in the documented literature. However, its miscibility in water and low K_{ow} suggest its adsorption to soil will be nominal (Lyman et al., 1982).

Octanol/water partition coefficient, log K_{ow}:
-0.53 (Collander, 1951)

Solubility in organics:
Miscible with acetone, alcohol, ether and liquid esters (Windholz et al., 1983).

Solubility in water:
Miscible (Price et al., 1974).

Vapor density:
3.68 g/L at 25 °C, 3.11 (air = 1)

Vapor pressure (mmHg):
3.8 at 20 °C, 7 at 30 °C (Verschueren, 1983)
5.63 at 25 °C (Banerjee et al., 1990)

Exposure Limits: NIOSH REL: TWA 0.5 ppm (1.8 mg/m^3), IDLH 500 ppm; OSHA PEL: TWA 200 ppm (740 mg/m^3); ACGIH TLV: TWA 5 ppm (19 mg/m^3).

Toxicity: Acute oral LD_{50} for guinea pigs 1,400 mg/kg, mice 2,451 mg/kg, rats 3,000 mg/kg, rabbits 3,100 mg/kg (RTECS, 1985).

Uses: Solvent for lacquers, varnishes and dopes, nitrocellulose, natural and synthetic resins; in cleaning solutions, varnish removers, dye baths; mutual solvent for formation of soluble oils; lacquer thinners; emulsion stabilizer; anti-icing additive for aviation fuels.

2-ETHOXYETHYL ACETATE

Synonyms: Acetic acid 2-ethoxyethyl ester; Cellosolve acetate; CSAC; Ekasolve EE acetate solvent; Ethoxyacetate; **2-Ethoxyethanol acetate**; Ethoxyethyl acetate; β-Ethoxyethyl acetate; Ethylene glycol ethyl ether acetate; Ethylene glycol monoethyl ether acetate; Glycol ether EE acetate; Glycol monoethyl ether acetate; Oxytol acetate; Poly-solv EE acetate; UN 1172.

$$H-\overset{\overset{\displaystyle H}{|}}{\underset{\underset{\displaystyle H}{|}}{C}}-\overset{\overset{\displaystyle H}{|}}{\underset{\underset{\displaystyle H}{|}}{C}}-O-\overset{\overset{\displaystyle H}{|}}{\underset{\underset{\displaystyle H}{|}}{C}}-\overset{\overset{\displaystyle H}{|}}{\underset{\underset{\displaystyle H}{|}}{C}}-O-\overset{\overset{\displaystyle O}{||}}{C}-\overset{\overset{\displaystyle H}{|}}{\underset{\underset{\displaystyle H}{|}}{C}}-H$$

CAS Registry No.: 111-15-9
DOT: 1172
DOT label: Flammable liquid
Molecular formula: $C_6H_{12}O_3$
Formula weight: 132.18
RTECS: KK8225000

Physical state, color and odor
Colorless liquid with a faint, pleasant odor. Odor threshold in air is 56 ppb (Amoore and Hautala, 1983).

Melting point (°C):
-61.7 (NIOSH, 1994)
-58 (Verschueren, 1983)

Boiling point (°C):
156 (NIOSH, 1994)

Density (g/cm³):
0.975 at 20/20 °C (Windholz et al., 1983)

Diffusivity in water (10^5 cm²/sec):
0.74 at 20 °C using method of Hayduk and Laudie (1974).

Flash point (°C):
51.5 (NIOSH, 1994)
56 (open cup, Windholz et al., 1983)

Lower explosive limit (%):
1.7 (NIOSH, 1994)

Henry's law constant (10^7 atm·m^3/mol):
9.07 at 25 °C (approximate - calculated from water solubility and vapor pressure)

Soil sorption coefficient, log K_{oc}:
Unavailable because experimental methods for estimation of this parameter for aliphatic amines are lacking in the documented literature. However, its high solubility in water suggests its adsorption to soil will be nominal (Lyman et al., 1982).

Octanol/water partition coefficient, log K_{ow}:
0.50 using method of Hansch et al. (1968).

Solubility in water:
230,000 mg/L at 20 °C (Verschueren, 1983)

Vapor density:
5.40 g/L at 25 °C, 4.56 (air = 1)

Vapor pressure (mmHg):
2 at 20 °C (NIOSH, 1994)
1.2 at 20 °C, 3.8 at 30 °C (Verschueren, 1983)

Exposure Limits: NIOSH REL: TWA 0.5 ppm (2.7 mg/m^3), IDLH 500 ppm; OSHA PEL: TWA 100 ppm (540 mg/m^3); ACGIH TLV: TWA 5 ppm (27 mg/m^3).

Toxicity: Acute oral LD$_{50}$ for guinea pigs 1,910 mg/kg, rats 2,900 mg/kg, rabbits 1,950 mg/kg (RTECS, 1985).

Uses: Automobile lacquers to reduce evaporation and to impart a high gloss; solvent for nitrocellulose, oils and resins; varnish removers; wood stains; textiles; leather.

ETHYL ACETATE

Synonyms: Acetic ether; **Acetic acid ethyl ester;** Acetidin; Acetoxy-ethane; Ethyl acetic ester; Ethyl ethanoate; RCRA waste number U112; UN 1173; Vinegar naphtha.

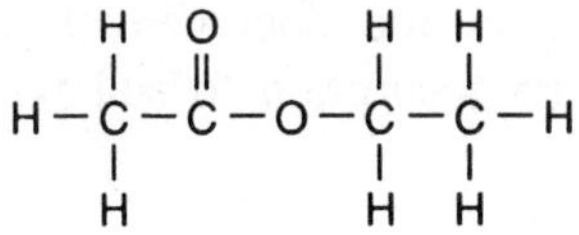

$$
\begin{array}{ccccccc}
 & H & O & & H & H & \\
 & | & \| & & | & | & \\
H- & C- & C- & O- & C- & C- & H \\
 & | & & & | & | & \\
 & H & & & H & H & \\
\end{array}
$$

CAS Registry No.: 141–78–6
DOT: 1173
DOT label: Flammable liquid
Molecular formula: $C_4H_8O_2$
Formula weight: 88.11
RTECS: AH5425000

Physical state, color and odor
Clear, colorless, mobile liquid with a pleasant fruity odor. Odor threshold in air is 3.9 ppm (Amoore and Hautala, 1983).

Melting point (°C):
-83.6 (Weast, 1986)

Boiling point (°C):
77.06 (Weast, 1986)

Density (g/cm³):
0.9003 at 20/4 °C (Weast, 1986)
0.8939 at 25/4 °C, 0.8876 at 30/4 °C (Abraham et al., 1971)

Diffusivity in water (10^5 cm²/sec at 25 °C):
1.12 (Hayduk and Laudie, 1974)

Flash point (°C):
-4.5 (NIOSH, 1994)
7.2 (open cup, Windholz et al., 1983)

Lower explosive limit (%):
2.0 (NFPA, 1984)

Upper explosive limit (%):
11.5 (NFPA, 1984)

474

Heat of fusion (kcal/mol):
2.505 (Dean, 1987)

Henry's law constant (10^4 atm·m^3/mol):
1.34 at 25 °C (Hine and Mookerjee, 1975)

Interfacial tension with water (dyn/cm at 25 °C):
6.8 (Donahue and Bartell, 1952)

Ionization potential (eV):
10.11 ± 0.02 (Franklin et al., 1969)
10.24 (Gibson et al., 1977)

Bioconcentration factor, log BCF:
4.13 (algae, Geyer et al., 1984)
3.52 (activated sludge), 2.28 (algae), 1.48 (golden ide) (Freitag et al., 1985)

Soil sorption coefficient, log K_{oc}:
Unavailable because experimental methods for estimation of this parameter for aliphatic esters are lacking in the documented literature.

Octanol/water partition coefficient, log K_{ow}:
0.66 (Collander, 1951)
0.73 (Hansch and Anderson, 1967)
0.81 (Kamlet et al., 1984)
0.68 (Tewari et al., 1982; Wasik et al., 1981)

Solubility in organics:
Miscible with acetone, alcohol, chloroform and ether (Windholz et al., 1983).

Solubility in water:
100 mL/L at 25 °C (Windholz et al., 1983)
79,000 mg/L at 20 °C, 74,000 mg/L at 35 °C (Verschueren, 1983)
80,000 mg/L at 25 °C (Banerjee, 1984)
In g/kg: 84.2 at 20 °C, 80.4 at 25 °C, 77.0 at 30 °C, 7.39 at 35 °C, 7.12 at 40 °C (Altshuller and Everson, 1953)
726 mmol/L at 25 °C (Tewari et al., 1982; Wasik et al., 1981)
85 g/L (Price et al., 1974)
731 mmol/L at 20 °C (Fühner, 1924)
9.50 wt % at 25 °C (Lo et al., 1986)
In wt %: 3.21 at 0 °C, 2.78 at 9.5 °C, 2.26 at 20.0 °C, 1.98 at 30.0 °C, 1.87 at 40.0 °C, 1.72 at 50.0 °C, 1.64 at 60.1 °C, 1.72 at 70.5 °C, 1.66 at 80.0 °C, 1.35 at 90.2 °C (Stephenson and Stuart, 1986)

Vapor density:
3.04 (air = 1) (Windholz et al., 1983)
3.60 g/L at 25 °C

Vapor pressure (mmHg):
72.8 at 20 °C, 115 at 30 °C (Verschueren, 1983)
94.5 at 25 °C (Abraham, 1984)

Environmental Fate
   Chemical/Physical. Hydrolyzes in water forming ethyl alcohol and acetic acid. The estimated hydrolysis half-life at 25 °C and pH 7 is 2.0 years (Mabey and Mill, 1978).

Exposure Limits: NIOSH REL: TWA 400 ppm (1,400 mg/m^3), IDLH 2,000 ppm; OSHA PEL: TWA 400 ppm.

Symptoms of Exposure: Inhalation of vapors may cause irritation of eyes, nose and throat (Patnaik, 1992).

Toxicity: Acute oral LD_{50} for mice 4,100 mg/kg, rats 5,620 mg/kg, rabbits 4,935 mg/kg; LC_{50} (inhalation) for rats 1,600 ppm/8-hour (RTECS, 1985).

Uses: Manufacture of smokeless powder, photographic film and plates, artificial leather and silk, perfumes; pharmaceuticals; in cleaning textiles; solvent for nitrocellulose, lacquers, varnishes and airplane dopes.

ETHYL ACRYLATE

Synonyms: Acrylic acid ethyl ester; Ethoxycarbonylethylene; Ethyl propenoate; Ethyl-2-propenoate; NCI-C50384; **2-Propenoic acid ethyl ester**; RCRA waste number U113; UN 1917.

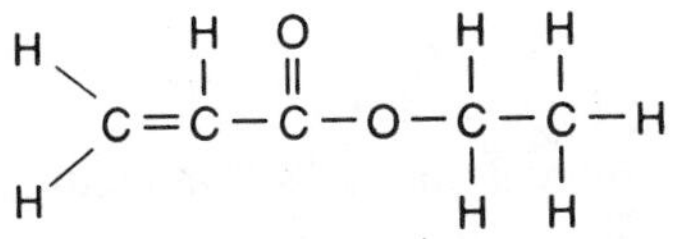

$$\underset{H}{\overset{H}{\diagdown}}C=\underset{}{\overset{H}{\underset{|}{C}}}-\overset{O}{\overset{||}{C}}-O-\underset{\underset{H}{|}}{\overset{\overset{H}{|}}{C}}-\underset{\underset{H}{|}}{\overset{\overset{H}{|}}{C}}-H$$

CAS Registry No.: 140-88-5
DOT: 1917
DOT label: Flammable liquid
Molecular formula: $C_5H_8O_2$
Formula weight: 100.12
RTECS: AT0700000

Physical state, color and odor
Colorless liquid with a sharp, penetrating odor. Odor threshold in air is 1.2 ppb (Amoore and Hautala, 1983).

Melting point (°C):
-71.2 (Weast, 1986)

Boiling point (°C):
99.8 (Weast, 1986)

Density (g/cm^3 at 20/4 °C):
0.9234 (Weast, 1986)
0.9405 (Windholz et al., 1983)

Diffusivity in water (10^5 cm^2/sec):
0.84 at 20 °C using method of Hayduk and Laudie (1974).

Flash point (°C):
8.9 (NIOSH, 1994)
15 (open cup, Windholz et al., 1983)
10 (open cup, NFPA, 1984)

Lower explosive limit (%):
1.4 (NIOSH, 1994)

Upper explosive limit (%):
14 (NFPA, 1984)

Henry's law constant (10^3 atm·m^3/mol):
2.25 at 20 °C (approximate – calculated from water solubility and vapor pressure)

Ionization potential (eV):
10.30 (NIOSH, 1994)

Soil sorption coefficient, log K_{oc}:
Unavailable because experimental methods for estimation of this parameter for aliphatic amines are lacking in the documented literature.

Octanol/water partition coefficient, log K_{ow}:
1.33 (Tanni et al., 1984)

Solubility in organics:
Soluble in alcohol, chloroform and ether (Weast, 1986).

Solubility in water:
20 g/L at 20 °C (Windholz et al., 1983)

Vapor density:
4.09 g/L at 25 °C, 3.45 (air = 1)

Vapor pressure (mmHg):
29 at 20 °C (NIOSH, 1994)
49 at 30 °C (Verschueren, 1983)

Environmental Fate
   Chemical/Physical. Polymerizes on standing and is catalyzed by heat, light and peroxides (Windholz et al., 1983). Slowly hydrolyzes in water forming ethyl alcohol and acrylic acid.

Exposure Limits: Potential occupational carcinogen. NIOSH REL: IDLH 300 ppm; OSHA PEL: TWA 25 ppm (100 mg/m^3); ACGIH TLV: TWA 5 ppm (25 mg/m^3).

Symptoms of Exposure: Strong irritant to eyes, skin and mucous membranes (Patnaik, 1992).

Toxicity: Acute oral LD$_{50}$ for rats is 800 mg/kg, rabbits 400 mg/kg, mice 1,799 mg/kg (RTECS, 1985).

Uses: Manufacture of water emulsion paints, textile and paper coatings, adhesives and leather finish resins.

ETHYLAMINE

Synonyms: Aminoethane; 1-Aminoethane; **Ethanamine**; Monoethylamine; UN 1036.

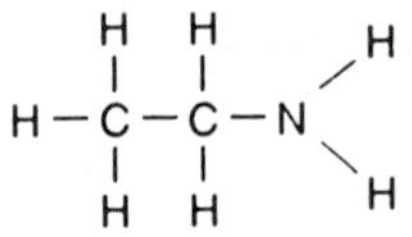

$$\begin{array}{c} \quad\; H \quad\; H \qquad\;\; H \\ \quad\; | \qquad | \qquad\; / \\ H-C-C-N \\ \quad\; | \qquad | \qquad\; \backslash \\ \quad\; H \quad\; H \qquad\;\; H \end{array}$$

CAS Registry No.: 75-04-7
DOT: 1036
DOT label: Flammable liquid
Molecular formula: C_2H_7N
Formula weight: 45.08
RTECS: KH2100000

Physical state, color and odor
Colorless liquid or gas with a strong ammonia-like odor. Odor threshold in air is 0.95 ppm (Amoore and Hautala, 1983).

Melting point (°C):
-81.1 (Weast, 1986)
-83 (Verschueren, 1983)

Boiling point (°C):
16.6 (Weast, 1986)

Density (g/cm³):
0.6829 at 20/4 °C (Weast, 1986)
0.662 at 20/4 °C (Sax and Lewis, 1987)
0.71 at 0/4 °C (Verschueren, 1983)

Diffusivity in water (10^5 cm²/sec):
1.13 at 20 °C using method of Hayduk and Laudie (1974).

Dissociation constant:
10.807 at 20 °C (Gordon and Ford, 1972)
10.63 at 25 °C (Dean, 1973)

Flash point (°C):
-17.4 (NIOSH, 1994)

Lower explosive limit (%):
3.5 (NIOSH, 1994)

Upper explosive limit (%):
14.0 (NIOSH, 1994)

Henry's law constant (10^5 atm·m^3/mol):
1.07 at 25 °C (Hine and Mookerjee, 1975)

Ionization potential (eV):
8.86 ± 0.02 (Franklin et al., 1969)
9.19 (Gibson et al., 1977)

Soil sorption coefficient, log K_{oc}:
Unavailable because experimental methods for estimation of this parameter for aliphatic amines are lacking in the documented literature.

Octanol/water partition coefficient, log K_{ow}:
-0.13, -0.30 (Sangster, 1989)

Solubility in organics:
Miscible with alcohol and ether (Hawley, 1981).

Solubility in water:
Miscible (NIOSH, 1994). A saturated solution in equilibrium with its own vapor had a concentration of 5,176 g/L at 25 °C (Kamlet et al., 1987).

Vapor density:
1.84 g/L at 25 °C, 1.56 (air = 1)

Vapor pressure (mmHg):
897 at 20 °C, 1,292 at 30 °C (Verschueren, 1983)

Environmental Fate
   Chemical/Physical. Reacts with hydroxyl radicals possibly forming acetaldehyde or acetamide (Atkinson et al., 1978). When ethylamine over kaolin is heated to 600 °C, hydrogen and acetonitrile formed as the major products. Trace amounts of ethylene, ammonia, hydrogen cyanide and methane were also produced. At 900 °C, however, acetonitrile was not produced (Hurd and Carnahan, 1930).
   Reacts with acid forming water-soluble salts.

Exposure Limits: NIOSH REL: TWA 10 ppm (18 mg/m^3), IDLH 600 ppm; OSHA PEL: TWA 10 ppm.

Symptoms of Exposure: Severe irritant to the eyes, skin and respiratory system (Patnaik, 1992).

Toxicity: Acute oral LD_{50} for rats 400 mg/kg (RTECS, 1985); LC_{50} (24-hour) for goldfish 190 mg/L at pH 10.1, LC_{50} (96-hour) for goldfish 170 mg/L at pH 10.1 (Verschueren, 1983).

Uses: Stabilizer for latex rubber; intermediate for dyestuffs and medicinals; resin and detergent manufacturing; solvent in petroleum and vegetable oil refining; starting material for manufacturing amides; plasticizer; stabilizer for rubber latex; in organic synthesis.

ETHYLBENZENE

Synonyms: EB; Ethylbenzol; NCI-C56393; Phenylethane; UN 1775.

CAS Registry No.: 100-41-4
DOT: 1175
DOT label: Flammable liquid
Molecular formula: C_8H_{10}
Formula weight: 106.17
RTECS: DA0700000

Physical state, color and odor
Clear, colorless liquid with a sweet, gasoline-like odor. Odor threshold in air is 2.3 ppm (Amoore and Hautala, 1983).

Melting point (°C):
-95.0 (Dean, 1973)
-94.4 (Huntress and Mulliken, 1941)

Boiling point (°C):
136.2 (Weast, 1986)

Density (g/cm³):
0.8670 at 20/4 °C (Weast, 1986)
0.86250 at 25/4 °C (Huntress and Mulliken, 1941)

Diffusivity in water (10^5 cm²/sec):
0.81 at 20 °C (Witherspoon and Bonoli, 1969)
0.90 (Hayduk and Laudie, 1974)

Dissociation constant:
>15 (Christensen et al., 1975)

Flash point (°C):
13 (NIOSH, 1994)
21 (NFPA, 1984)

Lower explosive limit (%):
0.8 (NIOSH, 1994)

Upper explosive limit (%):
6.7 (NIOSH, 1994)

Heat of fusion (kcal/mol):
2.195 (Dean, 1987)

Henry's law constant (10^3 atm·m^3/mol):
6.6 (Pankow and Rosen, 1988)
6.44 (Valsaraj, 1988)
8.68 at 25 °C (Hine and Mookerjee, 1975)
3.26, 4.51, 6.01, 7.88 and 10.5 at 10, 15, 20, 25 and 30 °C, respectively (Ashworth et al., 1988)
3.02, 4.22, 5.75, 7.84 and 10.3 at 10, 15, 20, 25 and 30 °C, respectively (Perlinger et al., 1993)
1.93, 2.05, 2.67, 5.02 and 6.62 at 2.0, 6.0, 10.0, 18.2 and 25.0 °C, respectively (Dewulf et al., 1995)

Interfacial tension with water (dyn/cm at 25 °C):
38.4 (Donahue and Bartell, 1952)

Ionization potential (eV):
8.76 ± 0.01 (Franklin et al., 1969)
9.12 (Yoshida et al., 1983)

Bioconcentration factor, log BCF:
1.19 (bluegill sunfish, Ogata et al., 1984)

Soil sorption coefficient, log K_{oc}:
1.98 (log K_{om} for a Woodburn silt loam soil, Chiou et al., 1983)
2.41 (Hodson and Williams, 1988; Vowles and Mantoura, 1987)

Octanol/water partition coefficient, log K_{ow}:
3.13 (Wasik et al., 1981, 1983; Yalkowsky et al., 1983)
3.15 (Campbell and Luthy, 1985; Hansch et al., 1968)

Solubility in organics:
Freely soluble in most solvents (U.S. EPA, 1985).

Solubility in water:
1.76 mmol/L at 25 °C (Wasik et al., 1981, 1983)
187 mg/L at 25 °C (Miller et al., 1985)
152 mg/kg at 25 °C (McAuliffe, 1966)
159 mg/kg at 25 °C (McAuliffe, 1963)

1.31 mmol/L at 25.0 °C (Andrews and Keefer, 1950a)

In mg/L: 219 at 0.4 °C, 213 at 5.2 °C, 207 at 20.7 °C, 207 at 21.2 °C, 208 mg/L at 25.0 °C, 209 at 25.6 °C, 211 at 30.2 °C, 221 at 34.9 °C, 231 at 42.8 °C (Bohon and Claussen, 1951)

0.014 wt % at 15 °C (Stephen and Stephen, 1963)

In mmol/L: 1.85 at 10.0 °C, 1.770 at 20.0 °C, 1.811 at 25.0 °C, 1.777 at 30.0 °C (Owens et al., 1986)

197 mg/kg at 0 °C, 177 mg/kg at 25 °C (Polak and Lu, 1973)

161.2 mg/kg at 25 °C, 111.0 mg/kg in artificial seawater at 25 °C (Sutton and Calder, 1975)

131.0 mg/kg at 25 °C (Price, 1976)

181 mg/L at 20 °C (Burris and MacIntyre, 1986)

77 mg/L in fresh water at 25 °C, 70 mg/L in salt water at 25 °C (Krasnoshchekova and Gubergrits, 1975)

2.00 mmol/L at 25 °C (Ben-Naim and Wilf, 1980)

1.12 mmol/L in 0.5 M NaCl at 25 °C (Wasik et al., 1984)

175 mg/L at 25 °C (Klevens, 1950)

140 mg/kg at 15 °C (Fühner, 1924)

147.7 mg/L at 30 °C (McNally and Grob, 1984)

1.51, 1.59, 1.65 and 1.83 mmol/L at 15, 25, 35 and 45 °C, respectively (Sanemasa et al., 1982)

172 mg/L (Coutant and Keigley, 1988)

1.55 mmol/kg at 25 °C (Morrison and Billett, 1952)

In mg/kg: 196, 192, 186, 187, 181, 183, 180, 184 and 180 at 4.5, 6.3, 7.1, 9.0, 11.8, 12.1, 15.1, 17.9 and 20.1 °C, respectively. In artificial seawater: 140, 133, 129, 125 and 122 at 0.19, 5.32, 10.05, 14.96 and 20.04 °C, respectively (Brown and Wasik, 1974)

1.91 mmol/kg at 25.0 °C (Vesala, 1974)

1.02 mM at 25.00 °C (Sanemasa et al., 1985)

1.37 mM at 25.0 °C (Sanemasa et al., 1987)

Vapor density:
4.34 g/L at 25 °C, 3.66 (air = 1)

Vapor pressure (mmHg):
12 at 30 °C (Verschueren, 1983)
7.08 at 20 °C (Burris and MacIntyre, 1986)
9.9 at 25 °C (Mackay et al., 1982)
9.6 at 25 °C (Banerjee et al., 1990)

Environmental Fate

Biological. Phenylacetic acid was reported to be the biooxidation product of ethylbenzene by *Nocardia* sp. in soil using *n*-hexadecane or *n*-octadecane as the substrate. In addition, *Methylosinus trichosporium* OB3b was reported to metabolize ethylbenzene to *o*- and *m*-hydroxybenzaldehyde with methane as the substrate (Keck et al., 1989). A culture of *Nocardia tartaricans* ATCC 31190, growing in a hexadecane medium, oxidized ethylbenzene to 1-phenethanol, which further oxidized to acetophenone (Cox and Goldsmith, 1979). When ethylbenzene (5 mg/L) was statically incubated in the dark at 25 °C with yeast extract and settled domestic wastewater inoculum, complete biodegradation with rapid acclimation was observed after 7 days. At a concentration of 10 mg/L, significant degradation occurred with gradual adaptation. Percent losses of 69, 78, 87 and 100 were obtained after 7, 14, 21 and 28-day incubation periods, respectively (Tabak et al., 1981).

Photolytic. Irradiation of ethylbenzene (λ <2537 Å) at low temperatures will form hydrogen, styrene and free radicals (Calvert and Pitts, 1966).

Exposure Limits: NIOSH REL: TWA 100 ppm (435 mg/m^3), STEL 125 ppm (545 mg/m^3), IDLH 800 ppm; OSHA PEL: TWA 100 ppm; ACGIH TLV: TWA 100 ppm, STEL 125 ppm.

Symptoms of Exposure: Narcotic at high concentrations. Irritant to the eyes, skin and nose (Patnaik, 1992).

Toxicity: Acute oral LD$_{50}$ for rats 3,500 mg/kg (RTECS, 1985).

Drinking Water Standard (final): MCLG: 0.7 mg/L; MCL: 0.7 mg/L (U.S. EPA, 1994).

Uses: Intermediate in production of styrene, acetophenone, ethylcyclohexane, benzoic acid, 1-bromo-1-phenylethane, 1-chloro-1-phenylethane, 2-chloro-1-phenylethane, *p*-chloroethylbenzene, *p*-chlorostyrene and many other compounds; solvent; in organic synthesis.

ETHYL BROMIDE

Synonyms: Bromic ether; **Bromoethane;** Halon 2001; Hydrobromic ether; Monobromoethane; NCI-C55481; UN 1891.

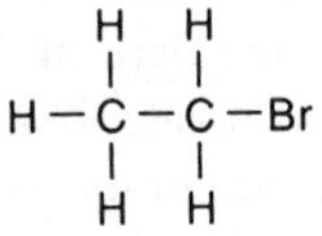

```
        H   H
        |   |
    H — C — C — Br
        |   |
        H   H
```

CAS Registry No.: 74-96-4
DOT: 1891
DOT label: Poison
Molecular formula: C_2H_5Br
Formula weight: 108.97
RTECS: KH6475000

Physical state, color and odor
Clear, colorless to yellow, volatile liquid with an ether-like odor. Odor threshold in air is 3.1 ppm (Amoore and Hautala, 1983).

Melting point (°C):
-119 (Windholz et al., 1983)

Boiling point (°C):
38.4 (Weast, 1986)

Density (g/cm³):
1.4604 at 20/4 °C (Weast, 1986)
1.4515 at 25/4 °C (Windholz et al., 1983)
1.4492 at 25/4 °C (Dreisbach, 1959)

Diffusivity in water (10^5 cm²/sec):
1.05 at 20 °C using method of Hayduk and Laudie (1974).

Flash point (°C):
<-20 (NIOSH, 1994)

Lower explosive limit (%):
6.8 (NFPA, 1984)

Upper explosive limit (%):
8.0 (NFPA, 1984)

Heat of fusion (kcal/mol):
1.4 (Dean, 1987)

Henry's law constant (10^3 atm·m^3/mol):
7.56 at 25 °C (Hine and Mookerjee, 1975)

Interfacial tension with water (dyn/cm at 25 °C):
31.3 (Donahue and Bartell, 1952)

Ionization potential (eV):
10.29 (Franklin et al., 1969)
10.46 (Gibson et al., 1977)

Soil sorption coefficient, log K_{oc}:
2.67 using method of Chiou et al. (1979).

Octanol/water partition coefficient, log K_{ow}:
1.61 (Hansch et al., 1975)

Solubility in organics:
Miscible with alcohol, chloroform and ether (Windholz et al., 1983).

Solubility in water:
In wt %: 0.965 at 10 °C, 0.914 at 20 °C, 0.896 at 30 °C (Windholz et al., 1983)
83 mmol/L at 20 °C (Fischer and Ehrenberg, 1948)
88.1 mmol/L at 17.5 °C (Fühner, 1924; Fischer and Ehrenberg, 1948)
1.47 x 10^{-3} at 25 °C (mole fraction, Li et al., 1993)

Vapor density:
4.05 g/L at 25 °C, 3.76 (air = 1)

Vapor pressure (mmHg):
375 (NIOSH, 1994)

Environmental Fate

Biological. A strain of *Acinetobacter* sp. isolated from activated sludge degraded ethyl bromide to ethanol and bromide ions (Janssen et al., 1987). When *Methanococcus thermolithotrophicus*, *Methanococcus deltae* and *Methanobacterium thermoautotrophicum*, were grown with H_2-CO_2 in the presence of ethyl bromide, methane and ethane were produced (Belay and Daniels, 1987).

Chemical/Physical. Hydrolyzes in water forming ethyl alcohol and bromide ions. The estimated hydrolysis half-life at 25 °C and pH 7 is 30 days (Mabey and Mill, 1978).

Groundwater under reducing conditions in the presence of hydrogen sulfide converted ethyl bromide to sulfur-containing products (Schwarzenbach et al., 1985).

Exposure Limits: NIOSH REL: IDLH 2,000 ppm; OSHA PEL: TWA 200 ppm (890 mg/m^3); ACGIH TLV: TWA 200 ppm, STEL 250 ppm (1,110 mg/m^3).

Symptoms of Exposure: Irritation of the respiratory system, eyes, pulmonary edema (RTECS, 1985).

Toxicity: Acute oral LD$_{50}$ for rats 1,350 mg/kg (RTECS, 1985).

Uses: In organic synthesis as an ethylating agent; refrigerant; solvent; grain and fruit fumigant; in medicine as an anesthetic.

ETHYLCYCLOPENTANE

Synonyms: None.

CAS Registry No.: 1640-89-7
Molecular formula: C_7H_{14}
Formula weight: 98.19
RTECS: GY4450000

Physical state and color
Colorless liquid.

Melting point (°C):
-138.4 (Weast, 1986)

Boiling point (°C):
103.5 (Weast, 1986)

Density (g/cm^3):
0.7665 at 20/4 °C (Weast, 1986)

Diffusivity in water (10^5 cm^2/sec):
0.76 at 20 °C using method of Hayduk and Laudie (1974).

Dissociation constant:
>14 (Schwarzenbach et al., 1993)

Flash point (°C):
15 (Aldrich, 1988)

Lower explosive limit (%):
1.1 (Hawley, 1981)

Upper explosive limit (%):
6.7 (Hawley, 1981)

Heat of fusion (kcal/mol):
1.642-1.889 (Dean, 1987)

Henry's law constant (10^3 atm·m^3/mol):
2.10 at 25 °C (approximate - calculated from water solubility and vapor pressure)

Soil sorption coefficient, log K_{oc}:
Unavailable because experimental methods for estimation of this parameter for alicyclic hydrocarbons are lacking in the documented literature.

Octanol/water partition coefficient, log K_{ow}:
1.90 using method of Hansch et al. (1968).

Solubility in organics:
Soluble in acetone, alcohol, benzene, ether and petroleum (Weast, 1986).

Solubility in water (mg/kg):
21.9 at 70.5 °C, 52.5 at 113 °C, 224 at 168.5 °C, 759 at 203 °C (Guseva and Parnov, 1964)

Vapor density:
4.01 g/L at 25 °C, 3.39 (air = 1)

Vapor pressure (mmHg):
39.9 at 25 °C (Wilhoit and Zwolinski, 1971)

Uses: Organic research.

ETHYLENE CHLOROHYDRIN

Synonyms: 2-Chloroethanol; δ-Chloroethanol; 2-Chloroethyl alcohol; β-Chloro-ethyl alcohol; Ethylene chlorhydrin; Glycol chlorohydrin; Glycol monochloro-hydrin; 2-Monochloroethanol; NCI-C50135; UN 1135.

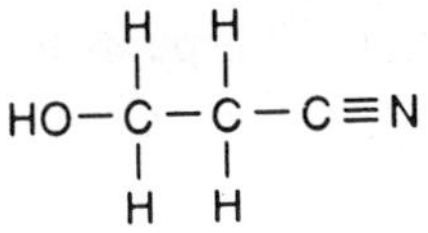

$$
\begin{array}{ccc}
 & H & H \\
 & | & | \\
HO- & C- & C-C\equiv N \\
 & | & | \\
 & H & H
\end{array}
$$

CAS Registry No.: 107-07-3
DOT: 1135
DOT label: Poison and flammable liquid
Molecular formula: C_2H_5ClO
Formula weight: 80.51
RTECS: KK0875000

Physical state, color and odor
Colorless liquid with a faint, ether-like odor. Odor threshold in air is 400 ppb (Keith and Walters, 1992).

Melting point (°C):
-89 (Aldrich, 1988)

Boiling point (°C):
128-130 (Windholz et al., 1983)

Density (g/cm^3 at 20/4 °C):
1.2003 (Weast, 1986)
1.121 (Verschueren, 1983)

Diffusivity in water (10^5 cm^2/sec):
1.12 at 20 °C using method of Hayduk and Laudie (1974).

Flash point (°C):
61 (NIOSH, 1994)
40 (open cup, Windholz et al., 1983)

Lower explosive limit (%):
4.9 (NIOSH, 1994)

Upper explosive limit (%):
15.9 (NIOSH, 1994)

Ionization potential (eV):
10.90 (NIOSH, 1994)

Soil sorption coefficient, log K_{oc}:
Unavailable because experimental methods for estimation of this parameter for chlorinated aliphatic alcohols are lacking in the documented literature. However, its high solubility in water suggests its adsorption to soil will be nominal (Lyman et al., 1982).

Solubility in organics:
Soluble in alcohol (Weast, 1986).

Solubility in water:
Miscible (Hawley, 1981).

Vapor density:
3.29 g/L at 25 °C, 2.78 (air = 1)

Vapor pressure (mmHg):
4.9 at 20 °C (Hawley, 1981)
8 at 25 °C (Nathan, 1978)

Environmental Fate
   Chemical/Physical. Reacts with aqueous sodium bicarbonate solutions at 105 °C producing ethylene glycol (Patnaik, 1992).

Exposure Limits: NIOSH REL: ceiling 1 ppm (3 mg/m^3), IDLH 7 ppm; OSHA PEL: TWA 5 ppm (16 mg/m^3); ACGIH TLV: ceiling 1 ppm.

Symptoms of Exposure: Respiratory distress, paralysis, brain damage, nausea and vomiting (Patnaik, 1992).

Toxicity: Acute oral LD_{50} for guinea pigs 110 mg/kg, mice 81 mg/kg, rats 71 mg/kg (RTECS, 1985).

Uses: Solvent for cellulose acetate, ethylcellulose; manufacturing insecticides, ethylene oxide and ethylene glycol; treating sweet potatoes before planting; organic synthesis (introduction of the hydroxyethyl group).

ETHYLENEDIAMINE

Synonyms: 1,2-Diaminoethane; Dimethylenediamine; **1,2-Ethanediamine;** 1,2-Ethylenediamine; NCI-C60402; UN 1604.

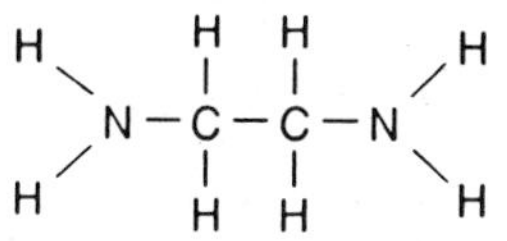

$$H-N-C-C-N-H$$

CAS Registry No.: 107-15-3
DOT: 1604
DOT label: Corrosive, flammable liquid
Molecular formula: $C_2H_8N_2$
Formula weight: 60.10
RTECS: KH8575000

Physical state, color and odor
Clear, colorless, volatile, slight viscous, hygroscopic liquid with an ammonia-like odor.

Melting point (°C):
8.5 (hydrated, Weast, 1986)
10 (anhydrous, Verschueren, 1983)

Boiling point (°C):
116.5 (Weast, 1986)
118 (hydrated, Verschueren, 1983)

Density (g/cm³):
0.8995 at 20/20 °C (Weast, 1986)
0.8994 at 20/4 °C (anhydrous), 0.963 at 21/4 °C (hydrated) (Verschueren, 1983)

Diffusivity in water (10^5 cm²/sec):
1.12 at 20 °C using method of Hayduk and Laudie (1974).

Dissociation constant (20 °C):
pK_1 = 10.075, pK_2 = 6.985 (Gordon and Ford, 1972)

Flash point (°C):
34.2 (NIOSH, 1994)
43 (Windholz et al., 1983)
66 (open cup, NFPA, 1984)

Lower explosive limit (%):
2.5 at 100 °C (NIOSH, 1994)
4.2 (NFPA, 1984)

Upper explosive limit (%):
12 at 100 °C (NIOSH, 1994)
11.4 (NFPA, 1984)

Henry's law constant (10^9 atm·m³/mol):
1.73 at 25 °C (Hine and Mookerjee, 1975)

Ionization potential (eV):
8.60 (NIOSH, 1994)

Bioconcentration factor, log BCF:
3.94 (activated sludge), 2.87 (algae) (hydrochloride salt, Freitag et al., 1985)

Soil sorption coefficient, log K_{oc}:
Unavailable because experimental methods for estimation of this parameter for diamines are lacking in the documented literature. However, its miscibility in water suggests its adsorption to soil will be nominal (Lyman et al., 1982).

Solubility in organics:
Soluble in alcohol; slightly soluble in benzene and ether (Windholz et al., 1983).

Solubility in water:
Miscible (Price et al., 1974).

Vapor density:
2.46 g/L at 25 °C, 2.07 (air = 1)

Vapor pressure (mmHg):
10.7 at 20 °C (Sax and Lewis, 1987)
116 at 20 °C (anhydrous), 9 at 20 °C, 16 at 30 °C (hydrated) (Verschueren, 1983)

Environmental Fate
   Chemical/Physical. Absorbs carbon dioxide forming carbonates (Patnaik, 1992; Windholz et al., 1983).

Exposure Limits: NIOSH REL: TWA 10 ppm (25 mg/m³), IDLH 1,000 ppm; OSHA PEL: TWA 10 ppm; ACGIH TLV: TWA 10 ppm.

Symptoms of Exposure: Severe skin irritant producing sensitization and blistering of the skin. Liquid splashed in eyes may cause injury (Patnaik, 1992). Inhalation may cause irritation of nose and respiratory system (NIOSH, 1994).

Toxicity: Acute oral LD_{50} for guinea pigs 470 mg/kg, rats 500 mg/kg (RTECS, 1985).

Uses: Stabilizing rubber latex; solvent for albumin, casein, shellac and sulfur; neutralizing oils; in antifreeze as a corrosion inhibitor; emulsifier; adhesives; textile lubricants; fungicides; manufacturing chelating agents such as EDTA (ethylenediaminetetraacetic acid); dimethylolethylene-urea resins; organic synthesis.

ETHYLENE DIBROMIDE

Synonyms: Acetylene dibromide; Bromofume; Celmide; DBE; Dibromoethane; **1,2-Dibromoethane;** *sym*-Dibromoethane; α,β-Dibromoethane; Dowfume 40; Dowfume EDB; Dowfume W-8; Dowfume W-85; Dow-fume W-90; Dowfume W-100; EDB; EDB-85; E-D-BEE; ENT 15349; Ethylene bromide; Ethylene bromide glycol dibromide; 1,2-Ethylene dibromide; Fumo-gas; Glycol bromide; Glycol dibromide; Iscobrome D; Kopfume; NCI-C00522; Nephis; Pestmaster; Pestmaster EDB-85; RCRA waste number U067; Soilbrom-40; Soilbrom-85; Soilbrom-90; Soilbrom-90EC; Soilbrom-100; Soilfume; UN 1605; Unifume.

$$\begin{array}{ccc} & H & H \\ & | & | \\ Br- & C - C & -Br \\ & | & | \\ & H & H \end{array}$$

CAS Registry No.: 106-93-4
DOT: 1605
DOT label: Poison
Molecular formula: $C_2H_4Br_2$
Formula weight: 187.86
RTECS: KH9275000

Physical state and odor
Colorless liquid with a sweet, chloroform-like odor. Odor threshold in air is 25 ppb (Keith and Walters, 1992).

Melting point (°C):
9.8 (Weast, 1986)

Boiling point (°C):
131.3 (Weast, 1986)
131.0 (Jones et al., 1977)

Density (g/cm^3):
2.1687 at 20/4 °C (Riddick et al., 1986)
2.1688 at 25/4 °C (Dreisbach, 1959)

Diffusivity in water (10^5 cm^2/sec):
0.96 at 20 °C using method of Hayduk and Laudie (1974).

Heat of fusion (kcal/mol):
2.62 (Dean, 1987)

496

Henry's law constant (10^4 atm·m³/mol):
7.06 at 25 °C (Hine and Mookerjee, 1975)
25.0 at 25 °C (Jafvert and Wolfe, 1987)
3.0, 4.8, 6.1, 6.5 and 8.0 at 10, 15, 20, 25 and 30 °C, respectively (Ashworth et al., 1988)

Interfacial tension with water (dyn/cm at 20 °C):
36.54 (Demond and Lindner, 1993)

Ionization potential (eV):
9.45 (Franklin et al., 1969)

Soil sorption coefficient, log K_{oc}:
1.64 (Kenaga and Goring, 1980)
1.56-2.21 (soil organic matter content 0.5-21.7%, Mingelgrin and Gerstl, 1983)
1.82 (average of 2 soil adsorbents, Rogers and MacFarlane, 1981)

Octanol/water partition coefficient, log K_{ow}:
1.76 (Rogers and MacFarlane, 1981)

Solubility in organics:
Soluble in acetone, alcohol, benzene and ether (Weast, 1986).

Solubility in water:
4,321 mg/L at 20 °C (Mackay and Yeun, 1983)
3,920 mg/kg at 15 °C, 4,310 mg/kg at 30 °C (Gross and Saylor, 1931)
2,910 mg/L at 25 °C (Jones et al., 1977)
4,200 mg/L at 25 °C (Dreisbach, 1952)
3.1 mg/g at 25 °C (Tokoro et al., 1986)
At 20 °C and vapor phase concentrations of 10.6, 19.7, 28.4, 38.1, 40.7 and 48.1 mg/m³, the concentrations of ethylene dibromide in water were 468, 816, 1,095, 1,516, 1,653, 1,930 mg/L, respectively (Call, 1957).

Vapor density:
7.68 g/L at 25 °C, 6.49 (air = 1)

Vapor pressure (mmHg):
11 at 20 °C (Mackay and Yeun, 1983)
11.4 at 25 °C (Hine and Mookerjee, 1975)
17.4 at 30 °C (Sax and Lewis, 1987)

Environmental Fate
Biological. Complete biodegradation by soil cultures resulted in the formation of

ethylene and bromide ions (Castro and Belser, 1968). A mutant of strain *Acinetobacter* sp. GJ70 isolated from activated sludge degraded ethylene dibromide to ethylene glycol and bromide ions (Janssen et al., 1987). When *Methanococcus thermolithotrophicus*, *Methanococcus deltae* and *Methanobacterium thermoautotrophicum* were grown with H_2-CO_2 in the presence of ethylene dibromide, methane and ethylene were produced (Belay and Daniels, 1987).

In a shallow aquifer material, ethylene dibromide aerobically degraded to carbon dioxide, microbial biomass and nonvolatile water-soluble compound(s) (Pignatello, 1987).

Soil. In both soil and water, chemical and biological mediated reactions can transform ethylene dibromide in the presence of hydrogen sulfides to ethyl mercaptan and other sulfur-containing compounds (Alexander, 1981).

Chemical/Physical. In an aqueous phosphate buffer solution (0.05 M) containing hydrogen sulfide ion, ethylene dibromide was transformed into 1,2-dithioethane and vinyl bromide. The hydrolysis half-lives for solutions with and without sulfides present ranged from 37-70 days and 0.8-4.6 years, respectively (Barbash and Reinhard, 1989).

Hydrolyzes in water to ethylene glycol and bromoethanol (Leinster et al., 1978). Dehydrobromination of ethylene dibromide to vinyl bromide was observed in various aqueous buffer solutions (pH 7 to 11) over the temperature range of 45 to 90 °C. The estimated hydrolysis half-life for this reaction at 25 °C and pH 7 was 2.5 years (Vogel and Reinhard, 1986). The hydrolysis rate constant for ethylene dibromide at pH 7 and 25 °C was determined to be 9.9 x 10^{-6}/hour, resulting in a half-life of 8.0 years (Ellington et al., 1988). In an earlier report, the hydrolysis half-lives at 30 °C and pH values of 5, 7 and 9 were reported to be 180, 410 and 170 days, respectively (Ellington et al., 1986).

Anticipated products from the reaction of ethylene dibromide with ozone or hydroxyl radicals in the atmosphere are bromoacetaldehyde, formaldehyde, bromoformaldehyde and bromide radicals (Cupitt, 1980). Emits toxic bromide fumes when heated to decomposition (Lewis, 1990).

Exposure Limits: Potential occupational carcinogen. NIOSH REL: TWA 45 ppb, 15-minute ceiling 130 ppb, IDLH 100 ppm; OSHA PEL: TWA 20 ppm, ceiling 30 ppm, 5-minute peak 50 ppm.

Symptoms of Exposure: Irritation of the respiratory system, eyes; dermatitis vesiculation (RTECS, 1985).

Toxicity: Acute oral LD_{50} for quail 130 mg/kg, rats 108 mg/kg, rabbits 55 mg/kg (RTECS, 1985); LC_{50} (48-hour) for bluegill sunfish 19 mg/L (Davis and Hardcastle, 1959).

Drinking Water Standard (final): MCLG: zero; MCL: 0.05 μg/L (U.S. EPA, 1994).

Uses: In anti-knock gasolines; grain and fruit fumigant; waterproofing preparations; insecticide; medicines; general solvent; organic synthesis.

ETHYLENIMINE

Synonyms: Aminoethylene; Azacyclopropane; Azirane; **Aziridine**; Dihydro-1*H*-azirine; Dihydroazirine; Dimethyleneimine; Dimethylenimine; EI; ENT 50324; Ethyleneimine; Ethylimine; RCRA waste number P054; TL 337; UN 1185.

CAS Registry No.: 151-56-4
DOT: 1185
DOT label: Flammable liquid and poison
Molecular formula: C_2H_5N
Formula weight: 43.07
RTECS: KX5075000

Physical state, color and odor
Colorless liquid with a very strong ammonia odor. Odor threshold in air is 1.5 ppm (Amoore and Hautala, 1983).

Melting point (°C):
-71.5 (Sax and Lewis, 1987)

Boiling point (°C):
57 (Hawley, 1981)

Density (g/cm³):
0.8321 at 20/4 °C (Weast, 1986)

Diffusivity in water (10^5 cm²/sec):
1.30 at 20 °C using method of Hayduk and Laudie (1974).

Dissociation constant:
8.04 (HSDB, 1989)

Flash point (°C):
-11.21 (NIOSH, 1994)

Lower explosive limit (%):
3.3 (NIOSH, 1994)
3.6 (NFPA, 1984)

Upper explosive limit (%):
54.8 (NIOSH, 1994)

Henry's law constant (10^7 atm·m^3/mol):
1.33 at 25 °C (Mercer et al., 1990)

Ionization potential (eV):
9.20 (NIOSH, 1994)
9.9 (Scala and Salomon, 1976)

Soil sorption coefficient, log K_{oc}:
0.11 (calculated, Mercer et al., 1990)

Octanol/water partition coefficient, log K_{ow}:
-1.01 (Mercer et al., 1990)

Solubility in organics:
Soluble in acetone, alcohol, benzene and ether (Weast, 1986).

Solubility in water:
Miscible (NIOSH, 1994).

Vapor density:
1.76 g/L at 25 °C, 1.49 (air = 1)

Vapor pressure (mmHg):
160 at 20 °C, 250 at 30 °C (Verschueren, 1983)

Environmental Fate
Photolytic. The vacuum UV photolysis (λ = 147 nm) and γ radiolysis of ethylenimine resulted in the formation of acetylene, methane, ethane, ethylene, hydrogen cyanide, methyl radicals and hydrogen (Scala and Salomon, 1976). Photolysis of ethylenimine vapor at krypton and xenon lines yielded ethylene, ethane, methane, acetylene, propane, butane, hydrogen, ammonia, ethylenimino radicals (Iwasaki et al., 1973).
Chemical/Physical. Polymerizes easily (Windholz et al., 1983). Hydrolyzes in water forming ethanolamine (HSDB, 1989). The estimated hydrolysis half-life in water at 25 °C and pH 7 is 154 days (Mabey and Mill, 1978).

Exposure Limits: Potential occupational carcinogen. NIOSH REL: IDLH 100 ppm; ACGIH TLV: TWA 0.5 ppm (1 mg/m^3).

Symptoms of Exposure: Severe irritation of the skin, eyes and mucous membranes. Eye contact with liquid may cause corneal opacity and loss of vision. Inhalation of vapors may cause eye, nose and throat irritation and breathing difficulties (Patnaik, 1992).

Toxicity: Acute oral LD_{50} for rats 15 mg/kg (RTECS, 1985).

Uses: Manufacture of triethylenemelamine and other amines; fuel oil and lubricant refining; ion exchange; protective coatings; adhesives; pharmaceuticals; polymer stabilizers; surfactants.

ETHYL ETHER

Synonyms: Aether; Anaesthetic ether; Anesthesia ether; Anesthetic ether; Diethyl ether; Diethyl oxide; Ether; Ethoxyethane; Ethyl oxide; **1,1′-Oxybis(ethane);** RCRA waste number U117; Solvent ether; Sulfuric ether; UN 1155.

$$\begin{array}{c} \quad H \quad H \qquad\quad H \quad H \\ \quad | \quad\; | \qquad\quad\; | \quad\; | \\ H-C-C-O-C-C-H \\ \quad | \quad\; | \qquad\quad\; | \quad\; | \\ \quad H \quad H \qquad\quad H \quad H \end{array}$$

CAS Registry No.: 60-29-7
DOT: 1155
DOT label: Flammable liquid
Molecular formula: $C_4H_{10}O$
Formula weight: 74.12
RTECS: KI5775000

Physical state, color and odor
Colorless, hygroscopic, volatile liquid with a sweet, pungent odor. Odor threshold in air is 330 ppb (Keith and Walters, 1992).

Melting point (°C):
–116.2 (stable form, Weast, 1986)
–123 (metastable form, Verschueren, 1983)

Boiling point (°C):
34.5 (Weast, 1986)
34.43 (Boublik et al., 1984)

Density (g/cm³):
0.7138 at 20/4 °C (Weast, 1986)
0.79125 at 15/4 °C, 0.70205 at 30/4 °C (Huntress and Mulliken, 1941)
0.71361 at 20/4 °C, 0.70782 at 25/4 °C (Riddick et al., 1986)

Diffusivity in water (10^5 cm²/sec):
0.86 at 20 °C using method of Hayduk and Laudie (1974).

Flash point (°C):
9.5 (NIOSH, 1994)

Lower explosive limit (%):
1.9 (NIOSH, 1994)

Upper explosive limit (%):
36.0 (NIOSH, 1994)

Heat of fusion (kcal/mol):
1.745 (Dean, 1987)

Henry's law constant (10^3 atm·m^3/mol):
1.28 at 25 °C (Hine and Mookerjee, 1975)

Interfacial tension with water (dyn/cm at 20 °C):
10.70 (Demond and Lindner, 1993)

Ionization potential (eV):
9.6 (Franklin et al., 1969)
9.53 (NIOSH, 1994)

Soil sorption coefficient, log K_{oc}:
Unavailable because experimental methods for estimation of this parameter for ethers are lacking in the documented literature. However, its high solubility in water suggests its adsorption to soil will be nominal (Lyman et al., 1982).

Octanol/water partition coefficient, log K_{ow}:
0.77 (Leo et al., 1971)
0.89 (Hansch et al., 1975)
0.83 (Collander, 1951)

Solubility in organics:
Soluble in acetone (Weast, 1986). Miscible with lower aliphatic alcohols, benzene, chloroform, petroleum ether and many oils (Windholz et al., 1983).

Solubility in water:
0.632, 1.010 and 1.2 mol/L at 38, 20 and 25 °C, respectively (Fischer and Ehrenberg, 1948)
6.80 wt % at 20 °C (Palit, 1947)
9.01, 7.95, 6.87 and 6.03 wt % at 10, 15, 20 and 25 °C, respectively (Bennett and Philip, 1928)
In wt %: 12.752 at -3.83 °C, 11.668 at 0 °C, 9.040 at 10 °C, 7.913 at 15 °C, 6.896 at 20 °C, 6.027 at 25 °C, 5.340 at 30 °C (Hill, 1923)
64 g/kg at 25 °C (Butler and Ramchandani, 1935)
0.80 mol/L at 25 °C (Hine and Weimar, 1965)

Vapor density:
3.03 g/L at 25 °C, 2.55 (air = 1)

Vapor pressure (mmHg):
442 at 20 °C (Verschueren, 1983)
439.8 at 20 °C (Windholz et al., 1983)
537 at 25 °C (Butler and Ramchandani, 1935)

Environmental Fate
   Chemical/Physical. The atmospheric oxidation of ethyl ether by hydroxyl radicals in the presence of nitric oxide yielded ethyl formate as the major product. Minor products included formaldehyde and nitrogen dioxide (Wallington and Japar, 1991).

Exposure Limits: NIOSH REL: IDLH 1,900 ppm; OSHA PEL: TWA 400 ppm (1,200 mg/m^3); ACGIH TLV: TWA 400 ppm, STEL 500 ppm (1,500 mg/m^3).

Symptoms of Exposure: Narcotic at high concentrations and a mild irritant to eyes, nose and skin (Patnaik, 1992).

Toxicity: Acute oral LD$_{50}$ for rats is 1,215 mg/kg (RTECS, 1985).

Uses: Solvent for oils, waxes, perfumes, alkaloids, fats and gums; organic synthesis (Grignard and Wurtz reactions); extractant; manufacture of gun powder, ethylene and other organic compounds; analytical chemistry; perfumery; alcohol denaturant; primer for gasoline engines; anesthetic.

ETHYL FORMATE

Synonyms: Areginal; Ethyl formic ester; Ethyl methanoate; **Formic acid ethyl ester;** Formic ether; UN 1190.

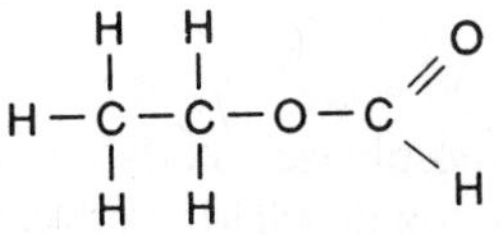

$$H-\underset{\underset{H}{|}}{\overset{\overset{H}{|}}{C}}-\underset{\underset{H}{|}}{\overset{\overset{H}{|}}{C}}-O-C\overset{\diagup O}{\underset{\diagdown H}{}}$$

CAS Registry No.: 109-94-4
DOT: 1190
DOT label: Flammable liquid
Molecular formula: $C_3H_6O_2$
Formula weight: 74.08
RTECS: LQ8400000

Physical state, color and odor
Colorless, clear liquid with a pleasant, fruity odor. Odor threshold in air is 31 ppm (Amoore and Hautala, 1983).

Melting point (°C):
–80.5 (Weast, 1986)
–79 (Verschueren, 1983)

Boiling point (°C):
54.5 (Weast, 1986)
53–54 (Windholz et al., 1983)

Density (g/cm^3):
0.9168 at 20/4 °C (Weast, 1986)
0.924 at 25/4 °C (Verschueren, 1983)

Diffusivity in water (10^5 cm^2/sec):
1.00 at 20 °C using method of Hayduk and Laudie (1974).

Flash point (°C):
–20 (Windholz et al., 1983)

Lower explosive limit (%):
2.8 (NFPA, 1984)

Upper explosive limit (%):
16.0 (NFPA, 1984)

Heat of fusion (kcal/mol):
2.20 (Dean, 1987)

Henry's law constant (10^4 atm·m^3/mol):
2.23 at 25 °C (Hine and Mookerjee, 1975)

Ionization potential (eV):
10.61 ± 0.01 (Franklin et al., 1969)

Soil sorption coefficient, log K_{oc}:
Unavailable because experimental methods for estimation of this parameter for esters are lacking in the documented literature. However, its high solubility in water suggests its adsorption to soil will be nominal (Lyman et al., 1982).

Octanol/water partition coefficient, log K_{ow}:
0.36 using method of Hansch et al. (1968).

Solubility in organics:
Miscible with alcohol, benzene and ether (Hawley, 1981).

Solubility in water:
9 wt % at 17.9 °C (NIOSH, 1994)
105.0 g/L at 20 °C, 118.0 g/L at 25 °C (Verschueren, 1983)

Vapor density:
3.03 g/L at 25 °C, 2.56 (air = 1)

Vapor pressure (mmHg):
192 at 20 °C, 300 at 30 °C (Verschueren, 1983)

Environmental Fate
Chemical/Physical. Slowly hydrolyzes in water forming formic acid and ethanol (Windholz et al., 1983).

Exposure Limits: NIOSH REL: TWA 100 ppm (300 mg/m^3), IDLH 1,500 ppm; OSHA PEL: TWA 100 ppm; ACGIH TLV: TWA 100 ppm.

Symptoms of Exposure: May irritate eyes and nose (Patnaik, 1992).

Toxicity: Acute oral LD$_{50}$ for guinea pigs 1,110 mg/kg, rats 1,850 mg/kg, rabbits 2,075 mg/kg (RTECS, 1985).

Uses: Solvent for nitrocellulose and cellulose acetate; artificial flavor for lemonades and essences; fungicide and larvacide for cereals, tobacco, dried fruits; acetone substitute; organic synthesis.

ETHYL MERCAPTAN

Synonyms: 2-Aminoethanethiol; **Ethanethiol;** Ethyl hydrosulfide; Ethyl sulf-hydrate; Ethyl thioalcohol; LPG ethyl mercaptan 1010; UN 2363; Thioethanol; Thioethyl alcohol.

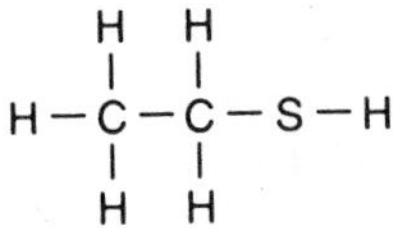

$$H-\underset{\underset{H}{|}}{\overset{\overset{H}{|}}{C}}-\underset{\underset{H}{|}}{\overset{\overset{H}{|}}{C}}-S-H$$

CAS Registry No.: 75-08-1
DOT: 2363
DOT label: Flammable liquid
Molecular formula: C_2H_6S
Formula weight: 62.13
RTECS: KI9625000

Physical state, color and odor
Colorless liquid with a strong skunk-like odor. Odor threshold in air is 0.76 ppb (Amoore and Hautala, 1983).

Melting point (°C):
-144.4 (Weast, 1986)
-147 (Sax and Lewis, 1987)

Boiling point (°C):
35 (Weast, 1986)

Density (g/cm³):
0.8391 at 20/4 °C (Weast, 1986)
0.83147 at 25/4 °C (Windholz et al., 1983)

Diffusivity in water (10^5 cm²/sec):
1.05 at 20 °C using method of Hayduk and Laudie (1974).

Dissociation constant:
10.50 at 20 °C (Dean, 1973)

Flash point (°C):
-48.7 (NIOSH, 1994)

Lower explosive limit (%):
2.8 (NIOSH, 1994)

Upper explosive limit (%):
18.0 (NIOSH, 1994)

Heat of fusion (kcal/mol):
1.189 (Dean, 1987)

Henry's law constant (10^3 atm·m^3/mol):
2.74 at 25 °C (Hine and Mookerjee, 1975)

Interfacial tension with water (dyn/cm at 20 °C):
26.12 (Demond and Lindner, 1993)

Ionization potential (eV):
9.285 ± 0.005 (Franklin et al., 1969)

Soil sorption coefficient, log K_{oc}:
Unavailable because experimental methods for estimation of this parameter for mercaptans are lacking in the documented literature. However, its high solubility in water suggests its adsorption to soil will be nominal (Lyman et al., 1982).

Octanol/water partition coefficient, log K_{ow}:
1.49 using method of Hansch et al. (1968).

Solubility in organics:
Soluble in acetone, alcohol and ether (Weast, 1986).

Solubility in water (20 °C):
0.7 wt % (NIOSH, 1994)
6.76 g/L (Windholz et al., 1983)

Vapor density:
2.54 g/L at 25 °C, 2.14 (air = 1)

Vapor pressure (mmHg):
440 at 20 °C, 640 at 30 °C (Verschueren, 1983)
527.2 at 25 °C (Wilhoit and Zwolinski, 1971)

Environmental Fate
   Chemical/Physical. In the presence of nitric oxide, ethyl mercaptan reacted with hydroxyl radical to give ethyl thionitrite (MacLeod et al., 1984).

Exposure Limits: NIOSH REL: 15-minute ceiling 0.5 ppm (1.3 mg/m^3), IDLH 500 ppm; OSHA PEL: ceiling 10 ppm (25 mg/m^3); ACGIH TLV: TWA 0.5 ppm.

Symptoms of Exposure: May produce irritation of the nose and throat, headache and fatigue (Patnaik, 1992).

Toxicity: Acute oral LD_{50} for rats 682 mg/kg (RTECS, 1985).

Uses: Odorant for natural gas; manufacturing of plastics, antioxidants, pesticides; adhesive stabilizer; chemical intermediate.

4-ETHYLMORPHOLINE

Synonym: *N*-Ethylmorpholine.

CAS Registry No.: 100-74-3
Molecular formula: $C_6H_{13}NO$
Formula weight: 115.18
RTECS: QE4025000

Physical state, color and odor
Colorless liquid with an ammonia-like odor. Odor threshold in air is 1.4 ppm (Amoore and Hautala, 1983).

Melting point (°C):
-65 to -63 (Verschueren, 1983)

Boiling point (°C):
138-139 at 763 mmHg (Weast, 1986)

Density (g/cm^3 at 20/4 °C):
0.9886 (Weast, 1986)
0.905 (Aldrich, 1988)

Diffusivity in water (10^5 cm^2/sec):
0.81 at 20 °C using method of Hayduk and Laudie (1974).

Flash point (°C):
32.5 (open cup, NIOSH, 1994)
27 (Aldrich, 1988)

Soil sorption coefficient, log K_{oc}:
Unavailable because experimental methods for estimation of this parameter for substituted morpholines are lacking in the documented literature. However, its high solubility in water suggests its adsorption to soil will be nominal (Lyman et al., 1982).

Solubility in organics:
Soluble in acetone, alcohol, benzene and ether (Weast, 1986).

Solubility in water:
Miscible (Hawley, 1981).

Vapor density:
4.71 g/L at 25 °C, 3.98 (air = 1)

Vapor pressure (mmHg at 20 °C):
6 (NIOSH, 1994)
6.1 (Verschueren, 1983)

Environmental Fate
   Chemical/Physical. Releases toxic nitrogen oxides when heated to decomposition (Sax and Lewis, 1987).

Exposure Limits: NIOSH REL: TWA 5 ppm (23 mg/m^3), IDLH 100 ppm; OSHA PEL: TWA 20 ppm (94 mg/m^3); ACGIH TLV: TWA 5 ppm.

Toxicity: Acute oral LD_{50} for rats 1,780 mg/kg, mice 1,200 mg/kg; LC_{50} (inhalation) for mice 18,000 mg/m^3/2-hour (RTECS, 1985).

Uses: Intermediate for pharmaceuticals, dyestuffs, emulsifying agents and rubber accelerators; solvent for dyes, resins and oils; catalyst for making polyurethane foams.

2-ETHYLTHIOPHENE

Synonyms: None.

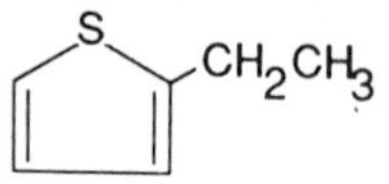

CAS Registry No.: 872-55-9
Molecular formula: C_6H_8S
Formula weight: 112.19

Physical state and odor
Liquid with a pungent odor.

Boiling point (°C):
104 (Weast, 1986)
134 (Wilhoit and Zwolinski, 1971)

Density (g/cm³):
0.9930 at 20/4 °C (Weast, 1986)

Diffusivity in water (10^5 cm²/sec):
0.82 at 20 °C using method of Hayduk and Laudie (1974).

Flash point (°C):
21 (Aldrich, 1988)

Ionization potential (eV):
8.8 ± 0.2 (Franklin et al., 1969)

Soil sorption coefficient, log K_{oc}:
Unavailable because experimental methods for estimation of this parameter for substituted thiophenes are lacking in the documented literature.

Octanol/water partition coefficient, log K_{ow}:
2.83 using method of Hansch et al. (1968).

Solubility in organics:
Soluble in alcohol and ether (Weast, 1986).

Solubility in water:
292 mg/kg at 25 °C (Price, 1976)

514

Vapor density:
4.59 g/L at 25 °C, 3.87 (air = 1)

Vapor pressure (mmHg):
60.9 at 60.3 °C, 92.2 at 70.3 °C, 136 at 80.3 °C, 197 at 90.3 °C, 280 at 100.3 °C (Eon et al., 1971)

Use: Ingredient in crude petroleum.

FLUORANTHENE

Synonyms: 1,2-Benzacenaphthene; Benzo[*jk*]fluorene; Idryl; 1,2-(1,8-Naphthyl-ene)benzene; 1,2-(1,8-Naphthalenediyl)benzene; RCRA waste number U120.

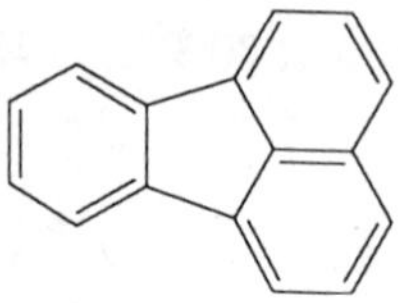

CAS Registry No.: 206-44-0
Molecular formula: $C_{16}H_{10}$
Formula weight: 202.26
RTECS: LL4025000

Physical state and color
Colorless to light yellow crystals.

Melting point (°C):
107 (Verschueren, 1983)
109-110 (Fluka, 1988)

Boiling point (°C):
375 (Weast, 1986)
384 (Aldrich, 1988)
367 (Sax, 1984)

Density (g/cm³):
1.252 at 0/4 °C (Weast, 1986)

Diffusivity in water (10^5 cm²/sec):
0.56 at 20 °C using method of Hayduk and Laudie (1974).

Dissociation constant:
>15 (Christensen et al., 1975)

Henry's law constant (10^6 atm·m³/mol):
16.0 at 25 °C (Hine and Mookerjee, 1975)
2.57, 6.32, 16.09, 23.49, 57.64 and 61.48 at 10.0, 20.0, 35.0, 40.1, 45.0 and 55.0 °C, respectively (ten Hulscher et al., 1992)

Ionization potential (eV):
8.54 (Franklin et al., 1969)

Bioconcentration factor, log BCF:
3.24 (*Daphnia magna*, Newsted and Giesy, 1987)
1.08 (*Polychaete* sp.), 0.76 (*Capitella capitata*) (Bayona et al., 1991)

Soil sorption coefficient, log K_{oc}:
4.62 (aquifer sands, Abdul et al., 1987)
6.38 (average, Kayal and Connell, 1990)

Octanol/water partition coefficient, log K_{ow}:
5.22 (Bruggeman et al., 1982)
5.20 (Yoshida et al., 1983)
5.148 (Brooke et al., 1990)
5.155 (Brooke et al., 1990; de Bruijn et al., 1989)

Solubility in organics:
Soluble in acetic acid, benzene, chloroform, carbon disulfide, ethanol and ether (U.S. EPA, 1985).

Solubility in water:
265 μg/L at 25 °C (Klevens, 1950; Harrison et al., 1975)
260 μg/L at 25 °C (Mackay and Shiu, 1977)
206 and 264 μg/kg at 25 and 29 °C, respectively (May et al., 1978a)
236 μg/L at 25 °C (Schwarz and Wasik, 1976)
240 μg/L at 27 °C (Davis et al., 1942)
133, 275 μg/L at 15 °C, 166 μg/L at 20 °C, 222, 373 μg/L at 25 °C (Kishi and Hashimoto, 1989)
199 μg/L at 25 °C (Walters and Luthy, 1984)
177 μg/L at 25 °C (Vadas et al., 1991)
1.4 μmol/L at 25 °C (Akiyoshi et al., 1987)

Vapor pressure (10^6 mmHg):
9.23 at 25 °C (Sonnefeld et al., 1983; Wasik et al., 1983)
50 at 25 °C (Bidleman, 1984)
71.6, 116 at 25 °C (Hinckley et al., 1990)

Environmental Fate
Photolytic. When an aqueous solution containing fluoranthene was photooxidized by UV light at 90-95 °C, 25, 50 and 75% degraded to carbon dioxide after 75.3, 160.6 and 297.4 hours, respectively (Knoevenagel and Himmelreich, 1976).
Chemical/Physical. 2-Nitrofluoranthene was the principal product formed from the gas-phase reaction of fluoranthene with hydroxyl radicals in a NO_x atmosphere. Minor products found include 7- and 8-nitrofluoranthene (Arey et al., 1986). The reaction of fluoranthene with NO_x to form 3-nitrofluoranthene was

reported to occur in urban air from St. Louis, MO (Randahl et al., 1982). Chlorination of fluoranthene in polluted, humus-poor lake water gave a large number of mono-, di- and trichlorofluoranthene derivatives (Johnsen et al., 1989). At pH <4, chlorination of fluoranthene produced 3-chlorofluoranthene as the major product (Oyler et al., 1983). It was suggested that the chlorination of fluoranthene in tap water accounted for the presence of chloro- and dichloro-fluoranthenes (Shiraishi et al., 1985).

Exposure Limits: Potential occupational carcinogen. No individual standards have been set, however, as a constituent in coal tar pitch volatiles, the following exposure limits have been established (mg/m^3): NIOSH REL: TWA 0.1 (cyclohex-ane-extractable fraction), IDLH 80; OSHA PEL: TWA 0.2 (benzene-soluble fraction); ACGIH TLV: TWA 0.2 (benzene solubles).

Toxicity: Acute oral LD_{50} for rats 2,000 mg/kg (RTECS, 1985).

Use: Research chemical.

FLUORENE

Synonyms: 2,3-Benzindene; *o*–Biphenylenemethane; *o*–Biphenylmethane; Diphen-ylenemethane; *o*–Diphenylenemethane; **9*H*-Fluorene**; 2,2′-Methylenebiphenyl.

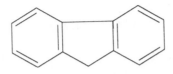

CAS Registry No.: 86-73-7
Molecular formula: $C_{13}H_{10}$
Formula weight: 166.22
RTECS: LL5670000

Physical state and color
Small white leaflets or flakes. Fluorescent when impure.

Melting point (°C):
116-117 (Weast, 1986)
114.0 (Pearlman et al., 1984)

Boiling point (°C):
298 (Aldrich, 1988)
294 (Huntress and Mulliken, 1941)

Density (g/cm^3):
1.203 at 0/4 °C (Weast, 1986)

Diffusivity in water (10^5 cm^2/sec):
0.61 at 20 °C using method of Hayduk and Laudie (1974).

Dissociation constant:
>15 (Christensen et al., 1975)

Heat of fusion (kcal/mol):
4.68 (Osborn and Douslin, 1975)
4.67 (Wauchope and Getzen, 1972)

Henry's law constant (10^5 atm·m^3/mol):
21 (Petrasek et al., 1983)
6.3 at 25 °C (Fendinger and Glotfelty, 1990)

Ionization potential (eV):
8.63 (Franklin et al., 1969)
8.56 (Yoshida et al., 1983)

Bioconcentration factor, log BCF:
2.70 (*Daphnia magna*, Newsted and Giesy, 1987)

Soil sorption coefficient, log K_{oc}:
3.70 (aquifer sands, Abdul et al., 1987)
3.85 (Meylan et al., 1992)
5.47 (average, Kayal and Connell, 1990)

Octanol/water partition coefficient, log K_{ow}:
4.12 (Chou and Jurs, 1979)
4.18 (Bruggeman et al., 1982; Hansch et al., 1972; Yalkowsky et al., 1983)

Solubility in organics:
Soluble in most solvents (U.S. EPA, 1985) including glacial acetic acid, carbon disulfide, ether and benzene (Windholz et al., 1983).

Solubility in water:
1.685 mg/kg at 25 °C (May et al., 1978a)
1.98 mg/L at 25 °C (Mackay and Shiu, 1977)
1.6622 mg/L at 25 °C (Sahyun, 1966)
1.68 mg/L at 25 °C (Wasik et al., 1983)
In mg/kg: 1.87, 1.88, 1.93 at 24.6 °C, 2.33, 2.34, 2.41 at 29.9 °C, 2.10, 2.23, 2.25 at 30.3 °C, 3.72, 3.73 at 38.4 °C, 3.84, 3.85, 3.88 at 40.1 °C, 5.59, 5.62, 5.68 at 47.5 °C, 6.31, 6.42, 6.54 at 50.1 °C, 6.31, 6.42, 6.54 at 50.1 °C, 6.27 at 50.2 °C, 8.31, 8.41, 8.56 at 54.7 °C, 10.5 at 59.2 °C, 10.7, 11.0, 11.6 at 60.5 °C, 14.1, 14.2, 14.2 at 65.1 °C, 18.5, 18.5, 18.9 at 70.7 °C, 18.8 at 71.9 °C, 21.5 at 73.4 °C (Wauchope and Getzen, 1972)
1.90 mg/L at 25 °C (Walters and Luthy, 1984)
2.23 mg/L at 25 °C (Vadas et al., 1991)
1.96 mg/L at 25 °C (Billington et al., 1988)
12.30 μmol/L at 30 °C (Yalkowsky et al., 1983)

Vapor pressure (10^4 mmHg):
1,130 at 75 °C (Osborn and Douslin, 1975)
1.64 at 33.30 °C, 1.95 at 34.85 °C, 2.5 at 37.20 °C, 2.81 at 38.45 °C, 3.43 at 40.30 °C, 5.43 at 45.00 °C, 7.08 at 47.75 °C, 8.18 at 49.25 °C, 8.33 at 49.55 °C (Bradley and Cleasby, 1953)
6.0 at 25 °C (Sonnefeld et al., 1983; Wasik et al., 1983)
45,100 at 127 °C (Eiceman and Vandiver, 1983)

29 at 25 °C (Bidleman, 1984)
35.5 at 25 °C (Hinckley et al., 1990)

Environmental Fate

Biological. Fluorene was statically incubated in the dark at 25 °C with yeast extract and settled domestic wastewater inoculum. Significant biodegradation with gradual adaptation was observed. At concentrations of 5 and 10 mg/L, bio-degradation yields at the end of 4 weeks of incubation were 77 and 45%, respectively (Tabak et al., 1981).

Chemical/Physical. Oxidation by ozone to fluorenone has been reported (Nikolaou, 1984). Chlorination of fluorene in polluted humus poor lake water gave a chlorinated derivative tentatively identified as 2-chlorofluorene (Johnsen et al., 1989). This compound was identified as a chlorination of fluorene at low pH (<4) (Oyler et al., 1983). It was suggested that the chlorination of fluoranthene in tap water accounted for the presence of chlorofluorene (Shiraishi et al., 1985).

Exposure Limits: Potential occupational carcinogen. No individual standards have been set, however, as a constituent in coal tar pitch volatiles, the following exposure limits have been established (mg/m^3): NIOSH REL: TWA 0.1 (cyclohex-ane-extractable fraction), IDLH 80; OSHA PEL: TWA 0.2 (benzene-soluble fraction); ACGIH TLV: TWA 0.2 (benzene solubles).

Drinking Water Standard: As of May 1994, no MCLGs or MCLs have been proposed (U.S. EPA, 1994).

Uses: Chemical intermediate in numerous applications and in the formation of polyradicals for resins; insecticides and dyestuffs. Derived from industrial and experimental coal gasification operations where maximum concentration detected in gas, liquid and coal tar streams were 9.1, 0.057 and 8.0 mg/m^3, respectively (Cleland, 1981).

FORMALDEHYDE

Synonyms: BFV; FA; Fannoform; Formalin; Formalin 40; Formalith; Formic aldehyde; Formol; Fyde; HOCH; Ivalon; Karsan; Lysoform; Methanal; Methyl aldehyde; Methylene glycol; Methylene oxide; Morbicid; NCI-C02799; Oxomethane; Oxymethylene; Paraform; Polyoxymethylene glycols; RCRA waste number U122; Superlysoform; UN 1198; UN 2209.

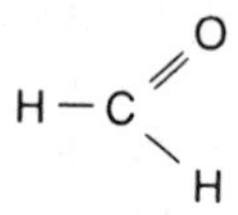

$$H-C{\overset{\displaystyle O}{\underset{\displaystyle H}{}}}$$

CAS Registry No.: 50-00-0
DOT: 1198
DOT label: Combustible liquid (aqueous solutions)
Molecular formula: CH_2O
Formula weight: 30.03
RTECS: LP8925000

Physical state, color and odor
Clear, colorless liquid with a pungent, suffocating odor. Burning taste. Odor threshold in air is 0.83 ppm (Amoore and Hautala, 1983).

Melting point (°C):
-92 (Weast, 1986)
-118 (Verschueren, 1983)

Boiling point (°C):
-21 (Weast, 1986)
-19.5 (Dean, 1987)

Density (g/cm³):
0.815 at -20/4 °C (Windholz et al., 1983)

Flash point (°C):
50 (37% aqueous solution, NFPA, 1984)

Lower explosive limit (%):
7.0 (NIOSH, 1994)

Upper explosive limit (%):
73 (NIOSH, 1994)

Henry's law constant (10^7 atm·m³/mol):
3.27 (Dong and Dasgupta, 1986)
1.67 (Gaffney et al., 1987)
34 at 25 °C (Zhou and Mopper, 1990)
3.37 at 25 °C (Betterton and Hoffmann, 1988)

Ionization potential (eV):
10.88 (Franklin et al., 1969)

Soil sorption coefficient, log K_{oc}:
0.56 (calculated, Mercer et al., 1990)

Octanol/water partition coefficient, log K_{ow}:
0.35 (Sangster, 1989)

Solubility in organics:
Soluble in acetone, alcohol, benzene and ether (Weast, 1986).

Solubility in water:
Miscible at 25 °C (Sax, 1984).

Vapor density:
1.067 (air = 1) (Windholz et al., 1983)
1.23 g/L at 25 °C

Vapor pressure (mmHg):
3,900 at 25 °C (Lide, 1990)

Environmental Fate
Biological. Biodegradation products reported include formic acid and ethanol, each of which can further degrade to carbon dioxide (Verschueren, 1983).

Photolytic. Major products reported from the photooxidation of formaldehyde with nitrogen oxides are carbon monoxide, carbon dioxide and hydrogen peroxide (Altshuller, 1983). In synthetic air, photolysis of formaldehyde yields hydrogen chloride and carbon monoxide (Su et al., 1979). Photooxidation of formaldehyde in nitrogen oxide-free air using radiation between 2900 and 3500 Å formed hydrogen peroxide, alkylhydroperoxides, carbon monoxide and lower molecular weight aldehydes. In the presence of NO_x, photooxidation products reported include ozone, hydrogen peroxide and peroxyacyl nitrates (Kopczynski et al., 1974).

Irradiation of gaseous formaldehyde containing an excess of nitrogen dioxide over chlorine yielded ozone, carbon monoxide, nitrogen pentoxide, nitryl chloride, nitric and hydrochloric acids. Peroxynitric acid was the major photolysis product when chlorine concentration exceeded the nitrogen dioxide concentration (Hanst and Gay, 1977).

Chemical/Physical. Oxidizes in air to formic acid. Trioxymethylene may precipitate under cold temperatures (Sax, 1984). Polymerizes easily (Windholz et al., 1983). Anticipated products from the reaction of formaldehyde with ozone or hydroxyl radicals in the atmosphere are carbon monoxide and carbon dioxide (Cupitt, 1980). Major products reported from the photooxidation of formaldehyde with nitrogen oxides are carbon monoxide, carbon dioxide and hydrogen peroxide (Altshuller, 1983).

Formaldehyde reacted with hydrogen chloride in moist air to form bis(chloromethyl)ether. This compound may also form from an acidic solution containing chloride ion and formaldehyde (Frankel et al., 1974). In an aqueous solution at 25 °C, nearly all the formaldehyde added is hydrated forming a gem-diol (Bell and McDougall, 1960).

Exposure Limits: Potential occupational carcinogen. NIOSH REL: TWA 16 ppb, 15-minute ceiling 100 ppb, IDLH 20 ppm; OSHA PEL: TWA 0.75 ppm, STEL 2 ppm; ACGIH TLV: TWA 1 ppm.

Symptoms of Exposure: Eye, nose and throat irritant; coughing, bronchospasm, pulmonary irritation, dermatitis, nausea, vomiting, loss of consciousness (NIOSH, 1994).

Toxicity: Acute oral LD_{50} for guinea pigs 260 mg/kg, mice 42 mg/kg, rats 800 mg/kg (RTECS, 1985).

Drinking Water Standard: As of May 1994, no MCLGs or MCLs have been proposed (U.S. EPA, 1994).

Uses: Manufacture of phenolic, melamine, urea and acetal resins; polyacetal and phenolic resins; pentaerythritol; fertilizers; dyes; hexamethylenetetramine; ethylene glycol; embalming fluids; textiles; fungicides; air fresheners; cosmetics; in medicine as a disinfectant and germicide; preservative; hardening agent; in oil wells as a corrosion inhibitor; industrial sterilant; reducing agent.

FORMIC ACID

Synonyms: Aminic acid; Formylic acid; Hydrogen carboxylic acid; Methanoic acid; RCRA waste number U123; UN 1779.

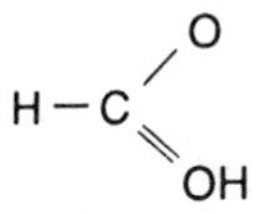

$$H-C\overset{O}{\underset{OH}{<}}$$

CAS Registry No.: 64-18-6
DOT: 1779
DOT label: Corrosive
Molecular formula: CH_2O_2
Formula weight: 46.03
RTECS: LQ4900000

Physical state, color and odor
Colorless, fuming liquid with a pungent, penetrating odor. Odor threshold in air is 49 ppm (Amoore and Hautala, 1983).

Melting point (°C):
8.1 (Barham and Clark, 1951)

Boiling point (°C):
100.7 (Weast, 1986)

Density (g/cm^3):
1.2200 at 20/4 °C (Barham and Clark, 1951)
1.2267 at 15/4 °C (Sax and Lewis, 1987)
1.21045 at 25/4 °C (Huntress and Mulliken, 1941)

Diffusivity in water (10^5 cm^2/sec):
1.57 at 20 °C using method of Hayduk and Laudie (1974).

Dissociation constant:
3.75 at 20 °C (Gordon and Ford, 1972)

Flash point (°C):
50.4 (90% solution, NIOSH, 1994)
69 (open cup, Hawley, 1981)

Lower explosive limit (%):
18 (90% solution, NIOSH, 1994)

Upper explosive limit (%):
57 (90% solution, NIOSH, 1994)

Heat of fusion (kcal/mol):
3.035 (Dean, 1987)

Henry's law constant (10^7 atm·m^3/mol):
1.67 at pH 4 (Gaffney et al., 1987)

Ionization potential (eV):
11.05 ± 0.01 (Franklin et al., 1969)
11.33 (Gibson et al., 1977)

Soil sorption coefficient, log K_{oc}:
Unavailable because experimental methods for estimation of this parameter for aliphatic carboxylic acids are lacking in the documented literature. However, its high solubility in water suggests its adsorption to soil will be nominal (Lyman et al., 1982).

Octanol/water partition coefficient, log K_{ow}:
-0.53 (Collander, 1951)

Solubility in organics:
Soluble in acetone and benzene (Weast, 1986). Miscible with alcohol, ether and glycerol (Windholz et al., 1983).

Solubility in water:
Miscible (Price et al., 1974).

Vapor density:
1.88 g/L at 25 °C, 1.59 (air = 1)

Vapor pressure (mmHg):
35 at 20 °C, 54 at 30 °C (Verschueren, 1983)
42.6 at 25 °C (Banerjee et al., 1990)

Environmental Fate
   Biological. Near Wilmington, NC, organic wastes containing formic acid (representing 11.4% of total dissolved organic carbon) were injected into an aquifer containing saline water to a depth of about 1,000 feet. The generation of gaseous components (hydrogen, nitrogen, hydrogen sulfide, carbon dioxide and methane) suggested that formic acid and possibly other waste constituents were anaerobically degraded by microorganisms (Leenheer et al., 1976).

Chemical/Physical. Pure formic acid slowly decomposes to give carbon monoxide and water. At 20 °C, 0.06 g of water would form in one year by 122 g formic acid. At standard temperature and pressure, this amount of formic acid would produce 0.15 mL of carbon monoxide per hour. Over time, the rate of decomposition will decrease because the water formed acts as a negative catalyst (Barham and Clark, 1951). Reacts with alkalies forming water-soluble salts. Slowly reacts with alcohols and anhydrides forming formate esters.

Exposure Limits: NIOSH REL: TWA 5 ppm (9 mg/m^3), IDLH 30 ppm; OSHA PEL: TWA 5 ppm.

Symptoms of Exposure: Corrosive to skin. Contact with liquid will burn eyes and skin (Patnaik, 1992).

Toxicity: Acute oral LD$_{50}$ for mice 700 mg/kg, dogs 4,000 mg/kg, rats 1,100 mg/kg (RTECS, 1985).

Uses: Chemical analysis; preparation of formate esters and oxalic acid; silvering glass; decalcifier; dehairing and plumping hides; manufacture of fumigants, insecticides, refrigerants, solvents for perfumes, lacquers; reducer in dyeing wool fast colors; ore flotation; electroplating; leather treatment; coagulating rubber latex; vinyl resin plasticizers.

FURFURAL

Synonyms: Artificial ant oil; Artificial oil of ants; Bran oil; Fural; 2-Furaldehyde; Furale; 2-Furanaldehyde; 2-Furancarbonal; **2-Furancarboxaldehyde**; Furfuraldehyde; Furfurol; Furfurole; Furole; α-Furole; 2-Furylmethanal; NCI-C56177; Pyromucic aldehyde; RCRA waste number U125; UN 1199.

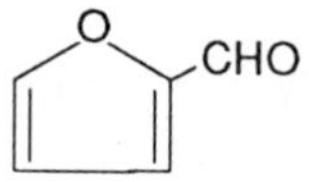

CAS Registry No.: 98-01-1
DOT: 1199
Molecular formula: $C_5H_4O_2$
Formula weight: 96.09
RTECS: LT7000000

Physical state, color and odor
Colorless to yellow liquid with an almond-like odor. Turns reddish brown on exposure to light and air. Odor and taste thresholds are 0.4 and 4 ppm, respectively (Keith and Walters, 1992).

Melting point (°C):
-38.7 (Weast, 1986)
-36.5 (Windholz et al., 1983)

Boiling point (°C):
161.7 (Weast, 1986)

Density (g/cm³):
1.1594 at 20/4 °C (Weast, 1986)
1.1563 at 25/4 °C (Windholz et al., 1983)

Diffusivity in water (10^5 cm²/sec at 25 °C):
1.12 (Hayduk and Laudie, 1974)

Flash point (°C):
61 (NIOSH, 1994)
68 (open cup, Windholz et al., 1983)

Lower explosive limit (%):
2.1 (NIOSH, 1994)

Upper explosive limit (%):
19.3 (NIOSH, 1994)

Henry's law constant (10^6 atm·m^3/mol):
1.52 at 20 °C (approximate - calculated from water solubility and vapor pressure)

Interfacial tension with water (dyn/cm at 25 °C):
4.7 (Demond and Lindner, 1993)

Ionization potential (eV):
9.21 ± 0.01 (Franklin et al., 1969)

Soil sorption coefficient, log K_{oc}:
Unavailable because experimental methods for estimation of this parameter for aldehydes are lacking in the documented literature. However, its high solubility in water suggests its adsorption to soil will be nominal (Lyman et al., 1982).

Octanol/water partition coefficient, log K_{ow}:
0.52 (Tewari et al., 1981, 1982)

Solubility in organics:
Soluble in acetone, alcohol, benzene, chloroform and ether (Weast, 1986).

Solubility in water:
83.0 g/L at 20 °C, 199.0 g/L at 90 °C (Verschueren, 1983)
0.81 M at 25.0 °C (Tewari et al., 1982; Wasik et al., 1981)
79.4 g/L at 20 °C, 84.0 g/L at 30 °C (Stephenson, 1993)
82.0 g/L at 40 °C (Jones, 1929)

Vapor density:
3.93 g/L at 25 °C, 3.32 (air = 1)

Vapor pressure (mmHg):
2 at 20 °C (NIOSH, 1994)
1 at 20 °C, 3 at 30 °C, 10 at 50 °C (Verschueren, 1983)
2.5 at 20 °C (Riddick et al., 1986)

Environmental Fate
   Biological. Under nitrate-reducing and methanogenic conditions, furfural biodegraded to methane and carbon dioxide (Knight et al., 1990).
   Chemical/Physical. Slowly resinifies at room temperature (Windholz et al., 1983). May polymerize on contact with strong acids or strong alkalies (NIOSH, 1994).

Exposure Limits: NIOSH REL: IDLH 100 ppm; OSHA PEL: TWA 5 ppm (20 mg/m^3); ACGIH TLV: TWA 2 ppm (8 mg/m^3).

Toxicity: Acute oral LD_{50} for dogs 950 mg/kg, guinea pigs 541 mg/kg, mice 400 mg/kg, rats 65 mg/kg (RTECS, 1985).

Uses: Synthesizing furan derivatives; adipic acid and adiponitrile; manufacture of furfural-phenol plastics; accelerator for vulcanizing rubber; solvent refining of petroleum oils; solvent for nitrated cotton, nitrocellulose, cellulose acetate, shoe dyes and gums; insecticide, fungicide and germicide; intermediate for tetrahydro-furan and furfuryl alcohol; weed killer; flavoring; wetting agent in manufacture of abrasive wheels and brake linings; road construction; refining of rare earths and metals; analytical chemistry.

FURFURYL ALCOHOL

Synonyms: 2-Furancarbinol; **2-Furanmethanol**; Furfural alcohol; Furyl alcohol; Furylcarbinol; 2-Furylcarbinol; α-Furylcarbinol; 2-Furylmethanol; 2-Hydroxy-meth-ylfuran; NCI-C56224; UN 2874.

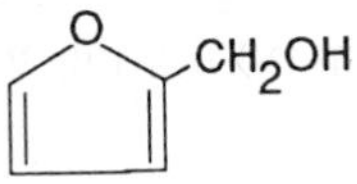

CAS Registry No.: 98-00-0
DOT: 2874
DOT label: Poison
Molecular formula: $C_5H_6O_2$
Formula weight: 98.10
RTECS: LU9100000

Physical state, color and odor
Colorless to yellow, mobile liquid with an irritating odor. Darkens on exposure to air. Odor threshold in air is 8 ppm (Keith and Walters, 1992).

Melting point (°C):
-14.6 (NIOSH, 1994)
-29 (metastable crystalline form, Verschueren, 1983)

Boiling point (°C):
170 (Windholz et al., 1983)

Density (g/cm^3):
1.1296 at 20/4 °C (Weast, 1986)

Diffusivity in water (10^5 cm^2/sec):
0.96 at 20 °C using method of Hayduk and Laudie (1974).

Flash point (°C):
66 (NIOSH, 1994)
75 (open cup, Windholz et al., 1983)

Lower explosive limit (%):
1.8 (NIOSH, 1994)

Upper explosive limit (%):
16.3 (NIOSH, 1994)

Heat of fusion (kcal/mol):
3.12 (Dean, 1987)

Soil sorption coefficient, log K_{oc}:
Unavailable because experimental methods for estimation of this parameter for unsaturated alicyclic alcohols are lacking in the documented literature. However, its high solubility in water suggests its adsorption to soil will be nominal (Lyman et al., 1982).

Solubility in organics:
Soluble in alcohol and ether (Weast, 1986).

Solubility in water:
Miscible (NIOSH, 1994).

Vapor density:
4.01 g/L at 25 °C, 3.39 (air = 1)

Vapor pressure (mmHg):
0.4 at 20 °C (Verschueren, 1983)
0.6 at 25 °C (NIOSH, 1994)

Environmental Fate
   Chemical/Physical. Easily resinified by acids (Windholz et al., 1983).

Exposure Limits: NIOSH REL: TWA 10 ppm (40 mg/m^3), STEL 15 ppm (60 mg/m^3), IDLH 75 ppm; OSHA PEL: TWA 50 ppm; ACGIH TLV: TWA 10 ppm, STEL 15 ppm.

Toxicity: Acute oral LD_{50} for mice 160 mg/kg, rats 88.3 mg/kg (RTECS, 1985).

Uses: Solvent for dyes and resins; preparation of furfuryl esters; furan polymers; solvent for textile printing; manufacturing wetting agents and resins; penetrant; flavoring; corrosion-resistant sealants and cements; viscosity reducer for viscous epoxy resins.

GLYCIDOL

Synonyms: Epihydric alcohol; Epihydrin alcohol; 2,3-Epoxypropanol; 2,3-Epoxy-1-propanol; Epoxypropyl alcohol; Glycide; Glycidyl alcohol; 3-Hydroxy-1,2-epoxypropane; Hydroxymethyl ethylene oxide; 2-Hydroxymethyloxiran; 3-Hydroxypropylene oxide; NCI-C55549; **Oxiranemethanol;** Oxiranylmethanol.

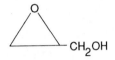

CAS Registry No.: 556-52-5
Molecular formula: $C_3H_6O_2$
Formula weight: 74.08
RTECS: UB4375000

Physical state, color and odor
Clear, colorless and odorless liquid.

Melting point (°C):
-45 (NIOSH, 1994)

Boiling point (°C):
Decomposes at 166-167 (Weast, 1986).
162 (Hawley, 1981)

Density (g/cm³):
1.1143 at 25/4 °C (Weast, 1986)
1.165 at 0/4 °C (Verschueren, 1983)

Diffusivity in water (10^5 cm²/sec):
0.23 at 25 °C using method of Hayduk and Laudie (1974).

Flash point (°C):
72.8 (NIOSH, 1994)
81 (Aldrich, 1988)

Soil sorption coefficient, log K_{oc}:
Unavailable because experimental methods for estimation of this parameter for epoxy aliphatic alcohols are lacking in the documented literature. However, its high solubility in water and low K_{ow} suggest its adsorption to soil will be nominal (Lyman et al., 1982).

Octanol/water partition coefficient, log K_{ow}:
-0.95 (Deneer et al., 1988)

Solubility in organics:
Soluble in acetone, alcohol, benzene, chloroform and ether (Weast, 1986).

Solubility in water:
Miscible (NIOSH, 1994).

Vapor density:
3.03 g/L at 25 °C, 2.56 (air = 1)

Vapor pressure (mmHg):
0.9 at 25 °C (Verschueren, 1983)

Environmental Fate
   Chemical/Physical. May hydrolyze in water forming glycerin (Lyman et al., 1982).

Exposure Limits: NIOSH REL: TWA 25 ppm (75 mg/m^3), IDLH 150 ppm; OSHA PEL: TWA 50 ppm (150 mg/m^3); ACGIH TLV: TWA 25 ppm.

Symptoms of Exposure: Eye, skin and lung irritant (Patnaik, 1992).

Toxicity: Acute oral LD_{50} for mice 431 mg/kg, rats 420 mg/kg (RTECS, 1985).

Uses: Demulsifier; dye-leveling agent; stabilizer for natural oils and vinyl polymers; organic synthesis.

HEPTACHLOR

Synonyms: Aahepta; Agroceres; Basaklor; 3-Chlorochlordene; Drinox; Drinox H-34; E 3314; ENT 15152; GPKh; H-34; Heptachlorane; 3,4,5,6,7,8,8-Heptachlorodicyclopentadiene; 3,4,5,6,7,8,8a-Heptachlorodicyclopentadiene; 1(3a),4,5,6,7,8,8-Heptachloro-3a(1),4,7,7a-tetrahydro-4,7-methanoindene; 1,4,5,6,7,8,8-Heptachloro-3a,4,7,7a-tetrahydro-4,7-methanoindene; **1,4,5,6,7,8,8-Heptachloro-3a,4,7,7a-tetrahydro-4,7-methanol-1*H*-indene;** 1,4,5,6,7,8,8-Heptachloro-3a,4,7,7a-tetrahydro-4,7-*endo*-methanoindene; 1,4,5,6,7,8,8a-Heptachloro-3a,4,7,7a-tetrahydro-4,7-methanoindene; 1,4,5,6,7,8,8-Heptachloro-3a,4,7,7a-tetrahydro-4,7-methyleneindene; 1,4,5,6,7,10,10-Heptachloro-4,7,8,9-tetrahydro-4,7-methyleneindene; 1,4,5,6,7,10,10-Heptachloro-4,7,8,9-tetrahydro-4,7-*endo*-methyleneindene; 3,4,5,6,7,8,8a-Heptachloro-α-dicyclopentadiene; Heptadichlorocyclopentadiene; Heptagran; Heptagranox; Heptamak; Heptamul; Heptasol; Heptox; NA 2761; NCI-C00180; Soleptax; RCRA waste number P059; Rhodiachlor; Velsicol 104; Velsicol heptachlor.

CAS Registry No.: 76-44-8
DOT: 2761
DOT label: Poison
Molecular formula: $C_{10}H_5Cl_7$
Formula weight: 373.32
RTECS: PC0700000

Physical state, color and odor
Crystalline white to light tan, waxy solid with a camphor-like odor.

Melting point (°C):
95-96 (Weast, 1986)
46-74 (Weiss, 1986)

Boiling point (°C):
Decomposes at 146.2 (NIOSH, 1994)
135-145 at 1-1.5 mmHg (IARC, 1979)

Density (g/cm³):
1.57 at 9/4 °C (Weast, 1986)

535

1.66 at 20/4 °C (Weiss, 1986)
1.65-1.67 at 25/4 °C (WHO, 1984)

Diffusivity in water (10^5 cm^2/sec):
0.46 at 20 °C using method of Hayduk and Laudie (1974).

Flash point (°C):
Noncombustible solid (NIOSH, 1994)

Henry's law constant (10^3 atm·m^3/mol):
2.3 (Petrasek et al., 1983)

Bioconcentration factor, log BCF:
4.14 (freshwater fish), 3.58 (fish, microcosm) (Garten and Trabalka, 1983)

Soil sorption coefficient, log K_{oc}:
4.38 (Jury et al., 1987)

Octanol/water partition coefficient, log K_{ow}:
5.44 (Travis and Arms, 1988)
5.5 (DeKock and Lord, 1987)

Solubility in organics:
At 27 °C (g/L): acetone (750), benzene (1,060), carbon tetrachloride (1,120), cyclohexanone (1,190), alcohol (45), xylene (1,020) (Windholz et al., 1983). Soluble in ether, kerosene and ligroin (U.S. EPA, 1985).

Solubility in water (μg/L):
100 at 15 °C, 180 at 25 °C, 315 at 35 °C, 490 at 45 °C (particle size ≤5 μ, Biggar and Riggs, 1974)
56 at 25-29 °C (Park and Bruce, 1968)

Vapor pressure (10^4 mmHg):
3 at 20 °C (Sims et al., 1988)
2.33 at 25 °C (Hinckley et al., 1990)

Environmental Fate
   Biological. Many soil microorganisms were found to oxidize heptachlor to heptachlor epoxide (Miles et al., 1969). In addition, hydrolysis produced hydroxychlordene with subsequent epoxidation yielding 1-hydroxy-2,3-epoxy-chlordene (Kearney and Kaufman, 1976). Heptachlor reacted with reduced hematin forming chlordene, which decomposed to hexachlorocyclopentadiene and cyclopentadiene (Baxter, 1990). In a model ecosystem containing plankton,

Daphnia magna, mosquito larva (*Culex pipiens quinquefasciatus*), fish (*Cambusia affinis*), alga (*Oedogonium cardiacum*) and snail (*Physa* sp.), heptachlor degraded to 1-hydroxychlordene, 1-hydroxy-2,3-epoxychlordene, hydroxychlordene epoxide, heptachlor epoxide and five unidentified compounds (Lu et al., 1975). In four successive 7-day incubation periods, heptachlor (5 and 10 mg/L) was recalcitrant to degradation in a settled domestic wastewater inoculum (Tabak et al., 1981).

Plant. On plant surfaces, heptachlor was oxidized to heptachlor epoxide (Gannon and Decker, 1958).

Photolytic. Sunlight and UV light convert heptachlor to photoheptachlor (Georgacakis and Khan, 1971). Eichelberger and Lichtenberg (1971) reported that heptachlor (10 μg/L) in river water, kept in a sealed jar under sunlight and fluorescent light, was completely converted to 1-hydroxychlordene. Under the same conditions, but in distilled water, 1-hydroxychlordene and heptachlor epoxide formed in yields of 60 and 40%, respectively (Eichelberger and Lichtenberg, 1971). The photolysis of heptachlor in various organic solvents afforded different photoproducts. Photolysis at 253.7 nm in hydrocarbon solvents yielded two olefinic monodechlorination isomers: 1,4,5,7,8,8-hexachloro-3a,4,7,7a-tetrahydro-4,7-methanoindene and 1,4,6,7,8,8-hexachloro-3a,4,7,7a-tetrahydro-4,7-methanoindene. Irradiation at 300 nm in acetone, 1,2,3,6,9,10,10-hepta-chloropentacyclo[$5.3.0.0^{2,5}.0^{3,9}.0^{4,8}$]decane is the only product formed. This compound and a C-1 cyclohexyl adduct are formed when heptachlor in a cyclohexane/acetone solvent system is irradiated at 300 nm (McGuire et al., 1972).

Heptachlor reacts with photochemically produced hydroxyl radicals in the atmosphere. At a concentration of 5×10^5 hydroxyl radicals/cm^3, the atmospheric half-life was estimated to be about 6 hours (Atkinson, 987).

Chemical/Physical. Slowly releases hydrogen chloride in aqueous media (Hartley and Kidd, 1987). The hydrolysis half-lives of heptachlor in a sterile 1% ethanol/water solution at 25 °C and pH values of 4.5, 5.0, 6.0, 7.0 and 8.0 were 0.77, 0.62, 0.64, 0.64 and 0.43 weeks, respectively (Chapman and Cole, 1982). Chemical degradation of heptachlor gave heptachlor epoxide (Newland et al., 1969). Heptachlor degraded in aqueous saturated calcium hypochlorite solution to 1-hydroxychlordene. Although further degradation occurred, no other metabolites were identified (Kaneda et al., 1974).

Heptachlor (1 mM) in ethyl alcohol (30 mL) underwent dechlorination in the presence of nickel boride (generated by the reacted of nickel chloride and sodium borohydride). The catalytic dechlorination of heptachlor by this method yielded a pentachloro derivative as the major product having the empirical formula $C_{10}H_9Cl_5$ (Dennis and Cooper, 1976). Emits toxic chloride fumes when heated to decomposition (Lewis, 1990).

Exposure Limits (mg/m^3): Potential occupational carcinogen. NIOSH REL: TWA 0.5, IDLH 35; OSHA PEL: TWA 0.5; ACGIH TLV: TWA 0.5.

Toxicity: LC_{50} (96-hour) for rainbow trout 7 μg/L, bluegill sunfish 26 μg/L, fathead minnow 78-130 μg/L (Hartley and Kidd, 1987), sheapshead minnow 2.7-8.8 μg/L and marine pin perch 0.20-4.4 μg/L (Verschueren, 1983); LC_{50} (48-hour) for red killifish 933 μg/L (Yoshioka et al., 1986); LC_{50} (24-hour) for sheapshead minnow 1.22-4.3 μg/L (Verschueren, 1983); acute oral LD_{50} for rats 40 mg/kg, guinea pigs 116 mg/kg, hamsters 100 mg/kg, mice 68 mg/kg (RTECS, 1985).

Drinking Water Standard (final): MCLG: zero; MCL: 0.4 μg/L (U.S. EPA, 1994).

Use: Insecticide for termite control.

HEPTACHLOR EPOXIDE

Synonyms: ENT 25584; Epoxy heptachlor; HCE; 1,4,5,6,7,8,8-Heptachloro-2,3-epoxy-2,3,3a,4,7,7a-hexahydro-4,7-methanoindene; 1,2,3,4,5,6,7,8,8-Heptachloro-2,3-epoxy-3a,4,7,7a-tetrahydro-4,7-methanoindene; **5a,6,6a-Hexahydro-2,5-methano-2H-indeno[1,2-b]-oxirene;** 2,3,4,5,6,7,7-Heptachloro-1a,1b,5,5a,6,6a-hexahydro-2,5-methano-2H-oxireno[a]indene; Velsicol 53-CS-17.

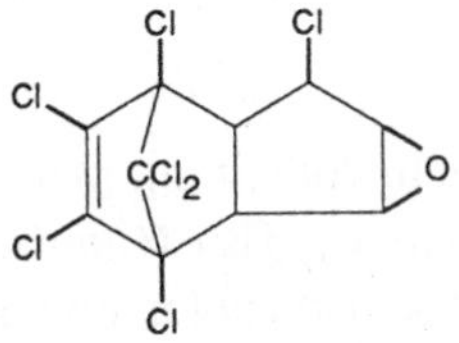

CAS Registry No.: 1024-57-3
DOT: 2761
DOT label: Poison
Molecular formula: $C_{10}H_5Cl_7O$
Formula weight: 389.32
RTECS: PB9450000

Physical state
Liquid.

Melting point (°C):
157-160 (Singh, 1969)

Diffusivity in water (10^5 cm^2/sec):
0.46 at 20 °C using method of Hayduk and Laudie (1974).

Henry's law constant (10^5 atm·m^3/mol):
3.2 at 25 °C (Warner et al., 1987)

Bioconcentration factor, log BCF:
2.90 (*B. subtilis*, Grimes and Morrison, 1975)
4.16 (freshwater fish), 3.69 (fish, microcosm) (Garten and Trabalka, 1983)

Soil sorption coefficient, log K_{oc}:
4.32 using method of Kenaga and Goring (1980).

Octanol/water partition coefficient, log K_{ow}:
5.40 (Travis and Arms, 1988)

Solubility in water (μg/L):
350 at 25-29 °C (Park and Bruce, 1968)

275 at 25 °C (Warner et al., 1987)
110 at 15 °C, 200 at 25 °C, 350 at 35 °C, 600 at 45 °C (particle size ≤5 μ, Biggar and Riggs, 1974)

Vapor pressure (10^6 mmHg):
2.6 at 20 °C (IARC, 1974)
300 at 30 °C (Nash, 1983)

Environmental Fate

Biological. In a model ecosystem containing plankton, *Daphnia magna*, mosquito larva (*Culex pipiens quinquefasciatus*), fish (*Cambusia affinis*), alga (*Oedogonium cardiacum*) and snail (*Physa* sp.), heptachlor epoxide degraded to hydroxychlordene epoxide (Lu et al., 1975). Using settled domestic wastewater inoculum, heptachlor epoxide (5 and 10 mg/L) did not degrade after 28 days of incubation at 25 °C (Tabak et al., 1981). This is consistent with the findings of Bowman et al. (1965). They reported that heptachlor epoxide was not significantly degraded in 7 air-dried soils that were incubated for 8 days.

Photolytic. Irradiation of heptachlor epoxide by a 450-W high-pressure mercury lamp gave two half-cage isomers, each containing a ketone functional group (Ivie et al., 1972).

Toxicity: Acute oral LD_{50} for rats is 47 mg/kg, mice 39 mg/kg (RTECS, 1985).

Drinking Water Standard (final): MCLG: zero; MCL: 0.2 μg/L (U.S. EPA, 1994).

Uses: Not known.

HEPTANE

Synonyms: Dipropylmethane; Gettysolve-C; *n*-Heptane; Heptyl hydride; UN 1206.

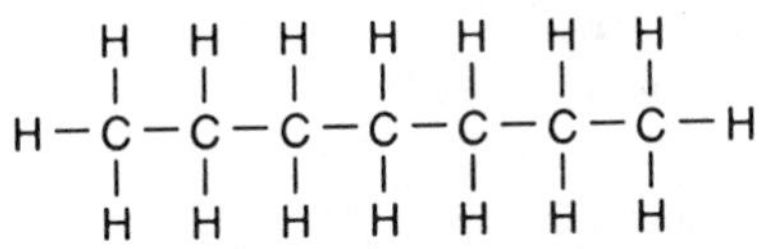

$$H-\overset{\displaystyle H}{\underset{\displaystyle H}{C}}-\overset{\displaystyle H}{\underset{\displaystyle H}{C}}-\overset{\displaystyle H}{\underset{\displaystyle H}{C}}-\overset{\displaystyle H}{\underset{\displaystyle H}{C}}-\overset{\displaystyle H}{\underset{\displaystyle H}{C}}-\overset{\displaystyle H}{\underset{\displaystyle H}{C}}-\overset{\displaystyle H}{\underset{\displaystyle H}{C}}-H$$

CAS Registry No.: 142-82-5
DOT: 1206
DOT label: Flammable liquid
Molecular formula: C_7H_{16}
Formula weight: 100.20
RTECS: MI7700000

Physical state, color and odor
Colorless liquid with a faint, pleasant odor. Odor threshold in air is 150 ppm (Amoore and Hautala, 1983).

Melting point (°C):
-90.6 (Weast, 1986)

Boiling point (°C):
98.4 (Weast, 1986)

Density (g/cm³):
0.6837 at 20/4 °C (Weast, 1986)

Diffusivity in water (10^5 cm²/sec):
0.71 at 20 °C using method of Hayduk and Laudie (1974).

Dissociation constant:
>14 (Schwarzenbach et al., 1993)

Flash point (°C):
-3.9 (NIOSH, 1994)
-1 (Affens and McLaren, 1972)

Lower explosive limit (%):
1.05 (NIOSH, 1994)

Upper explosive limit (%):
6.7 (NIOSH, 1994)

Heat of fusion (kcal/mol):
3.359 (Dean, 1987)

Henry's law constant (atm·m³/mol):
2.035 at 25 °C (Hine and Mookerjee, 1975)

Interfacial tension with water (dyn/cm at 25 °C):
50.2 (Donahue and Bartell, 1952)

Ionization potential (eV):
9.90 ± 0.05 (Franklin et al., 1969)
10.08 (HNU, 1986)

Soil sorption coefficient, log K_{oc}:
Unavailable because experimental methods for estimation of this parameter for aliphatic hydrocarbons are lacking in the documented literature.

Octanol/water partition coefficient, log K_{ow}:
4.66 (Tewari et al., 1982; Wasik et al., 1981)

Solubility in organics:
In methanol, g/L: 181 at 5 °C, 200 at 10 °C, 225 at 15 °C, 254 at 20 °C, 287 at 25 °C, 327 at 30 °C, 378 at 35 °C, 450 at 40 °C (Kiser et al., 1961).

Solubility in water:
In mg/kg: 2.24 at 25 °C, 2.63 at 40.1 °C, 3.11 at 55.7 °C (Price, 1976)
2.93 mg/kg at 25 °C (McAuliffe, 1963, 1966)
4.39 mg/kg at 0 °C, 3.37 mg/kg at 25 °C (Polak and Lu, 1973)
35.7 µmol/L at 25.0 °C (Tewari et al., 1982; Wasik et al., 1981)
2.19, 2.66 mg/L at 25 °C (Mackay and Shiu, 1981)
0.07 mL/L at 15.5 °C (Fühner, 1924)
2.9 mg/L (Coutant and Keigley, 1988)
70 mg/kg at 25 °C (Stearns et al., 1947)
Mole fraction x 10^7: 3.51, 3.63, 4.78, 4.07 and 4.32 at 4.3, 13.5, 25.0, 35.0 and 45.0 °C, respectively (Nelson and DeLigny, 1968)
5 x 10^{-7} in seawater at 25 °C (mole fraction, Krasnoshchekova and Gubergrits, 1973)

Vapor density:
4.10 g/L at 25 °C, 3.46 (air = 1)

Vapor pressure (mmHg):
35 at 20 °C, 58 at 30 °C (Verschueren, 1983)

46 at 25 °C (Milligan, 1924)
45.85 at 25 °C (Wilhoit and Zwolinski, 1971)
44.69 at 25.00 °C (Hussam and Carr, 1985)
14.8, 25.0, 45.7, 73.8 and 114.9 at 4.3, 13.5, 25.0, 35.0 and 45.0 °C, respectively
   (Nelson and DeLigny, 1968)

Environmental Fate
   Biological. Heptane may biodegrade in two ways. The first is the formation of
heptyl hydroperoxide, which decomposes to 1-heptanol followed by oxidation to
heptanoic acid. The other pathway involves dehydrogenation to 1-heptene, which
may react with water forming 1-heptanol (Dugan, 1972). Microorganisms can
oxidize alkanes under aerobic conditions (Singer and Finnerty, 1984). The most
common degradative pathway involves the oxidation of the terminal methyl group
forming the corresponding alcohol (1-heptanol). The alcohol may undergo a series
of dehydrogenation steps forming heptanal followed by oxidation forming
heptanoic acid. The acid may then be metabolized by β-oxidation to form the
mineralization products, carbon dioxide and water (Singer and Finnerty, 1984).
Hou (1982) reported hexanoic acid as a degradation product by the microorganism
Pseudomonas aeruginosa.

Exposure Limits: NIOSH REL: TWA 85 ppm (350 mg/m^3), 15-minute ceiling 440
ppm (1,800 mg/m^3), IDLH 750 ppm; OSHA PEL: TWA 500 ppm (2,000 mg/m^3);
ACGIH TLV: TWA 400 ppm, STEL 500 ppm.

Symptoms of Exposure: May cause nausea, dizziness and may impart a gasoline-
like taste (Patnaik, 1992).

Toxicity: LD$_{50}$ (intravenous) for mice 222 mg/kg (RTECS, 1985).

Uses: Standard in testing knock of gasoline engines and for octane rating
determinations; anesthetic; solvent; organic synthesis.

2-HEPTANONE

Synonyms: Amyl methyl ketone; *n*-Amyl methyl ketone; Methyl amyl ketone; Methyl *n*-amyl ketone; UN 1110.

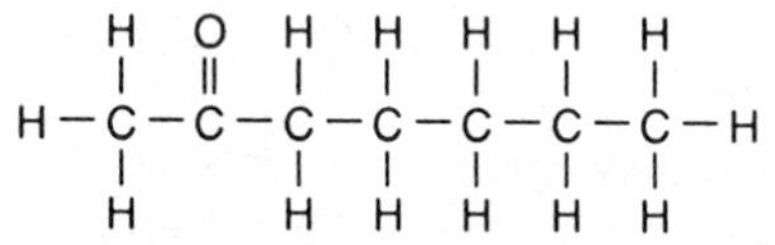

```
      H   O   H   H   H   H   H
      |   ||  |   |   |   |   |
  H — C — C — C — C — C — C — C — H
      |       |   |   |   |   |
      H       H   H   H   H   H
```

CAS Registry No.: 110-43-0
DOT: 1110
DOT label: Combustible liquid
Molecular formula: $C_7H_{14}O$
Formula weight: 114.19
RTECS: MJ5075000

Physical state, color and odor
Colorless liquid with a banana-like odor.

Melting point (°C):
-35.5 (Weast, 1986)
-27 (Verschueren, 1983)

Boiling point (°C):
151.4 (Weast, 1986)
150 (Verschueren, 1983)

Density (g/cm³):
0.8111 at 20/4 °C (Weast, 1986)
0.8197 at 15/4 °C (Sax and Lewis, 1987)
0.8115 at 25/4 °C (Ginnings et al., 1940)

Diffusivity in water (10^5 cm²/sec):
0.72 at 20 °C using method of Hayduk and Laudie (1974).

Flash point (°C):
39.2 (NIOSH, 1994)
47 (Aldrich, 1988)

Lower explosive limit (%):
1.1 at 66.7 °C (NIOSH, 1994)

544

Upper explosive limit (%):
7.9 at 122.1 °C (NIOSH, 1994)

Henry's law constant (10^4 atm·m^3/mol):
1.44 at 25 °C (Buttery et al., 1969)

Interfacial tension with water (dyn/cm at 25 °C):
12.4 (Donahue and Bartell, 1952)

Ionization potential (eV):
9.33 (NIOSH, 1994)

Soil sorption coefficient, log K_{oc}:
Unavailable because experimental methods for estimation of this parameter for aliphatic ketones are lacking in the documented literature.

Octanol/water partition coefficient, log K_{ow}:
2.03 (Sangster, 1989)
1.98 (Tewari et al., 1982; Wasik et al., 1981, 1983)

Solubility in organics:
Miscible with organic solvents (Hawley, 1981).

Solubility in water:
4,339 mg/L at 20 °C (Mackay and Yeun, 1983)
In wt %: 0.44 at 20 °C, 0.43 at 25 °C, 0.40 at 30 °C (Ginnings et al., 1940)
35.7 mmol/L at 25.0 °C (Tewari et al., 1982; Wasik et al., 1981, 1983)

Vapor density:
4.67 g/L at 25 °C, 3.94 (air = 1)

Vapor pressure (mmHg at 20 °C):
3 (NIOSH, 1994)
2.6 (Hawley, 1981)

Exposure Limits: NIOSH REL: TWA 100 ppm (465 mg/m^3), IDLH 800 ppm; OSHA PEL: TWA 100 ppm; ACGIH TLV: TWA 50 ppm.

Toxicity: Acute oral LD_{50} for mice 730 mg/kg, rats 1,670 mg/kg (RTECS, 1985).

Uses: Ingredient in artificial carnation oils; industrial solvent; synthetic flavoring; solvent for nitrocellulose lacquers; organic synthesis.

3-HEPTANONE

Synonyms: Butyl ethyl ketone; *n*-Butyl ethyl ketone; Ethyl butyl ketone; Heptan-3-one.

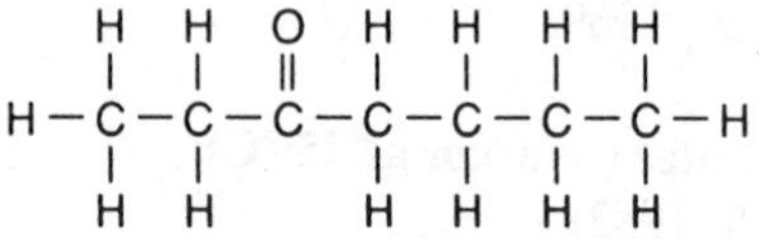

$$
\begin{array}{ccccccc}
H & H & O & H & H & H & H \\
| & | & \| & | & | & | & | \\
H-C-&C-&C-&C-&C-&C-&C-H \\
| & | & & | & | & | & | \\
H & H & & H & H & H & H
\end{array}
$$

CAS Registry No.: 106-35-4
Molecular formula: $C_7H_{14}O$
Formula weight: 114.19
RTECS: MJ5250000

Physical state, color and odor
Colorless liquid with a strong, fruity odor.

Melting point (°C):
-39 (Weast, 1986)

Boiling point (°C):
147.8 (Dean, 1987)

Density (g/cm^3):
0.8183 at 20/4 °C (Weast, 1986)

Diffusivity in water (10^5 cm^2/sec):
0.73 at 20 °C using method of Hayduk and Laudie (1974).

Flash point (°C):
46.5 (open cup, NIOSH, 1994)
41 (Aldrich, 1988)

Henry's law constant (10^5 atm·m^3/mol):
4.20 at 20 °C (approximate - calculated from water solubility and vapor pressure)

Ionization potential (eV):
9.02 (NIOSH, 1994)

Soil sorption coefficient, log K_{oc}:
Unavailable because experimental methods for estimation of this parameter for aliphatic ketones are lacking in the documented literature.

Octanol/water partition coefficient, log K_{ow}:
1.32 using method of Hansch et al. (1968).

1Solubility in organics:
Soluble in alcohol and ether (Weast, 1986).

Solubility in water:
14,300 mg/L at 20 °C (Verschueren, 1983)

Vapor density:
4.67 g/L at 25 °C, 3.94 (air = 1)

Vapor pressure (mmHg):
4 at 20 °C (NIOSH, 1994)
1.4 at 25 °C (Verschueren, 1983)

Exposure Limits: NIOSH REL: TWA 50 ppm (230 mg/m^3), IDLH 1,000 ppm; OSHA PEL: TWA 50 ppm; ACGIH TLV: TWA 50 ppm.

Symptoms of Exposure: May cause irritation of the eyes, skin and mucous membranes (Patnaik, 1992).

Toxicity: Acute oral LD_{50} for rats 2,760 mg/kg (RTECS, 1985).

Uses: Solvent mixtures for nitrocellulose and polyvinyl resins; in organic synthesis.

cis-2-HEPTENE

Synonyms: (Z)-2-Heptene; *cis*-2-Heptylene.

$$H-\overset{\displaystyle H}{\underset{\displaystyle H}{C}}-\overset{\displaystyle H}{\underset{\displaystyle H}{C}}-\overset{\displaystyle H}{\underset{\displaystyle H}{C}}-\overset{\displaystyle H}{\underset{\displaystyle H}{C}}\overset{\displaystyle H}{\diagdown}C=C\overset{\displaystyle H}{\diagup}\diagdown CH_3$$

CAS Registry No.: 6443-92-1
DOT: 2278
Molecular formula: C_7H_{14}
Formula weight: 98.19

Physical state and color
Colorless liquid.

Boiling point (°C):
98.5 (Weast, 1986)
98.41 (Wilhoit and Zwolinski, 1971)

Density (g/cm³):
0.708 at 20/4 °C (Weast, 1986)

Diffusivity in water (10^5 cm²/sec):
0.73 at 20 °C using method of Hayduk and Laudie (1974).

Dissociation constant:
>14 (Schwarzenbach et al., 1993)

Flash point (°C):
-6 (Aldrich, 1988)
-2.2 (commercial grade containing both *cis* and *trans* isomers, Hawley, 1981)

Henry's law constant (atm·m³/mol):
0.413 at 20 °C (approximate - calculated from water solubility and vapor pressure)

Soil sorption coefficient, log K_{oc}:
Unavailable because experimental methods for estimation of this parameter for aliphatic hydrocarbons are lacking in the documented literature.

Octanol/water partition coefficient, log K_{ow}:
2.88 using method of Hansch et al. (1968).

Solubility in organics:
Soluble in acetone, alcohol, benzene, chloroform and ether (Weast, 1986).

Solubility in water:
15 mg/kg at 25 °C (isomeric mixture, McAuliffe, 1966)
15.0 mg/kg at 23.5 °C (isomeric mixture, Schwarz, 1980)
271.6 μmol/L at 25 °C (Natarajan and Venkatachalam, 1972)

Vapor density:
4.01 g/L at 25 °C, 3.39 (air = 1)

Vapor pressure (mmHg):
48 at 25 °C (Wilhoit and Zwolinski, 1971)

Use: Organic synthesis.

trans-2-HEPTENE

Synonyms: (*E*)-2-Heptene; *trans*-2-Heptylene.

CAS Registry No.: 14686-13-6
DOT: 2278
Molecular formula: C_7H_{14}
Formula weight: 98.19

Physical state and color
Colorless liquid.

Melting point (°C):
-109.5 (Weast, 1986)

Boiling point (°C):
98 (Weast, 1986)

Density (g/cm^3):
0.7012 at 20/4 °C (Weast, 1986)

Diffusivity in water (10^5 cm^2/sec):
0.72 at 20 °C using method of Hayduk and Laudie (1974).

Dissociation constant:
>14 (Schwarzenbach et al., 1993)

Flash point (°C):
-1 (Aldrich, 1988)
-2.2 (commercial grade containing both isomers, Hawley, 1981)

Henry's law constant (atm·m^3/mol):
0.422 at 25 °C (approximate - calculated from water solubility and vapor pressure)

Soil sorption coefficient, log K$_{oc}$:
Unavailable because experimental methods for estimation of this parameter for aliphatic hydrocarbons are lacking in the documented literature.

Octanol/water partition coefficient, log K_{ow}:
2.88 using method of Hansch et al. (1968).

Solubility in organics:
Soluble in acetone, alcohol, benzene, chloroform and ether (Weast, 1986).

Solubility in water:
15 mg/kg at 25 °C (isomeric mixture, McAuliffe, 1966)
15.0 mg/kg at 23.5 °C (isomeric mixture, Schwarz, 1980)
271.6 μmol/L at 25 °C (isomeric mixture, Natarajan and Venkatachalam, 1972)

Vapor density:
4.01 g/L at 25 °C, 3.39 (air = 1)

Vapor pressure (mmHg):
49 at 25 °C (Wilhoit and Zwolinski, 1971)

Uses: Organic synthesis.

HEXACHLOROBENZENE

Synonyms: Amatin; Anticarie; Bunt-cure; Bunt-no-more; Co-op hexa; Granox NM; HCB; Hexa C.B.; Julin's carbon chloride; No bunt; No bunt 40; No bunt 80; No bunt liquid; Pentachlorophenyl chloride; Perchlorobenzene; Phenyl perchloryl; RCRA waste number U127; Sanocide; Smut-go; Snieciotox; UN 2729.

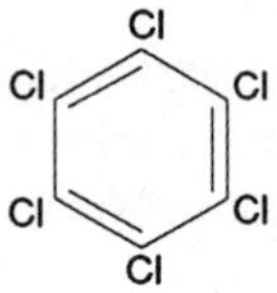

CAS Registry No.: 118-74-1
DOT: 2729
DOT label: Poison
Molecular formula: C_6Cl_6
Formula weight: 284.78
RTECS: DA2975000

Physical state and color
Monoclinic, white crystals.

Melting point (°C):
230 (Weast, 1986)
227 (Standen, 1964)

Boiling point (°C):
323–326 (Aldrich, 1988)

Density (g/cm^3):
1.5691 at 23.6/4 °C (Weast, 1986)
2.049 at 20/4 °C (Melnikov, 1971)

Diffusivity in water (10^5 cm^2/sec):
0.55 at 20 °C using method of Hayduk and Laudie (1974).

Flash point (°C):
242 (Hawley, 1981)

Heat of fusion (kcal/mol):
5.354 (Miller et al., 1984)
6.1 (Dean, 1987)
6.87 (Tsonopoulos and Prausnitz, 1971)

Henry's law constant (10^4 atm·m^3/mol):
17 at 25 °C (Warner et al., 1987)
At 25 °C: 13.1 and 17 at 23 °C in distilled water and seawater, respectively (Atlas et al., 1982)
71 at 20 °C (Oliver, 1985)
1.03, 1.97, 2.97, 3.44, 4.15, 4.73 and 4.78 at 14.8, 20.1, 22.1, 24.2, 34.8, 50.5 and 55 °C, respectively (ten Hulscher et al., 1992)
At 25 °C: 13.2 and 17.3 in distilled water and seawater, respectively (Brownawell et al., 1983)

Bioconcentration factor, log BCF:
4.27 (fathead minnow), 3.73 (rainbow trout), 4.34 (green sunfish) (Veith et al., 1979)
4.08-4.30 (rainbow trout, Oliver and Niimi, 1983)
4.54 (activated sludge), 3.37 (golden ide) (Freitag et al., 1985)
4.34 (fathead minnow, Carlson and Kosian, 1987)
4.39 (algae, Geyer et al., 1984)
4.09 (freshwater fish), 3.16 (fish, microcosm) (Garten and Trabalka, 1983)

Soil sorption coefficient, log K_{oc}:
3.59 (Kenaga, 1980)
4.49 (Briggs, 1981)
2.56, 2.70, 4.32, 4.54 (Hodson and Williams, 1988)
2.56, 2.70 (Rippen et al., 1982)
6.4 (average value using 10 suspended sediment samples collected from the St. Clair and Detroit Rivers, Lau et al., 1989)
3.23 (Captina silt loam), 4.73 (McLaurin sandy loam) (Walton et al., 1992)
K_d = 37.5 (Barcelona coastal sediments, Bayona et al., 1991)
4.98 (lake sediment, Schrap et al., 1994)
4.66 (fine sand, Enfield et al., 1989)

Octanol/water partition coefficient, log K_{ow}:
5.45 (Travis and Arms, 1988)
5.50 (Chiou et al., 1982)
5.66 (Hammers et al., 1982; Isnard and Lambert, 1988)
5.47 (Miller et al., 1984)
5.70, 5.79 (Garst and Wilson, 1984)
5.23 (Mackay, 1982)
5.31 (Watarai et al., 1982)
5.57 (Brooke et al., 1986)
5.20, 5.55 (Geyer et al., 1984)
5.44 (Briggs, 1981)
3.93 (Veith et al., 1980)

5.2 (Platford et al., 1982)
6.06 (Schwarzenbach et al., 1983)
5.68 at 13 °C, 5.70 at 19 °C, 5.58 at 28 °C, 5.17 at 33 °C (Opperhuizen et al., 1988)
6.22 (Rao and Davidson, 1980)
5.00 (Könemann et al., 1979)
6.86 (Burkhard et al., 1985a)
6.42 (DeKock and Lord, 1987)
5.731 (de Bruijn et al., 1989)

Solubility in organics:
Soluble in acetone, benzene, ether and chloroform (U.S. EPA, 1985).

Solubility in water (μg/L):
110 at 24 °C, 6 at 25 °C (Metcalf et al., 1973)
4.7 at 25 °C (Miller et al., 1985)
5 at 25 °C (Yalkowsky et al., 1979)
20 (Gile and Gillett, 1979)
40 at 20 °C (Riederer, 1990)
6.2 at 23.5 °C (Farmer et al., 1980)

Vapor pressure (10^6 mmHg):
10.89 at 20 °C (Isensee et al., 1976)
7.5 at 20 °C (Riederer, 1990)
912 at 25 °C (Bidleman, 1984)
18 at 25 °C (Banerjee et al., 1990)
4.2, 8.5 and 17.0 at 15, 20 and 25 °C, respectively (Farmer et al., 1980).
893, 1,190 at 25 °C (Hinckley et al., 1990)

Environmental Fate
    Biological. In activated sludge, only 1.5% of the applied hexachlorobenzene mineralized to carbon dioxide after 5 days (Freitag et al., 1985). Reductive monodechlorination occurred in an anaerobic sewage sludge yielding principally 1,3,5-trichlorobenzene. Other compounds identified include pentachlorobenzene, 1,2,3,5-tetrachlorobenzene and dichlorobenzenes (Fathepure et al., 1988). In a 5-day experiment, [^{14}C]hexachlorobenzene applied to soil water suspensions under aerobic and anaerobic conditions gave $^{14}CO_2$ yields of 0.4 and 0.2%, respectively (Scheunert et al., 1987).
    When hexachlorobenzene was statically incubated in the dark at 25 °C with yeast extract and settled domestic wastewater inoculum, no significant biodegradation was observed. At a concentration of 5 mg/L, percent losses after 7, 14, 21 and 28-day incubation periods were 56, 30, 8 and 5, respectively. At a concentration of 10 mg/L, only 21 and 3% losses were observed after the 7 and 14-day incubation periods, respectively. The decrease in concentration over time suggests bio-

degradation followed a deadaptive process (Tabak et al., 1981).

Photolytic. Solid hexachlorobenzene exposed to artificial sunlight for 5 months photolyzed at a very slow rate with no decomposition products identified (Plimmer and Klingebiel, 1976). The sunlight irradiation of hexachlorobenzene (20 g) in a 100-mL borosilicate glass-stoppered Erlenmeyer flask for 56 days yielded 64 ppm pentachlorobiphenyl (Uyeta et al., 1976). A carbon dioxide yield <0.1% was observed when hexachlorobenzene adsorbed on silica gel was irradiated with light (λ >290 nm) for 17 hours (Freitag et al., 1985).

Irradiation ($\lambda \geq 285$ nm) of hexachlorobenzene (1.1-1.2 mM/L) in an aceto-nitrile-water mixture containing acetone (0.553 mM/L) as a sensitizer gave the following products (% yield): pentachlorobenzene (71.0), 1,2,3,4-tetrachloro-benzene (0.6), 1,2,3,5-tetrachlorobenzene (2.2) and 1,2,4,5-tetrachlorobenzene (3.7) (Choudhry and Hutzinger, 1984). Without acetone, the identified photolysis products (% yield) included 1,2,3,4,5-pentachlorobenzene (76.8), 1,2,3,5-tetra-chlorobenzene (1.2), 1,2,4,5-tetrachlorobenzene (1.7), 1,2,4-trichlorobenzene (0.2) (Choudhry et al., 1984).

In another study, the irradiation (λ = 290-310 nm) of hexachlorobenzene in an aqueous solution gave only pentachlorobenzene and possibly pentachlorophenol as the transformation products. The photolysis rate increased with the addition of naturally occurring substances (tryptophan and pond proteins) and abiotic sensitizers (diphenylamine and skatole) (Hirsch and Hutzinger, 1989).

When an aqueous solution containing hexachlorobenzene (150 nM) and a nonionic surfactant micelle (0.50 M Brij 58, a polyoxyethylene cetyl ether) was illuminated by a photoreactor equipped with 253.7-nm monochromatic ultraviolet lamps, significant concentrations of pentachlorobenzene, all tetra-, tri- and dichloro-benzenes, chlorobenzene, benzene, phenol, hydrogen and chloride ions were formed. Two compounds, namely 1,2-dichlorobenzene and 1,2,3,4-tetrachloro-benzene, formed in minor amounts (<40 ppb). The half-life for this reaction, based on the first-order photodecomposition rate of 1.44×10^{-2}/sec, is 48 seconds (Chu and Jafvert, 1994).

Chemical/Physical. No hydrolysis was observed after 13 days at 85 °C and pH values of 3, 7 and 11 (Ellington et al., 1987).

Toxicity: Acute oral LD_{50} for cats 1,700 mg/kg, mice 4 gm/kg, rats 10,000 mg/kg, rabbits 2,600 mg/kg; LC_{50} (inhalation) for cats 1,600 mg/m^3, mice 4 gm/m^3, rats 3,600 mg/m^3, rabbits 1,800 mg/m^3; LC_{50} for 5 freshwater species 0.05-0.2 mg/L (Hartley and Kidd, 1987).

Drinking Water Standard: MCLG: zero; MCL: 1 μg/L (U.S. EPA, 1994).

Uses: Manufacture of pentachlorophenol; seed fungicide; wood preservative.

HEXACHLOROBUTADIENE

Synonyms: Dolen-pur; GP-40-66:120; HCBD; Hexachlorbutadiene; 1,1,2,3,4,4-Hexachlorobutadiene; 1,3-Hexachlorobutadiene; Hexachloro-1,3-butadiene; **1,1,2,3,4,4-Hexachloro-1,3-butadiene**; Perchlorobutadiene; RCRA waste number U128; UN 2279.

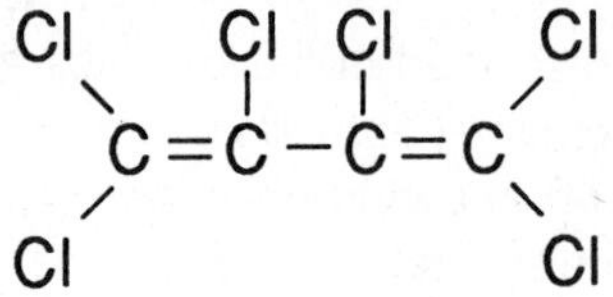

$$\underset{Cl}{\overset{Cl}{\diagdown}}C=\underset{}{\overset{Cl}{|}}C-\underset{}{\overset{Cl}{|}}C=\underset{Cl}{\overset{Cl}{\diagup}}C$$

CAS Registry No.: 87-68-3
DOT: 2279
DOT label: Poison
Molecular formula: C_4Cl_6
Formula weight: 260.76
RTECS: EJ0700000

Physical state, color and odor
Clear, yellowish-green liquid with a mild to pungent, turpentine-like odor. Odor threshold in air is 6 ppb (Keith and Walters, 1992).

Melting point (°C):
-21 (Weast, 1986)

Boiling point (°C):
215 (Weast, 1986)

Density (g/cm³):
1.6820 at 20/4 °C (Melnikov, 1971)

Diffusivity in water (10^5 cm²/sec):
0.65 at 20 °C using method of Hayduk and Laudie (1974).

Henry's law constant (10^3 atm·m³/mol):
26 (Pankow and Rosen, 1988)
1.02 at 25 °C (Warner et al., 1987)
7.8 (Warner et al., 1987)
4.3 at 20 °C (Oliver, 1985)

Soil sorption coefficient, log K_{oc}:
6.1 (average using 9 suspended sediment samples from the St. Clair and Detroit Rivers, Lau et al., 1989)

Octanol/water partition coefficient, log K_{ow}:
4.78 (Banerjee et al., 1980)
4.90 (Chiou, 1985)

Solubility in organics:
Soluble in ethanol and ether (U.S. EPA, 1985)

Solubility in water (mg/L):
3.23 at 25 °C (Banerjee et al., 1980)
4 at 20–25 °C (Geyer et al., 1980)
2.55 at 20 °C (Howard, 1989)

Vapor density:
10.66 g/L at 25 °C, 9.00 (air = 1)

Vapor pressure (mmHg):
0.15 at 20 °C (McConnell et al., 1975)

Environmental Fate
   Biological. In a model ecosystem containing plankton, *Daphnia magna*, mosquito larva (*Culex pipiens quinquefasciatus*), fish (*Cambusia affinis*), alga (*Oedogonium cardiacum*) and snail (*Physa* sp.), hexachlorocyclopentadiene degraded slightly, but no products were identified (Lu et al., 1975). When hexachlorobutadiene (5 and 10 mg/L) was statically incubated in the dark at 25 °C with yeast extract and settled domestic wastewater inoculum for 7 days, 100% biodegradation with rapid adaptation was observed (Tabak et al., 1981).

Exposure Limits: Potential occupational carcinogen. NIOSH REL: TWA 20 ppb (240 mg/m^3); ACGIH TLV: TWA 20 ppb (0.24 mg/m^3).

Toxicity: Acute oral LD$_{50}$ for rats 113 mg/kg; LD$_{50}$ (skin) for rabbits 430 mg/kg; LC$_{50}$ (inhalation) for rats 1,600 ppb/4-hour (RTECS, 1985).

Drinking Water Standard (tentative): MCLG: 1 μg/L; MCL: none proposed (U.S. EPA, 1994).

Uses: Solvent for elastomers, natural rubber, synthetic rubber; heat-transfer liquid; transformer and hydraulic fluid; wash liquor for removing C$_4$ and higher hydro-carbons; sniff gas recovery agent in chlorine plants; chemical intermediate for fluorinated lubricants and rubber compounds; fluid for gyroscopes; fumigant for grapes.

HEXACHLOROCYCLOPENTADIENE

Synonyms: C-56; Graphlox; HCCP; HCCPD; HCPD; Hex; **1,2,3,4,5,5-Hexachloro-1,3-cyclopentadiene**; HRS 1655; NCI-C55607; PCL; Perchlorocyclopentadiene; RCRA waste number U130; UN 2646.

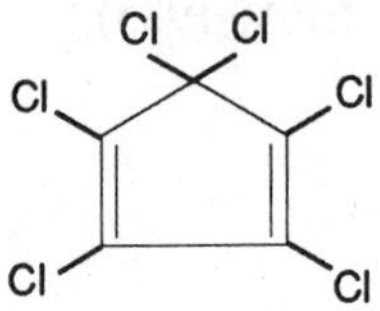

CAS Registry No.: 77-47-4
DOT: 2646
DOT label: Poison
Molecular formula: C_5Cl_6
Formula weight: 272.77
RTECS: GY1225000

Physical state, color and odor
Pale yellow to greenish-yellow liquid with a harsh, unpleasant odor. Odor threshold in air ranges from 1.4 to 1.6 μg/L (Keith and Walters, 1992).

Melting point (°C):
-9 (Weast, 1986)

Boiling point (°C):
236-238 (Melnikov, 1971)

Density (g/cm³):
1.7019 at 25/4 °C (Weast, 1986)
1.7119 at 20/4 °C (Ungnade and McBee, 1957)

Diffusivity in water (10^5 cm²/sec):
0.67 at 20 °C using method of Hayduk and Laudie (1974).

Flash point (°C):
Noncombustible liquid (NIOSH, 1994)

Lower explosive limit (%):
Noncombustible liquid (NIOSH, 1994)

Upper explosive limit (%):
Noncombustible liquid (NIOSH, 1994)

Henry's law constant (10^2 atm·m^3/mol):
1.6 (Pankow and Rosen, 1988)

Bioconcentration factor, log BCF:
3.04 (algae, Geyer et al., 1984)
3.38 (activated sludge), 3.09 (golden ide) (Freitag et al., 1985)

Soil sorption coefficient, log K_{oc}:
3.63 (Chou and Griffin, 1983)

Octanol/water partition coefficient, log K_{ow}:
5.00 (McDuffie, 1981)
5.51 (Mackay et al., 1982)
5.04 (Geyer et al., 1984)

Solubility in organics:
Based on structurally similar compounds, hexachlorocyclopentadiene is expected to
be soluble in benzene, ethanol, chloroform and other liquid halogenated solvents.

Solubility in water (mg/L):
0.805 (Lu et al., 1975)
1.8 at 25 °C (Zepp et al., 1979)
At 22 °C: 1.11 (distilled water), 1.14 (deionized water), 1.08 (tap water), 1.08 (Sugar
    Creek water) (Chou and Griffin, 1983)

Vapor density:
11.15 g/L at 25 °C, 9.42 (air = 1)

Vapor pressure (mmHg):
0.081 at 25 °C (Ungnade and McBee, 1957)
1 at 78-79 °C (Chou and Griffen, 1983)

Environmental Fate
   Biological. When hexachlorocyclopentadiene (5 and 10 mg/L) was statically
incubated in the dark at 25 °C with yeast extract and settled domestic wastewater
inoculum for 7 days, 100% biodegradation with rapid adaptation was observed
(Tabak et al., 1981).
   Photolytic. The major photolysis and hydrolysis products identified in distilled
water are pentachlorocyclopentenone and hexachlorocyclopentenone. In
mineralized water, the products identified include *cis*- and *trans*-pentachloro-
butadiene, tetrachlorobutenyne and pentachloropentadienoic acid (Chou and
Griffin, 1983). In a similar experiment, irradiation of hexachlorocyclopentadiene in
water by mercury-vapor lamps resulted in the formation of 2,3,4,4,5-pentachloro-

2-cyclopentenone. This was found to hydrolyze partially to hexachloroindenone (Butz et al., 1982). Other photodegradation products identified include hexachloro-2-cyclopentenone and hexachloro-3-cyclopentenone as major products. Secondary photodegradation products reported include pentachloro-*cis*-2,4-pentadienoic acid, *Z*- and *E*-pentachlorobutadiene and tetrachlorobutyne (Chou et al., 1987).

Chemical/Physical. Anticipated products from the reaction of hexachlorocyclopentadiene with ozone or hydroxyl radicals in the atmosphere are phosgene, diacylchlorides, ketones and chlorine radicals (Cupitt, 1980). Phosgene is hydrolyzed readily to hydrogen chloride and carbon dioxide (Morrison and Boyd, 1971).

Slowly reacts with water forming hydrochloric acid (NIOSH, 1994).

Exposure Limits: NIOSH REL: 10 ppb (100 mg/m^3); ACGIH TLV: TWA 10 ppb.

Drinking Water Standard (final): MCLG: 50 μg/L; MCL: 50 μg/L (U.S. EPA, 1994).

Uses: Intermediate in the synthesis of dyes, cyclodiene pesticides, (aldrin, dieldrin, endosulfan), fungicides and pharmaceuticals; manufacture of chlorendic anhydride and chlorendic acid.

HEXACHLOROETHANE

Synonyms: Avlothane; Carbon hexachloride; Carbon trichloride; Distokal; Distopan; Distopin; Egitol; Ethane hexachloride; Ethylene hexachloride; Falkitol; Fasciolin; HCE; 1,1,1,2,2,2-Hexachloroethane; Hexachloroethylene; Mottenhexe; NA 9037; NCI-C04604; Perchloroethane; Phenohep; RCRA waste number U131.

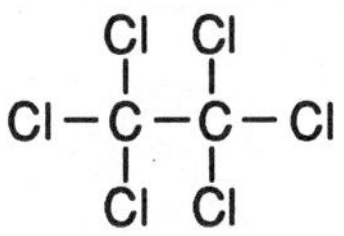

$$\underset{\underset{Cl}{|}}{\overset{\overset{Cl}{|}}{Cl-C}}-\underset{\underset{Cl}{|}}{\overset{\overset{Cl}{|}}{C}}-Cl$$

CAS Registry No.: 67-72-1
DOT: 9037
Molecular formula: C_2Cl_6
Formula weight: 236.74
RTECS: KI4025000

Physical state, color and odor
Rhombic, triclinic or cubic, colorless crystals with a camphor-like odor. Odor threshold in air is 0.15 ppm (Amoore and Hautala, 1983).

Melting point (°C):
186.6 (Weast, 1986)

Boiling point (°C):
187.5 (Dean, 1987)
190-195 (sublimes, Aldrich, 1988)

Density (g/cm³):
2.091 at 20/4 °C (Weast, 1986)

Diffusivity in water (10^5 cm²/sec):
0.63 at 20 °C using method of Hayduk and Laudie (1974).

Flash point (°C):
Noncombustible solid (NIOSH, 1994)

Heat of fusion (kcal/mol):
0.642 (Dean, 1987)

Henry's law constant (10^3 atm·m³/mol):
2.5 (Pankow and Rosen, 1988)

561

2.18 at 25 °C (Hine and Mookerjee, 1975)
10.3 at 30 °C (Jeffers et al., 1989)
5.93, 5.59, 5.91, 8.35 and 10.3 at 10, 15, 20, 25 and 30 °C, respectively (Ashworth et al., 1988)

Ionization potential (eV):
11.22 (NIOSH, 1994)
12.11 (Horvath, 1982)

Bioconcentration factor, log BCF:
2.14 (bluegill sunfish, Veith et al., 1980)

Soil sorption coefficient, log K_{oc}:
3.34 (Abdul et al., 1987)

Octanol/water partition coefficient, log K_{ow}:
4.62 (Abdul et al., 1987)
3.58 (Könemann et al., 1979)
3.93 (Veith et al., 1980)
4.14 (Chiou, 1985)

Solubility in organics:
Soluble in ethanol, benzene, chloroform and ether (U.S. EPA, 1985).

Solubility in water:
49.9 mg/L at 22 °C (Munz and Roberts, 1987)
27.2 mg/L at 25 °C (Lyman et al., 1982)
0.005 wt % at 22.3 °C (Stephen and Stephen, 1963)
50 ppm at 20 °C (Konietzko, 1984)

Vapor density:
6.3 kg/m^3 at the sublimation point and 750 mmHg (Konietzko, 1984).

Vapor pressure (mmHg):
0.8 at 30 °C (Verschueren, 1983)
0.18 at 20 °C (Munz and Roberts, 1987)

Environmental Fate
   Biological. Under aerobic conditions or in experimental systems containing mixed cultures, hexachloroethane was reported to degrade to tetrachloroethane (Vogel et al., 1987). In an uninhibited anoxic-sediment water suspension, hexachloroethane degraded to tetrachloroethylene. The reported half-life for this transformation was 19.7 minutes (Jafvert and Wolfe, 1987). When hexachloro-

ethane (5 and 10 mg/L) was statically incubated in the dark at 25 °C with yeast extract and settled domestic wastewater inoculum for 7 days, 100% biodegradation with rapid adaptation was observed (Tabak et al., 1981).

Photolytic. When an aqueous solution containing hexachloroethane was photooxidized by UV light at 90-95 °C, 25, 50 and 75% degraded to carbon dioxide after 25.2, 93.7 and 172.0 hours, respectively (Knoevenagel and Himmelreich, 1976).

Chemical/Physical. The reported hydrolysis half-life at 25 °C and pH 7 is 1.8 x 10^9 years (Jeffers et al., 1989). No hydrolysis was observed after 13 days at 85 °C and pH values of 3, 7 and 11 (Ellington et al., 1987).

The evaporation half-life of hexachloroethane (1 mg/L) from water at 25 °C using a shallow-pitch propeller stirrer at 200 rpm at an average depth of 6.5 cm is 40.7 minutes (Dilling, 1977).

Exposure Limits: Potential occupational carcinogen. NIOSH REL: TWA 1 ppm (10 mg/m^3), IDLH 300 ppm; OSHA PEL: TWA 1 ppm; ACGIH TLV: TWA 1 ppm.

Symptoms of Exposure: Vapors may cause irritation to the eyes and mucous membranes (Patnaik, 1992).

Toxicity: Acute oral LD_{50} for rats 4,460 mg/kg, guinea pigs 4,970 mg/kg (RTECS, 1985).

Drinking Water Standard: As of May 1994, no MCLGs or MCLs have been proposed although hexachloroethane has been listed for regulation (U.S. EPA, 1994).

Uses: Plasticizer for cellulose resins; moth repellant; camphor substitute in cellulose solvent; manufacturing of smoke candles and explosives; rubber vulcanization accelerator; insecticide; refining aluminum alloys.

HEXANE

Synonyms: Gettysolve-B; *n*-Hexane; Hexyl hydride; NCI-C60571; RCRA waste number U114; UN 1208.

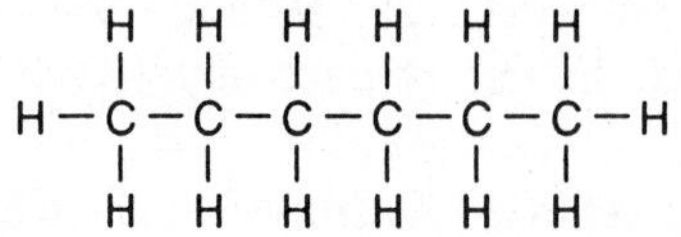

```
      H   H   H   H   H   H
      |   |   |   |   |   |
  H — C — C — C — C — C — C — H
      |   |   |   |   |   |
      H   H   H   H   H   H
```

CAS Registry No.: 110-54-3
DOT: 1208
DOT label: Flammable liquid
Molecular formula: C_6H_{14}
Formula weight: 86.18
RTECS: MN9275000

Physical state, color and odor
Clear, colorless liquid with a faint, gasoline-like odor. Odor threshold in air is 130 ppm (Amoore and Hautala, 1983).

Melting point (°C):
-95 (Weast, 1986)

Boiling point (°C):
69 (Weast, 1986)
68.74 (Wilhoit and Zwolinski, 1971)

Density (g/cm^3 at 20/4 °C):
0.6603 (Weast, 1986)
0.65937 (Hawley, 1981)

Diffusivity in water (10^5 cm^2/sec):
0.75 at 20 °C using method of Hayduk and Laudie (1974).

Dissociation constant:
>14 (Schwarzenbach et al., 1993)

Flash point (°C):
15 (Affens and McLaren, 1972)

Lower explosive limit (%):
1.1 (NIOSH, 1994)

Upper explosive limit (%):
7.5 (NIOSH, 1994)

Heat of fusion (kcal/mol):
3.126 (Dean, 1987)

Henry's law constant (atm·m³/mol):
1.184 at 25 °C (Hine and Mookerjee, 1975)
0.238, 0.413, 0.883, 0.768 and 1.56 at 10, 15, 20, 25 and 30 °C, respectively
  (Ashworth et al., 1988)

Interfacial tension with water (dyn/cm at 25 °C):
49.7 (Donahue and Bartell, 1952)

Ionization potential (eV):
10.18 (Franklin et al., 1969)

Soil sorption coefficient, log K_{oc}:
Unavailable because experimental methods for estimation of this parameter for
aliphatic hydrocarbons are lacking in the documented literature.

Octanol/water partition coefficient, log K_{ow}:
3.90 (Kamlet et al., 1984)
4.11 (Tewari et al., 1982; Wasik et al., 1981)

Solubility in organics:
Miscible with alcohol, chloroform and ether (Windholz et al., 1983).
In methanol, g/L: 324 at 5 °C, 370 g/L at 10 °C, 427 at 15 °C, 495 at 20 °C, 604 at
  25 °C, 830 at 30 °C. Miscible at higher temperatures (Kiser et al., 1961).

Solubility in water:
In mg/kg: 9.47 at 25 °C, 10.1 at 40.1 °C, 13.2 at 55.7 °C (Price, 1976)
9.5 mg/kg at 25 °C (McAuliffe, 1963, 1966)
16.5 mg/kg at 0 °C, 12.4 mg/kg at 25 °C (Polak and Lu, 1973)
13 mg/L at 20 °C, 75.5 mg/L in salt water at 20 °C (Verschueren, 1983)
12.3 mg/L at 25 °C. In NaCl solution (salinity, mol/L) at 25 °C, mg/L: 10.48 (0.31),
  8.06 (0.62), 7.54 (1.00), 4.88 (1.50), 3.75 (2.00), 2.55 (2.50) (Aquan-Yuen et al.,
  1979)
143 μmol/L at 25.0 °C (Tewari et al., 1982; Wasik et al., 1981)
9.52, 18.3 mg/L at 25 °C (Mackay and Shiu, 1981)
0.22 mL/L at 15.5 °C (Führer, 1924)
0.1882 mmol/L at 25 °C (Barone et al., 1967)
At 32 atmHg: 110, 150 and 260 mg/kg at 20, 37.8 and 71.4 °C, respectively

(Namiot and Beider, 1960)

120 mg/L at 25 °C (McBain and Lissant, 1951)

14 mg/L (Coutant and Keigley, 1988)

24 g/kg at 349.9 °C and 193.7 atmHg (DeLoos et al., 1982)

Mole fraction x 10^5: 3.42, 3.17, 3.83, 2.69, 4.64 and 4.42 at 4.0, 14.0, 25.0, 35.0, 45.0 and 55.0 °C, respectively (Nelson and DeLigny, 1968)

0.96 mmol/L at 25 °C (Barone et al., 1966)

2.8 and 13 μM at 25 °C in distilled water and seawater, respectively (Krasnoshchekova and Gubergrits, 1973)

12.3 mg/L at 25 °C (Leinonen and Mackay, 1973)

Vapor density:

3.52 g/L at 25 °C, 2.98 (air = 1)

Vapor pressure (mmHg):

150 at 25 °C (Milligan, 1924)

120 at 20 °C, 190 at 30 °C (Verschueren, 1983)

151.5 at 25 °C (Wilhoit and Zwolinski, 1971)

151.39 at 25.00 °C (Hussam and Carr, 1985)

55.94, 91.8, 151.2, 229.5, 337.9 and 483.3 at 4.0, 14.0, 25.0, 35.0, 45.0 and 55.0 °C, respectively (Nelson and DeLigny, 1968)

Environmental Fate

Biological. *n*-Hexane may biodegrade in two ways. The first is the formation of hexyl hydroperoxide, which decomposes to 1-hexanol followed by oxidation to hexanoic acid. The other pathway involves dehydrogenation to 1-hexene, which may react with water giving 1-hexanol (Dugan, 1972). Microorganisms can oxidize alkanes under aerobic conditions (Singer and Finnerty, 1984). The most common degradative pathway involves the oxidation of the terminal methyl group forming 1-hexanol. The alcohol may undergo a series of dehydrogenation steps forming a hexanal followed by oxidation to form hexanoic acid. The fatty acid may then be metabolized by β-oxidation to form the mineralization products, carbon dioxide and water (Singer and Finnerty, 1984).

Photolytic. An aqueous solution irradiated by UV light at 50 °C for 1 day resulted in a 50.51% yield of carbon dioxide (Knoevenagel and Himmelreich, 1976). Synthetic air containing gaseous nitrous acid and exposed to artificial sunlight (λ = 300-450 nm) photooxidized hexane into two isomers of hexyl nitrate and peroxyacetal nitrate (Cox et al., 1980).

Chemical/Physical. Complete combustion in air yields carbon dioxide and water vapor.

Exposure Limits: NIOSH REL: TWA 50 ppm (180 mg/m^3), IDLH 1,100 ppm; OSHA PEL: TWA 500 ppm (1,800 mg/m^3); ACGIH TLV: TWA 50 ppm.

Symptoms of Exposure: Irritation of respiratory tract. Narcotic at high concentrations (Patnaik, 1992).

Drinking Water Standard: As of May 1994, no MCLGs or MCLs have been proposed (U.S. EPA, 1994).

Toxicity: Acute oral LD_{50} for rats 28,710 mg/kg (RTECS, 1985).

Uses: Determining refractive index of minerals; paint diluent; dyed hexane is used in thermometers instead of mercury; polymerization reaction medium; calibrations; solvent for vegetable oils; alcohol denaturant; chief constituent of petroleum ether, rubber solvent and gasoline; in organic synthesis.

2-HEXANONE

Synonyms: Butyl ketone; Butyl methyl ketone; *n*-Butyl methyl ketone; Hexanone-2; MBK; Methyl *n*-butyl ketone; MNBK; Propylacetone.

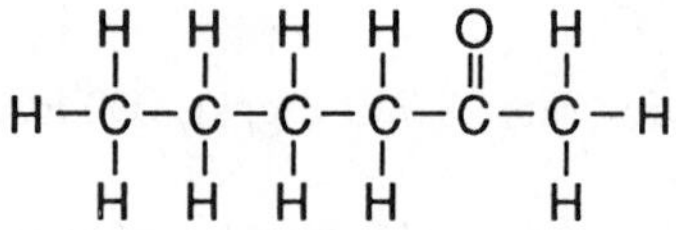

$$H-\underset{\underset{H}{|}}{\overset{\overset{H}{|}}{C}}-\underset{\underset{H}{|}}{\overset{\overset{H}{|}}{C}}-\underset{\underset{H}{|}}{\overset{\overset{H}{|}}{C}}-\underset{\underset{H}{|}}{\overset{\overset{H}{|}}{C}}-\overset{\overset{O}{\|}}{C}-\underset{\underset{H}{|}}{\overset{\overset{H}{|}}{C}}-H$$

CAS Registry No.: 591–78–6
Molecular formula: $C_6H_{12}O$
Formula weight: 100.16
RTECS: MP1400000

Physical state, color and odor
Clear, colorless liquid with an odor resembling acetone.

Melting point (°C):
-56.9 (Dean, 1973)
-55.8 (Riddick et al., 1986)

Boiling point (°C):
128 (Weast, 1986)

Density (g/cm^3 at 20/4 °C):
0.8113 (Weast, 1986)
0.8300 (Standen, 1967)

Diffusivity in water (10^5 cm^2/sec):
0.78 at 20 °C using method of Hayduk and Laudie (1974).

Flash point (°C):
-9 (NFPA, 1984)

Upper explosive limit (%):
8.0 (NIOSH, 1994)

Henry's law constant (10^3 atm·m^3/mol):
1.75 at 25 °C (approximate - calculated from water solubility and vapor pressure)

Interfacial tension with water (dyn/cm at 20 °C):
9.73 (Demond and Lindner, 1993)

Ionization potential (eV):
9.35 (Franklin et al., 1969)

Soil sorption coefficient, log K_{oc}:
2.13 using method of Kenaga and Goring (1980).

Octanol/water partition coefficient, log K_{ow}:
1.19 (Sangster, 1989)
1.38 (Leo et al., 1971)

Solubility in organics:
Soluble in acetone, ethanol and ether (Weast, 1986).

Solubility in water:
35,000 mg/L at 25 °C (Verschueren, 1983)
34,400 mg/L at 25 °C (Meites, 1963)
In wt %: 1.75 at 20 °C, 1.64 at 25 °C, 1.52 at 30 °C (Ginnings et al., 1940)

Vapor density:
4.09 g/L at 25 °C, 3.46 (air = 1)

Vapor pressure (mmHg):
2 at 20 °C (Verschueren, 1983)
3.8 at 25 °C (ACGIH, 1986)

Exposure Limits: NIOSH REL: TWA 1 ppm (4 mg/m^3), IDLH 1,600 ppm; OSHA PEL: TWA 100 ppm (410 mg/m^3); ACGIH TLV: TWA 5 ppm (20 mg/m^3).

Symptoms of Exposure: Vapor inhalation may cause muscle weakness, weakness in ankles and hand and difficulty in grasping objects (Patnaik, 1992).

Toxicity: Acute oral LD$_{50}$ for rats 2,590 mg/kg, mice 2,430 mg/kg (RTECS, 1985).

Uses: Solvent for paints, varnishes, nitrocellulose lacquers, oils, fats and waxes; denaturant for ethyl alcohol; in organic synthesis.

1-HEXENE

Synonyms: Butylethylene; Hex-1-ene; Hexylene; UN 2370.

$$H-\overset{\overset{\displaystyle H}{|}}{\underset{\underset{\displaystyle H}{|}}{C}}-\overset{\overset{\displaystyle H}{|}}{\underset{\underset{\displaystyle H}{|}}{C}}-\overset{\overset{\displaystyle H}{|}}{\underset{\underset{\displaystyle H}{|}}{C}}-\overset{\overset{\displaystyle H}{|}}{\underset{\underset{\displaystyle H}{|}}{C}}-\overset{\overset{\displaystyle H}{|}}{C}=C\overset{\displaystyle H}{\underset{\displaystyle H}{}}$$

CAS Registry No.: 592-41-6
DOT: 2370
Molecular formula: C_6H_{12}
Formula weight: 84.16
RTECS: MP6600100

Physical state and color
Colorless liquid.

Melting point (°C):
-139.8 (Weast, 1986)

Boiling point (°C):
63.3 (Weast, 1986)

Density (g/cm^3):
0.67317 at 20/4 °C, 0.66848 at 25/4 °C (Dreisbach, 1959)

Diffusivity in water (10^5 cm^2/sec):
0.77 at 20 °C using method of Hayduk and Laudie (1974).

Dissociation constant:
>14 (Schwarzenbach et al., 1993)

Flash point (°C):
-26.1 (Hawley, 1981)

Heat of fusion (kcal/mol):
2.234 (Dean, 1987)

Henry's law constant (atm·m^3/mol):
0.435 at 25 °C (Hine and Mookerjee, 1975)

Ionization potential (eV):
9.45 ± 0.02 (Franklin et al., 1969)

Soil sorption coefficient, log K_{oc}:
Unavailable because experimental methods for estimation of this parameter for aliphatic hydrocarbons are lacking in the documented literature.

Octanol/water partition coefficient, log K_{ow}:
2.25 (Coates et al., 1985)
2.70 (Hansch and Leo, 1979)
3.39 (Schantz and Martire, 1987)
3.47 (Wasik et al., 1981)

Solubility in organics:
Soluble in alcohol, benzene, chloroform, ether and petroleum (Weast, 1986).

Solubility in water:
49 mg/L at 23 °C (Coates et al., 1985)
50 mg/kg at 25 °C (McAuliffe, 1966)
In 1 mM nitric acid: 0.778, 0.643 and 0.501 mmol/L at 20, 25 and 30 °C, respectively (Natarajan and Venkatachalam, 1972)
82.8 mmol/L at 25.0 °C (Wasik et al., 1981)
55.4 mg/L at 25 °C (Leinonen and Mackay, 1973)

Vapor density:
3.44 g/L at 25 °C, 2.91 (air = 1)

Vapor pressure (mmHg):
176 at 23.7 °C (Forziati et al., 1950)
186.0 at 25 °C (Wilhoit and Zwolinski, 1971)

Environmental Fate
   Biological. Biooxidation of 1-hexene may occur yielding 5-hexen-1-ol, which may further oxidize to give 5-hexenoic acid (Dugan, 1972). Washed cell suspensions of bacteria belonging to the genera *Mycobacterium*, *Nocardia*, *Xanthobacter* and *Pseudomonas* and growing on selected alkenes metabolized 1-hexene to 1,2-epoxyhexane (van Ginkel et al., 1987).

Uses: Synthesis of perfumes, flavors, dyes and resins; polymer modifier; organic synthesis.

sec-HEXYL ACETATE

Synonyms: Acetic acid 1,3-dimethylbutyl ester; 1,3-Dimethylbutyl acetate; *sec*-Hexyl acetate; MAAC; Methylamyl acetate; Methylisoamyl acetate; Methyliso-butylcarbinol acetate; **4-Methyl-2-pentanol acetate**; 4-Methyl-2-pentyl acetate; UN 1233.

$$
\begin{array}{cccccc}
\text{H} & \text{H} & \text{H} & \text{H} & \text{H} & \text{H} \\
| & | & | & | & | & | \\
\text{H}-\text{C}-&\text{C}-&\text{C}-&\text{C}-&\text{C}-&\text{C}-\text{H} \\
| & | & | & | & | & | \\
\text{H} & \text{H} & \text{H} & \text{H} & \text{O} & \text{H}
\end{array}
$$

CAS Registry No.: 108-84-9
DOT: 1233
DOT label: Combustible liquid
Molecular formula: $C_8H_{16}O_2$
Formula weight: 144.21
RTECS: SA7525000

Physical state, color and odor
Colorless liquid with a fruity odor.

Melting point (°C):
-64.4 (NIOSH, 1994)

Boiling point (°C):
158 (Weast, 1986)
146.3 (Hawley, 1981)

Density (g/cm³):
0.8658 at 20/4 °C (Weast, 1986)

Diffusivity in water (10^5 cm²/sec):
0.65 at 20 °C using method of Hayduk and Laudie (1974).

Flash point (°C):
45 (NIOSH, 1994)

Henry's law constant (10^3 atm·m³/mol):
9.5 at 20 °C (approximate - calculated from water solubility and vapor pressure)

Soil sorption coefficient, log K_{oc}:
Unavailable because experimental methods for estimation of this parameter for aliphatic esters are lacking in the documented literature.

Octanol/water partition coefficient, log K_{ow}:
3.37 using method of Hansch et al. (1968).

Solubility in organics:
Soluble in alcohol and ether (Weast, 1986).

Solubility in water:
0.008 wt % at 20 °C (NIOSH, 1994)

Vapor density:
5.89 g/L at 25 °C, 4.98 (air = 1)

Vapor pressure (mmHg at 20 °C):
4 (NIOSH, 1994)
3 (Hawley, 1981)

Environmental Fate
   Chemical/Physical. Slowly hydrolyzes in water forming 4-methyl-2-pentanol and acetic acid.

Exposure Limits: NIOSH REL: TWA 50 ppm (300 mg/m^3), IDLH 500 ppm; OSHA PEL: TWA 50 ppm.

Toxicity: Acute oral LD_{50} for rats 6,160 mg/kg (RTECS, 1985).

Uses: Solvent for nitrocellulose and other lacquers.

HYDROQUINONE

Synonyms: Arctuvin; *p*-Benzenediol; **1,4-Benzenediol**; Benzohydroquinone; Benzoquinol; Black and white bleaching cream; Dihydroxybenzene; *p*-Dihydroxybenzene; 1,4-Dihydroxybenzene; *p*-Dioxobenzene; Eldopaque; Eldoquin; Hydroquinol; Hydroquinole; α-Hydroquinone; *p*-Hydroquinone; 4-Hydroxyphenol; *p*-Hydroxyphenol; NCI-C55834; Quinol; β-Quinol; Quinone; Tecquinol; Tenox HQ; Tequinol; UN 2662; USAF EK-356.

CAS Registry No.: 123-31-9
DOT: 2662
Molecular formula: $C_6H_6O_2$
Formula weight: 110.11
RTECS: MX3500000

Physical state, color and odor
Colorless to pale brown, odorless, hexagonal crystals.

Melting point (°C):
173–174 (Weast, 1986)
170–171 (Windholz et al., 1983)
174.00 (Martin et al., 1979)

Boiling point (°C):
285 at 750 mmHg (Weast, 1986)
285–287 (Windholz et al., 1983)

Density (g/cm³):
1.328 at 15/4 °C (Weast, 1986)
1.358 at 20/4 °C (Sax and Lewis, 1987)

Diffusivity in water (10^5 cm²/sec):
0.83 at 20 °C using method of Hayduk and Laudie (1974).

Dissociation constant (25 °C):
pK_1 = 10.0, pK_2 = 12.0 (Dean, 1973)

Flash point (°C):
166 (molten, NIOSH, 1994)

Henry's law constant (10^9 atm·m^3/mol):
<2.07 at 20 °C (approximate - calculated from water solubility and vapor pressure)

Ionization potential (eV):
7.95 (NIOSH, 1994)

Bioconcentration factor, log BCF:
1.54 (algae, Geyer et al., 1984)
2.94 (activated sludge), 1.60 (algae), 1.60 (golden ide) (Freitag et al., 1985)

Soil sorption coefficient, log K_{oc}:
0.98 using method of Kenaga and Goring (1980).

Octanol/water partition coefficient, log K_{ow}:
0.50, 0.59 (Leo et al., 1971)
0.55 (Geyer et al., 1984)
0.54 (Nahum and Horvath, 1980)
0.46 (Janini and Attari, 1983)

Solubility in organics:
Soluble in acetone, alcohol and ether (Weast, 1986). Slightly soluble in benzene (Windholz et al., 1983).

Solubility in water (mg/L):
59,000 at 15 °C, 70,000 at 25 °C, 94,000 at 28 °C (Verschueren, 1983)

Vapor pressure (mmHg):
<10^{-3} at 20 °C (Sax and Lewis, 1987)
4 at 150 °C (Verschueren, 1983)

Environmental Fate

Biological. In activated sludge, 7.5% mineralized to carbon dioxide after 5 days (Freitag et al., 1985). Under methanogenic conditions, inocula from a municipal sewage treatment plant digester degraded hydroquinone to phenol prior to being mineralized to carbon dioxide and methane (Young and Rivera, 1985). In various pure cultures, hydroquinone degraded to the following intermediates: benzoquinone, 2-hydroxy-1,4-benzoquinone and β-ketoadipic acid. Hydroquinone also degraded in activated sludge but no products were identified (Harbison and Belly, 1982).

Photolytic. A carbon dioxide yield of 53.7% was achieved when hydroquinone adsorbed on silica gel was irradiated with light (λ >290 nm) for 17 hours (Freitag et al., 1985).

Chemical/Physical. Ozonolysis products reported are *p*-quinone and dibasic acids (Verschueren, 1983). Moussavi (1979) studied the autoxidation of hydro-quinone in slightly alkaline (pH 7-9) aqueous solutions at room temperature. The oxidation of hydroquinone by oxygen followed first-order kinetics that yielded hydrogen peroxide and p-quinone as products. At pH values of 7.0, 8.0 and 9.0, the calculated half-lives of this reaction were 111, 41 and 0.84 hours, respectively (Moussavi, 1979).

Chlorine dioxide reacted with hydroquinone in an aqueous solution forming p-benzoquinone (Wajon et al., 1982). Kanno et al. (1982) studied the aqueous reaction of hydroquinone and other substituted aromatic hydrocarbons (aniline, toluidine, 1- and 2-naphthylamine, phenol, cresol, pyrocatechol, resorcinol and 1-naphthol) with hypochlorous acid in the presence of ammonium ion. They reported that the aromatic ring was not chlorinated as expected but was cleaved by chloramine forming cyanogen chloride. As the pH was lowered, the amount of cyanogen chloride formed increased (Kanno et al., 1982).

Exposure Limits (mg/m³): NIOSH REL: 15-minute ceiling 2, IDLH 50; OSHA PEL: TWA 2; ACGIH TLV: TWA 2.

Toxicity: Acute oral LD_{50} for cats 70 mg/kg, dogs 200 mg/kg, guinea pigs 550 mg/kg, mice 245 mg/kg, pigeons 300 mg/kg, rats 320 mg/kg (RTECS, 1985).

Uses: Antioxidant; photographic reducer and developer for black and white film; determination of phosphate; dye intermediate; medicine; in monomeric liquids to prevent polymerization; stabilizer in paints and varnishes; motor fuels and oils.

INDAN

Synonyms: 2,3-Dihydroindene; **2,3-Dihydro-1*H*-indene**; Hydrindene.

CAS Registry No.: 496-11-7
Molecular formula: C_9H_{10}
Formula weight: 118.18
RTECS: NK3750000

Physical state
Liquid.

Melting point (°C):
-51.4 (Weast, 1986)
-51.0 (Bjørrseth, 1983)

Boiling point (°C):
178 (Weast, 1986)
176.5 (Hawley, 1981)

Density (g/cm³):
0.9639 at 20/4 °C (Weast, 1986)

Diffusivity in water (10^5 cm²/sec):
0.71 at 20 °C using method of Hayduk and Laudie (1974).

Dissociation constant:
>14 (Schwarzenbach et al., 1993)

Flash point (°C):
50 (Aldrich, 1988)

Henry's law constant (10^3 atm·m³/mol):
2.14 at 25 °C (approximate - calculated from water solubility and vapor pressure)

Soil sorption coefficient, log K_{oc}:
2.48 using method of Karickhoff et al. (1979).

Octanol/water partition coefficient, log K$_{ow}$:
2.38 using method of Hansch et al. (1968).

Solubility in organics:
Soluble in alcohol and ether (Weast, 1986).

Solubility in water:
88.9 mg/kg at 25 °C (Price, 1976)
109.1 mg/kg at 25 °C (Mackay and Shiu, 1977)
6.93 mg/L (water soluble fraction of a 15-component simulated jet fuel mixture
    (JP8) containing 7.5 wt % indan, MacIntyre and deFur, 1985)

Vapor density:
4.83 g/L at 25 °C, 4.08 (air = 1)

Vapor pressure (mmHg):
1.5 at 25 °C (extrapolated, Ambrose and Sprake, 1975)

Use: Organic synthesis.

INDENO[1,2,3-*cd*]PYRENE

Synonyms: Indenopyren; IP; 1,10-(*o*-Phenylene)pyrene; 2,3-Phenylenepyrene; 2,3-*o*-Phenylenepyrene; 3,4-(*o*-Phenylene)pyrene; 2,3-Phenylene-*o*-pyrene; 1,10-(1,2-Phenylene)pyrene; *o*-Phenylpyrene; RCRA waste number U137.

CAS Registry No.: 193-39-5
Molecular formula: $C_{22}H_{12}$
Formula weight: 276.34
RTECS: NK9300000

Physical state
Solid.

Melting point (°C):
160-163 (Verschueren, 1983)

Boiling point (°C):
536 (Verschueren, 1983)

Diffusivity in water (10^5 cm^2/sec):
0.48 at 20 °C using method of Hayduk and Laudie (1974).

Dissociation constant:
>15 (Christensen et al., 1975)

Henry's law constant (10^7 atm·m^3/mol):
1.78, 2.86, 5.63, 6.02, 7.60 and 10.36 at 10.0, 20.0, 35.0, 40.1, 45.0 and 55.0 °C, respectively (ten Hulscher et al., 1992)

Soil sorption coefficient, log K_{oc}:
7.49 using method of Karickhoff et al. (1979).

Octanol/water partition coefficient, log K_{ow}:
5.97 (Sims et al., 1988)

Solubility in organics:
Soluble in most solvents (U.S. EPA, 1985).

Solubility in water:
62 μg/L (Sims et al., 1988)

Vapor pressure (mmHg):
1.01×10^{-10} at 25 °C (McVeety and Hites, 1988)

Exposure Limits: Potential occupational carcinogen. No individual standards have been set, however, as a constituent in coal tar pitch volatiles, the following exposure limits have been established (mg/m^3): NIOSH REL: TWA 0.1 (cyclohex-ane-extractable fraction), IDLH 80; OSHA PEL: TWA 0.2 (benzene-soluble fraction); ACGIH TLV: TWA 0.2 (benzene solubles).

Drinking Water Standard (proposed): MCLG: zero; MCL: 0.4 μg/L (U.S. EPA, 1994).

Uses: Derived from industrial and experimental coal gasification operations where the maximum concentration detected in coal tar streams was 1.7 mg/m^3 (Cleland, 1981).

INDOLE

Synonyms: 1-Azaindene; 1-Benzazole; Benzopyrrole; Benzo[*b*]pyrrole; 1-Benzo[*b*]pyrrole; 2,3-Benzopyrrole; **1*H*-Indole**; Ketole.

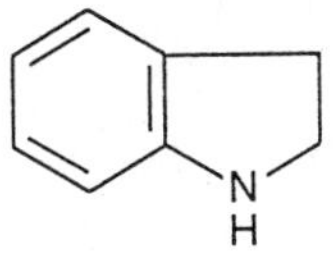

CAS Registry No.: 120-72-9
Molecular formula: C_8H_7N
Formula weight: 117.15
RTECS: NL2450000

Physical state, color and odor
Colorless to yellow scales with an unpleasant odor. Turns red on exposure to light and air.

Melting point (°C):
52.5 (Weast, 1986)

Boiling point (°C):
254 (Weast, 1986)

Density (g/cm^3):
1.22 (Weast, 1986)
1.643 (Dean, 1987)

Diffusivity in water (10^5 cm^2/sec):
0.76 at 20 °C using method of Hayduk and Laudie (1974).

Flash point (°C):
>110 (Aldrich, 1988)

Soil sorption coefficient, log K_{oc}:
1.69 using method of Kenaga and Goring (1980).

Octanol/water partition coefficient, log K_{ow}:
2.00, 2.25 (Leo et al., 1971)
2.14 (Hansch and Anderson, 1967)
1.81 (Eadsforth, 1986)

2.16 (Garst and Wilson, 1984)
2.28 (Rogers and Cammarata, 1969)

Solubility in organics:
Soluble in alcohol, benzene, ether and lignoin (Weast, 1986).

Solubility in water:
3,558 mg/kg at 25 °C (Price, 1976)

Environmental Fate
Biological. In 9% anaerobic municipal sludge, indole degraded to 1,3-dihydro-2H-indol-2-one (oxindole), which further degraded to methane and carbon dioxide (Berry et al., 1987).

Chemical/Physical. The aqueous chlorination of indole by hypochlorite/hypochlorous acid, chlorine dioxide and chloramines produced oxindole, isatin and possibly 3-chloroindole (Lin and Carlson, 1984).

Toxicity: Acute oral LD_{50} for rats 1,000 mg/kg (RTECS, 1985).

Uses: Chemical reagent; medicine; flavoring agent; perfumery; constituent of coal tar.

INDOLINE

Synonyms: 2,3-Dihydroindole; **2,3-Dihydro-1*H*-indole**.

CAS Registry No.: 496-15-1
Molecular formula: C_8H_9N
Formula weight: 119.17

Physical state and color
Dark brown liquid.

Boiling point (°C):
228–230 (Weast, 1986)
220–221 (Aldrich, 1988)

Density (g/cm^3):
1.069 at 20/4 °C (Weast, 1986)

Diffusivity in water (10^5 cm^2/sec):
0.83 at 20 °C using method of Hayduk and Laudie (1974).

Soil sorption coefficient, log K_{oc}:
1.42 using method of Kenaga and Goring (1980).

Octanol/water partition coefficient, log K_{ow}:
0.16 using method of Kenaga and Goring (1980).

Solubility in organics:
Soluble in acetone, benzene and ether (Weast, 1986).

Solubility in water:
10,800 mg/kg at 25 °C (Price, 1976)

Uses: Organic synthesis.

1-IODOPROPANE

Synonyms: Propyl iodide; *n*-Propyl iodide.

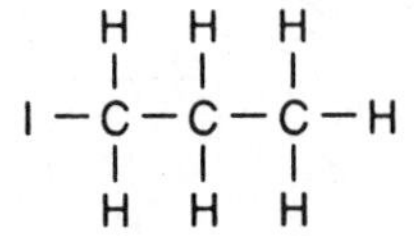

$$H \quad H \quad H$$
$$| \quad\; | \quad\; |$$
$$I - C - C - C - H$$
$$| \quad\; | \quad\; |$$
$$H \quad H \quad H$$

CAS Registry No.: 107-08-4
DOT: 2392
DOT label: Combustible liquid
Molecular formula: C_3H_7I
Formula weight: 169.99
RTECS: TZ4100000

Physical state and color
Colorless liquid.

Melting point (°C):
-101 (Weast, 1986)
≈ -98 (Windholz et al., 1983)

Boiling point (°C):
102.4 (Weast, 1986)

Density (g/cm³):
1.7489 at 20/4 °C (Weast, 1986)

Diffusivity in water (10^5 cm²/sec):
0.90 at 20 °C using method of Hayduk and Laudie (1974).

Flash point (°C):
None (Dean, 1987).

Henry's law constant (10^3 atm·m³/mol):
9.09 at 25 °C (Hine and Mookerjee, 1975)

Ionization potential (eV):
9.26 ± 0.01 (Franklin et al., 1969)

Soil sorption coefficient, log K_{oc}:
2.16 using method of Chiou et al. (1979).

Octanol/water partition coefficient, log K_{ow}:
2.49 using method of Hansch et al. (1968).

Solubility in organics:
Soluble in benzene and chloroform (Weast, 1986). Miscible with alcohol and ether (Windholz et al., 1983).

Solubility in water:
0.1065 wt % at 23.5 °C (Schwarz, 1980)
1,040 mg/kg at 30 °C (Gross and Saylor, 1931)
5.1 mmol/L at 20 °C (Fühner, 1924)
1.17 x 10^{-4} at 25 °C (mole fraction, Li et al., 1993)

Vapor density:
6.95 g/L at 25 °C, 5.87 (air = 1)

Vapor pressure (mmHg):
43.1 at 25 °C (Abraham, 1984)

Environmental Fate
Biological. A strain of *Acinetobacter* sp. isolated from activated sludge degraded 1-iodopropane to 1-propanol and iodide ions (Janssen et al., 1987).
Chemical/Physical. Slowly hydrolyzes in water forming 1-propanol and hydro-iodic acid.

Toxicity: LD_{50} (inhalation) for rats 73,000 mg/m^3/30-minute (RTECS, 1985).

Uses: Organic synthesis.

ISOAMYL ACETATE

Synonyms: Acetic acid isopentyl ester; Acetic acid 3-methylbutyl ester; Banana oil; Isoamyl ethanoate; Isopentyl acetate; Isopentyl alcohol acetate; **3-Methyl-1-butanol acetate;** 3-Methylbutyl acetate; 3-Methyl-1-butyl acetate; 3-Methylbutyl ethanoate; Pear oil.

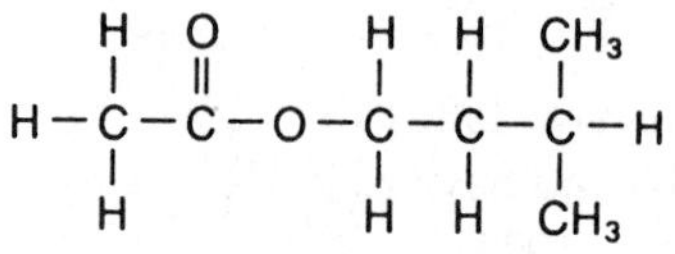

<pre>
      H   O   H   H CH₃
      | || | | |
 H — C — C — O — C — C — C — H
      | | | |
      H   H   H CH₃
</pre>

CAS Registry No.: 123-92-2
DOT: 1104
DOT label: Flammable liquid
Molecular formula: $C_7H_{14}O_2$
Formula weight: 130.19
RTECS: NS9800000

Physical state, color and odor
Clear, colorless liquid with a banana or pear-like odor. Odor threshold in air is 7 ppm (Keith and Walters, 1992).

Melting point (°C):
-78.5 (Weast, 1986)

Boiling point (°C):
142 (Weast, 1986)

Density (g/cm³):
0.8670 at 20/4 °C (Weast, 1986)
0.876 at 15/4 °C (Hawley, 1981)

Diffusivity in water (10^5 cm²/sec):
0.70 at 20 °C using method of Hayduk and Laudie (1974).

Flash point (°C):
25 (NIOSH, 1994)
33, 38 (open cup, Windholz et al., 1983)

Lower explosive limit (%):
1.0 at 100 °C (NIOSH, 1994)

586

Upper explosive limit (%):
7.5 (NIOSH, 1994)

Henry's law constant (10^2 atm·m^3/mol):
5.87 at 25 °C (Hine and Mookerjee, 1975)

Soil sorption coefficient, log K_{oc}:
1.95 using method of Chiou et al. (1979).

Octanol/water partition coefficient, log K_{ow}:
2.30 using method of Hansch et al. (1968).

Solubility in organics:
Miscible with alcohol, amyl alcohol, ether and ethyl acetate (Windholz et al., 1983).

Solubility in water (wt %):
0.340 at 0 °C, 0.265 at 9.1 °C, 0.212 at 19.4 °C, 0.208 at 30.3 °C, 0.184 at 39.7 °C, 0.174 at 50.0 °C, 0.152 at 60.1 °C, 0.203 at 70.2 °C, 0.182 at 80.3 °C, 0.205 at 90.7 °C (Stephenson and Stuart, 1986)

Vapor density:
5.32 g/L at 25 °C, 4.49 (air = 1)

Vapor pressure (mmHg):
4 at 20 °C (NIOSH, 1994)

Environmental Fate
  Chemical/Physical. Slowly hydrolyzes in water forming 3-methyl-1-butanol and acetic acid.

Exposure Limits: NIOSH REL: TWA 100 ppm (525 mg/m^3), IDLH 1,000 ppm; OSHA PEL: TWA 100 ppm; ACGIH TLV: TWA 100 ppm.

Symptoms of Exposure: Vapors may cause irritation to the eyes, nose and throat, fatigue, increased pulse rate and narcosis (Patnaik, 1992).

Toxicity: Acute oral LD$_{50}$ for rabbits 7.422 g/kg, rats 16.6 g/kg (RTECS, 1985).

Uses: Artificial pear flavor in mineral waters and syrups; dyeing and finishing textiles; solvent for tannins, lacquers, nitrocellulose, camphor, oil colors and celluloid; manufacturing of artificial leather, pearls or silk, photographic films, swelling bath sponges; celluloid cements, waterproof varnishes; bronzing liquids; metallic paints; perfumery; masking undesirable odors.

ISOAMYL ALCOHOL

Synonyms: Fermentation amyl alcohol; Fusel oil; Isoamylol; Isobutyl carbinol; Isopentanol; Isopentyl alcohol; 2-Methyl-4-butanol; 3-Methylbutanol; 3-Methyl-butan-1-ol; **3-Methyl-1-butanol**; Primary isoamyl alcohol; Primary isobutyl alcohol; UN 1105.

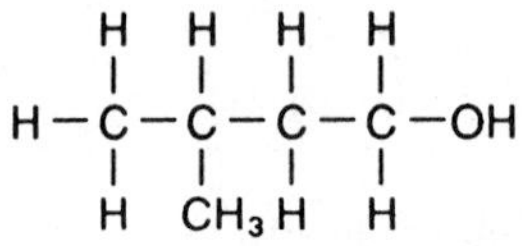

$$\begin{array}{ccccccc} & H & & H & & H & & H \\ & | & & | & & | & & | \\ H - & C & - & C & - & C & - & C & -OH \\ & | & & | & & | & & | \\ & H & & CH_3 & & H & & H \end{array}$$

CAS Registry No.: 123-51-3
DOT: 1105
DOT label: Combustible liquid
Molecular formula: $C_5H_{12}O$
Formula weight: 88.15
RTECS: EL5425000

Physical state, color and odor
Clear, colorless liquid with a pungent odor. Odor threshold in air is 22 ppm (Amoore and Hautala, 1983).

Melting point (°C):
-118.2 (NIOSH, 1994)

Boiling point (°C):
128.5 at 750 mmHg (Weast, 1986)
132 (Windholz et al., 1983)

Density (g/cm³):
0.8092 at 20/4 °C (Weast, 1986)
0.813 at 15/4 °C (Hawley, 1981)
0.8088 at 25/4 °C, 0.8046 at 30/4 °C, 0.7968 at 40/4 °C (Abraham et al., 1971)

Diffusivity in water (10^5 cm²/sec):
0.84 at 20 °C using method of Hayduk and Laudie (1974).

Dissociation constant:
>14 (Schwarzenbach et al., 1993)

Flash point (°C):
43.1 (NIOSH, 1994)
45, 55 (open cup, Windholz et al., 1983)

Lower explosive limit (%):
1.2 (NIOSH, 1994)

Upper explosive limit (%):
9.0 at 100 °C (NIOSH, 1994)

Henry's law constant (10^6 atm·m^3/mol):
8.89 at 20 °C (approximate – calculated from water solubility and vapor pressure)

Interfacial tension with water (dyn/cm at 25 °C):
4.8 (Donahue and Bartell, 1952)

Soil sorption coefficient, log K_{oc}:
Unavailable because experimental methods for estimation of this parameter for aliphatic esters are lacking in the documented literature.

Octanol/water partition coefficient, log K_{ow}:
1.16 (Leo et al., 1971)
1.42 (Sangster, 1989)

Solubility in organics:
Soluble in acetone (Weast, 1986). Miscible with alcohol, benzene, chloroform, ether, glacial acetic acid, oils and petroleum ether (Windholz et al., 1983).

Solubility in water:
20 g/L at 14 °C (Windholz et al., 1983)
30,000 mg/L at 20 °C, 26,720 mg/L at 22 °C (Verschueren, 1983)
3.18 wt % at 20 °C (Palit, 1947)
0.313 at 18 °C (Fühner, 1924)
In wt %: 3.73 at 0 °C, 3.14 at 10.1 °C, 2.64 at 19.8 °C, 2.29 at 30.2 °C, 2.18 at 40.0 °C, 2.03 at 49.9 °C, 2.19 at 59.8 °C, 2.11 at 70.0 °C, 2.20 at 80.0 °C, 2.27 at 90.0 °C (Stephenson et al., 1984)

Vapor density:
3.60 g/L at 25 °C, 3.04 (air = 1)

Vapor pressure (mmHg):
28 at 20 °C (NIOSH, 1994)
2.3 at 20 °C, 4.8 at 30 °C (Verschueren, 1983)

Exposure Limits: NIOSH REL: TWA 100 ppm (360 mg/m^3), IDLH 500 ppm; OSHA PEL: TWA 100 ppm; ACGIH TLV: TWA 100 ppm, STEL 125 ppm (450 mg/m^3).

Uses: Determining fat content in milk; solvent for alkaloids, fats, oils; manufacturing isovaleric acid, isoamyl or amyl compounds, esters, mercury fulminate, artificial silk, smokeless powders, lacquers, pyroxylin; photographic chemicals; pharmaceutical products; microscopy; in organic synthesis.

ISOBUTYL ACETATE

Synonyms: Acetic acid isobutyl ester; **Acetic acid 2-methylpropyl ester;** 2-Methylpropyl acetate; 2-Methyl-1-propyl acetate; β-Methylpropyl ethanoate; UN 1213.

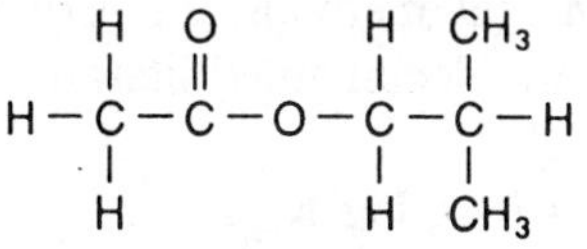

```
        H   O        H   CH₃
        |   ‖        |   |
  H  —  C — C — O — C — C — H
        |            |   |
        H            H   CH₃
```

CAS Registry No.: 110-19-0
DOT: 1213
DOT label: Combustible liquid
Molecular formula: $C_6H_{12}O_2$
Formula weight: 116.16
RTECS: AI4025000

Physical state, color and odor
Colorless liquid with a fruity odor. Odor threshold in air is 0.64 ppm (Amoore and Hautala, 1983).

Melting point (°C):
-98.58 (Weast, 1986)

Boiling point (°C):
117.2 (Weast, 1986)

Density (g/cm³):
0.8712 at 20/4 °C (Weast, 1986)

Diffusivity in water (10^5 cm²/sec):
0.75 at 20 °C using method of Hayduk and Laudie (1974).

Flash point (°C):
17.9 (NIOSH, 1994)

Lower explosive limit (%):
1.3 (NFPA, 1984)

Upper explosive limit (%):
10.5 (NFPA, 1984)

Henry's law constant (10^4 atm·m³/mol):
4.85 at 25 °C (approximate - calculated from water solubility and vapor pressure)

591

Ionization potential (eV):
9.97 (Franklin et al., 1969)

Soil sorption coefficient, log K_{oc}:
Unavailable because experimental methods for estimation of this parameter for aliphatic esters are lacking in the documented literature.

Octanol/water partition coefficient, log K_{ow}:
1.76 using method of Hansch et al. (1968).

Solubility in organics:
Miscible with most organic solvents (Patnaik, 1992).

Solubility in water:
6,300 mg/L at 25 °C (Verschueren, 1983)
658 mmol/L at 20 °C (Fühner, 1924)
In wt % (°C): 1.03 (0), 0.83 (10.0), 0.66 (19.7), 0.61 (29.9), 0.54 (39.7), 0.49 (50.0), 0.57 (60.5), 0.53 (70.1), 0.55 (80.2) (Stephenson and Stuart, 1986)

Vapor density:
4.75 g/L at 25 °C, 4.01 (air = 1)

Vapor pressure (mmHg):
13 at 20 °C (NIOSH, 1994)
10 at 16 °C, 20 at 25 °C (Verschueren, 1983)

Environmental Fate
  Chemical/Physical. Slowly hydrolyzes in water forming 2-methylpropanol and acetic acid.

Exposure Limits: NIOSH REL: TWA 150 ppm (700 mg/m^3), IDLH 1,300 ppm; OSHA PEL: TWA 150 ppm; ACGIH TLV: TWA 150 ppm, STEL 187 ppm (875 mg/m^3).

Symptoms of Exposure: Headache, drowsiness and irritation of upper respiratory tract (Patnaik, 1992).

Toxicity: Acute oral LD_{50} for rabbits 4,763 mg/kg, rats 13,400 mg/kg (RTECS, 1985).

Uses: Solvent for nitrocellulose; in thinners, sealers and topcoat lacquers; flavoring agent; perfumery.

ISOBUTYL ALCOHOL

Synonyms: Fermentation butyl alcohol; 1-Hydroxymethylpropane; IBA; Isobutanol; Isopropylcarbinol; 2-Methylpropanol; 2-Methylpropanol-1; **2-Methyl-1-propanol**; 2-Methyl-1-propan-1-ol; 2-Methylpropyl alcohol; RCRA waste number U140; UN 1212.

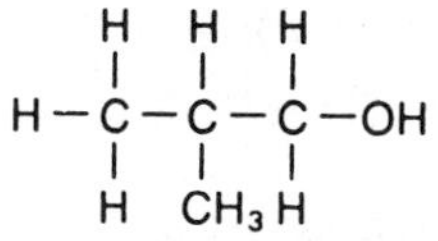

$$H-\overset{\overset{\displaystyle H}{|}}{\underset{\underset{\displaystyle H}{|}}{C}}-\overset{\overset{\displaystyle H}{|}}{\underset{\underset{\displaystyle CH_3}{|}}{C}}-\overset{\overset{\displaystyle H}{|}}{\underset{\underset{\displaystyle H}{|}}{C}}-OH$$

CAS Registry No.: 78-83-1
DOT: 1212
DOT label: Combustible liquid
Molecular formula: $C_4H_{10}O$
Formula weight: 74.12
RTECS: NP9625000

Physical state, color and odor
Colorless, oily liquid with a sweet, musty odor. Burning taste. Odor threshold in air is 1.6 ppm (Amoore and Hautala, 1983).

Melting point (°C):
-72.8 (NIOSH, 1994)

Boiling point (°C):
108.1 (Weast, 1986)
107 (Hawley, 1981)

Density (g/cm³):
0.8018 at 20/4 °C (Weast, 1986)
0.798 at 25/4 °C (Verschueren, 1983)
0.806 at 15/4 °C (Hawley, 1981)

Diffusivity in water (10^5 cm²/sec at 25 °C):
1.08 (Hayduk and Laudie, 1974)

Dissociation constant:
>14 (Schwarzenbach et al., 1993)

Flash point (°C):
28 (NIOSH, 1994)
37 (open cup, Hawley, 1981)

593

Lower explosive limit (%):
1.7 at 51 °C (NIOSH, 1994)

Upper explosive limit (%):
10.6 at 95 °C (NIOSH, 1994)

Henry's law constant (10^6 atm·m^3/mol):
9.79 at 25 °C (Snider and Dawson, 1985)

Interfacial tension with water (dyn/cm at 25 °C):
2.0 (Donahue and Bartell, 1952)

Ionization potential (eV):
10.12 (NIOSH, 1994)

Soil sorption coefficient, log K_{oc}:
Unavailable because experimental methods for estimation of this parameter for aliphatic alcohols are lacking in the documented literature.

Octanol/water partition coefficient, log K_{ow}:
0.65, 0.76, 0.83 (Sangster, 1989)

Solubility in organics:
Miscible with alcohol and ether (Windholz et al., 1983).

Solubility in water:
95,000 mg/L at 18 °C (Verschueren, 1983)
94.87 g/L at 20 °C (Mackay and Yeun, 1983)
85 g/L (Price et al., 1974)
1.351 M at 18 °C (Fühner, 1924)
In wt %: 11.60 at 0 °C, 10.05 at 9.8 °C, 8.84 at 19.7 °C, 7.87 at 30.6 °C, 7.30 at 40.4 °C, 7.08 at 50.1 °C, 7.05 at 60.2 °C, 6.93 at 70.3 °C, 7.31 at 80.5 °C, 7.71 at 90.7 °C (Stephenson and Stuart, 1986)

Vapor density:
3.03 g/L at 25 °C, 2.56 (air = 1)

Vapor pressure (mmHg at 20 °C):
9 (NIOSH, 1994)
10.0 (Mackay and Yeun, 1983)

Exposure Limits: NIOSH REL: TWA 50 ppm (150 mg/m^3), IDLH 1,600 ppm; OSHA PEL: TWA 100 ppm (300 mg/m^3); ACGIH TLV: TWA 50 ppm.

Symptoms of Exposure: Inhalation of vapors may cause eye and throat irritation and headache. Contact with skin may cause cracking (Patnaik, 1992).

Toxicity: Acute oral LD_{50} for rats 2,460 mg/kg (RTECS, 1985).

Uses: Preparation of esters for the flavoring industry; solvent for plastics, textiles, oils, perfumes, paint and varnish removers; intermediate for amino coating resins; liquid chromatography; fluorometric determinations; in organic synthesis.

ISOBUTYLBENZENE

Synonyms: (2-Methylpropyl)benzene; 2-Methyl-1-phenylpropane.

CAS Registry No.: 538-93-2
Molecular formula: $C_{10}H_{14}$
Formula weight: 134.22
RTECS: DA3550000

Physical state and color
Colorless liquid.

Melting point (°C):
-51.5 (Weast, 1986)

Boiling point (°C):
172.8 (Weast, 1986)
170.5 (Windholz et al., 1983)

Density (g/cm³ at 20/4 °C):
0.8532 (Weast, 1986)
0.8673 (Windholz et al., 1983)

Diffusivity in water (10^5 cm²/sec):
0.68 at 20 °C using method of Hayduk and Laudie (1974).

Dissociation constant:
>14 (Schwarzenbach et al., 1993)

Flash point (°C):
60 (Hawley, 1981)
55 (Aldrich, 1988)

Lower explosive limit (%):
0.8 (NFPA, 1984)

Upper explosive limit (%):
6.0 (NFPA, 1984)

Henry's law constant (10^2 atm·m^3/mol):
1.09 at 25 °C (Hine and Mookerjee, 1975)

Soil sorption coefficient, log K_{oc}:
3.90 using method of Karickhoff et al. (1979).

Octanol/water partition coefficient, log K_{ow}:
4.11 (Chiou et al., 1982)

Solubility in organics:
Soluble in acetone, alcohol, benzene, ether and petroleum hydrocarbons (Weast, 1986).

Solubility in water:
10.1 mg/kg at 25 °C (Price, 1976)
33.71 mg/L at 25 °C (Chiou et al., 1982)
70.2 μmol/L in 0.5 M NaCl at 25 °C (Wasik et al., 1984)

Vapor density:
5.49 g/L at 25 °C, 4.63 (air = 1)

Vapor pressure (mmHg):
2.06 at 25 °C (Mackay et al., 1982)

Environmental Fate
  Biological. Oxidation of isobutylbenzene by *Pseudomonas desmolytica* S44B1 and *Pseudomonas convexa* S107B1 yielded 3-isobutylcatechol and (+)-2-hydroxy-8-methyl-6-oxononanoic acid (Jigami et al., 1975).

Uses: Perfume synthesis; flavoring; pharmaceutical intermediate.

Drinking Water Standard: As of May 1994, no MCLGs or MCLs have been proposed although isophorone has been listed for regulation (U.S. EPA, 1994).

Uses: Solvent for paints, tin coatings, agricultural chemicals and synthetic resins; excellent solvent for vinyl resins, cellulose esters and ethers; pesticides; storing lacquers; pesticide manufacturing; intermediate in the manufacture of 3,5-xylenol, 2,3,5-trimethylcyclohexanol and 3,5-dimethylaniline.

ISOPROPYL ACETATE

Synonyms: Acetic acid isopropyl ester; **Acetic acid 1-methylethyl ester;** 2-Acetoxypropane; 2-Propyl acetate; UN 1220.

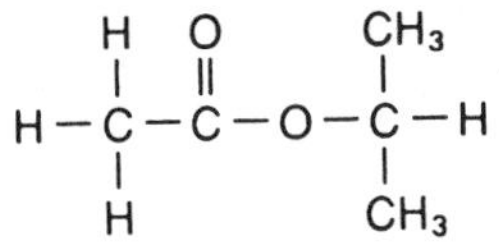

$$\begin{array}{ccccccc}
 & H & O & & CH_3 & & \\
 & | & \| & & | & & \\
H- & C- & C- & O- & C- & H \\
 & | & & & | & & \\
 & H & & & CH_3 & &
\end{array}$$

CAS Registry No.: 108-21-4
DOT: 1993
DOT label: Combustible liquid
Molecular formula: $C_5H_{10}O_2$
Formula weight: 102.13
RTECS: AI4930000

Physical state, color and odor
Colorless liquid with an aromatic odor. Odor threshold in air is 2.7 ppm (Amoore and Hautala, 1983).

Melting point (°C):
-73.4 (Weast, 1986)
-69.3 (Sax and Lewis, 1987)

Boiling point (°C):
90 (Weast, 1986)

Density (g/cm³):
0.8718 at 20/4 °C (Weast, 1986)
0.877 at 16/4 °C (Verschueren, 1983)
0.8690 at 25/4 °C (Hawley, 1981)

Diffusivity in water (10^5 cm²/sec):
0.80 at 20 °C using method of Hayduk and Laudie (1974).

Flash point (°C):
2.2 (NIOSH, 1994)
2-4 (open cup, Windholz et al., 1983)

Lower explosive limit (%):
1.8 at 38.1 °C (NIOSH, 1994)

Upper explosive limit (%):
8.0 (NIOSH, 1994)

601

ISOPROPYLAMINE

Synonyms: 2-Aminopropane; 1-Methylethylamine; Monoisopropylamine; **2-Propanamine;** *sec*-Propylamine; 2-Propylamine; UN 1221.

$$\begin{array}{ccc}
\text{H} & \text{CH}_3 & \quad \text{H} \\
| & | & \diagup \\
\text{H}-\text{C}-\text{C}-\text{N} & \\
| & | & \diagdown \\
\text{H} & \text{H} & \quad \text{H}
\end{array}$$

CAS Registry No.: 75-31-0
DOT: 1221
DOT label: Flammable/combustible liquid
Molecular formula: C_3H_9N
Formula weight: 59.11
RTECS: NT8400000

Physical state, color and odor
Colorless liquid with a penetrating, ammonia-like odor. Odor threshold in air is 1.2 ppm (Amoore and Hautala, 1983).

Melting point (°C):
-95.2 (Weast, 1986)
-101 (Windholz et al., 1983)

Boiling point (°C):
32.4 (Weast, 1986)
33-34 (Windholz et al., 1983)

Density (g/cm³):
0.6891 at 20/4 °C (Weast, 1986)
0.686 at 25/4 °C (Dean, 1987)
0.694 at 15/4 °C (Sax and Lewis, 1987)

Diffusivity in water (10^5 cm²/sec):
0.97 at 20 °C using method of Hayduk and Laudie (1974).

Dissociation constant:
10.53 at 25 °C (Dean, 1973)

Flash point (°C):
-37.5 (open cup, NIOSH, 1994)
-26 (open cup, Windholz et al., 1983)

Ionization potential (eV):
8.72 ± 0.03 (Franklin et al., 1969)
8.86 (Gibson et al., 1977)

Soil sorption coefficient, log K_{oc}:
Unavailable because experimental methods for estimation of this parameter for aliphatic amines are lacking in the documented literature. However, its high solubility in water and low K_{ow} suggest its adsorption to soil will be nominal (Lyman et al., 1982).

Octanol/water partition coefficient, log K_{ow}:
0.26 (Sangster, 1989)
-0.03 (Leo et al., 1971)

Solubility in organics:
Miscible with alcohol and ether (Windholz et al., 1983).

Solubility in water:
Miscible (NIOSH, 1994).

Vapor density:
2.42 g/L at 25 °C, 2.04 (air = 1)

Vapor pressure (mmHg):
460 at 20 °C (Verschueren, 1983)

Environmental Fate
   Chemical/Physical. Releases toxic nitrogen oxides when heated to decomposition (Sax and Lewis, 1987). Forms water-soluble salts with acids.

Exposure Limits: NIOSH REL: IDLH 750 ppm; OSHA PEL: TWA 5 ppm (12 mg/m^3).

Symptoms of Exposure: Strong irritation to the eyes, skin, throat and respiratory system; pulmonary edema. Skin contact may cause dermatitis and skin burns (Patnaik, 1992).

Toxicity: Acute oral LD_{50} for guinea pigs 2,700 mg/kg, mice 2,200 mg/kg, rats 820 mg/kg, rabbits 3,200 mg/kg (RTECS, 1985).

Uses: Intermediate in the synthesis of rubber accelerators, dyes, pharmaceuticals, insecticides, bactericides, textiles and surface-active agents; solvent; dehairing agent; solubilizer for 2,4-D.

1956)
170 mg/kg at 25 °C (Stearns et al., 1947)
9.80 x 10^{-6} at 25 °C (mole fraction, Li et al., 1993)

Vapor density:
4.91 g/L at 25 °C, 4.15 (air = 1)

Vapor pressure (mmHg):
3.2 at 20 °C (Verschueren, 1983)
4.6 at 25 °C (Mackay et al., 1982)

Environmental Fate

Biological. When isopropylbenzene was incubated with *Pseudomonas putida*, the substrate was converted to *ortho*-dihydroxy compounds in which the isopropyl part of the compound remained intact (Gibson, 1968). Oxidation of isopropylbenzene by *Pseudomonas desmolytica* S44B1 and *Pseudomonas convexa* S107B1 yielded 3-isopropylcatechol and a ring fission product, (+)-2-hydroxy-7-methyl-6-oxoocta-noic acid (Jigami et al., 1975).

Photolytic. Major products reported from the photooxidation of isopropyl-benzene with nitrogen oxides include nitric acid and benzaldehyde (Altshuller, 1983). An *n*-hexane solution containing isopropylbenzene and spread as a thin film (4 mm) on cold water (10 °C) was irradiated by a mercury medium pressure lamp. In 3 hours, 22% of the applied isopropylbenzene photooxidized into α,α-dimeth-ylbenzyl alcohol, 2-phenylpropionaldehyde and allylbenzene (Moza and Feicht, 1989).

Exposure Limits: NIOSH REL: TWA 50 ppm (245 mg/m^3), IDLH 900 ppm; OSHA PEL: TWA 50 ppm.

Symptoms of Exposure: Vapors may cause irritation to the eyes, skin and upper respiratory system. Narcotic at high concentrations (Patnaik, 1992).

Toxicity: Acute oral LD$_{50}$ for rats 1,400 mg/kg (RTECS, 1985).

Drinking Water Standard: As of May 1994, no MCLGs or MCLs have been proposed although isopropylbenzene has been listed for regulation (U.S. EPA, 1994).

Uses: Manufacture of acetone, acetophenone, diisopropylbenzene, α-methyl-styrene and phenol, polymerization catalysts; constituent of motor fuel, asphalt and naphtha; catalyst for acrylic and polyester-type resins; octane booster for gasoline; solvent.

ISOPROPYL ETHER

Synonyms: Diisopropyl ether; Diisopropyl oxide; DIPE; IPE; 2-Isopropoxy-propane; **2,2′-Oxybis(propane)**; UN 1159.

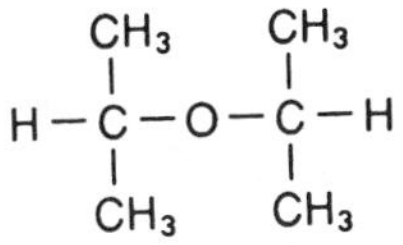

$$\begin{array}{ccc} CH_3 & & CH_3 \\ | & & | \\ H-C-O-C-H \\ | & & | \\ CH_3 & & CH_3 \end{array}$$

CAS Registry No.: 108-20-3
DOT: 1159
DOT label: Flammable liquid
Molecular formula: $C_6H_{14}O$
Formula weight: 102.18
RTECS: TZ5425000

Physical state, color and odor
Colorless liquid with a penetrating, sweet, ether-like odor. Odor threshold in air is 17 ppb (Amoore and Hautala, 1983).

Melting point (°C):
-85.9 (Weast, 1986)
-60 (Verschueren, 1983)
-85.5 (Riddick et al., 1986)

Boiling point (°C):
68 (Weast, 1986)
68.34 (Boublik et al., 1984)

Density (g/cm³):
0.7241 at 20/4 °C (Weast, 1986)
0.71854 at 25/4 °C (Riddick et al., 1986)

Diffusivity in water (10^5 cm²/sec):
0.72 at 20 °C using method of Hayduk and Laudie (1974).

Flash point (°C):
-28 (NIOSH, 1994)
-9 (open cup, Windholz et al., 1983)

Lower explosive limit (%):
1.4 (NIOSH, 1994)

609

KEPONE

Synonyms: Chlordecone; CIBA 8514; Compound 1189; 1,2,3,5,6,7,8,9,10,10-Decachloro[5.2.1.0^{2,6}.0^{3,9}.0^{5,8}]decano-4-one; Decachloroketone; Decachloro-1,3,4-metheno-2H-cyclobuta[cd]pentalen-2-one; Decachlorooctahydrokepone-2-one; Decachlorooctahydro-1,3,4-metheno-2H-cyclobuta[cd]pentalen-2-one; **1,1a,3,3a,4,5,5a,5b,6-Decachlorooctahydro-1,3,4-metheno-2H-cyclobuta[cd]pentalen-2-one;** Decachloropentacyclo[5.2.1.0^{2,6}.0^{3,9}.0^{5,8}]decan-3-one; Decachloropentacyclo[5.2.1.0^{2,6}.0^{4,10}.0^{5,9}]decan-3-one; Decachlorotetracyclodecanone; Decachlorotetrahydro-4,7-methanoindeneone; ENT 16391; GC-1189; General chemicals 1189; Merex; NA 2761; NCI-C00191; RCRA waste number U142.

CAS Registry No.: 143-50-0
DOT: 2761
DOT label: Poison
Molecular formula: $C_{10}Cl_{10}O$
Formula weight: 490.68
RTECS: PC8575000

Physical state, color and odor
Colorless to tan, odorless, crystalline solid.

Melting point (°C):
Sublimes at 352.8 (NIOSH, 1994).

Boiling point (°C):
Decomposes at 350 (Windholz et al., 1983).

Flash point (°C):
Noncombustible solid (NIOSH, 1994)

Henry's law constant (10^2 atm·m^3/mol):
3.11 at 25 °C (approximate - calculated from water solubility and vapor pressure)

Bioconcentration factor, log BCF:
4.00 (activated sludge), 2.65 (algae), 2.76 (golden ide) (Freitag et al., 1985)

Soil sorption coefficient, log K_{oc}:
4.74 (calculated, Mercer et al., 1990)

Octanol/water partition coefficient, log K_{ow}:
4.07 using method of Kenaga and Goring (1980).

Solubility in organics:
Soluble in acetic acid, alcohols and ketones (Windholz et al., 1983).

Solubility in water (mg/L):
7.600 at 24 °C (Hollifield, 1979)
2.7 at 20-25 °C (Kilzer et al., 1979)

Vapor pressure (mmHg):
2.25×10^{-7} at 25 °C (Kilzer et al., 1979)

Environmental Fate
  Photolytic. Kepone-contaminated soils from a site in Hopewell, VA were analyzed by GC/MS. 8-Chloro and 9-chloro homologs identified suggested these were photodegradation products of kepone (Borsetti and Roach, 1978). Products identified from the photolysis of kepone in cyclohexane were 1,2,3,4,6,7,9,10,10-nonachloro-5,5-dihydroxypentacyclo[$5.3.0.0^{2,6}.0^{3,9}.0^{4,8}$]decane for the hydrate and 1,2,3,4,6,7,9,10,10-nonachloro-5,5-dimethoxypentacyclo[$5.3.0.0^{2,6}.0^{3,9}.0^{4,8}$]decane (Alley et al., 1974).
  Chemical/Physical. Readily reacts with moisture forming hydrates (Hollifield, 1979). Decomposes at 350 °C (Windholz et al., 1983), probably emitting toxic chlorine fumes.

Exposure Limits: Potential occupational carcinogen. NIOSH REL: TWA 1 $\mu g/m^3$.

Symptoms of Exposure: May cause tremors (NIOSH, 1994).

Toxicity: LC_{50} (96-hour) for trout 0.02 ppm, bluegill sunfish 0.051, rainbow trout 0.036 ppm (Verschueren, 1983); LC_{50} (48-hour) for sunfish 0.27 ppm, trout 38 ppb (Verschueren, 1983); LC_{50} (24-hour) for bluegill sunfish 257 ppb, trout 66 ppb, rainbow trout 156 ppb (Verschueren, 1983); acute oral LD_{50} for dogs 250 mg/kg, quail 237 mg/kg, rats 95 mg/kg, rabbits 65 mg/kg (RTECS, 1985).

Uses: Not commercially produced in the United States. Insecticide; fungicide.

3.66 (Travis and Arms, 1988)
3.72 (Kurihara et al., 1973)
3.30 (Geyer et al., 1987)
3.20 (Geyer et al., 1984)
3.57 (Kishi and Hashimoto, 1989)
3.688 (de Bruijn et al., 1989)

Solubility in organics (20 °C):
In wt %: acetone (30.31), benzene (22.42), chloroform (19.35), ether (17.22), ethanol (6.02) (Windholz et al., 1983); 9.76 and 13.97 g/L in hexane at 10 and 20 °C, respectively (Mills and Biggar, 1969).

Solubility in water:
7.52 ppm at 25 °C (Masterton and Lee, 1972)
7.8 ppm at 25 °C (Weil et al., 1974)
7.3 ppm at 25 °C, 12 ppm at 35 °C, 14 ppm at 45 °C (Berg, 1983)
6.98 ppm (Caron et al., 1975)
7.5 g/m^3 at 25 °C (Spencer and Cliath, 1988)
7.87 mg/L at 24 °C (Chiou et al., 1986)
17.0 mg/L at 24 °C (Hollifield, 1979)
7.8 mg/L at 25 °C (Chiou et al., 1979)
5.75-7.40 ppm at 28 °C (Kurihara et al., 1973)
12 ppm at 26.5 °C (Bhavnagary and Jayaram, 1974)
In ppb: 2,150 at 15 °C, 6,800 at 25 °C, 11,400 at 35 °C, 15,200 at 45 °C (particle size ≤5 μ, Biggar and Riggs, 1974)
At 20-25 °C: 6,600 ppb (particle size ≤5 μ), 500 ppb (particle size ≤0.04 μ) (Robeck et al., 1975)
9.2 mg/L at 25 °C (Saleh et al., 1982)
8.5 mg/L at 20 °C (Mills and Biggar, 1969)

Vapor density (ng/L):
518 at 20 °C, 1,971 at 30 °C, 6,784 at 40 °C (Spencer and Cliath, 1970)

Vapor pressure (10^6 mmHg):
802.5 at 25 °C (Hinckley et al., 1990)
67 at 25 °C (Spencer and Cliath, 1988)
32.25 at 20 °C (Dobbs and Cull, 1982)
128 at 30 °C (Nash, 1983; Tinsley, 1979)
14.3 at 25 °C (Klöpffer et al., 1988)
491 at 25 °C (Bidleman, 1984)

Environmental Fate
Biological. In moist soils, lindane biodegraded to γ-pentachlorocyclohexene with

trace amounts of benzene and chlorobenzene as possible contaminants (Kearney and Kaufman, 1976). Under anaerobic conditions, degradation by soil bacteria yielded γ-3,4,5,6-tetrachloro-1-cyclohexane and α-BHC (Kobayashi and Rittman, 1982). Other reported biodegradation products include pentacyclohexane and tetrachlorocyclohexanes and pentachlorobenzene and tetrachlorobenzenes (Moore and Ramamoorthy, 1984).

Incubation of lindane for 6 weeks in a sandy loam soil under flooded conditions resulted in the formation of γ-3,4,5,6-tetrachlorocyclohexane, γ-2,3,4,5,6-penta-chlorocyclohex-1-ene and small amounts of 1,2,4-trichlorobenzene, 1,2,3,4-tetra-chlorobenzene, 1,2,3,5- and/or 1,2,4,5-tetrachlorobenzene (Mathur and Saha, 1975). Incubation of lindane in moist soil for 8 weeks yielded the following metabolites: γ-3,4,5,6-tetrachlorocyclohexene, γ-1,2,3,4,5-pentachlorocyclohex-1-ene, pentachlorobenzene, 1,2,3,4-tetrachlorobenzene, 1,2,3,5- and/or 1,2,4,5-tetrachlorobenzene, 1,2,4-trichlorobenzene, 1,3,5-trichlorobenzene, m- and/or p-dichlorobenzene (Mathur and Saha, 1977).

After a 30-day incubation period, the white rot fungus *Phanerochaete chrysosporium* converted to carbon dioxide. Mineralization began between the third and sixth day of incubation. The production of carbon dioxide was highest between 3-18 days of incubation, after which the rate of carbon dioxide produced decreased until the 30th day. It was suggested that the metabolism of lindane and other compounds, including *p,p'*-DDT, TCDD and benzo[*a*]pyrene, was dependent on the extracellular lignin-degrading enzyme system of this fungus (Bumpus et al., 1985).

Microorganisms isolated from a loamy sand soil degraded lindane and some of the metabolites identified were pentachlorobenzene, 1,2,4,5-tetrachlorobenzene, 1,2,3,5-tetrachlorobenzene, γ-2,3,4,5,6-pentachloro-1-cyclohexane (γ-PCCH), γ-3,4,5,6-tetrachloro-1-cyclohexane (γ-TCCH) and β-3,4,5,6-tetrachloro-1-cyclo-hexane (β-TCCH) (Tu, 1976). In a laboratory experiment, a *Pseudomonas* culture transformed lindane to γ-TCCH, γ-PCCH and α-BHC (Benezet and Matsumura, 1973). γ-TCCH was also reported as a product of lindane degradation by *Clostridium sphenoides* (Heritage and MacRae, 1977, 1977a; MacRae et al., 1969), an anaerobic bacterium isolated from flooded soils (MacRae et al., 1969; Sethunathan and Yoshida, 1973a). Evidence suggests that degradation of lindane in anaerobic cultures or flooded soils amended with lindane occurs via reductive dehalogenation, producing chlorine-free volatile metabolites (Sethunathan and Yoshida, 1973a). Indigenous microbes in soil partially degraded lindane to carbon dioxide (MacRae et al., 1967).

Beland et al. (1976) studied the degradation of lindane in sewage sludge under anaerobic conditions. Lindane underwent reductive hydrodechlorination forming γ-3,4,5,6-tetrachlorocyclohexene (γ-BTC). The amount of γ-BTC that formed reached a maximum concentration of 5% after two weeks. Further incubation with sewage sludge resulted in decreased concentrations. The evidence suggested that γ-BTC underwent further reduction affording benzene (Beland et al., 1976).

MALATHION

Synonyms: American Cyanamid 4049; S-1,2-Bis(carbethoxy)ethyl-O,O-dimethyl dithiophosphate; S-1,2-Bis(ethoxycarbonyl)ethyl-O,O-dimethyl phosphorodithioate; S-1,2-Bis(ethoxycarbonyl)ethyl-O,O-dimethyl thiophosphate; Calmathion; Carbethoxy malathion; Carbetovur; Carbetox; Carbofos; Carbophos; Celthion; Chemathion; Cimexan; Compound 4049; Cythion; Detmol MA; Detmol MA 96%; S-1,2-Dicarbethoxyethyl-O,O-dimethyl dithiophosphate; Dicarboethoxyethyl-O,O-dimethyl phosphorodithioate; 1,2-Di(ethoxycarbonyl)ethyl-O,O-dimethyl phosphorodithioate; S-1,2-Di(ethoxycarbonyl)ethyl dimethyl phosphorothiolo-thionate; Diethyl (dimethoxyphosphinothioylthio) butanedioate; Diethyl (dimethoxyphosphinothioylthio) succinate; Diethyl mercaptosuccinate, O,O-dimethyl phosphorodithioate; Diethyl mercaptosuccinate, O,O-dimethyl thiophosphate; Diethylmercaptosuccinic acid O,O-dimethyl phosphorodithioate; **[(Dimethoxyphosphinothioyl)thio]butanedioic acid diethyl ester;** O,O-Dimethyl-S-1,2-bis(ethoxy-carbonyl)ethyldithiophosphate; O,O-Dimethyl-S-(1,2-dicarbethoxyethyl)dithio-phosphate; O,O-Dimethyl-S-(1,2-dicarbethoxyethyl)phosphorodithioate; O,O-Di-methyl-S-(1,2-dicarbethoxyethyl)thiothionophosphate; O,O-Dimethyl-S-1,2-di-(ethoxycarbamyl)ethyl phosphorodithioate; O,O-Dimethyldithiophosphate dimeth-ylmercaptosuccinate; EL 4049; Emmatos; Emmatos extra; ENT 17034; Ethiolacar; Etiol; Experimental insecticide 4049; Extermathion; Formal; Forthion; Fosfothion; Fosfotion; Four thousand forty-nine; Fyfanon; Hilthion; Hilthion 25WDP; Insecticide 4049; Karbofos; Kop-thion; Kypfos; Malacide; Malafor; Malakill; Malagran; Malamar; Malamar 50; Malaphele; Malaphos; Malasol; Malaspray; Malathion E50; Malathion LV concentrate; Malathion ULV concentrate; Mala-thiozoo; Malathon; Malathyl LV concentrate & ULV concentrate; Malatol; Malatox; Maldison; Malmed; Malphos; Maltox; Maltox MLT; Mercaptosuccinic acid diethyl ester; Mercaptothion; MLT; Moscardia; NA 2783; NCI-C00215; Oleophosphothion; Orthomalathion; Phosphothion; Prioderm; Sadofos; Sadophos; SF 60; Siptox I; Sumitox; Tak; TM-4049; Vegfru malatox; Vetiol; Zithiol.

CAS Registry No.: 121-75-5
DOT: 2783
DOT label: Poison
Molecular formula: $C_{10}H_{19}O_6PS_2$
Formula weight: 330.36
RTECS: WM8400000

Physical state, color and odor
Clear yellow to brown liquid with a garlic odor.

Melting point (°C):
2.9 (Windholz et al., 1983)

Boiling point (°C):
156–157 at 0.7 mmHg (Windholz et al., 1983)
120 °C at 0.2 mmHg (Freed et al., 1977)

Density (g/cm^3):
1.23 at 25/4 °C (Windholz et al., 1983)

Diffusivity in water (10^5 cm^2/sec):
0.44 at 20 °C using method of Hayduk and Laudie (1974).

Flash point (°C):
>162.8 (open cup, Meister, 1988)

Henry's law constant (10^9 atm·m^3/mol):
4.89 at 25 °C (Fendinger and Glotfelty, 1990)

Bioconcentration factor, log BCF:
0.00 (fish, microcosm) (Garten and Trabalka, 1983)

Soil sorption coefficient, log K_{oc}:
2.61 (Lihue silty clay soil, Miles and Takashima, 1991)

Octanol/water partition coefficient, log K_{ow}:
2.89 (Chiou et al., 1977; Freed et al., 1979)
2.84 (Bowman and Sans, 1983a)
2.75 (Worthing and Hance, 1991)

Solubility in organics:
Miscible in most organic solvents (Meister, 1988).

Solubility in water (ppm):
300 at 30 °C (Freed et al., 1977)
141 at 10 °C, 145 at 20 °C, 164 at 30 °C (Bowman and Sans, 1985)
143 at 20 °C (Bowman and Sans, 1983; Miles and Takashima, 1991)

Vapor density:
13.50 g/L at 25 °C, 11.40 (air = 1)

diethyl fumarate, ethyl hydrogen fumarate and *O,O*-dimethyl phosphorodithioic acid. At pH 8, the reported half-lives at 0, 27 and 40 °C are 40 days, 36 hours and one hour, respectively (Wolfe et al., 1977a). However, under acidic conditions, it was reported that malathion degraded into diethyl thiomalate and *O,O*-dimethyl phosphorothionic acid (Wolfe et al., 1977a).

When applied as an aerial spray, malathion was converted to malaoxon and diethyl fumarate via oxidation and hydrolysis, respectively (Brown et al., 1993).

Emits toxic fumes of nitrogen and phosphorus oxides when heated to decomposition (Lewis, 1990). Products reported from the combustion of malathion at 900 °C include carbon monoxide, carbon dioxide, chlorine, sulfur oxides, nitrogen oxides, hydrogen sulfide and oxygen (Kennedy et al., 1972).

Exposure Limits (mg/m^3): NIOSH REL: TWA 10, IDLH 250; OSHA PEL: 15; ACGIH TLV: TWA 10.

Symptoms of Exposure: Miosis; eye and skin irritation; rhinorrhea; headache; tight chest, wheezing, laryngeal spasm; salivation; anorexia, nausea, vomiting, abdominal cramps; diarrhea, ataxia (NIOSH, 1994).

Toxicity: Acute oral LD$_{50}$ for wild birds 400 mg/kg, chickens 600 mg/kg, cattle 53 mg/kg, ducks 1.485 mg/kg, guinea pigs 570 mg/kg, mice 507 mg/kg, rats 370 mg/kg (RTECS, 1985); LC$_{50}$ (96-hour) for bluegill sunfish 100 μg/L (Hartley and Kidd, 1987), largemouth bass 285 μg/L (Worthing and Hance, 1991), coho salmon 100 μg/L, brown trout 200 μg/L, channel catfish 9.0 mg/L, channel black bullhead 12.9 mg/L, fathead minnow 8.7 mg/L, rainbow trout 170 μg/L, perch 260 μg/L (Macek and McAllister, 1970), green sunfish 120 μg/L (Verschueren, 1983); LC$_{50}$ (48-hour) for red killifish 3 mg/L (Yoshioka et al., 1986); LC$_{50}$ (24-hour) for bluegill sunfish 120 ppb and rainbow trout 100 ppb (Verschueren, 1983).

Drinking Water Standard: As of May 1994, no MCLGs or MCLs have been proposed (U.S. EPA, 1994).

Use: Insecticide for control of sucking and chewing insects and spider mites on vegetables, fruits, ornamentals, field crops, greenhouses, gardens and forestry.

MALEIC ANHYDRIDE

Synonyms: *cis*-Butenedioic anhydride; **2,5-Furanedione**; Maleic acid anhydride; RCRA waste number U147; Toxilic anhydride; UN 2215.

CAS Registry No.: 108-31-6
DOT: 2215
Molecular formula: $C_4H_2O_3$
Formula weight: 98.06
RTECS: ON3675000

Physical state and color
White crystals. Odor threshold in air is 0.32 ppm (Amoore and Hautala, 1983).

Melting point (°C):
60 (Weast, 1986)
52.8 (Windholz et al., 1983)

Boiling point (°C):
197–199 (Weast, 1986)
202 (Windholz et al., 1983)

Density (g/cm³):
1.314 at 60/4 °C (Weast, 1986)
1.48 at 20/4 °C (Sax and Lewis, 1987)

Diffusivity in water:
Not applicable - reacts with water.

Flash point (°C):
104.2 (NIOSH, 1994)
103 (Dean, 1987)

Lower explosive limit (%):
1.4 (NFPA, 1984)

Upper explosive limit (%):
7.1 (NIOSH, 1994)

Henry's law constant (10^6 atm·m^3/mol):
4.01 at 20 °C (approximate - calculated from water solubility and vapor pressure)

Ionization potential (eV):
9.08 ± 0.03 (Franklin et al., 1969)

Soil sorption coefficient, log K_{oc}:
Unavailable because experimental methods for estimation of this parameter for ketones are lacking in the documented literature.

Octanol/water partition coefficient, log K_{ow}:
1.25 using method of Hansch et al. (1968).

Solubility in organics:
Soluble in acetone, alcohol and ether (Weast, 1986).

Solubility in water (20 °C):
3 wt % (NIOSH, 1994)
28 g/L (Verschueren, 1983)

Vapor density:
4.01 g/L at 25 °C, 3.39 (air = 1)

Vapor pressure (mmHg):
8.7 at 20 °C (Verschueren, 1983)

Exposure Limits: NIOSH REL: TWA 10 ppm (40 mg/m^3), IDLH 1,400 ppm; OSHA PEL: TWA 25 ppm (100 mg/m^3); ACGIH TLV: TWA 15 ppm.

Symptoms of Exposure: Irritant to the eyes, skin and mucous membranes. Narcotic at high concentrations (Patnaik, 1992).

Toxicity: Acute oral LD_{50} for rats 1,120 mg/kg, mice 710 mg/kg (RTECS, 1985).

Uses: Solvent for nitrocellulose, gums and resins; roll-coating inks, varnishes, lacquers, stains and enamels; starting material for synthesizing methyl isobutyl ketone; insect repellent; ore flotation.

METHANOL

Synonyms: Carbinol; Colonial spirit; Columbian spirits; Columbian spirits (wood alcohol); Methyl alcohol; Methyl hydroxide; Methylol; Monohydroxymethane; NA 1230; Pyroxylic spirit; RCRA waste number U154; Wood alcohol; Wood naphtha; Wood spirit.

CAS Registry No.: 67-56-1
DOT: 1230
Molecular formula: CH_4O
Formula weight: 32.04
RTECS: PC1400000

Physical state, color and odor
Clear, colorless liquid with a characteristic odor. Odor threshold in air is 4.3 ppm (Keith and Walters, 1992).

Melting point (°C):
-93.9 (Weast, 1986)
-97.8 (Windholz et al., 1983)

Boiling point (°C):
65 (Weast, 1986)

Density (g/cm^3):
0.7914 at 20/4 °C (Weast, 1986)
0.7866 at 25/4 °C (Windholz et al., 1983)
0.796 at 15/4 °C (Verschueren, 1983)

Diffusivity in water (10^5 cm^2/sec):
1.50 at 20 °C using method of Hayduk and Laudie (1974).

Dissociation constant:
$\approx$ 16 (Gordon and Ford, 1972)

Flash point (°C):
16.9 (NIOSH, 1994)
12.2 (open cup, Hawley, 1981)

Lower explosive limit (%):
6.0 (NIOSH, 1994)

Upper explosive limit (%):
36 (NIOSH, 1994)

Heat of fusion (kcal/mol):
0.768 (Dean, 1987)

Henry's law constant (10^6 atm·m^3/mol):
4.44 at 25 °C (Snider and Dawson, 1985)
4.55 (Gaffney et al., 1987)

Ionization potential (eV):
10.84 (Franklin et al., 1969)

Bioconcentration factor, log BCF:
4.45 (algae, Geyer et al., 1984)
2.67 (activated sludge, Freitag et al., 1985)

Soil sorption coefficient, log K_{oc}:
0.44 (Gerstl and Helling, 1987)

Octanol/water partition coefficient, log K_{ow}:
-0.32, -0.52, -0.68, -0.71, -0.77, -0.82 (Sangster, 1989)
-0.66 (Hansch and Anderson, 1967)
-0.81 (Collander, 1951)
-0.70 (Geyer et al., 1984)

Solubility in organics:
Miscible with benzene, ethanol, ether, ketone and many other organic solvents (Windholz et al., 1983).

Solubility in water:
Miscible (Palit, 1947). A saturated solution in equilibrium with its own vapor had a concentration of 1,163 g/L at 25 °C (Kamlet et al., 1987).

Vapor density:
1.31 g/L at 25 °C, 1.11 (air = 1)

Vapor pressure (mmHg):
74.1 at 15 °C, 97.6 at 20 °C, 127.2 at 25 °C, 164.2 at 30 °C (Gibbard and Creek, 1974)

Environmental Fate

Biological. In a 5-day experiment, [^{14}C]methanol applied to soil water suspensions under aerobic and anaerobic conditions gave $^{14}CO_2$ yields of 53.4 and 46.3%, respectively (Scheunert et al., 1987).

Photolytic. Photooxidation of methanol in an oxygen-rich atmosphere (20%) yielded formaldehyde and hydroxy-peroxyl radicals. With chlorine, formaldehyde, carbon monoxide, hydrogen peroxide and formic acid were detected (Whitbeck, 1983).

Chemical/Physical. In a smog chamber, methanol reacted with nitrogen dioxide to give methyl nitrite and nitric acid (Takagi et al., 1986). The formation of these products was facilitated when this experiment was accompanied by UV light (Akimoto and Takagi, 1986).

Exposure Limits: NIOSH REL: TWA 200 ppm (260 mg/m^3), STEL 250 ppm (325 mg/m^3), IDLH 6,000 ppm; OSHA PEL: TWA 200 ppm.

Symptoms of Exposure: Ingestion may cause acidosis and blindness. Symptoms of poisoning include nausea, abdominal pain, headache, blurred vision, shortness of breath and dizziness (Patnaik, 1992).

Toxicity: Acute oral LD_{50} for monkeys 7 gm/kg, rats 5,628 mg/kg, mice 7,300 mg/kg; LC_{50} (inhalation) for rats 64,000 ppm/4-hour (RTECS, 1985).

Uses: Solvent for nitrocellulose, ethyl cellulose, polyvinyl butyral, rosin, shellac, manila resin, dyes; fuel for utility plants; home heating oil extender; preparation of methyl esters, formaldehyde, methacrylates, methylamines, dimethyl terephthalate, polyformaldehydes; methyl halides, ethylene glycol; in gasoline and diesel oil antifreezes; octane booster in gasoline; source of hydrocarbon for fuel cells; extractant for animal and vegetable oils; denaturant for ethanol; softening agent for certain plastics; dehydrator for natural gas.

METHOXYCHLOR

Synonyms: 2,2-Bis(*p*-anisyl)-1,1,1-trichloroethane; 1,1-Bis(*p*-methoxyphenyl)-2,2,2-trichloroethane; 2,2-Bis(*p*-methoxyphenyl)-1,1,1-trichloroethane; Chemform; 2,2-Di-*p*-anisyl-1,1,1-trichloroethane; Dimethoxy-DDT; *p,p'*-Dimethoxydiphenyltrichloroethane; Dimethoxy-DT; 2,2-Di(*p*-methoxyphenyl)-1,1,1-trichloroethane; Di(*p*-methoxyphenyl)trichloromethylmethane; DMDT; 4,4'-DMDT; *p,p'*-DMDT; DMTD; ENT 1716; Maralate; Marlate; Marlate 50; Methoxcide; Methoxo; 4,4'-Methoxychlor; *p,p'*-Methoxychlor; Methoxy-DDT; Metox; Moxie; NCI-C00497; RCRA waste number U247; 1,1,1-Trichloro-2,2-bis(*p*-anisyl)ethane; 1,1,1-Trichloro-2,2-bis(*p*-methoxyphenol)ethanol; 1,1,1-Trichloro-2,2-bis(*p*-methoxyphenyl)ethane; 1,1,1-Trichloro-2,2-di(4-methoxyphenyl)ethane; **1,1'-(2,2,2-Trichloroethylidene)bis(4-methoxybenzene).**

CAS Registry No.: 72-43-5
DOT: 2761
DOT label: Poison
Molecular formula: $C_{16}H_{15}Cl_3O_2$
Formula weight: 345.66
RTECS: KJ3675000

Physical state, color and odor
White, gray or pale yellow crystals or powder. May be dissolved in an organic solvent or petroleum distillate for application. Pungent to mild, fruity odor. Odor threshold in water is 4.7 mg/kg (Keith and Walters, 1992).

Melting point (°C):
98 (Verschueren, 1983)
92 (Kapoor et al., 1970)
77 (technical grade, Sunshine, 1969)

Boiling point (°C):
Decomposes (Weast, 1986)

Density (g/cm³):
1.41 at 25/4 °C (Verschueren, 1983)

Diffusivity in water (10^5 cm²/sec):
0.44 at 20 °C using method of Hayduk and Laudie (1974).

632

Flash point (°C):
Burns only at high temperatures (Weiss, 1986).

Lower explosive limit (%):
Not pertinent (Weiss, 1986)

Upper explosive limit (%):
Not pertinent (Weiss, 1986)

Henry's law constant (10^5 atm·m^3/mol):
1.58 at 25 °C (estimated, Howard, 1991)

Bioconcentration factor, log BCF:
3.92 (freshwater fish), 3.19 (fish, microcosm) (Garten and Trabalka, 1983)

Soil sorption coefficient, log K_{oc}:
4.90 (Kenaga, 1980)
4.95 (clay, Karickhoff et al., 1979)

Octanol/water partition coefficient, log K_{ow}:
4.30 (Mackay, 1982)
4.68 (Kenaga, 1980)
4.40 (Garten and Trabalka, 1983)
3.40 (Wolfe et al., 1977)

Solubility in organics:
Soluble in ethanol (Windholz et al., 1983), alcohol and chloroform (Meites, 1963).

Solubility in water:
40 µg/L at 24 °C (Hollifield, 1979)
100 µg/L at 25 °C (IARC, 1979)
620 ppb (Kapoor et al., 1970)
In ppb: 20 at 15 °C, 45 at 25 °C, 95 at 35 °C, 185 at 45 °C (particle size ≤5 µ)
   (Biggar and Riggs, 1974)
120 ppb at 25 °C (Zepp et al., 1976)

Environmental Fate

Biological. Degradation by *Aerobacter aerogenes* under aerobic or anaerobic conditions yielded 1,1-dichloro-2,2-bis(*p*-methoxyphenyl)ethylene and 1,1-dichloro-2,2-bis(*p*-methoxyphenyl)ethane (Kobayashi and Rittman, 1982).

In a model aquatic ecosystem, methoxychlor degraded to ethanol, dihydroxyethane, dihydroxyethylene and unidentified polar metabolites (Metcalf et al., 1971). Kapoor et al. (1970) also studied the biodegradation of methoxychlor in a model ecosystem containing snails, plankton, mosquito larvae, *Daphnia magna* and mosquito fish (*Gambusia affinis*). The following metabolites were identified: 2-(*p*-methoxyphenyl)-2-(*p*-hydroxyphenyl)-1,1,1-trichloroethane, 2,2-bis(*p*-hydroxyphenyl)-1,1,1-trichloroethane, 2,2-bis(*p*-hydroxyphenyl)-1,1,1-trichloroethylene and polar metabolites (Kapoor et al., 1970).

Photolytic. In air-saturated distilled water, direct photolysis of methoxychlor by >280 nm light produced 1,1-bis(*p*-methoxyphenyl)-2,2-dichloroethylene (DMDE), which photolyzed to *p*-methoxybenzaldehyde. The photolysis half-life was estimated to be 4.5 months (Zepp et al., 1976).

Methoxychlor-DDE and *p,p*-dimethoxybenzophenone were formed when methoxychlor in water was irradiated by UV light (Paris and Lewis, 1973). Compounds reported from the photolysis of methoxychlor in aqueous, alcoholic solutions were *p,p*-dimethoxybenzophenone, *p*-methoxybenzoic acid and 4-methoxyphenol (Wolfe et al., 1976). However, when methoxychlor in milk was irradiated by UV light (λ = 220 and 330 nm), 4-methoxyphenol, methoxychlor-DDE, *p,p*-dimethoxybenzophenone and 1,1,4,4-tetrakis(*p*-methoxyphenyl)-1,2,3-butatriene were formed (Li and Bradley, 1969).

Chemical/Physical. Hydrolysis at common aquatic pHs produced anisoin, anisil and 2,2-bis(*p*-methoxyphenyl)-1,1-dichloroethylene (Wolfe et al., 1977). Though no products were identified, the estimated hydrolysis half-life in water at 25 °C and pH 7.1 is 270 days (Mabey and Mill, 1978).

Decomposes in aqueous alkaline solutions forming diphenylethylene and hydrochloric acid (Hartley and Kidd, 1987). Emits toxic chloride fumes when heated to decomposition (Lewis, 1990).

Exposure Limits (mg/m^3): Potential occupational carcinogen. NIOSH REL: IDLH 5,000; OSHA PEL: TWA 15; ACGIH TLV: TWA 10.

Symptoms of Exposure: Slightly irritating to skin (NIOSH, 1994).

Toxicity: Acute oral LD_{50} for mice 1,850 mg/kg, rats 5,000 mg/kg (RTECS, 1985); LC_{50} (96-hour) for fathead minnow 7.5 μg/L, bluegill sunfish 62.0 μg/L, rainbow trout 62.6 μg/L, coho salmon 66.2 μg/L, chinook 27.9 μg/L, perch 20.0 μg/L (Verschueren, 1983); LC_{50} (24-hour) for bluegill sunfish 67 μg/L and rainbow trout 52 μg/L (Worthing and Hance, 1987).

Drinking Water Standard (final): MCLG: 40 μg/L; MCL: 40 μg/L (U.S. EPA, 1994).

Uses: Insecticide to control mosquito larvae and house flies; to control ectoparasites on cattle, sheep and goats; recommended for use in dairy barns.

METHYL ACETATE

Synonyms: Devoton; **Acetic acid methyl ester**; Methyl ethanoate; Tereton; UN 1231.

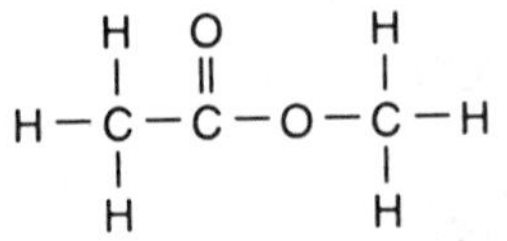

$$H-\overset{\overset{\displaystyle H}{|}}{\underset{\underset{\displaystyle H}{|}}{C}}-\overset{\overset{\displaystyle O}{\|}}{C}-O-\overset{\overset{\displaystyle H}{|}}{\underset{\underset{\displaystyle H}{|}}{C}}-H$$

CAS Registry No.: 79-20-9
DOT: 1231
DOT label: Flammable liquid
Molecular formula: $C_3H_6O_2$
Formula weight: 74.08
RTECS: AI9100000

Physical state, color and odor
Colorless liquid with a pleasant odor. Odor threshold in air is 4.6 ppm (Amoore and Hautala, 1983).

Melting point (°C):
-98.1 (Weast, 1986)
-99 (Verschueren, 1983)

Boiling point (°C):
57 (Weast, 1986)
54.05 (Hawley, 1981)

Density (g/cm^3):
0.9330 at 20/4 °C (Weast, 1986)
0.9342 at 20/4 °C, 0.9279 at 25/4 °C (Windholz et al., 1983)

Diffusivity in water (10^5 cm^2/sec):
1.01 at 20 °C using method of Hayduk and Laudie (1974).

Flash point (°C):
-5.6 (NIOSH, 1994)

Lower explosive limit (%):
3.1 (NIOSH, 1994)

Upper explosive limit (%):
16 (NIOSH, 1994)

Henry's law constant (10^5 atm·m^3/mol at 25 °C):
9.09 (Hine and Mookerjee, 1975)
11.5 (Buttery et al., 1969)

Ionization potential (eV):
10.27 ± 0.02 (Franklin et al., 1969)

Soil sorption coefficient, log K_{oc}:
Unavailable because experimental methods for estimation of this parameter for aliphatic esters are lacking in the documented literature.

Octanol/water partition coefficient, log K_{ow}:
0.17 (Collander, 1951)

Solubility in organics:
Soluble in acetone, benzene and chloroform (Weast, 1986). Miscible with alcohol and ether (Sax and Lewis, 1987).

Solubility in water:
25 wt % at 20 °C (NIOSH, 1994)
319,000 mg/L, 240,000 mg/L at 20 °C (Verschueren, 1983)
3.29 M at 20 °C (Fühner, 1924)

Vapor density:
3.03 g/L at 25 °C, 2.56 (air = 1)

Vapor pressure (mmHg):
173 at 20 °C (NIOSH, 1994)
170 at 20 °C, 235 at 25 °C, 255 at 30 °C (Verschueren, 1983)
216 at 25 °C (Abraham, 1984)

Environmental Fate
   Chemical/Physical. Slowly hydrolyzes in water forming methyl alcohol and acetic acid (NIOSH, 1994).

Exposure Limits: NIOSH REL: TWA 200 ppm (610 mg/m^3), STEL 250 ppm (760 mg/m^3), IDLH 3,100 ppm; OSHA PEL: TWA 200 ppm.

Symptoms of Exposure: Inflammation of the eyes, visual and nervous disturbances, tightness of the chest, drowsiness and narcosis (Patnaik, 1992).

Toxicity: Acute oral LD_{50} for rabbits 3,705 mg/kg, rats 5,450 mg/kg (RTECS, 1985).

Uses: Solvent for resins, lacquers, oils, acetylcellulose, nitrocellulose; paint removers; synthetic flavoring.

METHYL ACRYLATE

Synonyms: Acrylic acid methyl ester; Curithane 103; Methoxycarbonylethylene; Methyl propenate; Methyl propenoate; Methyl-2-propenoate; Propenoic acid methyl ester; **2-Propenoic acid methyl ester;** UN 1919.

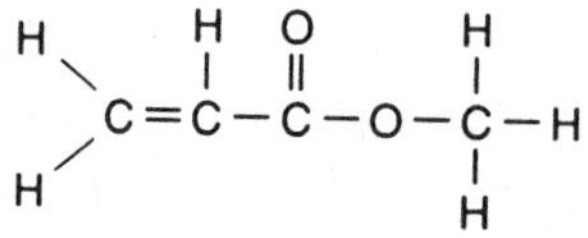

$$H_2C=CH-\overset{\overset{\displaystyle O}{\|}}{C}-O-CH_3$$

CAS Registry No.: 96-33-3
DOT: 1919
DOT label: Flammable liquid
Molecular formula: $C_4H_6O_2$
Formula weight: 86.09
RTECS: AT2800000

Physical state, color and odor
Clear, colorless liquid with a heavy, sweet odor. Odor threshold in air is 17 $\mu g/m^3$ (Keith and Walters, 1992).

Melting point (°C):
-77 (NIOSH, 1994)

Boiling point (°C):
80.5 (Weast, 1986)

Density (g/cm^3):
0.9561 at 20/4 °C (Windholz et al., 1983)

Diffusivity in water (10^5 cm^2/sec):
0.94 at 20 °C using method of Hayduk and Laudie (1974).

Flash point (°C):
-2.8 (NIOSH, 1994)
-3.8 (open cup, Hawley, 1981)

Lower explosive limit (%):
2.8 (NIOSH, 1994)

Upper explosive limit (%):
25.0 (NIOSH, 1994)

Henry's law constant (10^4 atm·m^3/mol):
1.3 at 20 °C (approximate - calculated from water solubility and vapor pressure)

Ionization potential (eV):
9.90 (NIOSH, 1994)
9.19 ± 0.05 (Franklin et al., 1969)

Soil sorption coefficient, log K_{oc}:
Unavailable because experimental methods for estimation of this parameter for aliphatic esters are lacking in the documented literature.

Octanol/water partition coefficient, log K_{ow}:
0.67 using method of Hansch et al. (1968).

Solubility in organics:
Soluble in acetone, alcohol, benzene and ether (Weast, 1986).

Solubility in water (g/L):
60 at 20 °C, 50 at 40 °C (Windholz et al., 1983)
52.0 g/L (Verschueren, 1983)

Vapor density:
3.52 g/L at 25 °C, 2.97 (air = 1)

Vapor pressure (mmHg):
70 at 20 °C, 110 at 30 °C (Verschueren, 1983)

Environmental Fate
    Photolytic. Polymerizes on standing and is accelerated by heat, light and peroxides (Windholz et al., 1983).
    Chemical/Physical. Begins to polymerize at 80.2 °C (Weast, 1986).

Exposure Limits: NIOSH REL: TWA 10 ppm (35 mg/m^3), IDLH 250 ppm; OSHA PEL: TWA 10 ppm.

Symptoms of Exposure: Lacrimation, irritation of respiratory tract, lethargy and convulsions (Patnaik, 1992).

Toxicity: Acute oral LD_{50} for rats 277 mg/kg, mice 827 mg/kg (RTECS, 1985).

Uses: Manufacturing plastic films, leather finish resins, textile and paper coatings; amphoteric surfactants; chemical intermediate.

METHYLAL

Synonyms: Anesthenyl; **Dimethoxymethane**; Formal; Formaldehyde dimethylacetal; Methyl formal; Methylene dimethyl ether; UN 1234.

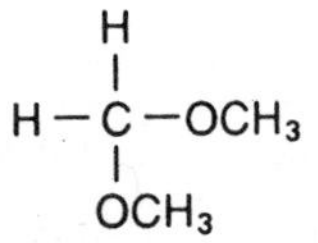

$$H-\underset{\underset{OCH_3}{|}}{\overset{\overset{H}{|}}{C}}-OCH_3$$

CAS Registry No.: 109-87-5
DOT: 1234
Molecular formula: $C_3H_8O_2$
Formula weight: 76.10
RTECS: PA8750000

Physical state, color and odor
Colorless liquid with a pungent, chloroform-like odor.

Melting point (°C):
-104.8 (Weast, 1986)

Boiling point (°C):
45.5 (Weast, 1986)
42.3 (Sax and Lewis, 1987)

Density (g/cm^3):
0.8593 at 20/4 °C (Weast, 1986)
0.86645 at 15/4 °C (Huntress and Mulliken, 1941)

Diffusivity in water (10^5 cm^2/sec):
0.95 at 20 °C using method of Hayduk and Laudie (1974).

Flash point (°C):
-17 (Dean, 1991)
-18 (Windholz et al., 1983)
-32 (open cup, NFPA, 1984)

Lower explosive limit (%):
2.2 (NFPA, 1984)

Upper explosive limit (%):
13.8 (NFPA, 1984)

Henry's law constant (10^4 atm·m^3/mol):
1.73 at 25 °C (Hine and Mookerjee, 1975)

Ionization potential (eV):
10.00 (NIOSH, 1994)

Soil sorption coefficient, log K_{oc}:
Unavailable because experimental methods for estimation of this parameter for aliphatic ethers are lacking in the documented literature. However, its high solubility in water and low K_{ow} suggest its adsorption to soil will be nominal (Lyman et al., 1982).

Octanol/water partition coefficient, log K_{ow}:
-0.01 (Collander, 1951)

Solubility in organics:
Soluble in acetone and benzene (Weast, 1986). Miscible with alcohol, ether and oils (Windholz et al., 1983).

Solubility in water:
330,000 mg/L (Verschueren, 1983)

Vapor density:
3.11 g/L at 25 °C, 2.63 (air = 1)

Vapor pressure (mmHg):
330 at 20 °C, 400 at 25 °C (Verschueren, 1983)

Exposure Limits: NIOSH REL: TWA 1,000 ppm (3,100 mg/m^3), IDLH 2,200 ppm; OSHA PEL: TWA 1,000 ppm.

Toxicity: Acute oral LD_{50} for rats 5,708 mg/kg; LC_{50} (inhalation) for rats 15,000 ppm (RTECS, 1985).

Uses: Artificial resins; perfumery; solvent; adhesives; protective coatings; special fuel; organic synthesis (Grignard and Reppe reactions).

METHYLAMINE

Synonyms: Aminomethane; Carbinamine; Mercurialin; **Methanamine**; Monomethylamine; UN 1061; UN 1235.

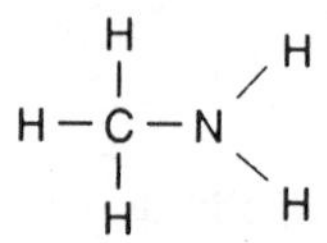

$$H-\underset{\underset{H}{|}}{\overset{\overset{H}{|}}{C}}-N\overset{H}{\underset{H}{\diagdown}}$$

CAS Registry No.: 74-89-5
DOT: 1061
DOT label: Flammable gas (anhydrous)/flammable liquid
Molecular formula: CH_5N
Formula weight: 31.06
RTECS: PF6300000

Physical state, color and odor
Colorless gas with a strong ammonia-like odor. Odor threshold in air is 3.2 ppm (Amoore and Hautala, 1983).

Melting point (°C):
-93.5 (Weast, 1986)

Boiling point (°C):
-6.3 (Weast, 1986)

Density (g/cm³):
0.6628 at 20/4 °C (Weast, 1986)
0.769 at -70/4 °C (Verschueren, 1983)

Diffusivity in water (10^5 cm²/sec):
1.38 at 20 °C using method of Hayduk and Laudie (1974).

Dissociation constant:
10.657 at 25 °C (Gordon and Ford, 1972)

Flash point (°C):
0 (Windholz et al., 1983)
-10 (liquid, NIOSH, 1994)

Lower explosive limit (%):
4.9 (NFPA, 1984)

Upper explosive limit (%):
20.7 (NFPA, 1984)

Heat of fusion (kcal/mol):
1.466 (Dean, 1987)

Henry's law constant (10^2 atm·m^3/mol):
1.81 at 25 °C (approximate - calculated from water solubility and vapor pressure)

Ionization potential (eV):
8.97 (Franklin et al., 1969)
9.18 (Gibson et al., 1977)

Soil sorption coefficient, log K_{oc}:
Unavailable because experimental methods for estimation of this parameter for aliphatic amines are lacking in the documented literature. However, its high solubility in water and low K_{ow} suggest its adsorption to soil will be nominal (Lyman et al., 1982).

Octanol/water partition coefficient, log K_{ow}:
-0.57 (Collander, 1951)

Solubility in organics:
Soluble in benzene (105 g/L at 25 °C) and miscible with ether (Windholz et al., 1983).

Solubility in water:
1,154 volumes at 12.5 °C, 959 volumes at 25 °C (Windholz et al., 1983)

Vapor density:
1.27 g/L at 25 °C, 1.07 (air = 1)

Vapor pressure (mmHg):
2,356 at 20 °C, 3,268 at 30 °C (Verschueren, 1983)

Environmental Fate
   Chemical/Physical. In an aqueous solution, chloramine reacted with methylamine to form *N*-chloromethylamine (Isaac and Morris, 1983).
   Reacts with acids forming water-soluble salts.

Exposure Limits: NIOSH REL: TWA 10 ppm (12 mg/m^3), IDLH 100 ppm; OSHA PEL: TWA 10 ppm.

Symptoms of Exposure: Severe irritant to eyes, skin and respiratory tract (Patnaik, 1992).

Toxicity: LC_{50} (inhalation) for mice 2,400 mg/kg/2-hour (RTECS, 1985).

Uses: Tanning; intermediate for accelerators, dyes, pharmaceuticals, insecticides, fungicides, tanning, surface active agents, fuel additive, dyeing of acetate textiles; polymerization inhibitor; ingredient in paint removers; photographic developer; solvent; rocket propellant; fuel additive; solvent; in organic synthesis.

METHYLANILINE

Synonyms: Anilinomethane; MA; (Methylamino)benzene; *N*-Methylaminoben-zene; *N*-Methylaniline; **N-Methylbenzenamine**; Methylphenylamine; *N*-Methyl-phenylamine; Monomethylaniline; *N*-Monomethylaniline; *N*-Phenylmethyl-amine; UN 2294.

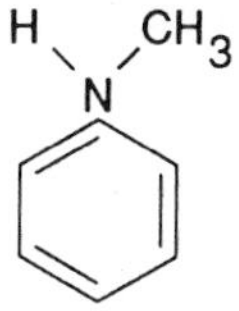

CAS Registry No.: 100-61-8
DOT: 2294
Molecular formula: C_7H_9N
Formula weight: 107.16
RTECS: BY4550000

Physical state, color and odor
Colorless to yellow to pale brown liquid with a faint, ammonia-like odor. Odor threshold in air is 1.7 ppm (Amoore and Hautala, 1983).

Melting point (°C):
-57 (Weast, 1986)

Boiling point (°C):
196.25 (Weast, 1986)
190-191 (Hawley, 1981)

Density (g/cm^3):
0.9891 at 20/4 °C (Weast, 1986)

Diffusivity in water (10^5 cm^2/sec):
0.84 at 20 °C using method of Hayduk and Laudie (1974).

Dissociation constant:
4.848 at 25 °C (Gordon and Ford, 1972)

Flash point (°C):
80.1 (NIOSH, 1994)

Henry's law constant (10^5 atm·m^3/mol):
1.19 at 25 °C (approximate - calculated from water solubility and vapor pressure)

Ionization potential (eV):
7.32 (Franklin et al., 1969)

Soil sorption coefficient, log K_{oc}:
2.28 (Meylan et al., 1992)

Octanol/water partition coefficient, log K_{ow}:
1.66, 1.82 (Leo et al., 1971)
1.40 (Johnson and Westall, 1990)

Solubility in organics:
Soluble in alcohol (Weast, 1986) and ether (Windholz et al., 1983).

Solubility in water:
5.624 g/L at 25 °C (Chiou et al., 1982)

Vapor density:
4.38 g/L at 25 °C, 3.70 (air = 1)

Vapor pressure (mmHg):
0.3 at 20 °C, 0.65 at 30 °C (Verschueren, 1983)

Environmental Fate
   Soil. Reacts slowly with humic acids or humates forming quinoidal structures (Parris, 1980).

Exposure Limits: NIOSH REL: TWA 0.5 ppm (2 mg/m^3), IDLH 100 ppm; OSHA PEL: TWA 2 ppm (9 mg/m^3); ACGIH TLV: TWA 0.5 ppm.

Toxicity: LC_{50} (48-hour) for red killifish 355 mg/L (Yoshioka et al., 1986).

Uses: Solvent; acid acceptor; organic synthesis.

2-METHYLANTHRACENE

Synonym: β-Methylanthracene.

CAS Registry No.: 613-12-7
Molecular formula: $C_{15}H_{12}$
Formula weight: 192.96
RTECS: CB0680000

Physical state
Solid.

Melting point (°C):
209 (Weast, 1986)

Boiling point (°C):
Sublimes (Weast, 1986)
358.5 (Wilhoit and Zwolinski, 1971)

Density (g/cm³):
1.165 using method of Lyman et al. (1982).

Diffusivity in water (10^5 cm²/sec):
0.56 at 20 °C using method of Hayduk and Laudie (1974).

Dissociation constant:
>14 (Schwarzenbach et al., 1993)

Soil sorption coefficient, log K_{oc}:
5.12 using method of Karickhoff et al. (1979).

Octanol/water partition coefficient, log K_{ow}:
5.52 using method of Yalkowsky and Valvani (1979).

Solubility in organics:
Soluble in benzene and chloroform (Weast, 1986).

Solubility in water:
In µg/kg: 7.06 at 6.3 °C, 8.48 at 9.1 °C, 9.43 at 10.8 °C, 11.1 at 13.9 °C, 14.5 at 18.3

°C, 19.1 at 23.1 °C, 24.2 at 27.0 °C, 32.1 at 31.1 °C (May et al., 1978)
21.3 μg/kg at 25 °C (May et al., 1978a)
39 μg/L at 25 °C (Mackay and Shiu, 1977)
In nmol/L: 39.2, 50.4, 63.9, 84.0 and 117 at 8.8, 12.9, 17.0, 21.1 and 25.3 °C, respectively. In seawater (salinity = 36.5 g/kg): 15.6, 19.5, 26.2, 35.7, 49.4 and 70.8 at 4.6, 8.8, 12.9, 17.0, 21.1 and 25.3 °C, respectively (Whitehouse, 1984).

Uses: Organic synthesis.

METHYL BROMIDE

Synonyms: Brom-o-gas; Brom-o-gaz; **Bromomethane**; Celfume; Dawson 100; Dowfume; Dowfume MC-2; Dowfume MC-2 soil fumigant; Dowfume MC-33; Edco; Embafume; Fumigant-1; Halon 1001; Iscobrome; Kayafume; MB; M-B-C Fumigant; MBX; MEBR; Metafume; Methogas; Monobromomethane; Pestmaster; Profume; R 40B1; RCRA waste number U029; Rotox; Terabol; Terr-o-gas 100; UN 1062; Zytox.

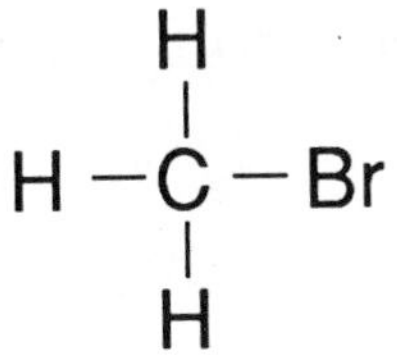

$$
\begin{array}{c}
\mathsf{H} \\
| \\
\mathsf{H-C-Br} \\
| \\
\mathsf{H}
\end{array}
$$

CAS Registry No.: 74-83-9
DOT: 1062
DOT label: Poison
Molecular formula: CH_3Br
Formula weight: 94.94
RTECS: PA4900000

Physical state, color and odor
Colorless liquid or gas with an odor similar to chloroform at high concentrations.

Melting point (°C):
-93.6 (Weast, 1986)

Boiling point (°C):
3.55 (Kudchadker et al., 1979)
4.5 (Worthing and Hance, 1991)

Density (g/cm^3):
1.6755 at 20/4 °C (Weast, 1986)
1.732 at 0/0 °C (Sax, 1984)

Diffusivity in water (10^5 cm^2/sec):
1.23 at 20 °C using method of Hayduk and Laudie (1974).

Flash point (°C):
Practically nonflammable (NFPA, 1984)

Lower explosive limit (%):
10 (NFPA, 1984)

649

Upper explosive limit (%):
16 (NFPA, 1984)

Heat of fusion (kcal/mol):
1.429 (Dean, 1987)

Henry's law constant (10^2 atm·m^3/mol):
62.3 (Glew and Moelyn-Hughes, 1953)
3.18 (low ionic strength, Jury et al., 1984)

Ionization potential (eV):
10.54 (NIOSH, 1994)
10.53 (Franklin et al., 1969)
10.69 (Gibson, 1982)

Soil sorption coefficient, log K_{oc}:
1.92 using method of Kenaga and Goring (1980).

Octanol/water partition coefficient, log K_{ow}:
1.00 (Mills et al., 1985)
1.19 (Hansch and Leo, 1979; Leo et al., 1975)

Solubility in organics:
Soluble in ethanol, ether (Weast, 1986), chloroform, carbon disulfide, carbon tetrachloride and benzene (ITII, 1986).

Solubility in water (mg/L):
13,200 at 25 °C (Gordon and Ford, 1972)
20,700 at 20 °C, 24,137 at 25 °C (Glew and Moelyn-Hughes, 1953)
17,500 at 20 °C under 748 mmHg atmosphere consisting of methyl bromide and water vapor (Standen, 1964)
In g/kg: 26.79, 18.30, 13.41 and 11.49 at 10, 17, 25 and 32 °C, respectively (Haight, 1951)

Vapor density:
3.88 g/L at 25 °C, 3.28 (air = 1)

Vapor pressure (mmHg):
1,633 at 25 °C (Howard, 1989)
1,420 at 20 °C (U.S. EPA, 1976)

Environmental Fate
   Chemical/Physical. Hydrolyzes in water forming methanol and hydrobromic

acid. The estimated hydrolysis half-life in water at 25 °C and pH 7 is 20 days (Mabey and Mill, 1978). Forms a voluminous crystalline hydrate at 0-5 °C (Keith and Walters, 1992).

When methyl bromide was heated to 550 °C in the absence of oxygen, methane, hydrobromic acid, hydrogen, bromine, ethyl bromide, anthracene, pyrene and free radicals were produced (Chaigneau et al., 1966). Emits toxic bromide fumes when heated to decomposition (Lewis, 1990).

Photolytic. When methyl bromide and bromine gas (concentration = 3%) were irradiated at 1850 Å, methane was produced (Kobrinsky and Martin, 1968).

Exposure Limits: Potential occupational carcinogen. NIOSH REL: IDLH 250 ppm; OSHA PEL: ceiling 20 ppm (80 mg/m^3); ACGIH TLV: TWA 5 ppm (20 mg/m^3).

Symptoms of Exposure: Inhalation may cause headache, visual disturbance, vertigo, nausea, vomiting, malaise, hand tremor, convulsions, eye and skin irritation (NIOSH, 1994).

Toxicity: Acute oral LD_{50} for rats is 100 mg/kg (Ashton and Monaco, 1991); LC_{50} (inhalation) for rats 302 ppm/8-hour, mice 1,540 mg/m^3/2-hour (RTECS, 1985).

Drinking Water Standard: As of May 1994, no MCLGs or MCLs have been proposed (U.S. EPA, 1994).

Uses: Soil, space and food fumigant; organic synthesis; fire extinguishing agent; refrigerant; disinfestation of potatoes, tomatoes and other crops; solvent for extracting vegetable oils.

2-METHYL-1,3-BUTADIENE

Synonyms: Hemiterpene; Isoprene; β-Methylbivinyl; 2-Methylbutadiene; UN 1218.

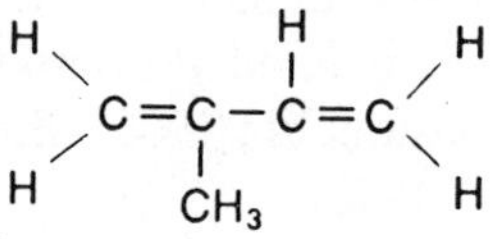

CAS Registry No.: 78-79-5
DOT: 1218 (inhibited)
Molecular formula: C_5H_8
Formula weight: 68.12
RTECS: NT4037000

Physical state and color
Colorless, volatile liquid.

Melting point (°C):
-146 (Weast, 1986)

Boiling point (°C):
34 (Weast, 1986)

Density (g/cm^3):
0.6810 at 20/4 °C (Weast, 1986)

Diffusivity in water (10^5 cm^2/sec):
0.88 at 20 °C using method of Hayduk and Laudie (1974).

Dissociation constant:
>14 (Schwarzenbach et al., 1993)

Flash point (°C):
-53.9 (Sax and Lewis, 1987)

Lower explosive limit (%):
1.5 (NFPA, 1984)

Upper explosive limit (%):
8.9 (NFPA, 1984)

Heat of fusion (kcal/mol):
1.115 (Dean, 1987)

Henry's law constant (10^2 atm·m^3/mol):
7.7 at 25 °C (Hine and Mookerjee, 1975)

Ionization potential (eV):
8.845 ± 0.005 (Franklin et al., 1969)

Soil sorption coefficient, log K_{oc}:
Unavailable because experimental methods for estimation of this parameter for aliphatic hydrocarbons are lacking in the documented literature.

Octanol/water partition coefficient, log K_{ow}:
1.76 using method of Hansch et al. (1968).

Solubility in organics:
Miscible with alcohol and ether (Windholz et al., 1983).

Solubility in water:
642 mg/kg at 25 °C (McAuliffe, 1966)

Vapor density:
2.78 g/L at 25 °C, 2.35 (air = 1)

Vapor pressure (mmHg):
493 at 20 °C, 700 at 30 °C (Verschueren, 1983)
550.1 at 25 °C (Wilhoit and Zwolinski, 1971)

Environmental Fate

Photolytic. Methyl vinyl ketone and methacrolein were reported as major photooxidation products for the reaction of 2-methyl-1,3-butadiene with hydroxyl radicals. Formaldehyde, nitrogen dioxide, nitric oxide and HO$_2$ were reported as minor products (Lloyd et al., 1983). Synthetic air containing gaseous nitrous acid and exposed to artificial sunlight (λ = 300-450 nm) photooxidized 2-methyl-1,3-butadiene into formaldehyde, methyl nitrate, peroxyacetal nitrate and a compound tentatively identified as methyl vinyl ketone (Cox et al., 1980).

Chemical/Physical. Slowly oxidizes and polymerizes in air (Huntress and Mulliken, 1941).

Toxicity: LC$_{50}$ (inhalation) for mice 139 gm/m^3/2-hour, rats 180 gm/m^3/4-hour (RTECS, 1985).

Uses: Manufacture of butyl and synthetic rubber; gasoline component; organic synthesis.

2-METHYLBUTANE

Synonyms: Ethyldimethylmethane; Isoamylhydride; Isopentane; UN 1265.

$$H-\underset{\underset{H}{|}}{\overset{\overset{H}{|}}{C}}-\underset{\underset{CH_3}{|}}{\overset{\overset{H}{|}}{C}}-\underset{\underset{H}{|}}{\overset{\overset{H}{|}}{C}}-\underset{\underset{H}{|}}{\overset{\overset{H}{|}}{C}}-H$$

CAS Registry No.: 78-78-4
DOT: 1265
Molecular formula: C_5H_{12}
Formula weight: 72.15
RTECS: EK4430000

Physical state, color and odor
Colorless liquid with a pleasant odor.

Melting point (°C):
-159.9 (Weast, 1986)

Boiling point (°C):
27.8 (Weast, 1986)
28.88 (Wilhoit and Zwolinski, 1971)

Density (g/cm^3):
0.61067 at 20/4 °C, 0.61462 at 25/4 °C (Dreisbach, 1959)
0.6201 at 20/4 °C (Weast, 1986)

Diffusivity in water (10^5 cm^2/sec):
0.81 at 20 °C using method of Hayduk and Laudie (1974).

Dissociation constant:
>14 (Schwarzenbach et al., 1993)

Flash point (°C):
-57 (Hawley, 1981)

Lower explosive limit (%):
1.4 (Sax and Lewis, 1987)

Upper explosive limit (%):
7.6 (Sax and Lewis, 1987)

Heat of fusion (kcal/mol):
1.231 (Dean, 1987)

Henry's law constant (atm·m^3/mol):
1.25 at 25 °C (approximate - calculated from water solubility and vapor pressure)

Interfacial tension with water (dyn/cm at 20 °C):
49.64 (Demond and Lindner, 1993)

Ionization potential (eV):
10.32 (Franklin et al., 1969)

Soil sorption coefficient, log K_{oc}:
Unavailable because experimental methods for estimation of this parameter for aliphatic hydrocarbons are lacking in the documented literature.

Octanol/water partition coefficient, log K_{ow}:
2.23 (Coates et al., 1985)

Solubility in organics:
Soluble in alcohol, ether (Weast, 1986), hydrocarbons and oils (Hawley, 1981).

Solubility in water (mg/kg):
51.8 at 23 °C (Coates et al., 1985)
48 at 25 °C (Price, 1976)
47.8 at 25 °C (McAuliffe, 1963, 1966)
72.4 at 0 °C, 49.6 at 25 °C (Polak and Lu, 1973)

Vapor density:
2.95 g/L at 25 °C, 2.49 (air = 1)

Vapor pressure (mmHg):
621 at 22.04 °C (Schumann et al., 1942)
628 at 22.44 °C (Willingham et al., 1945)

Environmental Fate
   Photolytic. Synthetic air containing gaseous nitrous acid and exposed to artificial sunlight (λ = 300-450 nm) photooxidized 2-methylbutane into acetone, acetalde- hyde, methyl nitrate, peroxyacetal nitrate, propyl nitrate and pentyl nitrate (Cox et al., 1980).

Symptoms of Exposure: May be narcotic at high concentrations (NIOSH, 1994; Patnaik, 1992).

Uses: Solvent; blowing agent for polystyrene; manufacturing chlorinated derivatives.

3-METHYL-1-BUTENE

Synonyms: Isopentene; Isopropylethylene; α-Isoamylene; UN 2561.

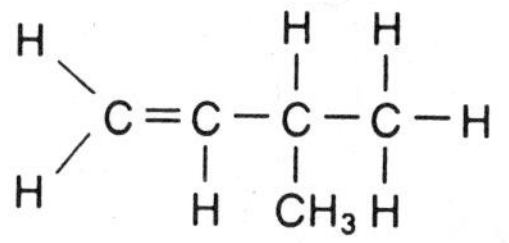

$$H_2C=CH-CH(CH_3)-CH_3$$

CAS Registry No.: 563-45-1
DOT: 2561
DOT label: Combustible liquid
Molecular formula: C_5H_{10}
Formula weight: 70.13
RTECS: EM7600000

Physical state, color and odor
Colorless liquid with a disagreeable odor.

Melting point (°C):
-168.5 (Weast, 1986)

Boiling point (°C):
20 (Weast, 1986)
21, 25 (Verschueren, 1983)

Density (g/cm³):
0.6272 at 20/4 °C, 0.6219 at 25/4 °C (Dreisbach, 1959)

Diffusivity in water (10^5 cm²/sec):
0.83 at 20 °C using method of Hayduk and Laudie (1974).

Dissociation constant:
>14 (Schwarzenbach et al., 1993)

Flash point (°C):
-57 (Hawley, 1981)

Lower explosive limit (%):
1.5 (NFPA, 1984)

Upper explosive limit (%):
9.1 (NFPA, 1984)

657

Heat of fusion (kcal/mol):
1.281 (Dean, 1987)

Henry's law constant (atm·m³/mol):
0.535 at 25 °C (Hine and Mookerjee, 1975)

Ionization potential (eV):
9.51 ± 0.03 (Franklin et al., 1969)

Soil sorption coefficient, log K_{oc}:
Unavailable because experimental methods for estimation of this parameter for aliphatic hydrocarbons are lacking in the documented literature.

Octanol/water partition coefficient, log K_{ow}:
2.30 using method of Hansch et al. (1968).

Solubility in organics:
Soluble in alcohol, benzene and ether (Weast, 1986).

Solubility in water:
130 mg/kg at 25 °C (McAuliffe, 1966)

Vapor density:
2.87 g/L at 25 °C, 2.42 (air = 1)

Vapor pressure (mmHg):
902.1 at 25 °C (Wilhoit and Zwolinski, 1971)

Uses: In high octane fuels; organic synthesis.

METHYL CELLOSOLVE

Synonyms: Dowanol EM; EGM; EGME; Ektasolve; Ethylene glycol methyl ether; Ethylene glycol monomethyl ether; Glycol ether EM; Glycol methyl ether; Glycol monomethyl ether; Jeffersol EM; MECS; **2-Methoxyethanol**; Methoxyhydroxyethane; Methyl ethoxol; Methyl glycol; Methyl oxitol; Poly-solv EM; Prist; UN 1188.

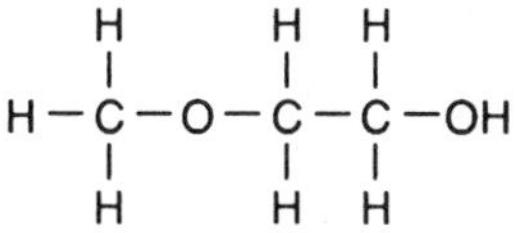

$$H-\underset{\underset{H}{|}}{\overset{\overset{H}{|}}{C}}-O-\underset{\underset{H}{|}}{\overset{\overset{H}{|}}{C}}-\underset{\underset{H}{|}}{\overset{\overset{H}{|}}{C}}-OH$$

CAS Registry No.: 109-86-4
DOT: 1188
DOT label: Flammable liquid
Molecular formula: $C_3H_8O_2$
Formula weight: 76.10
RTECS: KL5775000

Physical state, color and odor
Colorless liquid with a mild, ether-like odor. Odor threshold in air is 900 ppb (Keith and Walters, 1992).

Melting point (°C):
-85.1 (Weast, 1986)

Boiling point (°C):
124.6 (Dean, 1987)

Density (g/cm^3):
0.9647 at 20/4 °C (Weast, 1986)

Diffusivity in water (10^5 cm^2/sec):
1.02 at 20 °C using method of Hayduk and Laudie (1974).

Flash point (°C):
46.1 (open cup, Windholz et al., 1983)
39 (NFPA, 1984)

Lower explosive limit (%):
1.8 (NIOSH, 1994)
1.8 at 0 °C (NFPA, 1984)

659

Upper explosive limit (%):
14 (NIOSH, 1994)
14 at 0 °C (NFPA, 1984)

Henry's law constant (10^2 atm·m^3/mol):
4.41, 3.63, 11.6, 3.09 and 3.813 at 10, 15, 20, 25 and 30 °C, respectively (Ashworth et al., 1988).

Ionization potential (eV):
9.60 (NIOSH, 1994)

Soil sorption coefficient, log K_{oc}:
Unavailable because experimental methods for estimation of this parameter for methoxy alcohols are lacking in the documented literature. However, its miscibility in water suggests its adsorption to soil will be nominal (Lyman et al., 1982).

Octanol/water partition coefficient, log K_{ow}:
Unavailable because experimental methods for estimation of this parameter for methoxy alcohols are lacking in the documented literature.

Solubility in organics:
Very soluble in acetone, dimethylsulfoxide and 95% ethanol (Keith and Walters, 1992). Miscible with N,N-dimethylformamide, ether and glycerol (Windholz et al., 1983).

Solubility in water:
Miscible (Price et al., 1974).

Vapor density:
3.11 g/L at 25 °C, 2.63 (air = 1)

Vapor pressure (mmHg):
6.2 at 20 °C, 14 at 30 °C (Verschueren, 1983)
9.4 at 25 °C (Banerjee et al., 1990)

Exposure Limits: NIOSH REL: TWA 0.1 ppm (0.3 mg/m^3), IDLH 200 ppm; OSHA PEL: TWA 25 ppm (80 mg/m^3); ACGIH TLV: TWA 5 ppm (16 mg/m^3).

Symptoms of Exposure: Inhalation of vapors may cause headache, weakness, eye irritation, ataxia and tremor (Patnaik, 1992).

Toxicity: Acute oral LD$_{50}$ for guinea pigs 950 mg/kg, rats 2,460 mg/kg, rabbits 890 mg/kg (RTECS, 1985).

Uses: Solvent for natural and synthetic resins, cellulose acetate, nitrocellulose and some dyes; nail polishes; dyeing leather; sealing moisture-proof cellophane; lacquers, varnishes, enamels, wood stains; in solvent mixtures; perfume fixative; jet fuel deicing additive.

METHYL CELLOSOLVE ACETATE

Synonyms: Acetic acid 2-methoxyethyl ester; Ethylene glycol methyl ether acetate; Ethylene glycol monomethyl ether acetate; Glycol ether EM acetate; Glycol monomethyl ether acetate; **2-Methoxyethanol acetate**; 2-Methoxyethyl acetate; Methyl glycol acetate; Methyl glycol monoacetate; UN 1189.

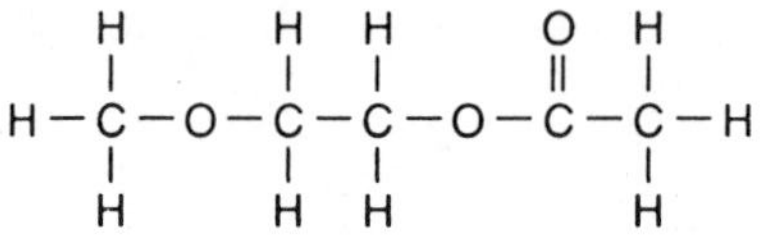

$$
\begin{array}{ccccccc}
H & & H & H & & O & H \\
| & & | & | & & \| & | \\
H-C-O-C-C-O-C-C-H \\
| & & | & | & & & | \\
H & & H & H & & & H \\
\end{array}
$$

CAS Registry No.: 110-49-6
DOT: 1189
DOT label: Combustible liquid
Molecular formula: $C_5H_{10}O_3$
Formula weight: 118.13
RTECS: KL5950000

Physical state, color and odor
Colorless liquid with a mild, ether-like odor.

Melting point (°C):
-65 (NIOSH, 1994)
-70 (Windholz et al., 1983)

Boiling point (°C):
144-145 (Weast, 1986)

Density (g/cm³):
1.0090 at 19/19 °C (Weast, 1986)

Diffusivity in water (10^5 cm²/sec):
0.81 at 19 °C using method of Hayduk and Laudie (1974).

Flash point (°C):
48.9 (NIOSH, 1987)
54 (Windholz et al., 1983)

Lower explosive limit (%):
1.7 (NFPA, 1984)

Upper explosive limit (%):
8.2 (NFPA, 1984)

Soil sorption coefficient, log K_{oc}:
Unavailable because experimental methods for estimation of this parameter for cellosolve esters are lacking in the documented literature. However, its miscibility in water suggests its adsorption to soil will be nominal (Lyman et al., 1982).

Octanol/water partition coefficient, log K_{ow}:
Unavailable because experimental methods for estimation of this parameter for cellosolve esters are lacking in the documented literature.

Solubility in organics:
Soluble in alcohol and ether (Weast, 1986).

Solubility in water:
Miscible (Lyman et al., 1982).

Vapor density:
4.83 g/L at 25 °C, 4.08 (air = 1)

Vapor pressure (mmHg at 20 °C):
2 (NIOSH, 1994)
7 (Verschueren, 1983)

Environmental Fate
   Chemical/Physical. Hydrolyzes in water forming methyl cellosolve and acetic acid.

Exposure Limits: NIOSH REL: TWA 0.1 ppm (0.5 mg/m^3), IDLH 200 ppm; OSHA PEL: TWA 25 ppm (120 mg/m^3); ACGIH TLV: TWA 5 ppm (24 mg/m^3).

Toxicity: Acute oral LD_{50} for guinea pigs 1,250 mg/kg, mice 3,390 mg/kg (RTECS, 1985).

Uses: Solvent for cellulose acetate, nitrocellulose, various gums, resins, waxes, oils; textile printing; lacquers; dopes; textile printing; photographic film.

METHYL CHLORIDE

Synonyms: Artic; **Chloromethane**; Monochloromethane; RCRA waste number U045; UN 1063.

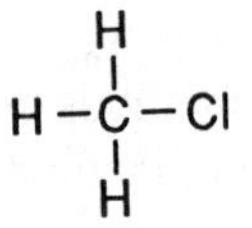

$$H-\overset{\displaystyle H}{\underset{\displaystyle H}{\overset{|}{\underset{|}{C}}}}-Cl$$

CAS Registry No.: 74-87-3
DOT: 1063
DOT label: Flammable gas
Molecular formula: CH_3Cl
Formula weight: 50.48
RTECS: PA6300000

Physical state, color and odor
Liquified compressed gas, colorless, odorless or sweet, ethereal odor.

Melting point (°C):
-97.1 (Weast, 1986)
-97.6 (McGovern, 1943)

Boiling point (°C):
-24.22 (Dreisbach, 1959)
-23.76 (McGovern, 1943)

Density (g/cm^3 at 20/4 °C):
0.9159 (Weast, 1986)
0.9214 (Riddick et al., 1986)

Diffusivity in water (10^5 cm^2/sec):
1.49 at 25 °C (Hayduk and Laudie, 1974)

Flash point (°C):
-50 (NFPA, 1984)

Lower explosive limit (%):
8.1 (NIOSH, 1994)

Upper explosive limit (%):
17.4 (NIOSH, 1994)

Heat of fusion (kcal/mol):
1.537 (Dean, 1987)

Henry's law constant (10^3 atm·m^3/mol):
8.82 (Gossett, 1987)
7.34 (McConnell et al., 1975)
9.41 (Glew and Moelyn-Hughes, 1953)
6.6 (low ionic strength, Pankow and Rosen, 1988)
10 at 25 °C (Hine and Mookerjee, 1975)
In seawater: 3.9, 4.6 and 5.3 at 0, 3 and 6 °C, respectively (Moore et al., 1995)

Ionization potential (eV):
11.26, 11.28 (Horvath, 1982)
11.33 (Gibson, 1977; Yoshida et al., 1983)
11.3 (Franklin et al., 1969)

Soil sorption coefficient, log K_{oc}:
1.40 using method of Chiou et al. (1979).

Octanol/water partition coefficient, log K_{ow}:
0.91 (Hansch et al., 1975)

Solubility in organics:
Miscible with chloroform, ether and glacial acetic acid (U.S. EPA, 1985).

Solubility in water:
7.4 g/kg at 30 °C (McGovern, 1943)
6,450–7,250 mg/L at 20 °C (Pearson and McConnell, 1975)
5,350 mg/L at 25 °C (Glew and Moelyn-Hughes, 1953)
0.648 wt % at 30 °C (Riddick et al., 1986)
4,800 mg/L at 25 °C (Standen, 1964)
0.46 wt % at 20 °C (Nathan, 1978)
In mmol/atm: 167 at 10.2 °C, 97.9 at 23.5 °C, 71.4 at 36.9 °C, 69.1 at 37.4 °C, 45.2
    at 59.2 °C (Boggs and Buck, 1958)

Vapor density:
2.06 g/L at 25 °C, 1.74 (air = 1)

Vapor pressure (mmHg):
3,756 at 20 °C (McConnell et al., 1975)
4,962 at 30 °C, 7,313 at 50 °C (Hsu et al., 1964)
4,309.7 at 30 °C (Howard, 1989)
4,028 at 25 °C (Nathan, 1978)

Environmental Fate

Biological. Enzymatic degradation of methyl chloride was reported to yield formaldehyde (Vogel et al., 1987).

Photolytic. Reported photooxidation products via hydroxyl radicals include formyl chloride, carbon monoxide, hydrogen chloride and phosgene (Spence et al., 1976). In the presence of water, formyl chloride hydrolyzes to hydrochloric acid and carbon monoxide, whereas phosgene hydrolyzes to hydrogen chloride and carbon monoxide (Morrison and Boyd, 1971).

Chemical/Physical. The estimated hydrolysis in water at 25 °C and pH 7 is 0.93 years (Mabey and Mill, 1978).

The evaporation half-life of methyl chloride (1 mg/L) from water at 25 °C using a shallow-pitch propeller stirrer at 200 rpm at an average depth of 6.5 cm is 27.6 minutes (Dilling, 1977).

Exposure Limits: Potential occupational carcinogen. NIOSH REL: IDLH 2,000 ppm; OSHA PEL: TWA 100 ppm, ceiling 200 ppm, 5-minute/3-hour peak 300 ppm; ACGIH TLV: TWA 50 ppm (105 mg/m^3), STEL 100 ppm (205 mg/m^3).

Symptoms of Exposure: Inhalation of vapors may cause headache, dizziness, drowsiness, nausea, vomiting, convulsions, coma and respiratory failure (Patnaik, 1992).

Toxicity: LC_{50} (inhalation) for mice 3,146 ppm/7-hour, rats 152,000 mg/m^3/30-minute (RTECS, 1985).

Drinking Water Standard: As of May 1994, no MCLGs or MCLs have been proposed although methyl chloride has been listed for regulation (U.S. EPA, 1994).

Uses: Coolant and refrigerant; herbicide and fumigant; organic synthesis-methylating agent; manufacturing of silicone polymers, pharmaceuticals, tetraethyl lead, synthetic rubber, methyl cellulose, agricultural chemicals and non-flammable films; preparation of methylene chloride, carbon tetrachloride, chloroform; low temperature solvent and extractant; catalytic carrier for butyl rubber polymerization; topical anesthetic; fluid for thermometric and thermostatic equipment.

METHYLCYCLOHEXANE

Synonyms: Cyclohexylmethane; Hexahydrotoluene; Sextone B; Toluene hexa-hydride; UN 2296.

CAS Registry No.: 108-87-2
DOT: 2296
Molecular formula: C_7H_{14}
Formula weight: 98.19
RTECS: GV6125000

Physical state and color
Colorless liquid. Odor threshold in air is 630 ppm (Amoore and Hautala, 1983).

Melting point (°C):
-126.6 (Weast, 1986)

Boiling point (°C):
100.9 (Weast, 1986)

Density (g/cm³):
0.7694 at 20/4 °C, 0.76506 at 25/4 °C (Dreisbach, 1959)
0.7864 at 0/4 °C (Sax and Lewis, 1987)

Diffusivity in water (10^5 cm²/sec):
0.77 at 20 °C using method of Hayduk and Laudie (1974).

Dissociation constant:
>14 (Schwarzenbach et al., 1993)

Flash point (°C):
-3.9 (NIOSH, 1994)

Lower explosive limit (%):
1.2 (NIOSH, 1994)

Upper explosive limit (%):
6.7 (NIOSH, 1994)

Heat of fusion (kcal/mol):
1.614 (Dean, 1987)

Henry's law constant (atm·m^3/mol):
0.435 at 25 °C (Hine and Mookerjee, 1975)

Ionization potential (eV):
9.85 ± 0.03 (Franklin et al., 1969)

Soil sorption coefficient, log K_{oc}:
Unavailable because experimental methods for estimation of this parameter for alicyclic hydrocarbons are lacking in the documented literature.

Octanol/water partition coefficient, log K_{ow}:
2.82 (Hansch and Leo, 1979)

Solubility in organics:
In methanol, g/L: 269 at 5 °C, 298 at 10 °C, 332 at 15 °C, 372 at 20 °C, 422 at 25 °C, 488 at 30 °C, 575 at 35 °C, 709 at 40 °C (Kiser et al., 1961).

Solubility in water:
In mg/kg: 16.0 at 25 °C, 18.0 at 40.1 °C, 18.9 at 55.7 °C, 33.8 at 99.1 °C, 79.5 at 120.0 °C, 139.0 at 137.3 °C, 244.0 at 149.5 °C (Price, 1976)
14.0 mg/kg at 25 °C (McAuliffe, 1963, 1966)
15.2 mg/L at 20 °C (Burris and MacIntyre, 1986)
At 25 °C: 16.7 mg/L (distilled water), 11.5 mg/L (3.3% NaCl) (Groves, 1988)

Vapor density:
4.01 g/L at 25 °C, 3.39 (air = 1)

Vapor pressure (mmHg):
47.7 at 24.46 °C (Willingham et al., 1945)
36.14 at 20 °C (Burris and MacIntyre, 1986)

Environmental Fate
  Biological. May be oxidized by microbes to 4-methylcyclohexanol, which may further oxidize to give 4-methylcycloheptanone (Dugan, 1972).

Exposure Limits: NIOSH REL: TWA 400 ppm (1,600 mg/m^3), IDLH 1,200 ppm; OSHA PEL: TWA 500 ppm (2,000 mg/m^3); ACGIH TLV: TWA 400 ppm.

Symptoms of Exposure: Vapors may irritate mucous membranes (NIOSH, 1994; Patnaik, 1992).

Toxicity: Acute oral LD_{50} for mice 2,250 mg/kg; LC_{50} (inhalation) for mice 41,500 mg/m^3/2-hour (RTECS, 1985).

Uses: Solvent for cellulose ethers and other organics; gasoline component; organic synthesis.

o-METHYLCYCLOHEXANONE

Synonyms: 2-Methylcyclohexanone; Tetrahydro-*o*-cresol.

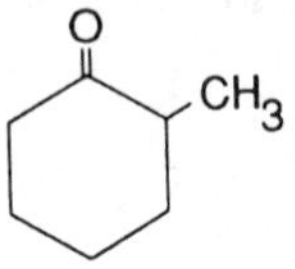

CAS Registry No.: 583-60-8
DOT: 2297
DOT label: Combustible liquid
Molecular formula: $C_7H_{12}O$
Formula weight: 112.17
RTECS: GW1750000

Physical state, color and odor
Colorless liquid with a weak, peppermint-like odor.

Melting point (°C):
-13.9 (Weast, 1986)
-19 (Verschueren, 1983)

Boiling point (°C):
165 at 757 mmHg (Weast, 1986)

Density (g/cm³):
0.9250 at 20/4 °C (Weast, 1986)

Diffusivity in water (10^5 cm²/sec):
0.79 at 20 °C using method of Hayduk and Laudie (1974).

Flash point (°C):
48.2 (NIOSH, 1994)

Solubility in organics:
Soluble in alcohol and ether (Weast, 1986).

Vapor density:
4.58 g/L at 25 °C, 3.87 (air = 1)

Vapor pressure (mmHg):
1 at 20 °C (NIOSH, 1994)

Exposure Limits: NIOSH REL: TWA 50 ppm (230 mg/m^3), STEL 75 ppm (345 mg/m^3), IDLH 600 ppm; OSHA PEL: TWA 100 ppm (460 mg/m^3); ACGIH TLV: TWA 50 ppm.

Toxicity: Acute oral LD$_{50}$ for rats 2,140 mg/kg, rabbits 1 gm/kg (RTECS, 1985).

Uses: Solvent; lacquers; organic synthesis.

1-METHYLCYCLOHEXENE

Synonyms: 1-Methyl-1-cyclohexene; 2,3,4,5-Tetrahydrotoluene.

CAS Registry No.: 591-49-1
Molecular formula: C_7H_{12}
Formula weight: 96.17

Physical state and color
Colorless liquid.

Melting point (°C):
-121 (Weast, 1986)

Boiling point (°C):
110 (Weast, 1986)

Density (g/cm³):
0.8102 at 20/4 °C, 0.8058 at 25/4 °C (Dreisbach, 1959)

Diffusivity in water (10^5 cm²/sec):
0.80 at 20 °C using method of Hayduk and Laudie (1974).

Dissociation constant:
>14 (Schwarzenbach et al., 1993)

Flash point (°C):
-3 (Dean, 1987)

Henry's law constant (10^2 atm·m³/mol):
7.45 at 25 °C (approximate - calculated from water solubility and vapor pressure)

Soil sorption coefficient, log K_{oc}:
Unavailable because experimental methods for estimation of this parameter for alicyclic hydrocarbons are lacking in the documented literature.

Octanol/water partition coefficient, log K_{ow}:
2.44 using method of Hansch et al. (1968).

Solubility in organics:
Soluble in benzene and ether (Weast, 1986).

Solubility in water:
52 mg/kg at 25 °C (McAuliffe, 1966)

Vapor density:
3.93 g/L at 25 °C, 3.32 (air = 1)

Vapor Pressure (mmHg):
30.6 at 25 °C (estimated using Antoine equation, Dreisbach, 1959)

Uses: Organic synthesis.

METHYLCYCLOPENTANE

Synonym: UN 2298.

CAS Registry No.: 96-37-7
DOT: 2298
DOT label: Combustible liquid
Molecular formula: C_6H_{12}
Formula weight: 84.16
RTECS: GY4640000

Physical state, color and odor
Colorless liquid with a sweetish odor.

Melting point (°C):
-142.4 (Weast, 1986)

Boiling point (°C):
71.8 (Weast, 1986)

Density (g/cm³):
0.74864 at 20/4 °C, 0.74394 at 25/4 °C (Dreisbach, 1959)

Diffusivity in water (10^5 cm²/sec):
0.85 at 20 °C (Witherspoon and Bonoli, 1969)
0.93 at 25 °C (Hayduk and Laudie, 1974)

Dissociation constant:
>14 (Schwarzenbach et al., 1993)

Flash point (°C):
<-7 (NFPA, 1984)

Lower explosive limit (%):
1.0 (NFPA, 1984)

Upper explosive limit (%):
8.35 (NFPA, 1984)

Heat of fusion (kcal/mol):
1.656 (Dean, 1987)

Henry's law constant (atm·m^3/mol):
0.362 at 25 °C (Hine and Mookerjee, 1975)

Soil sorption coefficient, log K_{oc}:
Unavailable because experimental methods for estimation of this parameter for alicyclic hydrocarbons are lacking in the documented literature.

Octanol/water partition coefficient, log K_{ow}:
3.37 (Sangster, 1989)

Solubility in organics:
In methanol, g/L: 380 at 5 °C, 415 at 10 °C, 500 at 15 °C, 595 at 20 °C, 740 at 25 °C, 1,100 at 30 °C. Miscible at higher temperatures (Kiser et al., 1961).

Solubility in water (mg/kg):
41.8 at 25 °C, 18.0 at 40.1 °C, 18.9 at 55.7 °C, 33.8 at 99.1 °C, 79.5 at 120.0 °C, 139.0 at 137.3 °C, 244.0 at 149.5 °C. In NaCl solution at 25 °C (salinity, g/kg): 38 (1.002), 36.3 (10.000), 29.2 (34.472), 27.0 (50.030), 12.7 (125.100), 5.72 (199.900), 3.36 (279.800), 1.89 (358.700) (Price, 1976)
42 at 25 °C (McAuliffe, 1966)
42.6 at 25 °C (McAuliffe, 1963)

Vapor density:
3.44 g/L at 25 °C, 2.91 (air = 1)

Vapor pressure (mmHg):
125.1 at 24.75 °C (Willingham et al., 1945)

Symptoms of Exposure: Vapors may irritate respiratory tract (Patnaik, 1992).

Uses: Extractive solvent; azeotropic distillation agent; organic synthesis.

METHYLENE CHLORIDE

Synonyms: Aerothene MM; DCM; **Dichloromethane**; Freon 30; Methane dichloride; Methylene bichloride; Methylene dichloride; Narcotil; NCI-C50102; RCRA waste number U080; Solaesthin; Solmethine; UN 1593.

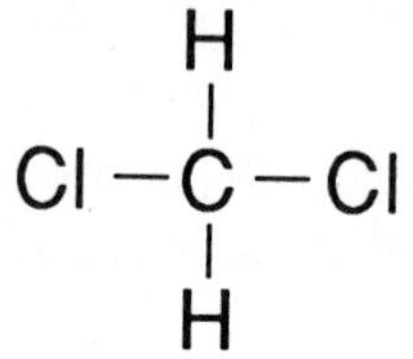

$$Cl - \underset{\underset{H}{|}}{\overset{\overset{H}{|}}{C}} - Cl$$

CAS Registry No.: 75-09-2
DOT: 1593
Molecular formula: CH_2Cl_2
Formula weight: 84.93
RTECS: PA8050000

Physical state, color and odor
Clear, colorless liquid with a sweet, penetrating ethereal odor. Odor threshold in air ranges from 205 to 307 ppm (Keith and Walters, 1992).

Melting point (°C):
-95.14 (Dreisbach, 1959)
-96.7 (Carlisle and Levine, 1932)
-94.92 (Riddick et al., 1986)

Boiling point (°C):
39.75 (Dreisbach, 1959)
40.1 (McConnell et al., 1975)

Density (g/cm³):
1.3361 at 20/4 °C (Carlisle and Levine, 1932)
1.3266 at 20/4 °C (Horvath, 1982)
1.3163 at 25/4 °C (Dreishbach, 1959)

Diffusivity in water (10^5 cm²/sec):
1.15 at 20 °C using method of Hayduk and Laudie (1974).

Flash point (°C):
≥30 (Kuchta et al., 1968)

Lower explosive limit (%):
13 (NIOSH, 1994)
15.1 at 103 °C (Coffee et al., 1972)

Upper explosive limit (%):
23 (NIOSH, 1994)
15.1 at 103 °C (Coffee et al., 1972)

Heat of fusion (kcal/mol):
1.1 (Dean, 1987)

Henry's law constant (10^3 atm·m^3/mol):
2.0 (Pankow and Rosen, 1988)
3.19 at 25 °C (Warner et al., 1987)
2.69 at 25 °C (Dilling, 1977)
2.18 at 25 °C (Hine and Mookerjee, 1975)
3.53 at 37 °C (Sato and Nakajima, 1979)
1.40, 1.69, 2.44, 2.96 and 3.61 at 10, 15, 20, 25 and 30 °C, respectively (Ashworth et al., 1988)
2.1, 3.1, 37 and 4.5 at 20, 30, 35 and 40 °C, respectively (Tse et al., 1992).

Interfacial tension with water (dyn/cm at 20 °C):
28.31 (Demond and Lindner, 1993)

Ionization potential (eV):
11.35 ± 0.01 (Franklin et al., 1969)
11.33 (Horvath, 1982)

Soil sorption coefficient, log K_{oc}:
0.94 (Schwille, 1988)
1.44 (Log K_{om}, Sabljić, 1984)

Octanol/water partition coefficient, log K_{ow}:
1.30 (Mills et al., 1985)
1.25 (Hansch et al., 1975)

Solubility in organics:
Miscible with ethanol and ether (U.S. EPA, 1985).

Solubility in water:
20,000 g/kg (Carlisle and Levine, 1932)
13,000 mg/L at 25 °C (Fluka, 1988; Haque, 1990)
19,400 mg/L at 25 °C (Dilling, 1977)
13,200 mg/L at 25 °C (Pearson and McConnell, 1975)
In mg/L: 23,600 at 0 °C, 21,200 at 10 °C, 19,700 at 30 °C (Standen, 1964)
1.96 wt % at 20 °C (Nathan, 1978)
18,000 mg/L at 17.5 °C, 17,200 mg/L at 26.8 °C (Stephenson, 1992)

22,700 mg/L at 1.5 °C (Suntio et al., 1988)
3.95 x 10^{-3} at 25 °C (mole fraction, Li et al., 1993)

Vapor density:
3.47 g/L at 25 °C, 2.93 (air = 1)

Vapor pressure (mmHg):
440 at 25 °C (ACGIH, 1986)
362.4 at 20 °C (McConnell et al., 1975)
455 at 25 °C (Valsaraj, 1988)
380 at 22 °C (Sax, 1984)
420 at 25 °C (Nathan, 1978)

Environmental Fate

Biological. Complete microbial degradation to carbon dioxide was reported under anaerobic conditions by mixed or pure cultures. Under enzymatic conditions formaldehyde was the only product reported (Vogel et al., 1987). In a static-culture-flask screening test, methylene chloride (5 and 10 mg/L) was statically incubated in the dark at 25 °C with yeast extract and settled domestic wastewater inoculum. After 7 days, 100% biodegradation with rapid adaptation was observed (Tabak et al., 1981).

Under aerobic conditions with sewage seed or activated sludge, complete biodegradation was observed between 6 hours to 1 week (Rittman and McCarty, 1980).

Photolytic. Reported photooxidation products via hydroxyl radicals include carbon dioxide, carbon monoxide, formyl chloride and phosgene (Spence et al., 1976). In the presence of water, phosgene hydrolyzes to hydrochloric acid and carbon dioxide whereas formyl chloride hydrolyzes to hydrogen chloride and carbon monoxide (Morrison and Boyd, 1971).

Chemical/Physical. Under laboratory conditions, methylene chloride hydrolyzed with subsequent oxidation and reduction to produce methyl chloride, methanol, formic acid and formaldehyde (Smith and Dragun, 1984). The experimental half-life for hydrolysis in water at 25 °C is approximately 18 months (Dilling et al., 1975).

The evaporation half-life of methylene chloride (1 mg/L) from water at 25 °C using a shallow-pitch propeller stirrer at 200 rpm at an average depth of 6.5 cm is 20.2 minutes (Dilling, 1977).

Exposure Limits: Potential occupational carcinogen. NIOSH REL: IDLH 2,300 ppm; OSHA PEL: TWA 500 ppm, ceiling 1,000 ppm, 5-minute/2-hour peak 2,000 ppm; ACGIH TLV: TWA 50 ppm (175 mg/m^3).

Symptoms of Exposure: May produce fatigue, weakness, headache, lightheadedness, euphoria, nausea and sleep. May be narcotic at high concentrations (Patnaik, 1992).

Toxicity: Acute oral LD_{50} for rats 2,136 mg/kg; LC_{50} (inhalation) for mice 14,400 ppm/7-hour, rats 88,000 mg/m^3/30-minute (RTECS, 1985).

Drinking Water Standard (final): MCLG: zero; MCL: 5 μg/L (U.S. EPA, 1994).

Uses: Low temperature solvent; ingredient in paint and varnish removers; cleaning, degreasing and drying metal parts; fumigant; manufacturing of aerosols; refrigerant; dewaxing; blowing agent in foams; solvent for cellulose acetate; organic synthesis.

METHYL FORMATE

Synonyms: Formic acid methyl ester; Methyl methanoate; UN 1243.

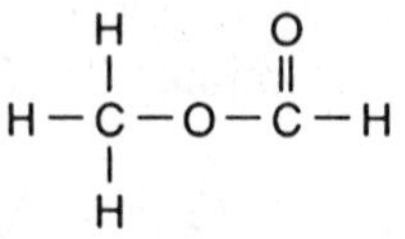

$$H-\underset{\underset{H}{|}}{\overset{\overset{H}{|}}{C}}-O-\overset{\overset{O}{\parallel}}{C}-H$$

CAS Registry No.: 107-31-3
DOT: 1243
DOT label: Flammable liquid
Molecular formula: $C_2H_4O_2$
Formula weight: 60.05
RTECS: LQ8925000

Physical state, color and odor
Clear, colorless, mobile liquid with a pleasant odor. Odor threshold in air is 600 ppm (Amoore and Hautala, 1983).

Melting point (°C):
-99 (Weast, 1986)

Boiling point (°C):
31.5 (Weast, 1986)

Density (g/cm^3):
0.9742 at 20/4 °C (Weast, 1986)
0.96697 at 25/4 °C (Huntress and Mulliken, 1941)

Diffusivity in water (10^5 cm^2/sec):
1.17 at 20 °C using method of Hayduk and Laudie (1974).

Flash point (°C):
-19 (NIOSH, 1994)

Lower explosive limit (%):
4.5 (NFPA, 1984)

Upper explosive limit (%):
23 (NFPA, 1984)

Heat of fusion (kcal/mol):
1.800 (Dean, 1987)

Henry's law constant (10^4 atm·m^3/mol):
2.23 at 25 °C (Hine and Mookerjee, 1975)

Ionization potential (eV):
10.815 ± 0.005 (Franklin et al., 1969)

Soil sorption coefficient, log K_{oc}:
Unavailable because experimental methods for estimation of this parameter for aliphatic esters are lacking in the documented literature. However, its high solubility in water suggests its adsorption to soil will not be nominal (Lyman et al., 1982).

Octanol/water partition coefficient, log K_{ow}:
-0.18 using method of Hansch et al. (1968).

Solubility in organics:
Soluble in ether (Weast, 1986). Miscible with alcohol (Windholz et al., 1983).

Solubility in water:
304 g/L at 20 °C (Verschueren, 1983)

Vapor density:
2.45 g/L at 25 °C, 2.07 (air = 1)

Vapor pressure (mmHg):
476 at 20 °C (NIOSH, 1994)
625 at 25 °C (Abraham, 1984)

Environmental Fate
   Photolytic. Methyl formate, formed from the irradiation of dimethyl ether in the presence of chlorine, degraded to carbon dioxide, water and small amounts of formic acid. Continued irradiation degraded formic acid to carbon dioxide, water and hydrogen chloride (Kallos and Tou, 1977).
   Chemical/Physical. Hydrolyzes slowly in water forming methyl alcohol and formic acid (NIOSH, 1994).

Exposure Limits: NIOSH REL: TWA 100 ppm (250 mg/m^3), STEL 150 ppm (375 mg/m^3), IDLH 4,500 ppm; OSHA PEL: TWA 100 ppm.

Symptoms of Exposure: May irritate eyes, nose and throat; inhalation may produce visual disturbances, narcotic effects and respiratory distress (Patnaik, 1992).

Toxicity: Acute oral LD_{50} for rabbits 1,622 mg/kg (RTECS, 1985).

Uses: Fumigant and larvacide for tobacco, cereals, dried fruits; cellulose acetate solvent; military poison gases; organic synthesis.

3-METHYLHEPTANE

Synonym: 5-Methylheptane.

$$H-\underset{\underset{H}{|}}{\overset{\overset{H}{|}}{C}}-\underset{\underset{H}{|}}{\overset{\overset{H}{|}}{C}}-\underset{\underset{CH_3}{|}}{\overset{\overset{H}{|}}{C}}-\underset{\underset{H}{|}}{\overset{\overset{H}{|}}{C}}-\underset{\underset{H}{|}}{\overset{\overset{H}{|}}{C}}-\underset{\underset{H}{|}}{\overset{\overset{H}{|}}{C}}-\underset{\underset{H}{|}}{\overset{\overset{H}{|}}{C}}-H$$

CAS Registry No.: 589-81-1
Molecular formula: C_8H_{18}
Formula weight: 114.23

Physical state and color
Colorless liquid.

Melting point (°C):
-120.5 (Weast, 1986)

Boiling point (°C):
119 (Weast, 1986)

Density (g/cm³):
0.70582 at 20/4 °C, 0.70175 at 25/4 °C (Dreisbach, 1959)

Diffusivity in water (10^5 cm²/sec):
0.67 at 20 °C using method of Hayduk and Laudie (1974).

Dissociation constant:
>14 (Schwarzenbach et al., 1993)

Heat of fusion (kcal/mol):
2.779 (Dean, 1987)

Henry's law constant (atm·m³/mol):
3.70 at 25 °C (approximate - calculated from water solubility and vapor pressure)

Soil sorption coefficient, log K_{oc}:
Unavailable because experimental methods for estimation of this parameter for aliphatic hydrocarbons are lacking in the documented literature.

Octanol/water partition coefficient, log K_{ow}:
3.97 using method of Hansch et al. (1968).

Solubility in organics:
In methanol, g/L: 154 at 5 °C, 170 at 10 °C, 190 at 15 °C, 212 at 20 °C, 242 at 25 °C, 274 at 30 °C, 314 at 35 °C, 365 at 40 °C (Kiser et al., 1961).

Solubility in water:
792 μg/kg at 25 °C (Price, 1976)

Vapor density:
4.67 g/L at 25 °C, 3.94 (air = 1)

Vapor pressure (mmHg):
19.5 at 25 °C (Wilhoit and Zwolinski, 1971)
8.260 at 10.000 °C, 11.146 at 15.000 °C (Osborn and Douslin, 1974)

Uses: Calibration; gasoline component; organic synthesis.

5-METHYL-3-HEPTANONE

Synonyms: Amyl ethyl ketone; EAK; Ethyl amyl ketone; Ethyl *sec*-amyl ketone; 3-Methyl-5-heptanone.

$$H-\underset{\underset{H}{|}}{\overset{\overset{H}{|}}{C}}-\underset{\underset{H}{|}}{\overset{\overset{H}{|}}{C}}-\overset{\overset{O}{||}}{C}-\underset{\underset{H}{|}}{\overset{\overset{H}{|}}{C}}-\underset{\underset{CH_3}{|}}{\overset{\overset{H}{|}}{C}}-\underset{\underset{H}{|}}{\overset{\overset{H}{|}}{C}}-\underset{\underset{H}{|}}{\overset{\overset{H}{|}}{C}}-H$$

CAS Registry No.: 541-85-5
DOT: 2271
DOT label: Combustible liquid
Molecular formula: $C_8H_{16}O$
Formula weight: 128.21
RTECS: MJ7350000

Physical state, color and odor
Colorless liquid with a fruity odor.

Melting point (°C):
-57.1 (NIOSH, 1994)

Boiling point (°C):
158.5 (NIOSH, 1987)
160, 162 (Verschueren, 1983)

Density (g/cm^3):
0.820–0.824 at 20/20 °C (Windholz et al., 1983)
0.85 at 0/4 °C (Verschueren, 1983)

Diffusivity in water (10^5 cm^2/sec):
0.68 at 20 °C using method of Hayduk and Laudie (1974).

Flash point (°C):
59 (NIOSH, 1994)

Henry's law constant (10^4 atm·m^3/mol):
1.30 at 20 °C (approximate - calculated from water solubility and vapor pressure)

Soil sorption coefficient, log K_{oc}:
Unavailable because experimental methods for estimation of this parameter for aliphatic hydrocarbons are lacking in the documented literature.

Octanol/water partition coefficient, log K_{ow}:
1.96 using method of Hansch et al. (1968).

Solubility in organics:
Mixes readily with alcohols, ether, ketones and other organic solvents (Patnaik, 1992).

Vapor density:
5.24 g/L at 25 °C, 4.43 (air = 1)

Vapor pressure (mmHg):
2 at 20 °C (NIOSH, 1987)
2 at 25 °C (Verschueren, 1983)

Exposure Limits: NIOSH REL: TWA 25 ppm (130 mg/m^3), IDLH 100 ppm; OSHA PEL: TWA 25 ppm.

Symptoms of Exposure: May irritate eyes, nose and throat. At high concentrations, ataxia, prostration, respiratory pain and narcosis may occur (Patnaik, 1992).

Toxicity: Acute oral LD_{50} for guinea pigs 2,500 mg/kg, mice 3,800 mg/kg, rats 3,500 mg/kg (RTECS, 1985).

Uses: Solvent for vinyl resins, nitrocellulose-alkyd and nitrocellulose-maleic acid resins.

2-METHYLHEXANE

Synonyms: Ethylisobutylmethane; Isoheptane.

$$H-\underset{\underset{H}{|}}{\overset{\overset{H}{|}}{C}}-\underset{\underset{CH_3}{|}}{\overset{\overset{H}{|}}{C}}-\underset{\underset{H}{|}}{\overset{\overset{H}{|}}{C}}-\underset{\underset{H}{|}}{\overset{\overset{H}{|}}{C}}-\underset{\underset{H}{|}}{\overset{\overset{H}{|}}{C}}-\underset{\underset{H}{|}}{\overset{\overset{H}{|}}{C}}-H$$

CAS Registry No.: 591-76-4
Molecular formula: C_7H_{16}
Formula weight: 100.20

Physical state and color
Colorless liquid.

Melting point (°C):
-118.3 (Weast, 1986)

Boiling point (°C):
90 (Weast, 1986)

Density (g/cm³):
0.67859 at 20/4 °C, 0.67439 at 25/4 °C (Dreisbach, 1959)

Diffusivity in water (10^5 cm²/sec):
0.70 at 20 °C using method of Hayduk and Laudie (1974).

Dissociation constant:
>14 (Schwarzenbach et al., 1993)

Flash point (°C):
<-17.7 (Hawley, 1981)

Heat of fusion (kcal/mol):
2.195 (Dean, 1987)

Henry's law constant (atm·m³/mol):
3.42 at 25 °C (approximate - calculated from water solubility and vapor pressure)

Soil sorption coefficient, log K_{oc}:
Unavailable because experimental methods for estimation of this parameter for aliphatic hydrocarbons are lacking in the documented literature.

687

Octanol/water partition coefficient, log K_{ow}:
3.30 (Coates et al., 1985)

Solubility in organics:
Soluble in acetone, alcohol, benzene, chloroform, lignoin and ether (Weast, 1986).

Solubility in water:
3.8 mg/L at 23 °C (Coates et al., 1985)
2.54 mg/kg at 25 °C (Price, 1976)

Vapor density:
4.10 g/L at 25 °C, 3.46 (air = 1)

Vapor pressure (mmHg):
65.9 at 25 °C (Wilhoit and Zwolinski, 1971)

Environmental Fate
   Biological. Riser-Roberts (1992) reported 2- and 5-methylhexanoic acids as metabolites by the microorganism *Pseudomonas aeruginosa.*

Use: Organic synthesis. Component of gasoline.

3-METHYLHEXANE

Synonym: 4-Methylhexane.

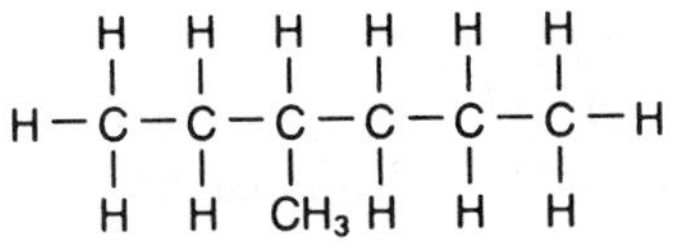

$$
\begin{array}{ccccccc}
 & H & H & H & H & H & H \\
 & | & | & | & | & | & | \\
H- & C- & C- & C- & C- & C- & C-H \\
 & | & | & | & | & | & | \\
 & H & H & CH_3 & H & H & H
\end{array}
$$

CAS Registry No.: 589-34-4
Molecular formula: C_7H_{16}
Formula weight: 100.20

Physical state and color
Colorless liquid.

Melting point (°C):
-119 (Weast, 1986)

Boiling point (°C):
91.85 (Wilhoit and Zwolinski, 1971)

Density (g/cm^3):
0.68713 at 20/4 °C, 0.68295 at 25/4 °C (Dreisbach, 1959)

Diffusivity in water (10^5 cm^2/sec):
0.71 at 20 °C using method of Hayduk and Laudie (1974).

Dissociation constant:
>14 (Schwarzenbach et al., 1993)

Flash point (°C):
-3.9 (Hawley, 1981)

Henry's law constant (atm·m^3/mol):
1.60 at 25 °C (approximate - calculated from water solubility and vapor pressure)

Soil sorption coefficient, log K_{oc}:
Unavailable because experimental methods for estimation of this parameter for aliphatic hydrocarbons are lacking in the documented literature.

Octanol/water partition coefficient, log K_{ow}:
3.41 (Coates et al., 1985)

Solubility in organics:
Soluble in acetone, alcohol, benzene, chloroform, lignoin and ether (Weast, 1986).

Solubility in water:
2.90 mg/L at 23 °C (Coates et al., 1985)
2.64 mg/kg at 25 °C (Price, 1976)
5.24 mg/kg at 0 °C, 4.95 mg/kg at 25 °C (Polak and Lu, 1973)

Vapor density:
4.10 g/L at 25 °C, 3.46 (air = 1)

Vapor pressure (mmHg):
61.6 at 25 °C (Wilhoit and Zwolinski, 1971)

Uses: Oil extender solvent; gasoline component; organic synthesis.

METHYLHYDRAZINE

Synonyms: Hydrazomethane; 1–Methylhydrazine; MMH; Monomethylhydrazine.

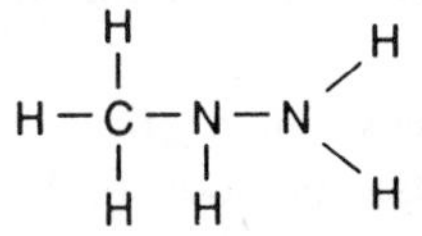

$$H-\underset{\underset{H}{|}}{\overset{\overset{H}{|}}{C}}-N-\underset{\underset{|}{H}}{N}\diagup \overset{H}{\diagdown H}$$

CAS Registry No.: 60-34-4
DOT: 1244
DOT label: Flammable liquid
Molecular formula: CH_6N_2
Formula weight: 46.07
RTECS: MV5600000

Physical state, color and odor
Fuming, clear, colorless liquid with an ammonia-like odor. Odor threshold in air ranges from 1 to 3 ppm (Keith and Walters, 1992).

Melting point (°C):
-52.4 (Weast, 1986)
-20.9 (Sax and Lewis, 1987)

Boiling point (°C):
87.5 (Weast, 1986)

Density (g/cm³):
0.874 at 25/4 °C (Weast, 1986)

Diffusivity in water (10^5 cm²/sec):
1.47 at 25 °C using method of Hayduk and Laudie (1974).

Flash point (°C):
-8.4 (NIOSH, 1994)

Lower explosive limit (%):
2.5 (NIOSH, 1994)

Upper explosive limit (%):
92 (NFPA, 1984)

Ionization potential (eV):
8.00 ± 0.06 (Franklin et al., 1969)

691

Soil sorption coefficient, log K_{oc}:
Unavailable because experimental methods for estimation of this parameter for hydrazines are lacking in the documented literature. However, its miscibility in water suggests its adsorption to soil will be nominal (Lyman et al., 1982).

Octanol/water partition coefficient, log K_{ow}:
Unavailable because experimental methods for estimation of this parameter for hydrazines are lacking in the documented literature.

Solubility in organics:
Soluble in alcohol and ether (Weast, 1986).

Solubility in water:
Miscible.

Vapor density:
1.88 g/L at 25 °C, 1.59 (air = 1)

Vapor pressure (mmHg):
38 at 20 °C (NIOSH, 1994)
49.6 at 25 °C (Sax, 1985)

Environmental Fate
   Biological. It was suggested that the rapid disappearance of methylhydrazine in sterile and nonsterile soil (Arrendondo fine sand) under aerobic conditions was due to chemical oxidation. Though the oxidation product was not identified, it biodegraded to carbon dioxide in the nonsterile soil. The oxidation product did not degrade further in the sterile soil (Ou and Street, 1988).

Exposure Limits: Potential occupational carcinogen. NIOSH REL: 2-hour ceiling 0.04 ppm (0.08 mg/m^3), IDLH 20 ppm; OSHA PEL: ceiling 0.2 ppm (0.35 mg/m^3).

Toxicity: Acute oral LD_{50} for hamsters 22 mg/kg, mice 29 mg/kg, rats 32 mg/kg (RTECS, 1985).

Uses: Rocket fuel; solvent; intermediate; organic synthesis.

METHYL IODIDE

Synonyms: Halon 10001; **Iodomethane**; RCRA waste number U138; UN 2644.

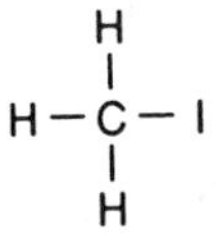

$$H-\overset{\displaystyle H}{\underset{\displaystyle H}{C}}-I$$

CAS Registry No.: 74-88-4
DOT: 2644
DOT label: Poison
Molecular formula: CH_3I
Formula weight: 141.94
RTECS: PA9450000

Physical state and color
Clear, colorless liquid which may become yellow, red or brown on exposure to light and moisture.

Melting point (°C):
-64.4 (Stull, 1947)

Boiling point (°C):
42.4 (Weast, 1986)

Density (g/cm^3):
2.279 at 20/4 °C (Weast, 1986)

Diffusivity in water (10^5 cm^2/sec):
1.17 at 20 °C using method of Hayduk and Laudie (1974).

Flash point (°C):
Noncombustible liquid (NIOSH, 1994).

Henry's law constant (10^3 atm·m^3/mol):
5.48 at 25 °C (Hine and Mookerjee, 1975)
5.87 at 25 °C (Liss and Slater, 1974)
5.26 at 25 °C (Hunter-Smith et al., 1983)
In seawater: 1.7, 3.2 and 5.4 at 0, 10 and 20 °C, respectively (Moore et al., 1995)

Ionization potential (eV):
9.54 (Franklin et al., 1969)

9.1 (Horvath, 1982)
9.86 (Gibson et al., 1977)

Soil sorption coefficient, log K_{oc}:
1.36 (calculated, Mercer et al., 1990)

Octanol/water partition coefficient, log K_{ow}:
1.69 at 19 °C (Collander, 1951)
1.51 (Hansch et al., 1975; Hansch and Leo, 1979)
1.69 (Leo et al., 1971)

Solubility in organics:
Soluble in acetone and benzene (Weast, 1986). Miscible with alcohol and ether (Windholz et al., 1983).

Solubility in water:
14,190 mg/L at 20 °C, 14,290 at 30 °C (Rex, 1906)
14 g/L at 20 °C (Verschueren, 1983)
18.217 mmol/L at 40.34 °C (Swain and Thornton, 1962)
95.9 mmol/L at 22 °C (Fühner, 1924)

Vapor density:
5.80 g/L at 25 °C, 4.90 (air = 1)

Vapor pressure (mmHg):
331 at 20 °C, 483 at 30 °C (Rex, 1906)
405 at 25 °C (calculated, Kudchadker et al., 1979)

Environmental Fate
   Chemical/Physical. Anticipated products from the reaction of methyl iodide with ozone or hydroxyl radicals in the atmosphere are formaldehyde, iodoform-aldehyde, carbon monoxide and iodine radicals (Cupitt, 1980). With hydroxyl radicals, CH_2, methyl radical, HOI and water are possible reaction products (Brown et al., 1990). The estimated half-life of methyl iodide in the atomosphere, based on a measuredrate constant for the vapor phase reaction with hydroxyl radicals, ranges from 535 hours to 32 weeks (Garraway and Donovan, 1979).
   Hydrolyzes in water forming methyl alcohol and hydriodic acid. The estimated half-life in water at 25 °C and pH 7 is 110 days (Mabey and Mill, 1978). May react with chlorides in seawater to form methyl chloride (Zafiriou, 1975).

Exposure Limits: Potential occupational carcinogen. NIOSH REL: TWA 2 ppm (10 mg/m³), IDLH 100 ppm; OSHA PEL: TWA 5 ppm (28 mg/m³); ACGIH TLV: TWA 2 ppm.

Toxicity: LD_{50} (intraperitoneal) for guinea pigs 51 mg/kg, mice 172 mg/kg, rats 101 mg/kg; LD_{50} (subcutaneous) for mice 110 mg/kg (RTECS, 1985).

Uses: Microscopy; medicine; testing for pyridine; methylating agent.

METHYL ISOCYANATE

Synonyms: Isocyanic acid methyl ester; **Isocyanatomethane**; RCRA waste number P064; TL 1450; UN 2480.

$$H-\underset{\underset{H}{|}}{\overset{\overset{H}{|}}{C}}-N=C=O$$

CAS Registry No.: 624-83-9
DOT: 2480
DOT label: Flammable liquid and poison
Molecular formula: C_2H_3NO
Formula weight: 57.05
RTECS: NQ9450000

Physical state, color and odor
Colorless liquid with a sharp, penetrating odor. Odor threshold in air is 2.1 ppm (Amoore and Hautala, 1983).

Melting point (°C):
–45 (Weast, 1986)

Boiling point (°C):
39.1–40.1 (Weast, 1986)

Density (g/cm³):
0.9230 at 27/4 °C (Weast, 1986)

Diffusivity in water (10^5 cm²/sec):
1.34 at 25 °C using method of Hayduk and Laudie (1974).

Flash point (°C):
-7 (NFPA, 1984)

Lower explosive limit (%):
5.3 (NIOSH, 1994)

Upper explosive limit (%):
26 (NIOSH, 1994)

Soil sorption coefficient, log K_{oc}:
Unavailable because experimental methods for estimation of this parameter for isocyanates are lacking in the documented literature.

Octanol/water partition coefficient, log K$_{ow}$:
Unavailable because experimental methods for estimation of this parameter for isocyanates are lacking in the documented literature.

Solubility in water:
10 wt % at 15 °C (NIOSH, 1994)

Vapor density:
2.33 g/L at 25 °C, 1.97 (air = 1)

Vapor pressure (mmHg):
348 at 20 °C (NIOSH, 1994)

Exposure Limits: NIOSH REL: TWA 0.02 ppm (0.05 mg/m^3), IDLH 3 ppm; OSHA PEL: TWA 0.02 ppm.

Symptoms of Exposure: Exposure to vapors may cause lacrimation, nose and throat irritation and respiratory difficulty. Oral intake or absorption through skin may produce asthma, chest pain, dyspnea, pulmonary edema, breathing difficulty and death (Patnaik, 1992).

Toxicity: In mice, the oral LD$_{50}$ is 69 mg/kg (Patnaik, 1992).

Uses: Manufacture of pesticides, e.g., aldicarb and carbaryl; chemical intermediate in production of plastics and polyurethane foams.

METHYL MERCAPTAN

Synonyms: Mercaptomethane; **Methanethiol;** Methyl sulfhydrate; Methyl thio-alcohol; RCRA waste number U153; Thiomethanol; Thiomethyl alcohol; UN 1064.

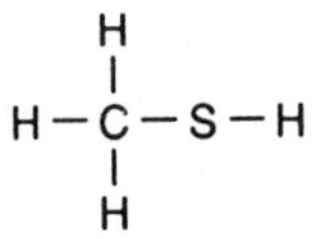

$$
\begin{array}{c}
\text{H} \\
| \\
\text{H} - \text{C} - \text{S} - \text{H} \\
| \\
\text{H}
\end{array}
$$

CAS Registry No.: 74-93-1
DOT: 1064
DOT label: Flammable gas
Molecular formula: CH_4S
Formula weight: 48.10
RTECS: PB4375000

Physical state, color and odor
Colorless gas with a garlic-like or rotten cabbage odor. Odor threshold in air is 1.6 ppb (Amoore and Hautala, 1983).

Melting point (°C):
-123 (Weast, 1986)
-121 (Hawley, 1981)

Boiling point (°C):
6.2 (Weast, 1986)
6-7.6 (Verschueren, 1983)
5.956 (Wilhoit and Zwolinski, 1971)

Density (g/cm³):
0.8665 at 20/4 °C (Weast, 1986)

Diffusivity in water (10^5 cm²/sec):
1.25 at 20 °C using method of Hayduk and Laudie (1974).

Dissociation constant (25 °C):
10.70 (Dean, 1973)

Flash point (°C):
-17.9 (open cup, NIOSH, 1994)

Lower explosive limit (%):
3.9 (NIOSH, 1994)

Upper explosive limit (%):
21.8 (NIOSH, 1994)

Henry's law constant (10^3 atm·m^3/mol):
3.01 at 25 °C (Hine and Mookerjee, 1975)

Ionization potential (eV):
9.440 ± 0.005 (Franklin et al., 1969)

Soil sorption coefficient, log K_{oc}:
Unavailable because experimental methods for estimation of this parameter for mercaptans are lacking in the documented literature.

Octanol/water partition coefficient, log K_{ow}:
Unavailable because experimental methods for estimation of this parameter for mercaptans are lacking in the documented literature.

Solubility in organics:
Soluble in alcohol, ether (Weast, 1986) and petroleum naphtha (Hawley, 1981).

Solubility in water:
23.30 g/L at 20 °C (Windholz et al., 1983)
0.330 mol/L at 25 °C (Hine and Weimar, 1965)

Vapor density:
1.97 g/L at 25 °C, 1.66 (air = 1)

Vapor pressure (mmHg):
1,292 at 20 °C (NIOSH, 1994)
1,516 at 25 °C (Wilhoit and Zwolinski, 1971)

Environmental Fate
Photolytic. Sunlight irradiation of a methyl mercaptan–nitrogen oxide mixture in an outdoor chamber yielded formaldehyde, sulfur dioxide, nitric acid, methyl nitrate, methanesulfonic acid and an inorganic sulfate (Grosjean, 1984).
Chemical/Physical. In the presence of nitric oxide, gaseous methyl mercaptan reacted with hydroxyl radicals to give methyl sulfenic acid and methyl thionitrite (MacLeod et al., 1984). Forms a crystalline hydrate with water (Patnaik, 1992).

Exposure Limits: NIOSH REL: 15-minute ceiling 0.5 ppm (1 mg/m^3), IDLH 150 ppm; OSHA PEL: TWA 10 ppm (20 mg/m^3); ACGIH TLV: TWA 0.5 ppm.

Symptoms of Exposure: Inhalation of vapors may cause headache, narcosis, nausea, pulmonary irritation and convulsions (Patnaik, 1992). Exposure of skin to liquid may cause frostbite (NIOSH, 1994).

Toxicity: LC_{50} (inhalation) for mice 6,530 $\mu g/m^3$/2-hour, rats 675 ppm (RTECS, 1985).

Uses: Synthesis of methionine; intermediate in the manufacture of pesticides, fungicides, jet fuels, plastics; catalyst; added to natural gas to give odor.

METHYL METHACRYLATE

Synonyms: Diakon; Methacrylic acid methyl ester; Methyl-α-methylacrylate; Methyl methacrylate monomer; Methyl-2-methyl-2-propenoate; MME; "Monocite" methacrylate monomer; NA 1247; NCI-C50680; **2-Propenoic acid 2-methyl methyl ester;** RCRA waste number U162; UN 1247.

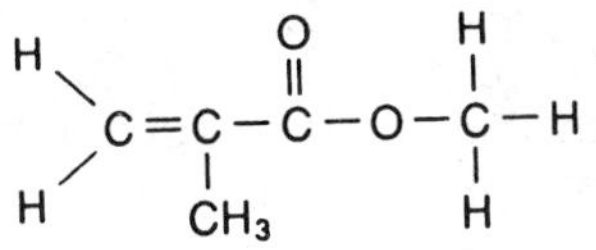

CAS Registry No.: 80-62-6
DOT: 1247
DOT label: Combustible liquid
Molecular formula: $C_5H_8O_2$
Formula weight: 100.12
RTECS: OZ5075000

Physical state, color and odor
Clear, colorless liquid with a penetrating, fruity odor. Odor threshold in air is 83 ppb (Amoore and Hautala, 1983).

Melting point (°C):
-48 (Weast, 1986)
-50 (Verschueren, 1983)

Boiling point (°C):
100-101 (Weast, 1986)

Density (g/cm³ at 20/4 °C):
0.9440 (Weast, 1986)
0.936 (Sax and Lewis, 1987)

Diffusivity in water (10^5 cm²/sec):
0.85 at 20 °C using method of Hayduk and Laudie (1974).

Flash point (°C):
10 (open cup, NIOSH, 1994)

Lower explosive limit (%):
1.7 (NIOSH, 1994)
2.1 (Sax and Lewis, 1987)

701

Upper explosive limit (%):
8.2 (NIOSH, 1994)
12.5 (Sax and Lewis, 1987)

Henry's law constant (10^4 atm·m^3/mol):
2.46 at 20 °C (approximate - calculated from water solubility and vapor pressure)

Soil sorption coefficient, log K_{oc}:
Unavailable because experimental methods for estimation of this parameter for unsaturated esters are lacking in the documented literature.

Octanol/water partition coefficient, log K_{ow}:
1.33 using method of Hansch et al. (1968).

Solubility in organics:
Soluble in acetone, alcohol, ether (Weast, 1986), methyl ethyl ketone, tetrahydrofuran, esters, aromatic and chlorinated hydrocarbons (Windholz et al., 1983).

Solubility in water:
1.5 wt % at 20 °C (NIOSH, 1994)

Vapor density:
4.09 g/L at 25 °C, 3.46 (air = 1)

Vapor pressure (mmHg):
29 at 20 °C (NIOSH, 1994)
28 at 20 °C, 40 at 26 °C, 49 at 30 °C (Verschueren, 1983)

Environmental Fate
   Chemical/Physical. Polymerizes easily (Windholz et al., 1983).

Exposure Limits: NIOSH REL: TWA 100 ppm (410 mg/m^3), IDLH 1,000 ppm; OSHA PEL: TWA 100 ppm.

Toxicity: Acute oral LD$_{50}$ for guinea pigs 6,300 mg/kg, mice 5,204 mg/kg, rats 7,872 mg/kg (RTECS, 1985).

Uses: Manufacturing methacrylate resins and plastics; impregnation of concrete.

2-METHYLNAPHTHALENE

Synonym: β-Methylnaphthalene.

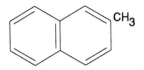

CAS Registry No.: 91-57-6
Molecular formula: $C_{11}H_{10}$
Formula weight: 142.20
RTECS: QJ9635000

Physical state
Solid.

Melting point (°C):
34.6 (Weast, 1986)

Boiling point (°C):
241.052 (Wilhoit and Zwolinski, 1971)

Density (g/cm³):
1.0058 at 20/4 °C (Weast, 1986)

Diffusivity in water (10^5 cm²/sec):
0.72 at 20 °C using method of Hayduk and Laudie (1974).

Dissociation constant:
>15 (Christensen et al., 1975)

Flash point (°C):
97 (Aldrich, 1988)

Heat of fusion (kcal/mol):
2.808 (Dean, 1987)

Heat of fusion (kcal/mol):
2.85 (Tsonopoulos and Prausnitz, 1971)

Henry's law constant (10^4 atm·m³/mol):
3.18 at 25 °C (Fendinger and Glotfelty, 1990)

Ionization potential (eV):
7.955 (HNU, 1986)
8.48 (Yoshida et al., 1983)

Bioconcentration factor, log BCF:
3.79 (Davies and Dobbs, 1984)

Soil sorption coefficient, log K_{oc}:
3.93 (Abdul et al., 1987)
3.87 (Hodson and Williams, 1988)
3.40 (estuarine sediment, Vowles and Mantoura, 1987)
3.66 (Tinker), 3.23 (Carswell), 2.96 (Barksdale), 3.29 (Blytheville), 3.83 (Traverse City), 3.51 (Borden), 3.40 (Lula) (Stauffer et al., 1989).

Octanol/water partition coefficient, log K_{ow}:
4.11 (Abdul et al., 1987)
3.86 (Hansch and Leo, 1979; Yoshida et al., 1983)
3.864 (Krishnamurthy and Wasik, 1978)

Solubility in organics:
Soluble in most solvents (U.S. EPA, 1985).

Solubility in water:
24.6 mg/kg at 25 °C (Eganhouse and Calder, 1976)
25.4 mg/L at 25 °C (Mackay and Shiu, 1977)
1.53 mg/L (water soluble fraction of a 15-component simulated jet fuel mixture
   (JP8) containing 6.3 wt % 2-methylnaphthalene (MacIntyre and deFur, 1985)
27.3 mg/L at 25 °C (Vadas et al., 1991)
20.0 mg/L at 25 °C (Vozňáková et al., 1978)

Vapor pressure (10^2 mmHg at 25 °C):
5.1 (calculated, Gherini et al., 1988)
5.4 (extrapolated, Mackay et al., 1982)

Environmental Fate
   Biological. 2-Naphthoic acid was reported as the biooxidation product of 2-methylnaphthalene by *Nocardia* sp. in soil using *n*-hexadecane as the substrate (Keck et al., 1989).
   Chemical/Physical. An aqueous solution containing chlorine dioxide in the dark for 3.5 days at room temperature oxidized 2-methylnaphthalene into the following: 1-chloro-2-methylnaphthalene, 3-chloro-2-methyl-naphthalene, 1,3-dichloro-2-methylnaphthalene, 3-hydroxymethylnaphthalene, 2-naphthaldehyde, 2-naphthoic acid and 2-methyl-1,4-naphthoquinone (Taymaz et al., 1979).

Noninhibited solutions will slowly hydrolyze in the presence of moisture forming methacrylic acid and methanol.

Toxicity: Acute oral LD_{50} for rats 1,630 mg/kg (RTECS, 1985).

Uses: Organic synthesis; insecticides; jet fuel component. Derived from industrial and experimental coal gasification operations where the maximum concentration detected in gas, liquid and coal tar streams were 2.1, 0.22 and 10 mg/m^3, respectively (Cleland, 1981).

4-METHYLOCTANE

Synonym: 5-Methyloctane.

$$
\begin{array}{ccccccccc}
 & H & H & H & H & H & H & H & H \\
 & | & | & | & | & | & | & | & | \\
H- & C- & C- & C- & C- & C- & C- & C- & C-H \\
 & | & | & | & | & | & | & | & | \\
 & H & H & H & CH_3 & H & H & H & H \\
\end{array}
$$

CAS Registry No.: 2216-34-4
Molecular formula: C_9H_{20}
Formula weight: 128.26

Physical state
Liquid.

Melting point (°C):
-113.2 (Weast, 1986)

Boiling point (°C):
142.4 (Weast, 1986)

Density (g/cm³):
0.7199 at 20/4 °C, 0.7169 at 25/4 °C (Dreisbach, 1959)

Diffusivity in water (10^5 cm²/sec):
0.63 at 20 °C using method of Hayduk and Laudie (1974).

Dissociation constant:
>14 (Schwarzenbach et al., 1993)

Henry's law constant (atm·m³/mol):
10.27 at 25 °C (approximate - calculated from water solubility and vapor pressure)

Soil sorption coefficient, log K_{oc}:
Unavailable because experimental methods for estimation of this parameter for aliphatic hydrocarbons are lacking in the documented literature.

Octanol/water partition coefficient, log K_{ow}:
4.69 using method of Hansch et al. (1968).

Solubility in organics:
Soluble in acetone, alcohol, benzene and ether (Weast, 1986).

Solubility in water:
115 μg/kg at 25 °C (Price, 1976)

Vapor density:
5.24 g/L at 25 °C, 4.43 (air = 1)

Vapor pressure (mmHg):
7 at 25 °C (Wilhoit and Zwolinski, 1971)

Uses: Gasoline component; organic synthesis.

2-METHYLPENTANE

Synonyms: Dimethylpropylmethane; Isohexane; UN 2462.

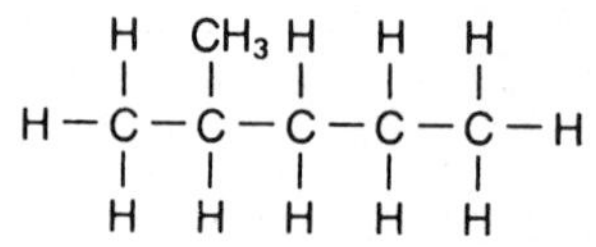

$$H-\overset{\overset{\displaystyle H}{|}}{\underset{\underset{\displaystyle H}{|}}{C}}-\overset{\overset{\displaystyle CH_3}{|}}{\underset{\underset{\displaystyle H}{|}}{C}}-\overset{\overset{\displaystyle H}{|}}{\underset{\underset{\displaystyle H}{|}}{C}}-\overset{\overset{\displaystyle H}{|}}{\underset{\underset{\displaystyle H}{|}}{C}}-\overset{\overset{\displaystyle H}{|}}{\underset{\underset{\displaystyle H}{|}}{C}}-H$$

CAS Registry No.: 107-83-5
DOT: 2462
DOT label: Flammable liquid
Molecular formula: C_6H_{14}
Formula weight: 86.18
RTECS: SA2995000

Physical state and color
Colorless liquid.

Melting point (°C):
-153.7 (Weast, 1986)

Boiling point (°C):
60.3 (Weast, 1986)

Density (g/cm^3):
0.63215 at 20/4 °C, 0.64852 at 25/4 °C (Dreisbach, 1959)

Diffusivity in water (10^5 cm^2/sec):
0.75 at 20 °C using method of Hayduk and Laudie (1974).

Dissociation constant:
>14 (Schwarzenbach et al., 1993)

Flash point (°C):
-23.3 (Hawley, 1981)
-7 (NFPA, 1984)

Lower explosive limit (%):
1.2 (NFPA, 1984)

Upper explosive limit (%):
7.0 (NFPA, 1984)

Heat of fusion (kcal/mol):
1.498 (Dean, 1987)

Henry's law constant (atm·m^3/mol):
1.732 at 25 °C (Hine and Mookerjee, 1975)
0.697, 0.694, 0.633, 0.825 and 0.848 at 10, 15, 20, 25 and 30 °C, respectively
   (Ashworth et al., 1988)

Ionization potential (eV):
10.12 (HNU, 1986)

Soil sorption coefficient, log K_{oc}:
Unavailable because experimental methods for estimation of this parameter for
aliphatic hydrocarbons are lacking in the documented literature.

Octanol/water partition coefficient, log K_{ow}:
2.77 (Coates et al., 1985)

Solubility in organics:
Soluble in acetone, alcohol, benzene, chloroform and ether (Weast, 1986).

Solubility in water:
14.0 mg/L at 23 °C (Coates et al., 1985)
In mg/kg: 13.0 at 25 °C, 13.8 at 40.1 °C, 15.7 at 55.7 °C, 27.1 at 99.1 °C, 44.9 at
   118.0 °C, 86.8 at 137.3 °C, 113.0 at 149.5 °C (Price, 1976)
13.8 mg/kg at 25 °C (McAuliffe, 1963, 1966)
19.45 mg/kg at 0 °C, 15.7 mg/kg at 25 °C (Polak and Lu, 1973)
14.2 mg/L at 25 °C (Leinonen and Mackay, 1973)
16.21 mg/L at 25 °C (Barone et al., 1966)

Vapor density:
3.52 g/L at 25 °C, 2.98 (air = 1)

Vapor pressure (mmHg):
211.8 at 25 °C (Wilhoit and Zwolinski, 1971)

Environmental Fate
   Photolytic. Synthetic air containing gaseous nitrous acid and exposed to artificial
sunlight (λ = 300-450 nm) photooxidized 2-methylpentane into acetone, propionaldehyde, peroxyacetal nitrate, peroxypropionyl nitrate and possibly two isomers of
hexyl nitrate and propyl nitrate (Cox et al., 1980).
   Chemical/Physical: Complete combustion is air yields carbon dioxide and water
vapor.

Symptoms of Exposure: Inhalation of vapors may cause irritation of respiratory tract (Patnaik, 1992).

Uses: Solvent; gasoline component; organic synthesis.

3-METHYLPENTANE

Synonyms: Diethylmethylmethane; UN 2462.

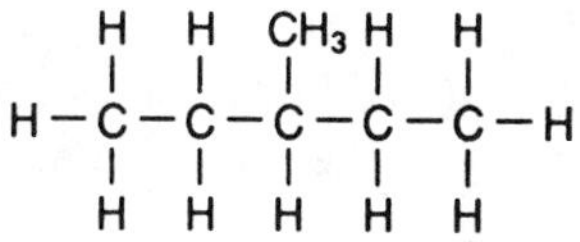

$$H-\overset{\overset{\displaystyle H}{|}}{\underset{\underset{\displaystyle H}{|}}{C}}-\overset{\overset{\displaystyle H}{|}}{\underset{\underset{\displaystyle H}{|}}{C}}-\overset{\overset{\displaystyle CH_3}{|}}{\underset{\underset{\displaystyle H}{|}}{C}}-\overset{\overset{\displaystyle H}{|}}{\underset{\underset{\displaystyle H}{|}}{C}}-\overset{\overset{\displaystyle H}{|}}{\underset{\underset{\displaystyle H}{|}}{C}}-H$$

CAS Registry No.: 96-14-0
DOT: 2462
DOT label: Combustible liquid
Molecular formula: C_6H_{14}
Formula weight: 86.18
RTECS: SA2995500

Physical state and color
Colorless liquid.

Melting point (°C):
-117.8 (Exxon Corp., 1985)

Boiling point (°C):
63.3 (Weast, 1986)

Density (g/cm^3):
0.66431 at 20/4 °C, 0.65976 at 25/4 °C (Dreisbach, 1959)

Diffusivity in water (10^5 cm^2/sec):
0.76 at 20 °C using method of Hayduk and Laudie (1974).

Dissociation constant:
>14 (Schwarzenbach et al., 1993)

Flash point (°C):
<-6.6 (Hawley, 1981)

Lower explosive limit (%):
1.2 (NFPA, 1984)

Upper explosive limit (%):
7.0 (NFPA, 1984)

Henry's law constant (atm·m^3/mol):
1.693 at 25 °C (Hine and Mookerjee, 1975)

711

Ionization potential (eV):
10.08 (HNU, 1986)

Soil sorption coefficient, log K_{oc}:
Unavailable because experimental methods for estimation of this parameter for aliphatic hydrocarbons are lacking in the documented literature.

Octanol/water partition coefficient, log K_{ow}:
2.88 (Coates et al., 1985)

Solubility in organics:
In methanol, g/L: 389 at 5 °C, 450 at 10 °C, 530 at 15 °C, 650 at 20 °C, 910 at 25 °C. Miscible at higher temperatures (Kiser et al., 1961).

Solubility in water:
10.5 mg/L at 23 °C (Coates et al., 1985)
13.1 mg/kg at 25 °C (Price, 1976)
12.8 mg/kg at 25 °C (McAuliffe, 1966)
21.5 mg/kg at 0 °C, 17.9 mg/kg at 25 °C (Polak and Lu, 1973)

Vapor density:
3.52 g/L at 25 °C, 2.98 (air = 1)

Vapor pressure (mmHg):
217.8 at 23.2 °C (Willimgham et al., 1945)

Symptoms of Exposure: Inhalation of vapors may cause irritation to respiratory tract. Narcotic at high concentrations (Patnaik, 1992).

Uses: Solvent; gasoline component; organic synthesis.

4-METHYL-2-PENTANONE

Synonyms: Hexanone; Hexone; Isobutyl methyl ketone; Isopropylacetone; Methyl isobutyl ketone; 2-Methyl-4-pentanone; MIBK; MIK; RCRA waste number U161; Shell MIBK; UN 1245.

$$
\begin{array}{c}
\quad\; \text{H}\quad\; \text{H}\quad\; \text{H}\quad\; \text{O}\quad\; \text{H} \\
\quad\; | \quad\; | \quad\; | \quad\; \| \quad\; | \\
\text{H} - \text{C} - \text{C} - \text{C} - \text{C} - \text{C} - \text{H} \\
\quad\; | \quad\; | \quad\; | \quad\quad\; | \\
\quad\; \text{H}\quad\; | \quad\; \text{H}\quad\quad\; \text{H} \\
\quad\quad\;\; \text{H} - \text{C} - \text{H} \\
\quad\quad\quad\quad\; | \\
\quad\quad\quad\quad\; \text{H}
\end{array}
$$

CAS Registry No.: 108-10-1
DOT: 1245
DOT label: Flammable liquid
Molecular formula: $C_6H_{12}O$
Formula weight: 100.16
RTECS: SA9275000

Physical state, color and odor
Clear, colorless, watery liquid with a mild, pleasant odor. Odor threshold in air is 100 ppb (Keith and Walters, 1992).

Melting point (°C):
-84.7 (Weast, 1986)

Boiling point (°C):
116.8 (Weast, 1986)

Density (g/cm³ at 20/4 °C):
0.7978 (Weast, 1986)
0.8008 (Huntress and Mulliken, 1941)

Diffusivity in water (10^5 cm²/sec):
0.77 at 20 °C using method of Hayduk and Laudie (1974).

Flash point (°C):
17.9 (NIOSH, 1994)

Lower explosive limit (%):
1.2 at 94 °C (NIOSH, 1994)

Upper explosive limit (%):
8.0 at 94 °C (NIOSH, 1994)

Henry's law constant (10^4 atm·m³/mol):
6.6, 3.7, 2.9, 3.9 and 6.8 at 10, 15, 20, 25 and 30 °C, respectively (Ashworth et al., 1988).

Interfacial tension with water (dyn/cm at 25 °C):
10.1 (Donahue and Bartell, 1952)

Ionization potential (eV):
9.30 (Franklin et al., 1969)

Soil sorption coefficient, log K_{oc}:
0.79 (estimated, Montgomery, 1989)

Octanol/water partition coefficient, log K_{ow}:
1.31 (Sangster, 1989)
1.09 (Hansch et al., 1968)

Solubility in organics:
Soluble in acetone, ethanol, benzene, chloroform, ether and many other solvents (U.S. EPA, 1985).

Solubility in water:
0.097 at 25 °C (mole fraction, Amidon et al., 1975)
In wt %: 2.04 at 20 °C, 1.91 at 25 °C, 1.78 at 30 °C (Ginnings et al., 1940)
19 g/L at 20 °C (Fluka, 1988)
1.77 wt % at 25 °C (Lo et al., 1986)

Vapor density:
4.09 g/L at 25 °C, 3.46 (air = 1)

Vapor pressure (mmHg):
16 at 20 °C (NIOSH, 1994)
14.5 at 20 °C (Howard, 1990)
19.9 at 25 °C (Banerjee et al., 1990)

Environmental Fate

Photolytic. Synthetic air containing gaseous nitrous acid and exposed to artificial sunlight (λ = 300-450 nm) converted 4-methyl-2-pentanone into acetone, peroxyacetal nitrate and methyl nitrate (Cox et al., 1980). In a subsequent experiment, the hydroxyl-initiated photooxidation of 4-methyl-2-pentanone in a smog chamber produced acetone (90% yield) and peroxyacetal nitrate (Cox et al., 1981). Irradiation at 3130 Å resulted in the formation of acetone, propyldiene and free radicals (Calvert and Pitts, 1966).

Exposure Limits: NIOSH REL: TWA 50 ppm (205 mg/m^3), STEL 75 ppm (300 mg/m^3), IDLH 500 ppm; OSHA PEL: TWA 100 ppm (410 mg/m^3); ACGIH TLV: TWA 50 ppm, STEL 75 ppm.

Symptoms of Exposure: Mild irritant and strong narcotic (Patnaik, 1992).

Toxicity: Acute oral LD$_{50}$ for mice 2,671 mg/kg, rats 2,080 mg/kg; LC$_{50}$ (inhalation) for rats 8,000 ppm/4-hour (RTECS, 1985).

Uses: Denaturant for ethyl alcohol; solvent for paints, varnishes, cellulose acetate, nitrocellulose lacquers, resins, fats, oils and waxes; preparation of methyl amyl alcohol; in hydraulic fluids and antifreeze; extraction of uranium from fission products; organic synthesis.

2-METHYL-1-PENTENE

Synonyms: 2-Methylpentene; 1-Methyl-1-propylethene; 1-Methyl-1-propyl-ethylene.

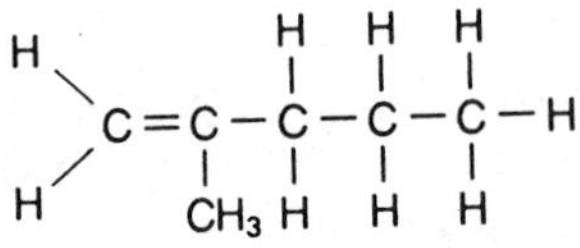

$$\begin{array}{c} H \\ \diagdown \\ H \diagup \end{array} C = C - \overset{\overset{\displaystyle H}{|}}{C} - \overset{\overset{\displaystyle H}{|}}{C} - \overset{\overset{\displaystyle H}{|}}{C} - H$$

CAS Registry No.: 763-29-1
DOT: 2288
Molecular formula: C_6H_{12}
Formula weight: 84.16
RTECS: SB2230000

Physical state and color
Colorless liquid.

Melting point (°C):
−135.7 (Weast, 1986)

Boiling point (°C):
60.7 (Weast, 1986)
61.5 (Verschueren, 1983)
62.6 (Hawley, 1981)
62.11 (Wilhoit and Zwolinski, 1971)

Density (g/cm³):
0.6799 at 20/4 °C, 0.6751 at 25/4 °C (Dreisbach, 1959)

Diffusivity in water (10^5 cm²/sec):
0.78 at 20 °C using method of Hayduk and Laudie (1974).

Dissociation constant:
>14 (Schwarzenbach et al., 1993)

Flash point (°C):
−26.1 (Dean, 1987)

Henry's law constant (atm·m³/mol):
0.277 at 25 °C (approximate - calculated from water solubility and vapor pressure)

Soil sorption coefficient, log K_{oc}:
Unavailable because experimental methods for estimation of this parameter for aliphatic hydrocarbons are lacking in the documented literature.

Octanol/water partition coefficient, log K_{ow}:
2.54 using method of Hansch et al. (1968).

Solubility in organics:
Soluble in alcohol, benzene, chloroform and petroleum (Weast, 1986).

Solubility in water:
78 mg/kg at 25 °C (McAuliffe, 1966)

Vapor density:
3.44 g/L at 25 °C, 2.91 (air = 1)

Vapor pressure (mmHg):
195.4 at 25 °C (Wilhoit and Zwolinski, 1971)

Toxicity: LC_{50} (inhalation) for mice 127 gm/m^3/2-hour, rats 115 $gm/^3$/4-hour (RTECS, 1985).

Uses: Flavors; perfumes; medicines; dyes; oils; resins; organic synthesis.

4-METHYL-1-PENTENE

Synonyms: 1-Isopropyl-2-methylethene; 1-Isopropyl-2-methylethylene.

$$\begin{array}{c} H \\ \diagdown \\ H \diagup \end{array} C = C - \overset{\overset{\displaystyle H}{|}}{C} - \overset{\overset{\displaystyle H}{|}}{C} - \overset{\overset{\displaystyle H}{|}}{C} - H$$

CAS Registry No.: 691-37-2
Molecular formula: C_6H_{12}
Formula weight: 84.16

Physical state and color
Colorless liquid.

Melting point (°C):
-153.6 (Weast, 1986)

Boiling point (°C):
53.9 (Weast, 1986)

Density (g/cm³):
0.6642 at 20/4 °C, 0.6594 at 25/4 °C (Dreisbach, 1959)

Diffusivity in water (10^5 cm²/sec):
0.77 at 20 °C using method of Hayduk and Laudie (1974).

Dissociation constant:
>14 (Schwarzenbach et al., 1993)

Flash point (°C):
-31.6 (Hawley, 1981)

Henry's law constant (atm·m³/mol):
0.615 at 25 °C (Hine and Mookerjee, 1975)

Soil sorption coefficient, log K_{oc}:
Unavailable because experimental methods for estimation of this parameter for aliphatic hydrocarbons are lacking in the documented literature.

Octanol/water partition coefficient, log K_{ow}:
2.70 using method of Hansch et al. (1968).

Solubility in organics:
Soluble in alcohol, benzene, chloroform and petroleum (Weast, 1986).

Solubility in water:
48 mg/kg at 25 °C (McAuliffe, 1966)

Vapor density:
3.44 g/L at 25 °C, 2.91 (air = 1)

Vapor pressure (mmHg):
270.8 at 25 °C (Wilhoit and Zwolinski, 1971)

Uses: Manufacture of plastics used in automobiles, laboratory ware and electronic components; organic synthesis.

1-METHYLPHENANTHRENE

Synonym: α–Methylphenanthrene.

CAS Registry No.: 832–69–9
Molecular formula: $C_{15}H_{12}$
Formula weight: 192.26
RTECS: SF7810000

Physical state and color
White powder or solid.

Melting point (°C):
123 (Weast, 1986)

Boiling point (°C):
358.6 (Wilhoit and Zwolinski, 1971)

Density (g/cm^3):
1.161 using method of Lyman et al. (1982).

Diffusivity in water (10^5 cm^2/sec):
0.55 at 20 °C using method of Hayduk and Laudie (1974).

Dissociation constant:
>14 (Schwarzenbach et al., 1993)

Soil sorption coefficient, log K_{oc}:
4.56 using method of Karickhoff et al. (1979).

Octanol/water partition coefficient, log K_{ow}:
5.27 using method of Yalkowsky and Valvani (1979).

Solubility in organics:
Soluble in alcohol (Weast, 1986).

Solubility in water (μg/L):
95.2 at 6.6 °C, 114 at 8.9 °C, 147 at 14.0 °C, 193 at 19.2 °C, 255 at 24.1 °C, 304 at

26.9 °C, 355 at 29.9 °C (May et al., 1978)
269 at 25 °C (May et al., 1978a)
173 at 25 °C, 300 in seawater at 22 °C (Verschueren, 1983)

Uses: Chemical research; organic synthesis.

2-METHYLPHENOL

Synonyms: 2-Cresol; *o*-Cresol; *o*-Cresylic acid; 1-Hydroxy-2-methylbenzene; 2-Hydroxytoluene; *o*-Hydroxytoluene; 2-Methylhydroxybenzene; *o*-Methylhydroxybenzene; *o*-Methylphenol; *o*-Methylphenylol; Orthocresol; *o*-Oxytoluene; RCRA waste number U052; 2-Toluol; *o*-Toluol; UN 2076.

CAS Registry No.: 95-48-7
DOT: 2076
DOT label: Corrosive material, poison
Molecular formula: C_7H_8O
Formula weight: 108.14
RTECS: GO6300000

Physical state, color and odor
Colorless solid or liquid with a phenolic odor; darkens on exposure to air.

Melting point (°C):
30.9 (Weast, 1986)

Boiling point (°C):
191.0 (Dean, 1973)

Density (g/cm³ at 20/4 °C):
1.0273 (Weast, 1986)
1.0465 (Standen, 1965)

Diffusivity in water (10^5 cm²/sec):
0.77 at 20 °C using method of Hayduk and Laudie (1974).

Dissociation constant:
10.26 at 25 °C (Dean, 1973)

Flash point (°C):
82 (NIOSH, 1994)

Lower explosive limit (%):
1.4 at 150 °C (NIOSH, 1994)

Henry's law constant (10^6 atm·m^3/mol):
1.23 at 25 °C (Hine and Mookerjee, 1975)
1.20 (Gaffney et al., 1987)

Ionization potential (eV):
8.93 (NIOSH, 1994)

Soil sorption coefficient, log K_{oc}:
1.70 (river sediment), 1.75 (coal wastewater sediment) (Kopinke et al., 1995)
1.34 (Brookston clay loam, Boyd, 1982)

Octanol/water partition coefficient, log K_{ow}:
1.93 (Howard, 1989)
1.95 (Leo et al., 1971)
1.99 (Dearden, 1985)
1.96 (Wasik et al., 1981)

Solubility in organics:
Miscible with ethanol, benzene, ether and glycerol (U.S. EPA, 1985).

Solubility in water:
31,000 mg/L at 40 °C, 56,000 mg/L at 100 °C (Verschueren, 1983)
In g/L: 13.0 at 0 °C, 29.0 at 46.2 °C (Standen, 1965)
30.8 g/L at 40 °C (Howard, 1989)
23 g/L at 8 °C, 26 g/L at 25 °C (Leuenberger et al., 1985)
23,000 mg/L at 23 °C (Pinal et al., 1990)
25.2 mmol/L at 25.0 °C (Wasik et al., 1981)

Vapor pressure (mmHg):
0.31 at 25 °C (Howard, 1989)
0.045 at 8 °C, 0.29 at 25 °C (Leuenberger et al., 1985)

Environmental Fate

Biological. Bacterial degradation of 2-methylphenol may introduce a hydroxyl group to produce *m*-methylcatechol (Chapman, 1972). In phenol-acclimated activated sludge, metabolites identified include 3-methylcatechol, 4-methyl-resorcinol and methylhydroquinone. Other metabolites identified include α-ketobutyric acid, dihydroxybenzaldehyde and trihydroxytoluene (Masunaga et al., 1986).

Groundwater contaminated with phenol and other phenols degraded in a methanogenic aquifer to methane and carbon dioxide. These results could not be duplicated in the laboratory utilizing an anaerobic digester (Godsy et al., 1983).

Chloroperoxidase, a fungal enzyme isolated from *Caldariomyces fumago*, reacted with 2-methylphenol forming 2-methyl-4-chlorophenol (38% yield) and 2-methyl-6-chlorophenol (Wannstedt et al., 1990).

Photolytic. Sunlight irradiation of 2-methylphenol and nitrogen oxides in air yielded the following gas-phase products: acetaldehyde, formaldehyde, pyruvic acid, peroxyacetylnitrate, nitrocresols and trace levels of nitric acid and methyl nitrate. Particulate phase products were also identified and these include 2-hydroxy-3-nitrotoluene, 2-hydroxy-5-nitrotoluene, 2-hydroxy-3,5-dinitrotoluene and tentatively identified nitrocresol isomers (Grosjean, 1984).

Chemical/Physical. Ozonation of an aqueous solution containing 2-methylphenol (200-600 mg/L) yielded formic, acetic, propionic, glyoxylic, oxalic and salicylic acids (Wang, 1990). In a different experiment, however, an aqueous solution containing 2-methylphenol (1 mM) reacted with ozone (11.7 mg/minute) forming 2-methylmuconic acid and hydrogen peroxide as end products. The proposed pathway of degradation involved electrophilic aromatic substitution by the first ozone molecule followed by a 1,3-dipolar addition of the second ozone molecule to the cleaved ring (Beltran et al., 1990).

In a smog chamber experiment, 2-methylphenol reacted with nitrogen oxides to form nitrocresols, dinitrocresols and hydroxynitrocresols (McMurry and Grosjean, 1985). Anticipated products from the reaction of 2-methylphenol with ozone or hydroxyl radicals in the atmosphere are hydroxynitrotoluenes and ring cleavage compounds (Cupitt, 1980).

Kanno et al. (1982) studied the aqueous reaction of *o*-cresol and other substituted aromatic hydrocarbons (toluidine, 1-naphthylamine, phenol, *m*- and *p*-cresol, pyrocatechol, resorcinol, hydroquinone and 1-naphthol) with hypochlorous acid in the presence of ammonium ion. They reported that the aromatic ring was not chlorinated as expected but was cleaved by chloramine forming cyanogen chloride. The amount of cyanogen chloride formed was increased as the pH was lowered (Kanno et al., 1982).

Exposure Limits: NIOSH REL: TWA 2.3 ppm (10 mg/m^3), IDLH 250 ppm; OSHA PEL: TWA 5 ppm (22 mg/m^3).

Symptoms of Exposure: May cause weakness, confusion, depression of central nervous system, dyspnea and respiratory failure. Eye and skin irritant. Contact with skin may cause burns and dermatitis. Chronic effects may include gastro-intestinal disorders, nervous disorders, tremor, confusion, skin eruptions, oliguria, jaundice and liver damage (Patnaik, 1992).

Toxicity: Acute oral LD_{50} for rats 121 mg/kg, mice 344 mg/kg; LD_{50} (skin) 620 mg/kg, rats 620 mg/kg, rabbits 890 mg/kg (RTECS, 1985).

Uses: Disinfectant; phenolic resins; tricresyl phosphate; ore flotation; textile scouring agent; organic intermediate; manufacturing salicylaldehyde, coumarin and herbicides; surfactant; synthetic food flavors (*para* isomer only); food anti-oxidant; dye, perfume, plastics and resins manufacturing.

4-METHYLPHENOL

Synonyms: 4-Cresol; *p*-Cresol; *p*-Cresylic acid; 1-Hydroxy-4-methylbenzene; *p*-Hydroxytoluene; 4-Hydroxytoluene; *p*-Kresol; 1-Methyl-4-hydroxybenzene; 4-Methylhydroxybenzene; *p*-Methylhydroxybenzene; *p*-Methylphenol; 4-Oxytoluene; *p*-Oxytoluene; Paracresol; Paramethylphenol; RCRA waste number U052; 4-Toluol; *p*-Toluol; *p*-Tolyl alcohol; UN 2076.

CAS Registry No.: 106-44-5
DOT: 2076
DOT label: Corrosive material, poison
Molecular formula: C_7H_8O
Formula weight: 108.14
RTECS: GO6475000

Physical state, color and odor
Colorless to pink crystals with a phenolic odor. Odor threshold in air is <1 ppm (Keith and Walters, 1992).

Melting point (°C):
34.8 (Weast, 1986)
36 (Huntress and Mulliken, 1941)

Boiling point (°C):
201.9 (Weast, 1986)

Density (g/cm³ at 20/4 °C):
1.0178 (Weast, 1986)
1.0341 (Standen, 1965)

Diffusivity in water (10⁵ cm²/sec):
0.77 at 20 °C using method of Hayduk and Laudie (1974).

Dissociation constant:
10.26 at 25 °C (Dean, 1973)

Flash point (°C):
87 (NIOSH, 1994)

Lower explosive limit (%):
1.1 at 151 °C (NIOSH, 1994)

Henry's law constant (10^7 atm·m³/mol):
7.92 at 25 °C (Hine and Mookerjee, 1975)
10 (Gaffney et al., 1987)

Ionization potential (eV):
8.97 (NIOSH, 1987)

Soil sorption coefficient, log K_{oc}:
1.69 (Brookston clay loam, Boyd, 1982)
2.06 (Dormont soil), 3.53 (Apison and Fullerton soils), 2.06 (Dormont soil) (Southworth and Keller, 1986)
2.81 (Coyote Creek sediments, Smith et al., 1978)
2.70 (Meylan et al., 1992)
K_d = 0.9 mL/g on a Cs^+-kaolinite (Haderlein and Schwarzenbach, 1993)

Octanol/water partition coefficient, log K_{ow}:
1.91, 1.92, 1.95, 1.99, 2.10 (Sangster, 1989)
1.67 (Neely and Blau, 1985)
3.01 (Mackay and Paterson, 1981)
1.92 (Leo et al., 1971)
1.94 (Campbell and Luthy, 1985; Fujita et al., 1964)
1.98 (Garst and Wilson, 1984)

Solubility in organics:
Miscible with ethanol, benzene, ether and glycerol (U.S. EPA, 1985).

Solubility in water:
24,000 mg/L at 40 °C, 53,000 mg/L at 100 °C (Verschueren, 1983)
22.10 mg/L at 30 °C, 54.00 mg/L at 105 °C, 164.0 g/L at 138 °C (Standen, 1965)
13 g/L at 8 °C, 18 g/L at 25 °C (Leuenberger et al., 1985)

Vapor pressure (10^2 mmHg):
4 at 20 °C (Verschueren, 1983)
8 at 25 °C (Valsaraj, 1988)
13 at 25 °C (Howard, 1989)
2.0 at 8 °C, 12 at 25 °C (Leuenberger et al., 1985)

Environmental Fate
  Biological. Protocatechuic acid (3,4-dihydroxybenzoic acid) is the central metabolite in the bacterial degradation of 4-methylphenol. Intermediate by-

products include 4-hydroxybenzyl alcohol, 4-hydroxybenzaldehyde and 4-hydroxybenzoic acid. In addition, 4-methylphenol may undergo hydroxylation to form 4-methylcatechol (Chapman, 1972). Chloroperoxidase, a fungal enzyme isolated from *Caldariomyces fumago*, reacted with 4-methylphenol forming 4-methyl-2-chlorophenol (Wannstedt et al., 1990). Under methanogenic conditions, inocula from a municipal sewage treatment plant digester degraded 4-methylphenol to phenol prior to being mineralized to carbon dioxide and methane (Young and Rivera, 1985).

A species of *Pseudomonas*, isolated from creosote-contaminated soil, degraded 4-methylphenol into 4-hydroxybenzaldehyde and 4-hydroxybenzoate. Both metabolites were then converted into protocatechuate (O'Reilly and Crawford, 1989).

Photolytic. Photooxidation products reported include 2,2'-dihydroxy-4,4'-dimethylbiphenyl, 2-hydroxy-3,4'-dimethylbiphenyl ether and 4-methyl catechol (Smith et al., 1978).

Chemical/Physical. Anticipated products from the reaction of 4-methylphenol with ozone or hydroxyl radicals in the atmosphere are hydroxynitrotoluene and ring cleavage compounds (Cupitt, 1980).

Kanno et al. (1982) studied the aqueous reaction of *p*-cresol and other substituted aromatic hydrocarbons (toluidine, 1-naphthylamine, phenol, *o*- and *m*-cresol, pyrocatechol, resorcinol, hydroquinone and 1-naphthol) with hypochlorous acid in the presence of ammonium ion. They reported that the aromatic ring was not chlorinated as expected but was cleaved by chloramine forming cyanogen chloride. The amount of cyanogen chloride formed was increased as the pH was lowered (Kanno et al., 1982).

Exposure Limits: NIOSH REL: TWA 2.3 ppm (10 mg/m^3), IDLH 250 ppm; OSHA PEL: TWA 5 ppm (22 mg/m^3).

Symptoms of Exposure: May cause weakness, confusion, depression of central nervous system, dyspnea, weak pulse and respiratory failure. May irritate eyes and mucous membranes. Contact with skin may cause burns and dermatitis. Chronic effects may include gastrointestinal disorders, nervous disorders, tremor, confusion, skin eruptions, oliguria, jaundice and liver damage (NIOSH, 1994; Patnaik, 1992).

Toxicity: Acute oral LD$_{50}$ for rats 207 mg/kg, mice 344 mg/kg; LD$_{50}$ (skin) for rats 750 mg/kg, rabbits 301 mg/kg (RTECS, 1985).

Uses: Disinfectant; phenolic resins; tricresyl phosphate; ore flotation; textile scouring agent; organic intermediate; manufacturing of salicylaldehyde, coumarin and herbicides; surfactant; synthetic food flavors.

2-METHYLPROPANE

Synonyms: Isobutane; Liquified petroleum gas; Trimethylmethane; UN 1075; UN 1969.

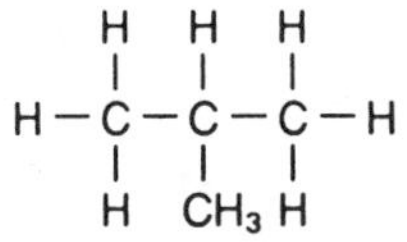

$$H-\underset{\underset{H}{|}}{\overset{\overset{H}{|}}{C}}-\underset{\underset{CH_3}{|}}{\overset{\overset{H}{|}}{C}}-\underset{\underset{H}{|}}{\overset{\overset{H}{|}}{C}}-H$$

CAS Registry No.: 75-28-5
DOT: 1011
DOT label: Flammable gas
Molecular formula: C_4H_{10}
Formula weight: 58.12
RTECS: TZ4300000

Physical state, color and odor
Colorless gas with a faint odor.

Melting point (°C):
-159.4 (Weast, 1986)
-145 (Verschueren, 1983)

Boiling point (°C):
-11.633 (Weast, 1986)
-11.83 (Wilhoit and Zwolinski, 1971)

Density (g/cm³):
0.5572 at 20/4 °C, 0.5510 at 25/4 °C (Dreisbach, 1959)
0.549 at 20/4 °C (Weast, 1986)

Diffusivity in water (10^5 cm²/sec):
0.85 at 20 °C using method of Hayduk and Laudie (1974).

Dissociation constant:
>14 (Schwarzenbach et al., 1993)

Flash point (°C):
-83 (Hawley, 1981)

Lower explosive limit (%):
1.6 (NIOSH, 1994)

Upper explosive limit (%):
8.4 (NFPA, 1984)

Heat of fusion (kcal/mol):
1.085 (Dean, 1987)

Henry's law constant (atm·m^3/mol):
1.171 at 25 °C (Hine and Mookerjee, 1975)

Ionization potential (eV):
10.57 (HNU, 1986)
10.74 (NIOSH, 1994)

Soil sorption coefficient, log K_{oc}:
Unavailable because experimental methods for estimation of this parameter for aliphatic hydrocarbons are lacking in the documented literature.

Octanol/water partition coefficient, log K_{ow}:
2.29 using method of Hansch et al. (1968).

Solubility in organics:
Mole fraction solubilities in 1-butanol: 0.0897, 0.0491 and 0.0308 at 25, 30 and 70 °C, respectively; chlorobenzene: 0.157, 0.0837 and 0.0542 at 25, 30 and 70 °C, respectively and octane: 0.301, 0.161 and 0.101 at 25, 30 and 70 °C, respectively (Hayduk et al., 1988).

Solubility in water:
48.9 mg/kg at 25 °C (McAuliffe, 1963, 1966)

Vapor density:
2.38 g/L at 25 °C, 2.01 (air = 1)

Vapor pressure (mmHg):
2,611 at 25 °C (Riddick et al., 1986)

Exposure Limits: NIOSH REL: TWA 800 ppm (1,900 mg/m^3)

Symptoms of Exposure: An asphyxiate. Inhalation of concentrations at 1% may cause narcosis and drowsiness (Patnaik, 1992).

Uses: Gasoline component; in liquified petroleum gas; organic synthesis.

2-METHYLPROPENE

Synonyms: γ-Butylene; *unsym*-Dimethylethylene; Isobutene; Isobutylene; Methyl-propene; **2-Methyl-1-propene**; 2-Methylpropylene.

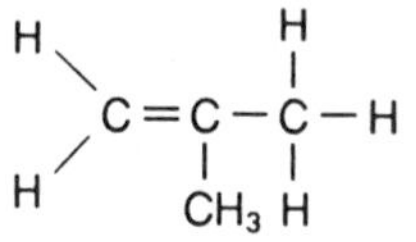

$$\begin{matrix} H & & & H \\ \diagdown & & & | \\ & C=C-C-H \\ \diagup & | & | \\ H & CH_3 & H \end{matrix}$$

CAS Registry No.: 115-11-7
DOT: 1055
DOT label: Flammable gas
Molecular formula: C_4H_8
Formula weight: 56.11
RTECS: UD0890000

Physical state, color and odor
Volatile liquid or colorless gas with a coal gas odor.

Melting point (°C):
-140.3 (Sax and Lewis, 1987)
-146.8 (McAuliffe, 1966)

Boiling point (°C):
-6.900 (Windholz et al., 1983)

Density (g/cm³):
0.5942 at 20/4 °C, 0.5879 at 25/4 °C, 0.5815 at 30/4 °C (Windholz et al., 1983)

Diffusivity in water (10^5 cm²/sec):
0.91 at 20 °C using method of Hayduk and Laudie (1974).

Dissociation constant:
>14 (Schwarzenbach et al., 1993)

Flash point (°C):
-76 (Hawley, 1981)

Lower explosive limit (%):
1.8 (Sax and Lewis, 1987)

Upper explosive limit (%):
9.6 (Sax and Lewis, 1987)

Heat of fusion (kcal/mol):
1.418 (Dean, 1987)

Henry's law constant (atm·m^3/mol):
0.21 at 25 °C (Hine and Mookerjee, 1975)

Ionization potential (eV):
9.23 (HNU, 1986)

Soil sorption coefficient, log K_{oc}:
Unavailable because experimental methods for estimation of this parameter for aliphatic hydrocarbons are lacking in the documented literature.

Octanol/water partition coefficient, log K_{ow}:
2.34, 2.40 (Sangster, 1989)

Solubility in organics (mole fraction):
In 1-butanol: 0.131, 0.0695 and 0.0458 at 25, 30 and 70 °C, respectively; chlorobenzene: 0.234, 0.132 and 0.0796 at 25, 30 and 70 °C, respectively; octane: 0.333, 0.184 and 0.119 at 25, 30 and 70 °C, respectively (Hayduk et al., 1988).

Solubility in water:
263 mg/kg at 25 °C (McAuliffe, 1966)

Vapor density:
2.29 g/L at 25 °C, 1.94 (air = 1)

Vapor pressure (mmHg):
1,976 at 20 °C, 2,736 at 30 °C (Verschueren, 1983)
2,270 at 25 °C (Wilhoit and Zwolinski, 1971)

Environmental Fate
   Chemical/Physical. Products identified from the photoirradiation of 1-butene with nitrogen dioxide in air are 2-butanone, 2-methylpropanal, acetone, carbon monoxide, carbon dioxide, methanol, methyl nitrate and nitric acid (Takeuchi et al., 1983).

Symptoms of Exposure: An asphyxiant (Patnaik, 1992).

Toxicity: LC$_{50}$ (inhalation) for mice 415 gm/m^3/2-hour, rats 620 gm/m^3/4-hour (RTECS, 1985).

Uses: Production of isooctane, butyl rubber, polyisobutene resins, high octane aviation fuels, *t*-butyl chloride, *t*-butyl methacrylates; copolymer resins with acrylonitrile, butadiene and other unsaturated hydrocarbons; organic synthesis.

α-METHYLSTYRENE

Synonyms: AMS; Isopropenylbenzene; **(1-Methylethenyl)benzene**; 1-Methyl-1-phenylethylene; 2-Phenylpropene; β-Phenylpropene; 2-Phenylpropylene; β-Phenylpropylene.

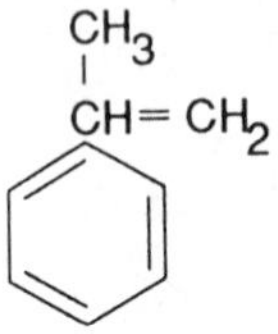

$$CH_3$$
$$CH = CH_2$$

CAS Registry No.: 98-83-9
DOT: 2303
Molecular formula: C_9H_{10}
Formula weight: 118.18
RTECS: WL5250000

Physical state, color and odor
Colorless liquid with a sharp aromatic odor. Odor threshold in air is 290 ppb (Amoore and Hautala, 1983).

Melting point (°C):
24.3 (Weast, 1986)
-96 (Sax and Lewis, 1987)
-23.21 (Hawley, 1981)

Boiling point (°C):
163-164 (Weast, 1986)
152.4 (Sax and Lewis, 1987)
165.38 °C (Hawley, 1981)

Density (g/cm³ at 20/4 °C):
0.9082 (Weast, 1986)
0.862 (Sax and Lewis, 1987)

Diffusivity in water (10^5 cm²/sec):
0.76 at 20 °C using method of Hayduk and Laudie (1974).

Dissociation constant:
>14 (Schwarzenbach et al., 1993)

Flash point (°C):
54.3 (NIOSH, 1994)

Lower explosive limit (%):
1.9 (NIOSH, 1994)

Upper explosive limit (%):
6.1 (NIOSH, 1994)

Ionization potential (eV):
8.35 ± 0.01 (Franklin et al., 1969)

Solubility in organics:
Soluble in benzene and chloroform (Weast, 1986). Miscible with alcohol and ether (Sax and Lewis, 1987).

Vapor density:
4.83 g/L at 25 °C, 4.08 (air = 1)

Vapor pressure (mmHg):
2 at 20 °C (NIOSH, 1994)

Environmental Fate
  Chemical/Physical. Polymerizes in the presence of heat or catalysts (Hawley, 1981).

Exposure Limits: NIOSH REL: TWA 50 ppm (240 mg/m^3), STEL 100 ppm (485 mg/m^3), IDLH 700 ppm; OSHA PEL: ceiling 100 ppm; ACGIH TLV: TWA 50 ppm.

Toxicity: Acute oral LD_{50} for mice 3,160 mg/kg, rats 4 gm/kg; LC_{50} (inhalation) for mice 3,020 mg/m^3 (RTECS, 1985).

Uses: Manufacture of polyesters.

MEVINPHOS

Synonyms: Apavinphos; 2-Butenoic acid 3-[(dimethoxyphosphinyl)oxy]methyl ester; 2-Carbomethoxy-1-methylvinyl dimethyl phosphate; α-Carbomethoxy-1-methylvinyl dimethyl phosphate; 2-Carbomethoxy-1-propen-2-yl dimethyl phosphate; CMDP; Compound 2046; **3-[(Dimethoxyphosphinyl)oxy]-2-butenoic acid methyl ester;** *O,O*-Dimethyl-*O*-(2-carbomethoxy-1-methylvinyl)phosphate; Dimethyl-1-carbomethoxy-1-propen-2-yl phosphate; *O,O*-Dimethyl 1-carbomethoxy-1-propen-2-yl phosphate; Dimethyl 2-methoxycarbonyl-1-methylvinyl phosphate; Dimethyl methoxycarbonylpropenyl phosphate; Dimethyl (1-methoxycarboxypropen-2-yl)phosphate; *O,O*-Dimethyl *O*-(1-methyl-2-carboxyvinyl)phosphate; Dimethyl phosphate of methyl-3-hydroxy-*cis*-crotonate; Duraphos; ENT 22324; Fosdrin; Gesfid; Gestid; 3-Hydroxycrotonic acid methyl ester dimethyl phosphate; Meniphos; Menite; 2-Methoxycarbonyl-1-methylvinyl dimethyl phosphate; *cis*-2-Methoxycarbonyl-1-methylvinyl dimethyl phosphate; 1-Methoxycarbonyl-1-propen-2-yl dimethyl phosphate; Methyl 3-(dimethoxyphosphinyloxy)crotonate; NA 2783; OS 2046; PD 5; Phosdrin; *cis*-Phosdrin; Phosfene; Phosphoric acid (1-methoxycarboxypropen-2-yl) dimethyl ester.

CAS Registry No.: 7786-34-7
DOT: 2783
DOT label: Poison
Molecular formula: $C_7H_{13}O_6P$
Formula weight: 224.16
RTECS: GQ5250000

Physical state, color and odor
Colorless to pale yellow liquid with a weak odor.

Melting point (°C):
6.7 (*trans*), 21.3 (*cis*), (NIOSH, 1994)

Boiling point (°C):
106–107.5 at 1 mmHg (Windholz et al., 1983)
99–103 at 0.03 mmHg (Verschueren, 1983)

Density (g/cm³):
1.25 at 20/4 °C (Windholz et al., 1983)

Diffusivity in water (10^5 cm^2/sec):
0.63 at 20 °C using method of Hayduk and Laudie (1974).

Flash point (°C):
176 (open cup, NIOSH, 1994)

Soil sorption coefficient, log K_{oc}:
Unavailable because experimental methods for estimation of this parameter for miscible insecticides are lacking in the documented literature. However, its miscibility in water suggests its adsorption to soil will be nominal (Lyman et al., 1982).

Octanol/water partition coefficient, log K_{ow}:
Unavailable because experimental methods for estimation of this parameter for miscible insecticides are lacking in the documented literature.

Solubility in organics:
Miscible with acetone, benzene, carbon tetrachloride, chloroform, ethanol, isopropanol, toluene and xylene. Soluble in carbon disulfide and kerosene (50 g/L) (Windholz et al., 1983).

Solubility in water:
Miscible (Gunther et al., 1968).

Vapor density:
9.16 g/L at 25 °C, 7.74 (air = 1)

Vapor pressure (10^3 mmHg):
3 at 20 °C (NIOSH, 1994)
2.2 at 20 °C, 5.7 at 29 °C (Freed et al., 1977)

Environmental Fate
  Plant. In plants, mevinphos is hydrolyzed to phosphoric acid dimethyl ester, phosphoric acid and other less toxic compounds (Hartley and Kidd, 1987). In one day, the compound is almost completely degraded in plants (Cremlyn, 1991). Casida et al. (1956) proposed two degradative pathways of mevinphos in bean plants and cabbage. In the first degradative pathway, cleavage of the vinyl phosphate bond affords methylacetoacetate and acetoacetic acid, which may be precursors to the formation of the end products dimethyl phosphoric acid, methanol, acetone and carbon dioxide. In the other degradative pathway, direct hydrolysis of the carboxylic ester would yield vinyl phosphates as intermediates. The half-life of mevinphos in bean plants was 0.5 days.
  Chemical/Physical. The reported hydrolysis half-lives of *cis*-mevinphos and

trans-mevinphos at pH 11.6 are 1.8 and 3.0 hours, respectively (Casida et al., 1956). Worthing and Hance (1991) reported that pH values of 6, 7, 9 and 11, the hydrolysis half-lives were 120 days, 35 days, 3.0 days and 1.4 hours, respectively (Worthing and Hance, 1991). Emits toxic phosphorus oxide fumes when heated to decomposition (Lewis, 1990).

Exposure Limits: NIOSH REL: TWA 0.01 ppm (0.1 mg/m^3), STEL 0.03 ppm (0.3 mg/m^3), IDLH 4 ppm; OSHA PEL: TWA 0.1 mg/m^3.

Toxicity: Acute oral LD$_{50}$ for wild birds 1.78 mg/kg, ducks 4.6 mg/kg, mice 4 mg/kg, pigeons 4.21 mg/kg, quail 23.7 mg/kg, rats 3 mg/kg (RTECS, 1985).

Uses: Insecticide and acaricide.

MORPHOLINE

Synonyms: Diethyleneimide oxide; Diethyleneimid oxide; Diethylene oximide; Diethylenimide oxide; 1-Oxa-4-azacyclohexane; Tetrahydro-1,4-isoxazine; Tetrahydro-1,4-oxazine; Tetrahydro-2H-1,4-oxazine; UN 2054.

CAS Registry No.: 110-91-8
DOT: 1760 (aqueous), 2054
DOT label: Flammable liquid
Molecular formula: C_4H_9NO
Formula weight: 87.12
RTECS: QD6475000

Physical state, color and odor
Colorless liquid with a weak ammonia-like odor. Hygroscopic. Odor threshold in air is 10 ppb (Amoore and Hautala, 1983).

Melting point (°C):
-4.7 (Weast, 1986)

Boiling point (°C):
128.3 (Weast, 1986)

Density (g/cm³):
1.0005 at 20/4 °C (Weast, 1986)

Diffusivity in water (10^5 cm²/sec):
0.96 at 20 °C using method of Hayduk and Laudie (1974).

Dissociation constant (25 °C):
8.33 (Gordon and Ford, 1972)

Flash point (°C):
37 (NIOSH, 1994)
38 (open cup, Windholz et al., 1983)

Lower explosive limit (%):
1.4 (NFPA, 1984)

739

Upper explosive limit (%):
11.2 (NFPA, 1984)

Ionization potential (eV):
8.88 (NIOSH, 1994)

Soil sorption coefficient, log K_{oc}:
Unavailable because experimental methods for estimation of this parameter for morpholines are lacking in the documented literature. However, its miscibility in water suggest its adsorption to soil will be nominal (Lyman et al., 1982).

Octanol/water partition coefficient, log K_{ow}:
-1.08 (Leo et al., 1971)

Solubility in organics:
Miscible with in acetone, benzene, castor oil, ethanol, ether, ethylene glycol, 2-hexanone, linseed oil, methanol, pine oil and turpentine (Windholz et al., 1983).

Solubility in water:
Miscible (NIOSH, 1994).

Vapor density:
3.56 g/L at 25 °C, 3.01 (air = 1)

Vapor pressure (mmHg):
6 at 20 °C (NIOSH, 1994)
4.3 at 10 °C, 8.0 at 20 °C, 13.4 at 25 °C (Verschueren, 1983)

Environmental Fate
  Chemical/Physical. In an aqueous solution, chloramine reacted with morpholine to form *N*-chloromorpholine (Isaac and Morris, 1983). The aqueous reaction of nitrogen dioxide (1-99 ppm) and morpholine yielded *N*-nitromorpholine and *N*-nitromorpholine (Cooney et al., 1987).

Exposure Limits: NIOSH REL: TWA 20 ppm (70 mg/m^3), STEL 30 ppm (105 mg/m^3), IDLH 1,400 ppm; OSHA PEL: TWA 20 ppm.

Symptoms of Exposure: Inhalation of vapors may cause visual disturbance, nasal irritation, coughing, and at high concentrations, respiratory distress (Patnaik, 1992).

Toxicity: Acute LD_{50} for mice 525 mg/kg, rats 1,050 mg/kg; LD_{50} (skin) for rabbits 500 mg/kg; LC_{50} (inhalation) for mice 1,320 mg/m^3/2-hour, rats 8,000 ppm/8-hour (RTECS, 1985).

Uses: Solvent for waxes, casein, dyes and resins; rubber accelerator; solvent; optical brightener for detergents; corrosion inhibitor; additive to boiler water; preservation of book paper; organic synthesis.

NALED

Synonyms: Arthodibrom; Bromchlophos; Bromex; Dibrom; 1,2-Dibromo-2,2-dichloroethyldimethyl phosphate; Dimethyl 1,2-dibromo-2,2-dichloroethyl phosphate; *O,O*-Dimethyl-*O*-(1,2-dibromo-2,2-dichloroethyl)phosphate; *O,O*-Dimethyl *O*-(2,2-dichloro-1,2-dibromoethyl)phosphate; ENT 24988; Hibrom; NA 2783; Ortho 4355; Orthodibrom; Orthodibromo; **Phosphoric acid 1,2-dibromo-2,2-dichloroethyl dimethyl ester**; RE-4355.

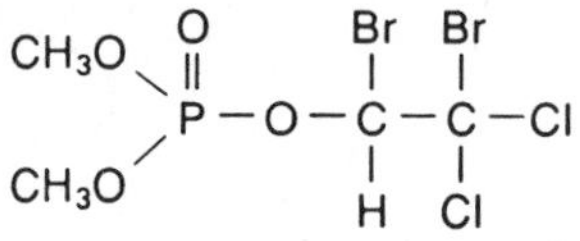

CAS Registry No.: 300-76-5
DOT: 2783
DOT label: Poison
Molecular formula: $C_4H_7Br_2Cl_2O_4P$
Formula weight: 380.79
RTECS: TB9450000

Physical state, color and odor
Colorless to pale yellow liquid or solid with a pungent odor.

Melting point (°C):
26.5-27.5 (Windholz et al., 1983)

Boiling point (°C):
110 at 0.5 mmHg (Windholz et al., 1983)

Density (g/cm³):
1.96 at 25/4 °C (Windholz et al., 1983)

Diffusivity in water (10^5 cm²/sec):
0.68 at 20 °C using method of Hayduk and Laudie (1974).

Flash point (°C):
Noncombustible solid (NIOSH, 1994).

Soil sorption coefficient, log K_{oc}:
Not applicable - reacts with water.

Octanol/water partition coefficient, log K_{ow}:
Not applicable - reacts with water.

Solubility in organics:
Freely soluble in ketone, alcohols, aromatic and chlorinated hydrocarbons but sparingly soluble in petroleum solvents and mineral oils (Windholz et al., 1983).

Vapor pressure (10^3 mmHg at 20 °C):
0.2 (NIOSH, 1994)
2 (Verschueren, 1983)

Environmental Fate
  Chemical/Physical. Completely hydrolyzed in water within 2 days (Windholz et al., 1983). In the presence of metals or reducing agents, dichlorvos is formed (Worthing and Hance, 1991). Emits toxic fumes of bromides, chlorides and phosphorus oxides when heated to decomposition (Lewis, 1990).

Exposure Limits (mg/m^3): NIOSH REL: TWA 3, IDLH 200; OSHA PEL: TWA 3.

Toxicity: Acute oral LD_{50} for ducks 52 mg/kg, mice 330 mg/kg, rats 250 mg/kg; LC_{50} (inhalation) for mice 156 mg/kg, rats 7.70 mg/kg (RTECS, 1985).

Uses: Not produced commercially in the United States. Insecticide; acaricide.

NAPHTHALENE

Synonyms: Camphor tar; Mighty 150; Mighty RD1; Moth balls; Moth flakes; Naphthalin; Naphthaline; Naphthene; NCI-C52904; RCRA waste number U165; Tar camphor; UN 1334; White tar.

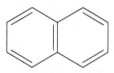

CAS Registry No.: 91-20-3
DOT: 1334 (crude/refined), 2304 (molten)
DOT label: Flammmable solid
Molecular formula: $C_{10}H_8$
Formula weight: 128.18
RTECS: QJ0525000

Physical state, color and odor
White, volatile, crystalline flakes with a strong aromatic odor resembling coal-tar or moth balls. Odor threshold in air is 3 ppb (Keith and Walters, 1992).

Melting point (°C):
80.5 (Weast, 1986)
80.28 (Fowler et al., 1968)

Boiling point (°C):
217.942 (Wilhoit and Zwolinski, 1971)

Density (g/cm³):
0.9625 at 100/4 °C (Weast, 1986)
1.162 at 20/4 °C (ACGIH, 1986)
1.145 at 20/4 °C (Weiss, 1986)
1.01813 at 30/4 °C, 0.9752 at 85/4 °C (Standen, 1967)

Diffusivity in water (10^5 cm²/sec):
0.70 at 20 °C using method of Hayduk and Laudie (1974).

Dissociation constant:
>15 (Christensen et al., 1975)

Flash point (°C):
79.5 (NIOSH, 1994)

Lower explosive limit (%):
0.9 (NIOSH, 1994)

Upper explosive limit (%):
5.9 (NIOSH, 1994)

Heat of fusion (kcal/mol):
4.54 (Tsonopoulos and Prausnitz, 1971)
4.56 (Wauchope and Getzen, 1972)
4.536 (Dean, 1987)

Henry's law constant (10^4 atm·m^3/mol):
4.6 (Pankow and Rosen, 1988)
4.8 (Valsaraj, 1988)
12.2 (Jury et al., 1984)
3.6 (Petrasek et al., 1983)
5.53 (Southworth, 1979)
7.34 at 25 °C (Fendinger and Glotfelty, 1990)

Ionization potential (eV):
8.26 (Yoshida et al., 1983)
8.12 (Franklin et al., 1969)

Bioconcentration factor, log BCF:
2.12 (*Daphnia pulex*, Southworth et al., 1978)
2.63 (fish, Veith et al., 1979)
1.64 (mussel, Lee et al., 1972)
3.00 (activated sludge), 1.48 (golden ide) (Freitag et al., 1985)
2.50 (bluegill sunfish, McCarthy and Jimenez, 1985)
2.11 (algae, Geyer et al., 1984)

Soil sorption coefficient, log K_{oc}:
2.74 (aquifer sands, Abdul et al., 1987)
3.11 (Karickhoff et al., 1979)
3.11, 3.52 (Chin et al., 1988)
2.96, 3.04 (Hodson and Williams, 1988)
3.11, 3.16, 3.21, 3.50 (Rippen et al., 1982)
2.72-3.32 (average = 3.04 for 10 Danish soils, Løkke, 1984)
2.62 (average for 5 soils, Briggs, 1981)
3.15 (Menlo Park soil), 2.76 (Eustis sand) (Podoll et al., 1989)
2.77 (Appalachee soil, Stauffer and MacIntyre, 1986)
2.93 (estuarine sediment, Vowles and Mantoura, 1987)
2.67 (average for 5 soils, Kishi et al., 1990)

3.02, 3.34 (Allerod), 3.10 (Borris), 3.67, 3.83 (Brande), 3.12 (Finderup), 3.12, 3.44 (Gunderup), 3.87 (Herborg), 3.71 (Rabis), 3.07, 3.21 (Tirstrup), 3.10 (Tylstrup), 3.10 (Vasby), 2.53, 3.56 (Vejen), 3.32, 3.69 (Vorbasse) (Larsen et al., 1992)

3.91 (Tinker), 3.24 (Carswell), 2.73 (Barksdale), 3.74 (Traverse City), 3.24 (Borden), 3.13 (Lula, Stauffer et al., 1989)

3.29 (Eustis fine sand, Wood et al., 1990)

5.00 (average, Kayal and Connell, 1990)

3.18 (Oshtemo, Sun and Boyd, 1993)

3.00 (river sediment), 3.08 (coal wastewater sediment) (Kopinke et al., 1995)

Octanol/water partition coefficient, log K_{ow}:

3.36, 3.40, 3.59 (Sangster, 1989)

3.36 (Karickhoff et al., 1979; Briggs, 1981)

3.59 (Mackay, 1982)

3.23, 3.24, 3.26, 3.28 (Brooke et al., 1986)

3.30 (Campbell and Luthy, 1985; Geyer et al., 1984)

3.31 (Kenaga and Goring, 1980)

3.01, 3.20, 3.45 (Leo et al., 1971)

3.37 (Hansch and Fujita, 1964)

3.35 (Bruggeman et al., 1982; Wasik et al., 1981, 1983)

3.29 (DeKock and Lord, 1987)

3.43 (Garst and Wilson, 1984)

3.395 (Krishnamurthy and Wasik, 1978)

Solubility in organics:

Soluble in methanol or ethanol (1 g/13 mL), benzene or toluene (1 g/3.5 mL), olive oil or turpentine (1 g/8 mL), chloroform or carbon tetrachloride (1 g/2 mL), carbon disulfide (1 g/2 mL) (Windholz et al., 1983).

66.2, 95.2, 97.8 and 334.0 g/L in methanol (23 °C), ethanol, acetone and propanol, respectively (Fu et al., 1986).

At 25.0 °C (mol/L): glycerol (0.01052), formamide (0.0539), ethylene glycol (0.0896), methanol (0.579), acetic acid (0.882), acetonitrile (1.715), dimethyl sulfoxide (2.02), acetone (2.75), N,N-dimethylformamide (3.12) (Van Meter and Neumann, 1976).

Solubility in water:

0.239 mmol/L at 25 °C (Wasik et al., 1981, 1983)

30 μg/L at 23 °C (Pinal et al., 1991)

In mg/L: 1.37 at 0 °C, 1.37 at 0.4 °C, 1.38 at 0.5 °C, 1.46 at 0.9 °C, 1.50 at 19 °C, 19.6 at 9.4 °C, 19.4 at 10.0 °C, 23.4 at 14.9 °C, 24.6 at 15.9 °C, 28.0 at 19.3 °C, 34.4 mg/L at 25 °C, 35.8 at 25.6 °C, 43.0 at 30.1 °C, 43.9 at 30.2 °C, 54.5 at 35.2 °C, 54.8 at 36.0 °C, 73.5 at 42.8 °C (Bohon and Claussen, 1951)

31.7 mg/L at 25 °C (Mackay and Shiu, 1977)

31.5 mg/kg at 25 °C (Andrews and Keefer, 1949)

22 mg/L at 25 °C (Schwarz and Wasik, 1976)

33.6 mg/L at 25 °C. In 35% NaCl solution: 23.6 mg/kg at 25 °C (Gordon and Thorne, 1967)

31.3 mg/kg at 25 °C. In artificial seawater (salinity = 35 g/kg): 22.0 mg/kg at 25 °C (Eganhouse and Calder, 1976)

In mg/kg: 28.8, 28.8, 29.1 at 22.2 °C, 30.1, 30.7, 30.8 at 24.5 °C, 38.1, 38.2, 38.3 at 29.9 °C, 37.6, 37.6, 38.1 at 30.3 °C, 43.8, 44.6 at 34.5 °C, 52.6, 52.8 at 39.2 °C, 54.8 at 40.1 °C, 65.3, 65.5, 66.0 at 44.7 °C, 78.6 at 50.2 °C, 106 at 55.6 °C, 151, 157, 166 at 64.5 °C, 240, 244, 247 at 73.4 °C (Wauchope and Getzen, 1972)

37.7 mg/L at 20-25 °C (Geyer et al., 1982)

20.315 mg/L at 25 °C (Sahyun, 1966)

0.22 mmol/L at 21 °C (Almgren et al., 1979)

157, 190 and 234 μmol/L at 12, 18 and 25 °C, respectively (Schwarz and Wasik, 1977)

In μmol/L: 140 at 8.4 °C, 149 at 11.1 °C, 166 at 14.0 °C, 188 at 17.5 °C, 207 at 20.2 °C, 222 at 23.2 °C, 236 at 25.0 °C, 248 at 26.3 °C, 268 at 29.2 °C, 283 at 31.8 °C. In 0.5 M NaCl: 84 at 8.4 °C, 92 at 11.1 °C, 109 at 14.0 °C, 123 at 17.1 °C, 137 at 20.0 °C, 158 at 23.0 °C, 173 at 25.0 °C, 222 at 31.8 °C (Schwarz, 1977)

In mg/L: 13.66 at 1 °C, 29.41 at 23 °C, 53.90 at 40 °C (Klöpffer et al., 1988)

In mg/L: 12.07 at 1.9 °C, 17.19 at 10.7 °C, 21.64 at 15.4 °C, 26.72 at 21.7 °C, 30.72 at 25.2 °C, 40.09 at 30.7 °C, 46.31 at 35.1 °C, 54.80 at 39.3 °C, 68.90 at 44.9 °C (Bennett and Canady, 1984)

30 mg/L at 23 °C (Pinal et al., 1990)

12.5 mg/L at 25 °C (Klevens, 1950)

32.9 mg/L at 25 °C (Walters and Luthy, 1984)

300 μmol/L at 25 °C (Edwards et al., 1991)

At 20 °C: 190, 134, 120 and 146 μmol/L in distilled water, Pacific seawater, artificial seawater and NaCl solution (35 wt %), respectively (Hashimoto et al., 1984)

In NaCl (g/kg) at 25 °C, mg/kg: 30.1 (12.40), 25.2 (25.31), 25.3 (30.59), 20.9 (43.70), 16.9 (61.63) (Paul, 1952)

31.69 mg/kg at 25 °C (May et al., 1978a)

30.6 mg/L at 25 °C (Vadas et al., 1991)

31.3 and 31.9 mg/L at 25 °C (Billington et al., 1988)

235 μmol/L at 25 °C (Akiyoshi et al., 1987)

251 μmol/kg at 25.0 °C (Vesala, 1974)

30.6 mg/L at 23 °C (Fu et al., 1986)

In mmol/L: 0.135 at 5.00 °C, 0.158 at 10.00 °C, 0.190 at 15.00 °C, 0.224 at 20.00 °C, 0.263 at 25.00 °C, 0.324 at 30.00 °C, 0.371 at 35.00 °C, 0.436 at 40.00 °C (Pérez-Tejeda et al., 1990)

254 μmol/L at 25.0 °C (Vesala and Lönnberg, 1980)

234 μmol/L at 25.0 °C (Van Meter and Neumann, 1976)

Vapor pressure (10^2 mmHg):
23 at 25 °C (Mackay and Wolkoff, 1973)
7.8 at 25 °C (Wasik et al., 1983)
17 at 25 °C (Hinckley et al., 1990)
1.32 at 7.15 °C, 2.32 at 12.80 °C, 4.19 at 18.40 °C, 4.45 at 18.85 °C, 9.44 at 26.40 °C
    (Macknick and Prausnitz, 1979)
0.122 at 6.70 °C, 0.141 at 8.10 °C, 0.222 at 12.30 °C, 0.235 at 12.70 °C, 0.263 at
    13.85 °C, 0.320 at 15.65 °C, 0.350 at 16.85 °C, 0.382 at 17.35 °C, 0.383 at 17.55
    °C, 0.438 at 18.70 °C, 0.534 at 20.70 °C (Bradley and Cleasby, 1953)
0.62 at 1 °C, 7.1 at 23 °C, 34 at 40 °C (Klöppfer et al., 1982)
35.4 at 40.33 °C (Fowler et al., 1968)
7.8 at 25 °C (Sonnefeld et al., 1983)
8.2 at 25 °C (Mackay et al., 1982)
21 at 25 °C (Bidleman, 1984)
8.5 at 25 °C (de Kruif, 1980)

Environmental Fate

Biological. In activated sludge, 9.0% of the applied amount mineralized to carbon dioxide after 5 days (Freitag et al., 1985). Under certain conditions, *Pseudomonas* sp. oxidized naphthalene to *cis*-1,2-dihydro-1,2-dihydroxynaphthalene (Dagley, 1972). This metabolite may be further oxidized by *Pseudomonas putida* to carbon dioxide and water (Jerina et al., 1971). Under aerobic conditions, *Cunninghamella elegans* biodegraded naphthalene to α-naphthol, β-naphthol, *trans*-1,2-dihydroxy-1,2-dihydronaphthalene, 4-hydroxy-1-tetralone and 1,4-naphthoquinone. Also under aerobic conditions, *Agnenellum, Oscillatoria* and *Anabaena* reportedly biodegraded naphthalene into 1-naphthol, *cis*-1,2-dihydroxyl-1,2-dihydronaphthalene and 4-hydroxy-1-tetralone (Kobayashi and Rittman, 1982; Riser-Roberts, 1992). *Candida lipolytica, Candida elegans* and species of *Cunninghamella, Syncephalastrum* and *Mucor* oxidized naphthalene to α-naphthol, β-naphthol, *trans*-1,2-dihydroxy-1,2-dihydronaphthalene, 4-hydroxy-1-tetralone, 1,2-naphthoquinone and 1,4-naphthoquinone (Cerniglia et al., 1978, 1980; Dodge and Gibson, 1980).

Cultures of *Bacillus* sp. oxidized naphthalene to (+)-*trans*-1,2-dihydro-1,2-dihydroxynaphthalene. In the presence of reduced nicotinamide adeninedinucleotide phosphate ($NADPH_2$) and ferrous ions, a cell extract oxidized naphthalene to *trans*-naphthalenediol (Gibson, 1968). Hydroxylation by pure microbial cultures yielded an unidentified phenol, 1- and 2-hydroxynaphthalene (Smith and Rosazza, 1974).

In a static-culture-flask screening test, naphthalene (5 and 10 mg/L) was statically incubated in the dark at 25 °C with yeast extract and settled domestic wastewater inoculum. After 7 days, 100% biodegradation with rapid adaptation was observed (Tabak et al., 1981). In freshwater sediments, naphthalene biodegraded to *cis*-1,2-dihydroxy-1,2-dihydronaphthalene, α-naphthol, salicylic acid and catechol.

Photolytic. Irradiation of naphthalene and nitrogen dioxide using a high pressure mercury lamp (λ >290 nm) yielded the following principal products: 1- and 2-hydroxynaphthalene, 1-hydroxy-2-nitronaphthalene, 1-nitronaphthalene, 2,3-dinitronaphthalene, phthalic anhydride, 1,3-, 1,5- and 1,8-dinitronaphthalene (Barlas and Parlar, 1987). A carbon dioxide yield of 30.0% was achieved when naphthalene adsorbed on silica gel was irradiated with light (λ >290 nm) for 17 hours (Freitag et al., 1985).

Fukuda et al. (1988) studied the photolysis of naphthalene in distilled water using a high pressure mercury lamp. After 96 hours of irradiation, a rate constant of 0.028/hour with a half-life of 25 hours was determined. When the experiment was replicated in the presence of various sodium chloride concentrations, they found that the rate of photolysis increased proportionately to the concentration of sodium chloride. The photolysis rates of naphthalene in sodium chloride concentrations of 0.2, 0.3, 0.4 and 0.5 M following 3 hours of irradiation were 33.3, 50.6, 91.6 and 99.2%, respectively.

Tuhkanen and Beltrán (1995) studied the decomposition of naphthalene in water using hydrogen peroxide and UV light via a low pressure mercury lamp (λ = 254 nm). Hydrogen peroxide alone did not cause any decrease in naphthalene concentration. However, UV light or UV light/hydrogen peroxide causes photolytic degradation of naphthalene. Intermediates identified from the direct photolysis of naphthalene in solution include 2,4-dimethyl-1,3-pentadiene, bicyclo[4.2.0]octa-1,3,5-triene, benzaldehyde, phenol, 2-hydroxybenzaldehyde, 2,3-dihydrobenzofuran, 2'-hydroxyacetophenone and 1(3H)isobenzofuranone. During the oxidation of naphthalene using UV/hydrogen peroxide the following intermediates formed: 2,4-dimethyl-1,3-pentadiene, bicyclo[4.2.0]octa-1,3,5-triene, 1,4-dimethylbenzene, phenol and 2-hydroxybenzaldehyde. The researchers concluded that naphthalene was first oxidized to naphthol and then to naphthoquinone, benzaldehyde, phthalic and benzoic acids. Continued irradiation with or without the presence of hydroxyl radicals would eventually result in the complete mineralization of naphthalene and its intermediates, i.e., carbon dioxide and water.

Chemical/Physical. An aqueous solution containing chlorine dioxide in the dark for 3.5 days oxidized naphthalene to chloronaphthalene, 1,4-dichloronaphthalene and methyl esters of phthalic acid (Taymaz et al., 1979). In the presence of bromide ions and a chlorinating agent (sodium hypochlorite), major products identified at various reaction times and pHs include 1-bromonaphthalene, dibromonaphthalene and 2-bromo-1,4-naphthoquinone. Minor products identified include chloronaphthalene, dibromonaphthalene, bromochloronaphthalene, bromonaphthol, dibromonaphthol, 2-bromonaphthoquinone, dichloronaphthalene and chlorodibromonaphthalene (Lin et al., 1984).

The gas-phase reaction of N_2O_5 and naphthalene in an environmental chamber at room temperature resulted in the formation of 1- and 2-nitronaphthalene with approximate yields of 18 and 7.5%, respectively (Pitts et al., 1985). The reaction of naphthalene with NO_x to form nitronaphthalene was reported to occur in urban air from St. Louis, MO (Randahl et al., 1982).

It was suggested that the chlorination of naphthalene in tap water accounted for the presence of chloro- and dichloronaphthalenes (Shiraishi et al., 1985). Kanno et al. (1982) studied the aqueous reaction of naphthalene and other aromatic hydrocarbons (benzene, toluene, *o*-, *m*- and *p*-xylene) with hypochlorous acid in the presence of ammonium ion. They reported that the aromatic ring was not chlorinated as expected but was cleaved by chloramine forming cyanogen chloride. The amount of cyanogen chloride increased at lower pHs (Kanno et al., 1982).

Exposure Limits: NIOSH REL: TWA 10 ppm (50 mg/m^3), STEL 15 ppm (75 mg/m^3), IDLH 250 ppm; OSHA PEL: TWA 10 ppm; ACGIH TLV: STEL 15 ppm.

Symptoms of Exposure: Inhalation of vapors may cause irritation of eyes, skin, respiratory tract, headache, nausea, confusion and excitement. Ingestion may cause gastrointestinal pain and kidney damage (Patnaik, 1992).

Toxicity: Acute oral LD_{50} for guinea pigs 1,200 mg/kg, mice 533 mg/kg, rats 1,250 mg/kg (RTECS, 1985).

Drinking Water Standard: As of May 1994, no MCLGs or MCLs have been proposed (U.S. EPA, 1994).

Uses: Intermediate for phthalic anhydride, naphthol, 1,4-napththoquinone, 1,4-dihydronaphthalene, 1,2,3,4-tetrahydronaphthalene (tetralin), decahydronaphthalene (decalin), 1-nitronaphthalene, halogenated naphthalenes, naphthyl, naphthol derivatives, dyes, explosives; mothballs manufacturing; preparation of pesticides, fungicides, detergents and wetting agents, synthetic resins, celluloids and lubricants; synthetic tanning; preservative; textile chemicals; emulsion breakers; scintillation counters; smokeless powders.

1-NAPHTHYLAMINE

Synonyms: 1-Aminonaphthalene; C.I. azoic diazo component 114; Fast garnet B base; Fast garnet base B; **1-Naphthalenamine**; Naphthalidam; Naphthalidine; α-Naphthylamine; RCRA waste number U167; UN 2077.

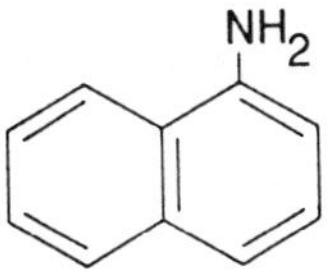

CAS Registry No.: 134-32-7
DOT: 2077
DOT label: Poison
Molecular formula: $C_{10}H_9N$
Formula weight: 143.19
RTECS: QM1400000

Physical state, color and odor
Colorless crystals or yellow, rhombic needles with an unpleasant odor. Becomes purplish-red in color on exposure to air. Odor threshold in air ranges from 140 to 290 $\mu g/m^3$ (Keith and Walters, 1992).

Melting point (°C):
50 (Weast, 1986)

Boiling point (°C):
Sublimes at 300.8 (Weast, 1986).

Density (g/cm³):
1.1229 at 25/25 °C (Weast, 1986)

Diffusivity in water (10^5 cm²/sec):
0.67 at 20 °C using method of Hayduk and Laudie (1974).

Dissociation constant (25 °C):
3.92 (Dean, 1973)

Flash point (°C):
158.5 (NIOSH, 1994)

Henry's law constant (10^{10} atm·m³/mol):
1.27 at 25 °C (Mercer et al., 1990)

Ionization potential (eV):
7.30 (NIOSH, 1994)

Soil sorption coefficient, log K_{oc}:
3.51 (average of 3 soils, Graveel et al., 1986)

Octanol/water partition coefficient, log K_{ow}:
2.27 (Sangster, 1989)

Solubility in organics:
Soluble in alcohol and ether (Weast, 1986).

Solubility in water:
1.6 g/L at 20 °C (Patnaik, 1992)
1,700 mg/L (Verschueren, 1983)

Vapor pressure (10^5 mmHg):
100,000 at 158.5 °C (NIOSH, 1994)
6.5 at 20–30 °C (Mercer et al., 1990)

Environmental Fate
   Biological. 1-Naphthylamine added to three different soils was incubated in the dark at 23 °C under a carbon dioxide-free atmosphere. After 308 days, 16.6 to 30.7% of the 1-naphthylamine added to soil biodegraded to carbon dioxide (Graveel et al., 1986).
   Chemical/Physical. Kanno et al. (1982) studied the aqueous reaction of 1-naphthylamine and other substituted aromatic hydrocarbons (aniline, toluidine, 2-naphthylamine, phenol, cresol, pyrocatechol, resorcinol, hydroquinone and 1-naphthol) with hypochlorous acid in the presence of ammonium ion. They reported that the aromatic ring was not chlorinated as expected but was cleaved by chloramine forming cyanogen chloride. The amount of cyanogen chloride that formed increased as the pH was lowered (Kanno et al., 1982).

Exposure Limits: Potential occupational carcinogen.

Symptoms of Exposure: Ingestion, skin contact or inhalation of vapors can cause acute hemorrhagic cystitis, dyspnea, ataxia, dysuria and hematuria (Patnaik, 1992).

Toxicity: Acute oral LD_{50} for rats 779 mg/kg; LD_{50} (intraperitoneal) for mice 96 mg/kg (RTECS, 1985); LC_{50} (48-hour) for red killifish 49 mg/L (Yoshioka et al., 1986).

Uses: Manufacture of dyes and dye intermediates; agricultural chemicals.

2-NAPHTHYLAMINE

Synonyms: 2-Aminonaphthalene; C.I. 37270; Fast scarlet base B; 2-Naphthal-amine; **2-Naphthalenamine;** β-Naphthylamine; 6-Naphthylamine; 2-Naphthyl-amine mustard; RCRA waste number U168; UN 1650; USAF CB-22.

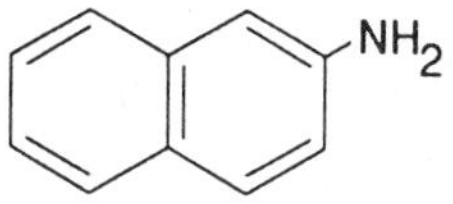

CAS Registry No.: 91-59-8
DOT: 1650
DOT label: Poison
Molecular formula: $C_{10}H_9N$
Formula weight: 143.19
RTECS: QM2100000

Physical state and color
Colorless to white crytals. Becomes purplish-red in color on exposure to air. Odor threshold in air ranges from 1.4 to 1.9 mg/m^3 (Keith and Walters, 1992).

Melting point (°C):
113 (Weast, 1986)
110.2 (Verschueren, 1983)

Boiling point (°C):
306.1 (Weast, 1986)

Density (g/cm^3):
1.0614 at 98/4 °C (Weast, 1986)

Diffusivity in water (10^5 cm^2/sec):
0.67 at 20 °C using method of Hayduk and Laudie (1974).

Dissociation constant (°C):
4.11 (Dean, 1973)

Flash point (°C):
158.5 (NIOSH, 1994)

Henry's law constant (10^9 atm·m^3/mol):
2.01 at 25 °C (Mercer et al., 1990)

Ionization potential (eV):
9.71 (NIOSH, 1994)

Soil sorption coefficient, log K_{oc}:
2.11 (calculated, Mercer et al., 1990)

Octanol/water partition coefficient, log K_{ow}:
2.40 (Sangster, 1989)

Solubility in organics:
Soluble in alcohol and ether (Weast, 1986).

Solubility in water:
0.002 wt % at 20 °C (NIOSH, 1994)

Vapor pressure (10^4 mmHg):
10,000 at 108.6 °C (NIOSH, 1994)
2.56 at 20–30 °C (Mercer et al., 1990)

Environmental Fate
   Chemical/Physical. Kanno et al. (1982) studied the aqueous reaction of 1-naphthylamine and other substituted aromatic hydrocarbons (aniline, toluidine, 2-naphthylamine, phenol, cresol, pyrocatechol, resorcinol, hydroquinone and 1-naphthol) with hypochlorous acid in the presence of ammonium ion. They reported that the aromatic ring was not chlorinated as expected but was cleaved by chloramine forming cyanogen chloride. At lower pHs, the amount of cyanogen chloride formed increased (Kanno et al., 1982).

Exposure Limits: Potential occupational carcinogen.

Symptoms of Exposure: Ingestion, skin contact or inhalation of vapors can cause acute hemorrhagic cystitis, respiratory stress and hematuria (Patnaik, 1992).

Toxicity: Acute oral LD_{50} for rats 727 mg/kg; LD_{50} (intraperitoneal) for mice 200 mg/kg (RTECS, 1985).

Uses: Manufacture of dyes and in rubber.

NITRAPYRIN

Synonyms: 2-Chloro-6-(trichloromethyl)pyridine; Dowco-163; N-serve; N-serve nitrogen stabilizer.

CAS Registry No.: 1929-82-4
Molecular formula: $C_6H_3Cl_4N$
Formula weight: 230.90
RTECS: US7525000

Physical state, color and odor
Colorless to white crystalline solid with a mild, sweet odor.

Melting point (°C):
62-63 (Verschueren, 1983)

Density (g/cm^3):
1.744 using method of Lyman et al. (1982).

Henry's law constant (10^3 atm·m^3/mol):
2.13 (approximate – calculated from water solubility and vapor pressure)

Soil sorption coefficient, log K_{oc}:
2.64 (Commerce soil), 2.68 (Tracy soil), 2.66 (Tracy soil) (McCall et al., 1981)
2.62 (Kenaga and Goring, 1980)
2.24 (average of 2 soils, Briggs, 1981)

Octanol/water partition coefficient, log K_{ow}:
3.41 (Kenaga and Goring, 1980)
3.02 (Briggs, 1981)

Solubility in organics:
Soluble in acetone, alcohol, benzene, chloroform, ether and lignoin (Weast, 1986).

Solubility in water:
40 mg/L (Kenaga and Goring, 1980)

Vapor pressure (mmHg):
2.8 x 10^{-3} at 20 °C (Verschueren, 1983)

Environmental Fate

   Biological. 6-Chloropicolinic acid and carbon dioxide were reported as bio-degradation products (Verschueren, 1983).

   Soil. Hydrolyzes in soil to 6-chloropyridine-2-carboxylic acid (Worthing and Hance, 1991).

   Photolytic. Photolysis of nitrapyrin in water yielded 6-chloropicolinic acid, 6-hydroxypicolinic acid and an unidentified polar material (Verschueren, 1983).

   Chemical/Physical. Emits toxic nitrogen oxides and chloride fumes when heated to decomposition (Lewis, 1990).

Exposure Limits (mg/m^3): NIOSH REL: TWA 10, STEL 20, IDLH 250 ppm; OSHA PEL: TWA 15 ppm.

Toxicity: Acute oral LD_{50} for chickens 235 mg/kg, mice 710 mg/kg, rats 940 mg/kg, rabbits 500 mg/kg (RTECS, 1985).

Uses: Not commercially produced in the United States. Nitrification inhibitor in ammonium fertilizers.

2-NITROANILINE

Synonyms: 1-Amino-2-nitrobenzene; Azoene fast orange GR base; Azoene fast orange GR salt; Azofix orange GR salt; Azogene fast orange GR; Azoic diazo component 6; Brentamine fast orange GR base; Brentamine fast orange GR salt; C.I. 37025; C.I. azoic diazo component 6; Devol orange B; Devol orange salt B; Diazo fast orange GR; Fast orange base GR; Fast orange base GR salt; Fast orange base JR; Fast orange GR base; Fast orange O base; Fast orange O salt; Fast orange salt JR; Hiltonil fast orange GR base; Hiltosal fast orange GR salt; Hindasol orange GR salt; Natasol fast orange GR salt; *o*-Nitraniline; *o*-Nitroaniline; **2-Nitrobenzenamine**; ONA; Orange base CIBA II; Orange base IRGA II; Orange GRS salt; Orange salt CIBA II; Orange salt IRGA II; Orthonitroaniline; UN 1661.

CAS Registry No.: 88-74-4
DOT: 1661
DOT label: Poison
Molecular formula: $C_6H_6N_2O_2$
Formula weight: 138.13
RTECS: BY6650000

Physical state, color and odor
Orange-yellow crystals with a musty odor.

Melting point (°C):
71.5 (Weast, 1986)
69.3 (Collett and Johnston, 1926)
69-70 (Dean, 1987)

Boiling point (°C):
284.1 (Dean, 1973)

Density (g/cm^3):
1.442 at 15/4 °C (Weast, 1986)
0.9015 at 25/4 °C (Sax, 1984)
1.44 at 20/4 °C (Weiss, 1986)

Diffusivity in water (10^5 cm^2/sec):
0.78 at 20 °C using method of Hayduk and Laudie (1974).

Dissociation constant (°C):
13.25 (Keith and Walters, 1992)

Flash point (°C):
168 (Hawley, 1981)

Lower explosive limit (%):
Not pertinent (Weiss, 1986)

Henry's law constant (10^5 atm·m^3/mol):
9.72 at 25 °C (approximate - calculated from water solubility and vapor pressure)

Ionization potential (eV):
8.66 (Franklin et al., 1969)

Soil sorption coefficient, log K_{oc}:
1.23–1.62 using method of Karickhoff et al. (1979).

Octanol/water partition coefficient, log K_{ow}:
1.44, 1.79, 1.83 (Leo et al., 1971)

Solubility in organics:
In g/kg: benzene (208 at 25 °C), chloroform (11.7 at 0 °C) and ethanol (278.7 at 25 °C) (Collett and Johnston, 1926).

Solubility in water:
1,260 mg/L at 25 °C (Verschueren, 1983)
1.47 g/L at 30 °C (Gross et al., 1933)
1.212 and 2.423 g/kg at 25 and 40 °C, respectively (Collett and Johnston, 1926)

Vapor pressure (mmHg):
8.1 at 25 °C (Mabey et al., 1982)

Environmental Fate
   Biological. Under aerobic and anaerobic conditions using a sewage inoculum, 2-nitroaniline degraded to 2-methylbenzimidazole and 2-nitroacetanilide (Hallas and Alexander, 1983). A *Pseudomonas* sp. strain P6, isolated from a Matapeake silt loam, did not grow on 2-nitroaniline as the sole source of carbon. However, in the presence of 4-nitroaniline, approximately 50% of the applied 2-nitroaniline metabolized to nonvolatile products which could not be identified by HPLC (Zeyer and Kearney, 1983).
   Plant. 2-Nitroaniline was degraded by tomato cell suspension cultures (*Lycopericon lycopersicum*). Transformation products identified included 2-nitro-

anilino-β-D-glucopyranoside, β-(2-amino-3-nitrophenyl)glucopyranoside and β-(4-amino-3-nitrophenyl)glucopyranoside (Pogány et al., 1990).

Toxicity: Acute LD_{50} for wild birds 750 mg/kg, guinea pigs 2,350 mg/kg, mice 1,070 mg/kg, quail 750 mg/kg, rats 1,600 mg/kg (RTECS, 1985).

Use: Organic synthesis.

3-NITROANILINE

Synonyms: Amarthol fast orange R base; 1-Amino-3-nitrobenzene; 3-Aminonitrobenzene; *m*-Aminonitrobenzene; Azobase MNA; C.I. 37030; C.I. azoic diazo component 7; Daito orange base R; Devol orange R; Diazo fast orange R; Fast orange base R; Fast orange M base; Fast orange MM base; Fast orange R base; Fast orange R salt; Hiltonil fast orange R base; MNA; Naphtolean orange R base; Nitranilin; *m*-Nitraniline; 3-Nitroaminobenzene; *m*-Nitroaminobenzene; *m*-Nitroaniline; **3-Nitrobenzenamine**; *m*-Nitrobenzenamine; *m*-Nitrophenylamine; Orange base IRGA I; UN 1661.

CAS Registry No.: 99-09-2
DOT: 1661
DOT label: Poison
Molecular formula: $C_6H_6N_2O_2$
Formula weight: 138.13
RTECS: BY6825000

Physical state and color
Yellow, rhombic crystals.

Melting point (°C):
114 (Weast, 1986)

Boiling point (°C):
306.4 (Dean, 1973)

Density (g/cm³):
0.9011 at 25/4 °C (Sax, 1984)

Diffusivity in water (10^5 cm²/sec):
0.78 at 20 °C using method of Hayduk and Laudie (1974).

Dissociation constant:
2.46 at 25 °C (Dean, 1973)

Henry's law constant (10^5 atm·m³/mol):
1.93 at 25 °C (approximate - calculated from water solubility and vapor pressure)

Ionization potential (eV):
8.80 (Franklin et al., 1969)

Soil sorption coefficient, log K_{oc}:
1.26 using method of Karickhoff et al. (1979).

Octanol/water partition coefficient, log K_{ow}:
1.37 (Fujita et al., 1964)

Solubility in organics:
In g/kg at 25 °C: benzene (27.18), chloroform (32.16) and ethanol (77.78) (Collett and Johnston, 1926).

Solubility in water:
890 mg/L at 25 °C (Verschueren, 1983)
0.910 and 1.785 g/kg at 25 and 40.1 °C, respectively (Collett and Johnston, 1926)
1,100 mg/L at 20 °C (Dean, 1973)
1.21 g/L at 30 °C (Gross et al., 1933)

Vapor pressure (mmHg):
9.56 x 10^{-5} at 25 °C (Banerjee et al., 1990)

Environmental Fate
  Biological. A bacterial culture isolated from the Oconee River in North Georgia degraded 3-nitroaniline to the intermediate 4-nitrocatechol (Paris and Wolfe, 1987). A *Pseudomonas* sp. strain P6, isolated from a Matapeake silt loam, did not grow on 3-nitroaniline as the sole source of carbon. However, in the presence of 4-nitroaniline, all of the applied 3-nitroaniline metabolized completely to carbon dioxide (Zeyer and Kearney, 1983).
  Chemical/Physical. Will react with acids forming water soluble salts.

Toxicity: Acute LD_{50} for guinea pigs 450 mg/kg, mice 308 mg/kg, quail 562 mg/kg, rats 535 mg/kg (RTECS, 1985).

Use: Organic synthesis.

4-NITROANILINE

Synonyms: 1-Amino-4-nitrobenzene; 4-Aminonitrobenzene; *p*-Aminonitroben-zene; Azoamine red ZH; Azofix Red GG salt; Azoic diazo component 37; C.I. 37035; C.I. azoic diazo component 37; C.I. developer 17; Developer P; Devol red GG; Diazo fast red GG; Fast red base GG; Fast red base 2J; Fast red 2G base; Fast red 2G salt; Fast red GG base; Fast red GG salt; Fast red MP base; Fast red P base; Fast red P salt; Fast red salt GG; Fast red salt 2J; IG base; Naphtolean red GG base; NCI-C60786; 4-Nitraniline; *p*-Nitraniline; Nitrazol 2F extra; *p*-Nitroaniline; **4-Nitrobenzenamine**; *p*-Nitrobenzenamine; *p*-Nitrophenylamine; PNA; RCRA waste number P077; Red 2G base; Shinnippon fast red GG base; UN 1661.

CAS Registry No.: 100-01-6
DOT: 1661
DOT label: Poison
Molecular formula: $C_6H_6N_2O_2$
Formula weight: 138.13
RTECS: BY7000000

Physical state, color and odor
Bright yellow crystalline powder with a faint, ammonia-like, slightly pungent odor.

Melting point (°C):
146 (Dean, 1987)
148-149 (Weast, 1986)

Boiling point (°C):
331.7 (Weast, 1986)
336 (Weiss, 1986)

Density (g/cm^3):
1.424 at 20/4 °C (Weast, 1986)

Diffusivity in water (10^5 cm^2/sec):
0.78 at 20 °C using method of Hayduk and Laudie (1974).

Dissociation constant:
0.99 at 25 °C (Dean, 1973)

Flash point (°C):
165 (Aldrich, 1988)
200.5 (NIOSH, 1994)

Heat of fusion (kcal/mol):
5.04 (Dean, 1987)

Henry's law constant (10^8 atm·m^3/mol):
1.14 at 25 °C (approximate - calculated from water solubility and vapor pressure)

Ionization potential (eV):
8.85 (NIOSH, 1994)

Soil sorption coefficient, log K_{oc}:
1.08 (estimated, Montgomery, 1989)

Octanol/water partition coefficient, log K_{ow}:
1.39 (Fujita et al., 1964)
1.40 (Campbell and Luthy, 1985)

Solubility in organics:
In g/kg at 25 °C: benzene (5.794), chloroform (9.290) and ethanol (60.48) (Collett and Johnston, 1926).

Solubility in water:
800 mg/L at 18.5 °C (Dean, 1973)
0.568 and 1.157 g/kg at 25 and 40 °C, respectively (Collett and Johnston, 1926).
728 mg/kg at 30 °C (Gross and Saylor, 1931)

Vapor pressure (10^3 mmHg):
1.5 at 20 °C, 7 at 30 °C (Verschueren, 1983)

Environmental Fate

Biological. A *Pseudomonas* sp. strain P6, isolated from a Matapeake silt loam, was grown using a yeast extract. After 8 days, 4-nitroaniline degraded completely to carbon dioxide (Zeyer and Kearney, 1983).

Chemical/Physical: Spacek et al. (1995) investigated the photodegradation of 4-nitroaniline using titanium dioxide-UV light and Fenton's reagent (hydrogen peroxide:substance - 10:1; Fe^{2+} 2.5 x 10^{-4} mol/L). Both experiments were carried out at 25 °C. The decomposition rate of 4-nitroaniline was very high by the photo-Fenton reaction in comparison to titanium dioxide-UV light (λ = 365 nm). Decomposition products identified in both reactions were nitrobenzene, *p*-benzoquinone, hydroquinone, oxalic acid and resorcinol. Oxalic acid, hydroquinone

and *p*-benzoquinone were identified as intermediate products using HPLC.
Will reacts with mineral acids forming water soluble salts.

Exposure Limits (mg/m^3): NIOSH REL: TWA 3; IDLH 300; OSHA PEL: TWA 6.

Toxicity: Acute LD_{50} for wild birds 75 mg/kg, guinea pigs 450 mg/kg, mice 810 mg/kg, quail 1,000 mg/kg, rats 750 mg/kg (RTECS, 1985); LD_{50} (intraperitoneal) for mice 250 mg/kg; LC_{50} (48-hour) for red killifish 363 mg/L (Yoshioka et al., 1986).

Uses: Intermediate for dyes and antioxidants; inhibits gum formation in gasoline; corrosion inhibiter; organic synthesis (preparation of *p*-phenylenediamine).

NITROBENZENE

Synonyms: Essence of mirbane; Essence of myrbane; Mirbane oil; NCI-C60082; Nitrobenzol; Oil of bitter almonds; Oil of mirbane; Oil of myrbane; RCRA waste number U169; UN 1662.

CAS Registry No.: 98-95-3
DOT: 1662
DOT label: Poison
Molecular formula: $C_6H_5NO_2$
Formula weight: 123.11
RTECS: DA6475000

Physical state, color and odor
Clear, light yellow to brown, oily liquid with an almond or shoe polish odor. Odor threshold in air is 18 ppb (Amoore and Hautala, 1983).

Melting point (°C):
5.7 (Weast, 1986)

Boiling point (°C):
210.8 (Weast, 1986)

Density (g/cm³):
1.2037 at 20/4 °C (Weast, 1986)
1.205 at 15/4 °C (Sax, 1984)
1.2125 at 10/4 °C, 1.205 at 18/4 °C, 1.1986 at 25/4 °C (Standen, 1967)

Diffusivity in water (10^5 cm²/sec):
0.87 at 20 °C using method of Hayduk and Laudie (1974).

Dissociation constant:
>15 (Christensen et al., 1975)

Flash point (°C):
88.5 (NIOSH, 1994)

Lower explosive limit (%):
1.8 at 100 °C (NIOSH, 1994)

Heat of fusion (kcal/mol):
2.780 (Dean, 1987)

Henry's law constant (10^5 atm·m^3/mol):
2.45 at 25 °C (Warner et al., 1987)

Interfacial tension with water (dyn/cm at 20 °C):
25.66 (Demond and Lindner, 1993)

Ionization potential (eV):
9.92 (Franklin et al., 1969)

Bioconcentration factor, log BCF:
2.41 (algae, Geyer et al., 1984)
1.60 (activated sludge), 1.30 (algae) (Freitag et al., 1985)

Soil sorption coefficient, log K_{oc}:
1.85 (river sediment), 1.95 (coal wastewater sediment) (Kopinke et al., 1995)
2.36 (Løkke, 1984)
1.49, 1.95, >2.01 (various Norwegian soils, Seip et al., 1986)
1.94 (Briggs, 1981)
1.95 (Captina silt loam), 2.02 (McLaurin sandy loam) (Walton et al., 1992)
K_d = 3.5 mL/g on a Cs^+-kaolinite (Haderlein and Schwarzenbach, 1993)

Octanol/water partition coefficient, log K_{ow}:
1.80 (Sangster, 1989)
1.85 (Briggs, 1981, Campbell and Luthy, 1985; Fujita et al., 1964; Walton et al., 1992; Wasik et al., 1981)
1.83 (Banerjee et al., 1980; Garst and Wilson, 1984)
1.84 (Geyer et al., 1984)
1.792 (Lu and Metcalf, 1975)
1.88 (Leo et al., 1971)
1.828 (Brooke et al., 1990; de Bruijn et al., 1989)
1.836 (Brooke et al., 1990)
1.70 (Hammers et al., 1982)

Solubility in organics:
Soluble in acetone, ethanol, benzene and ether (Weast, 1986).

Solubility in water:
1,900 mg/L at 20 °C, 8,000 mg/L at 80 °C (Verschueren, 1983)
2,000 mg/L at 25 °C (Warner et al., 1987)
2,090 mg/L at 25 °C (Banerjee et al., 1980)

1,780 mg/kg at 15 °C, 2,050 mg/kg at 30 °C (Gross and Saylor, 1931)
1,930 mg/L at 25 °C (Andrews and Keefer, 1950)
18.35 mmol/L at 35 °C (Hine et al., 1963)
31.1 mmol/L at 25 °C (Tewari et al., 1982; Wasik et al., 1981)

Vapor density:
5.03 g/L at 25 °C, 4.25 (air = 1)

Vapor pressure (mmHg):
0.15 at 20.0 °C, 0.35 at 30.0 °C (Verschueren, 1983)
0.28 at 25 °C (Warner et al., 1987)
0.600 at 35 °C (Hine et al., 1963)

Environmental Fate
Biological. In activated sludge, 0.4% of the applied nitrobenzene mineralized to carbon dioxide after 5 days (Freitag et al., 1985). Under anaerobic conditions using a sewage inoculum, nitrobenzene degraded to aniline (Hallas and Alexander, 1983). When nitrobenzene (5 and 10 mg/L) was statically incubated in the dark at 25 °C with yeast extract and settled domestic wastewater inoculum, complete biodegradation with rapid acclimation was observed after 7-14 days (Tabak et al., 1981).

Photolytic. Irradiation of nitrobenzene in the vapor phase produced nitrosobenzene and 4-nitrophenol (HSDB, 1989). Titanium dioxide suspended in an aqueous solution and irradiated with UV light (λ = 365 nm) converted nitrobenzene to carbon dioxide at a significant rate (Matthews, 1986). A carbon dioxide yield of 6.7% was achieved when nitrobenzene adsorbed on silica gel was irradiated with light (λ >290 nm) for 17 hours (Freitag et al., 1985).

An aqueous solution containing nitrobenzene (500 μM) and hydrogen peroxide (100 μM) was irradiated with UV light (λ = 285-360 nm). After 18 hours, 2% of the substrate was converted into *o*-, *m*- and *p*-nitrophenols having an isomer distribution of 50, 29.5 and 20.5%, respectively (Draper and Crosby, 1984).

Chemical/Physical. In an aqueous solution, nitrobenzene (100 μM) reacted with Fenton's reagent (35 μM). After 15 minutes, 2-, 3- and 4-nitrophenol were identified as products. After 6 hours, about 50% of the nitrobenzene was destroyed (Lipczynska-Kochany, 1991). When nitrobenzene, with nitrogen as a carrier, was passed through a quartz cell and irradiated by two 220-volt arcs, nitrosobenzene and *p*-nitrophenol formed as the major products (Hastings and Matsen, 1948).

Exposure Limits: NIOSH REL: TWA 1 ppm (5 mg/m^3), IDLH 200 ppm; OSHA PEL: TWA 1 ppm.

Symptoms of Exposure: Chronic exposure may cause anemia. Acute effects include headache, dizziness, nausea, vomiting and dyspnea (Patnaik, 1992).

Toxicity: Acute oral LD_{50} for mice 590 mg/kg, rats 489 mg/kg; LC_{50} (48-hour) for red killifish 275 mg/L (Yoshioka et al., 1986).

Uses: Solvent for cellulose ethers; modifying esterification of cellulose acetate; ingredient of metal polishes and shoe polishes; manufacture of aniline, benzidine, quinoline, azobenzene, drugs, photographic chemicals.

4-NITROBIPHENYL

Synonyms: **4-Nitro-1,1′-biphenyl**; 4-Nitrodiphenyl; *p*-Nitrobiphenyl; 4-Phenyl-nitrobenzene; *p*-Phenylnitrobenzene; PNB.

CAS Registry No.: 92-93-3
Molecular formula: $C_{12}H_9NO_2$
Formula weight: 199.21
RTECS: DV5600000

Physical state, color and odor
White to yellow crystals with a sweetish odor.

Melting point (°C):
114 (Weast, 1986)

Boiling point (°C):
340 (Weast, 1986)

Diffusivity in water (10^5 cm^2/sec):
0.59 at 20 °C using method of Hayduk and Laudie (1974).

Flash point (°C):
144.5 (NIOSH, 1994)

Lower explosive limit (%):
Not applicable (NFPA, 1984)

Upper explosive limit (%):
Not applicable (NFPA, 1984)

Solubility in organics:
Soluble in acetic acid, benzene, chloroform and ether (Weast, 1986).

Exposure Limits: Potential occupational carcinogen.

Toxicity: Acute oral LD_{50} for rats 2,230 mg/kg, rabbits 1,970 mg/kg (RTECS, 1985).

Uses: Organic synthesis.

NITROETHANE

Synonym: UN 2842.

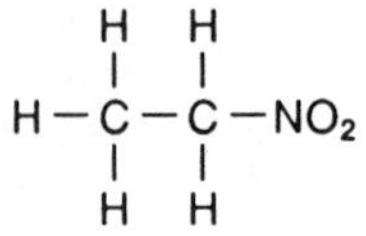

$$H-\underset{\underset{H}{|}}{\overset{\overset{H}{|}}{C}}-\underset{\underset{H}{|}}{\overset{\overset{H}{|}}{C}}-NO_2$$

CAS Registry No.: 79-24-3
DOT: 2842
DOT label: Combustible liquid
Molecular formula: $C_2H_5NO_2$
Formula weight: 75.07
RTECS: KI5600000

Physical state, color and odor
Colorless liquid with a fruity odor. Odor threshold in air is 2.1 ppm (Amoore and Hautala, 1983).

Melting point (°C):
-50 (Weast, 1986)

Boiling point (°C):
114.1 (Dean, 1987)
115 (Weast, 1986)

Density (g/cm^3):
1.0448 at 25/4 °C (Weast, 1986)

Diffusivity in water (10^5 cm^2/sec):
1.23 at 25 °C using method of Hayduk and Laudie (1974).

Dissociation constant:
8.5 (Gordon and Ford, 1972)
8.44 at 25 °C (Dean, 1973)

Flash point (°C):
28.0 (NIOSH, 1994)
41.11 (open cup, Windholz et al., 1983)

Lower explosive limit (%):
3.4 (NIOSH, 1994)

770

Henry's law constant (10^5 atm·m^3/mol):
4.66 at 25 °C (Hine and Mookerjee, 1975)
4.76 (Gaffney et al., 1987)

Ionization potential (eV):
10.88 ± 0.05 (Franklin et al., 1969)

Soil sorption coefficient, log K_{oc}:
Unavailable because experimental methods for estimation of this parameter for nitroaliphatics are lacking in the documented literature. However, its moderate solubility in water and low K_{ow} suggest its adsorption to soil will be low (Lyman et al., 1982).

Octanol/water partition coefficient, log K_{ow}:
0.18 (Hansch and Anderson, 1967)

Solubility in organics:
Soluble in acetone (Weast, 1986). Miscible with alcohol, chloroform and ether (Sax and Lewis, 1987).

Solubility in water:
45 mL/L at 20 °C (Windholz et al., 1983)

Vapor density:
3.07 g/L at 25 °C, 2.59 (air = 1)

Vapor pressure (mmHg):
21 at 25.2 °C (NIOSH, 1994)

Exposure Limits: NIOSH REL: TWA 100 ppm (310 mg/m^3), IDLH 1,000 ppm; OSHA PEL: TWA 100 ppm.

Toxicity: Acute oral LD_{50} for mice 860 mg/kg, rats 1,100 mg/kg (RTECS, 1985).

Uses: Solvent for nitrocellulose; cellulose acetate; cellulose acetobutyrate; cellulose acetopropionate, waxes, fats, dyestuffs, vinyl and alkyd resins; experimental propellant; fuel additive; organic synthesis (Friedel-Crafts reactions).

NITROMETHANE

Synonyms: Nitrocarbol; UN 1261.

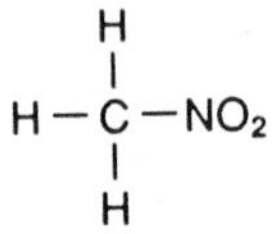

$$H-\overset{\overset{\displaystyle H}{|}}{\underset{\underset{\displaystyle H}{|}}{C}}-NO_2$$

CAS Registry No.: 75-52-5
DOT: 1261
DOT label: Combustible liquid
Molecular formula: CH_3NO_2
Formula weight: 61.04
RTECS: PA9800000

Physical state, color and odor
Colorless liquid with a strong, disagreeable odor. Odor threshold in air is 3.5 ppm (Amoore and Hautala, 1983).

Melting point (°C):
-17 (Weast, 1986)
-29 (Windholz et al., 1983)

Boiling point (°C):
100.8 (Weast, 1986)

Density (g/cm³):
1.1371 at 20/4 °C (Weast, 1986)
1.1322 at 25/4 °C (Windholz et al., 1983)

Diffusivity in water (10^5 cm²/sec):
1.27 at 20 °C using method of Hayduk and Laudie (1974).

Dissociation constant:
10.21 at 25 °C (Dean, 1973)

Flash point (°C):
35 (NIOSH, 1994)
44.4 (Windholz et al., 1983)

Lower explosive limit (%):
7.3 (NIOSH, 1994)

Heat of fusion (kcal/mol):
2.319 (Dean, 1987)

Henry's law constant (10^5 atm·m^3/mol):
2.86 (Gaffney et al., 1987)

Interfacial tension with water (dyn/cm at 25 °C):
9.5 (Donahue and Bartell, 1952)

Ionization potential (eV):
11.08 (NIOSH, 1994)

Soil sorption coefficient, log K_{oc}:
Unavailable because experimental methods for estimation of this parameter for nitroaliphatics are lacking in the documented literature. However, its high solubility in water and low K_{ow} suggest its adsorption to soil will be low (Lyman et al., 1982).

Octanol/water partition coefficient, log K_{ow}:
-0.33 (Hansch and Anderson, 1967)
-0.35 (Hansch and Leo, 1979)

Solubility in organics:
Soluble in acetone, alcohol, ether (Weast, 1986) and *N,N*-dimethylformamide (Windholz et al., 1983).

Solubility in water (at 20 °C):
10 wt % (NIOSH, 1994)
22 mL/L (Hawley, 1981)

Vapor density:
2.49 g/L at 25 °C, 2.11 (air = 1)

Vapor pressure (mmHg):
27.8 at 20 °C, 46 at 30 °C (Verschueren, 1983)
35.26 at 25.00 °C (Hussam and Carr, 1985)

Exposure Limits: NIOSH REL: IDLH 750 ppm; OSHA PEL: TWA 100 ppm (250 mg/m^3).

Toxicity: Acute oral LD_{50} for mice 950 mg/kg, rats 940 mg/kg; LD_{50} (intraperitoneal) for mice 110 mg/kg (RTECS, 1985).

Uses: Rocket fuel; coatings industry; solvent for cellulosic compounds, polymers, waxes, fats; gasoline additive; organic synthesis.

2-NITROPHENOL

Synonyms: 2-Hydroxynitrobenzene; *o*-Hydroxynitrobenzene; 2-Nitro-1-hydroxybenzene; *o*-Nitrophenol; ONP; UN 1663.

CAS Registry No.: 88-75-5
DOT: 1663
DOT label: Poison
Molecular formula: $C_6H_5NO_3$
Formula weight: 139.11
RTECS: SM2100000

Physical state, color and odor
Pale yellow crystals with an aromatic odor.

Melting point (°C):
45-46 (Weast, 1986)
44-45 (Standen, 1967)

Boiling point (°C):
216 (Weast, 1986)

Density (g/cm³):
1.485 at 14/4 °C (Weast, 1986)
1.495 at 20/4 °C (Sax, 1984)
1.2942 at 40/4 °C, 1.2712 at 60/4 °C, 1.2482 at 80/4 °C (Standen, 1967)

Diffusivity in water (10^5 cm²/sec):
0.81 at 20 °C using method of Hayduk and Laudie (1974).

Dissociation constant:
7.23 at 25 °C (Dean, 1973)

Flash point (°C):
73.5 (Sax, 1985)

Lower explosive limit (%):
Not pertinent (Weiss, 1986)

Upper explosive limit (%):
Not pertinent (Weiss, 1986)

Henry's law constant (10^6 atm·m^3/mol):
3.5 (Howard, 1989)

Bioconcentration factor, log BCF:
1.48 (activated sludge), 1.48 (algae), 1.60 (golden ide) (Freitag et al., 1985)

Soil sorption coefficient, log K_{oc}:
1.79 (river sediment), 2.32 (coal wastewater sediment) (Kopinke et al., 1995)
2.06 (Brookston clay loam, Boyd, 1982)
2.21 (coarse sand), 1.75 (loamy sand) (Kjeldsen et al., 1990)
$K_d = 35$ mL/g on a Cs^+-kaolinite (Haderlein and Schwarzenbach, 1993)

Octanol/water partition coefficient, log K_{ow}:
1.73 (Leo et al., 1971)
1.79 (Fujita et al., 1964)
1.89 (Schwarzenbach et al., 1988)

Solubility in organics:
In g/kg at 25 °C: benzene (1,472; 3,597 at 30 °C) and ethanol (460) (Windholz et al., 1983).
0.33 and 1,270 mmol/L at 25 °C in isooctane and butyl ether, respectively (Anderson et al., 1980).

Solubility in water:
3.30 g/kg at 40 °C (Palit, 1947)
2,100 mg/L at 20 °C, 10,800 mg/L at 100 °C (Verschueren, 1983)
2,000 mg/L at 25 °C (Morrison and Boyd, 1971)
3.89 g/L at 48 °C (Stephen and Stephen, 1963)
3,200 mg/L at 38 °C (Standen, 1967)
1,060 mg/L at 20 °C, 2,500 mg/L at 25 °C (Howard, 1989)
1.4 g/L at 20 °C (Leuenberger et al., 1985)
1,300 mg/L at 25 °C (Riederer, 1990)
1,079 mg/L at 20 °C (buffer solution at pH 1.5, Schwarzenbach et al., 1988)
At 20 °C: 10.0, 8.34, 8.70 and 8.55 mmol/L in doubly distilled water, Pacific seawater, artificial seawater and 35% NaCl, respectively (Hashimoto et al., 1984)

Vapor pressure (10^2 mmHg):
1.9 at 8 °C, 12 at 25 °C (Leuenberger et al., 1985)
9.3 at 25 °C (Sonnefeld et al., 1983)
8.9 at 25 °C (Riederer, 1990)

Environmental Fate

Biological. A microorganism, *Pseudomonas putida*, isolated from soil degraded *o*-nitrophenol to nitrite. Degradation by enzymatic mechanisms produced nitrite and catechol. Catechol subsequently degraded to β-ketoadipic acid (Zeyer and Kearney, 1984). When 2-nitrophenol was statically incubated in the dark at 25 °C with yeast extract and settled domestic wastewater inoculum, 100% biodegradation with rapid adaptation was achieved after 7 days (Tabak et al., 1981).

Chemical/Physical. Oxidation by Fenton's reagent (hydrogen peroxide and Fe^{3+}) produced nitrohydroquinone and 3-nitrocatechol (Andersson et al., 1986).

Toxicity: Acute oral LD_{50} for mice 1,300 mg/kg, rats 334 mg/kg (RTECS, 1985); LC_{50} (48-hour) for red killifish 275 mg/L (Yoshioka et al., 1986).

Uses: Indicator; preparation of *o*-nitroanisole and other organic compounds.

4-NITROPHENOL

Synonyms: 4-Hydroxynitrobenzene; *p*-Hydroxynitrobenzene; NCI-C55992; 4-Nitro-1-hydroxybenzene; *p*-Nitrophenol; PNP; RCRA waste number U170; UN 1663.

CAS Registry No.: 100-02-7
DOT: 1663
DOT label: Poison
Molecular formula: $C_6H_5NO_3$
Formula weight: 139.11
RTECS: SM2275000

Physical state, color and odor
Colorless to pale yellow, odorless crystals.

Melting point (°C):
112-114 (Dean, 1987)
114 (Standen, 1967)

Boiling point (°C):
Sublimes and decomposes at 279 (Weast, 1986).

Density (g/cm³):
1.479 at 20/4 °C, 1.270 at 120/4 °C (Standen, 1967)

Diffusivity in water (10^5 cm²/sec):
0.81 at 20 °C using method of Hayduk and Laudie (1974).

Dissociation constant:
7.15 at 25 °C (Dean, 1973)
7.08 at 21.5 °C (Schwarzenbach et al., 1988)

Flash point (°C):
Not pertinent (combustible solid, Weiss, 1986)

Henry's law constant (10^5 atm·m³/mol):
3.0 at 20 °C (calculated, Schwarzenbach et al., 1988)

Ionization potential (eV):
9.52 (Gordon and Ford, 1972)

Bioconcentration factor, log BCF:
1.48 (algae, Geyer et al., 1984)

Soil sorption coefficient, log K_{oc}:
1.75-2.73 (average = 2.33 for 10 Danish soils, Løkke, 1984)
1.74 (Brookstone clay loam, Boyd, 1982)
2.37 (Meylan et al., 1992)
1.94 (loamy sand, Kjeldsen et al., 1990)

Octanol/water partition coefficient, log K_{ow}:
1.91 (Campbell and Luthy, 1985; Leo et al., 1971)
1.85, 1.92 (Geyer et al., 1984)
2.04 (Schwarzenbach et al., 1988)
1.96 (Fujita et al., 1964; Garst and Wilson, 1984)

Solubility in organics:
Soluble in benzene (9.2 g/kg at 20 °C), ethanol (1,895 g/kg at 25 °C) and toluene (227 g/kg at 70 °C) (Palit, 1947). Also soluble in isooctane and *n*-butyl ether at 0.046 and 176.7 mg/L at 25 °C, respectively (Anderson et al., 1971)

Solubility in water:
32.8 g/kg at 40 °C (Palit, 1947)
16,000 mg/L at 25 °C, 269,000 mg/L at 90 °C (Smith et al., 1976)
16,900 mg/L at 25 °C (Morrison and Boyd, 1971)
In g/L: 8.04 at 15 °C, 16 at 25 °C, 29.1 at 90 °C (Standen, 1967)
11,300 mg/L at 20 °C, 25,000 mg/L at 25 °C (Howard, 1989)
11.57 g/L at 20 °C in a buffered solution (pH 1.5, Schwarzenbach et al., 1988)
0.1 M at 25 °C (Caturla et al., 1988)
14.746 g/L at 25 °C (Riederer, 1990)
At 20 °C: 97, 77.6, 78.4 and 77.6 mmol/L in distilled water, Pacific seawater, artificial seawater and 35 wt % NaCl solution, respectively (Hashimoto et al., 1984)

Vapor pressure (10^5 mmHg):
10 at 20 °C (Schwarzenbach et al., 1988)
30 at 30 °C (extrapolated, McCrady et al., 1985)
4.05 at 25 °C (Riederer, 1990)

Environmental Fate
   Biological. Under   anaerobic   conditions,   4-nitrophenol   may   undergo

nitroreduction producing 4-aminophenol (Kobayashi and Rittman, 1982). Estuarine sediment samples collected from the Mississippi River near Leeville, LA were used to study the mineralization of 4-nitrophenol under aerobic and anaerobic conditions. The rate of mineralization to carbon dioxide was found to be faster under aerobic conditions (1.04 x 10^{-3} μg/day/g dry sediment) than under anaerobic conditions (2.95 x 10^{-5} μg/day/g dry sediment) (Siragusa and DeLaune, 1986). In lake water samples collected from Beebe and Cayuga Lakes, Ithica, NY, 4-nitrophenol at 50, 75 and 100 μg/L was not mineralized after 7 days. When the lake water samples were inoculated with the microorganism *Corynebacterium* sp., extensive mineralization was observed. However, at a concentration of 26 μg/L, the extent of mineralization was much lower than at higher concentrations. The presence of a eucaryotic inhibitor (cycloheximide) also inhibited mineralization at the lower concentration but did not affect mineralization at the higher concentrations (Zaidi et al., 1989).

In activated sludge, 0.5% mineralized to carbon dioxide after 5 days (Freitag et al., 1985). Intermediate products include 4-nitrophenol, which further degraded to hydroquinone with lesser quantities of oxyhydroquinone (Nyholm et al., 1984). When 4-nitrophenol was statically incubated in the dark at 25 °C with yeast extract and settled domestic wastewater inoculum, 100% biodegradation with rapid adaptation was achieved after 7 days (Tabak et al., 1981).

Photolytic. An aqueous solution containing 200 ppm 4-nitrophenol exposed to sunlight for 1-2 months yielded hydroquinone, 4-nitrocatechol and an unidentified polymeric substance (Callahan et al., 1979). Under artificial sunlight, river water containing 2-5 ppm 4-nitrophenol photodegraded to produce trace amounts of 4-aminophenol (Mansour et al., 1989). A carbon dioxide yield of 39.5% was achieved when 4-nitrophenol adsorbed on silica gel was irradiated with light (λ >290 nm) for 17 hours (Freitag et al., 1985).

Chemical/Physical. Wet oxidation of 4-nitrophenol at 320 °C yielded formic and acetic acids (Randall and Knopp, 1980). Wet oxidation of 4-nitrophenol at an elevated pressure and temperature gave the following products: acetone, acetaldehyde, formic, acetic, maleic, oxalic and succinic acids (Keen and Baillod, 1985).

In an aqueous solution, 4-nitrophenol (100 μM) reacted with Fenton's reagent (35 μM). After 15 minutes into the reaction, the following products were identified: 1,2,4-trihydroxybenzene, hydroquinone, hydroxy-*p*-benzoquinone, *p*-benzoquinone and 4-nitro-catechol. After 3.5 hours, 90% of the 4-nitrophenol was destroyed (Lipczynska-Kochany, 1991). In a dilute aqueous solution at pH 6.0, 4-nitrophenol reacted with excess hypochlorous acid forming 2,6-dichlorobenzoquinone, 2,6-dichloro-4-nitrophenol and 2,3,4,6-tetrachlorophenol at yields of 20, 1 and 0.3%, respectively (Smith et al., 1976).

Toxicity: Acute oral LD_{50} for mice 380 mg/kg, rats 250 mg/kg (RTECS, 1985); LC_{50} (48-hour) for red killifish 100 mg/L (Yoshioka et al., 1986).

Drinking Water Standard: As of May 1994, no MCLGs or MCLs have been proposed (U.S. EPA, 1984).

Uses: Fungicide for leather; production of parathion; preparation of *p*-nitrophenyl acetate and other organic compounds.

1-NITROPROPANE

Synonym: UN 2608.

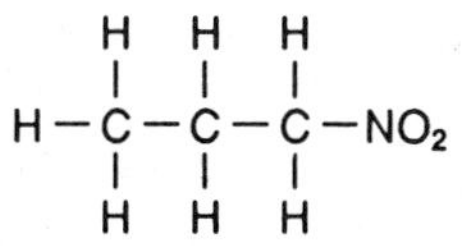

$$H - \underset{\underset{H}{|}}{\overset{\overset{H}{|}}{C}} - \underset{\underset{H}{|}}{\overset{\overset{H}{|}}{C}} - \underset{\underset{H}{|}}{\overset{\overset{H}{|}}{C}} - NO_2$$

CAS Registry No.: 108-03-2
DOT: 2608
DOT label: Combustible liquid
Molecular formula: $C_3H_7NO_2$
Formula weight: 89.09
RTECS: TZ5075000

Physical state, color and odor
Colorless, oily liquid with a mild, fruity odor. Odor threshold in air is 11 ppm (Amoore and Hautala, 1983).

Melting point (°C):
-108 (Weast, 1986)

Boiling point (°C):
130-131 (Weast, 1986)

Density (g/cm^3):
1.0081 at 24/4 °C (Weast, 1986)
0.9934 at 25/4 °C (Windholz et al., 1983)

Diffusivity in water (10^5 cm^2/sec):
1.08 at 25 °C using method of Hayduk and Laudie (1974).

Dissociation constant:
8.98 at 25 °C (Dean, 1987)

Flash point (°C):
35.8 (NIOSH, 1994)
34 (Windholz et al., 1983)

Lower explosive limit (%):
2.2 (NIOSH, 1994)

Henry's law constant (10^5 atm·m^3/mol):
8.68 at 25 °C (Hine and Mookerjee, 1975)

Ionization potential (eV):
10.81 ± 0.03 (Franklin et al., 1969)

Soil sorption coefficient, log K_{oc}:
Unavailable because experimental methods for estimation of this parameter for nitroaliphatics are lacking in the documented literature. However, its moderate solubility in water suggests its adsorption to soil will be low (Lyman et al., 1982).

Octanol/water partition coefficient, log K_{ow}:
0.65 (Hansch and Anderson, 1967)
0.87 (Hansch and Leo, 1979)

Solubility in organics:
Soluble in alcohol, chloroform and ether (Weast, 1986). Miscible with many organic solvents (Windholz et al., 1983).

Solubility in water:
14 mL/L (Windholz et al., 1983)

Vapor density:
3.64 g/L at 25 °C, 3.08 (air = 1)

Vapor pressure (mmHg):
8 at 20 °C (NIOSH, 1994)

Exposure Limits: NIOSH REL: TWA 25 ppm (90 mg/m^3), IDLH 1,000 ppm; OSHA PEL: TWA 25 ppm.

Toxicity: Acute oral LD_{50} for mice 800 mg/kg, rats 455 mg/kg (RTECS, 1985).

Uses: Solvent for cellulose acetate, lacquers, vinyl resins, fats, oils, dyes, synthetic rubbers; chemical intermediate; propellant; gasoline additive.

2-NITROPROPANE

Synonyms: Dimethylnitromethane; Isonitropropane; Nipar S-20 solvent; Nipar S-30 solvent; Nitroisopropane; 2-NP; RCRA waste number U171; UN 2608.

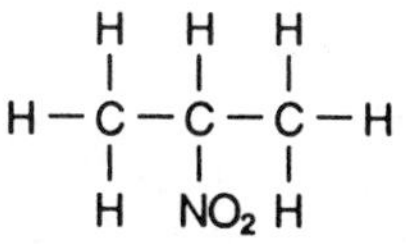

$$H-\underset{\underset{H}{|}}{\overset{\overset{H}{|}}{C}}-\underset{\underset{NO_2}{|}}{\overset{\overset{H}{|}}{C}}-\underset{\underset{H}{|}}{\overset{\overset{H}{|}}{C}}-H$$

CAS Registry No.: 79-46-9
DOT: 2608
DOT label: Combustible liquid
Molecular formula: $C_3H_7NO_2$
Formula weight: 89.09
RTECS: TZ5250000

Physical state, color and odor
Colorless liquid with a mild, fruity odor. Odor threshold in air is 70 ppm (Amoore and Hautala, 1983).

Melting point (°C):
-93 (Weast, 1986)

Boiling point (°C):
120 (Weast, 1986)

Density (g/cm³):
0.9876 at 20/4 °C (Weast, 1986)
0.9821 at 25/4 °C (Windholz et al., 1983)

Diffusivity in water (10^5 cm²/sec):
0.94 at 20 °C using method of Hayduk and Laudie (1974).

Dissociation constant:
7.675 at 25 °C (Dean, 1987)

Flash point (°C):
24.1 (NIOSH, 1994)

Lower explosive limit (%):
2.6 (NIOSH, 1994)

Upper explosive limit (%):
11.0 (NIOSH, 1994)

Henry's law constant (10^4 atm·m^3/mol):
1.23 at 25 °C (Hine and Mookerjee, 1975)

Ionization potential (eV):
10.71 ± 0.05 (Franklin et al., 1969)

Bioconcentration factor, log BCF:
1.85 (activated sludge), 1.30 (algae) (Freitag et al., 1985)

Soil sorption coefficient, log K_{oc}:
Unavailable because experimental methods for estimation of this parameter for nitroaliphatics are lacking in the documented literature. However, its moderate solubility in water suggests its adsorption to soil will be low (Lyman et al., 1982).

Octanol/water partition coefficient, log K_{ow}:
Unavailable because experimental methods for estimation of this parameter for nitroaliphatics are lacking in the documented literature.

Solubility in organics:
Miscible with many organic solvents (Windholz et al., 1983).

Solubility in water:
2 wt % at 20 °C (NIOSH, 1994)
17 mL/L (Windholz et al., 1983)

Vapor density:
3.64 g/L at 25 °C, 3.08 (air = 1)

Vapor pressure (mmHg):
13 at 20 °C (NIOSH, 1994)

Environmental Fate
   Photolytic. Anticipated products from the reaction of 2-nitropropane with ozone or hydroxyl radicals in the atmosphere are formaldehyde and acetaldehyde (Cupitt, 1980).

Exposure Limits: Potential occupational carcinogen. NIOSH REL: IDLH 100 ppm; OSHA PEL: TWA 25 ppm (90 mg/m^3); ACGIH TLV: TWA 10 ppm.

Toxicity: Acute oral LD$_{50}$ for rats 720 mg/kg (RTECS, 1985).

Uses: Solvent for cellulose acetate, lacquers, vinyl resins, fats, oils, dyes, synthetic rubbers; chemical intermediate; propellant; gasoline additive.

N-NITROSODIMETHYLAMINE

Synonyms: Dimethylnitrosamine; *N*-Dimethylnitrosamine; *N,N*-Dimethylnitros-amine; Dimethylnitrosomine; DMN; DMNA; **N-Methyl-*N*-nitrosomethanamine;** NDMA; Nitrous dimethylamide; RCRA waste number P082.

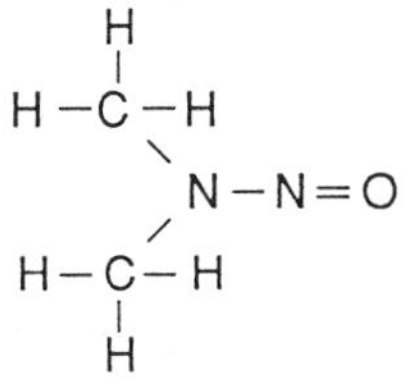

$$H-C-H$$

(structural diagram showing)

H
|
H—C—H
\\
N—N=O
/
H—C—H
|
H

CAS Registry No.: 62-75-9
DOT: 1955
DOT label: Poison
Molecular formula: $C_2H_6N_2O$
Formula weight: 74.09
RTECS: IQ0525000

Physical state, color and odor
Yellow, oily liquid with a faint, characteristic odor.

Boiling point (°C):
151-153 (Dean, 1987)
154 (Weast, 1986)

Density (g/cm^3):
1.0059 at 20/4 °C (Weast, 1986)
1.0049 at 18/4 °C (Weast and Astle, 1986)

Diffusivity in water (10^5 cm^2/sec):
1.06 at 20 °C using method of Hayduk and Laudie (1974).

Flash point (°C):
61 (Aldrich, 1988)

Henry's law constant (atm·m^3/mol):
0.143 at 25 °C (estimated using a solubility of 1,000 g/L)

Ionization potential (eV):
8.69 (NIOSH, 1994)

Soil sorption coefficient, log K_{oc}:
1.41 using method of Kenaga and Goring (1980).

Octanol/water partition coefficient, log K_{ow}:
0.06 (Radding et al., 1976)

Solubility in organics:
Soluble in solvents (U.S. EPA, 1985), including ethanol and ether (Weast, 1986).

Solubility in water:
Miscible (Mirvish et al., 1976).

Vapor pressure (mmHg):
8.1 at 25 °C (Mabey et al., 1982)
2.7 at 20 °C (Klein, 1982)

Environmental Fate

Biological. Two of seven microorganisms, *Escherichia coli* and *Pseudomonas fluorescens*, were capable of slowly degrading N-nitrosodimethylamine to dimethylamine (Mallik and Tesfai, 1981).

Photolytic. A Teflon bag containing air and N-nitrosodimethylamine was subjected to sunlight on two different days. On a cloudy day, half of the N-nitrosodimethylamine was photolyzed in 60 minutes. On a sunny day, half of the N-nitrosodimethylamine was photolyzed in 30 minutes. Photolysis products include nitric oxide, carbon monoxide, formaldehyde and an unidentified compound (Hanst et al., 1977). In a separate experiment, Tuazon et al. (1984) irradiated an ozone-rich atmosphere containing N-nitrosodimethylamine. Photolysis products identified include dimethylnitramine, nitromethane, formaldehyde, carbon monoxide, nitrogen dioxide, nitrogen pentoxide and nitric acid.

Exposure Limits: Potential occupational carcinogen.

Toxicity: Acute oral LD_{50} for hamsters 28 mg/kg, rats 45 mg/kg; LC_{50} (inhalation) for mice 57 ppm/4-hour, rats 78 ppm/4-hour (RTECS, 1985).

Uses: Rubber accelerator; solvent in fiber and plastic industry; rocket fuels; lubricants; condensers to increase dielectric constant; industrial solvent; antioxidant; nematocide; softener of copolymers; research chemical; plasticizer in acrylonitrile polymers; inhibit nitrification in soil; chemical intermediate for 1,1-dimethylhydrazine.

N-NITROSODIPHENYLAMINE

Synonyms: Benzenamine; Curetard A; Delac J; Diphenylnitrosamine; Diphenyl-*N*-nitrosamine; *N,N*-Diphenylnitrosamine; Naugard TJB; NCI-C02880; NDPA; NDPhA; Nitrosodiphenylamine; *N*-Nitroso-*n*-phenylamine; *N*-Nitroso-*n*-phenyl-benzenamine; Nitrous diphenylamide; Redax; Retarder J; TJB; Vulcalent A; Vulcatard; Vulcatard A; Vultrol.

CAS Registry No.: 86-30-6
Molecular formula: $C_{12}H_{10}N_2O$
Formula weight: 198.22
RTECS: JJ9800000

Physical state and color
Yellow to brown to orange power or flakes.

Melting point (°C):
66.5 (Weast, 1986)
65-66 (Fluka, 1988)

Diffusivity in water (10^5 cm²/sec):
0.58 at 20 °C using method of Hayduk and Laudie (1974).

Henry's law constant (10^8 atm·m³/mol):
2.33 at 25 °C (approximate - calculated from water solubility and vapor pressure)

Bioconcentration factor, log BCF:
2.34 (bluegill sunfish, Veith et al., 1980)

Soil sorption coefficient, log K_{oc}:
2.76 (estimated, Montgomery, 1989)

Octanol/water partition coefficient, log K_{ow}:
3.13 (Veith et al., 1980)
3.18 (Garst and Wilson, 1984)

Solubility in organics:
Soluble in ethanol, benzene, ethylene dichloride, gasoline, (Keith and Walters,

787

1992), ether, chloroform and slightly soluble in petroleum ether (Windholz et al., 1983).

Solubility in water:
35.1 mg/L at 25 °C (Banerjee et al., 1980)

Vapor pressure (mmHg):
0.1 at 25 °C (assigned by analogy, Mabey et al., 1982)

Environmental Fate
   Chemical/Physical. Above 85 °C, technical grades may decompose to nitrogen oxides (IARC, 1978).

Toxicity: Acute oral LD_{50} for mice 3,850 mg/kg, rats 1,650 mg/kg (RTECS, 1985).

Uses: Chemical intermediate for *N*-phenyl-*p*-phenylenediamine; rubber processing (vulcanization retarder).

N-NITROSODI-*n*-PROPYLAMINE

Synonyms: Dipropylnitrosamine; Di-*n*-propylnitrosamine; DPN; DPNA; NDPA; *N*-Nitrosodipropylamine; **N-Nitroso-*n*-propyl-1-propanamine**; RCRA waste number U111.

H H H
| | |
H—C—C—C
| | | \
H H H \
N—N=O
H H H /
| | | /
H—C—C—C
| | |
H H H

CAS Registry No.: 621-64-7
Molecular formula: $C_6H_{14}N_2O$
Formula weight: 130.19
RTECS: JL9700000

Physical state and color
Yellow to gold colored liquid.

Boiling point (°C):
205.9 (Dean, 1973)

Density (g/cm^3):
0.9160 at 20/4 °C (IARC, 1978)

Diffusivity in water (10^5 cm^2/sec):
0.72 at 20 °C using method of Hayduk and Laudie (1974).

Soil sorption coefficient, log K_{oc}:
1.01 (estimated, Montgomery, 1989)

Octanol/water partition coefficient, log K_{ow}:
1.31 (calculated, U.S. EPA, 1980)

Solubility in organics:
Very soluble in ethanol and ether (Keith and Walters, 1992).

Solubility in water:
9,900 mg/L at 25 °C (Mirvish et al., 1976)

Toxicity: Acute oral LD_{50} for rats 480 mg/kg (RTECS, 1985).

Use: Research chemical.

2-NITROTOLUENE

Synonyms: 1-Methyl-2-nitrobenzene; 2-Methylnitrobenzene; *o*-Methylnitroben-zene; *o*-Nitrotoluene; ONT; UN 1664.

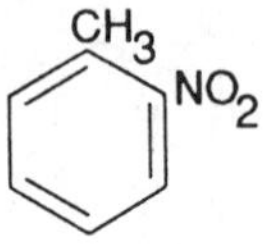

CAS Registry No.: 88-72-2
DOT: 1664
DOT label: Poison
Molecular formula: $C_7H_7NO_2$
Formula weight: 137.14
RTECS: XT3150000

Physical state, color and odor
Yellowish liquid with a faint, aromatic odor.

Melting point (°C):
-9.5 (needles), -2.5 °C (crystals) (Weast, 1986)
-4.1 (Stull, 1947)

Boiling point (°C):
221.7 (Weast, 1986)
225 (Gross et al., 1933)
220.4 (Hawley, 1981)

Density (g/cm³):
1.1629 at 20/4 °C (Weast, 1986)

Diffusivity in water (10^5 cm²/sec):
0.80 at 20 °C using method of Hayduk and Laudie (1974).

Flash point (°C):
107 (NIOSH, 1994)

Lower explosive limit (%):
2.2 (NIOSH, 1994)

Henry's law constant (10^5 atm·m³/mol):
4.51 at 20 °C (approximate - calculated from water solubility and vapor pressure)

Interfacial tension with water (dyn/cm at 20 °C):
27.19 (Demond and Lindner, 1993)

Ionization potential (eV):
9.43 (NIOSH, 1994)

Soil sorption coefficient, log K_{oc}:
Unavailable because experimental methods for estimation of this parameter for nitroaromatics are lacking in the documented literature.

Octanol/water partition coefficient, log K_{ow}:
2.30 (Leo et al., 1971)

Solubility in organics:
Soluble in alcohol, ether (Weast, 1986), benzene and petroleum ether (Windholz et al., 1983).

Solubility in water:
652 mg/kg at 30 °C (Gross et al., 1933)

Vapor density:
5.61 g/L at 25 °C, 4.73 (air = 1)

Vapor pressure (mmHg):
0.1 at 20 °C (NIOSH, 1987)
0.25 at 30 °C (Verschueren, 1983)

Exposure Limits: NIOSH REL: TWA 2 ppm (11 mg/m^3), IDLH 200 ppm; OSHA PEL: TWA 5 ppm (30 mg/m^3); ACGIH TLV: TWA 2 ppm.

Toxicity: Acute oral LD_{50} for mice 970 mg/kg, rats 891 mg/kg (RTECS, 1985); LC_{50} (48-hour) for red killifish 245 mg/L (Yoshioka et al., 1986).

Uses: Manufacture of dyes, nitrobenzoic acids, toluidines, etc.

3-NITROTOLUENE

Synonyms: 1-Methyl-3-nitrobenzene; 3-Methylnitrobenzene; *m*-Methylnitrobenzene; MNT; *m*-Nitrotoluene; 3-Nitrotoluol; UN 1664.

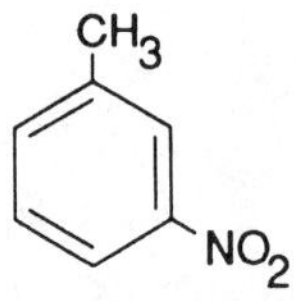

CAS Registry No.: 99-08-1
DOT: 1664
DOT label: Poison
Molecular formula: $C_7H_7NO_2$
Formula weight: 137.14
RTECS: XT2975000

Physical state, color and odor
Clear, yellowish liquid with an aromatic odor. Odor threshold in air is 45 ppb Odor threshold in air is 600 ppm (Amoore and Hautala, 1983).

Melting point (°C):
16 (Weast, 1986)

Boiling point (°C):
232.6 (Weast, 1986)

Density (g/cm³):
1.1571 at 20/4 °C (Weast, 1986)

Diffusivity in water (10^5 cm²/sec):
0.80 at 20 °C using method of Hayduk and Laudie (1974).

Flash point (°C):
107 (NIOSH, 1994)

Lower explosive limit (%):
1.6 (NIOSH, 1994)

Henry's law constant (10^5 atm·m³/mol):
5.41 at 20 °C (approximate - calculated from water solubility and vapor pressure)

Interfacial tension with water (dyn/cm at 20 °C):
27.68 (Demond and Lindner, 1993)

Ionization potential (eV):
9.48 (NIOSH, 1994)
9.65 ± 0.05 (Franklin et al., 1969)

Soil sorption coefficient, log K_{oc}:
Unavailable because experimental methods for estimation of this parameter for nitroaromatics are lacking in the documented literature.

Octanol/water partition coefficient, log K_{ow}:
2.40, 2.45 (Leo et al., 1971)
2.42 (Fujita et al., 1964)

Solubility in organics:
Soluble in alcohol, benzene and ether (Weast, 1986).

Solubility in water:
498 mg/kg at 30 °C (Gross et al., 1933)

Vapor density:
5.61 g/L at 25 °C, 4.73 (air = 1)

Vapor pressure (mmHg):
0.1 at 20 °C (NIOSH, 1987)
0.25 at 25 °C (Verschueren, 1983)

Environmental Fate
Biological. Under anaerobic conditions using a sewage inoculum, 3-nitrotoluene and 4-nitrotoluene both degraded to toluidine (Hallas and Alexander, 1983).

Exposure Limits: NIOSH REL: TWA 2 ppm (11 mg/m^3), IDLH 200 ppm; OSHA PEL: TWA 5 ppm (30 mg/m^3); ACGIH TLV: TWA 2 ppm.

Toxicity: Acute oral LD$_{50}$ for guinea pigs 3,600 mg/kg, mice 330 mg/kg, rats 1,072 mg/kg, rabbits 2,400 (RTECS, 1985).

Uses: Manufacture of dyes, nitrobenzoic acids, toluidines; organic synthesis.

4-NITROTOLUENE

Synonyms: 1-Methyl-4-nitrobenzene; 4-Methylnitrobenzene; *p*-Methylnitroben-zene; NCI-C60537; *p*-Nitrotoluene; 4-Nitrotoluol; PNT; UN 1664.

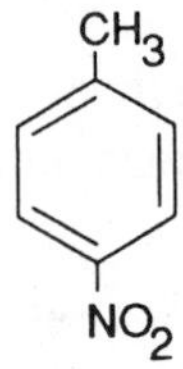

CAS Registry No.: 99-99-0
DOT: 1664
DOT label: Poison
Molecular formula: $C_7H_7NO_2$
Formula weight: 137.14
RTECS: XT3325000

Physical state, color and odor
Yellowish crystals with a weak, aromatic odor.

Melting point (°C):
54.5 (Weast, 1986)
51.3 (Verschueren, 1983)
51.9 (Stull, 1947)

Boiling point (°C):
238.3 (Weast, 1986)

Density (g/cm^3):
1.1038 at 75/4 °C (Weast, 1986)
1.286 at 20 °C (Verschueren, 1983)

Diffusivity in water (10^5 cm^2/sec):
0.81 at 20 °C using method of Hayduk and Laudie (1974).

Dissociation constant:
11.27 (Perrin, 1972)

Flash point (°C):
107 (NIOSH, 1994)

Lower explosive limit (%):
1.6 (NIOSH, 1994)

Henry's law constant (10^5 atm·m^3/mol):
5.0 at 25 °C using method of Hine and Mookerjee (1975).

Ionization potential (eV):
9.50 (NIOSH, 1994)
9.82 (Franklin et al., 1969)

Soil sorption coefficient, log K_{oc}:
Unavailable because experimental methods for estimation of this parameter for nitroaromatics are lacking in the documented literature.

Octanol/water partition coefficient, log K_{ow}:
2.37, 2.42 (Leo et al., 1971)

Solubility in organics:
Soluble in acetone, alcohol, benzene, ether (Weast, 1986) and chloroform (Windholz et al., 1983).

Solubility in water:
442 mg/kg at 30 °C (Gross et al., 1933)
At 20 °C: 210, 183, 187 and 177 μmol/L in distilled water, Pacific seawater, artificial seawater and 35 wt % NaCl, respectively (Hashimoto et al., 1984)

Vapor pressure (mmHg):
0.1 at 20 °C (NIOSH, 1994)
0.22 at 30 °C (Verschueren, 1983)
0.004633 at 23.9 °C, 0.005484 at 26.0 °C (Lenchitz and Velicky, 1970)

Environmental Fate
 Biological. Under anaerobic conditions using a sewage inoculum, 3- and 4-nitrotoluene both degraded to toluidine (Hallas and Alexander, 1983).

Exposure Limits: NIOSH REL: TWA 2 ppm (11 mg/m^3), IDLH 200 ppm; OSHA PEL: TWA 5 ppm (30 mg/m^3); ACGIH TLV: TWA 2 ppm.

Toxicity: Acute oral LD$_{50}$ for mice 1,231 mg/kg, rats 1,960 mg/kg (RTECS, 1985); LC$_{50}$ (48-hour) for red killifish 537 mg/L (Yoshioka et al., 1986).

Uses: Manufacture of dyes, nitrobenzoic acids, toluidines, etc.

NONANE

Synonyms: *n*-Nonane; Nonyl hydride; Shellsol 140; UN 1920.

$$H-\underset{\underset{H}{|}}{\overset{\overset{H}{|}}{C}}-\underset{\underset{H}{|}}{\overset{\overset{H}{|}}{C}}-\underset{\underset{H}{|}}{\overset{\overset{H}{|}}{C}}-\underset{\underset{H}{|}}{\overset{\overset{H}{|}}{C}}-\underset{\underset{H}{|}}{\overset{\overset{H}{|}}{C}}-\underset{\underset{H}{|}}{\overset{\overset{H}{|}}{C}}-\underset{\underset{H}{|}}{\overset{\overset{H}{|}}{C}}-\underset{\underset{H}{|}}{\overset{\overset{H}{|}}{C}}-\underset{\underset{H}{|}}{\overset{\overset{H}{|}}{C}}-H$$

CAS Registry No.: 111-84-2
DOT: 1920
Molecular formula: C_9H_{20}
Formula weight: 128.26
RTECS: RA6115000

Physical state, color and odor
Clear, colorless liquid with a gasoline-like odor. Odor threshold in air is 47 ppm (Amoore and Hautala, 1983).

Melting point (°C):
-53.7 (Stull, 1947)

Boiling point (°C):
150.8 (Weast, 1986)

Density (g/cm³):
0.71763 at 20/4 °C, 0.71381 at 25/4 °C (Dreisbach, 1959)

Diffusivity in water (10^5 cm²/sec):
0.63 at 20 °C using method of Hayduk and Laudie (1974).

Dissociation constant:
>14 (Schwarzenbach et al., 1993)

Flash point (°C):
.31.4 (NIOSH, 1994)
33 (Affens and McLaren, 1972)

Lower explosive limit (%):
0.8 (NIOSH, 1994)

Upper explosive limit (%):
2.9 (NIOSH, 1994)

Heat of fusion (kcal/mol):
3.72 (Dean, 1987)

Henry's law constant (atm·m³/mol):
0.400, 0.496, 0.332, 0.414 and 0.465 at 10, 15, 20, 25 and 30 °C, respectively (Ashworth et al., 1988).
2.32 at 25 °C (Jönsson et al., 1982)

Ionization potential (eV):
10.21 (NIOSH, 1994)

Soil sorption coefficient, log K_{oc}:
Unavailable because experimental methods for estimation of this parameter for aliphatic hydrocarbons are lacking in the documented literature.

Octanol/water partition coefficient, log K_{ow}:
4.67 using method of Hansch et al. (1968).

Solubility in organics:
In methanol, g/L: 84 at 5 °C, 95 at 10 °C, 105 at 15 °C, 116 at 20 °C, 129 at 25 °C, 142 at 30 °C, 155 at 35 °C, 170 at 40 °C (Kiser et al., 1961).

Solubility in water:
In mg/kg: 0.122 at 25 °C, 0.309 at 69.7 °C, 0.420 at 99.1 °C, 1.70 at 121.3 °C, 5.07 at 136.6 °C (Price, 1976)
0.07 mg/L at 20 °C, 0.43 mg/L in seawater at 20 °C (Verschueren, 1983)
220 μg/kg at 25 °C (McAuliffe, 1969)
1 x 10⁻⁸ at 25 °C (mole fraction, Krasnoshchekova and Gubergrits, 1973)
272 mg/L at 25 °C (Jönsson et al., 1982)

Vapor density:
5.24 g/L at 25 °C, 4.43 (air = 1)

Vapor pressure (mmHg):
3.22 at 20 °C (Verschueren, 1983)
4.35 at 25 °C (Wilhoit and Zwolinski, 1971)

Environmental Fate
   Biological. Nonane may biodegrade in two ways. This first is the formation of nonyl hydroperoxide, which decomposes to 1-nonanol followed by oxidation to nonanoic acid. The other pathway involves dehydrogenation to 1-nonene, which may react with water giving 1-nonanol (Dugan, 1972). Microorganisms can oxidize alkanes under aerobic conditions (Singer and Finnerty, 1984). The most common

degradative pathway involves the oxidation of the terminal methyl group forming the 1-nonanol. The alcohol may undergo a series of dehydrogenation steps forming nonanal then a fatty acid (nonanoic acid). The fatty acid may then be metabolized by β-oxidation to form the mineralization products, carbon dioxide and water (Singer and Finnerty, 1984).

   Chemical/Physical. Complete combustion in air yields carbon dioxide and water.

Exposure Limits: NIOSH REL: TWA 200 ppm (1,050 mg/m^3).

Toxicity: LC$_{50}$ (inhalation) for rats 3,200 ppm/4-hour (RTECS, 1985).

Uses: Solvent; standardized hydrocarbon; manufacturing paraffin products; biodegradable detergents; jet fuel research; rubber industry; paper processing industry; distillation chaser; component of gasoline and similar fuels; organic synthesis.

OCTACHLORONAPHTHALENE

Synonym: Halowax 1051.

CAS Registry No.: 2234-13-1
Molecular formula: $C_{10}Cl_8$
Formula weight: 403.73
RTECS: QK0250000

Physical state, color and odor
Waxy, light yellow solid with an aromatic odor.

Melting point (°C):
197-198 (Weast, 1986)

Boiling point (°C):
413 (NIOSH, 1994)

Density (g/cm³):
2.00 at 20/4 °C (NIOSH, 1994)

Diffusivity in water (10^5 cm²/sec):
0.45 at 20 °C using method of Hayduk and Laudie (1974).

Flash point (°C):
Noncombustible solid (NIOSH, 1994)

Bioconcentration factor, log BCF:
2.52 (rainbow trout, Oliver and Niimi, 1985)

Solubility in organics:
Soluble in benzene, chloroform and lignoin (Weast, 1986).

Vapor pressure (mmHg):
<1 at 20 °C (NIOSH, 1987)

Exposure Limits (mg/m³): NIOSH REL: TWA 0.1, STEL 0.3; OSHA PEL: TWA 0.1.

Uses: Chemical research; organic synthesis.

OCTANE

Synonyms: *n*-Octane; UN 1262.

$$H-\overset{\displaystyle H}{\underset{\displaystyle H}{C}}-\overset{\displaystyle H}{\underset{\displaystyle H}{C}}-\overset{\displaystyle H}{\underset{\displaystyle H}{C}}-\overset{\displaystyle H}{\underset{\displaystyle H}{C}}-\overset{\displaystyle H}{\underset{\displaystyle H}{C}}-\overset{\displaystyle H}{\underset{\displaystyle H}{C}}-\overset{\displaystyle H}{\underset{\displaystyle H}{C}}-\overset{\displaystyle H}{\underset{\displaystyle H}{C}}-H$$

CAS Registry No.: 111-65-9
DOT: 1262
DOT label: Flammable liquid
Molecular formula: C_8H_{18}
Formula weight: 114.23
RTECS: RG8400000

Physical state, color and odor
Colorless liquid with a gasoline-like odor. Odor threshold in air is 48 ppm (Amoore and Hautala, 1983).

Melting point (°C):
-56.8 (Weast, 1986)

Boiling point (°C):
125.67 (Dreisbach, 1959)
125.66 (Stephenson and Malanowski, 1987)

Density (g/cm^3):
0.70252 at 20/4 °C, 0.69849 at 25/4 °C (Dreisbach, 1959)
0.70267 at 20/4 °C, 0.68862 at 25/4 °C (Riddick et al., 1986)

Diffusivity in water (10^5 cm^2/sec):
0.66 at 20 °C using method of Hayduk and Laudie (1974).

Dissociation constant:
>14 (Schwarzenbach et al., 1993)

Flash point (°C):
13.4 (NIOSH, 1994)
48 (Affens and McLaren, 1972)

Lower explosive limit (%):
1.0 (NIOSH, 1994)

Upper explosive limit (%):
6.5 (NIOSH, 1994)
4.7 (Sax and Lewis, 1987)

Heat of fusion (kcal/mol):
4.957 (Dean, 1987)

Henry's law constant (atm·m^3/mol at 25 °C):
4.45 (Jönsson et al., 1982)
3.225 (Hine and Mookerjee, 1975)

Interfacial tension with water (dyn/cm at 25 °C):
50.2 (Donahue and Bartell, 1952)

Ionization potential (eV):
9.82 (NIOSH, 1994)

Soil sorption coefficient, log K_{oc}:
Unavailable because experimental methods for estimation of this parameter for aliphatic hydrocarbons are lacking in the documented literature.

Octanol/water partition coefficient, log K_{ow}:
4.00 (Coates et al., 1985)
5.18 (Tewari et al., 1982; Wasik et al., 1981)

Solubility in organics:
In methanol, g/L: 122 at 5 °C, 136 at 10 °C, 152 at 15 °C, 167 at 20 °C, 184 at 25 °C, 206 at 30 °C, 230 at 35 °C, 260 at 40 °C (Kiser et al., 1961)

Solubility in water:
In mg/kg: 0.431 at 25 °C, 0.524 at 40.1 °C, 0.907 at 69.7 °C, 1.12 at 99.1 °C, 4.62 at 121.3 °C, 8.52 at 136.6 °C, 11.8 at 149.5 °C (Price, 1976)
0.66 mg/kg at 25 °C (Coates et al., 1985; McAuliffe, 1963, 1966)
1.35 mg/kg at 0 °C, 0.85 mg/kg at 25 °C (Polak and Lu, 1973)
0.884 mg/L at 20 °C, 1.66 mg/L at 70 °C (Burris and MacIntyre, 1986)
9.66 μmol/L at 25.0 °C (Tewari et al., 1982; Wasik et al., 1981)
0.493 and 0.88 mg/L at 25 °C (Mackay and Shiu, 1981)
0.02 mL/L at 16 °C (Fühner, 1924)
0.090 mL/L (Baker, 1980)
1.25 mg/L (Coutant and Keigley, 1988)
Mole fraction x 10^7: 2.6, 1.4 and 2.9 at 5.0, 25.0 and 45.0 °C, respectively (Nelson and DeLigny, 1968)
10^{-7} at 25 °C (mole fraction, Krasnoshchekova and Gubergrits, 1973)

0.615 mg/L at 25 °C (Jönsson et al., 1982)
1.25 mg/L at 25 °C (Coutant and Keigley, 1988)

Vapor density:
4.67 g/L at 25 °C, 3.94 (air = 1)

Vapor pressure (mmHg):
11 at 20 °C, 18 at 30 °C (Verschueren, 1983)
14.14 at 25 °C (Wilhoit and Zwolinski, 1971)
10.37 at 20 °C, 118.4 at 70 °C (Burris and MacIntyre, 1986)
13.60 at 25.00 °C (Hussam and Carr, 1985)
4.2, 14.0 and 39.8 at 5.0, 25.0 and 45.0 °C, respectively (Nelson and DeLigny, 1968)

Environmental Fate
  Biological. *n*-Octane may biodegrade in two ways. This first is the formation of octyl hydroperoxide, which decomposes to 1-octanol followed by oxidation to octanoic acid. The other pathway involves dehydrogenation to 1-octene, which may react with water giving 1-octanol (Dugan, 1972). 1-Octanol was reported as the biodegradation product of octane by a *Pseudomonas* sp. (Riser-Roberts, 1992). Microorganisms can oxidize alkanes under aerobic conditions (Singer and Finnerty, 1984). The most common degradative pathway involves the oxidation of the terminal methyl group forming the corresponding alcohol (1-octanol). The alcohol may undergo a series of dehydrogenation steps forming an aldehyde (octanal) then a fatty acid (octanoic acid). The fatty acid may then be metabolized by β-oxidation to form the mineralization products, carbon dioxide and water (Singer and Finnerty, 1984).

Exposure Limits: NIOSH REL: TWA 75 ppm (350 mg/m^3), 15-minute ceiling 385 ppm (1,800 mg/m^3), IDLH 1,000 ppm; OSHA PEL: TWA 500 ppm (2,350 mg/m^3); ACGIH TLV: TWA 300 ppm.

Symptoms of Exposure: Irritates mucous membranes. Narcotic at high concentrations (Patnaik, 1992).

Uses: Solvent; rubber and paper industries; calibrations; azeotropic distillations; occurs in gasoline and petroleum naphtha; organic synthesis.

1-OCTENE

Synonyms: 1-Caprylene; 1-Octylene.

$$H_2C=CH-CH_2-CH_2-CH_2-CH_2-CH_2-CH_3$$

CAS Registry No.: 111-66-0
Molecular formula: C_8H_{16}
Formula weight: 112.22

Physical state and color
Colorless liquid.

Melting point (°C):
-101.7 (Weast, 1986)

Boiling point (°C):
121.3 (Weast, 1986)

Density (g/cm³):
0.71492 at 20/4 °C, 0.71085 at 25/4 °C (Dreisbach, 1959)

Diffusivity in water (10^5 cm²/sec):
0.68 at 20 °C using method of Hayduk and Laudie (1974).

Dissociation constant:
>14 (Schwarzenbach et al., 1993)

Flash point (°C):
21 (open cup, NFPA, 1984)

Heat of fusion (kcal/mol):
3.660 (Dean, 1987)

Henry's law constant (atm·m³/mol):
0.952 at 25 °C (Hine and Mookerjee, 1975)

Soil sorption coefficient, log K_{oc}:
Unavailable because experimental methods for estimation of this parameter for aliphatic hydrocarbons are lacking in the documented literature.

803

Octanol/water partition coefficient, log K_{ow}:
2.79 (Coates et al., 1985)
4.56 (Shantz and Martire, 1987)
4.88 (Wasik et al., 1981)

Solubility in organics:
Soluble in acetone, benzene and chloroform (Weast, 1986). Miscible with alcohol and ether (Windholz et al., 1983).

Solubility in water:
3.6 mg/L at 23 °C (Coates et al., 1985)
2.7 mg/kg at 25 °C (McAuliffe, 1966)
36.5 μmol/L at 25.0 °C (Wasik et al., 1981)
In 10^{-3} M nitric acid: 0.239, 0.197 and 0.153 mmol/L at 20, 25 and 30 °C, respectively (Natarajan and Venkatachalam, 1972)

Vapor density:
4.59 g/L at 25 °C, 3.87 (air = 1)

Vapor pressure (mmHg):
17.4 at 25 °C (Wilhoit and Zwolinski, 1971)

Environmental Fate
Biological. Biooxidation of 1-octene may occur yielding 7-octen-1-ol, which may further oxidize to 7-octenoic acid (Dugan, 1972).
Chemical/Physical. The reaction of ozone and hydroxyl radicals with 1-octene was studied in a flexible outdoor Teflon chamber (Paulson and Seinfeld, 1992). 1-Octene reacted with ozone producing heptanal, a thermally stabilized C_7 biradical and hexane at yields of 80, 10 and 1%, respectively. With hydroxyl radicals, only 15% of 1-octene was converted to heptanal. In both reactions, the remaining compounds were tentatively identified as alkyl nitrates (Paulson and Seinfeld, 1977).

Uses: Plasticizer; surfactants; organic synthesis.

OXALIC ACID

Synonyms: Ethanedioic acid; Ethanedionic acid; NCI-C55209.

CAS Registry No.: 144-62-7
DOT: 2449
Molecular formula: $C_2H_2O_4$
Formula weight: 90.04
RTECS: RO2450000

Physical state, color and odor
Colorless and odorless rhombic crystals. Hygroscopic.

Melting point (°C):
182-189.5 (anhydrous), 101.5 °C (hydrated) (Weast, 1986)

Boiling point (°C):
Sublimes at 157 (Weast, 1986).
Sublimes at 150 (Verschueren, 1983).

Density (g/cm^3):
1.895-1.900 at 17/4 °C (Weast, 1986)
1.653 at 18/4 °C (Windholz et al., 1983)

Diffusivity in water (10^5 cm^2/sec):
0.97 at 20 °C using method of Hayduk and Laudie (1974).

Dissociation constant (25 °C):
$pK_1 = 1.27$, $pK_2 = 4.27$ (Windholz et al., 1983)

Henry's law constant (10^{10} atm·m^3/mol):
1.43 at pH 4 (Gaffney et al., 1987)

Soil sorption coefficient, log K_{oc}:
0.89 using method of Kenaga and Goring (1980).

Octanol/water partition coefficient, log K_{ow}:
-0.43, -0.81 (Verschueren, 1983)

Solubility in organics:
In g/L: alcohol (40), ether (10), glycerol (181.8) (Windholz et al., 1983).

Solubility in water:
6.71 wt % at 15 °C, 8.34 wt % at 20 °C, 9.81 wt % at 25 °C (Windholz et al., 1983)
95,000 mg/L at 15 °C, 1,290 g/L at 90 °C (Verschueren, 1983)

Vapor pressure (mmHg):
3×10^{-4} at 30 °C (Verschueren, 1983)

Environmental Fate
   Chemical/Physical. Above 189.5 °C, decomposes to carbon dioxide, carbon monoxide, formic acid and water (Windholz et al., 1983). Ozonolysis of oxalic acid in distilled water at 25 °C under acidic conditions (pH 6.3) yielded carbon dioxide (Kuo et al., 1977).
   Absorbs moisture in air forming the dihydrate (Huntress and Mulliken, 1941).

Exposure Limits (mg/m^3): NIOSH REL: TWA 1, STEL 2, IDLH 500; OSHA PEL: TWA 1.

Symptoms of Exposure: Ingestion may cause vomiting, diarrhea, severe gastro-intestinal disorder, renal damage, shock, convulsions and coma (Patnaik, 1992).

Toxicity: Acute oral LD_{50} in rats 375 mg/kg (RTECS, 1985).

Uses: Calico printing and dyeing, analytical and laboratory reagent; bleaching agent; removing ink or rust stains, paint or varnish; stripping agent for permanent press resins; reducing agent; metal polishes; ceramics and pigments; cleaning wood; purifying methanol; cleanser in the metallurgical industry; in paper industry, photography and process engraving; producing glucose from starch; leather tanning; rubber manufacturing industry; automobile radiator cleanser; purifying agent and intermediate for many compounds; catalyst; rare earth processing.

PARATHION

Synonyms: AAT; AATP; AC 3422; ACC 3422; Alkron; Alleron; American Cyanamid 3422; Aphamite; B 404; Bay E-605; Bladan; Bladan F; Compound 3422; Corothion; Corthion; Corthione; Danthion; DDP; *O,O*-Diethyl-*O*-4-nitrophenyl phosphorothioate; *O,O*-Diethyl *O*-*p*-nitrophenyl phosphorothioate; Diethyl-4-nitrophenyl phosphorothionate; Diethyl-*p*-nitrophenyl thionophosphate; *O,O*-Diethyl-*O*-4-nitrophenyl thionophosphate; *O,O*-Diethyl-*O*-*p*-nitrophenyl thiono-phosphate; Diethyl-*p*-nitrophenyl thiophosphate; *O,O*-Diethyl-*O*-*p*-nitrophenyl thiophosphate; Diethylparathion; DNTP; DPP; Drexel parathion 8E; E 605; Ecatox; Ekatin WF & WF ULV; Ekatox; ENT 15108; Ethlon; Ethyl parathion; Etilon; Folidol; Folidol E605; Folidol E & E 605; Fosfermo; Fosferno; Fosfex; Fosfive; Fosova; Fostern; Fostox; Gearphos; Genithion; Kolphos; Kypthion; Lethalaire G-54; Lirothion; Murfos; NA 2783; NCI-C00226; Niran; Niran E-4; Nitrostigmine; Nitrostygmine; Niuif-100; Nourithion; Oleofos 20; Oleoparaphene; Oleoparathion; Orthophos; Pac; Panthion; Paradust; Paraflow; Paramar; Paramar 50; Paraphos; Paraspray; Parathene; Parathion-ethyl; Parawet; Penphos; Pestox plus; Pethion; Phoskil; Phosphemol; Phosphenol; **Phosphorothioic acid *O,O*-diethyl *O*-(4-nitrophenyl) ester;** Phosphostigmine; RB; RCRA waste number P089; Rhodiasol; Rhodiatox; Rhodiatrox; Selephos; Sixty-three special E.C. insecticide; SNP; Soprathion; Stabilized ethyl parathion; Stathion; Strathion; Sulphos; Super rodiatox; T-47; Thiofos; Tiophos; Thiophos 3422; Tox 47; Vapophos; Vitrex.

CAS Registry No.: 56-38-2
DOT: 2783 (liquid/dry), 2784 (flammable liquid)
DOT label: Poison
Molecular formula: $C_{10}H_{14}NO_5PS$
Formula weight: 291.27
RTECS: TF4550000

Physical state, color and odor
Pale yellow to dark brown liquid with a garlic-like odor.

Melting point (°C):
6.1 (Weast, 1986)

Boiling point (°C):
375 (Weast, 1986)

807

Density (g/cm^3):
1.2704 at 20/20 °C (Weast, 1986)
1.2683 at 25/4 °C (Williams, 1951)

Diffusivity in water (10^5 cm^2/sec):
0.51 at 20 °C using method of Hayduk and Laudie (1974).

Flash point (°C):
174 (Meister, 1988)
202 (open cup, NIOSH, 1994)

Henry's law constant (10^8 atm·m^3/mol):
8.56 at 25 °C (Fendinger and Glotfelty, 1990)

Soil sorption coefficient, log K_{oc}:
3.68 (Kenaga and Goring, 1980)
3.02 (Batcombe silt loam, Briggs, 1981)
2.50-4.20 (organic matter content 0.2-6.1%, Mingelgrin and Gerstl, 1983)
3.25 (Mediterranean soil), 3.43 (Terra rossa) and 3.79 (Dark rendzina) (Saltzman et al., 1972)
3.02-3.14 (average = 3.07 for 5 sterilized Iowa soils, Felsot and Dahm, 1979)
2.43 (Netanya soil), 2.25 (Mivtahim soil), 2.25 (Golan soil), 2.39 (Gilat soil), 2.31 (Shefer soil), 2.32 (Bet Dagan soil), 2.40 (Neve Yaar soil), 2.45 (Malkiya soil), 2.75 (Kinneret sediment), 2.67 (Kinneret-A sediment), 2.70 (Kinneret-F sediment), 2.72 (Kinneret-G sediment, Gerstl and Mingelgrin, 1984)

Octanol/water partition coefficient, log K_{ow}:
3.81 (Chiou et al., 1977; Freed et al., 1979a)
3.91 (Verschueren, 1983)
2.15 (Leo et al., 1971)
3.93 (Briggs, 1981)
3.76 (Bowman and Sans, 1983a)

Solubility in organics:
2,900 and 2,700 g/kg in petroleum ether and heptane, respectively (Williams, 1951)

Solubility in water:
6.54 ppm at 24 °C (Felsot and Dahm, 1979)
24 mg/L at 25 °C (Williams, 1951)
11.9 and 11 ppm at 20 and 40 °C, respectively (Freed et al., 1977)
10.3 and 15.2 mg/L at 10 and 30 °C, respectively (Bowman and Sans, 1985)
11 mg/L at 20 °C (Worthing and Hance, 1991)
12.9 mg/L at 20 °C (Bowman and Sans, 1977, 1985)

4.3 mg/L at 10 °C (Yaron and Saltzman, 1972)
12.4 mg/L at 20.0 °C (Bowman and Sans, 1979)

Vapor density:
11.91 g/L at 25 °C, 10.06 (air = 1)

Vapor pressure (10^5 mmHg):
0.98 at 25 °C (Kim et al., 1984)
(Bright et al., 1980)6 and 93 at 20.0, 26.8, 37.8, 49.0 and 60.0 °C, respectively

Environmental Fate

Biological. Parathion was reported to biologically hydrolyze to *p*-nitrophenol in different soils under flooded conditions (Sudhakar and Sethunathan, 1978). *p*-Nitrophenol also was identified as a hydrolysis product (Suffet et al., 1967).

When equilibrated with a prereduced pokkali soil, parathion instantaneously degraded to aminoparathion. The quick rate of reaction was reportedly due to soil enzymes and/or other heat labile substances (Wahid et al., 1980). Aminoparathion also was formed when parathion (500 ppm) was incubated in a flooded alluvial soil. The amount of parathion remaining after 6 and 12 days were 43.0 and 0.09%, respectively (Freed et al., 1979).

Initial hydrolysis products include diethyl-*O*-thiophosphoric acid, *p*-nitrophenol (Munnecke and Hsieh, 1976; Sethunathan, 1973, 1973a; Sethunathan et al., 1977; Verschueren, 1983) and the biodegradation products *p*-aminoparathion and *p*-aminophenol (Sethunathan, 1973). Mixed bacterial cultures were capable of growing on technical parathion as the sole carbon and energy source (Munnecke and Hsieh, 1976). Three oxidative pathways were reported. The primary degradative pathway is initial hydrolysis to yield *p*-nitrophenol and diethylthiophosphoric acid. The secondary pathway involves the formation of paraoxon (diethyl *p*-nitrophenyl phosphate), which subsequently undergoes hydrolysis to yield *p*-nitrophenol and diethylphosphoric acid. The third degradative pathway involved reduction of parathion under low oxygen conditions to yield *p*-aminoparathion followed by hydrolysis to *p*-aminophenol and diethylphosphoric acid. Other potential degradation products include hydroquinone, 2-hydroxyhydroquinone, ammonia and polymeric substances (Munnecke and Hsieh, 1976). The reported half-life in soil is 18 days (Jury et al., 1987).

A *Flavobacterium* sp. (ATCC 27551), isolated from rice paddy water, degraded parathion to *p*-nitrophenol. The microbial hydrolysis half-life of this reaction was <1 hour (Sethunathan and Yoshida, 1973). Sharmila et al. (1989) isolated a *Bacillus* sp. from a laterite soil which degraded parathion in the presence yeast extracts. At yeast concentrations of 0.05, 0.1 and 0.25%, parathion degraded via hydrolysis, hydrolysis and nitro group reduction, and exclusively by nitro group reduction, respectively. Aminoparathion and *p*-nitrophenol were the identified metabolites under these conditions. Rosenberg and Alexander (1979) demonstrated that two strains of *Pseudomonas* used parathion as the sole source of phosphorus. It was suggested that degradation of parathion resulted from an induced enzyme or enzyme system that catalytically hydrolyzed the aryl P-O bond, forming dimethyl phosphorothioate as the major product.

In both soils and water, chemical and biological mediated reactions transform parathion to paraoxon (Alexander, 1981). Parathion was reported to biologically hydrolyze to *p*-nitrophenol in different soils under flooded conditions (Ferris and Lichtenstein, 1980; Sudhakar-Barik and Sethunathan, 1978).

p-Nitrophenol, paraoxon and three unidentified metabolites were identified in a model ecosystem containing algae, *Daphnia magna*, fish, mosquito and snails (Yu and Sanborn, 1975).

Soil. A *Pseudomonas* sp. (ATCC 29354), isolated from parathion-amended treated soil, degraded *p*-nitrophenol to *p*-nitrocatechol, which was recalcitrant to further degradation. In an unsterilized soil, however, *p*-nitrocatechol was further degraded to nitrites and other unidentified compounds (Sudhakar-Barik et al., 1978a). *Pseudomonas* sp. and *Bacillus* sp., isolated from a parathion-amended flooded soil, degraded *p*-nitrophenol (parathion hydrolysis product) to nitrite ions (Siddaramappa et al., 1973; Sudhakar-Barik et al., 1976) and carbon dioxide (Sudhakar-Barik et al., 1976). When parathion was equilibrated with aerobic soils, virtually no degradation was observed (Adhya et al., 1981a). However, in flooded (anaerobic) acid sulfate soils or low sulfate soils, aminoparathion and desethyl aminoparathion formed as the major metabolites (Adhya et al., 1981). However under flooded acid conditions, parathion hydrolyzed to form *p*-nitrophenol (Sethunathan, 1973a). The rate of hydrolysis was found to be higher in soils containing higher organic matter content. The bacterium *Bacillus* sp., isolated from parathion-amended flooded alluvial soil, decomposed *p*-nitrophenol as a sole carbon source (Sethunathan, 1973a). In a similar study, Rajaram and Sethunathan (1975) were able to increase the rate of hydrolysis of parathion in soil by a variety of organic materials. They found that the rate of hydrolysis was as follows: glucose >rice straw >algal crust >farmyard manure >unamended soil. However, these organic materials inhibited the rate of hydrolysis when the soils were inoculated with a parathion-hydrolyzing bacterial culture.

p-Nitrophenol was identified as a hydrolysis product in soil (Camper, 1991; Miles et al., 1979; Somasundaram et al., 1991; Suffet et al., 1967). The rate of hydrolysis in soil is accelerated following repeated applications of parathion (Ferris and Lichtenstein, 1980) or the surface catalysis of clays (Mingelgrin et al., 1977). The reported hydrolysis half-lives at pH 7.4 and 20 and 37.5 °C were 130 and 26.8 days, respectively. At pH 6.1 and 20 °C, the hydrolysis half-life is 170 days (Freed et al., 1979). When equilibrated with a prereduced pokkali soil (acid sulfate), parathion instantaneously degraded to aminoparathion. The quick rate of reaction was reportedly due to soil enzymes and/or other heat labile substances. Desethyl aminoparathion was also identified as a metabolite in two separate studies (Wahid and Sethunathan, 1979; Wahid et al., 1980).

Aminoparathion also was formed when parathion (500 ppm) was incubated in a flooded alluvial soil. The amounts of parathion remaining after 6 and 12 days were 43.0 and 0.09%, respectively (Freed et al., 1979). It was observed that the degradation of parathion in flooded soil was more rapid in the presence of ferrous sulfate. Parathion degraded to aminoparathion and four unknown products (Rao and Sethunathan, 1979). In flooded alluvial soils, parathion degraded to amino-parathion via nitro group reduction. The rate of degradation remained constant despite variations in the redox potential of the soils (Adhya et al., 1981a). In soil, parathion may degrade via two oxidative pathways. The primary pathway is hydrolysis to p-nitrophenol (Sudhakar-Barik et al., 1979) and diethylthiophos-phoric acid (Miles et al., 1979). The other pathway involves oxidation to paraoxon, but aminoparathion is formed under anaerobic (Miles et al., 1979) and flooded conditions (Sudhakar-Barik et al., 1979). The degradation pathway as well as the rate of degradation of parathion in a flooded soil changed following each successive application of parathion (Sudhakar-Barik et al., 1979). After the first application, nitro group reduction gave aminoparathion as the major product. After the second application, both the hydrolysis product (p-nitrophenol) and aminoparathion were found. Following the third addition of parathion, p-nitrophenol was the only product detected. It was reported that the change from nitro group reduction to hydrolysis occurred as a result of rapid proliferation of parathion-hydrolyzing microorganisms that utilized p-nitrophenol as the carbon source (Sudhakar-Barik et al., 1979).

In a cranberry soil pretreated with p-nitrophenol, parathion was rapidly mineralized to carbon dioxide by indigenous microorganisms (Ferris and Lichtenstein, 1980). The half-lives of parathion (10 ppm) in a nonsterile sandy loam and a nonsterile organic soil were <1 and 1.5 weeks, respectively (Miles et al., 1979). Walker (1976) reported that 16-23% of parathion added to both sterile and nonsterile estuarine water was degraded after incubation in the dark for 40 days.

Plant. Oat plants were grown in two soils treated with [^{14}C]parathion. Less than 2% of the applied [^{14}C]parathion was translocated to the oat plant. Metabolites identified in both soils and leaves were paraoxon, aminoparaoxon, aminoparathion, p-nitrophenol and an aminophenol (Fuhremann and Lichtenstein, 1980).

The following metabolites were identified in a soil-oat system: paraoxon, aminoparathion, p-nitrophenol and p-aminophenol (Lichtenstein, 1980; Lichtenstein et al., 1982). Mick and Dahm (1970) reported that *Rhizobium* sp. converted 85% [^{14}C]parathion to aminoparathion in one day and 10% diethylphos-phorothioic acid.

One month after application of [^{14}C]parathion to cotton plants, 6.5-10.5% of the total reactivity was found to be unreacted [^{14}C]parathion. Photo-alteration products identified included S-ethyl parathion, S-phenyl parathion, paraoxon and p-nitrophenol (Joiner and Baetcke, 1973). Reddy and Sethunathan (1983) studied the mineralization of ring-labeled [2,6-^{14}C]parathion in the rhizosphere of rice seedlings under flooded and nonflooded soil conditions. In unplanted soil, only

5.5% of the ^{14}C in the parathion was evolved as $^{14}CO_2$ in 15 days under flooded and non-flooded conditions. However, in soils planted with rice, 9.2 and 22.6% of the ^{14}C in the parathion evolved as $^{14}CO_2$ under nonflooded and flooded conditions, respectively. In an earlier study, the presence of rice straw in a flooded alluvial soil inoculated with an enrichment culture greatly inhibited the hydrolysis of parathion to p-nitrophenol and O,O-diethylphosphorothioic acid. In uninoculated soils, however, rice straw enhanced the degradation of parathion via nitro group reduction to p-aminoparathion and a compound possessing a P=S bond (Sethunathan, 1973).

Photolytic. p-Nitrophenol and paraoxon were formed from the irradiation of parathion in water, aqueous methanol and aqueous n-propyl alcohol solutions by a low-pressure mercury lamp. Degradation was more rapid in water than in organic solvent/water mixtures with p-nitrophenol forming as the major product (Mansour et al., 1983). When parathion in aqueous tetrahydrofuran or ethanol solutions (80%) was irradiated at 2537 Å, O,O,S-triethylthiophosphate formed as the major product (Grunwell and Erickson, 1973). Minor photoproducts included O,O,O-triethylthiophosphate, triethylphosphate, paraoxon and traces of ethanethiol and p-nitrophenol.

Parathion degraded on both glass surfaces and on bean plant leaves. Metabolites reported were paraoxon, p-nitrophenol and a compound tentatively identified as s-ethyl parathion (El-Refai and Hopkins, 1966). Upon exposure to high intensity UV light, parathion was altered to the following photoproducts: paraoxon, O,S-diethyl O-4-nitrophenyl phosphorothioate, O,O-diethyl S-4-nitrophenyl phosphorothioate, O,O-bis(4-nitrophenyl) O-ethyl phosphorothioate, O,O-bis(4-nitrophenyl) O-ethyl phosphate, O,O-diethyl O-phenyl phosphorothioate and O,O-diethyl O-phenyl phosphate (Joiner et al., 1971).

When parathion was released in the atmosphere on a sunny day, it was rapidly converted to the photochemical paraoxon. The estimated photolytic half-life is 2 minutes (Woodrow et al., 1978). The reaction involving the oxidation of parathion to paraoxon is catalyzed in the presence of UV light, ozone, soil dust or clay minerals (Spencer et al., 1980, 1980a). Ozone and UV light alone could not oxide parathion. However, in the presence of ozone (300 ppb) and UV light, dry kaolinite clays were more effective than the montmorillonite clays in the oxidation of parathion. The Cu-saturated clays were most effective oxidizing parathion to paraoxon and the Ca-saturated clays were the least effective (Spencer et al., 1980).

When applied as thin films on leaf surfaces, parathion was converted to paraoxon and p-nitrophenol. The photodegradation half-life was reported to be 88 hours (HSDB, 1989).

When an aqueous solution containing parathion was photooxidized by UV light at 90-95 °C, 25, 50 and 75% degraded to carbon dioxide after 15.1, 57.6 and 148.8 hours, respectively (Knoevenagel and Himmelreich, 1976). p-Nitrophenol and paraoxon were the major products identified following the sunlight irradiation of parathion in distilled water and river water. The photolysis half-life of parathion in river water was 15 hours (Mansour et al., 1989). Kotronarou et al. (1992) studied the degradation of parathion-saturated deionized water solution at 30 °C by ultrasonic irradiation (sonolysis). After 2 hours of sonolysis, all the parathion degraded to the following final end products: sulfate ions, phosphate ions, nitrate ions, hydrogen ions and carbon dioxide. Precursors/intermediate compounds to the final end products included *p*-nitrophenol, diethylmonothiophosphoric acid, hydroquinone, benzoquinone, formic acid, oxalic acid, 4-nitro-catechol, nitrite ions and ethanol.

Chemical/Physical. The reported hydrolysis half-lives at pH 7.4 at 20 and 37.5 °C were 130 and 26.8 days, respectively (Freed et al., 1977). The hydrolysis half-lives of parathion in a sterile 1% ethanol/water solution at 25 °C and pH values of 4.5, 5.0, 6.0, 7.0 and 8.0, were 39, 43, 33, 24 and 15 weeks, respectively (Chapman and Cole, 1982).

Reported ozonation products of parathion in drinking water include sulfuric acid (Richard and Bréner, 1984), paraoxon, 2,4-dinitrophenol, picric and phosphoric acids (Laplanche et al., 1984).

Paraoxon was also found in fogwater collected near Parlier, CA (Glotfelty et al., 1990). It was suggested that parathion was oxidized in the atmosphere during daylight hours prior to its partitioning in the fog. On January 12, 1986, the distributions of parathion (9.4 ng/m^3) in the vapor phase, dissolved phase, air particles and water particles were 78, 10, 11 and 0.6%, respectively. For paraoxon (2.3 ng/m^3), the distribution in the vapor phase, dissolved phase, air particles and water particles were 7.8, 35.5, 56.7 and 0.09%, respectively (Glotfelty et al., 1990).

At 130 °C, parathion isomerizes to *O,S*-diethyl *O-p*-nitrophenyl phosphoro-thioate (Hartley and Kidd, 1987; Worthing and Hance, 1991). An 85% yield of this compound was reported when parathion was heated at 150 °C for 24 hours (Wolfe et al., 1976). Emits toxic oxides of nitrogen, sulfur and phosphorus when heated to decomposition (Lewis, 1990).

Exposure Limits (mg/m^3): NIOSH REL: TWA 0.05, IDLH 10; OSHA PEL: TWA 0.1.

Toxicity: Acute oral LD_{50} for wild birds 1,330 mg/kg, cats 930 μg/kg, ducks 2.1 mg/kg, dogs 3 mg/kg, guinea pigs 8 mg/kg, mice 6 mg/kg, pigeons 1.33 mg/kg, quail 4.04 mg/kg, rats 2 mg/kg, rabbits 10 mg/kg; LC_{50} (inhalation) for rats 84 mg/m^3/4-hour (RTECS, 1985); LC_{50} (48-hour) for red killifish 10 mg/L (Yoshioka et al., 1986).

Uses: Insecticide; acaricide.

PCB-1016

Synonyms: Arochlor 1016; **Aroclor 1016**; Chlorodiphenyl (41% Cl).

$$x + y = 0 \text{ thru } 6$$

CAS Registry No.: 12674-11-2
DOT: 2315
Molecular formula: Not definitive. PCB-1016 is a mixture of many biphenyls with varying degrees of chlorination. According to Hutzinger et al. (1974), the approximate composition of PCB-1016 by weight is as follows: biphenyls (<0.1%), chlorobiphenyls (1%), dichlorobiphenyls (20%), trichlorobiphenyls (57%), tetra-chlorobiphenyls (21%), pentachlorobiphenyls (1%), hexachlorobiphenyls (<0.1%) and heptachlorobiphenyls (0%).
Formula weight: 257.9 (average, Hutzinger et al., 1974)
RTECS: TQ1351000

Physical state, color and odor
Viscous, oily, light yellow liquid or white powder with a weak odor.

Boiling point (°C):
Distills at 325-356 (Monsanto, 1974)

Density (g/cm^3):
1.33 at 25/4 °C (Monsanto, 1974)

Diffusivity in water (10^5 cm^2/sec):
0.68 at 25 °C using method of Hayduk and Laudie (1974).

Flash point (°C):
Nonflammable (Sittig, 1985)

Henry's law constant (atm/mol fraction):
750 (Paris et al., 1978)

Bioconcentration factor, log BCF:
4.63 (freshwater fish, Garten and Trabalka, 1983)
3.81 (Doe Run pond bacteria), 3.73 (Hickory Hills pond bacteria) (Paris et al.,

1978)
4.69 (fish, Kenaga and Goring, 1980)

Soil sorption coefficient, log K$_{oc}$:
5.51 (Oconee River sediment), 5.21 (USDA Pond sediment), 4.99 (Doe Run Pond sediment), 4.41 (Hickory Hill Pond sediment) (Steen et al., 1978)

Octanol/water partition coefficient, log K$_{ow}$:
4.38 (Paris et al., 1978)
5.88 (Mackay, 1982)
5.580 (for 2,4,4'- and 2,2',5-trichlorobiphenyl, Kenaga and Goring, 1980)

Solubility in organics:
Soluble in most solvents (U.S. EPA, 1985).

Solubility in water (μg/L):
420 (Paris et al., 1978)
49 at 24 °C (Hollifield, 1979)
906 at 23 °C (Lee et al., 1979)

Vapor pressure (mmHg):
9.03 x 10^{-4} at 25 °C (Foreman and Bidleman, 1985)

Environmental Fate

Biological. Reported degradation products by the microorganism *Alcaligenes* BM-2 for a mixture of polychlorinated biphenyls include monohydroxychlorobiphenyl, 2-hydroxy-6-oxochlorophenylhexa-2,4-dieonic acid, chlorobenzoic acid, chlorobenzoylpropionic acid, chlorophenylacetic acid and 3-chlorophenyl-2-chloropropenic acid (Yagi and Sudo, 1980). When PCB-1016 was statically incubated in the dark at 25 °C with yeast extract and settled domestic wastewater inoculum, no significant biodegradation was observed. At a concentration of 5 mg/L, percent losses after 7, 14, 21 and 28-day incubation periods were 44, 47, 46 and 48, respectively. At a concentration of 10 mg/L, only 22, 46, 20 and 13% losses were observed after the 7, 14, 21 and 28-day incubation periods, respectively (Tabak et al., 1981).

Exposure Limits: Potential occupational carcinogen. NIOSH REL: TWA 1.0 μg/m^3, IDLH 5 mg/m^3.

Drinking Water Standard (final): For all PCBs, the MCLG and MCL are zero and 0.5 μg/L, respectively (U.S. EPA, 1994).

Toxicity: A teratogen having a low toxicity (Patnaik, 1992).

Uses: Insulator fluid for electric condensers; additive in high pressure lubricants.

PCB-1221

Synonyms: Arochlor 1221; **Aroclor 1221**; Chlorodiphenyl (21% Cl).

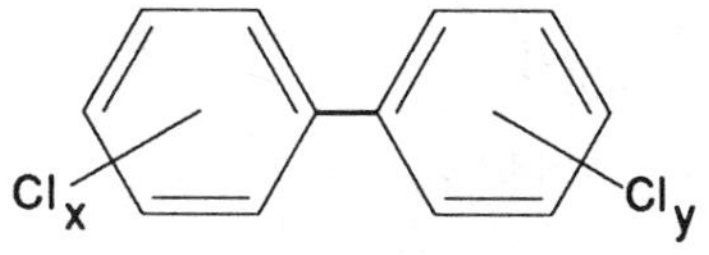

x + y = 0 thru 5

CAS Registry No.: 11104-28-2
DOT: 2315
Molecular formula: Not definitive. PCB-1221 is a mixture of many biphenyls with varying degrees of chlorination. According to Hutzinger et al. (1974), the approximate composition of PCB-1221 by weight is as follows: biphenyls (11%), chlorobiphenyls (51%), dichlorobiphenyls (32%), trichlorobiphenyls (4%), tetra-chlorobiphenyls (2%), pentachlorobiphenyls (<0.5%), hexachlorobiphenyls (0%) and heptachlorobiphenyls (0%).
Formula weight: 192 (average, Hutzinger et al., 1974)
RTECS: TQ1352000

Physical state, color and odor
Viscous, oily, colorless to light yellow liquid with a weak odor.

Melting point (°C):
1 (HSDB, 1989)

Boiling point (°C):
Distills at 275-320 (Monsanto, 1974)

Density (g/cm^3):
1.15 at 25/4 °C (Hutzinger et al., 1974)

Diffusivity in water (10^5 cm^2/sec):
0.75 at 25 °C using method of Hayduk and Laudie (1974).

Flash point (°C):
141-150 (Hutzinger et al., 1974)

Henry's law constant (10^4 atm·m^3/mol):
0.286, 0.452, 0.697, 1.06, 1.57, 2.28, 3.26, 4.83 and 6.68 at 0.0, 5.0, 10.0, 15.0, 20.0, 25.0, 30.0, 35.5 and 40.0 °C, respectively (predicted, Burkhard et al., 1985).

817

Soil sorption coefficient, log K_{oc}:
2.44 (estimated, Montgomery, 1989)

Octanol/water partition coefficient, log K_{ow}:
4.08 (Callahan et al., 1979)
2.8 (estimated, U.S. EPA, 1980)

Solubility in organics:
Soluble in most solvents (U.S. EPA, 1985).

Solubility in water (mg/L):
0.590 at 24 °C (Hollifield, 1979)
3.5 at 23 °C (Lee et al., 1979)
5.0 (Zitko, 1970)

Vapor pressure (mmHg):
7×10^{-3} at 25 °C (Pal et al., 1980)

Environmental Fate
   Biological. Reported degradation products by the microorganism *Alcaligenes* BM-2 for a mixture of polychlorinated biphenyls include monohydroxychlorobiphenyl, 2-hydroxy-6-oxochlorophenylhexa-2,4-dieonic acid, chlorobenzoic acid, chlorobenzoylpropionic acid, chlorophenylacetic acid and 3-chlorophenyl-2-chloropropenic acid (Yagi and Sudo, 1980).
   In sewage wastewater, *Pseudomonas* sp. 7509 degraded PCB-1221 into a yellow compound tentatively identified as a chlorinated derivative of α-hydroxymuconic acid (Liu, 1981). When PCB-1221 was statically incubated in the dark at 25 °C with yeast extract and settled domestic wastewater inoculum for 7 days, significant biodegradation with rapid adaptation was observed (Tabak et al., 1981).
   In activated sludge, 80.6% degraded after a 47-hour time period (Pal et al., 1980).
   Chemical/Physical. Zhang and Rusling (1993) evaluated the bicontinuous microemulsion of surfactant/oil/water as a medium for the dechlorination of polychlorinated biphenyls by electrochemical catalytic reduction. The microemulsion (20 mL) used was didodecyldimethylammonium bromide, dodecane and water at 21, 57 and 22 wt %, respectively. The catalyst used was zinc phthalocyanine (2.5 nM). When PCB-1221 (72 mg), the emulsion and catalyst were subjected to a current of mA/cm^2 on 11.2 cm^2 lead electrode for 10 hours, a dechlorination yield of 99% was achieved. Reaction products included monochlorobiphenyl (0.9 mg) and biphenyl and reduced alkylbenzene derivatives.

Exposure Limits: Potential occupational carcinogen. NIOSH REL: TWA 1.0 μg/m^3, IDLH 5 mg/m^3.

Drinking Water Standard (final): For all PCBs, the MCLG and MCL are zero and 0.5 μg/L, respectively (U.S. EPA, 1994).

Toxicity: A teratogen and suspected human carcinogen (Patnaik, 1992).

Uses: In polyvinyl acetate to improve fiber-tear properties; plasticizer for polystyrene; in epoxy resins to improve adhesion and resistance to chemical attack; as an insulator fluid for electric condensers and as an additive in very high pressure lubricants.

PCB-1232

Synonyms: Arochlor 1232; **Aroclor 1232**; Chlorodiphenyl (32% Cl).

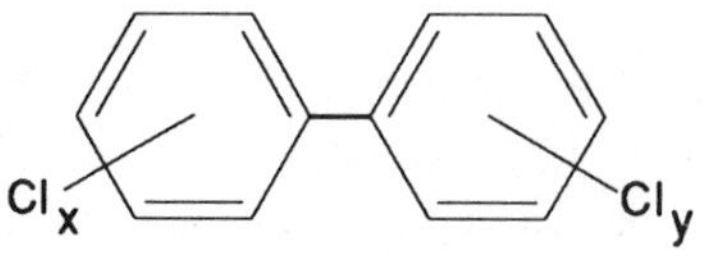

x + y = 0 thru 6

CAS Registry No.: 11141-16-5
DOT: 2315
Molecular formula: Not definitive. PCB-1232 is a mixture of many biphenyls with varying degrees of chlorination. According to Hutzinger et al. (1974), the approximate composition of PCB-1232 by weight is as follows: biphenyls (<0.1%), chlorobiphenyls (31%), dichlorobiphenyls (24%), trichlorobiphenyls (28%), tetra-chlorobiphenyls (12%), pentachlorobiphenyls (4%), hexachlorobiphenyls (<0.1%) and heptachlorobiphenyls (0%).
Formula weight: 221 (average, Hutzinger et al., 1974)
RTECS: TQ1354000

Physical state, color and odor
Viscous, oily, almost colorless to light yellow liquid with a weak odor.

Melting point (°C):
-35.5 (pour point, HSDB, 1989)

Boiling point (°C):
Distills at 290-325 (Monsanto, 1974)

Density (g/cm^3):
1.24 at 25/4 °C (Monsanto, 1974)
1.270-1.280 at 25/15.5 °C (Standen, 1964)

Diffusivity in water (10^5 cm^2/sec):
0.72 at 20 °C using method of Hayduk and Laudie (1974).

Flash point (°C):
152-154 (Hutzinger et al., 1974)

Henry's law constant (10^4 atm·m^3/mol):
8.64 (calculated, U.S. EPA, 1980)

820

Soil sorption coefficient, log K_{oc}:
2.83 (estimated, Montgomery, 1989)

Octanol/water partition coefficient, log K_{ow}:
3.2 (estimated, U.S. EPA, 1980)

Solubility in water:
1.45 mg/L at 25 °C (estimated, U.S. EPA, 1980)

Vapor density:
9.03 g/L at 25 °C, 7.63 (air = 1)

Vapor pressure (mmHg):
4.6 x 10^{-3} at 25 °C (estimated, U.S. EPA, 1980)

Environmental Fate

Biological. Reported degradation products by the microorganism *Alcaligenes* BM-2 for a mixture of polychlorinated biphenyls include monohydroxychlorobiphenyl, 2-hydroxy-6-oxochlorophenylhexa-2,4-dieonic acid, chlorobenzoic acid, chlorobenzoylpropionic acid, chlorophenylacetic acid and 3-chlorophenyl-2-chloropropenic acid (Yagi and Sudo, 1980).

When PCB-1232 was statically incubated in the dark at 25 °C with yeast extract and settled domestic wastewater inoculum for 7 days, significant biodegradation with rapid adaptation was observed (Tabak et al., 1981).

Photolytic. PCB-1232 in a 90% acetonitrile/water solution containing 0.2-0.3 M sodium borohydride and irradiated with UV light (λ = 254 nm) reacted to yield dechlorinated biphenyls. Without sodium borohydride, the reaction proceeded more slowly (Epling et al., 1988).

Chemical/Physical. Zhang and Rusling (1993) evaluated the bicontinuous microemulsion of surfactant/oil/water as a medium for the dechlorination of polychlorinated biphenyls by electrochemical catalytic reduction. The microemulsion (20 mL) used was didodecyldimethylammonium bromide, dodecane and water at concentrations of 21, 57 and 22 wt %, respectively. The catalyst used was zinc phthalocyanine (3.5 nM). When PCB-1232 (69 mg), the microemulsion and catalyst were subjected to an electrical current of mA/cm² on 11.2 cm² lead electrode for 12 hours, a dechlorination yield of >99.8% was achieved. Reaction products included minor amounts of mono- and dichloropbiphenyls (0.01 mg), biphenyl and reduced alkylbenzene derivatives.

Exposure Limits: NIOSH REL: TWA 1.0 μg/m³, IDLH 5 mg/m³.

Drinking Water Standard (final): For all PCBs, the MCLG and MCL are zero and 0.5 μg/L, respectively (U.S. EPA, 1994).

Toxicity: A teratogen and suspected human carcinogen having a low toxicity (Patnaik, 1992); acute oral LD_{50} for rats 4,470 mg/kg (RTECS, 1985).

Uses: In polyvinyl acetate to improve fiber-tear properties; as an insulator fluid for electric condensers and as an additive in very high pressure lubricants.

PCB-1242

Synonyms: Arochlor 1242; **Aroclor 1242**; Chlorodiphenyl (42% Cl).

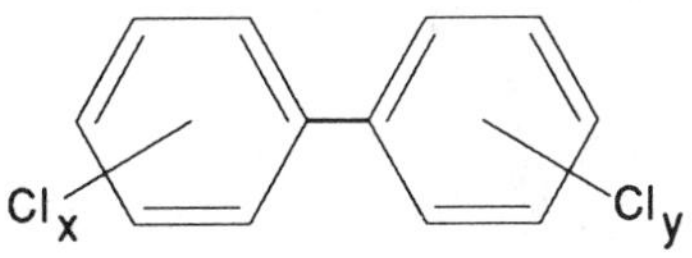

x + y = 0 thru 7

CAS Registry No.: 53469-21-9
DOT: 2315
Molecular formula: Not definitive. PCB-1242 is a mixture of many biphenyls with varying degrees of chlorination. According to Hutzinger et al. (1974), the approximate composition of PCB-1242 by weight is as follows: biphenyls (<0.1%), chlorobiphenyls (1%), dichlorobiphenyls (16%), trichlorobiphenyls (49%), tetra-chlorobiphenyls (25%), pentachlorobiphenyls (8%), hexachlorobiphenyls (1%) and heptachlorobiphenyls (<0.1%).
Formula weight ranges from 154 to 358 (Nisbet and Sarofim, 1972) with an average value of 261 (Hutzinger et al., 1974)
RTECS: TQ1356000

Physical state, color and odor
Colorless to light yellow, viscous, oily liquid with a weak, hydrocarbon odor.

Melting point (°C):
-19 (NIOSH, 1994)

Boiling point (°C):
Distills at 325-366 (Monsanto, 1974)

Density (g/cm^3):
1.392 at 15/4 °C, 1.381-1.392 at 25/15.5 °C (Standen, 1964)

Diffusivity in water (10^5 cm^2/sec):
0.61 at 20 °C using method of Hayduk and Laudie (1974).

Flash point (°C):
176-180 (Hutzinger et al., 1974)

Henry's law constant (10^4 atm·m^3/mol):
5.6 (Eisenreich et al., 1981)

823

2.8 at 20 °C (Murphy et al., 1987)

7.78 and 29.2 at 23 °C in distilled water and seawater, respectively (Atlas et al., 1972)

0.347, 0.575, 0.928, 1.47, 2.27, 3.43, 5.09, 7.60 and 10.8 at 0.0, 5.0, 10.0, 15.0, 20.0, 25.0, 30.0, 35.5 and 40.0 °C, respectively (predicted, Burkhard et al., 1985)

Soil sorption coefficient, log K_{oc}:
5.50 (Oconee River sediment), 5.13 (USDA Pond sediment), 4.94 (Doe Run Pond sediment), 4.35 (Hickory Hill Pond sediment) (Steen et al., 1978)

3.68, 4.37, 4.48, 4.51, 4.53, 4.75 (Paya-Perez et al., 1991)

Octanol/water partition coefficient, log K_{ow}:
4.11 (Paris et al., 1978)

Solubility in organics:
Soluble in most solvents (U.S. EPA, 1985).

Solubility in water:
100 μg/L at 24 °C (Hollifield, 1979)

240 μg/L at 25 °C (Monsanto, 1974; U.S. EPA, 1980)

340 μg/L (Paris et al., 1978)

200 μg/L at 20 °C (Nisbet and Sarofim, 1972)

132.9 and 18.6 ppb at 11.5 °C in distilled water and artificial seawater, respectively (Dexter and Pavlou, 1978)

277 μg/L at 20 °C (Murphy et al., 1987)

16 ppb (Watanabe et al., 1985)

703 μg/L at 23 °C (Lee et al., 1979)

Vapor density:
10.67 g/L at 25 °C, 9.01 (air = 1)

Vapor pressure (10^4 mmHg):
4.06 at 25 °C (Mackay and Wolkoff, 1973)

10 at 38 °C (Nisbet and Sarofim, 1972)

2.5 at 20 °C (Murphy et al., 1987)

5.74 at 25 °C (Foreman and Bidleman, 1985)

Environmental Fate
Biological. A strain of *Alcaligenes eutrophus* degraded 81% of the congeners by dechlorination under anaerobic conditions (Bedard et al., 1987). A bacterial culture isolated from Hamilton Harbour, Ontario was capable of degrading a commercial mixture of PCB-1242. The metabolites identified by GC/MS included isohexane, isooctane, ethylbenzene, isoheptane, isopropylbenzene, *n*-propylbenzene, iso-

butylbenzene, *n*-butylbenzene and isononane (Kaiser and Wong, 1974). A strain of *Pseudomonas*, isolated from activated sludge and grown with biphenyl as the sole carbon source, degraded 2,4′-dichlorobiphenyl yielding the following compounds: monochlorobenzoic acids, two monohydroxydichlorobiphenyls and the yellow compound hydroxyoxo(chlorophenyl)chlorohexadienoic acid. Irradiation of the mixture containing these compounds led to the formation of two monochloro-acetophenones and the disappearance of the yellow compound. Similar compounds were found when 2,4′-dichlorobiphenyl was replaced with a PCB-1242 mixture (Baxter and Sutherland, 1984).

Reported degradation products by the microorganism *Alcaligenes* BM-2 for a mixture of polychlorinated biphenyls include monohydroxychlorobiphenyl, 2-hydroxy-6-oxochlorophenylhexa-2,4-dieonic acid, chlorobenzoic acid, chloro-benzoylpropionic acid, chlorophenylacetic acid and 3-chlorophenyl-2-chloro-propenic acid (Yagi and Sudo, 1980).

When PCB-1242 was statically incubated in the dark at 25 °C with yeast extract and settled domestic wastewater inoculum, no significant biodegradation was observed. At a concentration of 5 mg/L, percent losses after 7, 14, 21 and 28-day incubation periods were 37, 41, 47 and 66, respectively. At a concentration of 10 mg/L, only 34, 33, 15 and 0% losses were observed after the 7, 14, 21 and 28-day incubation periods, respectively (Tabak et al., 1981).

Photolytic. PCB-1242 in a 90% acetonitrile/water solution containing 0.2-0.3 M sodium borohydride and irradiated with UV light (λ = 254 nm) reacted to yield dechlorinated biphenyls. Without sodium borohydride, the reaction proceeded much more slowly (Epling et al., 1988).

Chemical/Physical. When PCB-1242-contaminated sand was treated with a poly(ethylene glycol)/potassium hydroxide mixture at room temperature, 27% reacted after 2 weeks forming aryl poly(ethylene glycols) (Brunelle and Singleton, 1985).

When exposed to fire, black soot containing PCBs, polychlorinated dibenzo-furans and chlorinated dibenzo-*p*-dioxins is formed (NIOSH, 1994).

Exposure Limits: Potential occupational carcinogen. NIOSH REL: TWA 1.0 $\mu g/m^3$, IDLH 5 mg/m^3; OSHA PEL: TWA 1 mg/m^3.

Drinking Water Standard (final): For all PCBs, the MCLG and MCL are zero and 0.5 $\mu g/L$, respectively (U.S. EPA, 1994).

Toxicity: Acute oral LD$_{50}$ for rats 4250 mg/kg (RTECS, 1985).

Uses: Dielectric liquids; heat-transfer liquid widely used in transformers; swelling agents for transmission seals; ingredient in lubricants, oils and greases; plasticizers for cellulosics, vinyl and chlorinated rubbers; in polyvinyl acetate to improve fiber-tear properties.

PCB-1248

Synonyms: Arochlor 1248; **Aroclor 1248**; Chlorodiphenyl (48% Cl).

x + y = 2 thru 6

CAS Registry No.: 12672-29-6
DOT: 2315
Molecular formula: Not definitive. PCB-1248 is a mixture of many biphenyls with varying degrees of chlorination. According to Hutzinger et al. (1974), the approximate composition of PCB-1248 by weight is as follows: biphenyls (0%), chlorobiphenyls (0%), dichlorobiphenyls (2%), trichlorobiphenyls (18%), tetra-chlorobiphenyls (40%), pentachlorobiphenyls (36%), hexachlorobiphenyls (4%) and heptachlorobiphenyls (0%).
Formula weight: ranges from 222 to 358 (Nisbet and Sarofim, 1972) with an average value of 288 (Hutzinger et al., 1974)
RTECS: TQ1358000

Physical state, color and odor
Viscous, oily, light yellow liquid with a weak odor.

Melting point (°C):
-7 (pour point, HSDB, 1989)

Boiling point (°C):
Distills at 340-375 (Monsanto, 1974)

Density (g/cm^3):
1.41 at 25/4 °C (Monsanto, 1974)

Diffusivity in water (10^5 cm^2/sec):
0.66 at 20 °C using method of Hayduk and Laudie (1974).

Flash point (°C):
193-196 (Hutzinger et al., 1974)

Henry's law constant (10^3 atm·m^3/mol):
0.413, 0.696, 1.14, 1.83, 2.88, 4.40, 6.62, 9.91 and 14.3 at 0.0, 5.0, 10.0, 15.0, 20.0,

25.0, 30.0, 35.5 and 40.0 °C, respectively (predicted, Burkhard et al., 1985).
3.67 at 25 °C (Slinn et al., 1978)

Bioconcentration factor, log BCF:
4.85 (freshwater fish, Garten and Trabalka, 1983)
4.42 (bluegill sunfish, Stalling and Meyer, 1972)
5.08 (fathead minnow, DeFoe et al., 1978)

Soil sorption coefficient, log K_{oc}:
5.64 (estimated, Montgomery, 1989)

Octanol/water partition coefficient, log K_{ow}:
6.110 (for 2,2′,4,4′- and 2,2′,5,5′-tetrachlorobiphenyl, Kenaga and Goring, 1980)
6.0 (Mills et al., 1982)
6.11 (Chiou et al., 1977)

Solubility in organics:
Soluble in most solvents (U.S. EPA, 1985).

Solubility in water ($\mu g/L$):
54 (Monsanto, 1974)
50 at 20 °C (Nisbet and Sarofim, 1972)
60 at 24 °C (Hollifield, 1979)

Vapor pressure (10^4 mmHg):
4.94 at 25 °C (Mackay and Wolkoff, 1973)
0.6 at 38 °C (Nisbet and Sarofim, 1972)
1.835 at 25 °C (Foreman and Bidleman, 1985)

Environmental Fate
Biological. Reported degradation products by the microorganism *Alcaligenes* BM-2 for a mixture of polychlorinated biphenyls include monohydroxychlorobiphenyl, 2-hydroxy-6-oxochlorophenylhexa-2,4-dieonic acid, chlorobenzoic acid, chlorobenzoylpropionic acid, chlorophenylacetic acid and 3-chlorophenyl-2-chloropropenic acid (Yagi and Sudo, 1980). When PCB-1248 (5 and 10 mg/) was statically incubated in the dark at 25 °C with yeast extract and settled domestic wastewater inoculum, no biodegradation was observed (Tabak et al., 1981).

Chemical/Physical. Heating PCB-1248 in oxygen at 270-300 °C for 1 week resulted in the formation of mono-, di-, tri-, tetra- and pentachlorodibenzofurans (PCDFs). Above 330 °C, PCDFs decompose (Morita et al., 1978).

Exposure Limits: Potential occupational carcinogen. NIOSH REL: TWA 1.0 $\mu g/m^3$, IDLH 5 mg/m^3.

Drinking Water Standard (final): For all PCBs, the MCLG and MCL are zero and 0.5 μg/L, respectively (U.S. EPA, 1994).

Toxicity: Acute oral LD_{50} for rats 11 gm/kg (RTECS, 1985).

Uses: In epoxy resins to improve adhesion and resistance to chemical attack; as an insulator fluid for electric condensers and as an additive in very high pressure lubricants.

PCB-1254

Synonyms: Arochlor 1254; **Aroclor 1254**; Chlorodiphenyl (54% Cl); NCI-C02664.

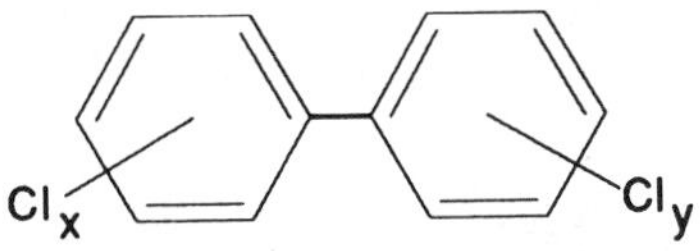

x + y = 0 thru 7

CAS Registry No.: 11097-69-1
DOT: 2315
Molecular formula: Not definitive. PCB-1254 is a mixture of many biphenyls with varying degrees of chlorination. According to Hutzinger et al. (1974), the approximate composition of PCB-1254 by weight is as follows: biphenyls (<0.1%), chlorobiphenyls (<0.1%), dichlorobiphenyls (0.5%), trichlorobiphenyls (1%), tetra-chlorobiphenyls (21%), pentachlorobiphenyls (48%), hexachlorobiphenyls (23%) and heptachlorobiphenyls (6%).
Formula weight: 327 (average, Hutzinger et al., 1974)
RTECS: TQ1360000

Physical state, color and odor
Light yellow, viscous, oily liquid with a weak odor.

Melting point (°C):
10 (NIOSH, 1994)

Boiling point (°C):
Distills at 365–390 (Monsanto, 1974)

Density (g/cm^3):
1.505 at 15.5/4 °C (Standen, 1964)
1.38 at 25/4 °C (NIOSH, 1994)

Diffusivity in water (10^5 cm^2/sec):
0.56 at 20 °C using method of Hayduk and Laudie (1974).

Henry's law constant (10^4 atm·m^3/mol):
27 (Eisenreich et al., 1981)
26.9 (Slinn et al., 1978)
23 (Petrasek et al., 1983)
1.9 at 20 °C (Murphy et al., 1987)

829

0.226, 0.394, 0.670, 1.11, 1.80, 2.83, 4.37, 6.65 and 9.81 at 0.0, 5.0, 10.0, 15.0, 20.0, 25.0, 30.0, 35.5 and 40.0 °C, respectively (predicted, Burkhard et al., 1985)

Bioconcentration factor, log BCF:
3.34 (guppies, Gooch and Hamdy, 1983)
4.70 (freshwater fish, Garten and Trabalka, 1983)

Soil sorption coefficient, log K_{oc}:
5.61 (Lake Michigan sediments, Voice and Weber, 1985)
4.628 (2,2′,4,5,5′-pentachlorobiphenyl, Kenaga and Goring, 1980)
4.88 (freshly amended Glendale soil), 4.73 (preconditioned Glendale soil), 4.47, 4.50 (Harvey soils), 4.40, 4.55 (Lea soils) (Fairbanks and O'Connor, 1984)

Octanol/water partition coefficient, log K_{ow}:
6.47 (Travis and Arms, 1988)
5.61 (Voice and Weber, 1985)

Solubility in organics:
Soluble in most solvents (U.S. EPA, 1985).

Solubility in water (ppb):
57 at 24 °C (Hollifield, 1979)
50 at 20 °C, $\approx$ 56 μg/L at 26 °C (Haque et al., 1974)
24.2 and 4.34 at 11.5 °C in distilled water and artificial seawater, respectively (Dexter and Pavlou, 1978)
43 at 20 °C (Murphy et al., 1987)
28.1 and 24.7 at 16.5 °C for membrane and carbon-filtered seawater, respectively (Wiese and Griffin, 1978)
$\approx$ 70 at 23 °C (Lee et al., 1979)

Vapor density:
13.36 g/L at 25 °C, 11.29 (air = 1)

Vapor pressure (10^5 mmHg):
7.71 at 25 °C (Mackay and Wolkoff, 1973)
2.2 at 20 °C (Murphy et al., 1987)
3.22 at 25 °C (Foreman and Bidleman, 1985)

Environmental Fate
Biological. A strain of *Alcaligenes eutrophus* degraded 35% of the congeners by dechlorination under anaerobic conditions (Bedard et al., 1987). Indigenous microbes in the Center Hill Reservoir, TN oxidized 2-chlorobiphenyl (a congener present in trace quantities) into chlorobenzoic acid and chlorobenzoylformic acid.

Biooxidation of the PCB mixture containing 54 wt % chlorine was not observed (Shiaris and Sayler, 1982).

When PCB-1254 was statically incubated in the dark at 25 °C with yeast extract and settled domestic wastewater inoculum, no significant biodegradation was observed. At a concentration of 5 mg/L, percent losses after 7, 14, 21 and 28-day incubation periods were 11, 42, 15 and 0, respectively. At a concentration of 10 mg/L, only 10 and 26% losses were observed after the 7 and 14-day incubation periods, respectively (Tabak et al., 1981).

Reported degradation products by the microorganism *Alcaligenes* BM-2 for a mixture of polychlorinated biphenyls include monohydroxychlorobiphenyl, 2-hydroxy-6-oxochlorophenylhexa-2,4-dieonic acid, chlorobenzoic acid, chloro-benzoylpropionic acid, chlorophenylacetic acid and 3-chlorophenyl-2-chloro-propenic acid (Yagi and Sudo, 1980).

Photolytic. PCB-1254 in a 90% acetonitrile/water solution containing 0.2-0.3 M sodium borohydride and irradiated with UV light (λ = 254 nm) reacted to yield dechlorinated biphenyls. After 16 hours, no chlorinated biphenyls were detected. Without sodium borohydride, only 25% of PCB-1254 were destroyed after 16 hours (Epling et al., 1988). In a similar experiment, PCB-1254 (1,000 mg/L) in an alkaline 2-propanol solution was exposed to UV light (λ = 254 nm). After 30 min, all of the PCB-1254 isomers were converted to biphenyl. When the radiation source was sunlight, only 25% was degraded after a 20-hour exposure. But when the sensitizer phenothiazine (5 mM) was added to the solution, photodechlorination of PCB-1254 was complete after 1 hour at 350 nm. In addition, when PCB-1254-contaminated soil was heated at about 80 °C in the presence of di-*t*-butyl peroxide, complete dechlorination to biphenyl was observed (Hawari et al., 1992).

When PCB-1254 in a methanol/water solution containing sodium methyl siliconate was irradiated with UV light (λ = 300 nm) for 5 hours, most of the congeners dechlorinated to biphenyl (Hawari et al., 1991). In a similar experiment, a dilute solution of PCB-1254 in cyclohexane was exposed to a Montreal sunlight in December and January for 55 days at 10 °C. Dechlorination of the higher chlorinated congeners at the *ortho* position was observed. In addition, two cyclohexyl adducts of polychlorinated biphenyls containing three and four chlorine atoms were also formed in small amounts (0.6% of the original PCB-1254 added) (Lépine et al., 1992).

Chemical/Physical. When PCB-1254-contaminated sand was treated with a poly(ethylene glycol)/potassium hydroxide mixture at room temperature, 81% reacted after 6 days forming aryl poly(ethylene glycols) (Brunelle and Singleton, 1985).

Using methanol, ethanol or 2-propanol in the presence of nickel chloride and sodium borohydride, dechlorination resulted in the formation of biphenyls with smaller quantities of mono- and dichlorobiphenyls (Dennis et al., 1979).

When exposed to fire, black soot containing PCBs, polychlorinated dibenzofurans and chlorinated dibenzo-*p*-dioxins is formed (NIOSH, 1994).

Exposure Limits: Potential occupational carcinogen. NIOSH REL: TWA 1 $\mu g/m^3$, IDLH 5 mg/m^3; OSHA PEL: TWA 0.5 mg/m^3.

Drinking Water Standard (final): For all PCBs, the MCLG and MCL are zero and 0.5 $\mu g/L$, respectively (U.S. EPA, 1994).

Toxicity: Acute oral LD_{50} for rats 1,010 mg/kg (RTECS, 1985).

Uses: Secondary plasticizer for polyvinyl chloride; co-polymers of styrene-butadiene and chlorinated rubber to improve chemical resistance to attack.

PCB-1260

Synonyms: Arochlor 1260; **Aroclor 1260**; Chlorodiphenyl (60% Cl); Clophen A60; Kanechlor; Phenoclor DP6.

$$x + y = 4 \text{ thru } 7$$

CAS Registry No.: 11096-82-5
DOT: 2315
Molecular formula: Not definitive. PCB-1260 is a mixture of many biphenyls with varying degrees of chlorination. According to Hutzinger et al. (1974), the approximate composition of PCB-1260 by weight is as follows: biphenyls (0%), chlorobiphenyls (0%), dichlorobiphenyls (0%), trichlorobiphenyls (0%), tetrachlorobiphenyls (1%), pentachlorobiphenyls (12%), hexachlorobiphenyls (38%) and heptachlorobiphenyls (41%).
Formula weight ranges from 324 to 460 (Nisbet and Sarofim, 1972) with an average value of 370 (Hutzinger et al., 1974)
RTECS: TQ1362000

Physical state, color and odor
Light yellow sticky, soft resin with a weak odor.

Melting point (°C):
31 (HSDB, 1989)

Boiling point (°C):
Distills at 385-420 (Monsanto, 1974)

Density (g/cm^3):
1.566 at 15.5/4 °C, 1.555-1.566 at 90/15.5 °C (Standen, 1964)
1.58 (Mills et al., 1982)

Diffusivity in water (10^5 cm^2/sec):
0.53 at 20 °C using method of Hayduk and Laudie (1974).

Flash point (°C):
None to the boiling point (Hutzinger et al., 1974).

Bioconcentration factor, log BCF:
5.29 (fathead minnow, Veith et al., 1979a)

Henry's law constant (10^4 atm·m^3/mol):
1.7 at 20 °C (Murphy et al., 1987)
0.244, 0.435, 0.754, 1.27, 2.10, 3.36, 5.27, 8.09 and 12.1 at 0.0, 5.0, 10.0, 15.0, 20.0, 25.0, 30.0, 35.5 and 40.0 °C, respectively (predicted, Burkhard et al., 1985)

Soil sorption coefficient, log K_{oc}:
6.42 (estimated, Montgomery, 1989)

Octanol/water partition coefficient, log K_{ow}:
6.91 (Mackay, 1982)
6.11 (Chiou et al., 1977)

Solubility in organics:
Soluble in most solvents (U.S. EPA, 1985).

Solubility in water (μg/L):
80 at 24 °C (Hollifield, 1979)
2.7 (Monsanto, 1974)
14.4 at 20 °C (Murphy et al., 1987)

Vapor pressure (10^7 mmHg):
405 at 25 °C (Mackay and Wolkoff, 1973)
2 at 38 °C (Nisbet and Sarofim, 1972)
63.1 at 20 °C (Murphy et al., 1987)
130.5 at 25 °C (Foreman and Bidleman, 1985)

Environmental Fate

Biological. Reported degradation products by the microorganism *Alcaligenes* BM-2 for a mixture of polychlorinated biphenyls include monohydroxychlorobiphenyl, 2-hydroxy-6-oxochlorophenylhexa-2,4-dieonic acid, chlorobenzoic acid, chlorobenzoylpropionic acid, chlorophenylacetic acid and 3-chlorophenyl-2-chloropropenic acid (Yagi and Sudo, 1980). When PCB-1260 was statically incubated in the dark at 25 °C with yeast extract and settled domestic wastewater inoculum, no significant biodegradation was observed (Tabak et al., 1981).

Photolytic. PCB-1260 in a 90% acetonitrile/water solution containing 0.2-0.3 M sodium borohydride and irradiated with UV light (λ = 254 nm) reacted to yield dechlorinated biphenyls. After 2 hours, about 75% of the congeners were destroyed. Without sodium borohydride, only 10% of the congeners had reacted. Products identified by GC include biphenyl, 2-, 3- and 4-chlorobiphenyl, six dichlorobiphenyls, three trichlorobiphenyls, 1-phenyl-1,4-cyclohexadiene and 1-

phenyl-3-cyclohexene (Epling et al., 1988).

Chemical/Physical. Zhang and Rusling (1993) evaluated the bicontinuous microemulsion of surfactant/oil/water as a medium for the dechlorination of polychlorinated biphenyls by electrochemical catalytic reduction. The microemulsion (20 mL) used was didodecyldimethylammonium bromide, dodecane and water at 21, 57 and 22 wt %, respectively. The catalyst used was zinc phthalocyanine (4.5 nM). When PCB-1260 (100 mg), the emulsion and catalyst were subjected to a current of mA/cm^2 on 11.2 cm^2 lead electrode for 18 hours, a dechlorination yield of >99.8 % was achieved. Reaction products included minor amounts of mono- and dichloropbiphenyls (0.02 mg), biphenyl and reduced alkylbenzene derivatives.

When PCB-1260-contaminated sand was treated with a poly(ethylene glycol)/potassium hydroxide mixture at room temperature, more than 99% reacted after 2 days forming aryl poly(ethylene glycols) (Brunelle and Singleton, 1985).

Drinking Water Standard (final): For all PCBs, the MCLG and MCL are zero and 0.5 μg/L, respectively (U.S. EPA, 1994).

Toxicity: Suspected human carcinogen. Acute oral LD$_{50}$ for rats 1,315 mg/kg (RTECS, 1985).

Uses: Secondary plasticizer for polyvinyl chloride; in polyester resins to increase strength of fiberglass; varnish formulations to improve water and alkali resistance; as an insulator fluid for electric condensers and as an additive in very high pressure lubricants.

PENTACHLOROBENZENE

Synonyms: QCB; RCRA waste number U183.

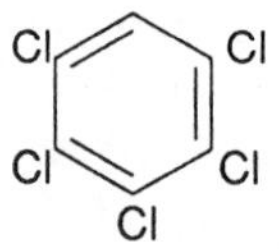

CAS Registry No.: 608-93-5
Molecular formula: C_6HCl_5
Formula weight: 250.34
RTECS: DA6640000

Physical state and color
White needles.

Melting point (°C):
86 (Weast, 1986)
82–85 (Verschueren, 1983)

Boiling point (°C):
277 (Weast, 1986)

Density (g/cm^3):
1.8342 at 16.5/4 °C (Weast, 1986)

Diffusivity in water (10^5 cm^2/sec):
0.59 at 20 °C using method of Hayduk and Laudie (1974).

Flash point (°C):
None (Dean, 1987)

Heat of fusion (kcal/mol):
4.92 (Miller et al., 1984)

Henry's law constant (10^4 atm·m^3/mol):
71 at 20 °C (Oliver, 1985)
3.69, 4.88, 6.72, 6.58, 12.25 and 27.26 at 14.8, 20.1, 22.1, 24.2, 34.8 and 50.5 °C, respectively (ten Hulscher et al., 1992)

Bioconcentration factor, log BCF:
3.53 (bluegill sunfish, Veith et al., 1980)

3.60 (algae, Geyer et al., 1984)
4.16 (activated sludge), 3.48 (golden ide) (Freitag et al., 1985)
3.84 (freshwater fish, Garten and Trabalka, 1983)

Soil sorption coefficient, log K_{oc}:
6.3 (average of 7 suspended sediment samples from the St. Clair and Detroit Rivers, Lau et al., 1989)
4.36, 4.52, 5.57 (Paya-Perez et al., 1991)
4.68 (lake sediment, Schrap et al., 1994)

Octanol/water partition coefficient, log K_{ow}:
5.17 (Watarai et al., 1982)
4.94 (Banerjee et al., 1980; Veith et al., 1980)
5.19 (Kenaga and Goring, 1980)
4.88 (Geyer et al., 1984; Könemann et al., 1979)
5.75 (DeKock and Lord, 1987)
5.20 at 13 °C, 5.05 at 19 °C, 4.70 at 28 °C, 4.66 at 33 °C (Opperhuizen et al., 1988)
5.183 (de Bruijn et al., 1989)
5.06 (Hammers et al., 1982)
5.03 (Miller et al., 1984)
5.69 (Bruggeman et al., 1982)

Solubility in organics:
Very soluble in ether (Sax and Lewis, 1987).

Solubility in water:
5.37 μmol/L at 20 °C (Veith et al., 1980)
5.32 μmol/L at 25 °C (Banerjee et al., 1980)
240 ppb at 22 °C (Verschueren, 1983)
3.32 μmol/L at 25 °C (Miller et al., 1984)
2.24 μmol/L at 25 °C (Yalkowsky et al., 1979)

Vapor pressure (mmHg):
6 x 10^{-3} at 20-30 °C (Mercer et al., 1990)

Environmental Fate
Biological. In activated sludge, <0.1% mineralized to carbon dioxide after 5 days (Freitag et al., 1985).
Photolytic. UV irradiation (λ = 2537 Å) of pentachlorobenzene in hexane solution for 3 hours produced a 50% yield of 1,2,4,5-tetrachlorobenzene and a 13% yield of 1,2,3,5-tetrachlorobenzene (Crosby and Hamadmad, 1971). Irradiation (λ $\geq$285 nm) of pentachlorobenzene (1.1-1.2 mM/L) in an acetonitrile-water mixture containing acetone (0.553 mM/L) as a sensitizer gave the following products (%

yield): 1,2,3,4-tetrachlorobenzene (6.6), 1,2,3,5-tetrachlorobenzene (52.8), 1,2,4,5-tetrachlorobenzene (15.1), 1,2,4-trichlorobenzene (1.9), 1,3,5-trichlorobenzene (5.3), 1,3-dichlorobenzene (0.9), 2,2',3,3',4,4',5,6,6'-nonachlorobiphenyl (2.08), 2,2',3,3',4,4',5,5',6-nonachlorobiphenyl (0.34), 2,2',3,3',4,5,5',6,6'-nonachloro-biphenyl (trace), one octachlorobiphenyl (0.53) and one heptachlorobiphenyl (0.49) (Choudhry and Hutzinger, 1984). Without acetone, the identified photolysis products (% yield) included 1,2,3,4-tetrachlorobenzene (3.7), 1,2,3,5-tetrachloro-benzene (13.5), 1,2,4,5-tetrachlorobenzene (2.8), 1,2,4-trichlorobenzene (12.7), 1,3,5-trichlorobenzene (1.0) and 1,4-dichlorobenzene (6.7) (Choudhry and Hutzinger, 1984).

A carbon dioxide yield of 2.0% was achieved when pentachlorobenzene adsorbed on silica gel was irradiated with light (λ >290 nm) for 17 hours (Freitag et al., 1985).

The experimental first-order decay rate for pentachlorobenzene in an aqueous solution containing a nonionic surfactant micelle (Brij 58, a polyoxyethylene cetyl ether) and illuminated by a photoreactor equipped with 253.7-nm monochromatic ultraviolet lamps, is 1.47 x 10^{-2}/sec. The corresponding half-life is 47 seconds. Photoproducts reported include, all tetra-, tri- and dichlorobenzenes, chloro-benzene, benzene, phenol, hydrogen and chloride ions (Chu and Jafvert, 1994).

Chemical/Physical. Based on an assumed base-mediated 1% disappearance after 16 days at 85 °C and pH 9.70 (pH 11.26 at 25 °C), the hydrolysis half-life was estimated to be >900 years (Ellington et al., 1988).

Toxicity: Acute oral LD_{50} for rats 1,080 mg/kg, mice 1,175 mg/kg (RTECS, 1985); LC_{50} (14-day) for guppies 178 ppb (Verschueren, 1983).

Uses: Chemical research; organic synthesis.

PENTACHLOROETHANE

Synonyms: Ethane pentachloride; NCI-C53894; Pentalin; RCRA waste number U184; UN 1669.

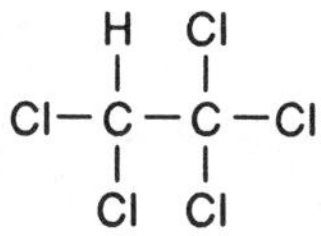

$$Cl-\underset{\underset{Cl}{|}}{\overset{\overset{H}{|}}{C}}-\underset{\underset{Cl}{|}}{\overset{\overset{Cl}{|}}{C}}-Cl$$

CAS Registry No.: 76-01-7
DOT: 1669
DOT label: Poison
Molecular formula: C_2HCl_5
Formula weight: 202.28
RTECS: KI6300000

Physical state, color and odor
Clear, colorless liquid with a sweetish, chloroform-like odor.

Melting point (°C):
-29.0 (Horvath, 1982)

Boiling point (°C):
162 (Weast, 1986)
159.72 (Boublik et al., 1973)
160.5 (Dean, 1987)

Density (g/cm^3):
1.6796 at 20/4 °C (Weast, 1986)
1.6808 at 20/4 °C, 1.6732 at 25/4 °C (Riddick et al., 1986)

Diffusivity in water (10^5 cm^2/sec):
0.79 at 20 °C using method of Hayduk and Laudie (1974).

Heat of fusion (kcal/mol):
2.7 (Dean, 1987)

Henry's law constant (10^3 atm·m^3/mol):
2.45 at 25 °C (Hine and Mookerjee, 1975)

Ionization potential (eV):
11.28 (NIOSH, 1994)

Bioconcentration factor, log BCF:
1.83 (bluegill sunfish, Veith et al., 1980)

Soil sorption coefficient, log K_{oc}:
3.28 (Mercer et al., 1990)

Octanol/water partition coefficient, log K_{ow}:
2.89 (Veith et al., 1980)

Solubility in organics:
Miscible with alcohol and ether (Windholz et al., 1983).

Solubility in water:
3.8 mmol/L at 20 °C (Veith et al., 1980)
470 mg/L at 25 °C (O'Connell, 1963)
500 and 769 mg/L at 20 and 25 °C, respectively (Mackay and Shiu, 1981)

Vapor density:
8.27 g/L at 25 °C, 6.98 (air = 1)

Vapor pressure (mmHg):
3.4 at 20 °C, 6 at 30 °C (Verschueren, 1983)
4.5 at 25 °C (Mackay and Shiu, 1981; Mackay et al., 1982)

Environmental Fate
   Chemical/Physical. At various pHs, pentachloroethane hydrolyzed to tetra-chloroethylene (Jeffers et al., 1989; Roberts and Gschwend, 1991). Dichloroacetic acid was also reported as a hydrolysis product. Reacts with alkalies and metals producing explosive chloroacetylenes (NIOSH, 1994). The reported hydrolysis half-life at 25 °C and pH 7 is 3.6 days (Jeffers et al., 1989).
   The evaporation half-life of pentachloroethane (1 mg/L) from water at 25 °C using a shallow-pitch propeller stirrer at 200 rpm at an average depth of 6.5 cm is 46.5 minutes (Dilling, 1977).

Drinking Water Standard: As of May 1994, no MCLGs or MCLs have been proposed (U.S. EPA, 1984).

Uses: Solvent for oil and grease in metal cleaning.

PENTACHLOROPHENOL

Synonyms: Acutox; Chempenta; Chemtol; Chlorophen; Cryptogil OL; Dowcide 7; Dowicide 7; Dowicide EC-7; Dowicide G; Dow pentachlorophenol DP-2 antimicrobial; Durotox; EP 30; Fungifen; Fungol; Glazd penta; Grundier arbezol; Lauxtol; Lauxtol A; Liroprem; Monsanto penta; Moosuran; NCI-C54933; NCI-C55378; NCI-C56655; PCP; Penchlorol; Penta; Pentachlorofenol; Pentachlorofenolo; Pentachlorophenate; Pentachlorphenol; 2,3,4,5,6-Pentachlorophenol; Pentacon; Penta-kil; Pentasol; Penwar; Peratox; Permacide; Permaguard; Permasan; Permatox DP-2; Permatox Penta; Permite; Priltox; RCRA waste number U242; Santobrite; Santophen; Santophen 20; Sinituho; Term-i-trol; Thompson's wood fix; Weedone; Witophen P.

CAS Registry No.: 87-86-5
DOT: 2020
Molecular formula: C_6HCl_5O
Formula weight: 266.34
RTECS: SM6300000

Physical state, color and odor
White to dark-colored flakes or crystalline solid with a phenolic odor. The medium odor threshold and upper taste threshold are both 857 ppb (Keith and Walters, 1992).

Melting point (°C):
191 (Weast, 1986)
174 (IARC, 1979)
188 (Weiss, 1986)

Boiling point (°C):
310 (Melnikov, 1971)
293 (Bailey and White, 1965)

Density (g/cm³):
1.978 at 22/4 °C (Weast, 1986)

Diffusivity in water (10^5 cm²/sec):
0.49 at 20 °C using method of Hayduk and Laudie (1974).

Dissociation constant:
4.74 (Mills et al., 1985)
5.3 (Eder and Weber, 1980)

Flash point (°C):
Noncombustible solid (NIOSH, 1994)

Henry's law constant (10^7 atm·m^3/mol):
2.8 (Eisenreich et al., 1981)
21 (Petrasek et al., 1983)

Bioconcentration factor, log BCF:
3.10 (algae, Geyer et al., 1984)
3.04 (activated sludge), 2.41 (golden ide) (Freitag et al., 1985)
2.89 (freshwater fish), 2.11 (fish, microcosm) (Garten and Trabalka, 1983)

Soil sorption coefficient, log K_{oc}:
2.95 (Kenaga, 1980)
2.96, 2.47, >2.68 (various Norwegian soils, Seip et al., 1986)
4.16 (Jury et al., 1987)
3.10-3.26 (calculated for dissociated species), 4.40 (calculated for undissociated species) (Lagas, 1988)
K_d = 1.73 (Eder and Weber, 1980)
2.76 (coarse sand), 2.48 (loamy sand) (Kjeldsen et al., 1990)
3.89 at pH 5.2 (Humaquept sand, Lafrance et al., 1994)

Octanol/water partition coefficient, log K_{ow}:
5.01 (Leo et al., 1971)
5.86 (Banerjee et al., 1984)
4.84 at pH 1.2, 4.72 at pH 2.4, 4.62 at pH 3.4, 4.57 at pH 4.7, 3.72 at pH 5.9, 3.56 at pH 6.5, 3.32 at pH 7.2, 3.20 at pH 7.8, 3.10 at pH 8.4, 2.75 at pH 8.9, 1.45 at pH 9.3, 1.36 at pH 9.8, 1.30 at pH 10.5, 2.42 at pH 10.5, 1.30 at pH 11.5, 1.67 at pH 12.5, 3.86 at pH 13.5 (Kaiser and Valdmanis, 1982)
3.69 (Geyer et al., 1982)
5.24 (Schellenberg et al., 1984)
3.807 (Lu and Metcalf, 1975)
5.0 (van Gestel and Ma, 1988)
4.16 (Rao and Davidson, 1980)
4.07 (Riederer, 1990)
5.11 (Garst and Wilson, 1984)

Solubility in organics:
Very soluble in ethanol and ether; slightly soluble in solvents and ligroin (U.S.

EPA, 1985). Also soluble in acetone, carbitol, cellosolve (ITII, 1986) and dimethylsulfoxide (Keith and Walters, 1992).

Solubility in water:
37.7 mmol/L at 22.7 °C and pH 9.46 (Arcand et al., 1995)
16 mg/L (Mills and Hoffman, 1993)
35 mg/L at 50 °C, 85 mg/L at 70 °C (Verschueren, 1983)
20 mg/L at 20 °C (Kearney and Kaufman, 1976)
14 ppm at 20 °C, 19 ppm at 30 °C (Bevenue and Beckman, 1967)
14 mg/L at 20 °C, 20 mg/L at 30 °C (Gunther et al., 1968)
2.0 mg/L (Gile and Gillett, 1979)
12 mg/L at 20 °C (Riederer, 1990)

Vapor pressure (10^5 mmHg):
17 at 20 °C (Melnikov, 1971)
2.14 at 20 °C (Dobbs and Cull, 1982)
11 at 20 °C, 1.7 at 23 °C, 0.76 at 30 °C, 26 at 40 °C (Klöpffer et al., 1988)
11 at 20 °C (Bevenue and Beckman, 1967)
1.7 at 20.25 °C (Gile and Gillett, 1979)

Environmental Fate
Biological. Under aerobic conditions, microbes in estuarine water partially dechlorinated pentachlorophenol to trichlorophenol. In distilled water, pentachlorophenol photolyzed to tetrachlorophenols, trichlorophenols, chlorinated dihydroxybenzenes and dichloromaleic acid (Hwang et al., 1986).

The disappearance of pentachlorophenol was studied in four aquaria, with and without mud, under aerobic and anaerobic conditions. Potential biological and/or chemical products identified include pentachloroanisole, 2,3,4,5-, 2,3,4,6- and 2,3,5,6-tetrachlorophenol (Boyle, 1980).

Pentachlorophenol degraded in anaerobic sludge to 3,4,5-trichlorophenol, which was further reduced to 3,5-dichlorophenol (Mikesell and Boyd, 1985). In activated sludge, only 0.2% of the applied amount was mineralized to carbon dioxide after 5 days (Freitag et al., 1985).

Pentachlorophenol was statically incubated in the dark at 25 °C with yeast extract and settled domestic wastewater inoculum. Significant biooxidation was observed but with a gradual adaptation over a 14-day period to achieve complete degradation at 5 mg/L substrate cultures. At a concentration of 10 mg/L, it took 28 days for pentachlorophenol to degrade completely (Tabak et al., 1981).

Pentachlorophenol was also subject to methylation by a culture medium containing *Trichoderma viride* affording pentachloroanisole (Cserjesi and Johnson, 1972).

Soil. Under anaerobic conditions, pentachlorophenol may undergo sequential dehalogenation to produce tetra-, tri-, di- and *m*-chlorophenol (Kobayashi and

Rittman, 1982). In aerobic and anaerobic soils, pentachloroanisole was the major metabolite, along with 2,3,6-trichlorophenol, 2,3,4,5- and 2,3,5,6-tetrachlorophenol (Murthy et al., 1979). Degradation was rapid in estuarine sediments having pH values of 6.5 and 8.0. Following a 17-day lag period, 70% of pentachlorophenol degraded (DeLaune, 1983). After 160 days in aerobic soil, only 6% biodegradation was observed (Baker and Mayfield, 1980).

Weiss et al. (1982) studied the fate of [^{14}C]pentachlorophenol added to flooded rice soil in a plant growth chamber. After one growing season, the following residues were observed (% of applied radioactivity): unidentified unextractable/bound compounds (28.61%), pentachlorophenol (0.51%), conjugated pentachlorophenol (0.61%), 2,3,6-, 2,4,5-, 2,4,6-, 2,3,4-, 2,3,5- and 3,4,5-trichlorophenols (1.27%), 2,3,4,5-tetrachlorophenol (0.38%), 2,3,4-, 2,3,6-, 2,4,6- and 3,4,5-trichloroanisoles (0.08%), 2,3,4,5-tetrachloroanisole (0.02%), pentachloroanisole (0.02%) and unidentified conversion products (including highly polar hydrolyzable and nonhydrolyzable compounds) (4.74%).

Metabolites identified in soil beneath a sawmill environment where pentachlorophenol was used as a wood preservative include pentachloroanisole, 2,3,4,6-tetrachloroanisole, tetrachlorocatechol, tetrachlorohydroquinone, 3,4,5-trichlorocatechol, 2,3,6-trichlorohydroquinone, 3,4,6-trichlorocatechol and 2,3,4,6-tetrachlorophenol (Knuutinen et al., 1990).

Photolytic. Wood treated with pure pentachlorophenol did not photolyze under natural sunlight or laboratory-induced UV radiation. However, in the presence of an antimicrobial (Dowcide EC-7), pure pentachlorophenol degraded to chlorinated dibenzo-*p*-dioxin. Wood containing composited technical grade pentachlorophenol yielded similar results (Lamparski et al., 1980).

Photodecomposition of pentachlorophenol was observed when an aqueous solution was exposed to sunlight for 10 days. The violet-colored solution contained 3,4,5-trichloro-6-(2'-hydroxy-3',4',5',6'-tetrachlorophenoxy)-*o*-benzoquinone as the major product. Minor photodecomposition products (% yield) included tetrachlororesorcinol (0.10%), 2,5-dichloro-3-hydroxy-6-pentachlorophenoxy-*p*-benzoquinone (0.16%) and 3,5-dichloro-2-hydroxy-5-2',4',5',6'-tetrachloro-3-hydroxyphenoxy-*p*-benzoquinone (0.08%) (Plimmer, 1970).

An aqueous solution containing pentachlorophenol and exposed to sunlight or laboratory UV light yielded tetrachlorocatechol, tetrachlororesorcinol and tetrachlorohydroquinone. These compounds were air-oxidized to chloranil, hydroxyquinones and 2,3-dichloromaleic acid. Other compounds identified include a cyclic dichlorodiketone, 2,3,5,6- and 2,3,4,6-tetrachlorophenol and trichlorophenols (Munakata and Kuwahara, 1969; Wong and Crosby, 1981).

UV irradiation (λ = 2537 Å) of pentachlorophenol in hexane solution for 32 hours produced a 30% yield of 2,3,5,6-tetrachlorophenol and about a 10% yield of a compound tentatively identified as an isomeric tetrachlorophenol (Crosby and Hamadmad, 1971). When pentachlorophenol in distilled water is exposed to UV irradiation, it is photolyzed to give tetrachlorophenols, trichlorophenols, chlorinated dihydroxybenzenes and dichloromaleic acid (Hwang et al., 1986).

A carbon dioxide yield of 50.0% was achieved when pentachlorophenol adsorbed on silica gel was irradiated with light (λ >290 nm) for 17 hours (Freitag et al., 1985). In soil, photodegradation was not significant (Baker et al., 1980).

When an aqueous solution containing pentachlorophenol (45 μM) and a suspension of titanium dioxide (2 g/L) was irradiated with UV light, carbon dioxide and hydrochloric acid formed in quantitative amounts. The half-life for this reaction at 45-50 °C is 8 minutes (Barbeni et al., 1985). When an aqueous solution containing pentachlorophenol was photooxidized by UV light at 90-95 °C, 25, 50 and 75% degraded to carbon dioxide after 31.7, 66.0 and 180.7 hours, respectively (Knoevenagel and Himmelreich, 1976). The photolysis half-lives of pentachlorophenol under sunlight irradiation in distilled water and river water were 27 and 53 hours, respectively (Mansour et al., 1989).

In a similar study, pentachlorophenol (47 μM) in an air-saturated solution containing titanium dioxide suspension was irradiated with UV light (λ = 330-370 nm). Major chemical intermediates included *p*-choranil, tetrachlorohydroquinone, hydrogen peroxide and *o*-chloranil. The intermediate compounds were attacked by hydroxyl radicals during the latter stages of irradiation forming HCO_2^-, acetate and formate ions, carbon dioxide and hydrochloric acid (Mills and Hoffman, 1993).

Petrier et al. (1992) studied the sonochemical degradation of pentachlorophenol in aqueous solutions saturated with different gases at 24 °C. Ultrasonic irradiation of solutions saturated with air or oxygen resulted in the liberation of chloride ions and mineralization of the parent compound to carbon dioxide. When the solution was saturated with argon, pentachlorophenol completely degraded to carbon monoxide and chloride ions.

Chemical/Physical. Wet oxidation of pentachlorophenol at 320 °C yielded formic and acetic acids (Randall and Knopp, 1980). In a dilute aqueous solution at pH 6.0, pentachlorophenol reacted with excess of hypochlorous acid forming 2,3,5,6-tetrachlorobenzoquinone (3% yield), 2,3,4,4,5,6-hexachlorobenzoquinone (20% yield) and two other chlorinated compounds (Smith et al., 1976).

Reacts with amines and alkali metals forming water-soluble salts (Sanborn et al., 1977). Hexachlorobenzene, octachlorodiphenylene dioxide and other polymeric compounds were formed when pentachlorophenol was heated to 300 °C for 24 hours (Sanderman et al., 1957).

Exposure Limits (mg/m^3): NIOSH REL: TWA 0.5, IDLH 2.5; OSHA PEL: TWA 0.5; ACGIH TLV: TWA 0.5.

Symptoms of Exposure: Ingestion may cause fluctuation in blood pressure, respiration, fever, urinary output, weakness, convulsions and possibly death (NIOSH, 1994).

Drinking Water Standard (final): MCLG: zero; MCL: 1 μg/L (U.S. EPA, 1994).

Toxicity: Acute oral LD_{50} for ducks 380 mg/kg, hamsters 168 mg/kg, mice 117 mg/kg, rats 27 mg/kg; LC_{50} (inhalation) for mice 225 mg/m^3, rats 355 mg/m^3 (RTECS, 1985); LC_{50} (48-hour) for rainbow trout 0.17 mg/L (sodium salt) (Hartley and Kidd, 1987); LC_{50} (24-hour) for goldfish 270 ppb (Verschueren, 1983).

Uses: Manufacture of insecticides (termite control), algicides, herbicides, fungicides and bactericides; wood preservative.

1,4-PENTADIENE

Synonyms: None.

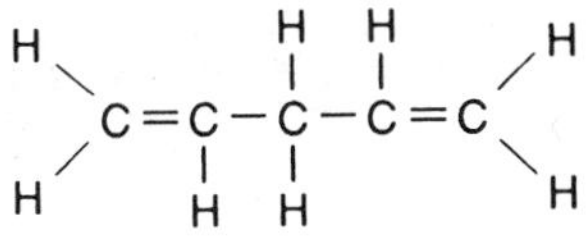

$$H \underset{H}{\overset{H}{\diagdown}} C = C - \overset{\overset{H}{|}}{\underset{\underset{H}{|}}{C}} - \overset{\overset{H}{|}}{\underset{\underset{H}{|}}{C}} = C \overset{H}{\underset{H}{\diagup}}$$

CAS Registry No.: 591-93-5
Molecular formula: C_5H_8
Formula weight: 68.12

Physical state
Liquid or gas.

Melting point (°C):
-148.3 (Weast, 1986)

Boiling point (°C):
26 (Weast, 1986)

Density (g/cm^3):
0.66076 at 20/4 °C, 0.65571 at 25/4 °C (Dreisbach, 1959)

Diffusivity in water (10^5 cm^2/sec):
0.87 at 20 °C using method of Hayduk and Laudie (1974).

Dissociation constant:
>14 (Schwarzenbach et al., 1993)

Flash point (°C):
<0 (Sax and Lewis, 1987)

Heat of fusion (kcal/mol):
1.468 (Dean, 1987)

Henry's law constant (atm·m^3/mol):
0.120 at 25 °C (Hine and Mookerjee, 1975)

Soil sorption coefficient, log K_{oc}:
Unavailable because experimental methods for estimation of this parameter for aliphatic hydrocarbons are lacking in the documented literature.

847

Octanol/water partition coefficient, log K_{ow}:
1.48 (Hansch and Leo, 1979)

Solubility in organics:
Soluble in acetone, alcohol, benzene and ether (Weast, 1986).

Solubility in water:
558 mg/kg at 25 °C (McAuliffe, 1966)

Vapor density:
2.78 g/L at 25 °C, 2.35 (air = 1)

Vapor pressure (mmHg):
734.6 at 25 °C (Wilhoit and Zwolinski, 1971)

Uses: Chemical research; organic synthesis.

PENTANE

Synonyms: Amyl hydride; Dimethylmethane; *n*-Pentane; UN 1265.

$$H-\overset{\overset{\displaystyle H}{|}}{\underset{\underset{\displaystyle H}{|}}{C}}-\overset{\overset{\displaystyle H}{|}}{\underset{\underset{\displaystyle H}{|}}{C}}-\overset{\overset{\displaystyle H}{|}}{\underset{\underset{\displaystyle H}{|}}{C}}-\overset{\overset{\displaystyle H}{|}}{\underset{\underset{\displaystyle H}{|}}{C}}-\overset{\overset{\displaystyle H}{|}}{\underset{\underset{\displaystyle H}{|}}{C}}-H$$

CAS Registry No.: 109-66-0
DOT: 1265
DOT label: Flammable liquid
Molecular formula: C_5H_{12}
Formula weight: 72.15
RTECS: RZ9450000

Physical state, color and odor
Clear, colorless, volatile liquid with an odor resembling gasoline. Odor threshold in air is 10 ppm (Keith and Walters, 1992).

Melting point (°C):
-130 (Weast, 1986)

Boiling point (°C):
36.1 (Weast, 1986)

Density (g/cm^3):
0.62624 at 20/4 °C, 0.62139 at 25/4 °C (Dreisbach, 1959)
0.6290 at 17.2/4 °C, 0.62139 at 25.00/4 °C (Curtice et al., 1972)

Diffusivity in water (10^5 cm^2/sec):
0.84 at 20 °C (Witherspoon and Bonoli, 1969)
0.97 at 25 °C (Hayduk and Laudie, 1974)

Dissociation constant:
>14 (Schwarzenbach et al., 1993)

Flash point (°C):
-49.8 (NIOSH, 1994)
-40 (Windholz et al., 1983)

Lower explosive limit (%):
1.5 (NIOSH, 1994)

Upper explosive limit (%):
7.8 (NIOSH, 1994)

Heat of fusion (kcal/mol):
2.008 (Dean, 1987)

Henry's law constant (atm·m^3/mol):
1.255 at 25 °C (Hine and Mookerjee, 1975)

Interfacial tension with water (dyn/cm at 25 °C):
49.0 (Donahue and Bartell, 1952)

Ionization potential (eV):
10.35 (HNU, 1986)

Soil sorption coefficient, log K_{oc}:
Unavailable because experimental methods for estimation of this parameter for aliphatic hydrocarbons are lacking in the documented literature.

Octanol/water partition coefficient, log K_{ow}:
3.23, 3.39 (Hansch and Leo, 1979)
3.62 (Schantz and Martire, 1987; Tewari et al., 1982; Wasik et al., 1981)

Solubility in organics:
In methanol: 620 and 810 g/L at 5 and 10 °C, respectively. Miscible at higher temperatures (Kiser et al., 1961).

Solubility in water:
In mg/kg: 39.5 at 25 °C, 39.8 at 40.1 °C, 41.8 at 55.7 °C, 69.4 at 99.1 °C. In NaCl
    solution at 25 °C (salinity, g/kg): 36.8 (1.002), 34.5 (10.000), 27.6 (34.472), 22.6
    (50.03), 10.9 (125.10), 5.91 (199.90), 2.64 (279.80), 2.01 (358.70) (Price, 1976)
38.5 mg/kg at 25 °C (McAuliffe, 1963, 1966)
65.7 mg/kg at 0 °C, 47.6 mg/kg at 25 °C (Polak and Lu, 1973)
39 mg/kg at 25 °C (Krzyzanowska and Szeliga, 1978)
360 mg/L at 16 °C (Fischer and Ehrenberg, 1948)
565 μmol/L at 25.0 °C (Tewari et al., 1982; Wasik et al., 1981)
40.0, 40.4 and 47.6 mg/L at 25 °C (Mackay and Shiu, 1981)
0.60 mL/L at 16 °C (Fühner, 1924)
0.11 g/kg at 20 °C and 32 atmHg (Namiot and Beider, 1960)
700 mg/L at 20 °C (Korenman and Aref'eva, 1977)
Mole fraction x 10^5: 1.02, 1.07, 0.98, 1.01 and 1.01 at 4.0, 10.0, 20.0, 25.0 and 30.0
    °C, respectively (Nelson and DeLigny, 1968)
1.03 mmol/L at 25 °C (Barone et al., 1966)

39.0 mg/L at 25 °C (Kryzanowska and Szeliga, 1978)
40.6 mg/L at 25 °C (Jönsson et al., 1982)

Vapor density:
2.95 g/L at 25 °C, 2.49 (air = 1)

Vapor pressure (mmHg):
512.8 at 25 °C (Wilhoit and Zwolinski, 1971)
433.67 at 20.572 °C, 525.95 at 25.698 °C (Osborn and Douslin, 1974)
516.65 at 25.00 °C (Hussam and Carr, 1985)
219.3, 283.7, 424.1, 511.3 and 614.8 at 4.0, 10.0, 20.0, 25.0 and 30.0 °C, respectively (Nelson and DeLigny, 1968)

Environmental Fate
Biological. n-Pentane may biodegrade in two ways. The first is the formation of pentyl hydroperoxide, which decomposes to 1-pentanol followed by oxidation to pentanoic acid. The other pathway involves dehydrogenation to 1-pentene, which may react with water giving 1-pentanol (Dugan, 1972). Microorganisms can oxidize alkanes under aerobic conditions (Singer and Finnerty, 1984). The most common degradative pathway involves the oxidation of the terminal methyl group forming 1-pentanol. The alcohol may undergo a series of dehydrogenation steps forming an aldehyde (valeraldehyde) then a fatty acid (valeric acid). The fatty acid may then be metabolized by β-oxidation to form the mineralization products, carbon dioxide and water (Singer and Finnerty, 1984). *Mycobacterium smegnatis* was capable of degrading pentane to 2-pentanone (Riser-Roberts, 1992).
Photolytic. Synthetic air containing gaseous nitrous acid and exposed to artificial sunlight (λ = 300-450 nm) photooxidized pentane into methyl nitrate, pentyl nitrate, peroxyacetal nitrate and peroxypropionyl nitrate (Cox et al., 1980).

Exposure Limits: NIOSH REL: TWA 120 ppm (350 mg/m^3), ceiling 610 ppm (1,800 mg/m^3), IDLH 1,500 ppm; OSHA PEL: TWA 1,000 ppm (2,950 mg/m^3); ACGIH TLV: TWA 600 ppm.

Symptoms of Exposure: Inhalation may cause narcosis and irritation of respiratory tract (Patnaik, 1992).

Toxicity: LD$_{50}$ (intravenous) for mice 466 mg/kg (RTECS, 1985).

Uses: Solvent recovery and extraction; blowing agent for plastic foams; low temperature thermometers; natural gas processing plants; production of olefin, hydrogen, ammonia; fuel production; pesticide; manufacture of artificial ice; organic synthesis.

2-PENTANONE

Synonyms: Ethyl acetone; Methyl propyl ketone; Methyl *n*-propyl ketone; MPK; Propyl methyl ketone; UN 1249.

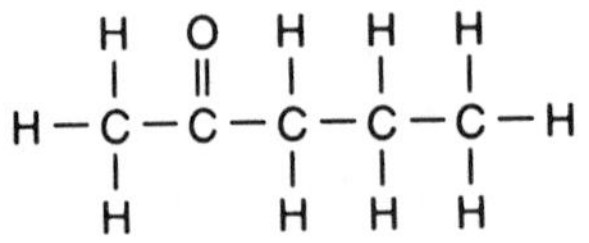

$$H-\overset{\displaystyle H}{\underset{\displaystyle H}{C}}-\overset{\displaystyle O}{C}-\overset{\displaystyle H}{\underset{\displaystyle H}{C}}-\overset{\displaystyle H}{\underset{\displaystyle H}{C}}-\overset{\displaystyle H}{\underset{\displaystyle H}{C}}-H$$

CAS Registry No.: 107–87–9
DOT: 1249
DOT label: Flammable liquid
Molecular formula: $C_5H_{10}O$
Formula weight: 86.13
RTECS: SA7875000

Physical state, color and odor
Colorless liquid with a characteristic, pungent odor. Odor threshold in air is 8 ppm (Keith and Walters, 1992).

Melting point (°C):
–77.8 (Weast, 1986)

Boiling point (°C):
102 (Weast, 1986)
103.3 (Stull, 1947)

Density (g/cm³):
0.8089 at 20/4 °C (Weast, 1986)
0.8018 at 25/4 °C (Ginnings et al., 1940)

Diffusivity in water (10^5 cm²/sec):
0.85 at 20 °C using method of Hayduk and Laudie (1974).

Flash point (°C):
7.3 (NIOSH, 1994)

Lower explosive limit (%):
1.5 (NIOSH, 1994)

Upper explosive limit (%):
8.2 (NIOSH, 1994)

Henry's law constant (10^5 atm·m^3/mol at 25 °C):
6.44 (Hine and Mookerjee, 1975)
6.36 (Buttery et al., 1969)
10.8 (28 °C, Nelson and Hoff, 1968)
11 (Hawthorne et al., 1985)

Ionization potential (eV):
9.37 ± 0.02 (Franklin et al., 1969)

Soil sorption coefficient, log K_{oc}:
Unavailable because experimental methods for estimation of this parameter for ketones are lacking in the documented literature.

Octanol/water partition coefficient, log K_{ow}:
0.78 (Sangster, 1989)
0.91 (Unger et al., 1978)

Solubility in organics:
Miscible with alcohol and ether (Windholz et al., 1983).

Solubility in water:
43,065 mg/L at 20 °C (Mackay and Yeun, 1983)
In wt %: 5.95 at 20 °C, 5.51 at 25 °C, 5.18 at 30 °C (Ginnings et al., 1940)
0.630 and 0.515 mol % at 30 and 50 °C, respectively (Palit, 1947)

Vapor density:
3.52 g/L at 25 °C, 2.97 (air = 1)

Vapor pressure (mmHg):
27 at 20 °C (NIOSH, 1994)
12 at 20 °C, 16 at 25 °C, 21 at 30 °C (Verschueren, 1983)

Exposure Limits: NIOSH REL: TWA 150 ppm (530 mg/m^3), IDLH 1,500 ppm; OSHA PEL: TWA 200 (700 mg/m^3).

Symptoms of Exposure: Inhalation of vapors may cause narcosis and irritation of eyes and respiratory tract (Patnaik, 1992).

Toxicity: Acute oral LD$_{50}$ for mice 2,205 mg/kg, rats 3,730 mg/kg (RTECS, 1985).

Uses: Solvent; substitute for 3-pentanone; flavoring.

1-PENTENE

Synonyms: α-n-Amylene; Propylethylene.

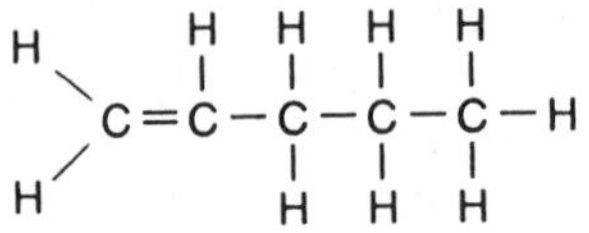

$$\begin{array}{c} H \\ \diagdown \\ \diagup \\ H \end{array} C = \overset{\displaystyle H}{\underset{\displaystyle H}{C}} - \overset{\displaystyle H}{\underset{\displaystyle H}{C}} - \overset{\displaystyle H}{\underset{\displaystyle H}{C}} - \overset{\displaystyle H}{\underset{\displaystyle H}{C}} - H$$

CAS Registry No.: 109-67-1
DOT: 1108
DOT label: Combustible liquid
Molecular formula: C_5H_{10}
Formula weight: 70.13

Physical state and color
Colorless liquid.

Melting point (°C):
-165.219 (Dreisbach, 1959)

Boiling point (°C):
30 (Weast, 1986)

Density (g/cm^3):
0.64050 at 20/4 °C, 0.63533 at 25/4 °C (Dreisbach, 1959)

Diffusivity in water (10^5 cm^2/sec):
0.84 at 20 °C using method of Hayduk and Laudie (1974).

Dissociation constant:
>14 (Schwarzenbach et al., 1993)

Flash point (°C):
-17.7 (open cup, NFPA, 1984)

Lower explosive limit (%):
1.5 (Sax and Lewis, 1987)

Upper explosive limit (%):
8.7 (Sax and Lewis, 1987)

Heat of fusion (kcal/mol):
1.388 (Dean, 1987)

Henry's law constant (atm·m³/mol):
0.406 at 25 °C (Hine and Mookerjee, 1975)

Ionization potential (eV):
9.50 (HNU, 1986)

Soil sorption coefficient, log K_{oc}:
Unavailable because experimental methods for estimation of this parameter for aliphatic hydrocarbons are lacking in the documented literature.

Octanol/water partition coefficient, log K_{ow}:
2.26 using method of Hansch et al. (1968).

Solubility in organics:
Miscible with alcohol, benzene and ether (Windholz et al., 1983).

Solubility in water:
148 mg/kg at 25 °C (McAuliffe, 1966)

Vapor density:
2.87 g/L at 25 °C, 2.42 (air = 1)

Vapor pressure (mmHg):
628.2 at 24.6 °C (Forziati, 1950)
637.7 at 25 °C (Wilhoit and Zwolinski, 1971)

Environmental Fate
Biological. Biooxidation of 1-pentene may occur yielding 4-penten-1-ol, which may further oxidize to give 4-pentenoic acid (Dugan, 1972). Washed cell suspensions of bacteria belonging to the genera *Mycobacterium*, *Nocardia*, *Xanthobacter* and *Pseudomonas* and growing on selected alkenes metabolized 1-pentene to 1,2-epoxypentane. *Mycobacterium* sp., growing on ethene, hydrolyzed 1,2-epoxypropane to 1,2-propanediol (van Ginkel et al., 1987).

Uses: Blending agent for high octane motor fuel; organic synthesis.

cis-2-PENTENE

Synonyms: *cis*-Pentene-2; **(Z)-2-Pentene.**

$$H_3C, CH_3$$

CAS Registry No.: 627-20-3
Molecular formula: C_5H_{10}
Formula weight: 70.13

Physical state and color
Colorless liquid.

Melting point (°C):
-151.4 (Weast, 1986)

Boiling point (°C):
36.9 (Weast, 1986)

Density (g/cm³):
0.6556 at 20/4 °C, 0.6504 at 25/4 °C (Dreisbach, 1959)
0.6614 at 16.50/4 °C, 0.65152 at 25.00/4 °C (Curtice et al., 1972)
0.6503 at 20/4 °C (Dean, 1987)

Diffusivity in water (10^5 cm²/sec):
0.85 at 20 °C using method of Hayduk and Laudie (1974).

Dissociation constant:
>14 (Schwarzenbach et al., 1993)

Flash point (°C):
-18 (Sax and Lewis, 1987)

Heat of fusion (kcal/mol):
1.700 (Dean, 1987)

Henry's law constant (atm·m³/mol):
0.225 at 25 °C (approximate - calculated from water solubility and vapor pressure)

Ionization potential (eV):
9.13 (HNU, 1986)

Soil sorption coefficient, log K$_{oc}$:
Unavailable because experimental methods for estimation of this parameter for aliphatic hydrocarbons are lacking in the documented literature.

Octanol/water partition coefficient, log K$_{ow}$:
2.15 using method of Hansch et al. (1968).

Solubility in organics:
Soluble in alcohol, benzene and ether (Weast, 1986).

Solubility in water:
203 mg/kg at 25 °C (McAuliffe, 1966)

Vapor density:
2.87 g/L at 25 °C, 2.42 (air = 1)

Vapor pressure (mmHg):
494.6 at 25 °C (Wilhoit and Zwolinski, 1971)

Uses: Polymerization inhibitor; organic synthesis.

trans-2-PENTENE

Synonyms: (*E*)-2-Pentene; *trans*-Pentene-2.

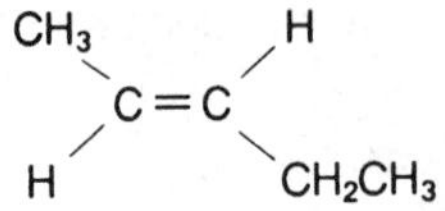

$$CH_3 \diagdown \qquad \diagup H$$
$$C = C$$
$$H \diagup \qquad \diagdown CH_2CH_3$$

CAS Registry No.: 646-04-8
Molecular formula: C_5H_{10}
Formula weight: 70.13

Physical state and color
Colorless liquid.

Melting point (°C):
-136 (Weast, 1986)
-140.2 (Dean, 1987)

Boiling point (°C):
36.3 (Weast, 1986)

Density (g/cm³):
0.6482 at 20/4 °C (Weast, 1986)

Diffusivity in water (10^5 cm²/sec):
0.84 at 20 °C using method of Hayduk and Laudie (1974).

Dissociation constant:
>14 (Schwarzenbach et al., 1993)

Flash point (°C):
<-20 (NFPA, 1984)

Heat of fusion (kcal/mol):
1.996 (Dean, 1987)

Henry's law constant (atm·m³/mol):
0.234 at 25 °C (Hine and Mookerjee, 1975)

Ionization potential (eV):
9.13 (HNU, 1986)

Soil sorption coefficient, log K_{oc}:
Unavailable because experimental methods for estimation of this parameter for aliphatic hydrocarbons are lacking in the documented literature.

Octanol/water partition coefficient, log K_{ow}:
2.15 using method of Hansch et al. (1968).

Solubility in organics:
Soluble in alcohol, benzene and ether (Weast, 1986).

Solubility in water:
203 mg/kg at 25 °C (McAuliffe, 1966)

Vapor density:
2.87 g/L at 25 °C, 2.42 (air = 1)

Vapor pressure (mmHg):
505.5 at 25 °C (Wilhoit and Zwolinski, 1971)

Uses: Polymerization inhibitor; organic synthesis.

PENTYLCYCLOPENTANE

Synonyms: None.

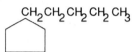

CAS Registry No.: 3741-00-2
Molecular formula: $C_{10}H_{20}$
Formula weight: 140.28

Physical state
Liquid.

Boiling point (°C):
180.6 (Wilhoit and Zwolinski, 1971)

Diffusivity in water (10^5 cm^2/sec):
0.57 at 20 °C using method of Hayduk and Laudie (1974).

Dissociation constant:
>14 (Schwarzenbach et al., 1993)

Soil sorption coefficient, log K_{oc}:
Unavailable because experimental methods for estimation of this parameter for alicyclic hydrocarbons are lacking in the documented literature.

Octanol/water partition coefficient, log K_{ow}:
4.90 using method of Hansch et al. (1968).

Solubility in water:
115 μg/kg at 25 °C (Price, 1976)

Vapor density:
5.73 g/L at 25 °C, 4.84 (air = 1)

Uses: Organic synthesis; gasoline component.

PHENANTHRENE

Synonyms: Phenanthren; Phenantrin.

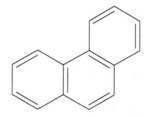

CAS Registry No.: 85-01-8
Molecular formula: $C_{14}H_{10}$
Formula weight: 178.24
RTECS: SF7175000

Physical state, color and odor
Colorless, monoclinic crystals with a faint, aromatic odor.

Melting point (°C):
100.5 (Dean, 1973)
98-100 (Fluka, 1988)

Boiling point (°C):
341.2 (Wilhoit and Zwolinski, 1971)

Density (g/cm³):
1.179 at 25/4 °C (Dean, 1987)

Diffusivity in water (10^5 cm²/sec):
0.59 at 20 °C using method of Hayduk and Laudie (1974).

Dissociation constant:
>15 (Christensen et al., 1975)

Flash point (°C):
171 (open cup, NFPA, 1984)

Heat of fusion (kcal/mol):
4.45 (Tsonopoulos and Prausnitz, 1971)
3.93 (Osborn and Douslin, 1975)
3.89 (Wauchope and Getzen, 1972)

Henry's law constant (10^5 atm·m³/mol):
3.9 (Mackay et al., 1979)

13 (Petrasek et al., 1983)
2.56 at 25 °C (Hine and Mookerjee, 1975)
5.48 (Southworth, 1979)
2.35 at 25 °C (Fendinger and Glotfelty, 1990)

Ionization potential (eV):
8.22 ± 0.29 (Franklin et al., 1969)
8.19 (Cavalieri and Rogan, 1985)

Bioconcentration factor, log BCF:
2.51 (*Daphnia pulex*, Southworth et al., 1978)
3.42 (fathead minnow, Veith et al., 1979)
2.51 (*Daphnia magna*, Newsted and Giesy, 1987)
0.77 (*Polychaete* sp.), 1.49 (*Capitella capitata*) (Bayona et al., 1991)
3.25 (algae, Geyer et al., 1984)
2.97 (activated sludge), 3.25 (golden ide) (Freitag et al., 1985)

Soil sorption coefficient, log K_{oc}:
3.72 (aquifer sands, Abdul et al., 1987)
4.36 (Karickhoff et al., 1979)
4.59 (Socha and Carpenter, 1987)
4.70 (Gauthier et al., 1986)
4.28 (estuarine sediment, Vowles and Mantoura, 1987)
4.07 (dark sand from Newfield, NY, Magee et al., 1991)
3.77 (Apison soil), 3.15 (Fullerton soil), 3.76 (Dormont soil) (Southworth and
    Keller, 1986)
4.42 (Eustis fine sand, Wood et al., 1990)
6.12 (average, Kayal and Connell, 1990)
4.12 (Oshtemo, Sun and Boyd, 1993)
4.60 (humic acid, Gauthier et al., 1986)
4.46 (Rotterdam Harbor sediment, Hegeman et al., 1995)

Octanol/water partition coefficient, log K_{ow}:
4.57 (Karickhoff et al., 1979)
4.46 (Hansch and Fujita, 1964)
4.52 (Yoshida et al., 1983)
4.517 (Kenaga and Goring, 1980)
4.16 (Landrum et al., 1984)
4.562 (Brooke et al., 1990; de Bruijn et al., 1989)
4.374 (Brooke et al., 1990)
4.54 (Hammers et al., 1982)
4.63 (Bruggeman et al., 1982)
4.45 (Wang et al., 1992)

Solubility in organics:
Soluble in toluene and carbon tetrachloride (1 g/2.4 mL) and anhydrous ether (1 g/3.3 mL) (Windholz et al., 1983).
24.5 and 31.7 g/L at 26 °C in methanol and ethanol, respectively (Fu et al., 1986).

Solubility in water:
At 25 °C: 0.85, 0.67, 0.48 and 0.33 mg/L in 0.50, 1.00, 1.50 and 2.00 M NaCl, respectively (Aquan-Yuen et al., 1979)
In mg/kg: 1.11-1.12 at 24.6 °C, 1.49 at 29.9 °C, 1.47-1.48 at 30.3 °C, 2.44-2.45 at 38.4 °C, 2.25-2.28 at 40.1 °C, 3.81-3.88 at 47.5 °C, 4.30-4.38 at 50.1 °C, 4.04-4.11 at 50.2 °C, 5.63-5.66 at 54.7 °C, 7.17-7.21 at 59.2 °C, 7.2-7.6 at 60.5 °C, 9.7-9.8 at 65.1 °C, 12.4-12.6 at 70.7 °C (Wauchope and Getzen, 1972)
1.29 mg/L at 25 °C (Mackay and Shiu, 1977; Walters and Luthy, 1984)
994 μg/L at 25 °C (Andrews and Keefer, 1949)
1.002 mg/kg at 25 °C, 1.220 mg/kg at 29 °C (May et al., 1978a)
1.07 mg/kg at 25 °C, 0.71 mg/kg in artificial seawater (salinity = 35 g/kg) at 25.0 °C (Eganhouse and Calder, 1976)
1.55-1.65 mg/L at 27 °C (Davis et al., 1942)
1.650 mg/L at 27 °C (Davis and Parke, 1942)
16 ng/L at 25 °C (Klevens, 1950)
710 μg/L at 25 °C (Sahyun, 1966)
In μg/kg: 423 at 8.5 °C, 468 at 10.0 °C, 512 at 12.5 °C, 601 at 15 °C, 816 at 21.0 °C, 995 at 24.3 °C, 1,277 at 29.9 °C (May et al., 1978)
In μmol/L: 2.81 at 8.4 °C, 3.09 at 11.1 °C, 3.59 at 14.0 °C, 4.40 at 17.5 °C, 4.94 at 20.2 °C, 6.09 at 23.3 °C, 6.46 at 25.0 °C, 7.7 at 29.3 °C, 9.13 at 31.8 °C (Schwarz, 1977)
235 ppb at 20-25 °C in Narragansett Bay water (pH 7.7, dissolved organic carbon 2.9 mg/L, salinity = 27 wt %) (Boehm and Quinn, 1973)
5.61 μmol/L at 25 °C (Wasik et al., 1983)
9.5 μmol/L at 29 °C (Stucki and Alexander, 1987)
9 μmol/L at 25 °C (Edwards et al., 1991)
In μmol/L: 2.01, 2.45, 3.12, 4.04, 4.94 and 6.16 at 4.6, 8.8, 12.9, 17.0, 21.1 and 25.3 °C, respectively. In seawater (salinity = 36.5 g/kg): 1.36, 1.75, 2.21, 2.81, 3.59 and 4.54 at 4.6, 8.8, 12.9, 17.0, 21.1 and 25.3 °C, respectively (Whitehouse, 1984)
At 20 °C: 6.2, 4.15, 4.01 and 4.14 μmol/L in distilled water, Pacific seawater and artificial seawater and 35% NaCl solution, respectively (Hashimoto et al., 1984)
1.00 mg/L at 25 °C (Vadas et al., 1991)
1.08 mg/L at 25 °C (Billington et al., 1988)
6.77 μmol/kg at 25.0 °C (Vesala, 1974)

Vapor pressure (10^4 mmHg):
1.21 at 25 °C (Sonnefeld et al., 1983; Wasik et al., 1983)
6.80 at 25 °C (Radding et al., 1976)

4.2, 8.3 at 25 °C (Hinckley et al., 1990)

34.9 at 51.60 °C (Macknick and Prausnitz, 1979)

0.64 at 36.70 °C, 0.81 at 39.15 °C, 0.89 at 39.85 °C, 1.11 at 42.10 °C, 8.1 at 42.60 °C, 1.45 at 44.62 °C, 1.85 at 46.70 °C (Bradley and Cleasby, 1953)

0.95 at 25 °C (McVeety and Hites, 1988)

15,500 at 127 °C (Eiceman and Vandiver, 1983)

5.16 at 25 °C (Bidleman, 1984)

Environmental Fate

Biological. Catechol is the central metabolite in the bacterial degradation of phenanthrene. Intermediate by-products include 1-hydroxy-2-napthoic acid, 1,2-dihydroxynaphthalene and salicylic acid (Chapman, 1972; Hou, 1982). It was reported that *Beijerinckia*, under aerobic conditions, degraded phenanthrene to *cis*-3,4-dihydroxy-3,4-dihydrophenanthracene (Kobayashi and Rittman, 1982).

In activated sludge, 39.6% mineralized to carbon dioxide (Freitag et al., 1985). When phenanthrene (5 and 10 mg/L) was statically incubated in the dark at 25 °C with yeast extract and settled domestic wastewater inoculum for 7 days, 100% bio-degradation with rapid adaptation was observed (Tabak et al., 1981).

Photolytic. A carbon dioxide yield of 24.2% was achieved when phenanthrene adsorbed on silica gel was irradiated with light (λ >290 nm) for 17 hours (Freitag et al., 1985). In a 2-week experiment, [^{14}C]phenanthrene applied to soil-water suspensions under aerobic and anaerobic conditions gave $^{14}CO_2$ yields of 7.2 and 6.3%, respectively (Scheunert et al., 1987).

Wang et al. (1995) investigated the photodegradation of phenanthrene in water using artificial light (λ >290 nm) in the presence of fulvic acids as sensitizers and hydrogen peroxide as an oxidant. The major photoproducts identified were 9,10-phenanthrenequinone, 1,3,4-trihydroxyphenanthrene, 9-hydroxyphenanthrene, 2,2'-biphenyldialdehyde, 2,2'-biphenyldicarbonic acid and 2-phenylbenz-aldehyde. The presence of fulvic acids/humic substances produced mixed results. The rate of photodegradation was retarded or accelerated depending upon the origins of the humic substances. Experimentally determined photolysis half-lives of phenanthrene in water using fulvic acids obtained from six different locations ranged from 1.30 to 5.78 hours. It was suggested that the formation of the photoproducts involved oxidation of phenanthrene via hydroxyl radicals generated from hydrogen peroxide.

Chemical/Physical. The aqueous chlorination of phenanthrene at pH <4 produced phenanthrene-9,10-dione and 9-chlorophenanthrene. At high pH (>8.8), phenanthrene-9,10-oxide, phenanthrene-9,10-dione and 9,10-dihydrophen-anthrenediol were identified as major products (Oyler et al., 1983). It was suggested that the chlorination of phenanthrene in tap water accounted for the presence of chloro- and dichlorophenanthrenes (Shiraishi et al., 1985).

Exposure Limits: Potential occupational carcinogen. No individual standards have been set, however, as a constituent in coal tar pitch volatiles, the following exposure limits have been established (mg/m^3): NIOSH REL: TWA 0.1 (cyclohex-ane-extractable fraction), IDLH 80; OSHA PEL: TWA 0.2 (benzene-soluble fraction); ACGIH TLV: TWA 0.2 (benzene solubles).

Toxicity: Acute oral LD$_{50}$ for mice 700 mg/kg; LD$_{50}$ (intraperitoneal) for mice 700 mg/kg (RTECS, 1985).

Drinking Water Standard: As of May 1994, no MCLGs or MCLs have been proposed (U.S. EPA, 1984).

Uses: Explosives; dyestuffs; biochemical research; synthesis of drugs; preparation of 9,10-phenanthrenequinone, 9,10-dihydrophenanthrene, 9-bromophenanthrene, 9,10-dibromo-9,10-dihydrophenanthrene and many other organic compounds.

PHENOL

Synonyms: Baker's P and S liquid and ointment; Benzenol; Carbolic acid; Hydroxybenzene; Monohydroxybenzene; NA 2821; NCI-C50124; Oxybenzene; Phenic acid; Phenyl hydrate; Phenyl hydroxide; Phenylic acid; Phenylic alcohol; RCRA waste number U188; UN 1671; UN 2312; UN 2821.

CAS Registry No.: 108–95–2
DOT: 1671 (solid); 2312 (molten); 2821 (solution)
DOT label: Poison
Molecular formula: C_6H_6O
Formula weight: 94.11
RTECS: SJ3325000

Physical state, color and odor
White crystals or light pink liquid which slowly turns brown on exposure to air. Sweet, tarry odor. Odor threshold in air is 16 ppb (Keith and Walters, 1992).

Melting point (°C):
43 (Weast, 1986)
40.9 (Weiss, 1986)

Boiling point (°C):
181.7 (Weast, 1986)
183 (Huntress and Mulliken, 1941)

Density (g/cm³):
1.0576 at 20/4 °C (Weast, 1986)
1.05760 at 41/4 °C (Standen, 1968)

Diffusivity in water (10^5 cm²/sec):
0.87 at 20 °C using method of Hayduk and Laudie (1974).

Dissociation constant:
9.99 at 25 °C (Dean, 1973)

Flash point (°C):
80 (NIOSH, 1994)

Lower explosive limit (%):
1.8 (NIOSH, 1994)

Upper explosive limit (%):
8.6 (NIOSH, 1994)

Heat of fusion (kcal/mol):
2.752 (Dean, 1987)

Henry's law constant (10^7 atm·m^3/mol):
2.7 (Petrasek et al., 1983)
1.71 (Jury et al., 1984)
39,900,000 kPa/unit mol at 27.0 °C (Abd-El-Bary et al., 1986)
3.97 at 25 °C (Hine and Mookerjee, 1975)
3.33 (Gaffney et al., 1987)

Ionization potential (eV):
8.51 (Franklin et al., 1969)

Soil sorption coefficient, log K_{oc}:
1.72 (Batcombe silt loam, Briggs, 1981)
3.46 (fine sediments), 3.49 (coarse sediments) (Isaacson and Frink, 1984)
1.21 (Brookston clay loam soil, Campbell and Luthy, 1985)
2.40 (Meylan et al., 1992)
1.74 (Apison soil), 2.85 (Fullerton soil), 0.845 (Dormont soil) (Southworth and Keller, 1986)
1.24 (Boyd, 1982)
1.4 (river sediment), 1.6 (coal wastewater sediment) (Kopinke et al., 1995)

Octanol/water partition coefficient, log K_{ow}:
1.37, 1.51, 1.54, 1.75 (Sangster, 1989)
1.48 (Leo et al., 1971)
1.46 (Briggs, 1981; Fujita et al., 1964; Leo et al., 1971)
1.39 (Riederer, 1990)
1.55 (Garst and Wilson, 1984)
1.45 (Mirrlees et al., 1976; Wasik et al., 1981)

Solubility in organics:
Soluble in carbon disulfide and chloroform; very soluble in ether; miscible with carbon tetrachloride, hot benzene (U.S. EPA, 1985) and alcohol (Meites, 1963).

Solubility in water:
82,000 mg/L at 15 °C (Verschueren, 1983)

93,000 mg/L at 25 °C (Morrison and Boyd, 1971)
67,000 mg/L at 25 °C (Warner et al., 1987)
8.2 wt % at 20 °C (Stephen and Stephen, 1963)
0.866 wt % at 25 °C (Riddick et al., 1986)
84,120 mg/L (Reinbold et al., 1979)
0.90 M at 25 °C (Caturla et al., 1988)
74 g/L at 8 °C, 82 g/L at 25 °C (Leuenberger et al., 1985)
670 mg/L at 16 °C (Meites, 1963)
84 g/L (Price et al., 1974)
0.813 M at 25.0 °C (Wasik et al., 1981)
At 20 °C: 1.6, 1.35, 1.39 and 1.33 M in distilled water, Pacific seawater, artificial seawater and 35 wt % NaCl, respectively (Hashimoto et al., 1984)

Vapor pressure (10^2 mmHg):
20 at 20 °C, 100 at 40 °C (Verschueren, 1983)
35 at 25 °C (ACGIH, 1986)
6.4 at 8 °C, 34 at 25 °C (Leuenberger et al., 1985)
62 at 25 °C (Riederer, 1990)

Environmental Fate

Biological. Under methanogenic conditions, inocula from a municipal sewage treatment plant digester degraded phenol to carbon dioxide and methane (Young and Rivera, 1985). In a methanogenic enrichment culture, phenol anaerobically biodegraded to carbon dioxide and methane (yield 70%) (Healy and Young, 1979). Chloroperoxidase, a fungal enzyme isolated from *Caldariomyces fumago*, reacted with phenol forming 2- and 4-chlorophenol, the latter in a 25% yield (Wannstedt et al., 1990). In activated sludge, 41.4% mineralized to carbon dioxide after 5 days (Freitag et al., 1985). When phenol was statically incubated in the dark at 25 °C with yeast extract and settled domestic wastewater inoculum, significant biodegradation with rapid adaptation was observed. At concentrations of 5 and 10 mg/L, 96 and 97% biodegradation, respectively, were observed after 7 days (Tabak et al., 1981). Phenol is rapidly degraded in aerobically incubated soil but is much slower under anaerobic conditions (Baker and Mayfield, 1980).

Photolytic. In an aqueous, oxygenated solution exposed to artificial light (λ = 234 nm), phenol photolyzed to hydroquinone, catechol, 2,2'-, 2,4'- and 4,4'-dihydroxybiphenyl (Callahan et al., 1979). Titanium dioxide suspended in an aqueous solution and irradiated with UV light (λ = 365 nm) converted phenol to carbon dioxide at a significant rate (Matthews, 1986). When an aqueous solution containing potassium nitrate (10 mM) and phenol (1 mM) was irradiated with UV light (λ = 290-350 nm) up to a conversion of 10%, the following products formed: hydroxyhydroquinone, hydroquinone, resorcinol, hydroxybenzoquinone, benzoquinone, catechol, nitrosophenol, 4-nitrocatechol, nitrohydroquinone, 2- and 4-nitrophenol (Niessen et al., 1988).

Irradiation of phenol with UV light (λ = 254 nm) in the presence of oxygen yielded substituted biphenyls, hydroquinone and catechol (Joschek and Miller, 1966). A carbon dioxide yield of 32.5% was achieved when phenol adsorbed on silica gel was irradiated with light (λ >290 nm) for 17 hours (Freitag et al., 1985). When an aqueous solution containing phenol was photooxidized by UV light at 50 °C, 10.96% degraded to carbon dioxide after 24 hours (Knoevenagel and Himmelreich, 1976).

Chemical/Physical. In an environmental chamber, nitrogen trioxide (10,000 ppb) reacted quickly with phenol (concentration 200-1400 ppb) to form phenoxy radicals and nitric acid (Carter et al., 1981). The phenoxy radicals may react with oxygen and nitrogen dioxide to form quinones and nitrohydroxy derivatives, respectively (Nielsen et al., 1983).

Anticipated products from the reaction of phenol with ozone or hydroxyl radicals in the atmosphere are dihydroxybenzenes, nitrophenols and ring cleavage products (Cupitt, 1980). Groundwater contaminated with various phenols degraded in a methanogenic aquifer. Similar results were obtained in the laboratory utilizing an anaerobic digester. Methane and carbon dioxide were reported as degradation products (Godsy et al., 1983).

Ozonization of phenol in water resulted in the formation of many oxidation products. The identified products in the order of degradation are catechol, hydroquinone, *o*-quinone, *cis,cis*-muconic acid, maleic (or fumaric) and oxalic acids (Eisenhauer, 1968). In addition, glyoxylic, formic and acetic acids also were reported as ozonization products prior to further oxidation to carbon dioxide (Kuo et al., 1977). Ozonation of an aqueous solution of phenol subjected to UV light (120 watt low pressure mercury lamp) gave glyoxal, glyoxylic, oxalic and formic acids as major products. Minor products included catechol, hydroquinone, muconic, fumaric and maleic acids (Takahashi, 1990). Wet oxidation of phenol at 320 °C yielded formic and acetic acids (Randall and Knopp, 1980).

Chlorination of water containing bromide ions converted phenol to 2,4,6-tribromophenol. Bromodichlorophenol, dibromochlorophenol and tribromophenol have also been reported to form from the chlorination of natural water under simulated conditions (Watanabe et al., 1984).

Wet oxidation of phenol at elevated pressure and temperature gave the following products: acetone, acetaldehyde, formic, acetic, maleic, oxalic and succinic acids (Keen and Baillod, 1985). Chlorine dioxide reacted with phenol in an aqueous solution forming *p*-benzoquinone and hypochlorous acid (Wajon et al., 1982).

Kanno et al. (1982) studied the aqueous reaction of phenol and other substituted aromatic hydrocarbons (aniline, toluidine, 1-naphthylamine, cresol, pyrocatechol, resorcinol, hydroquinone and 1-naphthol) with hypochlorous acid in the presence of ammonium ion. They reported that the aromatic ring was not chlorinated as expected but was cleaved by chloramine forming cyanogen chloride (Kanno et al., 1982). The amount of cyanogen chloride formed increased at lower pHs. At pH 6, the greatest amount of cyanogen chloride was formed when the reaction mixture contained ammonium ion and hypochlorous acid at a ratio of 2:3 (Kanno et al., 1982).

Spacek et al. (1995) investigated the photodegradation of phenol using titanium dioxide-UV light and Fenton's reagent (hydrogen peroxide:substance - 10:1; Fe^{2+} 2.5 x 10^{-4} mol/L) at 25 °C. The decomposition rate of 4-nitroaniline was very high by the photo-Fenton reaction in comparison to titanium dioxide-UV light (λ = 365 nm). Decomposition products identified in both reactions were p-benzoquinone, hydroquinone and oxalic acid.

Reacts with sodium and potassium hydroxide forming sodium and potassium phenolate, respectively.

Exposure Limits: NIOSH REL: TWA 5 ppm (19 mg/m^3), 15-minute ceiling 15.6 ppm (60 mg/m^3), IDLH 250 ppm; OSHA PEL: 5 ppm; ACGIH TLV: TWA 5 ppm (19 mg/m^3).

Symptoms of Exposure: Ingestion of 5-10 mg may cause death. Toxic symptoms include nausea, vomiting, weakness, cyanosis, tremor, convulsions, kidney and liver damage. Eye, nose and throat irritant. Burns skin on contact and may cause dermatitis (Patnaik, 1992).

Drinking Water Standard: As of May 1994, no MCLGs or MCLs have been proposed (U.S. EPA, 1984).

Toxicity: Acute oral LD_{50} for mice 270 mg/kg, rats 317 mg/kg; LC_{50} (inhalation) for mice 177 mg/m^3 (RTECS, 1985).

Uses: Antiseptic and disinfectant; pharmaceuticals; dyes; indicators; slimicide; phenolic resins; epoxy resins (bisphenol-A); nylon-6 (caprolactum); 2,4-D; solvent for refining lubricating oils; preparation of adipic acid, salicylic acid, phenol-phthalein, pentachlorophenol, acetophenetidin, picric acid, anisole, phenoxyacetic acid, phenyl benzoate, 2-phenolsulfonic acid, 4-phenolsulfonic acid, 2-nitro-phenol, 4-nitrophenol, 2,4,6-tribromophenol, 4-bromophenol, 4-t-butylphenol, salicylaldehyde and many other organic compounds; germicidal paints; laboratory reagent.

p-PHENYLENEDIAMINE

Synonyms: 4-Aminoaniline; *p*-Aminoaniline; BASF ursol D; **1,4-Benzenediamine**; *p*-Benzenediamine; Benzofur D; C.I. 76060; C.I. developer 13; C.I. oxidation base 10; Developer 13; Developer PF; 1,4-Diaminobenzene; *p*-Diaminobenzene; Durafur black R; Fouramine D; Fourrine 1; Fourrine D; Fur black 41867; Fur brown 41866; Furro D; Fur yellow; Futramine D; Nako H; Orsin; Oxidation base 10; Para; Pelagol D; Pelagol DR; Pelagol grey D; Peltol D; 1,4-Phenylenediamine; PPD; Renal PF; Santoflex IC; Tertral D; UN 1673; Ursol D; USAF EK-394; Vulkanox 4020; Zoba black D.

CAS Registry No.: 106-50-3
DOT: 1673
DOT label: Poison
Molecular formula: $C_6H_8N_2$
Formula weight: 108.14
RTECS: SS8050000

Physical state and color
White, red or brown crystals.

Melting point (°C):
140 (Weast, 1986)
145-147 (Windholz et al., 1983)

Boiling point (°C):
267 (Weast, 1986)

Diffusivity in water (10^5 cm^2/sec):
0.78 at 20 °C using method of Hayduk and Laudie (1974).

Dissociation constant (25 °C):
$pK_1 = 3.29$, $pK_2 = 6.08$ (Dean, 1973)

Flash point (°C):
156.8 (NIOSH, 1994)

Ionization potential (eV):
6.89 (NIOSH, 1994)

Bioconcentration factor, log BCF:
2.66 (activated sludge), 2.65 (algae) (hydrochloride, Freitag et al., 1985)

Soil sorption coefficient, log K_{oc}:
Unavailable because experimental methods for estimation of this parameter for aromatic amines are lacking in the documented literature.

Octanol/water partition coefficient, log K_{ow}:
Unavailable because experimental methods for estimation of this parameter for aromatic amines are lacking in the documented literature.

Solubility in organics:
Soluble in alcohol, chloroform and ether (Weast, 1986).

Solubility in water:
4.7 wt % at 20 °C (NIOSH, 1987)
38,000 mg/L at 24 °C, 6,690 g/L at 107 °C (Verschueren, 1983)

Vapor pressure (mmHg):
<1 (NIOSH, 1994)

Environmental Fate
   Biological. In activated sludge, 3.8% mineralized to carbon dioxide after 5 days (Freitag et al., 1985).
   Photolytic. A carbon dioxide yield of 53.7% was achieved when phenylenediamine (presumably an isomeric mixture) adsorbed on silica gel was irradiated with light (λ >290 nm) for 17 hours (Freitag et al., 1985).

Exposure Limits (mg/m^3): NIOSH REL: TWA 0.1, IDLH 25; OSHA PEL: TWA 0.1.

Symptoms of Exposure: May include vertigo, gastritis, jaundice, allergic asthma, dermatitis, cornea ulcer, eye burn (Patnaik, 1992).

Toxicity: Acute oral LD_{50} for wild birds 100 mg/kg, quail 100 mg/kg, rats 80 mg/kg (RTECS, 1985); LC_{50} (48-hour) for red killifish 186 mg/L (Yoshioka et al., 1986).

Uses: Manufacturing azo dyes, intermediates for antioxidants and accelerators for rubber; photochemical measurements; laboratory reagent; dyeing hair and fur.

PHENYL ETHER

Synonyms: Biphenyl ether; Biphenyl oxide; Diphenyl ether; Diphenyl oxide; Geranium crystals; **1,1-Oxybisbenzene**; Phenoxybenzene.

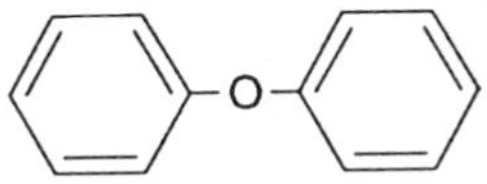

CAS Registry No.: 101-84-8
Molecular formula: $C_{12}H_{10}O$
Formula weight: 170.21
RTECS: KN8970000

Physical state, color and odor
Colorless solid or liquid with a geranium-like odor. Odor threshold in air is 1.2 ppb (Amoore and Hautala, 1983).

Melting point (°C):
28 (Verschueren, 1983)
26.87 (Riddick et al., 1986)

Boiling point (°C):
257.9 (Weast, 1986)

Density (g/cm³):
1.0748 at 20/4 °C (Weast, 1986)

Diffusivity in water (10^5 cm²/sec):
0.67 at 20 °C using method of Hayduk and Laudie (1974).

Flash point (°C):
116 (NIOSH, 1994)

Lower explosive limit (%):
0.7 (NIOSH, 1994)

Upper explosive limit (%):
6.0 (NIOSH, 1994)

Heat of fusion (kcal/mol):
4.115 (Dean, 1987)

Henry's law constant (10^4 atm·m^3/mol):
2.13 at 20 °C (approximate - calculated from water solubility and vapor pressure)

Ionization potential (eV):
8.82 ± 0.05 (Franklin et al., 1969)
8.09 (NIOSH, 1994)

Soil sorption coefficient, log K_{oc}:
Unavailable because experimental methods for estimation of this parameter for aromatic ethers are lacking in the documented literature.

Octanol/water partition coefficient, log K_{ow}:
4.21, 4.36 (Leo et al., 1971)
3.79 (Burkhard et al., 1985a)
4.08 (Banerjee et al., 1980)
4.20 (Chiou et al., 1977)

Solubility in organics:
Soluble in acetic acid, alcohol, benzene and ether (Weast, 1986).

Solubility in water:
106 μmol/L at 25 °C (Banerjee et al., 1980)
21 ppm at 25 °C (Chiou et al., 1977)

Vapor pressure (mmHg):
0.02 at 20 °C, 0.12 at 30 °C (Verschueren, 1983)

Exposure Limits: NIOSH REL: TWA 1 ppm (7 mg/m^3), IDLH 100 ppm; OSHA PEL: TWA 1 ppm.

Symptoms of Exposure: Mild skin irritant. Ingestion may cause liver and kidney damage (Patnaik, 1992).

Toxicity: Acute oral LD_{50} for rats 3,370 mg/kg (RTECS, 1985).

Uses: Heat transfer liquid; perfuming soaps; resins for laminated electrical insulation; organic synthesis.

PHENYLHYDRAZINE

Synonyms: Hydrazine-benzene; Hydrazinobenzene; UN 2562.

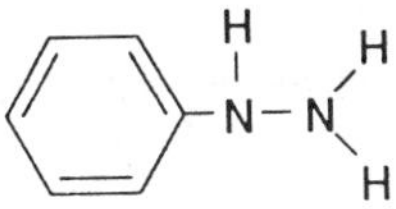

CAS Registry No.: 100-63-0
DOT: 2572
DOT label: Poison
Molecular formula: $C_6H_8N_2$
Formula weight: 108.14
RTECS: MV8925000

Physical state, color and odor
Yellow monoclinic crystals or oil with a faint, aromatic odor. Turns reddish-brown on exposure to air.

Melting point (°C):
19.8 (Weast, 1986)

Boiling point (°C):
243 (Weast, 1986) with decomposition (Windholz et al., 1983).

Density (g/cm³):
1.0986 at 20/4 °C (Weast, 1986)

Diffusivity in water (10^5 cm²/sec):
0.89 at 20 °C using method of Hayduk and Laudie (1974).

Dissociation constant:
8.79 at 15 °C (Windholz et al., 1983)
5.20 at 25 °C (Dean, 1973)

Flash point (°C):
88.5 (NIOSH, 1994)

Ionization potential (eV):
7.64 (NIOSH, 1994)

Soil sorption coefficient, log K_{oc}:
Unavailable because experimental methods for estimation of this parameter for hydrazines are lacking in the documented literature.

Octanol/water partition coefficient, log K_{ow}:
1.25 (Leo et al., 1971)

Solubility in organics:
Soluble in acetone (Weast, 1986). Miscible with alcohol, benzene, chloroform and ether (Windholz et al., 1983).

Vapor density:
4.42 g/L at 25 °C, 3.73 (air = 1)

Vapor pressure (mmHg):
0.04 at 25 °C (NIOSH, 1994)

Exposure Limits: Potential occupational carcinogen. NIOSH REL: 2-hour ceiling 0.14 ppm (0.6 mg/m^3), IDLH 15 ppm; OSHA PEL: TWA 5 ppm (22 mg/m^3); ACGIH TLV: TWA 5 ppm.

Symptoms of Exposure: Acute toxic symptoms include hematuria, vomiting, convulsions and respiratory arrest (Patnaik, 1992).

Toxicity: Acute oral LD_{50} for guinea pigs 80 mg/kg, rats 188 mg/kg, rabbits 80 mg/kg (RTECS, 1985).

Uses: Reagent for aldehydes, ketones, sugars; manufacturing dyes and antipyrine; organic synthesis.

PHTHALIC ANHYDRIDE

Synonyms: 1,2-Benzenedicarboxylic acid anhydride; 1,3-Dioxophthalan; ESEN; 1,3-Dihydro-1,3-dioxoisobenzofuran; **1,3-Isobenzofurandione;** NCI-C03601; Phthalandione; 1,3-Phthalandione; Phthalic acid anhydride; RCRA waste number U190; Retarder AK; Retarder ESEN; Retarder PD; UN 2214.

CAS Registry No.: 85-44-9
DOT: 2214
Molecular formula: $C_8H_4O_3$
Formula weight: 148.12
RTECS: TI3150000

Physical state, color and odor
White to pale cream crystals with a characteristic, choking odor. Odor threshold in air is 53 ppb (Amoore and Hautala, 1983).

Melting point (°C):
131.6 (Weast, 1986)
130.8 (Windholz et al., 1983)

Boiling point (°C):
Sublimes at 295 (Weast, 1986).

Density (g/cm³):
1.527 at 4/4 °C (Verschueren, 1983)

Flash point (°C):
152.9 (NIOSH, 1994)

Lower explosive limit (%):
1.7 (NIOSH, 1994)

Upper explosive limit (%):
10.5 (NIOSH, 1994)

Henry's law constant (10^9 atm·m³/mol):
6.29 at 20 °C (approximate - calculated from water solubility and vapor pressure)

Soil sorption coefficient, log K_{oc}:
1.90 using method of Chiou et al. (1979).

Octanol/water partition coefficient, log K_{ow}:
-0.62 (Kenaga and Goring, 1980)

Solubility in organics:
One part in 125 parts carbon disulfide (Windholz et al., 1983).
At 25 °C (mol/L): acetic acid (0.27), acetone (1.15), *t*-butyl alcohol (0.04), chloroform (0.78), cyclohexane (0.0064), decane (0.0049), di-*n*-propyl ether (0.04), dodecane (0.0048), ethyl ether (0.116), heptane (0.0049), hexane (0.0050), hexadecane (0.0047), isooctane (0.00422), pyridine (2.9) and tetrahydrofuran (1.2) (Fung and Higuchi, 1971)

Solubility in water:
0.6 wt % at 20 °C (NIOSH, 1994)

Vapor pressure (10^4 mmHg):
2 at 20 °C, 10 at 30 °C (Verschueren, 1983)

Environmental Fate
  Chemical/Physical. Reacts with water to form phthalic acid (Windholz et al., 1983). Pyrolysis of phthalic anhydride in the presence of polyvinyl chloride at 600 °C gave the following compounds: biphenyl, fluorene, benzophenone, 9-fluorenone, *o*-terphenyl, 9-phenylfluorene and three unidentified compounds (Bove and Dalven, 1984).

Exposure Limits: NIOSH REL: TWA 6 mg/m³ (1 ppm), IDLH 60 mg/m³; OSHA PEL: TWA 12 mg/m³ (2 ppm); ACGIH TLV: TWA 6 mg/m³.

Toxicity: Acute oral LD_{50} for mice 2 gm/kg, rats 4,020 mg/kg (RTECS, 1985).

Uses: Manufacturing phthalates, phthaleins, benzoic acid, synthetic indigo, pharmaceuticals, insecticides, chlorinated products and artificial resins.

PICRIC ACID

Synonyms: Carbazotic acid; C.I. 10305; 2-Hydroxy-1,3,5-trinitrobenzene; Lyddite; Melinite; Nitroxanthic acid; Pertite; Phenol trinitrate; Picronitric acid; Shimose; 1,3,5-Trinitrophenol; **2,4,6-Trinitrophenol**; UN 0154.

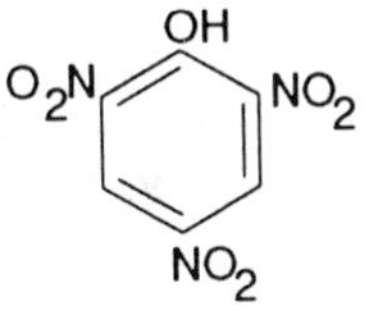

CAS Registry No.: 88-89-1
DOT: 1344
DOT label: Class A explosive
Molecular formula: $C_6H_3N_3O_7$
Formula weight: 229.11
RTECS: TJ7875000

Physical state, color and odor
Colorless to yellow, odorless liquid or crystals.

Melting point (°C):
122-123 (Weast, 1986)

Boiling point (°C):
Sublimes and explodes >300 °C (Weast, 1986).

Density (g/cm^3):
1.763 (Windholz et al., 1983)

Diffusivity in water (10^5 cm^2/sec):
0.72 at 20 °C using method of Hayduk and Laudie (1974).

Dissociation constant:
0.29 at 25 °C (Dean, 1973)

Flash point (°C):
151 (NIOSH, 1994)

Octanol/water partition coefficient, log K_{ow}:
2.03 (Verschueren, 1983)
1.34 (Hansch and Anderson, 1967)

Solubility in organics:
In g/L: alcohol (83.3), benzene (100), chloroform (28.57) and ether (15.38)

(Windholz et al., 1983).

At 25 °C (mmol/L): cyclohexane (0.41), decane (0.42), dodecane (0.33), heptane (0.33), hexane (0.30), hexadecane (0.39) and isooctane (0.25) (Fung and Higuchi, 1971).

Solubility in water (mg/L):
66,670 at 100 °C (Windholz et al., 1983)
14,000 at 20 °C, 68,000 at 100 °C (Verschueren, 1983)

Vapor pressure (mmHg):
1 at 197 °C (NIOSH, 1994)

Environmental Fate
   Chemical / Physical. Picric acid explodes when heated >300 °C (Weast, 1986). Shock sensitive! (Keith and Walters, 1992).

Exposure Limits (mg/m^3): NIOSH REL: TWA 0.1, STEL 0.3, IDLH 75; OSHA PEL: TWA 0.1.

Symptoms of Exposure: Contact with skin may cause sensitization dermatitis. Ingestion may cause severe poisoning. Toxic symptoms include headache, nausea, vomiting, abdominal pain and yellow coloration of skin (Patnaik, 1992).

Toxicity: In rabbits, the lethal dose is 120 mg/kg (RTECS, 1985); LC_{50} (48-hour) for red killifish 513 mg/L (Yoshioka et al., 1986).

Uses: Explosives, matches; electric batteries; in leather industry; manufacturing colored glass; etching copper; textile mordant; reagent.

PINDONE

Synonyms: Chemrat; **2-(2,2-Dimethyl-1-oxopropyl)-1H-indene-1,3(2H)-dione;** Pivacin; Pival; 2-Pivaloylindane-1,3-dione; 2-Pivaloyl-1,3-indanedione; Pivalyl; Pivalyl indandione; 2-Pivalyl-1,3-indandione; Pivalyl valone; Tri-Ban; UN 2472.

CAS Registry No.: 83-26-1
DOT: 2472
Molecular formula: $C_{14}H_{14}O_3$
Formula weight: 230.25
RTECS: NK6300000

Physical state and color
Bright yellow crystals.

Melting point (°C):
108-110 (Windholz et al., 1983)

Diffusivity in water (10^5 cm^2/sec):
0.51 at 20 °C using method of Hayduk and Laudie (1974).

Soil sorption coefficient, log K_{oc}:
2.95 using method of Kenaga and Goring (1980).

Octanol/water partition coefficient, log K_{ow}:
3.18 using method of Kenaga and Goring (1980).

Solubility in organics:
Soluble in most organic solvents (Worthing and Hance, 1991).

Solubility in water:
18 ppm at 25 °C (Gunther et al., 1968)

Exposure Limits (mg/m^3): NIOSH REL: TWA 0.1, IDLH 100; OSHA PEL: TWA 0.1.

Toxicity: Acute oral LD$_{50}$ for dogs 75 mg/kg, rats 280 mg/kg (RTECS, 1985).

Uses: Insecticide; rodenticide; intermediate in manufacturing pharmaceuticals.

PROPANE

Synonyms: Dimethylmethane; Propyl hydride; UN 1075; UN 1978.

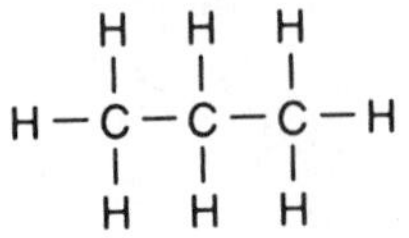

```
      H   H   H
      |   |   |
  H — C — C — C — H
      |   |   |
      H   H   H
```

CAS Registry No.: 74-98-6
DOT: 1978
DOT label: Flammable gas
Molecular formula: C_3H_8
Formula weight: 44.10
RTECS: TX2275000

Physical state, color and odor
Colorless gas. Odor threshold in air is 16,000 ppm (Amoore and Hautala, 1983).

Melting point (°C):
-189.7 (Weast, 1986)
-187.7 (Dean, 1987)

Boiling point (°C):
-42.1 (Weast, 1986)

Density (g/cm³):
0.5843 at -45/4 °C (Weast, 1986)

Diffusivity in water (10^5 cm²/sec):
1.16 at 25 °C (Hayduk and Laudie, 1974)

Dissociation constant:
≈ 44 (Gordon and Ford, 1972)

Flash point (°C):
-105 (Hawley, 1981)

Lower explosive limit (%):
2.1 (NIOSH, 1994)

Upper explosive limit (%):
9.5 (NIOSH, 1994)

Heat of fusion (kcal/mol):
0.842 (Dean, 1987)

Henry's law constant (atm·m³/mol):
0.706 at 25 °C (Hine and Mookerjee, 1975)

Ionization potential (eV):
11.07 (NIOSH, 1994)

Soil sorption coefficient, log K_{oc}:
Unavailable because experimental methods for estimation of this parameter for aliphatic hydrocarbons are lacking in the documented literature.

Octanol/water partition coefficient, log K_{ow}:
2.36 (Hansch et al., 1975)

Solubility in organics:
In vol %: alcohol (790 at 16.6 °C and 754 mmHg), benzene (1,452 at 21.5 °C and 757 mmHg), chloroform (1,299 at 21.6 °C and 757 mmHg), ether (926 at 16.6 °C and 757 mmHg) and turpentine (1,587 at 17.7 °C and 757 mmHg) (Windholz et al., 1983).

Solubility in water:
62.4 mg/kg at 25 °C (McAuliffe, 1963, 1966)
6.5 vol % at 17.8 °C and 753 mmHg (Windholz et al., 1983)
At 0 °C, 0.0394 volumes dissolved in a unit volume of water at 19.8 °C (Claussen and Polglase, 1952).
1.50 mmol/L at 25 °C (Barone et al., 1966)
3.46, 1.53 and 0.84 mmol/L at 4, 25 and 50 °C, respectively (Kresheck et al., 1965)

Vapor density:
2.0200 g/L at 0 °C, 1.8324 g/L at 25 °C (Windholz et al., 1983)
1.52 (air = 1)

Vapor pressure (mmHg):
6,384 at 21 °C (NIOSH, 1994)
6,460 at 20 °C, 8,360 at 30 °C (Verschueren, 1983)

Environmental Fate
Photolytic. Synthetic air containing gaseous nitrous acid and exposed to artificial sunlight (λ = 300-450 nm) photooxidized propane to acetone with a yield of 56% (Cox et al., 1980).
Biological. In the presence of methane, *Pseudomonas methanica* degraded

propane to 1-propanol, propionic acid and acetone (Leadbetter and Foster, 1959). The presence of carbon dioxide was required for "*Nocardia paraffinicum*" to degrade propane to propionic acid (MacMichael and Brown, 1987). Propane may biodegrade in two pathways. The first is the formation of propyl hydroperoxide, which decomposes to propanol followed by oxidation to propanoic acid. The other pathway involves dehydrogenation to 1-propene, which may react with water giving propanol (Dugan, 1972). Microorganisms can oxidize alkanes under aerobic conditions (Singer and Finnerty, 1984). The most common degradative pathway involves the oxidation of the terminal methyl group forming the corresponding alcohol (propyl alcohol). The alcohol may undergo a series of dehydrogenation steps forming an aldehyde (propionaldehyde), then a fatty acid (propionic acid). The fatty acid may then be metabolized by β-oxidation to form the mineralization products carbon dioxide and water (Singer and Finnerty, 1984).

Chemical/Physical. Incomplete combustion of propane in the presence of excess hydrogen chloride resulted in a high number of different chlorinated compounds including, but not limited, to alkanes, alkenes, monoaromatics, alicyclic hydrocarbons and polynuclear aromatic hydrocarbons. Without hydrogen chloride, 13 non-chlorinated polynuclear aromatic hydrocarbons were formed (Eklund et al., 1987).

Exposure Limits: NIOSH REL: TWA 1,000 ppm (1,800 mg/m^3), IDLH 2,100 ppm; OSHA PEL: TWA 1,000 ppm (1,800 mg/m^3).

Symptoms of Exposure: An asphyxiate. Narcotic at high concentrations (Patnaik, 1992).

Uses: Organic synthesis; refrigerant; fuel gas; manufacture of ethylene; solvent; extractant; aerosol propellant; mixture for bubble chambers.

1-PROPANOL

Synonyms: Ethyl carbinol; 1-Hydroxypropane; Optal; Osmosol extra; Propanol; 1-Propanol; *n*-Propanol; Propyl alcohol; 1-Propyl alcohol; *n*-Propyl alcohol; Propylic alcohol; UN 1274.

CAS Registry No.: 71-23-8
DOT: 1274
DOT label: Combustible liquid
Molecular formula: C_3H_8O
Formula weight: 60.10
RTECS: UH8225000

Physical state, color and odor
Colorless liquid with a mild, alcohol-like odor. Odor threshold in air is 2.6 ppm (Amoore and Hautala, 1983).

Melting point (°C):
-126.5 (Weast, 1986)

Boiling point (°C):
97.4 (Weast, 1986)

Density (g/cm^3):
0.8035 at 20/4 °C (Weast, 1986)
0.8053 at 20/4 °C, 0.8016 at 25/4 °C (Windholz et al., 1983)

Diffusivity in water (10^5 cm^2/sec):
1.12 at 25 °C (Hayduk and Laudie, 1974)

Dissociation constant:
>14 (Schwarzenbach et al., 1993)

Flash point (°C):
22 (NIOSH, 1994)
25 (open cup, Hawley, 1981)

Lower explosive limit (%):
2.2 (NIOSH, 1994)

Upper explosive limit (%):
13.7 (NIOSH, 1994)

Heat of fusion (kcal/mol):
1.242 (Dean, 1987)

Henry's law constant (10^6 atm·m^3/mol at 25 °C):
6.12 (Burnett, 1963)
6.85 (Butler et al., 1935)
6.74 (Hine and Mookerjee, 1975)
7.41 (Snider and Dawson, 1985)

Ionization potential (eV):
10.1 (Franklin et al., 1969)
10.15 (NIOSH, 1994)

Soil sorption coefficient, log K_{oc}:
0.48 (Gerstl and Helling, 1987)

Octanol/water partition coefficient, log K_{ow}:
0.34 (Hansch and Anderson, 1967)
0.25 (Hansch and Leo, 1979)

Solubility in organics:
Soluble in acetone and benzene (Weast, 1986). Miscible with alcohol and ether (Windholz et al., 1983).

Solubility in water:
Miscible (Palit, 1947). A saturated solution in equilibrium with its own vapor had a concentration of 250.4 g/L at 25 °C (Kamlet et al., 1987).

Vapor density:
2.46 g/L at 25 °C, 2.07 (air = 1)

Vapor pressure (mmHg):
14.5 at 20 °C, 20.8 at 25 °C, 27 at 30 °C (Verschueren, 1983)

Exposure Limits: NIOSH REL: TWA 200 ppm (500 mg/m^3), STEL 250 ppm (625 mg/m^3), IDLH 800 ppm; OSHA PEL: TWA 200 ppm.

Symptoms of Exposure: Ingestion causes headache, drowsiness, abdominal cramps, gastrointestinal pain, nausea and diarrhea. May irritate eyes on contact (Patnaik, 1992).

Toxicity: Acute oral LD_{50} for mice 6,800 mg/kg, rats 1,870 mg/kg (RTECS, 1985); LC_{50} for red killifish 83.2 g/L (Yoshioka et al., 1980).

Uses: Solvent for cellulose esters and resins; in manufacturing of printing inks, nail polishes, polymerization and spinning of acrylonitrile, dyeing wool, polyvinyl chloride adhesives, esters, waxes, vegetable oils; brake fluids; solvent degreasing; antiseptic; organic synthesis.

β-PROPIOLACTONE

Synonyms: Betaprone; BPL; Hydracrylic acid β-lactone; 3-Hydroxy-propionic acid lactone; β-Lactone; NSC 21626; **2-Oxetanone**; Propanolide; Propiolactone; 1,3-Propiolactone; 3-Propiolactone; β-Propionolactone; 3-Propionolactone; β-Proprolactone.

CAS Registry No.: 57-57-8
Molecular formula: C_3H_6O
Formula weight: 72.06
RTECS: RQ7350000

Physical state, color and odor
Colorless liquid with a sweet, but irritating odor.

Melting point (°C):
-33.4 (Weast, 1986)

Boiling point (°C):
Decomposes at 162 (Weast, 1986).
155 (commercial grade, James and Wellington, 1969)

Density (g/cm^3):
1.1460 at 20/5 °C (Weast, 1986)
1.1425 at 25/4 °C (Windholz et al., 1983)

Diffusivity in water (10^5 cm^2/sec):
1.16 at 20 °C using method of Hayduk and Laudie (1974).

Flash point (°C):
75 (NIOSH, 1994)
70 (Dean, 1987)

Lower explosive limit (%):
2.9 (NIOSH, 1994)

Henry's law constant (10^7 atm·m^3/mol):
7.6 at 25 °C (approximate - calculated from water solubility and vapor pressure)

Soil sorption coefficient, log K_{oc}:
Unavailable because experimental methods for estimation of this parameter for lactones are lacking in the documented literature.

Octanol/water partition coefficient, log K_{ow}:
Unavailable because experimental methods for estimation of this parameter for lactones are lacking in the documented literature.

Solubility in organics:
Miscible with acetone, alcohol, chloroform and ether (Windholz et al., 1983).

Solubility in water:
37 wt % at 25 °C (NIOSH, 1994)

Vapor density:
2.95 g/L at 25 °C, 2.49 (air = 1)

Vapor pressure (mmHg):
3 at 25 °C (NIOSH, 1994)

Environmental Fate
   Chemical/Physical. Slowly hydrolyzes to hyracrylic acid (Windholz et al., 1983). In a reactor heated to 250 °C and a pressure of 12 mmHg, β-propiolactone decomposed to give equal amounts of ethylene and carbon dioxide (James and Wellington, 1969).

Exposure Limits: Potential occupational carcinogen. ACGIH TLV: TWA 0.5 ppm.

Toxicity: LC_{50} (inhalation) for rats 25 ppm/6-hour (RTECS, 1985).

Uses: Organic synthesis; disinfectant; vapor sterilant.

n-PROPYL ACETATE

Synonyms: Acetic acid propyl ester; Acetic acid *n*-propyl ester; 1-Acetoxy-propane; Propyl acetate; 1-Propyl acetate; UN 1276.

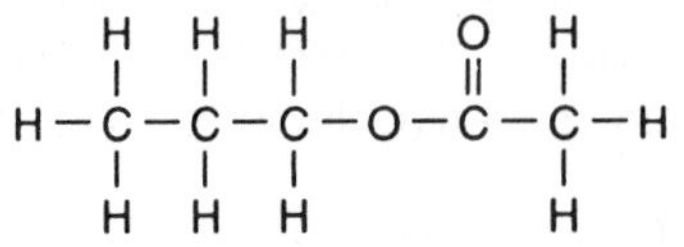

$$H-\overset{\displaystyle H}{\underset{\displaystyle H}{\overset{|}{\underset{|}{C}}}}-\overset{\displaystyle H}{\underset{\displaystyle H}{\overset{|}{\underset{|}{C}}}}-\overset{\displaystyle H}{\underset{\displaystyle H}{\overset{|}{\underset{|}{C}}}}-O-\overset{\displaystyle O}{\overset{||}{C}}-\overset{\displaystyle H}{\underset{\displaystyle H}{\overset{|}{\underset{|}{C}}}}-H$$

CAS Registry No.: 109-60-4
DOT: 1276
DOT label: Flammable liquid
Molecular formula: $C_5H_{10}O_2$
Formula weight: 102.12
RTECS: AJ3675000

Physical state, color and odor
Clear, colorless liquid with a pleasant, pear-like odor. Odor threshold in air is 670 ppb (Amoore and Hautala, 1983).

Melting point (°C):
-95 (Weast, 1986)
-92 (Verschueren, 1983)

Boiling point (°C):
101.6 (Weast, 1986)

Density (g/cm³ at 20/4 °C):
0.8878 (Weast, 1986)
0.836 (Windholz et al., 1983)

Diffusivity in water (10^5 cm²/sec):
0.81 at 20 °C using method of Hayduk and Laudie (1974).

Flash point (°C):
13 (NFPA, 1984)

Lower explosive limit (%):
1.7 at 38 °C (NFPA, 1984)

Upper explosive limit (%):
8 (NFPA, 1984)

Henry's law constant (10^4 atm·m^3/mol):
1.99 at 25 °C (Hine and Mookerjee, 1975)

Ionization potential (eV):
10.04 ± 0.03 (Franklin et al., 1969)

Soil sorption coefficient, log K_{oc}:
Unavailable because experimental methods for estimation of this parameter for aliphatic esters are lacking in the documented literature.

Octanol/water partition coefficient, log K_{ow}:
1.24 (Tewari et al., 1982; Wasik et al., 1981)

Solubility in organics:
Miscible with alcohol, ether (Windholz et al., 1983), hydrocarbons and ketones (Hawley, 1981).

Solubility in water:
18,900 mg/L at 20 °C, 26,000 mg/L at 20 °C (Verschueren, 1983)
16 g/L at 16 °C (Windholz et al., 1983)
0.200 M at 25.0 °C (Tewari et al., 1982; Wasik et al., 1981)
10.3 g/L in artificial seawater at 25 °C (Price et al., 1974)
0.185 M at 20 °C (Fühner, 1924)
In wt % (°C): 3.21 (0), 2.78 (9.5), 2.26 (20.0), 1.98 (30.0), 1.87 (40.0), 1.72 (50.0),
    1.66 (80.0), 1.35 (90.2) (Stephenson and Stuart, 1986)
22.6 g/kg at 25 °C (Butler and Ramchandani, 1935)

Vapor density:
4.17 g/L at 25 °C, 3.53 (air = 1)

Vapor pressure (mmHg):
25 at 20 °C, 35 at 25 °C, 42 at 30 °C (Verschueren, 1983)
33 at 25 °C (Butler and Ramchandani, 1935)

Environmental Fate
   Chemical/Physical. Slowly hydrolyzes in water forming acetic acid and propyl alcohol.

Exposure Limits: NIOSH REL: TWA 200 ppm (840 mg/m^3), STEL 250 ppm (1,050 mg/m^3), IDLH 1,700 ppm; OSHA PEL: TWA 200 ppm; ACGIH TLV: TWA 200 ppm, STEL 250 ppm.

Symptoms of Exposure: Vapors may irritate eyes, nose, throat and cause narcotic effects. Ingestion may cause narcotic action. At high concentrations death may occur (Patnaik, 1992).

Toxicity: Acute oral LD_{50} in mice 8,300 mg/kg, rats 9,370, rabbits 6,640 mg/kg (Patnaik, 1992).

Uses: Manufacture of flavors and perfumes; solvent for plastics, cellulose products and resins; lacquers, paints; natural and synthetic resins; lab reagent; organic synthesis.

n-PROPYLBENZENE

Synonyms: Isocumene; 1-Phenylpropane; **Propylbenzene**; UN 2364.

CAS Registry No.: 103-65-1
DOT: 2364
DOT label: Combustible liquid
Molecular formula: C_9H_{12}
Formula weight: 120.19
RTECS: DA8750000

Physical state and color
Colorless liquid.

Melting point (°C):
-99.5 (Weast, 1986)
-101.75 (Mackay et al., 1982)

Boiling point (°C):
159.2 (Weast, 1986)

Density (g/cm^3):
0.8620 at 20/4 °C (Weast, 1986)

Diffusivity in water (10^5 cm^2/sec):
0.7368 at 20 °C using method of Hayduk and Laudie (1974).

Dissociation constant:
>14 (Schwarzenbach et al., 1993)

Flash point (°C):
30 (Hawley, 1981)

Lower explosive limit (%):
0.8 (Sax and Lewis, 1987)

Upper explosive limit (%):
6 (Sax and Lewis, 1987)

Heat of fusion (kcal/mol):
2.03-2.215 (Dean, 1987)

Henry's law constant (10^3 atm·m^3/mol):
10 at 25 °C (Hine and Mookerjee, 1975)
5.68, 7.31, 8.81, 10.8 and 13.7 at 10, 15, 20, 25 and 30 °C, respectively (Ashworth et al., 1988)
4.35, 6.21, 8.37, 11.6 and 15.3 at 10, 15, 20, 25 and 30 °C, respectively (Perlinger et al., 1993)

Interfacial tension with water (dyn/cm at 20 °C):
38.5 (Demond and Lindner, 1993)

Ionization potential (eV):
9.14 (Yoshida et al., 1983)
8.72 (HNU, 1986)

Soil sorption coefficient, log K_{oc}:
2.87 (estuarine sediment, Vowles and Mantoura, 1987)

Octanol/water partition coefficient, log K_{ow}:
3.57 (Leo et al., 1971)
3.68 (Hansch et al., 1968)
3.60 (Galassi et al., 1988)
3.69 (Schantz and Martire, 1987; Tewari et al., 1982; Wasik et al., 1981, 1983)
3.72 (DeVoe et al., 1981; Camilleri et al., 1988)
3.44 (Nahum and Horvath, 1980)
3.71, 3.72, 3.73 at 23 °C (Wasik et al., 1981)

Solubility in organics:
Soluble in acetone and benzene (Weast, 1986). Miscible with alcohol and ether (Sax and Lewis, 1987).

Solubility in water:
360 μmol/L at 25 °C (Andrews and Keefer, 1950a)
60.24 mg/L at 25 °C (Chiou et al., 1982)
447, 435, 452, 443 and 437 μmol/L at 10.0, 15.0, 20.0, 25.0 and 30.0 °C, respectively (Owens et al., 1986)
434 μmol/L at 25.0 °C (Tewari et al., 1982; Wasik et al., 1981, 1983)
282 μmol/L in 0.5 M NaCl at 25 °C (Wasik et al., 1984)
388, 423, 455 and 528 μmol/L at 15, 25, 35 and 45 °C, respectively (Sanemasa et al., 1982)
426, 425, 427, 432 and 445 μmol/L at 15, 20, 23, 25 and 30 °C, respectively (Wasik

et al., 1981)
70 mg/L at 25 °C (Krasnoshchekova and Gubergrits, 1975)
120 mg/L at 25 °C (Klevens, 1950)
60 mg/kg at 15 °C (Fühner, 1924)
432 μmol/L at 25 °C (DeVoe et al., 1981)
110 mg/kg at 25 °C (Stearns et al., 1947)
239 μmol/L at 25.00 °C (Sanemasa et al., 1985)
415 μmol/L at 25.0 °C (Sanemasa et al., 1987)
7.35 x 10^{-6} at 25 °C (mole fraction, Li et al., 1993)

Vapor density:
4.91 g/L at 25 °C, 4.15 (air = 1)

Vapor pressure (mmHg):
2.5 at 20 °C (Verschueren, 1983)
3.43 at 25 °C (Mackay et al., 1982)

Environmental Fate
   Biological. A *Nocardia* sp., growing on *n*-octadecane, biodegraded *n*-propyl-benzene to phenyl acetic acid (Davis and Raymond, 1981).

Toxicity: Acute oral LD_{50} for rats 6,040 mg/kg (RTECS, 1985).

Drinking Water Standard: As of May 1994, no MCLGs or MCLs have been proposed (U.S. EPA, 1984).

Uses: In textile dyeing and printing; solvent for cellulose acetate; manufacturing methylstyrene.

PROPYLCYCLOPENTANE

Synonym: 1-Cyclopentylpropane.

CAS Registry No.: 2040-96-2
Molecular formula: C_8H_{16}
Formula weight: 112.22
RTECS: GY4700000

Physical state, color and odor
Colorless liquid with an ether-like odor.

Melting point (°C):
-117.3 (Weast, 1986)

Boiling point (°C):
131 (Weast, 1986)

Density (g/cm^3):
0.77633 at 20/4 °C, 0.77229 at 25/4 °C (Dreisbach, 1959)

Diffusivity in water (10^5 cm^2/sec):
0.71 at 20 °C using method of Hayduk and Laudie (1974).

Dissociation constant:
>14 (Schwarzenbach et al., 1993)

Heat of fusion (kcal/mol):
2.398 (Dean, 1987)

Henry's law constant (atm·m^3/mol):
0.890 at 25 °C (approximate - calculated from water solubility and vapor pressure)

Soil sorption coefficient, log K_{oc}:
Unavailable because experimental methods for estimation of this parameter for alicyclic hydrocarbons are lacking in the documented literature.

Octanol/water partition coefficient, log K_{ow}:
3.63 using method of Hansch et al. (1968).

Solubility in organics:
Soluble in acetone, alcohol, benzene and ether (Weast, 1986).

Solubility in water (25 °C):
2.04 mg/kg (Price, 1976)
1.77 mg/L (Kryzanowska and Szeliga, 1978)

Vapor density:
4.59 g/L at 25 °C, 3.87 (air = 1)

Vapor pressure (mmHg):
12.3 at 25 °C (Wilhoit and Zwolinski, 1971)

Uses: Organic synthesis; gasoline component.

PROPYLENE OXIDE

Synonyms: Epoxypropane; 1,2-Epoxypropane; Methyl ethylene oxide; **Methyloxirane**; NCI-C50099; Propene oxide; 1,2-Propylene oxide; UN 1280.

CAS Registry No.: 75-56-9
DOT: 1280
DOT label: Flammable liquid
Molecular formula: C_3H_6O
Formula weight: 58.08
RTECS: TZ2975000

Physical state, color and odor
Clear, colorless liquid with an agreeable, etheral odor. Odor threshold in air is 44 ppm (Amoore and Hautala, 1983).

Melting point (°C):
-113 (NIOSH, 1994)
-104.4 (Verschueren, 1983)

Boiling point (°C):
34.3 (Weast, 1986)

Density (g/cm³):
0.859 at 0/4 °C (Weast, 1986)

Flash point (°C):
-37.5 (NIOSH, 1994)
-35 (Windholz et al., 1983)

Lower explosive limit (%):
2.8 (Sax and Lewis, 1987)
2.3 (NFPA, 1984)

Upper explosive limit (%):
36 (NFPA, 1984)

Heat of fusion (kcal/mol):
1.561 (Dean, 1987)

Henry's law constant (10^5 atm·m^3/mol):
8.34 at 20 °C (approximate - calculated from water solubility and vapor pressure)

Ionization potential (eV):
10.22 ± 0.0 (Franklin et al., 1969)

Soil sorption coefficient, log K_{oc}:
Not applicable - reacts with water.

Octanol/water partition coefficient, log K_{ow}:
0.08 (Deneer et al., 1988)

Solubility in organics:
Miscible with alcohol and ether (Weast, 1986).

Solubility in water:
405,000 mg/L at 20 °C, 650,000 mg/L at 30 °C (Verschueren, 1983)
40 wt % at 20 °C (Gunther et al., 1968)

Vapor density:
2.37 g/L at 25 °C, 2.01 (air = 1)

Vapor pressure (mmHg):
445 at 20 °C (NIOSH, 1987)

Environmental Fate

Chemical/Physical. Anticipated products from the reaction of propylene oxide with ozone or hydroxyl radicals in the atmosphere are formaldehyde, pyruvic acid, $CH_3C(O)OCHO$ and $HC(O)OCHO$ (Cupitt, 1980). The reported hydrolysis half-life for the conversion of propylene oxide to 1,2-propanediol in water at 25 °C and pH 7 is 14.6 days (Mabey and Mill, 1978).

May polymerize ar high temperatures or on contact with alkalies, aqueous acids, amines and acid alcohols (NIOSH, 1994).

Exposure Limits: Potential occupational carcinogen. NIOSH REL: IDLH 400 ppm; OSHA PEL: TWA 100 ppm (240 mg/m^3); ACGIH TLV: TWA 20 ppm (50 mg/m^3).

Symptoms of Exposure: Vapors may irritate eyes, mucous membranes and skin. Inhalation may cause weakness and drowsiness (Patnaik, 1992). May cause blisters or burns (NIOSH, 1994).

Toxicity: Acute oral LD_{50} for guinea pigs 660 mg/kg, mice 630 mg/kg, rats 520 mg/kg; LD_{50} (skin) for rabbits 1,245 mg/kg; LC_{50} (inhalation) for mice 1,740 ppm/4-hour (RTECS, 1985).

Uses: Preparation of propylene and dipropylene glycols, lubricants, oil demulsifiers, surfactants, isopropanol amines, polyols for urethane foams; solvent; soil sterilant; fumigant.

n-PROPYL NITRATE

Synonyms: Nitric acid propyl ester; Propyl nitrate; UN 1865.

$$\underset{\underset{\displaystyle H}{|}}{\overset{\overset{\displaystyle H}{|}}{H-C}}-\underset{\underset{\displaystyle H}{|}}{\overset{\overset{\displaystyle H}{|}}{C}}-\underset{\underset{\displaystyle H}{|}}{\overset{\overset{\displaystyle H}{|}}{C}}-O-NO_2$$

CAS Registry No.: 627-13-4
DOT: 1865
Molecular formula: $C_3H_7NO_3$
Formula weight: 105.09
RTECS: UK0350000

Physical state, color and odor
Colorless to light yellow liquid with an ether-like odor. Odor threshold in air is 50 ppm (Amoore and Hautala, 1983).

Melting point (°C):
-101 (NIOSH, 1994)

Boiling point (°C):
110 at 762 mmHg (Weast, 1986)

Density (g/cm^3 at 20/4 °C):
1.0538 (Weast, 1986)
1.07 (Hawley, 1981)

Diffusivity in water (10^5 cm^2/sec):
0.88 at 20 °C using method of Hayduk and Laudie (1974).

Flash point (°C):
20 (NIOSH, 1994)

Lower explosive limit (%):
2 (NIOSH, 1994)

Upper explosive limit (%):
100 (NIOSH, 1994)

Ionization potential (eV):
11.07 (NIOSH, 1994)

Soil sorption coefficient, log K_{oc}:
Unavailable because experimental methods for estimation of this parameter for aliphatic nitrates are lacking in the documented literature.

Octanol/water partition coefficient, log K_{ow}:
Unavailable because experimental methods for estimation of this parameter for aliphatic nitrates are lacking in the documented literature.

Solubility in organics:
Soluble in alcohol and ether (Weast, 1986).

Vapor density:
4.30 g/L at 25 °C, 3.63 (air = 1)

Vapor pressure (mmHg):
18 at 20 °C (NIOSH, 1994)

Exposure Limits: NIOSH REL: TWA 25 ppm (109 mg/m^3), STEL 40 ppm (170 mg/m^3), IDLH 500 ppm; OSHA PEL: TWA 25 ppm; ACGIH TLV: TWA 25 ppm.

Uses: Rocket fuel formulations; organic synthesis.

PROPYNE

Synonyms: Allylene; Methyl acetylene; Propine; **1-Propyne**.

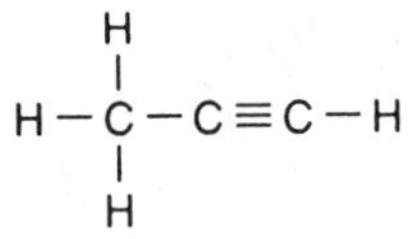

$$H-\overset{\displaystyle H}{\underset{\displaystyle H}{C}}-C\equiv C-H$$

CAS Registry No.: 74-99-7
Molecular formula: C_3H_4
Formula weight: 40.06
RTECS: UK4250000

Physical state, color and odor
Colorless gas with a sweet odor.

Melting point (°C):
-101.5 (Weast, 1986)

Boiling point (°C):
-23.2 (Weast, 1986)
-27.5 (Verschueren, 1983)

Density (g/cm³):
0.7062 at -50/4 °C (Weast, 1986)
0.678 at -27/4 °C (Verschueren, 1983)

Lower explosive limit (%):
1.7 (NIOSH, 1994)

Henry's law constant (atm·m³/mol):
0.11 at 25 °C (Hine and Mookerjee, 1975)

Ionization potential (eV):
10.36 ± 0.01 (Franklin et al., 1969)

Soil sorption coefficient, log K_{oc}:
Unavailable because experimental methods for estimation of this parameter for aliphatic hydrocarbons are lacking in the documented literature.

Octanol/water partition coefficient, log K_{ow}:
1.61 using method of Hansch et al. (1968).

Solubility in organics:
Soluble in alcohol, benzene and chloroform (Weast, 1986).

Solubility in water:
3,640 mg/L at 20 °C (Verschueren, 1983)

Vapor density:
1.64 g/L at 25 °C, 1.38 (air = 1)

Vapor pressure (mmHg):
3,952 at 20 °C, 5,244 at 30 °C (Verschueren, 1983)
4,310 at 25 °C (Wilhoit and Zwolinski, 1971)

Environmental Fate
   Chemical/Physical. When passed through a cold solution containing hydrobromite ions, 1-bromo-1-propyne was formed (Hatch and Kidwell, 1954).

Exposure Limits: NIOSH REL: TWA 1,000 ppm (1,650 mg/m^3), IDLH 1,700 ppm; OSHA PEL: TWA 1,000 ppm.

Symptoms of Exposure: An asphyxiant. Toxic at high concentrations (Patnaik, 1992).

Uses: Chemical intermediate; specialty fuel.

PYRENE

Synonyms: Benzo[*def*]phenanthrene; β-Pyrene; β-Pyrine.

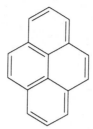

CAS Registry No.: 129-00-0
Molecular formula: $C_{16}H_{10}$
Formula weight: 202.26
RTECS: UR2450000

Physical state and color
Colorless solid (tetracene impurities impart a yellow color) or monoclinic prisms crystallized from alcohol. Solutions have a slight blue fluorescence.

Melting point (°C):
156 (Weast, 1986)
151 (Aldrich, 1988)
149 (Sims et al., 1988)

Boiling point (°C):
393 (Weast, 1986)
388 (Pearlman et al., 1984)
404 (Sax, 1984)

Density (g/cm^3):
1.271 at 23/4 °C (Weast, 1986)

Diffusivity in water (10^5 cm^2/sec):
0.56 at 20 °C using method of Hayduk and Laudie (1974).

Dissociation constant:
>15 (Christensen et al., 1975)

Henry's law constant (10^5 atm·m^3/mol):
1.09 (Mackay and Shiu, 1981)
1.87 (Southworth, 1979)

Ionization potential (eV):
7.70 ± 0.65 (Franklin et al., 1969)

7.55 (Yoshida et al., 1983)
7.50 (Cavalieri and Rogan, 1985)

Bioconcentration factor, log BCF:
3.43 (*Daphnia pulex*, Southworth et al., 1978)
3.43 (*Daphnia magna*, Newsted and Giesy, 1987)
2.66 (goldfish, Ogata et al., 1984)
0.72 (*Polychaete* sp.), 1.12 (*Capitella capitata*) (Bayona et al., 1991)

Soil sorption coefficient, log K_{oc}:
4.66 (Abdul et al., 1987)
4.92 (Schwarzenbach and Westall, 1981)
4.88 (Chin et al., 1988)
4.81 (Means et al., 1979)
4.80 (Hassett et al., 1980)
5.13 (Hodson and Williams, 1988; Vowles and Mantoura, 1987)
4.67 (Socha and Carpenter, 1987)
5.23 (Gauthier et al., 1987)
5.18 (Karickhoff et al., 1979)
6.51 (average, Kayal and Connell, 1990)
4.88 (fine sand, Enfield et al., 1989)

Octanol/water partition coefficient, log K_{ow}:
5.18 (Sangster, 1989)
4.88 (Chou and Jurs, 1979)
5.09 (Means et al., 1979)
5.52 (Burkhard et al., 1985a)
5.22 (Bruggeman et al., 1982)

Solubility in organics:
Soluble in most solvents (U.S. EPA, 1985).

Solubility in water:
160 μg/L at 26 °C, 32 μg/L at 24 °C (practical grade, Verschueren, 1983)
13 mg/kg at 25 °C. In seawater (salinity = 35 g/kg): 56, 71 and 89 μg/kg at 15, 20 and 25 °C, respectively (Rossi and Thomas, 1981)
13 μg/L at 25 °C (Miller et al., 1985)
135 μg/L at 25 °C (Mackay and Shiu, 1977)
135 μg/L at 24 °C (Means et al., 1979)
132 μg/kg at 25 °C, 162 μg/kg at 29 °C (May et al., 1978a)
171 μg/L at 25 °C (Schwarz and Wasik, 1976)
In mg/kg: 0.124, 0.128, 0.129 at 22.2 °C, 0.228, 0.235 at 34.5 °C, 0.395, 0.397, 0.405 at 44.7 °C, 0.556, 0.558, 0.576 at 50.1 °C, 0.75, 0.75, 0.77 at 55.6 °C, 0.74 at 56.0

°C, 0.90, 0.95, 0.96 at 60.7 °C, 1.27, 1.29 at 65.2 °C, 1.83, 1.86, 1.89 at 71.9 °C, 2.21 at 74.7 °C (Wauchope and Getzen, 1972)

160, 165 µg/L at 27 °C (Davis et al., 1942)

In nmol/L: 270 at 12.2 °C, 339 at 15.5 °C, 391 at 17.4 °C, 457 at 20.3 °C, 578 at 23.0 °C, 582 at 23.3 °C, 640 at 25.0 °C, 713 at 26.2 °C, 7.18 at 26.7 °C, 809 at 28.5 °C, 930 at 31.3 °C (Schwarz, 1977)

175 µg/L at 25 °C (Klevens, 1950)

133 µg/L at 25 °C (Walters and Luthy, 1984)

150 µg/L in Lake Michigan water at ≈ 25 °C (Eadie et al., 1990; Fatiadi, 1967)

1 µmol/L at 25 °C (Edwards et al., 1991)

At 20 °C: 470, 322, 318 and 311 nmol/L in doubly distilled water, Pacific seawater, artificial seawater and 35% NaCl, respectively (Hashimoto et al., 1984)

107 µg/L at 25 °C (Vadas et al., 1991)

118 µg/L at 25 °C (Billington et al., 1988)

Vapor pressure (10^7 mmHg):
6.85 at 25 °C (Radding et al., 1976)

25 at 25 °C (Mabey et al., 1982)

379 at 68.90 °C, 1,080 at 71.75 °C, 1,410 at 74.15 °C, 1,670 at 75.85 °C, 2,060 at 78.20 °C, 2,160 at 78.90 °C, 2,920 at 81.70 °C, 3,070 at 82.65 °C, 3,040 at 82.70 °C, 3,790 at 85.00 °C, 3,800 at 85.25 °C (Bradley and Cleasby, 1953)

45 at 25 °C (Sonnefeld et al., 1983; Wasik et al., 1983)

367 at 25 °C (Bidleman, 1984)

564, 848 at 25 °C (Hinckley et al., 1990)

Environmental Fate

Biological. When pyrene was statically incubated in the dark at 25 °C with yeast extract and settled domestic wastewater inoculum, complete degradation was demonstrated at the 5 mg/L substrate concentration after 2 weeks. At a substrate concentration of 10 mg/L, however, only 11 and 2% losses were observed after 7 and 14 days, respectively (Tabak et al., 1981).

Photolytic. Adsorption onto garden soil for 10 days at 32 °C and irradiated with UV light produced 1,1'-bipyrene, 1,6-pyrenedione, 1,8-pyrenedione and three unidentified compounds (Fatiadi, 1967). Microbial degradation by *Mycobacterium* sp. yielded the following ring-fission products: 4-phenanthroic acid, 4-hydroxy-perinaphthenone, cinnamic acid and phthalic acids. The compounds pyrenol and the *cis*- and *trans*-4,5-dihydrodiols of pyrene were identified as ring-oxidation products (Heitkamp et al., 1988).

Silica gel coated with pyrene and suspended in an aqueous solution containing nitrite ion and subjected to UV radiation yielded the tentatively identified product 1-nitropyrene (Suzuki et al., 1987). 1-Nitropyrene coated on glass surfaces and exposed to natural sunlight resulted in the formation of hydroxypyrene, possibly pyrene dione and dihydroxy pyrene and other unidentified compounds (Benson et al., 1985).

1-Nitropyrene also formed when pyrene deposited on glass filter paper containing sodium nitrite was irradiated with UV light at room temperature (Ohe, 1984). This compound was reported to have formed from the reaction of pyrene with NO_x in urban air from St. Louis, MO (Randahl et al., 1982).

Chemical/Physical. At room temperature, concentrated sulfuric acid will react with pyrene to form a mixture of disulfonic acids. In addition, an atmosphere containing 10% sulfur dioxide transformed pyrene into many sulfur compounds, including pyrene-1-sulfonic acid and pyrenedisulfonic acid (Nielsen et al., 1983).

2-Nitropyrene was the sole product formed from the gas-phase reaction of pyrene with hydroxyl radicals in a NO_x atmosphere (Arey et al., 1986). Pyrene adsorbed on glass fiber filters reacted rapidly with N_2O_5 to form 1-nitropyrene with a 60–70% yield (Pitts et al., 1985).

When pyrene, adsorbed from the vapor phase onto coal fly ash, silica and alumina, was exposed to nitrogen dioxide, no reaction occurred. However, in the presence of nitric acid, nitrated compounds were produced (Yokley et al., 1985). Ozonation of water containing pyrene (10–200 µg/L) yielded short-chain aliphatic compounds as the major products (Corless et al., 1990). A monochlorinated pyrene was the major product formed during the chlorination of pyrene in aqueous solutions. At pH 4, the reported half-lives at chlorine concentrations of 0.6 and 10 mg/L were 8.8 and <0.2 hours, respectively (Mori et al., 1991).

Exposure Limits: Potential occupational carcinogen. No individual standards have been set, however, as a constituent in coal tar pitch volatiles, the following exposure limits have been established (mg/m^3): NIOSH REL: TWA 0.1 (cyclohexane-extractable fraction), IDLH 80; OSHA PEL: TWA 0.2 (benzene-soluble fraction); ACGIH TLV: TWA 0.2 (benzene solubles).

Drinking Water Standard: As of May 1994, no MCLGs or MCLs have been proposed (U.S. EPA, 1984).

Toxicity: Acute oral LD_{50} for mice 800 mg/kg, rats 2,700 mg/kg; LC_{50} (inhalation) for rats 170 mg/m^3 (RTECS, 1985).

Use: Research chemical. Derived from industrial and experimental coal gasification operations where maximum concentration detected in gas and coal tar streams were 9.2 and 24 mg/m^3, respectively (Cleland, 1981).

PYRIDINE

Synonyms: Azabenzene; Azine; NCI-C55301; RCRA waste number U196; UN 1282.

CAS Registry No.: 110-86-1
DOT: 1282
DOT label: Flammable liquid
Molecular formula: C_5H_5N
Formula weight: 79.10
RTECS: UR8400000

Physical state, color and odor
Colorless liquid with a sharp, penetrating, fish-like odor. Odor threshold in air is 21 ppb (Keith and Walters, 1992).

Melting point (°C):
-42 (Weast, 1986)

Boiling point (°C):
115.5 (Weast, 1986)

Density (g/cm^3):
0.9819 at 20/4 °C (Weast, 1986)
0.9780 at 25/4 °C (Windholz et al., 1983)

Diffusivity in water (10^5 cm^2/sec):
1.00 at 20 °C using method of Hayduk and Laudie (1974).

Dissociation constant:
5.19 (Windholz et al., 1983)
5.21 at 25 °C (Gordon and Ford, 1972)

Flash point (°C):
20 (NIOSH, 1994)

Lower explosive limit (%):
1.8 (NIOSH, 1994)

909

Upper explosive limit (%):
12.4 (NIOSH, 1994)

Heat of fusion (kcal/mol):
1.979 (Dean, 1987)

Henry's law constant (10^6 atm·m^3/mol at 25 °C):
8.88 (Hine and Mookerjee, 1975)
12 (Hawthorne et al., 1985)

Ionization potential (eV):
9.32 (Franklin et al., 1969)
9.27 (NIOSH, 1994)

Soil sorption coefficient, log K_{oc}:
Unavailable because experimental methods for estimation of this parameter for pyridines are lacking in the documented literature. However, its miscibility in water and low K_{ow} suggest its adsorption to soil will be nominal (Lyman et al., 1982).

Octanol/water partition coefficient, log K_{ow}:
0.65 (Campbell and Luthy, 1985; Leo et al., 1971)
1.28 (Eadsforth, 1986)
0.63, 0.68, 0.72 (Garst and Wilson, 1984)
0.66 (Mirrlees et al., 1976)

Solubility in organics:
Soluble in acetone, alcohol, benzene and ether (Weast, 1986). Miscible with alcohol, ether, petroleum ether, oils and many other organic liquids (Windholz et al., 1983).

Solubility in water:
Miscible (Fischer and Ehrenberg, 1948). A saturated solution in equilibrium with its own vapor had a concentration of 233.4 g/L at 25 °C (Kamlet et al., 1987).

Vapor density:
3.23 g/L at 25 °C, 2.73 (air = 1)

Vapor pressure (mmHg):
18 at 20 °C (NIOSH, 1987)
14 at 20 °C, 20 at 25 °C, 26 at 30 °C (Verschueren, 1983)

Environmental Fate
   Photolytic. Irradiation of an aqueous solution at 50 °C for 24 hours resulted in a

23.06% yield of carbon dioxide (Knoevenagel and Himmelreich, 1976). When an aqueous pyridine solution in the presence of titanium dioxide was illuminated by UV light, ammonium and nitrate ions formed as the major products (Low et al., 1991).

Chemical/Physical. The gas-phase reaction of ozone with pyridine in synthetic air at 23 °C yielded a nitrated salt having the formula: $[C_6H_5NH]^+NO_3^-$ (Atkinson et al., 1987).

Ozonation of pyridine in aqueous solutions at 25 °C were studied with and without the addition of *t*-butyl alcohol (20 mM) as a radical scavenger. With *t*-butyl alcohol, ozonation of pyridine yielded mainly pyridine *N*-oxide (80% yield), which was very stable towards ozone. Without *t*-butyl alcohol, the heterocyclic ring is rapidly cleaved forming ammonia, nitrate and the amidic compound *N*-formyl oxamic acid (Andreozzi et al., 1991).

Forms water-soluble salts with strong acids.

Exposure Limits: NIOSH REL: TWA 5 ppm (15 mg/m³), IDLH 1,000 ppm; OSHA PEL: TWA 5 ppm; ACGIH TLV: TWA 5 ppm.

Symptoms of Exposure: Headache, dizziness, nervousness, nausea, insomnia, frequent urination and abdominal pain (Patnaik, 1992).

Toxicity: Acute oral LD_{50} for mice 1,500 mg/kg, rats 891 mg/kg (RTECS, 1985); LC_{50} for red killifish 9,330 mg/L (Yoshioka et al., 1986).

Uses: Organic synthesis (vitamins and drugs); analytical chemistry (cyanide analysis); solvent for anhydrous mineral salts; denaturant for alcohol; antifreeze mixtures; textile dyeing; waterproofing; fungicides; rubber chemicals.

p-QUINONE

Synonyms: 1,4-Benzoquine; Benzoquinone; 1,4-Benzoquinone; *p*-Benzoquinone; Chinone; Cyclohexadienedione; 1,4-Cyclohexadienedione; **2,5-Cyclohexadiene-1,4-dione**; 1,4-Cyclohexadiene dioxide; 1,4-Dioxybenzene; NCI-C55845; Quinone; 4-Quinone; RCRA waste number U197; UN 2587; USAF P-220.

CAS Registry No.: 106-51-4
DOT: 2587
DOT label: Poison
Molecular formula: $C_6H_4O_2$
Formula weight: 108.10
RTECS: DK2625000

Physical state, color and odor
Light yellow crystals with an acrid odor resembling chlorine. Odor threshold in air is 84 ppb (Amoore and Hautala, 1983).

Melting point (°C):
115–117 (Weast, 1986)

Boiling point (°C):
Sublimes (Weast, 1986)

Density (g/cm³):
1.318 at 20/4 °C (Weast, 1986)

Diffusivity in water (10^5 cm²/sec):
0.87 at 20 °C using method of Hayduk and Laudie (1974).

Flash point (°C):
38–94 (NIOSH, 1994)

Ionization potential (eV):
9.67 ± 0.02 (Franklin et al., 1969)

Soil sorption coefficient, log K_{oc}:
Unavailable because experimental methods for estimation of this parameter for quinones are lacking in the documented literature.

Octanol/water partition coefficient, log K_{ow}:
0.20 (Leo et al., 1971)

Solubility in organics:
Soluble in alcohol, ether (Weast, 1986) and hot petroleum ether (Windholz et al., 1983).

Vapor pressure (mmHg):
0.1 at 20 °C (NIOSH, 1994)

Exposure Limits: NIOSH REL: TWA 0.4 mg/m^3 (0.1 ppm), IDLH 100 mg/m^3; OSHA PEL: TWA 0.4 mg/m^3.

Symptoms of Exposure: Prolonged exposure may cause eye irritation. May damage cornea on contact. Contact with skin may cause irritation, ulceration and necrosis (Patnaik, 1992).

Toxicity: Acute oral LD$_{50}$ for rats 130 mg/kg (RTECS, 1985).

Uses: Oxidizing agent; manufacturing hydroquinone and dyes; in photography; strengthening animal fibers; tanning hides; analytical reagent; making gelatin insoluble; fungicides.

RONNEL

Synonyms: Dermaphos; Dimethyl trichlorophenyl thiophosphate; *O,O*-Dimethyl-*O*-2,4,5-trichlorophenyl phosphorothioate; *O,O*-Dimethyl *O*-(2,4,5-trichlorophenyl)thiophosphate; Dow ET 14; Dow ET 57; Ectoral; ENT 23284; ET 14; ET 57; Etrolene; Fenchlorfos; Fenchlorophos; Fenchchlorphos; Karlan; Korlan; Korlane; Nanchor; Nanker; Nankor; **Phosphorothioic acid *O,O*-dimethyl *O*-(2,4,5-trichlorophenyl)ester**; Trichlorometafos; Trolen; Trolene; Viozene.

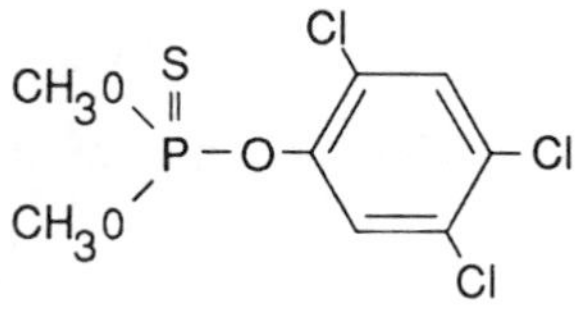

CAS Registry No.: 299-84-3
DOT: 2922
DOT label: Corrosive, poisonous
Molecular formula: $C_8H_8Cl_3O_3PS$
Formula weight: 321.57
RTECS: TG0525000

Physical state and color
White to light brown, waxy solid.

Melting point (°C):
41 (Windholz et al., 1983)
35–37 (Freed et al., 1977)

Boiling point (°C):
Decomposes (NIOSH, 1994)
97 at 0.01 mmHg (Freed et al., 1977)

Density (g/cm³):
1.48 at 25/4 °C (Verschueren, 1983)

Diffusivity in water (10^5 cm²/sec):
0.51 at 20 °C using method of Hayduk and Laudie (1974).

Flash point (°C):
Noncombustible solid (NIOSH, 1994)

Henry's law constant (10^6 atm·m³/mol):
8.46 at 25 °C (approximate - calculated from water solubility and vapor pressure)

Soil sorption coefficient, log K_{oc}:
2.76 using method of Kenaga and Goring (1980).

Octanol/water partition coefficient, log K_{ow}:
4.88 (Chiou et al., 1977; Freed et al., 1979a)
4.67 (Kenaga and Goring, 1980)
5.068 (de Bruijn et al., 1989)
4.81 (Bowman and Sans, 1983a)

Solubility in organics:
Soluble in acetone, carbon tetrachloride, ether, kerosene, methylene chloride and toluene (Windholz et al., 1983).

Solubility in water (mg/L):
1.08 at 20 °C (Chiou et al., 1977; Freed et al., 1977, 1979a)
40 at 25 °C (Windholz et al., 1983)
0.6 at 20.0 °C (Bowman and Sans, 1979)

Vapor pressure (10^5 mmHg):
80 at 25 °C (Windholz et al., 1983)
5.29 at 20 °C, 18.6 at 30 °C (Freed et al., 1977)

Environmental Fate

Chemical/Physical. Though no products were identified, the reported hydrolysis half-life at pH 7.4 and 70 °C using a 1:4 ethanol/water mixture is 10.2-10.4 hours (Freed et al., 1977). Ronnel decomposed at elevated temperatures on five clay surfaces, each treated with hydrogen, calcium, magnesium, aluminum and iron ions. At temperatures <950 °C (125, 300 and 750 °C), bentonite clays impregnated with technical ronnel (18.6 wt %) decomposed to 2,4,5-trichlorophenol and a rearrangement product tentatively identified as *O*-methyl *S*-methyl-*O*-(2,4,5-trichlorophenyl) phosphorothioate (Rosenfield and Van Valkenburg, 1965). At 950 °C, only the latter product formed. It was postulated that this compound resulted from an acid-catalyzed molecular rearrangement reaction. Ronnel also undergoes base-catalyzed hydrolysis at elevated temperatures. Products include methanol and a new compound that is formed via cleavage of a methyl group from one of the methoxy groups, which is then bonded to the sulfur atom (Rosenfield and Van Valkenburg, 1965).

Ronnel is stable to hydrolysis over the pH range of 5-6 (Mortland and Raman, 1967). However, in the presence of a Cu(II) salt (as cupric chloride) or when present as the exchangeable Cu(II) cation in montmorillonite clays, ronnel is completely hydrolyzed via first-order kinetics in <24 hours at 20 °C. The calculated half-life of ronnel at 20 °C for this reaction is 6.0 hours. It was suggested that decomposition in the presence of Cu(II) was a result of coordination of the copper atom through the oxygen or sulfur on the phosphorus atom, resulting in the cleavage of the side chain containing the phosphorus atom, forming *O,O*-ethyl-*O*-phosphorothioate and 1,2,4-trichlorobenzene (Mortland and Raman, 1967).

Emits very toxic fumes of chlorides, sulfur and phosphorus oxides when heated to decomposition (Lewis, 1990).

Exposure Limits (mg/m^3): NIOSH REL: TWA 10, IDLH 300; OSHA PEL: TWA 15; ACGIH TLV: TWA 10.

Symptoms of Exposure: In animals: cholinesterase inhibition; irritates eyes; liver and kidney damage.

Toxicity: Acute oral LD$_{50}$ for wild birds 80 mg/kg, chickens 6,375 mg/kg, ducks 3,500 mg/kg, dogs 500 mg/kg, guinea pigs 1,400 mg/kg, mice 2,000 mg/kg, rats 625 mg/kg, rabbits 420 mg/kg, turkeys 500 mg/kg (RTECS, 1985); LC$_{50}$ (96-hour) for fathead minnow 305 μg/L (Verschueren, 1983).

Uses: Not commercially produced in the United States. Insecticide.

STRYCHNINE

Synonyms: Certox; Dolco mouse cereal; Kwik-kil; Mole death; Mole-nots; Mouse-rid; Mouse-tox; Pied piper mouse seed; RCRA waste number P108; Rodex; Sanaseed; **Strychnidin-10-one**; Strychnos; UN 1692.

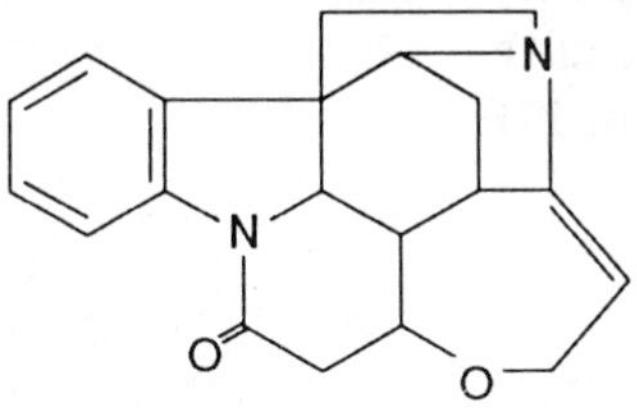

CAS Registry No.: 57-24-9
DOT: 1692
DOT label: Poison
Molecular formula: $C_{21}H_{22}N_2O_2$
Formula weight: 334.42
RTECS: WL2275000

Physical state, color and odor
Colorless to white, odorless crystals. Bitter taste.

Melting point (°C):
286-288 (Weast, 1986)
268 (Sax and Lewis, 1987)
284-286 (Dean, 1987)

Boiling point (°C):
270 at 5 mmHg (Weast, 1986)

Density (g/cm³):
1.36 at 20/4 °C (Weast, 1986)

Dissociation constant (20 °C):
$pK_1 = 6.0$, $pK_2 = 11.7$ (Windholz et al., 1983)

Soil sorption coefficient, log K_{oc}:
2.45 using method of Kenaga and Goring (1980).

Octanol/water partition coefficient, log K_{ow}:
1.93 (Mercer et al., 1990)

Solubility in organics:
In g/L: alcohol (6.67), amyl alcohol (4.55), benzene (5.56), chloroform (200),

glycerol (3.13), methanol (3.85), toluene (≈ 5) (Windholz et al., 1983), pyridine (>12.3 g/kg at 20-25 °C) (Dehn, 1917).

Solubility in water:
156.25 mg/L, 322.6 mg/L at 100 °C (Windholz et al., 1983)
143 ppm at 20-25 °C (Verschueren, 1983)
>0.2 g/kg at 20-25 °C (Dehn, 1917)

Environmental Fate
   Chemical/Physical. Reacts with water forming water-soluble salts (Worthing and Hance, 1991). Emits toxic nitrogen oxides when heated to decomposition (Lewis, 1990).

Exposure Limits (mg/m^3): NIOSH REL: TWA 0.15, IDLH 3; OSHA PEL: TWA 0.15.

Symptoms of Exposure: Powerful convulsant and toxic. Ingestion of 0.1 g may be fatal to humans (Patnaik, 1992).

Toxicity: Acute oral LD$_{50}$ for wild birds 4 mg/kg, cats 500 μg/kg, ducks 3 mg/kg, dogs 500 μg/kg, mice 2 mg/kg, quail 23 mg/kg, rats 16 mg/kg (RTECS, 1985).

Uses: Destroying rodents and predatory animals; trapping fur-bearing animals; medicine.

STYRENE

Synonyms: Cinnamene; Cinnamenol; Cinnamol; Diarex HF77; **Ethyenyl benzene;** NCI-C02200; Phenethylene; Phenylethene; Phenylethylene; Styrene monomer; Styrol; Styrolene; Styron; Styropol; Styropor; UN 2055; Vinyl benzene; Vinyl benzol.

CAS Registry No.: 100-42-5
DOT: 2055
DOT label: Flammable or combustible liquid
Molecular formula: C_8H_8
Formula weight: 104.15
RTECS: WL3675000

Physical state, color and odor
Colorless to light yellow, oily liquid with a sweet, penetrating odor. Odor threshold in air is 148 ppb (Keith and Walters, 1992).

Melting point (°C):
-30.6 (Weast, 1986)
-33 (Huntress and Mulliken, 1941)

Boiling point (°C):
145.2 (Weast, 1986)

Density (g/cm³):
0.9060 at 20/4 °C (Weast, 1986)
0.9237 at 0/4 °C, 0.9148 at 10/4 °C, 0.9059 at 20/4 °C, 0.8970 at 30/4 °C, 0.8880 at 40/4 °C, 0.8702 at 60/4 °C (Standen, 1969)

Diffusivity in water (10^5 cm²/sec):
0.81 at 20 °C using method of Hayduk and Laudie (1974).

Dissociation constant:
>14 (Schwarzenbach et al., 1993)

Flash point (°C):
31 (NIOSH, 1994)

919

Lower explosive limit (%):
0.9 (NIOSH, 1994)

Upper explosive limit (%):
6.8 (NIOSH, 1994)

Heat of fusion (kcal/mol):
2.617 (Dean, 1987)

Henry's law constant (10^3 atm·m^3/mol):
2.61 (calculated, U.S. EPA, 1980)

Interfacial tension with water (dyn/cm at 19 °C):
35.48 (Demond and Lindner, 1993)

Ionization potential (eV):
8.47 ± 0.02 (Franklin et al., 1969)
8.71 (Rav-Acha and Choshen, 1987)

Soil sorption coefficient, log K_{oc}:
2.96 (Meylan et al., 1992)
3.02 (aquifer sand), 4.25 (Lima loam), 3.65 (Edwards muck) (Fu and Alexander, 1992)

Octanol/water partition coefficient, log K_{ow}:
2.95 (Chou and Jurs, 1979)
3.16 (Banerjee et al., 1980)

Solubility in organics:
Soluble in acetone, ethanol, benzene, ether and carbon disulfide (Weast, 1986).

Solubility in water:
1.54 mmol/L at 25.0 °C (Banerjee et al., 1980)
2.22 mmol/L at 25 °C (Andrews and Keefer, 1950a)
9.6 g/kg at 60.3 °C (Fordyce and Chapin, 1947)
In wt % (°C): 0.032 (6), 0.066 (25), 0.084 (31), 0.101 (40), 0.123 (51) (Lane, 1946)
3.19 x 10^{-3} at 25 °C (mole fraction, Li et al., 1993)

Vapor density:
4.26 g/L at 25 °C, 3.60 (air = 1)

Vapor pressure (mmHg):
5 at 20 °C, 9.5 at 30 °C (Verschueren, 1983)
4.3 at 15 °C (ACGIH, 1987)

6.45 at 25 °C (WHO, 1983a)

Environmental Fate

Biological. Fu and Alexander (1992) observed that despite the high degree of adsorption onto soils, styrene was mineralized to carbon dioxide under aerobic conditions. Rates of mineralization from highest to lowest were sewage sludge, Lima soil (pH 7.23, 7.5% organic matter), groundwater (pH 8.25, 30.5 mg/L organic matter), Beebe Lake water from Ithica, NY (pH 7.5, 50-60 mg/L organic matter), aquifer sand (pH 6.95, 0.4% organic matter), Erie silt loam (pH 4.87, 5.74% organic matter). Styrene did not mineralize in sterile environmental samples.

Photolytic. Irradiation of styrene in solution forms polystyrene. In a benzene solution, irradiation of polystyrene will result in depolymerization to presumably styrene (Calvert and Pitts, 1966).

Chemical/Physical. In the dark, styrene reacted with ozone forming benzaldehyde, formaldehyde, benzoic acid and trace amounts of formic acid (Grosjean, 1985). Polymerizes readily in the presence of heat, light or a peroxide catalyst. Polymerization is exothermic and may become explosive (NIOSH, 1994).

Exposure Limits: NIOSH REL: TWA 50 ppm (215 mg/m^3), STEL 100 ppm (425 mg/m^3), IDLH 700 ppm; OSHA PEL: TWA 100 ppm, ceiling 200 ppm, 5-minute/3-hour peak 600 ppm; ACGIH TLV: TWA 50 ppm, STEL 100 ppm.

Symptoms of Exposure: Irritates, eye, skin and mucous membranes. Narcotic at high concentrations (Patnaik, 1992).

Toxicity: Acute oral LD_{50} for mice 316 mg/kg, rats 5,000 mg/kg; LC_{50} (inhalation) for mice 21,600 mg/m^3/2-hour, rats 24 gm/m^3/4-hour (RTECS, 1985).

Drinking Water Standard (final): MCLG: 0.1 mg/L; MCL: 0.1 mg/L (U.S. EPA, 1994).

Uses: Preparation of polystyrene, styrene oxide, ethylbenzene, ethylcyclo-hexane, benzoic acid, synthetic rubber, resins, protective coatings and insulators.

SULFOTEPP

Synonyms: ASP 47; Bay E-393; Bayer-E 393; Bis-*O,O*-diethylphosphorothionic anhydride; Bladafum; Bladafume; Bladafun; Dithio; Dithione; Dithiophos; Dithiophosphoric acid tetraethyl ester; Dithiotep; E 393; ENT 16273; Ethyl thiopyrophosphate; Lethalaire G-57; Pirofos; Plant dithio aerosol; Plantfume 103 smoke generator; Pyrophosphorodithioic acid tetraethyl ester; Pyrophosphorodithioic acid *O,O,O,O*-tetraethyl dithionopyrophosphate; RCRA waste number P109; Sulfatep; Sulfotep; TEDP; TEDTP; Tetraethyl dithionopyrophosphate; Tetraethyl dithiopyrophosphate; *O,O,O,O*-Tetraethyl dithiopyrophosphate; Thiodiphosphoric acid tetraethyl ester; **Thiophosphoric acid tetraethyl ester**; Thiopyrophosphoric acid tetraethyl ester; Thiotepp; UN 1704.

$$CH_3CH_2O \quad \overset{S}{\underset{\|}{}} \quad \overset{S}{\underset{\|}{}} \quad OCH_2CH_3$$
$$\diagdown \quad P-O-P \quad \diagup$$
$$CH_3CH_2O \diagup \qquad \diagdown OCH_2CH_3$$

CAS Registry No.: 3689-24-5
DOT: 1704
DOT label: Poison
Molecular formula: $C_8H_{20}O_5P_2S_2$
Formula weight: 322.30
RTECS: XN4375000

Physical state, color and odor
Pale yellow liquid with a garlic-like odor.

Boiling point (°C):
136-139 at 2 mmHg (Windholz et al., 1983)
Decomposes (NIOSH, 1994)

Density (g/cm³):
1.196 at 25/4 °C (Windholz et al., 1983)
1.796 at 20/4 °C (Worthing and Hance, 1991)

Diffusivity in water (10^5 cm²/sec):
0.63 at 20 °C using method of Hayduk and Laudie (1974).

Henry's law constant (10^6 atm·m³/mol):
2.88 at 20 °C (approximate - calculated from water solubility and vapor pressure)

Soil sorption coefficient, log K_{oc}:
2.87 using method of Kenaga and Goring (1980).

Octanol/water partition coefficient, log K_{ow}:
3.02 using method of Kenaga and Goring (1980).

Solubility in organics:
Miscible with most organic solvents (Windholz et al., 1983).

Solubility in water (mg/L):
25 (Windholz et al., 1983)
30 at 20 °C (Worthing and Hance, 1991)

Vapor density:
13.17 g/L at 25 °C, 11.13 (air = 1)

Vapor pressure (mmHg):
1.7 x 10^{-4} at 20 °C (Windholz et al., 1983)

Exposure Limits (mg/m^3): NIOSH REL: TWA 0.2, IDLH 10; OSHA PEL: TWA 0.2.

Environmental Fate
   Soil. Cleavage of the molecule yields diethyl phosphate, monoethyl phosphate and phosphoric acid (Hartley and Kidd, 1987).
   Chemical/Physical. Emits toxic oxides of sulfur and phosphorus when heated to decomposition (Lewis, 199).

Toxicity: Acute oral LD_{50} for wild birds 100 mg/kg, cats 3 mg/kg, chickens 25 mg/kg, dogs 5 mg/kg, mice 22 mg/kg, rats 5 mg/kg, rabbits 25 mg/kg (RTECS, 1985); LC_{50} (96-hour) for fathead minnow 178 μg/L, bluegill sunfish 1.6 μg/L and rainbow trout 18 μg/L (Verschueren, 1983).

Uses: Not commercially produced in the United States. Insecticide.

2,4,5-T

Synonyms: Amine 2,4,5-T for rice; BCF-bushkiller; Brush-off 445 low volatile brush killer; Brush-rhap; Brushtox; Dacamine; Debroussaillant concentre; Debroussaillant super concentre; Decamine 4T; Ded-weed brush killer; Ded-weed LV-6 brush-kil and T-5 brush-kil; Dinoxol; Envert-T; Estercide T-2 and T-245; Esteron; Esterone 245; Esteron 245 BE; Esteron brush killer; Farmco fence rider; Fence rider; Forron; Forst U 46; Fortex; Fruitone A; Inverton 245; Line rider; NA 2765; Phortox; RCRA waste number U232; Reddon; Reddox; Spontox; Super D weedone; Tippon; Tormona; Transamine; Tributon; **(2,4,5-Trichlorophenoxy)-acetic acid**; Trinoxol; Trioxon; Trioxone; U 46; Veon; Veon 245; Verton 2T; Visko-rhap low volatile ester; Weddar; Weedone; Weedone 2,4,5-T.

CAS Registry No.: 93-76-5
DOT: 2765
DOT label: Poison
Molecular formula: $C_8H_5Cl_3O_3$
Formula weight: 255.48
RTECS: AJ8400000

Physical state and color
Colorless to pale brown crystals. Odor threshold in water is 2.92 mg/kg (Keith and Walters, 1992). Metallic taste.

Melting point (°C):
153 (Windholz et al., 1983)

Density (g/cm³):
1.80 at 20/20 °C (Windholz et al., 1983)

Diffusivity in water (10^5 cm²/sec):
0.54 at 20 °C using method of Hayduk and Laudie (1974).

Dissociation constant:
2.80, 2.83 (Jafvert et al., 1990)
2.88 (Nelson and Faust, 1969)

Henry's law constant (10^8 atm·m³/mol):
4.87 at 25 °C (approximate - calculated from water solubility and vapor pressure)

Bioconcentration factor, log BCF:
1.36 (fish, microcosm) (Garten and Trabalka, 1983)

Soil sorption coefficient, log K_{oc}:
1.72 (Kenaga and Goring, 1980)
2.27 (Webster soil, Nkedi-Kizza et al., 1983)

Octanol/water partition coefficient, log K_{ow}:
3.31, 3.38 (Jafvert et al., 1990)
3.40 (Riederer, 1990)

Solubility in organics (°C):
Soluble in ethanol (548.2 mg/L), ether (243.2 mg/L), heptane (400 mg/L), xylenes (6.8 g/L), methanol (496 g/L), tolueme (7.32 g/L) (Keith and Walters, 1992).

Solubility in water:
278 ppm at 25 °C (Verschueren, 1983)
240 mg/L at 25 °C (Klöpffer et al., 1982)
220 mg/L at 20 °C (Riederer, 1990)
1.05 mmol/L at 25 °C (Gunther et al., 1968)
150 g/L at 25 °C (Lewis, 1989)

Vapor pressure (10^6 mmHg):
37.5 at 20 °C (Riederer, 1990)
6.46 at 25 °C (Lewis, 1989)

Environmental Fate

Biological. 2,4,5-T degraded in anaerobic sludge by reductive dechlorination to 2,4,5-trichlorophenol, 3,4-dichlorophenol and 4-chlorophenol (Mikesell and Boyd, 1985). An anaerobic methanogenic consortium, growing on 3-chlorobenzoate, metabolized 2,4,5-T to (2,5-dichlorophenoxy)acetic acid at a rate of 1.02×10^{-7} M/hr. The half-life was reported to be 2 days at 37 °C (Suflita et al., 1984). Under aerobic conditions, 2,4,5-T degraded to 2,4,5-trichlorophenol and 3,5-dichloro-catechol, which may further degrade to 4-chlorocatechol or *cis,cis*-2,4-dichloro-muconic acid, 2-chloro-4-carboxymethylenebut-2-enolide, chlorosuccinic acid and succinic acid (Byast and Hance, 1975). The cometabolic oxidation of 2,4,5-T by *Brevibacterium* sp. yielded a product tentatively identified as 3,5-dichlorocatechol (Horvath, 1970). The cometabolism of this compound by *Achromobacter* sp. gave 3,5-dichloro-2-hydroxymuconic semialdehyde (Horvath, 1970a).

Soil. 2,4,5-Trichlorophenol and 2,4,5-trichloroanisole were formed when 2,4,5-T was incubated in soil at 25 °C under aerobic conditions. The half-life under these conditions was 14 days (McCall et al., 1981). When 2,4,5-T (10 μg), in unsterilized tropical clay and silty clay soils, was incubated for 4 months, 5-35%

degradation yields were observed (Rosenberg and Alexander, 1980). Hydrolyzes in soil to 2,4,5-trichlorophenol (Somasundaram et al., 1989, 1991) and 2,4,5-trichloroanisole (Somasundaram et al., 1989). The rate of 2,4,5-T degradation in soil remained unchanged in a soil pretreated with its hydrolysis metabolite (2,4,5-trichlorophenol) (Somasundaram et al., 1989).

Photolytic. When 2,4,5-T (100 μM), in oxygenated water containing titanium dioxide (2 g/L) suspension, was irradiated by sunlight ($\lambda \geq 340$ nm), 2,4,5-trichlorophenol and 2,4,5-trichlorophenyl formate formed as major intermediates. Other compounds identified as intermediates included nine chlorinated aromatic hydrocarbons. Complete mineralization yielded hydrochloric acid, carbon dioxide and water (Barbeni et al., 1987).

Crosby and Wong (1973) studied the photolysis of 2,4,5-T in aqueous solutions (100 mg/L) under alkaline conditions (pH 8) using both outdoor sunlight and indoor irradiation ($\lambda = 300$-450 nm). 2,4,5-Trichlorophenol and 2-hydroxy-4,5-dichlorophenoxyacetic acid formed as major products. Minor photodecomposition products included 4,6-dichlororesorcinol, 4-chlororesorcinol, 2,5-dichlorophenol and a dark polymeric substance. The rate of photolysis increased 11-fold in the presence of sensitizers (acetone or riboflavin) (Crosby and Wong, 1973). The rate of photolysis of 2,4,5-T was also higher in natural waters containing fulvic acids when compared to distilled water. The major photoproduct found in the humic acid-induced reaction was 2,4,5-trichlorophenol. In addition, the presence of ferric ions and/or hydrogen peroxides may contribute to the sunlight-induced photolysis of 2,4,5-T in acidic, weakly-absorbing natural waters (Skurlatov et al., 1983).

Chemical/Physical. Carbon dioxide, chloride, dichloromaleic, oxalic and glycolic acids, were reported as ozonation products of 2,4,5-T in water at pH 8 (Struif et al., 1978). Reacts with alkali metals and amines forming water-soluble salts (Worthing and Hance, 1991). When 2,4,5-T was heated at 900 °C, carbon monoxide, carbon dioxide, chlorine, hydrochloric acid and oxygen were produced (Kennedy et al., 1972, 1972a).

Exposure Limits (mg/m^3): NIOSH REL: TWA 10, IDLH 250; OSHA PEL: TWA 10; ACGIH TLV: TWA 10.

Symptoms of Exposure: Skin irritation. May also cause eye, nose and throat irritation (NIOSH, 1987).

Toxicity: Acute oral LD_{50} for chickens 310 mg/kg, dogs 100 mg/kg, guinea pigs 381 mg/kg, hamsters 425 mg/kg, mice 389 mg/kg, rats 300 mg/kg (RTECS, 1985); LC_{50} (96-hour) for rainbow trout 350 mg/L and carp 355 mg/L (Hartley and Kidd, 1987).

Drinking Water Standard: As of May 1994, no MCLGs or MCLs have been proposed although 2,4,5-T has been listed for regulation (U.S. EPA, 1994).

Uses: Plant hormone; defoliant. Formerly used as a herbicide. Banned by the U.S. EPA.

TCDD

Synonyms: Dioxin; Dioxin (herbicide contaminant); Dioxine; NCI-C03714; TCDBD; 2,3,7,8-TCDD; 2,3,7,8-Tetrachlorodibenzodioxin; **2,3,7,8-Tetrachlorodibenzo[b,e][1,4]dioxin;** 2,3,7,8-Tetrachlorodibenzo-1,4-dioxin; 2,3,7,8-Tetrachlorodibenzo-*p*-dioxin; Tetradioxin.

CAS Registry No.: 1746-01-6
Molecular formula: $C_{12}H_4Cl_4O_2$
Formula weight: 321.98
RTECS: HP3500000

Physical state and color
Colorless to white needles.

Melting point (°C):
305.0 (Schroy et al., 1985)
303-305 (Crummett and Stehl, 1973)

Boiling point (°C):
421.2 (estimated, Schroy et al., 1985)
Begins to decompose at 500 (U.S. EPA, 1985).

Density (g/cm³):
1.827 at 25 °C (estimated, Schroy et al., 1985)

Diffusivity in water (10^5 cm²/sec):
0.49 at 20 °C using method of Hayduk and Laudie (1974).

Henry's law constant (10^{23} atm·m³/mol):
5.40 at 20 °C (approximate - calculated from water solubility and vapor pressure)

Ionization potential (eV):
9.148 (calculated, Koester and Hites, 1988)

Bioconcentration factor, log BCF:
3.73 (fish) (Kenega and Goring, 1980)
3.90 (fathead minnow, Adams et al., 1986)
3.97 (rainbow trout) (Branson et al., 1985)

4.43 (rainbow trout, Mehrle et al., 1988)
3.02 (Garten and Trabalka, 1983)
3.30-4.27 (algae), 3.30-4.45 (catfish), 3.89-4.68 (daphnids), 3.08-3.70 (duckweed), 3.00-4.80 (mosquito fish), 3.15-4.67 (snails) (Isensee and Jones, 1975)
3.02 (fish, microcosm) (Garten and Trabalka, 1983)

Soil sorption coefficient, log K_{oc}:
6.6 (Times Beach, MO soil, Walters and Guiseppi-Elie, 1988)
6.66 (Walters et al., 1989)

Octanol/water partition coefficient, log K_{ow}:
6.15 (Schroy et al., 1985; Travis and Arms, 1988)
6.64 (Marple et al., 1986)
6.20 (Doucette and Andren, 1988)
7.02 (Burkhard and Kuehl, 1986)
5.38 (Rappe et al., 1987)

Solubility in organics:
In mg/L: acetone (110), benzene (570), chloroform (370), o-dichlorobenzene (1,400), methanol (10) and octanol (50) (Arthur and Frea, 1989); benzene (570), tetrachloroethylene (680), chlorobenzene (720), lard oil (40), hexane (280) (Keith and Walters, 1992).

Solubility in water:
0.317 μg/L at 25 °C (Schroy et al., 1985)
0.2 ppb (Crummett and Stehl, 1973)
0.0193 ppb at 22 °C (Marple et al., 1986)
0.32 ppb at 25 °C (Rappe et al., 1987)
0.0129 and 0.483 ppb at 0.2 and 17.3 °C, respectively (Lodge, 1989)
7.91 ± 2.7 ng/L at 20-22 °C (Adams and Blaine, 1986)

Vapor pressure (10^{10} mmHg):
34.6 at 30.1 °C (Rodorf, 1986)
2.7 at 15 °C, 6.4 at 20 °C, 14 at 25 °C, 35 at 30 °C, 163 at 40 °C (Rappe et al., 1987)
7.2 at 25 °C (Podoll, 1986)
340 at 25 °C (Rodorf, 1985)

Environmental Fate
Biological. After a 30-day incubation period, the white rot fungus *Phanerochaete chrysosporium* was capable of oxidizing TCDD to carbon dioxide. Mineralization began between the third and sixth day of incubation. The production of carbon dioxide was highest between 3-18 days of incubation, after which the rate of a

carbon dioxide produced decreased until the 30th day. It was suggested that the metabolism of TCDD and other compounds, including p,p'-DDT, benzo[a]pyrene and lindane, was dependent on the extracellular lignin-degrading enzyme system of this fungus (Bumpus et al., 1985).

Photolytic. Pure TCDD did not photolyze under UV light. However, in aqueous solutions containing cationic (1-hexadecylpyridinium chloride), anionic (sodium dodecyl sulfate) and nonionic (methanol) surfactants, TCDD decomposed into the end product tentatively identified as 2-phenoxyphenol. The times required for total TCDD decomposition using the cationic, anionic and nonionic solutions were 4, 8 and 16 hours, respectively (Botré et al., 1978). TCDD photodegrades rapidly in alcoholic solutions by reductive dechlorination. In water, however, the reaction was very slow (Crosby et al., 1973). The major photodegradative pathway of TCDD involves a replacement of the chlorine atom by a hydrogen atom. The proposed degradative pathway is TCDD to 2,7,8-trichlorodibenzo[b,e][1,4]dioxin to 2,7-dichlorodibenzo[b,e][1,4]dioxin to 2-chlorodibenzo[b,e][1,4]dioxin to dibenzo[b,e]-[1,4]dioxin to 2-hydroxydiphenyl ether, which undergoes polymerization (Makino et al., 1992).

[^{14}C]TCDD on a silica plate was exposed to summer sunlight at Beltsville, MD for 20 hours. A polar product was formed which was not identified. An identical experiment using soil demonstrated that photodegradation occurred but, to a smaller extent. No photoproducts were identified (Plimmer, 1978).

When TCDD in isooctane (3.1 μM) was irradiated by UV light ($\lambda \leq 310$ nm) at 31-33 °C, $\approx$ 10% was converted to 2,3,7-trichlorodibenzo-p-dioxin. In addition, 4,4',5,5'-tetrachloro-2,2'-dihydroxybiphenyl was identified as a new photoproduct formed by the reductive rearrangement of TCDD (Kieatiwong et al., 1990).

Hilarides and others (1994) investigated the destruction of TCDD on artificially contaminated soils using ^{60}Co γ radiation. It appeared that TCDD underwent stepwise reduction dechlorination from tetra- to tri-, then di- to monochlordioxin, and then to presumably nonchlorinated dioxins and phenols. The investigators discovered that the greatest amount of TCDD destruction (92%) occurred when soils were amended with 25% water and 2% nonionic surfactant [alkoxylated fatty alcohol (Plurafac RA-40)]. Replicate experiments conducted without the surfactant lowered the rate of TCDD destruction.

Chemical/Physical. TCDD was dehalogenated by a solution composed of poly-(ethylene glycol), potassium carbonate and sodium peroxide. After 2 hours at 85 °C, >99.9% of the applied TCDD decomposed. Chemical intermediates identified include tri-, di- and monochloro[b,e]dibenzo[1,4]dioxin, dibenzodioxin, hydrogen, carbon monoxide, methane, ethylene and acetylene (Tundo et al., 1985).

Exposure Limits: An IDLH of 1 ppb was recommended by Schroy et al. (1985).

Symptoms of Exposure: Extremely toxic. Causes chromosome damage. May cause miscarriage, birth defects and fetotoxicity. Fatigue, headache, irritability, abdominal pain, blurred vision and ataxia have been observed. Contact with skin may cause dermatitis and allergic reactions (Patnaik, 1992).

Toxicity: Acute oral LD_{50} for dogs 1 μg/kg, frogs mg/kg, guinea pigs 500 ng/kg, hamsters 1.157 mg/kg, monkeys 2 μg/kg, mice 114 μg/kg, rats 20 μg/kg; LD_{50} (skin) for rabbits 275 μg/kg (RTECS, 1985); acute oral LD_{50} for guinea pigs 2 μg/kg, male rats 22 μg/kg, rabbits 115 μg/kg and mice 284 μg/kg (Kriebel, 1981).

Drinking Water Standard (final): MCLG: zero; MCL: 3×10^{-5} μg/L (U.S. EPA, 994).

Uses: Occurs as an impurity in the herbicide 2,4,5-T.

1,2,4,5-TETRABROMOBENZENE

Synonym: *sym*-Tetrabromobenzene.

CAS Registry No.: 636-28-2
Molecular formula: $C_6H_2Br_4$
Formula weight: 393.70

Physical state
Solid.

Melting point (°C):
189 (Weast, 1986)

Boiling point (°C):
329 (Weast, 1986)

Diffusivity in water (10^5 cm²/sec):
0.61 at 20 °C using method of Hayduk and Laudie (1974).

Bioconcentration factor, log BCF:
3.57–3.81 (rainbow trout, Oliver and Niimi, 1985)

Soil sorption coefficient, log K_{oc}:
4.82 using method of Chiou et al. (1979).

Octanol/water partition coefficient, log K_{ow}:
5.13 (Watarai et al., 1982)

Solubility in organics:
Soluble in alcohol, benzene and ether (Weast, 1986).

Solubility in water:
40 µg/L (Kim and Saleh, 1990)
104.7 nmol/L at 25 °C (Yalkowsky et al., 1979)

Environmental Fate
   Chemical/Physical. 1,2,4,5-Tetrabromobenzene (150 mg in 250-mL distilled

water) was stirred in the dark for 1 day. GC analysis of the solution showed 1,2,4-tribromobenzene as a major transformation product (Kim and Saleh, 1990).

Uses: Organic synthesis.

1,1,2,2-TETRABROMOETHANE

Synonyms: Acetylene tetrabromide; Muthmann's liquid; TBE; Tetrabromo-acetylene; Tetrabromoethane; *sym*-Tetrabromoethane; UN 2504.

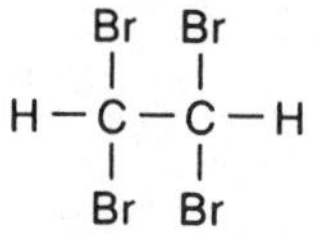

```
        Br  Br
        |   |
    H — C — C — H
        |   |
        Br  Br
```

CAS Registry No.: 79-27-6
DOT: 2504
Molecular formula: $C_2H_2Br_4$
Formula weight: 345.65
RTECS: KI8225000

Physical state, color and odor
Colorless to pale yellow liquid with a pungent odor resembling camphor and iodoform.

Melting point (°C):
0 (Weast, 1986)

Boiling point (°C):
243.5 (Weast, 1986)
239-242 (Verschueren, 1983)

Density (g/cm^3 at 20/4 °C):
2.8748 (Weast, 1986)
2.964 (Verschueren, 1983)
2.9529 (25/4 °C, Dean, 1987)

Diffusivity in water (10^5 cm^2/sec):
0.79 at 20 °C using method of Hayduk and Laudie (1974).

Flash point (°C):
Noncombustible liquid (NIOSH, 1994)

Henry's law constant (10^5 atm·m^3/mol):
6.40 at 20 °C (approximate - calculated from water solubility and vapor pressure)

Interfacial tension with water (dyn/cm at 20 °C):
38.82 (Demond and Lindner, 1993)

Soil sorption coefficient, log K_{oc}:
2.45 using method of Chiou et al. (1979).

Octanol/water partition coefficient, log K_{ow}:
2.91 using method of Hansch et al. (1968).

Solubility in organics:
Soluble in acetone and benzene (Weast, 1986). Miscible with acetic acid, alcohol, aniline, chloroform and ether (Windholz et al., 1983).

Solubility in water:
0.07 wt % at 20 °C (NIOSH, 1994)
651 mg/L at 30 °C (Verschueren, 1983)

Vapor density:
14.13 g/L at 25 °C, 11.93 (air = 1)

Vapor pressure (mmHg):
0.1 at 20 °C (Verschueren, 1983)

Exposure Limits: NIOSH REL: IDLH 8 ppm; OSHA PEL: 1 ppm (14 mg/m^3).

Toxicity: Acute oral LD_{50} for guinea pigs 400 mg/kg, mice 269 mg/kg, rats 1,100 mg/kg, rabbits 400 mg/kg (RTECS, 1985).

Uses: Solvent; in microscopy; separating minerals by density.

1,2,3,4-TETRACHLOROBENZENE

Synonym: 1,2,3,4-TCB.

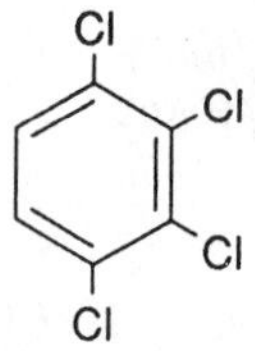

CAS Registry No.: 634-66-2
Molecular formula: $C_6H_2Cl_4$
Formula weight: 215.89
RTECS: DB9440000

Physical state and color
White crystals or needles.

Melting point (°C):
47.5 (Weast, 1986)
46.6 (Hawley, 1981)

Boiling point (°C):
254 (Weast, 1986)

Diffusivity in water (10^5 cm^2/sec):
0.63 at 20 °C using method of Hayduk and Laudie (1974).

Flash point (°C):
112 (Dean, 1987)

Heat of fusion (kcal/mol):
4.06 (Miller et al., 1984)

Henry's law constant (10^3 atm·m^3/mol):
69 at 20 °C (Oliver, 1985)
4.79, 5.13, 6.72, 7.00, 12.62 and 27.26 at 14.8, 20.1, 22.1, 24.2, 34.8 and 50.5 °C, respectively (ten Hulscher et al., 1992)

Bioconcentration factor, log BCF:
3.80–3.91 (rainbow trout, Oliver and Niimi, 1985)

Soil sorption coefficient, log K_{oc}:
5.4 (average of 5 suspended sediment samples from the St. Clair and Detroit Rivers

(Lau et al., 1989)
3.47 (average for 5 soils, Kishi et al., 1990)
K_d = 5.4 mL/g on a Cs^+-kaolinite (Haderlein and Schwarzenbach, 1993)
4.27, 4.35, 5.26 (Paya-Perez et al., 1991)
4.26 (lake sediment, Schrap et al., 1994)

Octanol/water partition coefficient, log K_{ow}:
4.37 (Watarai et al., 1982)
4.46 (Könemann et al., 1979)
4.83 at 13 °C, 4.61 at 19 °C, 4.37 at 28 °C, 4.25 at 33 °C (Opperhuizen et al., 1988)
4.635 (de Bruijn et al., 1989)
4.41 (Hammers et al., 1982)
4.55 (Miller et al., 1984)
4.60 (Chiou, 1985)
4.94 (Bruggeman et al., 1982)

Solubility in organics:
Soluble in acetic acid, ether and lignoin (Weast, 1986).

Solubility in water:
3.5 ppm at 22 °C (Verschueren, 1983)
3.27 mg/L (Kim and Saleh, 1990)
5.92 mg/L at 25 °C (Banerjee, 1984)
56.5 μmol/L at 25 °C (Miller et al., 1984)
20.0 μmol/L at 25 °C (Yalkowsky et al., 1979)

Vapor pressure (10^2 mmHg at 25 °C):
2.6 (Bidleman, 1984)
3.0 (Hinckley et al., 1990)

Environmental Fate
Biological. A mixed culture of soil bacteria or a *Pseudomonas* sp. transformed 1,2,3,4-tetrachlorobenzene to 2,3,4,5-tetrachlorophenol (Ballschiter and Scholz, 1980).
Photolytic. Irradiation ($\lambda \geq 285$ nm) of 1,2,3,4-tetrachlorobenzene (1.1-1.2 mM/L) in an acetonitrile-water mixture containing acetone (0.553 mM/L) as a sensitizer gave the following products (% yield): 1,2,3-trichlorobenzene (9.2), 1,2,4-trichlorobenzene (32.6), 1,3-dichlorobenzene (5.2), 1,4-dichlorobenzene (1.5), 2,2',3,3',4,4',5-heptachlorobiphenyl (2.52), 2,2',3,3',4,5,6'-heptachlorobiphenyl (1.22), 10 hexachlorobiphenyls (3.50), five pentachlorobiphenyls (0.87), dichlorophenyl cyanide, two trichloroacetophenones, trichlorocyanophenol, (trichlorophenyl)acetonitriles and 1-(trichlorophenyl)-2-propanone (Choudhry and Hutzinger, 1984). Without acetone, the identified photolysis products (% yield)

included 1,2,3-trichlorobenzene (7.8), 1,2,4-trichlorobenzene (26.8), 1,2-dichlorobenzene (0.5), 1,3-dichlorobenzene (0.7), 1,4-dichlorobenzene (30.4), 1,2,3,5-tetrachlorobenzene (2.26), 1,2,4,5-tetrachlorobenzene (0.72), 2,2′,3,3′,4,4′,5-heptachlorobiphenyl (<0.01) and 2,2′,3,3′,4,5,6′-heptachlorobiphenyl (<0.01) (Choudhry and Hutzinger, 1984). The sunlight irradiation of 1,2,3,4-tetrachloro-benzene (20 g) in a 100-mL borosilicate glass-stoppered Erlenmeyer flask for 56 days yielded 4,280 ppm heptachlorobiphenyl (Uyeta et al., 1976).

When an aqueous solution containing 1,2,3,4-tetrachlorobenzene and a nonionic surfactant micelle (Brij 58, a polyoxyethylene cetyl ether) was illuminated by a photoreactor equipped with 253.7-nm monochromatic ultraviolet lamps, no photo-isomerization was observed. However, based on photodechlorination of other polychlorobenzenes under similar conditions, it was suggested that tri- and dichlorobenzenes, chlorobenzene, benzene, phenol, hydrogen and chloride ions are likely to form. The photodecomposition half-life for this reaction, based on the first-order photodecomposition rate of 5.24×10^{-3}/sec, is 2.2 minutes (Chu and Jafvert, 1994).

Toxicity: Acute oral LD_{50} for rats 1,167 mg/kg (RTECS, 1985).

Uses: Dielectric fluids; organic synthesis.

1,2,3,5-TETRACHLOROBENZENE

Synonym: 1,2,3,5-TCB.

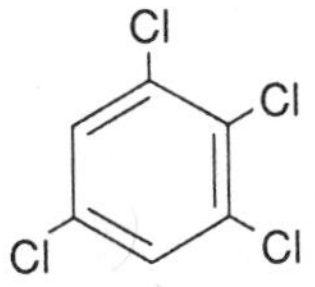

CAS Registry No.: 634-90-2
Molecular formula: $C_6H_2Cl_4$
Formula weight: 215.89
RTECS: DB9445000

Physical state
Solid.

Melting point (°C):
54.5 (Weast, 1986)
50-52 (Verschueren, 1983)

Boiling point (°C):
246 (Weast, 1986)

Diffusivity in water (10^5 cm^2/sec):
0.63 at 20 °C using method of Hayduk and Laudie (1974).

Heat of fusion (kcal/mol):
4.54 (Miller et al., 1984)

Henry's law constant (10^4 atm·m^3/mol):
15.8 at 25 °C (Hine and Mookerjee, 1975)
9.77 at 20.0 °C (ten Hulscher et al., 1992)

Bioconcentration factor, log BCF:
3.26 (bluegill sunfish, Veith et al., 1980)

Soil sorption coefficient, log K_{oc}:
6.0 (average using 2 suspended sediment samples from the St. Clair and Detroit
Rivers (Lau et al., 1989)
3.49 (Rothamsted Farm soil, Lord et al., 1980)
K_d = 7.2 mL/g on a Cs$^+$-kaolinite (Haderlein and Schwarzenbach, 1993)
4.23, 4.27, 5.19 (Paya-Perez et al., 1991)

939

Octanol/water partition coefficient, log K_{ow}:
4.46 (Veith et al., 1980)
4.658 (de Bruijn et al., 1989)
4.53 (Hammers et al., 1982)
4.65 (Miller et al., 1984)
4.50 (Könemann et al., 1979)
4.59 (Chiou, 1985)

Solubility in organics:
Soluble in alcohol, benzene, ether and lignoin (Weast, 1986).

Solubility in water:
19.055 μmol/L at 20 °C (Veith et al., 1980)
2.4 ppm at 22 °C (Verschueren, 1983)
2.79 mg/L (Kim and Saleh, 1990)
5.19 mg/L at 25 °C (Banerjee, 1984)
13.4 μmol/L at 25 °C (Miller et al., 1984)
16.2 μmol/L at 25 °C (Yalkowsky et al., 1979)

Vapor pressure (mmHg):
1 at 58.2 °C (estimated, Weast, 1986)
0.07 at 25 °C (extrapolated, Mackay et al., 1982)

Environmental Fate

Biological. A mixed culture of soil bacteria or a *Pseudomonas* sp. transformed 1,2,3,5-tetrachlorobenzene to 2,3,4,6-tetrachlorophenol (Ballschiter and Scholz, 1980).

Photolytic. Irradiation ($\lambda \geq 285$ nm) of 1,2,3,5-tetrachlorobenzene (1.1-1.2 mM/L) in an acetonitrile-water mixture containing acetone (0.553 mM/L) as a sensitizer gave the following products (% yield): 1,2,3-trichlorobenzene (5.3), 1,2,4-trichlorobenzene (4.9), 1,3,5-trichlorobenzene (49.3), 1,3-dichlorobenzene (1.8), 2,3,4,4',5,5',6-heptachlorobiphenyl (1.41), 2,2',3,4,4',6,6'-heptachlorobiphenyl (1.10), 2,2',3,3',4,5',6-heptachloro-biphenyl (4.50), four hexachlorobiphenyls (4.69), one pentachlorobiphenyl (0.64), trichloroacetophenone, 1-(trichlorophenyl)-2-propanone and (trichlorophenyl)acetonitrile (Choudhry and Hutzinger, 1984). Without acetone, the identified photolysis products (% yield) included 1,2,3-trichlorobenzene (trace), 1,2,4-trichlorobenzene (24.3), 1,3,5-trichlorobenzene (11.7), 1,3-dichlorobenzene (0.5), 1,4-dichlorobenzene (3.3), 1,2,3,4,5-pentachlorobenzene (1.43), 1,2,3,4-tetrachlorobenzene (5.99), two heptachlorobiphenyls (1.40), two hexachlorobiphenyls (<0.01) and one pentachlorobiphenyl (0.75) (Choudhry and Hutzinger, 1984).

When an aqueous solution containing 1,2,3,5-tetrachlorobenzene and a nonionic surfactant micelle (Brij 58, a polyoxyethylene cetyl ether) was illuminated by a photoreactor equipped with 253.7-nm monochromatic ultraviolet lamps, no photo-isomerization was observed. However, based on photodechlorination of other polychlorobenzenes under similar conditions, it was suggested that tri- and dichlorobenzenes, chlorobenzene, benzene, phenol, hydrogen and chloride ions are likely to form. The photodecomposition half-life for this reaction, based on the first-order photodecomposition rate of 2.67×10^{-3}/sec, is 4.3 minutes (Chu and Jafvert, 1994).

Toxicity: Acute oral LD_{50} for rats 1,727 mg/kg (RTECS, 1985).

Uses: Organic synthesis.

1,2,4,5-TETRACHLOROBENZENE

Synonyms: *sym*-Tetrachlorobenzene; RCRA waste number U207.

CAS Registry No.: 95-94-3
Molecular formula: $C_6H_2Cl_4$
Formula weight: 215.89
RTECS: DB9450000

Physical state, color and odor
White, odorless flakes or needles.

Melting point (°C):
139-140 (Weast, 1986)

Boiling point (°C):
243-246 (Weast, 1986)

Density (g/cm^3):
1.858 at 21/4 °C (Verschueren, 1983)
1.734 at 10/4 °C (Bailey and White, 1965)

Diffusivity in water (10^5 cm^2/sec):
0.63 at 20 °C using method of Hayduk and Laudie (1974).

Flash point (°C):
155 (Hawley, 1981)

Heat of fusion (kcal/mol):
5.76 (Miller et al., 1984)

Henry's law constant (10^2 atm·m^3/mol):
1.0 at 20 °C (Oliver, 1985)

Soil sorption coefficient, log K_{oc}:
3.72, 3.89, 3.91 (Schwarzenbach and Westall, 1981)
6.1 (average of 5 suspended sediment samples from the St. Clair and Detroit Rivers (Lau et al., 1989)
2.79 (McLaurin sandy loam, Walton et al., 1992)

Octanol/water partition coefficient, log K_{ow}:
4.56 (Watarai et al., 1982)
4.67 (Kenaga and Goring, 1980)
4.52 (Hammers et al., 1982; Könemann et al., 1979)
4.604 (de Bruijn et al., 1989)
4.51 (Miller et al., 1984)
4.70 (Chiou, 1985)

Solubility in organics:
Soluble in benzene, chloroform and ether (Weast, 1986).

Solubility in water:
10.9 μmol/L at 25 °C (Miller et al., 1984)
0.3 ppm at 22 °C (Verschueren, 1983)
0.56 mg/L (Kim and Saleh, 1990)
0.6 mg/L (Geyer et al., 1980)
0.465 mg/L at 25 °C (Banerjee, 1984)
2.75 μmol/L at 25 °C (Yalkowsky et al., 1979)

Vapor pressure (10^3 mmHg):
5 at 25 °C (extrapolated, Mackay et al., 1982)

Environmental Fate
    Biological. A mixed culture of soil bacteria or a *Pseudomonas* sp. transformed 1,2,4,5-tetrachlorobenzene to 2,3,5,6-tetrachlorophenol (Ballschiter and Scholz, 1980).
    Photolytic. Irradiation ($\lambda \geq 285$ nm) of 1,2,4,5-tetrachlorobenzene (1.1-1.2 mM/L) in an acetonitrile-water mixture containing acetone (0.553 mmol/L) as a sensitizer gave the following products (% yield): 1,2,4-trichlorobenzene (25.3), 1,3-dichlorobenzene (8.1), 1,4-dichlorobenzene (3.6), 2,2',3,4',5,5',6-heptachlorobiphenyl (4.19), four hexachlorobiphenyls (6.78), four pentachlorobiphenyls (2.33), one tetrachlorobiphenyl (0.32), 2,4,5-trichloroacetophenone and (2,4,5-trichlorophenyl)acetonitrile (Choudhry and Hutzinger, 1984). Without acetone, the identified photolysis products (% yield) included 1,2,4-trichlorobenzene (27.7), 1,3-dichlorobenzene (0.3), 1,4-dichlorobenzene (8.5), 1,2,3,4,5-pentachlorobenzene (trace), 1,2,3,4-tetrachlorobenzene (0.45), 1,2,3,5-tetrachlorobenzene (1.11), 2,2',3,4',5,5',6-heptachlorobiphenyl (1.24), three hexachlorobiphenyls (1.19) and four pentachlorobiphenyls (0.56) (Choudhry and Hutzinger, 1984). The sunlight irradiation of 1,2,4,5-tetrachlorobenzene (20 g) in a 100-mL borosilicate glass-stoppered Erlenmeyer flask for 28 days yielded 26 ppm heptachlorobiphenyl (Uyeta et al., 1976).
    When an aqueous solution containing 1,2,4,5-tetrachlorobenzene and a nonionic surfactant micelle (Brij 58, a polyoxyethylene cetyl ether) was illuminated by a

photoreactor equipped with 253.7-nm monochromatic ultraviolet lamps, no photo-isomerization was observed. However, based on photodechlorination of other polychlorobenzenes under similar conditions, it was suggested that tri- and dichlorobenzenes, chlorobenzene, benzene, phenol, hydrogen and chloride ions are likely to form. The photodecomposition half-life for this reaction, based on the first-order photodecomposition rate of 1.88×10^{-3}/sec, is 6.1 minutes (Chu and Jafvert, 1994).

Chemical/Physical. Based on an assumed base-mediated 1% disappearance after 16 days at 85 °C and pH 9.70 (pH 11.26 at °C), the hydrolysis half-life was estimated to be >900 years (Ellington et al., 1988).

Toxicity: Acute oral LD_{50} for rats 1,500 mg/kg, mice 1,035 mg/kg (RTECS, 1985).

Uses: Insecticides; intermediate for herbicides and defoliants; electrical insulation; impregnant for moisture resistance.

1,1,2,2-TETRACHLOROETHANE

Synonyms: Acetosol; Acetylene tetrachloride; Bonoform; Cellon; 1,1-Dichloro-2,2-dichloroethane; Ethane tetrachloride; NCI-C03554; RCRA waste number U208; TCE; Tetrachlorethane; Tetrachloroethane; *sym*-Tetrachloroethane; UN 1702; Westron.

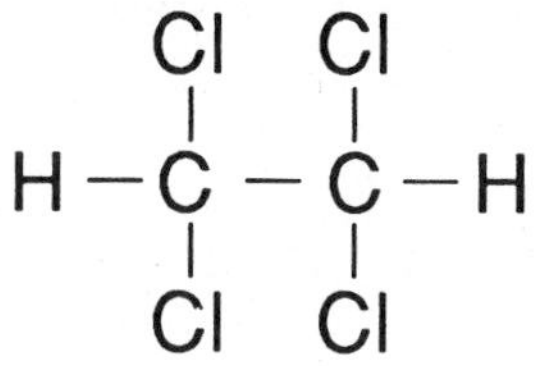

$$\underset{\underset{Cl}{|}}{\overset{\overset{Cl}{|}}{H-C}} - \underset{\underset{Cl}{|}}{\overset{\overset{Cl}{|}}{C}} - H$$

CAS Registry No.: 79-34-5
DOT: 1702
DOT label: Poison
Molecular formula: $C_2H_2Cl_4$
Formula weight: 167.85
RTECS: KI8575000

Physical state, color and odor
Colorless to pale yellow liquid with a sweet, chloroform-like odor. Odor threshold in air is 500 ppb (Keith and Walters, 1992).

Melting point (°C):
-36 (Weast, 1986)
-42.5 (Standen, 1964)

Boiling point (°C):
146.2 (Weast, 1986)
145.1 (Riddick et al., 1986)

Density (g/cm^3):
1.5953 at 20/4 °C (Weast, 1986)
1.60255 at 15/4 °C, 1.5869 at 25/4 °C, 1.57860 at 30/4 °C (Standen, 1964)

Diffusivity in water (10^5 cm^2/sec):
0.86 at 20 °C using method of Hayduk and Laudie (1974).

Flash point (°C):
Noncombustible liquid (NIOSH, 1994)

Henry's law constant (10^4 atm·m^3/mol):
3.8 (Pankow and Rosen, 1988)

4.56 at 25 °C (Hine and Mookerjee, 1975)
7.1 at 37 °C (Sato and Nakajima, 1979)
7 at 30 °C (Jeffers et al., 1989)
3.3, 2.0, 7.3, 2.5 and 7.0 at 10, 15, 20, 25 and 30 °C, respectively (Ashworth et al., 1988)
3, 5, 6 and 9 at 20, 30, 35 and 40 °C, respectively (Tse et al., 1992)

Ionization potential (eV):
11.10 (NIOSH, 1994)

Bioconcentration factor, log BCF:
0.90 (bluegill sunfish, Veith et al., 1980)

Soil sorption coefficient, log K_{oc}:
2.07 (Schwille, 1988)
1.663 (silt loam, Chiou et al., 1979)

Octanol/water partition coefficient, log K_{ow}:
2.56 (Mills et al., 1985)
2.39 (Banerjee et al., 1980; Yoshida et al., 1983)

Solubility in organics:
Soluble in acetone, ethanol, ether, benzene, carbon tetrachloride, petroleum ether, carbon disulfide, *N,N*-dimethylformamide and oils (U.S. EPA, 1985). Miscible with alcohol and chloroform (Meites, 1963).

Solubility in water:
2,970 mg/L at 25 °C (Banerjee et al., 1980)
0.287 wt % at 25 °C, 0.335 wt % at 55.6 °C (Stephen and Stephen, 1963)
3,230 mg/L at 20 °C (Chiou et al., 1979)
0.29 mass % at 20 °C (Konietzko, 1984)
2,910 mg/L at 20 °C, 2,920 mg/L at 29.7 °C (Stephenson, 1992)
0.296 wt % at 23.5 °C (Schwarz, 1980)
2,962 mg/L at 25 °C (Howard, 1990)
2,915 mg/L at 30 °C (McNally and Grob, 1984)

Vapor density:
5 kg/m^3 at the boiling point (Konietzko, 1984)
6.86 g/L at 25 °C, 5.79 (air = 1)

Vapor pressure (mmHg):
5 at 20 °C, 8.5 at 30 °C (Verschueren, 1983)
6.5 at 25 °C (Mackay et al., 1982)

Environmental Fate

Biological. Monodechlorination by microbes under laboratory conditions produced 1,1,2-trichloroethane (Smith and Dragun, 1984). In a static-culture-flask screening test, 1,1,2,2-tetrachloroethane (5 and 10 mg/L) was statically incubated in the dark at 25 °C with yeast extract and settled domestic wastewater inoculum. No significant degradation was observed after 28 days of incubation (Tabak et al., 1981).

Chemical/Physical. In an aqueous solution containing 0.100 M phosphate-buffered distilled water, 1,1,2,2-tetrachloroethane was abiotically transformed to 1,1,2-trichloroethane. This reaction was investigated over a temperature range of 30-95 °C at various pHs (5-9) (Cooper et al., 1987). Abiotic dehydrohalogenation of 1,1,2,2-tetrachloroethane was reported to yield 1,1,1-trichloroethylene. The half-life for this reaction at 20 °C was reported to be 0.8 years (Vogel et al., 1987). Under alkaline conditions, 1,1,2,2-tetrachloroethane dehydrohalogenated to trichloroethylene. The reported hydrolysis half-life of 1,1,2,2-tetrachloroethane in water at 25 °C and pH 7 is 146 days (Jeffers et al., 1989).

The evaporation half-life of 1,1,2,2-tetrachloroethane (1 mg/L) from water at 25 °C using a shallow-pitch propeller stirrer at 200 rpm at an average depth of 6.5 cm is 55.2 minutes (Dilling, 1977).

Exposure Limits: Potential occupational carcinogen. NIOSH REL: TWA 1 ppm (7 mg/m^3), IDLH 100 ppm; OSHA PEL: TWA 5 ppm (35 mg/m^3); ACGIH TLV: TWA 1 ppm.

Toxicity: Acute oral LD_{50} for rats 800 mg/kg (RTECS, 1985).

Drinking Water Standard: As of May 1994, no MCLGs or MCLs have been proposed although 1,1,2,2-tetrachloroethane has been listed for regulation (U.S. EPA, 1994).

Uses: Solvent for chlorinated rubber; insecticide and bleach manufacturing; paint, varnish and rust remover manufacturing; degreasing, cleansing and drying of metals; denaturant for ethyl alcohol; preparation of 1,1-dichloroethylene; extractant and solvent for oils and fats; insecticides; weed killer; fumigant; intermediate in the manufacturing of other chlorinated hydrocarbons; herbicide.

TETRACHLOROETHYLENE

Synonyms: Ankilostin; Antisol 1; Carbon bichloride; Carbon dichloride; Dee-Solv; Didakene; Dow-per; ENT 1860; Ethylene tetrachloride; Fedal-UN; NCI-C04580; Nema; PCE; PER; Perawin; PERC; Perchlor; Perchlorethylene; Perchloroethylene; Perclene; Perclene D; Percosolv; Perk; Perklone; Persec; RCRA waste number U210; Tetlen; Tetracap; Tetrachlorethylene; **Tetrachloroethene**; 1,1,2,2-Tetrachloroethylene; Tetraleno; Tetralex; Tetravec; Tetroguer; Tetropil; UN 1897.

$$\begin{array}{ccc} Cl & & Cl \\ \diagdown & & \diagup \\ & C = C & \\ \diagup & & \diagdown \\ Cl & & Cl \end{array}$$

CAS Registry No.: 127-18-4
DOT: 1897
Molecular formula: C_2Cl_4
Formula weight: 165.83
RTECS: KX3850000

Physical state, color and odor
Colorless liquid with a chloroform or sweet, ethereal odor. Odor threshold in air is 27 (Amoore and Hautala, 1983).

Melting point (°C):
-19 (Weast, 1986)
-22.35 (McGovern, 1943)

Boiling point (°C):
121.2 (Dean, 1973)

Density (g/cm^3):
1.623 at 20/4 °C (McGovern, 1943)
1.63109 at 15/4 °C, 1.62260 at 20/4 °C, 1.60640 at 30/4 °C (Standen, 1964)

Diffusivity in water (10^5 cm^2/sec):
0.87 at 20 °C using method of Hayduk and Laudie (1974).

Heat of fusion (kcal/mol):
2.5 (Dean, 1987)

Henry's law constant (10^3 atm·m^3/mol):
15.3 (Pankow and Rosen, 1988)

2.87 at 25 °C (Warner et al., 1987)
59.2 at 37 °C (Sato and Nakajima, 1979)
14.6 at 20 °C (Roberts et al., 1985)
24.5 at 30 °C (Jeffers et al., 1989)
8.46, 11.1, 14.1, 17.1 and 24.5 at 10, 15, 20, 25 and 30 °C, respectively (Ashworth et al., 1988)
3.85, 5.19, 6.27, 10.07 and 14.72 at 2.0, 6.0, 10.0, 18.2 and 25.0 °C, respectively (Dewulf et al., 1995)

Interfacial tension with water (dyn/cm at 20 °C):
47.48 (Demond and Lindner, 1993)

Ionization potential (eV):
9.32 (NIOSH, 1994)
9.71 (Yoshida et al., 1983)

Bioconcentration factor, log BCF:
1.69 (bluegill sunfish, Veith et al., 1980)

Soil sorption coefficient, log K_{oc}:
2.42 (Abdul et al., 1987)
2.322 (silt loam soil, Chiou et al., 1979)
2.25, 2.31, 2.54 (various Norwegian soils, Seip et al., 1986)
2.63 (Catlin soil, Roy et al., 1985)
2.65 (silty clay), 2.55, 2.99 (coarse sand) (Pavlostathis and Mathavan, 1992)

Octanol/water partition coefficient, log K_{ow}:
2.10 (Banerjee et al., 1980)
2.53 (Veith et al., 1980)

Solubility in organics:
Miscible with many other organic solvents (Keith and Walters, 1992).

Solubility in water:
150 mg/L at 20 °C (Pearson and McConnell, 1975)
2,200 mg/L at 20 °C (Chiou et al., 1977)
485 mg/L at 25 °C (Banerjee et al., 1980)
149 mg/L at 20 °C (Munz and Roberts, 1987)
240 mg/L at 23-24 °C (Broholm and Feenstra, 1995)
2.78 x 10^{-5} at 25 °C (mole fraction, Li et al., 1993)

Vapor density:
6.78 g/L at 25 °C, 5.72 (air = 1)

Vapor pressure (mmHg):
24 at 30 °C (Verschueren, 1983)
18.6 at 25 °C (Mackay et al., 1982)
20 at 25 °C (Valsaraj, 1988)
14 at 20 °C (McConnell et al., 1975)

Environmental Fate

Biological. Sequential dehalogenation by microbes under laboratory conditions produced trichloroethylene, *cis*-1,2-dichloroethylene, *trans*-1,2-dichloroethylene and vinyl chloride (Smith and Dragun, 1984). A microcosm composed of aquifer water and sediment collected from uncontaminated sites in the Everglades biotransformed tetrachloroethylene to *cis*- and *trans*-1,2-dichloroethylene (Parsons and Lage, 1985). Microbial degradation to trichloroethylene under anaerobic conditions or using mixed cultures was also reported (Vogel et al., 1987).

In a continuous-flow mixed-film methanogenic column study, tetrachloroethylene degraded to trichloroethylene with traces of vinyl chloride, dichloroethylene isomers and carbon dioxide (Vogel and McCarty, 1985). In a static-culture-flask screening test, tetrachloroethylene (5 and 10 mg/L) was statically incubated in the dark at 25 °C with yeast extract and settled domestic wastewater inoculum. Significant degradation with gradual adaptation was observed after 28 days of incubation. The amount lost due to volatilization after 10 days was 16-23% (Tabak et al., 1981).

Photolytic. Photolysis in the presence of nitrogen oxides yielded phosgene (carbonyl chloride) with minor amounts of carbon tetrachloride, dichloroacetyl chloride and trichloroacetyl chloride (Howard, 1990). In sunlight, photolysis products reported include chlorine, hydrogen chloride and trichloroacetic acid. Tetrachloroethylene reacts with ozone to produce a mixture of phosgene and trichloroacetyl chloride with a reported half-life of 8 days (Fuller, 1976). Reported photooxidation products in the troposphere include trichloroacetyl chloride and phosgene (Andersson et al., 1975; Gay et al., 1976; U.S. EPA, 1975). Phosgene is hydrolyzed readily to hydrogen chloride and carbon dioxide (Morrison and Boyd, 1971).

Chemical/Physical. The experimental half-life for hydrolysis of tetrachloroethylene in water at 25 °C is 8.8 months (Dilling et al., 1975).

The evaporation half-life of tetrachloroethylene (1 mg/L) from water at 25 °C using a shallow-pitch propeller stirrer at 200 rpm at an average depth of 6.5 cm is 25.4 minutes (Dilling, 1977).

At elevated temperatures, tetrachloroethylene decomposes to hydrogen chloride and phosgene (NIOSH, 1994).

Exposure Limits: Potential occupational carcinogen. NIOSH REL: IDLH 150 ppm; OSHA PEL: TWA 100 ppm, ceiling 200 ppm, 5-minute/3-hour peak 300 ppm; ACGIH TLV: TWA 50 ppm (335 mg/m^3), STEL 200 ppm (1,340 mg/m^3).

Symptoms of Exposure: May cause headache, dizziness, drowsiness, incoordination, irritation of eyes, nose and throat. Narcotic at high concentrations (Patnaik, 1992).

Drinking Water Standard (final): MCLG: zero; MCL: 5 μg/L (U.S. EPA, 1994).

Toxicity: Acute oral LD_{50} for mice 8,100 mg/kg, rats 3,005 mg/kg; LC_{50} for red killifish 263 mg/L (Yoshioka et al., 1986).

Uses: Dry cleaning fluid; degreasing and drying metals and other solids; solvent for waxes, greases, fats, oils, gums; manufacturing printing inks and paint removers; preparation of fluorocarbons and trichloroacetic acid; vermifuge; heat-transfer medium; organic synthesis.

TETRAETHYL PYROPHOSPHATE

Synonyms: Bis-*O,O*-diethylphosphoric anhydride; Bladan; **Diphosphoric acid tetraethyl ester;** ENT 18771; Ethyl pyrophosphate; Fosvex; Grisol; Hept; Hexamite; Killax; Kilmite 40; Lethalaire G-52; Lirohex; Mortopal; NA 2783; Nifos; Nifos T; Nifost; Pyrophosphoric acid tetraethyl ester; RCRA waste number P111; TEP; TEPP; Tetraethyl diphosphate; Tetraethyl pyrofosfaat; Tetrastigmine; Tetron; Tetron-100; Vapotone.

CAS Registry No.: 107-49-3
DOT: 2784
Molecular formula: $C_8H_{10}O_7P_2$
Formula weight: 290.20
RTECS: UX6825000

Physical state, color and odor
Colorless to amber liquid with an agreeable, fruity odor.

Melting point (°C):
0 (NIOSH, 1994)

Boiling point (°C):
Decomposes at 170-213 °C releasing ethylene (Windholz et al., 1983).
135-138 at 1 mmHg (Verschueren, 1983)

Density (g/cm³):
1.185 at 20/4 °C (Windholz et al., 1983)

Diffusivity in water:
Not applicable - reacts with water.

Flash point (°C):
Noncombustible liquid (NIOSH, 1994)

Soil sorption coefficient, log K_{oc}:
Not applicable - reacts with water.

Octanol/water partition coefficient, log K_{ow}:
Not applicable - reacts with water.

Solubility in organics:
Miscible with acetone, benzene, carbon tetrachloride, chloroform, ethanol, ethylene glycol, glycerol, methanol, propylene glycol, toluene and xylene (Windholz et al., 1983).

Solubility in water:
Miscible.

Vapor density:
11.86 g/L at 25 °C, 10.02 (air = 1)

Vapor pressure (10^4 mmHg):
4.7 at 30 °C (Windholz et al., 1983)
1.55 at 20 °C (Verschueren, 1983)

Environmental Fate
  Chemical/Physical. Tetraethyl pyrophosphate is quickly hydrolyzed by water forming pyrophosphoric acid (NIOSH, 1994). The reported hydrolysis half-life at 25 °C is 7 hours (Windholz et al., 1983). Decomposes at 170-213 °C releasing large amounts of ethylene (Hartley and Kidd, 1987).
  At 170-213 °C, tetraethyl pyrophosphate decomposes releasing copious amounts of ethylene (Keith and Walters, 1992).

Exposure Limits (mg/m^3): NIOSH REL: TWA 0.05, IDLH 5; OSHA PEL: TWA 0.05.

Toxicity: Acute oral LD_{50} for wild birds 1.30 mg/kg, ducks 3.56 mg/kg, guinea pigs 2.3 mg/kg, mice 7 mg/kg, rats 500 μg/kg (RTECS, 1985); LC_{50} (48-hour) for red killifish 1,020 mg/L (Yoshioka et al., 1986).

Uses: Insecticide for mites and aphids; rodenticide.

TETRAHYDROFURAN

Synonyms: Butylene oxide; Cyclotetramethylene oxide; Diethylene oxide; Furanidine; Hydrofuran; NCI-C60560; Oxacyclopentane; Oxolane; RCRA waste number U213; Tetramethylene oxide; THF; UN 2506.

CAS Registry No.: 109-99-9
DOT: 2056
DOT label: Flammable liquid
Molecular formula: C_4H_8O
Formula weight: 72.11
RTECS: LU5950000

Physical state, color and odor
Colorless liquid with an ether-like odor. Odor threshold in air is 2 ppm (Amoore and Hautala, 1983).

Melting point (°C):
-108 (Weast, 1986)
-65 (Hawley, 1981)

Boiling point (°C):
67 (Weast, 1986)
65.4 (Sax and Lewis, 1987)

Density (g/cm³):
0.8892 at 20/4 °C (Weast, 1986)

Diffusivity in water (10^5 cm²/sec):
1.00 at 20 °C using method of Hayduk and Laudie (1974).

Flash point (°C):
-14.6 (NIOSH, 1994)
-17.2 (Windholz et al., 1983)

Lower explosive limit (%):
2 (NIOSH, 1994)
1.8 (Sax and Lewis, 1987)

Upper explosive limit (%):
11.8 (NIOSH, 1994)

Henry's law constant (10^5 atm·m^3/mol):
7.06 at 25 °C (Hine and Mookerjee, 1975)

Ionization potential (eV):
9.45 (NIOSH, 1994)
9.54 (HNU, 1986)

Soil sorption coefficient, log K_{oc}:
1.37 (Captina silt loam), 1.26 (McLaurin sandy loam) (Walton et al., 1992)

Octanol/water partition coefficient, log K_{ow}:
0.46 (Hansch and Leo, 1979)

Solubility in organics:
Soluble in alcohols, ketones, esters, ethers and hydrocarbons (Windholz et al., 1983).

Solubility in water:
Miscible (Fischer and Ehrenberg, 1948; Palit, 1947)
4.2 M at 25 °C (Fischer and Ehrenberg, 1948)

Vapor density:
2.95 g/L at 25 °C, 2.49 (air = 1)

Vapor pressure (mmHg):
83.6 at 10 °C, 131.5 at 20 °C, 197.6 at 30 °C (Verschueren, 1983)
114 at 15 °C (Sax and Lewis, 1987)

Exposure Limits: NIOSH REL: TWA 200 ppm (590 mg/m^3), STEL 250 ppm (735 mg/m^3), IDLH 2,000 ppm; OSHA PEL: TWA 200 ppm; ACGIH TLV: TWA 200 ppm, STEL 250 ppm.

Symptoms of Exposure: Vapors may irritate respiratory tract and eyes. An anesthetic at high concentrations (Patnaik, 1992).

Toxicity: Acute oral LD$_{50}$ for rats 2,816 mg/kg (RTECS, 1985).

Uses: Solvent for uncured rubber and polyvinyl chlorides, vinyl chloride copolymers, vinylidene chloride copolymers, natural resins; topcoating solutions; cellophane; magnetic tapes; adhesives; printing inks; organic synthesis.

1,2,4,5-TETRAMETHYLBENZENE

Synonyms: Durene; Durol; *sym*-1,2,4,5-Tetramethylbenzene.

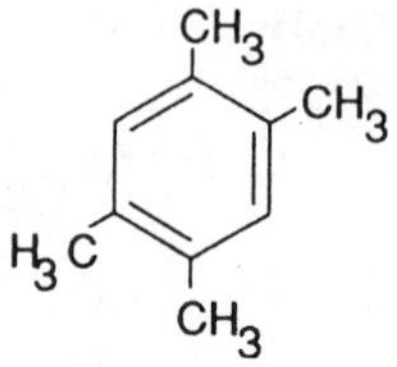

CAS Registry No.: 95-93-2
Molecular formula: $C_{10}H_{14}$
Formula weight: 134.22
RTECS: DC0500000

Physical state, color and odor
Colorless crystals or scales with a camphor-like odor.

Melting point (°C):
79.2 (Weast, 1986)
77 (Verschueren, 1983)

Boiling point (°C):
196.8 (Weast, 1986)
191–193 (Windholz et al., 1983)

Density (g/cm³):
0.8380 at 81/4 °C (Weast, 1986)

Diffusivity in water (10^5 cm²/sec):
0.62 at 20 °C using method of Hayduk and Laudie (1974).

Dissociation constant:
>14 (Schwarzenbach et al., 1993)

Flash point (°C):
54 (NFPA, 1984)
73 (Dean, 1987)

Heat of fusion (kcal/mol):
5.020 (Tsonopoulos and Prausnitz, 1971)

Henry's law constant (10^2 atm·m³/mol):
2.49 at 25 °C (approximate - calculated from water solubility and vapor pressure)

Soil sorption coefficient, log K_{oc}:
3.79 using method of Karickhoff et al. (1979).

Octanol/water partition coefficient, log K_{ow}:
4.00 (Camilleri et al., 1988)

Solubility in organics:
Soluble in acetone, alcohol, benzene and ether (Weast, 1986).

Solubility in water:
3.48 mg/kg at 25 °C (Price, 1976)

Vapor pressure (mmHg):
0.49 at 25 °C (Mackay et al., 1982)

Toxicity: Acute oral LD_{50} for rats 6,989 mg/kg (RTECS, 1985).

Uses: Plasticizers; polymers; fibers; organic synthesis.

TETRANITROMETHANE

Synonyms: NCI–C55947; RCRA waste number P112; Tetan; TNM; UN 1510.

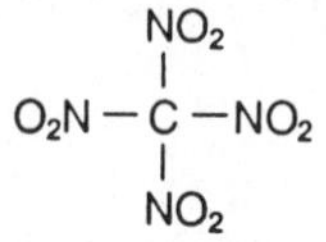

$$O_2N - \overset{\displaystyle NO_2}{\underset{\displaystyle NO_2}{\overset{|}{\underset{|}{C}}}} - NO_2$$

CAS Registry No.: 509-14-8
DOT: 1510
Molecular formula: CN_4O_8
Formula weight: 196.03
RTECS: PB4025000

Physical state, color and odor
Colorless to pale yellow liquid or solid with a pungent odor.

Melting point (°C):
14.2 (Weast, 1986)
12.5 (Hawley, 1981)

Boiling point (°C):
126 (Weast, 1986)

Density (g/cm³):
1.6380 at 20/4 °C (Weast, 1986)
1.6229 at 25/4 °C (Windholz et al., 1983)
1.650 at 13/4 °C (Hawley, 1981)

Diffusivity in water (10^5 cm²/sec):
0.79 at 20 °C using method of Hayduk and Laudie (1974).

Flash point (°C):
>112 (Dean, 1987)

Solubility in organics:
Miscible with alcohol and ether (Hawley, 1981).

Vapor density:
8.01 g/L at 25 °C, 6.77 (air = 1)

Vapor pressure (mmHg):
8 at 20 °C, 13 at 25 °C, 15 at 30 °C (Verschueren, 1983)

Exposure Limits: NIOSH REL: TWA 1 ppm (8 mg/m^3), IDLH 4 ppm; OSHA PEL: TWA 1 ppm.

Toxicity: Acute oral LD$_{50}$ for rats 130 mg/kg, mice 375 mg/kg; LC$_{50}$ (inhalation) for mice 54 ppm/4-hour, rats 18 ppm/4-hour (RTECS, 1985).

Drinking Water Standard: As of May 1994, no MCLGs or MCLs have been proposed (U.S. EPA, 1984).

Uses: Laboratory reagent for detecting double bonds in organic compounds; oxidizer in rocket propellants; diesel fuel booster.

TETRYL

Synonyms: *N*-Methyl-*N*,2,4,6-tetranitroaniline; **N-Methyl-N,2,4,6-tetranitro-benzenamine**; Nitramine; Picrylmethylnitramine; Picrylnitromethylamine; Tetralit; Tetralite; Tetril; 2,4,6-Tetryl; Trinitrophenylmethylnitramine; 2,4,6-Trinitrophenylmethylnitramine; 2,4,6-Trinitrophenyl-*N*-methylnitramine; UN 0208.

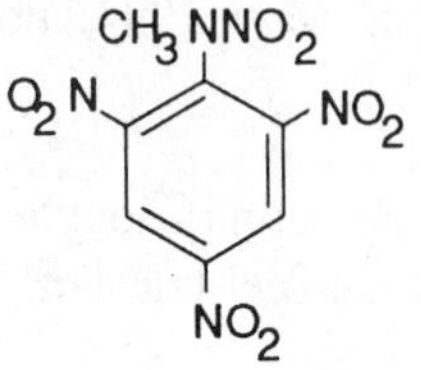

CAS Registry No.: 479-45-8
DOT: 0208
DOT label: Class A explosive
Molecular formula: $C_7H_5N_5O_8$
Formula weight: 287.15
RTECS: BY6300000

Physical state and color
Colorless to pale yellow crystals.

Melting point (°C):
131–132 (Weast, 1986)

Boiling point (°C):
Explodes at 187 (Weast, 1986).

Density (g/cm³):
1.57 at 10/4 °C (Weast, 1986)
1.57 at 19/4 °C (Verschueren, 1983)

Diffusivity in water (10^5 cm²/sec):
0.64 at 20 °C using method of Hayduk and Laudie (1974).

Flash point (°C):
Explodes (NIOSH, 1994)

Henry's law constant (10^3 atm·m³/mol):
<1.89 at 20 °C (approximate - calculated from water solubility and vapor pressure)

Soil sorption coefficient, log K_{oc}:
2.37 using method of Kenaga and Goring (1980).

Octanol/water partition coefficient, log K_{ow}:
2.04 using method of Kenaga and Goring (1980).

Solubility in organics:
Soluble in acetone, benzene (Weast, 1986), glacial acetic acid and ether (Windholz et al., 1983).

Solubility in water:
0.02 wt % at 20 °C (NIOSH, 1994)

Vapor pressure (mmHg):
<1 at 20 °C (NIOSH, 1994)

Environmental Fate
   Chemical/Physical. Produces highly toxic nitrogen oxides on decomposition (Lewis, 1990).

Exposure Limits (mg/m^3): NIOSH REL: 1.5, IDLH 750; OSHA PEL: TWA 1.5.

Uses: As an indicator in analytical chemistry; as a booster in artillery explosives.

THIOPHENE

Synonyms: CP 34; Divinylene sulfide; Huile H50; Huile HSO; Thiacyclopenta-diene; Thiaphene; Thiofuran; Thiofuram; Thiofurfuran; Thiole; Thiophen; Thio-tetrole; UN 2414; USAF EK-1860.

CAS Registry No.: 110-02-1
DOT: 2414
Molecular formula: C_4H_4S
Formula weight: 84.14
RTECS: XM7350000

Physical state, color and odor
Clear, colorless liquid with an aromatic odor resembling benzene.

Melting point (°C):
-38.2 (Weast, 1986)
-30 (Verschueren, 1983)

Boiling point (°C):
84.2 (Weast, 1986)

Density (g/cm³):
1.0649 at 20/4 °C (Weast, 1986)
1.0873 at 0/4 °C, 1.0573 at 25/4 °C (Windholz et al., 1983)

Diffusivity in water (10^5 cm²/sec):
1.01 at 20 °C using method of Hayduk and Laudie (1974).

Flash point (°C):
-1.1 (Hawley, 1981)

Heat of fusion (kcal/mol):
1.216 (Dean, 1987)

Henry's law constant (10^3 atm·m³/mol at 25 °C):
2.33 and 2.70 in distilled water and seawater, respectively (Przyjazny et al., 1983).

Ionization potential (eV):
8.860 (HNU, 1986)

Soil sorption coefficient, log K_{oc}:
1.73 using method of Kenaga and Goring (1980).

Octanol/water partition coefficient, log K_{ow}:
1.81 (Leo et al., 1971)
1.82 (Sangster, 1989)

Solubility in organics:
Miscible with carbon tetrachloride, heptane, pyrimidine, dioxane, toluene and many organic solvents (Keith and Walters, 1992).

Solubility in water:
3,015 mg/kg at 25 °C (Price, 1976)
3,600 mg/L at 18 °C (Verschueren, 1983)
2-5 mmol/L at 20 °C (Fischer and Ehrenberg, 1948)

Vapor density:
3.44 g/L at 25 °C, 2.90 (air = 1)

Vapor pressure (mmHg):
60 at 20 °C (Verschueren, 1983)
79.7 at 25 °C (Wilhoit and Zwolinski, 1971)
337 at 60.3 °C, 486 at 70.3 °C, 686 at 80.3 °C, 951 at 90.3 °C (Eon et al., 1971).

Toxicity: LD_{50} (intraperitoneal) for mice 100 mg/kg (RTECS, 1985).

Uses: Solvent; manufacturing resins, dyes and pharmaceuticals.

THIRAM

Synonyms: Aatack; Accelerator thiuram; Aceto TETD; Arasan; Arasan 70; Arasan 75; Arasan-M; Arasan 42-S; Arasan-SF; Arasan-SF-X; Aules; Bis(dimethylamino)carbonothioyl disulfide; Bis(dimethylthiocarbamoyl) disulfide; Bis(dimethylthiocarbamyl) disulfide; Chipco thiram 75; Cyuram DS; α,α′-Dithiobis(dimethylthio) formamide; N,N′-(Dithiodicarbonothioyl)bis(N-methylmethanamine); Ekagom TB; ENT 987; Falitram; Fermide; Fernacol; Fernasan; Fernasan A; Fernide; Flo pro T seed protectant; Hermal; Hermat TMT; Heryl; Hexathir; Kregasan; Mercuram; Methyl thiram; Methyl thiuramdisulfide; Methyl tuads; NA 2771; Nobecutan; Nomersan; Normersan; NSC 1771; Panoram 75; Polyram ultra; Pomarsol; Pomersol forte; Pomasol; Puralin; RCRA waste number U244; Rezifilm; Royal TMTD; Sadoplon; Spotrete; Spotrete-F; SQ 1489; Tersan; Tersan 75; Tetramethyldiurane sulphite; Tetramethylthiuram bisulfide; Tetramethylthiuram bisulphide; Tetramethylenethiuram disulfide; **Tetramethylthioperoxydicarbonic diamide**; Tetramethylthiocarbamoyl disulfide; Tetramethylthioperoxydicarbonic diamide; Tetramethylthiuram disulfide; Tetramethylthiuram disulphide; N,N-Tetramethylthiuram disulfide; N,N,N′,N′-Tetramethylthiuram disulfide; Tetramethylthiuran disulfide; Tetramethylthiurane disulphide; Tetramethylthiurum disulfide; Tetramethylthiurum disulphide; Tetrapom; Tetrasipton; Tetrathiuram disulfide; Tetrathiuram disulphide; Thillate; Thimer; Thiosan; Thiotex; Thiotox; Thiram 75; Thiramad; Thiram B; Thirasan; Thiulix; Thiurad; Thiuram; Thiuram D; Thiuramin; Thiuram M; Thiuram M rubber accelerator; Thiuramyl; Thylate; Tirampa; Tiuramyl; TMTD; TMTDS; Trametan; Tridipam; Tripomol; TTD; Tuads; Tuex; Tulisan; USAF B-30; USAF EK-2089; USAF P-5; Vancida TM-95; Vancide TM; Vulcafor TMTD; Vulkacit MTIC.

CAS Registry No.: 137-26-8
DOT: 2771
DOT label: Poison
Molecular formula: $C_6H_{12}N_2S_4$
Formula weight: 269.35
RTECS: JO1400000

Physical state and color
Colorless to white to cream colored crystals. May darken on exposure to air or light.

Melting point (°C):
155.6 (Weast, 1986)
146 (Sax and Lewis, 1987)

Boiling point (°C):
310-315 at 15 mmHg (Weast, 1986)
129 at 20 mmHg (Sax and Lewis, 1987)

Density (g/cm^3):
1.29 at 20/4 °C (Hawley, 1981)

Solubility in organics:
In wt %: acetone (1.2), alcohol (<0.2), benzene (2.5) and ether (<0.2) (Windholz et al., 1983).

Solubility in water:
30 ppm (Meister, 1988)

Vapor Pressure (mmHg):
8 x 10^{-6} at 20 °C (NIOSH, 1994)

Environmental Fate
Biological. Odeyemi and Alexander (1977) isolated three strains of *Rhizobium* sp. that degraded thiram. One of these strains, *Rhizobium meliloti*, metabolized thiram to yield dimethylamine (DMA) and carbon disulfide, which formed spontaneously from dimethyldithiocarbamate (DMDT). The conversion of DMDT to DMA and carbon disulfide occurred via enzymatic and nonenzymatic mechanisms.

Soil. In both soils and water, chemical and biological mediated reactions can transform thiram to compounds containing the mercaptan group (Alexander, 1981). Decomposes in soils to carbon disulfide and dimethylamine (Sisler and Cox, 1954; Kaars Sijpesteijn et al., 1977). When a spodosol (pH 3.8) pretreated with thiram was incubated for 24 days at 30 °C and relative humidity of 60-90%, dimethylamine formed as the major product. Minor degradative products included nitrite ions (nitration reduction) and dimethylnitrosamine (Ayanaba et al., 1973).

Plant. Major plant metabolites are ethylene thiourea, thiram monosulfide, ethylene thiram disulfide and sulfur (Hartley and Kidd, 1987).

Chemical/Physical. Emits toxic sulfur oxides when heated to decomposition (Lewis, 1990). Though no products were reported, the calculated hydrolysis half-life at 25 °C and pH 7 is 5.3 days (Ellington et al., 1988).

Exposure Limits (mg/m^3): NIOSH REL: TWA 5, IDLH 100; OSHA PEL: TWA 5.

Symptoms of Exposure: Irritates mucous membrane, dermatitis; with ethanol consumption: flush, erythema, pruritus, urticaria, headache, nausea, vomiting, diarrhea, weakness, dizziness, difficulty in breathing. Contact with skin may cause allergic reaction (NIOSH, 1994).

Toxicity: Acute oral LD_{50} for wild birds 300 mg/kg, mice 1,350 mg/kg, rats 560 mg/kg, rabbits 210 mg/kg (RTECS, 1985); LC_{50} (96-hour) for rainbow trout 0.13 mg/L, bluegill sunfish 0.23 mg/L and carp 4.0 mg/L (Hartley and Kidd, 1987).

Uses: Vulcanizer; seed disinfectant; rubber accelerator; animal repellant; fungicide; bacteriostat in soap.

TOLUENE

Synonyms: Antisal 1a; Methacide; **Methylbenzene**; Methylbenzol; NCI-C07272; Phenylmethane; RCRA waste number U220; Toluol; Tolu-sol; UN 1294.

CAS Registry No.: 108-88-3
DOT: 1294
DOT label: Flammable liquid
Molecular formula: C_7H_8
Formula weight: 92.14
RTECS: XS5250000

Physical state, color and odor
Colorless, water-white liquid with a pleasant odor similar to benzene. Odor threshold in air is 170 ppb (Keith and Walters, 1992).

Melting point (°C):
-95 (Weast, 1986)

Boiling point (°C):
110.6 (Weast, 1986)

Density (g/cm^3):
0.8669 at 20/4 °C (Weast, 1986)
0.86233 at 25/4 °C (Huntress and Mulliken, 1941)
0.8666 at 20/4 °C, 0.8573 at 30/4 °C, 0.8480 at 40/4 °C (Sumer et al., 1968)

Diffusivity in water (10^5 cm^2/sec):
0.85 at 20 °C (Witherspoon and Bonoli, 1969)
0.95 at 25 °C (Hayduk and Laudie, 1974)

Dissociation constant:
$\approx$ 35 (Gordon and Ford, 1972)

Flash point (°C):
4.5 (NIOSH, 1994)

Lower explosive limit (%):
1.1 (NIOSH, 1994)

Upper explosive limit (%):
7.1 (NIOSH, 1994)

Heat of fusion (kcal/mol):
1.586 (Dean, 1987)

Henry's law constant (10^3 atm·m^3/mol):
6.7 (Pankow and Rosen, 1988)
5.94 (Howard, 1990)
3.81, 4.92, 5.55, 6.42 and 8.08 at 10, 15, 20, 25 and 30 °C, respectively (Ashworth et al., 1988)
2.89, 3.85, 4.92, 6.51 and 8.27 at 10, 15, 20, 25 and 30 °C, respectively (Perlinger et al., 1993)
1.88, 2.14, 2.60, 4.29 and 5.49 at 2.0, 6.0, 10.0, 18.2 and 25.0 °C, respectively (Dewulf et al., 1995)

Interfacial tension with water (dyn/cm at 25 °C):
36.1 (Demond and Lindner, 1993)

Ionization potential (eV):
8.82 ± 0.01 (Franklin et al., 1969)

Bioconcentration factor, log BCF:
0.92 (goldfish, Ogata et al., 1984)
1.12 (eels, Ogata and Miyake, 1978)
0.62 (mussels, Geyer et al., 1982)
2.58 (algae, Geyer et al., 1984)
3.28 (activated sludge), 1.95 (golden ide) (Freitag et al., 1985)

Soil sorption coefficient, log K_{oc}:
2.06 (Abdul et al., 1987)
2.18 (Garbarini and Lion, 1986)
1.74, 1.97, 2.13 (various Norwegian soils, Seip et al., 1986)
1.66 (Vandreil sandy loam), 2.20 (Grimsby silt loam), 1.57 (Wendover silty loam) (Nathwani and Phillips, 1977)
2.25 (sandy soil, Wilson et al., 1981)
2.00 (estuarine sediment, Vowles and Mantoura, 1987)
1.99-3.05 (silty clay), 3.03, 3.39 (coarse sand) (Pavlostathis and Mathavan, 1992)
2.22 (Captina silt loam), 2.16 (McLaurin sandy loam) (Walton et al., 1992)
2.08 (river sediment), 2.03 (coal wastewater sediment) (Kopinke et al., 1995)

Octanol/water partition coefficient, log K_{ow}:
2.61 (Sangster, 1989)

2.65 (Tewari et al., 1982; Wasik et al., 1981, 1983)
2.69 (Hansch et al., 1968)
2.21 (Veith et al., 1980)
2.63 (Brooke et al., 1990; Yalkowsky et al., 1983)
2.50 (Walton et al., 1989)
2.11, 2.80 (Leo et al., 1971)
2.73 (Campbell and Luthy, 1985; Leo et al., 1971)
2.78 (Burkhard et al., 1985a)
2.79 (Fujita et al., 1964; Brooke et al., 1990; de Bruijn et al., 1989)
2.68 (Nahum and Horvath, 1980)
2.72 (Garst and Wilson, 1984)

Solubility in organics:
Soluble in acetone, carbon disulfide and ligroin; miscible with acetic acid, ethanol, benzene, ether, chloroform and other solvents (U.S. EPA, 1985).

Solubility in water:
5.58, 5.71, 5.88 and 6.28 mmol/L at 15, 25, 35 and 45 °C, respectively (Sanemasa et al., 1982)
519.5 mg/L at 25 °C (Mackay and Shiu, 1975)
563.3 μL/L at 25 °C (Sada et al., 1975)
500 mg/L at 25 °C (Klevens, 1950)
16.8 mmol/L at 25 °C (Banerjee et al., 1980)
524 mg/L at 25 °C (Banerjee, 1984)
490 mg/L at 25 °C (Meites, 1963)
506.7 mg/kg at 25 °C. In natural seawater: 410, 410 and 418.5 mg/kg at 15, 20 and 25 °C, respectively (Rossi and Thomas, 1981)
515 mg/kg at 25 °C (McAuliffe, 1966)
In mg/L: 658 at 0.4 °C, 646 at 3.6 °C, 628 at 10.0 °C, 624 at 11.2 °C, 623 at 14.9 °C, 621 at 15.9 °C, 627 mg/L at 25.0 °C, 625 at 25.6 °C, 640 at 30.0 °C, 642 at 30.2 °C, 657 at 35.2 °C, 701 at 42.8 °C, 717 at 45.3 °C (Bohon and Claussen, 1951).
530 mg/L at 25 °C (Andrews and Keefer, 1949)
570 mg/kg at 30 °C (Gross and Saylor, 1931)
724 mg/L at 0 °C (Brookman et al., 1985)
0.0368 wt % at 10 °C, 0.0492 wt % at 25 °C, 0.0344 vol % at 25 °C, 627 mg/L at 25 °C, 0.057 wt % at 30 °C (Stephen and Stephen, 1963)
724 mg/kg at 0 °C, 573 mg/kg at 25 °C (Polak and Lu, 1973)
534.8 mg/kg at 25 °C, 379.3 mg/kg in artificial seawater at 25 °C (Sutton and Calder, 1975)
554.0 mg/kg at 25 °C. In NaCl solution at 25 °C (salinity, g/kg), mg/kg: 526 (1.002), 490 (10.000), 402.0 (34.472), 359 (50.030), 182 (125.100), 106 (199.900), 53.8 (279.800), 37.2 (358.700) (Price, 1976)

6.28 mmol/L at 25 °C (Tewari et al., 1982; Wasik et al., 1981, 1983)

6.29 mmol/L at 25.00 °C (Keely et al., 1988)

220 mg/L in fresh water at 25 °C, 230 mg/L in salt water at 25 °C (Krasnoshchekova and Gubergrits, 1975)

660, 670 mg/kg at 23.5 °C (Schwarz, 1980)

470 mg/L at 30 °C (NAS, 1980)

6.69 mmol/L at 25 °C (Ben-Naim and Wilf, 1980)

392 mg/L in seawater at 25 °C (Bobra et al., 1979)

4.19 mmol/L in 0.5 M NaCl at 25 °C (Wasik et al., 1984)

7.13 mmol/L at 35 °C (Hine et al., 1963)

479 mg/kg at 25 °C (Chey and Calder, 1972)

5.1 mmol/L at 16 °C (Fühner, 1924)

26 mmol/L at 25 °C (Hogfeldt and Bolander, 1963)

466.9 mg/L at 30 °C (McNally and Grob, 1984)

0.052 wt % at 25 °C (Lo et al., 1986)

538 mg/L (Coutant and Keigley, 1988)

5.82 mmol/kg at 25 °C (Morrison and Billett, 1952)

In mg/kg: 612, 601, 586, 587, 573, 575, 569, 577 and 566 at 4.5, 6.3, 7.1, 9.0, 11.8, 12.1, 15.1, 17.9 and 20.1 °C, respectively. In artificial seawater: 449, 429, 416, 405 and 397 at 0.19, 5.32, 10.05, 14.96 and 20.04 °C, respectively (Brown and Wasik, 1974)

In wt % (°C): 0.823 (114), 1.640 (147), 2.387 (169), 2.790 (183), 4.113 (207), 5.072 (224) (Guseva and Parnov, 1963)

6.81 mmol/kg at 25.0 °C (Vesala, 1974)

3.43 mmol/L at 25.00 °C (Sanemasa et al., 1985)

5.65 mmol/L at 25.0 °C (Sanemasa et al., 1987)

Vapor density:
3.77 g/L at 25 °C, 3.18 (air = 1)

Vapor pressure (mmHg):
22 at 20 °C (Verschueren, 1983)

36.7 at 30 °C (Sax, 1984)

28.4 at 25 °C (Howard, 1990)

28.1 at 25 °C (Mackay et al., 1982)

46.7 at 35 °C (Chey and Calder, 1972)

27.82 at 25.00 °C (Hussam and Carr, 1985)

Environmental Fate
Biological. Toluene can undergo two types of microbial attack. The first type proceeds via immediate hydroxylation of the benzene ring, followed by ring cleavage. The second type of attack proceeds via oxidation of the methyl group followed by hydroxylation and ring cleavage (Fewson, 1981). A mutant of

Pseudomonas putida oxidized toluene to (+)-*cis*-2,3-dihydroxy-1-methylcyclo-hexa-1,4-diene (Dagley, 1972). Claus and Waker (1964) reported that *Pseudomonas* sp. and an *Achromobacter* sp. oxidized toluene to 3-methyl catechol. Other metabolites identified in the microbial degradation of toluene include *cis*-2,3-di-hydroxy-2,3-dihydrotoluene, 3-methyl catechol, benzyl alcohol, benzaldehyde, benzoic acid, catechol (Verschueren, 1983) and 1-hydroxy-2-naphthoic acid (Claus and Walker, 1964). In a methanogenic aquifer material, toluene degraded completely to carbon dioxide (Wilson et al., 1986). In activated sludge, 26.3% of the applied toluene mineralized to carbon dioxide after 5 days (Freitag et al., 1985).

In anoxic groundwater near Bemidji, MI, toluene anaerobically biodegraded to the intermediate benzoic acid (Cozzarelli et al., 1990). Methylmuconic acid was reported to be the biooxidation product of toluene by *Nocardia corallina* V-49, using *n*-hexadecane as the substrate. With methane as the substrate and *Methylosinus trichosporium* OB3b as the microorganism, *p*-hydroxytoluene and benzoic acid are the products of biooxidation. In addition, *Methyloccus capsulatus* was reported to bioxidize toluene to benzyl alcohol and cresol (Keck et al., 1989). When toluene (5 and 10 mg/L) was statically incubated in the dark at 25 °C with yeast extract and settled domestic wastewater inoculum for 7 days, 100% biodegradation with rapid adaptation was observed (Tabak et al., 1981). Pure microbial cultures isolated from soil hydroxylated toluene to 2- and 4-hydroxy-toluene (Smith and Rosazza, 1974). When toluene (5 and 10 mg/L) was statically incubated in the dark at 25 °C with yeast extract and settled domestic wastewater inoculum for 7 days, complete biodegradation with rapid acclimation was observed (Tabak et al., 1981).

Photolytic. Synthetic air containing gaseous nitrous acid and exposed to artificial sunlight (λ = 300-450 nm) photooxidized toluene into methyl nitrate, peroxyacetal nitrate and a nitro aromatic compound tentatively identified as a nitrophenol or nitrocresol (Cox et al., 1980). An *n*-hexane solution containing toluene and spread as a thin film (4 mm) on cold water (10 °C) was irradiated by a mercury medium pressure lamp. In 3 hours, 26% of the toluene photooxidized into benzaldehyde, benzyl alcohol, benzoic acid and *m*-cresol (Moza and Feicht, 1989). Methane and ethane were reported as products of the gas-phase photolysis of toluene at 2537 Å (Calvert and Pitts, 1966).

Irradiation of toluene (80 ppm) by UV light (λ = 200-300 nm) on titanium dioxide in the presence of oxygen (20%) and moisture resulted in the formation of benzaldehyde and carbon dioxide. Carbon dioxide concentrations increased linearly with the increase in relative humidity. However, the concentration of benz-aldehyde decreased with an increase in relative humidity. An identical experiment, but without moisture, resulted in the formation of benzaldehyde, carbon dioxide, hydrogen cyanide and nitrotoluenes. In an atmosphere containing moisture and nitrogen dioxide (80 ppm), cresols, benzaldehyde, carbon dioxide and nitrotoluenes were the photoirradiation products (Ibusuki and Takeuchi, 1986).

Irradiation of toluene in the presence of chlorine yielded benzyl hydroperoxide, benzaldehyde, peroxybenzoic acid, carbon monoxide, carbon dioxide and other unidentified products (Hanst and Gay, 1983). The photooxidation of toluene in the presence of nitrogen oxides (NO and NO_2) yielded small amounts of formaldehyde and traces of acetaldehyde or other low molecular weight carbonyls (Altshuller et al., 1970). Other photooxidation products not previously mentioned include phenol, phthalaldehydes and benzoyl alcohol (Altshuller, 1983). A carbon dioxide yield of 8.4% was achieved when toluene adsorbed on silica gel was irradiated with light (λ >290 nm) for 17 hours (Freitag et al., 1985).

Chemical/Physical. Products identified from the reaction of toluene with nitric oxide and hydroxyl radicals include benzaldehyde, benzyl alcohol, *m*-nitrotoluene, *p*-methylbenzoquinone, *o*-, *m*- and *p*-cresol (Kenley et al., 1978). Gaseous toluene reacted with nitrate radicals in purified air forming the following products: benzaldehyde, benzyl alcohol, benzyl nitrate, 2-, 3- and 4-nitrotoluene (Chiodini et al., 1993).

Under atmospheric conditions, the gas-phase reaction with hydroxyl radicals and nitrogen oxides resulted in the formation of benzaldehyde, benzyl nitrate, *m*-nitrotoluene, *o*-, *m*- and *p*-cresol (Atkinson, 1990).

Kanno et al. (1982) studied the aqueous reaction of toluene and other aromatic hydrocarbons (benzene, *o*-, *m*- and *p*-xylene and naphthalene) with hypochlorous acid in the presence of ammonium ion. Although chlorination of the aromatic was observed with the formation of 2- and 4-chlorotoluene, the major degradative pathway was cleavage of the aromatic ring by chloramine, forming cyanogen chloride (Kanno et al., 1982). The amount of cyanogen chloride formed was inversely proportional to the pH of the solution. At pH 6, the greatest amount of cyanogen chloride was formed when the reaction mixture contained ammonium ion and hypochlorous acid at a ratio of 2:3 (Kanno et al., 1982).

Exposure Limits: NIOSH REL: TWA 100 ppm (375 mg/m^3), STEL 150 ppm (560 mg/m^3), IDLH 500 ppm; OSHA PEL: TWA 200 ppm, ceiling 300 ppm, ceiling 10-minute peak 500 ppm; ACGIH TLV: TWA 100 ppm, STEL 150 ppm.

Symptoms of Exposure: May cause headache, dizziness, excitement, euphoria, hallucination, distorted perceptions and confusion. Narcotic at high concentrations (Patnaik, 1992).

Toxicity: Acute oral LD_{50} for rats 5,000 mg/kg; LC_{50} (inhalation) for mice 5,320 ppm/8-hour (RTECS, 1985).

Drinking Water Standard (final): MCLG: 1 mg/L; MCL: 1 mg/L (U.S. EPA, 1994).

Uses: Manufacture of caprolactum, saccharin, medicines, dyes, perfumes, benzoic acid, trinitrotoluene (TNT), *o*-nitrotoluene, *p*-nitrotoluene, *o*-toluenesulfonic acid, *p*-toluenesulfonic acid, *o*-xylene, *p*-xylene, benzyl chloride, benzal chloride, benzotrichloride, halogenated toluenes and many other organic compounds; solvent for paints and coatings, gums, resins, rubber, oils and vinyl compounds; adhesive solvent in plastic toys and model airplanes; diluent and thinner for nitrocellulose lacquers; detergent manufacturing; aviation gasoline and high-octane blending stock; preparation of toluene diisocyanate for polyurethane resins.

2,4-TOLUENE DIISOCYANATE

Synonyms: Desmodur T80; **2,4-Diisocyanato-1-methylbenzene**; Diisocyanatoluene; 2,4-Diisocyanotoluene; Isocyanic acid methylphenylene ester; Isocyanic acid 4-methyl-*m*-phenylene ester; Hylene T; Hylene TCPA; Hylene TLC; Hylene TM; Hylene TM-65; Hylene TRF; 4-Methylphenylene diisocyanate; 4-Methylphenylene isocyanate; Mondur TD; Mondur TD-80; Mondur TDS; Nacconate 100; NCI-C50533; Niax TDI; Niax TDI-P; RCRA waste number U223; Rubinate TDI 80/20; TDI; 2,4-TDI; TDI-80; Toluene 2,4-diisocyanate; Tolyene-2,4-diisocyanate; Tolylene-2,4-diisocyanate; 2,4-Tolylene diisocyanate; *m*-Tolylene diisocyanate.

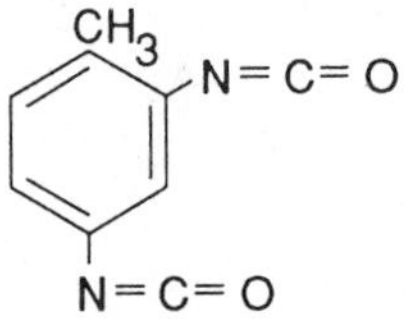

CAS Registry No.: 584-84-9
DOT: 2078
DOT label: Poison
Molecular formula: $C_9H_6N_2O_2$
Formula weight: 174.15
RTECS: CZ6300000

Physical state, color and odor
Clear, colorless to light yellow liquid with a pungent, fruity odor. Odor threshold in air ranges from 0.4 to 2.14 ppm (Keith and Walters, 1992).

Melting point (°C):
19.5-21.5 (Windholz et al., 1983)
20-21 (Dean, 1987)

Boiling point (°C):
251 (Windholz et al., 1983)

Density (g/cm³):
1.2244 at 20/4 °C (Windholz et al., 1983)

Flash point (°C):
127 (NFPA, 1984)

Lower explosive limit (%):
0.9 (NFPA, 1984)

Upper explosive limit (%):
9.5 (NIOSH, 1994)

Soil sorption coefficient, log K_{oc}:
Not applicable - reacts with water.

Octanol/water partition coefficient, log K_{ow}:
Not applicable - reacts with water.

Solubility in organics:
Miscible with acetone, alcohol (decomposes), benzene, carbon tetrachloride, diglycol monomethyl ether, ether, kerosene, olive oil (Windholz et al., 1983), chlorobenzene, and kerosene (Keith and Walters, 1992).

Solubility in water:
Not applicable - reacts with water.

Vapor density:
7.12 g/L at 25 °C, 6.01 (air = 1)

Vapor pressure (mmHg):
0.01 at 20 °C, 1 at 80 °C (Verschueren, 1983)

Environmental Fate
    Chemical/Physical. Slowly reacts with water forming carbon dioxide and polyureas (NIOSH, 1994; Windholz et al., 1983).

Exposure Limits: NIOSH REL: IDLH 2.5 ppm; OSHA PEL: ceiling 0.02 ppm (0.14 mg/m^3); ACGIH TLV: TWA 5 ppb.

Symptoms of Exposure: Vapors may cause bronchitis, headache, sleeplessness, pulmonary edema, wheezing, shortness of breath and chest congestion. Ingestion may cause coughing, vomiting and gastrointestinal pain. Contact with skin may cause nausea, vomiting, abdominal pain, dermatitis and skin sensitization (Patnaik, 1992).

Toxicity: Acute oral LD_{50} for rats 5,800 mg/kg, wild birds 100 mg/kg; LC_{50} (inhalation) for guinea pigs 13 ppm/4-hour, mice 10 ppm/4-hour, rats 14 ppm/4-hour (RTECS, 1985).

Uses: Manufacturing of polyurethane foams and other plastics; cross-linking agent for nylon 6.

o-TOLUIDINE

Synonyms: 1-Amino-2-methylbenzene; 2-Amino-1-methylbenzene; 2-Aminotoluene; *o*-Aminotoluene; C.I. 37077; 1-Methyl-2-aminobenzene; 2-Methyl-1-aminobenzene; 2-Methylaniline; *o*-Methylaniline; **2-Methylbenzenamine**; *o*-Methylbenzenamine; 2-Toluidine; *o*-Tolylamine; UN 1708.

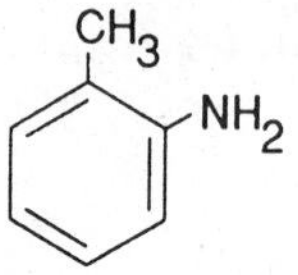

CAS Registry No.: 95-53-4
DOT: 1708
DOT label: Poison
Molecular formula: C_7H_9N
Formula weight: 107.16
RTECS: XU2975000

Physical state, color and odor
Colorless to pale yellow liquid with an aromatic, aniline-like odor. Becomes reddish-brown on exposure to air and light. Odor threshold in air is 250 ppb (Amoore and Hautala, 1983).

Melting point (°C):
-14.7 (Weast, 1986)

Boiling point (°C):
200.2 (Weast, 1986)

Density (g/cm³ at 20/4 °C):
0.9984 (Weast, 1986)
1.004 (Sax and Lewis, 1987)

Diffusivity in water (10^5 cm²/sec):
0.85 at 20 °C using method of Hayduk and Laudie (1974).

Dissociation constant:
4.45 at 25 °C (Gordon and Ford, 1972)

Flash point (°C):
86 (NIOSH, 1994)

Henry's law constant (10^3 atm·m³/mol):
3.01 at 25 °C (approximate - calculated from water solubility and vapor pressure)

976

Ionization potential (eV):
7.44 (NIOSH, 1994)

Soil sorption coefficient, log K$_{oc}$:
2.61 (calculated, Mercer et al., 1990)

Octanol/water partition coefficient, log K$_{ow}$:
1.29, 1.32 (Leo et al., 1971)
1.42 (Sangster, 1989)

Solubility in organics:
Soluble in alcohol and ether (Weast, 1986).

Solubility in water (g/L at 25 °C):
15 (Verschueren, 1983)
16.33 (Chiou et al., 1982)

Vapor density:
4.38 g/L at 25 °C, 3.70 (air = 1)

Vapor pressure (mmHg):
0.1 at 20 °C, 0.3 at 30 °C (Verschueren, 1983)
0.32 at 25 °C (Banerjee et al., 1990)

Environmental Fate
   Chemical/Physical. Kanno et al. (1982) studied the aqueous reaction of *o*-toluidine and other substituted aromatic hydrocarbons (aniline, toluidine, 1- and 2-naphthylamine, phenol, cresol, pyrocatechol, resorcinol, hydroquinone and 1-naphthol) with hypochlorous acid in the presence of ammonium ion. They reported that the aromatic ring was not chlorinated as expected but was cleaved by chloramine forming cyanogen chloride. As the pH was lowered, the amount of cyanogen chloride formed increased (Kanno et al., 1982).

Exposure Limits: Potential occupational carcinogen. NIOSH REL: IDLH 50 ppm; OSHA PEL: TWA 5 ppm (22 mg/m^3); ACGIH TLV: TWA 2 ppm (9 mg/m^3).

Symptoms of Exposure: May cause methemoglobinemia, anemia and reticulocytosis. Contact with skin may cause irritation and dermatitis (Patnaik, 1992).

Toxicity: Acute oral LD$_{50}$ in mice 520 mg/kg, rats 670 mg/kg, rabbits 840 mg/kg (RTECS, 1985).

Uses: Manufacture of dyes; vulcanization accelerator; organic synthesis.

TOXAPHENE

Synonyms: Agricide maggot killer (F); Alltex; Alltox; Attac 4-2; Attac 4-4; Attac 6; Attac 6-3; Attac 8; Camphechlor; Camphochlor; Camphoclor; Chemphene M5055; Chlorinated camphene; Chlorocamphene; Clor chem T-590; Compound 3956; Crestoxo; Crestoxo 90; ENT 9735; Estonox; Fasco-terpene; Geniphene; Gy-phene; Hercules 3956; Hercules toxaphene; Huilex; Kamfochlor; M 5055; Melipax; Motox; NA 2761; NCI-C00259; Octachlorocamphene; PCC; Penphene; Phenacide; Phenatox; Phenphane; Polychlorcamphene; Polychlorinated camphenes; Polychlorocamphene; RCRA waste number P123; Strobane-T; Strobane T-90; Synthetic 3956; Texadust; Toxakil; Toxon 63; Toxyphen; Vertac 90%; Vertac toxaphene 90.

Note: No definitive structure can be illustrated. Toxaphene is a complex mixture of at least 175 chlorinated camphenes. Of this amount, less than 10 structures are known (Casida et al., 1974).

CAS Registry No.: 8001-35-2
DOT: 2761
DOT label: Poison
Molecular formula: $C_{10}H_{10}Cl_8$
Formula weight: 413.82
RTECS: XW5250000

Physical state, color and odor
Yellow, waxy solid with a chlorine-like odor. Odor threshold in water is 140 μg/L (Keith and Walters, 1992).

Melting point (°C):
65-90 (IARC, 1979)
85 (Sims et al., 1988)

Boiling point (°C):
Decomposes at 120 (U.S. EPA, 1980).

Density (g/cm^3):
1.6 at 20/4 °C (Melnikov, 1971)
1.519-1.567 at 25/25 °C (Berg, 1983)
1.6 at 15/4 °C (Weiss, 1986)

Flash point (°C):
Nonflammable solid (NIOSH, 1994)
28.9 in 10% xylene (Keith and Walters, 1992)

Henry's law constant (10^2 atm·m^3/mol):
6.3 (Petrasek et al., 1983)

Bioconcentration factor, log BCF:
3.53 (*B. subtilis*, Paris et al., 1977)
3.81 (freshwater fish), 3.72 (fish, microcosm) (Garten and Trabalka, 1983)

Soil sorption coefficient, log K_{oc}:
3.18 using method of Kenaga and Goring (1980).

Octanol/water partition coefficient, log K_{ow}:
3.30 (Paris et al., 1977)
5.50 (Travis and Arms, 1988)
3.23 (Rao and Davidson, 1980)

Solubility in organics:
120 g/L in alcohol at 25–30 °C (Meites, 1963).

Solubility in water:
≈ 3 ppm at 25 °C (Brooks, 1974)
740 ppb at 25 °C (Weil et al., 1974)
1.75 mg/L at 25 °C (Warner et al., 1987)
400 ppb at 20–25 °C (Weber, 1972)
550 μg/L at 20 °C (Murphy et al., 1987)

Vapor pressure (10^6 mmHg):
33 at 20–25 °C (WHO, 1984a)
1 (Sims et al., 1988)
15.8 at 25 °C (2,2,5-*endo*-6-*exo*-8,9,10-heptachlorobornane, toxaphene component, Hinckley et al., 1990)

Environmental Fate
Soil. Under reduced soil conditions, about 50% of the C–Cl bonds were cleaved (dechlorinated) by Fe^{2+} porphyrins forming two major toxicants having the molecular formulas $C_{10}H_{10}Cl_8$ (Toxicant A) and $C_{10}H_{11}Cl_7$ (Toxicant B). Toxicant A reacted with reduced hematin yielding two reductive dechlorination products ($C_{10}H_{11}Cl_7$), two dehydrodechlorination products ($C_{10}H_9Cl_7$) and two other products ($C_{10}H_{10}Cl_6$). Similarly, products formed from the reaction of Toxicant B with reduced hematin included two reductive dechlorination products ($C_{10}H_{12}Cl_6$),

one dehydrochlorination product ($C_{10}H_{10}Cl_6$) and two products having the molecular formula $C_{10}H_{11}Cl_5$ (Khalifa et al., 1976).

Photolytic. Dehydrochlorination will occur after prolonged exposure to sunlight, releasing hydrochloric acid (HSDB, 1989). Two compounds isolated from toxaphene, 2-*exo*, 3-*exo*, 5,5,6-*endo*, 8,9,10,10-nonachloroborane and 2-*exo*, 3-*exo*, 5,5,6-*endo*, 8,10,10-octachloro-borane were irradiated with UV light (λ >290 nm) in a neutral aqueous solution and on a silica gel surface. Both compounds underwent reductive dechlorination, dehydrochlorination and/or oxidation to yield numerous products including bicyclo[2.1.1]hexane derivatives (Parlar, 1988).

Chemical/Physical. Saleh and Casida (1978) demonstrated that toxicant B (2,2,5-*endo*,6-*exo*,8,9,10-heptachlorobornane), the most active component of toxaphene, underwent reductive dechlorination at the geminal dichloro position, yielding 2-*endo*,5-*endo*,6-*exo*,8,9,10-hexachlorobornane and 2-*exo*,5-*endo*,6-*exo*,8,9,10-hexachlorobornane in various chemical, photochemical and metabolic systems. Emits toxic fumes of chlorides when heated to decomposition (Lewis, 1990). The hydrolysis rate constant for toxaphene at pH 7 and 25 °C was determined to be 8 x 10^{-6}/hour, resulting in a half-life of 9.9 years (Ellington et al., 1987).

Exposure Limits (mg/m^3): Potential occupational carcinogen. NIOSH REL: IDLH 200; OSHA PEL: TWA 0.5; ACGIH TLV: TWA 1.

Symptoms of Exposure: Nausea, confusion, agitation, tremors, convulsions, unconsciousness, dry red skin (NIOSH, 1987).

Toxicity: Acute oral LD_{50} for ducks 31 mg/kg, dogs 15 mg/kg, guinea pigs 250 mg/kg, mice 112 mg/kg, rats 55 mg/kg (RTECS, 1985); LC_{50} for young rainbow trout 0.2 mg/L, young pike 0.1 mg/L (Hartley and Kidd, 1987).

Drinking Water Standard (final): MCLG: zero; MCL: 3 μg/L (U.S. EPA, 1994).

Uses: Pesticide used primarily on cotton, lettuce, tomatoes, corn, peanuts, wheat and soybean.

1,3,5-TRIBROMOBENZENE

Synonym: *sym*-Tribromobenzene.

CAS Registry No.: 626-39-1
Molecular formula: $C_6H_3Br_3$
Formula weight: 314.80

Physical state
Solid.

Melting point (°C):
121-122 (Weast, 1986)

Boiling point (°C):
271 at 765 mmHg (Weast, 1986)

Diffusivity in water (10^5 cm^2/sec):
0.62 at 20 °C using method of Hayduk and Laudie (1974).

Bioconcentration factor, log BCF:
3.97-4.08 (rainbow trout, Oliver and Niimi, 1985)

Soil sorption coefficient, log K_{oc}:
4.05 using method of Chiou et al. (1979).

Octanol/water partition coefficient, log K_{ow}:
4.51 (Watarai et al., 1982)
5.26 (Gobas et al., 1978)

Solubility in organics:
Soluble in benzene, chloroform and ether (Weast, 1986).

Solubility in water:
2.51 μmol/L at 25 °C (Yalkowsky et al., 1979)

Uses: Organic synthesis.

TRIBUTYL PHOSPHATE

Synonyms: Celluphos 4; **Phosphoric acid tributyl ester;** TBP; Tri-*n*-butyl phosphate.

$$O = P \begin{array}{l} - O - CH_2CH_2CH_2CH_3 \\ - O - CH_2CH_2CH_2CH_3 \\ - O - CH_2CH_2CH_2CH_3 \end{array}$$

CAS Registry No.: 126-73-8
Molecular formula: $C_{12}H_{27}O_4P$
Formula weight: 266.32
RTECS: TC7700000

Physical state, color and odor
Colorless to pale yellow, odorless liquid.

Melting point (°C):
-81 (NIOSH, 1994)

Boiling point (°C):
289 (Weast, 1986) with decomposition (Windholz et al., 1983)

Density (g/cm³):
0.9727 at 25/4 °C (Weast, 1986)
0.982 at 20/4 °C (Sax and Lewis, 1987)

Diffusivity in water (10^5 cm²/sec):
0.49 at 20 °C using method of Hayduk and Laudie (1974).

Flash point (°C):
146 (open cup, NFPA, 1984)

Soil sorption coefficient, log K_{oc}:
2.29 (commercial mixture) using method of Kenaga and Goring (1980).

Octanol/water partition coefficient, log K_{ow}:
4.00 (commercial mixture, Saeger et al., 1979)

Solubility in organics:
Miscible with alcohol and ether (Sax and Lewis, 1987) and many other organic solvents (Dean, 1987).

Solubility in water:
≈ 0.61 vol % (Windholz et al., 1983)
280 ppm at 20–25 °C (commercial mixture, Saeger et al., 1979)

Vapor density:
10.89 g/L at 25 °C, 9.19 (air = 1)

Vapor pressure (10^3 mmHg):
4 at 25 °C (NIOSH, 1994)

Environmental Fate
  Biological. Indigenous microbes in Mississippi River water degraded tributyl phosphate to carbon dioxide. After 4 weeks, 90.8% of the theoretical carbon dioxide had evolved (Saeger et al., 1979).

Exposure Limits: NIOSH REL: TWA 0.2 ppm (2.5 mg/m^3), IDLH 30 mg/m^3; OSHA PEL: TWA 5 mg/m^3; ACGIH TLV: TWA 0.2 ppm.

Symptoms of Exposure: Depression of central nervous system and irritation of skin, eyes and respiratory tract (Patnaik, 1992).

Toxicity: Acute oral LD$_{50}$ for mice 1,189 mg/kg, rats 3,000 mg/kg (RTECS, 1985); LC$_{50}$ (48-hour) for red killifish 68 mg/L (Yoshioka et al., 1986).

Uses: Plasticizer for lacquers, plastics, cellulose esters and vinyl resins; heat-exchange liquid; solvent extraction of metal ions from solution of reactor products; pigment grinding assistant; antifoaming agent; solvent for nitrocellulose and cellulose acetate.

1,2,3-TRICHLOROBENZENE

Synonyms: 1,2,3-TCB; UN 2321.

CAS Registry No.: 87-61-6
DOT: 2321 (liquid)
Molecular formula: $C_6H_3Cl_3$
Formula weight: 181.45
RTECS: DC2095000

Physical state and color
White crystals or platelets.

Melting point (°C):
53–54 (Weast, 1986)

Boiling point (°C):
218–219 (Weast, 1986)

Density (g/cm^3):
1.69 (Windholz et al., 1983)

Diffusivity in water (10^5 cm^2/sec):
0.67 at 20 °C using method of Hayduk and Laudie (1974).

Flash point (°C):
113 (Windholz et al., 1983)

Heat of fusion (kcal/mol):
4.15 (Tsonopoulos and Prausnitz, 1971)
4.30 (Miller et al., 1984)

Henry's law constant (10^4 atm·m^3/mol at 20 °C):
89 (Oliver, 1985)
7.10 (ten Hulscher et al., 1992)

Soil sorption coefficient, log K_{oc}:
3.24 (average for 5 soils, Kishi et al., 1990)

3.97, 4.19, 5.09 (Paya-Perez et al., 1991)
3.81 (lake sediment, Schrap et al., 1994)

Octanol/water partition coefficient, log K_{ow}:
4.05 (Watarai et al., 1982)
4.11 (Könemann et al., 1979)
4.04 (Miller et al., 1984)
4.14 (Chiou, 1985)

Solubility in organics:
Soluble in benzene and ether (Weast, 1986).

Solubility in water:
67.6 μmol/L at 25 °C (Miller et al., 1984)
12 ppm at 22 °C (Verschueren, 1983)
18.0 mg/L at 25 °C (Banerjee, 1984; Chiou et al., 1986)
174 μmol/L at 25 °C (Yalkowsky et al., 1979)

Vapor pressure (mmHg):
2.1 at 25 °C (Banerjee et al., 1990)

Environmental Fate
Biological. Under aerobic conditions, soil microbes are capable of degrading 1,2,3-trichlorobenzene to 1,3- and 2,3-dichlorobenzene and carbon dioxide (Kobayashi and Rittman, 1982). A mixed culture of soil bacteria or a *Pseudomonas* sp. transformed 1,2,3-trichlorobenzene to 2,3,4-, 3,4,5- and 2,3,6-trichlorophenol (Ballschiter and Scholz, 1980).

Photolytic. The sunlight irradiation of 1,2,3-trichlorobenzene (20 g) in a 100-mL borosilicate glass-stoppered Erlenmeyer flask for 56 days yielded 32 ppm pentachlorobiphenyl (Uyeta et al., 1976).

When an aqueous solution containing 1,2,3-trichlorobenzene and a nonionic surfactant micelle (Brij 58, a polyoxyethylene cetyl ether) was illuminated by a photoreactor equipped with 253.7-nm monochromatic ultraviolet lamps, 1,2,4- and 1,3,5-trichlorobenzene formed as photoisomerization products. Continued irradiation of the solution would yield all dichlorobenzenes, chlorobenzene, benzene, phenol, hydrogen and chloride ions. The photodecomposition half-life for this reaction, based on the first-order photodecomposition rate of 1.10 x 10^{-3}/sec, is 10.5 minutes (Chu and Jafvert, 1994).

Chemical/Physical. At 70.0 °C and pH values of 3.07, 7.13 and 9.80, the hydrolysis half-lives were calculated to be 19.2, 15.0 and 34.4 days, respectively (Ellington et al., 1986).

Uses: The isomeric mixture is used to control termites; organic synthesis.

1,2,4-TRICHLOROBENZENE

Synonyms: 1,2,4-TCB; *unsym*-Trichlorobenzene; UN 2321.

CAS Registry No.: 120-82-1
DOT: 2321
Molecular formula: $C_6H_3Cl_3$
Formula weight: 181.45
RTECS: DC2100000

Physical state, color and odor
Colorless liquid with an odor similar to *o*-dichlorobenzene. Odor threshold in air is 1.4 (Amoore and Hautala, 1983).

Melting point (°C):
17 (Weast, 1986)

Boiling point (°C):
213.5 (Weast, 1986)
210 (Standen, 1964)

Density (g/cm³):
1.4542 at 20/4 °C (Weast, 1986)
1.4460 at 25/4 °C (Standen, 1964)

Diffusivity in water (10^5 cm²/sec):
0.78 at 20 °C using method of Hayduk and Laudie (1974).

Flash point (°C):
105 (NFPA, 1984)

Lower explosive limit (%):
2.5 at 150 °C (NFPA, 1984)

Upper explosive limit (%):
6.6 at 150 °C (NFPA, 1984)

Heat of fusion (kcal/mol):
3.70 (Tsonopoulos and Prausnitz, 1971)

Henry's law constant (10^3 atm·m^3/mol):
2.32 (Valsaraj, 1988)
1.2 (Oliver, 1985)
1.42 at 25 °C (Warner et al., 1987)
1.29, 1.05, 1.83, 19.2 and 29.7 at 10, 15, 20, 25 and 30 °C, respectively (Ashworth et al., 1988)
0.997 at 20.0 °C (ten Hulscher et al., 1992)

Bioconcentration factor, log BCF:
3.36-3.57 (fish tank), 3.08 (Lake Ontario) (rainbow trout, Oliver and Niimi, 1985)
2.40 (algae, Geyer et al., 1984)
3.15 (activated sludge), 2.69 (golden ide) (Freitag et al., 1985)

Soil sorption coefficient, log K_{oc}:
2.94 (Woodburn silt loam soil, Chiou et al., 1983)
3.98, 4.61 (lacustrine sediments, Chin et al., 1988)
3.09, 3.16 (Banerjee et al., 1985)
3.32 (Apison soil), 3.11 (Fullerton soil), 2.95 (Dormont soil) (Southworth and Keller, 1986)
4.08, 4.41, 5.11 (Paya-Perez et al., 1991)

Octanol/water partition coefficient, log K_{ow}:
4.02 (Chiou, 1985)
3.98 (Chin et al., 1986)
4.23 (Mackay, 1982)
4.11 (Hawker and Connell, 1988)
3.93 (Könemann et al., 1979)
4.176 (Kenaga and Goring, 1980)
4.12 (Anliker and Moser, 1987)
4.07 (DeKock and Lord, 1987)
4.050 (de Bruijn et al., 1989)
3.96 (Hammers et al., 1982)
3.63 (Wasik et al., 1981)

Solubility in organics:
Soluble in ether (Weast, 1986), other organic solvents and oils (ITII, 1986).

Solubility in water:
31.3 mg/L at 25 °C (Banerjee, 1984)
48.8 mg/L at 25 °C (Neely and Blau, 1985)
34.57 mg/L at 25 °C (Yalkowsky et al., 1979)
0.254 mmol/L at 25 °C (Miller et al., 1984)
64.5 mg/L at 30 °C (McNally and Grob, 1983, 1984)

Vapor density:
7.42 g/L at 25 °C, 6.26 (air = 1)

Vapor pressure (mmHg):
0.4 at 25 °C (Mackay et al., 1982; Neely and Blau, 1985)
0.29 at 25 °C (Warner et al., 1987)

Environmental Fate

Biological. Under aerobic conditions, biodegradation products may include 2,3-dichlorobenzene, 2,4-dichlorobenzene, 2,5-dichlorobenzene, 2,6-dichloro-benzene and carbon dioxide (Kobayashi and Rittman, 1982). A mixed culture of soil bacteria or a *Pseudomonas* sp. transformed 1,2,4-trichlorobenzene to 2,4,5- and 2,4,6-trichlorophenol (Ballschiter and Scholz, 1980). When 1,2,4-trichloro-benzene was statically incubated in the dark at 25 °C with yeast extract and settled domestic wastewater inoculum, significant biodegradation occurred, with gradual acclimation followed by a deadaptive process in subsequent subcultures. At a concentration of 5 mg/L, 54, 70, 59 and 24% losses were observed after 7, 14, 21 and 28-day incubation periods, respectively. At a concentration of 10 mg/L, only 43, 54, 14 and 0% were observed after 7, 14, 21 and 28-day incubation periods, respectively (Tabak et al., 1981). In activated sludge, <0.1% mineralized to carbon dioxide after 5 days (Freitag et al., 1985).

Photolytic. A carbon dioxide yield of 9.8% was achieved when 1,2,4-trichloro-benzene adsorbed on silica gel was irradiated with light (λ >290 nm) for 17 hours (Freitag et al., 1985).

The sunlight irradiation of 1,2,4-trichlorobenzene (20 g) in a 100-mL borosilicate glass-stoppered Erlenmeyer flask for 56 days yielded 9,770 ppm 2,4,5,2′,5′-pentachlorobiphenyl (Uyeta et al., 1976).

When an aqueous solution containing 1,2,4-trichlorobenzene (45 μM) and a nonionic surfactant micelle (Brij 58, a polyoxyethylene cetyl ether) was illuminated by a photoreactor equipped with 253.7-nm monochromatic ultraviolet lamps for 48 minutes, the chloride ion concentration increased from 9.4×10^{-7} to 1.1×10^{-4} M and the pH decreased from 6.9 to 4.0. Intermediate products identified during this reaction included all dichlorobenzenes and the photoisomerization products 1,2,3- and 1,3,5-trichlorobenzene. The photodecomposition half-life for this reaction, based on the first-order photodecomposition rate of 1.21×10^{-3}/sec, is 9.6 minutes (Chu and Jafvert, 1994).

Chemical/Physical. The hydrolysis half-life was estimated to be >900 years (Ellington et al., 1988). At 70.0 °C and pH values of 3.10, 7.11 and 9.77, the hydrolysis half-lives were calculated to be 18.4, 6.6 and 5.9 days, respectively (Ellington et al., 1986).

Exposure Limits: NIOSH REL: TWA ceiling 5 ppm (40 mg/m³); ACGIH TLV: TWA 5 ppm (40 mg/m³).

Drinking Water Standard (final): MCLG: 70 μg/L; MCL: 70 μg/L (U.S. EPA, 1994).

Toxicity: Acute oral LD_{50} for mice 300 mg/kg, rats 756 mg/kg (RTECS, 1985); LC_{50} for red killifish 65 mg/L (Yoshioka et al., 1986).

Uses: Solvent in chemical manufacturing; dyes and intermediates; dielectric fluid; synthetic transformer oils; lubricants; heat-transfer medium; insecticides; organic synthesis.

1,3,5-TRICHLOROBENZENE

Synonyms: 1,3,5-TCB; *sym*-Trichlorobenzene; UN 2321.

CAS Registry No.: 108-70-3
DOT: 2321 (liquid)
Molecular formula: $C_6H_3Cl_3$
Formula weight: 181.45
RTECS: DC2100100

Physical state
Crystals.

Melting point (°C):
63-64 (Weast, 1986)

Boiling point (°C):
208 at 763 mmHg (Weast, 1986)

Diffusivity in water (10^5 cm^2/sec):
0.67 at 20 °C using method of Hayduk and Laudie (1974).

Flash point (°C):
107 (Windholz et al., 1983)

Heat of fusion (kcal/mol):
4.49 (Miller et al., 1984)

Henry's law constant (10^3 atm·m^3/mol at 20 °C):
1.9 (Oliver, 1985)
1.89 (ten Hulscher et al., 1992)

Soil sorption coefficient, log K_{oc}:
2.85 (sand, Seip et al., 1986)
5.7 (average of 5 suspended sediment samples from the St. Clair and Detroit Rivers, Lau et al., 1989)
K_d = 1.3 mL/g on a Cs^+-kaolinite (Haderlein and Schwarzenbach, 1993)
4.18, 4.39, 5.23 (Paya-Perez et al., 1991)
3.96 (lake sediment, Schrap et al., 1994)

Octanol/water partition coefficient, log K_{ow}:
4.19 (Watarai et al., 1982)
4.15 (Könemann et al., 1979)
4.40 at 13 °C, 4.32 at 19 °C, 4.04 at 28 °C, 3.93 at 33 °C (Opperhuizen et al., 1988)
4.02 (Miller et al., 1984)
4.31 (Chiou, 1985)

Solubility in organics:
Soluble in benzene, ether, lignoin (Weast, 1986), glacial acetic acid, carbon disulfide and petroleum ether (Windholz et al., 1983).

Solubility in water:
2.7 µmol/L at 25 °C (Miller et al., 1984)
5.8 ppm at 20 °C (Verschueren, 1983)
6.01 mg/L at 25 °C (Banerjee, 1984)

Vapor pressure (mmHg):
0.58 at 25 °C (Mackay et al., 1982)

Environmental Fate
Biological. Under aerobic conditions, soil microbes degraded 1,3,5-trichlorobenzene to 1,4- and 2,4-dichlorobenzene and carbon dioxide (Kobayashi and Rittman, 1982). A mixed culture of soil bacteria or a *Pseudomonas* sp. transformed 1,3,5-trichlorobenzene to 2,4,6-trichlorophenol (Ballschiter and Scholz, 1980).
Photolytic. The sunlight irradiation of 1,3,5-trichlorobenzene (20 g) in a 100-mL borosilicate glass-stoppered Erlenmeyer flask for 56 days yielded 160 ppm pentachlorobiphenyl (Uyeta et al., 1976).

When an aqueous solution, containing 1,3,5-trichlorobenzene and a nonionic surfactant micelle (Brij 58, a polyoxyethylene cetyl ether), was illuminated by a photoreactor equipped with 253.7-nm monochromatic ultraviolet lamps, 1,2,4-trichlorobenzene formed as a result of photoisomerization. Based on photodechlorination of other polychlorobenzenes under similar conditions, it was suggested that dichlorobenzenes, chlorobenzene, benzene, phenol, hydrogen and chloride ions would form as photodegradation products. The photodecomposition half-life for this reaction, based on the first-order photodecomposition rate of 1.07×10^{-3}/sec, is 10.8 minutes (Chu and Jafvert, 1994).

Drinking Water Standard: As of May 1994, no MCLGs or MCLs have been proposed (U.S. EPA, 1984).

Uses: Organic synthesis.

1,1,1-TRICHLOROETHANE

Synonyms: Aerothene; Aerothene TT; Baltana; Chloroethene; Chloro-ethene NU; Chlorothane NU; Chlorothene; Chlorothene NU; Chlorothene VG; Chlorten; Genklene; Inhibisol; Methyl chloroform; Methyltrichloromethane; NCI-C04626; RCRA waste number U226; Solvent III; α-T; 1,1,1-TCA; 1,1,1-TCE; α-Trichloro-ethane; Tri-ethane; UN 2831.

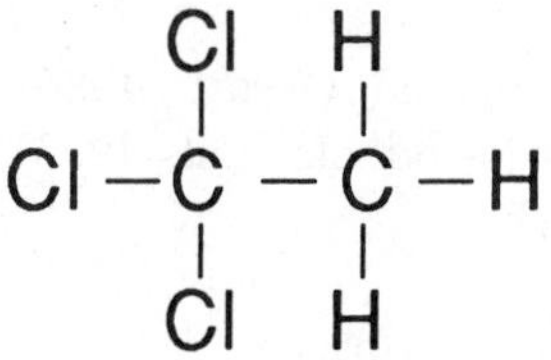

$$
\begin{array}{ccc}
\text{Cl} & & \text{H} \\
| & & | \\
\text{Cl}-\text{C}- & \text{C} & -\text{H} \\
| & & | \\
\text{Cl} & & \text{H}
\end{array}
$$

CAS Registry No.: 71-55-6
DOT: 2831
Molecular formula: $C_2H_3Cl_3$
Formula weight: 133.40
RTECS: KJ2975000

Physical state, color and odor
Colorless, watery liquid with an odor similar to chloroform. Odor threshold in air is 100 ppm (Keith and Walters, 1992).

Melting point (°C):
-30.6 (Stull, 1947)
-32.62 (Standen, 1964)

Boiling point (°C):
74.1 (Dreisbach, 1959)

Density (g/cm³):
1.3390 at 20/4 °C (Weast, 1986)
1.37068 at 0/4 °C, 1.34587 at 15/4 °C, 1.3296 at 30/4 °C (Standen, 1964)

Diffusivity in water (10^5 cm²/sec):
0.89 at 20 °C using method of Hayduk and Laudie (1974).

Flash point (°C):
None (NFPA, 1984)
≤25 (Kuchta et al., 1968)

Lower explosive limit (%):
7.5 (NFPA, 1984)

992

Upper explosive limit (%):
12.5 (NFPA, 1984)

Heat of fusion (kcal/mol):
0.45 (Dean, 1987)

Henry's law constant (10^2 atm·m^3/mol):
1.62 at 25 °C (Hine and Mookerjee, 1975)
1.5 at 20 °C (Roberts and Dändliker, 1983)
2.74 at 37 °C (Sato and Nakajima, 1979)
2.1 at 30 °C (Jeffers et al., 1989)
0.965, 1.15, 1.46, 1.74 and 2.11 at 10, 15, 20, 25 and 30 °C, respectively (Ashworth et al., 1988)
1.26, 2.00, 2.35 and 2.81 at 20, 30, 35 and 40 °C, respectively (Tse et al., 1992).
At 25 °C: 1.30 and 2.30 in distilled water and seawater, respectively (Hunter-Smith et al., 1983)
0.480, 0.635, 0.696, 1.127, and 1.490 at 2.0, 6.0, 10.0, 18.2 and 25.0 °C, respectively (Dewulf et al., 1995)

Ionization potential (eV):
10.82 (Horvath, 1982)
11.00 (NIOSH, 1994)

Bioconcentration factor, log BCF:
0.95 (bluegill sunfish, Veith et al., 1980)

Soil sorption coefficient, log K_{oc}:
2.017 (silt loam soil, Chiou et al., 1979)
2.15, 2.50 (Allerod), 2.40 (Borris), 3.18, 3.19 (Brande), 2.15, 2.85 (Finderup), 2.79 (Gunderup), 2.41 (Herborg), 3.01 (Rabis), 2.60, 2.63 (Tirstrup), 2.26 (Tylstrup), 3.24 (Vasby), 2.68, 2.80 (Vejen), 2.52, 3.29, 3.40 (Vorbasse) (Larsen et al., 1992)

Octanol/water partition coefficient, log K_{ow}:
2.18 (Mills et al., 1985)
2.48 (Neely and Blau, 1985)
2.49 (Hansch and Leo, 1979; Munz and Roberts, 1987)
2.47 (Veith et al., 1980)
2.17 (Schwarzenbach et al., 1983)

Solubility in organics:
Sparingly soluble in ethyl alcohol; freely soluble in carbon disulfide, benzene, ethyl ether, methanol, carbon tetrachloride (U.S. EPA, 1985), and many other organic solvents.

Solubility in water:
480 mg/L at 20 °C (Pearson and McConnell, 1975)
300 mg/L at 25 °C (IARC, 1979)
1,334 mg/L at 25 °C (Neely and Blau, 1985)
730 mg/L at 20 °C (Mackay and Shiu, 1981)
1,550 mg/L at 20 °C (Munz and Roberts, 1987)
0.132 wt % at 20 °C (Riddick et al., 1986)
1,360 mg/L at 20 °C (Chiou et al., 1979)
0.44 mass % at 20 °C (Konietzko, 1984)
0.1175 wt % 23.5 °C (Schwarz, 1980)
347 mg/L at 25 °C (Howard, 1990)
479.8 mg/L at 30 °C (McNally and Grob, 1984)
1,250 mg/L at 23-24 °C (Broholm and Feenstra, 1995)
1.69×10^{-4} at 25 °C (mole fraction, Li et al., 1993)

Vapor density:
5.45 g/L at 25 °C, 4.60 (air = 1)

Vapor pressure (mmHg):
62 at 10 °C, 100 at 20 °C, 150 at 30 °C (Anliker and Moser, 1987)
124 at 25 °C (Neely and Blau, 1985)
96 at 20 °C (U.S. EPA, 1980)

Environmental Fate
Biological. Microbial degradation by sequential dehalogenation under laboratory conditions produced 1,1-dichloroethane, *cis-* and *trans-*1,2-dichloroethylene, chloroethane and vinyl chloride. Hydrolysis products via dehydrohalogenation included acetic acid, 1,1-dichloroethylene (Dilling et al., 1975; Smith and Dragun, 1984) and hydrochloric acid (Dilling et al., 1975). The reported half-lives for this reaction at 20 and 25 °C are 0.5-2.5 (Vogel et al., 1987) and 1.1 years, respectively (ten Hulscher et al., 1992).

In an anoxic aquifer beneath a landfill in Ottawa, Ontario, Canada, there was evidence that 1,1,1-trichloroethane was biotransformed to 1,1-dichloroethane, 1,1-dichloro-ethylene and vinyl chloride (Lesage et al., 1990). In a similar study, 1,1,1-trichloroethane rapidly degraded in samples from an alluvial aquifer (Norman, OK) under both methanogenic and sulfate reducing conditions (KleČka et al., 1990). 1,1-Dichloroethane, ethyl chloride, possibly acetic acid and carbon dioxide (mineralization of acetic acid) were formed by biological processes, whereas 1,1-dichloroethylene and acetic acid were formed under abiotic conditions. The reported biological and abiotic half-lives were 46-206 and 1,155-1,390 days, respectively. Under aerobic or denitrifying conditions, 1,1,1-trichloroethane was recalcitrant to biodegradation (KleČka et al., 1990).

An anaerobic species of *Clostridium* biotransformed 1,1,1-trichloroethane to

1,1-dichloroethane, acetic acid and unidentified products (Gälli and McCarty, 1989). A microcosm composed of aquifer water and sediment collected from uncontaminated sites in the Everglades biotransformed 1,1,1-trichloroethane to 1,1-dichloroethylene (Parsons and Lage, 1985). In a static-culture-flask screening test, 1,1,1-trichloroethane was statically incubated in the dark at 25 °C with yeast extract and settled domestic wastewater inoculum. Percent degradation was 83 and 75% after 28 days at substrate concentrations of 5 and 10 mg/L, respectively. The amount lost due to volatilization was 7-27%, respectively (Tabak et al., 1981).

Photolytic. Reported photooxidation products include phosgene, chlorine, hydrochloric acid and carbon dioxide (McNally and Grob, 1984). Acetyl chloride (Christiansen et al., 1972) and trichloroacetaldehyde (U.S. EPA, 1975) have also been reported as photooxidation products. 1,1,1-Trichloroethane may react with hydroxyl radicals in the atmosphere producing chlorine atoms and chlorine oxides (McConnell and Schiff, 1978).

Chemical/Physical. The evaporation half-life of 1,1,1-trichloroethane (1 mg/L) from water at 25 °C using a shallow-pitch propeller stirrer at 200 rpm at an average depth of 6.5 cm is 18.7 minutes (Dilling, 1977).

The experimental half-life for hydrolysis of 1,1,1-trichloroetane in water at 25 °C is 6 months (Dilling et al., 1975).

Slowly reacts with water forming hydrochloric acid (NIOSH, 1994).

Exposure Limits: NIOSH REL: 15-minute ceiling 350 ppm (1,900 mg/m^3), IDLH 700 ppm; OSHA PEL: TWA 350 ppm; ACGIH TLV: TWA 350 ppm, STEL 450 ppm (2,450 mg/m^3).

Symptoms of Exposure: Vapors are irritating to eyes and mucous membranes (Patnaik, 1992).

Toxicity: Acute oral LD$_{50}$ for dogs 750 mg/kg, guinea pigs 9,470 mg/kg, mice 11,240 mg/kg, rats 10,300 rabbits 5,660 mg/kg (RTECS, 1985).

Drinking Water Standard (final): MCLG: 0.2 mg/L; MCL: 0.2 mg/L (U.S. EPA, 1994).

Uses: Organic synthesis; solvent for metal cleaning of precision instruments; textile processing; aerosol propellants; pesticide.

1,1,2-TRICHLOROETHANE

Synonyms: Ethane trichloride; NCI-C04579; RCRA waste number U227; β-T; 1,1,2-TCA; 1,2,2-Trichloroethane; β-Trichloroethane; Vinyl trichloride.

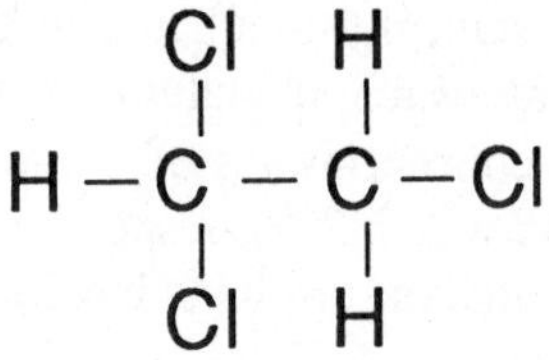

$$
\begin{array}{ccc}
& Cl & H \\
& | & | \\
H - & C - & C - Cl \\
& | & | \\
& Cl & H
\end{array}
$$

CAS Registry No.: 79-00-5
DOT: 2831
Molecular formula: $C_2H_3Cl_3$
Formula weight: 133.40
RTECS: KJ3150000

Physical state, color and odor
Colorless liquid with a pleasant, sweet, chloroform-like odor.

Melting point (°C):
-36.5 (Weast, 1986)
-37.0 (Standen, 1964)

Boiling point (°C):
113.8 (Weast, 1986)
111-114 (Fluka, 1988)

Density (g/cm³ at 20/4 °C):
1.4397 (Weast, 1986)
1.434 (Fluka, 1988)
1.4410 (Standen, 1964)

Diffusivity in water (10^5 cm²/sec):
0.92 at 20 °C using method of Hayduk and Laudie (1974).

Flash point (°C):
None (Dean, 1973)

Lower explosive limit (%):
6 (NIOSH, 1994)

Upper explosive limit (%):
15.5 (NIOSH, 1994)

Heat of fusion (kcal/mol):
2.7 (Dean, 1987)

Henry's law constant (10^4 atm·m³/mol):
7.4 (Pankow and Rosen, 1988)
9.09 at 25 °C (Hine and Mookerjee, 1975)
14.9 at 37 °C (Sato and Nakajima, 1979)
13.3 at 30 °C (Jeffers et al., 1989)
3.9, 6.3, 7.4, 9.1 and 13.3 at 10, 15, 20, 25 and 30 °C, respectively (Ashworth et al., 1988)
7, 11 and 17 at 20, 30 and 40 °C, respectively (Tse et al., 1992)

Interfacial tension with water (dyn/cm at 25 °C):
29.6 (Demond and Lindner, 1993)

Ionization potential (eV):
11.00 (NIOSH, 1994)

Soil sorption coefficient, log K_{oc}:
1.75 (Schwille, 1988)

Octanol/water partition coefficient, log K_{ow}:
2.18 (Mills et al., 1985)

Solubility in organics:
Soluble in ethanol and chloroform (U.S. EPA, 1985).

Solubility in water:
4,400 mg/L at 20 °C (Dean, 1987)
4,400 mg/L at 25 °C (McGovern, 1943)
3,704 mg/L at 25 °C (Van Arkel and Vles, 1936)
4,580 mg/L at 31.3 °C (Stephenson, 1992)
4,500 mg/L at 19.6 °C (Gladis, 1960)
0.45 wt % at 20 °C (Konietzko, 1984)
4,365.3 mg/L at 30 °C (McNally and Grob, 1984)

Vapor density:
4 kg/m³ at the boiling point (Konietzko, 1984).
5.45 g/L at 25 °C, 4.60 (air = 1)

Vapor pressure (mmHg):
19 at 20 °C, 32 at 30 °C (Verschueren, 1983)
30.3 at 25 °C (Mackay et al., 1982)

Environmental Fate

Biological. Vinyl chloride was reported to be a biodegradation product from an anaerobic digester at a wastewater treatment facility (Howard, 1990). Under aerobic conditions, *Pseudomonas putida* oxidized 1,1,2-trichloroethane to chloroacetic and glyoxylic acids. Simultaneously, 1,1,2-trichloroethane is reduced to vinyl chloride exclusively (Castro and Belser, 1990). In a static-culture-flask screening test, 1,1,2-trichloroethane was statically incubated in the dark at 25 °C with yeast extract and settled domestic wastewater inoculum. Biodegradative activity was slow to moderate, concomitant with a significant rate of volatilization (Tabak et al., 1981).

Chemical/Physical. If abiotic dehydrohalogenation occurs, the products would be 1,1-dichloroethylene and hydrochloric acid. The reported half-life for this reaction at 20 °C is 170 years (Vogel et al., 1987). Under alkaline conditions, 1,1,2-trichloroethane hydrolyzed to 1,2-dichloroethylene. The reported hydrolysis half-life in water at 25 °C and pH 7 is 139.2 years (Sata and Nakajima, 1979).

The evaporation half-life of 1,1,2-trichloroethane (1 mg/L) from water at 25 °C using a shallow-pitch propeller stirrer at 200 rpm at an average depth of 6.5 cm is 35.1 minutes (Dilling, 1977).

Exposure Limits: Potential occupational carcinogen. NIOSH REL: TWA 10 ppm (45 mg/m^3), IDLH 100 ppm; OSHA PEL: TWA 10 ppm.

Symptoms of Exposure: Irritant to eyes and mucous membranes. Ingestion may cause somnolence, nausea, vomiting, ulceration, hepatitis and necrosis (Patnaik, 1992).

Toxicity: Acute oral LD$_{50}$ for mice 378 mg/kg, rats 580 mg/kg (RTECS, 1985).

Drinking Water Standard (final): MCLG: 3 µg/L; MCL: 5 µg/L (U.S. EPA, 1994).

Uses: Solvent for fats, oils, resins, waxes, resins and other products; organic synthesis.

TRICHLOROETHYLENE

Synonyms: Acetylene trichloride; Algylen; Anamenth; Benzinol; Blacosolv; Blancosolv; Cecolene; Chlorilen; 1-Chloro-2,2-dichloroethylene; Chlorylea; Chlorylen; Circosolv; Crawhaspol; Densinfluat; 1,1-Dichloro-2-chloroethylene; Dow-tri; Dukeron; Ethinyl trichloride; Ethylene trichloride; Fleck-flip; Flock-flip; Fluate; Gemalgene; Germalgene; Lanadin; Lethurin; Narcogen; Narkogen; Narkosoid; NCI-C04546; Nialk; Perm-a-chlor; Perm-a-clor; Petzinol; Philex; RCRA waste number U228; TCE; Threthylen; Threthylene; Trethylene; Tri; Triad; Trial; Triasol; Trichloran; Trichloren; **Trichloroethene**; 1,1,2-Trichloroethene; 1,2,2-Trichloroethene; 1,1,2-Trichloroethylene; 1,2,2-Trichloroethylene; Tri-clene; Trielene; Trieline; Triklone; Trilen; Trilene; Triline; Trimar; Triol; Tri-plus; Tri-plus M; UN 1710; Vestrol; Vitran; Westrosol.

$$\underset{\displaystyle Cl}{\overset{\displaystyle Cl}{\diagdown}}C = C \underset{\displaystyle H}{\overset{\displaystyle Cl}{\diagup}}$$

CAS Registry No.: 79-01-6
DOT: 1710
Molecular formula: C_2HCl_3
Formula weight: 131.39
RTECS: KX4550000

Physical state, color and odor
Clear, colorless, watery-liquid with a chloroform-like odor. Odor threshold in air is 21.4 ppm (Keith and Walters, 1992).

Melting point (°C):
-86.4 (McGovern, 1943)
-87.1 (Standen, 1964)

Boiling point (°C):
87.2 (Dean, 1973)
86.7 (Standen, 1964)

Density (g/cm^3):
1.464 at 20/4 °C (McGovern, 1943)
1.461 at 20/4 °C (Fluka, 1988)
1.4996 at 0/4 °C, 1.4762 at 15/4 °C, 1.4514 at 30/4 °C (Carlisle and Levine, 1932a)

Diffusivity in water (10^5 cm^2/sec):
0.94 at 20 °C using method of Hayduk and Laudie (1974).

Flash point (°C):
32.2 (Weiss, 1986)

Lower explosive limit (%):
8 at 25 °C, 7.8 at 100 °C (NFPA, 1984)

Upper explosive limit (%):
10.5 at 25 °C, 52 at 100 °C (NFPA, 1984)

Henry's law constant (10^3 atm·m³/mol):
9.1 (Pankow and Rosen, 1988)
9.9 at 20 °C (Roberts and Dändliker, 1983)
19.6 at 37 °C (Sato and Nakajima, 1979)
10.1 at 20 °C (Roberts et al., 1985)
12.8 at 30 °C (Jeffers et al., 1989)
5.38, 6.67, 8.42, 10.2 and 12.8 at 10, 15, 20, 25 and 30 °C, respectively (Ashworth et al., 1988)
7.0, 11.4 and 17.3 at 20, 30 and 40 °C, respectively (Tse et al., 1992)
2.47, 3.06, 3.41, 6.22 and 8.60 at 2.0, 6.0, 10.0, 18.2 and 25.0 °C, respectively (Dewulf et al., 1995)

Interfacial tension with water (dyn/cm at 24 °C):
34.5 (Demond and Lindner, 1993)

Ionization potential (eV):
9.94 (Yoshida et al., 1983)
9.45 (Franklin et al., 1969)

Bioconcentration factor, log BCF:
1.23 (bluegill sunfish, Veith et al., 1980)
3.06 (algae, Geyer et al., 1984)
3.00 (activated sludge), 1.95 (golden ide) (Freitag et al., 1985)

Soil sorption coefficient, log K_{oc}:
1.81 (Abdul et al., 1987)
2.025 (Garbarini and Lion, 1986)
1.86, 1.98, 2.15 (various Norwegian soils, Seip et al., 1986)
1.79 (humic-coated alumina, Peterson et al., 1988)
1.66, 2.64, 2.83 (silty clay, Pavlostathis and Mathavan, 1992)

Octanol/water partition coefficient, log K_{ow}:
2.53 (Miller et al., 1985; Wasik et al., 1981)
2.29 (Schwarzenbach et al., 1983)

2.42 (Banerjee et al., 1980)
2.60 (Hawker and Connell, 1988)
3.24, 3.30 (Geyer et al., 1984)
2.37 (Green et al., 1983)
3.03 (Tewari et al., 1982)

Solubility in organics:
Soluble in acetone, ethanol, chloroform and ether (U.S. EPA, 1985).

Solubility in water:
1,100 mg/L at 20 °C (Pearson and McConnell, 1975)
1.1 g/kg at 25 °C (McGovern, 1943)
10.4 mmol/L at 25 °C (Tewari et al., 1982; Wasik et al., 1981)
1,080 mg/L at 20 °C (Munz and Roberts, 1987)
0.137 wt % at 25 °C (Riddick et al., 1986)
1,250 mg/L at 60 °C (Standen, 1964)
0.11 wt % at 25 °C (Nathan, 1978)
743.1 mg/L at 30 °C (McNally and Grob, 1984)
1,400 mg/L at 23-24 °C (Broholm and Feenstra, 1995)
1.14×10^{-4} at 25 °C (mole fraction, Li et al., 1993)

Vapor density:
5.37 g/L at 25 °C, 4.54 (air = 1)

Vapor pressure (mmHg):
74 at 25 °C (Mackay and Shiu, 1981)
100 at 32 °C (Sax, 1984)
53.5 at 20 °C (Munz and Roberts, 1987)
56.8 at 20 °C, 72.6 at 25 °C, 91.5 at 30 °C (Klöpffer et al., 1988)
69 at 25 °C (Howard, 1990)

Environmental Fate
   Biological. Microbial degradation of trichloroethylene via sequential dehalogenation produced *cis-* and *trans-*1,2-dichloroethylene and vinyl chloride (Smith and Dragun, 1984). Anoxic microcosms in sediment and water degraded trichloroethylene to 1,2-dichloroethylene and then to vinyl chloride (Barrio-Lage et al., 1986). Trichloroethylene in soil samples collected from Des Moines, IA anaerobically degraded to 1,2-dichloroethylene. The production of 1,1-dichloroethylene was not observed in this study (Kleopfer, 1985).
   In a methanogenic aquifer, trichloroethylene biodegraded to 1,2-dichloroethylene and vinyl chloride (Wilson et al., 1986). Dichloroethylene was reported as a biotransformation product under anaerobic conditions and in experimental systems using mixed or pure cultures. Under aerobic conditions, carbon dioxide

was the principal degradation product in experiments containing pure and mixed cultures (Vogel et al., 1987). A microcosm composed of aquifer water and sediment collected from uncontaminated sites in the Everglades biotransformed trichloroethylene to *cis*- and *trans*-1,2-dichloroethylene (Parsons and Lage, 1985). Trichloroethylene biodegraded to dichloroethylene in both anaerobic and aerobic soils of different soil types (Barrio-Lage et al., 1987). Under anaerobic conditions, nonmethanogenic fermenters and methanogens degraded trichloroethylene to chloroethane, methane, 1,1-dichloroethylene, 1,2-dichloroethylene and vinyl chloride (Baek and Jaffé, 1989). Titanium dioxide suspended in an aqueous solution and irradiated with UV light (λ = 365 nm) converted trichloroethylene to carbon dioxide at a significant rate (Matthews, 1986). In a static-culture-flask screening test, trichloroethylene was statically incubated in the dark at 25 °C with yeast extract and settled domestic wastewater inoculum. Percent degradation was 87 and 84% after 28 days at substrate concentrations of 5 and 10 mg/L, respectively. However, losses due to volatilization were 22-29% after 10 days (Tabak et al., 1981).

Photolytic. Under smog conditions, indirect photolysis via hydroxyl radicals will yield phosgene, dichloroacetyl chloride and formyl chloride (Howard, 1990). These compounds are readily hydrolyzed to hydrochloric acid, carbon monoxide, carbon dioxide and dichloroacetic acid (Morrison and Boyd, 1971). Though not identified, dichloroacetic acid and hydrogen chloride were reported to be aqueous photo-decomposition products (Dilling et al., 1975).

An aqueous solution containing 300 ng/μL trichloroethylene and colloidal platinum catalyst was irradiated with UV light. After 12 hours, 7.4 ng/μL trichloroethylene and 223.9 ng/μL ethane were detected. A duplicate experiment was performed but 1 g zinc was added to the system. After 5 hours, 259.9 ng/μL ethane was formed and trichloroethylene was non-detectable (Wang and Tan, 1988). Major products identified from the pyrolysis of trichloroethylene between 300-800 °C were carbon tetrachloride, tetrachloroethylene, hexachloroethane, hexachlorobutadiene and hexachlorobenzene (Yasuhara and Morita, 1990).

Chemical/Physical. Jacoby et al. (1994) studied the photocatalytic reaction of gaseous trichloroethylene in air in contact with UV-irradiated titanium dioxide catalyst. The UV radiation was kept below the maximum wavelength so that the catalyst could be excited by photons, i.e., λ <356 nm. Two reaction pathways were proposed. The first pathway includes the formation of the intermediate dichloro-acetyl chloride. This compound has a very short residence time and is quickly converted to the following compounds: phosgene, carbon dioxide, carbon monoxide, carbon dioxide and hydrogen chloride. The second pathway involves the formation of the final products without the formation of the intermediate.

The evaporation half-life of trichloroethylene (1 mg/L) from water at 25 °C using a shallow-pitch propeller stirrer at 200 rpm at an average depth of 6.5 cm is 18.5 minutes (Dilling, 1977).

The experimental hydrolysis half-life of trichloroethylene in water at 25 °C is

10.7 months (Dilling et al., 1975). The reported hydrolysis half-life at 25 °C and pH 7 is 1.3 x 10^{-6} years (Jeffers et al., 1989).

Exposure Limits: Potential occupational carcinogen. OSHA PEL: TWA 100 ppm, ceiling 200 ppm, 5-minute/2-hour peak 300 ppm; ACGIH TLV: TWA 50 ppm (270 mg/m^3), STEL 200 ppm (1,080 mg/m^3).

Symptoms of Exposure: Inhalation of vapors may cause dizziness, headache, fatigue and visual disturbances. Narcotic at high concentrations. Ingestion may cause nausea, vomiting, diarrhea and gastric disturbances (Patnaik, 1992).

Toxicity: Acute oral LD$_{50}$ for mice 2,402 mg/kg, rats 3,670 mg/kg (RTECS, 1985); LC$_{50}$ for red killifish 630 mg/L (Yoshioka et al., 1986).

Drinking Water Standard (final): MCLG: zero; MCL: 5 μg/L (U.S. EPA, 1994).

Uses: Dry cleaning fluid; degreasing and drying metals and electronic parts; extraction solvent for oils, waxes and fats; solvent for cellulose esters and ethers; removing caffeine from coffee; refrigerant and heat exchange liquid; fumigant; diluent in paints and adhesives; textile processing; aerospace operations (flushing liquid oxygen); anesthetic; medicine; organic synthesis.

TRICHLOROFLUOROMETHANE

Synonyms: Algofrene type 1; Arcton 9; Electro-CF 11; Eskimon 11; F 11; FC 11; Fluorocarbon 11; Fluorotrichloromethane; Freon 11; Freon 11A; Freon 11B; Freon HE; Freon MF; Frigen 11; Genetron 11; Halocarbon 11; Isceon 11; Isotron 11; Ledon 11; Monofluorotrichloromethane; NCI-C04637; RCRA waste number U121; Refrigerant 11; Trichloromonofluoromethane; Ucon 11; Ucon fluorocarbon 11; Ucon refrigerant 11.

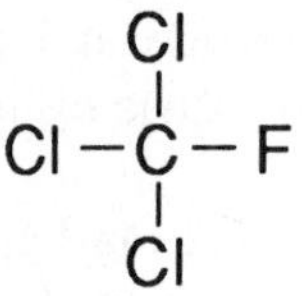

$$\underset{\displaystyle Cl}{\overset{\displaystyle Cl}{Cl - C - F}}$$

CAS Registry No.: 75-69-4
DOT: 1078
Molecular formula: CCl_3F
Formula weight: 137.37
RTECS: PB6125000

Physical state, color and odor
Colorless, odorless liquid.

Melting point (°C):
-111 (Windholz et al., 1983)

Boiling point (°C):
23.63 (Kudchadker et al., 1979)

Density (g/cm^3):
1.476 at 25/4 °C (Neely and Blau, 1985)
1.484 at 17.2/4 °C (Sax, 1984)
1.487 at 20/4 °C (Fluka, 1988)
1.467 at 25/4 °C (Horvath, 1982)

Diffusivity in water (10^5 cm^2/sec):
0.93 at 20 °C using method of Hayduk and Laudie (1974).

Flash point (°C):
Nonflammable gas (NIOSH, 1994)

Henry's law constant (10^3 atm·m^3/mol):
110 (Pankow and Rosen, 1988)

5.83 at 25 °C (Warner et al., 1987)
1,730 at 25 °C (Hine and Mookerjee, 1975)
120 at 25 °C (Liss and Slater, 1974)
5.38, 6.67, 8.42, 10.2 and 12.8 at 10, 15, 20, 25 and 30 °C, respectively (Ashworth et al., 1988)
At 25 °C: 88.2 and 123 in distilled water and seawater, respectively (Hunter-Smith et al., 1983)

Ionization potential (eV):
11.77 ± 0.02 (Franklin et al., 1969)

Soil sorption coefficient, log K_{oc}:
2.20 (Schwille, 1988)
2.13 (Neely and Blau, 1985)

Octanol/water partition coefficient, log K_{ow}:
2.53 (Hansch et al., 1975)

Solubility in organics:
Soluble in ethanol, ether and other solvents (U.S. EPA, 1985).

Solubility in water (mg/L):
1,240 at 25 °C (Neely and Blau, 1985)
1,100 at 20 °C (Du Pont, 1966; Pearson and McConnell, 1975)
1,080 at 30 °C (Horvath, 1982)

Vapor density:
5.85 g/L at 23.77 °C (Braker and Mossman, 1971)
5.61 g/L at 25 °C, 4.74 (air = 1)

Vapor pressure (mmHg):
687 at 20 °C, 980 at 30 °C (Verschueren, 1983)
792 at 25 °C (ACGIH, 1986)
667.4 at 20 °C (McConnell et al., 1975)
802.8 at 25 °C (Howard, 1990)

Environmental Fate
   Biological. In a static-culture-flask screening test, trichlorofluoromethane was statically incubated in the dark at 25 °C with yeast extract and settled domestic wastewater inoculum. No significant degradation was observed after 28 days of incubation. At substrate concentrations of 5 and 10 mg/L, percent losses due to volatilization were 58 and 37% after 10 days (Tabak et al., 1981).
   Chemical/Physical. When trichlorofluoromethane (50 µg/L) in an ultrasonicator

was exposed to 20-kHz ultrasound at 5 °C, nearly 100% degradation was achieved after 6 minutes. During sonication, the pH of the aqueous solution decreased, which is consistent with the formation of hydrochloric acid, hydrofluoric acid and acidic species from fluorine and chlorine. In this experiment <5% of trichlorofluoro-ethane was lost to volatilization (Cheung and Kurup, 1994).

Exposure Limits: NIOSH REL: ceiling 1,000 ppm (5,600 mg/m^3), IDLH 2,000 ppm; OSHA PEL: TWA 1,000 ppm; ACGIH TLV: ceiling 1,000 ppm.

Toxicity: LC$_{50}$ (inhalation) for guinea pigs 25 pph/30-minute, mice 10 pph/30-minute, rabbits 25 pph/30-minute (RTECS, 1985).

Uses: Aerosol propellant; refrigerant; solvent; blowing agent for polyurethane foams; fire extinguishing; chemical intermediate; organic synthesis.

2,4,5-TRICHLOROPHENOL

Synonyms: Collunosol; Dowicide 2; Dowicide B; NCI-C61187; Nurelle; Phenachlor; Preventol I; RCRA waste number U230; 2,4,5-TCP; 2,4,5-TCP-Dowicide 2.

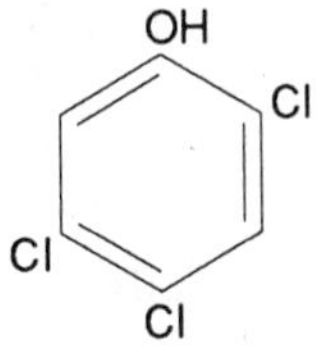

CAS Registry No.: 95-95-4
DOT: 2020
Molecular formula: $C_6H_3Cl_3O$
Formula weight: 197.45
RTECS: SN1400000

Physical state, color and odor
Colorless crystals or yellow to gray flakes with a strong, disinfectant or phenolic odor.

Melting point (°C):
68-70 (Weast, 1986)
61-63 (Sax, 1984)
57 (Weiss, 1986)

Boiling point (°C):
252 (Weast, 1986)

Density (g/cm³):
1.5 at 75/4 °C (Verschueren, 1983)
1.678 at 25/4 °C (Sax, 1984)

Dissociation constant:
7.37 at 25 °C (Dean, 1973)

Diffusivity in water (10^5 cm²/sec):
0.62 at 20 °C using method of Hayduk and Laudie (1974).

Flash point (°C):
Nonflammable (Weiss, 1986)

Lower explosive limit (%):
Nonflammable (Weiss, 1986)

Upper explosive limit (%):
Nonflammable (Weiss, 1986)

Henry's law constant (10^7 atm·m^3/mol):
1.76 at 25 °C (estimated, Leuenberger et al., 1985)

Soil sorption coefficient, log K_{oc}:
2.56 (Brookston clay loam, Boyd, 1982)
2.45 (river sediment, Eder and Weber, 1980)
2.55 (loamy sand, Kjeldsen et al., 1990)

Octanol/water partition coefficient, log K_{ow}:
3.72 (Leo et al., 1971)
4.19 (Schellenberg et al., 1984)
3.85 (Schultz et al., 1989)

Solubility in organics:
Soluble in ethanol and ligroin (U.S. EPA, 1985).

Solubility in water:
1,190 mg/kg at 25 °C (Verschueren, 1983)
1.2 g/L at 25 °C (Leuenberger et al., 1985)

Vapor pressure (10^3 mmHg):
3.5 at 8 °C, 22 at 25 °C (Leuenberger et al., 1985)

Environmental Fate
   Biological. Chloroperoxidase, a fungal enzyme isolated from *Caldariomyces fumago*, chlorinated 2,4,5-trichlorophenol to give 2,3,4,6-tetrachlorophenol (Wannstedt et al., 1990).
   Photolytic. When 2,4,5-trichlorophenol (100 μM) in an oxygenated, titanium dioxide (2 g/L) suspension was irradiated by sunlight ($\lambda \geq 340$ nm), complete mineralization to carbon dioxide and water and chloride ions was observed (Barbeni et al., 1987).

Toxicity: Acute oral LD_{50} for guinea pigs 1,000 mg/kg, mice 600 mg/kg, rats 820 mg/kg (RTECS, 1985); LC_{50} (48-hour) for red killifish 12 mg/L (Yoshioka et al., 1986).

Uses: Fungicide; bactericide; organic synthesis.

2,4,6-TRICHLOROPHENOL

Synonyms: Dowicide 2S; NCI-C02904; Omal; Phenachlor; RCRA waste number F027; 2,4,6-TCP; 2,4,6-TCP-Dowicide 25.

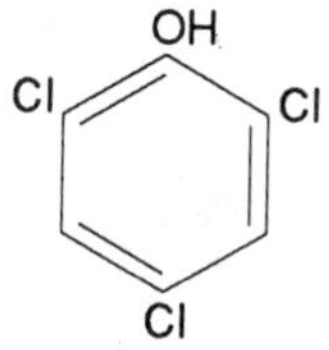

CAS Registry No.: 88-06-2
DOT: 2020
Molecular formula: $C_6H_3Cl_3O$
Formula weight: 197.45
RTECS: SN1575000

Physical state, color and odor
Colorless needles or yellow solid with a strong, phenolic odor.

Melting point (°C):
69.5 (Weast, 1986)

Boiling point (°C):
246 (Weast, 1986)

Density (g/cm^3):
1.4901 at 75/4 °C (Weast, 1986)

Diffusivity in water (10^5 cm^2/sec):
0.62 at 20 °C using method of Hayduk and Laudie (1974).

Dissociation constant:
7.42 at 25 °C (Howard, 1989)
6.0 (Eder and Weber, 1980)

Henry's law constant (10^8 atm·m^3/mol):
9.07 at 25 °C (estimated, Leuenberger et al., 1985)

Bioconcentration factor, log BCF:
1.71 (algae, Geyer et al., 1984)
1.78 (activated sludge), 2.49 (golden ide) (Freitag et al., 1985)

Soil sorption coefficient, log K_{oc}:
3.03 (Howard, 1989)
1.72 (river sediment, Eder and Weber, 1980)

Octanol/water partition coefficient, log K_{ow}:
2.80 (Geyer et al., 1982)
3.72 (Schellenberg et al., 1984)
3.06, 3.69 (Leo et al., 1971)

Solubility in organics:
Soluble in ethanol and ether (Weast, 1986).

Solubility in water (mg/L):
800 at 25 °C, 2,430 at 96 °C (Verschueren, 1983)
420 at 20-25 °C (Geyer et al., 1982)
900 at 20-25 °C (Kilzer et al., 1979)

Vapor pressure (10^3 mmHg):
8.4 at 24 °C (Howard, 1989)
2.5 at 8 °C, 17 at 25 °C (Leuenberger et al., 1985)

Environmental Fate

Biological. In activated sludge, only 0.3% mineralized to carbon dioxide after 5 days (Freitag et al., 1985). In anaerobic sludge, 2,4,6-trichlorophenol degraded to *p*-chlorophenol (Mikesell and Boyd, 1985). When 2,4,6-trichlorophenol was statically incubated in the dark at 25 °C with yeast extract and settled domestic wastewater inoculum, significant biodegradation with rapid adaptation was observed. At concentrations of 5 and 10 mg/L, 96 and 97% biodegradation, respectively, were observed after 7 days (Tabak et al., 1981).

Photolytic. Titanium dioxide suspended in an aqueous solution and irradiated with UV light (λ = 365 nm) converted 2,4,6-trinitrophenol to carbon dioxide at a significant rate (Matthews, 1986). A carbon dioxide yield of 65.8% was achieved when 2,4,6-trichlorophenol adsorbed on silica gel was irradiated with light (λ >290 nm) for 17 hours (Freitag et al., 1985).

Chemical/Physical. An aqueous solution containing chloramine reacted with 2,4,6-trichlorophenol to yield the following intermediate products after 2 hours at 25 °C: 2,6-dichloro-1,4-benzoquinone-4-(*N*-chloro)imine and 4,6-dichloro-1,2-benzoquinone-2-(*N*-chloro)imine (Maeda et al., 1987).

In a dilute aqueous solution at pH 6.0, 2,4,6-trichlorophenol reacted with an excess of hypochlorous acid forming 2,6-dichlorobenzoquinone (18% yield) and other chlorinated compounds (Smith et al., 1976). Based on an assumed 5% disappearance after 330 hours at 85 °C, the hydrolysis half-life was estimated to be >40 years (Ellington et al., 1988).

Drinking Water Standard: As of May 1994, no MCLGs or MCLs have been proposed although 2,4,6-trichlorophenol has been listed for regulation (U.S. EPA, 1994).

Toxicity: Acute oral LD_{50} for rats 820 mg/kg; LC_{50} for red killifish 7.6 mg/L (Yoshioka et al., 1986).

Uses: Manufacture of fungicides, bactericides, antiseptics, germicides; wood and glue preservatives; in textiles to prevent mildew; defoliant; disinfectant; organic synthesis.

1,2,3-TRICHLOROPROPANE

Synonyms: Allyl trichloride; Glycerin trichlorohydrin; Glycerol trichlorohydrin; Glyceryl trichlorohydrin; NCI-C60220; Trichlorohydrin.

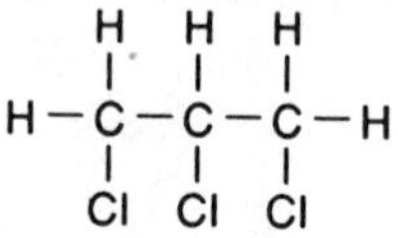

```
      H   H   H
      |   |   |
  H — C — C — C — H
      |   |   |
      Cl  Cl  Cl
```

CAS Registry No.: 96-18-4
Molecular formula: $C_3H_5Cl_3$
Formula weight: 147.43
RTECS: TZ9275000

Physical state, color and odor
Clear, colorless liquid with a strong, chloroform-like odor.

Melting point (°C):
-14.7 (Weast, 1986)

Boiling point (°C):
156.8 (Weast, 1986)

Density (g/cm³):
1.3889 at 20/4 °C (Horvath, 1982)
1.417 at 15/4 °C (Verschueren, 1983)

Diffusivity in water (10^5 cm²/sec):
0.85 at 20 °C using method of Hayduk and Laudie (1974).

Flash point (°C):
71.7 (NIOSH, 1994)
82.2 (open cup, Hawley, 1981)

Lower explosive limit (%):
3.2 at 121 °C (NIOSH, 1994)

Upper explosive limit (%):
12.6 at 151 °C (NIOSH, 1994)

Henry's law constant (10^5 atm·m³/mol):
9.79 at 25 °C (Tancréde and Yanagisawa, 1990)

Interfacial tension with water (dyn/cm at 20 °C):
38.50 (Demond and Lindner, 1993)

Solubility in organics:
Soluble in alcohol, chloroform and ether (Weast, 1986).

Solubility in water:
1,900 mg/L (Afghan and Mackay, 1980; Dilling, 1977)

Vapor density:
6.03 g/L at 25 °C, 5.09 (air = 1)

Vapor pressure (mmHg):
2 at 20 °C, 4 at 30 °C (Verschueren, 1983)
3.1 at 25 °C (Banerjee et al., 1990)

Environmental Fate

Chemical/Physical. The hydrolysis rate constant for 1,2,3-trichloropropane at pH 7 and 25 °C was determined to be 1.8×10^{-6}/hour, resulting in a half-life of 43.9 years (Ellington et al., 1988). The hydrolysis half-lives decrease at varying pHs and temperature. At 87 °C, the hydrolysis half-lives at pH values of 3.07, 7.12 and 9.71 were 21.1, 11.6 and 0.03 days, respectively (Ellington et al., 1986).

The evaporation half-life of 1,2,3-trichloropropane (1 mg/L) from water at 25 °C using a shallow-pitch propeller stirrer at 200 rpm at an average depth of 6.5 cm is 56.1 minutes (Dilling, 1977).

Exposure Limits: Potential occupational carcinogen. NIOSH REL: TWA 10 ppm (60 mg/m^3), IDLH 100 ppm; OSHA PEL: TWA 50 ppm (300 mg/m^3); ACGIH TLV: TWA 10 ppm.

Symptoms of Exposure: Inhalation of vapors may cause depression of central nervous system, narcosis and convulsions (Patnaik, 1992).

Toxicity: Acute oral LD$_{50}$ for rats 320 mg/kg; LC$_{50}$ (inhalation) for mice 3,400 mg/m^3/2-hour (RTECS, 1985).

Drinking Water Standard: As of May 1994, no MCLGs or MCLs have been proposed although 1,2,3-trichloropropane has been listed for regulation (U.S. EPA, 1994).

Uses: Solvent; degreaser; paint and varnish removers.

1,1,2-TRICHLOROTRIFLUOROETHANE

Synonyms: Fluorocarbon 113; Freon 113; Frigen 113 TR-T; Halocarbon 113; Kaiser chemicals 11; R 113; Refrigerant 113; 1,1,2-Trichloro-1,2,2-trifluoro-ethane; Trichlorotrifluoroethane; 1,1,2-Trifluoro-1,2,2-trichloroethane; TTE; Ucon 113; Ucon fluorocarbon 113; Ucon 113/halocarbon 113.

```
          Cl  F
          |   |
    F — C — C — F
          |   |
          Cl  Cl
```

CAS Registry No.: 76-13-1
DOT: 1078
Molecular formula: $C_2Cl_3F_3$
Formula weight: 187.38
RTECS: KJ4000000

Physical state, color and odor
Colorless liquid with a carbon tetrachloride-like odor at high concentrations.

Melting point (°C):
-36.4 (Weast, 1986)

Boiling point (°C):
47.7 (Weast, 1986)

Density (g/cm³):
1.5635 at 20/4 °C (Weast, 1986)
1.42 at 25/4 °C (Hawley, 1981)

Diffusivity in water (10^5 cm²/sec):
0.79 at 20 °C using method of Hayduk and Laudie (1974).

Flash point (°C):
Noncombustible at ordinary temperatures (NIOSH, 1994).

Henry's law constant (atm·m³/mol):
0.154, 0.215, 0.245, 0.319 and 0.321 at 10, 15, 20, 25 and 30 °C, respectively (Ashworth et al., 1988).

Ionization potential (eV):
11.99 ± 0.02 (Franklin et al., 1969)

Soil sorption coefficient, log K_{oc}:
2.59 using method of Chiou et al. (1979).

Octanol/water partition coefficient, log K_{ow}:
2.57 (Lesage et al., 1990)

Solubility in organics:
Soluble in alcohol, benzene and ether (Weast, 1986).

Solubility in water:
136 mg/L at 10 °C (Lesage et al., 1990)
170 mg/L at 25 °C (Du Pont, 1966; Jones et al., 1977)

Vapor density:
7.66 g/L at 25 °C, 6.47 (air = 1)

Vapor pressure (mmHg at 20 °C):
285 (NIOSH, 1994)
270 (Verschueren, 1983)

Environmental Fate
   Biological. In an anoxic aquifer beneath a landfill in Ottawa, Ontario, Canada, there was evidence to suggest that 1,1,2-trichlorotrifluoroethane underwent reductive dehalogenation to give 1,2-difluoro-1,1,2-trichloroethylene and 1,2-dichloro-1,1,2-trifluoroethane. It was proposed that the latter compound was further degraded via dehydrodehalogenation to give 1-chloro-1,1,2-trifluoroethylene (Lesage et al., 1990).

Exposure Limits: NIOSH REL: TWA 1,000 ppm (7,600 mg/m^3), STEL 1,250 ppm (9,500 mg/m^3), IDLH 2,000 ppm; OSHA PEL: TWA 1,000 ppm.

Symptoms of Exposure: May produce a weak narcotic effect, cardiac sensitization and irritation of respiratory passage (Patnaik, 1992).

Toxicity: Acute oral LD$_{50}$ for rats 43 mg/kg (RTECS, 1985).

Drinking Water Standard: As of May 1994, no MCLGs or MCLs have been proposed although 1,1,2-trichlorotrifluoroethane has been listed for regulation (U.S. EPA, 1994).

Uses: Fire extinguishers; dry-cleaning solvent; manufacture of chlorotrifluoroethylene; refrigerant; polymer intermediate; blowing agent; drying electronic parts; extraction solvent for analyzing hydrocarbons, oils and greases.

TRI-*o*-CRESYL PHOSPHATE

Synonyms: *o*-Cresyl phosphate; Phosflex 179-C; **Phosphoric acid tris(2-methyl-phenyl) ester;** Phosphoric acid tri-2-tolyl ester; Phosphoric acid tri-*o*-cresyl ester; TCP; TOCP; TOFK; *o*-Tolyl phosphate; TOTP; Tricresyl phosphate; Triorthocresyl phosphate; Tri-2-methylphenyl phosphate; Tris(*o*-cresyl)phosphate; Tris(*o*-meth-ylphenyl)phosphate; Tris(*o*-tolyl)phosphate; Tri-2-tolyl phosphate; Tri-*o*-tolyl phosphate.

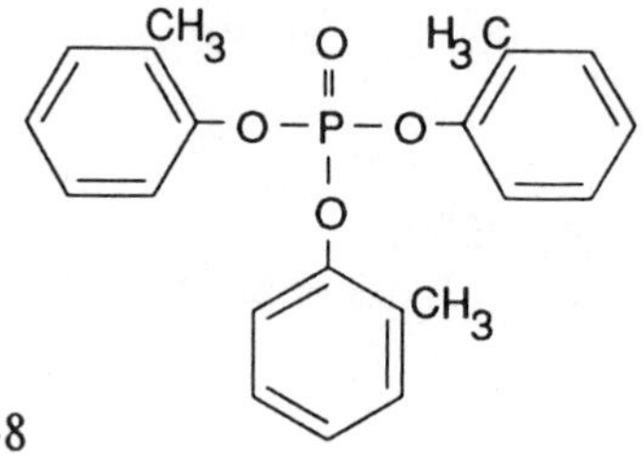

CAS Registry No.: 78-30-8
DOT: 2574
DOT label: Poison
Molecular formula: $C_{21}H_{21}O_4P$
Formula weight: 368.37
RTECS: TD0350000

Physical state, color and odor
Colorless to pale yellow, odorless liquid.

Melting point (°C):
11 (Weast, 1986)
-25, -30 (Verschueren, 1983)

Boiling point (°C):
410 (Weast, 1986)
420 (Verschueren, 1983)

Density (g/cm^3):
1.955 at 20/4 °C (Weast, 1986)
1.162 at 25/25 °C (Verschueren, 1983)

Diffusivity in water (10^5 cm^2/sec):
0.61 at 20 °C using method of Hayduk and Laudie (1974).

Flash point (°C):
227 (NIOSH, 1994)

Soil sorption coefficient, log K_{oc}:
3.37 using method of Kenaga and Goring (1980).

Octanol/water partition coefficient, log K_{ow}:
5.11 (commercial mixture containing tricresyl phosphates, Saeger et al., 1979)

Solubility in organics:
Soluble in acetic acid, alcohol, ether and toluene (Weast, 1986).

Solubility in water:
0.36 and 0.34 ppm at 20-25 and 25 °C, respectively (mixture containing tricresyl phosphates, Ofstad and Sletten, 1985; Saeger et al., 1979)
3.1 ppm at 25 °C ("Pliabrac 521", Ofstad and Sletten, 1985)

Vapor density:
15.06 g/L at 25 °C, 12.72 (air = 1)

Vapor pressure (10^5 mmHg):
2 at 25 °C (NIOSH, 1994)

Environmental Fate
  Biological. A commercial mixture containing tricresyl phosphates was completely degraded by indigenous microbes in Mississippi River water to carbon dioxide. After 4 weeks, 82.1% of the theoretical carbon dioxide had evolved (Saeger et al., 1979).
  Chemical/Physical. Tri-*o*-cresyl phosphate hydrolyzed rapidly in Lake Ontario water, presumably to di-*o*-cresyl phosphate (Howard and Deo, 1979). When an aqueous solution containing a mixture of isomers (0.1 mg/L) and chlorine (3-1,000 mg/L) was stirred in the dark at 20 °C for 24 hours, the benzene ring was substituted with 1-3 chlorine atoms (Ishikawa and Baba, 1988).

Exposure Limits (mg/m^3): NIOSH REL: TWA 0.1, IDLH 40; OSHA PEL: TWA 0.1.

Symptoms of Exposure: Ingestion may cause gastrointestinal pain, diarrhea, weakness, muscle pain, kidney damage, paralysis and death (Patnaik, 1992).

Toxicity: Acute oral LD_{50} for rats 160 mg/kg (RTECS, 1985).

Uses: Plasticizer in lacquers, varnishes, polyvinyl chloride, polystyrene, nitrocellulose; waterproofing agent; hydraulic fluid and heat exchange medium; fire retardant for plastics; solvent mixtures; synthetic lubricant; gasoline additive to prevent pre-ignition.

TRIETHYLAMINE

Synonyms: Diethylaminoethane; *N,N*-**Diethylethanamine**; TEN; UN 1296.

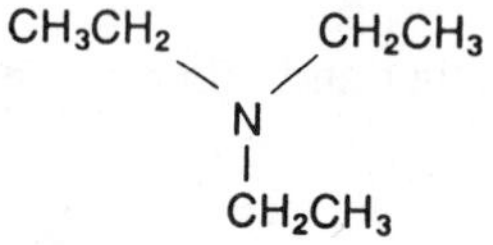

$$CH_3CH_2 \diagdown \quad \diagup CH_2CH_3$$
$$N$$
$$|$$
$$CH_2CH_3$$

CAS Registry No.: 121-44-8
DOT: 1296
DOT label: Flammable liquid
Molecular formula: $C_6H_{15}N$
Formula weight: 101.19
RTECS: YE0175000

Physical state, color and odor
Colorless liquid with a strong, ammonia-like odor. Odor threshold in air is 480 ppb (Amoore and Hautala, 1983).

Melting point (°C):
-114.7 (Weast, 1986)

Boiling point (°C):
89.3 (Weast, 1986)

Density (g/cm³):
0.7275 at 20/4 °C (Weast, 1986)
0.7255 at 25/4 °C (Windholz et al., 1983)

Diffusivity in water (10^5 cm²/sec):
0.73 at 20 °C using method of Hayduk and Laudie (1974).

Dissociation constant:
11.01 at 18 °C (Gordon and Ford, 1972)
10.72 at 25 °C (Dean, 1973)

Flash point (°C):
-6.7 (NIOSH, 1994)

Lower explosive limit (%):
1.2 (NIOSH, 1994)

Upper explosive limit (%):
8.0 (NIOSH, 1994)

Henry's law constant (10^4 atm·m^3/mol):
4.79 at 20 °C (approximate - calculated from water solubility and vapor pressure)

Ionization potential (eV):
7.50 ± 0.02 (Franklin et al., 1969)

Soil sorption coefficient, log K_{oc}:
Unavailable because experimental methods for estimation of this parameter for aliphatic amines are lacking in the documented literature.

Octanol/water partition coefficient, log K_{ow}:
1.15 (pH 9.4), 1.44, 1.45 (pH 13) (Sangster, 1989)

Solubility in organics:
Miscible with alcohol and ether (Windholz et al., 1983).

Solubility in water:
15,000 mg/L at 20 °C, 19,700 mg/L at 95 °C (Verschueren, 1983)
1.41 M at 20 °C (Fischer and Ehrenberg, 1948)
728 mmol/kg at 25.0 °C (Vesala, 1974)

Vapor density:
4.14 g/L at 25 °C, 3.49 (air = 1)

Vapor pressure (mmHg at 20 °C):
53 (NIOSH, 1994)
50 (Verschueren, 1983)

Environmental Fate
   Chemical/Physical. Triethylamine reacted with NO$_x$ in the dark to form diethyl-nitrosamine. In an outdoor chamber, photooxidation by natural sunlight yielded the following products: diethylnitramine, diethylformamide, diethylacet-amide, ethylacetamide, diethylhydroxylamine, ozone, acetaldehyde and peroxy-acetylnitrate (Pitts et al., 1978).

Exposure Limits: NIOSH REL: IDLH 200 ppm; OSHA PEL: TWA 25 ppm (100 mg/m^3); ACGIH TLV: TWA 10 ppm (40 mg/m^3), STEL 15 ppm (60 mg/m^3).

Toxicity: Acute oral LD$_{50}$ for mice 546 mg/kg, rats 460 mg/kg (RTECS, 1985).

Uses: Curing and hardening polymers; catalytic solvent in chemical synthesis; accelerator activators for rubber; wetting, penetrating and waterproofing agents of quaternary ammonium types; corrosion inhibitor; propellant.

TRIFLURALIN

Synonyms: Agreflan; Agriflan 24; Crisalin; Digermin; 2,6-Dinitro-*N*,*N*-dipropyl-4-trifluoromethylaniline; **2,6-Dinitro-*N*,*N*-dipropyl-4-(trifluoromethyl)benzenamine;** 2,6-Dinitro-*N*,*N*-di-*n*-propyl-α,α,α-trifluoro-*p*-toluidine; 4-(Di-*n*-propylamino)-3,5-dinitro-1-trifluoromethylbenzene; *N*,*N*-Di-*n*-propyl-2,6-dinitro-4-trifluoromethylaniline; *N*,*N*-Dipropyl-4-trifluoromethyl-2,6-dinitroaniline; Elancolan; L-36352; Lilly 36352; NCI-C00442; Nitran; Olitref; Trefanocide; Treficon; Treflam; Treflan; Treflanocide elancolan; Trifluoralin; α,α,α-Trifluoro-2,6-dinitro-*N*,*N*-dipropyl-*p*-toluidine; Trifluraline; Triflurex; Trikepin; Trim.

CAS Registry No.: 1582-09-8
DOT: 1609
Molecular formula: $C_{13}H_{16}F_3N_3O_4$
Formula weight: 335.29
RTECS: XU9275000

Physical state and color
Yellow to orange crystals.

Melting point (°C):
46–47 (Windholz et al., 1983)
48.5–49 (Verschueren, 1983)
43 (Bailey and White, 1965)

Boiling point (°C):
139–140 at 4.2 mmHg (Windholz et al., 1983)
96–97 at 0.18 mmHg (Probst et al., 1967)

Density (g/cm³):
1.294 at 25/4 °C (Keith and Walters, 1992)

Diffusivity in water (10^5 cm²/sec):
0.47 at 20 °C using method of Hayduk and Laudie (1974).

Henry's law constant (10^5 atm·m³/mol):
4.84 at 23 °C (Fendinger et al., 1989)

Bioconcentration factor, log BCF:
3.03 (freshwater fish), 2.45 (fish, microcosm) (Garten and Trabalka, 1983)

Soil sorption coefficient, log K_{oc}:
3.73 (Commerce soil), 3.67 (Tracy soil), 3.64 (Catlin soil) (McCall et al., 1981)
4.14 (Kenaga and Goring, 1980)
4.49 (Hickory Hill sediment, Brown and Flagg, 1981)
2.94 (average of 2 soil types, Kanazawa, 1989)
1.705 (Cecil sandy loam), 1.64 (Eustis fine sand), 2.25 (Glendale sandy clay loam),
   1.88 (Webster silty clay loam, Davidson et al., 1980)
3.70 (Mivtahim soil), 3.44 (Gilat soil), 3.46 (Neve Yaar soil), 3.63 (Malkiya soil),
   4.00 (Kinneret-A sediment), 3.97 (Kinneret-G sediment) (Gerstl and
   Mingelgrin, 1984)

Octanol/water partition coefficient, log K_{ow}:
5.34 (Kenaga and Goring, 1980)
5.28 (Brown and Flagg, 1981)
5.07 at pH 7 and 25 °C (Worthing and Hance, 1991)

Solubility in organics:
Freely soluble in Stoddard solvent (Windholz et al., 1983), chloroform, methanol
(Probst et al., 1967), acetone (400 g/L), xylene (580 g/L) (Worthing and Hance,
1991), ether and ethanol (Bailey and White, 1965).

Solubility in water:
240 mg/L (Bailey and White, 1965)
4 ppm at 27 °C (Verschueren, 1983)
3 μmol/L at 25 °C (LaFleur, 1979)

Vapor pressure (10^4 mmHg):
1.99 at 29.5 °C (Verschueren, 1983)
1.1 at 25 °C (Lewis, 1989)
2.42 at 30 °C (Nash, 1983)

Environmental Fate
   Biological. Laanio et al. (1973) incubated [$^{14}CF_3$]trifluralin with *Paecilomyces*,
Fusarium oxysporum, or *Aspergillus fumigatus* and reported that <1% was
converted to $^{14}CO_2$.
   Soil. Anaerobic degradation in a Crowley silt loam yielded α,α,α-trifluoro-
N^4,N^4-dipropyl-5-nitrotoluene-3,4-diamine and α,α,α-trifluoro-N^4,N^4-dipropyl-
toluene-3,4,5-triamine (Parr and Smith, 1973). Probst and Tepe (1969) reported
that trifluralin degradation in many soils was probably by chemical reduction of
the nitro groups into amino groups. Reported degradation products in aerobic soils

include α,α,α-trifluoro-2,6-dinitro-N-propyl-p-toluidine, α,α,α-trifluoro-2,6-di-nitro-p-toluidine, α,α,α-trifluoro-5-nitrotoluene-3,4-diamine and α,α,α-trifluoro-N,N-dipropyl-5-nitrotoluene-3,4-diamine. Anaerobic degradation products identified include α,α,α-trifluoro-N,N-dipropyl-5-nitrotoluene-3,4-diamine, α,α,α-trifluoro-N,N-dipropyltoluene-3,4,5-triamine, α,α,α-trifluorotoluene-3,4,5-triamine and α,α,α-trifluoro-N-propyltoluene-3,4,5-triamine. α,α,α-Trifluoro-5-nitro-N-propyltoluene-3,4-diamine was identified in both aerobic and anaerobic soils (Probst et al., 1967). The following compounds were reported as major soil metabolites: α,α,α-trifluoro-2,6-dinitro-N-propyl-p-toluidine, α,α,α-trifluoro-2,6-dinitro-p-toluidine, α,α,α-trifluoro-5-nitrotoluene-3,4-diamine, α,α,α-trifluorotoluene-3,4,5-triamine, 2-ethyl-7-nitro-1-propyl-5-(trifluoromethyl)benzimidazole, 2-ethyl-7-nitro-5-(trifluoromethyl)benzimidazole, 7-nitro-1-propyl-5-(trifluoromethyl)benzimidazole, 4-(dipropylamino)-3,5-dinitrobenzoic acid, 2,2'-azoxybis(α,α,α-trifluoro-6-nitro-N-propyl-p-toluidine), 2,2'-azobis(α,α,α-trifluoro-6-nitro-N-propyl-p-toluidine), 2,6-dinitro-N,N-dipropyl-4-(trifluoromethyl)-m-anisidine and α,α,α-trifluoro-2',6'-dinitro-N-propyl-p-propionotoluidine (Koskinen et al., 1984, 1985).

Golab et al. (1979) studied the degradation of trifluralin in soil over a 3-year period. They found that the herbicide undergoes N-dealkylation, reduction of nitro substituents, followed by the formation cyclized products. Of the 28 transformations products identified, none exceeded 3% of the applied amount. These compounds were: α,α,α-trifluoro-2,6-dinitro-N-propyl-p-toluidine, α,α,α-trifluoro-2,6-dinitro-p-toluidine, α,α,α-trifluoro-5-nitro-N^4,N^4-dipropyltoluene-3,4-diamine, α,α,α-trifluoro-5-nitro-N^4-propyltoluene-3,4-diamine, α,α,α-trifluoro-5-nitrotoluene-3,4-diamine, α,α,α-trifluoro-N^4,N^4-dipropyltoluene-3,4,5-triamine, α,α,α-trifluoro-N^4-propyltoluene-3,4,5-triamine, α,α,α-trifluorotoluene-3,4,5-triamine, α,α,α-trifluoro-2'-hydroxyamino-6'-nitro-N-propyl-p-propionotoluidide, 2-ethyl-7-nitro-1-propyl-5-(trifluoromethyl)benzimidazole 3-oxide, 2-ethyl-7-nitro-5-(trifluoromethyl)benzimidazole 3-oxide, 2-ethyl-7-nitro-1-propyl-5-(trifluoromethyl)benzimidazole, 7-amino-2-ethyl-1-propyl-5-(trifluoromethyl)benzimidazole, 2-ethyl-7-nitro-5-(trifluoromethyl)benzimidazole, 7-amino-2-ethyl-5-(trifluoromethyl)benzimidazole, 7-nitro-1-propyl-5-(trifluoromethyl)benzimidazole, 7-nitro-5-(trifluoromethyl)benzimidazole, 7-amino-5-(trifluoromethyl)benzimidazole, α,α,α-trifluoro-2,6-dinitro-p-cresol, 4-(dipropylamino)-3,5-dinitrobenzoic acid, 2,2'-azoxybis(α,α,α-trifluoro-6-nitro-N-propyl-p-toluidine), N-propyl-2,2'-azoxybis(α,α,α-trifluoro-6-nitro-p-toluidine), 2,2'-azoxybis(α,α,α-trifluoro-6-nitro-p-toluidine), 2,2'-azobis(α,α,α-trifluoro-6-nitro-N-propyl-p-toluidine), α,α,α-trifluoro-4,6-dinitro-5-(dipropylamino)-o-cresol, α,α,α-trifluoro-2-hydroxyamino-6-nitro-N,N-dipropyl-p-toluidine, α,α,α-trifluoro-2',6'-dinitro-N-propyl-p-propionotoluidide and α,α,α-trifluoro-2,6-dinitro-N-(propan-2-ol)-N-propyl-p-toluidine (Golab et al., 1979).

Zayed et al. (1983) studied the degradation of trifluralin by the microbes *Aspergillus carneus*, *Fusarium oxysporum* and *Trichoderma viride*. Following an inoculation and incubation period of 10 days in the dark at 25 °C, the following metabolites were identified: α,α,α-trifluoro-2,6-dinitro-N-propyl-p-toluidine, α,α,α-trifluoro-2,6-dinitro-p-toluidine, 2-amino-6-nitro-α,α,α-trifluoro-p-toluidine and 2,6-dinitro-4-trifluoromethyl phenol. The reported half-life in soil is 132 days (Jury et al., 1987).

Plant. Trifluralin was absorbed by carrot roots in greenhouse soils pretreated with the herbicide (0.75 lb/acre). The major metabolite formed was α,α,α-trifluoro-2,6-dinitro-N-(n-propyltoluene)-p-toluidine (Golab et al., 1967). Two metabolites of trifluralin that were reported in goosegrass (*Eleucine indica*) were 3-methoxy-2,6-dinitro-N,N'-dipropyl-4-(trifluoromethyl)benzenamine and N-(2,6-dinitro-4-(trifluoromethyl)phenyl)-N-propylpropanamide (Duke et al., 1991).

Photolytic. Irradiation of trifluralin in hexane by laboratory light produced α,α,α-trifluoro-2,6-dinitro-N-propyl-p-toluidine and α,α,α-trifluoro-2,6-dinitro-p-toluidine. The sunlight irradiation of trifluralin in water yielded α,α,α-trifluoro-N^4,N^4-dipropyl-5-nitrotoluene-3,4-diamine, α,α,α-trifluoro-N^4,N^4-dipropyl-toluene-3,4,5-triamine, 2-ethyl-7-nitro-5-trifluoromethylbenzimidazole, 2,3-dihydroxy-2-ethyl-7-nitro-1-propyl-5-trifluoromethylbenzimidazoline and 2-ethyl-7-nitro-5-trifluoromethylbenzimidazole. 2-Amino-6-nitro-α,α,α-trifluoro-p-toluidine and 2-ethyl-5-nitro-7-trifluoromethylbenzimidazole also were reported as major products under acidic and basic conditions, respectively (Crosby and Leitis, 1973). In a later study, Leitis and Crosby (1974) reported that trifluralin in aqueous solutions was very unstable to sunlight, especially in the presence of methanol. The photodecomposition of trifluralin involved oxidative N-dealkylation, nitro reduction and reductive cyclization. The principal photodecomposition products of trifluralin were 2-amino-6-nitro-α,α,α-trifluoro-p-toluidine, 2-ethyl-7-nitro-5-trifluoromethylbenzimidazole 3-oxide, 2,3-dihydroxy-2-ethyl-7-nitro-1-propyl-5-trifluoromethylbenzimidazole, two azoxybenzenes. Under alkaline conditions, the principal photodecomposition product was 2-ethyl-7-nitro-5-trifluoromethylbenzimidazole (Leitis and Crosby, 1974).

When trifluralin was released in the atmosphere on a sunny day, it was rapidly converted to the photochemical 2,6-dinitro-N-propyl-α,α,α-trifluoro-p-toluidine. The estimated half-life is 20 minutes (Woodrow et al., 1978). The vapor-phase photolysis of trifluralin was studied in the laboratory using a photoreactor, which simulated sunlight conditions (Soderquist et al., 1975). Vapor-phase photoproducts of trifluralin were identified as 2,6-dinitro-N-propyl-α,α,α-trifluoro-p-toluidine, 2,6-dinitro-α,α,α-trifluoro-p-toluidine, 2-ethyl-7-nitro-1-propyl-5-trifluoromethylbenzimidazole, 2-ethyl-7-nitro-5-trifluoromethylbenzimidazole and four benzimidazole precursors, reported by Leitis and Crosby (1974). Similar photoproducts were also identified in air above both bare surface treated soil and soil incorporated fields (Soderquist et al., 1975). Sullivan et al. (1980) studied the UV photolysis of trifluralin in anaerobic benzene solutions. Products identified included the three azoxybenzene derivatives N-propyl-2,2'-azoxybis(α,α,α-trifluoro-6-nitro-p-toluidine), 2,2'-azoxybis(α,α,α-trifluoro-6-nitro-N-propyl-p-

toluidine) and 2,2'-azoxybis(α,α,α-trifluoro-6-nitro-*p*-toluidine) and two azo-benzene derivatives, *N*-propyl-2,2'-azobis(α,α,α-trifluoro-6-nitro-*p*-toluidine) and 2,2'-azobis(α,α,α-trifluoro-6-nitro-*N*-propyl-*p*-toluidine) (Sullivan et al., 1980). [^{14}C]Trifluralin on a silica plate was exposed to summer sunlight for 7.5 hours. Although 52% of trifluralin was recovered, no photodegradation products were identified. In soil, no significant photodegradation of trifluralin was observed after 9 hours of irradiation (Plimmer, 1978). When trifluralin on glass plates was irradiated for 4-6 hours, it photodegraded to 2,3-dihydroxy-2-ethyl-7-nitro-1-propyl-5-trifluoromethylbenzimidazoline and 2-ethyl-7-nitro-5-trifluoromethyl-benzimidazole-3-oxide (Wright and Warren, 1965).

Chemical/Physical. Releases carbon monoxide, carbon dioxide and ammonia when heated to 900 °C (Kennedy, 1972, 1972a). Incineration may also release hydrofluoric acid in the off-gases (Sittig, 1985).

Toxicity: Acute oral LD_{50} for mice 5,000 mg/kg (RTECS, 1985); LC_{50} (96-hour) for young rainbow trout 10-40 μg/L and young bluegill sunfish 20-90 μg/L (Hartley and Kidd, 1987); LC_{50} (48-hour) for bluegill sunfish 19 ppb and rainbow trout 11 ppb (Verschueren, 1983).

Drinking Water Standard: As of May 1994, no MCLGs or MCLs have been proposed although trifluralin has been listed for regulation (U.S. EPA, 1994).

Use: Herbicide.

1,2,3-TRIMETHYLBENZENE

Synonym: Hemimellitene.

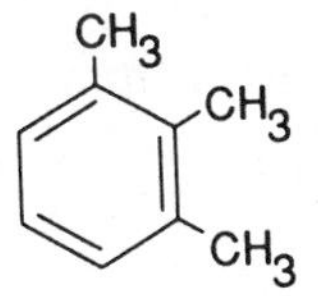

CAS Registry No.: 526-73-8
DOT: 2325
DOT label: Combustible liquid
Molecular formula: C_9H_{12}
Formula weight: 120.19
RTECS: DC3300000

Physical state, color and odor
Colorless liquid with an aromatic odor.

Melting point (°C):
-25.4 (Weast, 1986)

Boiling point (°C):
176.1 (Weast, 1986)

Density (g/cm³):
0.8944 at 20/4 °C (Weast, 1986)

Diffusivity in water (10^5 cm²/sec):
0.74 at 20 °C using method of Hayduk and Laudie (1974).

Dissociation constant:
>14 (Schwarzenbach et al., 1993)

Flash point (°C):
48 (Dean, 1987)
53 (90.5% solution, NFPA, 1984)

Lower explosive limit (%):
0.8 (NIOSH, 1994)

Upper explosive limit (%):
6.6 (NIOSH, 1994)

Heat of fusion (kcal/mol):
1.955 (Dean, 1987)

Henry's law constant (10^3 atm·m^3/mol):
3.18 at 25 °C (approximate - calculated from water solubility and vapor pressure)

Ionization potential (eV):
8.48 (NIOSH, 1994)

Soil sorption coefficient, log K_{oc}:
2.80 (Schwarzenbach and Westall, 1981)

Octanol/water partition coefficient, log K_{ow}:
3.55 (Tewari et al., 1982; Wasik et al., 1981)
3.66 (Camilleri et al., 1988; Hammers et al., 1982; Hansch and Leo, 1979)

Solubility in organics:
Soluble in acetone, alcohol, benzene and ether (Weast, 1986).

Solubility in water:
75.2 mg/kg at 25 °C, 48.6 mg/kg in artificial seawater (salinity = 34.5 g/kg) at 25.0 °C (Sutton and Calder, 1975)
545 μmol/L at 25.0 °C (Tewari et al., 1982; Wasik et al., 1981)
498, 520, 597 and 702 μmol/L at 15, 25, 35 and 45 °C, respectively (Sanemasa et al., 1982)

Vapor density:
4.91 g/L at 25 °C, 4.15 (air = 1)

Vapor pressure (mmHg):
1.51 at 25 °C (Mackay et al., 1982)

Environmental Fate
Photolytic. Glyoxal, methylglyoxal and biacetyl were produced from the photooxidation of 1,2,3-trimethylbenzene by hydroxyl radicals in air at 25 °C (Tuazon et al., 1986).

Exposure Limits: NIOSH REL: TWA 25 ppm (125 mg/m^3).

Uses: Organic synthesis.

1,2,4-TRIMETHYLBENZENE

Synonyms: *asym*-Trimethylbenzene; Pseudocumene; Pseudocumol.

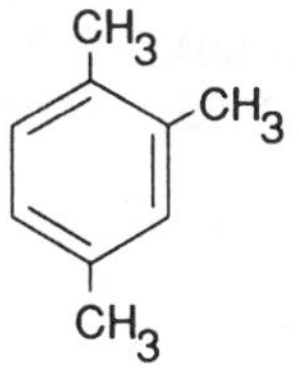

CAS Registry No.: 95-63-6
DOT: 2325
DOT label: Combustible liquid
Molecular formula: C_9H_{12}
Formula weight: 120.19
RTECS: DC3325000

Physical state, color and odor
Colorless liquid with a slight aromatic odor.

Melting point (°C):
-43.8 (Weast, 1986)

Boiling point (°C):
169.3 (Weast, 1986)

Density (g/cm³):
0.8758 at 20/4 °C (Weast, 1986)
0.888 at 4/4 °C (Sax and Lewis, 1987)

Diffusivity in water (10^5 cm²/sec):
0.73 at 20 °C using method of Hayduk and Laudie (1974).

Dissociation constant:
>14 (Schwarzenbach et al., 1993)

Flash point (°C):
54.4 (Sax and Lewis, 1987)
44 (NFPA, 1984)

Lower explosive limit (%):
0.9 (NFPA, 1984)

Upper explosive limit (%):
6.4 (NFPA, 1984)

1027

Heat of fusion (kcal/mol):
3.153 (Dean, 1987)

Henry's law constant (10^3 atm·m³/mol):
5.7 at 25 °C (Hine and Mookerjee, 1975)

Ionization potential (eV):
8.27 (NIOSH, 1994)

Soil sorption coefficient, log K_{oc}:
3.57 using method of Karickhoff et al. (1979).

Octanol/water partition coefficient, log K_{ow}:
3.65 (Mackay et al., 1980)
3.78 (Camilleri et al., 1988; Hammers et al., 1982)

Solubility in organics:
Soluble in acetone, alcohol, benzene and ether (Weast, 1986).

Solubility in water:
51.9 mg/kg at 25 °C (Price, 1976)
57 mg/kg at 25 °C (McAuliffe, 1966)
59.0 mg/kg at 25 °C (Sutton and Calder, 1975)
0.313 mmol/L in 0.5 M NaCl at 25 °C (Wasik et al., 1984)
435, 469, 514 and 571 μmol/L at 15, 25, 35 and 45 °C, respectively (Sanemasa et al., 1982)
3.47 mg/L (water soluble fraction of a 15-component simulated jet fuel mixture (JP8) containing 6.9 wt % 1,2,4-trimethylbenzene, MacIntyre and deFur, 1985)

Vapor density:
4.91 g/L at 25 °C, 4.15 (air = 1)

Vapor pressure (mmHg):
2.03 at 25 °C (Mackay et al., 1982)

Environmental Fate
Biological. In anoxic groundwater near Bemidji, MI, 1,2,4-trimethylbenzene anaerobically biodegraded to the intermediate 3,4-dimethylbenzoic acid and the tentatively identified compounds 2,4- and/or 2,5-dimethylbenzoic acid (Cozzarelli et al., 1990).
Photolytic. Glyoxal, methylglyoxal and biacetyl were produced from the photooxidation of 1,2,4-trimethylbenzene by hydroxyl radicals in air at 25 °C (Tuazon et al., 1986).

Exposure Limits: NIOSH REL: TWA 25 ppm (125 mg/m^3).

Toxicity: LC$_{50}$ (inhalation) for rats 18 gm/m^3/4-hour (RTECS, 1985).

Drinking Water Standard: As of May 1994, no MCLGs or MCLs have been proposed (U.S. EPA, 1984).

Uses: Manufacture of dyes, resins, perfumes, trimellitic anhydride, pseudo-cumidine.

1,3,5-TRIMETHYLBENZENE

Synonyms: Mesitylene; Fleet-X; TMB; *sym*-Trimethylbenzene; Trimethylbenzol; UN 2325.

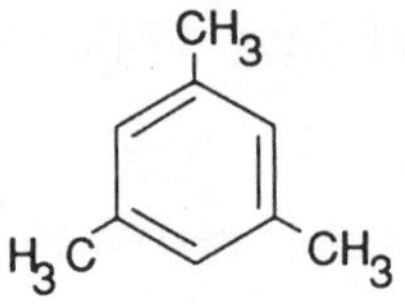

CAS Registry No.: 108-67-8
DOT: 2325
DOT label: Combustible liquid
Molecular formula: C_9H_{12}
Formula weight: 120.19
RTECS: OX6825000

Physical state, color and odor
Colorless liquid with a peculiar odor. Odor threshold in air is 550 ppb (Amoore and Hautala, 1983).

Melting point (°C):
-44.7 (Weast, 1986)
-52.7 (Verschueren, 1983)

Boiling point (°C):
164.7 (Weast, 1986)

Density (g/cm^3):
0.8652 at 20/4 °C (Weast, 1986)

Diffusivity in water (10^5 cm^2/sec):
0.73 at 20 °C using method of Hayduk and Laudie (1974).

Dissociation constant:
>14 (Schwarzenbach et al., 1993)

Flash point (°C):
44 (Dean, 1987)
50 (NFPA, 1984)

Heat of fusion (kcal/mol):
1.892-2.274 (Dean, 1987)

Henry's law constant (10^3 atm·m^3/mol):
4.03, 4.60, 5.71, 6.73 and 9.63 at 10, 15, 20, 25 and 30 °C, respectively (Ashworth et al., 1988).

Interfacial tension with water (dyn/cm at 20 °C):
38.70 (Demond and Lindner, 1993)

Ionization potential (eV):
8.39 (NIOSH, 1994)

Soil sorption coefficient, log K_{oc}:
2.82 (Schwarzenbach and Westall, 1981)

Octanol/water partition coefficient, log K_{ow}:
3.41 (Campbell and Luthy, 1985)
3.42 (Chiou et al., 1982; Hansch and Leo, 1979)
3.78 (Hammers et al., 1978)

Solubility in organics:
Miscible with alcohol, benzene and ether (Windholz et al., 1983).

Solubility in water:
48.2 mg/kg at 25 °C (Sutton and Calder, 1975)
640 µmol/L at 25.0 °C (Andrews and Keefer, 1950a)
97.7 mg/L at 25 °C (Chiou et al., 1982)
282 µmol/L in 0.5 M NaCl at 25 °C (Wasik et al., 1984)
383, 415, 455 and 485 µmol/L at 15, 25, 35 and 45 °C, respectively (Sanemasa et al., 1982)
<200 mg/L at 25 °C (Booth and Everson, 1948)
328 µmol/kg at 25.0 °C (Vesala, 1974)
8.55 x 10^{-6} at 25 °C (mole fraction, Li et al., 1993)

Vapor density:
4.91 g/L at 25 °C, 4.15 (air = 1)

Vapor pressure (mmHg):
2 at 20 °C (NIOSH, 1994)
2.42 at 25 °C (Mackay et al., 1982)

Environmental Fate
Biological. In anoxic groundwater near Bemidji, MI, 1,3,5-trimethylbenzene anaerobically biodegraded to the intermediate tentatively identified as 3,5-dimethylbenzoic acid (Cozzarelli et al., 1990).

Photolytic. Glyoxal, methylglyoxal and biacetyl were produced from the photooxidation of 1,3,5-trimethylbenzene by hydroxyl radicals in air at 25 °C (Tuazon et al., 1986).

Exposure Limits: NIOSH REL: TWA 25 ppm (125 mg/m^3).

Toxicity: LD$_{50}$ (inhalation) for rats 24 gm/m^3/4-hour (RTECS, 1985).

Drinking Water Standard: As of May 1994, no MCLGs or MCLs have been proposed (U.S. EPA, 1984).

Uses: UV oxidation stabilizer for plastics; manufacturing anthraquinone dyes.

1,1,3-TRIMETHYLCYCLOHEXANE

Synonym: 1,3,3-Trimethylcyclohexane.

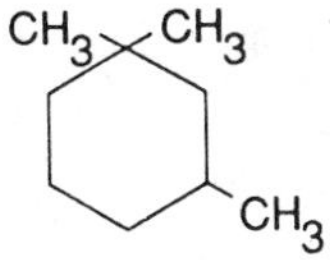

CAS Registry No.: 3073-66-3
Molecular formula: C_9H_{18}
Formula weight: 126.24
RTECS: GV7650000

Physical state
Liquid.

Boiling point (°C):
136.6 (Dean, 1987)
138–139 (Weast, 1986)

Density (g/cm³):
0.7664 at 20/4 °C (Weast, 1986)

Diffusivity in water (10^5 cm²/sec):
0.66 at 20 °C using method of Hayduk and Laudie (1974).

Dissociation constant:
>14 (Schwarzenbach et al., 1993)

Henry's law constant (atm·m³/mol):
1.04 at 25 °C (approximate – calculated from water solubility and vapor pressure)

Soil sorption coefficient, log K_{oc}:
Unavailable because experimental methods for estimation of this parameter for alicyclic hydrocarbons are lacking in the documented literature.

Octanol/water partition coefficient, log K_{ow}:
Unavailable because experimental methods for estimation of this parameter for alicyclic hydrocarbons are lacking in the documented literature.

Solubility in water:
1.77 mg/kg at 25 °C (Price, 1976)

Vapor density:
4.36 g/L at 25 °C, 3.68 (air = 1)

Vapor pressure (mmHg):
11.1 (estimated using Antoine equation, Dean, 1987)

Uses: Organic synthesis; gasoline component.

1,1,3-TRIMETHYLCYCLOPENTANE

Synonym: 1,3,3-Trimethylcyclopentane.

CAS Registry No.: 4516-69-2
Molecular formula: C_8H_{16}
Formula weight: 112.22

Physical state and color
Colorless liquid.

Melting point (°C):
-14.4 (Mackay et al., 1982)

Boiling point (°C):
104.89 °C (Wilhoit and Zwolinski, 1971)

Density (g/cm^3):
0.7703 at 20/4 °C (Weast, 1986)

Diffusivity in water (10^5 cm^2/sec):
0.71 at 20 °C using method of Hayduk and Laudie (1974).

Dissociation constant:
>14 (Schwarzenbach et al., 1993)

Henry's law constant (atm·m^3/mol):
1.57 at 25 °C (approximate - calculated from water solubility and vapor pressure)

Soil sorption coefficient, log K_{oc}:
Unavailable because experimental methods for estimation of this parameter for alicyclic hydrocarbons are lacking in the documented literature.

Octanol/water partition coefficient, log K_{ow}:
3.34 (calculated, Wang et al., 1992)

Solubility in water:
3.73 mg/kg at 25 °C (Price, 1976)

Vapor density:
4.59 g/L at 25 °C, 3.87 (air = 1)

Vapor pressure (mmHg):
39.7 at 25 °C (Wilhoit and Zwolinski, 1971)

Uses: Organic synthesis; gasoline component.

2,2,5-TRIMETHYLHEXANE

Synonym: 2,5,5-Trimethylhexane.

$$
\begin{array}{cccccc}
H & CH_3 & H & H & CH_3 & H \\
| & | & | & | & | & | \\
H-C- & C- & C- & C- & C- & C-H \\
| & | & | & | & | & | \\
H & CH_3 & H & H & H & H
\end{array}
$$

CAS Registry No.: 3522-94-9
Molecular formula: C_9H_{20}
Formula weight: 128.26

Physical state and color
Colorless liquid.

Melting point (°C):
-105.8 (Weast, 1986)

Boiling point (°C):
124 (Weast, 1986)

Density (g/cm³):
0.70721 at 20/4 °C, 0.70322 at 25/4 °C (Dreisbach, 1959)

Diffusivity in water (10^5 cm²/sec):
0.62 at 20 °C using method of Hayduk and Laudie (1974).

Dissociation constant:
>14 (Schwarzenbach et al., 1993)

Flash point (°C):
12.7 (Hawley, 1981)

Henry's law constant (atm·m³/mol):
2.42 at 25 °C (approximate - calculated from water solubility and vapor pressure)

Soil sorption coefficient, log K_{oc}:
Unavailable because experimental methods for estimation of this parameter for aliphatic hydrocarbons are lacking in the documented literature.

Octanol/water partition coefficient, log K_{ow}:
3.88 using method of Hansch et al. (1968).

Solubility in organics:
In methanol, g/L: 162 at 5 °C, 179 at 10 °C, 200 at 15 °C, 221 at 20 °C, 247 at 25 °C, 280 at 30 °C, 316 at 35 °C, 360 at 40 °C (Kiser et al., 1961).

Solubility in water (mg/kg):
0.79 at 0 °C, 0.54 at 25 °C (Polak and Lu, 1973)
1.15 at 25 °C (McAuliffe, 1966)

Vapor density:
5.24 g/L at 25 °C, 4.43 (air = 1)

Vapor pressure (mmHg):
16.5 at 25 °C (Wilhoit and Zwolinski, 1971)

Uses: Motor fuel additive; organic synthesis.

2,2,4-TRIMETHYLPENTANE

Synonyms: Isobutyltrimethylmethane; Isooctane; UN 1262.

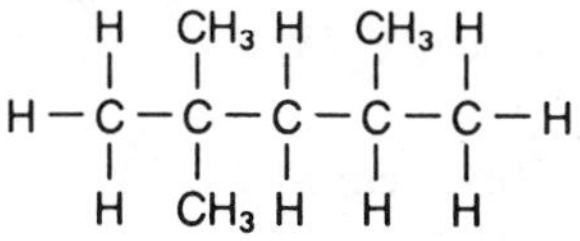

```
        H   CH₃ H   CH₃ H
        |   |   |   |   |
  H — C — C — C — C — C — H
        |   |   |   |   |
        H   CH₃ H   H   H
```

CAS Registry No.: 540-84-1
DOT: 1262
DOT label: Flammable liquid
Molecular formula: C_8H_{18}
Formula weight: 114.23
RTECS: SA3320000

Physical state, color and odor
Colorless liquid with a gasoline-like odor.

Melting point (°C):
-107.4 (Weast, 1986)
-116 (Sax and Lewis, 1987)

Boiling point (°C):
99.2 (Weast, 1986)

Density (g/cm³):
0.69192 at 20/4 °C, 0.68777 at 25/4 °C (Dreisbach, 1959)

Diffusivity in water (10^5 cm²/sec):
0.66 at 20 °C using method of Hayduk and Laudie (1974).

Dissociation constant:
>14 (Schwarzenbach et al., 1993)

Flash point (°C):
-12 (Windholz et al., 1983)

Lower explosive limit (%):
1.1 (Sax and Lewis, 1987)

Upper explosive limit (%):
6.0 (Sax and Lewis, 1987)

Heat of fusion (kcal/mol):
2.20 (Dean, 1987)

Henry's law constant (atm·m³/mol):
3.01 at 25 °C (Hine and Mookerjee, 1975)

Interfacial tension with water (dyn/cm at 25 °C):
50.1 (Demond and Lindner, 1993)

Ionization potential (eV):
9.86 (HNU, 1986)

Soil sorption coefficient, log K_{oc}:
Unavailable because experimental methods for estimation of this parameter for aliphatic hydrocarbons are lacking in the documented literature.

Octanol/water partition coefficient, log K_{ow}:
5.83 (Burkhard et al., 1985a)

Solubility in organics:
In methanol, g/L: 249 at 5 °C, 279 at 10 °C, 314 at 15 °C, 353 at 20 °C, 402 at 25 °C, 460 at 30 °C, 560 at 35 °C, 760 at 40 °C (Kiser et al., 1961).

Solubility in water:
1.14 mg/kg at 25 °C (Price, 1976)
2.46 mg/kg at 0 °C, 2.05 mg/kg at 25 °C (Polak and Lu, 1973)
0.56 mg/L at 25 °C (Verschueren, 1983)
2.44 mg/kg at 25 °C (McAuliffe, 1963, 1966)
0.900 and 1.296 mL/L in water and 0.1 wt % aqueous sodium naphthenate, respectively (Baker, 1980)

Vapor density:
4.67 g/L at 25 °C, 3.94 (air = 1)

Vapor pressure (mmHg):
47.8 at 24.4 °C (Willimgham et al., 1945)
49.3 at 25 °C (Wilhoit and Zwolinski, 1971)

Symptoms of Exposure: High concentrations may cause irritation of respiratory tract (Patnaik, 1992).

Uses: Determining octane numbers of fuels; solvent and thinner; in spectrophotometric analysis.

2,3,4-TRIMETHYLPENTANE

Synonyms: None.

$$
\begin{array}{ccccccc}
 & H & CH_3 & CH_3 & CH_3 & H & \\
 & | & | & | & | & | & \\
H - & C - & C - & C - & C - & C & - H \\
 & | & | & | & | & | & \\
 & H & H & H & H & H & \\
\end{array}
$$

CAS Registry No.: 565-75-3
Molecular formula: C_8H_{18}
Formula weight: 114.23

Physical state and color
Colorless liquid.

Melting point (°C):
−109.2 (Weast, 1986)

Boiling point (°C):
113.4 (Weast, 1986)

Density (g/cm³):
0.71906 at 20/4 °C, 0.71503 at 25/4 °C (Dreisbach, 1959)

Diffusivity in water (10⁵ cm²/sec):
0.67 at 20 °C using method of Hayduk and Laudie (1974).

Dissociation constant:
>14 (Schwarzenbach et al., 1993)

Heat of fusion (kcal/mol):
2.215 (Dean, 1987)

Henry's law constant (atm·m³/mol):
2.98 at 25 °C (approximate - calculated from water solubility and vapor pressure)

Soil sorption coefficient, log K_{oc}:
Unavailable because experimental methods for estimation of this parameter for aliphatic hydrocarbons are lacking in the documented literature.

Octanol/water partition coefficient, log K_{ow}:
3.78 using method of Hansch et al. (1968).

Solubility in organics:
Soluble in acetone, alcohol, benzene, chloroform and ether (Weast, 1986).

Solubility in water (mg/kg):
1.36 at 25 °C (Price, 1976)
2.34 at 0 °C, 2.30 at 25 °C (Polak and Lu, 1973)

Vapor density:
4.67 g/L at 25 °C, 3.94 (air = 1)

Vapor pressure (mmHg):
27.0 at 25 °C (Wilhoit and Zwolinski, 1971)

Uses: Organic synthesis.

2,4,6-TRINITROTOLUENE

Synonyms: Entsufon; 1-Methyl-2,4,6-trinitrobenzene; **2-Methyl-1,3,5-trinitro-benzene**; NCI-C56155; TNT; α-TNT; TNT-tolite; Tolit; Tolite; Trilit; Trinitro-toluene; *sym*-Trinitrotoluene; Trinitrotoluol; α-Trinitrotoluol; *sym*-Trinitrotoluol; Tritol; Triton; Trotyl; Trotyl oil; UN 0209.

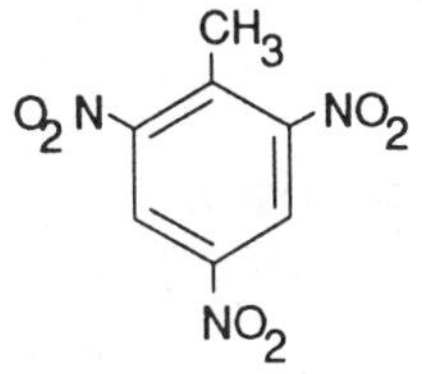

CAS Registry No.: 118-96-7
DOT: 1356
DOT label: Class A explosive
Molecular formula: $C_7H_5N_3O_6$
Formula weight: 227.13
RTECS: XU0175000

Physical state, color and odor
Colorless to light yellow, odorless monoclinic crystals.

Melting point (°C):
82 (Weast, 1986)
80.1 (Windholz et al., 1983)

Boiling point (°C):
Explodes at 240 (Weast, 1986)

Density (g/cm³):
1.654 at 20/4 °C (Weast, 1986)

Diffusivity in water (10^5 cm²/sec):
0.68 at 20 °C using method of Hayduk and Laudie (1974).

Flash point (°C):
Explodes (NIOSH, 1994)

Ionization potential (eV):
10.59 (NIOSH, 1994)

Soil sorption coefficient, log K_{oc}:
2.48 using method of Kenaga and Goring (1980).

Octanol/water partition coefficient, log K_{ow}:
2.25 using method of Kenaga and Goring (1980).

Solubility in organics:
Soluble in acetone, benzene, ether and pyrimidine (Weast, 1986).

Solubility in water:
0.013 wt % at 20 °C (NIOSH, 1987)
1.42 g/L at 100 °C (Windholz et al., 1983)
200 mg/L at 15 °C (Verschueren, 1983)

Vapor pressure (10^3 mmHg):
0.2 at 20 °C (NIOSH, 1994)
4.26 at 54.8 °C, 2.557 at 72.5 °C, 4.347 at 76.1 °C (Lenchitz and Velicky, 1970)

Environmental Fate
   Biological.   4-Amino-2,6-dinitrotoluene   and   2-amino-4,6-dinitrotoluene,
detected in contaminated groundwater beneath the Hawthorne Naval Ammunition
Depot, NV, were reported to have formed from the microbial degradation of
2,4,6-trinitrotoluene (Pereira et al., 1979).
   Will detonate upon heating (NIOSH, 1994).

Exposure Limits (mg/m^3): NIOSH 0.5, IDLH 500; OSHA PEL: TWA 1.5; ACGIH
TLV: TWA 0.5.

Symptoms of Exposure: May cause dermatitis, cyanosis, gastritis, sneezing, sore
throat, muscular pain, somnolence, tremor, convulsions and asplastic anemia
(Patnaik, 1992).

Toxicity: Acute oral LD_{50} in mice 660 mg/kg, rats 795 (RTECS, 1985).

Uses: High explosive; intermediate in dyestuffs and photographic chemicals.

TRIPHENYL PHOSPHATE

Synonyms: Celluflex TPP; **Phosphoric acid triphenyl ester;** Phenyl phosphate; TPP.

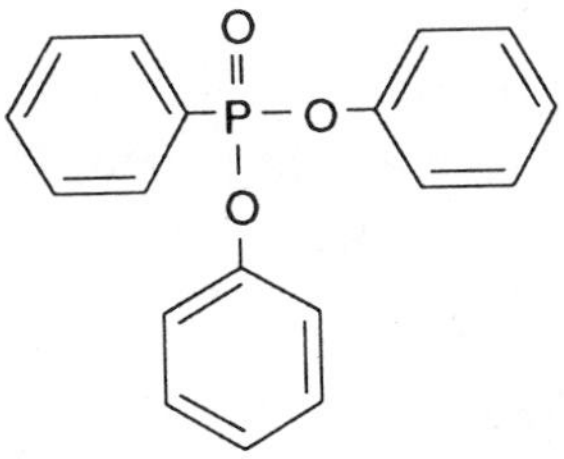

CAS Registry No.: 115-86-6
Molecular formula: $C_{18}H_{15}O_4P$
Formula weight: 326.29
RTECS: TC8400000

Physical state, color and odor
Colorless, solid with a faint, phenol-like odor.

Melting point (°C):
48.5 (Fordyce and Meyer, 1940)
50-51 (Weast, 1986)

Boiling point (°C):
245 at 11 mmHg (Weast, 1986)

Density (g/cm^3):
1.2055 at 50/4 °C (Weast, 1986)
1.268 at 60/4 °C (Sax and Lewis, 1987)

Diffusivity in water (10^5 cm^2/sec):
0.46 at 20 °C using method of Hayduk and Laudie (1974).

Flash point (°C):
222 (NIOSH, 1994)
223 (Dean, 1987)

Henry's law constant (10^2 atm·m^3/mol):
5.88 at 20 °C (approximate - calculated from water solubility and vapor pressure)

Soil sorption coefficient, log K_{oc}:
3.72 using method of Kenaga and Goring (1980).

Octanol/water partition coefficient, log K_{ow}:
5.27 using method of Saeger et al. (1979).

Solubility in organics:
Soluble in alcohol, benzene, chloroform and ether (Weast, 1986).

Solubility in water:
0.002 wt % at 54 °C (NIOSH, 1994)
0.73 mg/L at 24 °C (practical grade, Verschueren, 1983)

Vapor pressure (mmHg):
1 at 195 °C (NIOSH, 1994)
<0.1 at 20 °C (Verschueren, 1983)

Environmental Fate
  Chemical/Physical. When an aqueous solution containing triphenyl phosphate (0.1 mg/L) and chlorine (3-1,000 mg/L) was stirred in the dark at 20 °C for 24 hours, the benzene ring was substituted with 1-3 chlorine atoms (Ishikawa and Baba, 1988).

Exposure Limits (mg/m^3): NIOSH REL: TWA 3, IDLH 1,000; OSHA PEL: TWA 3.

Symptoms of Exposure: May cause depression of central nervous system and irritation of eyes, skin and respiratory tract (Patnaik, 1992).

Toxicity: Acute oral LD_{50} for mice 1,320 mg/kg, rats 3,500 mg/kg (RTECS, 1985).

Uses: Camphor substitute in celluloid; impregnating roofing paper; plasticizer in lacquers and varnishes; renders acetylcellulose, airplane "dope", nitrocellulose, stable and fireproof; gasoline additives; insecticides; floatation agents; antioxidants, stabilizers and surfactants.

VINYL ACETATE

Synonyms: Acetic acid ethenyl ester; Acetic acid ethylene ester; Acetic acid vinyl ester; 1-Acetoxyethylene; Ethenyl acetate; Ethenyl ethanoate; UN 1301; VAC; VAM; Vinyl acetate H.Q.

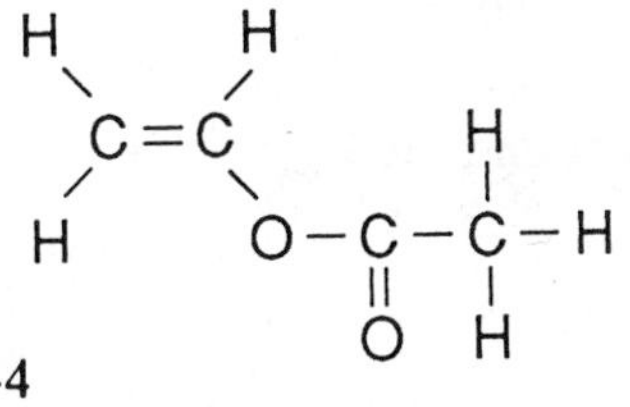

CAS Registry No.: 108-05-4
DOT: 1301
DOT label: Combustible liquid
Molecular formula: $C_4H_6O_2$
Formula weight: 86.09
RTECS: AK0875000

Physical state, color and odor
Colorless, watery liquid with a pleasant, fruity odor. Odor threshold in air is 500 ppb (Amoore and Hautala, 1983).

Melting point (°C):
-93.2 (Weast, 1986)

Boiling point (°C):
72.2 (Weast, 1986)

Density (g/cm³):
0.9317 at 20/4 °C (Weast, 1986)

Diffusivity in water (10^5 cm²/sec):
0.93 at 20 °C using method of Hayduk and Laudie (1974).

Flash point (°C):
-8 (NIOSH, 1994)

Lower explosive limit (%):
2.6 (NIOSH, 1994)

Upper explosive limit (%):
13.4 (NIOSH, 1994)

Henry's law constant (10^4 atm·m³/mol):
4.81 (calculated, Howard, 1989)

Ionization potential (eV):
9.19 ± 0.05 (Franklin et al., 1969)

Soil sorption coefficient, log K_{oc}:
0.45 (estimated, Montgomery, 1989)

Octanol/water partition coefficient, log K_{ow}:
0.60, 0.73 (Sangster, 1989)

Solubility in organics:
Soluble in acetone, ethanol, benzene, chloroform and ether (Weast, 1986).

Solubility in water (mg/L):
25,000 at 25 °C (Amoore and Hautala, 1983)
20,000 at 20 °C (Dean, 1973)

Vapor density:
3.52 g/L at 25 °C, 2.97 (air = 1)

Vapor pressure (mmHg):
83 at 20 °C, 115 at 25 °C, 140 at 30 °C (Verschueren, 1983)

Environmental Fate
   Chemical/Physical. Anticipated hydrolysis products would include acetic acid and vinyl alcohol. Slowly polymerizes in light to a colorless, transparent mass.

Exposure Limits: NIOSH REL: 15-minute ceiling 4 ppm (15 mg/m^3); ACGIH TLV: TWA 10 ppm (30 mg/m^3), STEL 20 ppm (60 mg/m^3).

Toxicity: Acute oral LD_{50} for mice 1,613 mg/kg, rats 2,920 mg/kg; LC_{50} (inhalation) for mice 1,550 ppm/4-hour, rats 4,000 ppm/2-hour, rabbits 2,500 ppm/4-hour (RTECS, 1985).

Uses: Manufacture of polyvinyl acetate, polyvinyl alcohol, polyvinyl chloride-acetate resins; used particularly in latex paint; paper coatings; adhesives; textile finishing; safety glass interlayers.

VINYL CHLORIDE

Synonyms: **Chlorethene;** Chlorethylene; Chloroethene; 1-Chloroethene; Chloro-ethylene; 1-Chloroethylene; Ethylene monochloride; Monochloroethene; Mono-chloroethylene; MVC; RCRA waste number U043; Trovidur; UN 1086; VC; VCM; Vinyl C monomer; Vinyl chloride monomer.

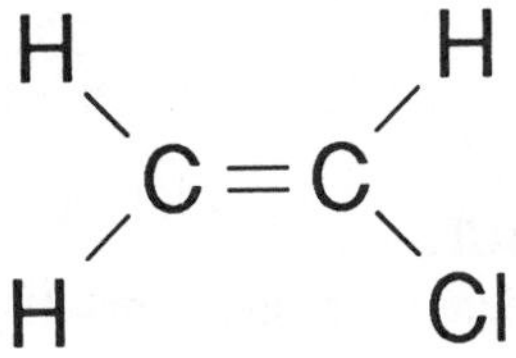

$$\begin{array}{ccc} H & & H \\ \diagdown & & \diagup \\ & C = C & \\ \diagup & & \diagdown \\ H & & Cl \end{array}$$

CAS Registry No.: 75-01-4
DOT: 1086
DOT label: Flammable gas
Molecular formula: C_2H_3Cl
Formula weight: 62.50
RTECS: KU9625000

Physical state, color and odor
Colorless, liquified compressed gas with a faint, sweetish odor.

Melting point (°C):
-153.8 (Weast, 1986)

Boiling point (°C):
-13.4 (Weast, 1986)

Density (g/cm^3):
0.9106 at 20/4 °C (Weast, 1986)
0.94 at 13.9/4 °C, 0.9121 at 20/4 °C (Standen, 1964)

Diffusivity in water (10^5 cm^2/sec):
1.34 at 25 °C (Hayduk and Laudie, 1974)

Lower explosive limit (%):
3.6 (NIOSH, 1994)

Upper explosive limit (%):
33 (NIOSH, 1994)

Henry's law constant (10^2 atm·m^3/mol):
278 (Gossett, 1987)

1049

2.2 (Pankow and Rosen, 1988)
122 at 10 °C (Dilling, 1977)
5.6 at 25 °C (Hine and Mookerjee, 1975)
1.50, 1.68, 2.17, 2.65 and 2.8 at 10, 15, 20, 25 and 30 °C, respectively (Ashworth et al., 1988)

Ionization potential (eV):
9.99 (Horvath, 1982)

Bioconcentration factor, log BCF:
3.04 (activated sludge), 1.60 (algae) (Freitag et al., 1985)

Soil sorption coefficient, log K_{oc}:
0.39 using method of Karickhoff et al. (1979).

Octanol/water partition coefficient, log K_{ow}:
0.60 (Radding et al., 1976)

Solubility in organics:
Soluble in ethanol, carbon tetrachloride and ether (U.S. EPA, 1985).

Solubility in water:
1,100 mg/L at 25 °C (Verschueren, 1983)
60 mg/L at 10 °C (Pearson and McConnell, 1975)
2,700 mg/L at 25 °C (Dilling, 1977)
In wt %: 0.95 at 15 °C, 0.995 at 16 °C, 0.915 at 20.5 °C, 0.88 at 26 °C, 0.89 at 29.5 °C, 0.94 at 35 °C, 0.89 at 41 °C, 0.88 at 46.5 °C, 0.95 at 55 °C, 0.92 at 65 °C, 0.98 at 72.5 °C, 1.00 at 80 °C, 1.12 at 85 °C (DeLassus and Schmidt, 1981)
14 mmol/L at 20 °C (Fischer and Ehrenberg, 1948)
0.68 wt % at 20 °C (Nathan, 1978)

Vapor density:
2.86 g/L at 0 °C (Hayduk and Laudie, 1974a)
2.55 g/L at 25 °C, 2.16 (air = 1)

Vapor pressure (mmHg):
2,531 at 20 °C, 3,428 at 30 °C (Standen, 1964)
2,320 at 20 °C (McConnell et al., 1975)
2,660 at 25 °C (Nathan, 1978)

Environmental Fate
   Biological. Under anaerobic or aerobic conditions, degradation to carbon dioxide was reported in experimental systems containing mixed or pure cultures (Vogel et

al., 1987). The anaerobic degradation of vinyl chloride dissolved in groundwater by static microcosms was enhanced by the presence of nutrients (methane, methanol, ammonium phosphate, phenol). Methane and ethylene were reported as the biodegradation end products (Barrio-Lage et al., 1990). When vinyl chloride (1 mM) was incubated with resting cells of *Pseudomonas* sp (0.1 g/L) in a 0.1 M phosphate buffer at pH 7.4, hydroxylation of the C-Cl bond occurred yielding acetaldehyde and chloride ions. Oxidation at both the methyl and carbonyl carbons produced acetic acid and hydroxyacetaldehyde, which underwent further oxidation to give glycolic acid (hydroxyacetic acid). The acid was further oxidized to carbon dioxide (Castro et al., 1992).

Surface Water. In natural surface waters, vinyl chloride was resistant to biological and chemical degradation (Hill et al., 1976).

Photolytic. Irradiation of vinyl chloride in the presence of nitrogen dioxide for 160 minutes produced formic acid, hydrochloric acid, carbon monoxide, formaldehyde, ozone and trace amounts of formyl chloride and nitric acid. In the presence of ozone, however, vinyl chloride photooxidized to carbon monoxide, formaldehyde, formic acid and small amounts of hydrochloric acid (Gay et al., 1976). Reported photooxidation products in the troposphere include hydrogen chloride and/or formyl chloride (U.S. EPA, 1985). In the presence of moisture, formyl chloride will decompose to carbon monoxide and hydrochloric acid (Morrison and Boyd, 1971).

Chemical/Physical. In a laboratory experiment, it was observed that the leaching of a vinyl chloride monomer from a polyvinyl chloride pipe into water reacted with chlorine to form chloroacetaldehyde, chloroacetic acid and other unidentified compounds (Ando and Sayato, 1984).

The evaporation half-life of vinyl chloride (1 mg/L) from water at 25 °C using a shallow-pitch propeller stirrer at 200 rpm at an average depth of 6.5 cm is 27.6 minutes (Dilling, 1977).

Exposure Limits: Potential occupational carcinogen. OSHA PEL: TWA 1 ppm, 15-minute ceiling 5 ppm; ACGIH TLV: TWA 5 ppm (10 mg/m^3).

Symptoms of Exposure: Narcotic at high concentrations (Patnaik, 1992).

Toxicity: Acute oral LD$_{50}$ for rats 500 mg/kg (RTECS, 1985).

Drinking Water Standard (final): MCLG: zero; MCL: 2 μg/L (U.S. EPA, 1994).

Uses: Manufacture of polyvinyl chloride and copolymers; adhesives for plastics; refrigerant; extraction solvent; organic synthesis.

WARFARIN

Synonyms: 3-(Acetonylbenzyl)-4-hydroxycoumarin; 3-(α-Acetonylbenzyl)-4-hydroxycoumarin; Athrombine-K; Athrombin-K; Brumolin; Compound 42; Corax; Coumadin; Coumafen; Coumafene; Cov-r-tox; D-con; Dethmor; Dethnel; Eastern states duocide; Fasco fascrat powder; 1-(4'-Hydroxy-3'-coumarinyl)-1-phenyl-3-butanone; **4-Hydroxy-3-(3-oxo-1-phenylbutyl)-2H-1-benzopyran-2-one;** 4-Hydroxy-3-(1-phenyl-3-oxobutyl)coumarin; Kumader; Kumadu; Kypfarin; Liqua-tox; Mar-frin; Martin's mar-frin; Maveran; Mouse pak; 3-(1'-Phenyl-2'-acetylethyl)-4-hydroxycoumarin; 3-α-Phenyl-β-acetylethyl-4-hydroxycoumarin; Prothromadin; Rat-a-way; Rat-b-gon; Rat-gard; Rat-kill; Rat & mice bait; Rat-mix; Rat-o-cide #2; Rat-ola; Ratorex; Ratox; Ratoxin; Ratron; Ratron G; Rats-no-more; Rat-trol; Rattunal; Rax; RCRA waste number P001; Rodafarin; Rodeth; Rodex; Rodex blox; Rosex; Rough & ready mouse mix; Solfarin; Spray-trol brand roden-trol; Temus W; Tox-hid; Twin light rat away; Vampirinip II; Vampirin III; Waran; W.A.R.F. 42; Warfarat; Warfarin plus; Warfarin Q; Warf compound 42; Warficide.

CAS Registry No.: 81-81-2
DOT: 3027
DOT label: Poison
Molecular formula: $C_{19}H_{16}O_4$
Formula weight: 308.33
RTECS: GN4550000

Physical state, color and odor
Colorless, odorless crystals.

Melting point (°C):
161 (Weast, 1986)
159-161 (Worthing and Hance, 1991)
159-165 (Berg, 1983)

Boiling point (°C):
Decomposes (NIOSH, 1994)

Diffusivity in water (10^5 cm^2/sec):

1052

0.44 at 20 °C using method of Hayduk and Laudie (1974).

Soil sorption coefficient, log K_{oc}:
2.96 using method of Kenaga and Goring (1980).

Octanol/water partition coefficient, log K_{ow}:
3.20 using method of Kenaga and Goring (1980).

Solubility in organics:
Soluble in alcohol, benzene, 1,4-dioxane (Weast, 1986) and acetone (Sax and Lewis, 1987). Moderately soluble in methanol, ethanol, isopropanol and some oils (Windholz et al., 1983).

Solubility in water (mg/L):
17 at 20 °C, 72 at 100 °C (Gunther et al., 1968)

Vapor pressure (mmHg):
0.09 at 22 °C (NIOSH, 1994)

Environmental Fate
Chemical/Physical. The hydrolysis half-lives at at 68.0 °C and pH values of 3.09, 7.11 and 10.18 were calculated to be 12.9, 57.4 and 23.9 days, respectively. At 25 °C and pH 7, the half-life was estimated to be 16 years (Ellington et al., 1986).

Exposure Limits (mg/m^3): NIOSH REL: TWA 0.1, IDLH 100; OSHA PEL: TWA 0.1.

Symptoms of Exposure: Hematuria, back pain, hematoma in arms, legs; epistaxis, bleeding lips, mucous membrane hemorrhage, abdominal pain, vomiting, fecal blood; petechial rash; abnormal hematology (NIOSH, 1987).

Toxicity: Acute oral LD_{50} for dogs 200 mg/kg, guinea pigs 182 mg/kg, mice 331 mg/kg, rats 3 mg/kg (RTECS, 1985).

Uses: Rodenticide.

o-XYLENE

Synonyms: 1,2-Dimethylbenzene; *o*-Dimethylbenzene; *o*-Methyltoluene; UN 1307; 1,2-Xylene; *o*-Xylol.

CAS Registry No.: 95-47-6
DOT: 1307
DOT label: Flammable liquid
Molecular formula: C_8H_{10}
Formula weight: 106.17
RTECS: ZE2450000

Physical state, color and odor
Clear, colorless liquid with an aromatic odor.

Melting point (°C):
-25.2 (Weast, 1986)

Boiling point (°C):
144.4 (Weast, 1986)

Density (g/cm³):
0.8802 at 20/4 °C (Weast, 1986)

Diffusivity in water (10^5 cm²/sec):
0.79 at 20 °C using method of Hayduk and Laudie (1974).

Dissociation constant:
>15 (Christensen et al., 1975)

Flash point (°C):
33 (NIOSH, 1994)
17 (Windholz et al., 1976)

Lower explosive limit (%):
0.9 (NIOSH, 1994)

Upper explosive limit (%):
6.7 (NIOSH, 1994)

Heat of fusion (kcal/mol):
3.25 (Dean, 1987)

Henry's law constant (10^3 atm·m^3/mol):
5.27 (Mackay and Wolkoff, 1973)
5.0 (Pankow and Rosen, 1988)
5.35 at 25 °C (Hine and Mookerjee, 1975)
2.85, 3.61, 4.74, 4.87 and 6.26 at 10, 15, 20, 25 and 30 °C, respectively (Ashworth et al., 1988)
1.47, 1.24, 1.62, 3.28 and 4.24 at 2.0, 6.0, 10.0, 18.2 and 25.0 °C, respectively (Dewulf et al., 1995)

Interfacial tension with water (dyn/cm at 25 °C):
37.2 (Donahue and Bartell, 1952)

Ionization potential (eV):
8.56 (Franklin et al., 1969)
9.04 (Yoshida et al., 1983)

Soil sorption coefficient, log K_{oc}:
2.11 (Abdul et al., 1987)
2.41 (Catlin soil, Roy et al., 1985)
2.35 (estuarine sediment, Vowles and Mantoura, 1987)
2.25 (Meylan et al., 1992)
2.03 (Grimsby silt loam), 1.92 (Vaudreil sandy loam), 2.07 (Wendover silty clay) (Nathwani and Phillips, 1977)
2.32 (river sediment), 2.40 (coal wastewater sediment) (Kopinke et al., 1995)

Octanol/water partition coefficient, log K_{ow}:
3.13 (Tewari et al., 1982; Wasik et al., 1981, 1983)
2.77 (Leo et al., 1971)
3.16 (Hodson and Williams, 1988)
3.08 (Galassi et al., 1988)
3.12 (Tuazon et al., 1986)
3.18 (Garst and Wilson, 1984)

Solubility in organics:
Soluble in acetone, ethanol, benzene and ether (Weast, 1986) and many other organic solvents.

Solubility in water:
152 mg/L at 20 °C (Mackay and Shiu, 1981)
204 mg/L at 25 °C (Andrews and Keefer, 1949)

175 mg/L at 25 °C (Mackay and Wolkoff, 1973)
175 mg/kg at 25 °C (McAuliffe, 1963)
2.08 mmol/L at 25 °C (Tewari et al., 1982; Wasik et al., 1981, 1983)
142 mg/L at 0 °C, 167 mg/L at 25 °C (Brookman et al., 1985)
142 mg/kg at 0 °C, 213 mg/kg at 25 °C (Polak and Lu, 1973)
170.5 mg/kg at 25 °C, 129.6 mg/kg in artificial seawater at 25 °C (Sutton and Calder, 1975)
176.2 mg/L at 25 °C (Hermann, 1972)
1.742 mmol/L at 20 °C (Galassi et al., 1988)
179 mg/L in seawater at 25 °C (Bobra et al., 1979)
1.58, 1.68, 1.85 and 2.00 mmol/L at 15, 25, 35 and 45 °C, respectively (Sanemasa et al., 1982)
167.0 mg/kg at 25 °C (Price, 1976)
135 ppm (Brooker and Ellison, 1974)
176 mg/L (Coutant and Keigley, 1988)
In wt % (°C): 0.047 (139), 0.093 (162), 0.407 (207), 0.960 (251) (Guseva and Parnov, 1963)
1.68 mmol/L at 25.0 °C (Sanemasa et al., 1987)
3.28×10^{-5} at 25 °C (mole fraction, Li et al., 1993)

Vapor density:
4.34 g/L at 25 °C, 3.66 (air = 1)

Vapor pressure (mmHg):
6.6 at 25 °C (Mackay and Wolkoff, 1973)

Environmental Fate
   Biological. Reported biodegradation products of the commercial product containing xylene include α-hydroxy-*p*-toluic acid, *p*-methylbenzyl alcohol, benzyl alcohol, 4-methylcatechol, *m*- and *p*-toluic acids (Fishbein, 1985). In anoxic groundwater near Bemidji, MI, *o*-xylene anaerobically biodegraded to the intermediate *o*-toluic acid (Cozzarelli et al., 1990).
   Photolytic. Synthetic air containing gaseous nitrous acid and exposed to artificial sunlight (λ = 300-450 nm) photooxidized *o*-xylene into biacetyl, peroxyacetal nitrate and methyl nitrate (Cox et al., 1980). An *n*-hexane solution containing *o*-xylene and spread as a thin film (4 mm) on cold water (10 °C) was irradiated by a mercury medium pressure lamp. In 3 hours, 13.6% of the *o*-xylene photooxidized into *o*-methylbenzaldehyde, *o*-benzyl alcohol, *o*-benzoic acid and *o*-methylaceto-phenone (Moza and Feicht, 1989). Irradiation of *o*-xylene at ≈ 2537 Å at 35 °C and 6 mmHg isomerizes to *m*-xylene (Calvert and Pitts, 1966). Glyoxal, methylglyoxal and biacetyl were produced from the photooxidation of *o*-xylene by hydroxyl radicals in air at 25 °C (Tuazon et al., 1986).
   Major products reported from the photooxidation of *o*-xylene with nitrogen

oxides not previously mentioned include formaldehyde, acetaldehyde, peroxyacetyl nitrate, glyoxal and methylglyoxal (Altshuller, 1983).

Chemical/Physical. Under atmospheric conditions, the gas-phase reaction of *o*-xylene with hydroxyl radicals and nitrogen oxides resulted in the formation of *o*-tolualdehyde, *o*-methylbenzyl nitrate, nitro-*o*-xylenes, 2,3-and 3,4-dimethylphenol (Atkinson, 1990). Kanno et al. (1982) studied the aqueous reaction of *o*-xylene and other aromatic hydrocarbons (benzene, toluene, *m*- and *p*-xylene and naphthalene) with hypochlorous acid in the presence of ammonium ion. They reported that the aromatic ring was not chlorinated as expected but was cleaved by chloramine forming cyanogen chloride. The amount of cyanogen chloride formed increased at lower pHs (Kanno et al., 1982). In the gas phase, *o*-xylene reacted with nitrate radicals in purified air forming the following products: 5-nitro-2-methyltoluene and 6-nitro-2-methyltoluene, 2-methylbenzaldehyde and an aryl nitrate (Chiodini et al., 1993).

Exposure Limits: NIOSH REL: 100 ppm (435 mg/m^3), STEL 150 ppm (655 mg/m^3), IDLH 900 ppm; OSHA PEL: TWA 100 ppm; ACGIH TLV: TWA 100 ppm, STEL 150 ppm.

Symptoms of Exposure: May cause irritation of eyes, nose and throat, headache, dizziness, excitement, drowsiness, nausea, vomiting, abdominal pain and dermatitis (Patnaik, 1992).

Toxicity: Acute oral LD$_{50}$ in rats is approximately 5 gm/kg (RTECS, 1992).

Drinking Water Standard (final): For all xylenes, the MCLG and MCL are both 10 mg/L (U.S. EPA, 1994).

Uses: Preparation of phthalic acid, phthalic anhydride, terephthalic acid, isophthalic acid; solvent for alkyd resins, lacquers, enamels, rubber cements; manufacture of dyes, pharmaceuticals and insecticides; motor fuels.

m-XYLENE

Synonyms: 1,3-Dimethylbenzene; *m*-Dimethylbenzene; *m*-Methyltoluene; UN 1307; 1,3-Xylene; *m*-Xylol.

CAS Registry No.: 108-38-3
DOT: 1307
DOT label: Flammable liquid
Molecular formula: C_8H_{10}
Formula weight: 106.17
RTECS: ZE2275000

Physical state, color and odor
Clear, colorless, watery liquid with a sweet, aromatic odor. Odor threshold in air is 1.1 ppm (Amoore and Hautala, 1983).

Melting point (°C):
-47.9 (Weast, 1986)
-47.40 (Martin et al., 1979)

Boiling point (°C):
139.1 (Weast, 1986)

Density (g/cm³):
0.8642 at 20/4 °C (Weast, 1986)
0.8684 at 25/4 °C (Hawley, 1981)
0.86407 at 20/4 °C, 0.85979 at 25/4 °C (Huntress and Mulliken, 1941)

Diffusivity in water (10^5 cm²/sec):
0.78 at 20 °C using method of Hayduk and Laudie (1974).

Dissociation constant:
>15 (Christensen et al., 1975)

Flash point (°C):
28 (NIOSH, 1994)

Lower explosive limit (%):
1.1 (NIOSH, 1994)

Upper explosive limit (%):
7.0 (NIOSH, 1994)

Heat of fusion (kcal/mol):
2.765 (Dean, 1987)

Henry's law constant (10^3 atm·m^3/mol):
7.0 (Pankow and Rosen, 1988)
6.3 at 25 °C (Hine and Mookerjee, 1975)
7.68 (Tuazon et al., 1986)
4.11, 4.96, 5.98, 7.44 and 8.87 at 10, 15, 20, 25 and 30 °C, respectively (Ashworth et al., 1988)
2.24, 2.15, 2.74, 4.78 and 6.08 at 2.0, 6.0, 10.0, 18.2 and 25.0 °C, respectively (Dewulf et al., 1995)

Interfacial tension with water (dyn/cm at 20 °C):
37.9 (Donahue and Bartell, 1952)

Ionization potential (eV):
8.58 (Franklin et al., 1969)
9.05 (Yoshida et al., 1983)

Soil sorption coefficient, log K_{oc}:
3.20 (Abdul et al., 1987)

Octanol/water partition coefficient, log K_{ow}:
3.20 (Leo et al., 1971)
3.13 (Wasik et al., 1981, 1983)
3.28 (Garst and Wilson, 1984)

Solubility in organics:
Soluble in acetone, ethanol, benzene and ether (Weast, 1986).

Solubility in water:
In mg/L: 209 at 0.4 °C, 201 at 5.2 °C, 192 at 14.9 °C, 196 at 21.0 °C, 196 at 25.0 °C, 196 at 25.6 °C, 198 at 30.3 °C, 203 at 34.9 °C, 218 at 39.6 °C (Bohon and Claussen, 1951)
173 mg/L at 25 °C (Andrews and Keefer, 1949)
1.51 mmol/L at 25 °C (Tewari et al., 1982; Wasik et al., 1981, 1983)
0.0146 wt % at 25 °C (Riddick et al., 1986)
196 mg/kg at 0 °C, 162 mg/kg at 25 °C (Polak and Lu, 1973)
146.0 mg/kg at 25 °C, 106.0 mg/kg in artificial seawater at 25 °C (Sutton and Calder, 1975)

157.0 mg/L at 25 °C (Hermann, 1972)
1.525 mmol/L at 20 °C (Galassi et al., 1988)
1.92 mmol/L at 35 °C (Hine et al., 1963)
1.49, 1.52, 1.57 and 1.73 mmol/L at 15, 25, 35 and 45 °C, respectively (Sanemasa et al., 1982)
134.0 mg/kg at 25 °C (Price, 1976)
In wt % (°C): 0.031 (127), 0.072 (149), 0.168 (187), 0.648 (239) (Guseva and Parnov, 1963)
1.94 mmol/kg at 25.0 °C (Vesala, 1974)
1.33 mmol/L at 25.0 °C (Sanemasa et al., 1987)
3.01×10^{-5} at 25 °C (mole fraction, Li et al., 1993)

Vapor density:
4.34 g/L at 25 °C, 3.66 (air = 1)

Vapor pressure (mmHg):
8.3 at 25 °C (Mackay et al., 1982)
15.2 at 35 °C (Hine et al., 1963)

Environmental Fate

Biological. Microbial degradation produced 3-methylbenzyl alcohol, 3-methylbenzaldehyde, *m*-toluic acid and 3-methyl catechol (Verschueren, 1983). *m*-Toluic acid was reported to be the biooxidation product of *m*-xylene by *Nocardia corallina* V-49 using *n*-hexadecane as the substrate (Keck et al., 1989). Reported biodegradation products of the commercial product containing xylene include α-hydroxy-*p*-toluic acid, *p*-methylbenzyl alcohol, benzyl alcohol, 4-methylcatechol, *m*- and *p*-toluic acids (Fishbein, 1985). In anoxic groundwater near Bemidji, MI, *m*-xylene anaerobically biodegraded to the intermediate *m*-toluic acid (Cozzarelli et al., 1990).

Photolytic. Synthetic air containing gaseous nitrous acid and exposed to artificial sunlight (λ = 300-450 nm) photooxidized *o*-xylene into biacetyl, peroxyacetal nitrate and methyl nitrate (Cox et al., 1980). An *n*-hexane solution containing *m*-xylene and spread as a thin film (4 mm) on cold water (10 °C) was irradiated by a mercury medium pressure lamp. In 3 hours, 25% of the *m*-xylene photooxidized into *m*-methylbenzaldehyde, *m*-benzyl alcohol, *m*-benzoic acid and *m*-methylacetophenone (Moza and Feicht, 1989). Irradiation of *m*-xylene isomerizes to *p*-xylene (Calvert and Pitts, 1966). Glyoxal, methylglyoxal and biacetyl were produced from the photooxidation of *m*-xylene by hydroxyl radicals in air at 25 °C (Tuazon et al., 1986). The photooxidation of *m*-xylene in the presence of nitrogen oxides (NO and NO_2) yielded small amounts of formaldehyde and a trace of acetaldehyde (Altshuller et al., 1970). *m*-Tolualdehyde and nitric acid also were identified as photooxidation products of *m*-xylene with nitrogen oxides (Altshuller, 1983).

Chemical/Physical. Under atmospheric conditions, the gas-phase reaction with hydroxyl radicals and nitrogen oxides resulted in the formation of *m*-tolualdehyde, *m*-methylbenzyl nitrate, nitro-*m*-xylenes, 2,4- and 2,6-dimethylphenol (Atkinson, 1990). Kanno et al. (1982) studied the aqueous reaction of *m*-xylene and other aromatic hydrocarbons (benzene, toluene, *o*- and *p*-xylene and naphthalene) with hypochlorous acid in the presence of ammonium ion. They reported that the aromatic ring was not chlorinated as expected but was cleaved by chloramine forming cyanogen chloride. The amount of cyanogen chloride formed increased at lower pHs (Kanno et al., 1982). In the gas phase, *m*-xylene reacted with nitrate radicals in purified air forming 3-methylbenzaldehyde, an aryl nitrate and trace amounts of 2,6-dimethylnitrobenzene, 2,4-dimethylnitrobenzene and 3,5-dimethylnitrobenzene (Chiodini et al., 1993).

Exposure Limits: NIOSH REL: 100 ppm (435 mg/m^3), STEL 150 ppm (655 mg/m^3), IDLH 900 ppm; OSHA PEL: TWA 100 ppm; ACGIH TLV: TWA 100 ppm, STEL 150 ppm.

Symptoms of Exposure: May cause irritation of eyes, nose and throat, headache, dizziness, excitement, drowsiness, nausea, vomiting, abdominal pain and dermatitis (Patnaik, 1992).

Toxicity: Acute oral LD$_{50}$ in rats 5 gm/kg (RTECS, 1985).

Drinking Water Standard (final): For all xylenes, the MCLG and MCL are both 10 mg/L (U.S. EPA, 1994).

Uses: Solvent; preparation of isophthalic acid, intermediate for dyes; insecticides; aviation fuel.

p-XYLENE

Synonyms: Chromar; 1,4-Dimethylbenzene; *p*-Dimethylbenzene; *p*-Methyl-toluene; Scintillar; UN 1307; 1,4-Xylene; *p*-Xylol.

CAS Registry No.: 106-42-3
DOT: 1307
DOT label: Flammable liquid
Molecular formula: C_8H_{10}
Formula weight: 106.17
RTECS: ZE2625000

Physical state, color and odor
Clear, colorless, watery liquid with a sweet odor.

Melting point (°C):
13.3 (Weast, 1986)
13.50 (Martin et al., 1979)

Boiling point (°C):
138.3 (Weast, 1986)

Density (g/cm³):
0.8811 at 20/4 °C (Weast, 1986)
0.86100 at 20/4 °C, 0.85665 at 25/4 °C (Huntress and Mulliken, 1941)
0.85655 at 25/4 °C (Kirchnerová and Cave, 1976)

Diffusivity in water (10⁵ cm²/sec):
0.79 at 20 °C using method of Hayduk and Laudie (1974).

Dissociation constant:
>15 (Christensen et al., 1975)

Flash point (°C):
27 (NIOSH, 1994)

Lower explosive limit (%):
1.1 (NIOSH, 1994)

Upper explosive limit (%):
7.0 (NIOSH, 1994)

Heat of fusion (kcal/mol):
4.090 (Tsonopoulos and Prausnitz, 1971)

Henry's law constant (10^3 atm·m³/mol):
7.1 (Pankow and Rosen, 1988)
6.3 at 25 °C (Hine and Mookerjee, 1975)
7.68 (Tuazon et al., 1986)
4.20, 4.83, 6.45, 7.44 and 9.45 at 10, 15, 20, 25 and 30 °C, respectively (Ashworth et al., 1988)
1.89, 1.67, 2.62, 4.73 and 5.68 at 2.0, 6.0, 10.0, 18.2 and 25.0 °C, respectively (Dewulf et al., 1995)

Interfacial tension with water (dyn/cm at 20 °C):
37.77 (Demond and Lindner, 1993)

Ionization potential (eV):
8.44 (Franklin et al., 1969)
8.99 (Yoshida et al., 1983)

Soil sorption coefficient, log K_{oc}:
2.31 (Abdul et al., 1987)
2.42 (estuarine sediment, Vowles and Mantoura, 1987)
2.72 (Captina silt loam), 2.87 (McLaurin sandy loam) (Walton et al., 1992)

Octanol/water partition coefficient, log K_{ow}:
3.18 (Tewari et al., 1982; Wasik et al., 1981, 1983)
3.15 (Campbell and Luthy, 1985; Leo et al., 1971)

Solubility in organics:
Soluble in acetone, ethanol and benzene (Weast, 1986).

Solubility in water:
200 mg/L at 25 °C (Andrews and Keefer, 1949)
2.02 mmol/L at 25 °C (Tewari et al., 1982; Wasik et al., 1981, 1983)
In mg/L: 156 at 0.4 °C, 188 at 10.0 °C, 195 at 14.9 °C, 197 at 21.0 °C, 198 at 25.0 °C, 199 at 25.6 °C, 201 at 30.2 °C, 204 at 30.3 °C, 207 at 34.9 °C, 207 at 35.2 °C, 222 at 42.8 °C (Bohon and Claussen, 1951)
180 mg/L at 25 °C (Banerjee, 1984)
0.0156 wt % at 20 °C (Riddick et al., 1986)
185 mg/kg at 25 °C (Polak and Lu, 1973)

156.0 mg/kg at 25 °C, 110.9 mg/kg in artificial seawater at 25 °C (Sutton and Calder, 1975)

163.3 mg/L at 25 °C (Hermann, 1972)

1.94 mmol/L at 35 °C (Hine et al., 1963)

1.48, 1.53, 1.61 and 1.66 mmol/L at 15, 25, 35 and 45 °C, respectively (Sanemasa et al., 1982)

0.019 wt % at 25 °C (Lo et al., 1986)

157.0 mg/kg at 25 °C (Price, 1976)

In wt % (°C): 0.049 (141), 0.096 (169), 0.231 (194), 0.607 (231), 1.283 (258) (Guseva and Parnov, 1963)

1.51 mmol/L at 25.0 °C (Sanemasa et al., 1987)

3.00×10^{-5} at 25 °C (mole fraction, Li et al., 1993)

Vapor density:
4.34 g/L at 25 °C, 3.66 (air = 1)

Vapor pressure (mmHg):
8.8 at 25 °C (Mackay et al., 1982)
15.8 at 35 °C (Hine et al., 1963)

Environmental Fate

Biological. Microbial degradation of *p*-xylene produced 4-methylbenzyl alcohol, 4-methylbenzaldehyde, *p*-toluic acid and 4-methyl catechol (Verschueren, 1983). Dimethyl-*cis,cis*-muconic acid and 2,3-dihydroxy-*p*-toluic acid were reported to be biooxidation products of *p*-xylene by *Nocardia corallina* V-49 using *n*-hexadecane as the substrate (Keck et al., 1989). Reported biodegradation products of the commercial product containing xylene include α-hydroxy-*p*-toluic acid, *p*-methylbenzyl alcohol, benzyl alcohol, 4-methylcatechol, *m*- and *p*-toluic acids (Fishbein, 1985). In anoxic groundwater near Bemidji, MI, *p*-xylene anaerobically biodegraded to the intermediate *p*-toluic acid (Cozzarelli et al., 1990).

Photolytic. An *n*-hexane solution containing *m*-xylene and spread as a thin film (4 mm) on cold water (10 °C) was irradiated by a mercury medium pressure lamp. In 3 hours, 18.5% of the *p*-xylene photooxidized into *p*-methylbenzaldehyde, *p*-benzyl alcohol, *p*-benzoic acid and *p*-methylacetophenone (Moza and Feicht, 1989). Glyoxal and methylglyoxal were produced from the photooxidation of *p*-xylene by hydroxyl radicals in air at 25 °C (Tuazon et al., 1986).

Chemical/Physical. Under atmospheric conditions, the gas-phase reaction with hydroxyl radicals and nitrogen oxides resulted in the formation of *p*-tolualdehyde (Atkinson, 1990). Kanno et al. (1982) studied the aqueous reaction of *p*-xylene and other aromatic hydrocarbons (benzene, toluene, *o*- and *m*-xylene and naphthalene) with hypochlorous acid in the presence of ammonium ion. They reported that the aromatic ring was not chlorinated as expected but was cleaved by chloramine forming cyanogen chloride. The amount of cyanogen chloride formed increased at lower pHs (Kanno et al., 1982).

In the gas phase, *p*-xylene reacted with nitrate radicals in purified air forming 3,6-dimethylnitrobenzene, 4-methylbenzaldehyde and an aryl nitrate (Chiodini et al., 1993).

Exposure Limits: NIOSH REL: 100 ppm (435 mg/m^3), STEL 150 ppm (655 mg/m^3), IDLH 900 ppm; OSHA PEL: TWA 100 ppm; ACGIH TLV: TWA 100 ppm, STEL 150 ppm.

Symptoms of Exposure: May cause irritation of eyes, nose and throat, headache, dizziness, excitement, drowsiness, nausea, vomiting, abdominal pain and dermatitis (Patnaik, 1992).

Toxicity: Acute oral LD$_{50}$ in rats 5 gm/kg; LC$_{50}$ (inhalation) for rats 4,550 ppm/4-hour (RTECS, 1985).

Drinking Water Standard (final): For all xylenes, the MCLG and MCL are both 10 mg/L (U.S. EPA, 1994).

Uses: Preparation of terephthalic acid for polyester resins and fibers (Dacron, Mylar and Terylene), vitamins, pharmaceuticals and insecticides.

References

Abd-El-Bary, M.F., M.F. Hamoda, S. Tanisho, and N. Wakao. "Henry's Constants for Phenol over its Diluted Aqueous Solution," *J. Chem. Eng. Data*, 31(2):229-230 (1986).

Abdelmagid, H.M. and M.A. Tabatabai. "Decomposition of Acrylamide in Soils," *J. Environ. Qual.*, 11(4):701-704 (1982).

Abdul, S.A., T.L. Gibson, and D.N. Rai. "Statistical Correlations for Predicting the Partition Coefficient for Nonpolar Organic Contaminants between Aquifer Organic Carbon and Water," *Haz. Waste Haz. Mater.*, 4(3):211-222 (1987).

Abraham, M.H. "Thermodynamics of Solution of Homologous Series of Solutes in Water," *J. Chem. Soc., Faraday Trans. 1*, 80:153-181 (1984).

Abraham, T., V. Bery, and A.P. Kudchadker. "Densities of Some Organic Substances," *J. Chem. Eng. Data*, 16(3):355-356 (1971).

ACGIH. 1986. Documentation of the Threshold Limit Values and Biological Exposure Indices (Cincinnati, OH: American Conference of Governmental Industrial Hygienists), 744 p.

Adams, A.J., G.M. DeGraeve, T.D. Sabourin, J.D. Cooney, and G.M. Mosher. "Toxicity and Bioconcentration of 2,3,7,8-TCDD to Fathead Minnows (*Pimephales promelas*)," *Chemosphere*, 15(9-12):1503-1511 (1986).

Adams, A.J. and K.M. Blaine. "A Water Solubility Determination of 2,3,7,8-TCDD," *Chemosphere*, 15(9-12):1397-1400 (1986).

Adewuyi, Y.G. and G.R. Carmichael. "Kinetics of Hydrolysis and Oxidation of Carbon Disulfide by Hydrogen Peroxide in Alkaline Medium and Application to Carbonyl Sulfide," *Environ. Sci. Technol.*, 21(2):170-177 (1987).

Adhya, T.K., Sudhakar-Barik, and N. Sethunathan. "Fate of Fenitrothion, Methyl Parathion, and Parathion in Anoxic Sulfur-Containing Soil Systems," *Pestic. Biochem. Physiol.*, 16(1):14-20 (1981).

Adhya, T.K., Sudhakar-Barik, and N. Sethunathan. "Stability of Commercial Formulation of Fenitrothion, Methyl Parathion, and Parathion in Anaerobic Soils," *J. Agric. Food Chem.*, 29(1):90-93 (1981a).

Adrian, P., E.S. Lahaniatis, F. Andreux, M. Mansour, I. Scheunert, and F. Korte. "Reaction of the Soil Pollutant 4-Chloroaniline with the Humic Acid Monomer Catechol," *Chemosphere*, 18(7/8):1599-1609 (1989).

Affens, W.A. and G.W. McLaren. "Flammability Properties of Hydrocarbon Solutions in Air," *J. Chem. Eng. Data*, 17(4):482-488 (1972).

Afghan, B.K. and D. Mackay, Eds. *Hydrocarbons and Halogenated Hydrocarbons in the Environment*, Volume 16, (New York: Plenum Press, 1990).

Ahel, M. and W. Giger. "Aqueous Solubility of Alkylphenols and Alkylphenol Polyethoxylates," *Environ. Sci. Technol.*, 26(8):1461-1470 (1993).

Ahmad, N., D.D. Walgenbach, and G.R. Sutter. "Degradation Rates of Technical Carbofuran and a Granular Formation in Four Soils with Known Insecticide Use History," *Bull. Environ. Contam. Toxicol.*, 23:572-574 (1979).

Akimoto, H. and H. Takagi. "Formation of Methyl Nitrite in the Surface Reaction of Nitrogen Dioxide and Methanol. 2. Photoenhancement," *Environ. Sci. Technol.*, 20(4):387-393 (1986).

Akiyoshi, M., T. Deguchi, and I. Sanemasa. "The Vapor Saturation Method for Preparing Aqueous Solutions of Solid Aromatic Hydrocarbons," *Bull. Chem. Soc. Jpn.*, 60(11):3935-3939 (1987).

Aksnes, G. and K. Sandberg. "On the Oxidation of Benzidine and *o*-Dianisidine with Hydrogen Peroxide and Acetylcholine in Alkaline Solution," *Acta Chem. Scand.*, 11:876-880 (1957).

Aldrich. 1990. Catalog Handbook of Fine Chemicals (Milwaukee, WI: Aldrich Chemical Co.), 2105 p.

Alexander, D.M. "The Solubility of Benzene in Water," *J. Phys. Chem.*, 63(6):1021-1022 (1959).

Alexander, M. "Biodegradation of Chemicals of Environmental Concern," *Science (Washington, DC)*, 211(4478):132-138 (1981).

Ali, S. "Degradation and Environmental Fate of Endosulfan Isomers and Endosulfan Sulfate in Mouse, Insect, and Laboratory Ecosystem," PhD Thesis, University of Illinois, Ann Arbor, MI (1978).

Alley, E.G., B.R. Layton, and J.P. Minyard, Jr. "Identification of the Photoproducts of the Insecticides Mirex and Kepone," *J. Agric. Food Chem.*, 22(3):442-445 (1974).

Almgren, M., F. Grieser, J.R. Powell, and J.K. Thomas. "A Correlation between the Solubility of Aromatic Hydrocarbons in Water and Micellar Solutions, with Their Normal Boiling Points," *J. Chem. Eng. Data*, 24(4):285-287 (1979).

Altshuller, A.P. "Measurements of the Products of Atmospheric Photochemical Reactions in Laboratory Studies and in Ambient Air-Relationships between Ozone and Other Products," *Atmos. Environ.*, 17(12):2383-2427 (1983).

Altshuller, A.P. and H.E. Everson. "The Solubility of Ethyl Acetate in Water," *J. Am. Chem. Soc.*, 75(7):1727 (1953).

Altshuller, A.P., S.L. Kopczynski, W.A. Lonneman, F.D. Sutterfield, and D.L. Wilson. "Photochemical Reactivities of Aromatic Hydrocarbon-Nitrogen Oxide and Related Systems," *Environ. Sci. Technol.*, 4(1):44-49 (1970).

Aly, O.M. and M.A. El-Dib. "Studies on the Persistence of Some Carbamate Insecticides in the Aquatic Environment-I. Hydrolysis of Sevin, Baygon, Pyrolam and Dimetilan in Waters," *Water Res.*, 5(12):1191-1205 (1971).

Ambrose, D. and H.S. Sprake. "The Vapor Pressure of Indane," *J. Chem. Thermodyn.*, 8:601-602 (1976).

Ambrose, D., J.H. Ellender, E.B. Lees, C.H.S. Sprake, and R. Townsend. "Thermodynamic Properties of Organic Compounds. XXXVIII. Vapor Pressures of Some Aliphatic Ketones," *J. Chem. Thermodyn.*, 7(5):453-472 (1975).

Amidon, G.L., S.H. Yalkowsky, S.T. Anik, and S.C. Valvani. "Solubility of Nonelectrolytes in Polar Solvents. V. Estimation of the Solubility of Aliphatic Monofunctional Compounds in Water Using a Molecular Surface Area Approach," *J. Phys. Chem.*, 79(21):2239-2246 (1975).

Amoore, J.E. and E. Hautala. "Odor as an Aide to Chemical Safety: Odor Thresholds Compared with Threshold Limit Values and Volatilities for 214 Industrial Chemicals in Air and Water Dilution," *J. Appl. Toxicol.*, 3(6):272-290 (1983).

Amoore, J.E. and R.G. Buttery. "Partition Coefficients and Comparative Olfactometry," *Chem. Sens. Flavour*, 3(1):57-70 (1978).

Anbar, M. and P. Neta. "A Compilation of Specific Bimolecular Rate and Hydroxyl Radical with Inorganic and Organic Compounds in Aqueous Solution," *Int. J. Appl. Rad. Isotopes*, 18:493-523 (1967).

Anderson, B.D., J.H. Rytting, and T. Higuchi. "Solubility of Polar Organic Solutes in Nonaqueous Systems: Role of Specific Interactions," *J. Pharm. Sci.*, 69(6):676-680 (1971).

Andersson, H.F., J.A. Dahlberg, and R. Wettstrom. "On the Formation of Phosgene and Trichloroacetylchloride in the Non-Sensitized Photooxidation of Perchloroethylene in Air," *Acta Chem. Scand.*, A29:473-474 (1975).

Andersson, J.T., R. Häussler, and K. Ballschmiter. "Chemical Degradation of Xenobiotics. III. Simulation of the Biotic Transformation of 2-Nitrophenol and 3-Nitrophenol," *Chemosphere*, 15(2):149-152 (1986).

Ando, M. and Y. Sayato. "Studies on Vinyl Chloride Migrating into Drinking Water from Polyvinyl Chloride Pipe and Reaction between Vinyl Chloride and Chlorine," *Water Res.*, 18(3):315-318 (1984).

Andon, R.J.L., D.P. Biddiscombe, F.D. Cox, R. Handley, D. Harrop, E.F.G. Herington, and J.F. Martin. "Thermodynamic Properties of Organic Oxygen Compounds. Part 1. Preparation of Physical Properties of Pure Phenol, Cresols and Xylenols," *J. Chem. Soc. (London)*, pp. 5246-5254, (1960).

Andreozzi, R., A. Insola, V. Caprio, and M.G. D'Amore. "Ozonation of Pyridine in Aqueous Solution: Mechanistic and Kinetic Aspects," *Water Res.*, 25(6):655-659 (1991).

Andrews, L.J. and R.M. Keefer. "Cation Complexes of Compounds Containing Carbon-Carbon Double Bonds. IV. The Argentation of Aromatic Hydrocarbons," *J. Am. Chem. Soc.*, 71(11):3644-3647 (1949).

Andrews, L.J. and R.M. Keefer. "Cation Complexes of Compounds Containing Carbon-Carbon Double Bonds. VI. The Argentation of Substituted Benzenes," *J. Am. Chem. Soc.*, 72(7):3113-3116 (1950).

Andrews, L.J. and R.M. Keefer. "Cation Complexes of Compounds Containing Carbon-Carbon Double Bonds. VII. Further Studies on the Argentation of Substituted Benzenes," *J. Am. Chem. Soc.*, 72(11):5034-5037 (1950a).

Anliker, R. and P. Moser. "The Limits of Bioaccumulation of Organic Pigments in Fish: Their Relation to the Partition Coefficient and the Solubility in Water and Octanol," *Ecotoxicol. Environ. Saf.*, 13(1):43-52 (1987).

Antelo, J.M., F. Arce, F. Barbadillo, J. Casado, and A. Varela. "Kinetics and Mechanism of Ethanolamine Chlorination," *Environ. Sci. Technol.*, 15(8):912-917 (1981).

Appleton, H.T. and H.C. Sikka. "Accumulation, Elimination, and Metabolism of

Dichlorobenzidine in the Bluegill Sunfish," *Environ. Sci. Technol.*, 14(1):50-54 (1980).

Aquan-Yuen, M., D. Mackay, and W.-Y. Shiu. "Solubility of Hexane, Phenanthrene, Chlorobenzene, and *p*-Dichlorobenzene in Aqueous Electrolyte Solutions," *J. Chem. Eng. Data*, 24(1):30-34 (1979).

Arcand, Y., J. Hawari, and S.R. Guiot. "Solubility of Pentachlorophenol in Aqueous Solutions: The pH Effect," *Water Res.*, 29(1):131-136 (1995).

Archer, T.E. "Malathion Residues on Ladino Clover Seed Screenings Exposed to Ultraviolet Irradiation," *Bull. Environ. Contam. Toxicol.*, 6(2):142-143 (1971).

Archer, T.E., I.K. Nazer, and D.G. Crosby. "Photodecomposition of Endosulfan and Related Products in Thin Films by Ultraviolet Light Irradiation," *J. Agric. Food Chem.*, 20(5):954-956 (1972).

Archer, T.E., J.D. Stokes, and R.S. Bringhurst. "Fate of Carbofuran and its Metabolites on Strawberries in the Environment," *J. Agric. Food Chem.*, 25(3):536-541 (1977).

Arey, J., B. Zielinska, R. Atkinson, A.M. Winer, T. Randahl, and J.N. Pitts, Jr. "The Formation of Nitro-PAH from the Gas-Phase Reactions of Fluoranthene and Pyrene with the OH Radical in the Presence of NO_x," *Atmos. Environ.*, 20(12):2339-2345 (1986).

Arthur, M.F. and J.I. Frea. "2,3,7,8-Tetrachloro-*p*-dioxin: Aspects of its Important Properties and its Potential Biodegradation in Soils," *J. Environ. Qual.*, 18(1):1-11 (1989).

Arunachalam, K.D. and M. Lakshmanan. "Microbial Uptake and Accumulation of (^{14}C-Carbofuran) 1,3-Dihydro-2,2-Dimethyl-7 Benzofuranylmethyl Carbamate in Twenty Fungal Strains Isolated by Miniecosystem Studies," *Bull. Environ. Contam. Toxicol.*, 41(1):127-134 (1988).

Ashton, F.M. and T.J. Monaco. *Weed Science: Principals and Practices* (New York: John Wiley & Sons, Inc., 1991), 466 p.

Ashworth, R.A., G.B. Howe, M.E. Mullins, and T.N. Rogers. "Air-Water Partitioning Coefficients of Organics in Dilute Aqueous Solutions," *J. Haz. Mater.*, 18:25-36 (1988).

Atkinson, R. "Kinetics and Mechanisms of the Gas-Phase Reactions of Hydroxyl Radical with Organic Compounds under Atmospheric Conditions," *Chem. Rev.*, 85:69-201 (1985).

Atkinson, R. "Structure-Activity Relationship for the Estimation of Rate Constants for the Gas Phase Reactions of OH Radicals with Organic Compounds," *Int. J. Chem. Kinetics*, 19:799-828 (1987).

Atkinson, R. "Gas-Phase Tropospheric Chemistry of Organic Compounds: A Review," *Atmos. Environ.*, 24A(1):1-41 (1990).

Atkinson, R., E.C. Tuazon, T.J. Wallington, S.M. Aschmann, J. Arey, A.M. Winer, and J.N. Pitts, Jr. "Atmospheric Chemistry of Aniline, *N,N*-Dimethylaniline, Pyridine, 1,3,5-Triazine, and Nitrobenzene," *Environ. Sci. Technol.*, 21(1):64-72 (1987).

Atkinson, R., R.A. Perry, and J.N. Pitts, Jr. "Rate Constants for the Reactions of the OH Radical with $(CH_3)_2NH$, $(CH_3)_3N$, and $C_2H_5NH_2$ Over the Temperature Range 298-426°K," *J. Chem. Phys.*, 68(4):1850-1853 (1978).

Atkinson, R. and W.P.L. Carter. "Kinetics and Mechanisms of Gas-Phase Ozone Reaction with Organic Compounds under Atmospheric Conditions," *Chem. Rev.*, 84:437-470 (1984).

Atlas, E., R. Foster, and C.S. Giam. "Air-Sea Exchange of High Molecular Weight Organic Pollutants: Laboratory Studies," *Environ. Sci. Technol.*, 16(5):283-286 (1982).

Ayanaba, A., W. Verstraete, and M. Alexander. "Formation of Dimethylnitros-amine, a Carcinogen and Mutagen, in Soils Treated with Nitrogen Compounds," *Soil Sci. Soc. Am. Proc.*, 37:565-568 (1973).

Babers, F.H. "The Solubility of DDT in Water Determined Radiometrically," *J. Am. Chem. Soc.*, 77(17):4666 (1955).

Bachmann, A., W.P. Wijnen, W. de Bruin, J.L.M. Huntjens, W. Roelofsen, and A.J.B. Zehnder. "Biodegradation of *Alpha*- and *Beta*-Hexachlorocyclohexane in a Soil Slurry under Different Redox Conditions," *Appl. Environ. Microbiol.*, 54(1):143-149 (1988).

Baek, N.H. and P.R. Jaffé. "The Degradation of Trichloroethylene in Mixed Methanogenic Cultures," *J. Environ. Qual.*, 18(4):515-518 (1989).

Bailey, G.W. and J.L. White. "Herbicides: A Compilation of Their Physical, Chemical, and Biological Properties," *Residue Rev.*, 10:97-122 (1965).

Baird, R., L. Carmona, and R.L. Jenkins. "Behavior of Benzidine and Other Aro-matic Amines in Aerobic Wastewater Treatment," *J. Water Pollut. Control Fed.*, 49(7):1609-1615 (1977).

Baker, E.G. "Origin and Migration of Oil," *Science (Washington, DC)*, 129(3353):871-874 (1959).

Baker, E.G. "A Hypothesis Concerning the Accumulation of Sediment Hydro-carbons to Form Crude Oil," *Geochim. Cosmochim. Acta*, 19:309-317 (1980).

Baker, M.D. and C.I. Mayfield. "Microbial and Nonbiological Decomposition of Chlorophenols and Phenols in Soil," *Water, Air, Soil Pollut.*, 13:411-424 (1980).

Baker, M.D., C.I. Mayfield, and W.E. Inniss. "Degradation of Chlorophenol in Soil, Sediment and Water at Low Temperature," *Water Res.*, 14:765-771 (1980).

Ballschiter, K. and C. Scholz. "Mikrobieller Abbau von Chlorier Aromaten: VI. Bildung von Dichlorphenolen in Mikromolarer Lösung durch *Pseudomonas* sp.," *Chemosphere*, 9(7/8):457-467 (1980).

Balson, E.W. "Studies in Vapour Pressure Measurement, Part III. - An Effusion Manometer Sensitive to 5 x 10^{-6} Millimetres of Mercury: Vapour Pressure of DDT and Other Slightly Volatile Substances," *Trans. Faraday Soc.*, 43:54-60 (1947).

Banerjee, P., M.D. Piwoni, and K. Ebeid. "Sorption of Organic Contaminants to a Low Carbon Subsurface Core," *Chemosphere*, 14(8):1057-1067 (1985).

Banerjee, S. "Solubility of Organic Mixtures in Water," *Environ. Sci. Technol.*,

18(8):587-591 (1984).

Banerjee, S., H.C. Sikka, R. Gray, and C.M. Kelly. "Photodegradation of 3,3'-Dichlorobenzidine," *Environ. Sci. Technol.*, 12(13):1425-1427 (1978).

Banerjee, S., P.H. Howard, A.M. Rosenburg, A.E. Dombrowski, H. Sikka, and D.L. Tullis. "Development of a General Kinetic Model for Biodegradation and its Application to Chlorophenols and Related Compounds," *Environ. Sci. Technol.*, 18(6):416-422 (1984).

Banerjee, S., P.H. Howard, and S.S. Lande. "General Structure-Vapor Pressure Relationships for Organics," *Chemosphere*, 21(10/11):1173-1180 (1990).

Banerjee, S., S.H. Yalkowsky, and S.C. Valvani. "Water Solubility and Octanol/Water Partition Coefficients of Organics. Limitations of the Solubility-Partition Coefficient Correlation," *Environ. Sci. Technol.*, 14(10):1227-1229 (1980).

Barbash, J.E. and M. Reinhard. "Abiotic Dehalogenation of 1,2-Dichloroethane and 1,2-Dibromoethane in Aqueous Solution Containing Hydrogen Sulfide," *Environ. Sci. Technol.*, 23(11):1349-1358 (1989).

Barbeni, M., E. Pramauro, and E. Pelizzetti. "Photodegradation of Pentachlorophenol Catalyzed by Semiconductor Particles," *Chemosphere*, 14(2):195-208 (1985).

Barbeni, M., E. Pramauro, E. Pelizzetti, M. Vincenti, E. Borgarello, and N. Serpone. "Sunlight Photodegradation of 2,4,5-Trichlorophenoxyacetic Acid and 2,4,5-Trichlorophenol on TiO_2. Identification of Intermediates and Degradation Pathway," *Chemosphere*, 16(6):1165-1179 (1987).

Barham, H.N. and L.W. Clark. "The Decomposition of Formic Acid at Low Temperatures," *J. Am. Chem. Soc.*, 73(10):4638-4640 (1951).

Barlas, H. and H. Parlar. "Reactions of Naphthaline with Nitrogen Dioxide in UV-Light," *Chemosphere*, 16(2/3):519-520 (1987).

Barone, G., V. Crescenzi, A.M. Liquori, and F. Quadrifoglio. "Solubilization of Polycyclic Aromatic Hydrocarbons in Poly(methacrylic acid) Aqueous Solutions," *J. Phys. Chem.*, 71(7):2341-2345 (1967).

Barone, G., V. Crescenzi, B. Pispisa, and F. Quadrifoglio. "Hydrophobic Interactions in Polyelectrolyte Solutions. II. Solubility of Some C_3-C_6 Alkanes in Poly(methacrylic acid) Aqueous Solutions," *J. Macromol. Chem.*, 1(4):761-771 (1966).

Barrio-Lage, G., F.Z. Parsons, R.S. Nassar, and P.A. Lorenzo. "Sequential Dehalogenation of Chlorinated Ethenes," *Environ. Sci. Technol.*, 20(1):96-99 (1986).

Barrio-Lage, G.A., F.Z. Parsons, R.M. Narbaitz, P.A. Lorenzo, and H.E. Archer. "Enhanced Anaerobic Biodegradation of Vinyl Chloride in Ground Water," *Environ. Toxicol. Chem.*, 9(1):403-415 (1990).

Barrio-Lage, G.A., F.Z. Parsons, R.S. Nassar, and P.A. Lorenzo. "Biotransformation of Trichloroethylene in a Variety of Subsurface Materials," *Environ. Toxicol. Chem.*, 6(8):571-578 (1987).

Baur, J.R. and R.W. Bovey. "Ultraviolet and Volatility Loss of Herbicides," *Arch.*

Environ. Contam., 2:275-288 (1974).

Baxter, R.M. "Reductive Dechlorination of Certain Chlorinated Organic Compounds by Reduced Hematin Compared with Their Behaviour in the Environment," *Chemosphere*, 21(4/5):451-458 (1990).

Baxter, R.M. and D.A. Sutherland. "Biochemical and Photochemical Processes in the Degradation of Chlorinated Biphenyls," *Environ. Sci. Technol.*, 18(8):608-610 (1984).

Bayona, J.M., P. Fernandez, C. Porte, I. Tolosa, M. Valls, and J. Albaiges. "Partitioning of Urban Wastewater Organic Microcontaminants Among Coastal Compartments," *Chemosphere*, 23(3):313-326 (1991).

Beard, J.E. and G.W. Ware. "Fate of Endosulfan on Plants and Glass," *J. Agric. Food Chem.*, 17(2):216-220 (1969).

Beck, J. and K.E. Hansen. "The Degradation of Quintozene, Pentachlorobenzene, Hexachlorobenzene and Pentachloroaniline in Soil," *Pestic. Sci.*, 5:41-48 (1974).

Bedard, D.L., R.E. Wagner, M.J. Brennan, M.L. Haberl, and J.F. Brown, Jr. "Extensive Degradation of Aroclors and Environmentally Transformed Polychlorinated Biphenyls by *Alcaligenes eutrophus* H850," *Appl. Environ. Microbiol.*, 53(5):1094-1102 (1987).

Beeman, R.W. and F. Matsumura. "Metabolism of *cis*- and *trans*-Chlordane by a Soil Microorganism," *J. Agric. Food Chem.*, 29(1):84-89 (1981).

Behymer, T.D. and R.A. Hites. "Photolysis of Polycyclic Aromatic Hydrocarbons Adsorbed on Simulated Atmospheric Particulates," *Environ. Sci. Technol.*, 19(10):1004-1006 (1985).

Beland, F.A., S.O. Farwell, A.E. Robocker, and R.D. Geer. "Electrochemical Reduction and Anaerobic Degradation of Lindane," *J. Agric. Food Chem.*, 24(4):753-756 (1976).

Belay, N. and L. Daniels. "Production of Ethane, Ethylene, and Acetylene from Halogenated Hydrocarbons by Methanogenic Bacteria," *Appl. Environ. Microbiol.*, 53(7):1604-1610 (1987).

Bell, G.R. "Photochemical Degradation of 2,4-Dichlorophenoxyacetic Acid and Structurally Related Compounds," *Botan. Gaz.*, 118:133-136 (1956).

Bell, R.P. and A.O. McDougall. "Hydration Equilibria of Some Aldehydes and Ketones," *Trans. Faraday Soc.*, 56:1281-1285 (1960).

Beltran, F.J., J.M. Encinar, and J.F. Garcia-Araya. "Ozonation of *o*-Cresol in Aqueous Solutions," *Water Res.*, 24(11):1309-1316 (1990).

Bender, M.E. "The Toxicity of the Hydrolysis and Breakdown Products of Malathion to the Fathead Minnow (*Pimephales Promelas*, Rafinesque)," *Water Res.*, 3(8):571-582 (1969).

Benezet, H.I. and F. Matsumura. "Isomerization of γ-BHC to a α-BHC in the Environment," *Nature (London)*, 243(5408):480-481 (1973).

Ben-Naim, A. and J. Wilf. "Solubilities and Hydrophobic Interactions in Aqueous Solutions of Monoalkylbenzene Molecules," *J. Phys. Chem.*, 84(6):583-586 (1980).

Bennett, D. and W.J. Canady. "Thermodynamics of Solution of Naphthalene in Various Water-Ethanol Mixtures," *J. Am. Chem. Soc.*, 106(4):910-915 (1984).

Bennett, G.M. and W.G. Philip. "The Influence of Structure on the Solubilities of Ethers. Part I. Aliphatic Ethers," *J. Chem. Soc. (London)*, 131:1930-1937 (1928).

Benson, J.M., A.L. Brooks, Y.S. Cheng, T.R. Henderson, and J.E. White. "Environmental Transformation of 1-Nitropyrene on Glass Surfaces," *Atmos. Environ.*, 19(7):1169-1174 (1985).

Benson, W.R. "Photolysis of Solid and Dissolved Dieldrin," *J. Agric. Food Chem.*, 19(1):66-72 (1971).

Berg, G.L., Ed. *The Farm Book* (Willoughby, OH: Meister Publishing Co., 1983), 440 p.

Berry, D.F., E.L. Madsen, and J.-M. Bollag. "Conversion of Indole to Oxindole under Methanogenic Conditions," *Appl. Environ. Microbiol.*, 53(1):180-182 (1987).

Betterton, E.A. and M.R. Hoffmann. "Henry's Law Constants of Some Environmentally Important Aldehydes," *Environ. Sci. Technol.*, 22(12):1415-11418 (1988).

Bevenue, A. and H. Beckman. "Pentachlorophenol: A Discussion of its Properties and its Occurrence as a Residue in Human and Animal Tissues," *Residue Rev.*, 19:83-134 (1967).

Bhavnagary, H.M. and M. Jayaram. "Determination of Water Solubilities of Lindane and Dieldrin at Different Temperatures," *Bull. Grain Technol.*, 12(2):95-99 (1974).

Bidleman, T.F. "Estimation of Vapor Pressures for Nonpolar Organic Compounds by Capillary Gas Chromatography," *Anal. Chem.*, 56(13):2490-2496 (1984).

Bielaszczyk, E., E. Czerwińska, Z. Janko, A. Kotarski, R. Kowalik, M. Kwiatkowski, and J. Żoledziowska. "Aerobic Reduction of Some Nitrochloro Substituted Benzene Compounds by Microorganisms," *Acta Microbiol. Pollut.*, 16:243-248 (1967).

Biggar, J.W., D.R. Nielson, and W.R. Tillotson. "Movement of DBCP in Laboratory Soil Columns and Field Soils to Groundwater," *Environ. Geol.*, 5(3):127-131 (1984).

Biggar, J.W. and I.R. Riggs. "Apparent Solubility of Organochlorine Insecticides in Water at Various Temperatures," *Hilgardia*, 42(10):383-391 (1974).

Billington, J.W., G.-L. Huang, F. Szeto, W.Y. Shiu, and D. Mackay. "Preparation of Aqueous Solutions of Sparingly Soluble Organic Substances: I. Single Component Systems," *Environ. Toxicol. Chem.*, 7:117-124 (1988).

Bjørseth, A. *Handbook of Polycyclic Aromatic Hydrocarbons* (New York: Marcel Dekker, 1983).

Blackman, G.E., M.H. Parke, and G. Garton. "The Physiological Activity of Substituted Phenols. I. Relationships between Chemical Structure and Physiological Activity," *Arch. Biochem. Biophys.*, 54(1):55-71 (1955).

Bobra, A., D. Mackay, and W.-Y. Shiu. "Distribution of Hydrocarbons Among Oil,

Water, and Vapor Phases During Oil Dispersant Toxicity Tests," *Bull. Environ. Contam. Toxicol.*, 23(4/5):558-565 (1979).

Boehm, P.D. and J.G. Quinn. "Solubilization of Hydrocarbons by the Dissolved Organic Matter in Sea Water," *Geochim. Cosmochim. Acta*, 37(11):2459-2477 (1973).

Boesten, J.J.T.I., L.J.T. van der Pas, M. Leistra, J.H. Smelt, and N.W.H. Houx. "Transformation of ^{14}C-Labelled 1,2-Dichloropropane in Water-Saturated Subsoil Materials," *Chemosphere*, 24(8):993-1011 (1992).

Boethling, R.S. and M. Alexander. "Effect of Concentration of Organic Chemicals on Their Biodegradation by Natural Microbial Communities," *Appl. Environ. Microbiol.*, 37(6):1211-1216 (1979).

Boggs, J.E. and A.E. Buck, Jr. "The Solubility of Some Chloromethanes in Water," *J. Phys. Chem.*, 62:1459-1461 (1958).

Bohon, R.L. and W.F. Claussen. "The Solubility of Aromatic Hydrocarbons in Water," *J. Am. Chem. Soc.*, 73(4):1571-1578 (1951).

Bollag, J.-M. and S.-Y. Liu. "Hydroxylations of Carbaryl by Soil Fungi," *Nature (London)*, 236(5343):177-178 (1972).

Booth, H.S. and H.E. Everson. "Hydrotrophic Solubilities. I. Solubilities in Forty Percent Xylenesulfonate," *Ind. Eng. Chem.*, 40:1491-1493 (1948).

Borello, R., C. Minero, E. Pramauro, E. Pelizzetti, N. Serpone, and H. Hidaka. "Photocatalytic Degradation of DDT Mediated in Aqueous Semiconductor Slurries by Simulated Sunlight," *Environ. Toxicol. Chem.*, 8(11):997-1002 (1989).

Borsetti, A.P. and J.A. Roach. "Identification of Kepone Alteration Products in Soil and Mullet," *Bull. Environ. Contam. Toxicol.*, 20(2):241-247 (1978).

Bossert, I.D. and R. Bartha. "Structure-Biodegradability Relationships of Polycyclic Aromatic-Hydrocarbon Systems in Soil," *Bull. Environ. Contam. Toxicol.*, 37:490-495 (1986).

Botré, C., A. Memoli, and F. Alhaique. "TCDD Solubilization and Photodecomposition in Aqueous Solutions," *Environ. Sci. Technol.*, 12(3):335-336 (1978).

Boublik, T., V. Fried, and E. Hala. *The Vapor Pressures of Pure Substances*, 2nd ed. (Amsterdam: Elsevier, 1973).

Boule, P., C. Guyon, and J. Lemaire. "Photochemistry and Environment. IV. Photochemical Behaviour of Monochlorophenols in Dilute Aqueous Solution," *Chemosphere*, 11(12):1179-1188 (1982).

Bouwer, E.J. and P.L. McCarty. "Transformations of 1- and 2-Carbon Halogenated Aliphatic Organic Compounds under Methanogenic Conditions," *Appl. Environ. Microbiol.*, 45:1286-1294 (1983).

Bove, J.L. and J. Arrigo. "Formation of Polycyclic Aromatic Hydrocarbon Via Gas Phase Benzyne," *Chemosphere*, 14(1):99-101 (1985).

Bove, J.L. and P. Dalven. "Pyrolysis of Phthalic Acid Esters: Their Fate," *Sci. Tot. Environ.*, 36:313-318 (1984).

Bowman, B.T. and W.W. Sans. "Adsorption of Parathion, Fenitrothion, Methyl Parathion, Aminoparathion and Paraoxon by Na^+, Ca^{2+}, and Fe^{3+}

Montmorillonite Suspensions," *Soil Sci. Soc. Am. J.*, 41:514-519 (1977).

Bowman, B.T. and W.W. Sans. "The Aqueous Solubility of Twenty-Seven Insect-icides and Related Compounds," *J. Environ. Sci. Health*, B14(6):625-634 (1979).

Bowman, B.T. and W.W. Sans. "Further Water Solubility Determininations of Insecticidal Compounds," *J. Environ. Sci. Health*, B18(2):221-227 (1983).

Bowman, B.T. and W.W. Sans. "Determination of Octanol-Water Partitioning Coef-ficients (K_{ow}) of 61 Organophosphorus and Carbamate Insecticides and Their Relationship to Respective Water Solubility (S) Values," *J. Environ. Sci. Health*, B18(6):667-683 (1983a).

Bowman, B.T. and W.W. Sans. "Effect of Temperature on the Water Solubility of Insecticides," *J. Environ. Sci. Health*, B20(6):625-631 (1985).

Bowman, M.C., F. Acree, Jr., and M.K. Corbett. "Solubility of Carbon-14 DDT in Water," *J. Agric. Food Chem.*, 8(5):406-408 (1960).

Bowman, M.C., M.S. Schecter, and R.L. Carter. "Behavior of Chlorinated Insecticides in a Broad Spectrum of Soil Types," *J. Agric. Food Chem.*, 13:360-365 (1965).

Boyd, S.A. "Adsorption of Substituted Phenols by Soil," *Soil Sci.*, 134(5):337-343 (1982).

Boyd, S.A., C.W. Kao, and J.M. Suflita. "Fate of 3,3'-Dichlorobenzidine in Soil: Persistence and Binding," *Environ. Toxicol. Chem.*, 3:201-208 (1984).

Boyd, S.A. and R. King. "Adsorption of Labile Organic Compounds by Soil," *Soil Sci.*, 137(2):115-119 (1984).

Boyle, T.P., E.F. Robinson-Wilson, J.D. Petty, and W. Weber. "Degradation of Pentachlorophenol in Simulated Lenthic Environment," *Bull. Environ. Contam. Toxicol.*, 24(2):177-184 (1980).

Bradley, R.S. and T.G. Cleasby. "The Vapour Pressure and Lattice Energy of Some Aromatic Ring Compounds," *J. Chem. Soc. (London)*, pp. 1690-1692, (1953).

Brady, A.P. and H. Huff. "The Vapor Pressure of Benzene Over Aqueous Detergent Solutions," *J. Phys. Chem.*, 62(6):644-649 (1958).

Braker, W. and A.L. Mossman. *Matheson Gas Data Book* (East Rutherford, NJ: Matheson Gas Products, 1971), 574 p.

Branson, D.R. "Predicting the Fate of Chemicals in the Aquatic Environment from Laboratory Data," in *Estimating the Hazard of Chemical Substances to Aquatic Life*, Cairns, J., Jr., K.L. Dickson, and A.W. Maki, Eds. (Philadelphia, PA: American Society for Testing and Materials, 1978), pp. 55-70.

Branson, D.R., I.T. Takahasi, W.M. Parker, and G.E. Blau. "Bioconcentration Kinetics of 2,3,7,8-Tetrachloro-*p*-dioxin in Rainbow Trout," *Environ. Toxicol. Chem.*, 4:779-788 (1985).

Bridie, A.L., C.J.M. Wolff, and M. Winter. "BOD and COD of Some Petro-chemicals," *Water Res.*, 13:627-630 (1979).

Briggs, G.G. "Theoretical and Experimental Relationships between Soil Adsorption, Octanol-Water Partition Coefficients, Water Solubilities, Bioconcentration Factors, and the Parachor," *J. Agric. Food Chem.*, 29(5):1050-1059 (1981).

Bright, N.F.H. "The Vapor Pressure of Diphenyl, Dibenzyl and Diphenylethane," *J. Chem. Soc. (London)*, pp. 624-626, (1951).

Bright, N.F.H., J.C. Cuthill, and N.H. Woodbury. "The Vapour Pressure of Parathion and Related Compounds," *J. Sci. Food Agric.*, 1:344-348 (1950).

Bringmann, G. and G. Kühn. "Comparison of the Toxicity Thresholds of Water Pollutants to Bacteria, Algae and Protozoa in the Cell Multiplication Inhibition Test," *Water Res.*, 14:231-241 (1980).

Broadhurst, M.G. "Use and Replaceability of Polychlorinated Biphenyls," *Environ. Health Perspect.*, (October 1972), pp. 81-102.

Brooke, D., I. Nielsen, J. de Bruijn, and J. Hermens. "An Interlaboratory Evaluation of the Stir-Flask Method for the Determination of Octanol-Water Partition Coefficients (Log P_{ow})," *Chemosphere*, 21(1/2):119-133 (1990).

Brooke, D.N., A.J. Dobbs, and N. Williams. "Octanol:Water Partition Coefficients (P): Measurement, Estimation, and Interpretation, Particularly for Chemicals with P >10^5," *Ecotoxicol. Environ. Saf.*, 11(3):251-260 (1986).

Brooker, P.J. and M. Ellison. "The Determination of the Water Solubility of Organic Compounds by a Rapid Turbidimetric Method," *Chem. Ind.*, (October 1974), pp. 785-787.

Brookman, G.T., M. Flanagan, and J.O. Kebe. "Literature Survey: Hydrocarbon Solubilities and Attenuation Mechanisms," API Publication 4414, (Washington, DC: American Petroleum Institute, 1985), 101 p.

Brooks, G.T. *Chlorinated Insecticides, Volume I, Technology and Applications* (Cleveland, OH: CRC Press, Inc., 1974), 249 p.

Brown, A.C., C.E. Canosa-Mas, and R.P. Wayne. "A Kinetic Study of the Reactions of OH with CH_3I and CF_3I," *Atmos. Environ.*, 24A(2):361-367 (1990).

Brown, D.S. and E.W. Flagg. "Empirical Prediction of Organic Pollutant Sorption in Natural Sediments," *J. Environ. Qual.*, 10(3):382-386 (1981).

Brown, L. and M. Rhead. "Liquid Chromatographic Determination of Acrylamide Monomer in Natural and Polluted Aqueous Environments," *Analyst*, 104(1238):391-399 (1979).

Brown, M.A., M.T. Petreas, H.S. Okamoto, T.M. Mischke, and R.D. Stephens. "Monitoring of Malathion and its Impurities and Environmental Transformation Products on Surfaces and in Air Following an Aerial Application," *Environ. Sci. Technol.*, 27(2):388-397 (1993).

Brown, R.L. and S.P. Wasik. "A Method of Measuring the Solubilities of Hydrocarbons in Aqueous Solutions," *J. Res. Nat. Bur. Stand.*, 78A(4):453-460 (1974).

Brownawell, B.J. "The Role of Colloid Organic Matter in the Marine Geochemistry of PCBs," Ph.D. Dissertation, Woods Hole-MIT Joint Program, Cambridge, MA (1986).

Bruggeman, W.A., J. Van Der Steen, and O. Hutzinger. "Reversed-Phase Thin-Layer Chromatography of Polynuclear Aromatic Hydrocarbons and Chlorinated Biphenyls. Relationship with Hydrophobicity as Measured by Aqueous Solubility and Octanol-Water Partition Coefficient," *J. Chromatogr.*, 238:335-

346 (1982).

Brunelle, D.J. and D.A. Singleton. "Chemical Reaction of Polychlorinated Biphenyls on Soils with Poly(ethylene glycol)/KOH," *Chemosphere*, 14(2):173-181 (1985).

Bufalini, J.J., B.W. Gay, and S.L. Kopczynski. "Oxidation of *n*-Butane by the Photolysis of NO_2," *Environ. Sci. Technol.*, 5(4):333-336 (1971).

Bull, D.L. and R.L. Ridgway. "Metabolism of Trichlorfon in Animals and Plants," *J. Agric. Food Chem.*, 27(2):268-272 (1979).

Bumpus, J.A. and S.D. Aust. "Biodegradation of DDT [1,1,1-Trichloro-2,2-bis(4-chlorophenyl)ethane] by the White Rot Fungus *Phanerochaete chrysosporium*," *Appl. Environ. Microbiol.*, 53(9):2001-2008 (1987).

Bumpus, J.A., M. Tien, D. Wright, and S.D. Aust. "Oxidation of Persistent Environmental Pollutants by a White Rot Fungus," *Science (Washington, DC)*, 228:1434-1436 (1985).

Bunton, C.A., N.A. Fuller, S.G. Perry, and V.J. Shiner. "The Hydrolysis of Carboxylic Anhydrides. Part III. Reactions in Initially Neutral Solution," *J. Chem. Soc. (London)*, pp. 2918-2926, (May 1963).

Burczyk, L., K. Walczyk, and R. Burczyk. "Kinetics of Reaction of Water Addition to the Acrolein Double-Bond in Dilute Aqueous Solution," *Przem. Chem.*, 47(10):625-627 (1968).

Burge, W.D. "Anaerobic Decomposition of DDT in Soil. Acceleration by Volatile Components of Alfalfa," *J. Agric. Food Chem.*, 19(2):375-378 (1971).

Burkhard, L.P., D.E. Armstrong, and A.W. Andren. "Henry's Law Constants for the Polychlorinated Biphenyls," *Environ. Sci. Technol.*, 19(7):590-596 (1985).

Burkhard, L.P. and D.W. Kuehl. "*n*-Octanol/Water Partition Coefficients by Reverse Phase Liquid Chromatography/Mass Spectrometry for Eight Tetrachlorinated Planar Molecules," *Chemosphere*, 15(2):163-167 (1986).

Burkhard, L.P., D.W. Kuehl, and G.D. Veith. "Evaluation of Reverse Phase Liquid Chromatography/Mass Spectrometry for Estimation of *n*-Octanol/Water Partition Coefficients for Organic Chemicals," *Chemosphere*, 14(10):1551-1560 (1985a).

Burlinson, N.E., L.A. Lee, and D.H. Rosenblatt. "Kinetics and Products of Hydrolysis of 1,2-Dibromo-3-chloropropane," *Environ. Sci. Technol.*, 16(9):627-632 (1982).

Burnett, M.G. "Determination of Partition Coefficients at Infinite Dilution by the Gas Chromatographic Analysis of the Vapor Above Dilute Solutions," *Anal. Chem.*, 35:1567-1570 (1963).

Burris, D.R. and W.G. MacIntyre. "A Thermodynamic Study of Solutions of Liquid Hydrocarbon Mixtures in Water," *Geochim. Cosmochim. Acta*, 50(7):1545-1549 (1986).

Burton, W.B. and G.E. Pollard. "Rate of Photochemical Isomerization of Endrin in Sunlight," *Bull. Environ. Contam. Toxicol.*, 12(1):113-116 (1974).

Butler, J.A.V. and C.N. Ramchandani. "The Solubility of Non-Electrolytes. Part II.

The Influence of the Polar Group on the Free Energy of Hydration of Aliphatic Compounds," *J. Chem. Soc. (London)*, pp. 952-955, (1935).

Butler, J.A.V., C.N. Ramchandani, and D.W. Thompson. "The Solubility of Non-Electrolytes. Part I. The Free Energy of Hydration of Some Aliphatic Alcohols," *J. Chem. Soc. (London)*, pp. 280-285, (1935).

Butler, J.A.V., D.W. Thomson, and W.H. Maclennan. "The Free Energy of the Normal Aliphatic Alcohols in Aqueous Solution. Part I. The Partial Vapour Pressures of Aqueous Solutions of Methyl, *n*-Propyl, and *n*-Butyl Alcohols. Part II. The Solubilities of Some Normal Aliphatic Alcohols in Water. Part III. The Theory of Binary Solutions, and its Application to Aqueous-Alcoholic Solutions," *J. Chem. Soc. (London)*, 136:674-686 (1933).

Butler, J.C. and W.P. Webb. "Upper Explosive Limits of Cumene," *J. Chem. Eng. Data*, 2(1):42-46 (1957).

Buttery, R.G., J.L. Bomben, D.G. Guadagni, and L.C. Ling. "Some Considerations of the Volatilities of Organic Flavor Compounds in Foods," *J. Agric. Food Chem.*, 19:1045-1048 (1971).

Buttery, R.G., L.C. Ling, and D.G. Guadagni. "Volatilities of Aldehydes, Ketones, and Esters in Dilute Water Solution," *J. Agric. Food Chem.*, 17(2):385-389 (1969).

Butz, R.G., C.C. Yu, and Y.H. Atallah. "Photolysis of Hexachlorocyclopentadiene in Water," *Ecotoxicol. Environ. Saf.*, 6(4):347-357 (1982).

Byast, T.H. and R.J. Hance. "Degradation of 2,4,5-T by South Vietnamese Soils Incubated in the Laboratory," *Bull. Environ. Contam. Toxicol.*, 14(1):71-76 (1975).

Call, F. "The Mechanism of Sorption of Ethylene Dibromide on Moist Soils," *J. Sci. Food Agric.*, 8:630-639 (1957).

Callahan, M.A., M.W. Slimak, N.W. Gable, I.P. May, C.F. Fowler, J.R. Freed, P. Jennings, R.L. Durfee, F.C. Whitmore, B. Maestri, W.R. Mabey, B.R. Holt, and C. Gould. "Water-Related Environmental Fate of 129 Priority Pollutants Volumes I and II," Office of Research and Development, U.S. EPA Report-440/4-79-029 (1979), 1160 p.

Calvert, J.G. and J.N. Pitts, Jr. *Photochemistry* (New York: John Wiley & Sons, Inc., 1966), 899 p.

Camilleri, P., S.A. Watts, and J.A. Boraston. "A Surface Area Approach to Determination of Partition Coefficients," *J. Chem. Soc., Perkin Trans. 2*, pp. 1699-1707, (September 1988).

Campbell, J.R. and R.G. Luthy. "Prediction of Aromatic Solute Partition Coefficients Using the UNIFAC Group Contribution Model," *Environ. Sci. Technol.*, 19(10):980-985 (1985).

Camper, N.D. "Effects of Pesticide Degradation Products on Soil Microflora" in *Pesticide Transformation Products. Fate and Significance in the Environment*, ACS Symposium Series 429, Somasundaram, L. and J.R. Coats, Eds., (New York: American Chemical Society, 1991), pp. 205-216.

Canton, J.H., G.J. van Esch, P.A. Greve, and A.B.A.M. avan Hellemond. "Accumulation and Elimination of Hexachlorocyclohexane (HCH) by the Marine Algae *Chlamydomonas* and *Dunaliella*," *Water Res.*, 11:111-115 (1977).

Carlisle, P.J. and A.A. Levine. "Stability of Chlorohydrocarbons," *Ind. Eng. Chem.*, 24(2):146-147 (1932).

Carlisle, P.J. and A.A. Levine. "Stability of Chlorohydrocarbons. II. Trichloroethylene," *Ind. Eng. Chem.*, 24(10):1164-1168 (1932a).

Carlson, A.R. and P.A. Kosian. "Toxicity of Chlorinated Benzenes to Fathead Minnows (*Pimephales promelas*)," *Arch. Environ. Contam. Toxicol.*, 16:129-135 (1987).

Caro, J.H., H.P. Freeman, D.E. Glotfelty, B.C. Turner, and W.M. Edwards. "Dissipation of Soil-Incorporated Carbofuran in the Field," *J. Sci. Food Agric.*, 21(6):1010-1015 (1973).

Caron, G., I.H. Suffet, and T. Belton. "Effect of Dissolved Organic Carbon on the Environmental Distribution of Nonpolar Organic Compounds," *Chemosphere*, 14(8):993-1000 (1985).

Carter, W.P.L., A.M. Winer, and J.N. Pitts, Jr. "Major Atmospheric Sink for Phenol and the Cresols. Reaction with the Nitrate Radical," *Environ. Sci. Technol.*, 15(7):829-831 (1981).

Casellato, F., C. Vecchi, A. Girelli, and B. Casu. "Differential Calorimetric Study of Polycyclic Aromatic Hydrocarbons," *Thermochim. Acta*, 6:361-368 (1973).

Casida, J.E., P.E. Gatterdam, L.W. Getzin, Jr., and R.K. Chapman. "Residual Properties of the Systemic Insecticide *O,O*-Dimethyl 1-Carbomethoxy-1-propen-2-yl Phosphate," *J. Agric. Food Chem.*, 4(3):236-243 (1956).

Casida, J.E., R.L. Holmstead, S. Khalifa, J.R. Knox, T. Ohsawa, K.J. Palmer, and R.Y. Wong. "Toxaphene Insecticide: A Complex Biodegradable Mixture," *Science (Washington, DC)*, 183(4124):520-521 (1974).

Castro, C.E. "The Rapid Oxidation of Iron(II) Porphyrins by Alkyl Halides. A Possible Mode of Intoxication of Organisms by Alkyl Halides," *J. Am. Chem. Soc.*, 86(2):2310-2311 (1964).

Castro, C.E. and N.O. Belser. "Biodehalogenation: Oxidative and Reductive Metabolism of 1,1,2-Trichloroethane by *Pseudomonas putida* - Biogeneration of Vinyl Chloride," *Environ. Toxicol. Chem.*, 9(6):707-714 (1990).

Castro, C.E. and N.O. Belser. "Biodehalogenation. Reductive Dehalogenation of the Biocides Ethylene Dibromide, 1,2-Dibromo-3-chloropropane, and 2,3-Dibromo-butane in Soil," *Environ. Sci. Technol.*, 2(10):779-783 (1968).

Castro, C.E. and N.O. Belser. "Hydrolysis of *cis*- and *trans*-1,3-Dichloropropene in Wet Soil," *J. Agric. Food Chem.*, 14(1):69-70 (1966).

Castro, C.E., R.S. Wade, D.M. Riebeth, E.W. Bartnicki, and N.O. Belser. "Biodehalogenation: Rapid Metabolism of Vinyl Chloride by a Soil *Pseudomonas* sp. Direct Hydrolysis of a Vinyl C-Cl Bond," *Environ. Toxicol. Chem.*, 11(6):757-764 (1992).

Castro, C.E. and T. Yoshida. "Degradation of Organochlorine Insecticides in

Flooded Soils in the Philippines," *J. Agric. Food Chem.*, 19:1168-1170 (1971).

Caswell, R.L., K.J. DeBold, and L.S. Gilbert, Eds. *Pesticide Handbook*, 29th ed. (College Park, MD: The Entomological Society of America, 1981), 286 p.

Caturla, F., J.M. Martin-Martinez, M. Molina-Sabio, F. Rodriguez-Reinoso, and R. Torregrosa. "Adsorption of Substituted Phenols on Activated Carbon," *J. Colloid Interface Sci.*, 124(2):528-534 (1988).

Cavalieri, E. and E. Rogan. "Role of Radical Cations in Aromatic Hydrocarbon Carcinogenesis," *Environ. Health Perspect.*, 64:69-84 (1985).

Cepeda, E.A., B. Gomez, and M. Diaz. "Solubility of Anthracene and Anthraquinone in Some Pure and Mixed Solvents," *J. Chem. Eng. Data*, 34(3):273-275 (1989).

Cerniglia, C.E. and D.T. Gibson. "Fungal Oxidation of Benzo[*a*]pyrene: Evidence for the Formation of a BP-7,8-Diol 9,10-Epoxide and BP-9,10-Diol 7,8-Epoxides," *Abstr. Ann. Mtg. Am. Soc. Microbiol.*, 1980, p. 138.

Cerniglia, C.E. and S.A. Crow. "Metabolism of Aromatic Hydrocarbons by Yeasts," *Arch. Microbiol.*, 129:9-13.

Chacko, C.I., J.L. Lockwood, and M. Zabik. "Chlorinated Hydrocarbon Pesticides: Degradation by Microbes," *Science (Washington, DC)*, 154(3751):893-895 (1966).

Chaigneau, M., G. LeMoan, and L. Giry. "Décomposition Pyrogénée du Bromurede Méthyle en L'absence D'oxygene et à 550 °C," *Comp. Rend. Ser. C.*, 263:259-261 (1966).

Chan, W.F. and R.A. Larson. "Formation of Mutagens from the Aqueous Reactions of Ozone and Anilines," *Water Res.*, 25(12):1529-1538 (1991).

Chan, W.F. and R.A. Larson. "Mechanisms and Products of Ozonolysis of Aniline in Aqueous Solution Containing Nitrite Ion," *Water Res.*, 25(12):1549-1544 (1991).

Chapman, P.J. "An Outline of Reaction Sequences Used for the Bacterial Degradation of Phenolic Compounds," in *Degradation of Synthetic Organic Molecules in the Biosphere: Natural, Pesticidal, and Various Other Man-Made Compounds* (Washington, DC: National Academy of Sciences, 1972), pp. 17-55.

Chapman, R.A. and C.M. Cole. "Observations on the Influence of Water and Soil pH on the Persistence of Insecticides," *J. Environ. Sci. Health*, B17(5):487-504 (1982).

Chen, P.N., G.A. Junk, and H.J. Svec. "Reactions of Organic Pollutants. I. Ozonation of Acenaphthylene and Acenaphthene," *Environ. Sci. Technol.*, 13(4):451-454 (1979).

Chen, P.R., W.P. Tucker, and W.C. Dauterman. "Structure of Biologically Produced Malathion Monoacid," *J. Agric. Food Chem.*, 17(1):86-90 (1969).

Cheung, H.M. and S. Kurup. "Sonochemical Destruction of CFC 11 and CFC 13 in Dilute Aqueous Solution," *Environ. Sci. Technol.*, 28(9):1619-1622 (1994).

Chey, W. and G.V. Calder. "Method for Determining Solubility of Slightly Soluble Organic Compounds," *J. Chem. Eng. Data*, 17(2):199-200 (1972).

Chin, Y.-P., C.S. Peven, and W.J. Weber. "Estimating Soil/Sediment Partition

Coefficients for Organic Compounds by High Performance Reverse Phase Liquid Chromatography," *Water Res.*, 22(7):873-881 (1988).

Chin, Y.-P., W.J. Weber, Jr., and T.C. Voice. "Determination of Partition Coefficients and Aqueous Solubilities by Reverse Phase Chromatography - II. Evaluation of Partitioning and Solubility Models," *Water Res.*, 20(11):1443-1451 (1986).

Chiodini, G., B. Rindone, F. Cariati, S. Polesello, G. Restilli, and J. Hjorth. "Comparison between the Gas-Phase and the Solution Reaction of the Nitrate Radical and Methylarenes," *Environ. Sci. Technol.*, 27(8):1659-1664 (1993).

Chiou, C.T. "Partition Coefficients of Organic Compounds in Lipid-Water Systems and Correlations with Fish Bioconcentration Factors," *Environ. Sci. Technol.*, 19(1):57-62 (1985).

Chiou, C.T., D.W. Schmedding, and M. Manes. "Partitioning of Organic Compounds in Octanol-Water Systems," *Environ. Sci. Technol.*, 16(1):4-10 (1982).

Chiou, C.T., L.J. Peters, and V.H. Freed. "A Physical Concept of Soil-Water Equilibria for Nonionic Organic Compounds," *Science (Washington, DC)*, 206(4420):831-832 (1979).

Chiou, C.T., P.E. Porter, and D.W. Schmedding. "Partition Equilibria of Nonionic Organic Compounds between Soil Organic Matter and Water," *Environ. Sci. Technol.*, 17(4):227-231 (1983).

Chiou, C.T., R.L. Malcolm, T.I. Brinton, and D.E. Kile. "Water Solubility Enhancement of Some Organic Pollutants and Pesticides by Dissolved Humic and Fulvic Acids," *Environ. Sci. Technol.*, 20(5):502-508 (1986).

Chiou, C.T., T.D. Shoup, and P.E. Porter. "Mechanistic Roles of Soil Humus and Minerals in the Sorption of Nonionic Organic Compounds from Aqueous and Organic Solutions," *Org. Geochem.*, 8(1):9-14 (1985).

Chiou, C.T., V.H. Freed, D.W. Schmedding, and R.L. Kohnert. "Partition Coefficients and Bioaccumulation of Selected Organic Chemicals," *Environ. Sci. Technol.*, 11(5):475-478 (1977).

Chopra, N.M. and A.M. Mahfouz. "Metabolism of Endosulfan I, Endosulfan II and Endosulfan sulfate in Tobacco Leaf," *J. Agric. Food Chem.*, 25(1):32-36 (1977).

Chou, J.T. and P.C. Jurs. "Computer Assisted Computation of Partition Coefficients from Molecular Structures Using Fragment Constants," *J. Chem. Info. Comp. Sci.*, 19:172-178 (1979).

Chou, S.-F.J., R.A. Griffin, M.-I.M. Chou, and R.A. Larson. "Products of Hexa-chlorocyclopentadiene (C-56) in Aqueous Solution," *Environ. Toxicol. Chem.*, 6(5):371-376 (1987).

Chou, S.F.J. and R.A. Griffin. "Soil, Clay and Caustic Soda Effects on Solubility, Sorption and Mobility of Hexachlorocyclopentadiene," Environmental Geology Notes 104, Illinois Department of Energy and Natural Resources (1983), 54 p.

Choudhry, G.G. and O. Hutzinger. "Acetone-Sensitized and Nonsensitized Photolyses of Tetra-, Penta-, and Hexachlorobenzenes in Acetonitrile-Water Mixtures: Photoisomerization and Formation of Several Products Including

Polychlorobiphenyls," *Environ. Sci. Technol.*, 18(4):235-241 (1984).

Choudhry, G.G., G. Sundström, L.O. Ruzo, and O. Hutzinger. "Photochemistry of Chlorinated Diphenyl Ethers," *J. Agric. Food Chem.*, 25(6):1371-1376 (1977).

Christensen, J.J., L.D. Hansen, and R.M. Izatt. *Handbook of Proton Ionization Heats* (New York: John Wiley & Sons, Inc., 1975), 269 p.

Christiansen, V.O., J.A. Dahlberg, and H.F. Andersson. "On the Nonsensitized Photooxidation of 1,1,1-Trichloroethane Vapor in Air," *Acta Chem. Scand.*, A26:3319-3324 (1972).

Chu, W. and C.T. Jafvert. "Photodechlorination of Polychlorobenzene Congeners in Surfactant Micelle Solutions," *Environ. Sci. Technol.*, 28(13):2415-2422 (1994).

Claus, D. and N. Walker. "The Decomposition of Toluene by Soil Bacteria," *J. Gen. Microbiol.*, 36:107-122 (1964).

Claussen, W.F. and M.F. Polglase. "Solubilities and Structures in Aqueous Aliphatic Hydrocarbon Solutions," *J. Am. Chem. Soc.*, 74(19):4817-4819 (1952).

Clayton, G.D. and F.E. Clayton, Eds. *Patty's Industrial Hygiene and Toxicology*, 3rd ed. (New York: John Wiley & Sons, Inc., 1981), 2878 p.

Cleland, J.G. "Project Summary - Environmental Hazard Rankings of Pollutants Generated in Coal Gasification Processes," Office of Research and Development, U.S. EPA Report-600/S7-81-101 (1981), 19 p.

Cleland, J.G. and G.L. Kingsbury. "Multimedia Environmental Goals for Environmental Assessment, Volume II. MEG Charts and Background Information," Office of Research and Development, U.S. EPA Report-600/7-77-136b (1977), 454 p.

Coates, M., D.W. Connell, and D.M. Barron. "Aqueous Solubility and Octan-1-ol to Water Partition Coefficients of Aliphatic Hydrocarbons," *Environ. Sci. Technol.*, 19(7):628-632 (1985).

Coffee, R.D., P.C. Vogel, Jr., and J.J. Wheeler. "Flammability Characteristics of Methylene Chloride (Dichloromethane)," *J. Chem. Eng. Data*, 17(1):89-93 (1972).

Collander, R. "The Partition of Organic Compounds between Higher Alcohols and Water," *Acta Chem. Scand.*, 5:774-780 (1951).

Collett, A.R. and J. Johnston. "Solubility Relations of Isomeric Organic Compounds. VI. Solubility of the Nitroanilines in Various Liquids," *J. Phys. Chem.*, 30(1):70-82 (1926).

Connors, T.F., J.D. Stuart, and J.B. Cope. "Chromatographic and Mutagenic Analyses of 1,2-Dichloropropane and 1,3-Dichloropropylene and Their Degradation Products," *Bull. Environ. Contam. Toxicol.*, 44(2):288-293 (1990).

Cooney, R.V., P.D. Ross, G.L. Bartolini, and J. Ramseyer. "*N*-Nitrosamine and *N*-Nitroamine Formation: Factors Influencing the Aqueous Reactions of Nitrogen Dioxide with Morpholine," *Environ. Sci. Technol.*, 21(1):77-83 (1987).

Cooper, W.J., M. Mehran, D.J. Riusech, and J.A. Joens. "Abiotic Transformations of Halogenated Organics. 1. Elimination Reaction of 1,1,2,2-Tetrachloroethane and Formation of 1,1,2-Trichloroethane," *Environ. Sci. Technol.*, 21(11):1112-1114 (1987).

Coover, M.P. and R.C.C. Sims. "The Effects of Temperature on Polycyclic Aromatic Hydrocarbon Persistence in an Unacclimated Agricultural Soil," *Haz. Waste Haz. Mater.*, 4:69-82 (1987).

Corbett, M.D. and B.R. Corbett. "Metabolism of 4-Chloronitro-benzene by the Yeast *Rhodosporidium* sp.," *Appl. Environ. Microbiol.*, 41(4):942-949 (1981).

Corby, T.C. and P.H. Elworthy. "The Solubility of Some Compounds in Hexadecylpolyoxyethylene Monoethers, Polyethylene Glycols, Water and Hexane," *J. Pharm. Pharmac.*, 23:39S-48S (1971).

Corless, C.E., G.L. Reynolds, N.J.D. Graham, and R. Perry. "Ozonation of Pyrene in Aqueous Solution," *Water Res.*, 24(9):1119-1123 (1990).

Coutant, R.W. and G.W. Keigley. "An Alternative Method for Gas Chromatographic Determination of Volatile Organic Compounds in Water," *Anal. Chem.*, 60:2536-2537 (1988).

Cowen, W.F. and R.K. Baynes. "Estimated Application of Gas Chromatographic Headspace Analysis for Priority Pollutants," *J. Environ. Sci. Health*, A15:413-427 (1980).

Cox, D.P. and C.D. Goldsmith. "Microbial Conversion of Ethylbenzene to 1-Phenethanol and Acetophenone by *Nocardia tartaricans* ATCC 31190," *Appl. Environ. Microbiol.*, 38(3):514-520 (1979).

Cox, R.A., K.F. Patrick, and S.A. Chant. "Mechanism of Atmospheric Photooxidation of Organic Compounds. Reactions of Alkoxy Radicals in Oxidation of *n*-Butane and Simple Ketones," *Environ. Sci. Technol.*, 15(5):587-592 (1981).

Cox, R.A., R.G. Derwent, and M.R. Williams. "Atmospheric Photooxidation Reactions. Rates, Reactivity, and Mechanism for Reaction of Organic Compounds with Hydroxyl Radicals," *Environ. Sci. Technol.*, 14(1):57-61 (1980).

Cozzarelli, I.M., R.P. Eganhouse, and M.J. Baedecker. "Transformation of Monoaromatic Hydrocarbons to Organic Acids in Anoxic Groundwater Environment," *Environ. Geol. Water Sci.*, 16(2):135-141 (1990).

Cremlyn, R.J. *Agrochemicals - Preparation and Mode of Action* (New York: John Wiley & Sons, Inc., 1991), 396 p.

Crosby, D.G. and A.S. Wong. "Photodecomposition of 2,4,5-Trichlorophenoxyacetic Acid (2,4,5-T) in Water," *J. Agric. Food Chem.*, 21(6):1052-1054 (1973).

Crosby, D.G. and E. Leitis. "The Photodecomposition of Trifluralin in Water," *Bull. Environ. Contam. Toxicol.*, 10(4):237-241 (1973).

Crosby, D.G. and H.O. Tutass. "Photodecomposition of 2,4-Dichlorophenoxyacetic Acid," *J. Agric. Food Chem.*, 14(6):596-599 (1966).

Crosby, D.G. and K.W. Moilanen. "Vapor-Phase Photodecomposition of Aldrin and Dieldrin," *Arch. Environ. Contam. Toxicol.*, 2(1):62-74 (1974).

Crosby, D.G., K.W. Moilanen, and A.S. Wong. "Environmental Generation and Degradation of Dibenzodioxins and Dibenzofurans," *Environ. Health Perspect.*, Experimental Issue No. 5, (September 1973), pp. 259-266.

Crosby, D.G. and N. Hamadmad. "The Photoreduction of Pentachlorobenzenes," *J. Agric. Food Chem.*, 19(6):1171-1174 (1971).

Crummett, W.B. and R.H. Stehl. "Determination of Chlorinated Dibenzo-*p*-dioxins and Dibenzofurans in Various Materials," *Environ. Health Perspect.* (September 1973), pp. 15-25.

Cserjsei, A.J. and E.L. Johnson. "Methylation of Pentachlorophenol by *Trichoderma Virgatum*," *Can. J. Microbiol.*, 18:45-49 (1972).

Cupitt, L.T. "Fate of Toxic and Hazardous Materials in the Air Environment," Office of Research and Development, U.S. EPA Report-600/3-80-084 (1980), 28 p.

Curtice, S., E.G. Felton, and H.W. Prengle, Jr., "Thermodynamics of Solutions. Low-Temperature Densities and Excess Volumes of *cis*-Pentene-2 and Mixtures," *J. Chem. Eng. Data*, 17(2):192-194 (1972).

Dagley, S. "Microbial Degradation of Stable Chemical Structures: General Features of Metabolic Pathways," in *Degradation of Synthetic Organic Molecules in the Biosphere: Natural, Pesticidal, and Various Other Man-Made Compounds* (Washington, DC: National Academy of Sciences, 1972), pp. 1-16.

Davidson, J.M., P.S.C. Rao, L.T. Ou, W.B. Wheeler, and D.F. Rothwell. "Adsorption, Movement, and Biological Degradation of Large Concentrations of Selected Pesticides in Soils," U.S. EPA Report-600/2-80-124 (1980), 110 p.

Davies, R.P. and A.J. Dobbs. "The Prediction of Bioconcentration in Fish," *Water Res.*, 18(10):1253-1262 (1984).

Davis, J.B. and R.L. Raymond. "Oxidation of Alkyl-Substituted Cyclic Hydrocarbons by a *Nocardia* During Growth on *n*-Alkanes," *Appl. Microbiol.*, 9:383-388 (1961).

Davis, J.T. and W.S. Hardcastle. "Biological Assay of Herbicides for Fish Toxicity," *Weeds*, 7:397-404 (1959).

Davis, W.W., M.E. Krahl, and G.H.A. Clowes. "Solubility of Carcinogenic and Related Hydrocarbons in Water," *J. Am. Chem. Soc.*, 64(1):108-110 (1942).

Davis, W.W. and T.V. Parke, Jr. "A Nephelometric Method for Determination of Solubilities of Extremely Low Order," *J. Am. Chem. Soc.*, 64(1):101-107 (1942).

Davis S.N. and R.J.M. DeWiest. *Hydrogeology* (New York: John Wiley & Sons, Inc., 1966).

Day, K.E. "Pesticide Transformation Products in Surface Waters" in *Pesticide Transformation Products, Fate and Significance in the Environment*, ACS Symposium Series 459, Somasundaram, L. and J.R. Coats, Eds. (Washington, DC: American Chemical Society, 1991), pp. 217-241.

de Bruijn, J., F. Busser, W. Seinen, and J. Hermens. "Determination of Octanol/Water Partition Coefficients for Hydrophobic Organic Chemicals with the 'Slow-Stirring' Method," *Environ. Toxicol. Chem.*, 8:499-512 (1989).

de Kruif, C.G. "Enthalpies of Sublimation and Vapour Pressures of 11 Polycyclic Hydrocarbons," *J. Chem. Thermodyn.*, 12:243-248 (1980).

Dean, J.A., Ed. *Lange's Handbook of Chemistry*, 11th ed. (New York: McGraw-Hill, Inc., 1973), 1570 p.

Dean, J.A. *Handbook of Organic Chemistry* (New York: McGraw-Hill Book Co.,

1987).

Dearden, J.C. "Partitioning and Lipophilicity in Quantitative Structure-Activity Relationships," *Environ. Health Perspect.* (September 1985), pp. 203-228.

DeFoe, D.L., G.W. Holcombe, D.E. Hammermeister, and K.E. Biesinger. "Solubility and Toxicity of Eight Phthalate Esters to Four Aquatic Organisms," *Environ. Toxicol. Chem.*, 9(5):623-636 (1990).

Dehn, W.M. "Comparative Solubilities in Water, in Pyridine and in Aqueous Pyridine," *J. Am. Chem. Soc.*, 39(7):1399-1404 (1917).

DeKock, A.C. and D.A. Lord. "A Simple Procedure for Determining Octanol-Water Partition Coefficients Using Reverse Phase High Performance Liquid Chromatography (RPHPLC)," *Chemosphere*, 16(1):133-142 (1987).

DeLassus, P.T. and D.D. Schmidt. "Solubilities of Vinyl Chloride and Vinylidene Chloride in Water," *J. Chem. Eng. Data*, 26(3):274-276 (1981).

DeLaune, R.D., R.P. Gambrell, and K.S. Reddy. "Fate of Pentachlorophenol in Estuarine Sediment," *Environ. Poll. Series B*, 6:297-308 (1983).

DeLoos, Th.W., W.G. Penders, and R.N. Lichtenthaler. "Phase Equilibria and Critical Phenomena in Fluid (*n*-Hexane + Water) at High Pressures and Temperatures," *J. Chem. Thermodyn.*, 14:83-91 (1982).

Deneer, J.W., T.L. Sinnige, W. Seinen, and J.L.M. Hermens. "A Quantitative Structure-Activity Relationship for the Acute Toxicity of Some Epoxy Compounds to the Guppy," *Aquat. Toxicol.*, 13(3):195-204 (1988).

Dennis, W.H., Jr. and W.J. Cooper. "Catalytic Dechlorination of Organochlorine Compounds. II. Heptachlor and Chlordane," *Bull. Environ. Contam. Toxicol.*, 16(4):425-430 (1976).

Dennis, W.H., Jr., Y.H. Chang, and W.J. Cooper. "Catalytic Dechlorination of Organochlorine Compounds-Aroclor 1254," *Bull. Environ. Contam. Toxicol.*, 22(6):750-753 (1979).

Derbyshire, M.K., J.S. Karns, P.C. Kearney, and J.O. Nelson. "Purification and Characterization of an *N*-Methylcarbamate Pesticide Hydrolyzing Enzyme," *J. Agric. Food Chem.*, 35(6):871-877 (1987).

DeVoe, H., M.M. Miller, and S.P. Wasik. "Generator Columns and High Pressure Liquid Chromatography for Determining Aqueous Solubilities and Octanol-Water Partition Coefficients of Hydrophobic Substances," *J. Res. Nat. Bur. Stand.*, 86(4):361-366 (1981).

Dewulf, J., D. Drijvers, and H. Van Langenhove. "Measurement of Henry's Law Constant as Function of Temperature and Salinity for the Low Temperature Range," *Atmos. Environ.*, 29(3):323-331 (1995).

Dexter, R.N. and S.P. Pavlou. "Mass Solubility and Aqueous Activity Coefficients of Stable Organic Chemicals in the Marine Environment: Polychlorinated Biphenyls," *Mar. Sci.*, 6:41-53 (1978).

Dickenson, W. "The Vapour Pressure of 1:1:*p.p'* Dichlorodiphenyl Trichloroethane (D.D.T.)," *Trans. Faraday Soc.*, pp. 31-35 (1956).

Dickhut, R.M., A.W. Andren, and D.E. Armstrong. "Aqueous Solubilities of Six

Polychlorinated Biphenyl Congeners at Four Temperatures," *Environ. Sci. Technol.*, 20(8):807–810 (1986).

DiGeronimo, M.J. and A.D. Antoine. "Metabolism of Acetonitrile and Propionitrile by *Nocardia rhodochrous* LL 100-21," *Appl. Environ. Microbiol.*, 31(6):900–906 (1976).

Dilling, W.L. "Interphase Transfer Processes. II. Evaporation Rates of Chloro Methanes, Ethanes, Ethylenes, Propanes, and Propylenes from Dilute Aqueous Solutions. Comparisons with Theoretical Predictions," *Environ. Sci. Technol.*, 11(4):405–409 (1977).

Dilling, W.L., L.C. Lickly, T.D. Lickly, P.G. Murphy, and R.L. McKellar. "Organic Photochemistry. 19. Quantum Yields for *O,O*-Diethyl *O*-(3,5,6-Tri-chloro-2-pyridinyl) Phosphorothioate (Chlorpyrifos) and 3,5,6-Trichloro-2-pyridinol in Dilute Aqueous Solutions and Their Environmental Phototransformation Rates," *Environ. Sci. Technol.*, 18(7):540–543 (1984).

Dilling, W.L., N.B. Tefertiller, and G.J. Kallos. "Evaporation Rates and Reactivities of Methylene Chloride, Chloroform, 1,1,1-Trichloroethane, Trichloroethylene, Tetrachloroethylene, and Other Chlorinated Compounds in Dilute Aqueous Solutions," *Environ. Sci. Technol.*, 9(9):833–837 (1975).

Dobbs, A.J. and M.R. Cull. "Volatilisation of Chemicals - Relative Loss Rates and the Estimation of Vapour Pressures," *Environ. Pollut. (Series B)*, 3(4):289–298 (1982).

Dodge R.H., C.E. Cerniglia, and D.T. Gibson. "Fungal Metabolism of Biphenyl," *Biochem. J.*, 178:223–230 (1979).

Dojlido, J.R. "Investigations of Biodegradability and Toxicity of Organic Compounds. Final Report 1975-79," Municipal Environmental Research Lab, U.S. EPA Report-600/2-79-163 (1980).

D'Oliveira, J.-C., G. Al-Sayyed, and P. Pichat. "Photodegradation of 2- and 3-Chlorophenol in TiO$_2$ Aqueous Suspensions," *Environ. Sci. Technol.*, 24(7):990–996 (1990).

Donahue, D.J. and F.E. Bartell. "The Boundary Tension at Water-Organic Liquid Interfaces," *J. Phys. Chem.*, 56(4):480–484 (1952).

Dong, S. and P.K. Dasgupta. "Solubility of Gaseous Formaldehyde in Liquid Water and Generation of Trace Standard Gaseous Formaldehyde," *Environ. Sci. Technol.*, 20(6):637–640 (1986).

Doucette, W.J. and A.W. Andren. "Estimation of Octanol/Water Partition Coefficients: Evaluation of Six Methods for Highly Hydrophobic Aromatic Hydrocarbons," *Chemosphere*, 17(2):345–359 (1988).

Doucette, W.J. and A.W. Andren. "Aqueous Solubility of Biphenyl, Furan, and Dioxin Congeners," *Chemosphere*, 17:243–252 (1988a).

Draper, W.M. and D.G. Crosby. "Solar Photooxidation of Pesticides in Dilute Hydrogen Peroxide," *J. Agric. Food Chem.*, 32(2):231–237 (1984).

Dreher, R.M. and B. Podratzki. "Development of an Enzyme Immunoassay for Endosulfan and its Degradation Products," *J. Agric. Food Chem.*, 36(5):1072-

1075 (1988).

Dreisbach, R.R. *Pressure-Volume-Temperature Relationships of Organic Compounds* (Sandusky, OH: Handbook Publishers, 1952), 349 p.

Dreisbach, R.R. *Physical Properties of Chemical Compounds II. Adv. Chem. Ser.* 22 (Washington, D.C.: American Chemical Society, 1959)

Dugan, P.R. *Biochemical Ecology of Water Pollution* (New York: Plenum Press, 1972), 159 p.

Duke, S.O., T.B. Moorman, and C.T. Bryson. "Phytotoxicity of Pesticide Degradation Products" in *Pesticide Transformation Products, Fate and Significance in the Environment*, ACS Symposium Series 459, Somasundaram, L. and J.R. Coats, Eds. (Washington, DC: American Chemical Society, 1991), pp. 188-204.

Dulfer, W.J., M.W.C. Bakker, and H.A.J. Govers. "Micellar Solubility and Micelle/Water Partitioning of Polychlorinated Biphenyls in Solutions of Sodium Dodecyl Sulfate," *Environ. Sci. Technol.*, 29(4):985-992 (1995).

Dulin, D., H. Drossman, and T. Mill. "Products and Quantum Yields for Photolysis of Chloroaromatics in Water," *Environ. Sci. Technol.*, 20(1):72-77 (1986).

Du Pont. 1966. *Solubility Relationships of the Freon Fluorocarbon Compounds*, Technical Bulletin B-7, (Wilmington, DE: Du Pont de Nemours & Co.)

Eadie, B.J., N.R. Morehead, and P.F. Landrum. "Three-Phase Partitioning of Hydrophobic Organic Compounds in Great Lakes Waters," *Chemosphere*, 20(1/2):161-178 (1990).

Eadsforth, C.V. "Application of Reverse-Phase H.P.L.C. for the Determination of Partition Coefficients," *Pestic. Sci.*, 17(3):311-325 (1986).

Eckert, J.W. "Fungistatic and Phytotoxic Properties of Some Derivatives of Nitrobenzene," *Phytopathology*, 52:642-650 (1962).

Eder, G. and K. Weber. "Chlorinated Phenols in Sediments and Suspended Matter of the Weser Estuary," *Chemosphere*, 9(2):111-118 (1980).

Edwards, D.A., R.G. Luthy, and Z. Liu. "Solubilization of Polycyclic Aromatic Hydrocarbons in Micellar Nonionic Surfactant Solutions," *Environ. Sci. Technol.*, 25(1):127-133 (1991).

Eganhouse, R.P. and J.A. Calder. "The Solubility of Medium Weight Aromatic Hydrocarbons and the Effect of Hydrocarbon Co-Solutes and Salinity," *Geochim. Cosmochim. Acta*, 40(5):555-561 (1976).

Eiceman, G.A. and V.J. Vandiver. "Adsorption of Polycyclic Aromatic Hydrocarbons on Fly Ash from a Municipal Incinerator and a Coal-Fired Power Plant," *Atmos. Environ.*, 17(3):461-465 (1983).

Eichelberger, J.W. and J.J. Lichtenberg. "Persistence of Pesticides in River Water," *Environ. Sci. Technol.*, 5(6):541-544 (1971).

Eisenhauer, H.R. "The Ozonization of Phenolic Wastes," *J. Water Pollut. Control Fed.*, 40(11):1887-1899 (1968).

Eisenreich, S.J., B.B. Looney, and J.D. Thornton. "Airborne Organic Contaminants in the Great Lakes Ecosystem," *Environ. Sci. Technol.*, 15(1):30-38 (1981).

Eklund, G., J.R. Pedersen, and B. Strömberg. "Formation of Chlorinated Organic Compounds During Combustion of Propane in the Presence of HCl," *Chemosphere*, 16(1):161–166 (1987).

Ellgehausen, H., C. D'Hondt, and R. Fuerer. "Reversed-Phase Chromatography as a General Method for Determining Octan-1-ol/Water Partition Coefficients," *Pestic. Sci.*, 12:219–227 (1981).

Elliott, S. "Effect of Hydrogen Peroxide on the Alkaline Hydrolysis of Carbon Disulfide," *Environ. Sci. Technol.*, 24(2):264–267 (1990).

Ellington, J.J., F.E. Stancil, Jr., and W.D. Payne. "Measurement of Hydrolysis Rate Constants for Evaluation of Hazardous Waste Land Disposal. Volume 1," Office of Research and Development, U.S. EPA Report-600/3-86/043 (1986), 122 p.

Ellington, J.J., F.E. Stancil, Jr., W.D. Payne, and C. Trusty. "Measurement of Hydrolysis Rate Constants for Evaluation of Hazardous Waste Land Disposal. Volume 2. Data on 54 Chemicals," Office of Research and Development, U.S. EPA Report-600/3-87/019 (1987), 152 p.

Ellington, J.J., F.E. Stancil, W.D. Payne, and C.D. Trusty. "Measurement of Hydrolysis Rate Constants for Evaluation of Hazardous Waste Land Disposal. Volume 3," Office of Research and Development, U.S. EPA Report-600/3-88/028 (1988), 23 p.

Elliott, S. "The Solubility of Carbon Disulfide Vapor in Natural Aqueous Systems," *Atmos. Environ.*, 23(9):1977–1980 (1989).

El-Refai, A. and T.L. Hopkins. "Parathion Absorption, Translocation, and Conversion to Paraoxon in Bean Plants," *J. Agric. Food Chem.*, 14(6):588–592 (1966).

Elsner, E., D. Bieniek, W. Klein, and F. Korte. "Verteilung und Umwandlumg von Aldrin, Heptachlor, und Lindan in der Grunalge *Chlorella pyrenoidosa*," *Chemosphere*, 1:247–250 (1972).

Enfield, C.G., G. Bengtsson, and R. Lindqvist. "Influence of Macromolecules on Chemical Transport," *Environ. Sci. Technol.*, 23(10):1278–1286 (1989).

Engelhardt, G., P.R. Wallnöfer, and O. Hutzinger. "The Microbial Metabolism of Di-*n*-butyl Phthalate and Related Dialkyl Phthalates," *Bull. Environ. Contam. Toxicol.*, 13(3):342–347 (1975).

Eon, C., C. Pommier, and G. Guiochon. "Vapor Pressures and Second Virial Coefficients of Some Five-Membered Heterocyclic Derivatives," *J. Chem. Eng. Data*, 16(4):408–410 (1971).

Epling, G.A., W.M. McVicar, and A. Kumar. "Borohydride-Enhanced Photodehalogenation of Aroclor 1232, 1242, 1254, and 1260," *Chemosphere*, 17(7):1355–1362 (1988).

Exxon. 1974. Tables of Useful Information (Houston, TX: Exxon Corp.), 70 p.

Fairbanks, B.C. and G.A. O'Connor. "Effect of Sewage Sludge on the Adsorption of Polychlorinated Biphenyls by Three New Mexico Soils," *J. Environ. Qual.*, 13(2):297–300 (1984).

Fairbanks, B.C., N.E. Schmidt, and G.A. O'Connor. "Butanol Degradation and Volatilization in Soils Amended with Spent Acid or Sulfuric Acid," *J. Environ.*

Qual., 14(1):83-86 (1985).

Farm Chemicals Handbook (Willoughby, OH: Meister Publishing Co., 1988).

Farmer, W.J., M.S. Yang, J. Letey, and W.F. Spencer. "Hexachlorobenzene: Its Vapor Pressure and Vapor Phase Diffusion in Soil," *Soil Sci. Soc. Am. J.*, 44:676-680 (1980).

Fathepure, B.Z., J.M. Tiedje, and S.A. Boyd. "Reductive Dechlorination of Hexachlorobenzene to Tri- and Dichlorobenzenes in Anaerobic Sewage Sludge," *Appl. Environ. Microbiol.*, 54(2):327-330 (1988).

Fatiadi, A.J. "Effects of Temperature and of Ultraviolet Radiation on Pyrene Adsorbed on Garden Soil," *Environ. Sci. Technol.*, 1(7):570-572 (1967).

Felsot, A.J., J.V. Maddox, and W. Bruce. "Dissipation of Pesticide from Soils with Histories of Furadan Use," *Bull. Environ. Contam. Toxicol.*, 26(6):781-788 (1981).

Felsot, A. and P.A. Dahm. "Sorption of Organophosphorus and Carbamate Insecticides by Soil," *J. Agric. Food Chem.*, 27(3):557-563 (1979).

Fendinger, N.J. and D.W. Glotfelty. "A Laboratory Method for the Experimental Determination of Air-Water Henry's Law Constants for Several Pesticides," *Environ. Sci. Technol.*, 22(11):1289-1293 (1988).

Fendinger, N.J., D.W. Glotfelty, and H.P. Freeman. "Comparison of Two Experimental Techniques for Determining Air/Water Henry's Law Constants," *Environ. Sci. Technol.*, 23(12):1528-1531 (1989).

Fendinger, N.J. and D.W. Glotfelty. "Henry's Law Constants for Selected Pesticides, PAHs and PCBs," *Environ. Toxicol. Chem.*, 9(6):731-735 (1990).

Ferris, I.G. and E.P. Lichtenstein. "Interactions between Agricultural Chemicals and Soil Microflora and Their Effects on the Degradation of [^{14}C]Parathion in a Cranberry Soil," *J. Agric. Food Chem.*, 28(5):1011-1019 (1980).

Feung, C.-S., R.H. Hamilton, and R.O. Mumma. "Metabolism of 2,4-Dichlorophenoxyacetic Acid. V. Identification of Metabolites in Soybean Callus Tissue Cultures," *J. Agric. Food Chem.*, 21(4):637-640 (1973).

Feung, C.S, R.H. Hamilton, and F.H. Witham. "Metabolism of 2,4-Dichlorophenoxyacetic Acid by Soybean Cotyledon Callus Tissue Cultures," *J. Agric. Food Chem.*, 19(3):475-479 (1971).

Feung, C.S., R.H. Hamilton, F.H. Witham, and R.O. Mumma. "The Relative Amounts and Identification of Some 2,4-Dichlorophenoxyacetic Acid Metabolites Isolated from Soybean Cotyledon Callus Cultures," *Plant Physiol.*, 50:80-86 (1972).

Fewson, C.A. "Biodegradation of Aromatics with Industrial Relevance," in *FEMS Symp. No. 12. Microbial Degradation of Xenobiotics and Recalcitrant Compounds*, 12:141-179 (1981).

Fischer, I. and L. Ehrenberg. "Studies of the Hydrogen Bond. II. Influence of the Polarizability of the Heteroatom," *Acta Chem. Scand.*, 2:669-677 (1948).

Fishbein, L. "Potential Halogenated Industrial Carcinogenic and Mutagenic Chemicals. III. Alkane Halides, Alkanols, and Ethers," *Sci. Tot. Environ.*, 11:223-

257 (1979).

Fishbein, L. "An Overview of Environmental and Toxicological Aspects of Aromatic Hydrocarbons. III. Xylene," *Sci. Tot. Environ.*, 43(1/2):165-183 (1985).

Fishbein, L. and P.W. Albro. "Chromatographic and Biological Aspects of the Phthalate Esters," *J. Chromatogr.*, 70(2):365-412 (1972).

Flathman, P.E. and J.R. Dahlgran. "Correspondence On: Anaerobic Degradation of Halogenated 1- and 2-Carbon Organic Compounds," *Environ. Sci. Technol.*, 16:130 (1982).

Fleeker, J. and R. Steen. "Hydroxylation of 2,4-D in Several Weed Species," *Weed Sci.*, 19:507 (1971).

Fluka. 1988. Fluka Catalog 1988/89 - Chemika-Biochemika (Ronkonkoma, NY: Fluka Chemical Corp.), 1536 p.

Fluka. 1990. Fluka Catalog 1989/90 - Chemika-Biochemika (Ronkonkoma, NY: Fluka Chemical Corp.), 1480 p.

Fordyce, C.R. and L.W.A. Meyer. "Plasticizers for Cellulose Acetate and Cellulose Acetate Butyrate," *Ind. Eng. Chem.*, 32(8):1053-1060 (1940).

Fordyce, R.G. and E.C. Chapin. "Copolymerization. I. The Mechanism of Emulsion Copolymerization of Styrene and Acrylonitrile," *J. Am. Chem. Soc.*, 69(3):581-583 (1947).

Foreman, W.T. and T.F. Bidleman. "Vapor Pressure Estimates of Individual Polychlorinated Biphenyls and Commercial Fluids Using Gas Chromatographic Retention Data," *J. Chromatogr.*, 330(2):203-216 (1985).

Forziati, A.F., W.R. Norris, and F.D. Rossini. "Vapor Pressures and Boiling Points of Sixty API-NBS Hydrocarbons," *J. Res. Nat. Bur. Stand.*, 43:555-563 (1949).

Fowler, L., W.N. Trump, and C.E. Vogler. "Vapor Pressure of Naphthalene - New Measurements between 40° and 180 °C," *J. Chem. Eng. Data*, 13(2):209-210 (1968).

Francis, A.W. and G.W. Robbins. "The Vapor Pressures of Propane and Propylene," *J. Am. Chem. Soc.*, 55(11):4339-4342 (1933).

Frank, R., H.E. Braun, M. Van Hove Holdrimet, G.J. Sirons, and B.D. Ripley. "Agriculture and Water Quality in the Canadian Great Lakes Basin: V. Pesticide Use in 11 Agricultural Watersheds and Presence in Stream Water, 1975-1977," *J. Environ. Qual.*, 11(3):497-505 (1982).

Frankel, L.S., K.S. McCallum, and L. Collier. "Formation of Bis(chloromethyl) Ether from Formaldehyde and Hydrogen Chloride," *Environ. Sci. Technol.*, 8(4):356-359 (1974).

Franklin, J.L., J.G. Dillard, H.M. Rosenstock, J.T. Herron, K. Draxl, and F.H. Field. "Ionization Potentials, Appearance Potentials and Heats of Formation of Gaseous Positive Ions," National Bureau of Standards Report NSRDS-NBS 26, U.S. Government Printing Office (1969), 289 p.

Franks, F. "Solute-Water Interactions and the Solubility Behaviour of Long-chain Paraffin Hydrocarbons," *Nature (London)*, 210(5031):87-88 (1966).

Franks, F., M. Gent, and H.H. Johnson. "The Solubility of Benzene in Water," *J.*

Chem. Soc. (London), pp. 2716-2723 (1963).

Freed, V.H., C.T. Chiou, and D.W. Schmedding. "Degradation of Selected Organo-phosphate Pesticides in Water and Soil," *J. Agric. Food Chem.*, 27(4):706-708 (1979).

Freed, V.H., C.T. Chiou, and R. Haque. "Chemodynamics: Transport and Behavior of Chemicals in the Environment - A Problem in Environmental Health," *Environ. Health Perspect.*, 20:55-70 (1977).

Freed, V.H., D. Schmedding, R. Kohnert, and R. Haque. "Physical Chemical Properties of Several Organophosphates: Some Implication in Environmental and Biological Behavior," *Pestic. Biochem. Physiol.*, 10:203-211 (1979a).

Freitag, D., I. Scheunert, W. Klein, and F. Korte. "Long-Term Fate of 4-Chloro-aniline-^{14}C in Soil and Plants under Outdoor Conditions. A Contribution to Terrestrial Ecotoxicology of Chemicals," *J. Agric. Food Chem.*, 32(2):203-207 (1984).

Freitag, D., L. Ballhorn, H. Geyer, and F. Korte. "Environmental Hazard Profile of Organic Chemicals," *Chemosphere*, 14(10):1589-1616 (1985).

Fries, G.F. "Degradation of Chlorinated Hydrocarbons under Anaerobic Conditions," in *Fate of Organic Pesticides in the Aquatic Environment, Advances in Chemistry Series*, R.F. Gould, Ed. (Washington, DC: American Chemical Society, 1972), pp. 256-270.

Fries, G.R., G.S. Marrow, and C.H. Gordon. "Metabolism of *o,p'*- and *p,p'*-DDT by Rumen Microorganisms," *J. Agric. Food Chem.*, 17(4):860-862 (1969).

Friesen, K.J., L.P. Sarna, and G.R.B. Webster. "Aqueous Solubility of Poly-chlorinated Dibenzo-*p*-Dioxins Determined by High Pressure Liquid Chromatography," *Chemosphere*, 14(9):1267-1274 (1985).

Fu, J.-K. and R.G. Luthy. "Aromatic Compound Solubility in Solvent/Water Mixtures," *J. Environ. Eng.*, 112(2):328-345 (1986).

Fu, M.H. and M. Alexander. "Biodegradation of Styrene in Samples of Natural Environments," *Environ. Sci. Technol.*, 26(8):1540-1544 (1992).

Fühner, H. "Water-Solubility in Homologous Series," *Ber. Dtsch. Chem. Ges.*, 57B:510-515 (1924).

Fuhremann, T.W. and E.P. Lichtenstein. "A Comparative Study of the Persistence, Movement and Metabolism of Six C^{14}-Insecticides in Soils and Plants," *J. Agric. Food Chem.*, 28:446-452 (1980).

Fujita, T., J. Iwasa, and C. Hansch. "A New Substituent Constant, π, Derived from Partition Coefficients," *J. Am. Chem. Soc.*, 86(23):5175-5180 (1964).

Fukano, I. and Y. Obata. "Solubility of Phthalates in Water," [Chemical abstract 120601u, 86(17):486 (1977)]: *Purasuchikkusu*, 27(7):48-49 (1976).

Fukuda, K., I. Yasushi, T. Maruyama, H.I. Kojima, and T. Yoshida. "On the Photolysis of Alkylated Naphthalenes in Aquatic Systems," *Chemosphere*, 17(4):651-659 (1988).

Fukui, S., T. Hirayama, H. Shindo, and M. Nohara. "Photochemical Reaction of Biphenyl (BP) and *o*-Phenylphenol (OPP) with Nitrogen Monoxide (1),"

Chemosphere, 9(12):771-775 (1980).

Fuller, B.B. *Air Pollution Assessment of Tetrachloroethylene* (McLean, VA: The Mitre Corp., 1976), 87 p.

Funaski, N., S. Hada, and S. Neya. "Partition Coefficients of Aliphatic Ethers-Molecular Surface Area Approach," *J. Phys. Chem.*, 89:3046-3049 (1985).

Fung, H.-L. and T. Higuchi. "Molecular Interactions and Solubility of Polar Nonelectrolytes in Nonpolar Solvents," *J. Pharm Sci.*, 60(12):1782-1788 (1971).

Gäb, S., H. Parlar, S. Nitz, K. Hustert, and F. Korte. "Beitrage zur Okologischen Chemie. LXXXI. Photochemischer Abbau von Aldrin, Dieldrin and Photodieldrin als Festkorper im Sauer-stoffstrom," *Chemosphere*, 3(5):183-186 (1974).

Gaffney, J.S., G.E. Streit, W.D. Spall, and J.H. Hall. "Beyond Acid Rain," *Environ. Sci. Technol.*, 21(6):519-524 (1987).

Galassi, S., M. Mingazzini, L. Viganò, D. Cesareo, and M.L. Tosato. "Approaches to Modeling Toxic Responses of Aquatic Organisms to Aromatic Hydrocarbons," *Ecotoxicol. Environ. Saf.*, 16(2):158-169 (1988).

Gälli, R. and P.L. McCarty. "Biotransformation of 1,1,1-Trichloroethane, Trichloromethane, and Tetrachloromethane by a *Clostridium* sp.," *Appl. Environ. Microbiol.*, 55(4):837-844 (1989).

Gambrell, R.P., B.A. Taylor, K.S. Reddy, and W.H. Patrick, Jr. "Fate of Selected Toxic Compounds under Controlled Redox Potential and pH Conditions in Soil and Sediment-Water Systems," U.S. EPA Report-600/3-83-018 (1984).

Gannon, N. and G.C. Decker. "The Conversion of Heptachlor to its Epoxide on Plants," *J. Econ. Entomol.*, 51(1):3-7 (1958).

Garbarini, D.R. and L.W. Lion. "Influence of the Nature of Soil Organics on the Sorption of Toluene and Trichloroethylene," *Environ. Sci. Technol.*, 20(12):1263-1269 (1986).

Garraway, J. and J. Donovan. "Gas Phase Reaction of Hydroxyl Radical with Alkyl Iodides," *J. Chem. Soc. Chem. Comm.*, 23:1108 (1979).

Garst, J.E. and W.C. Wilson. "Accurate, Wide-Range, Automated, High-Performance Liquid Chromatographic Method for the Estimation of Octanol/Water Partition Coefficients I: Effects of Chromatographic Conditions and Procedure Variables on Accuracy and Reproducibility of the Method," *J. Pharm. Sci.*, 73(11):1616-1622 (1984).

Garten, C.T. and J.R. Trabalka. "Evaluation of Models for Predicting Terrestrial Food Chain Behavior of Xenobiotics," *Environ. Sci. Technol.*, 17(10):590-595 (1983).

Gauthier, T.D., E.C. Shane, W.F. Guerin, W.R. Seitz, and C.L. Grant. "Fluorescence Quenching Method for Determining Equilibrium Constants for Polycyclic Aromatic Hydrocarbons Binding to Humic Materials," *Environ. Sci. Technol.*, 20(11):1162-1166 (1986).

Gauthier, T.D., W.R. Selfz, and C.L. Grant. "Effects of Structural and Compositional Variations of Dissolved Humic Materials on K_{oc} Values," *Environ. Sci. Technol.*, 21(5):243-248 (1987).

Gay, B.W., Jr., P.L. Hanst, J.J. Bufalini, and R.C. Noonan. "Atmospheric Oxidation of Chlorinated Ethylenes," *Environ. Sci. Technol.*, 10(1):58-67 (1976).

Georgacakis, E. and M.A.Q. Khan. "Toxicity of the Photoisomers of Cyclodiene Insecticides to Freshwater Animals," *Nature (London)*, 233(5315):120-121 (1971).

Gerstl, Z. and C.S. Helling. "Evaluation of Molecular Connectivity as a Predictive Method for the Adsorption of Pesticides in Soils," *J. Environ. Sci. Health*, B22(1):55-69 (1987).

Gerstl, Z. and U. Mingelgrin. "Sorption of Organic Substances by Soils and Sediments," *J. Environ. Sci. Health*, B19(3):297-312 (1984).

Getzin, L.W. and I. Rosefield. "Organophosphorus Insecticide Degradation by Heat-Labile Substances in Soil," *J. Agric. Food Chem.*, 16(4):598-601 (1968).

Geyer, H., A.S. Kraus, W. Klein, E. Richter, and F. Korte. "Relationship between Water Solubility and Bioaccumulation Potential of Organic Chemicals in Rats," *Chemosphere*, 9(5/6):277-294 (1980).

Geyer, H., G. Politzki, and D. Freitag. "Prediction of Ecotoxicological Behaviour of Chemicals: Relationship between *n*-Octanol/Water Partition Coefficient and Bioaccumulation of Organic Chemicals by Alga *Chlorella*," *Chemosphere*, 13(2):269-284 (1984).

Geyer, H., P. Sheehan, D. Kotzias, D. Freitag, and F. Korte. "Prediction of Ecotoxicological Behaviour of Chemicals: Relationship between Physico-Chemical Properties and Bioaccumulation of Organic Chemicals in the Mussell *Mytilus edulis*," *Chemosphere*, 11(11):1121-1134 (1982).

Geyer, H.J., I. Scheunert, and F. Korte. "Correlation between the Bioconcentration Potential of Organic Environmental Chemicals in Humans and Their *n*-Octanol/Water Partition Coefficients," *Chemosphere*, 16(1):239-252 (1987).

Gherini, S.A., K.V. Summers, R.K. Munson, and W.B. Mills. *Chemical Data for Predicting the Fate of Organic Compounds in Water, Volume 2: Database*, (Lafayette, CA: Tetra Tech, Inc., 1988), 433 p.

Giam, C.S., E. Atlas, H.S. Chan, and G.S. Neff. "Phthalate Esters, PCB and DDT Residues in the Gulf of Mexico Atmosphere," *Atmos. Environ.*, 14(1):65-69 (1980).

Gibbard, H.F. and J.L. Creek. "Vapor Pressure of Methanol from 288.15 to 337.65K," *J. Chem. Eng. Data*, 19(4):308-310 (1974).

Gibson, D.T. "Microbial Degradation of Aromatic Compounds," *Science (Washington, DC)*, 161(3846):1093-1097 (1968).

Gibson, D.T., V. Mahadevan, D.M. Jerina, H. Yagi, and H.J.C. Yeh. "Oxidation of the Carcinogens Benzo[*a*]pyrene and Benzo[*a*]anthracene to Dihydrols by a Bacterium," *Science (Washington, DC)*, 189:295-297 (1975).

Gibson, H.W. "Linear Free Energy Relationships. VIII. Ionization Potentials of Aliphatic Compounds," *Can. J. Chem.*, 55(14):2637-2641 (1977).

Gibson, W.P. and R.G. Burns. "The Breakdown of Malathion in Soil and Soil Components," *Microbial Ecol.*, 3:219-230 (1977).

Gile, J.D. and J.W. Gillett. "Fate of Selected Fungicides in a Terrestrial Laboratory

Ecosystem," *J. Agric. Food Chem.*, 27(6):1159-1164 (1979).

Ginnings, P.M., D. Plonk, and E. Carter. "Aqueous Solubilities of Some Aliphatic Ketones," *J. Am. Chem. Soc.*, 62(8):1923-1924 (1940).

Girifalco, L.A. and R.J. Good. "A Theory for the Estimation of Surface and Interfacial Energies: I. Derivation and Application to Interfacial Tension," *J. Phys. Chem.*, 61(7):904-909 (1957).

Gladis, G.P. "Effects of Moisture on Corrosion in Petrochemical Environments," *Chem. Eng. Prog.*, 56(10):43-51 (1960).

Gledhill, W.E., R.G. Kaley, W.J. Adams, O. Hicks, P.R. Michael, and V.W. Saeger. "An Environmental Safety Assessment of Butyl Benzyl Phthalate," *Environ. Sci. Technol.*, 14(3):301-305 (1980).

Glew, D.N. and E.A. Moelyn-Hughes. "Chemical Studies of the Methyl Halides in Water," *Disc. Faraday Soc.*, 15:150-161 (1953).

Glew, D.N. and R.E. Robertson. "The Spectrophotometric Determination of the Solubility of Cumene in Water by a Kinetic Method," *J. Phys. Chem.*, 60(3):332-337 (1956).

Gobas, F.A.P.C., J.M. Lahittete, G. Garofalo, W.Y. Shiu, and D. Mackay. "A Novel Method for Measuring Membrane-Water Partition Coefficients of Hydrophobic Organic Compounds: Comparison with 1-Octanol-Water Partitioning," *J. Pharm. Sci.*, 77(3):265-272 (1988).

Godsy, E.M., D.F. Goerlitz, and G.G. Ehrlich. "Methanogenesis of Phenolic Compounds by a Bacterial Consortium from a Contaminated Aquifer in St. Louis Park, Minnesota," *Bull. Environ. Contam. Toxicol.*, 30(3):261-268 (1983).

Golab, T., W.A. Althaus, and H.L. Wooten. "Fate of [^{14}]Trifluralin in Soil," *J. Agric. Food Chem.*, 27(1):163-179 (1979).

Golab, T., R.J. Herberg, S.J. Parka, and J.B. Tepe. "Metabolism of Carbon-14 Trifluralin in Carrots," *J. Agric. Food Chem.*, 15(4):638-641 (1967).

González, V., J.H. Ayala, and A.M. Afonso. "Degradation of Carbaryl in Natural Waters: Enhanced Hydrolysis Rate in Micellar Solution," *Bull. Environ. Contam. Toxicol.*, 42(2):171-178 (1992).

Gooch, J.A. and M.K. Hamdy. "Uptake and Concentration Factor of Arochlor 1254 in Aquatic Organisms," *Bull. Environ. Contam. Toxicol.*, 31:445-452 (1983).

Gorder, G.W., P.A. Dahm, and J.J. Tollefson. "Carbofuran Persistence in Cornfield Soils," *J. Econ. Entomol.*, 75(4):637-642 (1982).

Gordon, A.J. and R.A. Ford. *The Chemist's Companion* (New York: John Wiley & Sons, Inc., 1972), 551 p.

Gordon, J.E. and R.L. Thorne. "Salt Effects on the Activity Coefficient of Naphthalene in Mixed Aqueous Electrolyte Solutions. I. Mixtures of Two Salts," *J. Phys. Chem.*, 71(13):4390-4399 (1967).

Gossett, J.M. "Measurement of Henry's Law Constants for C_1 and C_2 Chlorinated Hydrocarbons," *Environ. Sci. Technol.*, 21(2):202-208 (1987).

Goursot, P., H.L. Girdhar, and E.F. Westrum, Jr. "Thermodynamics of Polynuclear Aromatic Molecules. III. Heat Capacities and Enthalpies of Fusion of

Anthracene," *J. Phys. Chem.*, 74:2538-2541 (1970).

Gowda, T.K.S. and N. Sethunathan. "Persistence of Endrin in Indian Rice Soils under Flooded Conditions," *J. Agric. Food Chem.*, 24(4):750-753 (1976).

Grant, B.F. "Endrin Toxicity and Distribution in Freshwater: A Review," *Bull. Environ. Contam. Toxicol.*, 15(3):283-290 (1976).

Graveel, J.G., L.E. Sommers, and D.W. Nelson. "Decomposition of Benzidine, α-Naphthylamine, and *p*-Toluidine in Soils," *J. Environ. Qual.*, 15(1):53-59 (1986).

Grayson, B.T. and L.A. Fosbraey. "Determination of Vapor Pressures of Pesticides," *Pestic. Sci.*, 13:269-278 (1982).

Green, W.J., G.F. Lee, R.A. Jones, and T. Palit. "Interaction of Clay Soils with Water and Organic Solvents: Implications for the Disposal of Hazardous Wastes," *Environ. Sci. Technol.*, 17(5):278-282 (1983).

Greene, S., M. Alexander, and D. Leggett. "Formation of *N*-Nitrosodimethylamine During Treatment of Municipal Waste Water by Simulated Land Application," *J. Environ. Qual.*, 10(3):416-421 (1981).

Greenhalgh, R. and A. Belanger. "Persistence and Uptake of Carbofuran in a Humic Mesisol and the Effects of Drying and Storing Soil Samples on Residue Levels," *J. Agric. Food Chem.*, 29:231-235 (1981).

Grimes, D.J. and S.M. Morrison. "Bacterial Bioconcentration of Chlorinated Hydrocarbon Insecticides from Aquatic Systems," *Microbial Ecol.*, 2:43-59 (1975).

Grisak, G.E., J.F. Pickens, and J.A. Cherry. "Solute Transport Through Fractured Media. 2. Column Study of Fractured Till," *Water Resourc. Res.*, 16(4):731-739 (1980).

Groenewegen, D. and H. Stolp. "Microbial Breakdown of Polycyclic Aromatic Hydrocarbons," *Zentralbl. Bakteriol. Parasitenkd. Infekitionskr. Hyg. Abt., Abt. 1:Orig., Reihe, B.*, 162:225-232 (1976).

Grosjean, D. "Atmospheric Reactions of Ortho Cresol: Gas Phase and Aerosol Products," *Atmos. Environ.*, 19(8):1641-1652 (1984).

Grosjean, D. "Atmospheric Reactions of Styrenes and Peroxybenzoyl Nitrate," *Sci. Tot. Environ.*, 50:41-59 (1985).

Grosjean, D. "Photooxidation of Methyl Sulfide, Ethyl Sulfide, and Methanethiol," *Environ. Sci. Technol.*, 18(6):460-468 (1984).

Gross, F.C. and J.A. Colony. "The Ubiquitous Nature and Objectionable Characteristics of Phthalate Esters in Aerospace Industry," *Environ. Health Perspect.* (January 1973), pp. 37-48.

Gross, P. "The Determination of the Solubility of Slightly Soluble Liquids in Water and the Solubilities of the Dichloro- Ethanes and Propanes," *J. Am. Chem. Soc.*, 51(8):2362-2366 (1929).

Gross, P.M. and J.H. Saylor. "The Solubilities of Certain Slightly Soluble Organic Compounds in Water," *J. Am. Chem. Soc.*, 53(5):1744-1751 (1931).

Gross, P.M., J.H. Saylor, and M.A. Gorman. "Solubility Studies. IV. The Solubilities of Certain Slightly Soluble Organic Compounds in Water," *J. Am. Chem. Soc.*, 55(2):650-652 (1933).

Groves, F., Jr. "Solubility of Cycloparaffins in Distilled Water and Salt Water," *J. Chem. Eng. Data*, 33(2):136-138 (1988).

Grunwell, J.R. and R.H. Erickson. "Photolysis of Parathion (*O,O*-Diethyl-*O*-(4-nitrophenyl)thio-phosphate). New Products," *J. Agric. Food Chem.*, 21(5):929-931 (1973).

Guenzi, W.D. and W.E. Beard. "Anaerobic Biodegradation of DDT to DDD in Soil," *Science (Washington, DC)*, 156(3778):1116-1117 (1967).

Gunther, F.A., W.E. Westlake, and P.S. Jaglan. "Reported Solubilities of 738 Pesticide Chemicals in Water," *Residue Rev.*, 20:1-148 (1968).

Guseva, A.N. and E.I. Parnov. "The Solubility of Aromatic Hydrocarbons in Water," *Vestn. Mosk. Univ., Ser. II Khim.*, 18(1):76-79 (1963).

Guseva, A.N. and E.I. Parnov. "Solubility of Cyclohexane in Water," *Zh. Fiz. Khim.*, 37(12):2763 (1963a).

Guseva, A.N. and E.I. Parnov. "Isothermal Cross-Sections of the Systems Cyclanes-Water," *Vestn. Mosk. Univ., Ser. II Khim.*, 19:77 (1964).

Haderlein. S.B. and R.P. Schwarzenbach. "Adsorption of Substituted Nitrobenzenes and Nitrophenols to Mineral Surfaces," *Environ. Sci. Technol.*, 27(2):316-326 (1993).

Hafkenscheid, T.L. and E. Tomlinson. "Estimation of Aqueous Solubilities of Organic Non-Electrolytes Using Liquid Chromatographic Retention Data," *J. Chromatogr.*, 218:409-425 (1981).

Hafkenscheid, T.L. and E. Tomlinson. "Isocratic Chromatographic Retention Data for Estimating Aqueous Solubilities of Acidic, Basic and Neutral Drugs," *Int. J. Pharm.*, 17:1-21 (1983).

Hageman, W.J.M., C.H. Van der Weijden, and J.P.G. Loch. "Sorption of Benzo[*a*]pyrene and Phenanthrene on Suspended Harbor Sediment as a Function of Suspended Sediment Concentration and Salinity: A Laboratory Study Using the Cosolvent Partition Coefficient," *Environ. Sci. Technol.*, 29(2):363-371 (1995).

Hagin, R.D., D.L. Linscott, and J.E. Dawson. "2,4-D Metabolism in Resistant Grasses," *J. Agric. Food Chem.*, 18:848-850 (1970).

Haider, K., G. Jagnow, R. Kohnen, and S.U. Lim. "Degradation of Chlorinated Benzenes, Phenols, and Cyclohexane Derivatives by Benzene- and Phenol-Utilizing Bacteria under Aerobic Conditions," in *Decomposition of Toxic and Nontoxic Organic Compounds in Soil*, V.R. Overcash, Ed. (Ann Arbor, MI: Ann Arbor Science Publishers, 1981), pp. 207-223.

Haight, G.P., Jr. "Solubility of Methyl Bromide in Water and in Some Fruit Juices," *Ind. Eng. Chem.*, 43(8):1827-1828 (1951).

Haigler, B.E., S.F. Nishino, and J.C. Spain. "Degradation of 1,2-Dichlorobenzene by a *Pseudomonas* sp.," *Appl. Environ. Microbiol.*, 54(2):294-301 (1988).

Hakuta, T., A. Negishi, T. Goto, J. Kato, and S. Ishizaka. "Vapor-Liquid Equilibria of Some Pollutants in Aqueous and Saline Solutions," *Desalination*, 21(1):11-21 (1977).

Hallas, L.E. and M. Alexander. "Microbial Transformation of Nitroaromatic Compounds in Sewage Effluent," *Appl. Environ. Microbiol.*, 45(4):1234-1241 (1983).

Hammers, W.E., G.J. Meurs, and C.L. de Ligny. "Correlations between Liquid Chromatographic Capacity Ratio Data on Lichrosorb RP-18 and Partition Coefficients in the Octanol-Water System," *J. Chromatogr.*, 247:1-13 (1982).

Hansch, C. and A. Leo. *Substituent Constants for Correlation Analysis in Chemistry and Biology* (New York: John Wiley & Sons, Inc., 1979), 339 p.

Hansch, C., A. Leo, and D. Nikaitani. "On the Additive-Constitutive Character of Partition Coefficients," *J. Org. Chem.*, 37(20):3090-3092 (1972).

Hansch, C., A. Vittoria, C. Silipo, and P.Y.C. Jow. "Partition Coefficients and the Structure-Activity Relationship of the Anesthetic Gases," *J. Med. Chem.*, 18(6):546-548 (1975).

Hansch, C., J.E. Quinlan, and G.L. Lawrence. "The Linear Free-Energy Relationship between Partition Coefficients and Aqueous Solubility of Organic Liquids," *J. Org. Chem.*, 33(1):347-350 (1968).

Hansch, C. and S.M. Anderson. "The Effect of Intramolecular Hydrophobic Bonding on Partition Coefficients," *J. Org. Chem.*, 32(8):2583-2586 (1967).

Hansch, C. and T. Fujita. "p-σ-π Analysis. A Method for the Correlation of Biological Activity and Chemical Structure," *J. Am. Chem. Soc.*, 86(8):1616-1617 (1964).

Hanst, P.L. and B.W. Gay, Jr. "Atmospheric Oxidation of Hydrocarbons: Formation of Hydroperoxides and Peroxyacids," *Atmos. Environ.*, 17(11):2259-2265 (1983).

Hanst, P.L. and B.W. Gay, Jr. "Photochemical Reactions Among Formaldehyde, Chlorine, and Nitrogen Dioxide in Air," *Environ. Sci. Technol.*, 11(12):1105-1109 (1977).

Hanst, P.L., J.W. Spence, and M. Miller. "Atmospheric Chemistry of N-Nitrosodimethylamine," *Environ. Sci. Technol.*, 11(4):403-405 (1977).

Haque, R., D.W. Schmedding, and V.H. Freed. "Aqueous Solubility, Adsorption, and Vapor Behavior of Polychlorinated Biphenyl Aroclor 1254," *Environ. Sci. Technol.*, 8(2):139-142 (1974).

Harbison, K.G. and R.T. Belly. "The Biodegradation of Hydroquinone," *Environ. Toxicol. Chem.*, 1(1):9-15 (1982).

Harkins, W.D., G.L. Clark, and L.E. Roberts. "The Orientation of Molecules in Surfaces, Surface Energy, Adsorption, and Surface Catalysis. V. The Adhesional Work between Organic Liquids and Water," *J. Am. Chem. Soc.*, 42(4):700-713 (1920).

Harris, C.I., R.A. Chapman, C. Harris, and C.M. Tu. "Biodegradation of Pesticides in Soil: Rapid Induction of Carbamate Degrading Factors After Carbofuran Treatment," *J. Environ. Sci. Health*, B19:1-11 (1984).

Harrison, R.M., R. Perry, and R.A. Wellings. "Polynuclear Aromatic Hydrocarbons in Raw, Potable and Waste Waters," *Water Res.*, 9(4):331-346 (1975).

Hartley D. and H. Kidd, Eds. *The Agrochemicals Handbook*, 2nd ed. (England: Royal Society of Chemistry, 1987).

Hashimoto, Y., K. Tokura, H. Kishi, and W.M.J. Strachan. "Prediction of Seawater Solubility of Aromatic Compounds," *Chemosphere*, 13(8):881-888 (1984).

Hassett, J.J., J.C. Means, W.L. Banwart, and S.G. Wood. "Sorption Properties of Sediments and Energy-Related Pollutants," Office of Research and Development, U.S. EPA Report-600/3-80-041 (1980), 150 p.

Hastings, S.H. and F.A. Matsen. "The Photodecomposition of Nitrobenzene," *J. Am. Chem. Soc.*, 70(10):3514-3515 (1948).

Hatakeyama, S., M. Ohno, J. Weng, H. Takagi, and H. Akimoto. "Mechanism for the Formation of Gaseous and Particulate Products from Ozone-Cycloalkene Reactions in Air," *Environ. Sci. Technol.*, 21(1):52-57 (1987).

Hatch, L.F. and L.E. Kidwell, Jr. "Preparation and Properties of 1-Bromo-1-propyne, 1,3-Dibromopropyne and 1-Bromo-3-chloro-1-propyne," *J. Am. Chem. Soc.*, 76(1):289-290 (1954).

Hattula, M.L., V.-M. Wasenius, H. Reunanen, and A.U. Arstila. "Acute Toxicity of Some Chlorinated Phenols, Catechols and Cresols in Trout," *Bull. Environ. Contam. Toxicol.*, 26:295-298 (1981).

Hawari, J., A. Demeter, and R. Samson. "Sensitized Photolysis of Polychlorobiphenyls in Alkaline 2-Propanol: Dechlorination of Aroclor 1254 in Soil Samples by Solar Radiation," *Environ. Sci. Technol.*, 26(10):2022-2027 (1992).

Hawari, J., J. Tronczynski, A. Demeter, R. Samson, and D. Mourato. "Photodechlorination of Chlorobiphenyls by Sodium Methyl Siliconate," *Chemosphere*, 22(1/2):189-199 (1991).

Hawker, D.W. and D.W. Connell. "Influence of Partition Coefficient of Lipophilic Compounds on Bioconcentration Kinetics with Fish," *Water Res.*, 22(6):701-707 (1988).

Hawley, G.G. *The Condensed Chemical Dictionary* (New York: Van Nostrand Reinhold Co., 1981), 1135 p.

Hawthorne, S.B., R.E. Sievers, and R.M. Barkley. "Organic Emissions from Shale Oil Wastewaters and Their Implications for Air Quality," *Environ. Sci. Technol.*, 19(10):992-997 (1985).

Hayduk, W. and B.S. Minhas. "Diffusivity and Solubility of 1,3-Butadiene in Heptane and Other Properties of Heptane-Styrene and Aqueous Potassium Laurate Solutions," *J. Chem. Eng. Data*, 32(3):285-290 (1987).

Hayduk, W., H. Asatani, and Y. Miyano. "Solubilities of Propene, Butane, Isobutane and Isobutene Gases in *n*-Octane, Chlorobenzene and *n*-Butanol Solvents," *Can. J. Chem. Eng.*, 66:466-473 (1988).

Hayduk, W. and H. Laudie. "Prediction of Diffusion Coefficients for Nonelectrolytes in Dilute Aqueous Solution," *Am. Inst. Chem. Eng.*, 20(3):611-615 (1974).

Hayduk, W. and H. Laudie. "Vinyl Chloride Gas Compressibility and Solubility in Water and Aqueous Potassium Laurate Solutions," *J. Chem. Eng. Data*, 19(3):253-257 (1974a).

Hayes, W.J. *Pesticides Studied in Man* (Baltimore, MD: The Williams and Wilkens Co., 1982), pp. 234-247.

Healy, J.B. and L.Y. Young. "Anaerobic Biodegradation of Eleven Aromatic Compounds to Methane," *Appl. Environ. Microbiol.*, 38(1):85–89 (1979).

Heitkamp, M.A., J.P. Freeman, and C.E. Cerniglia. "Naphthalene Biodegradation in Environmental Microcosms: Estimates of Degradation Rates and Characterization of Metabolites," *Appl. Environ. Microbiol.*, 53(1):129–136 (1987).

Heitkamp, M.A., J.P. Freeman, D.W. Miller, and C.E. Cerniglia. "Pyrene Degradation by a *Mycobacterium* sp.: Identification of Ring Oxidation and Ring Fission Products," *Appl. Environ. Microbiol.*, 54(10):2556–2565 (1988).

Henderson, C., Q.H. Pickering, and C.M. Tarzwell. "Relative Toxicity of Ten Chlorinated Hydrocarbon Insecticides to Four Species of Fish," *Trans. Am. Fish Soc.*, 88(1):23–32 (1959).

Henderson, G.L. and D.G. Crosby. "The Photodecomposition of Dieldrin Residues in Water," *Bull. Environ. Contam. Toxicol.*, 3(3):131–134 (1968).

Heritage, A.D. and I.C. MacRae. "Identification of Intermediates Formed During the Degradation of Hexachlorocyclohexanes by *Clostridium sphenoides*," *Appl. Environ. Microbiol.*, 33(6):1295–1297 (1977).

Heritage, A.D. and I.C. MacRae. "Degradation of Lindane by Cell-Free Preparations of *Clostridium sphenoides*," *Appl. Environ. Microbiol.*, 34(2):222–224 (1977a).

Hermann, R.B. "Theory of Hydrophobic Bonding. II. The Correlation of Hydrocarbon Solubility in Water with Solvent Cavity Surface Area," *J. Phys. Chem.*, 76(19):2754–2759 (1972).

Herzel, F. and A.S. Murty. "Do Carrier Solvents Enhance the Water Solubility of Hydrophobic Compounds?," *Bull. Environ. Contam. Toxicol.*, 32(1):53–58 (1984).

Hilarides, R.J., K.A. Gray, J. Guzzetta, N. Cortellucci, and C. Sommer. "Radiolytic Degradation of 2,3,7,8-TCDD in Artificially Contaminated Soils," *Environ. Sci. Technol.*, 28(13):2249–2258 (1994).

Hill, A.E. "The System, Silver Perchlorate-Water-Benzene," *J. Am. Chem. Soc.*, 44(6):1163–1193 (1922).

Hill, A.E. "The Mutual Solubility of Liquids. I. The Mutual Solubility of Ethyl Ether and Water. II. The Solubility of Water in Benzene," *J. Am. Chem. Soc.*, 46(5):1143–1155 (1924).

Hill, A.E. and R. Macy. "Ternary Systems. II. Silver Perchlorate, Aniline and Water," *J. Am. Chem. Soc.*, 46:1132–1143 (1924).

Hill, J., H.P. Kollig, D.F. Paris, N.L. Wolfe, and R.G. Zepp. "Dynamic Behavior of Vinyl Chloride in Aquatic Ecosystems," U.S. EPA Report-600/3-76-001 (1976), 59 p.

Hinckley, D.A., T.F. Bidleman, W.T. Foreman, and J.R. Tuschall. "Determination of Vapor Pressures for Nonpolar and Semipolar Organic Compounds from Gas Chromatographic Retention Data," *J. Chem. Eng. Data*, 35(3):232–237 (1990).

Hine, J., H.W. Haworth, and O.B. Ramsey. "Polar Effects on Rates and Equilibria. VI. The Effect of Solvent on the Transmission of Polar Effects," *J. Am. Chem. Soc.*, 85(10):1473–1475 (1963).

Hine, J. and P.K. Mookerjee. "The Intrinsic Hydrophilic Character of Organic Compounds. Correlations in Terms of Structural Contributions," *J. Org. Chem.*, 40(3):292-298 (1975).

Hine, J. and R.D. Weimar, Jr. "Carbon Basicity," *J. Am. Chem. Soc.*, 87(15):3387-3396 (1965).

Hinga, K.R. and M.E.Q. Pilson. "Persistence of Benz[a]anthracene Degradation Products in an Enclosed Marine Ecosystem," *Environ. Sci. Technol.*, 21(7):648-653 (1987).

Hinga, K.R., M.E.Q. Pilson, R.F. Lee, J.W. Farrington, K. Tjessem, and A.C. Davis. "Biogeochemistry of Benzanthracene in an Enclosed Marine Ecosystem," *Environ. Sci. Technol.*, 14(9):1136-1143 (1980).

Hirsch, M. and O. Hutzinger. "Naturally Occurring Proteins from Pond Water Sensitize Hexachlorobenzene Photolysis," *Environ. Sci. Technol.*, 23(10):1306-1307 (1989).

Hirzy, J.W., W.J. Adams, W.E. Gledhill, and J.P. Mieure. "Phthalate Esters: The Environmental Issues," unpublished seminar document, Monsanto Industrial Chemicals Co. (1978), 54 p.

HNU. 1986. Instruction Manual - Model ISP1 101: Intrinsically Safe Portable Photoionization Analyzer (Newton, MA: HNU Systems, Inc.), 86 p.

Hodgman, C.D., R.C. Weast, R.S. Shankland, and S.M. Selby. *Handbook of Chemistry and Physics* (Cleveland, OH: Chemical Rubber Publishing Co., 1961).

Hodson, J. and N.A. Williams. "The Estimation of the Adsorption Coefficient (K_{oc}) for Soils by High Performance Liquid Chromatography," *Chemosphere*, 19(1):67-77 (1988).

Hogfeldt, E. and B. Bolander. "On the Extraction of Water and Nitric Acid by Aromatic Hydrocarbons," *Arkiv Kemi*, 21(16):161-186 (1963).

Hollifield, H.C. "Rapid Nephelometric Estimate of Water Solubility of Highly Insoluble Organic Chemicals of Environmental Interest," *Bull. Environ. Contam. Toxicol.*, 23(4/5):579-586 (1979).

Hommelen, J.R. "The Elimination of Errors Due to Evaporation of the Solute in the Determination of Surface Tensions," *J. Colloid Sci.*, 34:385-400 (1959).

Horvath, A.L. *Halogenated Hydrocarbons. Solubility-Miscibility with Water* (New York: Marcel Dekker, Inc., 1982), 889 p.

Horvath, R.S. "Microbial Cometabolism of 2,4,5-Trichlorophenoxyacetic Acid," *Bull. Environ. Contam. Toxicol.*, 5(6):537-541 (1970).

Horvath, R.S. "Co-metabolism of Methyl- and Chloro-substituted Catechols by an *Achromobacter* sp. Possessing a New *meta*-Cleavage Oxygenase," *Biochem. J.*, 119:871-876 (1970a).

Hou, C.T. "Microbial Transformation of Important Industrial Hydrocarbons," in *Microbial Transformations of Bioactive Compounds*, J.P. Rosazza, Ed. (Boca Raton, FL: CRC Press, Inc., 1982), pp. 81-107.

Houser, J.J. and B.A. Sibbio. "Liquid-Phase Photolysis of Dioxane," *J. Org. Chem.*, 42(12):2145-2151 (1977).

Howard, P.H. and P.G. Deo. "Degradation of Aryl Phosphates in Aquatic Environments," *Bull. Environ. Contam. Toxicol.*, 22(3):337-344 (1979).

Howard, P.H. *Handbook of Environmental Fate and Exposure Data for Organic Chemicals - Volume I. Large Production and Priority Pollutants* (Chelsea, MI: Lewis Publishers, Inc., 1989), 574 p.

Howard, P.H. *Handbook of Environmental Fate and Exposure Data for Organic Chemicals - Volume II. Solvents* (Chelsea, MI: Lewis Publishers, Inc., 1990), 546 p.

Howard, P.H. *Handbook of Environmental Fate and Exposure Data for Organic Chemicals - Volume III. Pesticides* (Chelsea, MI: Lewis Publishers, Inc., 1989), 684 p.

Howard, P.H. and P.R. Durkin. "Sources of Contamination, Ambient Levels, and Fate of Benzene in the Environment," Office of Toxic Substances, U.S. EPA Report-560/5-75-005 (1974), 73 p.

Howard, P.H., S. Banerjee, and K.H. Robillard. "Measurement of Water Solubilities, Octanol/Water Partition Coefficients and Vapor Pressures of Commercial Phthalate Esters," *Environ. Toxicol. Chem.*, 4:653-661 (1985).

Hoy, K.L. "New Values of the Solubility Parameters from Vapor Pressure Data," *J. Paint Technol.*, 42(541):76-118 (1970).

HSDB. 1989. U.S. Department of Health and Human Services. *Hazardous Substances Data Bank*. National Library of Medicine, Toxnet File (Bethedsa, MD: National Institute of Health).

Hsu, C.C. and J.J. McKetta. "Pressure-Volume-Temperature Properties of Methyl Chloride," *J. Chem. Eng. Data*, 9(1):45-51 (1964).

Humburg, N.E., S.R. Colby, R.G. Lym, E.R. Hill, W.J. McAvoy, L.M. Kitchen, and R. Prasad, Eds. *Herbicide Handbook of the Weed Science Society of America*, 6th ed. (Champaign, IL: Weed Science Society of America, 1989), 301 p.

Hunter-Smith, R.J., P.W. Balls, and P.S. Liss. "Henry's Law Constants and the Air-Sea Exchange of Various Low Molecular Weight Halocarbon Gases," *Tellus*, 35B:170-176 (1983).

Huntress, E.H. and S.P. Mulliken. *Identification of Pure Organic Compounds - Tables of Data on Selected Compounds of Order I* (New York: John Wiley & Sons, Inc., 1941), 691 p.

Hurd, C.D. and F.L. Carnahan. "The Action of Heat on Ethylamine and Benzylamine," *J. Am. Chem. Soc.*, 52(10):4151-414158 (1930).

Hussam, A. and P.W. Carr. "Rapid and Precise Method for the Measurement of Vapor/Liquid Equilibria by Headspace Gas Chromatography," *Anal. Chem.*, 57(4):793-801 (1985).

Hustert, K., D. Kotzias, and F. Korte. "Photokatalytischer Abbau von Chlornitrobenzolen an TiO_2 in Wäßriger Phase," *Chemosphere*, 16(4):809-812 (1987).

Hustert, K., M. Mansour, H. Parlar, and F. Korte. "The EPA Test - A Method to Determine the Photochemical Degradation of Organic Compounds in Aquatic

Systems," *Chemosphere*, 10:995-998 (1981).

Hustert, K. and P.N. Moza. "Photokatalytischer Abbau von Phthalaten an Titandioxid in Wässriger Phase," *Chemosphere*, 17(9):1751-1754 (1988).

Hutzinger, O., S. Safe, and V. Zitko. *The Chemistry of PCB's* (Boca Raton, FL: CRC Press, Inc., 1974), 269 p.

Hwang, H.-M., R.E. Hodson, and R.F. Lee. "Degradation of Phenol and Chlorophenols by Sunlight and Microbes in Estuarine Water," *Environ. Sci. Technol.*, 20(10):1002-1007 (1986).

IARC. 1973. IARC Monographs on the Evaluation of Carcinogenic Risk of Chemicals to Man. Certain Polycyclic Aromatic Hydrocarbons and Heterocyclic Compounds, Volume 3 (Lyon, France: International Agency for Research on Cancer), 271 p.

IARC. 1974. IARC Monographs on the Evaluation of Carcinogenic Risk of Chemicals to Man. Some Organochlorine Pesticides, Volume 5 (Lyon, France: International Agency for Research on Cancer), 241 p.

IARC. 1978. IARC Monographs on the Evaluation of Carcinogenic Risk of Chemicals to Man. Some *N*-Nitroso Compounds Volume 17 (Lyon, France: International Agency for Research on Cancer), 365 p.

IARC. 1979. IARC Monographs on the Evaluation of Carcinogenic Risk of Chemicals to Man. Some Halogenated Hydrocarbons, Volume 20 (Lyon, France: International Agency for Research on Cancer), 609 p.

IARC. 1983. IARC Monographs on the Evaluation of the Carcinogenic Risk of Chemicals to Humans. Polynuclear Aromatic Compounds, Part 1, Chemical, Environmental and Experimental Data, Volume 32 (Lyon, France: International Agency for Research on Cancer), 477 p.

Ibusuki, T. and K. Takeuchi. "Toluene Oxidation on U.V.-Irradiated Titanium Dioxide With and Without O_2, NO_2 or H_2O at Ambient Temperature," *Atmos. Environ.*, 29(9):1711-1715 (1986).

Inokuchi, H., S. Shiba, T. Handa, and H. Akamatsu. "Heats of Sublimation of Condensed Polynuclear Aromatic Hydrocarbons," *Bull. Chem. Soc. Jpn.*, 25:299-302 (1952).

Isaac, R.A. and J.C. Morris. "Transfer of Active Chlorine from Chloramine to Nitrogenous Organic Compounds. 1. Kinetics," *Environ. Sci. Technol.*, 17(12):738-742 (1983).

Isaacson, P.J. and C.R. Frink. "Nonreversible Sorption of Phenolic Compounds by Sediment Fractions: The Role of Sediment Organic Matter," *Environ. Sci. Technol.*, 18(1):43-48 (1984).

Isensee, A.R., E.R. Holden, E.A. Woolson, and G.E. Jones. "Soil Persistence and Aquatic Bioaccumulation Potential of Hexachlorobenzene," *J. Agric. Food Chem.*, 24(6):1210-1214 (1976).

Isensee, A.R. and G.E. Jones. "Distribution of 2,3,7,8-Tetrachlorodibenzo-*p*-dioxin (TCDD) in Aquatic Model Ecosystem," *Environ. Sci. Technol.*, 9(7):668-672 (1975).

Ishikawa, S. and K. Baba. "Reaction of Organic Phosphate Esters with Chlorine in Aqueous Solution," *Bull. Environ. Contam. Toxicol.*, 41(1):143-150 (1988).

Isnard, S. and S. Lambert. "Estimating Bioconcentration Factors from Octanol-Water Partition Coefficient and Aqueous Solubility," *Chemosphere*, 17(1):21-34 (1988).

ITII. 1986. *Toxic and Hazardous Industrial Chemicals Safety Manual for Handling and Disposal with Toxicity and Hazard Data* (Tokyo, Japan: International Technical Information Institute), 700 p.

Ivie, G.W. and J.E. Casida. "Enhancement of Photoalteration of Cyclodiene Insecticide Chemical Residues by Rotenone," *Science (Washington, DC)*, 167(3935):1620-1622 (1970).

Ivie, G.W. and J.E. Casida. "Sensitized Photodecomposition and Photosensitizer Activity of Pesticide Chemicals Exposed to Sunlight of Silica Gel Chromatoplates," *J. Agric. Food Chem.*, 19(3):405-409 (1971).

Ivie, G.W. and J.E. Casida. "Photosensitizers for the Accelerated Degradation of Chlorinated Cyclodienes and Other Insecticide Chemicals Exposed to Sunlight on Bean Leaves," *J. Agric. Food Chem.*, 19(3):410-416 (1971a).

Ivie, G.W., J.R. Knox, S. Khalifa, I. Yamamoto, and J.E. Casida. "Novel Photoproducts of Heptachlor Epoxide, *Trans*-Chlordane, and *Trans*-Nonachlor," *Bull. Environ. Contam. Toxicol.*, 7(6):376-383 (1972).

Iwasaki, M., T. Ibuki, and Y. Takezaki. "Primary Processes of the Photolysis of Ethylenimine at Xe and Kr Resonance Lines," *J. Chem. Phys.*, 59(12):6321-6327 (1973).

Jacobson, S.N. and M. Alexander. "Enhancement of the Microbial Dehalogenation of a Model Chlorinated Compound," *Appl. Environ. Microbiol.*, 42(6):1062-1066 (1981).

Jacoby, W.A., M.R. Nimios, D.M. Blake, R.D. Noble, and C.A. Koval. "Products, Intermediates, Mass Balances, and Reaction Pathways for the Oxidation of Trichloroethylene in Air via Heterogeneous Photocatalysts," *Environ. Sci. Technol.*, 28(9):1661-1668 (1994).

Jafvert, C.T., J.C. Westall, E. Grieder, and R.P. Schwarzenbach. "Distribution of Hydrophobic Ionogenic Organic Compounds between Octanol and Water: Organic Acids," *Environ. Sci. Technol.*, 24(12):1795-1803 (1990).

Jafvert, C.T. and N.L. Wolfe. "Degradation of Selected Halogenated Ethanes in Anoxic Sediment-Water Systems," *Environ. Toxicol. Chem.*, 6(11):827-837 (1987).

James, T.L. and C.A. Wellington. "Thermal Decomposition of β-Propiolactone in the Gas Phase," *J. Am. Chem. Soc.*, 91(27):7743-7746 (1969).

Janini, G.M. and S.A. Attari. "Determination of Partition Coefficient of Polar Organic Solutes in Octanol/Micellar Solutions," *Anal. Chem.*, 55(4):659-661 (1983).

Janssen, D.B., D. Jager, and B. Wilholt. "Degradation of *n*-Haloalkanes and α,ω-Dihaloalkanes by Wild Type and Mutants of *Acinetobacter* sp. Strain GJ70,"

Appl. Environ. Microbiol., 53(3):561-566 (1987).

Jayasinghe, D.S., B.J. Brownawell, H. Chen, and J.C. Westall. "Determination of Henry's Law Constants of Organic Compounds of Low Volatility," *Environ. Sci. Technol.*, 26(11):2275-2281 (1992).

Jeffers, P.M., L.M. Ward, L.M. Woytowitch, and N.L. Wolfe. "Homogeneous Hydrolysis Rate Constants for Selected Chlorinated Methanes, Ethanes, Ethenes, and Propanes," *Environ. Sci. Technol.*, 23(8):965-969 (1989).

Jensen, S., R. Göthe, and M.-O. Kindstedt. "Bis(*p*-chlorophenyl)acetonitrile (DDN), a New DDT Derivative Formed in Anaerobic Digested Sewage Sludge and Lake Sediment," *Nature (London)*, 240(5381):421-422 (1972).

Jerina, D.M., J.W. Daly, A.M. Jeffrey, and D.T. Gibson. "*cis*-1,2-Dihydroxy-1,2-dihydronaphthalene: A Bacterial Metabolite from Naphthalene," *Arch. Biochem. Biophys.*, 142:394-396 (1971).

Jewett, D. and J.G. Lawless. "Formate Esters of 1,2-Ethanediol: Major Decomposition Products of *p*-Dioxane During Storage," *Bull. Environ. Contam. Toxicol.*, 25(1):118-121 (1980).

Jigami, Y., T. Omori, and Y. Minoda. "The Degradation of Isopropylbenzene and Isobutylbenzene by *Pseudomonas* sp.," *Agric. Biol. Chem.*, 39(9):1781-1788 (1975).

Johnsen, R.E. "DDT Metabolism in Microbial Systems," *Residue Rev.*, 61:1-28 (1976).

Johnsen, S., I.S. Gribbestad, and S. Johansen. "Formation of Chlorinated PAH - A Possible Health Hazard from Water Chlorination," *Sci. Tot. Environ.*, 81/82:231-238 (1989).

Johnson, C.A. and J.C. Westall. "Effect of pH and KCl Concentration on the Octanol-Water Distribution of Methylanilines," *Environ. Sci. Technol.*, 24(12):1869-1875 (1990).

Johnson-Logan, L.R., R.E. Broshears, and S.J. Klaine. "Partitioning Behavior and the Mobility of Chlordane in Groundwater," *Environ. Sci. Technol.*, 26(11):2234-2239 (1992).

Johnston, H.S. and J. Heicklen. "Photochemical Oxidations. IV. Acetaldehyde," *J. Am. Chem Soc.*, 86(20):4254-4256 (1964).

Joiner, R.L. and K.P. Baetcke. "Parathion: Persistence on Cotton and Identification of its Photoalteration Products," *J. Agric. Food Chem.*, 21(3):391-396 (1973).

Joiner, R.L., H.W. Chambers, and K.P. Baetcke. "Toxicity of Parathion and Several of its Photoalteration Products to Boll Weevils," *Bull. Environ. Contam. Toxicol.*, 6(3):220-224 (1971).

Jones, C.J., B.C. Hudson, and A.J. Smith. "The Leaching of Some Halogenated Organic Compounds from Domestic Waste," *J. Haz. Mater.*, 2:227-233 (1977).

Jones, D.C. "The Systems of *n*-Butyl Alcohol-Water and *n*-Butyl Alcohol-Water-Acetone," *J. Chem. Soc. (London)*, pp. 799-816 (1929).

Jönsson, J.A., J. Vejrosta, and J. Novak. "Air/Water Partition Coefficients for Normal Alkanes (*n*-Pentane to *n*-Nonane)," *Fluid Phase Equil.*, 9:279-286

(1982).

Jordan, T.E. *Vapor Pressures of Organic Compounds* (New York: Interscience Publishers, Inc., 1954), 266 p.

Joschek, H.I. and S.T. Miller. "Photooxidation of Phenol, Cresols, and Dihydroxy-benzenes," *J. Am. Chem. Soc.*, 88:3273-3281 (1966).

Jury, W.A., D.D. Focht, and W.J. Farmer. "Evaluation of Pesticide Pollution Potential from Standard Indices of Soil-Chemical Adsorption and Bio-degradation," *J. Environ. Qual.*, 16(4):422-428 (1987).

Jury, W.A., W.F. Spencer, and W.J. Farmer. "Use of Models for Assessing Relative Volatility, Mobility, and Persistence of Pesticides and Other Trace Organics in Soil Systems," in *Hazard Assessment of Chemicals, Volume 2*, J. Saxena, Ed. (New York: Academic Press, Inc., 1983), pp. 1-43.

Jury, W.A., W.F. Spencer, and W.J. Farmer. "Behavior Assessment Model for Trace Organics in Soil: IV. Review of Experimental Evidence," *J. Environ. Qual.*, 13(4):580-586 (1984).

Kaars Sijpesteijn, A., H.M. Dekhuijzen, and J.W. Vonk. "Biological Conversion of Fungicides in Plants and Microorganisms in *Antifungal Compounds*", (New York: Marcel Dekker, 1977), pp. 91-147.

Kaiser, K.L. and P.T.S. Wong. "Bacterial Degradation of Polychlorinated Biphenyls. I. Identification of Some Metabolic Products from Aroclor 1242," *Bull. Environ. Contam. Toxicol.*, 11(3):291-296 (1974).

Kaiser, K.L.E. and I. Valdmanis. "Apparent Octanol/Water Partition Coefficients of Pentachlorophenol as a Function of pH," *Can. J. Chem.*, 60(16):2104-2106 (1982).

Kale, S.P., N.B.K. Murthy, and K. Raghu. "Effect of Carbofuran, Carbaryl, and Their Metabolites on the Growth of *Rhizobium* sp. and *Azotobacter chroococcum*," *Bull. Environ. Contam. Toxicol.*, 42(5):769-772 (1989).

Kallos, G.J. and J.C. Tou. "Study of Photolytic Oxidation and Chlorination Re-actions of Dimethyl Ether and Chlorine in Ambient Air," *Environ. Sci. Technol.*, 11(12):1101-1105 (1977).

Kamlet, M.J., M.H. Abraham, R.M. Doherty, and R.W. Taft. "Solubility Properties in Polymers and Biological Media. 4. Correlation of Octanol/Water Partition Coefficients with Solvatochromic Parameters," *J. Am. Chem. Soc.*, 106(2):464-466 (1984).

Kamlet, M.J., R.M. Doherty, M.H. Abraham, P.W. Carr, R.F. Doherty, and R.W. Taft. "Linear Solvation Energy Relationships. 41. Important Differences between Aqueous Solubility Relationships for Aliphatic and Aromatic Solutes," *J. Phys. Chem.*, 91(7):1996-2004 (1987).

Kanazawa, J. "Relationship between the Soil Sorption Constants for Pesticides and Their Physicochemical Properties," *Environ. Toxicol. Chem.*, 8(6):477-484 (1989).

Kaneda, Y., K. Nakamura, H. Nakahara, and M. Iwaida. "Degradation of Organo-chlorine Pesticides with Calcium Hypochlorite," [Chemical Abstracts 82:94190e] *I. Eisei Kagaku*, 20:296-299 (1974).

Kanno, S. and K. Nojima. "Studies on Photochemistry of Aromatic Hydrocarbons. V. Photochemical Reaction of Chlorobenzene with Nitrogen Oxides in Air," *Chemosphere*, 8(4):225-232 (1979).

Kanno, S., K. Nojima, and T. Ohya. "Formation of Cyanide Ion or Cyanogen Chloride through the Cleavage of Aromatic Rings by Nitrous Acid or Chlorine. IV.," *Chemosphere*, 11(7):663-667 (1982).

Kapoor, I.P., R.L. Metcalf, A.S. Hirwe, J.R. Coats, and M.S. Khalsa. "Structure Activity Correlations of Biodegradability of DDT Analogs," *J. Agric. Food Chem.*, 21(2):310-315 (1973).

Kapoor, I.P., R.L. Metcalf, R.F. Nystrom, and G.K. Sanghua. "Comparative Metabolism of Methoxychlor, Methiochlor, and DDT in Mouse, Insects, and in a Model Ecosystem," *J. Agric. Food Chem.*, 18(6):1145-1152 (1970).

Kappeler, T. and K. Wuhrmann. "Microbial Degradation of the Water-Soluble Fraction of Gas Oil - II. Bioassays with Pure Strains," *Water Res.*, 12:335-342 (1978).

Karickhoff, S.W. "Correspondence - On the Sorption of Neutral Organic Solutes in Soils," *J. Agric. Food Chem.*, 29(2):425-426 (1981).

Karickhoff, S.W., D.S. Brown, and T.A. Scott. "Sorption of Hydrophobic Pollutants on Natural Sediments," *Water Res.*, 13(3):241-248 (1979).

Karickhoff, S.W. and K.R. Morris. "Sorption Dynamics of Hydrophobic Pollutants in Sediment Suspensions," *Environ. Toxicol. Chem.*, 4:469-479 (1985).

Karinen, J.F., J.G. Lamberton, N.E. Stewart, and L.C. Terriere. "Persistence of Carbaryl in the Marine Estuarine Environment. Chemical and Biological Stability in Aquarium Systems," *J. Agric. Food Chem.*, 15(1):148-156 (1967).

Katz, M. "Acute Toxicity of Some Organic Insecticides to Three Species of Salmonids and to Threespine Stickleback," *Trans. Am. Fish Soc.*, 90(3):264-268 (1961).

Kawaguchi, H. and M. Furuya. "Photodegradation of Monochlorobenzene in Titanium Dioxide Aqueous Suspensions," *Chemosphere*, 21(12):1435-1440 (1990).

Kawamoto, K. and K. Urano. "Parameters for Predicting Fate of Organochlorine Pesticides in the Environment (I) Octanol-Water and Air-Water Partition Coefficients," *Chemosphere*, 18(9/10):1987-1996 (1989).

Kawasaki, M. "Experiences with the Test Scheme under the Chemical Control Law of Japan: An Approach to Structure-Activity Correlations," *Ecotoxicol. Environ. Saf.*, 4:444-454 (1980).

Kayal, S.I. and D.W. Connell. "Partitioning of Unsubstituted Polycyclic Aromatic Hydrocarbons between Surface Sediments and the Water Column in the Brisbane River Estuary," *Aust. J. Mar. Freshwater Res.*, 41:443-456 (1990).

Kazano, H., P.C. Kearney, and D.D. Kaufman. "Metabolism of Methylcarbamate Insecticides in Soils," *J. Agric. Food Chem.*, 20(5):975-979 (1972).

Kearney, P.C. and D.D. Kaufman. *Herbicides: Chemistry, Degradation and Mode of Action* (New York: Marcel Dekker, Inc., 1976), 1036 p.

Keck, J., R.C. Sims, M. Coover, K. Park, and B. Symons. "Evidence for

Cooxidation of Polynuclear Aromatic Hydrocarbons in Soil," *Water Res.*, 23(12):1467–1476 (1989).

Keely, D.F., M.A. Hoffpauir, and J.R. Meriwether. "Solubility of Aromatic Hydrocarbons in Water and Sodium Chloride Solutions of Different Ionic Strengths: Benzene and Toluene," *J. Chem. Eng. Data*, 33(2):87–89 (1988).

Keen, R. and C.R. Baillod. "Toxicity to *Daphnia* of the End Products of Wet Oxidation of Phenol and Substituted Phenols," *Water Res.*, 19(6):767–772 (1985).

Keith, L.H. and D.B. Walters. *The National Toxicology Program's Chemical Data Compendium - Volume II. Chemical and Physical Properties* (Chelsea, MI: Lewis Publishers, Inc., 1992), 1642 p.

Kelley, J.D. and F.O. Rice. "The Vapor Pressures of Some Polynuclear Aromatic Hydrocarbons," *J. Phys. Chem.*, 68(12):3794–3796 (1964).

Kenaga, E.E. *Environmental Dynamics of Pesticides* (New York: Plenum Press, 1975), 243 p.

Kenaga, E.E. "Correlation of Bioconcentration Factors of Chemicals in Aquatic and Terrestrial Organisms with Their Physical and Chemical Properties," *Environ. Sci. Technol.*, 14(5):553–556 (1980).

Kenaga, E.E. "Predicted Bioconcentration Factors and Soil Sorption Coefficients of Pesticides and Other Chemicals," *Ecotoxicol. Environ. Saf.*, 4(1):26–38 (1980).

Kenaga, E.E. and C.A.I. Goring. "Relationship between Water Solubility, Soil Sorption, Octanol-Water Partitioning and Concentration of Chemicals in Biota," in *Aquatic Toxicology, ASTM STP 707*, Eaton, J.G., P.R. Parrish, and A.C. Hendricks, Eds. (Philadelphia, PA: American Society for Testing and Materials, 1980), pp. 78–115.

Kenley, R.A., J.E. Davenport, and D.G. Hendry. "Hydroxyl Radical Reactions in the Gas Phase. Products and Pathways for the Reaction of OH with Toluene," *J. Phys. Chem.*, 82(9):1095–1096 (1978).

Kennedy, M.V., B.J. Stojanovic, and F.L. Shuman, Jr. "Analysis of Decomposition Products of Pesticides," *J. Agric. Food Chem.*, 20(2):341–343 (1972).

Kennedy, M.V., B.J. Stojanovic, and F.L. Shuman, Jr. "Chemical and Thermal Aspects of Pesticide Disposal," *J. Environ. Qual.*, 1(1):63–65 (1972a).

Khalifa, S., R.L. Holmstead, and J.E. Casida. "Toxaphene Degradation by Iron(II) Protoporphyrin Systems," *J. Agric. Food Chem.*, 24(2):277–282 (1976).

Khalil, M.A.K. and R.A. Rasmussen. "Global Sources, Lifetimes and Mass Balances of Carbonyl Sulfide (OCS) and Carbon Disulfide (CS_2) in the Earth's Atmosphere," *Atmos. Environ.*, 18(9):1805–1813 (1984).

Kieatiwong, S., L.V. Nguyen, V.R. Hebert, M. Hackett, G.C. Miller, M.J. Miille, and R. Mitzel. "Photolysis of Chlorinated Dioxins in Organic Solvents and on Soils," *Environ. Sci. Technol.*, 24(10):1575–1580 (1990).

Kile, D.E., C.T. Chiou, H. Zhou, H. Li, and O. Zu. Partition of Nonpolar Organic Pollutants from Water to Soil and Sediment Organic Matters," *Environ. Sci. Technol.*, 29(5):1401–1406 (1995).

Kilzer, L., I. Scheunert, H. Geyer, W. Klein, and F. Korte. "Laboratory Screening of the Volatilization Rates of Organic Chemicals from Water and Soil," *Chemosphere*, 8(10):751-761 (1979).

Kim, I.-Y. and F.Y. Saleh. "Aqueous Solubilities and Transformations of Tetrahalogenated Benzenes and Effects of Aquatic Fulvic Acids," *Bull. Environ. Contam. Toxicol.*, 44(6):813-818 (1990).

Kim, Y.-H., J.E. Woodrow, and J.N. Seiber. "Evaluation of a Gas Chromatographic Method for Calculating Vapor Pressures with Organophosphorus Pesticides," *J. Chromatogr.*, 314:37-53 (1984).

Kincannon, D.F and Y.S. Lin. "Microbial Degradation of Hazardous Wastes by Land Treatment," *Proc. Indust. Waste Conf.*, 40:607-619 (1985).

Kirchnerová, J. and G.C.B. Cave. "The Solubility of Water in Low-Dielectric Solvents," *Can. J. Chem.*, 54(24):3909-3916 (1976).

Kiser, R.W., G.D. Johnson, and M.D. Shetlar. "Solubilities of Various Hydrocarbons in Methanol," *J. Chem. Eng. Data*, 6(3):338-341 (1961).

Kishi, H., N. Kogure, and Y. Hashimoto. "Contribution of Soil Constituents in Adsorption Coefficient of Aromatic Compounds, Halogenated Alicyclic and Aromatic Compounds to Soil," *Chemosphere*, 21(7):867-876 (1990).

Kishi, H. and Y. Hashimoto. "Evaluation of the Procedures for the Measurement of Water Solubility and *n*-Octanol/Water Partition Coefficient of Chemicals. Results of a Ring Test in Japan," *Chemosphere*, 18(9/10):1749-1759 (1989).

Kishino, T. and K. Kobayashi. "Relation between the Chemical Structures of Chlorophenols and Their Dissociation Constants and Partition Coefficients in Several Solvent-WaterSystems," *Wat. Res.*, 7:1547-1552 (1994).

Kjeldsen, P., J. Kjølholt, B. Schultz, T.H. Christensen, and J.C. Tjell. "Sorption and Degradation of Chlorophenols, Nitrophenols and Organophosphorus Pesticides in the Subsoil under Landfills," *J. Contam. Hydrol.*, 6(2):165-184 (1990).

Klečka, G.M., S.J. Gonsior, and D.A. Markham. "Biological Transformations of 1,1,1-Trichloroethane in Subsurface Soils and Groundwater," *Environ. Toxicol. Chem.*, 9(12):1437-1451 (1990).

Klein, A.W., M. Harnish, H.J. Porenski, and F. Schmidt-Bleek. "OECD Chemicals Testing Programme Physico-Chemical Tests," *Chemosphere*, 10:153-207 (1981).

Klein, R.G. "Calculations and Measurements on the Volatility of *N*-Nitrosoamines and Their Aqueous Solutions," *Toxicology*, 23:135-147 (1982).

Klein, W., J. Kohli, I. Weisgerber, and F. Korte. "Fate of Aldrin-^{14}C in Potatoes and Soil under Outdoor Conditions," *J. Agric. Food Chem.*, 21(2):152-156 (1973).

Klein, W., W. Kördel, M. Weiβ, and H.J. Poremski. "Updating of the OECD Test Guideline 107 'Partition Coefficient *n*-Octanol/Water': OECD Laboratory Intercomparison Test of the HPLC Method," *Chemosphere*, 17(2):361-386 (1988).

Kleopfer, R.D., D.M. Easley, B.B. Haas, Jr., T.G. Deihl, D.E. Jackson, and C.J. Wurrey. "Anaerobic Degradation of Trichloroethylene in Soil," *Environ. Sci. Technol.*, 19(3):277-280 (1985).

Klevens, H.B. "Solubilization of Polycyclic Hydrocarbons," *J. Phys. Colloid Chem.*,

54(2):283-298 (1950).

Klöpffer, W., F. Haag, E.-G. Kohl, and R. Frank. "Testing of the Abiotic Degradation of Chemicals in the Atmosphere: The Smog Chamber Approach," *Ecotoxicol. Environ. Saf.*, 15(3):298-319 (1988).

Klöpffer, W., G. Kaufman, G. Rippen, and H.-P. Poremski. "A Laboratory Method for Testing the Volatility from Aqueous Solution: First Results and Comparison with Theory," *Ecotoxicol. Environ. Saf.*, 6(6):545-559 (1982).

Knight, E.V., N.J. Novick, D.L. Kaplan, and J.R. Meeks. "Biodegradation of 2-Furaldehyde under Nitrate-Reducing and Methanogenic Conditions," *Environ. Toxicol. Chem.*, 9(6):725-730 (1990).

Knoevenagel, K. and R. Himmelreich. "Degradation of Compounds Containing Carbon Atoms by Photo-Oxidation in the Presence of Water," *Arch. Environ. Contam. Toxicol.*, 4:324-333 (1976).

Knuutinen, J., H. Palm, H. Hakala, J. Haimi, V. Huhta, and J. Salminen. "Polychlorinated Phenols and Their Metabolites in Soil and Earthworms of Sawmill Environment," *Chemosphere*, 20(6):609-623 (1990).

Kobayashi, H. and B.E. Rittman. "Microbial Removal of Hazardous Organic Compounds," *Environ. Sci. Technol.*, 16(3):170A-183A (1982).

Kobrinsky, P.C. and M.R. Martin. "High-Energy Methyl Radicals; the Photolysis of Methyl Bromide at 1850 Å," *J. Chem. Phys.*, 48(12):5728-5729 (1968).

Koester, C.J. and R.A. Hites. "Calculated Physical Properties of Polychlorinated Dibenzo-*p*-dioxins and Dibenzofurans," *Chemosphere*, 17(12):2355-2362 (1988).

Kohring, G.-W., J.E. Rogers, and J. Wiegel. "Anaerobic Biodegradation of 2,4-Dichlorophenol in Freshwater Lake Sediments at Different Temperatures," *Appl. Environ. Microbiol.*, 55(2):348-353 (1989).

Kollig, H.P., J.J. Ellington, E.J. Weber, and N.L. Wolfe. "Environmental Research Brief - Pathway Analysis of Chemical Hydrolysis for 14 RCRA Chemicals," Office of Research and Development, U.S. EPA Report-600/M-89/009 (1990), 6 p.

Könemann, H. "Quantitative Structure-Activity Relationships in Fish Toxicity Studies," *Toxicology*, 19:209-221 (1981).

Könemann, H., R. Zelle, and F. Busser. "Determination of Log P_{oct} Values of Chloro-Substituted Benzenes, Toluenes, and Anilines by High-Performance Liquid Chromatography on ODS-Silica," *J. Chromatogr.*, 178:559-565 (1979).

Konietzko, H. "Chlorinated Ethanes: Sources, Distribution, Environmental Impact, and Health Effects," in *Hazard Assessment of Chemicals, Volume 3*, J. Saxena, Ed. (New York: Academic Press, Inc., 1984), pp. 401-448.

Konrad, J.G., G. Chesters, and D.E. Armstrong. "Soil Degradation of Malathion, a Phosphorothioate Insecticide," *Soil Sci. Soc. Am. Proc.*, 33:259-262 (1969).

Kopczynski, S.L., A.P. Altshuller, and F.D. Sutterfield. "Photochemical Reactivities of Aldehyde-Nitrogen Oxide Systems," *Environ. Sci. Technol.*, 8(10):909-918 (1974).

Kopinke, F.-D., J. Pörschmann, and U. Stottmeister. "Sorption of Organic

Pollutants on Anthropogenic Humic Matter," *Environ. Sci. Technol.*, 29(4):941-950 (1995).

Korenman, I.M. and R.P. Aref'eva. "Determination of the Solubility of Liquid Hydrocarbons in Water," *Otkrytiya Izobret. Prom. Obraztsy Tovarnye Znaki*, 54(13):101 (1977).

Korenman, Ya I., T.E. Nefedova, and R.I. Byukova. "Extraction of Picramic Acid," *Int. J. Pharmacol. Ther. Toxicol.*, 51:734-735 (1977).

Korfmacher, W.A., E.L. Wehry, G. Mamantov, and D.F.S. Natusch. "Resistance to Photochemical Decomposition of Polycyclic Aromatic Hydrocarbons Vapor-Adsorbed on Coal Fly Ash," *Environ. Sci. Technol.*, 14(9):1094-1099 (1980).

Koskinen, W.C., H.R. Leffler, J.E. Oliver, P.C. Kearney, and C.G. McWhorter. "Effect of Trifluralin Soil Metabolites on Cotton Boll Components and Fiber and Seed Properties," *J. Agric. Food Chem.*, 33(5):958-961 (1985).

Koskinen, W.C., J.E. Oliver, P.C. Kearney, and C.G. McWhorter. "Effect of Trifluralin Soil Metabolites on Cotton Growth and Yield," *J. Agric. Food Chem.*, 32(6):1246-1248 (1984).

Kotronarou, A., G. Mills, and M.R. Hoffmann. "Decomposition of Parathion in Aqueous Solution by Ultrasonic Irradiation," *Environ. Sci. Technol.*, 26(7):1460-1462 (1992).

Krasnoshchekova, R.Ya. and M. Gubergrits. "Solubility of Paraffin Hydrocarbons in Fresh and Salt Water," *Neftekhimiya*, 13(6):885-888 (1973).

Krasnoshchekova, R.Ya. and M. Gubergrits. "Solubility of *n*-Alkylbenzene in Fresh and Salt Waters," *Vodn. Resur.*, 2:170-173 (1975).

Kresheck, G.C., H. Schneider, and H.A. Scheraga. "The Effect of D_2O on the Thermal Stability of Proteins. Thermodynamic Parameters for the Transfer of Model Compounds from H_2O to D_2O," *J. Phys. Chem.*, 69(9):3132-3144 (1965).

Kretschmer, C.B. and R. Wiebe. "Solubility of Gaseous Paraffins in Methanol and Isopropyl Alcohol," *J. Am. Chem. Soc.*, 74(5):1276-1277 (1952).

Kriegman-King, M.R. and M. Reinhard. "Transformation of Carbon Tetrachloride in the Presence of Sulfide, Biotite, and Vermiculite," *Environ. Sci. Technol.*, 26(11):2198-2206 (1992).

Krijgsheld, K.R. and A. van der Gen. "Assessment of the Impact of the Emission of Certain Organochlorine Compounds on the Aquatic Environment," *Chemosphere*, 15(7):861-880 (1986).

Krijgsheld, K.R. and A. van der Gen. "Assessment of the Impact of the Emission of Certain Organochlorine Compounds on the Aquatic Environment - Part I: Monochlorophenols and 2,4-Dichlorophenol," *Chemosphere*, 15(7):825-860 (1986).

Krishnamurthy, T. and S.P. Wasik. "Fluorometric Determination of Partition Coefficients of Naphthalene Homologues in Octanol-Water Mixtures," *J. Environ. Sci. Health*, A13(8):595-602 (1978).

Kronberger, H. and J. Weiss. "Formation and Structure of Some Organic Molecular Compounds. III. The Dielectric Polarization of Some Solid Crystalline Molecular

Compounds," *J. Chem. Soc. (London)*, pp. 464-469, (1944).

Krzyzanowska, T. and J. Szeliga. "A Method for Determining the Solubility of Individual Hydrocarbons," *Nafta*, 28:414-417 (1978).

Kuchta, J.M., A.L. Furno, A. Bartkowiak, and G.H. Martindill. "Effect of Pressure and Temperature on Flammability Limits of Chlorinated Hydrocarbons in Oxygen-Nitrogen and Nitrogen Tetroxide-Nitrogen Atmospheres," *J. Chem. Eng. Data*, 13(3):421-428 (1968).

Kudchadker, A.P., S.A. Kudchadker, R.P. Shukla, and P.R. Patnaik. "Vapor Pressures and Boiling Points of Selected Halomethanes," *J. Phys. Chem. Ref. Data*, 8(2):499-517 (1979).

Kuhr, R.J. "Metabolism of Methylcarbamate Insecticide Chemicals in Plants," *J. Agric. Food Chem. Suppl.*, 44-49 (1968).

Kuo, P.P.K., E.S.K. Chian, and B.J. Chang. "Identification of End Products Resulting from Ozonation and Chlorination of Organic Compounds Commonly Found in Water," *Environ. Sci. Technol.*, 11(13):1177-1181 (1977).

Kurihara, N., M. Uchida, T. Fujita, and M. Nakajima. "Studies on BHC Isomers and Related Compounds. V. Some Physicochemical Properties of BHC Isomers (1)," *Pestic. Biochem. Physiol.*, 2(4):383-390 (1973).

Laanio, T.L., P.C. Kearney, and D.D. Kaufman. "Microbial Metabolism of Dinitramine," *Pestic. Biochem. Physiol.*, 3:271-277 (1973).

Lacorte, S., S.B. Lartiges, P. Garrigues, and D. Barceló. "Degradation of Organophosphorus Pesticides and Their Transformation Products in Estuarine Waters," *Environ. Sci. Technol.*, 29(2):431-438 (1995).

LaFleur, K.S. "Sorption of Pesticides by Model Soils and Agronomic Soils: Rates and Equilibria," *Soil Sci.*, 127(2):94-101 (1979).

Lafrance, P., L. Marineau, L. Perreault, and J.-P. Villenueve. "Effect of Natural Dissolved Organic Matter Found in Groundwater on Soil Adsorption and Transport of Pentachlorophenol," *Environ. Sci. Technol.*, 28(13):2314-2320 (1994).

Lagas, P. "Sorption of Chlorophenols in the Soil," *Chemosphere*, 17(2):205-216 (1988).

Laha, S. and R.G. Luthy. "Oxidation of Aniline and Other Primary Aromatic Amines by Manganese Dioxide," *Environ. Sci. Technol.*, 24(3):363-373 (1990).

Lamparski, L.L., R.H. Stehl, and R.L. Johnson. "Photolysis of Pentachlorophenol-Treated Wood. Chlorinated Dibenzo-*p*-Dioxin Formation," *Environ. Sci. Technol.*, 14(2):196-200 (1980).

Lande, S.S., D.F. Hagen, and A.E. Seaver. "Computation of Total Molecular Surface Area from Gas Phase Ion Mobility Data and its Correlation with Aqueous Solubilities of Hydrocarbons," *Environ. Toxicol. Chem.*, 4:325-334 (1985).

Landrum, P.F., S.R. Nihart, B.J. Eadie, and W.S. Gardner. "Reverse-Phase Separation Method for Determining Pollutant Binding to Aldrich Humic Acid and Dissolved Organic Carbon of Natural Waters," *Environ. Sci. Technol.*,

19(3):187-192 (1984).

Lane, W.H. "Determination of the Solubility of Styrene in Water and of Water in Styrene," *Ind. Eng. Chem.*, 18:295-296 (1946).

Laplanche, A., G. Martin, and F. Tonnard. "Ozonation Schemes of Organophosphorous Pesticides. Application in Drinking Water Treatment," *Ozone: Sci. Eng.*, 6:207-219 (1984).

Larsen, T., P. Kjeldsen, and T.H. Christensen. "Correlation of Benzene, 1,1,1-Trichloroethane, and Naphthalene Distribution Coefficients to the Characteristics of Aquifer Materials with Low Organic Carbon Content," *Chemosphere*, 24(8):979-991 (1992).

Lau, Y.L., B.G. Oliver, and B.G. Krishnappan. "Transport of Some Chlorinated Contaminants by the Water, Suspended Sediments, and Bed Sediments in the St. Clair and Detroit Rivers," *Environ. Toxicol. Chem.*, 8(4):293-301 (1989).

Leadbetter, E.R. and J.W. Foster. "Oxidation Products formed from Gaseous Alkanes by the Bacterium *Pseudomonas methanica*," *Arch. Biochem. Biophys.*, 82:491-492 (1959).

Leavitt, D.D. and M.A. Abraham. "Acid-Catalyzed Oxidation of 2,4-Dichlorophenoxyacetic Acid by Ammonium Nitrate in Aqueous Solution," *Environ. Sci. Technol.*, 24(4):566-571 (1990).

Lee, M.C., E.S.K. Chian, and R.A. Griffin. "Solubility of Polychlorinated Biphenyls and Capacitor Fluid in Water," *Water Res.*, 13(12):1249-1258 (1979).

Lee, R.F., R. Sauerheber, and A.A. Benson. "Petroleum Hydrocarbons: Uptake and Discharge by the Marine Mussel *Mytilus edulis*," *Science (Washington, DC)*, 177:344-345 (1972).

Leenheer, J.A., R.L. Malcolm, and W.R. White. "Investigation of the Reactivity and Fate of Certain Organic Components of an Industrial Waste after Deep-Well Injection," *Environ. Sci. Technol.*, 10(5):445-451 (1976).

Leffingwell, J.T. "The Photolysis of DDT in Water," Ph.D. Thesis, University of California, Davis, CA (1975).

Leinon, P.J. and D. Mackay. "The Multicomponent Solubility of Hydrocarbons in Water," *Can. J. Chem. Eng.*, 51:230-233 (1973).

Leinster, P., R. Perry, and R.J. Young. "Ethylenebromide in Urban Air," *Atmos. Environ.*, 12:2382-2398 (1978).

Leitis, E. and D.G. Crosby. "Photodecomposition of Trifluralin," *J. Agric. Food Chem.*, 22(5):842-848 (1974).

Leland, H.V., W.N. Bruce, and N.F. Shimp. "Chlorinated Hydrocarbon Insecticides in Sediments in Southern Lake Michigan," *Environ. Sci. Technol.*, 7(9):833-838 (1973).

Lenchitz, C. and R.W. Velicky. "Vapor Pressure and Heat of Sublimation of Three Nitrotoluenes," *J. Chem. Eng. Data*, 15(3):401-403 (1970).

Leo, A., C. Hansch, and D. Elkins. "Partition Coefficients and Their Uses," *Chem. Rev.*, 71(6):525-616 (1971).

Leo, A., P.Y.C. Jow, C. Silipo, and C. Hansch. "Calculation of Hydrophobic

Constant from π and f Constants," *J. Med. Chem.*, 18(9):865-868 (1975).

Lépine, F., S. Milot, and N. Vincent. "Formation of Toxic PCB Congeners and PCB-Solvent Adducts in a Sunlight Irradiated Cyclohexane Solution of Aroclor 1254," *Bull. Environ. Contam. Toxicol.*, 48(1):152-156 (1992).

Lesage, S., R.E. Jackson, M.W. Priddle, and P.G. Riemann. "Occurrence and Fate of Organic Solvent Residues in Anoxic Groundwater at the Gloucester Landfill, Canada," *Environ. Sci. Technol.*, 24(4):559-566 (1990).

Leuenberger, C., M.P. Ligocki, and J.F. Pankow. "Trace Organic Compounds in Rain. 4. Identities, Concentrations, and Scavenging Mechanisms for Phenols in Urban Air and Rain," *Environ. Sci. Technol.*, 19(11):1053-1058 (1985).

Lewis Publishers. 1989. Drinking Water Health Advisory - Pesticides (Chelsea, MI: Lewis Publishers, Inc.), 819 p.

Lewis, R.J., Sr. *Rapid Guide to Hazardous Chemicals in the Workplace* (New York: Van Nostrand Reinhold, 1990), 286 p.

Leyder, F. and P. Boulanger. "Ultraviolet Absorption, Aqueous Solubility and Octanol-Water Partition for Several Phthalates," *Bull. Environ. Contam. Toxicol.*, 30(2):152-157 (1983).

Li, A. and A.W. Andren. "Solubility of Polychlorinated Biphenyls in Water/Alcohol Mixtures. 1. Experimental Data," *Environ. Sci. Technol.*, 28(1):47-52 (1994).

Li, A. and W.J. Doucette. "The Effect of Cosolutes on the Aqueous Solubilities and Octanol/Water Partition Coefficients of Selected Polychlorinated Biphenyl Congeners," *Environ. Toxicol. Chem.*, 12:2031-2035 (1993).

Li, C.F. and R.L. Bradley. "Degradation of Chlorinated Hydrocarbon Pesticides in Milk and Butter Oil by Ultraviolet Energy," *J. Dairy Sci.*, 52:27-30 (1969).

Li, J., A.J. Dallas, D.I. Eikens, P.W. Carr, D.L. Bergmann, M.J. Hait, and C.A. Eckert. "Measurement of Large Infinite Dilution Activity Coefficients of Nonelectrolytes in Water by Inert Gas Stripping and Gas Chromatography," *Anal. Chem.*, 65(22):3212-3218 (1993).

Lichtenstein, E.P. "'Bound' Residues in Soils and Transfer of Soil Residues in Crops," *Residue Rev.*, 76:147-153 (1980).

Lichtenstein, E.P., T.T. Liang, and M.K. Koeppe. "Effects of Fertilizers, Captafol, and Atrazine on the Fate and Translocation of [^{14}C]Fonofos and [^{14}C]Parathion in a Soil-Plant Microcosm," *J. Agric. Food Chem.*, 30(5):871-878 (1982).

Lichtenstein, E.P., T.W. Fuhremann, and K.R. Schulz. "Persistence and Vertical Distribution of DDT, Lindane, and Aldrin Residues, 10 and 15 Years after a Single Soil Application," *J. Agric. Food Chem.*, 19(4):718-721 (1971).

Lieberman, M.T. and M. Alexander. "Microbial and Nonenzymatic Steps in the Decomposition of Dichlorvos (2,2-Dichlorovinyl *O,O*-Dimethyl Phosphate)," *J. Agric. Food Chem.*, 31(2):265-267 (1983).

Lide, C.R. *CRC Handbook of Chemistry and Physics*, 70th ed. (Boca Raton, FL: CRC Press, Inc., 1990).

Lin, S., M.T. Lukasewycz, R.J. Liukkonen, and R.M. Carlson. "Facile Incorporation of Bromine into Aromatic Systems under Conditions of Water Chlorination," *Environ. Sci. Technol.*, 18(12):985-986 (1984).

Lin, S. and R.M. Carlson. "Susceptibility of Environmentally Important Heterocycles to Chemical Disinfection: Reactions with Aqueous Chlorine, Chlorine Dioxide, and Chloramine," *Environ. Sci. Technol.*, 18(10):743-748 (1984).

Lin, W.S. and M. Kapoor. "Induction of Aryl Hydrocarbon Hydroxylase in *Neurospora crassa* by Benzo[*a*]pyrene," *Curr. Microbiol.*, 3:177-180 (1979).

Linscott, D.L. and R.D. Hagin. "Additions to the Aliphatic Moiety of Chlorophenoxy Compounds," *Weed Sci.*, 18:197-198 (1970).

Lipczynska-Kochany, E. "Degradation of Aqueous Nitrophenols and Nitrobenzene by Means of the Fenton Reaction," *Chemosphere*, 22(5/6):529-536 (1991).

Liss, P.S. and P.G. Slater. "Flux of Gases across the Air-Sea Interface," *Nature (London)*, 247(5438):181-184 (1974).

Liu, D. "Biodegradation of Aroclor 1221 Type PCBs in Sewage Wastewater," *Bull. Environ. Contam. Toxicol.*, 27(5):695-703 (1981).

Liu, S.-Y. and J.M. Bollag. "Carbaryl Decomposition to 1-Naphthyl Carbamate by *Aspergillus terreus*," *Pestic. Biochem. Physiol.*, 1:366-372 (1971).

Liu, S.-Y. and J.M. Bollag. "Metabolism of Carbaryl by a Soil Fungus," *J. Agric. Food Chem.*, 19(3):487-490 (1971a).

Liu, D., W.M.J. Strachan, K. Thomson, and K. Kwasniewska. "Determination of the Biodegradability of Organic Compounds," *Environ. Sci. Technol.*, 15(7):788-793 (1981).

Lloyd, A.C., R. Atkinson, F.W. Lurmann, and B. Nitta. "Modeling Potential Ozone Impacts from Natural Hydrocarbons - I. Development and Testing of a Chemical Mechanism for the NO_x-Air Photooxidations of Isoprene and α-Pinene under Ambient Conditions," *Atmos. Environ.*, 17(10):1931-1950 (1983).

Lo, J.M., C.L. Tseng, and J.Y. Yang. "Radiometric Method for Determining Solubility of Organic Solvents in Water," *Anal. Chem.*, 58(7):1596-1597 (1986).

Lodge, K.B. "Solubility Studies Using a Generator Column for 2,3,7,8-Tetrachlorodibenzo-*p*-dioxin," *Chemosphere*, 18(1-6):933-940 (1989).

Løkke, H. "Sorption of Selected Organic Pollutants in Danish Soils," *Ecotoxicol. Environ. Saf.*, 8(5):395-409 (1984).

Lombardo, P., I.H. Pomerantz, and I.J. Egry. "Identification of Photoaldrin Chlorohydrin as a Photoalteration Product of Dieldrin," *J. Agric. Food Chem.*, 20(6):1278-1279 (1972).

Lopez-Gonzales, J.De D. and C. Valenzuela-Calahorro. "Associated Decomposition of DDT to DDE in the Diffusion of DDT on Homoionic Clays," *J. Agric. Food Chem.*, 18(3):520-523 (1970).

Lord, K.A., G.G. Briggs, M.C. Neale, and R. Manlove. "Uptake of Pesticides from Water and Soil by Earthworms," *Pestic Sci.*, 11(4):401-408 (1980).

Low, G.K.-C., S.R. McEvoy, and R.W. Matthews. "Formation of Nitrate and Ammonium Ions in Titanium Dioxide Mediated Photocatalytic Degradation of Organic Compounds Containing Nitrogen Atoms," *Environ. Sci. Technol.*, 25(3):460-467 (1991).

Lu, P.-Y. and R.L. Metcalf. "Environmental Fate and Biodegradability of Benzene Derivatives as Studied in a Model Ecosystem," *Environ. Health Perspect.*, 10:269-284 (1975).

Lu, P.-Y., R.L. Metcalf, A.S. Hirwe, and J.W. Williams. "Evaluation of Environmental Distribution and Fate of Hexachlorocyclopentadiene, Chlordene, Heptachlor, and Heptachlor Epoxide in a Laboratory Model Ecosystem," *J. Agric. Food Chem.*, 23(5):967-973 (1975).

Lu, P.-Y., R.L. Metcalf, N. Plummer, and D. Mandel. "The Environmental Fate of Three Carcinogens: Benzo-[a]-Pyrene, Benzidine, and Vinyl Chloride Evaluated in Laboratory Model Ecosystems," *Arch. Environ. Contam. Toxicol.*, 6:129-142 (1977).

Lyman, W.J., W.F. Reehl, and D.H. Rosenblatt. *Handbook of Chemical Property Estimation Methods: Environmental Behavior of Organic Compounds* (New York: McGraw-Hill, Inc., 1982).

Lyons, C.D., S. Katz, and R. Bartha. "Mechanisms and Pathways of Aniline Elimination from Aquatic Environments," *Appl. Environ. Microbiol.*, 48(3):491-496 (1984).

Lyons, C.D., S.E. Katz, and R. Bartha. "Fate of Herbicide-Drived Aniline Residues During Ensilage," *Bull. Environ. Contam. Toxicol.*, 35(5):704-710 (1985).

Mabey, W. and T. Mill. "Critical Review of Hydrolysis of Organic Compounds in Water under Environmental Conditions," *J. Phys. Chem. Ref. Data*, 7(2):383-415 (1978).

Mabey, W.R., J.H. Smith, R.T. Podoll, H.L. Johnson, T. Mill, T.-W. Chou, J. Gates, I.W. Partridge, H. Jaber, and D. Vandenberg. "Aquatic Fate Process Data for Organic Priority Pollutants - Final Report," Office of Regulations and Standards, U.S. EPA Report-440/4-81-014 (1982), 407 p.

Macek, K.J. and W.A. McAllister. "Insecticide Susceptibility of Some Common Fish Family Representatives," *Trans. Am. Fish Soc.*, 99(1):20-27 (1970).

MacIntyre, W.G. and P.O. deFur. "The Effect of Hydrocarbon Mixtures on Adsorption of Substituted Naphthalenes by Clay and Sediment from Water," *Chemosphere*, 14(1):103-111 (1985).

Mackay, D. "Correlation of Bioconcentration Factors," *Environ. Sci. Technol.*, 16(5):274-278 (1982).

Mackay, D., A. Bobra, D.W. Chan, and W.Y. Shiu. "Vapor Pressure Correlations for Low-Volatility Environmental Chemicals," *Environ. Sci. Technol.*, 16(10):645-649 (1982).

Mackay, D., A. Bobra, W.-Y. Shiu, and S.H. Yalkowsky. "Relationships between Aqueous Solubility and Octanol-Water Partition Coefficients," *Chemosphere*, 9:701-711 (1980).

Mackay, D. and A.T.K. Yeun. "Mass Transfer Coefficient Correlations for Volatilization of Organic Solutes from Water," *Environ. Sci. Technol.*, 17(4):211-217 (1983).

Mackay, D. and A.W. Wolkoff. "Rate of Evaporation of Low-Solubility Contaminants from Water Bodies to Atmosphere," *Environ. Sci. Technol.*, 7(7):611-614 (1973).

Mackay, D. and P.J. Leinonen. "Notes - Rate of Evaporation of Low-Solubility Contaminants from Water Bodies to Atmosphere," *Environ. Sci. Technol.*, 9(13):1178-1180 (1975).

Mackay, D. and S. Paterson. "Calculating Fugacity," *Environ. Sci. Technol.*, 15(9):1006-1014 (1981).

Mackay, D. and W.-Y. Shiu. "The Determination of the Solubility of Hydrocarbons in Aqueous Sodium Chloride Solutions," *Can. J. Chem. Eng.*, 53:239-242 (1975).

Mackay, D. and W.-Y. Shiu. "Aqueous Solubility of Polynuclear Aromatic Hydrocarbons," *J. Chem. Eng. Data*, 22(4):399-402 (1977).

Mackay, D., W.-Y. Shiu, and R.P. Sutherland. "Determination of Air-Water Henry's Law Constants for Hydrophobic Pollutants," *Environ. Sci. Technol.*, 13(3):333-337 (1979).

Mackay, D. and W.-Y. Shiu. "A Critical Review of Henry's Law Constants for Chemicals of Environmental Interest," *J. Phys. Chem. Ref. Data*, 10(4):1175-1199 (1981).

Macknick, A.B. and J.M. Prausnitz. "Vapor Pressures of High Molecular Weight Hydrocarbons," *J. Chem. Eng. Data*, 24(3):175-178 (1979).

MacLeod, H., L. Jourdain, G. Poulet, and G. LeBras. "Kinetic Study of Reactions of Some Organic Sulfur Compounds with OH Radicals," *Atmos. Environ.*, 18(12):2621-2626 (1984).

MacMichael, G.J. and L.R. Brown. "Role of Carbon Dioxide in Catabolism of Propane by *"Nocardia paraffinicum"* (*Rhodococcus rhodochrous*)," *Appl. Environ. Microbiol.*, 53(1):65-69 (1987).

MacRae, I.C. "Microbial Metabolism of Pesticides and Structurally Related Compounds," *Rev. Environ. Contam. Toxicol.*, 109:2-87 (1989).

MacRae, I.C., K. Raghu, and E.M. Bautista. "Anaerobic Degradation of the Insecticide Lindane by *Clostridium* sp.," *Nature (London)*, 221(5183):859-860 (1969).

MacRae, I.C., K. Raghu, and T.F. Castro. "Persistence and Biodegradation of Four Common Isomers of Benzene Hexachloride in Submerged Soils," *J. Agric. Food Chem.*, 15(5):911-914 (1967).

Madhun, Y.A. and V.H. Freed. "Degradation of the Herbicides Bromacil, Diuron and Chlortoluron in Soil," *Chemosphere*, 16(5):1003-1011 (1987).

Maeda, N., T. Ohya, K. Nojima, and S. Kanno. "Formation of Cyanide Ion or Cyanogen Chloride through the Cleavage of Aromatic Rings by Nitrous Acid or Chlorine. IX. On the Reactions of Chlorinated, Nitrated, Carboxylated or Methylated Benzene Derivatives with Hypochlorous Acid in the Presence of Ammonium Ion," *Chemosphere*, 16(10-12):2249-2258 (1987).

Magee, B.R., L.W. Lion, and A.T. Lemley. "Transport of Dissolved Organic Macromolecules and Their Effect on the Transport of Phenanthrene in Porous Media," *Environ. Sci. Technol.*, 25(2):323-331 (1991).

Mahaffey, W.R., D.T. Gibson, and C.E. Cerniglia. "Bacterial Oxidation of Chemical Carcinogens: Formation of Polycyclic Aromatic Acids from Benz[*a*]anthracene," *Appl. Environ. Microbiol.*, 54(10):2415-2423 (1988).

Mailhot, H. and R.H. Peters. "Empirical Relationships between the 1-Octanol/Water Partition Coefficient and Nine Physiochemical Properties," *Environ. Sci. Technol.*, 22(12):1479-1488 (1988).

Makino, M., M. Kamiya, and H. Matsushita. "Photodegradation Mechanism of 2,3,7,8-Tetrachlorodibenzo-*p*-dioxin as Studied by the PM3-MNDO Method," *Chemosphere*, 24(3):291-307 (1992).

Maksimov, Y.Y. "Vapor Pressures of Aromatic Nitro Compounds at Various Temperatures," *Zh. Fiz. Khim.*, 42(11):2921-2925 (1968).

Mallik, M.A.B. and K. Tesfai. "Transformation of Nitrosamines in Soil and *in Vitro* by Soil Micro-Organisms," *Bull. Environ. Contam. Toxicol.*, 27(1):115-121 (1981).

Mallon, B.J. and F.L. Harrison. "Octanol-Water Partition Coefficient of Benzo[*a*]pyrene: Measurement, Calculation and Environmental Implications," *Bull. Environ. Contam. Toxicol.*, 32(3):316-323 (1984).

Mansour, M., E. Feicht, and P. Méallier. "Improvement of the Photostability of Selected Substances in Aqueous Medium," *Toxicol. Environ. Chem.*, 20-21:139-147 (1989).

Mansour, M., S. Thaller, and F. Korte. "Action of Sunlight on Parathion," *Bull. Environ. Contam. Toxicol.*, 30(3):358-364 (1983).

Marple, L., B. Berridge, and L. Throop. "Measurement of the Water-Octanol Partition Coefficient of 2,3,7,8-Tetrachlorodibenzo-*p*-Dioxin," *Environ. Sci. Technol.*, 20(4):397-399 (1986).

Marple, L., R. Brunck, and L. Throop. "Water Solubility of 2,3,7,8-Tetrachloro-dibenzo-*p*-dioxin," *Environ. Sci. Technol.*, 20(2):180-182 (1986).

Martens, R. "Degradation of Endosulfan-8,9-^{14}C in Soil under Different Conditions," *Bull. Environ. Contam. Toxicol.*, 17(4):438-446 (1977).

Martin, E., S.H. Yalkowsky, and J.E. Wells. "Fusion of Disubstituted Benzenes," *J. Pharm. Sci.*, 68(5):565-568 (1979).

Martin, G., A. Laplanche, J. Morvan, Y. Wei, and C. LeCloirec. "Action of Ozone on Organo-Nitrogen Products," in *Proceedings Symposium on Ozonization: Environmental Impact and Benefit* (Paris, France: International Ozone Association, 1983), pp. 379-393.

Martin, H., Ed. *Pesticide Manual*, 3rd ed. (Worcester, England: British Crop Protection Council, 1972).

Masterton, W.L. and T.P. Lee. "Effect of Dissolved Salts on Water Solubility of Lindane," *Environ. Sci. Technol.*, 6(10):919-921 (1972).

Masunaga, S., Y. Urushigawa, and Y. Yonezawa. "Biodegradation Pathway of *o*-Cresol By Heterogeneous Culture," *Water Res.*, 20(4):477-484 (1986).

Matheson, L.J. and P.G. Tratnyek. "Reductive Dehalogenation of Chlorinated Methanes by Iron Metal," *Environ. Sci. Technol.*, 28(12):2045-2053 (1994).

Mathur, S.P. and J.G. Saha. "Microbial Degradation of Lindane-C^{14} in a Flooded

Sandy Loam Soil," *Soil Sci.*, 120(4):301-307 (1975).

Mathur, S.P. and J.G. Saha. "Degradation of Lindane-^{14}C in a Mineral Soil and in an Organic Soil," *Bull. Environ. Contam. Toxicol.*, 17(4):424-430 (1977).

Mathur, S.P. and J.W. Rouatt. "Utilization of the Pollutant Di-2-Ethylhexyl Phthalate by a Bacterium," *J. Environ. Qual.*, 4(2):273-275 (1975).

Matsumura, F. and G.M. Boush. "Malathion Degradation by *Trichoderma viride* and a *Pseudomonas* Species," *Science (Washington, DC)*, 153(3741):1278-1280 (1966).

Matsumura, F. and G.M. Boush. "Degradation of Insecticides by a Soil Fungus, *Trichoderma viride*," *J. Econ. Entomol.*, 61:610-612 (1968).

Matthews, R.W. "Photo-Oxidation of Organic Material in Aqueous Suspensions of Titanium Dioxide," *Water Res.*, 20(5):569-578 (1986).

Mattson, V.R., J.W. Arthur, and C.T. Walbridge. "Acute Toxicity of Selected Organic Compounds to Fathead Minnows," Office of Research and Development, U.S. EPA Report-600/3-76-097 (1976).

Maule, A., S. Plyte, and A.V. Quirk. "Dehalogenation of Organochlorine Insecticides by Mixed Anaerobic Microbial Populations," *Pestic. Biochem. Physiol.*, 277:229-236 (1987).

May, W.E., S.P. Wasik, and D.H. Freeman. "Determination of the Aqueous Solubility of Polynuclear Aromatic Hydrocarbons by a Coupled Column Liquid Chromatographic Technique," *Anal. Chem.*, 50(1):175-179 (1978).

May, W.E., S.P. Wasik, and D.H. Freeman. "Determination of the Solubility Behavior of Some Polycyclic Aromatic Hydrocarbons in Water," *Anal. Chem.*, 50(7):997-1000 (1978).

Mazzocchi, P.H. and M.W. Bowen. "Photolysis of Dioxane," *J. Org. Chem.*, 40(18):2689-2690 (1975).

McAuliffe, C. "Solubility in Water of C_1-C_9 Hydrocarbons," *Nature (London)*, 200(4911):1092-1093 (1963).

McAuliffe, C. "Solubility in Water of Paraffin, Cycloparaffin, Olefin, Acetylene, Cycloolefin, and Aromatic Compounds," *J. Phys. Chem.*, 70(4):1267-1275 (1966).

McAuliffe, C. "Solubility in Water of Normal C_9 and C_{10} Alkane Hydrocarbons," *Science (Washington, DC)*, 163(3866):478-479 (1969).

McBain, J.W. and K.J. Lissant. "The Solubilization of Four Typical Hydrocarbons in Aqueous Solution by Three Typical Detergents," *J. Am. Chem. Soc.*, 55:655-662 (1951).

McCall, P.J., R.L. Swann, D.A. Laskowski, S.A. Vrona, S.M. Unger, and H.J. Dishburger. "Prediction of Chemical Mobility in Soil from Sorption Coefficients," in *Aquatic Toxicology and Hazard Assessment, Fourth Conference, ASTM STP 737*, Branson, D.R. and K.L. Dickson, Eds. (Philadelphia, PA: American Society for Testing and Materials, 1981), pp. 49-58.

McCall, P.J., S.A. Vrona, and S.S. Kelley. "Fate of Uniformly Carbon-14 Ring Labeled 2,4,5-Trichlorophenoxyacetic Acid and 2,4-Dichlorophenoxy Acid," *J. Agric. Food Chem.*, 29(1):100-107 (1981).

McCarthy, J.F. "Role of Particulate Organic Matter in Decreasing Accumulation of Polynuclear Aromatic Hydrocarbons by *Daphnia magna*," *Arch. Environ. Contam. Toxicol.*, 12:559-568 (1983).

McCarthy, J.F. and B.D. Jimenez. "Reduction in Bioavailability to Bluegills of Polycyclic Aromatic Hydrocarbons Bound to Dissolved Humic Material," *Environ. Toxicol. Chem.*, 4:511-521 (1985).

McConnell, G., D.M. Ferguson, and C.R. Pearson. "Chlorinated Hydrocarbons and the Environment," *Endeavour*, 34(121):13-18 (1975).

McConnell, G. and H.I. Schiff. "Methyl Chloroform: Impact on Stratospheric Ozone," *Science (Washington, DC)*, 199:174-177 (1978).

McCrady, J.K., D.E. Johnson, and L.W. Turner. "Volatility of Ten Priority Pollutants from Fortified Avian Toxicity Test Diets," *Bull. Environ. Contam. Toxicol.*, 34(5):634-644 (1985).

McDevit, W.F. and F.A. Long. "The Activity Coefficient of Benzene in Aqueous Salt Solutions," *J. Am. Chem. Soc.*, 74(7):1773-1777 (1952).

McDuffie, B. "Estimation of Octanol/Water Partition Coefficients for Organic Pollutants using Reversed Phase HPLC," *Chemosphere*, 10(1):73-78 (1981).

McGuire, R.R., M.J. Zabik, R.D. Schuetz, and R.D. Flotard. "Photochemistry of Bioactive Compounds. Photochemical Reactions of Heptachlor: Kinetics and Mechanisms," *J. Agric. Food Chem.*, 20(4):856-861 (1972).

McGovern, E.W. "Chlorohydrocarbon Solvents," *Ind. Eng. Chem.*, 35(12):1230-1239 (1943).

McMurry, P.H. and D. Grosjean. "Photochemical Formation of Organic Aerosols: Growth Laws and Mechanisms," *Atmos. Environ.*, 19(9):1445-1451 (1985).

McNally, M.E. and R.L. Grob. "Determination of the Solubility Limits of Organic Priority Pollutants by Gas Chromatographic Headspace Analysis," *J. Chromatogr.*, 260:23-32 (1983).

McNally, M.E. and R.L. Grob. "Headspace Determination of Solubility Limits of the Base Neutral and Volatile Organic Components from the Environmental Protection Agency's List of Priority Pollutants," *J. Chromatogr.*, 284:105-116 (1984).

McVeety, B.D. and R.A. Hites. "Atmospheric Deposition of Polycyclic Aromatic Hydrocarbons to Water Surfaces: A Mass Balance Approach," *Atmos. Environ.*, 22(3):511-536 (1988).

Means, J.C., J.J. Hassett, S.G. Wood, and W.L. Banwart. "Sorption Properties of Energy-Related Pollutants and Sediments," in *Polynuclear Aromatic Hydrocarbons, Third International Symposium on Chemistry, Biology, Carcinogenesis and Mutagenesis*, Jones, P.W. and P. Leber, Eds. (Ann Arbor, MI: Ann Arbor Science Publishers, Inc., 1979), pp. 327-340.

Means, J.C., S.G. Wood, J.J. Hassett, and W.L. Banwart. "Sorption of Polynuclear Aromatic Hydrocarbons by Sediments and Soils," *Environ. Sci. Technol.*, 14(2):1524-1528 (1980).

Medley, D.R. and E.L. Stover. "Effects of Ozone on the Biodegradability of

Biorefractory Pollutants," *J. Water Pollut. Control Fed.*, 55(5):489-494 (1983).

Meikle, R.W. and C.R. Youngson. "The Hydrolysis Rate of Chlorpyrifos, *O,O*-Diethyl *O*-(3,5,6-trichloro-2-pyridyl) Phosphorothioate, and its Dimethyl Analog, Chloropyrifos-Methyl in Dilute Aqueous Solution," *Arch. Environ. Contam. Toxicol.*, 7(1):13-22 (1978).

Meites, L., Ed. *Handbook of Analytical Chemistry*, 1st ed. (New York: McGraw-Hill, Inc., 1963), 1782 p.

Melnikov, N.N. *Chemistry of Pesticides* (New York: Springer-Verlag, Inc., 1971), 480 p.

Mehrle, P.M., D.R. Buckler, E.E. Little, L.M. Smith, J.D. Petty, P.H. Peterman, D.L. Stalling, G.M. DeGraeve, J.J. Coyle, and W.J. Adams. "Toxicity and Bioconcentration of 2,3,7,8-Tetrachlorodibenzodioxin and 2,3,7,8-Tetrachlorodibenzofuran in Rainbow Trout," *Environ. Toxicol. Chem.*, 7:47-62 (1988).

Menges, R.A., G.L. Bertrand, and D.W. Armstrong. "Direct Measurement of Octanol-Water Partition Coefficient Chromatograph with a Back-Flushing Technique," *J. Liq. Chromatogr.*, 13(15):3061-3077 (1990).

Mercer, J.W., D.C. Skipp, and D. Giffin. "Basics of Pump-and-Treat Ground-Water Remediation Technology," U.S. EPA Report-600/8-90-003 (1990), 60 p.

Metcalf, R.L. "A Century of DDT," *J. Agric. Food Chem.*, 21(4):511-519 (1973).

Metcalf, R.L., I.P. Kapoor, P.-Y. Lu, C.K. Schuth, and P. Sherman. "Model Ecosystem Studies of the Environmental Fate of Six Organochlorine Pesticides," *Environ. Health Perspect.*, (June 1973), pp. 35-44.

Metcalf, R.L., G.K. Sangha, and I.P. Kapoor. "Model Ecosystem for the Evaluation of Pesticide Biodegradability and Ecological Magnification," *Environ. Sci. Technol.*, 5(8):709-713 (1971).

Metcalf, R.L., J.R. Sanborn, P.-Y. Lu, and D. Nye. "Laboratory Model Ecosystem Studies of the Degradation and Fate of Radiolabeled Tri-, Tetra-, and Pentachlorobiphenyl Compared with DDE," *Arch. Environ. Contam. Toxicol.*, 3(2):151-165 (1975).

Metcalf, R.L., T.R. Fukuto, C. Collins, K. Borck, S. Abd El-Aziz, R. Munoz, and C.C. Cassil. "Metabolism of 2,2-Dimethyl-2,3-dihydrobenzofuranyl-7 *N*-Methylcarbamate (Furadan) in Plants, Insects, and Mammals," *J. Agric. Food Chem.*, 16(2):300-311 (1968).

Meylan, W., P.H. Howard, and R.S. Boethling. "Molecular Topology/Fragment Contribution Method for Predicting Soil Sorption Coefficients," *Environ. Sci. Technol.*, 26(8):1560-1567 (1992).

Mick, D.L. and P.A. Dahm. "Metabolism of Parathion by Two Species of *Rhizobium*," *J. Econ. Entomol.*, 63:1155-1159 (1970).

Mihelcic, J.R. and R.G. Luthy. "Microbial Degradation of Acenaphthene and Naphthalene under Denitrification Conditions in Soil-Water Systems," *Appl. Environ. Microbiol.*, 54(5):1188-1198 (1988).

Mikami, Y., Y. Fukunaga, M. Arita, Y. Obi, and T. Kisaki. "Preparation of Aroma Compounds by Microbial Transformation of Isophorone by *Aspergillus niger*,"

Agric. Biol. Chem., 45(3):791–793 (1981).

Mikesell, M.D. and S.A. Boyd. "Reductive Dechlorination of the Pesticides 2,4-D, 2,4,5-T, and Pentachlorophenol in Anaerobic Sludges," *J. Environ. Qual.*, 14(3):337–340 (1985).

Milano, J.C., A. Guibourg, and J.L. Vernet. "Nonbiological Evolution, in Water, of Some Three- and Four-Carbon Atoms Organohalogenated Compounds: Hydrolysis and Photolysis," *Water Res.*, 22(12):1553–1562 (1988).

Miles, C.J. and S. Takashima. "Fate of Malathion and O,O,S-Trimethyl Phosphorothioate By-Product in Hawaiian Soil and Water," *Arch. Environ. Contam. Toxicol.*, 20(3):325–329 (1991).

Miles, C.J., M.L. Trehy, and R.A. Yost. "Degradation of N-Methylcarbamate and Carbamoyl Oxime Pesticides in Chlorinated Water," *Bull. Environ. Contam. Toxicol.*, 41(6):838–843 (1988).

Miles, J.R.W., C.M. Tu, and C.R. Harris. "Metabolism of Heptachlor and its Degradation Products by Soil Microorganisms," *J. Econ. Entomol.*, 62:1334 (1969).

Miles, J.R.W. and P. Moy. "Degradation of Endosulfan and its Metabolites by a Mixed Culture of Soil Microorganisms," *Bull. Environ. Contam. Toxicol.*, 23(1/2):13–19 (1979).

Mill, T., W.R. Mabey, B.Y. Lan, and A. Baraze. "Photolysis of Polycyclic Aromatic Hydrocarbons in Water," *Chemosphere*, 10(11/12):1281–1290 (1981).

Miller, G.C. and D.G. Crosby. "Photooxidation of 4-Chloroaniline and N-(4-Chlorophenyl)benzenesulfonamide to Nitroso- and Nitro-Products," *Chemosphere*, 12(9/10):1217–1227 (1983).

Miller, G.C. and R.G. Zepp. "Photoreactivity of Aquatic Pollutants Sorbed on Suspended Sediments," *Environ. Sci. Technol.*, 13(7):860–863 (1979).

Miller, M.M., S. Ghodbane, S.P. Wasik, Y.B. Tewari, and D.E. Martire. "Aqueous Solubilities, Octanol/Water Partition Coefficients, and Entropies of Melting of Chlorinated Benzenes and Biphenyls," *J. Chem. Eng. Data*, 29(2):184–190 (1984).

Miller, M.M., S.P. Wasik, G.-L. Huang, W.-Y. Shiu, and D. Mackay. "Relationships between Octanol-Water Partition Coefficient and Aqueous Solubility," *Environ. Sci. Technol.*, 19(6):522–529 (1985).

Milligan, L.H. "The Solubility of Gasoline (Hexane and Heptane) in Water at 25 °C," *J. Phys. Chem.*, 28(5):495–497 (1924).

Mills, A.C. and J.W. Biggar. "Solubility-Temperature Effect on the Adsorption of Gamma- and Beta-BHC from Aqueous and Hexane Solutions by Soil Materials," *Soil Sci. Soc. Am. Proc.*, 33:210–216 (1969).

Mills, A.L. and M. Alexander. "Factors Affecting Dimethylnitrosamine Formation in Samples of Soil and Water," *J. Environ. Qual.*, 5(4):437–440 (1976).

Mills, G. and M.R. Hoffman. "Photocatalytic Degradation of Pentachlorophenol on TiO_2 Particles: Identification of Intermediates and Mechanism of Reaction," *Environ. Sci. Technol.*, 27(8):1681–1689 (1993).

Mills, W.B., D.B. Porcella, M.J. Ungs, S.A. Gherini, K.V. Summers, L. Mok, G.L.

Rupp, and G.L. Bowie. "Water Quality Assessment: A Screening Procedure for Toxic and Conventional Pollutants in Surface and Groundwater - Part I," Office of Research and Development, U.S. EPA Report-600/6-85-002a (1985), 638 p.

Minard, R.D., S. Russel, and J.-M. Bollag. "Chemical Transformation of 4-Chloroaniline to a Triazene in a Bacterial Culture Medium," *J. Agric. Food Chem.*, 25(4):841-844 (1977).

Mingelgrin, U., S. Saltzman, and B. Yaron. "A Possible Model for the Surface-Induced Hydrolysis of Organophosphorus Pesticides on Kaolinite Clays," *Soil Sci. Soc. Am. J.*, 41:519-523 (1977).

Mingelgrin, U. and Z. Gerstl. "Reevaluation of Partitioning as a Mechanism of Nonionic Chemicals Adsorption in Soils," *J. Environ. Qual.*, 12(1):1-11 (1983).

Mirrlees, M.S., S.J. Moulton, C.T. Murphy, and P.J. Taylor. "Direct Measurement of Octanol-Water Partition Coefficient by High-Pressure Liquid Chromatography," *J. Med. Chem.*, 19(5):615-619 (1976).

Mirvish, S.S., P. Issenberg, and H.C. Sornson. "Air-Water and Ether-Water Distribution of *N*-Nitroso Compounds: Implications for Laboratory Safety, Analytic Methodology, and Carcinogenicity for the Rat Esophagus, Nose, and Liver," *J. Nat. Cancer Instit.*, 56(6):1125-1129 (1976).

Miyano, Y. and W. Hayduk. "Solubility of Butane in Several Polar and Nonpolar Solvents and in an Acetone-Butanol Solvent Solution," *J. Chem. Eng. Data*, 31(1):77-80 (1986).

Moilanen, K.W., D.G. Crosby, J.R. Humphrey, and J.W. Giles. "Vapor-Phase Photodecomposition of Chloropicrin (Trichloronitromethane)," *Tetrahedron*, 34(22):3345-3349 (1978).

Monsanto Industrial Chemicals Co. "PCBs-Aroclors," Technical Bulletin O/PL 306A, (1974).

Montgomery, J.H. Unpublished results (1989).

Moore, J.W. and S. Ramamoorthy. *Organic Chemicals in Natural Waters - Applied Monitoring and Impact Assessment* (New York: Springer-Verlag, Inc., 1984), 289 p.

Moore, R.M., C.E. Green, and V.K. Tait. "Determination of Henry's Law Constants for a Suite of Naturally Occurring Halogenated Methanes in Seawater," *Chemosphere*, 30(6):1183-1191 (1995).

Moreale, A. and R. Van Bladel. "Soil Interactions of Herbicide-Derived Aniline Residues: A Thermodynamic Approach," *Soil Sci.*, 127(1):1-9 (1979).

Morehead, N.R., B.J. Eadie, B. Lake, P.F. Landrum, and D. Berner. "The Sorption of PAH onto Dissolved Organic Matter in Lake Michigan Waters," *Chemosphere*, 15(4):403-412 (1986).

Mori, Y., S. Goto, S. Onodera, S. Naito, and H. Matsushita. "Aqueous Chlorination of Tetracyclic Aromatic Hydrocarbons: Reactivity and Product Distribution," *Chemosphere*, 22(5/6):495-501 (1991).

Moriguchi, I. "Quantitative Structure-Activity Studies on Parameters Related to Hydrophobicity," *Chem. Pharm. Bull.*, 23:247-257 (1975).

Morita, M., J. Nakagawa, and C. Rappe. "Polychlorinated Dibenzofuran (PCDF) Formation from PCB Mixture by Heat and Oxygen," *Bull. Environ. Contam. Toxicol.*, 19(6):665-670 (1978).

Morris, D.A. and A.I. Johnson. "Summary of Hydrologic and Physical Properties of Rocks and Soil Materials as Analyzed by the Hydrologic Laboratory of the U.S. Geological Survey, 1948-1960," U.S. Geological Survey Water-Supply Paper 1839-D (1967), 42 p.

Morrison, R.T. and R.N. Boyd. *Organic Chemistry* (Boston, MA: Allyn & Bacon, Inc., 1971), 1258 p.

Morrison, T.J. and F. Billett. "The Salting-Out of Non-Electrolytes. Part II. The Effect of Variation in Non-Electrolyte," *J. Chem. Soc. (London)*, pp. 3819-3822, (1952).

Mortland, M.M. and K.V. Raman. "Catalytic Hydrolysis of Some Organic Phosphate Pesticides by Copper(II)," *J. Agric. Food Chem.*, 15(1):163-167 (1967).

Mosher, D.R. and A.M. Kadoum. "Effects of Four Lights on Malathion Residues on Glass Beads, Sorghum Grain, and Wheat Grain," *J. Econ. Entomol.*, 65:847-850 (1972).

Mosier, A.R., W.D. Guenzi, and L.L. Miller. "Photochemical Decomposition of DDT by a Free-Radical Mechanism," *Science (Washington, DC)*, 164(3883):1083-1085 (1969).

Mostafa, I.Y., M.R.E. Bahig, I.M.I. Fakhr, and Y. Adam. "Metabolism of Organophosphorus Pesticides. XIV. Malathion Breakdown by Soil Fungi," *Z. Naturforsch. B.*, 27:1115-1116 (1972).

Moussavi, M. "Effect of Polar Substituents on Autoxidation of Phenols," *Water Res.*, 13(12):1125-1128 (1979).

Moza, P.N. and E. Feicht. "Photooxidation of Aromatic Hydrocarbons as Liquid Film on Water," *Toxicol. Environ. Chem.*, 20-21:135-138 (1989).

Moza, P.N., K. Fytianos, V. Samanidou, and F. Korte. "Photodecomposition of Chlorophenols in Aqueous Medium in Presence of Hydrogen Peroxide," *Bull. Environ. Contam. Toxicol.*, 41(5):678-682 (1988).

Mulla, M.S., L.S. Mian, and J.A. Kawecki. "Distribution, Transport and Fate of the Insecticides Malathion and Parathion in the Environment," *Residue Rev.*, 81:116-125 (1981).

Mulley, B.A. and A.D. Metcalf. "Solubilizationof Phenols by Non-Ionic Surface Active Reagents," *Sci. Pharm.*, 6:481-488 (1966).

Munakata, K. and M. Kuwahara. "Photochemical Degradation Products of Pentachlorophenol," *Residue Rev.*, 25:13-23 (1969).

Munnecke, D.M. and D.P.H. Hsieh. "Pathways of Microbial Metabolism of Parathion," *Appl. Environ. Microbiol.*, 31(1):63-69 (1976).

Munshi, H.B., K.V.S. Rama Rao, and R.M. Iyer. "Rate Constants of the Reactions of Ozone with Nitriles, Acrylates and Terpenes in Gas Phase," *Atmos. Environ.*, 23(9):1971-1976 (1989).

Munshi, H.B., K.V.S. Rama Rao, and R.M. Iyer. "Characterization of Products of

Ozonolysis of Acrylonitrile in Liquid Phase," *Atmos. Environ.*, 23(9):1945-1948 (1989).

Munz, C. and P.V. Roberts. "Air-Water Phase Equilibria of Volatile Organic Solutes," *J. Am. Water Works Assoc.*, 79(5):62-69 (1987).

Murphy, T.J., M.D. Mullin, and J.A. Meyer. "Equilibrium of Polychlorinated Biphenyls and Toxaphene with Air and Water," *Environ. Sci. Technol.*, 21(2):155-162 (1987).

Murray, J.M., R.F. Pottie, and C. Pupp. "The Vapor Pressures and Enthalpies of Sublimation of Five Polycyclic Aromatic Hydrocarbons," *Can. J. Chem.*, 52(4):557-563 (1974).

Murthy, N.B.K., D.D. Kaufman, and G.F. Fries. "Degradation of Pentachlorophenol (PCP) in Aerobic and Anaerobic Soil," *J. Environ. Sci. Health*, B14(1):1-14 (1979).

Musoke, G.M.S., D.J. Roberts, and M. Cooke. "Heterogeneous Hydrodechlorination of Chlordan," *Bull. Environ. Contam. Toxicol.*, 28(4):467-472 (1982).

Nahum, A. and C. Horvath. "Evaluation of Octanol-Water Partition Coefficients by Using High-Performance Liquid Chromatography," *J. Chromatogr.*, 192:315-322 (1980).

Naik, M.N., R.B. Jackson, J. Stokes, and R.J. Swabay. "Microbial Degradation and Phytotoxicity of Picloram and Other Substituted Pyridines," *Soil Biol. Biochem.*, 4:13-23 (1972).

Nakajima, S., N. Nakato, and T. Tani. "Microbial Transformation of 2,4-D and its Analog," *Chem. Pharm. Bull.*, 21:671-673 (1973).

Namiot, Yu. and S.Ya. Beider. "The Water Solubility of *n*-Pentane and *n*-Hexane," *Khim. Tekhnol. Topl. Masel*, 5(7):52-55 (1960).

NAS. 1977. Drinking Water and Health (Washington, DC: National Academy of Sciences), 939 p.

NAS. 1980. Drinking Water and Health (Washington, DC: National Academy of Sciences), 415 p.

Nash, R.G. "Comparative Volatilization and Dissipation Rates of Several Pesticides from Soil," *J. Agric. Food Chem.*, 31(2):210-217 (1983).

Natarajan, G.S. and K.A. Venkatachalam. "Solubilities of Some Olefins in Aqueous Solutions," *J. Chem. Eng. Data*, 17(3):328-329 (1972).

Nathan, M.F. "Choosing a Process for Chloride Removal," *Chem. Eng.*, 85:93-100 (1978).

Nathwani, J.S. and C.R. Phillips. "Adsorption-Desorption of Selected Hydrocarbons in Crude Oil on Soils," *Chemosphere*, 6(4):157-162 (1977).

Neely, W.B., D.R. Branson, and G.E. Blau. "Partition Coefficient to Measure Bioconcentration Potential of Organic Pesticides in Fish," *Environ. Sci. Technol.*, 6(7):629-632 (1974).

Neely, W.B. and G.E. Blau, Eds. *Environmental Exposure from Chemicals. Volume 1* (Boca Raton, FL: CRC Press, Inc. 1985), 245 p.

Nelson, H.D. and C.L. DeLigny. "The Determination of the Solubilities of Some *n*-

Alkanes in Water at Different Temperatures by Means of Gas Chromatography," *Rec. Trav. Chim. Pays-Bays*, 87(6):528-544 (1968).

Nelson, N.H. and S.D. Faust. "Acidic Dissociation Constants of Selected Aquatic Herbicides," *Environ. Sci. Technol.*, 3(11):1186-1188 (1969).

Nelson, P.E. and J.E. Hoff. "Food Volatiles: Gas Chromatographic Determination of Partition Coefficient in Water-Lipid Systems," *J. Food Sci.*, 33:479-482 (1968).

Neudorf, S. and M.A.Q. Khan. "Pick Up and Metabolism of DDT, Dieldrin, and Photodieldrin by a Fresh Water Alga (*Ankistrodesmus amalloides*) and a Microcrustacean (*Daphnia pulex*)," *Bull. Environ. Contam. Toxicol.*, 13:443-450 (1975).

Newland, L.W., G. Chesters, and G.B. Lee. "Degradation of γ-BHC in Simulated Lake Impoundments as Affected by Aeration," *J. Water Pollut. Control Fed.*, 41(5):R174-R188 (1969).

Newsted, J.L. and G.P. Giesy. "Predictive Models for Photoinduced Acute Toxicity of Polycyclic Aromatic Hydrocarbons to *Daphnia magna*, Strauss (*Cladocera, Crustacea*)," *Environ. Toxicol. Chem.*, 6:445-461 (1987).

NFPA. 1984. Fire Protection Guide on Hazardous Materials (Quincy, MA: National Fire Protection Association), 443 p.

Ngabe, B., T.F. Bidleman, and R.L. Falconer. "Base Hydrolysis of α- and β-Hexachlorocyclohexanes," *Environ. Sci. Technol.*, 27(9):1930-1933 (1993).

Nicholson, B.C., B.P. Maguire, and D.B. Bursill. "Henry's Law Constants for the Trihalomethanes: Effects of Water Composition and Temperature," *Environ. Sci. Technol.*, 18(7):518-521 (1984).

Nielsen, T., T. Ramdahl, and A. Bjørseth. "The Fate of Airborne Polycyclic Organic Matter," *Environ. Health Perspect.*, 47:103-114 (1983).

Niessen, R., D. Lenoir, and P. Boule. "Phototransformation of Phenol Induced by Excitation of Nitrate Ions," *Chemosphere*, 17(10):1977-1984 (1988).

Nikolaou, K., P. Masclet, and G. Mouvier. "Sources and Chemical Reactivity of Polynuclear Aromatic Hydrocarbons in the Atmosphere - A Critical Review," *Sci. Tot. Environ.*, 32(2):103-132 (1984).

NIOSH. 1994. "NIOSH Pocket Guide to Chemical Hazards," U.S. Department of Health and Human Services, U.S. Government Printing Office, 398 p.

Nisbet, I.C. and A.F. Sarofim. "Rates and Routes of Transport of PCBs in the Environment," *Environ. Health Perspect.*, (April 1972), pp. 21-38.

Nkedi-Kizza, P., P.S.C. Rao, and A.G. Hornsby. "Influence of Organic Cosolvents on Sorption of Hydrophobic Organic Chemicals by Soils," *Environ. Sci. Technol.*, 19(10):975-979 (1985).

Nkedi-Kizza, P., P.S.C. Rao, and J.W. Johnson. "Adsorption of Diuron and 2,4,5-T on Soil Particle-Size Separates," *J. Environ. Qual.*, 12(2):195-197 (1983).

Nojima, K. and S. Kanno. "Studies on Photochemistry of Aromatic Hydrocarbons. VII. Photochemical Reaction of *p*-Dichlorobenzene with Nitrogen Oxides in Air," *Chemosphere*, 9(7/8):437-440 (1980).

Nojima, K., T. Ikarigawa, and S. Kanno. "Studies on Photochemistry of Aromatic Hydrocarbons. VI. Photochemical Reaction of Bromobenzene with Nitrogen Oxides in Air," *Chemosphere*, 9(7/8):421–436 (1980).

Nyholm, N., P. Lindgaard-Jørgensen, and N. Hansen. "Biodegradation of 4-Nitrophenol in Standardized Aquatic Degradation Tests," *Ecotoxicol. Environ. Saf.*, 8(6):451–470 (1984).

Nyssen, G.A., E.T. Miller, T.F. Glass, C.R. Quinn II, J. Underwood, and D.J. Wilson. "Solubilities of Hydrophobic Compounds in Aqueous-Organic Solvent Mixtures," *Environ. Monit. Assess.*, 9(1):1–11 (1987).

O'Connell, W.L. "Properties of Heavy Liquids," *Trans. Am. Inst. Mech. Eng.*, 226(2):126–132 (1963).

Odeyemi, O. and M. Alexander. "Resistance of *Rhizobium* Strains to Phygon, Spergon, and Thiram," *Appl. Environ. Microbiol.*, 33(4):784–790 (1977).

Ofstad, E.B. and T. Sletten. "Composition and Water Solubility Determination of a Commercial Tricresylphosphate," *Sci. Tot. Environ.*, 43(3):233–241 (1985).

Ogata, M., K. Fujisawa, Y. Ogino, and E. Mano. "Partition Coefficients as a Measure of Bioconcentration Potential of Crude Oil Compounds in Fish and Shellfish," *Bull. Environ. Contam. Toxicol.*, 33:561–567 (1988).

Ogata, M. and Y. Miyake. "Disappearance of Aromatic Hydrocarbons and Organic Sulfer Compounds from Fish Flesh Reared in Crude Oil Suspension," *Water Res.*, 12:1041–1044 (1978).

Ohe, T. "Mutagenicity of Photochemical Reaction Products of Polycyclic Aromatic Hydrocarbons with Nitrite," *Sci. Tot. Environ.*, 39(1/2):161–175 (1984).

Oliver, B.G. "Desorption of Chlorinated Hydrocarbons from Spiked and Anthropogenically Contaminated Sediments," *Chemosphere*, 14(8):1087–1106 (1985).

Oliver, B.G. and A.J. Niimi. "Bioconcentration of Chlorobenzenes from Water by Rainbow Trout: Correlations with Partition Coefficients and Environmental Residues," *Environ. Sci. Technol.*, 17:287–291 (1983).

Oliver, B.G. and A.J. Niimi. "Bioconcentration Factors of Some Halogenated Organics for Rainbow Trout: Limitations in Their Use for Prediction of Environmental Residues," *Environ. Sci. Technol.*, 19(9):842–849 (1985).

Oliver, B.G. and J.H. Carey. "Photochemical Production of Chlorinated Organics in Aqueous Solutions," *Environ. Sci. Technol.*, 11(9):893–895 (1977).

Oliver, B.G. and M.N. Charlton. "Chlorinated Organic Contaminants on Settling Particulates in the Niagara River Vicinity of Lake Ontario," *Environ. Sci. Technol.*, 18(12):903–908 (1984).

Onitsuka, S., Y. Kasai, and K. Yoshimura. "Quantitative Structure-Toxicity Activity Relationship of Fatty Acids and the Sodium Salts to Aquatic Organisms," *Chemosphere*, 18(7/8):1621–1631 (1989).

Opperhuizen, A., P. Serné, and J.M.D. Van der Steen. "Thermodynamics of Fish/Water and Octan-1-ol/Water Partitioning of Some Chlorinated Benzenes," *Environ. Sci. Technol.*, 22(3):286–298 (1988).

Opperhuizen, A., F.A.P.C. Gobas, J.M.D. Vander Steen, and O. Hutzinger.

"Aqueous Solubility of Polychlorinated Biphenyls Related to Molecular Structure," *Environ. Sci. Technol.*, 22(6):638-646 (1988).

O'Reilly, K.T. and R.L. Crawford. "Kinetics of *p*-Cresol Degradation by an Immobilized *Pseudomonas* sp.," *Appl. Environ. Microbiol.*, 55(4):866-870 (1989).

Osborn, A.G. and D.R. Douslin. "Vapor-Pressure Relationships for 15 Hydrocarbons," *J. Chem. Eng. Data*, 19(2):114-117 (1974).

Osborn, A.G. and D.R. Douslin. "Vapor Pressures and Derived Enthalpies of Vaporization of Some Condensed-Ring Hydrocarbons," *J. Chem. Eng. Data*, 20(3):229-231 (1975).

OSHA. 1976. Criteria for a Recommended Standard. . .Occupational Exposure to Phenol (Cincinnati, OH: National Institute for Occupational Safety and Health), 167 p.

OSHA. 1982. "General Industry Standards for Toxic and Hazardous Substances," U.S. Code of Federal Regulations 1910, Subpart Z, Section 1910.1000.

Ou, L.-T., D.H. Gancarz, W.B. Wheeler, P.S.C. Rao, and J.M. Davidson. "Influence of Soil Temperature and Soil Moisture on Degradation and Metabolism of Carbofuran in Soils," *J. Environ. Qual.*, 11(2):293-298 (1982).

Ou, L.-T. and J.J. Street. "Monomethylhydrazine Degradation and its Effect on Carbon Dioxide Evolution and Microbial Populations in Soil," *Bull. Environ. Contam. Toxicol.*, 41(3):454-460 (1988).

Owens, J.W., S.P. Wasik, and H. DeVoe. "Aqueous Solubilities and Enthalpies of Solution of *n*-Alkylbenzenes," *J. Chem. Eng. Data*, 31(1):47-51 (1986).

Oyler, A.R., R.J. Llukkonen, M.T. Lukasewycz, K.E. Heikkila, D.A. Cox, and R.M. Carlson. "Chlorine 'Disinfection' Chemistry of Aromatic Compounds. Polynuclear Aromatic Hydrocarbons: Rates, Products, and Mechanisms," *Environ. Sci. Technol.*, 17(6):334-342 (1983).

Pal, D., J.B. Weber, and M.R. Overcash. "Fate of Polychlorinated Biphenyls (PCBs) in Soil Plant Systems," *Pest. Rev.*, 74:45-98 (1980).

Palit, S.R. "Electronic Interpretations of Organic Chemistry. II. Interpretation of the Solubility of Organic Compounds," *J. Phys. Chem.*, 51(3):837-857 (1947).

Pankow, J.F. and M.E. Rosen. "Determination of Volatile Compounds in Water by Purging Directly to a Capillary Column with Whole Column Cryotrapping," *Environ. Sci. Technol.*, 22(4):398-405 (1988).

Paris, D.F. and D.L. Lewis. "Chemical and Microbial Degradation of Ten Selected Pesticides in Aquatic Systems," *Residue Rev.*, 45:95-124 (1973).

Paris, D.F., D.L. Lewis, and J.T. Barnett. "Bioconcentration of Toxaphene by Microorganisms," *Bull. Environ. Contam. Toxicol.*, 17(5):564-572 (1977).

Paris, D.F. and N.L. Wolfe. "Relationship between Properties of a Series of Anilines and Their Transformation by Bacteria," *Appl. Environ. Microbiol.*, 53(5):911-916 (1987).

Paris, D.F., W.C. Steen, and G.L. Baughman. "Role of the Physico-Chemical Properties of Aroclors 1016 and 1242 in Determining Their Fate and Transport in Aquatic Environments," *Chemosphere*, 7(4):319-325 (1978).

Park, K.S. and W.N. Bruce. "The Determination of the Water Solubility of Aldrin, Dieldrin, Heptachlor, and Heptachlor Epoxide," *J. Econ. Entomol.*, 61(3):770-774 (1968).

Park, K.S., R.C. Sims, R.R. Dupont, W.J. Doucette, and J.E. Matthews. "Fate of PAH Compounds in Two Soil Types: Influence of Volatilization, Abiotic Loss and Biological Activity," *Environ. Toxicol. Chem.*, 9:187-195 (1990).

Parks, G.S. and H.M. Huffman. "Some Fusion and Transition Data for Hydrocarbons," *Ind. Eng. Chem.*, 23(10):1138-1139 (1931).

Parlar, H. "Photoinduced Reactions of Two Toxaphene Compounds in Aqueous Medium and Adsorbed on Silica Gel," *Chemosphere*, 17(11):2141-2150 (1988).

Parr, J.F. and S. Smith. "Degradation of Trifluralin under Laboratory Conditions and Soil Anaerobiosis," *Soil Sci.*, 115(1):55-63 (1973).

Parr, J.F. and S. Smith. "Degradation of DDT in an Everglades Muck as Affected by Lime, Ferrous Ion, and Anaerobiosis," *Soil Sci.*, 118(1):45-52 (1974).

Parris, G.E. "Covalent Binding of Aromatic Amines to Humates. 1. Reactions with Carbonyls and Quinones," *Environ. Sci. Technol.*, 14(9):1099-1106 (1980).

Parsons, F. and G.B. Lage. "Chlorinated Organics in Simulated Groundwater Environments," *J. Am. Water Works Assoc.*, 75(5):52-59 (1985).

Patil, K.C. and F. Matsumura. "Degradation of Endrin, Aldrin, and DDT by Soil Microorganisms," *Appl. Microbiol.*, 19:879-881 (1970).

Patil, K.C., F. Matsumura, and G.M. Boush. "Metabolic Transformation of DDT, Dieldrin, Aldrin, and Endrin by Marine Microorganisms," *Environ. Sci. Technol.*, 6(7):629-632 (1972).

Patnaik, P. *A Comprehensive Guide to the Hazardous Properties of Chemical Substances* (New York: Van Nostrand Reinhold, 1992), 763 p.

Paul, M.A. "The Solubilities of Naphthalene and Biphenyl in Aqueous Solutions of Electrolytes," *J. Am. Chem. Soc.*, 74(21):5274-5277 (1952).

Paulson, S.E. and J.H. Seinfeld. "Atmospheric Photochemical Oxidation of 1-Octene: OH, O_3, and $O(^3P)$ Reactions," *Environ. Sci. Technol.*, 26(6):1165-1173 (1992).

Pavlostathis, S.G. and G.N. Mathavan. "Desorption Kinetics of Selected Volatile Organic Compounds from Field Contaminated Soils," *Environ. Sci. Technol.*, 26(3):532-538 (1992).

Paya-Perez, A.B., M. Riaz, and B.R. Larsen. "Soil Sorption of 20 PCB Congeners and Six Chlorobenzenes," *Ecotoxicol. Environ. Saf.*, 21(1):1-17 (1991).

Pearlman, R.S., S.H. Yalkowsky, and S. Banerjee. "Water Solubilities of Polynuclear Aromatic and Heteroaromatic Compounds," *J. Phys. Chem. Ref. Data*, 13(2):555-562 (1984).

Pearson, C.R. and G. McConnell. "Chlorinated C_1 and C_2 Hydrocarbons in the Marine Environment," *Proc. R. Soc. London*, B189(1096):305-332 (1975).

Pereira, W.E., C.E. Rostad, C.T. Chiou, T.I. Brinton, L.B. Barber, II, D.K. Demcheck, and C.R. Demas. "Contamination of Estuarine Water, Biota and Sediment by Halogenated Organic Compounds: A Field Study," *Environ. Sci.*

Technol., 22:772-778 (1988).

Pereira, W.E., D.L. Short, D.B. Manigold, and P.K. Roscio. "Isolation and Characterization of TNT and its Metabolites in Groundwater by Gas Chromatograph-Mass Spectrometer-Computer Techniques," *Bull. Environ. Contam. Toxicol.*, 21(4/5):554-562 (1979).

Pérez-Tejeda, P., C. Yanes, and A. Maestre. "Solubility of Naphthalene in Water + Alcohol Solutions at Various Temperatures," *J. Chem. Eng. Data*, 35(3):244-246 (1990).

Perlinger, J.A., S.J. Eisenreich, and P.D. Capel. "Application of Headspace Analysis to the Study of Sorption of Hydrophobic Organic Compounds ti α-Al$_2$O$_3$," *Environ. Sci. Technol.*, 27(5):928-937 (1993).

Perrin, D.D. *Dissociation Constants of Organic Bases in Aqueous Solutions*, IUPAC Chemical Data Series; Supplement (London: Butterworth, 1972).

Perscheid, M., H. Schlueter, and K. Ballschmiter. "Aerober Abbau von Endosulfan Durch Bodenmikroorganismen," *Zeitschrift Naturforsch*, 28:761-763 (1973).

Peterson, M.S., L.W. Lion, and C.A. Shoemaker. "Influence of Vapor-Phase Sorption and Dilution on the Fate of Trichloroethylene in an Unsaturated Aquifer System," *Environ. Sci. Technol.*, 22(5):571-578 (1988).

Petrasek, A.C., I.J. Kugelman, B.M. Austern, T.A. Pressley, L.A. Winslow, and R.H. Wise. "Fate of Toxic Organic Compounds in Wastewater Treatment Plants," *J. Water Pollut. Control Fed.*, 55(10):1286-1296 (1983).

Petrier, C., M. Micolle, G. Merlin, J.-L. Luche, and G. Reverdy. "Characteristics of Pentachlorophenate Degradation in Aqueous Solution by Means of Ultrasound," *Environ. Sci. Technol.*, 26(8):1639-1642 (1992).

Peyton, T.O., R.V. Steel, and W.R. Mabey. "Carbon Disulfide, Carbonyl Sulfide: Literature Review and Environmental Assessment," U.S. EPA Report PB-257-947/2 (1976), 64 p.

Pfaender, F.K. and M. Alexander. "Extensive Microbial Degradation of DDT *in Vitro* and DDT Metabolism by Natural Communities," *J. Agric. Food Chem.*, 20(4):842-846 (1972).

Pfaender, F.K. and M. Alexander. "Effect of Nutrient Additions on the Apparent Cometabolism of DDT," *J. Agric. Food Chem.*, 21(3):397-399 (1973).

Piacente, V., P. Scardala, D. Ferro, and R. Gigli. "Vaporization Study of *o*-, *m*-, and *p*-Chloroaniline by Torsion-Weighing Effusion Vapor Pressure Measurements," *J. Chem. Eng. Data*, 30(4):372-376 (1985).

Pierce, R.H., C.E. Olney, and G.T. Felbeck. "*p,p'*-DDT Adsorption to Suspended Particulate Matter in Sea Water," *Geochim. Cosmochim. Acta*, 38(7):1061-1073 (1974).

Pierotti, C., C. Deal, and E. Derr. "Activity Coefficient and Molecular Structure," *Ind. Chem. Eng. Fundam.*, 51:95-101 (1959).

Pignatello, J.J. "Microbial Degradation of 1,2-Dibromoethane in Shallow Aquifer Materials," *J. Environ. Qual.*, 16(4):307-312 (1987).

Pillai, P., C.S. Helling, and J. Dragun. "Soil-Catalyzed Oxidation of Aniline,"

Chemosphere, 11(3):299-317 (1982).

Pinal, R., L.S. Lee, P.S.C. Rao. "Prediction of the Solubility of Hydrophobic Compounds in Nonideal Solvent Mixtures," *Chemosphere*, 22(9/10):939-951 (1991).

Pinal, R., P.S.C. Rao, L.S. Lee, P.V. Cline, and S.H. Yalkowsky. "Cosolvency of Partially Miscible Organic Solvents on the Solubility of Hydrophobic Organic Chemicals," *Environ. Sci. Technol.*, 24(5):639-647 (1990).

Pitts, J.N., Jr., B. Zielinska, J.A. Sweetman, R. Atkinson, and A.M. Winer. "Reactions of Adsorbed Pyrene and Perylene with Gaseous N_2O_5 under Simulated Atmospheric Conditions," *Atmos. Environ.*, 19(6):911-915 (1985).

Pitts, J.N., Jr., D. Grosjean, K.V. Cauwenberghe, J.P. Schmid, and D.R. Fitz. "Photooxidation of Aliphatic Amines under Simulated Atmospheric Conditions: Formation of Nitrosamines, Nitramines, Amides, and Photochemical Oxidant," *Environ. Sci. Technol.*, 12(8):946-953 (1978).

Pitts, J.N., Jr., D.M. Lokensgard, P.S. Ripley, K.A. Van Cauwenberghe, L. Van Vaeck, S.D. Schaffer, A.J. Thill, and W.L. Belser, Jr. "'Atmospheric' Epoxidation of Benzo[a]pyrene by Ozone: Formation of the Metabolite Benzo[a]pyrene-4,5-oxide," *Science (Washington, DC)*, 210(4476):1347-1349 (1980).

Pitts, J.N., Jr., R. Atkinson, J.A. Sweetman, and B. Zielinska. "The Gas-Phase Reaction of Naphthalene with N_2O_5 to Form Nitronaphthalenes," *Atmos. Environ.*, 19(5):701-705 (1985).

Platford, R.F., J.H. Carey, and E.J. Hale. "The Environmental Significance of Surface Films: Part 1. Octanol-Water Partition Coefficients for DDT and Hexachlorobenzene," *Environ. Pollut. (Series B)*, 3(2):125-128 (1982).

Plimmer, J.R. "The Photochemistry of Halogenated Herbicides," *Residue Rev.*, 33:47-74 (1970).

Plimmer, J.R. "Photolysis of TCDD and Trifluralin on Silica and Soil," *Bull. Environ. Contam. Toxicol.*, 20(1):87-92 (1978).

Plimmer, J.R., P.C. Kearney, and D.W. Von Endt. "Mechanism of Conversion of DDT to DDD by *Aerobacter aerogenes*," *J. Agric. Food Chem.*, 16(4):594-597 (1968).

Plimmer, J.R. and U.J. Klingebiel. "Photolysis of Hexachlorobenzene," *J. Agric. Food Chem.*, 24(4):721-723 (1976).

Plimmer, J.R., U.J. Klingbiel, and B.E. Hummer. "Photooxidation of DDT and DDE," *Science (Washington, DC)*, 167(3914):67-69 (1970).

Podoll, R.T., H.M. Jaber, and T. Mill. "Tetrachlorodioxin: Rates of Volatilization and Photolysis in the Environment," *Environ. Sci. Technol.*, 20(5):490-492 (1986).

Podoll, R.T., K.C. Irwin, and H.J. Parish. "Dynamic Studies of Naphthalene Sorption on Soil from Aqueous Solution," *Chemosphere*, 18(11/12):2399-2412 (1989).

Pogány, E., P.R. Wallnöfer, W. Ziegler, and W. Mücke. "Metabolism of *o*-Nitroaniline and Di-*n*-butyl Phthalate in Cell Suspension Cultures of Tomatoes," *Chemosphere*, 21(4/5):557-562 (1990).

Polak, J. and B.C.-Y. Lu. "Mutual Solubilities of Hydrocarbons and Water at 0 and 25 °C," *Can. J. Chem.*, 51(24):4018-4023 (1973).

Politzki, G.R., D. Bieniek, E.S. Lahaniatis, I. Scheunert, I. Klein, and F. Korte. "Determination of Vapor Pressures of Nine Organic Chemicals on Silica Gel," *Chemosphere*, 11(12):1217-1229 (1982).

Pollero, R. and S.C. dePollero. "Degradation of DDT by a Soil Amoeba," *Bull. Environ. Contam. Toxicol.*, 19(3):345-350 (1978).

Price, K.S., G.T. Waggy, and R.A. Conway. "Brine Shrimp Bioassay and Seawater BOD of Petro-Chemicals," *J. Water Pollut. Control Fed.*, 46(1):63-77 (1974).

Price, L.C. "Aqueous Solubility of Petroleum as Applied to its Origin and Primary Migration," *Am. Assoc. Pet. Geol. Bull.*, 60(2):213-244 (1976).

Probst, G.W. and J.B. Tepe. "Trifluralin and Related Compounds" in *Degradation of Herbicides*, Kearney, P.C. and D.D. Kaufman, Eds. (New York: Marcel Dekker, Inc., 1969), pp. 255-282.

Probst, G.W., T. Golab, R.J. Herberg, F.J. Holzer, S.J. Parka, C. van der Schans, and J.B. Tepe. "Fate of Trifluralin in Soils and Plants," *J. Agric. Food Chem.*, 15(4):592-599 (1967).

Pruden, A.L. and D.F. Ollis. "Degradation of Chloroform by Photoassisted Heterogeneous Catalysis in Dilute Aqueous Suspensions of Titanium Dioxide," *Environ. Sci. Technol.*, 17(10):628-631 (1983).

Przyjazny, A., W. Janicki, W. Chrzanowski, and R. Staszewki. "Headspace Gas Chromatographic Determination of Distribution Coefficients of Selected Organosulphur Compounds and Their Dependence of Some Parameters," *J. Chromatogr.*, 280:249-260 (1983).

Pupp, C., R.C. Lao, J.J. Murray, and R.F. Pottie. "Equilibrium Vapor Concentrations of Some Polycyclic Hydrocarbons, Arsenic Trioxide (As_4O_6) and Selenium Dioxide and the Collection Efficiencies of These Air Pollutants," *Atmos. Environ.*, 8:915-925 (1974).

Pussemier, L., G. Szabo, and R.A. Bulman. "Prediction of the Soil Adsorption Coefficient K_{oc} for Aromatic Pollutants," *Chemosphere*, 21(10-11):1199-1212 (1990).

Putnam, T.B., D.D. Bills, and L.M. Libbey. "Identification of Endosulfan Based on the Products of Laboratory Photolysis," *Bull. Environ. Contam. Toxicol.*, 13(6):662-665 (1975).

Que Hee, S.S. and R.G. Sutherland. *The Phenoxyalkanoic Herbicides. Vol. 1. Chemistry, Analysis, and Environmental Pollution* (Boca Raton, FL: CRC Press, Inc, 1981), 321 p.

Quirke, J.M.E., A.S.M. Marei, and G. Eglinton. "The Degradation of DDT and its Degradative Products by Reduced Iron (III) Porphyrins and Ammonia," *Chemosphere*, 8(3):151-155 (1979).

Radchenko, L.G. and A.I. Kitiagorodskii. "Vapor Pressure and Heat of Sublimation of Naphthalene, Biphenyl, Octafluoronaphthalene, Decafluorobiphenyl, Acenaphthene and α-Nitronaphthalene," *Zhur. Fiz. Khim.*, 48:2702-2704 (1974).

Radding, S.B., T. Mill, C.W. Gould, D.H. Lia, H.L. Johnson, D.S. Bomberger, and C.V. Fojo. "The Environmental Fate of Selected Polynuclear Aromatic Hydrocarbons," Office of Toxic Substances, U.S. EPA Report-560/5-75-009 (1976), 122 p.

Raghu, K. and I.C. MacRae. "Biodegradation of the Gamma Isomer of Benzene Hexachloride in Submerged Soils," *Science (Washington, DC)*, 154(3746):263-264 (1966).

Raha, P. and A.K. Das. "Photodegradation of Carbofuran," *Chemosphere*, 21(1/2):99-106 (1990).

Rajagopal, B.S., G.P. Brahmaprakash, B.R. Reddy, U.D. Singh, and N. Sethunathan. "Effect and Persistence of Selected Carbamate Pesticides in Soil," *Residue Rev.*, 93:1-199 (1984).

Rajagopal, B.S., K. Chendrayan, B.R. Reddy, and N. Sethunathan. "Persistence of Carbaryl in Flooded Soils and its Degradation by Soil Enrichment Cultures," *Plant Soil*, 73(1):35-45 (1983).

Rajagopal, B.S., S. Panda, and N. Sethunathan. "Accelerated Degradation of Carbaryl and Carbofuran in a Flooded Soil Pretreated with Hydrolysis Products, 1-Naphthol and Carbofuran Phenol," *Bull. Environ. Contam. Toxicol.*, 36(6):827-832 (1986).

Rajagopal, B.S., V.R. Rao, G. Nagendrappa, and N. Sethunathan. "Metabolism of Carbaryl and Carbofuran by Soil-Enrichment and Bacterial Cultures," *Can. J. Microbiol.*, 30(12):1458-1466 (1984a).

Rajagopalan, R., K.G. Vohra, and A.M. Mohan Rao. "Studies on Oxidation of Benzo[a]pyrene by Sunlight and Ozone," *Sci. Tot. Environ.*, 27(1):33-42 (1983).

Rajaram, K.P. and N. Sethunathan. "Effect of Organic Sources on the Degradation of Parathion in Flooded Alluvial Soil," *Soil Sci.*, 119(4):296-300 (1975).

Ralston, A.W. and C.W. Hoerr. "The Solubilities of the Normal Saturated Fatty Acids," *J. Org. Chem.*, 7:546-555 (1942).

Ramamoorthy, S. "Competition of Fate Processes in the Bioconcentration of Lindane," *Bull. Environ. Contam. Toxicol.*, 34(3):349-358 (1985).

Ramanand, K., S. Panda, M. Sharmila, T.K. Adhya, and N. Sethunathan. "Development and Acclimatization of Carbofuran-Degrading Soil Enrichment Cultures at Different Temperatures," *J. Agric. Food Chem.*, 36(1):200-205 (1988).

Ramanand, K., M. Sharmila, and N. Sethunathan. "Mineralization of Carbofuran by a Soil Bacterium," *Appl. Environ. Microbiol.*, 54(8):2129-2133 (1988a).

Ramdahl, T., G. Becher, and A. Bjørseth. "Nitrated Polycyclic Aromatic Hydrocarbons in Urban Air Particles," *Environ. Sci. Technol.*, 16(12):861-865 (1982).

Randall, T.L. and P.V. Knopp. "Detoxification of Specific Organic Substances by Wet Oxidation," *J. Water Pollut. Control Fed.*, 52(8):2117-2130 (1980).

Rao, P.S.C. and J.M. Davidson. "Estimation of Pesticide Retention and Transformation Parameters Required in Nonpoint Source Pollution Models," in *Environmental Impact of Nonpoint Source Pollution*, Overcash, M.R. and J.M.

Davidson, Eds. (Ann Arbor, MI: Ann Arbor Science Publishers, Inc., 1980), pp. 23-67.

Rao, P.S.C. and N. Sethunathan. "Effect of Ferrous Sulfate on Parathion Degradation in Flooded Soil," *J. Environ. Sci. Health*, B14(3):335-351 (1979).

Rapaport, R.A. and S.J. Eisenreich. "Chromatographic Determination of Octanol-Water Partition Coefficients (K_{ow}'s) for 58 Polychlorinated Biphenyl Congeners," *Environ. Sci. Technol.*, 18(3):163-170 (1984).

Rappe, C., G. Choudhary, and L.H. Keith. *Chlorinated Dioxins and Dibenzofurans in Perspective* (Chelsea, MI: Lewis Publishers, Inc., 1987), 570 p.

Rav-Acha, C. and E. Choshen. "Aqueous Reactions of Chlorine Dioxide with Hydrocarbons," *Environ. Sci. Technol.*, 21(11):1069-1074 (1987).

Reddy, B.R. and N. Sethunathan. "Mineralization of Parathion in the Rice Rhizophere," *Appl. Environ. Microbiol.*, 45(3):826-829 (1983).

Reed, C.D. and J.J. McKetta. "Solubility of 1,3-Butadiene in Water," *J. Chem. Eng. Data*, 4(4):294-295 (1959).

Reinbold, K.A., J.J. Hassett, J.C. Means, and W.L. Banwart. "Adsorption of Energy-Related Organic Pollutants: A Literature Review," Office of Research and Development, U.S. EPA Report-600/3-79-086 (1979), 180 p.

Reinert, K.H. and J.H. Rodgers. "Fate and Persistence of Aquatic Herbicides," *Rev. Environ. Contam. Toxicol.*, 98:61-98 (1987).

Renner, G. "Gas Chromatographic Studies of Chlorinated Phenols, Chlorinated Anisoles, and Chlorinated Phenylacetates," *Toxicol. Environ. Chem.*, 27:217-224 (1990).

Reuber, M.D. "Carcinogenicity of Lindane," *Environ. Res.*, 19:460-491 (1979).

Rex, A. "Über die Löslichkeit der Halogenderivate der Kohlenwasserstoffe in Wasser," *Z. Phys. Chem.*, 55:355-370 (1906).

Rhodes, R.C. "Metabolism of [2-^{14}C]Terbacil in Alfalfa," *J. Agric. Food Chem.*, 25(5):1066-1068 (1977).

Richard, Y. and L. Bréner. "Removal of Pesticides from Drinking Water by Ozone," in *Handbook of Ozone Technology and Applications, Volume II. Ozone for Drinking Water Treatment*, Rice, A.G. and A. Netzer, Eds. (Montvale, MA: Butterworth Publishers, 1984), pp. 77-97.

Riddick, J.A., W.B. Bunger, and T.K. Sakano. *Organic Solvents - Physical Properties and Methods of Purification. Volume II* (New York: John Wiley & Sons, Inc., 1986), 1325 p.

Riederer, M. "Estimating Partitioning and Transport of Organic Chemicals in the Foliage/Atmosphere System: Discussion of a Fugacity-Based Model," *Environ. Sci. Technol.*, 24(6):829-837 (1990).

Rippen, G., M. Ilgenstein, and W. Klöpffer. "Screening of the Adsorption Behavior of New Chemicals: Natural Soils and Model Adsorbents," *Ecotoxicol. Environ. Saf.*, 6(3):236-245 (1982).

Riser-Roberts, E. *Bioremediation of Petroleum Contaminated Sites* (Boca Raton, FL: CRC Press, Inc., 1992), 496 p.

Rittman, B.E. and P.L. McCarty. "Utilization of Dichloromethane by Suspended and Fixed Film Bacteria," *Appl. Environ. Microbiol.*, 39:1225-1226 (1980).

Roark, R.C. "A Digest of Information on Chlordane," Bureau of Entomology and Plant Quarantine, U.S. Dept. of Agric. Report E-817 (1951), 132 p.

Robb, I.D. "Determination of the Aqueous Solubility of Fatty Acids and Alcohols," *Aust. J. Chem.*, 18:2281-2285 (1966).

Robeck, G.G., K.A. Dostal, J.M. Cohen, and J.F. Kreissl. "Effectiveness of Water Treatment Processes in Pesticide Removal," *J. Am. Water Works Assoc.*, 57(2):181-200 (1965).

Roberts, A.L. and P.M. Gschwend. "Mechanism of Pentachloroethane Dehydrochlorination to Tetrachloroethylene," *Environ. Sci. Technol.*, 25(1):76-86 (1991).

Roberts, M.S., R.A. Anderson, and J. Swarbrick. "Permeability of Human Epidermis to Phenolic Compounds," *J. Pharm. Pharmac.*, 29:677-683 (1977).

Roberts, P.V. "Nature of Organic Contaminants in Groundwater and Approaches to Treatment," in *Proceedings, AWWA Seminar on Organic Chemical Contaminants in Groundwater: Transport and Removal* (St. Louis, MO: American Water Works Association, 1981), pp. 47-65.

Roberts, P.V. and P.G. Dändliker. "Mass Transfer of Volatile Organic Contaminants from Aqueous Solution to the Atmosphere During Surface Aeration," *Environ. Sci. Technol.*, 17(8):484-489 (1983).

Roberts, P.V., C. Munz, and P. Dändliker. "Modeling Volatile Organic Solute Removal by Surface and Bubble Aeration," *J. Water Pollut. Control Fed.*, 56(2):157-163 (1985).

Roberts, T.R. and G. Stoydin. "The Degradation of (*Z*)- and (*E*)-1,3-Dichloropropenes and 1,2-Dichloropropane in Soil," *Pestic. Sci.*, 7:325-335 (1976).

Robertson, R.E. and S.E. Sugamori. "The Hydrolysis of Dimethyl Sulfate and Diethyl Sulfate in Water," *Can. J. Chem.*, 44(14):1728-1730 (1966).

Robinson, J., A. Richardson, B. Bush, and K.E. Elgar. "A Photo-Isomerization Product of Dieldrin," *Bull. Environ. Contam. Toxicol.*, 1(4):127-132 (1966).

Rodorf, B.F. "Thermodynamic and Thermal Properties of Polychlorinated Compounds: The Vapor Pressures and Flow Tube Kinetics of Ten Dibenzo-*para*-Dioxins," *Chemosphere*, 14(7):885-892 (1985).

Roeder, K.D. and E.A. Weiant. "The Site of Action of DDT in the Cockroach," *Science (Washington, DC)*, 103(2671):304-305 (1946).

Rogers, K.S. and A. Cammarata. "Superdelocalizability and Charge Density. A Correlation with Partition Coefficients," *J. Med. Chem.*, 12(4):692-693 (1969).

Rogers, R.D. and J.C. MacFarlane. "Sorption of Carbon Tetrachloride, Ethylene Dibromide, and Trichloroethylene on Soil and Clay," *Environ. Monitor. Assess.*, 1(2):155-162 (1981).

Rogers, R.D., J.C. McFarlane, and A.J. Cross. "Adsorption and Desorption of Benzene in Two Soils and Montmorillonite Clay," *Environ. Sci. Technol.*, 14(4):457-461 (1980).

Rondorf, B. "Thermal Properties of Dioxins, Furans and Related Compounds,"

Chemosphere, 15:1325-1332 (1986).

Rontani, J.F., E. Rambeloarisoa, J.C. Bertrand, and G. Giusti. "Favourable Interaction between Photooxidation and Bacterial Degradation of Anthracene in Sea Water," *Chemosphere*, 14(11/12):1909-1912 (1985).

Rordorf, B.F. "Prediction of Vapor Pressures, Boiling Points and Enthalpies of Fusion for Twenty-Nine Halogenated Dibenzo-*p*-Dioxins and Fifty-Five Dibenzofurans by a Vapor Pressure Correlation Method," *Chemosphere*, 18(1-6):783-788 (1989).

Rosen, J.D. and D.J. Sutherland. "The Nature and Toxicity of the Photoconversion Products of Aldrin," *Bull. Environ. Contam. Toxicol.*, 2(1):1-9 (1967).

Rosen, J.D. and W.F. Carey. "Preparation of the Photoisomers of Aldrin and Dieldrin," *J. Agric. Food Chem.*, 16(3):536-537 (1968).

Rosen, J.D., J. Sutherland, and G.R. Lipton. "The Photochemical Isomerization of Dieldrin and Endrin and Effects on Toxicity," *Bull. Environ. Contam. Toxicol.*, 1(4):133-140 (1966).

Rosenberg, A. and M. Alexander. "Microbial Cleavage of Various Organophosphorus Insecticides," *Appl. Environ. Microbiol.*, 37(5):886-891 (1979).

Rosenberg, A. and M. Alexander. "Microbial Metabolism of 2,4,5-Trichlorophenoxyacetic Acid in Soil, Soil Suspensions, and Axenic Culture," *J. Agric. Food Chem.*, 28:297-302 (1980).

Rosenfield, C. and W. Van Valkenburg. "Decomposition of (*O,O*-Dimethyl-*O*-2,4,5-trichlorophenyl) Phosphorothioate (Ronnel) Adsorbed on Bentonite and Other Clays," *J. Agric. Food Chem.*, 13(1):68-72 (1972).

Ross, R.D. and D.G. Crosby. "The Photooxidation of Aldrin in Water," *Chemosphere*, 4(5):277-282 (1975).

Ross, R.D. and D.G. Crosby. "Photooxidant Activity in Natural Waters," *Environ. Toxicol. Chem.*, 4(6):773-778 (1985).

Rossi, S.S. and W.H. Thomas. "Solubility Behavior of Three Aromatic Hydrocarbons in Distilled Water and Natural Seawater," *Environ. Sci. Technol.*, 15(6):715-716 (1981).

Rothman, A.M. "Low Vapor Pressure Determination by the Radiotracer Transpiration Method," *Anal. Chem.*, 28(6):1225-1228 (1980).

Roy, W.R., C.C. Ainsworth, S.F.J. Chou, R.A. Griffin, and I.G. Krapac. "Development of Standardized Batch Adsorption Procedures: Experimental Considerations," in *Proceedings of the Eleventh Annual Research Symposium of the Solid and Hazardous Waste Research Division, April 29-May 1, 1985* (Cincinnati, OH: U.S. Environmental Protection Agency, 1985), U.S. EPA Report-600/9-85-013, pp. 169-178.

RTECS, 1985. "Registry of Toxic Effects of Chemical Substances," U.S. Department of Health and Human Services, National Institute of Occupational Safety and Health, 2050 p.

Ruepert, C., A. Grinwis, and H. Govers. "Prediction of Partition Coefficients of Unsubstituted Polycyclic Aromatic Hydrocarbons from C_{18} Chromatographic and Structural Properties," *Chemosphere*, 14(3/4):279-291 (1985).

Russell, D.J. and B. McDuffie. "Chemodynamic Properties of Phthalate Esters: Partitioning and Soil Migration," *Chemosphere*, 15(8):1003-1021 (1986).

Sabljić, A. "Predictions of the Nature and Strength of Soil Sorption of Organic Pollutants by Molecular Topology," *J. Agric. Food Chem.*, 32(2):243-246 (1984).

Sada, E., S. Kito, and Y. Ito. "Solubility of Toluene in Aqueous Salt Solutions," *J. Chem. Eng. Data*, 20(4):373-375 (1975).

Saeger, V.W., O. Hicks, R.G. Kaley, P.R. Michael, J.P. Mieure, and E.S. Tucker. "Environmental Fate of Selected Phosphate Esters," *Environ. Sci. Technol.*, 13(7):840-844 (1979).

Sahu, S.K., K.K. Patnaik, and N. Sethunathan. "Dehydrochlorination of δ-Isomer of Hexachlorocyclohexane by a Soil Bacterium, *Pseudomonas* sp.," *Bull. Environ. Contam. Toxicol.*, 48(2):265-268 (1992).

Sahyun, M.R.V. "Binding of Aromatic Compounds to Bovine Serum Albumin," *Nature (London)*, 209(5023):613-614 (1966).

Saleh, M.A. and J.E. Casida. "Reductive Dechlorination of the Toxaphene Component 2,2,5-*endo*,6-*exo*,8,9,10-Heptachlorobornane in Various Chemical, Photochemical, and Metabolic Systems," *J. Agric. Food Chem.*, 26(3):583-590 (1978).

Saleh, F.Y., K.L. Dickson, and J.H. Rodgers, Jr. "Fate of Lindane in the Aquatic Environment: Rate Constants of Physical and Chemical Processes," *Environ. Toxicol. Chem.*, 1(4):289-297 (1982).

Saltzman, S., L. Kliger, and B. Yaron. "Adsorption-Desorption of Parathion as Affected by Soil Organic Matter," *J. Agric. Food Chem.*, 20(6):1224-1226 (1972).

Sanborn, J.R., B.M. Francis, and R.L. Metcalf. "The Degradation of Selected Pesticides in Soil: A Review of the Published Literature," Office of Research and Development, U.S. EPA Report-600/9-77-022 (1977), 616 p.

Sanborn, J.R., R.L. Metcalf, W.N. Bruce, and P.-Y. Lu. "The Fate of Chlordane and Toxaphene in a Terrestrial-Aquatic Model Ecosystem," *Environ. Entomol.*, 5(3):533-538 (1976).

Sanderman, W., H. Stockmann, and R. Casten. "Über die Pyrolyse des Pentachloro-phenols," *Chem. Ber.*, 90:690-692 (1957).

Sanders, H.O. and O.B. Cope. "The Relative Toxicities of Several Pesticides to Naiads of Three Species of Stoneflies," *Limnol. Oceanogr.*, 13(1):112-117 (1968).

Sanemasa, I., S. Arakawa, C.Y. Piao, and T. Deguchi. "Equilibrium Solubilities of Benzene, Toluene, Ethylbenzene, and Propylbenzene in Silver Nitrate Solutions," *Bull. Chem. Soc. Jpn.*, 58(10):3033-3034 (1985).

Sanemasa, I., M. Araki, T. Deguchi, and H. Nagai. "Solubility Measurements of Benzene and the Alkylbenzenes in Water by Making Use of Solute Vapor," *Bull. Chem. Soc. Jpn.*, 55(4):1054-1062 (1982).

Sanemasa, I., Y. Miyazaki, S. Arakawa, M. Kumamaru, and T. Deguchi. "The Solubility of Benzene-Hydrocarbon Binary Mixtures in Water," *Bull. Chem. Soc. Jpn.*, 60(2):517-523 (1987).

Sangster, J. "Octanol-Water Partition Coefficients of Simple Organic Compounds," *J. Phys. Chem. Ref. Data*, 18(3):1111-1229 (1989).

Saravanja-Bozanic, V., S. Gäb, K. Hustert, and F. Korte. "Reacktionen von Aldrin, Chlorden, und 2,2-Dichlorobiphenyl mit O(^{3}P)," *Chemosphere*, 6(1):21-26 (1977).

Sato, A. and T. Nakajima. "A Structure-Activity Relationship of Some Chlorinated Hydrocarbons," *Arch. Environ. Health*, 34(2):69-75 (1979).

Sax, N.I. and R.J. Lewis, Sr. *Hazardous Chemicals Desk Reference* (New York: Van Nostrand Reinhold, 1987), 1084 p.

Sax, N.I. *Dangerous Properties of Industrial Materials* (New York: Van Nostrand Reinhold Co., 1984), 3124 p.

Sax, N.I., Ed. *Dangerous Properties of Industrial Materials Report* (New York: Van Nostrand Reinhold Co., 1985), 5(4): 94 p.

Sax, N.I., Ed. *Dangerous Properties of Industrial Materials Report* (New York: Van Nostrand Reinhold Co., 1985), 5(3): 88 p.

Sax, N.I., Ed. *Dangerous Properties of Industrial Materials Report* (New York: Van Nostrand Reinhold Co., 1984), 4(1): 112 p.

Sax, N.I., Ed. *Dangerous Properties of Industrial Materials Report* (New York: Van Nostrand Reinhold Co., 1984), 4(3): 93 p.

Sax, N.I., Ed. *Dangerous Properties of Industrial Materials Report* (New York: Van Nostrand Reinhold Co., 1984), 4(6): 105 p.

Saylor, J.H., J.M. Stuckey, and P.M. Gross. "Solubility Studies. V. The Validity of Henry's Lay for the Calculation of Vapor Solubilities," *J. Am. Chem. Soc.*, 60(2):373-376 (1938).

Scala, A.A. and D. Salomon. "The Gas Phase Photolysis and γ Radiolysis of Ethylenimine," *J. Chem. Phys.*, 65(11):4455-4461 (1976).

Schantz, M.M. and D.E. Martire. "Determination of Hydrocarbon-Water Partition Coefficients from Chromatographic Data and Based on Solution Thermodynamics and Theory," *J. Chromatogr.*, 391(1):35-51 (1987).

Schellenberg, K., C. Leuenberger, and R.P. Schwarzenbach. "Sorption of Chlorinated Phenols by Natural Sediments and Aquifer Materials," *Environ. Sci. Technol.*, 18(9):652-657 (1984).

Scheunert, I., D. Vockel, J. Schmitzer, and F. Korte. "Biomineralization Rates of ^{14}C-Labelled Organic Chemicals in Aerobic and Anaerobic Suspended Soil," *Chemosphere*, 16(5):1031-1041 (1987).

Schrap, S.M., P.J. de Vries, and A. Opperhuizen. "Experimental Problems in Determining Sorption Coefficients of Organic Chemicals; An Example of Chlorobenzenes," *Chemosphere*, 28(5):931-945 (1994).

Schroy, J.M., F.D. Hileman, and S.C. Cheng. "Physical/Chemical Properties of 2,3,7,8-TCDD," *Chemosphere*, 14(6/7):877-880 (1985).

Schultz, T.W., S.K. Wesley, and L.L. Baker. "Structure-Activity Relationships for Di and Tri Alkyl and/or Halogen Substituted Phenols," *Bull. Environ. Contam. Toxicol.*, 43(2):192-198 (1989).

Schumacher, H.G., H. Parlar, W. Klein, and F. Korte. "Beitrage zur Okologischen Chemie. LV. Photochemische Reaktion von Endosulfan," *Chemosphere*, 3(2):65-70 (1974).

Schumann, S.C., J.S. Aston, and M. Sagenkahn. "The Heat Capacity and Entropy, Heats of Fusion and Vaporization and the Vapor Pressures of Isopentane," *J. Am. Chem. Soc.*, 64:1039-1043 (1942).

Schwarz, F.P. "Determination of Temperature Dependence of Solubilities of Polycyclic Aromatic Hydrocarbons in Aqueous Solutions by a Fluorescence Method," *J. Chem. Eng. Data*, 22(3):273-277 (1977).

Schwarz, F.P. "Measurement of the Solubilities of Slightly Soluble Organic Liquids in Water by Elution Chromatography," *Anal. Chem.*, 52(1):10-15 (1980).

Schwarz, F.P. and S.P. Wasik. "A Fluorescence Method for the Measurement of the Partition Coefficients of Naphthalene, 1-Methylnaphthalene, and 1-Ethylnaphthalene in Water," *J. Chem. Eng. Data*, 22(3):270-273 (1977).

Schwarz, F.P. and S.P. Wasik. "Fluorescence Measurements of Benzene, Naphthalene, Anthracene, Pyrene, Fluoranthene, and Benzo[e]pyrene in Water," *Anal. Chem.*, 48(3):524-528 (1976).

Schwarzenbach, R.P. and J. Westall. "Transport of Nonpolar Organic Compounds from Surface Water to Groundwater. Laboratory Sorption Studies," *Environ. Sci. Technol.*, 15(11):1360-1367 (1981).

Schwarzenbach, R.P., P.M. Gschwend, and D.M. Imboden. *Environmental Organic Chemistry* (New York: John Wiley & Sons, Inc., 1993), 681 p.

Schwarzenbach, R.P., R. Stierli, B.R. Folsom, and J. Zeyer. "Compound Properties Relevant for Assessing the Environmental Partitioning of Nitrophenols," *Environ. Sci. Technol.*, 22(1):83-92 (1988).

Schwarzenbach, R.P., W. Giger, C. Schaffner, and O. Wanner. "Groundwater Contamination by Volatile Halogenated Alkanes: Abiotic Formation of Volatile Sulfur Compounds under Anaerobic Conditions," *Environ. Sci. Technol.*, 19(4):322-327 (1985).

Schwarzenbach, R.P., W. Giger, E. Hoehn, and J.K. Schneider. "Behavior of Organic Compounds during Infiltration of River Water to Groundwater. Field Studies," *Environ. Sci. Technol.*, 17(8):472-479 (1983).

Schwille, F. *Dense Chlorinated Solvents* (Chelsea, MI: Lewis Publishers, Inc., 1988), 146 p.

Schmidt-Bleek, F., W. Haberland, A.W. Klein, and S. Caroli. "Steps Towards Environment Assessment of New Chemicals," *Chemosphere*, 11:383-415 (1982).

Sears, G.W. and E.R. Hopke. "Vapor Pressures of Naphthalene, Anthracene, and Hexachlorobenzene in a Low Pressure Region," *J. Am. Chem. Soc.*, 71(5):1632-1634 (1949).

Sedlak, D.L. and A.W. Andren. "Oxidation of Chlorobenzene with Fenton's Reagent," *Environ. Sci. Technol.*, 25(4):777-782 (1991).

Seip, H.M., J. Alstad, G.E. Carlberg, K. Martinsen, and R. Skaane. "Measurement of Mobility of Organic Compounds in Soils," *Sci. Tot. Environ.*, 50:87-101

(1986).

Sethunathan, N., R. Siddaramappa, K.P. Rajaram, S. Barik, and P.A. Wahid. "Parathion: Residues in Soil and Water," *Residue Rev.*, 68:91-122 (1977).

Sethunathan, N. and T. Yoshida. "A *Flavobacterium* sp. that Degrades Diazinon and Parathion," *Can. J. Microbiol.*, 19(5):873-875 (1973).

Sethunathan, N. and T. Yoshida. "Degradation of Chlorinated Hydrocarbons by *Clostridium* sp. Isolated from Lindane-Amended, Flooded Soil," *Plant Soil*, 38:663-666 (1973a).

Sharma, R.K. and H.B. Palmer. "Vapor Pressure of Biphenyl Near Fusion Temperature," *J. Chem. Eng. Data*, 19(1):6-8 (1974).

Sharmila, M., K. Ramanand, and N. Sethunathan. "Effect of Yeast Extract on the Degradation of Organophosphorus Insecticides by Soil Enrichment and Bacterial Cultures," *Can. J. Microbiol.*, 35(12):1105-1110 (1989).

Sharom, M.S., J.R.W. Miles, J.W. Harris, and F.L. McEwen. "Persistence of 12 Insecticides in Water," *Water Res.*, 14(8):1089-1093 (1980).

Sharom, M.S., J.R.W. Miles, C.R. Harris, and F.L. McEwen. "Behavior of 12 Pesticides in Soil and Aqueous Suspensions of Soil and Sediment," *Water Res.*, 14:1095-1100 (1980a).

Shelton, D.R., S.A. Boyd, and J.M. Tiedje. "Anaerobic Biodegradation of Phthalic Acid Esters in Sludge," *Environ. Sci. Technol.*, 18(2):93-97 (1984).

Shevchenko, M.A., P.N. Taran, and P.V. Marchenko. "Technology of Water Treatment and Demineralization," *Soviet J. Water Chem. Technol.*, 4(4):53-71 (1982).

Shiaris, M.P. and G.S. Sayler. "Biotransformation of PCB by Natural Assemblages of Freshwater Microorganisms," *Environ. Sci. Technol.*, 16(6):367-369 (1982).

Shinozuka, N., C. Lee, and S. Hayano. "Solubilizing Action of Humic Acid from Marine Sediment," *Sci. Tot. Environ.*, 62:311-314 (1987).

Shiraishi, H., N.H. Pilkington, A. Otsuki, and K. Fuwa. "Occurrence of Chlorinated Polynuclear Aromatic Hydrocarbons in Tap Water," *Environ. Sci. Technol.*, 19(7):585-590 (1985).

Shriner, C.R., J.S. Drury, A.S. Hammons, L.E. Towill, E.B. Lewis, and D.M. Opresko. "Reviews of the Environmental Effects of Pollutants: II. Benzidine," Office of Research and Development, U.S. EPA Report-600/1-78-024 (1978), 157 p.

Siddaramappa, R., K.P. Rajaram, and N. Sethunathan. "Degradation of Parathion by Bacteria Isolated from Flooded Soil," *Appl. Microbiol.*, 26(6):846-849 (1973).

Sims, G.K. and L.E. Sommers. "Degradation of Pyridine Derivatives in Soil," *J. Environ. Qual.*, 14:580-584 (1985).

Sims, R.C., W.C. Doucette, J.E. McLean, W.J. Grenney, and R.R. DuPont. "Treatment Potential for 56 EPA Listed Hazardous Chemicals in Soil," U.S. EPA Report-600/6-88-001 (1988), 105 p.

Singer, M.E. and W.R. Finnerty. "Microbial Metabolism of Straight-Chain and Branched Alkanes," in *Petroleum Microbiology*, Atlas, R.M., Ed. (New York: Macmillan Publishing Co., 1984), pp. 1-59.

Singh, A.K. and P.K. Seth. "Degradation of Malathion by Microorganisms Isolated from Industrial Effluents," *Bull. Environ. Contam. Toxicol.*, 43(1):28-35 (1989).

Singh, J. "Conversion of Heptachloride to its Epoxide," *Bull. Environ. Contam. Toxicol.*, 4(2):77-79 (1969).

Singmaster, J.A., III. "Environmental Behavior of Hydrophobic Pollutants in Aqueous Solutions," PhD Thesis, University of California, Davis CA (1975).

Sinke, G.C. "A Method for Measurement of Vapor Pressures of Organic Compounds Below 0.1 Torr, Naphthalene as a Reference Substance," *J. Chem. Thermodyn.*, 6:311-316 (1974).

Siragusa, G.R. and D.D. DeLaune. "Mineralization and Sorption of *p*-Nitrophenol in Estuarine Sediment," *Environ. Toxicol. Chem.*, 5(1):175-178 (1986).

Sisler, H.D. and C.E. Cox. "Effects of Tetramethyl Thiuram Disulfide on Metabolism of *Fusarium roseum*," *Am. J. Bot.*, 41:338 (1954).

Sittig, M. *Handbook of Toxic and Hazardous Chemicals and Carcinogens* (Park Ridge, NJ: Noyes Publications, 1985), 950 p.

Skurlatov, Y.I., R.G. Zepp, and G.L. Baughman. "Photolysis Rates of (2,4,5-Trichlorophenoxy)acetic Acid and 4-Amino-3,5,6-trichloropicolinic Acid in Natural Waters," *J. Agric. Food Chem.*, 31(5):1065-1071 (1983).

Slater, R.M. and D.J. Spedding. "Transport of Dieldrin between Air and Water," *Arch. Environ. Contam. Toxicol.*, 10(1):25-33 (1981).

Slinn, W.G.N., L. Hasse, B.B. Hicks, A.W. Hogan, D. Lal, P.S. Liss, K.O. Munnich, G.A. Sehmel, and O. Vittori. "Some Aspects of the Transfer of Atmospheric Trace Constituents Past the Air-Sea Interface," *Atmos. Environ.*, 12:2055-2087 (1978).

Smith, A.E. "Identification of 2,4-Dichloroanisole and 2,4-Dichlorophenol as Soil Degradation Products of Ring-Labelled [^{14}C]2,4-D," *Bull. Environ. Contam. Toxicol.*, 34(2):150-157 (1985).

Smith, J.G., S.-F. Lee, and A. Netzer. "Model Studies in Aqueous Chlorination: The Chlorination of Phenols in Dilute Aqueous Solutions," *Water Res.*, 10(11):985-990 (1976).

Smith, J.H., W.R. Mabey, N. Bohonos, B.R. Holt, and S.S. Lee. "Environmental Pathways of Selected Chemicals in Freshwater Systems. Part II. Laboratory Studies," Environmental Research Lab, Athens, GA., U.S. EPA Report-600/7-78-074 (1978), 432 p.

Smith, L.R. and J. Dragun. "Degradation of Volatile Chlorinated Aliphatic Priority Pollutants in Groundwater," *Environ. Int.*, 19(4):291-298 (1984).

Smith, R.V. and J.P. Rosazza. "Microbial Models of Mammalian Metabolism. Aromatic Hydroxylation," *Arch. Biochem. Biophys.*, 161(2):551-558 (1974).

Snider, E.H. and F.C. Alley. "Kinetics of the Chlorination of Biphenyl under Conditions of Waste Treatment Processes," *Environ. Sci. Technol.*, 13(10):1244-1248 (1979).

Snider, J.R. and G.A. Dawson. "Tropospheric Light Alcohols, Carbonyls, and Acetonitrile: Concentrations in the Southwestern United States and Henry's Law

Data," *J. Geophys. Res., D: Atmos.*, 90(D2):3797–3805 (1985).

Socha, S.B. and R. Carpenter. "Factors Affecting the Pore Water Hydrocarbon Concentrations in Puget Sound Sediments," *Geochim. Cosmochim. Acta*, 51(5):1273–1284 (1987).

Soderquist, C.J., D.G. Crosby, K.W. Moilanen, J.N. Seiber, and J.E. Woodrow. "Occurrence of Trifluralin and its Photoproducts in Air," *J. Agric. Food Chem.*, 23(2):304–309 (1975).

Solon, J.M. and J.H. Nair. "The Effect of a Sublethal Concentration of LAS on the Acute Toxicity of Various Phosphate Pesticides to the Fathead Minnow *Pimephales promelas* Rafinesque," *Bull. Environ. Contam. Toxicol.*, 5(5):408–413 (1970).

Somasundaram, L. and J.R. Coats. "Interactions between Pesticides and Their Major Degradation Products," in *Pesticide Transformation Products. Fate and Significance in the Environment*, ACS Symposium Series 459, Somasundaram, L. and J.R. Coats, Eds. (Washington, DC: American Chemical Society, 1991), pp. 162–171.

Somasundaram, L., J.R. Coats, and K.D. Racke. "Degradation of Pesticides in Soil as Influenced by the Presence of Hydrolysis Metabolites," *J. Environ. Sci. Health*, B24(5):457–478 (1989).

Somasundaram, L., J.R. Coats, K.D. Racke, and V.M. Shanbhag. "Mobility of Pesticides and Their Metabolites in Soil," *Environ. Toxicol. Chem.*, 10(2):185–194 (1991).

Sonnefeld, W.J., W.H. Zoller, and W.E. May. "Dynamic Coupled-Column Liquid Chromatographic Determination of Ambient Temperature Vapor Pressures of Polynuclear Aromatic Hydrocarbons," *Anal. Chem.*, 55(2):275–280 (1983).

Sonobe, H., L.R. Kamps, E.P. Mazzola, and J.A.G. Roach. "Isolation and Identification of a New Conjugated Carbofuran Metabolite in Carrots: Angelic Acid Ester of 3-Hydroxycarbofuran," *J. Agric. Food Chem.*, 29(6):1125–1129 (1981).

Southworth, G.R. "Transports and Transformation of Anthracene in Natural Waters," in *Aquatic Toxicology, ASTM STP 667*, Marking, L.L. and R.A. Kimerle, Eds. (Philadelphia, PA: American Society for Testing and Materials, 1977), pp. 359–380.

Southworth, G.R. "The Role of Volatilization in Removing Polycyclic Aromatic Hydrocarbons from Aquatic Environments," *Bull. Environ. Contam. Toxicol.*, 21(4/5):507–514 (1979).

Southworth, G.R. and J.L. Keller. "Hydrophobic Sorption of Polar Organics by Low Organic Carbon Soils," *Water, Air, Soil Pollut.*, 28(3–4):239–248 (1986).

Southworth, G.R., J.J. Beauchamp, and P.K. Schmieders. "Bioaccumulation Potential of Polycyclic Aromatic Hydrocarbons in *Daphnia pulex*," *Water Res.*, 12:973–977 (1978).

Spacek, W., R. Bauer, and G. Heisler. "Heterogeneous and Homogeneous Wastewater Treatment – Comparison between Photodegradation with TiO$_2$ and Photo-

Fenton Reaction," *Chemosphere*, 30(3):477–484 (1995).

Spacie, A., R.F. Lundrum, and G.L. Leversee. "Uptake, Depuration and Bio-transformation of Anthracene and Benzo[*a*]pyrene in Bluegill Sunfish," *Ecotoxicol. Environ. Saf.*, 7:330–341 (1983).

Special Occupational Hazard Review for DDT (Rockville, MD: National Institute for Occupational Safety and Health, 1978), 205 p.

Spence, J.W., P.L. Hanst, and B.W. Gay, Jr. "Atmospheric Oxidation of Methyl Chloride, Methylene Chloride, and Chloroform," *J. Air. Pollut. Control Assoc.*, 26(10):994–996 (1976).

Spencer, W.F. and M.M. Cliath. "Vapor Density and Apparent Vapor Pressure of Lindane (γ-BHC)," *J. Agric. Food Chem.*, 18(3):529–530 (1970).

Spencer, W.F. and M.M. Cliath. "Vapor Density of Dieldrin," *Environ. Sci. Technol.*, 3(7):670–674 (1969).

Spencer, W.F. and M.M. Cliath. "Volatility of DDT and Related Compounds," *J. Agric. Food Chem.*, 20(3):645–649 (1972).

Spencer, W.F., J.D. Adams, T.D. Shoup, and R.C. Spear. "Conversion of Parathion to Paraoxon on Soil Dusts and Clay Minerals as Affected by Ozone and UV Light," *J. Agric. Food Chem.*, 28(2):366–371 (1980).

Spencer, W.F., M.M. Cliath, W.A. Jury, and L.-Z. Zhang. "Volatilization of Organic Chemicals from Soil as Related to Their Henry's Law Constants," *J. Environ. Qual.*, 17(3):504–509 (1988).

Stalling, D.L. and F.L. Meyer. "Toxicities of PCBs to Fish and Environmental Residues," *Environ. Health Perspect.*, 1:159–164 (1972).

Standen, A., Ed. *Kirk-Othmer Encyclopedia of Chemical Technology, Volume 1*, 2nd ed. (New York: John Wiley & Sons, Inc., 1963), 990 p.

Standen, A., Ed. *Kirk-Othmer Encyclopedia of Chemical Technology, Volume 3*, 2nd ed. (New York: John Wiley & Sons, Inc., 1964), 927 p.

Standen, A., Ed. *Kirk-Othmer Encyclopedia of Chemical Technology, Volume 5*, 2nd ed. (New York: John Wiley & Sons, Inc., 1965), 884 p.

Standen, A., Ed. *Kirk-Othmer Encyclopedia of Chemical Technology, Volume 12*, 2nd ed. (New York: John Wiley & Sons, Inc., 1967), 905 p.

Standen, A., Ed. *Kirk-Othmer Encyclopedia of Chemical Technology, Volume 15*, 2nd ed. (New York: John Wiley & Sons, Inc., 1968), 923 p.

Standen, A., Ed. *Kirk-Othmer Encyclopedia of Chemical Technology, Volume 19*, 2nd ed. (New York: John Wiley & Sons, Inc., 1969), 839 p.

Standen, A., Ed. *Kirk-Othmer Encyclopedia of Chemical Technology, Volume 21*, 2nd ed. (New York: John Wiley & Sons, Inc., 1970), 707 p.

Stauffer, T.B. and W.G. MacIntyre. "Sorption of Low-Polarity Organic Compounds on Oxide Minerals and Aquifer Material," *Environ. Toxicol. Chem.*, 5(11):949–955 (1986).

Stauffer, T.B., W.G. MacIntyre, and D.C. Wickman. "Sorption of Nonpolar Organic Chemicals on Low-Carbon-Content Aquifer Materials," *Environ. Toxicol. Chem.*, 8:845–852 (1989).

Stearns, R.S., H. Oppenheimer, E. Simon, and W.D. Harkins. "Solubilization by Solutions of Long-Chain Colloidal Electrolytes," *J. Chem. Phys.*, 15(7):496-507 (1947).

Steen, W.C., D.F. Paris, and G.L. Baughman. "Partitioning of Selected Polychlorinated Biphenyls to Natural Sediments," *Water Res.*, 12(9):655-657 (1978).

Steen, W.C. and S.W. Karickhoff. "Biosorption of Hydrophobic Organic Pollutants by Mixed Microbial Populations," *Chemosphere*, 10(1):27-32 (1981).

Steenson, T.I. and N. Walker. "The Pathway of Breakdown of 2,4-Dichloro- and 4-Chloro-2-methylphenoxyacetic Acid by Bacteria," *J. Gen. Microbiol.*, 16:146-155 (1957).

Steinwandter, H. "Experiments on Lindane Metabolism in Plants III. Formation of β-HCH," *Bull. Environ. Contam. Toxicol.*, 20(4):535-536 (1978).

Stephen, H. and T. Stephen. *Solubilities of Inorganic and Organic Compounds - Part 1, Volume 1* (London: Pergamon Printing & Art Services, Ltd., 1963), 960 p.

Stephenson, R. "Mutual Solubility of Water and Aldehydes," *J. Chem. Eng. Data*, 37(1):80-95 (1992).

Stephenson, R. and J. Stuart. "Mutual Binary Solubilities: Water-Alcohols and Water Esters," *J. Chem. Eng. Data*, 31(1):56-70 (1986).

Stephenson, R., J. Stuart, and M. Tabak. "Mutual Solubility of Water and Aliphatic Alcohols," *J. Chem. Eng. Data*, 29(3):287-290 (1984).

Stephenson, R.M. and S. Malanowski. *Handbook of the Thermodynamic of Organic Compounds* (New York: Elsevier Science Publishing Co., 1987).

Stevens, A.A., C.J. Slocum, D.R. Seeger, and G.G. Robeck. "Chlorination of Organics in Drinking Water," *J. Am. Water Works Assoc.*, 68(11):615-620 (1976).

Stewart, D.K.R. and K.G. Cairns. "Endosulfan Persistence in Soil and Uptake by Potato Tubers," *J. Agric. Food Chem.*, 22(6):984-986 (1974).

Stockdale, M. and M.J. Selwyn. "Effects of Rings Substituents on the Activity of Phenols as Inhibitors and Uncouplers of Mitochondrial Respiration," *Eur. J. Biochem.*, 21:565-574 (1971).

Stott, D.E., J.P. Martin, D.D. Focht, and K. Haider. "Biodegradation, Stabilization in Humus, and Incorporation into Soil Biomass of 2,4-D and Chlorocatechol Carbons," *Soil Sci. Soc. Am. J.*, 47:66-70 (1983).

Streitwieser, A., Jr. and L.L. Nebenzahl. "Carbon Acidity. LII. Equilibrium Acidity of Cyclopentadiene in Water and in Cyclohexylamine," *J. Am. Chem. Soc.*, 98(8):2188-2190 (1976).

Struif, B., L. Weil, and K.-E. Quentin. "VerhaltenHerbizider Phenoxyalkancarbonäuren bei der Wasseraufbereitung mit Ozon," *Zeit. fur Wasser und Abwasser-Forschung*, 3/4:118-127 (1978).

Stucki, G. and M. Alexander. "Role of Dissolution Rate and Solubility in Biodegradation of Aromatic Compounds," *Appl. Environ. Microbiol.*, 53(2):292-297 (1987).

Stull, D.R. "Vapor Pressures of Pure Organic Compounds," *Ind. Eng. Chem.*,

39(4):517-560 (1947).

Su, F., J.G. Calvert, and J.H. Shaw. "Mechanism of the Photooxidation of Gaseous Formaldehyde," *J. Phys. Chem.*, 83(25):3185-3191 (1979).

Subba-Rao, R.V., H.E. Rubin, and M. Alexander. "Kinetics and Extent of Mineralization of Organic Chemicals at Trace Levels in Freshwater and Sewage," *Appl. Environ. Microbiol.*, 43:1139-1150 (1982).

Subba-Rao, R.V. and M. Alexander. "Effect of DDT Metabolites on Soil Respiration and on an Aquatic Alga," *Bull. Environ. Contam. Toxicol.*, 25(2):215-220 (1980).

Sud, R.K., A.K. Sud, and K.G. Gupta. "Degradation of Sevin (1-Naphthyl *N*-methylcarbamate) by *Achromobacter* sp.," *Arch. Microbiol.*, 87:353-358 (1972).

Sudhakar-Barik and N. Sethunathan. "Biological Hydrolysis of Parathion in Natural Ecosystems," *J. Environ. Qual.*, 7(3):346-348 (1978).

Sudhakar-Barik, P.A. Wahid, C. Ramakrishna, and N. Sethunathan. "A Change in the Degradation Pathway of Parathion after Repeated Applications to Flooded Soil," *J. Agric. Food Chem.*, 27(6):1391-1392 (1979).

Sudhakar-Barik, R. Siddaramappa, and N. Sethunathan. "Metabolism of Nitrophenols by Bacteria Isolated from Parathion-Amended Flooded Soil," *Antonie van Leeuwenhoek*, 42(4):461-470 (1976).

Sudhakar-Barik, R. Siddaramappa, P.A. Wahid, and N. Sethunathan. "Conversion of *p*-Nitrophenol to 4-Nitrocatechol by a *Pseudomonas* sp.," *Antonie van Leeuwenhoek*, 44(2):171-176 (1978a).

Suffet, I.H., S.D. Faust, and W.F. Carey. "Gas-Liquid Chromatographic Separation of Some Organophosphate Pesticides, Their Hydrolysis Products, and Oxons," *Environ. Sci. Technol.*, 1(8):639-643 (1967).

Suflita, J.M., J. Stout, and J.M. Tiedje. "Dechlorination of (2,4,5-Trichlorophenoxy)acetic acid by Anaerobic Microorganisms," *J. Agric. Food Chem.*, 32(2):218-221 (1984).

Sugiura, K., T. Washino, M. Hattori, E. Sato, and M. Goto. "Accumulation of Organochlorine Compounds in Fishes. Difference of Accumulation Factors by Fishes," *Chemosphere*, 8:359-364 (1979).

Sullivan, R.G., H.W. Knoche, and J.C. Markle. "Photolysis of Trifluralin: Characterization of Azobenzene and Azoxybenzene Photodegradation Products," *J. Agric. Food Chem.*, 28(4):746-755 (1980).

Sumer, K.M. and A.R. Thompson. "Refraction, Dispersion, and Densities of Benzene, Toluene, and Xylene Mixtures," *J. Chem. Eng. Data*, 13(1):30-34 (1968).

Summers, L. "The *alpha*-Haloalkyl Ethers," *Chem. Rev.*, 55:301-353 (1955).

Sun, S. and S.A. Boyd. "Sorption of Nonionic Organic Compounds in Soil-Water Systems Containing Petroleum Sulfonate-Oil Surfactants," *Environ. Sci. Technol.*, 27(7):1340-1348 (1993).

Sun, Y. and J.J. Pignatello. "Photochemical Reactions Involved in the Total Mineralization of 2,4-D by $Fe^{3+}/H_2O_2/UV$," *Environ. Sci. Technol.*, 27(2):304-

310 (1993).

Sunshine, I., Ed. *Handbook of Analytical Toxicology* (Cleveland, OH: The Chemical Rubber Co., 1969), 1081 p.

Suntio, L.R., W.-Y. Shiu, and D. Mackay. "A Review of the Nature and Properties of Chemicals Present in Pulp Mill Effluents," *Chemosphere*, 17(7):1249-1290 (1988).

Sutton, C. and J.A. Calder. "Solubility of Alkylbenzenes in Distilled Water and Seawater at 25 °C," *J. Chem. Eng. Data*, 20(3):320-322 (1975).

Sutton, C. and J.A. Calder. "Solubility of Higher-Molecular-Weight *n*-Paraffins in Distilled Water and Seawater," *Environ. Sci. Technol.*, 8(7):654-657 (1974).

Suzuki, J., T. Hagino, and S. Suzuki. "Formation of 1-Nitropyrene by Photolysis of Pyrene in Water Containing Nitrite Ion," *Chemosphere*, 16(4):859-867 (1987).

Swain, C.G. and E.R. Thornton. "Initial-State and Transition-State Isotope Effects of Methyl Halides in Light and Heavy Water," *J. Am. Chem. Soc.*, 84:822-826 (1962).

Swann, R.L., D.A. Laskowski, P.J. McCall, K. Vander Kuy, and H.J. Dishburger. "A Rapid Method for the Estimation of the Environmental Parameters Octanol/Water Partition Coefficient, Soil Sorption Constant, Water to Air Ratio, and Water Solubility," *Residue Rev.*, 85:17-28 (1983).

Tabak, H.H., S.A. Quave, C.I. Mashni, and E.F. Barth. "Biodegradability Studies with Organic Priority Pollutant Compounds," *J. Water Pollut. Control Fed.*, 53(10):1503-1518 (1981).

Taha, A.A., R.D. Grigsby, J.R. Johnson, S.D. Christian, and H.E. Affsprung. "Manometric Apparatus for Vapor and Solution Studies," *J. Chem. Educ.*, 43(8):432-43435 (1966).

Takagi, H., S. Hatakeyama, H. Akimoto, and S. Koda. "Formation of Methyl Nitrite in the Surface Reaction of Nitrogen Dioxide and Methanol. 1. Dark Reaction," *Environ. Sci. Technol.*, 20(4):387-393 (1986).

Takahashi, N. "Ozonation of Several Organic Compounds having Low Molecular Weight under Ultraviolet Irradiation," *Ozone Sci. Eng.*, 12(1):1-18 (1990).

Takeuchi, K., T. Yazawa, and T. Ibusuki. "Heterogeneous Photocatalytic Effect of Zinc Oxide on Photochemical Smog Formation Reaction of C_4H_8-NO_2-Air," *Atmos. Environ.*, 17(11):2253-2258 (1983).

Tan, C.K. and T.C. Wang. "Reduction of Trihalomethanes in a Water-Photolysis System," *Environ. Sci. Technol.*, 21(5):508-511 (1987).

Tancréde, M.V. and Y. Yanagisawa. "An Analytical Method to Determine Henry's Law Constant for Selected Volatile Organic Compounds at Concentrations and Temperatures Corresponding to Tap Water Use," *J. Air Waste Manage. Assoc.*, 40:1658-1663 (1990).

Tanni, H., H. Tsuji, and K. Hashimoto. "Structure-Acivity Relationship of Monoketones," *Toxicol. Letters*, 30:13-17 (1984).

Tanni, H. and K. Hashimoto. "Studies on the Mechanism of Acute Toxicity of Nitriles in Mice," *Arch. Toxicol.*, 55:47-54 (1984).

Tate, R.L. III and M. Alexander. "Microbial Formation and Degradation of Dimethylamine," *Appl. Environ. Microbiol.*, 31:399-403 (1976).

Taymaz, K., D.T. Williams, and F.M. Benoit. "Chlorine Dioxide Oxidation of Aromatic Hydrocarbons Commonly Found in Water," *Bull. Environ. Contam. Toxicol.*, 23(3):398-404 (1979).

ten Hulscher, Th.E.M., L.E. er Velde, and W.A. Bruggeman. "Temperature Dependence of Henry's Law Constants for Selected Chlorobenzenes, Polychlorinated Biphenyls and Polycyclic Aromatic Hydrocarbons," *Environ. Toxicol. Chem.*, 11(11):1595-1603 (1992).

Tewari, Y.B., M.M. Miller, S.P. Wasik, and D.E. Martire. "Aqueous Solubility and Octanol/Water Partition Coefficient of Organic Compounds at 25.0 °C," *J. Chem. Eng. Data*, 27(4):451-454 (1982).

Tinsley, L.J. *Chemical Concepts in Pollutant Behavior* (New York: John Wiley & Sons, Inc., 1979), 265 p.

Tipker, J., C.P. Groen, J.K. Van Den Bergh-Swart, and J.H.M. Van Den Berg. "Contribution of Electronic Effects to the Lipophilicity Determined by Comparison of Values of Log *P* Obtained by High-Performance Liquid Chromatography and Calculation," *J. Chromatogr.*, 452:227-239 (1988).

Tokoro, R., R. Bilcovicz, and J. Osteryoung. "Polarographic Reduction and Determination of 1,2-Dibromoethane in Aqueous Solutions," *Anal. Chem.*, 58(9):1964-1969 (1986).

Tomkiewicz, M.A., A. Groen, and M. Cocivera. "Electron Paramagnetic Resonance Spectra of Semiquinone Intermediates Observed during the Photooxidation of Phenol in Water," *J. Am. Chem. Soc.*, 93(25):7102-7103 (1971).

Tomlinson, E. "Chromatographic Hydrophobic Parameters in Correlation Analysis of Structure-Activity Relationships," *J. Chromatogr.*, 113:1-45 (1975).

Tou, J.C. and G.J. Kallos. "Study of Aqueous HCl and Formaldehyde Mixtures for Formation of Bis(chloromethyl)ether," *J. Am. Ind. Hyg. Assoc.*, 35(7):419-422 (1974).

Travis, C.C. and A.D. Arms. "Bioconcentration of Organics in Beef, Milk and Vegetation," *Environ. Sci. Technol.*, 22(3):271-274 (1988).

Tse, G., H. Orbey, and S.I. Sandler. "Infinite Dilution Activity Coefficients and Henry's Law Coefficients of Some Priority Water Pollutants Determined by a Relative Gas Chromatographic Method," *Environ. Sci. Technol.*, 25(10):2017-2022 (1992).

Tu, C.M. "Utilization and Degradation of Lindane by Soil Microorganisms," *Arch. Microbiol.*, 108(3):259-263 (1976).

Tuazon, E.C., A.M. Winer, R.A. Graham, J.P. Schmid, and J.N. Pitts, Jr. "Fourier Transform Infrared Detection of Nitramines in Irradiated Amine-NO$_x$ Systems," *Environ. Sci. Technol.*, 12(8):954-958 (1978).

Tuazon, E.C., H.M. Leod, R. Atkinson, and W.P.L. Carter. "α-Dicarbonyl Yields from the NO$_x$-Air Photooxidations of a Series of Aromatic Hydrocarbons in Air," *Environ. Sci. Technol.*, 20(4):383-387 (1986).

Tuazon, E.C., R. Atkinson, A.M. Winer, and J.N. Pitts, Jr. "A Study of the Atmospheric Reactions of 1,3-Dichloropropene and Other Selected Organochlorine Compounds," *Arch. Environ. Contam. Toxicol.*, 13(6):691-700 (1984).

Tuazon, E.C., W.P.L. Carter, R. Atkinson, A.M. Winer, and J.N. Pitts, Jr. "Atmospheric Reactions of *N*-Nitrosodimethylamine and Dimethylnitramine," *Environ. Sci. Technol.*, 18(1):49-54 (1984).

Tuhkanen, T.A. and F.J. Beltrán. "Intermediates in the Oxidation of Naphthalene in Water with the Combination of Hydrogen Peroxide and UV Radiation," *Chemosphere*, 30(8):1463-1475 (1995).

Tundo, P., S. Facchetti, W. Tumiatti, and U.G. Fortunati. "Chemical Degradation of 2,3,7,8-TCDD by Means of Polyethylene-Glycols in the Presence of Weak Bases and an Oxidant," *Chemosphere*, 14(5):403-410 (1985).

Uchrin, C.G. and G. Michaels. "Chloroform Sorption to New Jersey Coastal Plain Ground Water Aquifer Solids," *Environ. Toxicol. Chem.*, 5(4):339-343 (1986).

Unger, S.H., J.R. Cook, and J.S. Hollenberg. "Simple Procedure for Determining Octanol-Aqueous Partition, Distribution, and Ionization Coefficients by Reversed-Phase High-Pressure Liquid Chromatography," *J. Pharm. Sci.*, 67(10):1364-1366 (1978).

Ungnade, H.E. and E.T. McBee. "The Chemistry of Perchlorocyclopentadienes and Cyclopentadienes," *Chem. Rev.*, 58:249-320 (1957).

Ursin, C. "Degradation of Organic Chemicals at Trace Levels in Seawater and Marine Sediment. The Effect of Concentration on the Initial Turnover Rate," *Chemosphere*, 14(10):1539-1550 (1985).

U.S. EPA. 1975. "Preliminary Study of Selected Potential Environmental Contaminants - Optical Brighteners, Methyl Chloroform, Trichloroethylene, Tetrachloroethylene, Ion Exchange Resins," Office of Toxic Substances, Report-560/2-75-002, 286 p.

U.S. EPA. 1975a. "Report on the Problem of Halogenated Air Pollutants and Stratigraphic Ozone," Office of Research and Development, Report-600/9-75-008, 55 p.

U.S. EPA. 1976. "A Literature Survey Oriented Towards Adverse Environmental Effects Resultant from the Use of Azo Compounds, Brominated Hydrocarbons, EDTA, Formaldehyde Resins, and *o*-Nitrochlorobenzene," Office of Toxic Substances, Report-560/2-76-005, 480 p.

U.S. EPA. 1980. "Treatability Manual - Volume 1: Treatability Data," Office of Research and Development, Report-600/8-80-042a, 1035 p.

U.S. EPA. 1982. "Aquatic Fate Process Data for Organic Priority Pollutants," Office of Water Regulations and Standards, Report-440/4-81-014, 407 p.

U.S. EPA. 1985. "Chemical, Physical, and Biological Properties of Compounds Present at Hazardous Waste Sites," Report-530/SW-89-010, 619 p.

U.S. EPA. 1994. "Drinking Water Regulations and Health Advisories," Report-822-R-94-001, 11 p.

Uyeta, M., S. Taue, K. Chikasawa, and M. Mazaki. "Photoformation of Polychlorinated Biphenyls from Chlorinated Benzenes," *Nature (London)*, 264(5586):583-584 (1976).

Vadas, G.G., W.G. MacIntyre, and D.R. Burris. "Aqueous Solubility of Liquid Hydrocarbon Mixtures Containing Dissolved Solid Components," *Environ. Toxicol. Chem.*, 10:633-639 (1991).

Vaishnav, D.D. and L. Babeu. Comparison of Occurrence and Rates of Chemical Biodegradation in Natural Waters," *Bull. Environ. Contam. Toxicol.*, 39:237-244 (1987).

Valsaraj, K.T. "On the Physio-Chemical Aspects of Partitioning of Non-Polar Hydrophobic Organics at the Air-Water Interface," *Chemosphere*, 17(5):875-887 (1988).

Valvani, S.C., S.H. Yalkowsky, and T.J. Roseman. "Solubility and Partitioning IV: Aqueous Solubility and Octanol-Water Partition Coefficients of Liquid Non-electrolytes," *J. Pharm. Sci.*, 70(5):502-506 (1981).

Van Arkel, A.E. and S.E. Vles. "Löslichkeit von Organishen Verbindungen in Wasser," *Recl. Trav. Chim. Pays-Bas*, 55:407-411 (1936).

Van Bladel, M. and A. Moreale. "Adsorption of Herbicide-Derived *p*-Chloroaniline Residues in Soils: A Predictive Equation," *J. Soil Sci.*, 28:93-102 (1977).

Van der Linden, A.C. "Degradation of Oil in the Marine Environment," *Dev. Biodegradation Hydrocarbons*, 1:165-200 (1978).

van Gestel, C.A.M. and W.-C. Ma. "Toxicity and Bioaccumulation of Chlorophenols in Earthworms, in Relation to Bioavailability in Soil," *Ecotoxicol. Environ. Saf.*, 15(3):289-297 (1988).

van Ginkel, C.G., H.G.J. Welten, and J.A.M. de Bont. "Oxidation of Gaseous and Volatile Hydrocarbons by Selected Alkene-Utilizing Bacteria," *Appl. Environ. Microbiol.*, 53(12):2903-2907 (1987).

Van Meter, F.M. and H.M. Neumann. "Solvation of the Tris(1,10-phenanthroline)iron(II) Cation as Measured by Solubility and Nuclear Magnetic Resonance Shifts," *J. Am. Chem. Soc.*, 98(6):1382-1388 (1976).

Veith, G.D., D.J. Call, and L.T. Brooke. "Structure-Toxicity Relationships for the Fathead Minnow *Pimephales promelas*: Narcotic Industrial Chemicals," *Can. J. Fish. Aquat. Sci.*, 40(6):743-748 (1983).

Veith, G.D., D.L. DeFoe, and B.V. Bergstedt. "Measuring and Estimating the Bioconcentration Factor of Chemicals in Fish," *J. Fish Res. Board Can.*, 26:1040-1048 (1979).

Veith, G.D., K.J. Macek, S.R. Petrocelli, and J. Carroll. "An Evaluation of Using Partition Coefficients and Water Solubility to Estimate Bioconcentration Factors for Organic Chemicals in Fish," in *Aquatic Toxicology, ASTM STP 707*, Eaton, J.G., P.R. Parrish, and A.C. Hendricks, Eds. (Philadelphia, PA: American Society for Testing and Materials, 1980), pp. 116-129.

Veith, G.D., N.M. Austin, and R.T. Morris. "A Rapid Method for Estimating Log P for Organic Chemicals," *Water Res.*, 13(1):43-47 (1979a).

Venkateswarlu, K., K. Chendrayan, and N. Sethunathan. "Persistence and Biodegradation of Carbaryl in Soils," *J. Environ. Sci. Health*, B15(4):421-429 (1980).

Venkateswarlu, K. and N. Sethunathan. "Metabolism of Carbofuran in Rice Straw-Amended and Unamended Rice Soils," *J. Environ. Qual.*, 8(3):365-368 (1979).

Venkateswarlu, K. and N. Sethunathan. "Degradation of Carbofuran by *Azospirillium lipoferum* and *Streptomyces* spp. Isolated from Flooded Alluvial Soil," *Bull. Environ. Contam. Toxicol.*, 33(5):556-560 (1984).

Venkateswarlu, K., T.K.S. Gowda, and N. Sethunathan. "Persistence and Biodegradation of Carbofuran in Flooded Soils," *J. Agric. Food Chem.*, 25(3):533-536 (1977).

Verschueren, K. *Handbook of Environmental Data on Organic Chemicals* (New York: Van Nostrand Reinhold Co., 1983), 1310 p.

Vesala, A. "Thermodynamics of Transfer of Nonelectrolytes from Light to Heavy Water. I. Linear Free Energy Correlations of Free Energy of Transfer with Solubility and Heat of Melting of a Nonelectrolyte," *Acta Chem. Scand.*, A28:839-845 (1974).

Vesala, A. and H. Lönnberg. "Salting-In Effects of Alkyl-Substituted Pyridinium Chlorides on Aromatic Compounds," *Acta Chem. Scand.*, A34:187-192 (1980).

Vogel, T.M. and M. Reinhard. "Reaction Products and Rates of Disappearance of Simple Bromoalkanes, 1,2-Dibromopropane, and 1,2-Dibromoethane in Water," *Environ. Sci. Technol.*, 20(10):992-997 (1986).

Vogel, T.M. and P.L. McCarty. "Biotransformation of Tetrachloroethylene to Trichloroethylene, Dichloroethylene, Vinyl Chloride, and Carbon Dioxide under Methanogenic Conditions," *Appl. Environ. Microbiol.*, 49(5):1080-1083 (1985).

Vogel, T.M., C.S. Criddle, and P.L. McCarty. "Transformations of Halogenated Aliphatic Compounds," *Environ. Sci. Technol.*, 21(8):722-736 (1987).

Voice, T.C. and W.J. Weber, Jr. "Sorbent Concentration Effects in Liquid/Solid Partitioning," *Environ. Sci. Technol.*, 19(9):789-796 (1985).

Vontor, T., J. Socha, and M. Vecera. "Kinetics and Mechanism of Hydrolysis of 1-Naphthyl, *N*-Methylcarbamate and *N,N*-Dimethyl Carbamates," *Collect. Czech. Chem. Commun.*, 37:2183-2196 (1972).

Vowles, P.D. and R.F.C. Mantoura. "Sediment-Water Partition Coefficients and HPLC Retention Factors of Aromatic Hydrocarbons," *Chemosphere*, 16(1):109-116 (1987).

Vozňáková, Z., M. Popl, and M. Berka. "Recovery of Aromatic Hydrocarbons from Water," *J. Chromatogr. Sci.*, 16:123-127 (1978).

Wahid, P.A. and N. Sethunathan. "Sorption-Desorption of *Alpha*, *Beta* and *Gamma* Isomers of BHC in Soils," *J. Agric. Food Chem.*, 27(5):1050-1053 (1979).

Wahid, P.A., C. Ramakrishna, and N. Sethunathan. "Instantaneous Degradation of Parathion in Anaerobic Soils," *J. Environ. Qual.*, 9(1):127-130 (1980).

Wajon, J.E., D.H. Rosenblatt, and E.P. Burrows. "Oxidation of Phenol and Hydroquinone by Chlorine Dioxide," *Environ. Sci. Technol.*, 16(7):396-402 (1982).

Walker, W.W. "Chemical and Microbiological Degradation of Malathion and Parathion in an Estuarine Environment," *J. Environ. Qual.*, 5(2):210-216 (1976).

Walker, W.W. and B.J. Stojjanovic. "Malathion Degradation by an *Arthrobacter* sp.," *J. Environ. Qual.*, 3(1):4-10 (1974).

Wallington, T.J. and S.M. Japar. "Atmospheric Chemistry of Diethyl Ether and Ethyl *tert*-Butyl Ether," *Environ. Sci. Technol.*, 25(3):410-415 (1991).

Walters, R.W. and A. Guiseppi-Elie. "Sorption of 2,3,7,8-Tetrachlorodibenzo-*p*-dioxin to Soils from Water/Methanol Mixtures," *Environ. Sci. Technol.*, 22(7):819-825 (1988).

Walters, R.W., S.A. Ostazeski, and A. Guiseppi-Elie. "Sorption of 2,3,7,8-Tetrachlorodibenzo-*p*-Dioxin from Water by Surface Soils," *Environ. Sci. Technol.*, 23(4):480-484 (1989).

Walton, B.T., M.S. Hendricks, T.A. Anderson, W.H. Griest, R. Merriweather, J.J. Beauchamp, and C.W. Francis. "Soil Sorption of Volatile and Semivolatile Organic Compounds in a Mixture," *J. Environ. Qual.*, 21(4):552-558 (1992).

Walton, B.T., T.A. Anderson, M.S. Hendricks, and S.S. Talmage. "Physicochemical Properties as Predictors of Organic Chemical Effects on Soil Microbial Respiration," *Environ. Toxicol. Chem.*, 8(1):53-63 (1989).

Walton, W.C. *Practical Aspects of Ground Water Modeling* (Worthington, OH: National Water Well Association, 1985), 587 p.

Wams, T.J. "Diethylhexylphthalate as an Environmental Contaminant - A Review," *Sci. Tot. Environ.*, 66:1-16 (1987).

Wang, C.X., A. Yediler, A. Peng, and A. Kettrup. "Photodegradation of Phenanthrene in the Presence of Humic Substances and Hydrogen Peroxide," *Chemosphere*, 30(3):501-510 (1995).

Wang, L., L. Kong, and C. Chang. "Photodegradation of 17 PAHs in Methanol (or Acetonitrile)-Water Solution," *Environ. Chem.*, 10(2):15-20 (1991).

Wang, L., Y. Zhao, and G. Hong. "Predicting Aqueous Solubility and Octanol/Water Partition Coefficients of Organic Chemicals from Molecular Volume," *Environ. Chem.*, 11:55-70 (1992).

Wang, L., R. Govind, and R.A. Dobbs. "Sorption of Toxic Organic Compounds on Wastewater Solids: Mechanism and Modeling," *Environ. Sci. Technol.*, 27(1):152-158 (1992).

Wang, T.C. and C.K. Tan. "Enhanced Degradation of Halogenated Hydrocarbons in a Water-Photolysis System," *Bull. Environ. Contam. Toxicol.*, 40(1):60-65 (1988).

Wang, T.C., C.K. Tan, and M.C. Liou. "Degradation of Bromoform and Chlorodibromomethane in a Catalyzed H_2-Water System," *Bull. Environ. Contam. Toxicol.*, 41(4):563-568 (1988).

Wang, Y.-S., R.V. Subba-Rao, and M. Alexander. "Effect of Substrate Concentration and Organic and Inorganic Compounds on the Occurrence and Rate of Mineralization and Cometabolism," *Appl. Environ. Microbiol.*, 47:1195-1200 (1984).

Wang, Y.-T. "Methanogenic Degradation of Ozonation Products of Biorefractory or Toxic Aromatic Compounds," *Water Res.*, 24(2):185-190 (1990).

Wannstedt, C., D. Rotella, and J.F. Siuda. "Chloroperoxidase Mediated Halogenation of Phenols," *Bull. Environ. Contam. Toxicol.*, 44(2):282-287 (1990).

Warneck, P. and T. Zerbach. "Synthesis of Peroxyacetyl Nitrate in Air by Acetone Photolysis," *Environ. Sci. Technol.*, 26(1):74-79 (1992).

Warner, H.P., J.M. Cohen, and J.C. Ireland. "Determination of Henry's Law Constants of Selected Priority Pollutants," Office of Science and Development, U.S. EPA Report-600/D-87/229 (1987), 14 p.

Wasik, S.P., F.P. Schwarz, Y.B. Tewari, and M.M. Miller. "A Head-Space Method for Measuring Activity Coefficients, Partition Coefficients, and Solubilities of Hydrocarbons in Saline Solutions," *J. Res. Nat. Bur. Stand.*, 89(3):273-277 (1984).

Wasik, S.P., M.M. Miller, Y.B. Tewari, W.E. May, W.J. Sonnefeld, H. DeVoe, and W.H. Zoller. "Determination of the Vapor Pressure, Aqueous Solubility, and Octanol/Water Partition Coefficient of Hydrophobic Substances by Coupled Generator Column/Liquid Chromatographic Methods," *Residue Rev.*, 85:29-42 (1983).

Wasik, S.P., Y.B. Tewari, M.M. Miller, and D.E. Martire. "Octanol/Water Partition Coefficients and Aqueous Solubilities of Organic Compounds," U.S. EPA Report PB82-141797 (1981), 56 p.

Watanabe, I., T. Kashimoto, and R. Tatsukawa. "Brominated Phenol Production from the Chlorination of Wastewater Containing Bromide Ions," *Bull. Environ. Contam. Toxicol.*, 33(4):395-399 (1984).

Watanabe, N., E. Sato, and Y. Ose. "Adsorption and Desorption of Polydimethylsiloxane, Cadmium Nitrate, Copper Sulfate, Nickel Sulfate and Zinc Nitrate by River Surface Sediments," *Sci. Tot. Environ.*, 41(2):163-161 (1985).

Watarai, H., M. Tanaka, and N. Suzuki. "Determination of Partition Coefficients of Halobenzenes in Heptane/Water and 1-Octanol/Water Systems and Comparison with the Scaled Particle Calculation," *Anal. Chem.*, 54(4):702-705 (1982).

Watson, E. and C.F. Parrish. "Laser Induced Decomposition of 1,4-Dioxane," *J. Chem. Phys.*, 54(3):1427-1428 (1971).

Wauchope, R.D. and F.W. Getzen. "Temperature Dependence of Solubilities in Water and Heats of Fusion of Solid Aromatic Hydrocarbons," *J. Chem. Eng. Data*, 17(1):38-41 (1972).

Wauchope, R.D. and R. Haque. "Effects of pH, Light and Temperature on Carbaryl in Aqueous Media," *Bull. Environ. Contam. Toxicol.*, 9(5):257-261 (1973).

Weast, R.C. and M.J. Astle, Eds. *CRC Handbook of Data on Organic Compounds - 2 Volumes* (Boca Raton, FL: CRC Press, Inc. 1986).

Weast, R.C., Ed. *CRC Handbook of Chemistry and Physics*, 67th ed. (Boca Raton, FL: CRC Press, Inc., 1986), 2406 p.

Webb, R.F., A.J. Duke, and L.S.A. Smith. "Acetals and Oligoacetals. Part I. Preparation and Properties of Reactive Oligoformals," *J. Chem. Soc. (London)*, pp. 4307-4319, (1962).

Weber, J.B. "Interaction of Organic Pesticides with Particulate Matter in Aquatic and Soil Systems," in *Fate of Organic Pesticides in the Aquatic Environment,*

Advances in Chemistry Series, R.F. Gould, Ed. (Washington, DC: American Chemical Society, 1972), pp. 55-120.

Webster, G.R.B., K.J. Friesen, L.P. Sarna, and D.C.G. Muir. "Environmental Fate Modelling of Chlorodioxins: Determination of Physical Constants," *Chemosphere*, 14(6/7):609-622 (1985).

Wedemeyer, G.A. "Dechlorination of DDT by *Aerobacter aerogenes*," *Science (Washington, DC)*, 152(3722):647 (1966).

Weil, L., G. Dure, and K.E. Quentin. "Solubility in Water of Insecticide Chlorinated Hydrocarbons and Polychlorinated Biphenyls in View of Water Pollution," *Z. Wasser Forsch.*, 7(6):169-175 (1974).

Weiss, G. *Hazardous Chemicals Data Book* (Park Ridge, NJ: Noyes Data Corp., 1986), 1069 p.

Weiss, U.M., I. Scheunert, W. Klein, and F. Korte. "Fate of Pentachlorophenol-^{14}C in Soil under Controlled Conditions," *J. Agric. Food Chem.*, 30(6):1191-1194 (1982).

Wescott, J.W. and T.F. Bidleman. "Determination of Polychlorinated Biphenyl Vapor Pressures by Capillary Gas Chromatography," *J. Chromatogr.*, 210(2):331-336 (1981).

Westcott, J.W., C.G. Simon, and T.F. Bidleman. "Determination of Polychlorinated Biphenyl Vapor Pressures by a Semimicro Gas Saturation Method," *Environ. Sci. Technol.*, 15(11):1375-1378 (1981).

Whitbeck, M. "Photo-Oxidation of Methanol," *Atmos. Environ.*, 17(1):121-126 (1983).

Whitehouse, B.G. "The Effects of Temperature and Salinity on the Aqueous Solubility of Polynuclear Aromatic Hydrocarbons," *Mar. Chem.*, 14:319-332 (1984).

WHO. 1979. Environmental Health Criteria 10: Carbon disulfide (Geneva: World Health Organization), 100 p.

WHO. 1983. Environmental Health Criteria 28: Acrylonitrile (Geneva: World Health Organization), 125 p.

WHO. 1983a. Environmental Health Criteria 26: Styrene (Geneva: World Health Organization), 123 p.

WHO. 1984. Environmental Health Criteria 38: Heptachlor (Geneva: World Health Organization), 81 p.

WHO. 1984a. Environmental Health Criteria 45: Camphechlor (Geneva: World Health Organization), 66 p.

Wiese, C.S. and D.A. Griffin. "The Solubility of Aroclor 1254 in Seawater," *Bull. Environ. Contam. Toxicol.*, 19(4):403-411 (1978).

Wilhoit, R.C. and B.J. Zwolinski. *Handbook of Vapor Pressure and Heats of Vaporization of Hydrocarbons and Related Compounds*, Publication 101 (College Station, TX: Thermodynamics Research Station, 1971), 329 p.

Williams, E.F. "Properties of *O,O*-Diethyl *O-p*-Nitrophenyl Thiophosphate and *O,O*-Diethyl *O-p*-Nitrophenyl Phosphate," *Ind. Eng. Chem.*, 43(4):950-954

(1951).

Willingham, C.B., W.J. Taylor, J.M. Pignocco, and F.D. Rossinni. "Vapor Pressure and Boiling Points of Some Paraffin, Akylcyclopentane, Akylcyclohexane, and Akylbenzene Hydrocarbons," *J. Res. Nat. Bur. Stand.*, 34:219 (1945).

Wilson, B.H., G.B. Smith, and J.F. Rees. "Biotransformations of Selected Alkyl-benzenes and Halogenated Aliphatic Hydrocarbons in Methanogenic Aquifer Material: A Microcosm Study," *Environ. Sci. Technol.*, 20(10):997–1002 (1986).

Wilson, J.T., C.G. Enfield, W.J. Dunlap, R.L. Cosby, D.A. Foster, and L.B. Baskin. "Transport and Fate of Selected Organic Pollutants in a Sandy Soil," *J. Environ. Qual.*, 10(4):501–506 (1981).

Windholz, M., S. Budavari, L.S. Stroumtsos, and M.N. Fertig, Eds. *The Merck Index*, 9th ed. (Rahway, NJ: Merck & Co., 1976), 1313 p.

Windholz, M., S. Budavari, R.F. Blumetti, and E.S. Otterbein, Eds. *The Merck Index*, 10th ed. (Rahway, NJ: Merck & Co., 1983), 1463 p.

Wiseman, A., T.-K. Lim, and L.F.J. Woods. "Regulation of the Biosynthesis of Cytochrome P-450 in Brewer's Yeast. Role of Cyclic AMP," *Biochim. Biophys. Acta*, 544:615–623 (1978).

Witherspoon, P.A. and L. Bonoli. "Correlation of Diffusion Coefficients for Paraffin, Aromatic, and Cycloparaffin Hydrocarbons in Water," *Ind. Eng. Chem. Fundam.*, 8(3):589–591 (1969).

Wolfe, N.L., R.G. Zepp, G.L. Baughman, R.C. Fincher, and J.A. Gordon. "Chemical and Photochemical Transformations of Selected Pesticides in Aquatic Systems," U.S. EPA Report 600/3-76-067 (1976), 141 p.

Wolfe, N.L., R.G. Zepp, D.F. Paris, G.L. Baughman, and R.C. Hollis. "Methoxy-chlor and DDT Degradation in Water: Rates and Products," *Environ. Sci. Technol.*, 11(12):1077–1081 (1977).

Wolfe, N.L., R.G. Zepp, J.A. Gordon, G.L. Baughman, and D.M. Cline. "Kinetics of Chemical Degradation of Malathion in Water," *Environ. Sci. Technol.*, 11(1):88–93 (1977a).

Wolfe, N.L., R.G. Zepp, and D.F. Paris. "Use of Structure-Reactivity Relationships to Estimate Hydrolytic Persistence of Carbamate Pesticides," *Water Res.*, 12(8):561–563 (1978).

Wolfe, N.L., R.G. Zepp, and D.F. Paris. "Carbaryl, Phospham and Chloropropham: A Comparison of the Rates of Hydrolysis and Photolysis with the Rate of Biolysis," *Water Res.*, 12(8):565–571 (1978a).

Wolfe, N.L., W.C. Steen, and L.A. Burns. "Phthalate Ester Hydrolysis: Linear Free Energy Relationships," *Chemosphere*, 9(7/8):403–408 (1980).

Wong, A.S. and D.G. Crosby. "Photodecomposition of Pentachlorophenol in Water," *J. Agric. Food Chem.*, 29(1):125–130 (1981).

Wood, A.L., D.C. Bouchard, M.L. Brusseau, and P.S.C. Rao. "Cosolvent Effects on Sorption and Mobility of Organic Contaminants in Soils," *Chemosphere*, 21(4/5):575–587 (1990).

Woodburn, K.B., W.J. Doucette, and A.W. Andren. "Generator Column Determination of Octanol/Water Partition Coefficients for Selected Polychlorinated Biphenyl Congeners," *Environ. Sci. Technol.*, 18(6):457-459 (1984).

Woodrow, J.E., D.G. Crosby, T. Mast, K.W. Moilanen, and J.N. Seiber. "Rates of Transformation of Trifluralin and Parathion Vapors in Air," *J. Agric. Food Chem.*, 26(6):1312-1316 (1978).

Worley, J.D. "Benzene as a Solute in Water," *Can. J. Chem.*, 45:2465-2467 (1967).

Worthing, C.R. and R.J. Hance, Eds. *The Pesticide Manual - A World Compendium*, 9th ed. (Great Britain: British Crop Protection Council, 1991), 1141 p.

Xie, T.M. "Determination of Trace Amounts of Chlorophenols and Chloroguaicols in Sediment," *Chemosphere*, 12(9/10):1183-1191 (1983).

Yagi, O. and R. Sudo. "Degradation of Polychlorinated Biphenyls by Micro-organisms," *J. Water Pollut. Control Fed.*, 52(5):1035-1043 (1980).

Yalkowsky, S.H. and S.C. Valvani. "Solubilities and Partitioning 2. Relationships between Aqueous Solubilities, Partition Coefficients, and Molecular Surface Areas of Rigid Aromatic Hydrocarbons," *J. Chem. Eng. Data*, 24(2):127-129 (1979).

Yalkowsky, S.H., R.J. Orr, and S.C. Valvani. "Solubility and Partitioning. 3. The Solubility of Halobenzenes in Water," *Indust. Eng. Chem. Fundam.*, 18(4):351-353 (1979).

Yalkowsky, S.H., S.C. Valvani, and D. Mackay. "Estimation of the Aqueous Solubility of Some Aromatic Compounds," *Residue Rev.*, 85:43-55 (1983).

Yalkowsky, S.H., S.C. Valvani, and T.J. Roseman. "Solubility and Partitioning VI. Octanol Solubility and Octanol-Water Partition Coefficients," *J. Pharm. Sci.*, 72(8):866-870 (1983).

Yaron, B. and S. Saltzman. "Influence of Water and Temperature on Adsorption of Parathion by Soils," *Soil Sci. Soc. Am. Proc.*, 36:583-586 (1972).

Yasuhara, A. and M. Morita. "Formation of Chlorinated Compounds in Pyrolysis of Trichloroethylene," *Chemosphere*, 21(4/5):479-486 (1990).

Yokley, R.A., A.A. Garrison, G. Mamantov, and E.L. Wehry. "The Effect of Nitrogen Dioxide on the Photochemical and Nonphotochemical Degradation of Pyrene and Benzo[*a*]pyrene Adsorbed on Coal Fly Ash," *Chemosphere*, 14(11/12):1771-1778 (1985).

Yoshida, K., S. Tadayoshi, and F. Yamauchi. "Relationship between Molar Refraction and *n*-Octanol/Water Partition Coefficient," *Ecotoxicol. Environ. Saf.*, 7(6):558-565 (1983).

Yoshida, K., S. Tadayoshi, and F. Yamauchi. "Relationship between Molar Refraction and *n*-Octanol/Water Partition Coefficient," *Ecotoxicol. Environ. Saf.*, 7(6):558-565 (1983).

Yoshida, K., T. Shigeoka, and F. Yamauchi. "Non-Steady-State Equilibrium Model for the Preliminary Prediction of the Fate of Chemicals in the Environment," *Ecotoxicol. Environ. Saf.*, 7(2):179-190 (1983).

Yoshioka, Y., T. Mizuno, Y. Ose, and T. Sato. "The Estimation of Toxicity of

Chemicals on Fish by Physico-Chemical Properties," *Chemosphere*, 15(2):195-203 (1986).

Youezawa, Y. and Y. Urushigawa. "Chemico-Biological Interactions in Biological Purification Systems. V. Relation between Biodegradation Rate Constants of Aliphatic Alcohols by Activated Sludge and Their Partition Coefficients in a 1-Octanol-Water System," *Chemosphere*, 8(3):139-142 (1979).

Young, L.Y. and M.D. Rivera. "Methanogenic Degradation of Four Phenolic Compounds," *Water Res.*, 19(10):1325-1332 (1985).

Yu, C.-C., D.J. Hansen, and G.M. Booth. "Fate of Dicamba in a Model Ecosystem," *Bull. Environ. Contam. Toxicol.*, 13(3):280-283 (1975).

Yule, W.N., M. Chiba, and H.V. Morley. "Fate of Insecticide Residues. Decomposition of Lindane in Soil," *J. Agric. Food Chem.*, 15(6):1000-1004 (1967).

Zabik, M.J., R.D. Schuetz, W.L. Burton, and B.E. Pape. "Photochemistry of Bioreactive Compounds. Studies of a Major Product of Endrin," *J. Agric. Food Chem.*, 19(2):308-313 (1971).

Zafiriou, O.C. "Reaction of Methyl Halides with Seawater and Marine Aerosols," *J. Mar. Res.*, 33(1):75-81 (1975).

Zaidi, B.R., Y. Murakami, and M. Alexander. "Predation and Inhibitors in Lake Water Affect the Success of Inoculation to Enhance Biodegradation of Organic Chemicals," *Environ. Sci. Technol.*, 23(7):859-863 (1989).

Zayed, S.M.A.D., I.Y. Mostafa, M.M. Farghaly, H.S.H. Attaby, Y.M. Adam, and F.M. Mahdy. "Microbial Degradation of Trifluralin by *Aspergillus carneus*, *Fusarium oxysporum* and *Trichoderma viride*," *J. Environ. Sci. Health*, B18(2):253-267 (1983).

Zepp, R.G. and P.F. Scholtzhauer. "Photoreactivity of Selected Aromatic Hydrocarbons in Water," in *Polynuclear Aromatic Hydrocarbons, 3rd International Symposium on Chemistry, Biology, Carcinogenesis and Mutagenesis*, Jones, P.W. and P. Leber, Eds. (Ann Arbor, MI: Ann Arbor Science Publishers, Inc., 1979), pp. 141-158.

Zepp, R.G., N.L. Wolfe, G.L. Baughman, P.F. Schlotzhauer, and J.N. MacAllister. "Dynamics of Processes Influencing the Behavior of Hexachlorocyclopentadiene in the Aquatic Environment," 178th Meeting of the American Chemical Society, Washington, DC (September 1979).

Zepp, R.G., N.L. Wolfe, J.A. Gordon, and R.C. Fincher. "Light-Induced Transformations of Methoxychlor in Aquatic Systems," *J. Agric. Food Chem.*, 24(4):727-733 (1976).

Zeyer, J. and P.C. Kearney. "Degradation of *o*-Nitrophenol and *m*-Nitrophenol by a *Pseudomonas putida*," *J. Agric. Food Chem.*, 32(2):238-242 (1984).

Zeyer, J. and P.C. Kearney. "Microbial Metabolism of [14C]Nitroanilines to [14C]Carbon Dioxide," *J. Agric. Food Chem.*, 31(2):304-308 (1983).

Zhang, S. and J.F. Rusling. "Dechlorination of Polychlorinated Biphenyls by Electrochemical Catalysis in a Biocontinuous Microemulsion," *Environ. Sci.*

Technol., 27(7):1375–1380 (1993).

Zhou, X. and K. Mopper. "Apparent Partition Coefficients of 15 Carbonyl Compounds between Air and Seawater and between Air and Freshwater; Implications for Air-Sea Exchange," *Environ. Sci. Technol.*, 24(12):1864–1869 (1990).

Zitko, V. "Polychlorinated Biphenyls (PCBs) Solubilized in Water by Monoionic Surfactants for Study of Toxicity to Aquatic Animals," *Bull. Environ. Contam. Toxicol.*, 5(3):279–285 (1970).

Zoro, J.A., J.M. Hunter, and G.C. Ware. "Degradation of *p,p'*-DDT in Reducing Environments," *Nature (London)*, 247(5438):235–237 (1974).

Appendix A. Conversion Factors between Various Concentration Units

To obtain	From	Compute
g/L of solution	molality	$g/L = 1000dm(MW_1)/[1000+m(MW_1)]$
g/L of solution	molarity	$g/L = M(MW_1)$
g/L of solution	mole fraction	$g/L = 1000dn(MW_1)/[n(MW_1)+(1-n)MW_2]$
g/L of solution	wt %	$g/L = 10d(\text{wt }\%)$
molality	molarity	$m = 1000M/[1000d-M(MW_1)]$
molality	mole fraction	$m = 1000n/[MW_2-n(MW_2)]$
molality	g/L of solution	$m = 1000G/[MW_1(1000d-G)]$
molality	wt % of solute	$m = 1000(\text{wt }\%)/[MW_1(100-\text{wt }\%)]$
molarity	g/L of solution	$M = G/MW_1$
molarity	molality	$M = 1000dm/[1000+m(MW_1)]$
molarity	mole fraction	$M = 1000dn/[n(MW_1)+(1-n)MW_2]$
molarity	wt % of solute	$M = 10d(\text{wt }\%)/MW_1$
mole fraction	g/L of solution	$n = G(MW_2)/[G(MW_2-MW_1)+1000d(MW_1)]$
mole fraction	molality	$n = m(MW_2)/[m(MW_2)+1000]$
mole fraction	molarity	$n = M(MW_2)/[M(MW_2-MW_1)+1000d]$
mole fraction	wt % of solute	$n = [(\text{wt }\%)/MW_1]/[(\text{wt }\%/MW_1)+$ $(100-\text{wt }\%)MW_2]$
wt % of solute	g/L of solution	$\text{wt }\% = G/10d$
wt % of solute	molality	$\text{wt }\% = 100m(MW_1)/[1000+m(MW_1)]$
wt % of solute	molarity	$\text{wt }\% = M(MW_1)/10d$
wt % of solute	mole fraction	$\text{wt }\% = 100n(MW_1)/[n(MW_1)+(1-n)MW_2]$

d = density of solution (g/L); G = g of solute/L of solution; m = molality; M = mass of solute; MW_1 = formula weight of solute; MW_2 = formula weight of solvent; n = mole fraction; wt % = weight percent of solute

Appendix B. Conversion Factors

To convert	Into	Multiply by
	A	
acre-feet	feet^3	4.356 x 10^4
acre-feet	gallons (U.S.)	3.529 x 10^5
acre-feet	inches^3	7.527 x 10^7
acre-feet	liters	1.233 x 10^6
acre-feet	meters^3	1,233
acre-feet	yards^3	1,613
acre-feet/day	feet^3/second	5.042 x 10^{-1}
acre-feet/day	gallons (U.S.)/minute	226.3
acre-feet/day	liters^3/second	14.28
acre-feet/day	meters^3/day	1,234
acre-feet/day	meters^3/second	1.428 x 10^{-2}
acres	feet^2	43,560
acres	hectares	4.047 x 10^{-1}
acres	inches^2	6.273 x 10^6
acres	kilometers^2	4.047 x 10^{-3}
acres	meters^2	4,047
acres	miles^2	1.563 x 10^{-3}
angstroms	inches	3.937 x 10^{-6}
angstroms	meters	10^{-10}
angstroms	microns	10^{-4}
atmospheres	bars	1.01325
atmospheres	inches of Hg	29.92126
atmospheres	millibars	1013.25
atmospheres	millimeters of Hg	760
atmospheres	millimeters of water	1.033227 x 10^4
atmospheres	pascals	1.01325 x 10^5
atmospheres	pounds/inch^2	14.70
atmospheres	torrs	760
	B	
barrels (petroleum)	liters	159.0
bars	atmospheres	9.869 x 10^{-1}
bars	pounds/inch^2	14.50
British thermal unit (BTU)	joules	1.055 x 10^{-3}
British thermal unit (BTU)	kilowatt-hour	2.928 x 10^{-4}
British thermal unit (BTU)	watts	2.931 x 10^{-1}
	C	
calories	joules	4.187

To convert	Into	Multiply by
Celsius (°C)	Fahrenheit (°F)	$(1.8 \times °C)+32$
Celsius (°C)	Kelvin (°K)	$°C+273.15$
centimeters	feet	3.291×10^{-2}
centimeters	inches	3.937×10^{-1}
centimeters	miles	6.214×10^{-6}
centimeters	millimeters	10
centimeters	mils	393.7
centimeters	yards	1.094×10^{-2}
centimeters^2	feet^2	1.076×10^{-3}
centimeters^2	inches^2	1.55×10^{-1}
centimeters^2	meters^2	1×10^{-4}
centimeters^2	yards^2	1.196×10^{-4}
centimeters^3	fluid ounces	3.381×10^{-2}
centimeters^3	feet^3	3.5314×10^{-5}
centimeters^3	inches^3	6.102×10^{-2}
centimeters^3	liters	1×10^{-3}
centimeters^3	ounces (U.S., fluid)	3.381×10^{-2}
centimeters/second	feet/day	2,835
centimeters/second	feet/minute	1.968
centimeters/second	feet/second	3.281×10^{-2}
centimeters/second	kilometers/hour	3.6×10^{-2}
centimeters/second	liters/meter/second	9.985
centimeters/second	meters/minute	6×10^{-1}
centimeters/second	miles/hour	2.237×10^{-2}
centimeters/second	miles/minute	3.728×10^{-4}

D

To convert	Into	Multiply by
Darcy	centimeters/second	9.66×10^{-4}
Darcy	feet/second	3.173×10^{-5}
Darcy	liters/meter/second	8.58×10^{-3}
days	seconds	86,400.0
dynes/centimeter^2	atmospheres	9.869×10^{-7}
dynes/centimeter^2	inches of Hg at 0 °C	2.953×10^{-5}
dynes/centimeter^2	inches of water at 4 °C	4.015×10^{-4}

F

To convert	Into	Multiply by
Fahrenheit (°F)	Celsius (°C)	$5(°F-32)/9$
Fahrenheit (°F)	Kelvin (°K)	$5(°F+459.67)/9$
Fahrenheit (°F)	Rankine (°R)	$°F+459.67$
feet	centimeters	30.48

To convert	Into	Multiply by
feet	inches	12
feet	kilometers	3.048×10^{-4}
feet	meters	3.048×10^{-1}
feet	miles	1.894×10^{-4}
feet	millimeters	304.8
feet	yards	3.33×10^{-1}
feet/day	feet/second	1.157×10^{-5}
feet/day	meters/second	3.528×10^{-6}
feet^2	acres	2.296×10^{-5}
feet^2	hectares	9.29×10^{-9}
feet^2	inches^2	144
feet^2	kilometers^2	9.29×10^{-8}
feet^2	meters^2	9.29×10^{-2}
feet^2	miles^2	3.587×10^{-8}
feet^2	yards^2	1.111×10^{-1}
feet^2/day	meters^2/day	9.290×10^{-2}
feet^2/sec	meters^2/day	8,027
feet^3	acre-feet	2.296×10^{-5}
feet^3	gallons (U.S.)	7.481
feet^3	inches^3	1,728
feet^3	liters	28.32
feet^3	meters^3	2.832×10^{-2}
feet^3	yards^3	3.704×10^{-2}
feet^3/foot/day	gallons/foot/day	7.48052
feet^3/foot/day	liters/meter/day	92.903
feet^3/foot/day	meters^3/meter/day	9.29×10^{-2}
feet^3/foot/day	feet^3/foot^2/minute	6.944×10^{-4}
feet^3/foot/day	gallons/foot^2/day	7.4805
feet^3/foot/day	inches^3/inch^2/hour	5×10^{-1}
feet^3/foot/day	liters/meter^2/day	304.8
feet^3/foot/day	meters^3/meter^2/day	3.048×10^{-1}
feet^3/foot/day	millimeters^3/inch^2/hour	5×10^{-1}
feet^3/foot^2/day	millimeters^3/millimeter^2/hour	25.4
feet/mile	meters/kilometer	1.894×10^{-1}
feet/second	centimeters/second	5.080×10^{-1}
feet/second	feet/day	86,400
feet/second	feet/hour	3,600
feet/second	gallons (U.S.)/foot^2/day	5.737×10^{5}
feet/second	kilometers/hour	1.097
feet/second	meters/second	3.048×10^{-1}

To convert	Into	Multiply by
feet/second	miles/hour	6.818×10^{-1}
feet/year	centimeters/second	9.665×10^{-7}
feet^3/second	acre-feet/day	1.983
feet^3/second	feet^3/minute	60.0
feet^3/second	gallons (U.S.)/minute	448.8
feet^3/second	liters/second	28.32
feet^3/second	meters^3/second	2.832×10^{-2}
feet^3/second	meters^3/day	2,447
	G	
gallons (U.K.)	gallons (U.S.)	1.200
gallons (U.S.)	acre-feet	3.068×10^{-6}
gallons (U.S.)	feet^3	1.337×10^{-1}
gallons (U.S.)	fluid ounces	128.0
gallons (U.S.)	liters	3.785
gallons (U.S.)	meters^3	3.785×10^{-3}
gallons (U.S.)	yards^3	4.951×10^{-3}
gallons/day	acre-feet/year	1.12×10^{-3}
gallons/foot/day	feet^3/foot/day	1.3368×10^{-1}
gallons/foot/day	liters/meter/day	12.42
gallons/foot/day	meters^3/meter/day	1.242×10^{-2}
gallons (U.S.)/foot^2/day	centimeters/second	4.717×10^{-5}
gallons (U.S.)/foot^2/day	Darcy	5.494×10^{-2}
gallons (U.S.)/foot^2/day	feet/day	1.3368×10^{-1}
gallons (U.S.)/foot^2/day	feet/second	1.547×10^{-6}
gallons (U.S.)/foot^2/day	gallons/foot^2/minute	6.944×10^{-4}
gallons (U.S.)/foot^2/day	liters/meter^2/day	40.7458
gallons (U.S.)/foot^2/day	meters/day	4.07458×10^{-2}
gallons (U.S.)/foot^2/day	meters/minute	2.83×10^{-5}
gallons (U.S.)/foot^2/day	meters/second	4.716×10^{-7}
gallons (U.S.)/foot^2/minute	meters/day	58.67
gallons (U.S.)/foot^2/minute	meters/second	6.791×10^{-2}
gallons (U.S.)/minute	acre-feet/day	4.419×10^{-3}
gallons (U.S.)/minute	feet^3/second	2.228×10^{-3}
gallons (U.S.)/minute	feet^3/hour	8.0208
gallons (U.S.)/minute	liters/second	6.309×10^{-2}
gallons (U.S.)/minute	meters^3/day	5.30
gallons (U.S.)/minute	meters^3/second	6.309×10^{-5}
grams/centimeter^3	kilograms/meter^3	1,000
grams/centimeter^3	pounds/feet^3	62.428

To convert	Into	Multiply by
grams/centimeter^3	pounds/gallon (U.S.)	8.345
grams	kilograms	1×10^{-3}
grams	ounces (avoirdupois)	3.527×10^{-2}
grams	pounds	2.2046×10^{-3}
grams/liter	grains/gallon (U.S.)	58.4178
grams/liter	grams/centimeter^3	1×10^{-3}
grams/liter	kilograms/meter^3	1
grams/liter	pounds/feet^3	6.24×10^{-2}
grams/liter	pounds/inch^3	3.61×10^{-5}
grams/liter	pounds/gallon (U.S.)	8.35×10^{-3}
grams/meter^3	grains/feet^3	4.37×10^{-1}
grams/meter^3	milligrams/liter	1.0
grams/meter^3	pounds/gallon (U.S.)	8.345×10^{-5}
grams/meter^3	pounds/inch^3	7.433×10^{-3}
	H	
hectares	acres	2.471
hectares	feet^2	1.076×10^5
hectares	inches^2	1.55×10^7
hectares	kilometers^2	1×10^{-2}
hectares	meters^2	10,000
hectares	miles^2	3.861×10^{-3}
hectares	yards^2	11,959.90
horsepower	kilowatts/hour	7.457×10^{-1}
	I	
inches	centimeters	2.540
inches	feet	8.333×10^{-1}
inches	kilometers	2.54×10^{-5}
inches	meters	2.54×10^{-2}
inches	miles	1.578×10^{-5}
inches	millimeters	25.4
inches	yards	2.778×10^{-2}
inches of Hg	pascals	3,386
inches^2	acres	1.594×10^{-8}
inches^2	centimeters^2	6.4516
inches^2	feet^2	6.944×10^{-3}
inches^2	hectares	6.452×10^{-8}
inches^2	kilometers^2	6.452×10^{-10}
inches^2	meters^2	6.452×10^{-4}

To convert	Into	Multiply by
inches^2	millimeters^2	645.16
inches^3	acre-feet	1.329×10^{-8}
inches^3	centimeters^3	16.39
inches^3	feet^3	5.787×10^{-4}
inches^3	gallons (U.S.)	4.329×10^{-3}
inches^3	liters	1.639×10^{-2}
inches^3	meters^3	1.639×10^{-5}
inches^3	milliliters	16.387
inches^3	yards^3	2.143×10^{-5}
	K	
Kelvin (°K)	Celsius	°K - 273.15
Kelvin (°K)	Fahrenheit	1.8(°K) - 459.67
kilograms	grams	1,000
kilograms	milligrams	1×10^6
kilograms	ounces (avoirdupois)	35.28
kilograms	pounds	2.205
kilograms	tons (long)	9.842×10^{-4}
kilograms	tons (metric)	1×10^{-3}
kilograms/meter^3	grams/centimeter^3	1×10^{-3}
kilograms/meter^3	grams/liter	1.0
kilograms/meter^3	pounds/inch^3	3.613×10^{-5}
kilograms/meter^3	pounds/feet^3	0.0624
kilometers	feet	3,281
kilometers	inches	39,370
kilometers	meters	1,000
kilometers	miles	6.214×10^{-1}
kilometers	millimeters	1×10^6
kilometers^2	acres	247.1
kilometers^2	hectares	100
kilometers^2	inches^2	1.55×10^9
kilometers^2	meters^2	1×10^6
kilometers^2	miles^2	0.3861
kilometers/hour	feet/day	78,740
kilometers/hour	feet/second	9.113×10^{-1}
kilometers/hour	meters/second	2.778×10^{-1}
kilometers/hour	miles/hour	6.214×10^{-1}
	L	
liters	acre-feet	8.106×10^{-7}

To convert	Into	Multiply by
liters	feet^3	3.531×10^{-2}
liters	fluid ounces	33.814
liters	gallons (U.S.)	2.642×10^{-1}
liters	inches^3	61.02
liters	meters^3	10^{-3}
liters	yards^3	1.308×10^{-3}
liters/second	acre-feet/day	7.005×10^{-2}
liters/second	feet^3/second	3.531×10^{-2}
liters/second	gallons (U.S.)/minute	15.85
liters/second	meters^3/day	86.4

M

To convert	Into	Multiply by
meters	feet	3.28084
meters	inches	39.3701
meters	kilometers	1×10^{-3}
meters	miles	6.214×10^{-4}
meters	millimeters	1,000
meters	yards	1.0936
meters^2	acres	2.471×10^{-4}
meters^2	feet^2	10.76
meters^2	hectares	1×10^{-4}
meters^2	inches^2	1.550
meters^2	kilometers	1×10^{-6}
meters^2	miles^2	3.861×10^{-7}
meters^2	yards^2	1.196
meters^3	acre-feet	8.106×10^{-4}
meters^3	feet^3	35.31
meters^3	gallons (U.S.)	264.2
meters^3	inches^3	6.102×10^4
meters^3	liters	1,000
meters^3	yards^3	1.308
meters/second	feet/minute	196.8
meters/second	feet/day	283,447
meters/second	kilometers/hour	3.6
meters/second	feet/second	3.281
meters/second	miles/hour	2.237
meters^3/day	acre-feet/day	6.051×10^6
meters^3/day	feet^3/second	3.051×10^6
meters^3/day	gallons (U.S.)/minute	1.369×10^9

To convert	Into	Multiply by
meters^3/day	liters/second	8.64 x 10^7
miles	feet	5,280
miles	kilometers	1.609
miles	meters	1,609
miles	millimeters	1.609 x 10^6
miles^2	acres	640
miles^2	feet^2	2.778 x 10^7
miles^2	hectares	259
miles^2	inches^2	4.014 x 10^9
miles^2	kilometers^2	2.590
miles^2	meters^2	2.59 x 10^6
miles/hour	feet/day	1.267 x 10^5
miles/hour	kilometers/hour	1.609
miles/hour	meters/second	4.47 x 10^{-1}
millibars	pascals	100.0
milligrams/liter	grams/meter^3	1.0
milligrams/liter	parts/million	1.0
milligrams/liter	pounds/feet^3	6.2428 x 10^{-5}
milliliters	centimeters^3	1.0
milliliters	liters	1 x 10^{-3}
millimeters	centimeters	1 x 10^{-1}
millimeters	feet	3.281 x 10^{-3}
millimeters	inches	0.03937
millimeters	kilometers	1.0 x 10^{-6}
millimeters	meters	1.0 x 10^{-3}
millimeters	miles	6.214 x 10^{-7}
millimeters^2	centimeters^2	1 x 10^{-2}
millimeters^2	inches^2	1.55 x 10^{-3}
millimeters^3	centimeters^3	1 x 10^{-3}
millimeters^3	inches^3	6.102 x 10^{-5}
millimeters^3	liters	10^{-6}
millimeters of Hg	atmospheres	1.316 x 10^{-3}
millimeters of Hg	pascals	133.3224
millimeters of Hg	torrs	1.0

O

ounces (avoirdupois)	grams	28.35
ounces (avoirdupois)	kilograms	2.8355 x 10^{-2}
ounces (avoirdupois)	pounds	6.25 x 10^{-2}

To convert	Into	Multiply by
	P	
pascals	atmospheres	9.869×10^{-6}
pascals	millimeters of Hg	7.501×10^{-3}
pounds (avoirdupois)	kilograms	4.535×10^{-1}
pounds (avoirdupois)	ounces (avoirdupois)	16
pounds/centimeter^3	pounds/inch^3	3.61×10^{-2}
pounds/inch^2	pascals	6,895
pounds/inch^2	kilograms/centimeter^2	7.31×10^{-2}
pounds/feet^3	grams/centimeter^3	1.6×10^{-2}
pounds/feet^3	grams/liter	27,680
pounds/feet^3	pounds/inch^3	5.787×10^{-4}
pounds/feet^3	pounds/gallon (U.S.)	1.337×10^{-1}
pounds/foot^2	pascals	47.88
pounds/inch^3	pounds/centimeter^3	27.68
pounds/inch^3	pounds/feet^3	1,728
pounds/inch^3	pounds/gallon (U.S.)	231
pounds/gallon (U.S.)	grams/centimeter^3	1.198×10^{-1}
pounds/gallon (U.S.)	grams/liter	119.8
pounds/gallon (U.S.)	pounds/feet^3	7.481
pounds/gallon (U.S.)	pounds/inch^3	4.329×10^{-3}
	Q	
quarts	liters	9.464×10^{-1}
	T	
torrs	atmospheres	1.316×10^{-3}
torrs	millimeters of Hg	1.0
torrs	pascals	133.322
	W	
watts	BTU/hour	3.4129
	Y	
yards	centimeters	91.44
yards	fathoms	5×10^{-1}
yards	feet	3.0
yards	inches	36.0
yards	meters	9.144×10^{-1}
yards	miles	5.682×10^{-4}

To convert	Into	Multiply by
yards	feet^3	27
yards	gallons (U.S.)	202
yards	inches^3	4.666×10^4
yards	liters	764.6
yards	meters^3	7.646×10^{-1}
yards^2	meters^2	8.361×10^{-1}
yards^3	meters^3	7.646×10^{-1}
years (normal calendar)	hours	8,760
years (normal calendar)	minutes	5.256×10^5
years (normal calendar)	seconds	3.1536×10^7
years (normal calendar)	weeks	52.1428

Appendix C. U.S. EPA Approved Test Methods

The numbers that follow each compound name below are the U.S. EPA-approved test methods based on information obtained from the "Guide to Environmental Analytical Methods" (Schenectady, NY: Genium Publishing Corp., 1992) and the U.S. EPA's Sampling and Analysis Methods Database. The database information is available in hard copy from Keith, L.H. *Compilation of EPA's Sampling and Analysis Methods* (Chelsea, MI: Lewis Publishers, Inc., 1991), 803 p.

Chemical Name	Test Methods
Acrolein	8030, 8240
Acrylonitrile	8030, 8240
Alachlor	525
Aldrin	508, 525, 608, 625, SM-6410, SM-6630, 8080, 8270
Anilazine	8270
Atrazine	525
Azinphos-methyl	8140, 8270
α-BHC	508, 608, 625, SM-6410, SM-6630, 8080, 8270
β-BHC	508, 608, 625, SM-6410, SM-6630, 8080, 8270
δ-BHC	508, 608, 625, SM-6410, SM-6630, 8080, 8270
Bis(2-chloroethyl)ether	625, SM-6410, 8270
Bis(2-chloroisopropyl)ether	625, SM-6410, 8270
Bromoxynil	8270
Carbaryl	8270
Carbofuran	8270
Carbon disulfide	8240
Carbon tetrachloride	502.1, 502.2, 601, 524.1, 524.2, 624, SM-6210 SM-6230, 8010, 8240
Carbophenothion	8270
Chloramben	515.1
Chlordane (technical)	508, 608, SM-6630, 8080, 8270
Chlorobenzilate	508, 8270
Chlorothalonil	508
Chlorpyrifos	508, 8140
Crotoxyphos	8270
2,4-D	515.1, 8150
Dalapon	8150
p,p'-DDD	508, 608, SM-6630, 8080, 8270
p,p'-DDE	508, 608, SM-6630, 8080, 8270
p,p'-DDT	508, 608, SM-6630, 8080, 8270
Diallate	8270
Diazinon	8140
1,2-Dibromo-3-chloropropane	502.2, 504, 524.1, 524.2, SM-6210, 8240

1171

Chemical Name	Test Methods
Di-*n*-butyl phthalate	525, 625, SM-6410, 8060, 8270
Dicamba	515.1, 8150
Dichlone	8270
1,2-Dichloropropane	502.1, 502.2, 524.1, 524.2, 601, 624, SM-6210, SM-6230 8010, 8240
cis-1,3-Dichloropropylene	502.1, 502.2, 524.1, 601, 624, SM-6210, SM-6230 8010, 8240
trans-1,3-Dichloropropylene	502.1, 502.2, 524.1, 601, 624, SM-6210, SM-6230 8010, 8240
Dichlorvos	8140, 8270
Dicrotophos	8270
Dieldrin	508, 608, SM-6630, 8080, 8270
Dimethoate	8270
Dimethyl phthalate	525, 625, SM-6410, 8060, 8270
4,6-Dinitro-*o*-cresol	8040
Dinoseb	515.1, 8150, 8270
Disulfoton	8140, 8270
α-Endosulfan	508, 608, 625, SM-6410, SM-6630, 8080, 8270
β-Endosulfan	508, 608, 625, SM-6410, SM-6630, 8080, 8270
Endosulfan sulfate	508, 608, 625, SM-6410, SM-6630, 8080, 8270
Endrin	508, 525, 608, 625, SM-6410, SM-6630, 8080, 8270
Endrin aldehyde	508, 608, 625, SM-6410, SM-6630, 8080, 8270
EPN	8270
Ethion	8270
Ethoprop	8140
Ethylene dibromide	502.1, 502.2, 504, 524.1, 524.2, SM-6210, 8240
Fensulfothion	8140, 8270
Fenthion	8140, 8270
Heptachlor	508, 525, 608, 625, SM-6410, SM-6630, 8080, 8270
Heptachlor epoxide	508, 525, 608, 625, SM-6410, SM-6630, 8080, 8270
Hexachlorobenzene	508, 525, 625, SM-6410, 8120, 8270
Kepone	8270
Lindane	508, 525, 608, SM-6630, 8080, 8270
Malathion	8270
MCPA	8150
Methoxychlor	508, 525, SM-6630, 8080, 8270
Methyl bromide	502.1, 502.2, 524.1, 524.2, 601, 624, SM-6210, SM-6230 8010, 8240
Mevinphos	8140, 8270
Monocrotophos	8270

Chemical Name	Test Methods
Naled	8140, 8270
Parathion	8270
Pentachlorobenzene	8270
Pentachlorophenol	525, 625, SM-6410, 8040, 8270
Phorate	8140, 8270
Phosalone	8270
Phosmet	8270
Phosphamidon	8270
Picloram	515.1
Propachlor	508
Propyzamide	8270
Ronnel	8140
Simazine	525
Strychnine	525
Sulprofos	8140
2,4,5-T	515.1, 8150, SM-6640
Terbufos	8270
Tetraethyl pyrophosphate	8270
Toxaphene	508, 525, 608, 625, SM-6630, 8080, 8270
Trifluralin	508, 8270

NOTE: Methods beginning with the prefix "SM-" refer to standard methods found in the 17th edition of the "Standard Methods for the Examination of Water and Waste Water", published by the American Water Works Association in 1989.

Appendix D. CAS Registry Number Index

50-00-0	Formaldehyde
50-29-3	*p,p'*-DDT
50-32-8	Benzo[*a*]pyrene
51-28-5	2,4-Dinitrophenol
53-70-3	Dibenz[*a,h*]anthracene
53-96-3	2-Acetylaminofluorene
56-23-5	Carbon tetrachloride
56-38-2	Parathion
56-55-3	Benzo[*a*]anthracene
57-14-7	1,1-Dimethylhydrazine
57-24-9	Strychnine
57-57-8	*β*-Propiolactone
57-74-9	Chlordane
58-89-9	Lindane
59-50-7	*p*-Chloro-*m*-cresol
60-11-7	*p*-Dimethylaminoazobenzene
60-29-7	Ethyl ether
60-34-4	Methylhydrazine
60-57-1	Dieldrin
62-53-3	Aniline
62-73-7	Dichlorvos
62-75-9	*N*-Nitrosodimethylamine
63-25-2	Carbaryl
64-18-6	Formic acid
64-19-7	Acetic acid
65-85-0	Benzoic acid
67-56-1	Methanol
67-64-1	Acetone
67-66-3	Chloroform
67-72-1	Hexachloroethane
68-12-2	*N,N*-Dimethylformamide
71-23-8	1-Propanol
71-36-3	1-Butanol
71-43-2	Benzene
71-55-6	1,1,1-Trichloroethane
72-20-8	Endrin
72-43-5	Methoxychlor
72-54-8	*p,p'*-DDD
72-55-9	*p,p'*-DDE
74-83-9	Methyl bromide
74-87-3	Methyl chloride
74-88-4	Methyl iodide

74-89-5	Methylamine
74-93-1	Methyl mercaptan
74-96-4	Ethyl bromide
74-97-5	Bromochloromethane
74-98-6	Propane
74-99-7	Propyne
75-00-3	Chloroethane
75-01-4	Vinyl chloride
75-04-7	Ethylamine
75-05-5	Acetonitrile
75-07-0	Acetaldehyde
75-08-1	Ethyl mercaptan
75-09-2	Methylene chloride
75-15-0	Carbon disulfide
75-25-2	Bromoform
75-27-4	Bromodichloromethane
75-28-5	2-Methylpropane
75-31-0	Isopropylamine
75-34-3	1,1-Dichloroethane
75-35-4	1,1-Dichloroethylene
75-43-4	Dichlorofluoromethane
75-52-5	Nitromethane
75-56-9	Propylene oxide
75-61-6	Dibromodifluoromethane
75-63-8	Bromotrifluoromethane
75-65-0	*tert*-Butyl alcohol
75-69-4	Trichlorofluoromethane
75-71-8	Dichlorodifluoromethane
75-83-2	2,2-Dimethylbutane
76-01-7	Pentachloroethane
76-06-2	Chloropicrin
76-11-9	1,1-Difluorotetrachloroethane
76-12-0	1,2-Difluorotetrachloroethane
76-13-1	1,1,2-Trichlorotrifluoroethane
76-22-2	Camphor
76-44-8	Heptachlor
77-47-4	Hexachlorocyclopentadiene
77-78-1	Dimethyl sulfate
78-30-8	Tri-*o*-cresyl phosphate
78-59-1	Isophorone
78-78-4	2-Methylbutane
78-79-5	2-Methyl-1,3-butadiene
78-83-1	Isobutyl alcohol

78-87-5	1,2-Dichloropropane
78-92-2	*sec*-Butyl alcohol
78-93-3	2-Butanone
79-00-5	1,1,2-Trichloroethane
79-01-6	Trichloroethylene
79-06-1	Acrylamide
79-20-9	Methyl acetate
79-24-3	Nitroethane
79-27-6	1,1,2,2-Tetrabromoethane
79-29-8	2,3-Dimethylbutane
79-34-5	1,1,2,2-Tetrachloroethane
79-46-9	2-Nitropropane
80-62-6	Methyl methacrylate
81-81-2	Warfarin
83-26-1	Pindone
83-32-9	Acenaphthene
84-66-2	Diethyl phthalate
84-74-2	Di-*n*-butyl phthalate
85-01-8	Phenanthrene
85-44-9	Phthalic anhydride
85-68-7	Benzyl butyl phthalate
86-30-6	*N*-Nitrosodiphenylamine
86-73-7	Fluorene
86-88-4	ANTU
87-61-6	1,2,3-Trichlorobenzene
87-68-3	Hexachlorobutadiene
87-86-5	Pentachlorophenol
88-06-2	2,4,6-Trichlorophenol
88-72-2	2-Nitrotoluene
88-74-4	2-Nitroaniline
88-75-5	2-Nitrophenol
88-89-1	Picric acid
90-04-0	*o*-Anisidine
91-17-8	Decahydronaphthalene
91-20-3	Naphthalene
91-57-6	2-Methylnaphthalene
91-58-7	2-Chloronaphthalene
91-59-8	2-Naphthylamine
91-94-1	3,3'-Dichlorobenzidine
92-52-4	Biphenyl
92-67-1	4-Aminobiphenyl
92-87-5	Benzidine
92-93-3	4-Nitrobiphenyl

93-37-8	2,7-Dimethylquinoline
93-76-5	2,4,5-T
94-75-7	2,4-D
95-47-6	*o*-Xylene
95-48-7	2-Methylphenol
95-50-1	1,2-Dichlorobenzene
95-53-4	*o*-Toluidine
95-57-8	2-Chlorophenol
95-63-6	1,2,4-Trimethylbenzene
95-93-2	1,2,4,5-Tetramethylbenzene
95-94-3	1,2,4,5-Tetrachlorobenzene
95-95-4	2,4,5-Trichlorophenol
96-12-8	1,2-Dibromo-3-chloropropane
96-14-0	3-Methylpentane
96-18-4	1,2,3-Trichloropropane
96-33-3	Methyl acrylate
96-37-7	Methylcyclopentane
98-00-0	Furfuryl alcohol
98-01-1	Furfural
98-06-6	*tert*-Butylbenzene
98-82-8	Isopropylbenzene
98-83-9	α-Methylstyrene
98-95-3	Nitrobenzene
99-08-1	3-Nitrotoluene
99-09-2	3-Nitroaniline
99-65-0	1,3-Dinitrobenzene
99-99-0	4-Nitrotoluene
100-00-5	*p*-Chloronitrobenzene
100-01-6	4-Nitroaniline
100-02-7	4-Nitrophenol
100-25-4	1,4-Dinitrobenzene
100-37-8	2-Diethylaminoethanol
100-41-4	Ethylbenzene
100-42-5	Styrene
100-44-7	Benzyl chloride
100-51-6	Benzyl alcohol
100-61-8	Methylaniline
100-63-0	Phenylhydrazine
100-74-3	4-Ethylmorpholine
101-55-3	4-Bromophenyl phenyl ether
101-84-8	Phenyl ether
103-65-1	*n*-Propylbenzene
104-51-8	*n*-Butylbenzene

104-94-9	*p*-Anisidine
105-46-4	*sec*-Butyl acetate
105-67-9	2,4-Dimethylphenol
106-35-4	3-Heptanone
106-37-6	1,4-Dibromobenzene
106-42-3	*p*-Xylene
106-44-5	4-Methylphenol
106-46-7	1,4-Dichlorobenzene
106-47-8	4-Chloroaniline
106-50-3	*p*-Phenylenediamine
106-51-4	*p*-Quinone
106-89-8	Epichlorohydrin
106-92-3	Allyl glycidyl ether
106-93-4	Ethylene dibromide
106-97-8	Butane
106-98-9	1-Butene
106-99-0	1,3-Butadiene
107-02-8	Acrolein
107-05-1	Allyl chloride
107-06-2	1,2-Dichloroethane
107-07-3	Ethylene chlorohydrin
107-08-4	1-Iodopropane
107-13-1	Acrylonitrile
107-15-3	Ethylenediamine
107-18-6	Allyl alcohol
107-20-0	Chloroacetaldehyde
107-31-3	Methyl formate
107-49-3	Tetraethyl pyrophosphate
107-83-5	2-Methylpentane
107-87-9	2-Pentanone
108-03-2	1-Nitropropane
108-05-4	Vinyl acetate
108-08-7	2,4-Dimethylpentane
108-10-1	4-Methyl-2-pentanone
108-18-9	Diisopropylamine
108-20-3	Isopropyl ether
108-21-4	Isopropyl acetate
108-24-7	Acetic anhydride
108-31-6	Maleic anhydride
108-38-3	*m*-Xylene
108-60-1	Bis(2-chloroisopropyl)ether
108-67-8	1,3,5-Trimethylbenzene
108-70-3	1,3,5-Trichlorobenzene

108-83-8	Diisobutyl ketone
108-84-9	*sec*-Hexyl acetate
108-86-1	Bromobenzene
108-87-2	Methylcyclohexane
108-88-3	Toluene
108-90-7	Chlorobenzene
108-93-0	Cyclohexanol
108-94-1	Cyclohexanone
108-95-2	Phenol
109-60-4	*n*-Propyl acetate
109-66-0	Pentane
109-67-1	1-Pentene
109-73-9	*n*-Butylamine
109-79-5	*n*-Butyl mercaptan
109-86-4	Methyl cellosolve
109-87-5	Methylal
109-89-7	Diethylamine
109-94-4	Ethyl formate
109-99-9	Tetrahydrofuran
110-02-1	Thiophene
110-19-0	Isobutyl acetate
110-43-0	2-Heptanone
110-49-6	Methyl cellosolve acetate
110-54-3	Hexane
110-75-8	2-Chloroethyl vinyl ether
110-80-5	2-Ethoxyethanol
110-82-7	Cyclohexane
110-83-8	Cyclohexene
110-86-1	Pyridine
110-91-8	Morpholine
111-15-9	2-Ethoxyethyl acetate
111-44-4	Bis(2-chloroethyl)ether
111-65-9	Octane
111-66-0	1-Octene
111-76-2	2-Butoxyethanol
111-84-2	Nonane
111-91-1	Bis(2-chloroethoxy)methane
112-40-3	Dodecane
115-11-7	2-Methylpropene
115-86-6	Triphenyl phosphate
117-81-7	Bis(2-ethylhexyl)phthalate
117-84-0	Di-*n*-octyl phthalate
118-52-5	1,3-Dichloro-5,5-dimethylhydantoin

118-74-1	Hexachlorobenzene
118-96-7	2,4,6-Trinitrotoluene
120-12-7	Anthracene
120-72-9	Indole
120-82-1	1,2,4-Trichlorobenzene
120-83-2	2,4-Dichlorophenol
121-14-2	2,4-Dinitrotoluene
121-44-8	Triethylamine
121-69-7	Dimethylaniline
121-75-5	Malathion
122-66-7	1,2-Diphenylhydrazine
123-31-9	Hydroquinone
123-42-2	Diacetone alcohol
123-51-3	Isoamyl alcohol
123-86-4	*n*-Butyl acetate
123-91-1	1,4-Dioxane
123-92-2	Isoamyl acetate
124-18-5	Decane
124-40-3	Dimethylamine
124-48-1	Dibromochloromethane
126-73-8	Tributyl phosphate
126-99-8	Chloroprene
127-18-4	Tetrachloroethylene
127-19-5	*N,N*-Dimethylacetamide
129-00-0	Pyrene
131-11-3	Dimethyl phthalate
132-64-9	Dibenzofuran
134-32-7	1-Naphthylamine
135-98-8	*sec*-Butylbenzene
137-26-8	Thiram
140-88-5	Ethyl acrylate
141-43-5	Ethanolamine
141-78-6	Ethyl acetate
141-79-7	Mesityl oxide
142-29-0	Cyclopentene
142-82-5	Heptane
143-50-0	Kepone
144-62-7	Oxalic acid
151-56-4	Ethylenimine
156-60-5	*trans*-1,2-Dichloroethylene
191-24-2	Benzo[*ghi*]perylene
192-97-2	Benzo[*e*]pyrene
193-39-5	Indeno[1,2,3-*cd*]pyrene

205-99-2	Benzo[b]fluoranthene
206-44-0	Fluoranthene
207-08-9	Benzo[k]fluoranthene
208-96-8	Acenaphthylene
218-01-9	Chrysene
287-92-3	Cyclopentane
291-64-5	Cycloheptane
299-84-3	Ronnel
300-76-5	Naled
309-00-2	Aldrin
319-84-6	α-BHC
319-85-7	β-BHC
319-86-8	δ-BHC
330-54-1	Diuron
463-82-1	2,2-Dimethylpropane
479-45-8	Tetryl
496-11-7	Indan
496-15-1	Indoline
504-29-0	2-Aminopyridine
509-14-8	Tetranitromethane
526-73-8	1,2,3-Trimethylbenzene
528-29-0	1,2-Dinitrobenzene
532-27-4	α-Chloroacetophenone
534-52-1	4,6-Dinitro-o-cresol
538-93-2	Isobutylbenzene
540-84-1	2,2,4-Trimethylpentane
540-88-5	tert-Butyl acetate
541-73-1	1,3-Dichlorobenzene
541-85-5	5-Methyl-3-heptanone
542-88-1	sym-Dichloromethyl ether
542-92-7	Cyclopentadiene
556-52-5	Glycidol
562-49-2	3,3-Dimethylpentane
563-45-1	3-Methyl-1-butene
565-59-3	2,3-Dimethylpentane
565-75-3	2,3,4-Trimethylpentane
583-60-8	o-Methylcyclohexanone
584-84-9	2,4-Toluene diisocyanate
589-34-4	3-Methylhexane
589-81-1	3-Methylheptane
591-49-1	1-Methylcyclohexene
591-76-4	2-Methylhexane
591-78-6	2-Hexanone

591-93-5	1,4-Pentadiene
592-41-6	1-Hexene
600-25-9	1-Chloro-1-nitropropane
606-20-2	2,6-Dinitrotoluene
608-93-5	Pentachlorobenzene
613-12-7	2-Methylanthracene
621-64-7	*N*-Nitrosodi-*n*-propylamine
624-83-9	Methyl isocyanate
626-38-0	*sec*-Amyl acetate
626-39-1	1,3,5-Tribromobenzene
627-13-4	*n*-Propyl nitrate
627-20-3	*cis*-2-Pentene
628-63-7	*n*-Amyl acetate
634-66-2	1,2,3,4-Tetrachlorobenzene
634-90-2	1,2,3,5-Tetrachlorobenzene
636-28-2	1,2,4,5-Tetrabromobenzene
646-04-8	*trans*-2-Pentene
691-37-2	4-Methyl-1-pentene
763-29-1	2-Methyl-1-pentene
832-69-9	1-Methylphenanthrene
872-55-9	2-Ethylthiophene
959-98-8	α-Endosulfan
1024-57-3	Heptachlor epoxide
1031-07-8	Endosulfan sulfate
1563-66-2	Carbofuran
1582-09-8	Trifluralin
1640-89-7	Ethylcyclopentane
1746-01-6	TCDD
1929-82-4	Nitrapyrin
2040-96-2	Propylcyclopentane
2104-64-5	EPN
2207-01-4	*cis*-1,2-Dimethylcyclohexane
2216-34-4	4-Methyloctane
2234-13-1	Octachloronaphthalene
2698-41-1	*o*-Chlorobenzylidenemalononitrile
2921-88-2	Chlorpyrifos
3073-66-3	1,1,3-Trimethylcyclohexane
3522-94-9	2,2,5-Trimethylhexane
3689-24-5	Sulfotepp
3741-00-2	Pentylcyclopentane
4170-30-3	Crotonaldehyde
4516-69-2	1,1,3-Trimethylcyclopentane
5103-71-9	*trans*-Chlordane

5103-74-2	*cis*-Chlordane
6443-92-1	*cis*-2-Heptene
6876-23-9	*trans*-1,2-Dimethylcyclohexane
7005-72-3	4-Chlorophenyl phenyl ether
7421-93-4	Endrin aldehyde
7664-41-7	Ammonia
7786-34-7	Mevinphos
8001-35-2	Toxaphene
10061-01-5	*cis*-1,3-Dichloropropylene
10061-02-6	*trans*-1,3-Dichloropropylene
11096-82-5	PCB-1260
11097-69-1	PCB-1254
11104-28-2	PCB-1221
11141-16-5	PCB-1232
12672-29-6	PCB-1248
12674-11-2	PCB-1016
14686-13-6	*trans*-2-Heptene
33213-65-9	β-Endosulfan
53469-21-9	PCB-1242

Appendix E. RTECS Number Index

AB1000000	Acenaphthene
AB1254000	Acenaphthylene
AB1925000	Acetaldehyde
AB2450000	Chloroacetaldehyde
AB7700000	*N,N*-Dimethylacetamide
AB9450000	2-Acetylaminofluorene
AF1225000	Acetic acid
AF7350000	*n*-Butyl acetate
AF7380000	*sec*-Butyl acetate
AF7400000	*tert*-Butyl acetate
AG6825000	2,4-D
AH5425000	Ethyl acetate
AI4025000	Isobutyl acetate
AI4930000	Isopropyl acetate
AI9100000	Methyl acetate
AJ1925000	*n*-Amyl acetate
AJ2100000	*sec*-Amyl acetate
AJ3675000	*n*-Propyl acetate
AJ8400000	2,4,5-T
AK0875000	Vinyl acetate
AK1925000	Acetic anhydride
AL3150000	Acetone
AL7700000	Acetonitrile
AM6300000	α-Chloroacetophenone
AS1050000	Acrolein
AS3325000	Acrylamide
AT0700000	Ethyl acrylate
AT2800000	Methyl acrylate
AT5250000	Acrylonitrile
BA5075000	Allyl alcohol
BO0875000	Ammonia
BW6650000	Aniline
BX0700000	4-Chloroaniline
BX4725000	Dimethylaniline
BX7350000	*p*-Dimethylaminoazobenzene
BY4550000	Methylaniline
BY6300000	Tetryl
BY6650000	2-Nitroaniline
BY6825000	3-Nitroaniline
BY7000000	4-Nitroaniline
BZ5410000	*o*-Anisidine
BZ5450000	*p*-Anisidine

CA9350000	Anthracene
CB0680000	2-Methylanthracene
CU1400000	Benzo[b]fluoranthene
CV9275000	Benzo[a]anthracene
CY1400000	Benzene
CY9000000	Bromobenzene
CY9070000	n-Butylbenzene
CY9100000	sec-Butylbenzene
CY9120000	tert-Butylbenzene
CZ0175000	Chlorobenzene
CZ1050000	p-Chloronitrobenzene
CZ1791000	1,4-Dibromobenzene
CZ4499000	1,3-Dichlorobenzene
CZ4500000	1,2-Dichlorobenzene
CZ4550000	1,4-Dichlorobenzene
CZ6300000	2,4-Toluene diisocyanate
CZ7350000	1,3-Dinitrobenzene
CZ7450000	1,2-Dinitrobenzene
CZ7525000	1,4-Dinitrobenzene
DA0700000	Ethylbenzene
DA2975000	Hexachlorobenzene
DA3550000	Isobutylbenzene
DA6475000	Nitrobenzene
DA6640000	Pentachlorobenzene
DA8750000	n-Propylbenzene
DB9440000	1,2,3,4-Tetrachlorobenzene
DB9445000	1,2,3,5-Tetrachlorobenzene
DB9450000	1,2,4,5-Tetrachlorobenzene
DC0500000	1,2,4,5-Tetramethylbenzene
DC2095000	1,2,3-Trichlorobenzene
DC2100000	1,2,4-Trichlorobenzene
DC2100100	1,3,5-Trichlorobenzene
DC3300000	1,2,3-Trimethylbenzene
DC3325000	1,2,4-Trimethylbenzene
DC9625000	Benzidine
DD0525000	3,3'-Dichlorobenzidine
DF6350000	Benzo[k]fluoranthene
DG0875000	Benzoic acid
DI6200500	Benzo[ghi]perylene
DJ3675000	Benzo[a]pyrene
DJ4200000	Benzo[e]pyrene
DK2625000	p-Quinone
DN3150000	Benzyl alcohol

DU8925000	4-Aminobiphenyl
DV5600000	4-Nitrobiphenyl
EI9275000	1,3-Butadiene
EI9625000	Chloroprene
EJ0700000	Hexachlorobutadiene
EJ4200000	Butane
EJ9300000	2,2-Dimethylbutane
EJ9350000	2,3-Dimethylbutane
EK4430000	2-Methylbutane
EK6300000	*n*-Butyl mercaptan
EL5425000	Isoamyl alcohol
EL6475000	2-Butanone
EM7600000	3-Methyl-1-butene
EO1400000	1-Butanol
EO1750000	*sec*-Butyl alcohol
EO1925000	*tert*-Butyl alcohol
EO2975000	*n*-Butylamine
EX1225000	Camphor
FB9450000	Carbofuran
FC5950000	Carbaryl
FF6650000	Carbon disulfide
FG4900000	Carbon tetrachloride
FS9100000	Chloroform
GC0700000	Chrysene
GN4550000	Warfarin
GO6300000	2-Methylphenol
GO6475000	4-Methylphenol
GO7100000	*p*-Chloro-*m*-cresol
GO9625000	4,6-Dinitro-*o*-cresol
GP9499000	Crotonaldehyde
GQ5250000	Mevinphos
GR8575000	Isopropylbenzene
GU3140000	Cycloheptane
GU6300000	Cyclohexane
GV3500000	α-BHC
GV4375000	β-BHC
GV4550000	δ-BHC
GV4900000	Lindane
GV6125000	Methylcyclohexane
GV7650000	1,1,3-Trimethylcyclohexane
GV7875000	Cyclohexanol
GW1050000	Cyclohexanone
GW1750000	*o*-Methylcyclohexanone

GW2500000	Cyclohexene
GW7700000	Isophorone
GY1000000	Cyclopentadiene
GY1225000	Hexachlorocyclopentadiene
GY2390000	Cyclopentane
GY4450000	Ethylcyclopentane
GY4640000	Methylcyclopentane
GY4700000	Propylcyclopentane
GY5950000	Cyclopentene
HD6550000	Decane
HN2625000	Dibenz[*a,h*]anthracene
HP3500000	TCDD
HZ8750000	Diethylamine
IM4025000	Diisopropylamine
IO1575000	Endrin
IO1750000	Dieldrin
IO2100000	Aldrin
IP8750000	Dimethylamine
IQ0525000	*N*-Nitrosodimethylamine
JG8225000	1,4-Dioxane
JJ9800000	*N*-Nitrosodiphenylamine
JL9700000	*N*-Nitrosodi-*n*-propylamine
JO1400000	Thiram
JR2125000	Dodecane
KH2100000	Ethylamine
KH6475000	Ethyl bromide
KH7525000	Chloroethane
KH8575000	Ethylenediamine
KH9275000	Ethylene dibromide
KI0175000	1,1-Dichloroethane
KI0525000	1,2-Dichloroethane
KI0700000	*p,p'*-DDD
KI1420000	1,2-Difluorotetrachloroethane
KI1425000	1,1-Difluorotetrachloroethane
KI4025000	Hexachloroethane
KI5600000	Nitroethane
KI5775000	Ethyl ether
KI6300000	Pentachloroethane
KI8225000	1,1,2,2-Tetrabromoethane
KI8575000	1,1,2,2-Tetrachloroethane
KI9625000	Ethyl mercaptan
KJ2975000	1,1,1-Trichloroethane
KJ3150000	1,1,2-Trichloroethane

KJ3325000	*p,p'*-DDT
KJ3675000	Methoxychlor
KJ4000000	1,1,2-Trichlorotrifluoroethane
KJ5775000	Ethanolamine
KK0875000	Ethylene chlorohydrin
KK5075000	2-Diethylaminoethanol
KK8050000	2-Ethoxyethanol
KK8225000	2-Ethoxyethyl acetate
KL5775000	Methyl cellosolve
KL5950000	Methyl cellosolve acetate
KN0875000	Bis(2-chloroethyl)ether
KN1575000	*sym*-Dichloromethyl ether
KN1750000	Bis(2-chloroisopropyl)ether
KN6300000	2-Chloroethyl vinyl ether
KN8970000	Phenyl ether
KU9625000	Vinyl chloride
KV9275000	1,1-Dichloroethylene
KV9400000	*trans*-1,2-Dichloroethylene
KV9450000	*p,p'*-DDE
KX3850000	Tetrachloroethylene
KX4550000	Trichloroethylene
KX5075000	Ethylenimine
LL4025000	Fluoranthene
LL5670000	Fluorene
LP8925000	Formaldehyde
LQ2100000	*N,N*-Dimethylformamide
LQ4900000	Formic acid
LQ8400000	Ethyl formate
LQ8925000	Methyl formate
LT7000000	Furfural
LU5950000	Tetrahydrofuran
LU9100000	Furfuryl alcohol
MI7700000	Heptane
MJ5075000	2-Heptanone
MJ5250000	3-Heptanone
MJ5775000	Diisobutyl ketone
MJ7350000	5-Methyl-3-heptanone
MN9275000	Hexane
MP1400000	2-Hexanone
MP6600100	1-Hexene
MU0700000	1,3-Dichloro-5,5-dimethylhydantoin
MV2450000	1,1-Dimethylhydrazine
MV5600000	Methylhydrazine

MV8925000	Phenylhydrazine
MW2625000	1,2-Diphenylhydrazine
MX3500000	Hydroquinone
NK3750000	Indan
NK6300000	Pindone
NK9300000	Indeno[1,2,3-*cd*]pyrene
NL2450000	Indole
NP9625000	Isobutyl alcohol
NQ9450000	Methyl isocyanate
NS9800000	Isoamyl acetate
NT4037000	2-Methyl-1,3-butadiene
NT8400000	Isopropylamine
NU8050000	Biphenyl
ON3675000	Maleic anhydride
OO3675000	*o*-Chlorobenzylidenemalononitrile
OX6825000	1,3,5-Trimethylbenzene
OZ5075000	Methyl methacrylate
PA3675000	Bis(2-chloroethoxy)methane
PA4900000	Methyl bromide
PA5250000	Bromochloromethane
PA5310000	Bromodichloromethane
PA5425000	Bromotrifluoromethane
PA6300000	Methyl chloride
PA6360000	Dibromochloromethane
PA7525000	Dibromodifluoromethane
PA8050000	Methylene chloride
PA8200000	Dichlorodifluoromethane
PA8400000	Dichlorofluoromethane
PA8750000	Methylal
PA9450000	Methyl iodide
PA9800000	Nitromethane
PB4025000	Tetranitromethane
PB4375000	Methyl mercaptan
PB5600000	Bromoform
PB6125000	Trichlorofluoromethane
PB6300000	Chloropicrin
PB9450000	Heptachlor epoxide
PB9705000	*trans*-Chlordane
PB9800000	Chlordane
PC0175000	*cis*-Chlordane
PC0700000	Heptachlor
PC1400000	Methanol
PC8575000	Kepone

PF6300000	Methylamine
QD6475000	Morpholine
QE4025000	4-Ethylmorpholine
QJ0525000	Naphthalene
QJ2275000	2-Chloronaphthalene
QJ3150000	Decahydronaphthalene
QJ9635000	2-Methylnaphthalene
QK0250000	Octachloronaphthalene
QM1400000	1-Naphthylamine
QM2100000	2-Naphthylamine
RA6115000	Nonane
RB9275000	α-Endosulfan, β-Endosulfan
RG8400000	Octane
RO2450000	Oxalic acid
RQ7350000	β-Propiolactone
RR0875000	Allyl glycidyl ether
RZ9450000	Pentane
SA2995000	2-Methylpentane
SA2995500	3-Methylpentane
SA3320000	2,2,4-Trimethylpentane
SA7525000	sec-Hexyl acetate
SA7875000	2-Pentanone
SA9100000	Diacetone alcohol
SA9275000	4-Methyl-2-pentanone
SB2230000	2-Methyl-1-pentene
SB4200000	Mesityl oxide
SF7175000	Phenanthrene
SF7810000	1-Methylphenanthrene
SJ3325000	Phenol
SK2625000	2-Chlorophenol
SK8575000	2,4-Dichlorophenol
SL2800000	2,4-Dinitrophenol
SM2100000	2-Nitrophenol
SM2275000	4-Nitrophenol
SM6300000	Pentachlorophenol
SN1400000	2,4,5-Trichlorophenol
SN1575000	2,4,6-Trichlorophenol
SS8050000	p-Phenylenediamine
TB1925000	EPN
TB9450000	Naled
TC0350000	Dichlorvos
TC7700000	Tributyl phosphate
TC8400000	Triphenyl phosphate

TD0350000	Tri-*o*-cresyl phosphate
TF4550000	Parathion
TF6300000	Chlorpyrifos
TG0525000	Ronnel
TH9990000	Benzyl butyl phthalate
TI0350000	Bis(2-ethylhexyl)phthalate
TI0875000	Di-*n*-butyl phthalate
TI1050000	Diethyl phthalate
TI1575000	Dimethyl phthalate
TI1925000	Di-*n*-octyl phthalate
TI3150000	Phthalic anhydride
TJ7875000	Picric acid
TQ1351000	PCB-1016
TQ1352000	PCB-1221
TQ1354000	PCB-1232
TQ1356000	PCB-1242
TQ1358000	PCB-1248
TQ1360000	PCB-1254
TQ1362000	PCB-1260
TX2275000	*n*-Propane
TX4900000	Epichlorohydrin
TX5075000	1-Chloro-1-nitropropane
TX8750000	1,2-Dibromo-3-chloropropane
TX9625000	1,2-Dichloropropane
TY1190000	2,2-Dimethylpropane
TZ2925000	Propylene oxide
TZ4100000	1-Iodopropane
TZ4300000	2-Methylpropane
TZ5075000	1-Nitropropane
TZ5250000	2-Nitropropane
TZ5425000	Isopropyl ether
TZ9275000	1,2,3-Trichloropropane
UB4375000	Glycidol
UC7350000	Allyl chloride
UC8320000	*trans*-1,3-Dichloropropylene
UC8325000	*cis*-1,3-Dichloropropylene
UD0890000	2-Methylpropene
UH8225000	1-Propanol
UK0350000	*n*-Propyl nitrate
UK4250000	Propyne
UR2450000	Pyrene
UR8400000	Pyridine
US1575000	2-Aminopyridine

US7525000	Nitrapyrin
UX6825000	Tetraethyl pyrophosphate
WL2275000	Strychnine
WL3675000	Styrene
WL5075300	α-Methylstyrene
WM8400000	Malathion
WS8225000	Dimethyl sulfate
XM7350000	Thiophene
XN4375000	Sulfotepp
XS5250000	Toluene
XS8925000	Benzyl chloride
XT1575000	2,4-Dinitrotoluene
XT1925000	2,6-Dinitrotoluene
XT2975000	3-Nitrotoluene
XT3150000	2-Nitrotoluene
XT3325000	4-Nitrotoluene
XU0175000	2,4,6-Trinitrotoluene
XU2975000	o-Toluidine
XU9275000	Trifluralin
XW5250000	Toxaphene
YE0175000	Triethylamine
YS8925000	Diuron
ZE2275000	m-Xylene
ZE2450000	o-Xylene
ZE2625000	p-Xylene
ZE5600000	2,4-Dimethylphenol

Appendix F. Empirical Formula Index

$CBrF_3$	Bromotrifluoromethane
CBr_2F_2	Dibromodifluoromethane
CCl_2F_2	Dichlorodifluoromethane
CCl_3F	Trichlorofluoromethane
CCl_3NO_2	Chloropicrin
CCl_4	Carbon tetrachloride
$CHBrCl_2$	Bromodichloromethane
$CHBr_2Cl$	Dibromochloromethane
$CHBr_3$	Bromoform
$CHCl_2F$	Dichlorofluoromethane
$CHCl_3$	Chloroform
CH_2BrCl	Bromochloromethane
CH_2Cl_2	Methylene chloride
CH_2O	Formaldehyde
CH_2O_2	Formic acid
CH_3Br	Methyl bromide
CH_3Cl	Methyl chloride
CH_3I	Methyl iodide
CH_3NO_2	Nitromethane
CH_4O	Methanol
CH_4S	Methyl mercaptan
CH_5N	Methylamine
CH_6N_2	Methylhydrazine
CN_4O_8	Tetranitromethane
CS_2	Carbon disulfide
$C_2Cl_3F_3$	1,1,2-Trichlorotrifluoroethane
C_2Cl_4	Tetrachloroethylene
$C_2Cl_4F_2$	1,1-Difluorotetrachloroethane
	1,2-Difluorotetrachloroethane
C_2Cl_6	Hexachloroethane
C_2HCl_3	Trichloroethylene
C_2HCl_5	Pentachloroethane
$C_2H_2Br_4$	1,1,2,2-Tetrabromoethane
$C_2H_2Cl_2$	1,1-Dichloroethylene
	trans-1,2-Dichloroethylene
$C_2H_2Cl_4$	1,1,2,2-Tetrachloroethane
$C_2H_2O_4$	Oxalic acid
C_2H_3Cl	Vinyl chloride
C_2H_3ClO	Chloroacetaldehyde
$C_2H_3Cl_3$	1,1,1-Trichloroethane
	1,1,2-Trichloroethane
C_2H_3N	Acetonitrile

C_2H_3NO	Methyl isocyanate
$C_2H_4Br_2$	Ethylene dibromide
$C_2H_4Cl_2$	1,1-Dichloroethane
	1,2-Dichloroethane
$C_2H_4Cl_2O$	sym-Dichloromethyl ether
C_2H_4O	Acetaldehyde
$C_2H_4O_2$	Acetic acid
	Methyl formate
C_2H_5Br	Ethyl bromide
C_2H_5Cl	Chloroethane
C_2H_5ClO	Ethylene chlorohydrin
C_2H_5N	Ethylenimine
$C_2H_5NO_2$	Nitroethane
$C_2H_6O_3$	Acetic anhydride
$C_2H_6O_4S$	Dimethyl sulfate
C_2H_6S	Ethyl mercaptan
C_2H_7N	Dimethylamine
	Ethylamine
C_2H_7NO	Ethanolamine
$C_2H_8N_2$	1,1-Dimethylhydrazine
	Ethylenediamine
C_3H_3N	Acrylonitrile
C_3H_4	Propyne
$C_3H_4Cl_2$	cis-1,3-Dichloropropylene
	trans-1,3-Dichloropropylene
C_3H_4O	Acrolein
$C_3H_5Br_2Cl$	1,2-Dibromo-3-chloropropane
C_3H_5Cl	Allyl chloride
C_3H_5ClO	Epichlorohydrin
$C_3H_5Cl_3$	1,2,3-Trichloropropane
C_3H_5NO	Acrylamide
$C_3H_6ClNO_2$	1-Chloro-1-nitropropane
$C_3H_6Cl_2$	1,2-Dichloropropane
$C_3H_6N_2O$	N-Nitrosodimethylamine
$C_3H_6O_2$	Vinyl acetate
C_3H_6O	Acetone
	Allyl alcohol
	β-Propiolactone
	Propylene oxide
$C_3H_6O_2$	Ethyl formate
	Glycidol
	Methyl acetate
C_3H_7I	1-Iodopropane

C_3H_7NO	*N*,*N*-Dimethylformamide
$C_3H_7NO_2$	1-Nitropropane
	2-Nitropropane
$C_3H_7NO_3$	*n*-Propyl nitrate
C_3H_8	Propane
C_3H_8O	1-Propanol
$C_3H_8O_2$	Methylal
	Methyl cellosolve
C_3H_9N	Isopropylamine
C_4Cl_6	Hexachlorobutadiene
$C_4H_2O_3$	Maleic anhydride
C_4H_4NO	*N*,*N*-Dimethylacetamide
C_4H_4S	Thiophene
C_4H_5Cl	Chloroprene
C_4H_6	1,3-Butadiene
C_4H_6O	Crotonaldehyde
$C_4H_6O_2$	Methyl acrylate
$C_4H_7Br_2Cl_2O_4P$	Naled
C_4H_7ClO	2-Chloroethyl vinyl ether
$C_4H_7Cl_2O_4P$	Dichlorvos
C_4H_8	1-Butene
	2-Methylpropene
$C_4H_8Cl_2O$	Bis(2-chloroethyl)ether
C_4H_8O	2-Butanone
	Tetrahydrofuran
$C_4H_8O_2$	1,4-Dioxane
	Ethyl acetate
C_4H_9NO	Morpholine
C_4H_{10}	Butane
	2-Methylpropane
$C_4H_{10}O$	1-Butanol
	sec-Butyl alcohol
	tert-Butyl alcohol
	Ethyl ether
	Isobutyl alcohol
$C_4H_{10}O_2$	2-Ethoxyethanol
$C_4H_{10}S$	*n*-Butyl mercaptan
$C_4H_{11}N$	*n*-Butylamine
	Diethylamine
C_5Cl_6	Hexachlorocyclopentadiene
$C_5H_4O_2$	Furfural
C_5H_5N	Pyridine
C_5H_6	Cyclopentadiene

$C_5H_6Cl_2N_2O_2$	1,3-Dichloro-5,5-dimethylhydantoin
$C_5H_6N_2$	2-Aminopyridine
$C_5H_6O_2$	Furfuryl alcohol
C_5H_8	Cyclopentene
	2-Methyl-1,3-butadiene
	1,4-Pentadiene
$C_5H_8O_2$	Ethyl acrylate
	Methyl methacrylate
C_5H_{10}	Cyclopentane
	3-Methyl-1-butene
	1-Pentene
	cis-2-Pentene
	trans-2-Pentene
$C_5H_{10}Cl_2O_2$	Bis(2-chloroethoxy)methane
	Bis(2-chloroisopropyl)ether
$C_5H_{10}O$	2-Pentanone
$C_5H_{10}O_2$	Isopropyl acetate
	n-Propyl acetate
$C_5H_{10}O_3$	Methyl cellosolve acetate
C_5H_{12}	2,2-Dimethylpropane
	2-Methylbutane
	Pentane
$C_5H_{12}O$	Isoamyl alcohol
C_6Cl_6	Hexachlorobenzene
C_6HCl_5	Pentachlorobenzene
C_6HCl_5O	Pentachlorophenol
$C_6H_2Br_4$	1,2,4,5-Tetrabromobenzene
$C_6H_2Cl_4$	1,2,3,4-Tetrachlorobenzene
	1,2,3,5-Tetrachlorobenzene
	1,2,4,5-Tetrachlorobenzene
$C_6H_3Br_3$	1,3,5-Tribromobenzene
$C_6H_3Cl_3$	1,2,3-Trichlorobenzene
	1,2,4-Trichlorobenzene
	1,3,5-Trichlorobenzene
$C_6H_3Cl_3O$	2,4,5-Trichlorophenol
	2,4,6-Trichlorophenol
$C_6H_3Cl_4N$	Nitrapyrin
$C_6H_3N_3O_7$	Picric acid
$C_6H_4Br_2$	1,4-Dibromobenzene
$C_6H_4ClNO_2$	*p*-Chloronitrobenzene
$C_6H_4Cl_2$	1,2-Dichlorobenzene
	1,3-Dichlorobenzene
	1,4-Dichlorobenzene

$C_6H_4Cl_2O$	2,4-Dichlorophenol
$C_6H_4N_2O_5$	2,4-Dinitrophenol
$C_6H_4N_2O_4$	1,2-Dinitrobenzene
	1,3-Dinitrobenzene
	1,4-Dinitrobenzene
$C_6H_4O_2$	*p*-Quinone
C_6H_5Br	Bromobenzene
C_6H_5Cl	Chlorobenzene
C_6H_5ClO	2-Chlorophenol
$C_6H_5NO_2$	Nitrobenzene
$C_6H_5NO_3$	2-Nitrophenol
	4-Nitrophenol
C_6H_6	Benzene
C_6H_6ClN	4-Chloroaniline
$C_6H_6Cl_6$	α-BHC
	β-BHC
	δ-BHC
	Lindane
$C_6H_6N_2O_2$	2-Nitroaniline
	3-Nitroaniline
	4-Nitroaniline
C_6H_6O	Phenol
$C_6H_6O_2$	Hydroquinone
C_6H_7N	Aniline
$C_6H_8N_2$	*p*-Phenylenediamine
	Phenylhydrazine
C_6H_8S	2-Ethylthiophene
C_6H_{10}	Cyclohexene
$C_6H_{10}O$	Cyclohexanone
	Mesityl oxide
$C_6H_{10}O_2$	Allyl glycidyl ether
C_6H_{12}	Cyclohexane
	1-Hexene
	Methylcyclopentane
	2-Methyl-1-pentene
	4-Methyl-1-pentene
$C_6H_{12}N_2S_4$	Thiram
$C_6H_{12}O$	Cyclohexanol
	2-Hexanone
	4-Methyl-2-pentanone
$C_6H_{12}O_2$	*n*-Butyl acetate
	sec-Butyl acetate
	tert-Butyl acetate

	Diacetone alcohol
	Isobutyl acetate
$C_6H_{12}O_3$	2-Ethoxyethyl acetate
$C_6H_{13}NO$	4-Ethylmorpholine
C_6H_{14}	2,2-Dimethylbutane
	2,3-Dimethylbutane
	Hexane
	2-Methylpentane
	3-Methylpentane
$C_6H_{14}N_2O$	N-Nitrosodi-n-propylamine
$C_6H_{14}O$	Isopropyl ether
$C_6H_{14}O_2$	2-Butoxyethanol
$C_6H_{15}N$	Diisopropylamine
	Triethylamine
$C_6H_{15}NO$	2-Diethylaminoethanol
$C_7H_5N_3O_6$	2,4,6-Trinitrotoluene
$C_7H_5N_5O_8$	Tetryl
$C_7H_6N_2O_4$	2,4-Dinitrotoluene
	2,6-Dinitrotoluene
$C_7H_6N_2O_5$	4,6-Dinitro-o-cresol
$C_7H_6O_2$	Benzoic acid
C_7H_7Cl	Benzyl chloride
C_7H_7ClO	p-Chloro-m-cresol
$C_7H_7NO_2$	2-Nitrotoluene
	3-Nitrotoluene
	4-Nitrotoluene
C_7H_8	Toluene
C_7H_8O	Benzyl alcohol
	2-Methylphenol
	4-Methylphenol
C_7H_9N	Methylaniline
	o-Toluidine
C_7H_9NO	o-Anisidine
	p-Anisidine
C_7H_{12}	1-Methylcyclohexene
$C_7H_{12}O$	o-Methylcyclohexanone
$C_7H_{13}O_6P$	Mevinphos
C_7H_{14}	Cycloheptane
	Ethylcyclopentane
	cis-2-Heptene
	$trans$-2-Heptene
C_7H_{14}	Methylcyclohexane
$C_7H_{14}O$	2-Heptanone

	3-Heptanone
$C_7H_{14}O_2$	*n*-Amyl acetate
	sec-Amyl acetate
	Isoamyl acetate
C_7H_{16}	2,3-Dimethylpentane
	2,4-Dimethylpentane
	3,3-Dimethylpentane
	Heptane
	2-Methylhexane
	3-Methylhexane
$C_8H_4O_3$	Phthalic anhydride
$C_8H_5Cl_3O_3$	2,4,5-T
$C_8H_6Cl_2O_3$	2,4-D
C_8H_7ClO	α-Chloroacetophenone
C_8H_7N	Indole
C_8H_8	Styrene
$C_8H_8Cl_3O_3PS$	Ronnel
C_8H_9N	Indoline
C_8H_{10}	Ethylbenzene
	m-Xylene
	o-Xylene
	p-Xylene
$C_8H_{10}O$	2,4-Dimethylphenol
$C_8H_{10}O_7P_2$	Tetraethyl pyrophosphate
$C_8H_{11}N$	Dimethylaniline
C_8H_{16}	*cis*-1,2-Dimethylcyclohexane
	trans-1,4-Dimethylcyclohexane
	1-Octene
	Propylcyclopentane
	1,1,3-Trimethylcyclopentane
$C_8H_{16}O$	5-Methyl-3-heptanone
$C_8H_{16}O_2$	*sec*-Hexyl acetate
C_8H_{18}	3-Methylheptane
	Octane
	2,2,4-Trimethylpentane
	2,3,4-Trimethylpentane
$C_8H_{20}O_5P_2S_2$	Sulfotepp
$C_9H_6Cl_6O_3S$	α-Endosulfan
	β-Endosulfan
$C_9H_6Cl_6O_4S$	Endosulfan sulfate
$C_9H_6N_2O_2$	2,4-Toluene diisocyanate
C_9H_{10}	Indan
	α-Methylstyrene

$C_9H_{10}Cl_2N_2O$	Diuron
$C_9H_{11}Cl_3NO_3PS$	Chlorpyrifos
C_9H_{12}	Isopropylbenzene
	n-Propylbenzene
	1,2,3-Trimethylbenzene
	1,2,4-Trimethylbenzene
	1,3,5-Trimethylbenzene
$C_9H_{14}O$	Isophorone
C_9H_{18}	1,1,3-Trimethylcyclohexane
$C_9H_{18}O$	Diisobutyl ketone
C_9H_{20}	4-Methyloctane
	Nonane
	2,2,5-Trimethylhexane
$C_{10}Cl_8$	Octachloronaphthalene
$C_{10}Cl_{10}O$	Kepone
$C_{10}H_5ClN_2$	o-Chlorobenzylidenemalononitrile
$C_{10}H_5Cl_7$	Heptachlor
$C_{10}H_5Cl_7O$	Heptachlor epoxide
$C_{10}H_6Cl_8$	Chlordane
	cis-Chlordane
	trans-Chlordane
$C_{10}H_7Cl$	2-Chloronaphthalene
$C_{10}H_8$	Naphthalene
$C_{10}H_9N$	1-Naphthylamine
	2-Naphthylamine
$C_{10}H_{10}Cl_8$	Toxaphene
$C_{10}H_{10}O_4$	Dimethyl phthalate
$C_{10}H_{14}$	n-Butylbenzene
	sec-Butylbenzene
	tert-Butylbenzene
	Isobutylbenzene
	1,2,4,5-Tetramethylbenzene
$C_{10}H_{14}NO_5PS$	Parathion
$C_{10}H_{16}O$	Camphor
$C_{10}H_{18}$	Decahydronaphthalene
$C_{10}H_{19}O_6PS_2$	Malathion
$C_{10}H_{20}$	Pentylcyclopentane
$C_{10}H_{22}$	Decane
$C_{11}H_{10}$	2-Methylnaphthalene
$C_{11}H_{10}N_2S$	ANTU
$C_{11}H_{11}N$	2,7-Dimethylquinoline
$C_{12}H_4Cl_4O_2$	TCDD
$C_{12}H_8$	Acenaphthylene

$C_{12}H_8Cl_6$	Aldrin
$C_{12}H_8Cl_6O$	Dieldrin
	Endrin
$C_{12}H_8O$	Dibenzofuran
$C_{12}H_9BrO$	4-Bromophenyl phenyl ether
$C_{12}H_9ClO$	4-Chlorophenyl phenyl ether
$C_{12}H_9NO_2$	4-Nitrobiphenyl
$C_{12}H_{10}$	Acenaphthene
	Biphenyl
$C_{12}H_{10}Cl_2N_2$	3,3'-Dichlorobenzidine
$C_{12}H_{10}N_2O$	N-Nitrosodiphenylamine
$C_{12}H_{10}O$	Phenyl ether
$C_{12}H_{11}N$	4-Aminobiphenyl
$C_{12}H_{11}NO_2$	Carbaryl
$C_{12}H_{12}$	Benzidine
$C_{12}H_{12}N_2$	1,2-Diphenylhydrazine
$C_{12}H_{14}O_4$	Diethyl phthalate
$C_{12}H_{15}NO$	Carbofuran
$C_{12}H_{26}$	Dodecane
$C_{12}H_{27}O_4P$	Tributyl phosphate
$C_{13}H_{10}$	Fluorene
$C_{13}H_{16}F_3N_3O_4$	Trifluralin
$C_{14}H_8Cl_4$	p,p'-DDE
$C_{14}H_9Cl_5$	p,p'-DDT
$C_{14}H_{10}$	Anthracene
	Phenanthrene
$C_{14}H_{10}Cl_4$	p,p'-DDD
$C_{14}H_{14}NO_4PS$	EPN
$C_{14}H_{14}O_3$	Pindone
$C_{14}H_{15}N_3$	p-Dimethylaminoazobenzene
$C_{15}H_{12}$	2-Methylanthracene
	1-Methylphenanthrene
$C_{15}H_{13}NO$	2-Acetylaminofluorene
$C_{16}H_{10}$	Fluoranthene
	Pyrene
$C_{16}H_{15}Cl_3O_2$	Methoxychlor
$C_{16}H_{22}O_4$	Di-n-butyl phthalate
$C_{18}H_{12}$	Benzo[a]anthracene
	Chrysene
$C_{18}H_{15}O_4P$	Triphenyl phosphate
$C_{19}H_{16}O_4$	Warfarin
$C_{19}H_{20}O_4$	Benzyl butyl phthalate
$C_{20}H_{12}$	Benzo[b]fluoranthene

	Benzo[k]fluoranthene
	Benzo[a]pyrene
	Benzo[e]pyrene
$C_{21}H_{21}O_4P$	Tri-o-cresyl phosphate
$C_{21}H_{22}N_2O_2$	Strychnine
$C_{22}H_{12}$	Benzo[ghi]perylene
	Indeno[1,2,3-cd]pyrene
$C_{22}H_{14}$	Dibenzo[a,h]anthracene
$C_{24}H_{38}O_4$	Bis(2-ethylhexyl)phthalate
	Di-n-octyl phthalate
H_3N	Ammonia

Appendix G. Typical Bulk Density Values for Selected Soils and Rocks

Soil/Rock Type	Bulk Density, B
Silt	1.38
Clay	1.49
Loess	1.45
Sand, dune	1.58
Sand, fine	1.55
Sand, medium	1.69
Sand, coarse	1.73
Gravel, fine	1.76
Gravel, medium	1.85
Gravel, coarse	1.93
Till, predominantly silt	1.78
Till, predominantly sand	1.88
Till, predominantly gravel	1.91
Glacial drift, predominantly silt	1.38
Glacial drift, predominantly sand	1.55
Glacial drift, predominantly gravel	1.60
Siltstone	1.61
Claystone	1.51
Sandstone, fine-grained	1.76
Sandstone, medium-grained	1.68
Limestone	1.94
Dolomite	2.02
Schist	1.76
Basalt	2.53
Shale	2.53
Gabbro, weathered	1.73
Granite, weathered	1.50

Morris and Johnson (1967)

1205

Appendix H. Ranges of Porosity Values
for Selected Soils and Rocks

Soil/Rock Type	Porosity, n
Peat	0.60–0.80
Silt	0.34–0.61
Clay	0.34–0.57
Loess	0.40–0.57
Sand, dune	0.35–0.51
Sand, fine	0.25–0.55
Sand, medium	0.28–0.49
Sand, coarse	0.30–0.46
Gravel, fine	0.25–0.40
Gravel, medium	0.24–0.44
Gravel, coarse	0.25–0.35
Sand + gravel	0.20–0.35
Till, predominantly silt	0.30–0.41
Till, predominantly sand	0.22–0.37
Till, predominantly gravel	0.22–0.30
Till, clay-loam	0.30–0.35
Glacial drift, predominantly silt	0.38–0.59
Glacial drift, predominantly sand	0.36–0.48
Glacial drift, predominantly gravel	0.35–0.42
Siltstone	0.20–0.41
Claystone	0.41–0.45
Sandstone, fine-grained	0.14–0.49
Sandstone, medium-grained	0.30–0.44
Volcanic, dense	0.01–0.10
Volanic, pumice	0.80–0.90
Volcanic, vesicular	0.10–0.50
Volcanic, tuff	0.10–0.40
Limestone	0.05–0.56
Dolomite	0.19–0.33
Schist	0.05–0.55
Basalt	0.03–0.35
Shale	0.01–0.10
Igneous, dense metamorphic and plutonic	0.01–0.05
Igneous, weathered metamorphic and plutonic	0.34–0.55

Davis and DeWiest (1966); Grisak et al. (1980); Morris and Johnson (1967)

Appendix I. Solubility Data of Miscellaneous Compounds

Compound	CAS Registry No.	Temp. °C	Log S mol/L	Reference
Acetal	105-57-7	25.0	-0.12	1
Acetophenone.	98-86-2	25.0	-1.34	2
Allyl bromide	106-95-6	25.0	-1.50	1
Anisole	100-66-3	25.00	-1.85	3
		25.0	-1.89	2
Atrazine	1912-24-9	22	-3.80	4
2,3-Benzanthracene	92-24-0	25	-8.60	5
		27	-8.36	6
1,2-Benzofluorene	238-84-6	25	-6.68	7
2,3-Benzofluorene	243-17-4	25	-7.73	8
		25	-8.03	7
Benzonitrile	100-47-0	25	-1.38	9
Benzyl acetate	140-11-4	25.0	-1.77	10
Bibenzyl	103-29-7	20	-5.62	11
		25.0	-4.80	12
Bis(2-ethylhexyl)isophthalate	137-89-3	24	-7.55	13
Borneol	464-45-9	25	-2.32	14
4-Bromobiphenyl	92-66-0	25	-5.55	15
1-Bromobutane	109-65-9	25.0	-2.20	1
		30	-2.35	16
4-Bromo-1-butene	5162-44-7	25.0	-2.25	1
2-Bromochlorobenzene	694-80-4	25	-3.19	17
3-Bromochlorobenzene	108-37-2	25	-3.21	17
4-Bromochlorobenzene	106-39-8	25	-3.63	17
1-Bromo-2-chloroethane	107-04-0	30.00	-1.32	18
1-Bromo-3-chloropropane	109-70-6	25.0	-1.85	1
2-Bromoethylacetate	927-68-4	25.0	-0.67	1
1-Bromoheptane	629-04-9	25.0	-4.43	1
1-Bromohexane	111-25-1	25.0	-3.81	1
4-Bromoiodobenzene	589-87-7	25	-4.56	17
1-Bromooctane	111-83-1	25.0	-5.06	1
1-Bromopentane	110-53-2	25.0	-3.08	1
n-Butyl butyrate	109-21-7	25.0	-2.37	10
n-Butyl formate	592-84-7	25.0	-0.97	10
n-Butyl isobutyrate	97-87-0	25.0	-2.27	10
n-Butyl lactate	138-22-7	25	-0.53	19
4-sec-Butyl-2-nitrophenol	3555-18-8	20.0	-3.84	20
n-Butyl propionate	590-01-2	25.0	-1.85	10
Carbazole	86-74-8	20	-5.14	21
Carbon tetrabromide	558-13-4	30	-3.14	16
		30.00	-2.71	18
2-Chlorobiphenyl	2051-60-7	25	-4.57	22
4-Chlorobiphenyl	2051-62-9	25	-5.15	23
		25	-5.15	24
1-Chlorobutane	109-69-3	25.0	-2.03	1

Compound	CAS Registry No.	Temp. °C	Log S mol/L	Reference
2-Chlorodibenzo-*p*-dioxin	39227-540-8	25	-5.84	15
1-Chloroheptane	629-06-1	25.0	-4.00	1
2-Chloroiodobenzene	615-41-8	25	-3.54	17
3-Chloroiodobenzene	625-99-0	25	-3.55	17
4-Chloroiodobenzene	637-87-6	25	-4.03	17
2-Chloronitrobenzene	88-73-3	20	-2.55	25
3-Chloronitrobenzene	121-73-3	20	-2.76	25
4-Chloronitrobenzene	100-00-5	20	-2.85	26
		20	-2.54	25
4-Chloro-2-nitrophenol	89-64-5	20.0	-3.09	20
4-Chlorophenol	106-48-9	20	-0.99	26
6-Chloro-3-nitrotoluene	7147-89-9	20.0	-4.39	20
4-Chlorotoluene	106-43-4	20	-2.97	26
		20	-3.08	27
Chlorprothixene	113-59-7	20	-4.41	26
Chlorpyrifos	2921-88-2	20	-8.00	11
Cholanthrene	479-23-2	27	-7.86	6
Cinnamic acid	621-82-9	30.00	-2.39	18
Coronene	191-07-1	25	-9.48	8
		25	-9.33	5
Cycloheptanol	5-2-41-0	25.0	-0.89	10
Cycloheptatriene	544-25-2	25	-2.17	28
Cycloheptene	628-92-2	25	-3.16	28
1,4-Cyclohexadiene	628-41-1	25	-2.06	28
Cyclohexyl acetate	622-45-7	25.0	-1.71	10
Cyclohexyl formate	4351-54-6	25.0	-1.19	10
Cyclohexylmethanol	100-49-2	25.0	-1.14	10
Cyclohexyl propionate	6222-35-1	25.0	-2.31	10
Cyclooctane	292-64-8	25	-4.15	28
Cyclooctanol	696-71-9	25.0	-1.30	10
Cyclopentanol	96-41-3	25.0	0.07	10
Decachlorobiphenyl	2051-24-3	22.0	-10.38	29
		25	-11.89	30
		25	-10.83	22
Decanoic acid	334-48-5	20.0	-3.06	31
1-Decanol	112-30-1	20.0	-3.57	32
		25	-3.50	33
		29.6	-2.88	10
2-Decanone	693-54-9	25.0	-3.30	1
1-Decene	872-05-9	25	-4.39	34
n-Decylbenzene	104-72-3	25.0	-7.94	35
Dially phthalate	131-17-9	20	-3.13	36
Diazinon	333-41-5	25	-2.47	37
1,2,3,4-Dibenzanthracene	215-58-7	25	-8.24	8
Dibenzo-*p*-dioxin	262-12-4	25	-5.31	15
1,2-Dibromobenzene	583-53-9	25	-3.50	17
1,3-Dibromobenzene	108-36-1	35.0	-3.54	38
		25	-3.38	17

Compound	CAS Registry No.	Temp. °C	Log S mol/L	Reference
Dibromomethane	74-95-3	30	-1.16	16
1,3-Dibromopropane	109-64-8	30.00	-2.08	18
Dicapthon	2463-84-5	20	-4.68	39
3,5-Dichloroaniline	626-43-7	23	-2.37	40
2,2'-Dichlorobiphenyl	13029-08-8	25	-5.27	41
		25	-5.45	41
2,4-Dichlorobiphenyl	33284-50-3	25	-5.29	41
2,5-Dichlorobiphenyl	34883-39-1	25	-5.30	41
		25	-5.06	22
		25	-5.59	41
2,6-Dichlorobiphenyl	33146-45-1	22.0	-5.62	29
		25	-4.97	41
		25	-5.21	22
3,3'-Dichlorobiphenyl	2050-67-1	25	-5.80	41
3,4-Dichlorobiphenyl	2974-92-7	25	-7.45	41
3,5-Dichlorobiphenyl	31883-41-5	25	-6.37	42
4,4'-Dichlorobiphenyl	2050-08-2	25	-6.56	39
		25	-6.32	42
2,8-Dichlorodibenzofuran	5409-83-6	25	-7.21	15
2,3-Dichloronitrobenzene	3209-22-1	20	-3.49	25
2,5-Dichloronitrobenzene	89-61-2	20	-3.32	25
		25	-6.79	41
		25	-6.60	41
3,4-Dichloronitrobenzene	99-54-7	20	-3.20	25
Dicofol	115-32-2	24	-5.49	13
Dicyclohexyl phthalate	84-61-7	24	-4.92	13
Di-n-decyl phthalate	2432-90-8	24	-6.18	13
Diethyl carbonate	105-58-8	25.0	-0.70	10
Diethyl oxalate	95-92-1	25.0	-0.59	10
1,2-Difluorobenzene	367-11-3	25	-2.00	17
1,3-Difluorobenzene	372-18-9	25	-2.00	17
1,4-Difluorobenzene	540-36-3	25	-1.97	17
1,2-Diiodobenzene	615-42-9	25	-4.24	17
1,4-Diiodobenzene	624-38-4	25	-5.25	17
		25.0	-5.50	2
Diiodomethane	75-11-6	30	-2.33	16
Diisobutyl phthalate	84-69-5	20	-4.14	36
		24	-4.65	13
Diisopropyl phthalate	605-45-8	20	-2.88	36
1,3-Dimethoxybenzene	151-10-1	25	-2.06	9
Dimethyl adipate	627-93-0	25.0	-0.77	10
9,10-Dimethylanthracene	781-43-1	25	-6.57	7
7,12-Dimethylbenz[a]anthracene	57-97-6	25	-7.02	43
		25	-6.62	5
9,10-Dimethylbenz[a]anthracene	58429-99-5	24	-6.67	13
		27	-6.78	6
4,4'-Dimethylbiphenyl	613-33-2	25	-6.02	15
2,2-Dimethyl-1-butanol	1185-33-7	25	-1.13	44

Compound	CAS Registry No.	Temp. °C	Log S mol/L	Reference
2,3-Dimethyl-2-butanol	594-60-5	25	-0.34	44
3,3-Dimethyl-2-butanol	464-07-3	25	-0.62	44
		25.0	-0.68	10
3,3-Dimethyl-2-butanone	75-97-8	25	-0.72	45
Dimethyl carbonate	616-38-6	25.0	0.03	10
5,6-Dimethylchrysene	3697-27-6	27	-7.01	6
2,6-Dimethylcyclohexanol	5337-72-4	25.0	-1.46	10
Dimethyl glutarate	1119-40-0	25.0	-0.43	10
2,6-Dimethyl-4-heptanol	108-82-7	25.0	-2.35	10
Dimethyl maleate	624-48-6	25.0	-0.27	10
1,3-Dimethylnaphthalene	575-41-7	25	-3.79	5
1,4-Dimethylnaphthalene	571-58-4	25	-4.29	5
1,5-Dimethylnaphthalene	571-61-9	25.0	-4.74	46
		25	-4.14	5
2,3-Dimethylnaphthalene	581-40-8	25.0	-4.89	46
		25	-4.66	5
2,6-Dimethylnaphthalene	581-42-0	25.0	-5.20	47
		25.0	-5.08	46
		25	-4.72	5
2,2-Dimethylpentane	590-35-2	25	-4.36	48
2,2-Dimethyl-3-pentanol	3970-62-5	25	-1.15	49
2,3-Dimethyl-2-pentanol	4911-70-0	25	-0.88	49
2,3-Dimethyl-3-pentanol	595-41-5	25	-0.85	49
2,4-Dimethyl-2-pentanol	625-06-9	25	-0.94	50
2,4-Dimethyl-3-pentanol	600-36-2	25	-1.22	50
		25	-1.24	51
2,4-Dimethyl-3-pentanone	565-80-0	25	-1.30	45
2,2-Dimethyl-1-propanol	75-84-3	24.4	-0.44	51
2,7-Dimethylquinoline	93-37-8	25	-1.94	48
Dimethyl sulfide	75-18-3	25	-0.45	52
2,5-Dinitrophenol	329-71-5	20.0	-2.68	20
Di-n-pentyl phthalate	131-18-0	20	-5.84	36
1,1-Diphenylethylene	530-48-3	25.0	-4.52	12
Diphenyl phosphate	838-85-7	24	-2.97	13
Diphenyl phthalate	84-62-8	24	-6.59	13
Diphenylamine	122-39-4	20	-3.55	21
Diphenylmethane	101-81-5	24	-4.75	13
		25.0	-5.08	53
Diphenylmethyl phosphate	115-89-9	24	-5.44	13
Di-n-propyl phthalate	131-16-8	20	-3.36	36
Ditridecyl phthalate	119-06-2	24	-6.19	13
		25	-0.40	54
Dodecanoic acid	143-07-7	20.0	-3.56	31
		25	-4.58	55
1-Dodecanol	112-53-8	25	-4.64	55
		29.5	-2.67	10
Eicosane	112-95-8	25.0	-8.18	56
Ethyl adipate	141-28-6	20	-1.68	58

Compound	CAS Registry No.	Temp. °C	Log S mol/L	Reference
		25.0		10
		30	-1.68	18
2-Ethylanthracene	52251-71-5	25.0	-6.78	47
		25.3	-6.89	59
10-Ethylbenz[a]anthracene	3697-30-1	27	-6.81	60
Ethyl benzoate	93-89-0	25.0	-2.40	2
		25.0	-2.26	10
2-Ethyl-1-butanol	97-95-0	25.0	-1.06	10
Ethyl butyrate	105-54-4	25.0	-1.26	10
Ethyl chloroacetate	105-39-5	25.0	-0.80	10
Ethyl decanoate	110-38-3	20	-4.13	58
Ethyl heptanoate	106-30-9	20	-2.74	58
2-Ethyl-1,3-hexanediol	94-96-2	25.0	-2.81	1
Ethyl hexanoate	123-66-0	20	-2.36	58
		25.0	-2.26	10
2-Ethyl-1-hexanol	104-76-7	20.0	-2.17	32
		24.7	-2.07	61
Ethyl hydrocinnamate	2021-28-5	25.0	-3.01	12
Ethyl iodide	75-03-6	30	-1.59	16
Ethyl isobutyrate	97-62-1	25.0	-1.27	10
Ethyl isopentanoate	108-64-5	25.0	-1.87	10
Ethyl malonate	105-53-3	20	-0.89	58
1-Ethylnaphthalene	1127-26-0	25	-4.19	62
		25.0	-4.19	63
		25	-4.16	5
2-Ethylnaphthalene	939-27-5	25.0	-4.29	46
Ethyl octanoate	123-29-5	20	-3.79	58
3-Ethyl-3-pentanol	597-49-9	25	-0.84	50
Ethyl propionate	105-37-3	25.0	-0.83	1
		25.0	-0.69	10
Ethyl salicylate	119-61-6	25.0	-2.65	10
Ethyl succinate	123-25-1	20	-0.96	58
Ethyl succinate	123-25-1	25.0	-0.93	10
2-Ethylthiophene	872-55-9	25	-2.58	48
2-Ethyltoluene	611-14-3	25.0	-3.21	1
Ethyl trimethylacetate	3938-95-2	25.0	-1.38	10
Fluorobenzene	462-06-6	25.0	-1.87	2
		30.00	-1.79	18
m-Fluorobenzyl chloride	456-42-8	25.0	-2.54	1
o-Fluorobenzyl chloride	345-35-7	25.0	-2.54	1
4-Formyl-2-nitrophenol	3011-34-5	20.0	-2.95	20
2,2',3,3',4,4',6-Heptachlorobiphenyl	52663-71-5	25	-8.26	22
2,2',3,4,4',5,5'-Heptachlorobiphenyl	35065-29-3	25	-8.71	42
1,2,3,4,6,7,8-Heptachlorodibenzo-p-dioxin	35822-46-9	20.0	-11.25	64
		26.0	-11.22	23
1,2,3,4,6,7,8-Heptachlorodibenzofuran	67462-39-4	22.7	-11.48	64
Heptadecanoic acid	506-12-7	20.0	-4.81	31
1,6-Heptadiene	3070-53-9	25	-3.34	28

Compound	CAS Registry No.	Temp. °C	Log S mol/L	Reference
1,6-Heptadiyne	2396-63-6	25	-1.75	28
Heptanoic acid	111-14-8	20.0	-1.73	31
		30	-1.85	65
1-Heptanol	111-70-6	20.0	-1.84	66
		25	-1.89	23
		25.0	-1.95	1
		25.4	-1.84	61
2-Heptanol	543-49-7	25.1	-1.48	61
3-Heptanol	589-82-2	20.0	-1.39	32
		25.0	-1.40	10
4-Heptanol	589-55-9	20.0	-1.39	32
		25.0	-1.42	10
1-Heptene	592-76-7	25	-3.55	34
		25.0	-3.73	1
n-Heptyl acetate	112-06-1	25.0	-2.88	10
n-Heptyl formate	112-23-2	25.0	-1.98	10
1-Heptyne	628-71-7	25	-3.01	28
2,2',3,3',4,4'-Hexachlorobiphenyl	38380-07-3	25	-9.01	41
		25	-9.11	22
		25	-8.87	41
		25	-8.41	42
2,2',3,3',4,5-Hexachlorobiphenyl	55215-18-4	25	-7.79	41
		25	-8.59	41
2,2',3,3',6,6-Hexachlorobiphenyl	38411-22-2	25	-7.90	30
		25	-7.78	22
		25	-8.35	42
2,2',4,4',5,5'-Hexachlorobiphenyl	35065-27-1	22.0	-8.50	29
		24	-8.58	39
		25	-7.63	15
		25	-8.62	41
2,2',4,4',6,6'-Hexachlorobiphenyl	33979-03-2	22.0	-8.52	29
		25	-8.20	41
		25.0	-8.04	67
		25	-8.10	23
		25	-8.04	24
		25	-8.95	22
		25	-8.56	41
		25	-8.46	42
1,2,3,4,7,8-Hexachlorodibenzo-p-dioxin	39227-28-6	20.0	-10.95	64
		26.0	-10.69	23
1,2,3,4,7,8-Hexachlorodibenzofuran	70658-26-9	22.7	-10.66	68
Hexacosane	630-01-3	25.0	-8.33	56
Hexadecane	544-76-3	25.0	-8.40	56
		25	-7.56	69
Hexadecanoic acid	57-10-3	20.0	-4.55	31
		25	-5.47	55
1-Hexadecanol	36653-82-4	25	-6.77	55
1,5-Hexadiene	592-42-7	25	-2.69	28

Compound	CAS Registry No.	Temp. °C	Log S mol/L	Reference
Hexanoic acid	142-62-1	20.0	-1.08	31
		30	-1.19	65
1-Hexanol	111-27-3	20.0	-1.22	32
		24.9	-1.22	61
		25	-1.28	23
		25.0	-1.38	1
2-Hexanol	623-93-7	20.0	-0.99	32
		24.9	-0.92	61
		25	-0.87	44
3-Hexanol	623-37-0	25	-0.80	44
3-Hexanone	589-38-8	25	-0.83	45
2-Hexene	592-43-8	25	-3.10	34
1-Hexen-3-ol	4798-44-1	25	-0.60	70
4-Hexen-3-ol	4798-58-7	25	-0.42	70
n-Hexyl acetate	142-92-7	25.0	-2.48	10
Hexylbenzene	1077-16-3	25.0	-5.20	1
		25.0	-5.25	72
n-Hexyl formate	629-33-4	25.0	-1.97	10
n-Hexyl propionate	2445-76-3	25.0	-1.94	10
Hexyne	693-02-7	25.0	-2.08	1
		25	-2.36	28
4-Hydroxybenzoic acid	99-96-7	20	-1.46	71
1-Iodoheptane	4282-40-0	25.0	-4.81	1
4-Isopropyltoluene	99-87-6	25.0	-3.76	73
Iodobenzene	591-50-4	25.00	-2.95	3
		25.0	-3.11	2
		25.0	-3.01	1
		30.00	-2.78	18
Isobutyl butyrate	97-85-8	25.0	-2.46	10
Isononyl acetate	40379-24-6	25.0	-2.99	10
Isopentyl acetate	123-92-2	25.0	-1.79	10
Isopentyl butyrate	106-27-4	25.0	-2.80	10
Isopentyl propionate	105-68-0	25.0	-2.37	10
Isopropyl butyrate	638-11-9	25.0	-0.68	10
Leptophos	21609-90-5	20	-7.94	39
Methane	74-82-8	25	-2.79	28
		25.0	-2.79	74
4-Methoxy-2-nitrophenol	1568-70-3	20.0	-2.84	20
9-Methylanthracene	779-02-2	25.0	-5.56	47
		25	-5.87	5
1-Methylbenz[a]anthracene	2498-77-3	27	-6.64	6
9-Methylbenz[a]anthracene	2381-16-0	24	-6.82	13
		27	-6.56	6
10-Methylbenz[a]anthracene	2381-15-9	24	-7.34	13
		27	-6.64	6
Methyl benzoate	93-58-3	25.0	-1.74	10
4-Methylbiphenyl	644-08-6	25	-4.62	15
		25	-5.15	67

Compound	CAS Registry No.	Temp. °C	Log S mol/L	Reference
2-Methyl-1-butanol	137-32-6	24.6	-0.52	61
		25	-0.47	54
2-Methyl-2-butanol	75-85-4	25	0.10	54
		25.2	0.08	61
3-Methyl-2-butanol	598-75-4	25	-0.20	54
		25.0	-0.20	61
3-Methyl-2-butanone	563-80-4	25	-0.15	45
2-Methyl-2-butene	513-35-9	25	-2.34	34
Methyl butyrate	623-42-7	25.0	-0.76	10
Methyl chloroacetate	96-34-4	25.0	-0.33	10
Methyl 2-chlorobutyrate	3153-37-5	26.0	-1.09	10
Methyl 2-chloropropionate	17639-93-9	25.0	-0.79	10
3-Methylcholanthrene	56-49-5	25	-7.92	43
		25	-7.97	5
5-Methylchrysene	3697-24-3	27	-5.69	6
6-Methylchrysene	1705-85-7	27	-6.57	6
2-Methylcyclohexanol	583-59-5	25.0	-0.83	10
3-Methylcyclohexanol	591-23-1	25.0	-0.93	10
4-Methylcyclohexanol	589-91-3	25.0	-0.97	10
1-Methylcyclohexene	591-47-9	25	-3.27	28
Methyl decanoate	110-42-9	25.0	-4.69	1
Methyl dichloroacetate	116-54-1	25.0	-1.64	10
6-Methyl-2,4-dinitrophenol	534-52-1	20.0	-3.00	20
Methyl enanthate	106-73-0	25.0	-2.20	10
1-Methylfluorene	1730-37-6	25	-4.96	8
Methyl hexanoate	106-70-7	25.0	-1.89	10
2-Methyl-2-hexanol	625-23-0	25	-1.08	50
3-Methyl-3-hexanol	597-96-6	25	-0.99	50
5-Methyl-2-hexanol	627-59-8	25.0	-1.38	10
Methyl isobutyrate	547-63-7	25.0	-0.75	10
Methyl p-methoxybenzoate	121-98-2	20	-2.41	71
1-Methylnaphthalene	90-12-0	25.0	-3.74	46
		25	-3.68	62
		25.0	-3.85	63
		25	-3.65	5
3-Methyl-2-nitrophenol	4920-77-8	20.0	-1.64	20
3-Methyl-4-nitrophenol	2581-34-2	20.0	-2.11	20
4-Methyl-2-nitrophenol	119-33-5	20.0	-2.55	20
5-Methyl-2-nitrophenol	700-838-9	20.0	-2.75	20
Methyl nonanoate	1731-84-6	25.0	-3.88	1
Methyl pentanoate	624-24-8	25.0	-6.05	48
2-Methyl-1-pentanol	105-30-6	25.2	-1.37	10
2-Methyl-2-pentanol	590-36-3	25	-0.50	44
2-Methyl-3-pentanol	565-67-3	25	-0.71	44
3-Methyl-1-pentanol	589-35-5	25.0	-1.04	10
3-Methyl-2-pentanol	565-60-6	25	-0.72	44
3-Methyl-3-pentanol	77-74-7	24.7	-0.52	61
		25	-0.38	44

Compound	CAS Registry No.	Temp. °C	Log S mol/L	Reference
4-Methyl-2-pentanol	108-11-2	25	-0.79	44
		25.0	-0.81	61
2-Methyl-3-Pentanone	565-69-5	25	-0.82	45
3-Methyl-2-pentanone	565-61-7	25	-0.68	45
4-Methyl-1-penten-3-ol	4798-45-2	25	-0.51	70
3-Methylphenol	108-39-4	25.0	-1.59	1
Methyl propionate	554-12-1	25.0	-0.17	10
Methyl salicylate	119-36-8	25.0	-2.10	10
Methyl trichloroacetate	598-99-2	25.0	-2.29	10
Methyl trimethylacetate	598-98-1	25.0	-1.74	10
Mirex	2385-85-5	24	-6.44	13
1-Naphthol	90-15-3	23	-2.23	40
2-Nitroanisole	91-23-6	30.00	-1.96	18
4-Nitroanisole	100-02-7	30.00	-2.41	18
3-Nitrophenol	554-84-7	20	-1.08	76
2,2',3,3',4,4',5,5',6-Nonachlorobiphenyl	40186-72-9	22.0	-9.77	29
		25	-10.26	30
2,2',3,3',4,5,5',6,6'-Nonachlorobiphenyl	52663-77-1	25	-10.41	22
1,8-Nonadiyne	2396-65-8	25	-2.98	28
Nonanoic acid	112-05-0	20.0	-2.78	31
1-Nonanol	143-08-8	20.0	-3.03	32
		25.0	-3.13	1
		25.0	-2.68	10
1-Nonene	124-11-8	25.0	-5.05	1
4-Nonylphenol	104-40-5	20.5	-4.61	77
1-Nonyne	3452-09-3	25	-4.24	28
2-Nonanol	628-99-9	25.0	-2.59	10
2-Nonanone	821-55-6	25.0	-2.58	1
2,2',3,3',4,4',5,5'-Octachlorobiphenyl	35694-08-7	22.0	-9.54	29
2,2',3,3',5,5',6,6'-Octachlorobiphenyl	2136-99-4	25	-9.47	30
		25	-9.04	22
Octachlorodibenzo-p-dioxin	3268-87-9	20.0	-12.06	64
		20	-12.06	78
Octadecane	593-45-3	25.0	-8.08	56
Octadecanoic acid	57-11-4	20.0	-4.99	31
		25	-5.70	55
Octanoic acid	127-04-2	20.0	-2.33	31
		30	-2.46	65
1-Octanol	111-87-5	20.0	-2.43	32
		25.0	-2.39	79
		25	-2.49	23
		25	-2.35	80
		25.6	-2.36	61
2-Octanol	123-96-6	20.0	-2.07	32
		25.0	-2.07	10
		25	-2.01	14
2-Octanone	111-13-7	25.0	-2.05	1
Octanol	589-98-0	25.0	-1.97	10

Compound	CAS Registry No.	Temp. °C	Log S mol/L	Reference
2-Octylbenzene	777-22-0	25	-5.80	81
n-Octyl formate	112-32-3	25.0	-2.48	10
1-Octyne	629-05-0	25	-3.66	28
2,2',4,5,5'-Pentachlorobiphenyl	37680-72-3	20	-7.91	11
2,2',4,6,6'-Pentachlorobiphenyl	56558-16-8	25	-7.32	41
2,3,4,5,6-Pentachlorobiphenyl	18259-05-7	22.0	-7.38	29
		25	-7.91	41
		25	-7.77	22
		25	-7.68	41
		25	-7.91	42
1,2,3,4,7-Pentachloro-p-dioxin	39227-61-7	20.0	-9.47	64
		26.0	-9.33	23
2,3,4,7,8-Pentachlorobibenzofuran	57117-31-4	22.7	-9.16	68
Pentachloronitrobenzene	82-68-8	20	-5.82	25
Pentadecanoic acid	1002-84-2	20.0	-4.30	31
1-Pentadecanol	629-76-5	25	-6.35	55
Pentanoic acid	109-52-4	30	-0.60	65
1-Pentanol	71-41-0	25	-0.64	23
		25	-0.60	54
		25.0	-0.88	1
		25	-0.60	80
		25.4	-0.61	61
2-Pentanol	6032-29-7	25	-0.30	54
		25.1	-0.31	61
3-Pentanol	584-02-1	25	-0.23	54
		25.0	-0.20	61
3-Pentanone	96-22-0	25	-0.25	45
		25.0	-0.28	1
1-Penten-3-ol	616-25-1	25	-0.02	70
3-Penten-2-ol	1569-50-2	25	0.02	70
4-Penten-1-ol	821-09-0	25	-0.18	70
n-Pentyl acetate	628-63-7	25.0	-1.84	10
Pentylbenzene	528-68-1	25.0	-4.59	1
		25.0	-4.64	72
tert-Pentylbenzene	2049-95-8	25.0	-4.25	12
n-Pentyl butyrate	540-18-1	25.0	-2.80	10
1-Pentyne	627-19-0	25.0	-1.81	1
		25	-1.64	28
Perylene	198-55-0	25	-8.80	5
Phenetole	103-73-1	25.00	-2.33	3
		25	-2.35	9
2-Phenyldecane	4537-13-7	25.0	-7.59	82
3-Phenyldecane	4621-36-7	25.0	-7.42	82
4-Phenyldecane	4537-12-6	25.0	-7.44	82
5-Phenyldecane	4537-11-5	25.0	-7.46	82
2-Phenyldodecane	2719-61-1	25.0	-8.40	82
3-Phenyldodecane	2400-00-2	25.0	-8.15	82
4-Phenyldodecane	2719-64-4	25.0	-8.30	82

Compound	CAS Registry No.	Temp. °C	Log S mol/L	Reference
5-Phenyldodecane	2719-63-3	25.0	-8.30	82
6-Phenyldodecane	2719-62-2	25.0	-8.40	82
2-Phenylethanol	60-12-8	25.0	-0.67	10
Phenylbutazone	50-33-9	20	-4.08	26
4-Phenyl-2-nitrophenol	885-82-5	20.0	-4.41	20
1-Phenyl-1-propanol	93-54-9	25.0	-1.12	10
2-Phenyltetradecane	4534-59-2	25.0	-8.40	35
3-Phenyltetradecane	4534-58-1	25.0	-8.30	35
4-Phenyltetradecane	4534-57-0	25.0	-8.40	35
5-Phenyltetradecane	4534-56-9	25.0	-8.30	35
6-Phenyltetradecane	4534-55-8	25.0	-8.40	35
2-Phenyltridecane	4534-53-6	25.0	-8.40	35
3-Phenyltridecane	4534-52-5	25.0	-8.40	35
4-Phenyltridecane	4534-51-4	25.0	-8.40	35
5-Phenyltridecane	4534-50-3	25.0	-8.40	35
6-Phenyltridecane	4534-49-0	25.0	-8.40	35
2-Phenylundecane	4536-88-3	25.0	-8.10	35
3-Phenylundecane	4536-87-2	25.0	-7.92	35
4-Phenylundecane	4536-86-1	25.0	-8.05	35
5-Phenylundecane	4537-15-9	25.0	-8.00	35
6-Phenylundecane	4537-14-8	25.0	-7.96	35
Phosalone	2310-17-0	20	-5.23	39
Prometryn	7287-19-6	25	-4.68	37
Propene	115-07-1	25	-2.32	28
n-Propyl bromide	106-94-5	30	-1.73	16
n-Propyl butyrate	105-66-8	25.0	-1.88	10
n-Propyl ether	111-43-3	25	-1.61	75
n-Propyl propionate	106-36-5	25.0	-1.29	10
Propyne	115-07-1	25	-1.06	28
Quinoline	91-22-5	23	-1.28	40
trans-Stilbene	103-03-0	25.0	-5.93	12
Terbufos	13071-79-9	20	-4.72	83
m-Terphenyl	92-06-8	25.0	-5.18	84
o-Terphenyl	84-15-1	25.0	-5.27	84
p-Terphenyl	92-94-4	25.0	-7.11	84
2,2',3,3'-Tetrachlorobiphenyl	38444-93-8	25	-7.27	41
		25	-7.01	42
2,2',4,4'-Tetrachlorobiphenyl	2437-79-8	22.0	-6.73	29
2,2',4,5-Tetrachlorobiphenyl	41464-40-8	25	-7.25	22
2,2',5,5'-Tetrachlorobiphenyl	35693-99-3	22.0	-7.28	29
		25	-6.43	41
2,2',5,6'-Tetrachlorobiphenyl	41464-41-9	25	-6.79	41
2,2',6,6'-Tetrachlorobiphenyl	15968-05-5	22.0	-8.03	29
		25	-7.39	41
2,3,4,5-Tetrachlorobiphenyl	33284-53-6	25	-7.32	41
		25.0	-7.33	67
		25	-7.33	24
		25	-7.14	22

Compound	CAS Registry No.	Temp. °C	Log S mol/L	Reference
		25	−7.18	41
		25	−7.26	42
2,4,4′,6-Tetrachlorobiphenyl	32598−12−2	25	−6.51	41
3,3′,4,4′-Tetrachlorobiphenyl	32598−13−3	22.0	−8.21	29
		25	−8.73	41
		25	−8.71	30
		25	−8.59	41
3,3′,5,5′-Tetrachlorobiphenyl	33284−52−5	25	−8.37	41
1,2,3,4-Tetrachloro-p-dioxin	30746−58−8	25	−8.84	15
1,2,3,7-Tetrachloro-p-dioxin	67028−18−6	20.0	−8.87	64
		26.0	−8.65	23
1,3,6,8-Tetrachlorodibenzo-p-dioxin	33423−92−6	20.0	−9.00	64
		20	−9.00	78
2,3,7,8-Tetrachlorodibenzofuran	51207−31−9	22.7	−8.86	68
Tetradecane	629−59−4	23	−8.78	57
		25.0	−7.96	56
		25	−7.46	69
Tetradecanoic acid	544−63−8	20.0	−4.06	31
		25	−5.43	55
1-Tetradecanol	112−72−1	25	−5.84	55
1,2,3,4-Tetrahydronaphthalene	119−64−2	20.0	−3.49	85
2,3,4,5-Tetrachloronitrobenzene	879−39−0	20	−4.55	25
2,3,5,6-Tetrachloronitrobenzene	117−18−0	20	−5.10	25
Thioanisole	110−68−5	25	−2.39	52
Thiophene	110−02−1	25	−1.45	48
Thiophenol	108−98−5	25	−2.12	52
p-Toluidine	106−49−0	20	−1.21	76
p-Tolunitrile	104−85−8	20	−1.89	9
1,2,4-Tribromobenzene	615−54−3	25	−4.50	17
Tri-n-butylamine	102−82−9	25.00	−3.12	3
2,2′,5-Trichlorobiphenyl	37680−65−2	25	−5.70	41
		25	−5.60	41
2,3′,5-Trichlorobiphenyl	38444−81−4	25	−6.01	41
2,4,4′-Trichlorobiphenyl	7012−37−5	22.0	−6.58	29
		25	−6.34	41
		25	−6.35	86
		25	−6.00	41
2,4,5-Trichlorobiphenyl	15862−07−4	25	−6.20	22
		25	−6.08	42
2,4,6-Trichlorobiphenyl	35693−92−6	25	−6.14	15
		25	−6.01	41
		25	−6.03	67
		25	−6.06	22
		25	−5.83	42
2,3,4-Trichloronitrobenzene	17700−09−3	20	−3.94	25
2,4,5-Trichloronitrobenzene	89−69−0	20	−3.89	25
n-Tridecanoic acid	638−53−9	20.0	−3.81	31
4-Trifluoromethyl-2-nitrophenol	400−99−7	20.0	−2.50	20

Compound	CAS Registry No.	Temp. °C	Log S mol/L	Reference
2,3,3-Trimethyl-2-butanol	594-83-2	40	-0.72	50
1,1,3-Trimethylcyclopentane	4516-69-2	25	-4.48	48
3,3,5-Trimethyl-1-hexanol	1484-87-3	25.0	-1.99	10
		20.0	-2.51	32
		25.0	-2.46	10
1,4,5-Trimethylnaphthalene	213-41-1	25	-4.91	5
2,2,3-Trimethyl-3-pentanol	7294-83-2	25	-1.28	87
Tri-n-propylamine	102-69-2	25.00	-2.28	3
Triphenylene	217-59-4	25.0	-6.74	84
		25	-6.73	7
		25	-6.73	88
		27	-6.78	6
Tris(1,3-dichloroisopropyl) phosphate	13674-87-8	24	-4.79	13
Tris(2-ethylhexyl) phosphate	126-72-7	24	-4.94	13
1-Undecanol	112-42-5	25.0	-2.63	10
Undecylbenzene	6742-54-7	25.0	-9.05	35
4-Vinyl-1-cyclohexene	100-40-3	25	-3.33	28

1) Tewari et al., 1982; 2) Andrews and Keefer, 1950; 3) Vesala, 1974; 4) Mills and Thurman, 1994; 5) Mackay and Shiu, 1977; 6) Davis et al., 1942; 7) Sutton and Calder, 1975; 8) Billington et al., 1988; 9) McGowan et al., 1966; 10) Stephenson and Stuart, 1986; 11) Swann et al., 1983; 12) Andrews and Keefer, 1950a; 13) Hollifield, 1979; 14) Mitchell, 1926; 15) Doucette and Andren, 1988a; 16) Gross and Saylor, 1931; 17) Yalkowsky et al., 1979; 18) Gross et al., 1933; 19) Rehberg and Dixon, 1950; 20) Schwarzenbach et al., 1988; 21) Hashimoto et al., 1982; 22) Miller et al., 1984; 23) Li and Andren, 1994; 24) Li and Doucette, 1993; 25) Eckert, 1962; 26) Hafkenscheid and Tomlinson, 1983; 27) Hafkenscheid and Tomlinson, 1981; 28) McAuliffe, 1966; 29) Opperhuizen et al., 1988; 30) Dickhut et al., 1986; 31) Ralston and Hoerr, 1942; 32) Hommelen, 1959; 33) Stearns et al., 1947; 34) Natarajan and Venkatachalam, 1972; 35) Sherblom et al., 1992; 36) Leyder and Boulanger, 1983; 37) Somasundaram et al., 1991; 38) Hine et al., 1963; 39) Chiou et al., 1977; 40) Fu et al., 1986; 41) Dunnivant and Elzerman, 1988; 41) Weil et al., 1974; 42) Dulfer et al., 1995; 43) Means et al., 1980; 44) Ginnings and Webb, 1938; 45) Ginnings et al., 1940; 46) Eganhouse and Calder, 1976; 47) Vadas et al., 1991; 48) Price, 1976; 49) Ginnings et al., 1938; 50) Ginnings and Hauser, 1938; 51) Stephenson and Stuart, 1984; 52) Hine and Weimar, 1965; 53) Andrews and Keefer, 1949; 54) Ginnings and Baum, 1937; 55) Robb, 1966; 56) Sutton and Calder, 1974; 57) Coates et al., 1985; 58) Sobotka and Kahn, 1931; 59) Whitehouse, 1984; 60) Davis and Parke, 1942; 61) Stephenson et al., 1984; 62) Schwarz and Wasik, 1977; 63) Schwarz, 1977; 64) Friesen et al., 1985; 65) Bell, 1971; 66) Hommelen, 1979; 67) Li et al., 1992; 68) Friesen et al., 1990; 69) Franks, 1966; 70) Ginnings et al., 1939; 71) Corvy and Elworthy, 1971; 72) Owens et al., 1986; 73) Banerjee et al., 1980; 74) McAuliffe, 1963; 75) Bennett and Philip, 1928; 76) Hashimoto et al., 1984; 77) Ahel and Giger, 1993; 78) Webster et al., 1985; 79) Sanemassa et al., 1987; 80) Butler et al., 1933; 81) Deno and Berkheimer, 1960; 82) Sherblom et al., 1982; 83) Bowman and Sans, 1982; 84) Akiyoshi et al., 1987; 85) Burris and MacIntyre, 1986; 86) Chiou et al., 1986; 87) Ginnings and Coltrane, 1939; 88) Klevens, 1950.

Appendix J. Synonym Index

A 1068, see Chlordane
AA, see Allyl alcohol
AAF, see 2-Acetylaminofluorene
2-AAF, see 2-Acetylaminofluorene
Aahepta, see Heptachlor
Aalindan, see Lindane
AAT, see Parathion
Aatack, see Thiram
AATP, see Parathion
AC 3422, see Parathion
ACC 3422, see Parathion
Accelerator thiuram, see Thiram
Acenaphthalene, see Acenaphthene
Acenaphthene
Acenaphthylene
Acetaldehyde
2-Acetamidofluorene, see 2-Acetylaminofluorene
2-Acetaminofluorene, see 2-Acetylaminofluorene
Acetdimethylamide, see *N,N*-Dimethylacetamide
Acetic acid
Acetic acid amyl ester, see *n*-Amyl acetate
Acetic acid anhydride, see Acetic anhydride
Acetic acid (aqueous soln), see Acetic acid
Acetic acid 2-butoxy ester, see *sec*-Butyl acetate
Acetic acid butyl ester, see *n*-Butyl acetate
Acetic acid *sec*-butyl ester, see *sec*-Butyl acetate
Acetic acid *tert*-butyl ester, see *tert*-Butyl acetate
Acetic acid dimethylamide, see *N,N*-Dimethylacetamide
Acetic acid 1,3-dimethylbutyl ester, see *sec*-Hexyl acetate
Acetic acid 1,1-dimethylethyl ester, see *tert*-Butyl acetate
Acetic acid ethenyl ester, see Vinyl acetate
Acetic acid 2-ethoxyethyl ester, see 2-Ethoxyethyl acetate
Acetic acid ethyl ester, see Ethyl acetate
Acetic acid ethylene ester, see Vinyl acetate
Acetic acid isobutyl ester, see Isobutyl acetate
Acetic acid isopentyl ester, see Isoamyl acetate
Acetic acid isopropyl ester, see Isopropyl acetate
Acetic acid 2-methoxyethyl ester, see Methyl cellosolve acetate
Acetic acid 3-methylbutyl ester, see Isoamyl acetate
Acetic acid methyl ester, see Methyl acetate
Acetic acid 1-methylethyl ester, see Isopropyl acetate
Acetic acid 1-methylpropyl ester, see *sec*-Butyl acetate

1223

Acetic acid 2-methylpropyl ester, see Isobutyl acetate
Acetic acid pentyl ester, see *n*-Amyl acetate
Acetic acid propyl ester, see *n*-Propyl acetate
Acetic acid *n*-propyl ester, see *n*-Propyl acetate
Acetic acid vinyl ester, see Vinyl acetate
Acetic aldehyde, see Acetaldehyde
Acetic anhydride
Acetic ether, see Ethyl acetate
Acetic oxide, see Acetic anhydride
Acetidin, see Ethyl acetate
Acetoaminofluorene, see 2-Acetylaminofluorene
Acetone
Acetonitrile
3-(Acetonylbenzyl)-4-hydroxycoumarin, see Warfarin
3-(α-Acetonylbenzyl)-4-hydroxycoumarin, see Warfarin
Acetosol, see 1,1,2,2-Tetrachloroethane
Aceto TETD, see Thiram
Acetoxyethane, see Ethyl acetate
1-Acetoxyethylene, see Vinyl acetate
2-Acetoxypentane, see *sec*-Amyl acetate
1-Acetoxypropane, see *n*-Propyl acetate
2-Acetoxypropane, see Isopropyl acetate
2-Acetylaminofluorene
2-(Acetylamino)fluorene, see 2-Acetylaminofluorene
N-Acetyl-2-aminofluorene, see 2-Acetylaminofluorene
Acetyl anhydride, see Acetic anhydride
Acetylene dibromide, see Ethylene dibromide
Acetylene dichloride, see *trans*-1,2-Dichloroethylene
trans-Acetylene dichloride, see *trans*-1,2-Dichloroethylene
Acetylene tetrabromide, see 1,1,2,2-Tetrabromoethane
Acetylene tetrachloride, see 1,1,2,2-Tetrachloroethane
Acetylene trichloride, see Trichloroethylene
Acetyl ether, see Acetic anhydride
Acetyl oxide, see Acetic anhydride
Acquinite, see Chloropicrin
Acraldehyde, see Acrolein
Acritet, see Acrylonitrile
Acrolein
Acrylaldehyde, see Acrolein
Acrylamide
Acrylamide monomer, see Acrylamide
Acrylic acid ethyl ester, see Ethyl acrylate
Acrylic acid methyl ester, see Methyl acrylate

Acrylic aldehyde, see Acrolein
Acrylon, see Acrylonitrile
Acrylonitrile
Acrylonitrile monomer, see Acrylonitrile
Acutox, see Pentachlorophenol
Acylic amide, see Acrylamide
Adakane 12, see Dodecane
Adronal, see Cyclohexanol
Aether, see Ethyl ether
Aerothene, see 1,1,1-Trichloroethane
Aerothene MM, see Methylene chloride
Aerothene TT, see 1,1,1-Trichloroethane
Aethylis, see Chloroethane
Aethylis chloridum, see Chloroethane
AF 101, see Diuron
Aficide, see Lindane
AGE, see Allyl glycidyl ether
Agreflan, see Trifluralin
Agricide maggot killer (F), see Toxaphene
Agriflan 24, see Trifluralin
Agrisol G-20, see Lindane
Agritan, see *p,p'*-DDT
Agroceres, see Heptachlor
Agrocide, see Lindane
Agrocide 2, see Lindane
Agrocide 6G, see Lindane
Agrocide 7, see Lindane
Agrocide III, see Lindane
Agrocide WP, see Lindane
Agronexit, see Lindane
Agrotect, see 2,4-D
Aldehyde, see Acetaldehyde
Aldifen, see 2,4-Dinitrophenol
Aldrec, see Aldrin
Aldrex, see Aldrin
Aldrex 30, see Aldrin
Aldrin
Aldrite, see Aldrin
Aldrosol, see Aldrin
Algofrene type 1, see Trichlorofluoromethane
Algofrene type 2, see Dichlorodifluoromethane
Algofrene type 5, see Dichlorofluoromethane
Algylen, see Trichloroethylene

Alkron, see Parathion
Alleron, see Parathion
Alltex, see Toxaphene
Alltox, see Toxaphene
Allyl al, see Allyl alcohol
Allyl alcohol
Allyl aldehyde, see Acrolein
Allyl chloride
Allylene, see Propyne
Allyl-2,3-epoxypropyl ether, see Allyl glycidyl ether
Allyl glycidyl ether
Allylic alcohol, see Allyl alcohol
1-Allyloxy-2,3-epoxypropane, see Allyl glycidyl ether
Allyl trichloride, see 1,2,3-Trichloropropane
Altox, see Aldrin
Alvit, see Dieldrin
Amarthol fast orange R base, see 3-Nitroaniline
Amatin, see Hexachlorobenzene
Ameisenatod, see Lindane
Ameisenmittel merck, see Lindane
American Cyanamid 3422, see Parathion
American Cyanamid 4049, see Malathion
Am-fol, see Ammonia
Amidox, see 2,4-D
Amine 2,4,5-T for rice, see 2,4,5-T
Aminic acid, see Formic acid
4-Aminoaniline, see *p*-Phenylenediamine
p-Aminoaniline, see *p*-Phenylenediamine
2-Aminoanisole, see *o*-Anisidine
4-Aminoanisole, see *p*-Anisidine
o-Aminoanisole, see *o*-Anisidine
p-Aminoanisole, see *p*-Anisidine
Aminobenzene, see Aniline
4-Aminobiphenyl
p-Aminobiphenyl, see 4-Aminobiphenyl
1-Aminobutane, see *n*-Butylamine
1-Amino-4-chloroaniline, see 4-Chloroaniline
1-Amino-4-chlorobenzene, see 4-Chloroaniline
1-Amino-*p*-chlorobenzene, see 4-Chloroaniline
4-Aminochlorobenzene, see 4-Chloroaniline
p-Aminochlorobenzene, see 4-Chloroaniline
4-Aminodiphenyl, see 4-Aminobiphenyl
p-Aminodiphenyl, see 4-Aminobiphenyl

Amyl methyl ketone, see 2-Heptanone
n-Amyl methyl ketone, see 2-Heptanone
An, see Acrylonitrile
Anaesthetic ether, see Ethyl ether
Anamenth, see Trichloroethylene
Anesthenyl, see Methylal
Anesthesia ether, see Ethyl ether
Anesthetic ether, see Ethyl ether
Anhydrous ammonia, see Ammonia
Aniline
Aniline oil, see Aniline
Anilinobenzene, see 4-Aminobiphenyl
Anilinomethane, see Methylaniline
4-Anisidine, see *p*-Anisidine
***o*-Anisidine**
***p*-Anisidine**
2-Anisylamine, see *o*-Anisidine
p-Anisylamine, see *p*-Anisidine
Ankilostin, see Tetrachloroethylene
Annulene, see Benzene
Anodynon, see Chloroethane
Anofex, see *p,p'*-DDT
Anol, see Cyclohexanol
Anone, see Cyclohexanone
Anozol, see Diethyl phthalate
Anthracene
Anthracin, see Anthracene
Anticarie, see Hexachlorobenzene
Antinonin, see 4,6-Dinitro-*o*-cresol
Antinonnon, see 4,6-Dinitro-*o*-cresol
Antisal 1a, see Toluene
Antisol 1, see Tetrachloroethylene
ANTU
Anturat, see ANTU
Anyvim, see Aniline
Aparasin, see Lindane
Apavap, see Dichlorvos
Apavinphos, see Mevinphos
Aphamite, see Parathion
Aphtiria, see Lindane
Aplidal, see Lindane
Aptal, see *p*-Chloro-*m*-cresol
Aqua-kleen, see 2,4-D

Aqualin, see Acrolein
Aqualine, see Acrolein
Arasan, see Thiram
Arasan 70, see Thiram
Arasan 75, see Thiram
Arasan-M, see Thiram
Arasan 42-S, see Thiram
Arasan-SF, see Thiram
Arasan-SF-X, see Thiram
Arbitex, see Lindane
Arborol, see 4,6-Dinitro-*o*-cresol
Arcton 6, see Dichlorodifluoromethane
Arcton 7, see Dichlorofluoromethane
Arcton 9, see Trichlorofluoromethane
Arctuvin, see Hydroquinone
Areginal, see Ethyl formate
Arkotine, see *p,p'*-DDT
Arochlor 1221, see PCB-1221
Arochlor 1232, see PCB-1232
Arochlor 1242, see PCB-1242
Arochlor 1248, see PCB-1248
Arochlor 1254, see PCB-1254
Arochlor 1260, see PCB-1260
Aroclor 1016, see PCB-1016
Aroclor 1221, see PCB-1221
Aroclor 1232, see PCB-1232
Aroclor 1242, see PCB-1242
Aroclor 1248, see PCB-1248
Aroclor 1254, see PCB-1254
Aroclor 1260, see PCB-1260
Arthodibrom, see Naled
Artic, see Methyl chloride
Artificial ant oil, see Furfural
Artificial oil of ants, see Furfural
Arylam, see Carbaryl
ASP 47, see Sulfotepp
Aspon-chlordane, see Chlordane
Astrobot, see Dichlorvos
Atgard, see Dichlorvos
Atgard C, see Dichlorvos
Atgard V, see Dichlorvos
Athrombine-K, see Warfarin
Athrombin-K, see Warfarin

Attac 4-2, see Toxaphene
Attac 4-4, see Toxaphene
Attac 6, see Toxaphene
Attac 6-3, see Toxaphene
Attac 8, see Toxaphene
Atul fast yellow R, see *p*-Dimethylaminoazobenzene
Aules, see Thiram
Avlothane, see Hexachloroethane
Avolin, see Dimethyl phthalate
Azabenzene, see Pyridine
Azacyclopropane, see Ethylenimine
1-Azaindene, see Indole
Azine, see Pyridine
Azirane, see Ethylenimine
Aziridine, see Ethylenimine
Azoamine Red ZH, see 4-Nitroaniline
Azobase MNA, see 3-Nitroaniline
Azoene fast orange GR base, see 2-Nitroaniline
Azoene fast orange GR salt, see 2-Nitroaniline
Azofix orange GR salt, see 2-Nitroaniline
Azofix red GG salt, see 4-Nitroaniline
Azogene fast orange GR, see 2-Nitroaniline
Azoic diazo component 6, see 2-Nitroaniline
Azoic diazo component 37, see 4-Nitroaniline
Azoic diazo component 112, see Benzidine
Azotox, see *p,p'*-DDT
B 404, see Parathion
BA, see Benzo[*a*]anthracene
B(*a*)A, see Benzo[*a*]anthracene
Baker's P and S liquid and ointment, see Phenol
Baktol, see *p*-Chloro-*m*-cresol
Baktolan, see *p*-Chloro-*m*-cresol
Baltana, see 1,1,1-Trichloroethane
Banana oil, see *n*-Amyl acetate, Isoamyl acetate
Bantu, see ANTU
Basaklor, see Heptachlor
BASF ursol D, see *p*-Phenylenediamine
Bay 19149, see Dichlorvos
Bay 70143, see Carbofuran
Bay E-393, see Sulfotepp
Bay E-605, see Parathion
Bayer-E 393, see Sulfotepp
BBC 12, see 1,2-Dibromo-3-chloropropane

BBP, see Benzyl butyl phthalate
BBX, see Lindane
BCEXM, see Bis(2-chloroethoxy)methane
BCF-bushkiller, see 2,4,5-T
BCIE, see Bis(2-chloroisopropyl)ether
BCME, see *sym*-Dichloromethyl ether
BCMEE, see Bis(2-chloroisopropyl)ether
BDCM, see Bromodichloromethane
Belt, see Chlordane
Benfos, see Dichlorvos
Ben-hex, see Lindane
Bentox 10, see Lindane
Benxole, see Benzene
1,2-Benzacenaphthene, see Fluoranthene
Benz[*e*]acephenanthrylene, see Benzo[*b*]fluoranthene
3,4-Benz[*e*]acephenanthrylene, see Benzo[*b*]fluoranthene
Benzal alcohol, see Benzyl alcohol
Benzanthracene, see Benzo[*a*]anthracene
Benz[*a*]anthracene, see Benzo[*a*]anthracene
1,2-Benzanthracene, see Benzo[*a*]anthracene
1,2-Benz[*a*]anthracene, see Benzo[*a*]anthracene
2,3-Benzanthracene, see Benzo[*a*]anthracene
1,2:5,6-Benzanthracene, see Dibenz[*a,h*]anthracene
Benzanthrene, see Benzo[*a*]anthracene
1,2-Benzanthrene, see Benzo[*a*]anthracene
1-Benzazole, see Indole
Benzenamine, see Aniline, *N*-Nitrosodiphenylamine
Benzene
Benzeneazodimethylaniline, see *p*-Dimethylaminoazobenzene
Benzene carbinol, see Benzyl alcohol
Benzenecarboxylic acid, see Benzoic acid
Benzene chloride, see Chlorobenzene
1,4-Benzenediamine, see *p*-Phenylenediamine
p-Benzenediamine, see *p*-Phenylenediamine
1,2-Benzenedicarboxylic acid anhydride, see Phthalic anhydride
1,2-Benzenedicarboxylic acid bis(2-ethylhexyl)ester, see Bis(2-ethylhexyl)phthalate
1,2-Benzenedicarboxylic acid butyl phenylmethyl ester, see Benzyl butyl phthalate
1,2-Benzenedicarboxylic acid dibutyl ester, see Di-*n*-butyl phthalate
o-Benzenedicarboxylic acid dibutyl ester, see Di-*n*-butyl phthalate
Benzene-*o*-dicarboxylic acid di-*n*-butyl ester, see Di-*n*-butyl phthalate
1,2-Benzenedicarboxylic acid diethyl ester, see Diethyl phthalate
1,2-Benzenedicarboxylic acid dimethyl ester, see Dimethyl phthalate
1,2-Benzenedicarboxylic acid dioctyl ester, see Di-*n*-octyl phthalate

1,2-Benzenedicarboxylic acid di-*n*-octyl ester, see Di-*n*-octyl phthalate
o-Benzenedicarboxylic acid dioctyl ester, see Di-*n*-octyl phthalate
1,4-Benzenediol, see Hydroquinone
p-Benzenediol, see Hydroquinone
Benzeneformic acid, see Benzoic acid
Benzene hexachloride, see Lindane
Benzene hexachloride-α-isomer, see α-BHC
Benzene-*cis*-hexachloride, see β-BHC
Benzene-γ-hexachloride, see Lindane
α-Benzene hexachloride, see α-BHC
β-Benzene hexachloride, see β-BHC
δ-Benzene hexachloride, see δ-BHC
γ-Benzene hexachloride, see Lindane
trans-α-Benzene hexachloride, see β-BHC
Benzene hexahydride, see Cyclohexane
Benzene methanoic acid, see Benzoic acid
Benzene methanol, see Benzyl alcohol
Benzene tetrahydride, see Cyclohexene
Benzenol, see Phenol
2,3-Benzfluoranthene, see Benzo[*b*]fluoranthene
3,4-Benzfluoranthene, see Benzo[*b*]fluoranthene
8,9-Benzfluoranthene, see Benzo[*k*]fluoranthene
Benzidam, see Aniline
Benzidene base, see Benzidine
Benzidine
p-Benzidine, see Benzidine
2,3-Benzindene, see Fluorene
Benzinoform, see Carbon tetrachloride
Benzinol, see Trichloroethylene
Benzoanthracene, see Benzo[*a*]anthracene
Benzo[*a*]anthracene
1,2-Benzoanthracene, see Benzo[*a*]anthracene
Benzoate, see Benzoic acid
Benzo[*d,e,f*]chrysene, see Benzo[*a*]pyrene
Benzoepin, see α-Endosulfan, β-Endosulfan
Benzo[*b*]fluoranthene
2,3-Benzofluoranthene, see Benzo[*b*]fluoranthene
3,4-Benzo[*b*]fluoranthene, see Benzo[*b*]fluoranthene
11,12-Benzofluoranthene, see Benzo[*k*]fluoranthene
Benzo[*e*]fluoranthene, see Benzo[*b*]fluoranthene
Benzo[*k*]fluoranthene
11,12-Benzo[*k*]fluoranthene, see Benzo[*k*]fluoranthene
3,4-Benzofluoranthene, see Benzo[*b*]fluoranthene

Benzyl alcohol
Benzyl butyl phthalate
Benzyl *n*-butyl phthalate, see Benzyl butyl phthalate
Benzyl chloride
Benzyl chloride anhydrous, see Benzyl chloride
3,4-Benzypyrene, see Benzo[*a*]pyrene
Beosit, see α-Endosulfan, β-Endosulfan
Betaprone, see β-Propiolactone
Bexol, see Lindane
B(*b*)F, see Benzo[*b*]fluoranthene
B(*k*)F, see Benzo[*k*]fluoranthene
BFV, see Formaldehyde
BHC, see Lindane
α-BHC
β-BHC
δ-BHC
γ-BHC, see Lindane
BH 2,4-D, see 2,4-D
4,4'-Bianiline, see Benzidine
N,N'-Bianiline, see 1,2-Diphenylhydrazine
p,p'-Bianiline, see Benzidine
Bibenzene, see Biphenyl
Bibesol, see Dichlorvos
Bicarburet of hydrogen, see Benzene
1,2-Bichloroethane, see 1,2-Dichloroethane
Bicyclo[4.4.0]decane, see Decahydronaphthalene
Biethylene, see 1,3-Butadiene
Bihexyl, see Dodecane
Biisopropyl, see 2,3-Dimethylbutane
2,3,1',8'-Binaphthylene, see Benzo[*k*]fluoranthene
Binitrobenzene, see 1,3-Dinitrobenzene
Bio 5462, see α-Endosulfan, β-Endosulfan
Biocide, see Acrolein
Bioflex 81, see Bis(2-ethylhexyl)phthalate
Bioflex DOP, see Bis(2-ethylhexyl)phthalate
Biphenyl
1,1'-Biphenyl, see Biphenyl
Biphenylamine, see 4-Aminobiphenyl
4-Biphenylamine, see 4-Aminobiphenyl
p-Biphenylamine, see 4-Aminobiphenyl
[1,1'-Biphenyl]-4-amine, see 4-Aminobiphenyl
(1,1'-Biphenyl)-4,4'-diamine, see Benzidine
4,4'-Biphenyldiamine, see Benzidine

Bivinyl, see 1,3-Butadiene
Black and white bleaching cream, see Hydroquinone
Blacosolv, see Trichloroethylene
Bladafum, see Sulfotepp
Bladafume, see Sulfotepp
Bladafun, see Sulfotepp
Bladan, see Parathion, Tetraethyl pyrophosphate
Bladan F, see Parathion
Blancosolv, see Trichloroethylene
Blue oil, see Aniline
Bonoform, see 1,1,2,2-Tetrachloroethane
Borer sol, see 1,2-Dichloroethane
2-Bornanone, see Camphor
Bosan Supra, see *p,p'*-DDT
Bovidermol, see *p,p'*-DDT
BP, see Benzo[*a*]pyrene
3,4-BP, see Benzo[*a*]pyrene
B(*a*)P, see Benzo[*a*]pyrene
B(*e*)P, see Benzo[*e*]pyrene
B(*ghi*)P, see Benzo[*ghi*]perylene
BPL, see β-Propiolactone
Bran oil, see Furfural
Brentamine fast orange GR base, see 2-Nitroaniline
Brentamine fast orange GR salt, see 2-Nitroaniline
Brevinyl, see Dichlorvos
Brevinyl E50, see Dichlorvos
Brilliant fast oil yellow, see *p*-Dimethylaminoazobenzene
Brilliant fast spirit yellow, see *p*-Dimethylaminoazobenzene
Brilliant fast yellow, see *p*-Dimethylaminoazobenzene
Brilliant oil yellow, see *p*-Dimethylaminoazobenzene
Brocide, see 1,2-Dichloroethane
Brodan, see Chlorpyrifos
Bromchlophos, see Naled
Bromex, see Naled
Bromic ether, see Ethyl bromide
Bromobenzene
Bromochloromethane
Bromodichloromethane
4-Bromodiphenyl ether, see 4-Bromophenyl phenyl ether
p-Bromodiphenyl ether, see 4-Bromophenyl phenyl ether
Bromoethane, see Ethyl bromide
Bromofluoroform, see Bromotrifluoromethane
Bromoform

Bromofume, see Ethylene dibromide
Brom-o-gaz, see Methyl bromide
Bromomethane, see Methyl bromide
1-Bromo-4-phenoxybenzene, see 4-Bromophenyl phenyl ether
1-Bromo-*p*-phenoxybenzene, see 4-Bromophenyl phenyl ether
4-Bromophenyl ether, see 4-Bromophenyl phenyl ether
p-Bromophenyl ether, see 4-Bromophenyl phenyl ether
4-Bromophenyl phenyl ether
p-Bromophenyl phenyl ether, see 4-Bromophenyl phenyl ether
Bromotrifluoromethane
Brumolin, see Warfarin
Brush-off 445 low volatile brush killer, see 2,4,5-T
Brush-rhap, see 2,4-D, 2,4,5-T
Brushtox, see 2,4,5-T
B-Selektonon, see 2,4-D
Bunt-cure, see Hexachlorobenzene
Bunt-no-more, see Hexachlorobenzene
Butadiene, see 1,3-Butadiene
1,3-Butadiene
α,γ-Butadiene, see 1,3-Butadiene
Buta-1,3-diene, see 1,3-Butadiene
1-Butanamine, see *n*-Butylamine
Butane
n-Butane, see Butane
Butanethiol, see *n*-Butyl mercaptan
1-Butanethiol, see *n*-Butyl mercaptan
n-Butanethiol, see *n*-Butyl mercaptan
Butanol-2, see *sec*-Butyl alcohol
1-Butanol
2-Butanol, see *sec*-Butyl alcohol
n-Butanol, see 1-Butanol
sec-Butanol, see *sec*-Butyl alcohol
t-Butanol, see *tert*-Butyl alcohol
tert-Butanol, see *tert*-Butyl alcohol
Butan-1-ol, see 1-Butanol
Butan-2-ol, see *sec*-Butyl alcohol
1-Butanol acetate, see *n*-Butyl acetate
2-Butanol acetate, see *sec*-Butyl acetate
Butanone, see 2-Butanone
2-Butanone
2-Butenal, see Crotonaldehyde
1-Butene
cis-Butenedioic anhydride, see Maleic anhydride

2-Butenoic acid 3-[(dimethoxyphosphinyl)oxy]methyl ester, see Mevinphos
2-Butoxyethanol
Butter yellow, see *p*-Dimethylaminoazobenzene
Butyl acetate, see *n*-Butyl acetate
1-Butyl acetate, see *n*-Butyl acetate
2-Butyl acetate, see *sec*-Butyl acetate
***n*-Butyl acetate**
***sec*-Butyl acetate**
t-Butyl acetate, see *tert*-Butyl acetate
***tert*-Butyl acetate**
Butyl alcohol, see 1-Butanol
n-Butyl alcohol, see 1-Butanol
***sec*-Butyl alcohol**
t-Butyl alcohol, see *tert*-Butyl alcohol
***tert*-Butyl alochol**
2-Butyl alcohol, see *sec*-Butyl alcohol
sec-Butyl alcohol acetate, see *sec*-Butyl acetate
Butylamine, see *n*-Butylamine
***n*-Butylamine**
Butylbenzene, see *n*-Butylbenzene
***n*-Butylbenzene**
***sec*-Butylbenzene**
t-Butylbenzene, see *tert*-Butylbenzene
***tert*-Butylbenzene**
Butyl benzyl phthalate, see Benzyl butyl phthalate
n-Butyl benzyl phthalate, see Benzyl butyl phthalate
Butyl cellosolve, see 2-Butoxyethanol
α-Butylene, see 1-Butene
γ-Butylene, see 2-Methylpropene
Butylene hydrate, see *sec*-Butyl alcohol
Butylene oxide, see Tetrahydrofuran
Butyl ethanoate, see *n*-Butyl acetate
n-Butyl ethanoate, see *n*-Butyl acetate
Butylethylene, see 1-Hexene
Butyl ethyl ketone, see 3-Heptanone
n-Butyl ethyl ketone, see 3-Heptanone
Butyl glycol, see 2-Butoxyethanol
Butyl glycol ether, see 2-Butoxyethanol
Butyl hydroxide, see 1-Butanol
t-Butyl hydroxide, see *tert*-Butyl alcohol
Butyl ketone, see 2-Hexanone
Butyl mercaptan, see *n*-Butyl mercaptan
***n*-Butyl mercaptan**

Butyl methyl ketone, see 2-Hexanone
n-Butyl methyl ketone, see 2-Hexanone
Butyl oxitol, see 2-Butoxyethanol
Butyl phenylmethyl 1,2-Benzenedicarboxylate, see Benzyl butyl phthalate
Butyl phthalate, see Di-*n*-butyl phthalate
n-Butyl phthalate, see Di-*n*-butyl phthalate
n-Butyl thioalcohol, see *n*-Butyl mercaptan
Butyric alcohol, see 1-Butanol
C-56, see Hexachlorocyclopentadiene
CAF, see α-Chloroacetophenone
Calmathion, see Malathion
2-Camphanone, see Camphor
Camphechlor, see Toxaphene
Camphochlor, see Toxaphene
Camphoclor, see Toxaphene
Camphor
Camphor-natural, see Camphor
Camphor-synthetic, see Camphor
Camphor tar, see Naphthalene
Candaseptic, see *p*-Chloro-*m*-cresol
Canogard, see Dichlorvos
CAP, see α-Chloroacetophenone
1-Caprylene, see 1-Octene
Capsine, see 4,6-Dinitro-*o*-cresol
Carbacryl, see Acrylonitrile
Carbamine, see Carbaryl
Carbaryl
Carbatox, see Carbaryl
Carbatox 60, see Carbaryl
Carbatox 75, see Carbaryl
Carbazotic acid, see Picric acid
Carbethoxy malathion, see Malathion
Carbetovur, see Malathion
Carbetox, see Malathion
Carbinamine, see Methylamine
Carbinol, see Methanol
Carbofos, see Malathion
Carbofuran
Carbolic acid, see Phenol
2-Carbomethoxy-1-methylvinyl dimethyl phosphate, see Mevinphos
α-Carbomethoxy-1-methylvinyl dimethyl phosphate, see Mevinphos
2-Carbomethoxy-1-propen-2-yl dimethyl phosphate, see Mevinphos
Carbona, see Carbon tetrachloride

Carbon bichloride, see Tetrachloroethylene
Carbon bisulfide, see Carbon disulfide
Carbon bisulphide, see Carbon disulfide
Carbon chloride, see Carbon tetrachloride
Carbon dichloride, see Tetrachloroethylene
Carbon disulfide
Carbon disulphide, see Carbon disulfide
Carbon hexachloride, see Hexachloroethane
Carbon oil, see Benzene
Carbon sulfide, see Carbon disulfide
Carbon sulphide, see Carbon disulfide
Carbon tet, see Carbon tetrachloride
Carbon tetrachloride
Carbon trichloride, see Hexachloroethane
Carbophos, see Malathion
Carboxybenzene, see Benzoic acid
Carpolin, see Carbaryl
Carylderm, see Carbaryl
CB, see Bromochloromethane
CBM, see Bromochloromethane
CCS 203, see 1-Butanol
CCS 301, see sec-Butyl alcohol
CD 68, see Chlordane
CDBM, see Dibromochloromethane
Cecolene, see Trichloroethylene
Cekiuron, see Diuron
Cekubaryl, see Carbaryl
Cekusan, see Dichlorvos
Celanex, see Lindane
Celfume, see Methyl bromide
Cellon, see 1,1,2,2-Tetrachloroethane
Cellosolve, see 2-Ethoxyethanol
Cellosolve acetate, see 2-Ethoxyethyl acetate
Cellosolve solvent, see 2-Ethoxyethanol
Celluflex DOP, see Di-n-octyl phthalate
Celluflex DPB, see Di-n-butyl phthalate
Celluflex TPP, see Triphenyl phosphate
Celluphos 4, see Tributyl phosphate
Celmide, see Ethylene dibromide
Celthion, see Malathion
Cerasine yellow CG, see p-Dimethylaminoazobenzene
Certox, see Strychnine
Chelen, see Chloroethane

Chemform, see Methoxychlor
Chemathion, see Malathion
Chemical 109, see ANTU
Chemical mace, see α-Chloroacetophenone
Chemox PE, see 2,4-Dinitrophenol
Chempenta, see Pentachlorophenol
Chemphene, see Toxaphene
Chemrat, see Pindone
Chemsect DNOC, see 4,6-Dinitro-*o*-cresol
Chemtol, see Pentachlorophenol
Chevron acetone, see Acetone
Chinone, see *p*-Quinone
Chipco thiram 75, see Thiram
Chipco turf herbicide "D", see 2,4-D
Chlorallylene, see Allyl chloride
Chloran, see Lindane
4-Chloraniline, see 4-Chloroaniline
p-Chloraniline, see 4-Chloroaniline
Chlorbenzene, see Chlorobenzene
Chlorbenzol, see Chlorobenzene
p-Chlor-*m*-cresol, see *p*-Chloro-*m*-cresol
Chlordan, see Chlordane
α-Chlordan, see *trans*-Chlordane
cis-Chlordan, see *trans*-Chlordane
γ-Chlordan, see Chlordane
Chlordane
α-Chlordane, see *cis*-Chlordane, *trans*-Chlordane
α(*cis*)-Chlordane, see *trans*-Chlordane
β-Chlordane, see *cis*-Chlordane
***cis*-Chlordane**
γ-Chlordane, see *trans*-Chlordane
***trans*-Chlordane**
Chlordecone, see Kepone
Chloresene, see Lindane
Chlorethene, see Vinyl chloride
Chlorethyl, see Chloroethane
Chlorethylene, see Vinyl chloride
2-Chlorethyl vinyl ether, see 2-Chloroethyl vinyl ether
Chlorex, see Bis(2-chloroethyl)ether
Chloridan, see Chlordane
Chloridum, see Chloroethane
Chlorilen, see Trichloroethylene
Chlorinated camphene, see Toxaphene

Chlorinated hydrochloric ether, see 1,1-Dichloroethane

Chlorindan, see Chlordane

Chlor kil, see Chlordane

Chloroacetaldehyde

2-Chloroacetaldehyde, see Chloroacetaldehyde

Chloroacetaldehyde monomer, see Chloroacetaldehyde

2-Chloroacetophenone, see α-Chloroacetophenone

α-Chloroacetophenone

Ω-Chloroacetophenone, see α-Chloroacetophenone

4-Chloroaniline

p-Chloroaniline, see 4-Chloroaniline

Chloroben, see 1,2-Dichlorobenzene

2-Chlorobenzalmalononitrile, see *o*-Chlorobenzylidenemalononitrile

o-Chlorobenzalmalononitrile, see *o*-Chlorobenzylidenemalononitrile

4-Chlorobenzamine, see 4-Chloroaniline

p-Chlorobenzamine, see 4-Chloroaniline

Chlorobenzene

Chlorobenzol, see Chlorobenzene

2-Chlorobenzylidenemalononitrile, see *o*-Chlorobenzylidenemalononitrile

o-Chlorobenzylidenemalononitrile

2-Chlorobmn, see *o*-Chlorobenzylidenemalononitrile

Chlorobromomethane, see Bromochloromethane

Chlorobutadiene, see Chloroprene

2-Chlorobutadiene-1,3, see Chloroprene

2-Chlorobuta-1,3-diene, see Chloroprene

2-Chloro-1,3-butadiene, see Chloroprene

Chlorocamphene, see Toxaphene

3-Chlorochlordene, see Heptachlor

1-Chloro-2-(β-chloroethoxy)ethane, see Bis(2-chloroethyl)ether

1-Chloro-2-(β-chloroisopropoxy)propane, see Bis(2-chloroisopropyl)ether

Chloro(chloromethoxy)methane, see *sym*-Dichloromethyl ether

Chlorocresol, see *p*-Chloro-*m*-cresol

4-Chlorocresol, see *p*-Chloro-*m*-cresol

p-Chlorocresol, see *p*-Chloro-*m*-cresol

4-Chloro-*m*-cresol, see *p*-Chloro-*m*-cresol

6-Chloro-*m*-cresol, see *p*-Chloro-*m*-cresol

p-Chloro-m-cresol

Chlorodane, see Chlordane

Chloroden, see 1,2-Dichlorobenzene

Chlorodibromomethane, see Dibromochloromethane

1-Chloro-2,3-dibromopropane, see 1,2-Dibromo-3-chloropropane

3-Chloro-1,2-dibromopropane, see 1,2-Dibromo-3-chloropropane

1-Chloro-2,2-dichloroethylene, see Trichloroethylene

Chlorodiphenyl (21% Cl), see PCB-1221
Chlorodiphenyl (32% Cl), see PCB-1232
Chlorodiphenyl (41% Cl), see PCB-1016
Chlorodiphenyl (42% Cl), see PCB-1242
Chlorodiphenyl (48% Cl), see PCB-1248
Chlorodiphenyl (54% Cl), see PCB-1254
Chlorodiphenyl (60% Cl), see PCB-1260
4-Chlorodiphenyl ether, see 4-Chlorophenyl phenyl ether
p-Chlorodiphenyl ether, see 4-Chlorophenyl phenyl ether
1-Chloro-2,3-epoxypropane, see Epichlorohydrin
3-Chloro-1,2-epoxypropane, see Epichlorohydrin
2-Chloroethanal, see Chloroacetaldehyde
2-Chloro-1-ethanal, see Chloroacetaldehyde
Chloroethane
2-Chloroethanol, see Ethylene chlorohydrin
δ-Chloroethanol, see Ethylene chlorohydrin
Chloroethene, see Vinyl chloride
1-Chloroethene, see Vinyl chloride
Chloroethene NU, see 1,1,1-Trichloroethane
(2-Chloroethoxy)ethene, see 2-Chloroethyl vinyl ether
2-Chloroethyl alcohol, see Ethylene chlorohydrin
β-Chloroethyl alcohol, see Ethylene chlorohydrin
Chloroethylene, see Vinyl chloride
1-Chloroethylene, see Vinyl chloride
Chloroethyl ether, see Bis(2-chloroethyl)ether
2-Chloroethyl ether, see Bis(2-chloroethyl)ether
(β-Chloroethyl)ether, see Bis(2-chloroethyl)ether
2-Chloroethyl vinyl ether
Chloroform
1-Chloro-2-hydroxybenzene, see 2-Chlorophenol
1-Chloro-o-hydroxybenzene, see 2-Chlorophenol
2-Chlorohydroxybenzene, see 2-Chlorophenol
3-Chloro-4-hydroxychlorobenzene, see 2,4-Dichlorophenol
4-Chloro-1-hydroxy-3-methylbenzene, see p-Chloro-m-cresol
2-Chlorohydroxytoluene, see p-Chloro-m-cresol
2-Chloro-5-hydroxytoluene, see p-Chloro-m-cresol
4-Chloro-3-hydroxytoluene, see p-Chloro-m-cresol
6-Chloro-3-hydroxytoluene, see p-Chloro-m-cresol
2-Chloroisopropyl ether, see Bis(2-chloroisopropyl)ether
β-Chloroisopropyl ether, see Bis(2-chloroisopropyl)ether
Chloromethane, see Methyl chloride
(Chloromethyl)benzene, see Benzyl chloride
Chloromethyl ether, see sym-Dichloromethyl ether

(Chloromethyl)ethylene oxide, see Epichlorohydrin

(2-Chloro-1-methylethyl)ether, see Bis(2-chloroisopropyl)ether

(Chloromethyl)oxirane, see Epichlorohydrin

2-(Chloromethyl)oxirane, see Epichlorohydrin

4-Chloro-3-methylphenol, see *p*-Chloro-*m*-cresol

p-Chloro-3-methylphenol, see *p*-Chloro-*m*-cresol

Chloromethyl phenyl ketone, see α-Chloroacetophenone

2-Chloronaphthalene

β-Chloronaphthalene, see 2-Chloronaphthalene

4-Chloronitrobenzene, see *p*-Chloronitrobenzene

_p_-Chloronitrobenzene

1-Chloro-4-nitrobenzene, see *p*-Chloronitrobenzene

4-Chloro-1-nitrobenzene, see *p*-Chloronitrobenzene

Chloronitropropane, see 1-Chloro-1-nitropropane

1-Chloro-1-nitropropane

Chlorophen, see Pentachlorophenol

2-Chlorophenol

o-Chlorophenol, see 2-Chlorophenol

Chlorophenothan, see *p,p'*-DDT

Chlorophenothane, see *p,p'*-DDT

Chlorophenotoxum, see *p,p'*-DDT

1-Chloro-4-phenoxybenzene, see 4-Chlorophenyl phenyl ether

1-Chloro-*p*-phenoxybenzene, see 4-Chlorophenyl phenyl ether

4-Chlorophenylamine, see 4-Chloroaniline

p-Chlorophenylamine, see 4-Chloroaniline

4-Chlorophenyl chloride, see 1,4-Dichlorobenzene

p-Chlorophenyl chloride, see 1,4-Dichlorobenzene

2-Chloro-1-phenylethanone, see α-Chloroacetophenone

4-Chlorophenyl ether, see 4-Chlorophenyl phenyl ether

p-Chlorophenyl ether, see 4-Chlorophenyl phenyl ether

Chlorophenylmethane, see Benzyl chloride

[(2-Chlorophenyl)methylene]propanedinitrile,   see   *o*-Chlorobenzylidenemalononi-
    trile

4-Chlorophenyl phenyl ether

p-Chlorophenyl phenyl ether, see 4-Chlorophenyl phenyl ether

Chlor-o-pic, see Chloropicrin

Chloropicrin

Chloroprene

3-Chloroprene, see Allyl chloride

β-Chloroprene, see Chloroprene

3-Chloroprene-1, see Allyl chloride

1-Chloropropene-2, see Allyl chloride

1-Chloro-2-propene, see Allyl chloride

3-Chloropropene, see Allyl chloride
3-Chloro-1-propene, see Allyl chloride
3-Chloropropylene, see Allyl chloride
α-Chloropropylene, see Allyl chloride
1-Chloropropylene-2, see Allyl chloride
1-Chloro-2-propylene, see Allyl chloride
3-Chloro-1-propylene, see Allyl chloride
2-Chloropropylene oxide, see Epichlorohydrin
γ-Chloropropylene oxide, see Epichlorohydrin
3-Chloro-1,2-propylene oxide, see Epichlorohydrin
Chlorothane NU, see 1,1,1-Trichloroethane
Chlorothene, see 1,1,1-Trichloroethane
Chlorothene NU, see 1,1,1-Trichloroethane
Chlorothene VG, see 1,1,1-Trichloroethane
α-Chlorotoluene, see Benzyl chloride
Ω-Chlorotoluene, see Benzyl chloride
2-Chloro-6-(trichloromethyl)pyridine, see Nitrapyrin
Chloroxone, see 2,4-D
Chlorpyrifos
Chlorpyrifos-ethyl, see Chlorpyrifos
Chlorten, see 1,1,1-Trichloroethane
Chlorthiepin, see α-Endosulfan, β-Endosulfan
Chlortox, see Chlordane
Chloryl, see Chloroethane
Chloryl anesthetic, see Chloroethane
Chlorylea, see Trichloroethylene
Chlorylen, see Trichloroethylene
Chromar, see p-Xylene
Chrysene
C.I. 10305, see Picric acid
C.I. 11020, see p-Dimethylaminoazobenzene
C.I. 23060, see 3,3′-Dichlorobenzidine
C.I. 37025, see 2-Nitroaniline
C.I. 37030, see 3-Nitroaniline
C.I. 37035, see 4-Nitroaniline
C.I. 37077, see o-Toluidine
C.I. 37225, see Benzidine
C.I. 37270, see 2-Naphthylamine
C.I. 76000, see Aniline
C.I. 76060, see p-Phenylenediamine
C.I. azoic diazo component 6, see 2-Nitroaniline
C.I. azoic diazo component 7, see 3-Nitroaniline
C.I. azoic diazo component 37, see 4-Nitroaniline

C.I. azoic diazo component 112, see Benzidine
C.I. azoic diazo component 114, see 1-Naphthylamine
CIBA 8514, see Kepone
C.I. developer 13, see *p*-Phenylenediamine
C.I. developer 17, see 4-Nitroaniline
Cimexan, see Malathion
Cinnamene, see Styrene
Cinnamenol, see Styrene
Cinnamol, see Styrene
C.I. oxidation base 1, see Aniline
C.I. oxidation base 10, see *p*-Phenylenediamine
Circosolv, see Trichloroethylene
C.I. solvent yellow 2, see *p*-Dimethylaminoazobenzene
Citox, see *p,p'*-DDT
Clofenotane, see *p,p'*-DDT
Clophen A60, see PCB-1260
Clor chem T-590, see Toxaphene
Cloroben, see 1,2-Dichlorobenzene
CMDP, see Mevinphos
CN, see α-Chloroacetophenone
Coal naphtha, see Benzene
Coal tar naphtha, see Benzene
Codechine, see Lindane
Colamine, see Ethanolamine
Collunosol, see 2,4,5-Trichlorophenol
Colonial spirit, see Methanol
Columbian spirits, see Methanol
Columbian spirits (wood alcohol), see Methanol
Compound 42, see Warfarin
Compound 118, see Aldrin
Compound 269, see Endrin
Compound 497, see Dieldrin
Compound 889, see Bis(2-ethylhexyl)phthalate
Compound 1189, see Kepone
Compound 2046, see Mevinphos
Compound 3422, see Parathion
Compound 3956, see Toxaphene
Compound 4049, see Malathion
Co-op hexa, see Hexachlorobenzene
Corax, see Warfarin
Corodane, see Chlordane
Corothion, see Parathion
Corthion, see Parathion

Corthione, see Parathion
Cortilan-neu, see Chlordane
Coumadin, see Warfarin
Coumafen, see Warfarin
Coumafene, see Warfarin
Cov-r-tox, see Warfarin
CP 34, see Thiophene
Crag sevin, see Carbaryl
Crawhaspol, see Trichloroethylene
2-Cresol, see 2-Methylphenol
4-Cresol, see 4-Methylphenol
o-Cresol, see 2-Methylphenol
p-Cresol, see 4-Methylphenol
Crestoxo, see Toxaphene
Crestoxo 90, see Toxaphene
o-Cresylic acid, see 2-Methylphenol
p-Cresylic acid, see 4-Methylphenol
o-Cresyl phosphate, see Tri-o-cresyl phosphate
Crisalin, see Trifluralin
Crisulfan, see α-Endosulfan, β-Endosulfan
Crisuron, see Diuron
Crolean, see Acrolein
Crop rider, see 2,4-D
Crotenaldehyde, see Crotonaldehyde
Crotilin, see 2,4-D
Crotonal, see Crotonaldehyde
Crotonaldehyde
Crotonic aldehyde, see Crotonaldehyde
Cryptogil OL, see Pentachlorophenol
CS, see o-Chlorobenzylidenemalononitrile
CSAC, see 2-Ethoxyethyl acetate
Cumene, see Isopropylbenzene
Cumol, see Isopropylbenzene
Curetard A, see N-Nitrosodiphenylamine
Curithane 103, see Methyl acrylate
Curithane C126, see 3,3'-Dichlorobenzidine
Cyanoethylene, see Acrylonitrile
Cyanol, see Aniline
Cyanomethane, see Acetonitrile
Cyanophos, see Dichlorvos
Cyclodan, see α-Endosulfan, β-Endosulfan
Cycloheptane
Cyclohexadienedione, see p-Quinone

1,2-DCB, see 1,2-Dichlorobenzene
1,3-DCB, see 1,3-Dichlorobenzene
1,4-DCB, see 1,4-Dichlorobenzene
o-DCB, see 1,2-Dichlorobenzene
m-DCB, see 1,3-Dichlorobenzene
p-DCB, see 1,4-Dichlorobenzene
1,1-DCE, see 1,1-Dichloroethylene
1,2-DCE, see 1,2-Dichloroethane
DCEE, see Bis(2-chloroethyl)ether
DCM, see Methylene chloride
DCMU, see Diuron
D-con, see Warfarin
DCP, see 2,4-Dichlorophenol
2,4-DCP, see 2,4-Dichlorophenol
DDD, see p,p'-DDD
4,4'-DDD, see p,p'-DDD
p,p'-DDD
DDE, see p,p'-DDE
p,p'-DDE
DDH, see 1,3-Dichloro-5,5-dimethylhydantoin
DDP, see Parathion
DDT, see p,p'-DDT
4,4'-DDT, see p,p'-DDT
DDT dihydrochloride, see p,p'-DDE
p,p'-DDT
DDVF, see Dichlorvos
DDVP, see Dichlorvos
DEAE, see 2-Diethylaminoethanol
Debroussaillant 600, see 2,4-D
Debroussaillant concentre, see 2,4,5-T
Debroussaillant super concentre, see 2,4,5-T
Dec, see Decahydronaphthalene
1,2,3,5,6,7,8,9,10,10-Decachloro[5.2.1.0^{2,6}.0^{3,9}.0^{5,8}]decano-4-one, see Kepone
Decachloroketone, see Kepone
Decachloro-1,3,4-metheno-2H-cyclobuta[cd]pentalen-2-one, see Kepone
Decachlorooctahydrokepone-2-one, see Kepone
Decachlorooctahydro-1,3,4-metheno-2H-cyclobuta[cd]pentalen-2-one, see Kepone
1,1a,3,3a,4,5,5a,5b,6-Decachlorooctahydro-1,3,4-metheno-2H-cyclobuta[cd]penta-
    len-2-one, see Kepone
Decachloropentacyclo[5.2.1.0^{2,6}.0^{3,9}.0^{5,8}]decan-3-one, see Kepone
Decachloropentacyclo[5.2.1.0^{2,6}.0^{4,10}.0^{5,9}]decan-3-one, see Kepone
Decachlorotetracyclodecanone, see Kepone
Decachlorotetrahydro-4,7-methanoindeneone, see Kepone

Decahydronaphthalene
Decalin, see Decahydronaphthalene
Decalin solvent, see Decahydronaphthalene
Decamine, see 2,4-D
Decamine 4T, see 2,4,5-T
Decane
n-Decane, see Decane
Decyl hydride, see Decane
Dedelo, see p,p'-DDT
Dedevap, see Dichlorvos
Ded-weed, see 2,4-D
Ded-weed brush killer, see 2,4,5-T
Ded-weed LV-69, see 2,4-D
Ded-weed LV-6 brush-kil and T-5 brush-kil, see 2,4,5-T
Dee-Solv, see Tetrachloroethylene
Degrassan, see 4,6-Dinitro-o-cresol
DEHP, see Bis(2-ethylhexyl)phthalate
Dekalin, see Decahydronaphthalene
Dekrysil, see 4,6-Dinitro-o-cresol
Delac J, see N-Nitrosodiphenylamine
Denapon, see Carbaryl
Densinfluat, see Trichloroethylene
Deoval, see p,p'-DDT
DEP, see Diethyl phthalate
Deriban, see Dichlorvos
Dermaphos, see Ronnel
Derribante, see Dichlorvos
Desmodur T80, see 2,4-Toluene diisocyanate
Desormone, see 2,4-D
Destruxol borer-sol, see 1,2-Dichloroethane
Detal, see 4,6-Dinitro-o-cresol
Dethmor, see Warfarin
Dethnel, see Warfarin
Detmol-extrakt, see Lindane
Detmol MA, see Malathion
Detmol MA 96%, see Malathion
Detmol U.A., see Chlorpyrifos
Detox, see p,p'-DDT
Detox 25, see Lindane
Detoxan, see p,p'-DDT
Developer 13, see p-Phenylenediamine
Developer P, see 4-Nitroaniline
Developer PF, see p-Phenylenediamine

Devicarb, see Carbaryl
Devikol, see Dichlorvos
Devol orange B, see 2-Nitroaniline
Devol orange R, see 3-Nitroaniline
Devol orange salt B, see 2-Nitroaniline
Devol red GG, see 4-Nitroaniline
Devoran, see Lindane
Devoton, see Methyl acetate
Diacetone, see Diacetone alcohol
Diacetone alcohol
Diacetonyl alcohol, see Diacetone alcohol
Diakon, see Methyl methacrylate
1,4-Diaminobenzene, see *p*-Phenylenediamine
p-Diaminobenzene, see *p*-Phenylenediamine
4,4′-Diaminobiphenyl, see Benzidine
p,p′-Diaminobiphenyl, see Benzidine
4,4′-Diamino-1,1′-biphenyl, see Benzidine
4,4′-Diamino-3,3′-dichlorobiphenyl, see 3,3′-Dichlorobenzidine
4,4′-Diaminodiphenyl, see Benzidine
p-Diaminodiphenyl, see Benzidine
p,p′-Diaminodiphenyl, see Benzidine
1,2-Diaminoethane, see Ethylenediamine
4,4′-Dianiline, see Benzidine
p,p′-Dianiline, see Benzidine
2,2-Di-*p*-anisyl-1,1,1-trichloroethane, see Methoxychlor
Diarex HF77, see Styrene
Diater, see Diuron
Diazo fast orange GR, see 2-Nitroaniline
Diazo fast orange R, see 3-Nitroaniline
Diazo fast red GG, see 4-Nitroaniline
1,2:5,6-Dibenzanthracene, see Dibenz[*a,h*]anthracene
1,2:5,6-Dibenz[*a,h*]anthracene, see Dibenz[*a,h*]anthracene
Dibenz[*a,h*]anthracene
1,2:5,6-Dibenzoanthracene, see Dibenz[*a,h*]anthracene
Dibenzo[*a,h*]anthracene, see Dibenz[*a,h*]anthracene
Dibenzo[*b,jk*]fluorene, see Benzo[*k*]fluoranthene
Dibenzofuran
1,2-Dibenzonaphthalene, see Chrysene
1,2,5,6-Dibenzonaphthalene, see Chrysene
DIBK, see Diisobutyl ketone
Dibovan, see *p,p′*-DDT
Dibrom, see Naled
1,4-Dibromobenzene

asym-Dichloroethane, see 1,1-Dichloroethane
sym-Dichloroethane, see 1,2-Dichloroethane
1,1-Dichloroethene, see 1,1-Dichloroethylene
1,2-Dichloroethene, see *trans*-1,2-Dichloroethylene
(*E*)-1,2-Dichloroethene, see *trans*-1,2-Dichloroethylene
1,2-*trans*-Dichloroethene, see *trans*-1,2-Dichloroethylene
2,2-Dichloroethenyl dimethyl phosphate, see Dichlorvos
1,1'-(Dichloroethenylidene)bis(4-chlorobenzene), see *p,p'*-DDE
2,2-Dichloroethenyl phosphoric acid dimethyl ester, see Dichlorvos
Dichloroether, see Bis(2-chloroethyl)ether
Dichloroethylene, see 1,2-Dichloroethane
1,1-Dichloroethylene
asym-Dichloroethylene, see 1,1-Dichloroethylene
sym-Dichloroethylene, see *trans*-1,2-Dichloroethylene
trans-Dichloroethylene, see *trans*-1,2-Dichloroethylene
***trans*-1,2-Dichloroethylene**
1,2-*trans*-Dichloroethylene, see *trans*-1,2-Dichloroethylene
Dichloroethyl ether, see Bis(2-chloroethyl)ether
Di(2-chloroethyl)ether, see Bis(2-chloroethyl)ether
Di(*β*-chloroethyl)ether, see Bis(2-chloroethyl)ether
2,2'-Dichloroethyl ether, see Bis(2-chloroethyl)ether
α,α'-Dichloroethyl ether, see Bis(2-chloroethyl)ether
sym-Dichloroethyl ether, see Bis(2-chloroethyl)ether
Di(*β*-chloroethyl)ether, see Bis(2-chloroethyl)ether
Di-2-(chloroethyl)ether, see Bis(2-chloroethyl)ether
Dichloroethyl formal, see Bis(2-chloroethoxy)methane
Di-2-chloroethyl formal, see Bis(2-chloroethoxy)methane
1,1'-(2,2-Dichloroethylidene)bis(4-chlorobenzene), see *p,p'*-DDD
Dichloroethyl oxide, see Bis(2-chloroethyl)ether
Dichlorofluoromethane
2,4-Dichlorohydroxybenzene, see 2,4-Dichlorophenol
4,6-Dichlorohydroxybenzene, see 2,4-Dichlorophenol
Dichloroisopropyl ether, see Bis(2-chloroisopropyl)ether
2,2'-Dichloroisopropyl ether, see Bis(2-chloroisopropyl)ether
Dichloromethane, see Methylene chloride
Dichloromethyl ether, see *sym*-Dichloromethyl ether
***sym*-Dichloromethyl ether**
Dichloromonofluoromethane, see Dichlorofluoromethane
2,4-Dichlorophenol
3-(3,4-Dichlorophenol)-1,1-dimethylurea, see Diuron
Dichlorophenoxyacetic acid, see 2,4-D
(2,4-Dichlorophenoxy)acetic acid, see 2,4-D
3-(3,4-Dichlorophenyl)-1,1-dimethylurea, see Diuron

2-*N*-Diethylaminoethanol, see 2-Diethylaminoethanol
2-Diethylaminoethyl alcohol, see 2-Diethylaminoethanol
β-Diethylaminoethyl alcohol, see 2-Diethylaminoethanol
Diethyl (dimethoxyphosphinothioylthio) butanedioate, see Malathion
Diethyl (dimethoxyphosphinothioylthio) succinate, see Malathion
Diethylene dioxide, see 1,4-Dioxane
1,4-Diethylene dioxide, see 1,4-Dioxane
Diethylene ether, see 1,4-Dioxane
Diethylene oxide, see 1,4-Dioxane, Tetrahydrofuran
Diethylene oximide, see Morpholine
Diethyleneimid oxide, see Morpholine
Diethyleneimide oxide, see Morpholine
Diethylenimide oxide, see Morpholine
N,N-Diethylethanamine, see Triethylamine
Diethylethanolamine, see 2-Diethylaminoethanol
N,N-Diethylethanolamine, see 2-Diethylaminoethanol
Diethyl ether, see Ethyl ether
Di(2-ethylhexyl)orthophthalate, see Bis(2-ethylhexyl)phthalate
Di(2-ethylhexyl)phthalate, see Bis(2-ethylhexyl)phthalate
Diethyl-(2-hydroxyethyl)amine, see 2-Diethylaminoethanol
N,N-Diethyl-*N*-(*β*-hydroxyethyl)amine, see 2-Diethylaminoethanol
Diethyl mercaptosuccinic acid *O,O*-dimethyl phosphorodithioate, see Malathion
Diethyl mercaptosuccinate, *O,O*-dimethyl phosphorodithioate, see Malathion
Diethyl mercaptosuccinate, *O,O*-dimethyl thiophosphate, see Malathion
Diethylmethane, see Pentane
Diethylmethylmethane, see 3-Methylpentane
Diethyl-4-nitrophenyl phosphorothionate, see Parathion
O,O-Diethyl-*O*-4-nitrophenyl phosphorothioate, see Parathion
O,O-Diethyl *O*-*p*-nitrophenyl phosphorothioate, see Parathion
Diethyl-*p*-nitrophenyl thionophosphate, see Parathion
O,O-Diethyl-*O*-4-nitrophenyl thionophosphate, see Parathion
O,O-Diethyl-*O*-*p*-nitrophenyl thionophosphate, see Parathion
Diethyl-*p*-nitrophenyl thiophosphate, see Parathion
O,O-Diethyl-*O*-*p*-nitrophenyl thiophosphate, see Parathion
Diethyl oxide, see Ethyl ether
Diethylparathion, see Parathion
Diethyl phthalate
Diethyl-*o*-phthalate, see Diethyl phthalate
O,O-Diethyl-*O*-3,5,6-trichloro-2-pyridyl phosphorothioate, see Chlorpyrifos
Difluorodibromomethane, see Dibromodifluoromethane
Difluorodichloromethane, see Dichlorodifluoromethane
1,1-Difluorotetrachloroethane
1,1-Difluoro-1,2,2,2-tetrachloroethane, see 1,1-Difluorotetrachloroethane

1,2-Difluorotetrachloroethane
1,2-Difluoro-1,1,2,2-tetrachloroethane, see 1,2-Difluorotetrachloroethane
2,2-Difluoro-1,1,1,2-tetrachloroethane, see 1,1-Difluorotetrachloroethane
Digermin, see Trifluralin
Dihexyl, see Dodecane
1,2-Dihydroacenaphthylene, see Acenaphthene
Dihydroazirine, see Ethylenimine
Dihydro-1*H*-azirine, see Ethylenimine
2,3-Dihydro-2,2-dimethyl-7-benzofuranol methyl carbamate, see Carbofuran
1,3-Dihydro-1,3-dioxoisobenzofuran, see Phthalic anhydride
2,3-Dihydroindene, see Indan
2,3-Dihydro-1*H*-indene, see Indan
2,3-Dihydroindole, see Indoline
2,3-Dihydro-1*H*-indole, see Indoline
Dihydroxybenzene, see Hydroquinone
1,4-Dihydroxybenzene, see Hydroquinone
p-Dihydroxybenzene, see Hydroquinone
Diisobutyl ketone
Diisocyanatoluene, see 2,4-Toluene diisocyanate
2,4-Diisocyanotoluene, see 2,4-Toluene diisocyanate
2,4-Diisocyanato-1-methylbenzene, see 2,4-Toluene diisocyanate
Diisopropyl, see 2,3-Dimethylbutane
sym-Diisopropylacetone, see Diisobutyl ketone
Diisopropylamine
Diisopropyl ether, see Isopropyl ether
Diisopropylmethane, see 2,4-Dimethylpentane
Diisopropyl oxide, see Isopropyl ether
Diketone alcohol, see Diacetone alcohol
Dilantin DB, see 1,2-Dichlorobenzene
Dilatin DB, see 1,2-Dichlorobenzene
Dilene, see *p,p'*-DDD
Dimazine, see 1,1-Dimethylhydrazine
Dimethoxy-DDT, see Methoxychlor
p,p'-Dimethoxydiphenyltrichloroethane, see Methoxychlor
Dimethoxy-DT, see Methoxychlor
Dimethoxymethane, see Methylal
2,2-Di-(*p*-methoxyphenyl)-1,1,1-trichloroethane, see Methoxychlor
Di(*p*-methoxyphenyl)trichloromethyl methane, see Methoxychlor
[(Dimethoxyphosphinothioyl)thio]butanedioic acid diethyl ester, see Malathion
3-[(Dimethoxyphosphinyl)oxy]-2-butenoic acid methyl ester, see Mevinphos
Dimethylacetamide, see *N,N*-Dimethylacetamide
N,N-**Dimethylacetamide**
Dimethylacetone amide, see *N,N*-Dimethylacetamide

Dimethyl-1,1'-dichloroether, see *sym*-Dichloromethyl ether
1,1-Dimethyl-3-(3,4-dichlorophenyl)urea, see Diuron
Dimethyl dichlorovinyl phosphate, see Dichlorvos
Dimethyl-2,2-dichlorovinyl phosphate, see Dichlorvos
O,O-Dimethyl-*O*-(2,2-dichlorovinyl)phosphate, see Dichlorvos
O,O-Dimethyl-*S*-1,2-di(ethoxycarbamyl)ethyl phosphorodithioate, see Malathion
2,2-Dimethyl-2,3-dihydro-7-benzofuranyl-*N*-methyl carbamate, see Carbofuran
O,O-Dimethyldithiophosphate dimethylmercaptosuccinate, see Malathion
Dimethylenediamine, see Ethylenediamine
Dimethyleneimine, see Ethylenimine
Dimethylenimine, see Ethylenimine
1,1-Dimethylethanol, see *tert*-Butyl alcohol
(1,1-Dimethylethyl)benzene, see *tert*-Butylbenzene
unsym-Dimethylethylene, see 2-Methylpropene
Dimethylformaldehyde, see Acetone
Dimethylformamide, see Dimethylformamide
N,N-Dimethylformamide
2,6-Dimethylheptan-4-one, see Diisobutyl ketone
2,6-Dimethyl-4-heptanone, see Diisobutyl ketone
1,1-Dimethylhydrazine
asym-Dimethylhydrazine, see 1,1-Dimethylhydrazine
N,N-Dimethylhydrazine, see 1,1-Dimethylhydrazine
unsym-Dimethylhydrazine, see 1,1-Dimethylhydrazine
Dimethylketal, see Acetone
Dimethyl ketone, see Acetone
Dimethylmethane, see Propane
Dimethyl 2-methoxycarbonyl-1-methylvinyl phosphate, see Mevinphos
Dimethyl methoxycarbonylpropenyl phosphate, see Mevinphos
Dimethyl (1-methoxycarboxypropen-2-yl)phosphate, see Mevinphos
O,O-Dimethyl *O*-(1-methyl-2-carboxyvinyl)phosphate, see Mevinphos
Dimethyl monosulfate, see Dimethyl sulfate
Dimethylnitromethane, see 2-Nitropropane
Dimethylnitrosamine, see *N*-Nitrosodimethylamine
N-Dimethylnitrosamine, see *N*-Nitrosodimethylamine
N,N-Dimethylnitrosamine, see *N*-Nitrosodimethylamine
Dimethylnitrosomine, see *N*-Nitrosodimethylamine
2-(2,2-Dimethyl-1-oxopropyl)-1*H*-indene-1,3(2*H*)-dione, see Pindone
2,3-Dimethylpentane
2,4-Dimethylpentane
3,3-Dimethylpentane
3,4-Dimethylpentane, see 2,3-Dimethylpentane
2,4-Dimethylphenol
4,6-Dimethylphenol, see 2,4-Dimethylphenol

Dimethylphenylamine, see Dimethylaniline

N,N-Dimethylphenylamine, see Dimethylaniline

N,N-Dimethyl-4-(phenylazo)benzamine, see *p*-Dimethylaminoazobenzene

N,N-Dimethyl-*p*-(phenylazo)benzamine, see *p*-Dimethylaminoazobenzene

N,N-Dimethyl-4-(phenylazo)benzenamine, see *p*-Dimethylaminoazobenzene

N,N-Dimethyl-*p*-(phenylazo)benzenamine, see *p*-Dimethylaminoazobenzene

Dimethyl phosphate of methyl-3-hydroxy-*cis*-crotonate, see Mevinphos

Dimethyl phthalate

2,2-Dimethylpropane

Dimethylpropylmethane, see 2-Methylpentane

2,7-Dimethylquinoline

Dimethyl sulfate

O,O-Dimethyl-*O*-2,4,5-trichlorophenyl phosphorothioate, see Ronnel

Dimethyl trichlorophenyl thiophosphate, see Ronnel

O,O-Dimethyl *O*-(2,4,5-trichlorophenyl)thiophosphate, see Ronnel

Dimethyl yellow, see *p*-Dimethylaminoazobenzene

Dimethyl yellow analar, see *p*-Dimethylaminoazobenzene

Dimethyl yellow *N,N*-dimethylaniline, see *p*-Dimethylaminoazobenzene

1,2-Dinitrobenzene

1,3-Dinitrobenzene

1,4-Dinitrobenzene

2,4-Dinitrobenzene, see 1,3-Dinitrobenzene

m-Dinitrobenzene, see 1,3-Dinitrobenzene

o-Dinitrobenzene, see 1,2-Dinitrobenzene

p-Dinitrobenzene, see 1,4-Dinitrobenzene

1,3-Dinitrobenzol, see 1,3-Dinitrobenzene

o-Dinitrobenzol, see 1,2-Dinitrobenzene

Dinitrocresol, see 4,6-Dinitro-*o*-cresol

Dinitro-*o*-cresol, see 4,6-Dinitro-*o*-cresol

2,4-Dinitro-*o*-cresol, see 4,6-Dinitro-*o*-cresol

3,5-Dinitro-*o*-cresol, see 4,6-Dinitro-*o*-cresol

4,6-Dinitro-*o*-cresol

Dinitrodendtroxal, see 4,6-Dinitro-*o*-cresol

2,6-Dinitro-*N,N*-dipropyl-4-trifluoromethylaniline, see Trifluralin

2,6-Dinitro-*N,N*-dipropyl-4-(trifluoromethyl)benzenamine, see Trifluralin

2,6-Dinitro-*N,N*-di-*n*-propyl-α,α,α-trifluoro-*p*-toluidine, see Trifluralin

3,5-Dinitro-2-hydroxytoluene, see 4,6-Dinitro-*o*-cresol

Dinitrol, see 4,6-Dinitro-*o*-cresol

2,4-Dinitromethylbenzene, see 2,4-Dinitrotoluene

2,6-Dinitromethylbenzene, see 2,6-Dinitrotoluene

Dinitromethyl cyclohexyltrienol, see 4,6-Dinitro-*o*-cresol

2,4-Dinitro-2-methylphenol, see 4,6-Dinitro-*o*-cresol

2,4-Dinitro-6-methylphenol, see 4,6-Dinitro-*o*-cresol

Diphenyl ether, see Phenyl ether
1,2-Diphenylhydrazine
sym-Diphenylhydrazine, see 1,2-Diphenylhydrazine
N,N'-Diphenylhydrazine, see 1,2-Diphenylhydrazine
Diphenylnitrosamine, see *N*-Nitrosodiphenylamine
Diphenyl-*N*-nitrosamine, see *N*-Nitrosodiphenylamine
N,N-Diphenylnitrosamine, see *N*-Nitrosodiphenylamine
Diphenyl oxide, see Phenyl ether
Diphenyltrichloroethane, see *p,p'*-DDT
Diphosphoric acid tetraethyl ester, see Tetraethyl pyrophosphate
4-(Di-*n*-propylamino)-3,5-dinitro-1-trifluoromethylbenzene, see Trifluralin
N,N-Di-*n*-propyl-2,6-dinitro-4-trifluoromethylaniline, see Trifluralin
Dipropylmethane, see Heptane
Dipropylnitrosamine, see *N*-Nitrosodi-*n*-propylamine
Di-*n*-propylnitrosamine, see *N*-Nitrosodi-*n*-propylamine
N,N-Dipropyl-4-trifluoromethyl-2,6-dinitroaniline, see Trifluralin
Direx 4L, see Diuron
Distokal, see Hexachloroethane
Distopan, see Hexachloroethane
Distopin, see Hexachloroethane
Dithane A-4, see 1,4-Dinitrobenzene
Dithio, see Sulfotepp
α,α'-Dithiobis(dimethylthio)formamide, see Thiram
Dithiocarbonic anhydride, see Carbon disulfide
N,N'-(Dithiodicarbonothioyl)bis(*N*-methylmethanamine), see Thiram
Dithione, see Sulfotepp
Dithiophos, see Sulfotepp
Dithiophosphoric acid tetraethyl ester, see Sulfotepp
Dithiotep, see Sulfotepp
Diurex, see Diuron
Diurol, see Diuron
Diuron
Divinylene sulfide, see Thiophene
Divinyl erythrene, see 1,3-Butadiene
Divipan, see Dichlorvos
Dizene, see 1,2-Dichlorobenzene
DMA, see *N,N*-Dimethylacetamide, Dimethylamine
DMA-4, see 2,4-D
DMAB, see *p*-Dimethylaminoazobenzene
DMAC, see *N,N*-Dimethylacetamide
DMDT, see Methoxychlor
4,4'-DMDT, see Methoxychlor
p,p'-DMDT, see Methoxychlor

DMF, see *N,N*-Dimethylformamide
DMFA, see *N,N*-Dimethylformamide
DMK, see Acetone
DMN, see *N*-Nitrosodimethylamine
DMNA, see *N*-Nitrosodimethylamine
DMP, see Dimethyl phthalate
2,4-DMP, see 2,4-Dimethylphenol
DMTD, see Methoxychlor
DMS, see Dimethyl sulfate
DMU, see Diuron
DN, see 4,6-Dinitro-*o*-cresol
DNC, see 4,6-Dinitro-*o*-cresol
DN-dry mix no. 2, see 4,6-Dinitro-*o*-cresol
DNOC, see 4,6-Dinitro-*o*-cresol
DNOP, see Di-*n*-octyl phthalate
DNP, see 2,4-Dinitrophenol
2,4-DNP, see 2,4-Dinitrophenol
DNT, see 2,4-Dinitrotoluene
2,4-DNT, see 2,4-Dinitrotoluene
2,6-DNT, see 2,6-Dinitrotoluene
DNTP, see Parathion
Dodat, see *p,p'*-DDT
Dodecane
n-Dodecane, see Dodecane
Dolco mouse cereal, see Strychnine
Dolen-pur, see Hexachlorobutadiene
Dol granule, see Lindane
Dolochlor, see Chloropicrin
DOP, see Bis(2-ethylhexyl)phthalate, Di-*n*-octyl phthalate
Dormone, see 2,4-D
Dowanol EB, see 2-Butoxyethanol
Dowanol EE, see 2-Ethoxyethanol
Dowanol EM, see Methyl cellosolve
Dowcide 7, see Pentachlorophenol
Dowco-163, see Nitrapyrin
Dowco-179, see Chlorpyrifos
Dow ET 14, see Ronnel
Dow ET 57, see Ronnel
Dowfume, see Methyl bromide
Dowfume 40, see Ethylene dibromide
Dowfume EDB, see Ethylene dibromide
Dowfume MC-2, see Methyl bromide
Dowfume MC-2 soil fumigant, see Methyl bromide

Dowfume MC-33, see Methyl bromide
Dowfume W-8, see Ethylene dibromide
Dowfume W-85, see Ethylene dibromide
Dowfume W-90, see Ethylene dibromide
Dowfume W-100, see Ethylene dibromide
Dowicide 2, see 2,4,5-Trichlorophenol
Dowicide 2S, see 2,4,6-Trichlorophenol
Dowicide 7, see Pentachlorophenol
Dowicide B, see 2,4,5-Trichlorophenol
Dowicide EC-7, see Pentachlorophenol
Dowicide G, see Pentachlorophenol
Dowklor, see Chlordane
Dow pentachlorophenol DP-2 antimicrobial, see Pentachlorophenol
Dow-per, see Tetrachloroethylene
Dowtherm E, see 1,2-Dichlorobenzene
Dow-tri, see Trichloroethylene
DPH, see 1,2-Diphenylhydrazine
DPN, see N-Nitrosodi-n-propylamine
DPNA, see N-Nitrosodi-n-propylamine
DPP, see Parathion
Dracylic acid, see Benzoic acid
Drexel parathion 8E, see Parathion
Drinox, see Aldrin and Heptachlor
Drinox H-34, see Heptachlor
Dukeron, see Trichloroethylene
Duodecane, see Dodecane
Duo-kill, see Dichlorvos
Durafur black R, see p-Phenylenediamine
Duraphos, see Mevinphos
Duravos, see Dichlorvos
Durene, see 1,2,4,5-Tetramethylbenzene
Durol, see 1,2,4,5-Tetramethylbenzene
Durotox, see Pentachlorophenol
Dursban, see Chlorpyrifos
Dursban F, see Chlorpyrifos
Dutch liquid, see 1,2-Dichloroethane
Dutch oil, see 1,2-Dichloroethane
Dykol, see p,p'-DDT
Dynex, see Diuron
E 393, see Sulfotepp
E 605, see Parathion
E 3314, see Heptachlor
EAK, see 5-Methyl-3-heptanone

Eastern states duocide, see Warfarin
EB, see Ethylbenzene
Ecatox, see Parathion
ECH, see Epichlorohydrin
Ectoral, see Ronnel
EDB, see Ethylene dibromide
EDB-85, see Ethylene dibromide
E-D-BEE, see Ethylene dibromide
EDC, see 1,2-Dichloroethane
Edco, see Methyl bromide
Effusan, see 4,6-Dinitro-*o*-cresol
Effusan 3436, see 4,6-Dinitro-*o*-cresol
Egitol, see Hexachloroethane
EGM, see Methyl cellosolve
EGME, see Methyl cellosolve
EI, see Ethylenimine
Ekagom TB, see Thiram
Ekasolve EE acetate solvent, see 2-Ethoxyethyl acetate
Ekatin WF & WF ULV, see Parathion
Ekatox, see Parathion
Ektasolve, see 2-Butoxyethanol, Methyl cellosolve
Ektasolve EE, see 2-Ethoxyethanol
EL 4049, see Malathion
Elancolan, see Trifluralin
Elaol, see Di-*n*-butyl phthalate
Eldopaque, see Hydroquinone
Eldoquin, see Hydroquinone
Electro-CF 11, see Trichlorofluoromethane
Electro-CF 12, see Dichlorodifluoromethane
Elgetol, see 4,6-Dinitro-*o*-cresol
Elgetol 30, see 4,6-Dinitro-*o*-cresol
Elipol, see 4,6-Dinitro-*o*-cresol
Embafume, see Methyl bromide
Emmatos, see Malathion
Emmatos extra, see Malathion
Emulsamine BK, see 2,4-D
Emulsamine E-3, see 2,4-D
Endocel, see α-Endosulfan, β-Endosulfan
Endosol, see α-Endosulfan, β-Endosulfan
Endosulfan, see α-Endosulfan, β-Endosulfan
Endosulfan I, see α-Endosulfan
Endosulfan II, see β-Endosulfan
α-Endosulfan

β-Endosulfan

Endosulfan sulfate

Endosulphan, see α-Endosulfan, β-Endosulfan

Endrex, see Endrin

Endrin

Endrin aldehyde

Enial yellow 2G, see p-Dimethylaminoazobenzene

ENT 54, see Acrylonitrile

ENT 154, see 4,6-Dinitro-o-cresol

ENT 262, see Dimethyl phthalate

ENT 987, see Thiram

ENT 1506, see p,p'-DDT

ENT 1656, see 1,2-Dichloroethane

ENT 1716, see Methoxychlor

ENT 1860, see Tetrachloroethylene

ENT 4225, see p,p'-DDD

ENT 4504, see Bis(2-chloroethyl)ether

ENT 4705, see Carbon tetrachloride

ENT 7796, see Lindane

ENT 8538, see 2,4-D

ENT 9232, see α-BHC

ENT 9233, see β-BHC

ENT 9234, see δ-BHC

ENT 9735, see Toxaphene

ENT 9932, see Chlordane

ENT 15108, see Parathion

ENT 15152, see Heptachlor

ENT 15349, see Ethylene dibromide

ENT 15406, see 1,2-Dichloropropane

ENT 15949, see Aldrin

ENT 16225, see Dieldrin

ENT 16273, see Sulfotepp

ENT 16391, see Kepone

ENT 17034, see Malathion

ENT 17251, see Endrin

ENT 17798, see EPN

ENT 18771, see Tetraethyl pyrophosphate

ENT 20738, see Dichlorvos

ENT 22324, see Mevinphos

ENT 23284, see Ronnel

ENT 23969, see Carbaryl

ENT 23979, see α-Endosulfan, β-Endosulfan

ENT 24988, see Naled

ENT 25552-X, see Chlordane
ENT 25584, see Heptachlor epoxide
ENT 27164, see Carbofuran
ENT 27311, see Chlorpyrifos
ENT 50324, see Ethylenimine
Entomoxan, see Lindane
Entsufon, see 2,4,6-Trinitrotoluene
Envert 171, see 2,4-D
Envert DT, see 2,4-D
Envert-T, see 2,4,5-T
EP 30, see Pentachlorophenol
Epichlorohydrin
α-Epichlorohydrin, see Epichlorohydrin
(*dl*)-α-Epichlorohydrin, see Epichlorohydrin
Epichlorophydrin, see Epichlorohydrin
Epihydric alcohol, see Glycidol
Epihydrin alcohol, see Glycidol
EPN
EPN 300, see EPN
1,2-Epoxy-3-allyloxypropane, see Allyl glycidyl ether
1,2-Epoxy-3-chloropropane, see Epichlorohydrin
Epoxy heptachlor, see Heptachlor epoxide
Epoxypropane, see Propylene oxide
1,2-Epoxypropane, see Propylene oxide
2,3-Epoxypropanol, see Glycidol
2,3-Epoxy-1-propanol, see Glycidol
Epoxypropyl alcohol, see Glycidol
2,3-Epoxypropyl chloride, see Epichlorohydrin
Equigard, see Dichlorvos
Equigel, see Dichlorvos
Eradex, see Chlorpyrifos
Ergoplast FDO, see Bis(2-ethylhexyl)phthalate
ESEN, see Phthalic anhydride
Eskimon 11, see Trichlorofluoromethane
Eskimon 12, see Dichlorodifluoromethane
Essence of mirbane, see Nitrobenzene
Essence of myrbane, see Nitrobenzene
Estercide T-2 and T-245, see 2,4,5-T
Esteron, see 2,4-D, 2,4,5-T
Esteron 44 weed killer, see 2,4-D
Esteron 76 BE, see 2,4-D
Esteron 99, see 2,4-D
Esteron 99 concentrate, see 2,4-D

Esteron 245 BE, see 2,4,5-T
Esteron brush killer, see 2,4-D, 2,4,5-T
Esterone 4, see 2,4-D
Esterone 245, see 2,4,5-T
Estol 1550, see Diethyl phthalate
Estonate, see *p,p'*-DDT
Estone, see 2,4-D
Estonox, see Toxaphene
Estrosel, see Dichlorvos
Estrosol, see Dichlorvos
ET 14, see Ronnel
ET 57, see Ronnel
Ethanal, see Acetaldehyde
Ethanamine, see Ethylamine
1,2-Ethanediamine, see Ethylenediamine
Ethane dichloride, see 1,2-Dichloroethane
Ethanedioic acid, see Oxalic acid
1,2-Ethanediol dipropanoate, see Crotonaldehyde
Ethanedionic acid, see Oxalic acid
Ethane hexachloride, see Hexachloroethane
Ethanenitrile, see Acetonitrile
Ethane pentachloride, see Pentachloroethane
Ethane tetrachloride, see 1,1,2,2-Tetrachloroethane
Ethanethiol, see Ethyl mercaptan
Ethane trichloride, see 1,1,2-Trichloroethane
Ethanoic acid, see Acetic acid
Ethanoic anhydrate, see Acetic anhydride
Ethanoic anhydride, see Acetic anhydride
Ethanolamine
β-Ethanolamine, see Ethanolamine
Ethene dichloride, see 1,2-Dichloroethane
Ethenyl acetate, see Vinyl acetate
Ethenyl ethanoate, see Vinyl acetate
Ether, see Ethyl ether
Ether chloratus, see Chloroethane
Ether hydrochloric, see Chloroethane
Ether muriatic, see Chloroethane
Ethinyl trichloride, see Trichloroethylene
Ethiolacar, see Malathion
Ethlon, see Parathion
Ethoxyacetate, see 2-Ethoxyethyl acetate
Ethoxycarbonylethylene, see Ethyl acrylate
Ethoxyethane, see Ethyl ether

2-Ethoxyethanol
2-Ethoxyethanol acetate
Ethoxyethyl acetate, see 2-Ethoxyethyl acetate
2-Ethoxyethyl acetate, see 2-Ethoxyethyl acetate
β-Ethoxyethyl acetate, see 2-Ethoxyethyl acetate
Ethoxy-4-nitrophenoxy phenylphosphine sulfide, see EPN
Ethyenyl benzene, see Styrene
Ethyl acetate
Ethyl acetic ester, see Ethyl acetate
Ethyl acetone, see 2-Pentanone
Ethyl acrylate
Ethylaldehyde, see Acetaldehyde
Ethylamine
Ethyl amyl ketone, see 5-Methyl-3-heptanone
Ethyl sec-amyl ketone, see 5-Methyl-3-heptanone
Ethylbenzene
Ethylbenzol, see Ethylbenzene
Ethyl bromide
Ethyl butyl ketone, see 3-Heptanone
Ethyl carbinol, see 1-Propanol
Ethyl cellosolve, see 2-Ethoxyethanol
Ethyl chloride, see Chloroethane
Ethylcyclopentane
Ethyldimethylmethane, see 2-Methylbutane
Ethylene aldehyde, see Acrolein
Ethylene bromide, see Ethylene dibromide
Ethylene bromide glycol dibromide, see Ethylene dibromide
Ethylene carboxamide, see Acrylamide
Ethylene chlorhydrin, see Ethylene chlorohydrin
Ethylene chloride, see 1,2-Dichloroethane
Ethylene chlorohydrin
Ethylenediamine
1,2-Ethylenediamine, see Ethylenediamine
Ethylene dibromide
1,2-Ethylene dibromide, see Ethylene dibromide
Ethylene dichloride, see 1,2-Dichloroethane
1,2-Ethylene dichloride, see 1,2-Dichloroethane
Ethylene dipropionate, see Crotonaldehyde
Ethylene glycol dipropionate, see Crotonaldehyde
Ethylene glycol ethyl ether, see 2-Ethoxyethanol
Ethylene glycol ethyl ether acetate, see 2-Ethoxyethyl acetate
Ethylene glycol methyl ether, see Methyl cellosolve
Ethylene glycol methyl ether acetate, see Methyl cellosolve acetate

Ethylene glycol monobutyl ether, see 2-Butoxyethanol
Ethylene glycol mono-*n*-butyl ether, see 2-Butoxyethanol
Ethylene glycol monoethyl ether, see 2-Ethoxyethanol
Ethylene glycol monoethyl ether acetate, see 2-Ethoxyethyl acetate
Ethylene glycol monomethyl ether, see Methyl cellosolve
Ethylene glycol monomethyl ether acetate, see Methyl cellosolve acetate
Ethylene hexachloride, see Hexachloroethane
Ethylenimine
Ethylene monochloride, see Vinyl chloride
Ethylenenaphthalene, see Acenaphthene
1,8-Ethylenenaphthalene, see Acenaphthene
Ethylene propionate, see Crotonaldehyde
Ethylene tetrachloride, see Tetrachloroethylene
Ethylene trichloride, see Trichloroethylene
N-Ethylethanamine, see Diethylamine
Ethyl ethanoate, see Ethyl acetate
Ethyl ether
Ethylethylene, see 1-Butene
Ethyl formate
Ethyl formic ester, see Ethyl formate
Ethylhexyl phthalate, see Bis(2-ethylhexyl)phthalate
2-Ethylhexyl phthalate, see Bis(2-ethylhexyl)phthalate
Ethyl hydrosulfide, see Ethyl mercaptan
Ethylic acid, see Acetic acid
Ethylidene chloride, see 1,1-Dichloroethane
Ethylidene dichloride, see 1,1-Dichloroethane
1,1-Ethylidene dichloride, see 1,1-Dichloroethane
Ethylimine, see Ethylenimine
Ethylisobutylmethane, see 2-Methylhexane
Ethyl mercaptan
Ethyl methanoate, see Ethyl formate
Ethylmethyl carbinol, see *sec*-Butyl alcohol
Ethyl methyl ketone, see 2-Butanone
4-Ethylmorpholine
N-Ethylmorpholine, see 4-Ethylmorpholine
Ethyl nitrile, see Acetonitrile
Ethyl *p*-nitrophenyl benzenethionophosphate, see EPN
Ethyl *p*-nitrophenyl benzenethiophosphonate, see EPN
Ethyl *p*-nitrophenyl ester, see EPN
Ethyl *p*-nitrophenyl phenylphosphonothioate, see EPN
O-Ethyl *O*-4-nitrophenyl phenylphosphonothioate, see EPN
O-Ethyl *O*-*p*-nitrophenyl phenylphosphonothioate, see EPN
Ethyl *p*-nitrophenyl thionobenzenephosphate, see EPN

Ethyl *p*-nitrophenyl thionobenzenephosphonate, see EPN
Ethylolamine, see Ethanolamine
Ethyl oxide, see Ethyl ether
Ethyl parathion, see Parathion
O-Ethyl phenyl *p*-nitrophenyl phenylphosphorothioate, see EPN
O-Ethyl phenyl *p*-nitrophenyl thiophosphonate, see EPN
Ethyl phthalate, see Diethyl phthalate
Ethyl propenoate, see Ethyl acrylate
Ethyl-2-propenoate, see Ethyl acrylate
Ethyl pyrophosphate, see Tetraethyl pyrophosphate
Ethyl sulfhydrate, see Ethyl mercaptan
Ethyl thioalcohol, see Ethyl mercaptan
2-Ethylthiophene
Ethyl thiopyrophosphate, see Sulfotepp
Etilon, see Parathion
Etiol, see Malathion
Etrolene, see Ronnel
Eviplast 80, see Bis(2-ethylhexyl)phthalate
Eviplast 81, see Bis(2-ethylhexyl)phthalate
Evola, see 1,4-Dichlorobenzene
Exagama, see Lindane
Experimental insecticide 269, see Endrin
Experimental insecticide 4049, see Malathion
Experimental insecticide 7744, see Carbaryl
Extermathion, see Malathion
Extrar, see 4,6-Dinitro-*o*-cresol
F 11, see Trichlorofluoromethane
F 12, see Dichlorodifluoromethane
F 112, see 1,2-Difluorotetrachloroethane
F 13B1, see Bromotrifluoromethane
FA, see Formaldehyde
FAA, see 2-Acetylaminofluorene
2-FAA, see 2-Acetylaminofluorene
Falitram, see Thiram
Falkitol, see Hexachloroethane
Fannoform, see Formaldehyde
Farmco, see 2,4-D
Farmco diuron, see Diuron
Farmco fence rider, see 2,4,5-T
Fasciolin, see Carbon tetrachloride and Hexachloroethane
Fasco fascrat powder, see Warfarin
Fasco-terpene, see Toxaphene
Fast corinth base B, see Benzidine

Fast garnet base B, see 1-Naphthylamine
Fast garnet B base, see 1-Naphthylamine
Fast oil yellow B, see *p*-Dimethylaminoazobenzene
Fast orange base GR, see 2-Nitroaniline
Fast orange base GR salt, see 2-Nitroaniline
Fast orange base JR, see 2-Nitroaniline
Fast orange base R, see 3-Nitroaniline
Fast orange GR base, see 2-Nitroaniline
Fast orange M base, see 3-Nitroaniline
Fast orange MM base, see 3-Nitroaniline
Fast orange O base, see 2-Nitroaniline
Fast orange O salt, see 2-Nitroaniline
Fast orange R base, see 3-Nitroaniline
Fast orange R salt, see 3-Nitroaniline
Fast orange salt JR, see 2-Nitroaniline
Fast red 2G base, see 4-Nitroaniline
Fast red base GG, see 4-Nitroaniline
Fast red base 2J, see 4-Nitroaniline
Fast red 2G salt, see 4-Nitroaniline
Fast red GG base, see 4-Nitroaniline
Fast red GG salt, see 4-Nitroaniline
Fast red MP base, see 4-Nitroaniline
Fast red P base, see 4-Nitroaniline
Fast red P salt, see 4-Nitroaniline
Fast red salt GG, see 4-Nitroaniline
Fast red salt 2J, see 4-Nitroaniline
Fast scarlet base B, see 2-Naphthylamine
Fast yellow, see *p*-Dimethylaminoazobenzene
Fat yellow, see *p*-Dimethylaminoazobenzene
Fat yellow A, see *p*-Dimethylaminoazobenzene
Fat yellow AD OO, see *p*-Dimethylaminoazobenzene
Fat yellow ES, see *p*-Dimethylaminoazobenzene
Fat yellow ES extra, see *p*-Dimethylaminoazobenzene
Fat yellow extra conc., see *p*-Dimethylaminoazobenzene
Fat yellow R, see *p*-Dimethylaminoazobenzene
Fat yellow R (8186), see *p*-Dimethylaminoazobenzene
FC 11, see Trichlorofluoromethane
FC 12, see Dichlorodifluoromethane
Fecama, see Dichlorvos
Fedal-UN, see Tetrachloroethylene
Fence rider, see 2,4,5-T
Fenchchlorphos, see Ronnel
Fenchlorfos, see Ronnel

Fenchlorophos, see Ronnel

Fenoxyl carbon n, see 2,4-Dinitrophenol

Fermentation amyl alcohol, see Isoamyl alcohol

Fermentation butyl alcohol, see Isobutyl alcohol

Fermide, see Thiram

Fermine, see Dimethyl phthalate

Fernacol, see Thiram

Fernasan, see Thiram

Fernasan A, see Thiram

Fernesta, see 2,4-D

Fernide, see Thiram

Fernimine, see 2,4-D

Fernoxone, see 2,4-D

Ferxone, see 2,4-D

Fleck-flip, see Trichloroethylene

Fleet-X, see 1,3,5-Trimethylbenzene

Fleximel, see Bis(2-ethylhexyl)phthalate

Flexol DOP, see Bis(2-ethylhexyl)phthalate

Flexol plasticizer DOP, see Bis(2-ethylhexyl)phthalate

Flock-flip, see Trichloroethylene

Flo pro T, seed protectant, see Thiram

Fluate, see Trichloroethylene

Flukoids, see Carbon tetrachloride

Fluoranthene

N-Fluoren-2-acetylacetamide, see 2-Acetylaminofluorene

Fluorene

9H-Fluorene, see Fluorene

2-Fluorenylacetamide, see 2-Acetylaminofluorene

N-2-Fluorenylacetamide, see 2-Acetylaminofluorene

N-Fluorenyl-2-acetamide, see 2-Acetylaminofluorene

N-9H-Fluoren-2-ylacetamide, see 2-Acetylaminofluorene

Fluorocarbon 11, see Trichlorofluoromethane

Fluorocarbon 12, see Dichlorodifluoromethane

Fluorocarbon 21, see Dichlorofluoromethane

Fluorocarbon 113, see 1,1,2-Trichlorotrifluoroethane

Fluorodichloromethane, see Dichlorofluoromethane

Fluorotrichloromethane, see Trichlorofluoromethane

Fly-die, see Dichlorvos

Fly fighter, see Dichlorvos

FMC 5462, see α-Endosulfan, β-Endosulfan

Folidol, see Parathion

Folidol E605, see Parathion

Folidol E & E 605, see Parathion

Foredex 75, see 2,4-D
Forlin, see Lindane
Formal, see Malathion, Methylal
Formaldehyde
Formaldehyde bis(β-chloroethylacetal), see Bis(2-chloroethoxy)methane
Formaldehyde dimethylacetal, see Methylal
Formalin, see Formaldehyde
Formalin 40, see Formaldehyde
Formalith, see Formaldehyde
Formic acid
Formic acid ethyl ester, see Ethyl formate
Formic acid methyl ester, see Methyl formate
Formic aldehyde, see Formaldehyde
Formic ether, see Ethyl formate
Formol, see Formaldehyde
Formosa camphor, see Camphor
Formula 40, see 2,4-D
N-Formyldimethylamine, see N,N-Dimethylformamide
Formylic acid, see Formic acid
Formyl trichloride, see Chloroform
Forron, see 2,4,5-T
Forst U 46, see 2,4,5-T
Fortex, see 2,4,5-T
Forthion, see Malathion
Fosdrin, see Mevinphos
Fosfermo, see Parathion
Fosferno, see Parathion
Fosfex, see Parathion
Fosfive, see Parathion
Fosfothion, see Malathion
Fosfotion, see Malathion
Fosova, see Parathion
Fostern, see Parathion
Fostox, see Parathion
Fosvex, see Tetraethyl pyrophosphate
Fouramine D, see p-Phenylenediamine
Fourrine 1, see p-Phenylenediamine
Fourrine D, see p-Phenylenediamine
Four thousand forty-nine, see Malathion
Freon 10, see Carbon tetrachloride
Freon 11, see Trichlorofluoromethane
Freon 11A, see Trichlorofluoromethane
Freon 11B, see Trichlorofluoromethane

Freon 12, see Dichlorodifluoromethane
Freon 12-B2, see Dibromodifluoromethane
Freon 13-B1, see Bromotrifluoromethane
Freon 20, see Chloroform
Freon 21, see Dichlorofluoromethane
Freon 30, see Methylene chloride
Freon 112, see 1,2-Difluorotetrachloroethane
Freon 112A, see 1,1-Difluorotetrachloroethane
Freon 113, see 1,1,2-Trichlorotrifluoroethane
Freon 150, see 1,2-Dichloroethane
Freon F-12, see Dichlorodifluoromethane
Freon HE, see Trichlorofluoromethane
Freon MF, see Trichlorofluoromethane
Frigen 11, see Trichlorofluoromethane
Frigen 12, see Dichlorodifluoromethane
Frigen 113 TR-T, see 1,1,2-Trichlorotrifluoroethane
Fruitone A, see 2,4,5-T
Fumagon, see 1,2-Dibromo-3-chloropropane
Fumazone, see 1,2-Dibromo-3-chloropropane
Fumazone 86, see 1,2-Dibromo-3-chloropropane
Fumazone 86E, see 1,2-Dibromo-3-chloropropane
Fumigant-1, see Methyl bromide
Fumigrain, see Acrylonitrile
Fumo-gas, see Ethylene dibromide
Fungifen, see Pentachlorophenol
Fungol, see Pentachlorophenol
Furadan, see Carbofuran
Fural, see Furfural
2-Furaldehyde, see Furfural
Furale, see Furfural
2-Furanaldehyde, see Furfural
2-Furancarbinol, see Furfuryl alcohol
2-Furancarbonal, see Furfural
2-Furancarboxaldehyde, see Furfural
2,5-Furanedione, see Maleic anhydride
Furanidine, see Tetrahydrofuran
2-Furanmethanol, see Furfuryl alcohol
Fur black 41867, see p-Phenylenediamine
Fur brown 41866, see p-Phenylenediamine
Furfural
Furfural alcohol, see Furfuryl alcohol
Furfuraldehyde, see Furfural
Furfurol, see Furfural

Furfurole, see Furfural
Furfuryl alcohol
Furole, see Furfural
α-Furole, see Furfural
Furro D, see *p*-Phenylenediamine
Fur yellow, see *p*-Phenylenediamine
Furyl alcohol, see Furfuryl alcohol
Furylcarbinol, see Furfuryl alcohol
2-Furylcarbinol, see Furfuryl alcohol
α-Furylcarbinol, see Furfuryl alcohol
2-Furylmethanal, see Furfural
2-Furylmethanol, see Furfuryl alcohol
Fusel oil, see Isoamyl alcohol
Futramine D, see *p*-Phenylenediamine
Fyde, see Formaldehyde
Fyfanon, see Malathion
G 25, see Chloropicrin
Gallogama, see Lindane
Gamacid, see Lindane
Gamaphex, see Lindane
Gamene, see Lindane
Gamiso, see Lindane
Gammahexa, see Lindane
Gammalin, see Lindane
Gammexene, see Lindane
Gammopaz, see Lindane
Gamonil, see Carbaryl
GC-1189, see Kepone
Gearphos, see Parathion
Gemalgene, see Trichloroethylene
General chemicals 1189, see Kepone
Genetron 11, see Trichlorofluoromethane
Genetron 12, see Dichlorodifluoromethane
Genetron 21, see Dichlorofluoromethane
Genetron 112, see 1,2-Difluorotetrachloroethane
Geniphene, see Toxaphene
Genithion, see Parathion
Genitox, see *p,p'*-DDT
Genklene, see 1,1,1-Trichloroethane
Geranium crystals, see Phenyl ether
Germain's, see Carbaryl
Germalgene, see Trichloroethylene
Gesafid, see *p,p'*-DDT

Gesapon, see *p,p'*-DDT
Gesarex, see *p,p'*-DDT
Gesarol, see *p,p'*-DDT
Gesfid, see Mevinphos
Gestid, see Mevinphos
Gettysolve-B, see Hexane
Gettysolve-C, see Heptane
Gexane, see Lindane
Glacial acetic acid, see Acetic acid
Glazd penta, see Pentachlorophenol
Glycol dichloride, see 1,2-Dichloroethane
Glycerin trichlorohydrin, see 1,2,3-Trichloropropane
Glycerol epichlorohydrin, see Epichlorohydrin
Glycerol trichlorohydrin, see 1,2,3-Trichloropropane
Glyceryl trichlorohydrin, see 1,2,3-Trichloropropane
Glycide, see Glycidol
Glycidol
Glycidyl alcohol, see Glycidol
Glycidyl allyl ether, see Allyl glycidyl ether
Glycinol, see Ethanolamine
Glycol bromide, see Ethylene dibromide
Glycol chlorohydrin, see Ethylene chlorohydrin
Glycol dibromide, see Ethylene dibromide
Glycol ether EE, see 2-Ethoxyethanol
Glycol ether EE acetate, see 2-Ethoxyethyl acetate
Glycol ether EM, see Methyl cellosolve
Glycol ether EM acetate, see Methyl cellosolve acetate
Glycol ethylene ether, see 1,4-Dioxane
Glycol methyl ether, see Methyl cellosolve
Glycol monobutyl ether, see 2-Butoxyethanol
Glycol monochlorohydrin, see Ethylene chlorohydrin
Glycol monoethyl ether, see 2-Ethoxyethanol
Glycol monoethyl ether acetate, see 2-Ethoxyethyl acetate
Glycol monomethyl ether, see Methyl cellosolve
Glycol monomethyl ether acetate, see Methyl cellosolve acetate
Good-rite GP 264, see Bis(2-ethylhexyl)phthalate
GP-40-66:120, see Hexachlorobutadiene
GPKh, see Heptachlor
Granox NM, see Hexachlorobenzene
Graphlox, see Hexachlorocyclopentadiene
Grasal brilliant yellow, see *p*-Dimethylaminoazobenzene
Green oil, see Anthracene
Grisol, see Tetraethyl pyrophosphate

Grundier arbezol, see Pentachlorophenol
Guesapon, see *p,p'*-DDT
Gum camphor, see Camphor
Gy-phene, see Toxaphene
Gyron, see *p,p'*-DDT
H-34, see Heptachlor
Halane, see 1,3-Dichloro-5,5-dimethylhydantoin
Hallucinogen, see *N,N*-Dimethylacetamide
Halocarbon 11, see Trichlorofluoromethane
Halocarbon 112, see 1,2-Difluorotetrachloroethane
Halocarbon 112a, see 1,1-Difluorotetrachloroethane
Halocarbon 113, see 1,1,2-Trichlorotrifluoroethane
Halocarbon 13B1, see Bromotrifluoromethane
Halon, see Dichlorodifluoromethane
Halon 21, see Dichlorofluoromethane
Halon 104, see Carbon tetrachloride
Halon 122, see Dichlorodifluoromethane
Halon 1001, see Methyl bromide
Halon 1011, see Bromochloromethane
Halon 1202, see Dibromodifluoromethane
Halon 1301, see Bromotrifluoromethane
Halon 2001, see Ethyl bromide
Halon 10001, see Methyl iodide
Halowax 1051, see Octachloronaphthalene
Hatcol DOP, see Bis(2-ethylhexyl)phthalate
Havero-extra, see *p,p'*-DDT
HCB, see Hexachlorobenzene
HCBD, see Hexachlorobutadiene
HCCH, see Lindane
HCCP, see Hexachlorocyclopentadiene
HCCPD, see Hexachlorocyclopentadiene
HCE, see Heptachlor epoxide, Hexachloroethane
HCH, see Lindane
α-HCH, see α-BHC
β-HCH, see β-BHC
δ-HCH, see δ-BHC
γ-HCH, see Lindane
HCPD, see Hexachlorocyclopentadiene
HCS 3,260, see Chlordane
Heclotox, see Lindane
Hedolit, see 4,6-Dinitro-*o*-cresol
Hedolite, see 4,6-Dinitro-*o*-cresol
Hedonal, see 2,4-D

Hemimellitene, see 1,2,3-Trimethylbenzene

Hemiterpene, see 2-Methyl-1,3-butadiene

HEOD, see Dieldrin

Hept, see Tetraethyl pyrophosphate

Heptachlor

Heptachlorane, see Heptachlor

Heptachlor epoxide

3,4,5,6,7,8,8-Heptachlorodicyclopentadiene, see Heptachlor

3,4,5,6,7,8,8a-Heptachlorodicyclopentadiene, see Heptachlor

3,4,5,6,7,8,8a-Heptachloro-α-dicyclopentadiene, see Heptachlor

1,4,5,6,7,8,8-Heptachloro-2,3-epoxy-2,3,3a,4,7,7a-hexahydro-4,7-methanoindene, see Heptachlor epoxide

1,2,3,4,5,6,7,8,8-Heptachloro-2,3-epoxy-3a,4,7,7a-tetrahydro-4,7-methanoindene, see Heptachlor epoxide

2,3,4,5,6,7,7-Heptachloro-1a,1b,5,5a,6,6a-hexahydro-2,5-methano-2*H*-indeno-[*1,2-b*]oxirene, see Heptachlor epoxide

2,3,4,5,6,7,7-Heptachloro-1a,1b,5,5a,6,6a-hexahydro-2,5-methano-2*H*-oxireno[*a*]-indene, see Heptachlor epoxide

1(3a),4,5,6,7,8,8-Heptachloro-3a(1),4,7,7a-tetrahydro-4,7-methanoindene, see Heptachlor

1,4,5,6,7,8,8-Heptachloro-3a,4,7,7a-tetrahydro-4,7-methanoindene, see Heptachlor

1,4,5,6,7,8,8a-Heptachloro-3a,4,7,7a-tetrahydro-4,7-methanoindene, see Heptachlor

1,4,5,6,7,8,8-Heptachloro-3a,4,7,7a-tetrahydro-4,7-*endo*-methanoindene, see Heptachlor

1,4,5,6,7,10,10-Heptachloro-4,7,8,9-tetrahydro-4,7-methanoindene, see Heptachlor

1,4,5,6,7,8,8-Heptachloro-3a,4,7,7a-tetrahydro-4,7-methanol-1*H*-indene, see Heptachlor

1,4,5,6,7,8,8-Heptachloro-3a,4,7,7a-tetrahydro-4,7-methyleneindene, see Heptachlor

1,4,5,6,7,10,10-Heptachloro-4,7,8,9-tetrahydro-4,7-*endo*-methyleneindene, see Heptachlor

Heptadichlorocyclopentadiene, see Heptachlor

Heptagran, see Heptachlor

Heptagranox, see Heptachlor

Heptamak, see Heptachlor

Heptamethylene, see Cycloheptane

Heptamul, see Heptachlor

Heptane

n-Heptane, see Heptane

2-Heptanone

3-Heptanone

Heptan-3-one, see 3-Heptanone
Heptasol, see Heptachlor
cis-2-Heptene
(*E*)-2-Heptene, see *cis*-2-Heptene
***trans*-2-Heptene**
(*Z*)-2-Heptene, see *cis*-2-Heptene
Heptox, see Heptachlor
cis-2-Heptylene, see *cis*-2-Heptene
trans-2-Heptylene, see *trans*-2-Heptene
Heptyl hydride, see Heptane
Herbatox, see Diuron
Herbidal, see 2,4-D
Hercoflex 260, see Bis(2-ethylhexyl)phthalate
Hercules 3956, see Toxaphene
Hercules toxaphene, see Toxaphene
Herkal, see Dichlorvos
Herkol, see Dichlorvos
Hermal, see Thiram
Hermat TMT, see Thiram
Heryl, see Thiram
Hex, see Hexachlorocyclopentadiene
Hexa, see Lindane
Hexa C.B., see Hexachlorobenzene
γ-Hexachlor, see Lindane
Hexachloran, see Lindane
α-Hexachloran, see α-BHC
γ-Hexachloran, see Lindane
Hexachlorane, see Lindane
α-Hexachlorane, see α-BHC
γ-Hexachlorane, see Lindane
Hexachlorbutadiene, see Hexachlorobutadiene
α-Hexachlorcyclohexane, see α-BHC
Hexachlorobenzene
β-Hexachlorobenzene, see β-BHC
γ-Hexachlorobenzene, see Lindane
α,β-1,2,3,7,7-Hexachlorobicyclo[2.2.1]-2-heptane-5,6-bisoxymethylene sulfite, see
    α-Endosulfan, β-Endosulfan
1,2,3,7,7-Hexachlorobicyclo[2.2.1]-2-heptene-5,6-bisoxymethylene   sulfite,   see
    α-Endosulfan, β-Endosulfan
Hexachlorobutadiene
1,3-Hexachlorobutadiene, see Hexachlorobutadiene
Hexachloro-1,3-butadiene, see Hexachlorobutadiene
1,1,2,3,4,4-Hexachlorobutadiene, see Hexachlorobutadiene

1,1,2,3,4,4-Hexachloro-1,3-butadiene, see Hexachlorobutadiene

1,2,3,4,5,6-Hexachlorocyclohexane, see Lindane

1α,2α,3β,4α,5α,6β-Hexachlorocyclohexane, see Lindane

1,2,3,4,5,6-Hexachloro-α-cyclohexane, see α-BHC

1,2,3,4,5,6-Hexachloro-β-cyclohexane, see β-BHC

1,2,3,4,5,6-Hexachloro-δ-cyclohexane, see δ-BHC

1,2,3,4,5,6-Hexachloro-γ-cyclohexane, see Lindane

1,2,3,4,5,6-Hexachloro-*trans*-cyclohexane, see β-BHC

α-Hexachlorocyclohexane, see α-BHC

β-Hexachlorocyclohexane, see β-BHC

δ-Hexachlorocyclohexane, see δ-BHC

α-1,2,3,4,5,6-Hexachlorocyclohexane, see α-BHC

β-1,2,3,4,5,6-Hexachlorocyclohexane, see β-BHC

δ-1,2,3,4,5,6-Hexachlorocyclohexane, see δ-BHC

γ-Hexachlorocyclohexane, see Lindane

δ-(aeeeee)-1,2,3,4,5,6-Hexachlorocyclohexane, see δ-BHC

γ-1,2,3,4,5,6-Hexachlorocyclohexane, see Lindane

1α,2α,3α,4β,5β,6β-Hexachlorocyclohexane, see δ-BHC

1α,2α,3β,4α,5β,6β-Hexachlorocyclohexane, see α-BHC

1α,2β,3α,4β,5α,6β-Hexachlorocyclohexane, see β-BHC

Hexachlorocyclopentadiene

1,2,3,4,5,5-Hexachloro-1,3-cyclopentadiene, see Hexachlorocyclopentadiene

2,2a,3,3,4,7-Hexachlorodecahydro-1,2,4-methenocyclopenta[*c,d*]pentalene-5-carboxaldehyde, see Endrin aldehyde

Hexachloroepoxyoctahydro-*endo,endo*-dimethanonaphthalene, see Endrin

Hexachloroepoxyoctahydro-*endo,exo*-dimethanonaphthalene, see Dieldrin

1,2,3,4,10,10-Hexachloro-6,7-epoxy-1,4,4a,5,6,7,8,8a-octahydro-1,4-*endo,exo*-5,8-dimethanonaphthalene, see Dieldrin

1,2,3,4,10,10-Hexachloro-6,7-epoxy-1,4,4a,5,6,7,8,8a-octahydro-*endo,endo*-1,4:5,8-dimethanonaphthalene, see Endrin

Hexachloroethane

1,1,1,2,2,2-Hexachloroethane, see Hexachloroethane

Hexachloroethylene, see Hexachloroethane

Hexachlorohexahydro-*endo,exo*-dimethanonaphthalene, see Aldrin

1,2,3,4,10,10-Hexachloro-1,4,4a,5,8,8a-hexahydro-1,4:5,8-dimethanonaphthalene, see Aldrin

1,2,3,4,10,10-Hexachloro-1,4,4a,5,8,8a-hexahydro-1,4-*endo,exo*-5,8-dimethano naphthalene, see Aldrin

1,2,3,4,10,10-Hexachloro-1,4,4a,5,8,8a-hexahydro-*exo*-1,4-*endo*-5,8-*endo*-dimethanonaphthalene, see Aldrin

6,7,8,9,10,10-Hexachloro-1,5,5a,6,9,9a-hexahydro-3,3-dioxide, see Endosulfan sulfate

(3α,5aα,6β,9β,9aα)-6,7,8,9,10,10-Hexachloro-1,5,5a,6,9,9a-hexahydro-6,9-methano-2,4,3-benzodioxathiepin-3-oxide, see β-Endosulfan

(3α,5aβ,6α,9α,9aβ)-6,7,8,9,10,10-Hexachloro-1,5,5a,6,9,9a-hexahydro-6,9-meth-
ano-2,4,3-benzodioxathiepin-3-oxide, see α-Endosulfan

Hexachlorohexahydromethano-2,4,3-benzodioxathiepin-3-oxide, see α-Endosulfan, β-Endosulfan

1,4,5,6,7,7-Hexachloro-5-norborene-2,3-dimethanol cyclic sulfite, see α-Endosulfan, β-Endosulfan

3,4,5,6,9,9-Hexachloro-1a,2,2a,3,6,6a,7,7a-octahydro-2,7:3,6-dimethanonaphth-
[2,3b]oxirene, see Dieldrin

Hexadrin, see Endrin

Hexahydrobenzene, see Cyclohexane

1,4,4a,5,8,8a-Hexahydro-1,4-*endo,exo*-5,8-dimethanonaphthalene, see Aldrin

Hexahydrophenol, see Cyclohexanol

Hexahydrotoluene, see Methylcyclohexane

cis-1,2-Hexahydroxylene, see *cis*-1,2-Dimethylcyclohexane

Hexalin, see Cyclohexanol

Hexamethylene, see Cyclohexane

Hexamite, see Tetraethyl pyrophosphate

Hexanaphthene, see Cyclohexane

Hexane

n-Hexane, see Hexane

Hexanon, see Cyclohexanone

Hexanone, see 4-Methyl-2-pentanone

Hexanone-2, see 2-Hexanone

2-Hexanone

Hexaplas M/B, see Di-*n*-butyl phthalate

Hexatox, see Lindane

Hexaverm, see Lindane

Hexathir, see Thiram

Hexavin, see Carbaryl

1-Hexene

Hex-1-ene, see 1-Hexene

Hexicide, see Lindane

Hexone, see 4-Methyl-2-pentanone

Hexyclan, see Lindane

sec-**Hexyl acetate**

Hexylene, see 1-Hexene

Hexyl hydride, see Hexane

HGI, see Lindane

HHDN, see Aldrin

Hibrom, see Naled

Hildan, see α-Endosulfan, β-Endosulfan

Hilthion, see Malathion

4-Hydroxy-1,3-dimethylbenzene, see 2,4-Dimethylphenol
1-Hydroxy-2,4-dinitrobenzene, see 2,4-Dinitrophenol
3-Hydroxy-1,2-epoxypropane, see Glycidol
Hydroxy ether, see 2-Ethoxyethanol
2-Hydroxyethylamine, see Ethanolamine
β-Hydroxyethylamine, see Ethanolamine
4-Hydroxy-2-keto-4-methylpentane, see Diacetone alcohol
1-Hydroxy-2-methylbenzene, see 2-Methylphenol
1-Hydroxy-4-methylbenzene, see 4-Methylphenol
Hydroxymethyl ethylene oxide, see Glycidol
2-Hydroxymethylfuran, see Furfuryl alcohol
2-Hydroxymethyloxiran, see Glycidol
4-Hydroxy-4-methylpentanone-2, see Diacetone alcohol
4-Hydroxy-4-methylpentan-2-one, see Diacetone alcohol
4-Hydroxy-4-methyl-2-pentanone, see Diacetone alcohol
1-Hydroxymethylpropane, see Isobutyl alcohol
2-Hydroxynitrobenzene, see 2-Nitrophenol
4-Hydroxynitrobenzene, see 4-Nitrophenol
o-Hydroxynitrobenzene, see 2-Nitrophenol
p-Hydroxynitrobenzene, see 4-Nitrophenol
4-Hydroxy-3-(3-oxo-1-phenylbutyl)-2H-1-benzopyran-2-one, see Warfarin
4-Hydroxyphenol, see Hydroquinone
p-Hydroxyphenol, see Hydroquinone
4-Hydroxy-3-(1-phenyl-3-oxobutyl)coumarin, see Warfarin
1-Hydroxypropane, see 1-Propanol
3-Hydroxypropene, see Allyl alcohol
3-Hydroxypropionic acid lactone, see β-Propiolactone
3-Hydroxypropylene oxide, see Glycidol
2-Hydroxytoluene, see 2-Methylphenol
4-Hydroxytoluene, see 4-Methylphenol
o-Hydroxytoluene, see 2-Methylphenol
p-Hydroxytoluene, see 4-Methylphenol
α-Hydroxy toluene, see Benzyl alcohol
2-Hydroxytriethylamine, see 2-Diethylaminoethanol
2-Hydroxy-1,3,5-trinitrobenzene, see Picric acid
Hylene T, see 2,4-Toluene diisocyanate
Hylene TCPA, see 2,4-Toluene diisocyanate
Hylene TLC, see 2,4-Toluene diisocyanate
Hylene TM, see 2,4-Toluene diisocyanate
Hylene TM-65, see 2,4-Toluene diisocyanate
Hylene TRF, see 2,4-Toluene diisocyanate
Hytrol O, see Cyclohexanone
IBA, see Isobutyl alcohol

Idryl, see Fluoranthene
IG base, see 4-Nitroaniline
Illoxol, see Dieldrin
Indan
Indenopyren, see Indeno[1,2,3-*cd*]pyrene
Indeno[1,2,3-*cd*]pyrene
Indole
1*H*-Indole, see Indole
Indoline
Inexit, see Lindane
Inhibisol, see 1,1,1-Trichloroethane
Insecticide 497, see Dieldrin
Insecticide 4049, see Malathion
Insectophene, see α-Endosulfan, β-Endosulfan
Inverton 245, see 2,4,5-T
Iodomethane, see Methyl iodide
1-Iodopropane
IP, see Indeno[1,2,3-*cd*]pyrene
Ipaner, see 2,4-D
IPE, see Isopropyl ether
Isceon 11, see Trichlorofluoromethane
Isceon 122, see Dichlorodifluoromethane
Iscobrome, see Methyl bromide
Iscobrome D, see Ethylene dibromide
Isoacetophorone, see Isophorone
Isoamyl acetate
Isoamyl alcohol
α-Isoamylene, see 3-Methyl-1-butene
Isoamyl ethanoate, see Isoamyl acetate
Isoamylhydride, see 2-Methylbutane
Isoamylol, see Isoamyl alcohol
1,3-Isobenzofurandione, see Phthalic anhydride
Isobutane, see 2-Methylpropane
Isobutanol, see Isobutyl alcohol
Isobutene, see 2-Methylpropene
Isobutenyl methyl ketone, see Mesityl oxide
Isobutyl acetate
Isobutyl alcohol
Isobutylbenzene
Isobutyl carbinol, see Isoamyl alcohol
Isobutylene, see 2-Methylpropene
Isobutyl ketone, see Diisobutyl ketone
Isobutyl methyl ketone, see 4-Methyl-2-pentanone

Isobutyltrimethylmethane, see 2,2,4-Trimethylpentane
Isocumene, see *n*-Propylbenzene
Isocyanatomethane, see Methyl isocyanate
Isocyanic acid methyl ester, see Methyl isocyanate
Isocyanic acid methylphenylene ester, see 2,4-Toluene diisocyanate
Isocyanic acid 4-methyl-*m*-phenylene ester, see 2,4-Toluene diisocyanate
Isodrin epoxide, see Endrin
Isoforon, see Isophorone
Isoforone, see Isophorone
Isoheptane, see 2-Methylhexane
Isohexane, see 2-Methylpentane
β-Isomer, see β-BHC
γ-Isomer, see Lindane
Isonitropropane, see 2-Nitropropane
Isooctane, see 2,2,4-Trimethylpentane
Isooctaphenone, see Isophorone
Isopentane, see 2-Methylbutane
Isopentanol, see Isoamyl alcohol
Isopentene, see 3-Methyl-1-butene
Isopentyl acetate, see Isoamyl acetate
Isopentyl alcohol, see Isoamyl alcohol
Isopentyl alcohol acetate, see Isoamyl acetate
Isophoron, see Isophorone
Isophorone
Isoprene, see 2-Methyl-1,3-butadiene
Isopropenylbenzene, see α-Methylstyrene
2-Isopropoxypropane, see Isopropyl ether
Isopropyl acetate
Isopropylacetone, see 4-Methyl-2-pentanone
Isopropylamine
Isopropylbenzene
Isopropylbenzol, see Isopropylbenzene
Isopropylcarbinol, see Isobutyl alcohol
Isopropyldimethylmethane, see 2,3-Dimethylbutane
Isopropyl ether
Isopropylethylene, see 3-Methyl-1-butene
Isopropylidene acetone, see Mesityl oxide
1-Isopropyl-2-methylethene, see 4-Methyl-1-pentene
1-Isopropyl-2-methylethylene, see 4-Methyl-1-pentene
Isotox, see Lindane
Isotron 2, see Dichlorodifluoromethane
Isotron 11, see Trichlorofluoromethane
Isovalerone, see Diisobutyl ketone

Ivalon, see Formaldehyde
Ivoran, see *p,p'*-DDT
Ixodex, see *p,p'*-DDT
Jacutin, see Lindane
Japan camphor, see Camphor
Jeffersol EB, see 2-Butoxyethanol
Jeffersol EE, see 2-Ethoxyethanol
Jeffersol EM, see Methyl cellosolve
Julin's carbon chloride, see Hexachlorobenzene
K III, see 4,6-Dinitro-*o*-cresol
K IV, see 4,6-Dinitro-*o*-cresol
Kaiser chemicals 11, see 1,1,2-Trichlorotrifluoroethane
Kaiser chemicals 12, see Dichlorodifluoromethane
Kamfochlor, see Toxaphene
Kanechlor, see PCB-1260
Karbaspray, see Carbaryl
Karbatox, see Carbaryl
Karbofos, see Malathion
Karbosep, see Carbaryl
Karlan, see Ronnel
Karmex, see Diuron
Karmex diuron herbicide, see Diuron
Karmex DW, see Diuron
Karsan, see Formaldehyde
Kayafume, see Methyl bromide
Kelene, see Chloroethane
Kepone
Ketohexamethylene, see Cyclohexanone
Ketole, see Indole
Ketone propane, see Acetone
β-Ketopropane, see Acetone
2-Keto-1,7,7-trimethylnorcamphane, see Camphor
Killax, see Tetraethyl pyrophosphate
Kilmite 40, see Tetraethyl pyrophosphate
Kodaflex DOP, see Bis(2-ethylhexyl)phthalate
Kokotine, see Lindane
Kolphos, see Parathion
Kopfume, see Ethylene dibromide
Kopsol, see *p,p'*-DDT
KOP-thiodan, see *α*-Endosulfan, *β*-Endosulfan
Kop-thion, see Malathion
Korax, see 1-Chloro-1-nitropropane
Korlan, see Ronnel

Korlane, see Ronnel
Krecalvin, see Dichlorvos
Kregasan, see Thiram
Kresamone, see 4,6-Dinitro-*o*-cresol
p-Kresol, see 4-Methylphenol
Krezotol 50, see 4,6-Dinitro-*o*-cresol
Krotiline, see 2,4-D
Krysid, see ANTU
Krystallin, see Aniline
Kumader, see Warfarin
Kumadu, see Warfarin
Kwell, see Lindane
Kwik-kil, see Strychnine
Kypchlor, see Chlordane
Kyanol, see Aniline
Kypfarin, see Warfarin
Kypfos, see Malathion
Kypthion, see Parathion
L-36352, see Trifluralin
β-Lactone, see β-Propiolactone
Lanadin, see Trichloroethylene
Lanstan, see 1-Chloro-1-nitropropane
Larvacide 100, see Chloropicrin
Laurel camphor, see Camphor
Lauxtol, see Pentachlorophenol
Lauxtol A, see Pentachlorophenol
Lawn-keep, see 2,4-D
Ledon 11, see Trichlorofluoromethane
Ledon 12, see Dichlorodifluoromethane
Lemonene, see Biphenyl
Lendine, see Lindane
Lentox, see Lindane
Lethalaire G-52, see Tetraethyl pyrophosphate
Lethalaire G-54, see Parathion
Lethalaire G-57, see Sulfotepp
Lethurin, see Trichloroethylene
Lidenal, see Lindane
Lilly 36352, see Trifluralin
Lindafor, see Lindane
Lindagam, see Lindane
Lindagrain, see Lindane
Lindagranox, see Lindane
Lindan, see Dichlorvos

Lindane
α-Lindane, see α-BHC
β-Lindane, see β-BHC
δ-Lindane, see δ-BHC
γ-Lindane, see Lindane
Lindapoudre, see Lindane
Lindatox, see Lindane
Lindosep, see Lindane
Line rider, see 2,4,5-T
Lintox, see Lindane
Lipan, see 4,6-Dinitro-*o*-cresol
Liroprem, see Pentachlorophenol
Liqua-tox, see Warfarin
Liquified petroleum gas, see 2-Methylpropane
Lirohex, see Tetraethyl pyrophosphate
Lirothion, see Parathion
Lorexane, see Lindane
Lorsban, see Chlorpyrifos
LPG ethyl mercaptan 1010, see Ethyl mercaptan
Lyddite, see Picric acid
Lysoform, see Formaldehyde
M 140, see Chlordane
M 410, see Chlordane
M 5055, see Toxaphene
MA, see Methylaniline
MAAC, see *sec*-Hexyl acetate
Mace, see α-Chloroacetophenone
Macrondray, see 2,4-D
Mafu, see Dichlorvos
Mafu strip, see Dichlorvos
Magnacide, see Acrolein
Malacide, see Malathion
Malafor, see Malathion
Malagran, see Malathion
Malakill, see Malathion
Malamar, see Malathion
Malamar 50, see Malathion
Malaphele, see Malathion
Malaphos, see Malathion
Malasol, see Malathion
Malaspray, see Malathion
Malathion
Malathion E50, see Malathion

Malathion LV concentrate, see Malathion
Malathion ULV concentrate, see Malathion
Malathiozoo, see Malathion
Malathon, see Malathion
Malathyl LV concentrate & ULV concentrate, see Malathion
Malatol, see Malathion
Malatox, see Malathion
Maldison, see Malathion
Maleic acid anhydride, see Maleic anhydride
Maleic anhydride
Malix, see α-Endosulfan, β-Endosulfan
Malmed, see Malathion
Malphos, see Malathion
Maltox, see Malathion
Maltox MLT, see Malathion
Maralate, see Methoxychlor
Mar-frin, see Warfarin
Marlate, see Methoxychlor
Marlate 50, see Methoxychlor
Marmer, see Diuron
Maroxol-50, see 2,4-Dinitrophenol
Martin's mar-frin, see Warfarin
Marvex, see Dichlorvos
Matricaria camphor, see Camphor
Maveran, see Warfarin
MB, see Methyl bromide
M-B-C fumigant, see Methyl bromide
MBK, see 2-Hexanone
MBX, see Methyl bromide
MCB, see Chlorobenzene
ME-1700, see *p,p′*-DDD
MEA, see Ethanolamine
MEBR, see Methyl bromide
MECS, see Methyl cellosolve
Meetco, see 2-Butanone
MEK, see 2-Butanone
Melinite, see Picric acid
Melipax, see Toxaphene
Mendrin, see Endrin
Meniphos, see Mevinphos
Menite, see Mevinphos
1-Mercaptobutane, see *n*-Butyl mercaptan
Mercaptomethane, see Methyl mercaptan

Mercaptosuccinic acid diethyl ester, see Malathion
Mercaptothion, see Malathion
Mercuram, see Thiram
Mercurialin, see Methylamine
Merex, see Kepone
Mesitylene, see 1,3,5-Trimethylbenzene
Mesityl oxide
Metafume, see Methyl bromide
Methacide, see Toluene
Methacrylic acid methyl ester, see Methyl methacrylate
Methanal, see Formaldehyde
Methanamine, see Methylamine
Methanecarbonitrile, see Acetonitrile
Methanecarboxylic acid, see Acetic acid
Methane dichloride, see Methylene chloride
Methane tetrachloride, see Carbon tetrachloride
Methanethiol, see Methyl mercaptan
Methane trichloride, see Chloroform
6,9-Methano-2,4,3-benzodioxathiepin, see Endosulfan sulfate
Methanoic acid, see Formic acid
Methanol
Methenyl chloride, see Chloroform
Methenyl tribromide, see Bromoform
Methenyl trichloride, see Chloroform
Methogas, see Methyl bromide
Methoxcide, see Methoxychlor
Methoxo, see Methoxychlor
2-Methoxy-1-aminobenzene, see *o*-Anisidine
4-Methoxy-1-aminobenzene, see *p*-Anisidine
2-Methoxyaniline, see *o*-Anisidine
4-Methoxyaniline, see *p*-Anisidine
o-Methoxyaniline, see *o*-Anisidine
p-Methoxyaniline, see *p*-Anisidine
2-Methoxybenzenamine, see *o*-Anisidine
4-Methoxybenzenamine, see *p*-Anisidine
p-Methoxybenzenamine, see *p*-Anisidine
Methoxycarbonylethylene, see Methyl acrylate
2-Methoxycarbonyl-1-methylvinyl dimethyl phosphate, see Mevinphos
cis-2-Methoxycarbonyl-1-methylvinyl dimethyl phosphate, see Mevinphos
1-Methoxycarbonyl-1-propen-2-yl dimethyl phosphate, see Mevinphos
Methoxychlor
4,4′-Methoxychlor, see Methoxychlor
p,p′-Methoxychlor, see Methoxychlor

Methoxy-DDT, see Methoxychlor
2-Methoxyethanol, see Methyl cellosolve
2-Methoxyethanol acetate, see Methyl cellosolve acetate
2-Methoxyethyl acetate, see Methyl cellosolve acetate
Methoxyhydroxyethane, see Methyl cellosolve
4-Methoxyphenylamine, see *p*-Anisidine
o-Methoxyphenylamine, see *o*-Anisidine
p-Methoxyphenylamine, see *p*-Anisidine
Methyl acetate
Methyl acetone, see 2-Butanone
Methyl acetylene, see Propyne
β-Methylacrolein, see Crotonaldehyde
Methyl acrylate
Methylal
Methyl alcohol, see Methanol
Methyl aldehyde, see Formaldehyde
Methylamine
(Methylamino)benzene, see Methylaniline
1-Methyl-2-aminobenzene, see *o*-Toluidine
2-Methyl-1-aminobenzene, see *o*-Toluidine
N-Methylaminobenzene, see Methylaniline
Methylamyl acetate, see *sec*-Hexyl acetate
Methyl amyl ketone, see 2-Heptanone
Methyl *n*-amyl ketone, see 2-Heptanone
Methylaniline
2-Methylaniline, see *o*-Toluidine
N-Methylaniline, see Methylaniline
o-Methylaniline, see *o*-Toluidine
2-Methylanthracene
β-Methylanthracene, see 2-Methylanthracene
2-Methylbenzenamine, see *o*-Toluidine
N-Methylbenzenamine, see Methylaniline
o-Methylbenzenamine, see *o*-Toluidine
Methylbenzene, see Toluene
Methylbenzol, see Toluene
β-Methylbivinyl, see 2-Methyl-1,3-butadiene
Methyl bromide
2-Methylbutadiene, see 2-Methyl-1,3-butadiene
2-Methyl-1,3-butadiene
2-Methylbutane
2-Methyl-4-butanol, see Isoamyl alcohol
3-Methylbutanol, see Isoamyl alcohol
3-Methyl-1-butanol, see Isoamyl alcohol

3-Methylbutan-1-ol, see Isoamyl alcohol
3-Methyl-1-butanol acetate, see Isoamyl acetate
3-Methyl-1-butene
1-Methylbutyl acetate, see *sec*-Amyl acetate
3-Methylbutyl acetate, see Isoamyl acetate
3-Methyl-1-butyl acetate, see Isoamyl acetate
3-Methylbutyl ethanoate, see Isoamyl acetate
Methyl *n*-butyl ketone, see 2-Hexanone
Methyl carbamate-1-naphthalenol, see Carbaryl
Methyl carbamate-1-naphthol, see Carbaryl
Methyl carbamic acid 2,3-dihydro-2,2-dimethyl-7-benzofuranyl ester, see Carbo-
    furan
Methyl carbamic acid 1-naphthyl ester, see Carbaryl
Methyl cellosolve
Methyl cellosolve acetate
Methyl chloride
Methyl chloroform, see 1,1,1-Trichloroethane
3-Methyl-4-chlorophenol, see *p*-Chloro-*m*-cresol
Methyl cyanide, see Acetonitrile
Methylcyclohexane
2-Methylcyclohexanone, see *o*-Methylcyclohexanone
o-**Methylcyclohexanone**
1-Methylcyclohexene
1-Methyl-1-cyclohexene, see 1-Methylcyclohexene
Methylcyclopentane
Methyl 3-(dimethoxyphosphinyloxy)crotonate, see Mevinphos
1-Methyl-2,4-dinitrobenzene, see 2,4-Dinitrotoluene
2-Methyl-1,3-dinitrobenzene, see 2,6-Dinitrotoluene
2-Methyl-4,6-dinitrophenol, see 4,6-Dinitro-*o*-cresol
6-Methyl-2,4-dinitrophenol, see 4,6-Dinitro-*o*-cresol
Methylene bichloride, see Methylene chloride
Methylenebiphenyl, see Fluorene
2,2'-Methylenebiphenyl, see Fluorene
1,1'-[Methylenebis(oxy)]bis(2-chloroethane), see Bis(2-chloroethoxy)methane
1,1'-[Methylenebis(oxy)]bis(2-chloroformaldehyde), see Bis(2-chloroethoxy)meth-
    ane
Methylene chloride
Methylene chlorobromide, see Bromochloromethane
Methylene dichloride, see Methylene chloride
Methylene dimethyl ether, see Methylal
Methylene glycol, see Formaldehyde
Methylene oxide, see Formaldehyde
Methyl ethanoate, see Methyl acetate

(1-Methylethenyl)benzene, see α-Methylstyrene
Methyl ethoxol, see Methyl cellosolve
1-Methylethylamine, see Isopropylamine
(1-Methylethyl)benzene, see Isopropylbenzene
Methylethylcarbinol, see *sec*-Butyl alcohol
Methyl ethylene oxide, see Propylene oxide
Methyl ethyl ketone, see 2-Butanone
Methylethylmethane, see Butane
N-(1-Methylethyl)-2-propanamine, see Diisopropylamine
Methyl formal, see Methylal
Methyl formate
Methyl glycol, see Methyl cellosolve
Methyl glycol acetate, see Methyl cellosolve acetate
Methyl glycol monoacetate, see Methyl cellosolve acetate
3-Methylheptane
5-Methylheptane, see 3-Methylheptane
3-Methyl-5-heptanone, see 5-Methyl-3-heptanone
5-Methyl-3-heptanone
2-Methylhexane
3-Methylhexane
4-Methylhexane, see 3-Methylhexane
Methylhydrazine
1-Methylhydrazine, see Methylhydrazine
Methyl hydroxide, see Methanol
1-Methyl-4-hydroxybenzene, see 4-Methylphenol
2-Methylhydroxybenzene, see 2-Methylphenol
4-Methylhydroxybenzene, see 4-Methylphenol
o-Methylhydroxybenzene, see 2-Methylphenol
p-Methylhydroxybenzene, see 4-Methylphenol
Methyl iodide
Methylisoamyl acetate, see *sec*-Hexyl acetate
Methyl isobutenyl ketone, see Mesityl oxide
Methylisobutylcarbinol acetate, see *sec*-Hexyl acetate
Methyl isobutyl ketone, see 4-Methyl-2-pentanone
Methyl isocyanate
Methyl ketone, see Acetone
Methyl mercaptan
Methyl methacrylate
Methyl methacrylate monomer, see Methyl methacrylate
N-Methylmethanamine, see Dimethylamine
Methyl methanoate, see Methyl formate
Methyl-α-methylacrylate, see Methyl methacrylate
Methyl-2-methyl-2-propenoate, see Methyl methacrylate

2-Methylnaphthalene
β-Methylnaphthalene, see 2-Methylnaphthalene
N-Methyl-1-naphthyl carbamate, see Carbaryl
N-Methyl-α-naphthyl carbamate, see Carbaryl
N-Methyl-α-naphthylurethan, see Carbaryl
1-Methyl-2-nitrobenzene, see 2-Nitrotoluene
1-Methyl-3-nitrobenzene, see 3-Nitrotoluene
1-Methyl-4-nitrobenzene, see 4-Nitrotoluene
2-Methylnitrobenzene, see 2-Nitrotoluene
3-Methylnitrobenzene, see 3-Nitrotoluene
4-Methylnitrobenzene, see 4-Nitrotoluene
m-Methylnitrobenzene, see 3-Nitrotoluene
o-Methylnitrobenzene, see 2-Nitrotoluene
p-Methylnitrobenzene, see 4-Nitrotoluene
n-Methyl-N-nitrosomethanamine, see N-Nitrosodimethylamine
4-Methyloctane
5-Methyloctane, see 4-Methyloctane
Methylol, see Methanol
Methylolpropane, see 1-Butanol
Methyloxirane, see Propylene oxide
Methyl oxitol, see Methyl cellosolve
2-Methylpentane
3-Methylpentane
4-Methyl-2-pentanol acetate, see sec-Hexyl acetate
2-Methyl-2-pentanol-4-one, see Diacetone alcohol
2-Methyl-4-pentanone, see 4-Methyl-2-pentanone
4-Methyl-2-pentanone
2-Methylpentene, see 2-Methyl-1-pentene
2-Methyl-1-pentene
4-Methyl-1-pentene
2-Methyl-2-penten-4-one, see Mesityl oxide
4-Methyl-2-pentene-2-one, see Mesityl oxide
4-Methyl-2-pentyl acetate, see sec-Hexyl acetate
1-Methylphenanthrene
α-Methylphenanthrene, see 1-Methylphenanthrene
2-Methylphenol
4-Methylphenol
o-Methylphenol, see 2-Methylphenol
p-Methylphenol, see 4-Methylphenol
Methylphenylamine, see Methylaniline
N-Methylphenylamine, see Methylaniline
4-Methylphenylene diisocyanate, see 2,4-Toluene diisocyanate
4-Methylphenylene isocyanate, see 2,4-Toluene diisocyanate

1-Methyl-1-phenylethylene, see α-Methylstyrene
o-Methylphenylol, see 2-Methylphenol
2-Methyl-1-phenylpropane, see Isobutylbenzene
2-Methyl-2-phenylpropane, see *tert*-Butylbenzene
Methyl phthalate, see Dimethyl phthalate
2-Methylpropane
2-Methylpropanol, see Isobutyl alcohol
2-Methylpropanol-1, see Isobutyl alcohol
2-Methyl-1-propanol, see Isobutyl alcohol
2-Methyl-1-propan-1-ol, see Isobutyl alcohol
2-Methyl-2-propanol, see *tert*-Butyl alcohol
Methyl propenate, see Methyl acrylate
Methylpropene, see 2-Methylpropene
2-Methylpropene
2-Methyl-1-propene, see 2-Methylpropene
Methyl propenoate, see Methyl acrylate
Methyl-2-propenoate, see Methyl acrylate
1-Methylpropyl acetate, see *sec*-Butyl acetate
2-Methylpropyl acetate, see Isobutyl acetate
2-Methyl-1-propyl acetate, see Isobutyl acetate
2-Methylpropyl alcohol, see Isobutyl alcohol
(1-Methylpropyl)benzene, see *sec*-Butylbenzene
(2-Methylpropyl)benzene, see Isobutylbenzene
2-Methylpropylene, see 2-Methylpropene
1-Methyl-1-propylethene, see 2-Methyl-1-pentene
1-Methyl-1-propylethylene, see 2-Methyl-1-pentene
β-Methylpropyl ethanoate, see Isobutyl acetate
Methyl propyl ketone, see 2-Pentanone
Methyl *n*-propyl ketone, see 2-Pentanone
α-Methylstyrene
Methyl sulfate, see Dimethyl sulfate
Methyl sulfhydrate, see Methyl mercaptan
N-Methyl-N,2,4,6-tetranitroaniline, see Tetryl
N-Methyl-N,2,4,6-tetranitrobenzenamine, see Tetryl
Methyl thioalcohol, see Methyl mercaptan
Methyl thiram, see Thiram
Methyl thiuramdisulfide, see Thiram
m-Methyltoluene, see m-Xylene
o-Methyltoluene, see o-Xylene
p-Methyltoluene, see p-Xylene
Methyl tribromide, see Bromoform
Methyl trichloride, see Chloroform
Methyltrichloromethane, see 1,1,1-Trichloroethane

1-Methyl-2,4,6-trinitrobenzene, see 2,4,6-Trinitrotoluene
2-Methyl-1,3,5-trinitrobenzene, see 2,4,6-Trinitrotoluene
Methyl tuads, see Thiram
Methyl yellow, see *p*-Dimethylaminoazobenzene
Metox, see Methoxychlor
Mevinphos
MIBK, see 4-Methyl-2-pentanone
Microlysin, see Chloropicrin
Mighty 150, see Naphthalene
Mighty RD1, see Naphthalene
MIK, see 4-Methyl-2-pentanone
Mil-B-4394-B, see Bromochloromethane
Milbol 49, see Lindane
Miller's fumigrain, see Acrylonitrile
Mineral naphthalene, see Benzene
Mipax, see Dimethyl phthalate
Miracle, see 2,4-D
Mirbane oil, see Nitrobenzene
MLT, see Malathion
MME, see Methyl methacrylate
MMH, see Methylhydrazine
MNA, see 3-Nitroaniline
MNBK, see 2-Hexanone
MNT, see 3-Nitrotoluene
Mole death, see Strychnine
Mole-nots, see Strychnine
Mollan 0, see Bis(2-ethylhexyl)phthalate
Mondur TD, see 2,4-Toluene diisocyanate
Mondur TD-80, see 2,4-Toluene diisocyanate
Mondur TDS, see 2,4-Toluene diisocyanate
Monobromobenzene, see Bromobenzene
Monobromoethane, see Ethyl bromide
Monobromomethane, see Methyl bromide
Monobromotrifluoromethane, see Bromotrifluoromethane
Monobutylamine, see *n*-Butylamine
Mono-*n*-butylamine, see *n*-Butylamine
Monochlorbenzene, see Chlorobenzene
Monochlorethane, see Chloroethane
Monochloroacetaldehyde, see Chloroacetaldehyde
Monochlorobenzene, see Chlorobenzene
Monochlorodiphenyl oxide, see 4-Chlorophenyl phenyl ether
Monochloroethane, see Chloroethane
2-Monochloroethanol, see Ethylene chlorohydrin

Monochloroethene, see Vinyl chloride
Monochloroethylene, see Vinyl chloride
Monochloromethane, see Methyl chloride
"Monocite" methacrylate monomer, see Methyl methacrylate
Monoethanolamine, see Ethanolamine
Monoethylamine, see Ethylamine
Monofluorotrichloromethane, see Trichlorofluoromethane
Monohydroxybenzene, see Phenol
Monohydroxymethane, see Methanol
Monoisopropylamine, see Isopropylamine
Monomethylamine, see Methylamine
Monomethylaniline, see Methylaniline
N-Monomethylaniline, see Methylaniline
Monomethylhydrazine, see Methylhydrazine
Monosan, see 2,4-D
Monsanto penta, see Pentachlorophenol
Moosuran, see Pentachlorophenol
Mopari, see Dichlorvos
Morbicid, see Formaldehyde
Morpholine
Mortopal, see Tetraethyl pyrophosphate
Moscardia, see Malathion
Moth balls, see Naphthalene
Moth flakes, see Naphthalene
Motor benzol, see Benzene
Motox, see Toxaphene
Mottenhexe, see Hexachloroethane
Mouse pak, see Warfarin
Mouse-rid, see Strychnine
Mouse-tox, see Strychnine
Moxie, see Methoxychlor
Moxone, see 2,4-D
MPK, see 2-Pentanone
Mszycol, see Lindane
Murfos, see Parathion
Muriatic ether, see Chloroethane
Muthmann's liquid, see 1,1,2,2-Tetrabromoethane
Mutoxin, see *p,p'*-DDT
MVC, see Vinyl chloride
NA 1120, see 1-Butanol
NA 1230, see Methanol
NA 1247, see Methyl methacrylate
NA 1583, see Chloropicrin

β-Naphthylamine, see 2-Naphthylamine
2-Naphthylamine mustard, see 2-Naphthylamine
1,2-(1,8-Naphthylene)benzene, see Fluoranthene
1-Naphthyl methyl carbamate, see Carbaryl
1-Naphthyl-*N*-methyl carbamate, see Carbaryl
α-Naphthyl-*N*-methyl carbamate, see Carbaryl
α-Naphthylthiocarbamide, see ANTU
1-(1-Naphthyl)-2-thiourea, see ANTU
N-1-Naphthylthiourea, see ANTU
α-Naphthylthiourea, see ANTU
Narcogen, see Trichloroethylene
Narcotil, see Methylene chloride
Narcotile, see Chloroethane
Narkogen, see Trichloroethylene
Narkosoid, see Trichloroethylene
Natasol fast orange GR salt, see 2-Nitroaniline
Naugard TJB, see *N*-Nitrosodiphenylamine
Naxol, see Cyclohexanol
NBA, see 1-Butanol
NCI-C00044, see Aldrin
NCI-C00099, see Chlordane
NCI-C00113, see Dichlorvos
NCI-C00124, see Dieldrin
NCI-C00157, see Endrin
NCI-C00180, see Heptachlor
NCI-C00191, see Kepone
NCI-C00204, see Lindane
NCI-C00215, see Malathion
NCI-C00226, see Parathion
NCI-C00259, see Toxaphene
NCI-C00442, see Trifluralin
NCI-C00464, see *p,p*′-DDT
NCI-C00475, see *p,p*′-DDD
NCI-C00497, see Methoxychlor
NCI-C00500, see 1,2-Dibromo-3-chloropropane
NCI-C00511, see 1,2-Dichloroethane
NCI-C00522, see Ethylene dibromide
NCI-C00533, see Chloropicrin
NCI-C00555, see *p,p*′-DDE
NCI-C00566, see α-Endosulfan, β-Endosulfan
NCI-C01854, see 1,2-Diphenylhydrazine
NCI-C01865, see 2,4-Dinitrotoluene
NCI-C02039, see 4-Chloroaniline

NCI-C02200, see Styrene
NCI-C02664, see PCB-1254
NCI-C02686, see Chloroform
NCI-C02799, see Formaldehyde
NCI-C02880, see *N*-Nitrosodiphenylamine
NCI-C02904, see 2,4,6-Trichlorophenol
NCI-C03054, see 1,3-Dichloro-5,5-dimethylhydantoin
NCI-C03361, see Benzidine
NCI-C03554, see 1,1,2,2-Tetrachloroethane
NCI-C03601, see Phthalic anhydride
NCI-C03689, see 1,4-Dioxane
NCI-C03714, see TCDD
NCI-CO3736, see Aniline
NCI-C04535, see 1,1-Dichloroethane
NCI-C04546, see Trichloroethylene
NCI-C04579, see 1,1,2-Trichloroethane
NCI-C04580, see Tetrachloroethylene
NCI-C04591, see Carbon disulfide
NCI-C04604, see Hexachloroethane
NCI-C04615, see Allyl chloride
NCI-C04626, see 1,1,1-Trichloroethane
NCI-C04637, see Trichlorofluoromethane
NCI-C06111, see Benzyl alcohol
NCI-C06224, see Chloroethane
NCI-C06360, see Benzyl chloride
NCI-C07272, see Toluene
NCI-C50044, see Bis(2-chloroisopropyl)ether
NCI-C50099, see Propylene oxide
NCI-C50102, see Methylene chloride
NCI-C50124, see Phenol
NCI-C50135, see Ethylene chlorohydrin
NCI-C50384, see Ethyl acrylate
NCI-C50533, see 2,4-Toluene diisocyanate
NCI-C50602, see 1,3-Butadiene
NCI-C50680, see Methyl methacrylate
NCI-C52733, see Bis(2-ethylhexyl)phthalate
NCI-C52904, see Naphthalene
NCI-C53894, see Pentachloroethane
NCI-C54262, see 1,1-Dichloroethylene
NCI-C54375, see Benzyl butyl phthalate
NCI-C54853, see 2-Ethoxyethanol
NCI-C54886, see Chlorobenzene
NCI-C54933, see Pentachlorophenol

NCI-C54944, see 1,2-Dichlorobenzene
NCI-C54955, see 1,4-Dichlorobenzene
NCI-C55005, see Cyclohexanone
NCI-C55107, see α-Chloroacetophenone
NCI-C55118, see o-Chlorobenzylidenemalononitrile
NCI-C55130, see Bromoform
NCI-C55141, see 1,2-Dichloropropane
NCI-C55209, see Oxalic acid
NCI-C55243, see Bromodichloromethane
NCI-C55254, see Dibromochloromethane
NCI-C55276, see Benzene
NCI-C55301, see Pyridine
NCI-C55345, see 2,4-Dichlorophenol
NCI-C55367, see tert-Butyl alcohol
NCI-C55378, see Pentachlorophenol
NCI-C55481, see Ethyl bromide
NCI-C55492, see Bromobenzene
NCI-C55549, see Glycidol
NCI-C55607, see Hexachlorocyclopentadiene
NCI-C55618, see Isophorone
NCI-C55834, see Hydroquinone
NCI-C55845, see p-Quinone
NCI-C55947, see Tetranitromethane
NCI-C55992, see 4-Nitrophenol
NCI-C56155, see 2,4,6-Trinitrotoluene
NCI-C56177, see Furfural
NCI-C56224, see Furfuryl alcohol
NCI-C56279, see Crotonaldehyde
NCI-C56326, see Acetaldehyde
NCI-C56393, see Ethylbenzene
NCI-C56428, see Dimethylaniline
NCI-C56655, see Pentachlorophenol
NCI-C56666, see Allyl glycidyl ether
NCI-C60048, see Diethyl phthalate
NCI-C60082, see Nitrobenzene
NCI-C60220, see 1,2,3-Trichloropropane
NCI-C60402, see Ethylenediamine
NCI-C60537, see 4-Nitrotoluene
NCI-C60560, see Tetrahydrofuran
NCI-C60571, see Hexane
NCI-C60786, see 4-Nitroaniline
NCI-C60822, see Acetonitrile
NCI-C60866, see n-Butyl mercaptan

Niax TDI-P, see 2,4-Toluene diisocyanate
Nicochloran, see Lindane
Nifos, see Tetraethyl pyrophosphate
Nifos T, see Tetraethyl pyrophosphate
Nifost, see Tetraethyl pyrophosphate
Nipar S-20 solvent, see 2-Nitropropane
Nipar S-30 solvent, see 2-Nitropropane
Niran, see Chlordane, Parathion
Niran E-4, see Parathion
Nitrador, see 4,6-Dinitro-*o*-cresol
Nitramine, see Tetryl
Nitran, see Trifluralin
Nitranilin, see 3-Nitroaniline
4-Nitraniline, see 4-Nitroaniline
m-Nitraniline, see 3-Nitroaniline
o-Nitraniline, see 2-Nitroaniline
p-Nitraniline, see 4-Nitroaniline
Nitrapyrin
Nitration benzene, see Benzene
Nitrazol 2F extra, see 4-Nitroaniline
Nitric acid propyl ester, see *n*-Propyl nitrate
Nitrile, see Acrylonitrile
3-Nitroaminobenzene, see 3-Nitroaniline
m-Nitroaminobenzene, see 3-Nitroaniline
2-Nitroaniline
3-Nitroaniline
4-Nitroaniline
m-Nitroaniline, see 3-Nitroaniline
o-Nitroaniline, see 2-Nitroaniline
p-Nitroaniline, see 4-Nitroaniline
2-Nitrobenzenamine, see 2-Nitroaniline
3-Nitrobenzenamine, see 3-Nitroaniline
m-Nitrobenzenamine, see 3-Nitroaniline
4-Nitrobenzenamine, see 4-Nitroaniline
p-Nitrobenzenamine, see 4-Nitroaniline
Nitrobenzene
Nitrobenzol, see Nitrobenzene
4-Nitrobiphenyl
4-Nitro-1,1'-biphenyl, see 4-Nitrobiphenyl
p-Nitrobiphenyl, see 4-Nitrobiphenyl
Nitrocarbol, see Nitromethane
4-Nitrochlorobenzene, see *p*-Chloronitrobenzene
p-Nitrochlorobenzene, see *p*-Chloronitrobenzene

No bunt 40, see Hexachlorobenzene
No bunt 80, see Hexachlorobenzene
No bunt liquid, see Hexachlorobenzene
Nogos, see Dichlorvos
Nogos 50, see Dichlorvos
Nogos G, see Dichlorvos
Nomersan, see Thiram
Nonane
n-Nonane, see Nonane
Nonyl hydride, see Nonane
No-pest, see Dichlorvos
No-pest strip, see Dichlorvos
Norcamphor, see Camphor
Normersan, see Thiram
Norvalamine, see *n*-Butylamine
Nourithion, see Parathion
Novigam, see Lindane
2-NP, see 2-Nitropropane
NSC 423, see 2,4-D
NSC 1532, see 2,4-Dinitrophenol
NSC 1771, see Thiram
NSC 3138, see *N,N*-Dimethylacetamide
NSC 5536, see *N,N*-Dimethylformamide
NSC 6738, see Dichlorvos
NSC 8819, see Acrolein
NSC 21626, see *β*-Propiolactone
N-serve, see Nitrapyrin
N-serve nitrogen stabilizer, see Nitrapyrin
NTM, see Dimethyl phthalate
Nuoplaz DOP, see Bis(2-ethylhexyl)phthalate
Nurelle, see 2,4,5-Trichlorophenol
Nuva, see Dichlorvos
Nuvan, see Dichlorvos
Nuvan 100EC, see Dichlorvos
OCBM, see *o*-Chlorobenzylidenemalononitrile
Octachlor, see Chlordane
1,2,4,5,6,7,8,8-Octachlor-2,3,3a,4,7,7a-hexahydro-4,7-methanoindene, see Chlordane
Octachlorocamphene, see Toxaphene
Octachlorodihydrodicyclopentadiene, see Chlordane
1,2,4,5,6,7,8,8-Octachloro-2,3,3a,4,7,7a-hexahydro-4,7-methylene indane, see Chlordane
Octachloro-4,7-methanohydroindane, see Chlordane

Octachloro-4,7-methanotetrahydroindane, see Chlordane
1,2,4,5,6,7,8,8-Octachloro-4,7-methano-3a,4,7,7a-tetrahydroindane, see Chlordane
Octachloronaphthalene
1,2,4,5,6,7,8,8-Octachloro-3a,4,7,7a-tetrahydro-4,7-methanoindan, see Chlordane, *trans*-Chlordane
α-1,2,4,5,6,7,8,8-Octachloro-3a,4,7,7a-tetrahydro-4,7-methanoindan, see Chlordane
1,2,4,5,6,7,8,8-Octachloro-3a,4,7,7a-tetrahydro-4,7-methanoindane, see Chlordane
1,2,4,5,6,7,8,8-Octachloro-3a,4,7,7a-tetrahydro-4,7-methyleneindane, see Chlordane
Octaklor, see Chlordane
Octalene, see Aldrin
Octalox, see Dieldrin
Octane
n-Octane, see Octane
Octaterr, see Chlordane
1-Octene
Octoil, see Bis(2-ethylhexyl)phthalate
1-Octylene, see 1-Octene
Octyl phthalate, see Bis(2-ethylhexyl)phthalate, Di-*n*-octyl phthalate
n-Octyl phthalate, see Di-*n*-octyl phthalate
ODB, see 1,2-Dichlorobenzene
ODCB, see 1,2-Dichlorobenzene
Oil of bitter almonds, see Nitrobenzene
Oil of mirbane, see Nitrobenzene
Oil of myrbane, see Nitrobenzene
Oil yellow, see *p*-Dimethylaminoazobenzene
Oil yellow 20, see *p*-Dimethylaminoazobenzene
Oil yellow 2625, see *p*-Dimethylaminoazobenzene
Oil yellow 7463, see *p*-Dimethylaminoazobenzene
Oil yellow BB, see *p*-Dimethylaminoazobenzene
Oil yellow D, see *p*-Dimethylaminoazobenzene
Oil yellow DN, see *p*-Dimethylaminoazobenzene
Oil yellow FF, see *p*-Dimethylaminoazobenzene
Oil yellow FN, see *p*-Dimethylaminoazobenzene
Oil yellow G, see *p*-Dimethylaminoazobenzene
Oil yellow G-2, see *p*-Dimethylaminoazobenzene
Oil yellow 2G, see *p*-Dimethylaminoazobenzene
Oil yellow GG, see *p*-Dimethylaminoazobenzene
Oil yellow GR, see *p*-Dimethylaminoazobenzene
Oil yellow II, see *p*-Dimethylaminoazobenzene
Oil yellow N, see *p*-Dimethylaminoazobenzene
Oil yellow PEL, see *p*-Dimethylaminoazobenzene

Oko, see Dichlorvos
Olamine, see Ethanolamine
Oleal yellow 2G, see *p*-Dimethylaminoazobenzene
Oleofos 20, see Parathion
Oleoparaphene, see Parathion
Oleoparathion, see Parathion
Oleophosphothion, see Malathion
Olitref, see Trifluralin
Omal, see 2,4,6-Trichlorophenol
Omchlor, see 1,3-Dichloro-5,5-dimethylhydantoin
OMS-14, see Dichlorvos
OMS-29, see Carbaryl
OMS-570, see α-Endosulfan, β-Endosulfan
OMS-971, see Chlorpyrifos
Omnitox, see Lindane
ONA, see 2-Nitroaniline
ONP, see 2-Nitrophenol
ONT, see 2-Nitrotoluene
Optal, see 1-Propanol
Orange base CIBA II, see 2-Nitroaniline
Orange base IRGA I, see 3-Nitroaniline
Orange base IRGA II, see 2-Nitroaniline
Orange GRS salt, see 2-Nitroaniline
Orange salt CIBA II, see 2-Nitroaniline
Orange salt IRGA II, see 2-Nitroaniline
Organol yellow ADM, see *p*-Dimethylaminoazobenzene
Orient oil yellow GG, see *p*-Dimethylaminoazobenzene
Orsin, see *p*-Phenylenediamine
Ortho 4355, see Naled
Orthocresol, see 2-Methylphenol
Orthodibrom, see Naled
Orthodibromo, see Naled
Orthodichlorobenzene, see 1,2-Dichlorobenzene
Orthodichlorobenzol, see 1,2-Dichlorobenzene
Orthoklor, see Chlordane
Orthomalathion, see Malathion
Orthonitroaniline, see 2-Nitroaniline
Orthophos, see Parathion
Orvinylcarbinol, see Allyl alcohol
OS 1987, see 1,2-Dibromo-3-chloropropane
OS 2046, see Mevinphos
Osmosol extra, see 1-Propanol
Ottafact, see *p*-Chloro-*m*-cresol

Ovadziak, see Lindane
Owadziak, see Lindane
1-Oxa-4-azacyclohexane, see Morpholine
Oxacyclopentane, see Tetrahydrofuran
Oxalic acid
2-Oxetanone, see β-Propiolactone
Oxidation base 10, see p-Phenylenediamine
Oxiranemethanol, see Glycidol
Oxiranylmethanol, see Glycidol
Oxitol, see 2-Ethoxyethanol
2-Oxobornane, see Camphor
Oxolane, see Tetrahydrofuran
Oxomethane, see Formaldehyde
Oxybenzene, see Phenol
1,1'-Oxybisbenzene, see Phenyl ether
1,1'-Oxybis(2-chloroethane), see Bis(2-chloroethyl)ether
Oxybis(chloromethane), see sym-Dichloromethyl ether
2,2'-Oxybis(1-chloropropane), see Bis(2-chloroisopropyl)ether
1,1'-Oxybis(ethane), see Ethyl ether
2,2'-Oxybis(propane), see Isopropyl ether
Oxy DBCP, see 1,2-Dibromo-3-chloropropane
Oxymethylene, see Formaldehyde
Oxytol acetate, see 2-Ethoxyethyl acetate
o-Oxytoluene, see 2-Methylphenol
p-Oxytoluene, see 4-Methylphenol
Pac, see Parathion
Palatinol A, see Diethyl phthalate
Palatinol AH, see Bis(2-ethylhexyl)phthalate
Palatinol BB, see Benzyl butyl phthalate
Palatinol C, see Di-n-butyl phthalate
Palatinol M, see Dimethyl phthalate
Panam, see Carbaryl
Panoram 75, see Thiram
Panoram D-31, see Dieldrin
Panthion, see Parathion
Para, see p-Phenylenediamine
Parachlorocidum, see p,p'-DDT
Paracide, see 1,4-Dichlorobenzene
Para-cresol, see 4-Methylphenol
Para crystals, see 1,4-Dichlorobenzene
Paradi, see 1,4-Dichlorobenzene
Paradichlorobenzene, see 1,4-Dichlorobenzene
Paradichlorobenzol, see 1,4-Dichlorobenzene

Paradow, see 1,4-Dichlorobenzene
Paradust, see Parathion
Paraflow, see Parathion
Paraform, see Formaldehyde
Paramar, see Parathion
Paramar 50, see Parathion
Paramethylphenol, see 4-Methylphenol
Paraminodiphenyl, see 4-Aminobiphenyl
Paramoth, see 1,4-Dichlorobenzene
Paranaphthalene, see Anthracene
Paranuggetts, see 1,4-Dichlorobenzene
Paraphos, see Parathion
Paraspray, see Parathion
Parathene, see Parathion
Parathion
Parathion-ethyl, see Parathion
Parawet, see Parathion
Parazene, see 1,4-Dichlorobenzene
Parmetol, see p-Chloro-m-cresol
Parodi, see 1,4-Dichlorobenzene
Parol, see p-Chloro-m-cresol
PCB-1016
PCB-1221
PCB-1232
PCB-1242
PCB-1248
PCB-1254
PCB-1260
PCC, see Toxaphene
PCE, see Tetrachloroethylene
PCL, see Hexachlorocyclopentadiene
PCMC, see p-Chloro-m-cresol
PCNB, see p-Chloronitrobenzene
PCP, see Pentachlorophenol
PD 5, see Mevinphos
PDAB, see p-Dimethylaminoazobenzene
PDB, see 1,4-Dichlorobenzene
PDCB, see 1,4-Dichlorobenzene
Pear oil, see Isoamyl acetate, n-Amyl acetate
PEB1, see p,p'-DDT
Pedraczak, see Lindane
Pelagol D, see p-Phenylenediamine
Pelagol DR, see p-Phenylenediamine

n-Pentyl acetate, see *n*-Amyl acetate
Pentylcyclopentane
Penwar, see Pentachlorophenol
PER, see Tetrachloroethylene
Peratox, see Pentachlorophenol
Perawin, see Tetrachloroethylene
PERC, see Tetrachloroethylene
Perchlor, see Tetrachloroethylene
Perchlorethylene, see Tetrachloroethylene
Perchlorobenzene, see Hexachlorobenzene
Perchlorobutadiene, see Hexachlorobutadiene
Perchlorocyclopentadiene, see Hexachlorocyclopentadiene
Perchloroethane, see Hexachloroethane
Perchloroethylene, see Tetrachloroethylene
Perchloromethane, see Carbon tetrachloride
Perclene, see Tetrachloroethylene
Perclene D, see Tetrachloroethylene
Percosolv, see Tetrachloroethylene
Perhydronaphthalene, see Decahydronaphthalene
Periethylenenaphthalene, see Acenaphthene
Peritonan, see *p*-Chloro-*m*-cresol
Perk, see Tetrachloroethylene
Perklone, see Tetrachloroethylene
Perm-a-chlor, see Trichloroethylene
Permacide, see Pentachlorophenol
Perm-a-clor, see Trichloroethylene
Permaguard, see Pentachlorophenol
Permasan, see Pentachlorophenol
Permatox DP-2, see Pentachlorophenol
Permatox Penta, see Pentachlorophenol
Permite, see Pentachlorophenol
Persec, see Tetrachloroethylene
Persia-Perazol, see 1,4-Dichlorobenzene
Pertite, see Picric acid
Pestmaster, see Ethylene dibromide, Methyl bromide
Pestmaster EDB-85, see Ethylene dibromide
Pestox plus, see Parathion
Pethion, see Parathion
Petrol yellow WT, see *p*-Dimethylaminoazobenzene
Petzinol, see Trichloroethylene
Pflanzol, see Lindane
Phenachlor, see 2,4,5-Trichlorophenol, 2,4,6-Trichlorophenol
Phenacide, see Toxaphene

Phenylformic acid, see Benzoic acid
Phenyl hydrate, see Phenol
Phenylhydrazine
Phenyl hydride, see Benzene
Phenyl hydroxide, see Phenol
Phenylic acid, see Phenol
Phenylic alcohol, see Phenol
Phenylmethane, see Toluene
Phenyl methanol, see Benzyl alcohol
Phenyl methyl alcohol, see Benzyl alcohol
N-Phenylmethylamine, see Methylaniline
4-Phenylnitrobenzene, see 4-Nitrobiphenyl
p-Phenylnitrobenzene, see 4-Nitrobiphenyl
Phenyl perchloryl, see Hexachlorobenzene
Phenyl phosphate, see Triphenyl phosphate
Phenylphosphonothioic acid O-ethyl O-p-nitrophenyl ester, see EPN
1-Phenylpropane, see n-Propylbenzene
2-Phenylpropane, see Isopropylbenzene
2-Phenylpropene, see α-Methylstyrene
β-Phenylpropene, see α-Methylstyrene
2-Phenylpropylene, see α-Methylstyrene
β-Phenylpropylene, see α-Methylstyrene
o-Phenylpyrene, see Indeno[1,2,3-cd]pyrene
Philex, see Trichloroethylene
Phortox, see 2,4,5-T
Phosdrin, see Mevinphos
cis-Phosdrin, see Mevinphos
Phosfene, see Mevinphos
Phosflex 179-C, see Tri-o-cresyl phosphate
Phoskil, see Parathion
Phosphemol, see Parathion
Phosphenol, see Parathion
Phosphonothioic acid O,O-diethyl O-(3,5,6-trichloro-2-pyridinyl) ester, see EPN
Phosphoric acid 1,2-dibromo-2,2-dichloroethyl dimethyl ester, see Naled
Phosphoric acid 2,2-dichloroethenyl dimethyl ester, see Dichlorvos
Phosphoric acid 2,2-dichlorovinyl dimethyl ester, see Dichlorvos
Phosphoric acid (1-methoxycarboxypropen-2-yl) dimethyl ester, see Mevinphos
Phosphoric acid tributyl ester, see Tributyl phosphate
Phosphoric acid tri-o-cresyl ester, see Tri-o-cresyl phosphate
Phosphoric acid triphenyl ester, see Triphenyl phosphate
Phosphoric acid tris(2-methylphenyl)ester, see Tri-o-cresyl phosphate
Phosphoric acid tri-2-tolyl ester, see Tri-o-cresyl phosphate
Phosphorothioic acid O,O-diethyl O-(4-nitrophenyl)ester, see Parathion

Phosphorothioic acid *O,O*-dimethyl *O*-(2,4,5-trichlorophenyl)ester, see Ronnel
Phosphorothionic acid *O,O*-diethyl *O*-(3,5,6-trichloro-2-pyridyl)ester, see Chlor-
  pyrifos
Phosphostigmine, see Parathion
Phosphothion, see Malathion
Phosvit, see Dichlorvos
Phthalandione, see Phthalic anhydride
1,3-Phthalandione, see Phthalic anhydride
Phthalic acid anhydride, see Phthalic anhydride
Phthalic acid bis(2-ethylhexyl) ester, see Bis(2-ethylhexyl)phthalate
Phthalic acid dibutyl ester, see Di-*n*-butyl phthalate
Phthalic acid dimethyl ester, see Dimethyl phthalate
Phthalic acid dioctyl ester, see Bis(2-ethylhexyl)phthalate
Phthalic acid methyl ester, see Dimethyl phthalate
Phthalic anhydride
Phthalol, see Diethyl phthalate
Pic-clor, see Chloropicrin
Picfume, see Chloropicrin
Picric acid
Picride, see Chloropicrin
Picronitric acid, see Picric acid
Picrylmethylnitramine, see Tetryl
Picrylnitromethylamine, see Tetryl
Pied piper mouse seed, see Strychnine
Pielik, see 2,4-D
Pimelic ketone, see Cyclohexanone
Pin, see EPN
Pindone
Pirofos, see Sulfotepp
Pittsburgh PX-138, see Bis(2-ethylhexyl)phthalate
Pivacin, see Pindone
Pival, see Pindone
2-Pivaloylindane-1,3-dione, see Pindone
2-Pivaloyl-1,3-indanedione, see Pindone
Pivalyl, see Pindone
Pivalyl indandione, see Pindone
2-Pivalyl-1,3-indandione, see Pindone
Pivalyl valone, see Pindone
Placidol E, see Diethyl phthalate
Planotox, see 2,4-D
Plant dithio aerosol, see Sulfotepp
Plantfume 103 smoke generator, see Sulfotepp
Plantgard, see 2,4-D

Propylcyclopentane
Propylene aldehyde, see Crotonaldehyde
Propylene chloride, see 1,2-Dichloropropane
Propylene dichloride, see 1,2-Dichloropropane
α,β-Propylene dichloride, see 1,2-Dichloropropane
Propylene oxide
1,2-Propylene oxide, see Propylene oxide
Propylethylene, see 1-Pentene
Propyl hydride, see Propane
Propylic alcohol, see 1-Propanol
Propyl iodide, see 1-Iodopropane
n-Propyl iodide, see 1-Iodopropane
Propylmethanol, see 1-Butanol
Propyl methyl ketone, see 2-Pentanone
Propyl nitrate, see n-Propyl nitrate
n-Propyl nitrate
Propyne
1-Propyne, see Propyne
Prothromadin, see Warfarin
PS, see Chloropicrin
Pseudobutylbenzene, see tert-Butylbenzene
Pseudocumene, see 1,2,4-Trimethylbenzene
Pseudocumol, see 1,2,4-Trimethylbenzene
Puralin, see Thiram
PX 104, see Di-n-butyl phthalate
PX 138, see Di-n-octyl phthalate
Pyranton, see Diacetone alcohol
Pyranton A, see Diacetone alcohol
Pyrene
β-Pyrene, see Pyrene
2-Pyridinamine, see 2-Aminopyridine
α-Pyridinamine, see 2-Aminopyridine
Pyridine
2-Pyridylamine, see 2-Aminopyridine
α-Pyridylamine, see 2-Aminopyridine
β-Pyrine, see Pyrene
Pyrinex, see Chlorpyrifos
Pyroacetic acid, see Acetone
Pyroacetic ether, see Acetone
Pyrobenzol, see Benzene
Pyrobenzole, see Benzene
Pyroligneous acid, see Acetic acid
Pyromucic aldehyde, see Furfural

Pyropentylene, see Cyclopentadiene

Pyrophosphoric acid tetraethyl ester, see Tetraethyl pyrophosphate

Pyrophosphorodithioic acid *O,O,O,O*-tetraethyl dithionopyrophosphate, see Sulfotepp

Pyrophosphorodithioic acid tetraethyl ester, see Sulfotepp

Pyroxylic spirit, see Methanol

Pyrrolylene, see 1,3-Butadiene

QCB, see Pentachlorobenzene

Quellada, see Lindane

Quinol, see Hydroquinone

β-Quinol, see Hydroquinone

Quinone, see Hydroquinone, *p*-Quinone

4-Quinone, see *p*-Quinone

p-**Quinone**

Quintox, see Dieldrin

R 10, see Carbon tetrachloride

R 12, see Dichlorodifluoromethane

R 20, see Chloroform

R 20 (refrigerant), see Chloroform

R 113, see 1,1,2-Trichlorotrifluoroethane

R 717, see Ammonia

Rafex, see 4,6-Dinitro-*o*-cresol

Rafex 35, see 4,6-Dinitro-*o*-cresol

Raphatox, see 4,6-Dinitro-*o*-cresol

Raschit, see *p*-Chloro-*m*-cresol

Raschit K, see *p*-Chloro-*m*-cresol

Rasenanicon, see *p*-Chloro-*m*-cresol

Rat-a-way, see Warfarin

Rat-b-gon, see Warfarin

Rat-gard, see Warfarin

Rat-kill, see Warfarin

Rat & mice bait, see Warfarin

Rat-mix, see Warfarin

Rat-o-cide #2, see Warfarin

Rat-ola, see Warfarin

Ratorex, see Warfarin

Ratox, see Warfarin

Ratoxin, see Warfarin

Ratron, see Warfarin

Ratron G, see Warfarin

Rats-no-more, see Warfarin

Rattrack, see ANTU

Rat-trol, see Warfarin

Rattunal, see Warfarin
Ravyon, see Carbaryl
Rax, see Warfarin
RB, see Parathion
RC plasticizer DOP, see Bis(2-ethylhexyl)phthalate
RCRA waste number F027, see 2,4,6-Trichlorophenol
RCRA waste number P001, see Warfarin
RCRA waste number P003, see Acrolein
RCRA waste number P004, see Aldrin
RCRA waste number P005, see Allyl alcohol
RCRA waste number P016, see *sym*-Dichloromethyl ether
RCRA waste number P022, see Carbon disulfide
RCRA waste number P023, see Chloroacetaldehyde
RCRA waste number P024, see 4-Chloroaniline
RCRA waste number P028, see Benzyl chloride
RCRA waste number P037, see Dieldrin
RCRA waste number P047, see 4,6-Dinitro-*o*-cresol
RCRA waste number P048, see 2,4-Dinitrophenol
RCRA waste number P050, see α-Endosulfan, β-Endosulfan
RCRA waste number P051, see Endrin
RCRA waste number P054, see Ethylenimine
RCRA waste number P059, see Heptachlor
RCRA waste number P064, see Methyl isocyanate
RCRA waste number P077, see 4-Nitroaniline
RCRA waste number P082, see *N*-Nitrosodimethylamine
RCRA waste number P089, see Parathion
RCRA waste number P108, see Strychnine
RCRA waste number P109, see Sulfotepp
RCRA waste number P111, see Tetraethyl pyrophosphate
RCRA waste number P112, see Tetranitromethane
RCRA waste number P123, see Toxaphene
RCRA waste number U001, see Acetaldehyde
RCRA waste number U002, see Acetone
RCRA waste number U003, see Acetonitrile
RCRA waste number U005, see 2-Acetylaminofluorene
RCRA waste number U007, see Acrylamide
RCRA waste number U009, see Acrylonitrile
RCRA waste number U012, see Aniline
RCRA waste number U018, see Benzo[*a*]anthracene
RCRA waste number U019, see Benzene
RCRA waste number U021, see Benzidine
RCRA waste number U022, see Benzo[*a*]pyrene
RCRA waste number U024, see Bis(2-chloroethoxy)methane

RCRA waste number U025, see Bis(2-chloroethyl)ether
RCRA waste number U027, see Bis(2-chloroisopropyl)ether
RCRA waste number U028, see Bis(2-ethylhexyl)phthalate
RCRA waste number U029, see Methyl bromide
RCRA waste number U031, see 1-Butanol
RCRA waste number U036, see Chlordane
RCRA waste number U037, see Chlorobenzene
RCRA waste number U039, see *p*-Chloro-*m*-cresol
RCRA waste number U041, see Epichlorohydrin
RCRA waste number U042, see 2-Chloroethyl vinyl ether
RCRA waste number U043, see Vinyl chloride
RCRA waste number U044, see Chloroform
RCRA waste number U045, see Methyl chloride
RCRA waste number U047, see 2-Chloronaphthalene
RCRA waste number U048, see 2-Chlorophenol
RCRA waste number U050, see Chrysene
RCRA waste number U052, see 2-Methylphenol
RCRA waste number U053, see Crotonaldehyde
RCRA waste number U055, see Isopropylbenzene
RCRA waste number U056, see Cyclohexane
RCRA waste number U057, see Cyclohexanone
RCRA waste number U060, see *p,p'*-DDD
RCRA waste number U061, see *p,p'*-DDT
RCRA waste number U063, see Dibenz[*a,h*]anthracene
RCRA waste number U066, see 1,2-Dibromo-3-chloropropane
RCRA waste number U067, see Ethylene dibromide
RCRA waste number U069, see Di-*n*-butyl phthalate
RCRA waste number U070, see 1,2-Dichlorobenzene
RCRA waste number U071, see 1,3-Dichlorobenzene
RCRA waste number U072, see 1,4-Dichlorobenzene
RCRA waste number U073, see 3,3'-Dichlorobenzidine
RCRA waste number U075, see Dichlorodifluoromethane
RCRA waste number U076, see 1,1-Dichloroethane
RCRA waste number U077, see 1,2-Dichloroethane
RCRA waste number U078, see 1,1-Dichloroethylene
RCRA waste number U080, see Methylene chloride
RCRA waste number U081, see 2,4-Dichlorophenol
RCRA waste number U083, see 1,2-Dichloropropane
RCRA waste number U088, see Diethyl phthalate
RCRA waste number U092, see Dimethylamine
RCRA waste number U093, see *p*-Dimethylaminoazobenzene
RCRA waste number U101, see 2,4-Dimethylphenol
RCRA waste number U102, see Dimethyl phthalate

RCRA waste number U103, see Dimethyl sulfate
RCRA waste number U105, see 2,4-Dinitrotoluene
RCRA waste number U106, see 2,6-Dinitrotoluene
RCRA waste number U107, see Di-*n*-octyl phthalate
RCRA waste number U108, see 1,4-Dioxane
RCRA waste number U109, see 1,2-Diphenylhydrazine
RCRA waste number U111, see *N*-Nitrosodi-*n*-propylamine
RCRA waste number U112, see Ethyl acetate
RCRA waste number U113, see Ethyl acrylate
RCRA waste number U114, see Hexane
RCRA waste number U117, see Ethyl ether
RCRA waste number U120, see Fluoranthene
RCRA waste number U121, see Trichlorofluoromethane
RCRA waste number U122, see Formaldehyde
RCRA waste number U123, see Formic acid
RCRA waste number U125, see Furfural
RCRA waste number U127, see Hexachlorobenzene
RCRA waste number U128, see Hexachlorobutadiene
RCRA waste number U129, see Lindane
RCRA waste number U130, see Hexachlorocyclopentadiene
RCRA waste number U131, see Hexachloroethane
RCRA waste number U137, see Indeno[1,2,3-*cd*]pyrene
RCRA waste number U138, see Methyl iodide
RCRA waste number U140, see Isobutyl alcohol
RCRA waste number U142, see Kepone
RCRA waste number U147, see Maleic anhydride
RCRA waste number U153, see Methyl mercaptan
RCRA waste number U154, see Methanol
RCRA waste number U159, see 2-Butanone
RCRA waste number U161, see 4-Methyl-2-pentanone
RCRA waste number U162, see Methyl methacrylate
RCRA waste number U165, see Naphthalene
RCRA waste number U167, see 1-Naphthylamine
RCRA waste number U168, see 2-Naphthylamine
RCRA waste number U169, see Nitrobenzene
RCRA waste number U170, see 4-Nitrophenol
RCRA waste number U171, see 2-Nitropropane
RCRA waste number U183, see Pentachlorobenzene
RCRA waste number U184, see Pentachloroethane
RCRA waste number U188, see Phenol
RCRA waste number U190, see Phthalic anhydride
RCRA waste number U196, see Pyridine
RCRA waste number U197, see *p*-Quinone

Rhodiasol, see Parathion
Rhodiatox, see Parathion
Rhodiatrox, see Parathion
Rhothane, see *p,p'*-DDD
Rhothane D-3, see *p,p'*-DDD
Rodafarin, see Warfarin
Ro-deth, see Warfarin
Rodex, see Strychnine, Warfarin
Rodex blox, see Warfarin
Ronnel
Rosex, see Warfarin
Rothane, see *p,p'*-DDD
Rotox, see Methyl bromide
Rough & ready mouse mix, see Warfarin
Royal TMTD, see Thiram
R-pentine, see Cyclopentadiene
Rubinate TDI 80/20, see 2,4-Toluene diisocyanate
Rukseam, see *p,p'*-DDT
Rylam, see Carbaryl
S 1, see Chloropicrin
Sadofos, see Malathion
Sadophos, see Malathion
Sadoplon, see Thiram
Salvo, see 2,4-D
Salvo liquid, see Benzoic acid
Salvo powder, see Benzoic acid
Sanaseed, see Strychnine
Sandolin, see 4,6-Dinitro-*o*-cresol
Sandolin A, see 4,6-Dinitro-*o*-cresol
Sanocide, see Hexachlorobenzene
Santicizer 160, see Benzyl butyl phthalate
Santobane, see *p,p'*-DDT
Santobrite, see Pentachlorophenol
Santochlor, see 1,4-Dichlorobenzene
Santoflex IC, see *p*-Phenylenediamine
Santophen, see Pentachlorophenol
Santophen 20, see Pentachlorophenol
Santox, see EPN
SBA, see *sec*-Butyl alcohol
Scintillar, *p*-Xylene
Sconatex, see 1,1-Dichloroethylene
SD 1750, see Dichlorvos
SD 1897, see 1,2-Dibromo-3-chloropropane

SD 5532, see Chlordane
Seedrin, see Aldrin
Seedrin liquid, see Aldrin
Seffein, see Carbaryl
Selephos, see Parathion
Selinon, see 4,6-Dinitro-*o*-cresol
Septene, see Carbaryl
Sevimol, see Carbaryl
Sevin, see Carbaryl
Sextone, see Cyclohexanone
Sextone B, see Methylcyclohexane
SF 60, see Malathion
Shell MIBK, see 4-Methyl-2-pentanone
Shell SD-5532, see Chlordane
Shellsol 140, see Nonane
Shimose, see Picric acid
Shinnippon fast red GG base, see 4-Nitroaniline
Sicol 150, see Bis(2-ethylhexyl)phthalate
Sicol 160, see Benzyl butyl phthalate
Silotras yellow T2G, see *p*-Dimethylaminoazobenzene
Silvanol, see Lindane
Sinituho, see Pentachlorophenol
Sinox, see 4,6-Dinitro-*o*-cresol
Siptox I, see Malathion
Sixty-three special E.C. insecticide, see Parathion
Slimicide, see Acrolein
Smut-go, see Hexachlorobenzene
Snieciotox, see Hexachlorobenzene
SNP, see Parathion
Soilbrom-40, see Ethylene dibromide
Soilbrom-85, see Ethylene dibromide
Soilbrom-90, see Ethylene dibromide
Soilbrom-90EC, see Ethylene dibromide
Soilbrom-100, see Ethylene dibromide
Soilfume, see Ethylene dibromide
Sok, see Carbaryl
Solaesthin, see Methylene chloride
Soleptax, see Heptachlor
Solfarin, see Warfarin
Solfo black B, see 2,4-Dinitrophenol
Solfo black 2B supra, see 2,4-Dinitrophenol
Solfo black BB, see 2,4-Dinitrophenol
Solfo black G, see 2,4-Dinitrophenol

Solfo black SB, see 2,4-Dinitrophenol
Solmethine, see Methylene chloride
Solvanol, see Diethyl phthalate
Solvanom, see Dimethyl phthalate
Solvarone, see Dimethyl phthalate
Solvent ether, see Ethyl ether
Solvent III, see 1,1,1-Trichloroethane
Somalia yellow A, see *p*-Dimethylaminoazobenzene
Soprathion, see Parathion
Special termite fluid, see 1,2-Dichlorobenzene
Spirit of Hartshorn, see Ammonia
Spontox, see 2,4,5-T
Spotrete, see Thiram
Spotrete-F, see Thiram
Spray-trol brand roden-trol, see Warfarin
Spritz-hormin/2,4-D, see 2,4-D
Spritz-hormit/2,4-D, see 2,4-D
Spritz-rapidin, see Lindane
Spruehpflanzol, see Lindane
SQ 1489, see Thiram
Stabilized ethyl parathion, see Parathion
Staflex DBP, see Di-*n*-butyl phthalate
Staflex DOP, see Bis(2-ethylhexyl)phthalate
Stathion, see Parathion
Stear yellow JB, see *p*-Dimethylaminoazobenzene
Strathion, see Parathion
Strobane-T, see Toxaphene
Strobane T-90, see Toxaphene
Struenex, see Lindane
Strychnidin-10-one, see Strychnine
Strychnine
Strychnos, see Strychnine
Styrene
Styrene monomer, see Styrene
Styrol, see Styrene
Styrolen, see Styrene
Styron, see Styrene
Styropol, see Styrene
Styropor, see Styrene
Suberane, see Cycloheptane
Sudan GG, see *p*-Dimethylaminoazobenzene
Sudan yellow, see *p*-Dimethylaminoazobenzene
Sudan yellow 2G, see *p*-Dimethylaminoazobenzene

Sudan yellow 2GA, see *p*-Dimethylaminoazobenzene
Sulfatep, see Sulfotepp
Sulfotepp
Sulfuric acid dimethyl ester, see Dimethyl sulfate
Sulfuric ether, see Ethyl ether
Sulphocarbonic anhydride, see Carbon disulfide
Sulphos, see Parathion
Sumitox, see Malathion
Super D weedone, see 2,4-D, 2,4,5-T
Superlysoform, see Formaldehyde
Super rodiatox, see Parathion
Sup'r flo, see Diuron
Synklor, see Chlordane
Synthetic 3956, see Toxaphene
Synthetic camphor, see Camphor
Szklarniak, see Dichlorvos
α-T, see 1,1,1-Trichloroethane
β-T, see 1,1,2-Trichloroethane
T-47, see Parathion
2,4,5-T
Tak, see Malathion
Tap 85, see Lindane
Tap 9VP, see Dichlorvos
Tar camphor, see Naphthalene
Task, see Dichlorvos
Task tabs, see Dichlorvos
Tat chlor 4, see Chlordane
TBA, see *tert*-Butyl alcohol
TBE, see 1,1,2,2-Tetrabromoethane
TBH, see Lindane, α-BHC, β-BHC, δ-BHC
TBP, see Tributyl phosphate
1,1,1-TCA, see 1,1,1-Trichloroethane
1,1,2-TCA, see 1,1,2-Trichloroethane
1,2,3-TCB, see 1,2,3-Trichlorobenzene
1,2,4-TCB, see 1,2,4-Trichlorobenzene
1,2,3,4-TCB, see 1,2,3,4-Tetrachlorobenzene
1,2,3,5-TCB, see 1,2,3,5-Tetrachlorobenzene
1,3,5-TCB, see 1,3,5-Trichlorobenzene
TCDBD, see TCDD
TCDD
2,3,7,8-TCDD, see TCDD
TCE, see 1,1,2,2-Tetrachloroethane, Trichloroethylene
1,1,1-TCE, see 1,1,1-Trichloroethane

TCM, see Chloroform
TCP, see Tri-*o*-cresyl phosphate
2,4,5-TCP, see 2,4,5-Trichlorophenol
2,4,5-TCP-Dowicide 2, see 2,4,5-Trichlorophenol
2,4,6-TCP, see 2,4,6-Trichlorophenol
2,4,6-TCP-Dowicide 25, see 2,4,6-Trichlorophenol
TDE, see *p,p'*-DDD
4,4'-TDE, see *p,p'*-DDD
p,p'-TDE, see *p,p'*-DDD
TDI, see 2,4-Toluene diisocyanate
TDI-80, see 2,4-Toluene diisocyanate
2,4-TDI, see 2,4-Toluene diisocyanate
Tear gas, see α-Chloroacetophenone
Tecquinol, see Hydroquinone
TEDP, see Sulfotepp
TEDTP, see Sulfotepp
Telvar, see Diuron
Telvar diuron weed killer, see Diuron
Temus W, see Warfarin
TEN, see Triethylamine
Tenac, see Dichlorvos
Tenn-plas, see Benzoic acid
Tenox HQ, see Hydroquinone
TEP, see Tetraethyl pyrophosphate
TEPP, see Tetraethyl pyrophosphate
Tequinol, see Hydroquinone
Terabol, see Methyl bromide
Tercyl, see Carbaryl
Tereton, see Methyl acetate
Termitkil, see 1,2-Dichlorobenzene
Term-i-trol, see Pentachlorophenol
Terr-o-gas, see Methyl bromide
Tersan, see Thiram
Tersan 75, see Thiram
Tertral D, see *p*-Phenylenediamine
Tetan, see Tetranitromethane
Tetlen, see Tetrachloroethylene
Tetrabromoacetylene, see 1,1,2,2-Tetrabromoethane
1,2,4,5-Tetrabromobenzene
Tetrabromoethane, see 1,1,2,2-Tetrabromoethane
1,1,2,2-Tetrabromoethane
sym-Tetrabromoethane, see 1,1,2,2-Tetrabromoethane
Tetracap, see Tetrachloroethylene

Tetrachloormetaan, see Carbon tetrachloride
Tetrachlorethane, see 1,1,2,2-Tetrachloroethane
Tetrachlorethylene, see Tetrachloroethylene
1,2,3,4-Tetrachlorobenzene
1,2,3,5-Tetrachlorobenzene
1,2,4,5-Tetrachlorobenzene
sym-Tetrabromobenzene, see 1,2,4,5-Tetrachlorobenzene
sym-Tetrachlorobenzene, see 1,2,4,5-Tetrachlorobenzene
Tetrachlorocarbon, see Carbon tetrachloride
2,3,7,8-Tetrachlorodibenzodioxin, see TCDD
2,3,7,8-Tetrachlorodibenzo-1,4-dioxin, see TCDD
2,3,7,8-Tetrachlorodibenzo-*p*-dioxin, see TCDD
2,3,7,8-Tetrachlorodibenzo[*b,e*][1,4]dioxin, see TCDD
1,1,1,2-Tetrachloro-2,2-difluoroethane, see 1,1-Difluorotetrachloroethane
1,1,2,2-Tetrachloro-1,2-difluoroethane, see 1,2-Difluorotetrachloroethane
Tetrachlorodiphenylethane, see *p,p'*-DDD
Tetrachloroethane, see 1,1,2,2-Tetrachloroethane
1,1,2,2-Tetrachloroethane
sym-Tetrachloroethane, see 1,1,2,2-Tetrachloroethane
Tetrachloroethene, see Tetrachloroethylene
Tetrachloroethylene
1,1,2,2-Tetrachloroethylene, see Tetrachloroethylene
Tetrachloromethane, see Carbon tetrachloride
Tetradioxin, see TCDD
Tetraethyl diphosphate, see Tetraethyl pyrophosphate
Tetraethyl dithionopyrophosphate, see Sulfotepp
Tetraethyl dithiopyrophosphate, see Sulfotepp
O,O,O,O-Tetraethyl dithiopyrophosphate, see Sulfotepp
Tetraethyl pyrofosfaat, see Tetraethyl pyrophosphate
Tetraethyl pyrophosphate
Tetrafinol, see Carbon tetrachloride
Tetraform, see Carbon tetrachloride
Tetrahydrobenzene, see Cyclohexene
1,2,3,4-Tetrahydrobenzene, see Cyclohexene
Tetrahydro-*o*-cresol, see *o*-Methylcyclohexanone
Tetrahydro-1,4-dioxin, see 1,4-Dioxane
Tetrahydro-*p*-dioxin, see 1,4-Dioxane
Tetrahydrofuran
Tetrahydro-1,4-isoxazine, see Morpholine
Tetrahydro-1,4-oxazine, see Morpholine
Tetrahydro-2*H*-1,4-oxazine, see Morpholine
2,3,4,5-Tetrahydrotoluene, see 1-Methylcyclohexene
Tetraleno, see Tetrachloroethylene

Tetralex, see Tetrachloroethylene

Tetralit, see Tetryl

Tetralite, see Tetryl

1,2,4,5-Tetramethylbenzene

sym-1,2,4,5-Tetramethylbenzene, see 1,2,4,5-Tetramethylbenzene

Tetramethyldiurane sulphite, see Thiram

Tetramethylene oxide, see Tetrahydrofuran

Tetramethylenethiuram disulfide, see Thiram

Tetramethylmethane, see 2,2-Dimethylpropane

Tetramethylthiocarbamoyl disulfide, see Thiram

Tetramethylthioperoxydicarbonic diamide, see Thiram

Tetramethylthiuram bisulfide, see Thiram

Tetramethylthiuram bisulphide, see Thiram

Tetramethylthiuram disulfide, see Thiram

N,N-Tetramethylthiuram disulfide, see Thiram

N,N,N',N'-Tetramethylthiuram disulfide, see Thiram

Tetramethylthiuran disulfide, see Thiram

Tetramethylthiurane disulphide, see Thiram

Tetramethylthiurum disulfide, see Thiram

Tetramethylthiurum disulphide, see Thiram

Tetranitromethane

Tetra olive N2G, see Anthracene

Tetraphene, see Benzo[*a*]anthracene

Tetrapom, see Thiram

Tetrasipton, see Thiram

Tetrasol, see Carbon tetrachloride

Tetrastigmine, see Tetraethyl pyrophosphate

Tetrasulphur black PB, see 2,4-Dinitrophenol

Tetrathiuram disulfide, see Thiram

Tetrathiuram disulphide, see Thiram

Tetravec, see Tetrachloroethylene

Tetravos, see Dichlorvos

Tetril, see Tetryl

Tetroguer, see Tetrachloroethylene

Tetron, see Tetraethyl pyrophosphate

Tetron-100, see Tetraethyl pyrophosphate

Tetropil, see Tetrachloroethylene

Tetrosulphur PBR, see 2,4-Dinitrophenol

Tetryl

2,4,6-Tetryl, see Tetryl

Texaco lead appreciator, see *tert*-Butyl acetate

Texadust, see Toxaphene

THF, see Tetrahydrofuran

Thiacyclopentadiene, see Thiophene
Thiaphene, see Thiophene
Thifor, see α-Endosulfan, β-Endosulfan
Thillate, see Thiram
Thimer, see Thiram
Thimul, see α-Endosulfan, β-Endosulfan
Thiobutyl alcohol, see *n*-Butyl mercaptan
Thiodan, see α-Endosulfan, β-Endosulfan
Thiodiphosphoric acid tetraethyl ester, see Sulfotepp
Thioethanol, see Ethyl mercaptan
Thioethyl alcohol, see Ethyl mercaptan
Thiofalco M-50, see Ethanolamine
Thiofor, see α-Endosulfan, β-Endosulfan
Thiofos, see Parathion
Thiofuram, see Thiophene
Thiofuran, see Thiophene
Thiofurfuran, see Thiophene
Thiole, see Thiophene
Thiomethanol, see Methyl mercaptan
Thiomethyl alcohol, see Methyl mercaptan
Thiomul, see α-Endosulfan, β-Endosulfan
Thionex, see α-Endosulfan, β-Endosulfan
Thiophen, see Thiophene
Thiophene
Thiophos 3422, see Parathion
Thiophosphoric acid tetraethyl ester, see Sulfotepp
Thiopyrophosphoric acid tetraethyl ester, see Sulfotepp
Thiosan, see Thiram
Thiosulfan, see α-Endosulfan, β-Endosulfan
Thiotepp, see Sulfotepp
Thiotetrole, see Thiophene
Thiotex, see Thiram
Thiotox, see Thiram
Thiram
Thiram 75, see Thiram
Thiramad, see Thiram
Thiram B, see Thiram
Thirasan, see Thiram
Thiulix, see Thiram
Thiurad, see Thiram
Thiuram, see Thiram
Thiuram D, see Thiram
Thiuram M, see Thiram

Thiuram M rubber accelerator, see Thiram
Thiuramin, see Thiram
Thiuramyl, see Thiram
Thompson's wood fix, see Pentachlorophenol
Threthylen, see Trichloroethylene
Threthylene, see Trichloroethylene
Thylate, see Thiram
Tionel, see α-Endosulfan, β-Endosulfan
Tiophos, see Parathion
Tiovel, see α-Endosulfan, β-Endosulfan
Tippon, see 2,4,5-T
Tirampa, see Thiram
Tiuramyl, see Thiram
TJB, see *N*-Nitrosodiphenylamine
TL 314, see Acrylonitrile
TL 337, see Ethylenimine
TL 1450, see Methyl isocyanate
TLA, see *tert*-Butyl acetate
TM-4049, see Malathion
TMB, see 1,3,5-Trimethylbenzene
TMTD, see Thiram
TMTDS, see Thiram
TNM, see Tetranitromethane
TNT, see 2,4,6-Trinitrotoluene
α-TNT, see 2,4,6-Trinitrotoluene
TNT-tolite, see 2,4,6-Trinitrotoluene
TOCP, see Tri-*o*-cresyl phosphate
TOFK, see Tri-*o*-cresyl phosphate
Tolit, see 2,4,6-Trinitrotoluene
Tolite, see 2,4,6-Trinitrotoluene
Toluene
Toluene 2,4-diisocyanate, see 2,4-Toluene diisocyanate
2,4-Toluene diisocyanate
Toluene hexahydride, see Methylcyclohexane
α-Toluenol, see Benzyl alcohol
2-Toluidine, see *o*-Toluidine
o-**Toluidine**
Toluol, see Toluene
2-Toluol, see 2-Methylphenol
4-Toluol, see 4-Methylphenol
o-Toluol, see 2-Methylphenol
p-Toluol, see 4-Methylphenol
Tolu-sol, see Toluene

Toluene-2,4-diisocyanate, see 2,4-Toluene diisocyanate
p-Tolyl alcohol, see 4-Methylphenol
o-Tolylamine, see *o*-Toluidine
Tolyl chloride, see Benzyl chloride
Tolylene-2,4-diisocyanate, see 2,4-Toluene diisocyanate
2,4-Tolylene diisocyanate, see 2,4-Toluene diisocyanate
m-Tolylene diisocyanate, see 2,4-Toluene diisocyanate
o-Tolyl phosphate, see Tri-*o*-cresyl phosphate
Topanel, see Crotonaldehyde
Topichlor 20, see Chlordane
Topiclor, see Chlordane
Topiclor 20, see Chlordane
Tormona, see 2,4,5-T
TOTP, see Tri-*o*-cresyl phosphate
Tox 47, see Parathion
Toxakil, see Toxaphene
Toxan, see Carbaryl
Toxaphene
Tox-hid, see Warfarin
Toxichlor, see Chlordane
Toxilic anhydride, see Maleic anhydride
Toxon 63, see Toxaphene
Toyo oil yellow G, see *p*-Dimethylaminoazobenzene
Toxyphen, see Toxaphene
TPP, see Triphenyl phosphate
Trametan, see Thiram
Transamine, see 2,4-D, see 2,4,5-T
Trefanocide, see Trifluralin
Treficon, see Trifluralin
Treflam, see Trifluralin
Treflan, see Trifluralin
Treflanocide elancolan, see Trifluralin
Trethylene, see Trichloroethylene
Tri, see Trichloroethylene
Tri-6, see Lindane
Triad, see Trichloroethylene
Trial, see Trichloroethylene
Triasol, see Trichloroethylene
Tri-Ban, see Pindone
1,3,5-Tribromobenzene
sym-Tribromobenzene, see 1,3,5-Tribromobenzene
Tribromomethane, see Bromoform
Tributon, see 2,4-D, 2,4,5-T

Tributyl phosphate
Tri-*n*-butyl phosphate, see Tributyl phosphate
Tricarnam, see Carbaryl
Trichloran, see Trichloroethylene
Trichloren, see Trichloroethylene
1,2,3-Trichlorobenzene
1,2,4-Trichlorobenzene
unsym-Trichlorobenzene, see 1,2,4-Trichlorobenzene
1,3,5-Trichlorobenzene
sym-Trichlorobenzene, see 1,3,5-Trichlorobenzene
1,1,1-Trichloro-2,2-bis(*p*-anisyl)ethane, see Methoxychlor
1,1,1-Trichlorobis(4-chlorophenyl)ethane, see *p,p'*-DDT
1,1,1-Trichlorobis(*p*-chlorophenyl)ethane, see *p,p'*-DDT
1,1,1-Trichloro-2,2-bis(*p*-chlorophenyl)ethane, see *p,p'*-DDT
1,1,1-Trichloro-2,2-bis(*p*-methoxyphenol)ethanol, see Methoxychlor
1,1,1-Trichloro-2,2-bis(*p*-methoxyphenyl)ethane, see Methoxychlor
1,1,1-Trichloro-2,2-di(4-chlorophenyl)ethane, see *p,p'*-DDT
1,1,1-Trichloro-2,2-di(*p*-chlorophenyl)ethane, see *p,p'*-DDT
1,1,1-Trichloro-2,2-di(4-methoxyphenyl)ethane, see Methoxychlor
β-Trichloroethane, see 1,1,2-Trichloroethane
1,1,1-Trichloroethane
1,1,2-Trichloroethane
1,2,2-Trichloroethane, see 1,1,2-Trichloroethane
α-Trichloroethane, see 1,1,1-Trichloroethane
Trichloroethene, see Trichloroethylene
1,1,2-Trichloroethene, see Trichloroethylene
1,2,2-Trichloroethene, see Trichloroethylene
Trichloroethylene
1,1,2-Trichloroethylene, see Trichloroethylene
1,2,2-Trichloroethylene, see Trichloroethylene
1,1'-(2,2,2-Trichloroethylidene)bis(4-chlorobenzene), see *p,p'*-DDT
1,1'-(2,2,2-Trichloroethylidene)bis(4-methoxybenzene), see Methoxychlor
Trichlorofluoromethane
Trichloroform, see Chloroform
Trichlorohydrin, see 1,2,3-Trichloropropane
Trichlorometafos, see Ronnel
Trichloromethane, see Chloroform
Trichloromonofluoromethane, see Trichlorofluoromethane
Trichloronitromethane, see Chloropicrin
2,4,5-Trichlorophenol
2,4,6-Trichlorophenol
(2,4,5-Trichlorophenoxy)acetic acid, see 2,4,5-T
1,2,3-Trichloropropane

Trichlorotrifluoroethane, see 1,1,2-Trichlorotrifluoroethane
1,1,2-Trichlorotrifluoroethane
1,1,2-Trichloro-1,2,2-trifluoroethane, see 1,1,2-Trichlorotrifluoroethane
Tri-clene, see Trichloroethylene
Tri-clor, see Chloropicrin
Tricresyl phosphate, see Tri-*o*-cresyl phosphate
Tri-*o*-cresyl phosphate
Tridipam, see Thiram
Trielene, see Trichloroethylene
Trieline, see Trichloroethylene
Tri-ethane, see 1,1,1-Trichloroethane
Triethylamine
Trifina, see 4,6-Dinitro-*o*-cresol
Trifluoralin, see Trifluralin
Trifluorobromomethane, see Bromotrifluoromethane
α,α,α-Trifluoro-2,6-dinitro-*N,N*-dipropyl-*p*-toluidine, see Trifluralin
Trifluoromonobromomethane, see Bromotrifluoromethane
1,1,2-Trifluorotrichloroethane, see 1,1,2-Trichlorotrifluoroethane
Trifluralin
Trifluraline, see Trifluralin
Triflurex, see Trifluralin
Trifocide, see 4,6-Dinitro-*o*-cresol
Trikepin, see Trifluralin
Triklone, see Trichloroethylene
Trilen, see Trichloroethylene
Trilene, see Trichloroethylene
Triline, see Trichloroethylene
Trilit, see 2,4,6-Trinitrotoluene
Trim, see Trifluralin
Trimar, see Trichloroethylene
1,2,3-Trimethylbenzene
1,2,4-Trimethylbenzene
1,3,5-Trimethylbenzene
asym-Trimethylbenzene, see 1,2,4-Trimethylbenzene
sym-Trimethylbenzene, see 1,3,5-Trimethylbenzene
Trimethylbenzol, see 1,3,5-Trimethylbenzene
1,7,7-Trimethylbicyclo[2.2.1]heptan-2-one, see Camphor
Trimethylcarbinol, see *tert*-Butyl alcohol
1,1,3-Trimethylcyclohexane
1,3,3-Trimethylcyclohexane, see 1,1,3-Trimethylcyclohexane
1,1,3-Trimethyl-3-cyclohexene-5-one, see Isophorone
Trimethylcyclohexenone, see Isophorone
3,5,5-Trimethyl-2-cyclohexen-1-one, see Isophorone

1,1,3-Trimethylcyclopentane
1,3,3-Trimethylcyclopentane, see 1,1,3-Trimethylcyclopentane
2,2,5-Trimethylhexane
2,5,5-Trimethylhexane, see 2,2,5-Trimethylhexane
Trimethylmethane, see 2-Methylpropane
Trimethyl methanol, see *tert*-Butyl alcohol
2,2,4-Trimethylpentane
2,3,4-Trimethylpentane
Trimethylphenylmethane, see *tert*-Butylbenzene
Tri-2-methylphenyl phosphate, see Tri-*o*-cresyl phosphate
1,3,5-Trinitrophenol, see Picric acid
2,4,6-Trinitrophenol, see Picric acid
Trinitrophenylmethylnitramine, see Tetryl
2,4,6-Trinitrophenylmethylnitramine, see Tetryl
2,4,6-Trinitrophenyl-*N*-methylnitramine, see Tetryl
Trinitrotoluene, see 2,4,6-Trinitrotoluene
2,4,6-Trinitrotoluene
sym-Trinitrotoluene, see 2,4,6-Trinitrotoluene
Trinitrotoluol, see 2,4,6-Trinitrotoluene
α-Trinitrotoluol, see 2,4,6-Trinitrotoluene
sym-Trinitrotoluol, see 2,4,6-Trinitrotoluene
Trinoxol, see 2,4-D, 2,4,5-T
Triol, see Trichloroethylene
Triorthocresyl phosphate, see Tri-*o*-cresyl phosphate
Trioxon, see 2,4,5-T
Trioxone, see 2,4,5-T
Triphenyl phosphate
Tri-plus, see Trichloroethylene
Tri-plus M, see Trichloroethylene
Tripomol, see Thiram
Tris(*o*-cresyl)phosphate, see Tri-*o*-cresyl phosphate
Tris(*o*-methylphenyl)phosphate, see Tri-*o*-cresyl phosphate
Tris(*o*-tolyl)phosphate, see Tri-*o*-cresyl phosphate
Tritol, see 2,4,6-Trinitrotoluene
Tri-2-tolyl phosphate, see Tri-*o*-cresyl phosphate
Tri-*o*-tolyl phosphate, see Tri-*o*-cresyl phosphate
Triton, see 2,4,6-Trinitrotoluene
Trolen, see Ronnel
Trolene, see Ronnel
Trotyl, see 2,4,6-Trinitrotoluene
Trotyl oil, see 2,4,6-Trinitrotoluene
Trovidur, see Vinyl chloride
Truflex DOP, see Bis(2-ethylhexyl)phthalate

TTD, see Thiram
TTE, see 1,1,2-Trichlorotrifluoroethane
Tuads, see Thiram
Tuex, see Thiram
Tulisan, see Thiram
Twin light rat away, see Warfarin
U 46, see 2,4-D, 2,4,5-T
U 46DP, see 2,4-D
U 4224, see *N,N*-Dimethylformamide
U 5043, see 2,4-D
U 5954, see *N,N*-Dimethylacetamide
UC 7744, see Carbaryl
Ucon 11, see Trichlorofluoromethane
Ucon 12, see Dichlorodifluoromethane
Ucon 113, see 1,1,2-Trichlorotrifluoroethane
Ucon fluorocarbon 11, see Trichlorofluoromethane
Ucon fluorocarbon 113, see 1,1,2-Trichlorotrifluoroethane
Ucon 12/halocarbon 12, see Dichlorodifluoromethane
Ucon 113/halocarbon 113, see 1,1,2-Trichlorotrifluoroethane
Ucon refrigerant 11, see Trichlorofluoromethane
UDMH, see 1,1-Dimethylhydrazine
UDVF, see Dichlorvos
UN 0154, see Picric acid
UN 0208, see Tetryl
UN 0209, see 2,4,6-Trinitrotoluene
UN 1005, see Ammonia
UN 1009, see Bromotrifluoromethane
UN 1011, see Butane
UN 1028, see Dichlorodifluoromethane
UN 1029, see Dichlorofluoromethane
UN 1032, see Dimethylamine
UN 1036, see Ethylamine
UN 1037, see Chloroethane
UN 1061, see Methylamine
UN 1062, see Methyl bromide
UN 1063, see Methyl chloride
UN 1064, see Methyl mercaptan
UN 1075, see 2-Methylpropane, Propane
UN 1086, see Vinyl chloride
UN 1089, see Acetaldehyde
UN 1090, see Acetone
UN 1092, see Acrolein
UN 1093, see Acrylonitrile

UN 1098, see Allyl alcohol
UN 1100, see Allyl chloride
UN 1104, see *n*-Amyl acetate
UN 1105, see Isoamyl alcohol
UN 1110, see 2-Heptanone
UN 1114, see Benzene
UN 1120, see 1-Butanol, *tert*-Butyl alcohol
UN 1123, see *sec*-Butyl acetate, *n*-Butyl acetate, *tert*-Butyl acetate
UN 1125, see *n*-Butylamine
UN 1131, see Carbon disulfide
UN 1134, see Chlorobenzene
UN 1135, see Ethylene chlorohydrin
UN 1145, see Cyclohexane
UN 1146, see Cyclopentane
UN 1147, see Decahydronaphthalene
UN 1148, see Diacetone alcohol
UN 1154, see Diethylamine
UN 1155, see Ethyl ether
UN 1157, see Diisobutyl ketone
UN 1158, see Diisopropylamine
UN 1159, see Isopropyl ether
UN 1160, see Dimethylamine
UN 1165, see 1,4-Dioxane
UN 1171, see 2-Ethoxyethanol
UN 1172, see 2-Ethoxyethyl acetate
UN 1173, see Ethyl acetate
UN 1184, see 1,2-Dichloroethane
UN 1185, see Ethylenimine
UN 1188, see Methyl cellosolve
UN 1189, see Methyl cellosolve acetate
UN 1190, see Ethyl formate
UN 1193, see 2-Butanone
UN 1198, see Formaldehyde
UN 1199, see Furfural
UN 1206, see Heptane
UN 1208, see 2,2-Dimethylbutane, Hexane
UN 1212, see Isobutyl alcohol
UN 1213, see Isobutyl acetate
UN 1218, see 2-Methyl-1,3-butadiene
UN 1220, see Isopropyl acetate
UN 1221, see Isopropylamine, Isopropylbenzene
UN 1229, see Mesityl oxide
UN 1231, see Methyl acetate

UN 1232, see 2-Butanone
UN 1233, see *sec*-Hexyl acetate
UN 1234, see Methylal
UN 1235, see Methylamine
UN 1243, see Methyl formate
UN 1245, see 4-Methyl-2-pentanone
UN 1247, see Methyl methacrylate
UN 1249, see 2-Pentanone
UN 1261, see Nitromethane
UN 1262, see Octane, 2,2,4-Trimethylpentane
UN 1265, see 2,2-Dimethylpropane, 2-Methylbutane, Pentane
UN 1274, see 1-Propanol
UN 1276, see *n*-Propyl acetate
UN 1280, see Propylene oxide
UN 1282, see Pyridine
UN 1294, see Toluene
UN 1296, see Triethylamine
UN 1301, see Vinyl acetate
UN 1307, see *m*-Xylene, *o*-Xylene, *p*-Xylene
UN 1334, see Naphthalene
UN 1510, see Tetranitromethane
UN 1547, see Aniline
UN 1578, see *p*-Chloronitrobenzene
UN 1580, see Chloropicrin
UN 1591, see 1,2-Dichlorobenzene, 1,3-Dichlorobenzene
UN 1592, see 1,4-Dichlorobenzene
UN 1593, see Methylene chloride
UN 1595, see Dimethyl sulfate
UN 1597, see 1,2-Dinitrobenzene, 1,3-Dinitrobenzene, 1,2-Dinitrobenzene
UN 1604, see Ethylenediamine
UN 1605, see Ethylene dibromide
UN 1648, see Acetonitrile
UN 1650, see 2-Naphthylamine
UN 1661, see 2-Nitroaniline, 3-Nitroaniline, 4-Nitroaniline
UN 1662, see Nitrobenzene
UN 1663, see 2-Nitrophenol, 4-Nitrophenol
UN 1664, see 2-Nitrotoluene, 3-Nitrotoluene, 4-Nitrotoluene
UN 1669, see Pentachloroethane
UN 1671, see Phenol
UN 1673, see *p*-Phenylenediamine
UN 1692, see Strychnine
UN 1697, see α-Chloroacetophenone
UN 1702, see 1,1,2,2-Tetrachloroethane

UN 2253, see Dimethylaniline
UN 2256, see Cyclohexene
UN 2265, see *N,N*-Dimethylformamide
UN 2279, see Hexachlorobutadiene
UN 2294, see Methylaniline
UN 2296, see Methylcyclohexane
UN 2298, see Methylcyclopentane
UN 2312, see Phenol
UN 2321, see 1,2,3-Trichlorobenzene, 1,3,5-Trichlorobenzene
UN 2321, see 1,2,4-Trichlorobenzene
UN 2325, see 1,3,5-Trimethylbenzene
UN 2347, see *n*-Butyl mercaptan
UN 2362, see 1,1-Dichloroethane
UN 2363, see Ethyl mercaptan
UN 2364, see *n*-Propylbenzene
UN 2370, see 1-Hexene
UN 2414, see Thiophene
UN 2431, see *o*-Anisidine
UN 2457, see 2,3-Dimethylbutane
UN 2462, see 2-Methylpentane, 3-Methylpentane
UN 2472, see Pindone
UN 2480, see Methyl isocyanate
UN 2490, see Bis(2-chloroisopropyl)ether
UN 2491, see Ethanolamine
UN 2504, see 1,1,2,2-Tetrabromoethane
UN 2506, see Tetrahydrofuran
UN 2514, see Bromobenzene
UN 2515, see Bromoform
UN 2561, see 3-Methyl-1-butene
UN 2562, see Phenylhydrazine
UN 2587, see *p*-Quinone
UN 2608, see 1-Nitropropane, 2-Nitropropane
UN 2644, see Methyl iodide
UN 2646, see Hexachlorocyclopentadiene
UN 2662, see Hydroquinone
UN 2671, see 2-Aminopyridine
UN 2686, see 2-Diethylaminoethanol
UN 2709, see *n*-Butylbenzene, *tert*-Butylbenzene
UN 2717, see Camphor
UN 2729, see Hexachlorobenzene
UN 2789, see Acetic acid
UN 2790, see Acetic acid
UN 2821, see Phenol

UN 2831, see 1,1,1-Trichloroethane
UN 2842, see Nitroethane
UN 2872, see 1,2-Dibromo-3-chloropropane
UN 2874, see Furfuryl alcohol
Unidron, see Diuron
Unifos, see Dichlorvos
Unifos 50 EC, see Dichlorvos
Unifume, see Ethylene dibromide
Unimoll BB, see Benzyl butyl phthalate
Union Carbide 7744, see Carbaryl
Univerm, see Carbon tetrachloride
Urox D, see Diuron
Ursol D, see p-Phenylenediamine
USAF B-30, see Thiram
USAF CB-22, see 2-Naphthylamine
USAF EK-338, see p-Dimethylaminoazobenzene
USAF EK-356, see Hydroquinone
USAF EK-394, see p-Phenylenediamine
USAF EK-488, see Acetonitrile
USAF EK-1597, see Ethanolamine
USAF EK-1860, see Thiophene
USAF EK-2089, see Thiram
USAF KF-11, see o-Chlorobenzylidenemalononitrile
USAF P-5, see Thiram
USAF P-7, see Diuron
USAF P-220, see p-Quinone
USAF XR-42, see Diuron
VAC, see Vinyl acetate
Valerone, see Diisobutyl ketone
VAM, see Vinyl acetate
Vampirin III, see Warfarin
Vampirinip II, see Warfarin
Vancida TM-95, see Thiram
Vancide TM, see Thiram
Vapona, see Dichlorvos
Vaponite, see Dichlorvos
Vapophos, see Parathion
Vapora II, see Dichlorvos
Vapotone, see Tetraethyl pyrophosphate
VC, see 1,1-Dichloroethylene
VC, see Vinyl chloride
VCM, see Vinyl chloride
VCN, see Acrylonitrile

VDC, see 1,1-Dichloroethylene
Vegfru malatox, see Malathion
Velsicol 104, see Heptachlor
Velsicol 1068, see Chlordane
Velsicol 53-CS-17, see Heptachlor epoxide
Velsicol heptachlor, see Heptachlor
Ventox, see Acrylonitrile
Veon, see 2,4,5-T
Veon 245, see 2,4,5-T
Verdican, see Dichlorvos
Verdipor, see Dichlorvos
Vergemaster, see 2,4-D
Vermoestricid, see Carbon tetrachloride
Versneller NL 63/10, see Dimethylaniline
Vertac 90%, see Toxaphene
Vertac toxaphene 90, see Toxaphene
Verton, see 2,4-D
Verton D, see 2,4-D
Verton 2D, see 2,4-D
Verton 2T, see 2,4,5-T
Vertron 2D, see 2,4-D
Vestinol 80, see Bis(2-ethylhexyl)phthalate
Vestrol, see Trichloroethylene
Vetiol, see Malathion
Vidon 638, see 2,4-D
Vinegar acid, see Acetic acid
Vinegar naphtha, see Ethyl acetate
Vinicizer 85, see Di-*n*-octyl phthalate
Vinyl acetate
Vinyl acetate H.Q., see Vinyl acetate
Vinyl alcohol 2,2-dichlorodimethyl phosphate, see Dichlorvos
Vinyl A Monomer, see Vinyl acetate
Vinyl benzene, see Styrene
Vinyl benzol, see Styrene
Vinyl carbinol, see Allyl alcohol
Vinyl chloride
Vinyl chloride monomer, see Vinyl chloride
Vinyl 2-chloroethyl ether, see 2-Chloroethyl vinyl ether
Vinyl β-chloroethyl ether, see 2-Chloroethyl vinyl ether
Vinyl C monomer, see Vinyl chloride
Vinyl cyanide, see Acrylonitrile
Vinylethylene, see 1,3-Butadiene
Vinylofos, see Dichlorvos

Vinylophos, see Dichlorvos
Vinyl trichloride, see 1,1,2-Trichloroethane
Vinylidene chloride, see 1,1-Dichloroethylene
Vinylidene chloride (II), see 1,1-Dichloroethylene
Vinylidene dichloride, see 1,1-Dichloroethylene
Vinylidine chloride, see 1,1-Dichloroethylene
Viozene, see Ronnel
Visko-rhap, see 2,4-D
Visko-rhap drift herbicides, see 2,4-D
Visko-rhap low volatile ester, see 2,4,5-T
Visko-rhap low volatile 4L, see 2,4-D
Viton, see Lindane
Vitran, see Trichloroethylene
Vitrex, see Parathion
Vonduron, see Diuron
Vulcafor TMTD, see Thiram
Vulcalent A, see N-Nitrosodiphenylamine
Vulcatard, see N-Nitrosodiphenylamine
Vulcatard A, see N-Nitrosodiphenylamine
Vulkacit MTIC, see Thiram
Vulkanox 4020, see p-Phenylenediamine
Vultrol, see N-Nitrosodiphenylamine
VyAc, see Vinyl acetate
W.A.R.F. 42, see Warfarin
Waran, see Warfarin
Warfarat, see Warfarin
Warfarin
Warfarin plus, see Warfarin
Warfarin Q, see Warfarin
Warf compound 42, see Warfarin
Warficide, see Warfarin
Waxoline yellow AD, see p-Dimethylaminoazobenzene
Waxoline yellow ADS, see p-Dimethylaminoazobenzene
Weddar, see 2,4,5-T
Weddar-64, see 2,4-D
Weddatul, see 2,4-D
Weedar, see 2,4-D
Weed-b-gon, see 2,4-D
Weedez wonder bar, see 2,4-D
Weedone, see Pentachlorophenol, 2,4-D, 2,4,5-T
Weedone LV4, see 2,4-D
Weedone 2,4,5-T, see 2,4,5-T
Weed-rhap, see 2,4-D

Weed tox, see 2,4-D
Weedtrol, see 2,4-D
Weeviltox, see Carbon disulfide
Westron, see 1,1,2,2-Tetrachloroethane
Westrosol, see Trichloroethylene
White tar, see Naphthalene
Winterwash, see 4,6-Dinitro-*o*-cresol
Witicizer 300, see Di-*n*-butyl phthalate
Witicizer 312, see Bis(2-ethylhexyl)phthalate
Witophen P, see Pentachlorophenol
Wood alcohol, see Methanol
Wood naphtha, see Methanol
Wood spirit, see Methanol
Xenylamine, see 4-Aminobiphenyl
1,2-Xylene, see *o*-Xylene
1,3-Xylene, see *m*-Xylene
1,4-Xylene, see *p*-Xylene
***o*-Xylene**
***m*-Xylene**
***p*-Xylene**
1,3,4-Xylenol, see 2,4-Dimethylphenol
2,4-Xylenol, see 2,4-Dimethylphenol
m-Xylenol, see 2,4-Dimethylphenol
o-Xylol, see *o*-Xylene
p-Xylol, see *o*-Xylene
Yellow G soluble in grease, see *p*-Dimethylaminoazobenzene
Zeidane, see *p,p′*-DDT
Zerdane, see *p,p′*-DDT
Zeset T, see Vinyl acetate
Zithiol, see Malathion
Zoba black D, see *p*-Phenylenediamine
Zytox, see Methyl bromide